Mathematical Symbols and Constants

$=$	equal to	$\gg$		much greater than
$\neq$	not equal to	$\ll$		much less than
$\equiv$	defined as	$\approx$		approximately equal to
$\propto$	proportional to	$\lvert X \rvert$		the magnitude of X
$>$	greater than	π		3.1415926536
$<$	less than	e		2.7182818285

Fundamental SI Units

Quantity	Name	Symbol
Length	meter	m
Mass	kilogram	kg
Time	second	s
Temperature	Kelvin	K
Amount of substance	mole	mol
Electric current	ampere	A
Luminous intensity	candela*	cd

* Not used in this text.

Derived SI Units (Common)

Quantity	Unit Name	Symbol	Expressed in Fundamental Units	Expressed in Other SI Units
Capacitance	farad	F	$A^2 \cdot s^4 / (kg \cdot m^2)$	
Electric charge	coulomb	C	$A \cdot s$	
Electric potential	volt	V	$kg \cdot m^2 / A \cdot s^3$	J/C
Electric resistance	ohm	Ω	$kg \cdot m^2 / A^2 \cdot s^3$	V/A
Force	newton	N	$kg \cdot m/s^2$	
Frequency	hertz	Hz	s^{-1}	
Inductance	henry	H	$kg \cdot m^2 / (A^2 \cdot s^2)$	$V \cdot s / A$
Magnetic induction	tesla	T	$kg / (A \cdot s^2)$	$N \cdot s / C \cdot m$
Power	watt	W	$kg \cdot m^2 / s^3$	J/s
Pressure	pascal	Pa	$kg / (m \cdot s^2)$	N/m^2
Viscosity	poiseuille	Pl	$kg / (m \cdot s)$	$N \cdot s / m^2$
Work and energy	joule	J	$kg \cdot m^2 / s^2$	$N \cdot m$

Symbols for Acceptable Non-SI Units Often Used*

Symbol	Meaning	Quantity
atm†	atmosphere	pressure
Btu†	British thermal unit	heat energy
cal†	calorie	heat energy
°C	degree Celsius	temperature
Ci	curie	radioactivity
db	decibel	logarithmic measure
eV	electronvolt	energy
R†	roentgen	radiation exposure
rad	rad	absorbed dose of radiation
rem	roentgen equivalent man	radiation dose equivalent
torr†	millimeter of mercury	pressure
u	atomic mass unit (unified)	mass

* The roentgen, rad, and rem are to be replaced by their SI equivalents and will eventually disappear.

† Not compatible with SI.

The Greek Alphabet

alpha	A	α		nu	N	ν
beta	B	β		xi	Ξ	ξ
gamma	Γ	γ		omicron	O	o
delta	Δ	δ		pi	Π	π
epsilon	E	ϵ		rho	P	ρ
zeta	Z	ζ		sigma	Σ	σ
eta	H	η		tau	T	τ
theta	Θ	θ		upsilon	Y	υ
iota	I	ι		phi	Φ	ϕ
kappa	K	κ		chi	X	χ
lambda	Λ	λ		psi	Ψ	ψ
mu	M	μ		omega	Ω	ω

COLLEGE PHYSICS

COLLEGE PHYSICS

Gary L. Buckwalter
David M. Riban
Indiana University of Pennsylvania

McGraw-Hill Book Company
New York   St. Louis   San Francisco   Auckland   Bogotá   Hamburg
London   Madrid   Mexico   Milan   Montreal   New Delhi
Panama   Paris   São Paulo   Singapore   Sydney   Tokyo   Toronto

COLLEGE PHYSICS

Copyright © 1987 by McGraw-Hill, Inc. All rights reserved. Printed in the United States of America. Except as permitted under the United States Copyright Act of 1976, no part of this publication may be reproduced or distributed in any form or by any means, or stored in a data base or retrieval system, without the prior written permission of the publisher.

2 3 4 5 6 7 8 9 0   VNHVNH   8 9 4 3 2 1 0 9 8 7

ISBN 0-07-052142-5

This book was set in Berkeley Old Style by Progressive Typographers Inc. The editors were Irene Nunes, Karen S. Misler, and Steven Tenney; the production supervisor was Phil Galea; the design was done by Caliber Design Planning. The drawings were done by Felix Cooper. Von Hoffmann Press, Inc., was printer and binder.

Library of Congress Cataloging-in-Publication Data

Buckwalter, Gary L.
   College physics.

  Includes index.
  1. Physics. I. Riban, David. M. II. Title.
QC21.2.B814   1987   530   86-20160
ISBN 0-07-052142-5

Contents in Brief

Preface *xiii*

1. Preliminaries and Definitions   1
2. Linear Motion   13
3. Vectors and Multidimensional Motion   35
4. Newton's Laws of Motion   55
5. Rotational Kinematics and Gravity   79
6. Equilibrium and Torques   105
7. Work and Energy   127
8. Impulse and Linear Momentum   155
9. Rotational Dynamics   179
10. Simple Harmonic Motion   205
11. Mechanical Waves and Sound   233
12. Some Properties of Materials   267
13. Mechanics of Fluids   287
14. Temperature, Gases, and Kinetic Theory   315
15. Heat and Heat Transfer   341
16. Thermodynamics   371
17. Electrostatic Forces   403
18. Electrostatic Energy and Capacitance   423
19. Electric Current, Resistance, EMF   451
20. Direct Current Circuits   471
21. Magnetic Phenomena   497
22. Inductance, Motors, and Generators   525
23. Alternating Current and Electrical Safety   551
24. Light and Geometric Optics   581
25. Lenses and Optical Instruments   611
26. Physical Optics   641
27. Theory of Relativity   673
28. Birth of Quantum Physics   701
29. Atomic Physics   723
30. Quantum Mechanics   745
31. The Nucleus   771
32. Ionizing Radiation, Safety, and Nuclear Medicine   795
33. Nuclear Fission and Fusion   815

Appendix A Mathematical Review   A1
Appendix B Table of Selected Isotopes   A11
Appendix C Periodic Table of the Elements   A15
Appendix D Answers to Odd-Numbered Problems   A16

Index *I1*

Contents

Preface   xiii

1 Preliminaries and Definitions   1

1.1 Introduction   1
1.2 What Is Physics?   2
1.3 Measurement   2
1.4 Dimensions, Systems of Units   3
1.5 Derived Units   5
1.6 Unit Conversion   5
1.7 Significant Figures   7
1.8 Scientific Notation and Powers of 10   8

Minimum Learning Objectives   9
Problems   10

2 Linear Motion   13

2.1 Introduction   13
2.2 Distance and Displacement   14
2.3 Speed and Velocity   16
2.4 Uniform Acceleration   21
2.5 The Acceleration Due to Gravity   24
Special Topic: Problem Solving in Physics   27

Minimum Learning Objectives   30
Problems   30

3 Vectors and Multidimensional Motion   35

3.1 Introduction   35
3.2 Resolution of Vectors   38
3.3 Velocity in Two Dimensions   42
3.4 Uniform Acceleration in Two Dimensions   44
3.5 Projectile Motion   45
3.6 Relative Velocities in Two Dimensions   49

Minimum Learning Objectives   51
Problems   51

4 Newton's Laws of Motion   55

4.1 Introduction   55
4.2 Newton's First Law of Motion   56
4.3 Newton's Second Law of Motion   59
4.4 Weight Distinguished from Mass   62
4.5 Newton's Third Law of Motion   63
4.6 Tension   65
4.7 Friction   68

Minimum Learning Objectives   73
Problems   73

5 Rotational Kinematics and Gravity   79

5.1 Introduction   79
5.2 Angular Measure   80
5.3 Angular Velocity   81
5.4 Uniform Rotational Acceleration   88
5.5 Circular Motion and Centripetal Acceleration   90
5.6 Newton's Law of Universal Gravitation   95

Minimum Learning Objectives   99
Problems   100

6 Equilibrium and Torques 105

6.1 Introduction 105
6.2 Torques 108
6.3 Second Condition of Equilibrium 113
6.4 Center of Gravity and Center of Mass 114
6.5 Application: Equilibrium and the Human Body 118

Minimum Learning Objectives 121
Problems 121

7 Work and Energy 127

7.1 Introduction 127
7.2 A Sample of Accomplishment 128
7.3 Work 130
7.4 Power 136
7.5 Kinetic Energy 138
7.6 Potential Energy 139
7.7 Work and Energy 141
7.8 Conservation of Mechanical Energy 143
7.9 An Application of Mechanical Energy 145
7.10 Conservation of Total Energy 147

Minimum Learning Objectives 149
Problems 149

8 Impulse and Linear Momentum 155

8.1 Introduction 155
8.2 Impulse 156
8.3 Collisions 159
8.4 Elastic Collisions in One Dimension 161
8.5 Elastic Collisions in Two Dimensions 164
8.6 Inelastic Collisions 166
Special Topic: The Rocket 171

Minimum Learning Objectives 174
Problems 174

9 Rotational Dynamics 179

9.1 Introduction 179
9.2 Torques and Moments of Inertia 180
9.3 Work and Rotational Kinetic Energy 187
9.4 Translational and Rotational Energy Combined 189
9.5 Angular Momentum 192
9.6 Directional Property of Torque and Angular Momentum 196

Minimum Learning Objectives 198
Problems 198

10 Simple Harmonic Motion 205

10.1 Introduction 205
10.2 Producing Simple Harmonic Motion 206
10.3 A Description of Simple Harmonic Motion 209
10.4 The Reference Circle 210
10.5 Mechanical Energy in Simple Harmonic Motion 216
10.6 Some Other Simple Harmonic Motion Systems 219
10.7 Damped and Driven Harmonic Motion 223

Minimum Learning Objectives 227
Problems 227

11 Mechanical Waves and Sound 233

11.1 Introduction 233
11.2 Traveling Waves 235
11.3 Speed of Wave Propagation 241
11.4 Reflections of Waves and Superposition 243
11.5 Standing Waves on a String 245
11.6 Standing Longitudinal Waves 251
11.7 Beats 255
11.8 Sound Intensity Level 256
11.9 The Doppler Effect 259

Minimum Learning Objectives 261
Problems 261

12 Some Properties of Materials 267

12.1 Introduction 267
12.2 Stress and Strain 268
12.3 Elastic Moduli 271
12.4 Applications and Extensions 278

12.5 Strength of Materials 278
Special Topic: Strength and Scaling Laws 280

Minimum Learning Objectives 282
Problems 282

13 Mechanics of Fluids 287

13.1 Introduction 287
13.2 Density 288
13.3 Pressure and Pascal's Principle 289
13.4 Measurement of Pressure 292
13.5 Buoyancy and Archimedes' Principle 294
13.6 Hydrodynamics and Continuity 296
13.7 Work-Energy and Bernoulli's Equation 297
13.8 Applications of Fluid Flow Equations 299
13.9 Viscosity 302
13.10 Stokes' Law 305
13.11 Turbulence 306
Special Topic: Measuring Blood Pressure 308

Minimum Learning Objectives 308
Problems 309

14 Temperature, Gases, and Kinetic Theory 315

14.1 Introduction 315
14.2 Temperature 316
14.3 Temperature Scales 316
14.4 Thermal Expansion 319
14.5 Gas Laws 320
14.6 Ideal Gases 323
14.7 Absolute Zero 324
14.8 Kinetic Theory of Gases 326
Special Topic I: How Hot Are You? 331
Special Topic II: States of a Real Gas 332

Minimum Learning Objectives 335
Problems 336

15 Heat and Heat Transfer 341

15.1 Introduction 341
15.2 Specific Heat Capacity 343

15.3 Latent Heat of Phase Change 344
15.4 Conduction 348
15.5 Convection 352
15.6 Radiation 355
15.7 Evaporation and Humidity 357
Special Topic: Keeping Your Cool 362

Minimum Learning Objectives 365
Problems 366

16 Thermodynamics 371

16.1 Introduction 371
16.2 The First Law of Thermodynamics 373
16.3 Molar Heat Capacity 374
16.4 Reversible Thermodynamic Processes 375
16.5 Heat Engines and Refrigerators 382
16.6 A Practical Cycle: The Otto Cycle 384
16.7 An Ideal Cycle: The Carnot Cycle 386
16.8 The Second Law of Thermodynamics 388
16.9 Application: The Refrigerator 392

Minimum Learning Objectives 396
Problems 396

17 Electrostatic Forces 403

17.1 Introduction 403
17.2 Electric Charge 403
17.3 The Electroscope 406
17.4 Coulomb's Law 407
17.5 The Electric Field 412
17.6 Electric Field Lines of Force 415
17.7 Electric Dipoles 416

Minimum Learning Objectives 419
Problems 419

18 Electrostatic Energy and Capacitance 423

18.1 Introduction 423
18.2 Potential Energy and Potential 424
18.3 "Absolute" Potential 428
18.4 Potential Difference and Electric Field 431

18.5 Capacitors 435
18.6 Capacitors and Energy 437
18.7 Dielectric Coefficient 439
18.8 Application: Painting Pictures with Electricity 441

Minimum Learning Objectives 446
Problems 447

19 Electric Current, Resistance, EMF 451

19.1 Introduction 451
19.2 Electric Current 452
19.3 Resistance and Ohm's Law 455
19.4 Voltage Sources; EMFs 457
19.5 Terminal Potential Difference 459
19.6 Resistivity 460
19.7 Energy and Power in Electric Circuits 463

Minimum Learning Objectives 465
Problems 465

20 Direct-Current Circuits 471

20.1 Introduction 471
20.2 Batteries in Series and Parallel 472
20.3 Resistors in Series and Parallel 473
20.4 Kirchhoff's Rules of Electric Circuits 477
20.5 Measuring Direct Currents and Voltages 483
20.6 Capacitors in Parallel and Series 486
20.7 Resistors and Capacitors in dc Circuits 488

Minimum Learning Objectives 491
Problems 492

21 Magnetic Phenomena 497

21.1 Introduction 497
21.2 Magnetic Force on Moving Charges 498
21.3 Magnetic Fields Produced by Moving Charges 504
21.4 Magnetic Force on Currents 507
21.5 Magnetic Properties of Materials 509
21.6 Induced Voltages 513
Special Topic: Magnetic Field of the Earth 516

Minimum Learning Objectives 518
Problems 519

22 Inductance, Motors, and Generators 525

22.1 Introduction 525
22.2 Mutual Inductance and Self-Inductance 526
22.3 Inductors in Electric Circuits 530
22.4 Torque on a Current Loop in a Magnetic Field 534
22.5 Direct-Current Motors 536
22.6 Electric Generators 538
22.7 Alternating-Current Motors 540
22.8 Transformers 541

Minimum Learning Objectives 545
Problems 545

23 Alternating Current and Electrical Safety 551

23.1 Introduction 551
23.2 Phasor Diagrams 552
23.3 Simple ac Circuits 553
23.4 *RCL* Alternating-Current Circuits 562
23.5 Power and RMS Values in ac Circuits 567
23.6 ac Meters and Signal Detection 570
23.7 Electrical Safety 573

Minimum Learning Objectives 576
Problems 576

24 Light and Geometric Optics 581

24.1 Introduction 581
24.2 Nature of Electromagnetic Waves 582
24.3 Light Wavefronts and Rays 585
24.4 Reflection 587
24.5 Refraction 589

24.6 Dispersion   597
24.7 Mirrors and Reflected Images   598
Special Topic: Fiber Optics   604

Minimum Learning Objectives   606
Problems   606

25 Lenses and Optical Instruments   611

25.1 Introduction   611
25.2 Types of Lenses   612
25.3 Thin Lenses   614
25.4 Lens Aberrations   620
25.5 Single-Lens Applications   621
25.6 Lens Combinations   625
25.7 Optical Instruments   629
25.8 The Human Eye   633

Minimum Learning Objectives   636
Problems   637

26 Physical Optics   641

26.1 Introduction   641
26.2 Development of Physical Optics   642
26.3 Young's Double-Slit Experiment   643
26.4 Thin-Film Interference   647
26.5 Diffraction (Single-Aperture)   650
26.6 The Plane Diffraction Grating   654
26.7 Polarized Light   659
Special Topic: Liquid Crystal Displays: A Pervasive Use of Polarization   666

Minimum Learning Objectives   668
Problems   668

27 Theory of Relativity   673

27.1 Introduction   673
27.2 The Ether: Historical Precursor to Relativity   673
27.3 Frames of Reference   675
27.4 Michelson-Morley and Other Ether Experiments   679

27.5 Einstein's Postulates and the Special Theory   683
27.6 Velocities and Masses in Relativity   690
27.7 Work, Energy, and Momentum in Special Relativity   693
27.8 Additional Thoughts on Relativity   695

Minimum Learning Objectives   696
Problems   697

28 Birth of Quantum Physics   701

28.1 Introduction   701
28.2 The Nature of Radiant Energy   701
28.3 Blackbody Radiation   704
28.4 The Photoelectric Effect   707
28.5 Diffraction of X-rays   711
28.6 The Compton Effect   714
28.7 Ordering Atomic Spectra   716

Minimum Learning Objectives   719
Problems   719

29 Atomic Physics   723

29.1 Introduction   723
29.2 The Atomic Model   724
29.3 The Bohr Atom   726
29.4 Energy Levels in the Bohr Atom and Beyond   731
29.5 A Modification to the Bohr Model   732
29.6 The Pauli Exclusion Principle   734
29.7 Application: The Laser   736

Minimum Learning Objectives   740
Problems   741

30 Quantum Mechanics   745

30.1 Introduction   745
30.2 The de Broglie Hypothesis and Matter Waves   746
30.3 Wave Mechanics   751
30.4 Matrix Mechanics and the Uncertainty Principle   755

xi

30.5 Probability, Duality, and Complementarity 758
30.6 Solid-State Band Theory and Semiconductors 759
Special Topic: "Seeing" with Electrons 765

Minimum Learning Objectives 767
Problems 768

31 The Nucleus 771

31.1 Introduction 771
31.2 The Composition and Properties of the Nucleus 773
31.3 Nuclear Forces and Binding Energy 777
31.4 Beta Decay and Neutrinos 780
31.5 Radioactive Decay 782
31.6 Detection of Ionizing Radiation 785
31.7 Elementary Particles 787

Minimum Learning Objectives 791
Problems 791

32 Ionizing Radiation, Safety, and Nuclear Medicine 795

32.1 Introduction 795
32.2 Cosmic Rays 796
32.3 Other Natural Background Radiation 798
32.4 Units of Ionizing Radiation 799
32.5 Levels of Human Exposure 802
32.6 Biological Effects of Ionizing Radiation 804
32.7 Radiation Safety and Standards for Protection 806
32.8 Radiation in Medicine 809

Minimum Learning Objectives 812
Problems 812

33 Nuclear Fission and Fusion 815

33.1 Introduction 815
33.2 The Discovery of Nuclear Fission 816
33.3 Nuclear Fission and Its Applications 818
33.4 Electricity Generated from Nuclear Power 822
33.5 Breeder Reactors 829
33.6 Controlled Fusion 830

Minimum Learning Objectives 837
Problems 837

Appendix A Mathematical Review A1

Appendix B Table of Selected Isotopes A11

Appendix C Periodic Table of the Elements A15

Appendix D Answers to Odd-Numbered Problems A16

Index I1

Preface

One of the distinctions of a course in college physics is that very few of the students taking it are doing so by choice. Physics is the base for much of modern science and technology, and a knowledge of it is required in many fields of study. It is usually the case, however, that students who major in the physical sciences and engineering begin by taking a physics course that is based on a knowledge of calculus. This leaves the student clientele for the physics course based on algebra and trigonometry — "college" physics — a captured audience that, once the course is finished, will likely not be seen again by the instructor or the physics department. This situation makes for poor feedback on the effectiveness of a textbook unless instructors are constantly trying to assess whether the presentation really works with the students. Much of the motivation for our writing this book grows from this type of assessment.

Physics is a difficult subject. There is no way to gloss over that fact, and any student wise enough to gather comments from other students who have taken the course will know that already. An ancient prince fascinated by mathematics despaired at the time it took to understand things. He was informed that "there is no royal road to geometry." Likewise, there is no way through physics without an honest effort. *Because the subject is quantitative and cumulative, you will need to do your best from the very start or else, in order to pass later tests, go back and relearn things after having already been tested on them.* This is not new or different, and you are not being discriminated against. All who have passed the course have "paid their dues" in intensive study and long sessions of solving problems.

There is no virtue in making something that is relatively simple appear harder than it is. There is a very considerable danger, however, in making something inherently complex look much simpler than it is. In writing this text, we were faithful to a simple set of ideas. First and foremost, *the physics had to be correct*. Nothing horrifies an instructor more than finding a section of a text that is just plain wrong. Should such error slip past the initial class, there is no telling what damage and unlearning might have to occur if the area is one in which a student will need to be knowledgeable later on. Of course, saying it correctly bears a penalty at times, but we have tried to work our way to the careful statement of the physics throughout the book. Normally, the introductory portion of a chapter explores phenomena in an intuitive way to allow the reader to develop a feeling for the area to be considered. Later, precise definitions are developed and used. Usually this approach is good for students but frustrating for instructors, more used to a direct mathematical plunge into the subject. Thus, for example, this book takes several pages and runs through many examples to get to the point of defining the important concept of work. Since instructors are well familiar with the applications and intuitively accept what the mathematical definition is stating, many might be happy with a cursory introduction to new concepts and a plunge into a derivation. You, the student, will benefit greatly if you patiently work through the introductory section to get to the real takeoff point for the chapter.

The second guiding principle in writing this text was that the book be *readable*. If a text is unreadable or hard to read, it usually isn't read. This tends to be a prescription for disaster for the student. Physics has great subtlety. Doing well frequently depends not only on what you know but also on whether you understand the subject. Many courses that you have had, even quite difficult ones, put a premium on memory. Physics may represent a novel task for you in that memory is not nearly so important as understanding is. In writing the text, we have been very careful to state clearly just why or when a particular idea is valid and may be used. Merely to know the idea may not be enough to apply it correctly on a test.

All physical laws have conditions under which they apply and limitations to their general applicability. The fewer the conditions, the more general the law. Normally, it is a great mistake to search for and use indiscriminately any equation that just possibly could be right because it involves the appropriate variables that are given in a problem.

A textbook is an organized guide to a subject that facilitates your learning. It should be clear, complete, and correct, but the textbook cannot learn for you. At best, it is a vehicle to optimize your ability to learn — to let you get the most understanding possible for the time expenditure you must make. Several devices have been used in the text to aid you in using the book.

Worked Examples

In each chapter, you will find many worked example problems. These are illustrations that show how to use new concepts in attacking problems like those found at the end of the chapter. We have tried to make the explanations complete and to show all the essential steps in the solution. Also, even though space in a book is always at a premium, we have selected some problems for examples that are at least as long and involved as anything found in the problem sets. Careful study of these examples can provide you with a good base for doing homework problems.

Problems

Well over 1200 multiple-part problems are presented at the end of the chapters. Throughout the writing of the text, we took a great deal of time to develop problems without reference to any text in print. That does not say that every one represents a new and unique problem not found in any other text, as authors have memories, too. It does imply, though, that each came from an attempt to use the chapter materials as they had been written as the source for the physics needed to work the problems. It will be little consolation to a student laboring over them, but many reviewers have specifically noted the problem sets as unique and a particularly strong feature of the text. The proof of the merit of these problems, however, is in how well they teach you physics. In every set, the physics of the chapter has been applied to some outside subject areas. Physics is physics no matter what the area of application. It is well to remember that the professionals in your own area of study feel that a knowledge of physics is important enough to require it of all the students with your major. Clearly, they believe it can be applied in your area of specialization.

The problems are divided into three types. Problem numbers printed in normal type indicate relatively easy, straightforward problems that should not present a great deal of difficulty to the average student. Problem numbers printed in **boldface type** indicate problems of a higher level of difficulty, usually involving more than the simple application of an idea freshly presented or more than routine mathematical manipulation. Problem numbers printed in color designate the most difficult problems in the sets. All physics problems may be hard, however, if you don't know what you're doing; most may be easy if you do. In any case, the levels grade continuously into each other. Thus, a problem marked less difficult in one chapter may well give more of a challenge than another marked more difficult in another chapter. Problems from areas that "click" for you might prove easy. On the other hand, those at the end of a chapter from which a fundamental idea really didn't enter your consciousness may give you trouble from the start. Each problem is an opportunity to gain in the ability to work with physics, and even the hardest problems in this book have been judged by the authors to be workable by a reasonable student trying hard to do well in this course.

The answers to the odd-numbered problems are provided at the back of the book. It becomes very rewarding to find that you have the right answer the first time you work a problem. If you don't have the right answer, the answers list allows you to find out promptly so that you can reexamine your approach. Although only the numerical answers with units are given, this is usually enough to show if you are on the right track or completely off the mark.

After gross content errors, there is probably nothing more distracting in a text than the presence of numerous erroneous solutions in the answers. Certainly nothing undermines student confidence faster than to find that the number she or he has been struggling for an hour to obtain is simply wrong. This goes beyond being distracting to become counterproductive with students, encouraging them to quit early when the going gets tough. It is also a familiar reason for instructors to drop a text from consideration for use, even if it has other meritorious points. Everything reasonable has been done to ensure that the answers presented in this book are correct. Each problem was worked completely and independently for the answers section by Dr. Dana Klinck of Hillsboro Community College, Tampa, Florida, and by Dr. Alvin Rusk of Southwest State University, Marshall, Minnesota. Their full solutions were then compared with those of the authors and merged by us to generate the answers. The final typed version was then rechecked against the originals to eliminate typographical errors. While no procedure is a complete guarantee that a text or its problem solutions will be error-free, everything reasonable has been done to minimize the probability of error.

One final note. Chapter 2 presents a long section on how to solve a physics problem. You may find this of help since problem solving will invariably occupy a great deal of your study time for this course. There is no way to bypass such practice and still do well in physics.

Minimum Learning Objectives

It is well to remember that, following the study of a chapter in this book, you will be expected to be able to do something as a result. At the end of each chapter, you will find a set of minimum learning objectives for the chapter. (These are at the end because you might not understand the vocabulary of the objectives before reading the chapter, but you are welcome to try.) The first objective gives the new terms you should be able to define and use after reading the chapter. Definitions in physics are expected to be precise, not vague — so practice. At the same time, not all definitions are printed in boldface in the text because some items do not lend themselves to a one-sentence definition. If you cannot give a crisp, clean definition of a term, reread the appropriate portion of the text. It is very important that you do achieve this first objective, however, because it will be difficult to work the problems if you don't really understand what the words mean. The later objectives in the list stipulate exactly what you should be able to do after mastering the material presented. Furthermore, each of the objectives stated is reflected in one or more of the problems of the following problems set.

Be warned again, however, that *physics is a cumulative subject.* It is completely appropriate that problems in a later chapter require you to know the content of an earlier chapter. This is one of the key reasons for you to determine to do as well as you possibly can from the very start of the course. Any deficiency you develop in earlier materials must be made up if you are to do well later. It really is much less painful in the long run not to allow a lack of understanding to develop in the first place.

Marginal Notes

Many times, to emphasize a particular point or to restate something in more intuitive terms, a marginal note is included alongside a discussion. Many of these notes are tips based on our experience in teaching this course for many years. We hope the suggestions in the notes are of help to you. Some try to clear up points that students overlook or get incorrect time after time even though the points are stated correctly in the text and even emphasized. We have avoided using an overabundance of these marginal comments to make sure they are noted when they are presented.

Appendices

In addition to the inside front cover, which contains frequently used constants and conversion factors, there are four appendices at the back of the book. These appendices contain information referenced less often but needed nonetheless. Some students will find it quite helpful to work through Appendix A, which reviews necessary mathematical operations and procedures. Students frequently find that their mathematical skills have become rusty if they haven't used them recently.

Study Guide

An accompanying study guide is available for the course, written by Dr. Marllin Simon of Auburn University. Dr. Simon began working from the earliest drafts of our manuscript to develop materials based directly on the text treatment of the subject. Many students will profit from the detailed practice a study guide can provide.

Acknowledgments

In his office, a very famous author keeps a painting that shows a turtle perched atop a wooden fence. This, he says, is to remind him that "if you see a turtle sitting on a fence post, you know it had some help getting there." Such a painting would be especially appropriate for the authors/turtles of this textbook, and we wish to acknowledge the help, direct and indirect, we received during the book's development.

We have benefited from the comments of the large number of scholars who read and reviewed the manuscript at one or more of its various stages of completion. We found the advice and suggestions of the following to be quite useful and wish to express our sincere appreciation: Dr. John Paul Barach, Vanderbilt University; Dr. Bennet B. Brabson, Indiana University; Dr. Fred E. Domann, University of Wisconsin; Dr. David J. Ernst, Texas A & M University; Dr. Richard E. Garrett, University of Florida; Dr. Philip Hetland, Concordia College; Dr. Verner Jensen, University of Northern Iowa; Dr. Dana S. Klinck, Hillsborough Community College; Dr. David Markowitz, University of Connecticut; Dr. Konrad Mauersberger, University of Minnesota; Dr. Joseph L. McCauley Jr., University of Houston; Dr. Robert Merlino, University of Iowa; Dr. William Riley, Ohio State University; Dr. Marllin L. Simon, Auburn University; Dr. Charles W. Smith, University of Maine at Orono; Dr. Thor Stromberg, New Mexico State University; and Dr. Gordon Wiseman, University of Kansas.

Daniel G. Reiber, Professor Emeritus, Indiana University of Pennsylvania, read the entire manuscript in some of its earliest drafts. He provided many valuable recommendations for which we are most grateful.

Many people on the editorial, design, and production staffs of McGraw-Hill as well as several free-lancers have contributed to the final product. There is a danger in attempting to list all these people by name, in that some could be slighted by unintentional omissions. Thus we simply recognize the support of the entire College Division of McGraw-Hill Book Company. We do, however, extend a special thank you to our basic book editor, Irene Nunes, whose ability and dedication are well known but who, more than once, made the extra effort to snatch victory from the jaws of defeat, as it were.

We are deeply indebted to Mary Buckwalter, who prepared each of the many versions of the manuscript, correcting many errors and deciphering the nearly illegible handwritings of both authors in the process. Her efficiency and dedication all but eliminated one of our major concerns — manuscript preparation.

All the constructive criticism and other assistance we received combined to eliminate errors and provide a superior textbook. Thus the dearth of errors in this text is testimony to the quality of this assistance. Of course, any errors that remain are the sole responsibility of the authors.

Finally, we wish to express our fond appreciation to our wives, Mary Buckwalter and Kathleen Riban, who during this project provided the support and encouragement we could not obtain elsewhere. Further, they uncomplainingly endured the neglect that is the inevitable result of fully employed professors taking their discretionary time to produce this kind of textbook.

Gary L. Buckwalter
David M. Riban

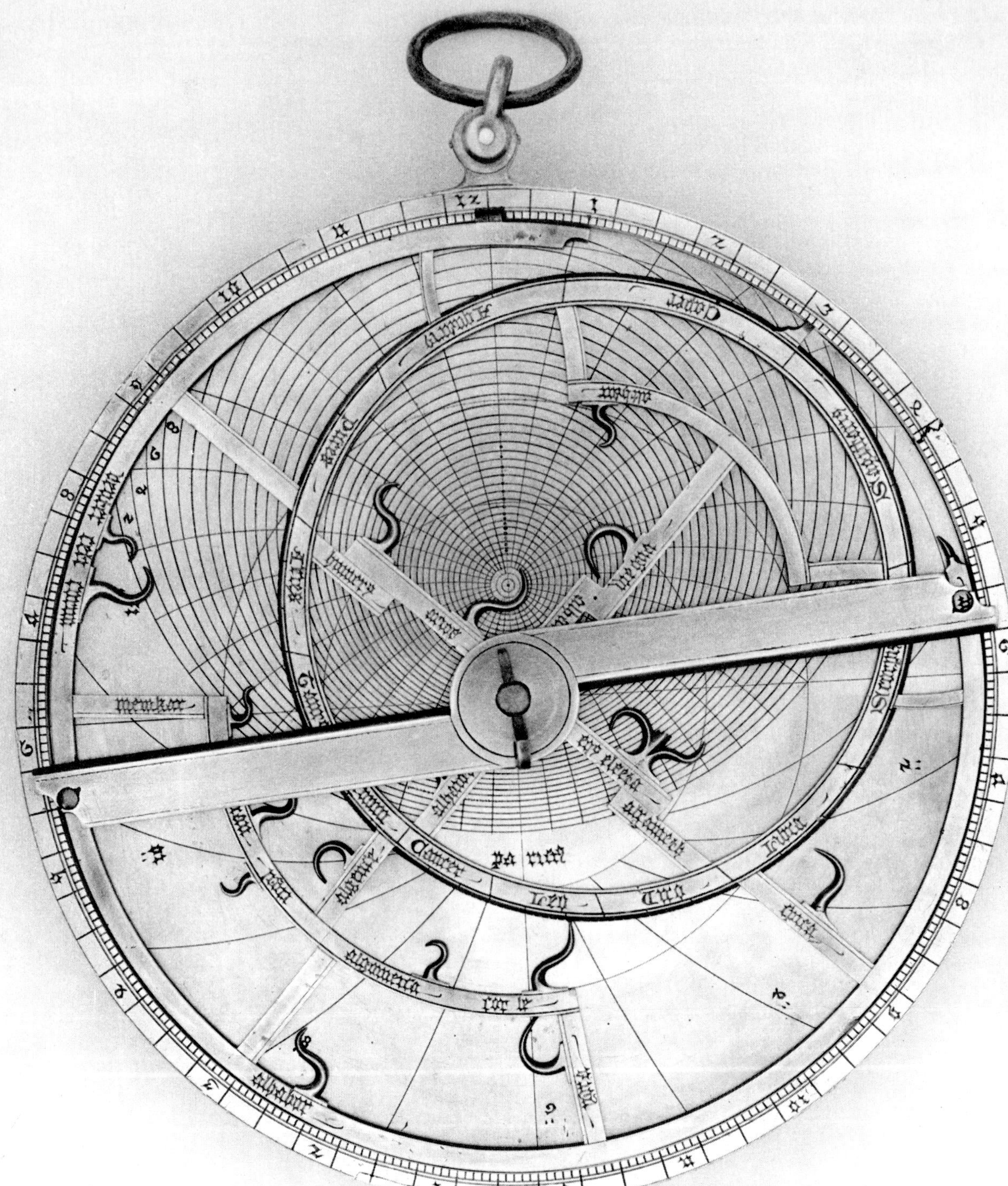

1 Preliminaries and Definitions

1-1 Introduction

The typical student begins a first college physics course wondering, "Is there anything in this for me or is this just another hurdle which I have to get over?" Although some study physics because they are interested in the laws of nature, most students in this course are here because it is required of their majors. However, just knowing that experts in a student's chosen field have agreed that a knowledge of physics is an essential component of the student's education does not make the course more interesting or easier. In fact, many of those experts might agree that there is no way to make the study of physics easy. However, it is possible to make physics palatable — agreeable to the mind — so that the student, having conquered the principles, will feel deep satisfaction in having made the necessary effort.

In addition, there are so many interesting facets of physics that each of us can take delight in understanding some physical principle which explains some occurrence in our daily lives. Physical principles help explain the rocket, why a table tennis ball may curve, how a refrigerator works, and why so little power is required for a liquid-crystal display. These topics and hundreds more will be encountered in this text. First, however, we must learn the language.

1-2 What Is Physics?

Perhaps a more fundamental question at this point is "What is science?" A good dictionary definition is:

Science: Knowledge as of facts, phenomena, laws, and proximate causes gained and verified by exact observation, organized experiment, and correct thinking.

It is important to recognize that the definition hinges on experiment and observation. The results of science must be objective. That is, whether a particular phenomenon or prediction made by science will take place is *not* a matter of opinion. A ripe apple falls from the branch whether or not anyone likes the idea. Hydrogen burns whether public opinion is for it or against it. Weed seeds germinate even if no one votes in their favor.

Physics is a particular branch of science with the following dictionary definition:

Physics: The science that treats of matter and energy and of the laws governing their reciprocal interplay under conditions susceptible to precise observation, experimental control, and exact measurement.

Again we have the words *experiment* and *observation*. As we will see, the ultimate test of any physical theory is whether it conforms to objective reality. This leads us to the requirement that experiments be designed with objectivity, measurements be objective, and results communicated objectively. This objectivity is essential if the predictions, results, and conclusions are to be communicated to someone else (another scientist, industrial firm, etc.) who may wish to duplicate an experiment or make some other use of the information. If agreement of experimental results is to be reached between people separated by time and/or distance, there must be prior agreement of definition of quantities observed. Further, in order to provide objectivity devoid of subjective language, results of experiments most often rely on reporting quantities as numbers. This necessitates attaching a representative number to a physical quantity. We are thus led to measurement.

1-3 Measurement

We often make a measurement — the length of this textbook, for example — by comparing that measurement to some standard (a ruler?). The standard may be crude ("My room is a stone's throw from the student union building") or rather precise ("He ran the length of the football field in 9.82 seconds"). In any case, there is always a comparison between an unknown quantity to be measured (distance to the union, time to run the length of the field) and a standard quantity which is more or less well known (stone's throw, second). In science we rarely accept the stone's throw, but we do require standards of measure. What standards of measure and how many standards will depend on just what we are going to measure.

A carpenter needs a standard measure of length, while a butcher might think a standard of weight is more important. Length and weight are just two of a very large number of possible **physical quantities** — physical properties which are measurable. In a physics text, we must deal with all of them: mass, length, time, weight, speed, acceleration, torque, temperature, energy—just to name a few.

Some of these quantities are combinations of other quantities. For instance, we might recognize speed as being length or distance divided by time. Because of the physical and mathematical relationships which exist among quantities, we arbitrarily choose certain ones as being **fundamental quantities** and call the rest **derived quantities.** Units are agreed upon and assigned to the fundamental quantities, while the units of the derived quantities depend on physical and/or mathematical definitions. The system of units used by the scientific community (and adopted by most industrial countries) will be described in the next section. Two of the fundamental quantities in this system are the *meter* for length and the *second* for time. In this system, speed is measured in meters per second and is a derived quantity.

1-4 Dimensions, Systems of Units

The system of units most commonly used throughout the world and used primarily in this text is called the International System of units, abbreviated **SI**, from the French phrase *Système International d'Unites*. This system was recommended at the General Conference on Weights and Measures of the International Academy of Sciences in 1960 and adopted by the U.S. National Bureau of Standards in 1964. There are seven fundamental quantities in the SI. The last four of these will be introduced and described as they are required later in this text. In the meantime, we will define the first three fundamental quantities because they are required immediately.

A fundamental standard should have several properties if it is to be most useful:

1. It should be *practical* for the measurement under consideration.
2. It should be widely *accessible,* either the **primary standard** or a duplicate called a **secondary standard.**
3. It should stay *constant in time*.
4. It must define something which is *reproducible*.
5. There must be *agreement* among users and potential users that this is the standard.

The three fundamental quantities in the SI which we must specify are length, mass, and time. We have general ideas of what these quantities are. We will often mention in this text that the language of science, in general, and physics, in particular, is very precise. In order that there be agreement on scientific principles, we must be sure that our defining words mean the same thing to everyone. Thus physicists and physics professors are often very picky about precision of language. Sometimes, however, when we first introduce a topic or concept, we find it useful to present a vague general meaning for a word and refine the definition as we proceed. Such is the case with these three quantities. For the present, **length** can be considered to be a physical distance; **mass,** a measure of the quantity of matter; and **time,** a measure of the duration of interval which lapses between successive events.

The choice of which physical quantities to define as fundamental is dictated somewhat by convenience and is fairly arbitrary. Once these are defined, however, the others are derived automatically. If we choose to measure length in miles and time in hours, speed is measured in the derived unit "miles per hour" (mi/h), with no new definition of this unit necessary.

Length

The fundamental SI standard of length is the meter. A unit system which is based on the meter as a standard of measurement is called a **metric** system. Hence, the SI is a particular metric system. Proposed as the length standard by the French Academy of Sciences nearly 200 years ago, the meter was originally intended to be one ten-millionth (1/10,000,000) the distance between the earth's equator and the north pole measured along the meridian passing through Paris. A standard for comparison was constructed of a platinum-iridium alloy bar with two marks, one near each end of the bar, separated by exactly one meter. This became the world's primary standard and remained so even after it became known that the distance originally postulated was not exactly 10,000,000 meters. The 1960 General Conference on Weights and Measures redefined the standard meter as 1,650,763.73 wavelengths in a vacuum of a particular orange-red spectral line of the krypton 86 atom. This redefinition was made in the interest of increasing the accessibility and precision of the standard as well as of enhancing the reproducibility. Though an improvement over the previous standard, this definition stood for only 23 years before it was superseded. The 1983 General Conference on Weights and Measures defined the **meter** as the distance light travels in a vacuum during 1/299,792,458 of a second. In addition to being more precise for a standard than the distance between scratches on the original metal bar, the standard meter now possesses all the desirable qualities of a fundamental unit. Furthermore, it fixes the speed of light with whatever precision time can be measured (currently, several parts in 1000 billion).

Mass

The fundamental SI standard of mass, the **kilogram,** was adopted by the General Conference on Weights and Measures in 1901. It is the quantity of matter equivalent to that contained in the primary standard, a small platinum-iridium alloy cylinder kept at very controlled temperature and humidity conditions in Sevres, France.

Time

The fundamental standard of time is the second, originally defined as 1/86,400 part of a mean solar day. To allow for greater accuracy when scientists discovered that the average time interval of the solar day is decreasing by about 0.001 second per century, the second was defined as 1/86,400 of the mean solar day of the tropical year 1900. In 1967, the present atomic standard was accepted. The **second** is now defined to be equal to the duration of 9,192,631,770 oscillations of the radiation associated with a certain electronic transition of the cesium 133 atom.

    Of course, the primary standards (particularly that of mass) are not always accessible to everyone. Thus many secondary standards are made. Copies or reasonable facsimiles of the secondary standards are mass-produced and are then available to everyone. Examples are the metersticks we have all used, sets of masses for laboratory and demonstration use, and crystal-controlled clocks and oscillators. Most measurements have depended and will depend on the copies of our secondary standards.

When we report a physical quantity from some measurement, it is absolutely mandatory that we state both a number and unit. For example, we might say that the dean's desk is 1.524 meters wide and 0.78 meters high. It would make no sense at all to report that the desk was 1.524 wide and 0.78 high. We should also note that for both convenience and brevity we will use the abbreviations of the units. That is, one meter becomes 1 m, one kilogram becomes 1 kg, and one second becomes 1 s.

There are several other systems of units. You are familiar with at least some of the quantities in the British engineering system — the foot, inch, pound, mile, and so on. Much of the system is in wide use in the United States but in no other industrialized country. It was even abandoned by Great Britain. SI is now making inroads in the United States, and no doubt will eventually be adopted by everyone. Although we will occasionally do some converting between systems to give a more intuitive feeling for certain quantities, we will try to use SI almost exclusively throughout the text.

1-6 Unit Conversion

A system is a *decimal* system if it defines a unit, then has multiple (or submultiple) units which are 10, 100, 1000, etc., times as large. The mks (meter-kilogram-second) and cgs (centimeter-gram-second) systems are metric decimal systems, as are SI units. SI differs from the older mks system only in units which we will define much later. In either, the *meter* is the *only* unit of length, others being decimal multiples.

1-5 Derived Units

Most of the quantities we use in physics have units which are combinations of those of the fundamental units. We have already mentioned speed as having units of length divided by time (meters per second in SI). Consider a typical bedroom in someone's residence. It may be 4 m long by 3 m wide by 2.5 m high. This gives it a floor area of 4 m times 3 m, or 12 m^2. The units of area (a derived quantity) in SI are meters times meters, or square meters (m^2). The volume of the room is 4 m times 3 m times 2.5 m, or 30 m^3. Thus the units of volume in SI are cubic meters (m^3).

It should be becoming clear to you that derived quantities and their units are necessary to describe the things with which science is concerned. Further, we find we may treat the abbreviations of units as algebraic symbols. The units of speed arise quite naturally from the units of length (distance) and time and the definition of speed as the distance advanced per unit time. Likewise, the units of area and volume are derived from the units of length and the definitions of area and volume. Most of the quantities with which we deal will be derived quantities. Each of these derived quantities will have units which are some combination of the fundamental quantities. However, we will often use a new single name for the derived unit either to simplify our discussion or to honor a scientist or both. A useful problem-solving-check technique is to investigate the units of each term in an equation to see if they agree. We may often discover errors or we may reinforce our belief in the correctness of a solution or equation by investigating the units of each term.

Many practical problems and some of those invented by your professor will have quantities whose units are given in different systems. When you encounter such a problem, you may have to convert some of the quantities to SI.

1-6 Unit Conversion

It is always possible to convert a quantity given in one unit to the same quantity in another unit if you know a relationship between the two units. For instance, if you know that a football field is 100 yards long and you know that 1 yard equals

3 feet, you can state that the field is 300 feet long. Although this is a rather simple example, it is instructive to ascertain a formal method for this conversion. We write

$$3 \text{ feet} = 1 \text{ yard}$$

Recognizing this as an equation and realizing we may multiply or divide at random and retain the equality *provided* we conduct the operation on both sides of the equation, we can divide both sides by 1 yard:

$$\frac{3 \text{ feet}}{1 \text{ yard}} = \frac{1 \text{ yard}}{1 \text{ yard}} = 1$$

Next we recall that we can always multiply or divide anything by 1 without changing it. Thus we write

$$100 \text{ yards} \times \frac{3 \text{ feet}}{1 \text{ yard}} = 300 \text{ feet}$$

Note we have treated the units as algebraic quantities and the yards canceled, leaving the desired units, feet.

More often we find we wish to convert between systems. For example, we may wish to convert the length of the field given in British units to SI units. From the Conversion Table on the inside front cover, we see that 1 yard = 0.9144 m. This gives us

$$1 = \frac{0.9144 \text{ m}}{1 \text{ yard}}$$

and

$$100 \text{ yards} \times \frac{0.9144 \text{ m}}{1 \text{ yard}} = 91.44 \text{ m}$$

In any conversion calculation, we will simply be multiplying by 1. We may do this once or a number of times.

Example 1-1 Convert 25 miles/hour to meters per second.

Solution We must know that

$$1 \text{ mile} = 5280 \text{ feet} \qquad \frac{5280 \text{ feet}}{1 \text{ mile}} = 1$$

$$1 \text{ foot} = 0.3048 \text{ m} \qquad \frac{0.3048 \text{ m}}{1 \text{ foot}} = 1$$

$$1 \text{ hour} = 3600 \text{ s} \qquad \frac{1 \text{ hour}}{3600 \text{ s}} = 1$$

Thus

$$25 \frac{\text{miles}}{\text{hour}} \times 1 \times 1 \times 1 = 25 \frac{\text{miles}}{\text{hour}} \times \frac{5280 \text{ feet}}{1 \text{ mile}} \times \frac{0.3048 \text{ m}}{1 \text{ foot}} \times \frac{1 \text{ hour}}{3600 \text{ s}}$$

$$= 11.176 \frac{\text{m}}{\text{s}}$$

Note that in Example 1-1 we were careful to place the proper units of the conversion factors in the numerators and denominators so we could cancel algebraically all unwanted units and retain only meters in the numerator and seconds in the denominator. This technique is worth remembering and practicing.

Once in a while, you may find yourself with an SI quantity for which you have no intuitive feeling—forces in newtons, volumes in liters, and so on. On these occasions you certainly could convert to a system with which you are more comfortable (clearly, the British engineering system). If you do this often enough, you will eventually be comfortable with SI as well.

Example 1-2 The winner of the short sprint in a track meet averaged 10 m/s over the course. Calculate his speed in miles per hour.

Solution This is the reverse of Example 1-1, and we may utilize the reciprocal of each of those conversion factors. Thus

$$10 \frac{m}{s} \times \frac{1 \text{ ft}}{0.3048 \text{ m}} \times \frac{1 \text{ mi}}{5280 \text{ ft}} \times \frac{3600 \text{ s}}{1 \text{ h}} = 22.37 \frac{\text{mi}}{\text{h}}$$

1-7 Significant Figures

A very important characteristic of measurement is the precision (or degree to which we may trust the number) obtained. Some measurements result in a number which is only slightly better than a rough estimate, while others yield numbers which we may report with confidence to one part in a billion. The fineness of the resulting number is an indication of the precision of the measurement.

We account for the difference in precision of numbers by using the concept of **significant figures.** The number of significant figures in a reported measurement may be determined by locating the first nonzero digit on the left and counting this and each of the following digits (including zeros) as significant. As examples, the number 160 has three significant figures, 357.1300 has seven significant figures, and 0.008612 has four.

Significant figures are crucial when the results of measurement must be manipulated mathematically. For instance, suppose the length and width of a bedroom are measured to be 4.1 m and 3.0 m. We might wish to report the floor area as 12.3 m². This would be wrong, since we cannot gain precision (number of significant figures) by multiplying. Thus we should drop the last digit and report the area as 12 m². This is part of a technique called **rounding off.** You must often round off the numbers from your calculator, since most operate as though every number entered has at least as many significant figures as the number of digits in the display (usually 8 or 10). Round off to a given number of significant figures as follows: Look at the digit immediately to the right of the last digit you wish to keep. If it is less than 5, retain the previous numbers without change when you drop the unwanted digits. If it is 5 or greater, increase the last retained digit by 1. Consider the number 61.453719. If we round off to seven significant figures, it becomes 61.45372; to six significant figures, it becomes 61.4537; to five, 61.454; to four, 61.45; to three, 61.5; to two, 61.

The following rules may be applied for consistency in reporting the number of significant figures resulting from mathematical manipulation:

> The only way to obtain more significant figures in a result is to perform better measurements. Mathematical manipulation can only decrease the number of significant figures, not increase it.

1. In multiplying or dividing, the product or quotient should have the same number of significant figures as the number in the least precise factor.
2. In addition or subtraction, the result should not be carried beyond the first column having a doubtful figure.

The area of the bedroom floor is an example of the first rule, while we note that $212 + 4.367 = 216$ (*not* 216.367) is an example of the second rule.

Because the number of significant figures is limited ultimately by experiment, we must pay strict attention to this aspect in the laboratory. We must neither report more precision than our data allow (by reporting too many significant figures) nor hide important physics by not reporting all the significant figures available. Thus in the laboratory, proper use of significant figures is intertwined with and part of the primary goals. In lecture or recitation, where we may postulate as many significant figures as we like, the goals are to impart an understanding of concepts and to help the student acquire problem-solving skills. While laboratory and lecture goals are not in conflict, it is important that the student's attention remain focused properly. Therefore, throughout *most* of this text, except for those few specific instances where three significant figures are inappropriate, data and answers will be given to three significant figures. On those occasions where you find the answer to a problem you have worked disagrees in the last significant figure with the text's answer, it will most often (though not always) be due to a difference in rounding off intermediate answers.

> If we are told that a room is 7 m × 7 m, each measurement only assures us that the length was closer to 7 than to either 6 or 8. Thus the real length could be anything from just over 6.5 to almost 7.5. Reporting the area of such a room as 49 m² exceeds the knowledge of our measurements, since the room could be anything from 42 m² up to 56 m² (that is, 6.5 × 6.5 up to 7.5 × 7.5 m²).

1-8 Scientific Notation and Powers of 10

Scientists and engineers work with numbers and measurements ranging from very tiny to very large. For this reason, a system of recording numbers called standard **scientific notation** was devised. In this system, we express a number as the product of a number between 1 and 9.999... with a power of 10. For example, we may wish to express the average distance s from the earth to the moon (384,000,000 m) to three significant figures as

$$s = 3.84 \times 10^8 \text{ m}$$

or we may wish to express the radius of the hydrogen atom to two significant figures. Instead of 0.000 000 000 053 m, we write

$$R_H = 5.3 \times 10^{-11} \text{ m}$$

We can see that this is somewhat more practical, providing both a standard and a more compact way to express all numbers. Furthermore, it helps to avoid possible ambiguity in the number of significant figures.

To express any number in scientific notation, we simply move the decimal place to obtain a number between 1 and 10 and multiply by the appropriate power of 10. Because the advantages of using powers of 10 and dividing quantities into 10 parts and each of those parts into 10 smaller parts was obvious, it became the basis of the metric system which evolved to SI. Table 1-1 lists the typical powers of 10 which are used enough to have acquired common names. The names in boldface type are used often enough in this course that it would be wise to know them.

Example 1-3 Write the average distance from the earth to the moon in megameters, centimeters, and picometers.

TABLE 1-1 Powers of 10 Used in SI

Number	Equals	Prefix	Symbol
0.000 000 000 001	10^{-12}	pico	p
0.000 000 001	10^{-9}	nano	n
0.000 001	10^{-6}	**micro**	μ
0.001	10^{-3}	**milli**	m
0.01	10^{-2}	**centi**	c
0.1	10^{-1}	deci	d
1	10^{0}		
10	10^{1}	deka	da
100	10^{2}	hecto	h
1000	10^{3}	**kilo**	k
1 000 000	10^{6}	**mega**	M
1 000 000 000	10^{9}	giga	G
1 000 000 000 000	10^{12}	tera	T

Solution   The prefixes mega (M), centi (c), and pico (p) mean $\times 10^{6}$, $\times 10^{-2}$, and $\times 10^{-12}$, respectively. Thus

$$3.84 \times 10^{8} \text{ m} = 3.84 \times 10^{2} \times 10^{6} \text{ m} = 3.84 \times 10^{2} \text{ Mm}$$

Similarly,

$$3.84 \times 10^{8} \text{ m} = 3.84 \times 10^{10} \times 10^{-2} \text{ m} = 3.84 \times 10^{10} \text{ cm}$$

$$3.84 \times 10^{8} \text{ m} = 3.84 \times 10^{20} \times 10^{-12} \text{ m} = 3.84 \times 10^{20} \text{ pm}$$

In concluding this chapter, we call your attention to a feature which has been incorporated into each chapter, the section just before the problems called Minimum Learning Objectives. That section always serves as a chapter summary, but it is much more. As you conclude each chapter, you should be able to assess your understanding and the knowledge you have acquired by reading just what the Minimum Learning Objectives of the chapter are and determine how well you are able to meet those objectives.

Minimum Learning Objectives

After studying this chapter, you should be able to:
1. Define:
   - derived quantity
   - fundamental quantity
   - kilogram [unit]
   - mass
   - meter [unit]
   - metric system
   - physical quantity
   - physics
   - primary standard
   - rounding off
   - science
   - scientific notation
   - second [unit]
   - secondary standard
   - SI
   - significant figure
2. Convert the values of physical quantity from one unit system to another.
3. Specify the correct number of significant figures in a given number.
4. Report the result of any arithmetic operation on measured numbers to the correct number of significant figures.
5. Round off the result of a calculation to the appropriate number of significant figures.
6. Write a decimal number in scientific notation, and vice versa.
7. Convert a number expressed in multiples and/or submultiples of a fundamental unit to some other multiple or submultiple unit by employing the correct prefix and adjusting the power of 10.

Problems

Throughout this text, the problems are of three levels of difficulty: Standard, somewhat difficult (indicated by a boldface black number preceding the problem), and most difficult (indicated by a colored number).

1-1 What is the SI unit of (a) area, (b) volume flow rate (volume per unit time), and (c) mass density (mass per unit volume)?

1-2 Using the given unit definitions, determine the SI unit for (a) linear momentum—mass times speed, (b) moment of inertia—mass times distance squared, (c) kinetic energy—mass times speed squared, and (d) sound intensity—energy per unit time per unit area.

1-3 Using the given unit definitions, determine the SI unit for (a) acceleration—change in speed per unit time, (b) force—mass times acceleration, (c) pressure—force per unit area, and (d) work—force times distance.

1-4 Convert (a) 55.0 mi/h to m/s, (b) 1500 m to mi, (c) 2000 ft² to m², and (d) 30.0 m³ to ft³.

1-5 Express the speed of sound (760 mi/h) in (a) ft/s, (b) m/s, and (c) km/h.

1-6 Convert (a) 300 Mm/s to mi/s, (b) 24.6 m/s to mi/h, (c) 183 cm to m, and (d) 186 cm to ft and in.

1-7 Determine the conversion factor between (a) cubic yards and cubic feet, (b) cubic meters and cubic feet, and (c) cubic inches and liters.

1-8 Convert (a) 4.00 mi/h to m/s, (b) 9.80 m/s² to ft/s², and (c) 11.2 km/s to mi/h.

1-9 How many seconds are there in (a) a day, (b) a week, and (c) a leap year?

1-10 A long-distance runner completed a marathon race (26 miles 385 yards) in 2 hours 22 minutes and 45 seconds. Determine (a) the average speed in miles per hour, (b) the average time in minutes to complete 1 mile, (c) the total distance in meters, and (d) the average speed in meters per second.

1-11 While on a trip, you see a road sign at 1:40 P.M. showing that your destination is 276 km away. (a) What distance in miles will your odometer show you have traveled to reach your destination? (b) If you average 45.0 mi/h, will you make your destination by 5:00 P.M.?

1-12 Your refrigerator is exactly 0.800 m wide, 0.700 m deep, and 1.70 m high. (a) If your basement doorway is 30.0 in wide by 6.00 ft high, can the refrigerator be moved to your basement without dismantling anything? (b) Determine the width and height in inches of the smallest opening through which the refrigerator can be passed.

1-13 A physics student is exactly 5 feet 7 inches tall. What is that in (a) centimeters and (b) meters?

1-14 (a) How many square feet of carpeting are needed for a dormitory lounge 25 ft by 18 ft? (b) How many square yards? (c) How many square meters? (d) How much will it cost at $18.00 per square yard?

1-15 A 1-in cube of gold has a mass of 0.316 kg. Calculate its mass density (mass per unit volume) in kilograms per cubic meter.

1-16 A suburban house lot is 150 ft by 120 ft. Calculate the area of the lot in (a) square feet and (b) acres.

1-17 The living room in a sorority house measures 6.256 m wide by 8.416 m long by 3.212 m high. To three significant figures, determine (a) the area of the floor in square meters and square yards and (b) the volume of the room in cubic meters. (c) What is the largest number of significant figures to which the area and volume should be reported?

1-18 Round off the number 4.8162593 to (a) three, (b) four, (c) five, (d) six, and (e) seven significant figures.

1-19 A piece of notebook paper is 28.23 cm by 22.8 cm. To the proper number of significant figures, determine its (a) perimeter and (b) area.

1-20 Add the following and report the sum to the proper number of significant figures: 348.62 m, 17.435 m, 4626.23 m, 8.1076 m.

1-21 Determine the number of significant figures in (a) 43.171, (b) 3.0026, (c) 0.430, and (d) 0.00631.

1-22 Determine the number of significant figures in (a) 74.01, (b) 9400, (c) 48.2, and (d) 0.00317.

1-23 A bedspread is measured to be 66.02 in by 107.79 in. Determine to the proper number of significant figures its (a) area and (b) perimeter.

1-24 To four significant figures, the average radius of the earth is 6.376×10^6 m. How should this value be reported in standard scientific notation in (a) kilometers, (b) millimeters, (c) nanometers, and (d) gigameters?

1-25 Write the following quantities in standard scientific notation: (a) 5216 m, (b) 0.00417 kg, (c) 11.21 s, (d) 0.00000213 s, and (e) 299,792,458 m/s.

1-26 Add the following lengths and report the answer in standard scientific notation and to the proper number of significant figures: 1.47 m, 62.9 m, 121.46 m, 3.516 m, 0.08 m.

1-27 Evaluate the quantity 4.00×10^{-3} m times 0.0025 m times 36.14×10^{-4} m divided by 271 s. Report your answer in standard scientific notation to the proper number of significant figures.

1-28 Assume the hydrogen atom is a sphere of radius 5.29×10^{-11} m. Calculate the volume it occupies, and report your answer to the proper number of significant figures and in standard scientific notation.

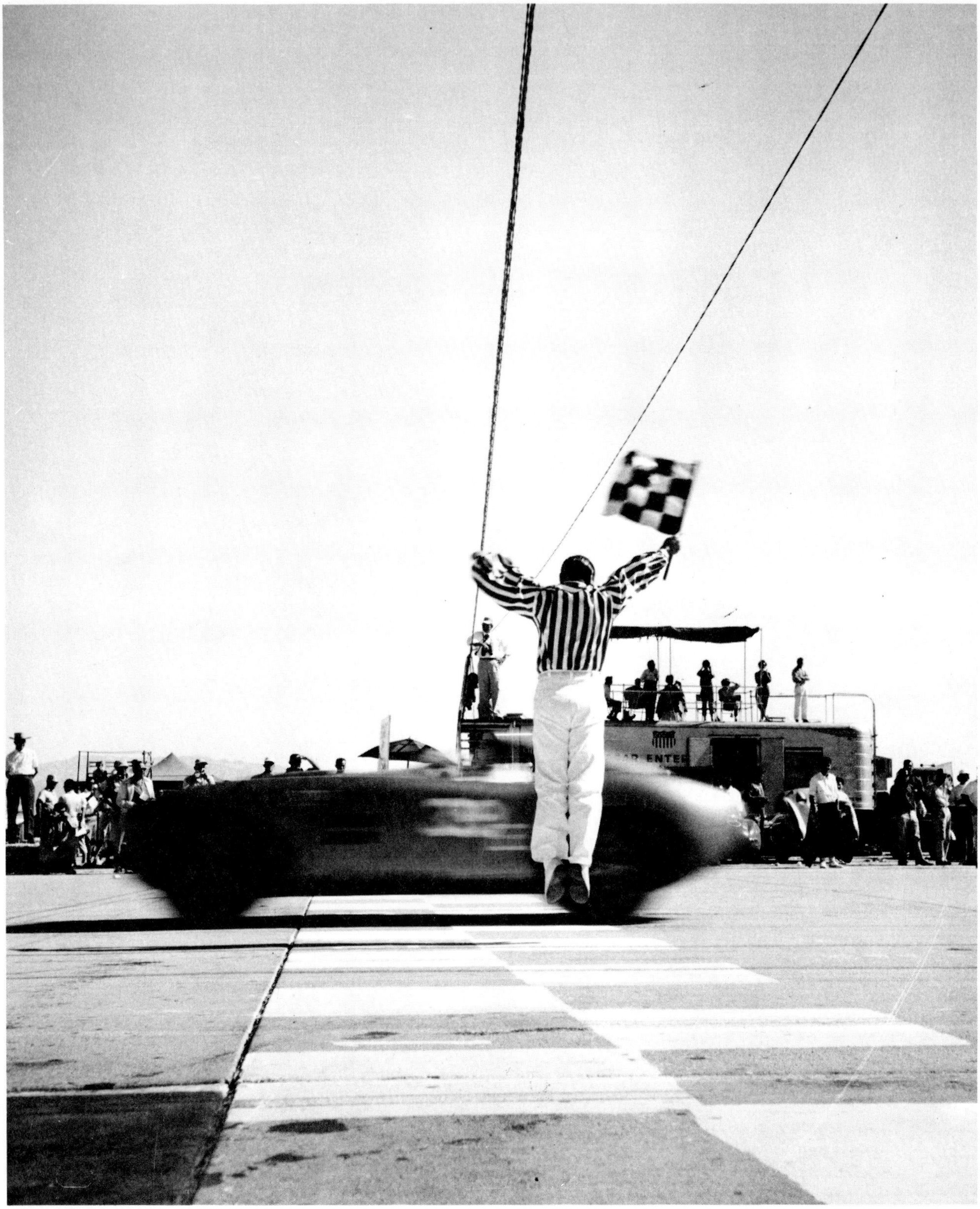

2 Linear Motion

2-1 Introduction

Modern science began in the 1500s with the systematic interpretation of motion, as can be seen in an example. In the 1530s the Italian scientist Tartaglia was studying the motion of cannonballs fired from a cannon. The old ideas of motion gave him no ability to predict the trajectory of such a projectile, and Tartaglia was forced to lengthy experiments in which he carefully fired a cannon at varying angles of elevation (Figure 2-1). He concluded that the cannon attains its maximum range when fired at an inclination of 45° from the horizontal. Tilt the cannon more or tilt it less and the ball strikes nearer to the cannon than a shot at 45°. Tartaglia could not explain why this result should be correct, merely that it was an experimental fact that nature was operating in this fashion.

Galileo Galilei (1564–1642) was born in the next generation, and his work began the modern study of motion. He asserted that in the absence of forces, motions persist and objects move at constant speed. He further held that falling objects increase their downward speed at a constant rate. From these two ideas, he showed that the 45° rule of Tartaglia was an inevitable consequence. Galileo had developed the ability to *describe* motion. However, using a new and untried cannon even with a known charge of gunpowder and a cannonball of

tion is chosen, however, the displacement of one point with respect to another has its sign fixed. In equation form, the displacement of point C in Figure 2-3 from point A is

$$\Delta s = s_2 - s_1 \quad (2\text{-}1)$$

where s = general symbol used to indicate the displacement of a point from the origin
s_1 = coordinate of the first point, A in this case
s_2 = coordinate of the second point, C in this case

and the Greek letter Δ (delta) before any variable means a change of or difference in that variable. For example, the expression Δx, read as "the change in x," may always be replaced by the quantity "$x - x_0$," where x is some final value of x for the interval considered and x_0 is some initial or starting value.

2-3 Speed and Velocity

Average Speed and Velocity

Figure 2-4 shows that a moving object is changing its position in space as it changes its location in time. To predict a body's position or displacement after some time interval, we must know the rate at which that body's position is changing with time. **Speed** is defined as the rate at which a body is changing its distance along its path. Similarly, we define **velocity** as the time rate at which a body is changing its displacement. Consider an automobile trip from Pittsburgh to Chicago, some 800 km apart. Along the straight line connecting the two cities, we choose Pittsburgh as the origin and the direction toward Chicago as positive. If you drove a scenic route with some side trips, you could find that the distance registered on your trip odometer was 1200 km when you reached Chicago exactly 40.0 h (1.44×10^5 s) after you left Pittsburgh. Your average speed for the trip is 30.0 km/h, or 8.33 m/s, but your average velocity for the trip is only +16.4 km/h, or +4.56 m/s. We found the **average speed** by dividing the total distance traveled by the total time of travel. The **average velocity** is the change in displacement divided by the total time of travel, or

$$v_{av} = \frac{s_2 - s_1}{t_2 - t_1} = \frac{\Delta s}{\Delta t} \quad (2\text{-}2)$$

We have used Equation 2-1 for the difference in displacement and recognized that the time interval $t_2 - t_1$ rather than a specific time is the most important

Mathematically, the word *rate* means the quotient of two changes. For example, the rate at which a tree is growing is $\Delta h/\Delta t$, the change in height Δh divided by the change in time Δt during which this change in height was measured. If the denominator of the rate is time, it is a **time rate of change**.

Figure 2-4 A moving object is located at a different point in space at each instant of time. We completely describe the motion of a point whenever we can specify its displacement at any time during the motion. (D. Riban)

quantity. After all, you probably did not start your watch just as you left Pittsburgh. You found the elapsed time by subtracting your reading when you left Pittsburgh from that when you reached Chicago. Note also that we need not have chosen the origin of our coordinates to be at Pittsburgh. We could have picked our origin at any arbitrary point along the line, since we are concerned only with the *difference* in displacement between Pittsburgh and Chicago.

Although Equation 2-2, in which subscripts are used to indicate particular points, is perfectly valid, we often write the definition of average velocity in more general symbols as

$$v_{av} = \frac{s - s_0}{t - t_0} \tag{2-3}$$

where s represents any displacement that the moving object has undergone at a time t (any time after the clock is started) and s_0 is the displacement at a time t_0 (when our measurement was started). We often set t_0 equal to zero by starting the clock as we begin our measurement. Then Equation 2-3 becomes

$$v_{av} = \frac{s - s_0}{t} \tag{2-4}$$

We may find our displacement s at any time after we begin by solving Equation 2-4 for s as a function of time:

$$s = v_{av}t + s_0 \tag{2-5}$$

If v_{av} is a constant, this equation is identical in form to the slope-intercept form of the equation of a straight line, $y = mx + b$. Thus, if we graph s on the y axis and t on the x axis, we get a straight line with a slope of value v_{av} and a y intercept of value s_0, as shown in Figure 2-5.

Example 2-1 An automobile traveling along a straight, level highway passes a sign at a service station (Figure 2-6), and 8.00 s later it passes a billboard 300 m from the service station. It arrives at a point opposite an entrance sign to a restaurant 250 m beyond the billboard exactly 20.0 s after it passes the billboard. Determine the magnitude of the average velocity of the automobile (*a*) between the service station and the billboard, (*b*) between the billboard and the restaurant, and (*c*) between the service station and the restaurant.

Solution For convenience, we label the positions in order as 1, 2, and 3 and choose the original direction of motion as positive. We may use Equation 2-2 for

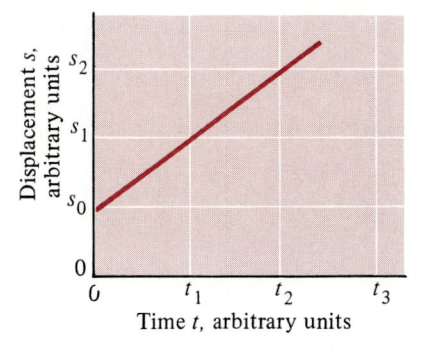

Figure 2-5 The slope of a displacement vs. time curve gives the speed of the motion. Here the curve is a straight line of constant slope, indicating a constant-speed motion.

2-3 Speed and Velocity

17

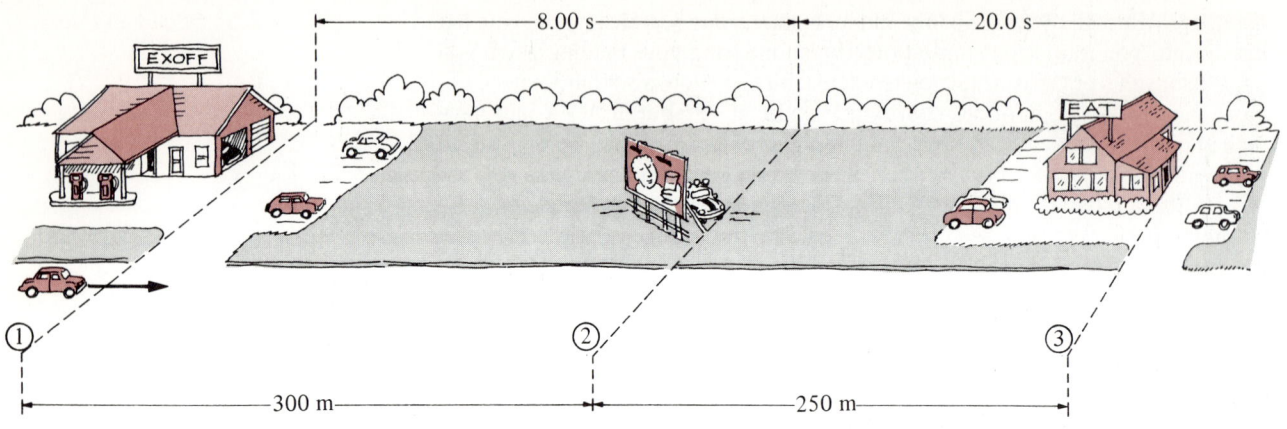

Figure 2-6 Example 2-1.

each part of the problem, since we have displacements and times and are looking for average velocities. We have

(a) From 1 → 2: $\quad v_{av} = \dfrac{\Delta s}{\Delta t} = \dfrac{+300 \text{ m}}{8.00 \text{ s}} = 37.5 \dfrac{\text{m}}{\text{s}}$

(b) From 2 → 3: $\quad v_{av} = \dfrac{+250 \text{ m}}{20.0 \text{ s}} = +12.5 \dfrac{\text{m}}{\text{s}}$

(c) From 1 → 3: $\quad v_{av} = \dfrac{+550 \text{ m}}{28.0 \text{ s}} = +19.6 \dfrac{\text{m}}{\text{s}}$

where we have rounded our answers to three significant figures.

Example 2-2 Suppose the driver of the automobile in Example 2-1 stops at the restaurant, turns around, and returns to the service station 110 s after he passed it originally. For that 110-s interval, determine (a) the average velocity and (b) the average speed.

Solution (a) Since his total displacement during the time interval is zero (he's back at the service station), his average velocity for the interval is zero.
(b) The average speed is simply the total distance traveled, 1100 m, divided by the total time elapsed, 110 s:

$$v_{av} = \dfrac{1100 \text{ m}}{110 \text{ s}} = 10.0 \dfrac{\text{m}}{\text{s}}$$

Instantaneous Speed and Velocity

So far we have considered only average speeds and average velocities, although either can vary during the course of an object's motion (Figure 2-7). We often wish to know more than average values of these quantities. We would like to know their values at a specific instant of time. The police officer is interested not in your average speed since you left the house but in your measured speed at the instant you passed the radar trap.

The **instantaneous velocity** of an object is defined as the time rate of change of the object's displacement at a particular instant. **Instantaneous**

18

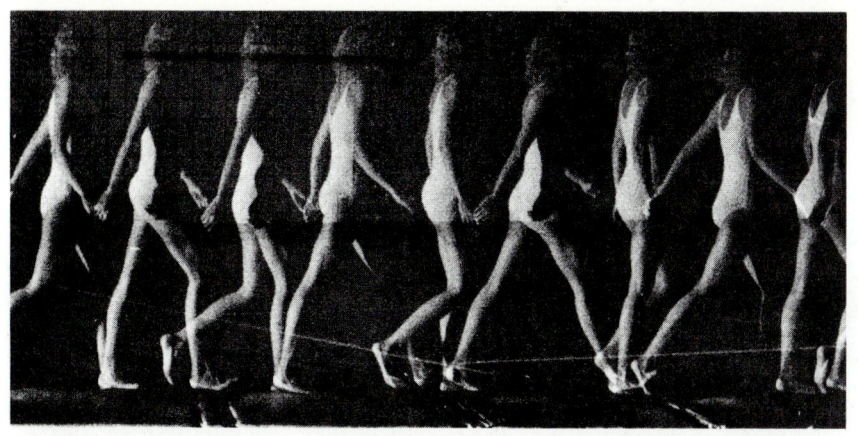

Figure 2-7 Constant-velocity motion is infrequent. A walking motion appears to be smooth, but this stroboscopic photograph shows that the velocity of the various body parts changes frequently. Each limb slows down and speeds up. (D. Riban)

speed is simply the magnitude of the instantaneous velocity. Thus, if we say that an automobile has an instantaneous velocity of 30 m/s to the west, we mean that if its velocity did not change, it would acquire an additional 30 m displacement to the west for every second of travel. Real objects do not often move at constant velocities; therefore we need a method for determining instantaneous velocities.

Figure 2-8 is a photocomposite of a tennis ball thrown upward by a very happy player who has just won her match. The position of the ball is shown for nine equally spaced time intervals. Displacement above the ground from the lines on the court can be measured on the photograph. Note that the motion of the ball is almost purely vertical, or essentially a one-dimensional motion. Choosing upward as positive, we see that the ball started with a large upward velocity and returned to hit the court with a large negative velocity. It must have slowed, stopped, and changed direction between those two points. In fact, the velocity of the ball was continually changing during the flight. We can see this by observing that the difference in displacement between any two photographs is different in each case. The photocomposite mixes the situation for nine separate instants in time. We may print each of these separate instants as an independent photograph and space them equally along the x axis to allow us to visualize more easily the height, or s, variation (Figure 2-9a). From this it is a small step to abstract the displacement and time variations from the photographs, as in Table 2-1, and produce a graph of height vs. time, as in Figure 2-9b. Suppose that we want the instantaneous velocity at time $t = 0.55$ s. We can obtain easily the average velocity for various time intervals. For example, the average velocity between position 1 (release of the ball) and position 5 (near peak point in flight) can be determined from photograph, graph, or table to be

$$v_{av} = \frac{\Delta s}{\Delta t} = \frac{s_5 - s_1}{t_5 - t_1} = \frac{(4.7 - 1.6) \text{ m}}{(0.84 - 0) \text{ s}} = 3.7 \, \frac{\text{m}}{\text{s}}$$

This is the average velocity for the entire upward motion. We have already noted that the ball begins with a high positive velocity and slows to a stop during this motion. Thus, the ball begins with a greater value than this and ends with a lower value. There is only one instant in the entire motion where the

Figure 2-8 Photocomposite of a tennis ball being thrown upward. The motion is essentially vertical. The decreasing separation of successive images after the ball leaves the player's hand indicates that its upward velocity is decreasing as the ball rises to its peak position. (D. Riban)

19

TABLE 2-1 Data from Figure 2-9a

Photo	Time, s	Height, m
1	0.0	1.7
2	0.21	3.1
3	0.42	4.0
4	0.63	4.6
5	0.84	4.7
6	1.05	4.3
7	1.26	3.6
8	1.47	2.4
9	1.68	0.9

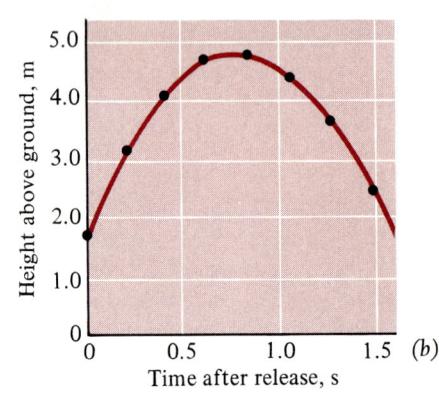

Figure 2-9 (a) The photographs used to produce Figure 2-8 spaced evenly along the x axis, allowing variation in vertical displacement to be viewed more easily. (D. Riban) (b) From these photographs, it is easy to abstract the information in the form of a graph of s vs. t. Displacements are measured directly from the photographs to two significant figures. The origin is chosen to be the line at the player's feet. Time intervals are 0.20 s.

instantaneous velocity of the ball is identical with this average velocity. We have no reason to believe that the instant we are interested in would happen to be the one and only time where $v_{av} = v_{inst}$; in fact, we have reason to believe this will *not* be the case. The instant considered is not the midpoint in time or in space of this upward motion and is rather closer to the end than the beginning. Our instant occurs between positions 3 and 4 in the photographs. We may obtain an average velocity for this interval of

$$v_{av} = \frac{\Delta s}{\Delta t} = \frac{s_4 - s_3}{t_4 - t_3} = \frac{(4.6 - 4.0) \text{ m}}{(0.63 - 0.42) \text{ s}} = 2.9 \frac{\text{m}}{\text{s}}$$

This value is quite different from that for the entire first half of the motion. Furthermore, since it considers only the motion during a small duration about the time of the instant being considered, we have reason to believe it is much closer to the value of the velocity at that instant — closer, but still not *the* instantaneous velocity for $t = 0.55$ s. However, comparison of these two averaged velocities gives us the clue we need. As the time interval Δt about the point we are considering decreases, our confidence that v_{av} is close to the value we seek increases. As Δt shrinks, Δs does also, and the value of the quotient $\Delta s/\Delta t$ as Δt approaches zero is the *instantaneous velocity* at our point in time. This may be done in steps, as we have just begun above, until Δt is small enough that we are confident that we are close enough to the value of v_{inst} for our purposes. It may also be done mathematically, where it is known as a limiting process and is one of the primary concerns of beginning calculus.

There is, however, a very practical way to obtain v_{inst} from our graph in Figure 2-9b. The slope of a line m is defined as $m = \Delta y/\Delta x$. Our graph has s graphed on the y axis and t graphed on the x axis, so its slope is $m = \Delta s/\Delta t$, which is just the expression for v_{av}. Of course, our graph is not a straight line and does not have a constant slope. How could it? We already know that v is constantly changing; thus the slope of the graph must be constantly changing. At any point the graph has a slope which is the same as a straight line tangent to the curve at that point. Figure 2-10 shows the curve of Figure 2-9b with a line constructed tangent to the point on the curve where $t = 0.55$ s. In the tiny zone about $t = 0.55$ s, the curve is growing as though it were a part of this line. Thus the slope of this line is the slope of the curve at $t = 0.55$ s. Taking the two arbitrary points a and b on the straight line, we may calculate its slope from their coordinates:

$$m = \frac{\Delta y}{\Delta x} = \frac{\Delta s}{\Delta t} = \frac{s_b - s_a}{t_b - t_a} = \frac{2.0 \text{ m}}{0.80 \text{ s}} = 2.5 \frac{\text{m}}{\text{s}} = v$$

This is the value of v_{inst} at $t = 0.55$ s, since it is the slope of the one and only line tangent to the curve at the instant considered. Note that the value is indeed closer to v_{av} calculated between points 3 and 4 than to the value calculated for the entire upward motion. Note also that while close to these values, it will not be the same as v_{av} for any finite interval except by chance.

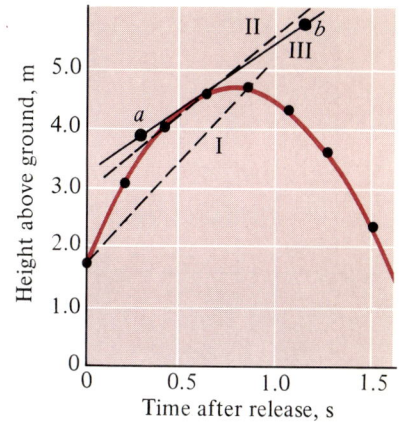

Figure 2-10 To determine v_{inst}, we obtain the slopes of three lines. The slope of line I is the average speed between the start and the peak of the motion v_{av}. The slope of line II is v_{av} for the interval 3–4. The only line yielding v_{inst} for $t = 0.55$ s is line III, the line tangent to the graph at $t = 0.55$ s, and we use points a and b on this line to determine its slope v.

Example 2-3 Table 2-2 shows the displacement of a racing stock car at various times after it left the starting line. Calculate its instantaneous velocity and speed at $t = 2.00$ s.

Solution We calculate the average velocity between 2.00 and 3.00 s, then between 2.00 and 2.20 s, then between 2.00 and 2.10, and so on (Table 2-3). We hope to be able to determine the value we are approaching as the time interval beyond 2.00 s approaches zero.

We can be fairly certain that the instantaneous velocity at $t = 2.00$ s is $+20.0$ m/s, since the values of v_{inst} approach closer and closer to this number as we decrease the size of the time interval about $t = 2.00$ s being considered. The instantaneous speed of the car at $t = 2.00$ s would be 20.0 m/s.

2-4 Uniform Acceleration

Objects in motion frequently have changing velocities; such objects are said to be **accelerating.** The definition of **acceleration** is the time rate of change of velocity:

$$a = \frac{v_2 - v_1}{t_2 - t_1} = \frac{\Delta v}{\Delta t} \qquad (2\text{-}6)$$

where a is the average acceleration of an object during the time interval $t_2 - t_1 = \Delta t$, and v_1 and v_2 are the instantaneous velocities at t_1 and t_2, respectively. (We are still dealing with motion along a straight line and will take direction into account with positive and negative signs.)

Consider an automobile which increases its velocity smoothly from 4.00 to 20.0 m/s during the 8.00-s interval immediately after the driver depresses the accelerator pedal. Its instantaneous velocity is plotted vs. time in Figure 2-11.

TABLE 2-2 Data for Example 2-3

Time, s	Displacement, m
2.00	+10.00
2.01	+10.20
2.02	+10.40
2.10	+12.05
2.20	+14.20
3.00	+35.00

TABLE 2-3

Time Interval	Δs, m	Δt, s	v, m/s
2.00 to 3.00	+25.00	1.00	+25.0
2.00 to 2.20	+4.20	0.20	+21.0
2.00 to 2.10	+2.05	0.10	+20.5
2.00 to 2.02	+0.40	0.02	+20.0
2.00 to 2.01	+0.20	0.01	+20.0

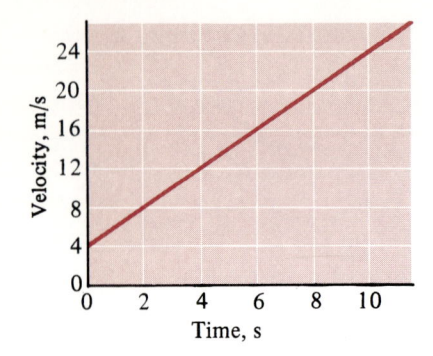

Figure 2-11 Velocity of an accelerating automobile vs. time.

We calculate the acceleration by using two corresponding pairs of coordinates from the graph. Choosing $t_1 = 0$ s and $t_2 = 8.00$ s, then $v_1 = +4.00$ m/s and $v_2 = 20.0$ m/s. Now Equation 2-6 yields

$$a = \frac{\Delta v}{\Delta t} = \frac{(20.0 - 4.00) \text{ m/s}}{(8.00 - 0) \text{ s}} = 2.00 \frac{\text{m}}{\text{s}^2}$$

In order to calculate the acceleration, we had to select two specific times with two specific associated velocities. The units of acceleration are those of velocity (length per unit time) divided by time. The automobile accelerating with a magnitude of 2.00 m/s each second is changing its velocity by 2.00 m/s for every second the acceleration continues. This could be written as 2.00 m/s/s, but it is common practice to write it as 2.00 m/s^2. Although Equation 2-6 defines only average acceleration, very often acceleration is *uniform* (constant), and thus instantaneous acceleration has the same value as the average acceleration. Unless otherwise stated, you may assume accelerations encountered in this book are uniform.

Problems involving motion contain various known quantities and require us to obtain different unknown values. It is useful to have a complete set of equations to cover any eventuality. Let us use the definitions of velocity and acceleration to obtain equations which describe the motion of a body moving on a straight line with a constant acceleration. Suppose we rewrite Equation 2-6 to be a little more general as

$$a = \frac{v - v_0}{t - t_0} \quad (2\text{-}7)$$

where v is now a variable velocity corresponding to time t and v_0 is the velocity when we started our measurement at t_0. We may also agree to start our clock at the beginning of the measurement, insisting that $t_0 = 0$. Then Equation 2-7 becomes

$$a = \frac{v - v_0}{t} \quad (2\text{-}8)$$

Solving for v, we have

$$v = v_0 + at \quad (2\text{-}9)$$

It is useful to recognize from the defining equations that the relationship between acceleration and velocity is mathematically the same as the relationship between velocity and displacement. Therefore, just as the velocity was equal to the slope of the displacement vs. time curve, acceleration is equal to the slope of the velocity vs. time curve. Note that Equation 2-9 is of the form $y = mx + b$. When the acceleration is constant, the graph of velocity vs. time is a straight line. Further, the average velocity of the object between time $t = 0$ when its velocity is v_0 and any time t when its velocity is v is

$$v_{av} = \frac{v + v_0}{2} \tag{2-10}$$

(You should be able to convince yourself that this is correct.)

Rearranging Equation 2-5 we have

$$s - s_0 = \Delta s = v_{av} t \tag{2-11}$$

Substituting for v_{av} from Equation 2-10, we obtain

$$\Delta s = \frac{v + v_0}{2} t$$

Substituting for v from Equation 2-9 gives us

$$\Delta s = \frac{(v_0 + at) + v_0}{2} t = \frac{2v_0 + at}{2} t$$

or

$$\Delta s = v_0 t + \tfrac{1}{2} a t^2 \tag{2-12}$$

If instead of substituting for v we substituted $(v - v_0)/a$ for t from Equation 2-8, then Equation 2-11 would become

$$\Delta s = \frac{v + v_0}{2} t = \frac{v + v_0}{2} \frac{v - v_0}{a}$$

A little algebra yields

$$2a\,\Delta s = v^2 - v_0^2 \tag{2-13}$$

These equations governing straight-line motion with constant acceleration are so useful that we collect them in Table 2-4.

2-4 Uniform Acceleration

TABLE 2-4 Equations Describing Motion at Constant Acceleration

$v = v_0 + at$	(2-9)
$v_{av} = \dfrac{v + v_0}{2}$	(2-10)
$\Delta s = v_{av} t$	(2-11)
$\Delta s = v_0 t + \tfrac{1}{2} a t^2$	(2-12)
$2a\,\Delta s = v^2 - v_0^2$	(2-13)

Example 2-4 A subway train is traveling at a speed of 30.0 m/s. Just as it passes a station mark, the brake is applied and the train slows with an acceleration of magnitude 4.00 m/s². Calculate (a) the time required to stop, (b) the distance the train travels from the station mark until it stops, and (c) the train's velocity 2.00 s after the brake is applied.

Solution If we choose the forward direction of the train as positive, then $v_0 = +30.0$ m/s, $v = 0$, and $a = -4.00$ m/s².
(a) Since we know v_0, v, and a and are looking for the value of t, we use Equation 2-9:

$$v = v_0 + at$$

or

$$t = \frac{v - v_0}{a} = \frac{0 - (+30.0) \text{ m/s}}{-4.00 \text{ m/s}^2} = 7.50 \text{ s}$$

Note that a negative acceleration in one-dimensional motion can imply *either* that the object is slowing while moving in the direction defined as positive *or* that the speed is increasing in the negative direction.

(b) We are seeking Δs and we know the values of v_0, v, and a, so we may apply Equation 2-13 as follows:

$$2a\,\Delta s = v^2 - v_0^2$$

23

$$\Delta s = \frac{v^2 - v_0^2}{2a} = \frac{0 - (30.0 \text{ m/s})^2}{2(-4.00 \text{ m/s}^2)} = \frac{-900 \text{ m}^2/\text{s}^2}{-8.00 \text{ m/s}^2} = 113 \text{ m}$$

We could also have used Equation 2-12. Can you think of any reason for choosing Equation 2-13 instead?

(c) Now we must apply Equation 2-9 between the time when the brake is applied and 2 s later. We still have $v_0 = +30.0$ m/s and $a = -4.00$ m/s², but we are trying to determine v at $t = 2.00$ s. Thus

$$v = v_0 + at = 30.0 \text{ m/s} + (-4.00 \text{ m/s}^2)(2.00 \text{ s}) = 22.0 \text{ m/s}$$

2-5 The Acceleration due to Gravity

Galileo maintained that objects fall with constant acceleration. To test Galileo's assertion, we can drop an object to measure its motion. Figure 2-12 is a photocomposite of a bowling ball being dropped from a seventh-floor window to the ground below. The ball has white stripes painted on it for improved visibility but is otherwise standard. A repeating camera capable of taking about three photographs per second recorded the individual pictures shown combined in Figure 2-12. Each photograph clearly defines the position of the ball for that instant. We may measure the motion of the ball by counting the number of bricks the ball has traveled between pictures. Alternatively, we may measure the displacement of the ball on the photograph with a ruler and translate this into meters, given that the distance between the windowsills of adjacent floors is 3.15 m. You are encouraged to verify the numbers in Table 2-5 from the photograph.

To determine the duration between photographs to three significant figures, a precision phonograph turntable adjusted to $33\frac{1}{3}$ revolutions per minute (r/min) was photographed sequentially with the same camera settings (Figure 2-13). The movement of the arrow on the turntable defines its rotation between pictures. Comparison of the first and seventh images indicates that the turntable moves through slightly more than one rotation (366°) in the six intervals represented. Since $33\frac{1}{3}$ turns per minute is 1.80 s per turn, the interval between photographs may be calculated as 0.305 s. For convenience we choose the origin of coordinates at the point of release and choose downward as positive.

Table 2-5 gives us the time from release t and the change in displacement Δs from the starting position. We select Equation 2-12 and divide all terms by t. This yields

$$\frac{\Delta s}{t} = v_0 + \tfrac{1}{2}at$$

But $\Delta s/t$ is, by definition, v_{av}. Thus we have

$$v_{av} = \tfrac{1}{2}at + v_0$$

This has the form $y = mx + b$, so that a graph of average velocity as the y variable vs. time as the x variable will yield a straight-line graph provided that the acceleration is constant. If so, the slope of this graph should be one-half the acceleration of the bowling ball. To calculate v_{av} from the data in Table 2-5, we simply divide the total displacement shown in a particular photograph by the

Figure 2-12 A photocomposite allows the motion of a falling ball to be studied in great detail. *(D. Riban)*

TABLE 2-5 Data Obtained from Figure 2-12

Photo	Time, s	Δs, m
1	0.000	0.00
2	0.305	+0.61
3	0.610	+1.89
4	0.915	+4.11
5	1.22	+7.15
6	1.53	+11.2
7	1.83	+16.4

Figure 2-13 A series of photographs of a phonograph turntable turning at a rate of $33\frac{1}{3}$ r/min made with the camera producing Figure 2-12. The motion of the arrow allows us to determine the time between photographs. (*D. Riban*)

elapsed time for that photograph. Figure 2-14 graphs these values of v_{av} vs. the elapsed time. The curve is indeed a line with constant slope. The value of this slope is calculated on the graph to be 4.90 m/s². Since the slope is one-half the acceleration, we may conclude that striped bowling balls fall from dormitories in western Pennsylvania at a constant acceleration of 9.80 m/s². Of course, we could just as easily have chosen to drop an anvil, a 2-lb box of lollipops, or some other object. Figure 2-15 shows the simultaneous drop of the same bowling ball and a smaller ball in the laboratory. This suggests that there is nothing unusual in the motion of the bowling ball. We conclude that unsupported objects near the earth's surface fall with a constant acceleration of 9.80 m/s². We call this acceleration the **acceleration due to gravity,** indicated by the symbol g, and have measured it to be downward and of magnitude 9.80 m/s².

Example 2-5 A ball is thrown vertically upward from the edge of a tower 50.0 m above the earth with an initial speed of 30.0 m/s. It slows, reaches a high point, stops, and returns, just missing the tower, and finally falls to earth. Assuming upward is positive and $a = -g = -9.80$ m/s², calculate (*a*) the time required for the ball to reach its highest point, (*b*) its highest displacement above the earth, (*c*) the velocity when it strikes the ground, and (*d*) the total time elapsed as it strikes the ground.

Solution (*a*) Since we know the ball's initial velocity (at $t = 0$), the acceleration, and its velocity at the time in which we are interested (certainly $v = 0$ at

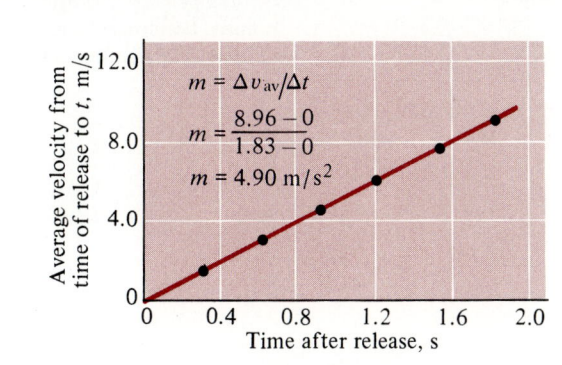

Figure 2-14 The average velocity of the falling bowling ball vs. time calculated from the data table for Figure 2-12. The graph is indeed a straight line, indicating that the acceleration was constant. The slope of the graph is measured to be 4.90 m/s².

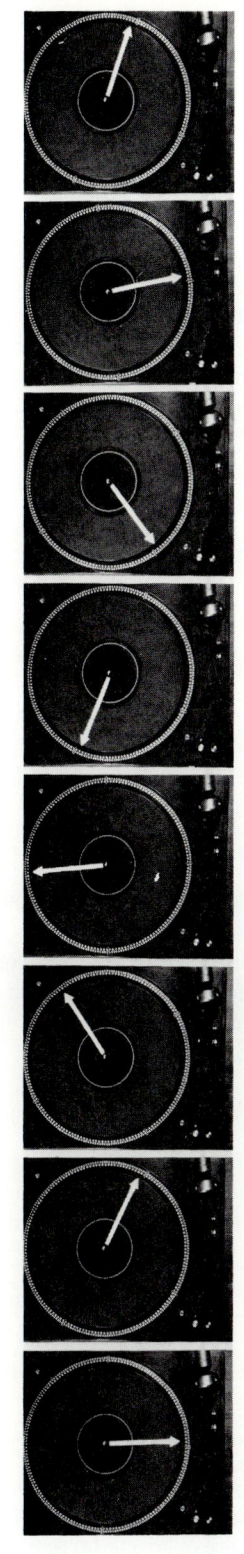

$$m = \Delta v_{av}/\Delta t$$
$$m = \frac{8.96 - 0}{1.83 - 0}$$
$$m = 4.90 \text{ m/s}^2$$

25

the highest point!), we should apply Equation 2-9 between the point of release and the highest point. Thus

$$v = v_0 + at$$

$$t = \frac{v - v_0}{a} = \frac{0 - (+30.0 \text{ m/s})}{-9.80 \text{ m/s}^2} = 3.06 \text{ s}$$

(b) We could use either Equation 2-12, since we now know the time to the highest point, or Equation 2-13 to determine Δs. It is customary to use the equation or method which depends on the fewest calculated quantities (none if possible) and which relies more on measured or given values. This way, an error made in an early calculation does not make all later answers to parts of the problem incorrect also. Thus we choose to apply Equation 2-13 between the release and the highest point, obtaining

$$\Delta s = \frac{v^2 - v_0^2}{2a} = \frac{0 - (30.0 \text{ m/s})^2}{2(-9.80 \text{ m/s}^2)} = 45.9 \text{ m}$$

Thus the highest displacement above the earth is

$$50.0 \text{ m} + 45.9 \text{ m} = 95.9 \text{ m}$$

(c) Application of Equation 2-13 again is appropriate since we know v_0, a, and the final value of Δs, -50.0 m, as the ball hits the ground. Rearranging, we obtain

$$v^2 = 2a\,\Delta s + v_0^2 = 2(-9.80 \text{ m/s}^2)(-50.0 \text{ m}) + (30.0 \text{ m/s})^2 = 1880 \text{ m}^2/\text{s}^2$$

Now we note that both $+43.4$ and -43.4 are valid mathematical solutions for v. Of course, we will choose

$$v = -43.4 \text{ m/s}$$

as the velocity when it strikes the ground, since we know it is headed in the negative direction at impact.[1]

(d) In keeping with the attempt to use only measured or given information, we select Equation 2-12 and apply it between the point of release and the point where the ball strikes the earth to determine the total time:

$$\Delta s = v_0 t + \tfrac{1}{2} a t^2$$

$$-50 \text{ m} = (30 \text{ m/s})t + \tfrac{1}{2}(-9.8 \text{ m/s}^2)t^2$$

Dropping units and rearranging to emphasize the form of the equation, we see that we have a quadratic equation with t as the variable: $4.9t^2 - 30t - 50 = 0$. This is in the form which can be solved by utilizing the quadratic formula (see Appendix A):

$$t = \frac{+(30 \text{ m/s}) \pm \sqrt{(-30 \text{ m/s})^2 - 4(4.9 \text{ m/s}^2)(-50 \text{ m})}}{2(4.9 \text{ m/s}^2)}$$

$$= +7.49 \text{ s} \quad \text{or} \quad -1.37 \text{ s}$$

Figure 2-15 When two objects of different mass are dropped together, they appear to have the same displacement vs. time behavior. Despite the difference in mass, both objects accelerate downward at the same rate. (C. Riban)

---

[1] It is interesting to note, however, that $+43.4$ m/s has possible physical meaning also. It is the initial velocity that the ball would have needed in order to reach the same height had it been projected upward from the ground. It is often true that both roots of a quadratic equation have *possible* physical meaning. When that happens, one must consider the real physics of the situation in order to select the correct solution.

Note that each time an addition or subtraction was called for, the terms being combined had identical units. Also, the final answer is in seconds, which are indeed units of time. Again, since we are looking for time *after* we started the clock, we select the positive root, $t = +7.49$. (Can you think of what possible physical meaning $t = -1.36$ s has?)

For practice, and to verify these answers, try to solve (c) and (d) by applying one or more of the other equations.

SPECIAL TOPIC
Problem Solving in Physics

A class in physics inevitably means spending considerable time solving problems. Experience indicates that this is the most effective way to learn a quantitative subject like physics. Furthermore, problem solving is almost the only valid way to demonstrate that you really have learned the subject. Since this problem solving is inevitable, it is a good idea to begin correctly, with good habits and strategies to improve your likelihood of attaining the correct answer.

We *assume* that you have read the relevant material and are familiar with the salient definitions and relationships. Lacking this knowledge, even if *by chance* you stumble on the correct equations and apply them to yield the appropriate answer, you will have gained little, if any, understanding from working the problem. That, in turn, will be very little help on a test of the material.

Standard Steps to Solving a Physics Problem

1. *Read the problem* very carefully for a good understanding. Most instructors are amazed by the number of problems that students miss because they have not read key conditions of the problem.
2. *Make a sketch* of the physical situation if at all applicable. Include a coordinate system of your choice, selecting both a convenient origin and positive directions for coordinates.
3. *Select symbols* for every quantity you know. Place them on your sketch if applicable and alongside otherwise. Do not forget any known quantities which apply to the problem but are unstated.
4. *Convert units* to a single system. Mixed units like centimeters and kilometers in the same problem can cause difficulties.
5. *Select symbols* for unknown quantities which you are trying to find. Place them on your sketch with a question mark.
6. *Determine* the definitions, laws, and relationships which apply to this problem. Write them down in equation form if necessary.
7. *Use the equations* and your knowledge of algebra and trigonometry to solve for unknowns. This can, on occasion, require much work — for example, obtaining two equations in two unknowns and solving them simultaneously.
8. *Check your answer* for arithmetic and substitutions, for whether the answer is reasonable, and for the appropriate *unit* resulting from the calculation.

Solving a Problem: An Example

Let us apply these steps to a problem appropriate for this chapter.

Problem The Sears Tower in Chicago, the world's tallest office building, is 320 m high. The observatory elevators take visitors from ground level to the top floor in 90.0 s. Assume the elevator accelerates through the first 30.0 m of the climb then moves at constant speed until it slows with the same magnitude acceleration (but in the opposite direction) through the last 30.0 m to come to rest at the top. (a) What is the magnitude of either acceleration of the elevator? (b) What is its maximum speed?

Please note that this is a difficult problem. We have specifically chosen it to illustrate as many phases of problem solving as is convenient.

A reading of the problem (step 1) yields the height of the building, the distance through which final and initial accelerations act, and the total time of the trip. We can see that there are three separate phases of the motion: start-up, coasting at constant v, slowdown. We choose upward as positive for each of the three phases.

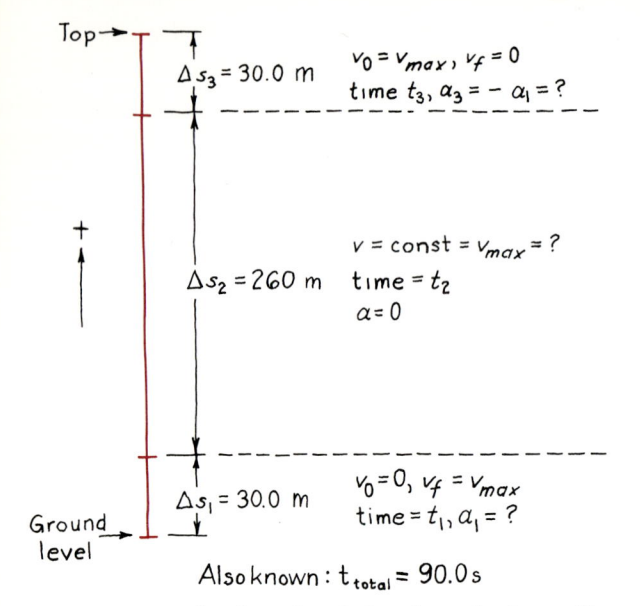

Figure 2-16 A sketch made to help solve a physics problem.

This means that a_1 is a positive number and $a_3 = -a_1$ is negative. We could draw the situation as in Figure 2-16 (step 2). We have labeled the figure with known information (step 3). The problem states that the elevator speeds up for 30.0 m, slows down at the top for 30.0 m, and that the total height is 320 m. Therefore the constant-velocity phase is $320 - 60$ m, or 260 m, which we indicated on the drawing. All values are given in standard SI units, so no conversions are required (step 4).

We are looking for the acceleration during the initial part of the motion, which we have labeled a_1. The second unknown we are seeking is the maximum speed, which we label v_{max}. We note that this is reached at the end of the initial acceleration phase and is the speed at which the central portion of the motion is conducted, and we so label the figure (step 5).

Each of the three phases is one-dimensional motion with constant acceleration. Therefore we recognize that any or all of Equations 2-9 through 2-13, previously reproduced in Table 2-4, are fair game for the solution of this problem and may be applied in any combination to each phase of the problem (step 6). Now on to the solution (step 7). Examination of the conditions known and the relationships reveals no strategy for a single-step solution of the problem. We note that one more bit of information is known. The total time is 90.0 s, and thus we write

$$t_1 + t_2 + t_3 = 90.0 \text{ s} \quad (1)$$

Working Strategy If we are able to obtain expressions for each of these three times in terms of the magnitude of the initial acceleration a_1, we will, in principle, be able to solve Equation 1 for a_1.

Applying Equation 2-12 to the start-up phase, we have (since $v_0 = 0$)

$$\Delta s_1 = 0 + \tfrac{1}{2} a_1 t_1^2$$

or since $\Delta s_1 = 30.0$ m, we obtain

$$t_1^2 = \frac{2(30.0 \text{ m})}{a_1}$$

or

$$t_1 = \sqrt{\frac{60.0 \text{ m}}{a_1}} \quad (2)$$

Now during the constant-velocity phase, we have (since the acceleration during this phase is zero)

$$\Delta s_2 = v_2 t_2 = v_{max} t_2$$

or

$$t_2 = \frac{\Delta s_2}{v_{max}} = \frac{260 \text{ m}}{v_{max}} \quad (3)$$

This must be further reduced because we do not yet know the value of v_{max}. What we do know is that v_{max} is the final velocity of the first phase of motion. Thus an expression for v_{max} in terms of a_1 may be obtained by applying Equation 2-13 to the first phase. Since $v_0 = 0$ for the first phase, we have

$$2a_1 \Delta s_1 = v_{max}^2 - 0$$

or

$$v_{max} = \sqrt{2a_1 \Delta s_1} = \sqrt{2a_1(30.0 \text{ m})}$$
$$= \sqrt{a_1(60.0 \text{ m})}$$

Substituting this equation into Equation 3 yields

$$t_2 = \frac{260 \text{ m}}{\sqrt{a_1(60.0 \text{ m})}} \quad (4)$$

Now applying Equation 2-9 for the acceleration in the slowdown phase gives

$$0 = v_{max} + a_3 t_3$$

or

$$a_3 = -\frac{v_{max}}{t_3}$$

but since $a_3 = -a_1$, we have

$$-a_1 = \frac{-v_{max}}{t_3}$$

or

$$t_3 = \frac{v_{max}}{a_1}$$

28

But we found the expression for v_{max} above to be $\sqrt{a_1(60.0 \text{ m})}$. Thus

$$t_3 = \frac{\sqrt{a_1(60.0 \text{ m})}}{a_1} = \sqrt{\frac{60.0 \text{ m}}{a_1}} \quad (5)$$

That this expression is identical to that for t_1 (Equation 2) is not really surprising. After all, the two accelerations were equal in magnitude. The magnitude of the change in velocity in bringing the elevator up to a speed v_{max} from rest is equal to the magnitude of the change in velocity in slowing the elevator from v_{max} to rest. Further, the distance traveled in the speeding up was identical to that in the slowing down. Thus Equation 2-13 dictates that the magnitudes of the accelerations were identical. The symmetry further dictates that t_3 is equal to t_1.

Now substituting the expressions from Equations 2, 4, and 5 into Equation 1 yields

$$\sqrt{\frac{60.0 \text{ m}}{a_1}} + \frac{260 \text{ m}}{\sqrt{a_1(60.0 \text{ m})}} + \sqrt{\frac{60.0 \text{ m}}{a_1}} = 90.0 \text{ s} \quad (6)$$

These fractions may be added if we find a common denominator, $\sqrt{a_1(60.0 \text{ m})}$ in this case. Multiplying the first and third terms by $\sqrt{60.0 \text{ m}}/\sqrt{60.0 \text{ m}}$, we obtain

$$\frac{60.0 \text{ m}}{\sqrt{a_1(60.0 \text{ m})}} + \frac{260 \text{ m}}{\sqrt{a_1(60.0 \text{ m})}} + \frac{60.0 \text{ m}}{\sqrt{a_1(60.0 \text{ m})}}$$

$$= \frac{380 \text{ m}}{\sqrt{a_1(60.0 \text{ m})}}$$

$$= 90.0 \text{ s}$$

thus

$$\sqrt{a_1(60.0 \text{ m})} = \frac{380 \text{ m}}{90.0 \text{ s}} = 4.22 \text{ m/s}$$

$$a_1(60.0 \text{ m}) = (4.22 \text{ m/s})^2$$

$$a_1 = \frac{1}{60.0 \text{ m}}(4.22 \text{ m/s})^2 = 0.297 \text{ m/s}^2 \quad (7)$$

For part (b), we use the intermediate step above that

$$v_{max} = \sqrt{a_1(60.0 \text{ m})}$$
$$= \sqrt{(0.297 \text{ m/s}^2)(60.0 \text{ m})} = 4.22 \text{ m/s} \quad (8)$$

Again, be aware that this is a very difficult problem for a beginner. You should also be aware, however, that this detailed analysis is necessary even for the experienced physicist to work an involved problem.

Checking Your Answer

At this point, we have the responsibility of convincing ourselves that the answers are likely to be correct. Are the answers consistent and/or reasonable? There are at least four checks we could make.

1. Go back through and check *substitutions and arithmetic*. Many students have an urge to kick themselves after most tests for something careless in this line done in the haste of the moment.

2. *Think* — Use your head! Do you remember anything in your experience that leads you to believe that your answer is possible (or ridiculous)?

   For example, suppose you had misplaced a decimal point in the calculation and obtained an acceleration for the elevator of 29.7 m/s². Would you recognize this as being ridiculous? Remember the acceleration due to gravity is only about 9.80 m/s². Thus if 29.7 m/s² were the correct answer, the elevator would have to accelerate at a rate more than three times that due to gravity. The elevator would have to apply very large forces on its contents. Consider the downward trip. If you were a passenger, something in the elevator (the roof of the cab?) would have to actively push you downward during the acceleration with a force equal to more than twice your weight. Would you ride that elevator again? Survive?

3. Are the dimensions of your quantities consistent? Make sure the units of the answer are consistent with the quantity it represents. Make sure the units of quantities in your substitution actually do yield the appropriate units of the answer.

   The correct units for the unknown quantity must emerge from our equations, or we have done something wrong. For example, in Equation 6 above, every term on the left has units of $m/\sqrt{m}$ or $\sqrt{m}$. Solving for a in Equation 7 yielded units of m/s² only after squaring both sides of the equation. Those indeed were acceleration units emerging from the substitutions, which increases our confidence in the procedures used and thus in our answer.

4. Finally, consider what would happen if you changed some important quantities in the problem. It frequently helps to take the extreme case and have some quantity approach zero or infinity to see if the mathematics and your answer behave as expected.

   For example, extreme values are not appropriate to this problem, but some speculation can convince us that the answer is reasonable. What if the elevator accelerated halfway up, then decelerated the rest of the way? It could certainly use a smaller value of accelera-

29

tion to accomplish the same trip in the same total time. We recognize that the elevator would again start from rest and accelerate for half the distance, 160 m, for half the total time, 45.0 s. Thus, applying Equation 2-12 for the first half of the journey yields

$$a = \frac{2(160 \text{ m})}{(45 \text{ s})^2} = 0.158 \frac{\text{m}}{\text{s}^2}$$

Since this is smaller than our value but near the same magnitude, our confidence in our answer is increased.

Remember, problem solving is a skill which must be developed. Practice is the only method to develop and improve a skill. Thus, you should work on as many problems as time will permit. No one denies that problem solving can be very, very frustrating at times — your instructor, your authors, your fellow students have all been there. Yet, emerging with the correct solution can also be very personally rewarding, and it really is the only way to master a subject.

Minimum Learning Objectives

After studying this chapter, you should be able to:
1. Define:
   - acceleration
   - acceleration due to gravity
   - average speed
   - average velocity
   - displacement
   - distance traveled
   - dynamics
   - instantaneous speed
   - instantaneous velocity
   - kinematics
   - speed
   - time rate of change
   - velocity
2. Explain what a unit like meters per second squared (m/s²) means.
3. Explain how to determine the instantaneous velocity from a graph of displacement vs. time.
4. Determine the position of an object at each instant, given its initial velocity and acceleration.
5. Calculate change in displacement during a time interval by using a given initial velocity and constant acceleration.
6. Find how long an object has to move to arrive at a given position, knowing starting position, velocity, and acceleration.
7. Determine the acceleration of an object, given its position in space at specified instants in time.
8. Read a situation involving motion stated in normal English, and translate the information given into mathematical conditions and statements (a critical skill underlying all future success in physics).

Problems

Throughout this text, the problems are of three levels of difficulty: Standard, somewhat difficult (indicated by a boldface black number preceding the problem), and most difficult (indicated by a colored number).

2-1 An outstanding marathon runner may complete the course of 26 mi 385 yd in 2 h 30 min. Determine the runner's average speed in meters per second.

2-2 In what time interval must a runner complete a 10,000-m race to accomplish an average speed of 7.50 km/h?

2-3 A radioactive fluid is injected into a tree trunk at ground level. In exactly 2 h 6 min a detector 35.0 m up the trunk detects the first arrival of radioactive material. What is the average speed of fluid moving up the tree?

2-4 The professor eats a double onion sandwich and goes to his lecture. At 1.40 min after he enters the room, a student in the back row some 22.6 m from the lecture podium detects the odor of onions. What is the speed of diffusion of onion molecules in the air?

2-5 There are 45 million children in the United States under the age of 13 years. The average distance between children is 480 m. Santa Claus is to visit all these children in the time between 6 P.M. and 6 A.M. EST. Assuming Santa spends zero time at each destination, what must be his average speed?

2-6 Light travels at a speed of about 3.00×10^8 m/s. At that rate, how far will light travel (a) in 1 day and (b) in 1 year? (c) How long would it take a radio signal (traveling at the speed of light) to reach a spaceship which was dispatched to Alpha Centauri, the star nearest to the sun, about 3.97×10^{16} m away? Convert your answer to days and years.

2-7 The earth turns on its axis once each 24 h. The circumference of the earth at the latitude of Cape Canaveral is 3.53×10^4 km. (a) What is the speed, due to the earth's rotation, of a stationary object at Cape Canaveral? (b) To attain a circular orbit, an object at 180 km above the surface must attain a speed of 2.81×10^4 km/h perpendicular to the line joining it to the center of the earth. Why are all the rockets launched from Cape Canaveral launched to the east? (Two reasons, one of which would not be true of a launch site in California.)

2-8 One day at noon, four vacationing students touring the United States find themselves at Wichita, Kansas. Their map shows that this is exactly 2.10×10^3 km east of San Francisco, California, which is their destination for a concert at noon 3 days later. The map shows the road distance to San Francisco to be 2.75×10^3 km. (a) Determine the minimum average velocity which they must travel in order to be in the audience at the start of the concert. (b) Determine their minimum average speed. (c) If they can average 45.0 km/h on the road, when will they arrive?

2-9 During a speedway race, car 1 makes a pit stop, and upon completion the driver finds that she is four laps behind the leader, car 2. The laps are 0.750 km. Car 2 maintains an average speed of 120 km/h, and car 1 can maintain an average speed of 135 km/h. (a) How long will it take car 1 to overtake car 2? (b) What distance will each car travel during that time interval?

2-10 What is the average acceleration in meters per second squared of an automobile which increases its speed from 5.00 to 20.0 m/s in (a) 10.0 s, (b) 5.00 s, and (c) 3.00 s?

2-11 An airplane increases its speed from 350 to 600 km/h in 40.0 s. (a) Calculate the magnitude of its average acceleration. (b) How far does the airplane travel while it is accelerating?

2-12 The TGV train in France can increase its speed from 90.0 to 180 km/h in 30.0 s. Calculate (a) the average speed during this 30.0 s if the acceleration is constant, and (b) the train's speed at 15.0 s.

2-13 An astronaut wishes to slow the vehicle from 200 to 10.0 m/s in order to make a soft landing on Mars. The retro-rockets can impart a net acceleration of magnitude 7.50 m/s². (a) How long should the retro-rockets be fired? (b) How far will the vehicle travel during this time?

2-14 A cosmic ray moving at a speed of 2.00×10^8 m/s is brought to rest in a large tank of liquid in 5.00×10^{-2} m. Calculate (a) the magnitude of acceleration and (b) the time required to stop.

2-15 A student riding a bicycle passes the entrance to the Student Union building at a speed of 12.0 m/s. As she passes the Science building 180 m further on, her speed is 6.00 m/s. Calculate (a) her average acceleration, (b) the time required if her acceleration was constant, and (c) the average speed by two different methods.

2-16 Measurements indicate that the northern Baltic Sea has risen 100 m in the last 6000 years, after the removal of the weight of a glacier several miles thick by melting. The current rate of uplift is 110 cm/century. Could the uplift have occurred at a constant rate over the six millenia?

2-17 The archaeological team excavating the El Kibur mound in the Middle East has a working season of 3 months. The layer of particular interest to the team dates to 1800 B.C. and is located 6.40 m below the surface. Their untrained

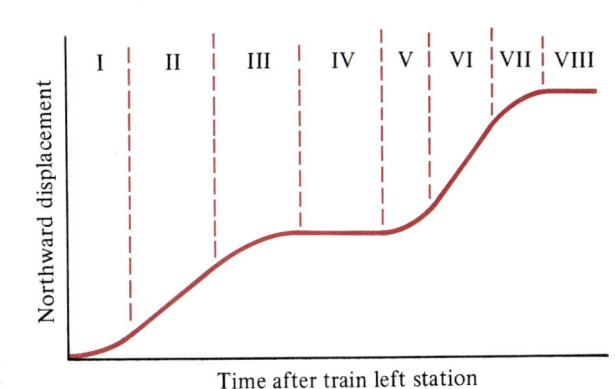

Figure 2-17 Problem 2-18.

team can carefully excavate at an average rate of 1.00 cm/day initially. At what rate must the training accelerate the average excavation rate over the first 2 months to leave the last full month available for detailed attention to the layers of interest?

2-18 Figure 2-17 represents the northward horizontal displacement of a train. In which segment(s) of the curve is the train (a) increasing speed, (b) decreasing speed, (c) stopped, and (d) moving with constant speed? (e) Which segment shows the greatest speed?

2-19 Astronauts who have landed on the moon must hope that their rockets will give them an acceleration of 10.8 m/s² for 220 s in order that they escape from the moon. (a) What is the escape velocity of the moon? (b) How high above the moon's surface will they be when the rockets cease burning?

2-20 A speed of 110 m/s is required in order that a large jet-propelled passenger aircraft be able to take off. During the takeoff phase, a large passenger plane accelerates at an average rate of 3.80 m/s². Calculate (a) the minimum length that the runway must be and (b) the minimum time from the start of the aircraft's motion until it is airborne.

2-21 An airplane landing on an aircraft carrier touches down at a speed of 60.0 m/s and is brought to rest in 120 m. Calculate (a) the average acceleration and (b) the time required to stop.

2-22 A block of wood is released from rest at one end of a wooden surface which is inclined at an angle to the horizontal. The surface is 4.00 m long, and the block moves with a constant acceleration of 0.250 m/s². Calculate (a) the time required for the block to complete the 4.00 m and (b) the speed of the block at the bottom of the incline.

2-23 During the Late Jurassic Period of geological time some 100 million years ago, South America was connected to Africa. Currently, South America is moving 5.60 cm/yr due westward from Africa. (a) If the change occurred at the end of the Jurassic Period and South America has maintained a constant speed since then, what should the average width of the Atlantic Ocean be? (b) If South America began accelerating uniformly away from Africa at the end of the

Jurassic Period and maintained this acceleration to the present, how wide would the Atlantic Ocean be? (c) The South Atlantic Ocean is 6000 km wide on average. Which of the two situations above is more likely?

2-24 An automobile starts from rest on a straight level road and accelerates to a speed of 30.0 m/s in 8.00 s. (a) Calculate the magnitude of its acceleration. (b) If its acceleration were 5.00 m/s², how many seconds would it take to reach a speed of 30.0 m/s? (c) How far would the automobile travel in each case while accelerating to 30.0 m/s?

2-25 A certain subway train typically leaves a station with an acceleration of 1.20 m/s² until it reaches a speed of 30.0 m/s. It holds this speed for 75.0 s and then begins to slow down with an acceleration of −1.20 m/s² until it stops at the next station, where it remains loading and unloading for 25.0 s. If it loads and unloads an average of 165 people at each stop, how many people does this train service in an hour?

2-26 An automobile starts at an intersection with a constant acceleration of 2.50 m/s². At the same time, a truck at a constant velocity of 15.0 m/s passes the automobile. Calculate (a) the time when the automobile will catch the truck and (b) the distance from the point where the automobile started in which it will catch the truck.

2-27 With maximum braking on a dry concrete roadway, an automobile can attain a stopping acceleration of 4.50 m/s² (opposite to the direction of its velocity). Assume you are traveling at 25.0 m/s (55.9 mi/h) and suddenly notice a stationary automobile 80.0 m ahead of you. (a) Assuming your reaction time is 0.300 s, can you stop before you collide? If so, how far from the stationary automobile will you stop? (b) Suppose your speed is 40.0 m/s (89.5 mi/h) and the stationary automobile is 160 m away when you spot it. Will you now be able to stop before you collide?

2-28 Figure 2-18 shows the height of the surface of North Bay, Lake Huron, vs. time from about 8600 B.C. to the present. The origin of coordinates is taken to be approximately 10,500 years ago at 1000 ft below the present surface level. Use data from this graph (read points directly from the curve), and assume a constant acceleration. (Be careful:

The rate of change of velocity of the surface is negative; thus the acceleration is a negative number.) Calculate (a) the average velocity of the surface during the first few hundreds of years, (b) the average velocity during the last few thousands of years, and (c) the acceleration of the surface.

2-29 A ball is dropped from the top of a building 25.0 m high. Calculate (a) the instantaneous speed with which the ball strikes the ground and (b) the time from the instant of release until the ball strikes the ground.

2-30 How far will an object fall in a vacuum near the earth's surface in (a) 1.00 s, (b) 2.00 s, and (c) 10.0 s?

2-31 An air rifle is fired upward, and the pellet has an initial speed of 60.0 m/s. Neglecting air resistance, calculate the pellet's (a) position at 2.00 s, (b) maximum height above the rifle, and (c) time to reach its maximum height.

2-32 Pacific salmon are capable of a leap of 2.50 m straight up out of the water. Calculate (a) the initial velocity necessary for this performance and (b) the time a salmon must spend out of the water.

2-33 The archvillain Zarg drops a vial of nitroglycerin from a balcony 18.0 m above a crowded street. If Superhero is 12.0 m from the point of impact, how fast will he have to run to catch the vial before it hits?

2-34 A softball is thrown straight upward and returns to its original position in exactly 4.00 s. Determine (a) its initial velocity and (b) the maximum height it attained.

2-35 Galileo rolled balls down ramps to weaken the acceleration due to gravity. Figure 2-19 is a stroboscopic photo of such an experiment. The ball is shown every 0.0650 s, and the scale shows the distance the ball travels. Use this photograph to build a table of observed and calculated data. Head columns "distance traveled," "time," and "average speed." Note that these are from the first point considered and the ball was not necessarily at rest. That is, v_0 is not necessarily zero. (a) Plot average speed vs. time. You should obtain a straight line. (b) From the slope and intercept of the line, determine the acceleration and the initial speed of the ball. Knowing the ramp is at an angle of 10.0°, can you determine the acceleration due to gravity?

2-36 A student is in an elevator which is moving upward at a speed of 10.0 m/s. Her calculator slides off the top of her physics book which is being held 1.30 m above the floor. Calculate the distance the elevator moves before the calculator hits the floor.

2-37 A stone is dropped from rest down an elevator shaft from the tenth floor of a building exactly 34.0 m above the top of the elevator, which is in the basement. At the same instant, the elevator begins accelerating upward from rest at 3.00 m/s². Calculate (a) the time at which the stone hits the top of the elevator, (b) the distance the elevator ascends before impact, and (c) the speed of both the stone and elevator when they collide.

2-38 Two physics students are tossing stones off a very high bridge. At one point, one student drops a stone. Exactly 2.00 s later, the other student throws a second stone down-

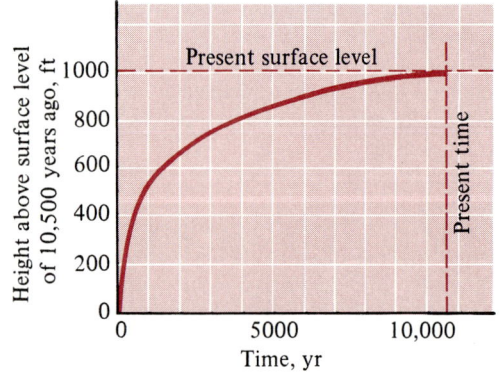

Figure 2-18 Problem 2-28.

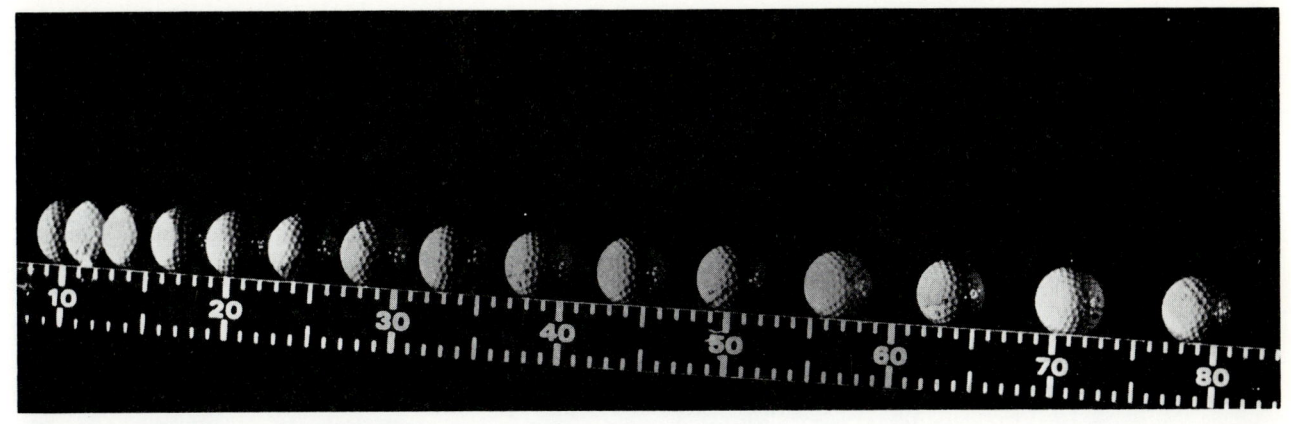

Figure 2-19 Problem 2-35. *(C. Riban)*

ward with an initial speed of 30.0 m/s. The two stones strike the water below at the same instant. Calculate (a) the time each stone required to reach the water and (b) the height of the bridge above the water.

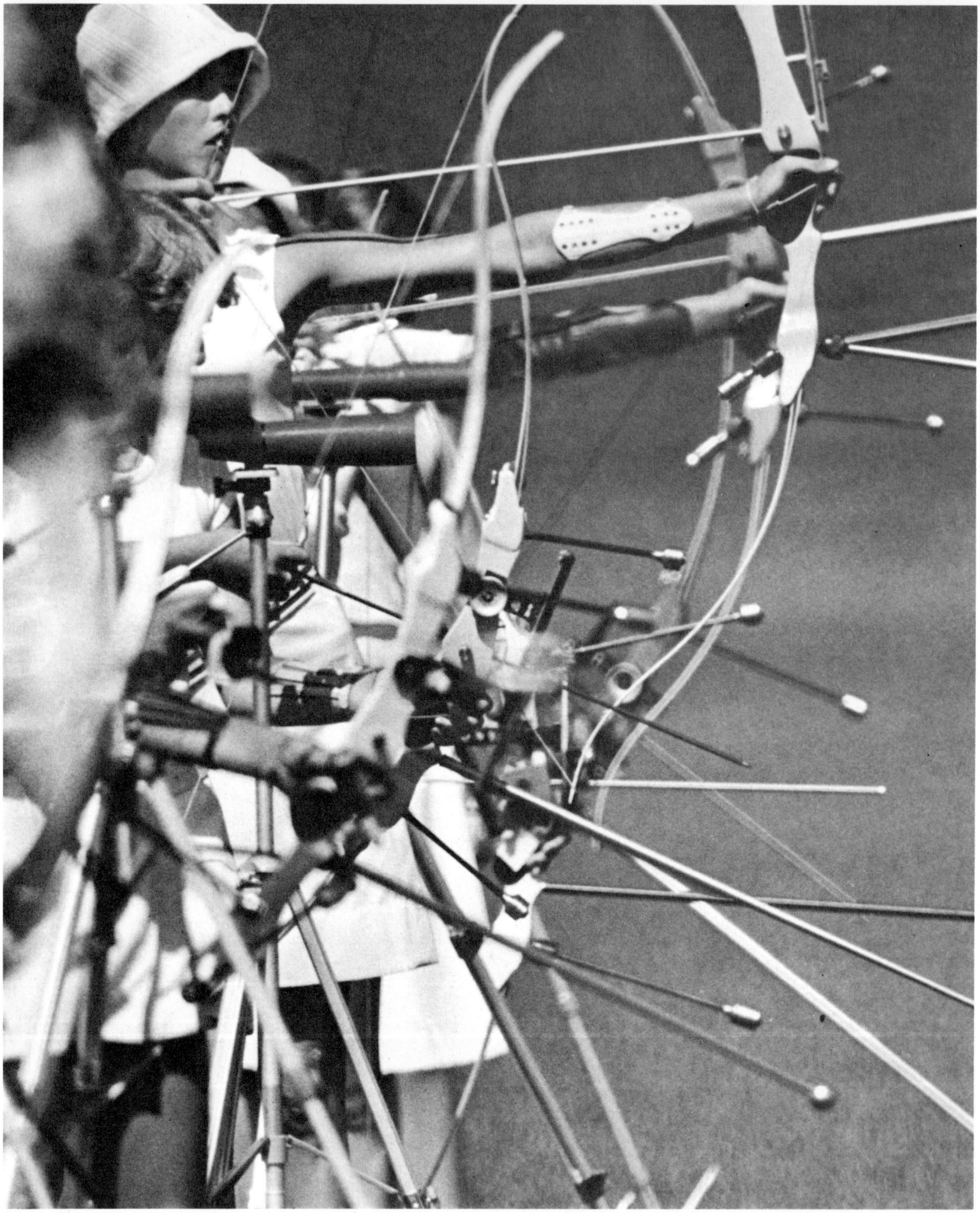

3 Vectors and Multidimensional Motion

3-1 Introduction

An ant crawling about on a tabletop is moving in two dimensions, while a fly buzzing around moves in three dimensions. To treat situations such as these, it is necessary to describe directional properties in more than one dimension. Physical quantities are generally divided into two classifications:

1. **Scalar quantities** are specified by stating a single number or magnitude (in addition to their units). Scalars have magnitude only, as, for example, mass, distance, time, and temperature.
2. **Vector quantities** have direction as well as magnitude; thus we must state both a magnitude and a direction for each vector. Examples are displacement, velocity, and acceleration.

   Neglecting the vector nature of a quantity can lead to serious complications. Walking three blocks north has very different physical consequences than walking three blocks east if there is any objective to the walk other than getting exercise. When a body is moving in one dimension, we can specify the direction simply by designating one direction as positive and the opposite direction as negative. It is not quite this easy when the object moves in two or three dimensions. We have to learn a few rules to handle vectors.

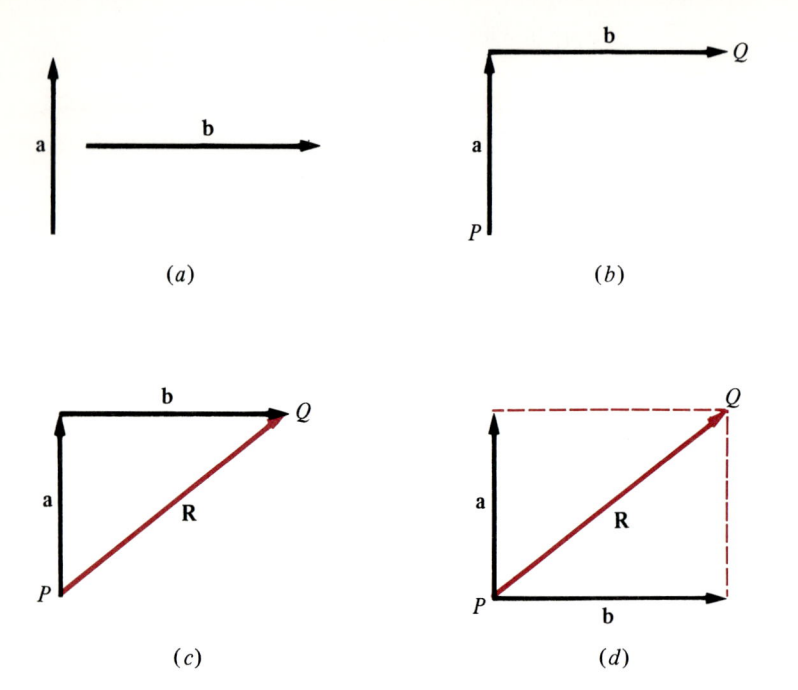

Figure 3-1 (a) Two vectors **a** and **b** representing three blocks north and four blocks east, respectively. (b) To apply the vectors sequentially from P, we draw **a** with its tail at P and preserve its magnitude and direction. Then we draw **b** beginning where **a** stopped and again preserving magnitude and direction to arrive at Q. (c) The vector **R** also begins at P and ends at Q and has the same result as the sequential addition of **a** and **b**. (d) Two vectors being added are sometimes placed tail to tail at the starting point. Dashed lines run from the tip of each to an intersection determine the resultant **R**.

We illustrate vectors by drawing a straight line with an arrowhead at one end. The arrow points in the direction of the vector, and the length is chosen to a proper scale to represent the magnitude of the vector quantity. Consider the problem "We walk three blocks due north and then turn and walk four blocks due east. Where are we from our starting position?" The vectors represented in Figure 3-1a could be the displacements of three blocks in the northerly direction and four blocks to the east. Vectors added graphically are added tail to head in sequence, much as you would execute this walk at the real scale. Starting at point P, we place the tail of the first vector on point P (Figure 3-1b). At the arrow point representing the end of this vector, we then begin the tail of the second vector, in both cases preserving the direction the vector points on the page. The head of the second vector ends up at point Q. The result of beginning at point P, walking three blocks north, and then four blocks east is to end at point Q. We now draw another vector which begins at P and ends at Q. This new vector is said to be the **resultant** of the first two vectors. We can express this in equation form as

$$\mathbf{a} + \mathbf{b} = \mathbf{R} \tag{3-1}$$

where the letters representing vectors are printed in **boldface** type. This is the convention we will use throughout this text. Here **a** represents the three blocks to the north, **b** is four blocks to the east, and **R** is the result of adding these vectors. You will note that **a** and **b** are the two sides of a 3-4-5 right triangle, and thus the magnitude of **R** is a length of five blocks. Trigonometry tells us that the direction of **R** is 36.9° N of E. Thus we could write **R** = 5 blocks at 36.9°N of E. Finally, we may note that when only two vectors are being added, as here, a frequently seen scheme is to place the vectors tail to tail, radiating from point P, as in Figure 3-1d, and complete the parallelogram shown by dashed lines. This gives the same result since opposite sides of the parallelogram are identical, and

Vectors are indicated on paper or on chalkboards by placing an arrow over the symbol for the vector: $\vec{a} + \vec{b} = \vec{R}$. In printing, vectors are indicated by boldface type: **a** + **b** = **R**. However, if a vector quantity is referred to without being in boldface, we are considering its magnitude only and neglecting its direction for the moment.

the method becomes equivalent to the tail-to-head method by simply moving one of the vectors parallel to itself.

It is easy to see that we could have arrived at point Q by first moving four blocks east and then three blocks north. In other words, the order in which we add the two vectors doesn't matter, and we say

$$\mathbf{b} + \mathbf{a} = \mathbf{R} \tag{3-2}$$

and

$$\mathbf{b} + \mathbf{a} = \mathbf{a} + \mathbf{b} \tag{3-3}$$

Although we have deduced this result for our special case, Equation 3-3 is quite general for any two vectors.

Suppose we wished to add a third vector of two blocks at 45° south of east (S of E) to the first two vectors. This could be done graphically, as in Figure 3-2. Our final displacement from P is the vector **r** beginning at P and ending at S.

If the scale and directions are accurate, we can determine the resultant displacement from measurement. In this case, the distance from P to S is compared to the scale to determine the magnitude of **r**. The direction is determined by using a protractor to measure the angle **r** makes with any arbitrary direction, say east. We find **r** to be about 5.6 blocks at about 15° N of E.

Suppose that we had added the three vectors in a different order. Consider Figure 3-3. We see that if we begin at point P and undertake the three displacements in any order, we always arrive at point S and the resultant displacement is always **r**. Thus we may add any given number of vectors in any order to achieve the same result.

Figure 3-2 Graphical addition of a third vector **c** to **a** and **b** from Figure 3-1 to yield a new resultant **r**.

Figure 3-3 The three vectors of Figure 3-2 could be added in any of six different orders. Note that all produce *the same resultant* vector; thus the order in which vectors are added does not affect the result.

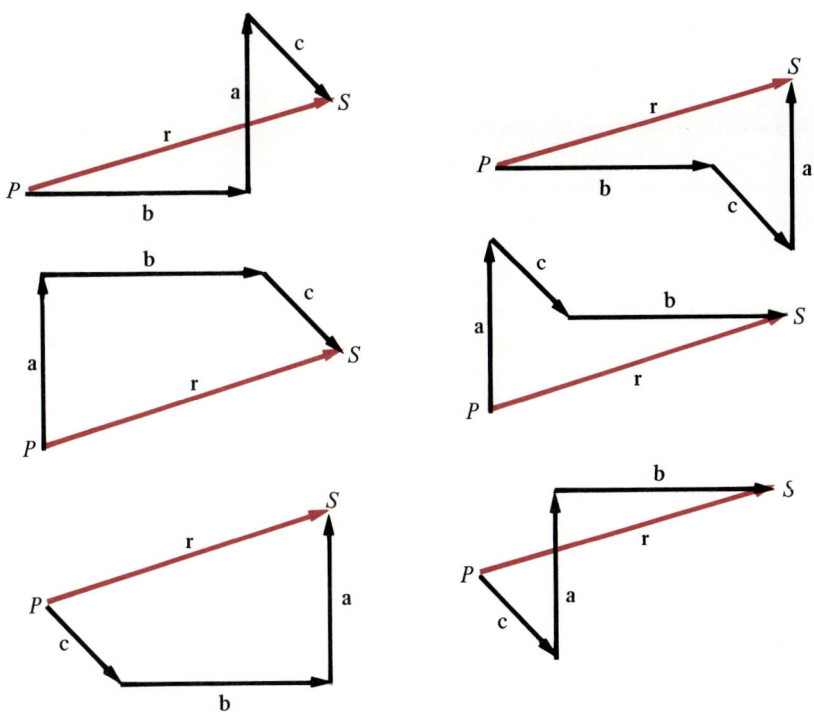

37

We must often subtract one vector from another. To do this we must define vector subtraction. This is similar to subtracting real numbers. For instance, we say that $3 = 5 - 2$ can be written as $3 = 5 + (-2)$, where the number (-2) is that number which when added to 2 gives zero, or $2 + (-2) = 0$. The vector analogy is that $\mathbf{a} - \mathbf{b} = \mathbf{a} + (-\mathbf{b})$, where the vector $(-\mathbf{b})$ is defined as that vector which when added to $\mathbf{b}$ equals zero. A little reflection should convince you that $-\mathbf{b}$ must have the same magnitude as $\mathbf{b}$ but be directed exactly opposite the direction of $\mathbf{b}$ (Figure 3-4). The only vector which completely cancels the effects of a walk three blocks north is one of three blocks south. Thus whenever we wish to subtract $\mathbf{b}$ from $\mathbf{a}$, we do so by adding $-\mathbf{b}$ to $\mathbf{a}$.

Figure 3-4 The negative of a vector has equal-length magnitude but is in an opposite direction. As with numbers, $\mathbf{b} + (-\mathbf{b}) = 0$.

Example 3-1 Show graphically $\mathbf{a} - \mathbf{b}$, where $\mathbf{a}$ is 3 m to the north and $\mathbf{b}$ is 4 m to the east.

Solution From the definition, we know that $-\mathbf{b}$ is 4 m to the west. Thus we have in Figure 3-5 that the vector connecting P to T is $\mathbf{a} - \mathbf{b}$.

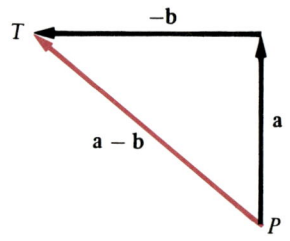

Figure 3-5 The graphical solution of a minus b, using the same vectors as before, adds $(-\mathbf{b})$ to $\mathbf{a}$. The resultant is $\mathbf{a} - \mathbf{b}$.

Graphical addition of vectors works fairly well, but analytical methods yield more precision. Graphing does have the advantage of helping us to visualize a situation. Thus, it often helps to draw the vectors being added even if only in a free-hand sketch.

3-2 Resolution of Vectors

When we add two vectors, we know that the resultant is equivalent to the two which were added and can be used to replace them. For instance, in our original example of Figure 3-1, instead of using vector $\mathbf{a}$ and then vector $\mathbf{b}$, we can use $\mathbf{R}$ and obtain the same result. Furthermore, if it were useful, we could replace $\mathbf{R}$ with $\mathbf{a}$ and $\mathbf{b}$. The vectors $\mathbf{a}$ and $\mathbf{b}$ are said to be **vector components** of $\mathbf{R}$. For example, if P in Figure 3-1b represented your house and Q represented the post office, the sum determined by Equation 3-1 would give the correct displacement of the post office from your house. If a passerby asked you where the nearest post office was, you would be very unlikely to respond with the resultant "It's five blocks at an angle of 36.9° north of east." You would be much more likely to give the vector components "It's three blocks north and four blocks east." Actually there are an infinite number of pairs of vectors which may be added vectorially to yield $\mathbf{R}$. This just implies that for our example there are many possible paths from our house to the post office. Some of these may be practical routes and some not, but each set of displacements would result in the same total displacement.

If the directions along which we would like to obtain the components of a vector are already chosen, the magnitudes of the components are uniquely determined. A particularly useful set of directions are those of a *system of rectangular coordinates*. We may **resolve** a vector into two mutually perpendicular components by projecting the vector successively onto any convenient set of rectangular coordinates. Figure 3-6 shows the vector $\mathbf{R}$ and its rectangular

Just as two or more vectors may be replaced by their resultant, any single vector, say $\mathbf{A}$, may be replaced by any two or more vectors which add up to $\mathbf{A}$. Thus if $\mathbf{A} = \mathbf{B} + \mathbf{C}$, then $(\mathbf{B} + \mathbf{C})$ may be substituted for $\mathbf{A}$ where convenient to do so.

components R_x and R_y. It is customary to specify the angle that the vector makes with the x axis by using the Greek letter θ (theta). We can determine the magnitude of the components from trigonometry:

$$R_x = R \cos \theta \qquad R_y = R \sin \theta \qquad (3\text{-}4)$$

where we have dropped the boldface vector designation since we are only interested in the magnitudes. Note that R_x and R_y have their directions already specified by their subscripts as being in the x and y directions, respectively. Thus, a statement like $R_x = -100$ m specifies the entire vector, with a magnitude of 100 m and in the negative x direction.

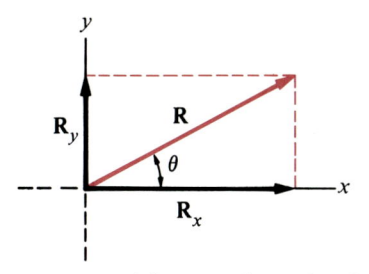

Figure 3-6 While **R** may be replaced by any set of vectors which adds up to **R**, a frequent and useful substitution is to replace a vector with *rectangular components*. The component of **R** along the defined x axis is named R_x, and the perpendicular component in the y direction is R_y. Clearly, $\mathbf{R} = \mathbf{R}_x + \mathbf{R}_y$.

Example 3-2 A cannonball leaves the cannon at a speed of 500 m/s at an angle of 31.5° to the horizontal. At what rate is the cannonball rising as it leaves the muzzle of the cannon? At what rate is it moving horizontally?

Solution The velocity vector is specified since both its magnitude (the speed) and its direction are given. Since **v** is a vector, we may resolve it into horizontal and vertical components by the methods just discussed. We take the vertical direction to be our y coordinate and the horizontal direction to be the x coordinate. Then

$$v_v = v \sin \theta = (500 \text{ m/s})(\sin 31.5°) = (500 \text{ m/s})(0.522) = 261 \text{ m/s}$$
$$v_h = v \cos \theta = (500 \text{ m/s})(\cos 31.5°) = (500 \text{ m/s})(0.853) = 426 \text{ m/s}$$

As it leaves the muzzle of the cannon, the cannonball is rising at a rate of 261 m/s and moving 426 m/s horizontally.

Sometimes, because the coordinate directions are already selected for us or because the physical situation indicates a particular choice of coordinates, one or both vector components may be negative.

Example 3-3 Resolve the vector **v** shown in Figure 3-7 into its rectangular components on the x and y axes.

Solution

$$v_x = v \cos \theta = (10.0 \text{ m/s})(\cos 60°)$$
$$= (10.0 \text{ m/s})(0.500) = 5.00 \text{ m/s} \quad \text{in negative } x \text{ direction}$$

$$v_y = v \sin \theta = (10.0 \text{ m/s})(\sin 60°)$$
$$= (10.0 \text{ m/s})(0.866) = 8.66 \text{ m/s} \quad \text{in positive } y \text{ direction}$$

Note how it simplifies the answer to say

$$v_x = -5.00 \text{ m/s} \qquad \text{and} \qquad v_y = +8.66 \text{ m/s}$$

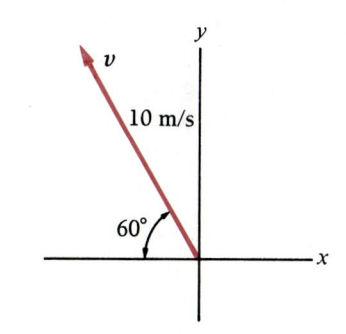

Figure 3-7 Example 3-3.

Resolving vectors along the axes of a rectangular coordinate system allows us to add vectors algebraically without requiring precision graphing. Suppose we wish to add two vectors: **a** is 3.00 m long at an angle of 60.0° to the positive x axis, while **b** is 3.20 m long at an angle of 21.0° to the same axis. These

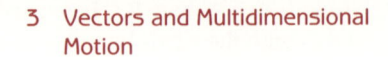

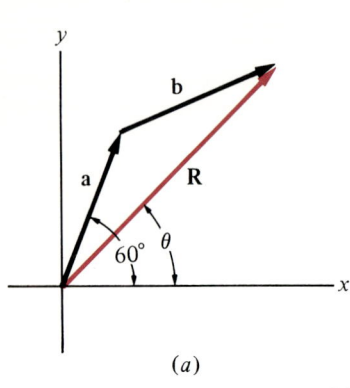

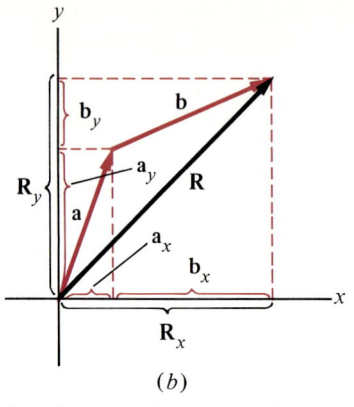

Figure 3-8 (a) The vectors **a** and **b** are added graphically to produce the resultant **R**. (b) If each vector is resolved into x and y components, it can be seen that $R_x = a_x + b_x$ and $R_y = a_y + b_y$, which is a perfectly general result.

are added tail to head in Figure 3-8a; however, this presents a problem. To determine the magnitude and direction of **R** to the three significant figures of the data is a difficult drawing task. Have you ever worked with or seen a protractor calibrated to a tenth of a degree? Alternatively, we may resolve **a** and **b** into their rectangular components:

$$a_x = a \cos \theta_a = (3.00 \text{ m})(\cos 60.0°) = 1.50 \text{ m}$$
$$a_y = a \sin \theta_a = (3.00 \text{ m})(\sin 60.0°) = 2.60 \text{ m}$$
$$b_x = b \cos \theta_b = (3.20 \text{ m})(\cos 21.0°) = 2.99 \text{ m}$$
$$b_y = b \sin \theta_b = (3.20 \text{ m})(\sin 21.0°) = 1.15 \text{ m}$$

The two components a_y and b_y are in the y direction; therefore no part of them can contribute to R_x; hence, $R_x = a_x + b_x$. Similarly, $R_y = a_y + b_y$, as can be seen in Figure 3-8b. Thus

$$R_x = a_x + b_x = 1.50 \text{ m} + 2.99 \text{ m} = 4.49 \text{ m}$$
$$R_y = a_y + b_y = 2.60 \text{ m} + 1.15 \text{ m} = 3.75 \text{ m}$$

The magnitude of **R** can be found from the pythagorean theorem:

$$R = \sqrt{R_x^2 + R_y^2} = \sqrt{(4.49 \text{ m})^2 + (3.75 \text{ m})^2} = 5.85 \text{ m}$$

The direction of **R** may be found from the definition of the tangent of an angle:

$$\tan \theta = \frac{R_y}{R_x} = \frac{3.75 \text{ m}}{4.49 \text{ m}} = 0.835$$

thus
$$\theta = \tan^{-1} 0.835 = 39.9°$$

Recall, always, that when vectors are added or subtracted, they must represent the same physical quantity and be expressed in the same units.

The expression $\theta = \tan^{-1} B$ may be read as "theta is the angle whose tangent is B." This way of expressing $\tan \theta = B$ is particularly handy for the hand calculator. Key B, then press INV, then press tan to have the value of θ displayed.

3-2 Resolution of Vectors

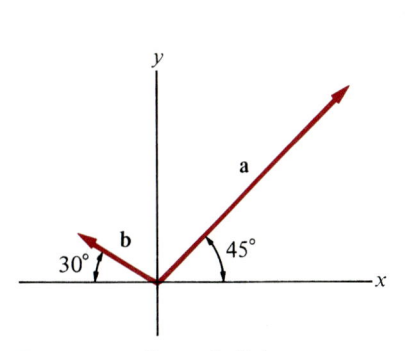

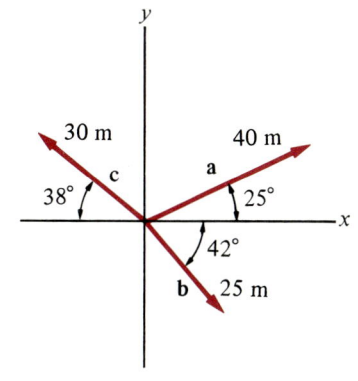

Figure 3-9 Example 3-4.

Figure 3-10 Example 3-5.

Example 3-4 Determine the resultant of the two vectors shown in Figure 3-9, where **a** is 12.0 m at an angle of 45.0° above the positive x axis and **b** is 4.00 m at an angle of 30.0° above the negative x axis.

Solution Note that we have drawn both the vectors originating at the origin. We could equally well have drawn them tail to head, but this way allows us to assign the signs of the components by inspection. We will first find the components of **a** and **b** and then add like components to obtain the components of **R**. We can then find the magnitude of **R** and the angle **R** makes with the x axis.

$$a_x = +a \cos 45.0° = +(12.0 \text{ m})(0.707) = +8.49 \text{ m}$$
$$a_y = +a \sin 45.0° = +(12.0 \text{ m})(0.707) = +8.49 \text{ m}$$
$$b_x = -b \cos 30.0° = -(4.00 \text{ m})(0.866) = -3.46 \text{ m}$$
$$b_y = +b \sin 30.0° = +(4.00 \text{ m})(0.500) = +2.00 \text{ m}$$
$$R_x = a_x + b_x = +5.03 \text{ m}$$
$$R_y = a_y + b_y = +10.5 \text{ m}$$
$$R = \sqrt{R_x^2 + R_y^2} = \sqrt{(5.03)^2 + (10.5)^2} = 11.6 \text{ m}$$
$$\tan \theta = \frac{R_y}{R_x} = \frac{10.5 \text{ m}}{5.03 \text{ m}} \qquad \theta = 64.4°$$

Example 3-5 Add the three vectors shown in Figure 3-10 to obtain the magnitude and direction of the resultant.

Solution The resultant **R** has $R_x = a_x + b_x + c_x$, $R_y = a_y + b_y + c_y$, and $\tan \theta = R_y/R_x$:

$$a_x = (+40.0 \text{ m})(\cos 25°) = (+40.0 \text{ m})(0.906) = +36.2 \text{ m}$$
$$a_y = (+40.0 \text{ m})(\sin 25°) = (+40.0 \text{ m})(0.423) = +16.9 \text{ m}$$
$$b_x = (+25.0 \text{ m})(\cos 42°) = (+25.0 \text{ m})(0.743) = +18.6 \text{ m}$$
$$b_y = (-25.0 \text{ m})(\sin 42°) = (-25.0 \text{ m})(0.669) = -16.7 \text{ m}$$

41

$$c_x = (-30.0 \text{ m})(\cos 38°) = (-30.0 \text{ m})(0.788) = -23.6 \text{ m}$$
$$c_y = (+30.0 \text{ m})(\sin 38°) = (+30.0 \text{ m})(0.616) = +18.5 \text{ m}$$
$$R_x = +31.2 \text{ m} \qquad R_y = +18.7 \text{ m}$$
$$R = \sqrt{R_x^2 + R_y^2} = \sqrt{(31.2 \text{ m})^2 + (18.7 \text{ m})^2} = 36.4 \text{ m}$$
$$\tan \theta = \frac{18.7 \text{ m}}{31.2 \text{ m}} = 0.599 \qquad \theta = 30.9° \text{ above the } +x \text{ axis}$$

3-3 Velocity in Two Dimensions

A complex motion can be described completely *only* if it is done so in terms of vectors. Even when we described motion in a straight line, we had to take the directional properties (the vector nature) of the displacement, velocity, and acceleration into account with plus and minus signs. Motion in more than one dimension requires that we make explicit use of the vector formulation we have developed.

Consider Figure 3-11. The jogger starts at the gate and runs clockwise, passing points A and B at times t_1 and t_2, respectively. Her displacements from the starting point are given by the vectors $\mathbf{s}_1$ and $\mathbf{s}_2$ as she passes those points. The change in displacement between A and B is $\Delta \mathbf{s} = \mathbf{s}_2 - \mathbf{s}_1$. It is useful to note from the figure that $\mathbf{s}_2 = \mathbf{s}_1 + \Delta \mathbf{s}$. This is the equation for $\Delta \mathbf{s}$ rewritten by adding $\mathbf{s}_1$ to both sides of the equation. In words it would read that "the second displacement is just the first displacement plus the change in displacement." It often makes the concept of change in a quantity more meaningful if we state it in this alternative fashion.

The jogger's average velocity between A and B is

$$\mathbf{v}_{av} = \frac{\mathbf{s}_2 - \mathbf{s}_1}{t_2 - t_1} = \frac{\Delta \mathbf{s}}{\Delta t} \tag{3-5}$$

Both sides of a vector equation must agree in direction; thus the average velocity from t_1 to t_2 is in the same direction as the *change* in displacement. We empha-

Figure 3-11 Top view of a running track. A jogger begins at the gate, which is taken as the origin, and follows the dashed path. At time t_1 she is at A with displacement $\mathbf{s}_1$. At time t_2 she is at B with displacement $\mathbf{s}_2$. The change in displacement from A to B is $\Delta \mathbf{s}$.

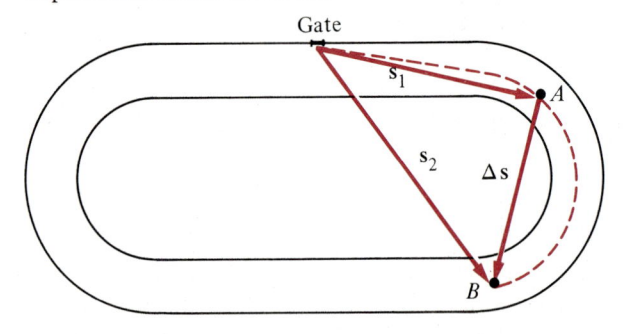

Figure 3-12 To determine the instantaneous velocity at A, we take successively smaller time intervals with correspondingly smaller changes in displacement; $\Delta \mathbf{s}_1, \Delta \mathbf{s}_2, \ldots$. As the time interval decreases, the direction of $\Delta \mathbf{s}$ approaches the line tangent to the jogger's path at A. Thus, in the limit as Δt approaches zero, $\Delta \mathbf{s}$ (hence v) is along her actual path.

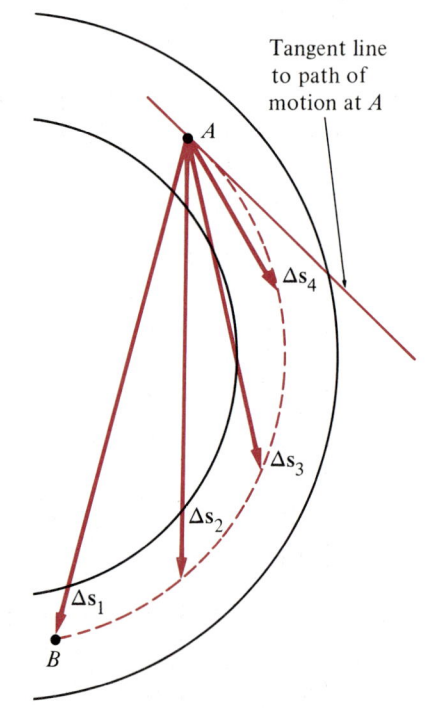

size that this is the *average* velocity. You can see from Figure 3-11 that the jogger could have been moving in the direction of the average velocity over only a small part of the track between A and B. As with movement in one dimension, we need a way to find instantaneous velocities.

The dashed line of Figure 3-11 shows the actual path that the jogger followed in going from A to B. **Instantaneous velocity in two dimensions** is defined as it was in one dimension, that is, as the value of $\Delta \mathbf{s}/\Delta t$ as Δt approaches zero. We determined $\Delta \mathbf{s}$, call it $\Delta \mathbf{s}_1$, for the entire interval from A to B. We now take a succession of increasingly smaller intervals from A toward B, as in Figure 3-12, and calculate the new magnitude and direction of $\Delta \mathbf{s}$ for each. In the limit as Δt approaches zero, $\mathbf{v}_{av}$ approaches $\mathbf{v}$ (the instantaneous velocity). We see from Figure 3-12 that the direction of $\mathbf{v}_{av}$ for increasingly smaller intervals of time approaches the straight line which is tangent to the path of motion at point A. The *direction of the instantaneous velocity* $\mathbf{v}$ *at any point* is that of a tangent to the path of the motion at that point.

Just as with displacements, it is often convenient to resolve velocities into components in a rectangular coordinate system. Thus we may write

$$v_x = v \cos \theta \qquad v_y = v \sin \theta$$

$$v = \sqrt{v_x^2 + v_y^2} \qquad \tan \theta = \frac{v_y}{v_x}$$

where θ is the angle between $\mathbf{v}$ and the positive x axis and v is the magnitude of $\mathbf{v}$.

The advantage of resolving velocities into mutually perpendicular components can be considerable. When the two components of a vector are along the x and y axes, for example, a change may occur in the x component while the y component is constant, or vice versa. We saw in the bowling ball drop of Chapter 2 that a falling object has a constant acceleration in the downward direction. What effect does this vertical acceleration have on the ball's horizontal motion? Figure 3-13 is a stroboscopic photograph of two balls falling from a

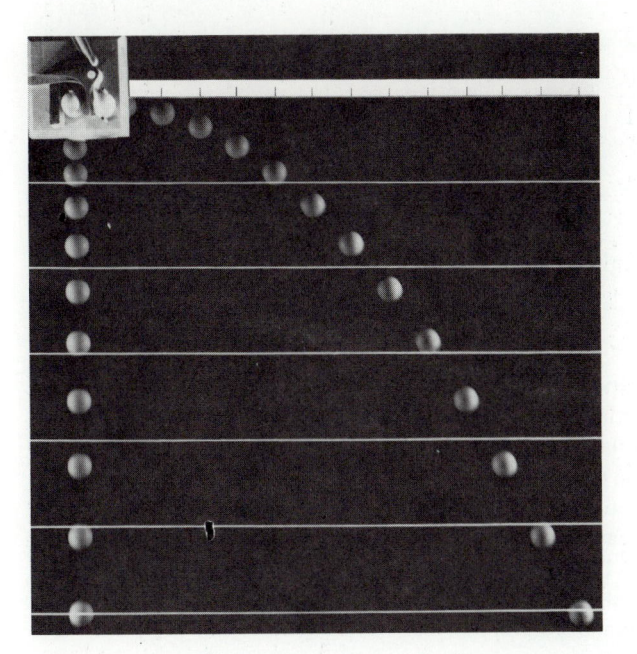

3-3 Velocity in Two Dimensions

In Section 2-5, we measured the acceleration due to gravity with the bowling ball drop and found it to be $g = 9.80$ m/s². This value varies somewhat with position on the earth and height above the surface. Usually we ignore this small variation.

Location	Value of g, m/s²
Equator	9.780
New York City	9.803
Denver	9.796
North pole	9.832
Standard value (by international agreement)	9.80665

Figure 3-13 A composite of photographs of two falling balls, one dropped and the other projected horizontally at the same instant. Each photograph shows the horizontally moving ball to have fallen the same vertical distance as the dropped ball, while marks along the top reveal it to have moved equal distances horizontally in the equal time intervals between photographs. (PSSC Physics/D.C. Heath/EDC)

common release height. The ball on the left was simply dropped, while the ball on the right had a horizontal velocity at the instant of release. The set of black marks at the top of the photograph was added to mark the horizontal position of the center of the ball in each successive photograph. The photograph shows that both balls are accelerating downward at the same rate. However, we also note that the horizontal displacements of the second ball during the constant time intervals between photographs are constant. Thus we conclude that the horizontal component of the second ball's velocity was constant. The purely vertical acceleration acting on this second ball has no apparent effect on its horizontal motion. We say that these components of velocity are independent.

From this we may appreciate the value of treating the motion of the ball by vector components. If we choose a horizontal (x) axis and a vertical (y) axis to analyze the ball's motion, the analysis is greatly simplified. The motion in the x direction is characterized by constant speed, which we already know how to analyze. Likewise, the motion in the y direction is characterized by constant acceleration, which has also already been studied. Thus, the complex two-dimensional motion of the second ball can be resolved into two one-dimensional motions occurring simultaneously.

3-4 Uniform Acceleration in Two Dimensions

As we saw in Figure 3-13, accelerations may occur in a direction which is different from that of the original velocity of the object. In that case we must develop kinematic equations to describe the motion. The definition of acceleration becomes

$$\mathbf{a} = \frac{\mathbf{v} - \mathbf{v}_0}{t - t_0} = \frac{\Delta \mathbf{v}}{\Delta t} \qquad (3\text{-}6)$$

or, choosing to start our clock at $t_0 = 0$ when $\mathbf{v} = \mathbf{v}_0$, we have

$$\mathbf{a} = \frac{\mathbf{v} - \mathbf{v}_0}{t} \qquad (3\text{-}7)$$

We may write the equation for velocity as a function of the acceleration and time, beginning at $t_0 = 0$, in the form of Equation 2-9 as

$$\mathbf{v} = \mathbf{v}_0 + \mathbf{a}t \qquad (3\text{-}8)$$

This is a vector equation and corresponds to the two scalar equations which must be true simultaneously:

$$v_x = v_{0x} + a_x t \quad \text{and} \quad v_y = v_{0y} + a_y t$$

We may use the independence of motion in mutually perpendicular directions to write a set of one-dimensional equations for each direction. Assuming that we have a constant acceleration a (hence constant components a_x and a_y along each axis of a rectangular coordinate system), we may use the equations from Section 2-4 with very little change. We may even choose our coordinates so that the acceleration is only along one axis. This won't cause even the slightest loss of generality. If the acceleration is constant, it must have constant direction as well as constant magnitude. Just as we are free to pick the origin of our coordinate system to be at any convenient point, we may always orient it in any direction we wish. It is most convenient to orient the axes such that one of them is in the same direction as the constant acceleration, because the component of acceleration along the other axis is then zero, thus simplifying the component equations.

Let us assume the acceleration is along the y axis, as in Figure 3-13. Then v_x is constant and equal to the original value of the velocity in the x direction, v_{0x}. Thus we know $v_x = v_{0x}$ and $a_x = 0$, so our equations in the x direction reduce to

$$\Delta x = v_{0x} t \tag{3-9}$$

The set of equations of motion from Chapter 2, 2-9 through 2-13, may be rewritten specifically for the motion in the y direction:

$$v_y = v_{0y} + a_y t \tag{3-10}$$

$$v_{y,\,av} = \frac{v_y + v_{0y}}{2} \tag{3-11}$$

$$\Delta y = v_{y,\,av} t \tag{3-12}$$

$$\Delta y = v_{0y} t + \tfrac{1}{2} a_y t^2 \tag{3-13}$$

$$2 a_y \Delta y = v_y^2 - v_{0y}^2 \tag{3-14}$$

The rewritten equations use subscripts to remind us that they are components along the axis of the acceleration and independent of any motion along the perpendicular axis. These equations can be applied to the motion of any object in two dimensions provided the acceleration is constant and in the y direction. We now apply them to a general case.

3-5 Projectile Motion

Figure 3-14 shows the partial path of an object which has been launched (or projected; hence we call the object a **projectile**) near the surface of the earth. Our coordinates are chosen so that the x axis is horizontal along the constant horizontal component of **v** and the y axis is vertical along the direction of the

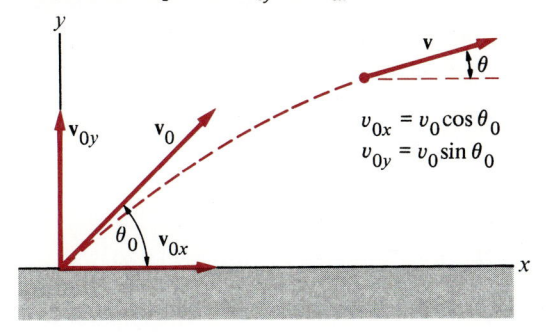

Figure 3-14 A portion of the trajectory of a projectile with an initial velocity $\mathbf{v}_0$ at an angle of θ_0 to the horizontal. At any later point, the velocity is **v** at an angle to the horizontal of θ. The initial velocity may be resolved into the initial vertical and horizontal components v_{0y} and v_{0x}.

$$v_{0x} = v_0 \cos \theta_0$$
$$v_{0y} = v_0 \sin \theta_0$$

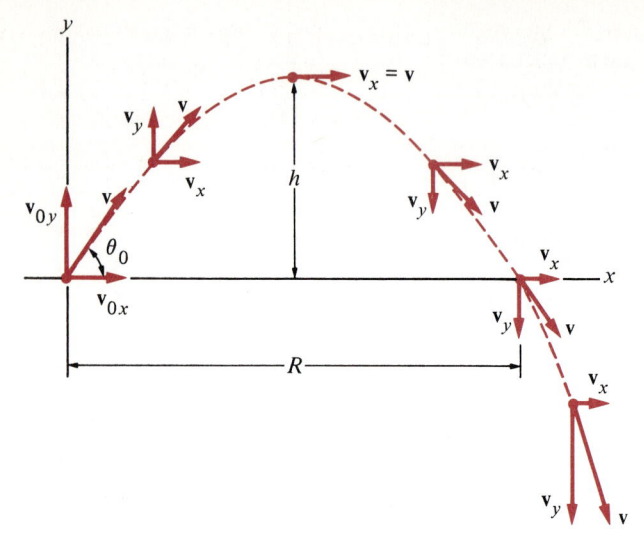

Figure 3-15 The trajectory of a projectile with the velocity at several points resolved into horizontal (v_x) and vertical (v_y) components. Note that v_x is constant and that v_y is constantly decreasing (increasing in the negative direction).

constant acceleration. For convenience we choose upward as positive and select the point at which the projectile was launched as the origin. The initial velocity v_0 resolves into a horizontal component v_{0x} and a vertical component v_{0y}, as shown. We may now apply Equations 3-9 through 3-14 directly to the object to determine times and/or displacements at any point in the motion.

We call the path followed by the projectile its **trajectory** (Figure 3-15). The maximum height of the trajectory above the projectile's initial position is designated h. When the projectile is at that point, it has stopped its upward motion so that $y = h$ and $v_y = 0$. The symbol R is used to designate the **range of the projectile,** which is the horizontal displacement of the object after it has returned to its original height ($y = 0$).

Because of our choice of coordinates, there is no acceleration in the x direction; thus, for an original angle of projection θ_0,

$$v_x = v_{0x} = v_0 \cos \theta_0$$

for the whole trajectory. Also, since $a_y = -g$,

$$v_y = v_{0y} - gt = v_0 \sin \theta_0 - gt$$

from Equation 3-10.

At any time t we can find the x and y positions of the object from Equations 3-9 and 3-13 as

$$x = (v_0 \cos \theta_0)t \tag{3-15}$$

and
$$y = (v_0 \sin \theta_0)t - \tfrac{1}{2}gt^2 \tag{3-16}$$

If we wish, we may even obtain a single equation for every point on the trajectory by solving Equation 3-15 for t and substituting into Equation 3-16. Thus

$$t = \frac{x}{v_0 \cos \theta_0}$$

and

$$y = (v_0 \sin \theta_0) \frac{x}{v_0 \cos \theta_0} - \tfrac{1}{2}g \left(\frac{x}{v_0 \cos \theta_0}\right)^2$$

This yields

$$y = (\tan \theta_0)x - \frac{gx^2}{2v_0^2 \cos^2 \theta_0} \qquad (3\text{-}17)$$

All the terms of Equation 3-17 are constant except x and y, and the equation of the trajectory is of the form

$$y = c_1 x - c_2 x^2$$

where c_1 and c_2 are positive constants. You might perhaps recognize this as the equation of a parabola. Thus the trajectory is parabolic and is symmetric about a vertical line through the point where the object attains its maximum height.

There are a number of ways we can determine R, h, and t_R (the time of flight for the projectile to reach R). If we apply Equation 3-16 for $y = 0$, then we should be able to find t_R. Thus

$$0 = (v_0 \sin \theta_0)t_R - \tfrac{1}{2}g t_R^2$$

This is satisfied when either $t_R = 0$ (note again this makes sense physically) or

$$t_R = \frac{2v_0 \sin \theta_0}{g} \qquad (3\text{-}18)$$

We may now apply Equation 3-15 at the same point, substituting the known expression for t_R. Thus

$$R = (v_0 \cos \theta_0)t_R = \frac{v_0^2}{g}(2 \sin \theta_0 \cos \theta_0)$$

$$= \frac{v_0^2 \sin 2\theta_0}{g} \qquad (3\text{-}19)$$

since $\sin 2\theta_0 = 2 \sin \theta_0 \cos \theta_0$.

The maximum height can be found by applying Equation 3-14 between the point of launch ($y = 0$) and the highest point in the trajectory ($y = h$). Thus

$$2(-g)h = 0 - (v_0 \sin \theta_0)^2$$

or

$$h = \frac{v_0^2 \sin^2 \theta_0}{2g} \qquad (3\text{-}20)$$

These equations for time of flight, range, and maximum height in terms of initial conditions could have been derived from other starting points. (For example, see Problem 3-24.)

Stop for a moment and consider Equation 3-19. Recall from Chapter 2 that Tartaglia found that the maximum range occurred when a cannon was fired at an angle of 45° to the horizontal. We asserted that Galileo could arrive at the same conclusion without firing the cannon. Galileo's assumption was that objects have a constant downward acceleration. We measured this in Chapter 2 and found it to be correct. Using this physical fact, we arrived at Equation 3-19 for the range of a projectile. What does this equation imply about the maximum range? Clearly, the larger v_0, the greater the range, but v_0 is fixed by the amount of gunpowder and the mass of the cannonball used. Range would also be changed if g could be adjusted, but it, too, is fixed. This leaves the angle of elevation θ_0 as the quantity which can be varied to obtain the maximum range. Since the largest value of a sine occurs for an angle of 90°, $\sin 2\theta_0$ is a maximum

---

The path and time relationships of a projectile are both symmetric, and this is sometimes useful. It takes just as long for it to rise to its peak as to return from its peak to the original height. For any height, the vertical velocity on the way up is of the same magnitude as the vertical velocity on the way down at that height.

when $2\theta_0 = 90°$, or $\theta_0 = 45°$. By applying physical principles, we have come to Tartaglia's conclusion, as did Galileo, without having fired a cannon.

Example 3-6 A golfer strikes a ball which leaves the tee at an angle of 35° to the horizontal with an initial speed of 40.0 m/s. Assume the course from tee to green is level and determine (a) the time of flight, (b) the range, and (c) the maximum height of the ball.

Solution Let us use the tee as the origin of coordinates. The y direction is vertically upward, and x is along the ground. The acceleration after the ball leaves the tee is in the y direction and is $a_y = -g = -9.80$ m/s².
(a) Applying Equation 3-13 between the point where the ball is struck and the point where it lands, we have

$$y = 0 = v_{0y}t_R + \tfrac{1}{2}a_y t_R^2$$

$$= (40.0 \text{ m/s})(\sin 35°)t_R + \tfrac{1}{2}(-9.80 \text{ m/s}^2)t_R^2$$

$$= 22.9 \text{ m/s} - 4.90 t_R$$

$$t_R = 4.68 \text{ s}$$

(b) Applying Equation 3-9 between the same two points, $x = R = v_{0x}t$ and

$$R = (40.0 \text{ m/s})(\cos 35°)(4.68 \text{ s}) = 153 \text{ m}$$

(c) We may find the maximum height by applying Equation 3-14 between the point where the ball is struck and the point of maximum height. Thus

$$2a_y \Delta y = 2(-9.80 \text{ m/s}^2)h = 0^2 - v_{0y}^2$$

or

$$h = 0 - \frac{[(40.0 \text{ m/s})(\sin 35°)]^2}{-19.6 \text{ m/s}^2}$$

$$= 26.9 \text{ m}$$

Example 3-7 A baseball thrown horizontally from the edge of the top of a building 20.0 m above a level field lands 45.0 m away from the building. Calculate the initial speed.

Solution Consider Figure 3-16. In the vertical (y) direction, we know the distance the ball moves, the acceleration, and the initial velocity (zero). In the horizontal (x) direction, we know the final distance only. Perusing Equations 3-9 through 3-14, we see that there is no one equation which allows us to solve for the velocity in one step. However, we could use Equation 3-9 if we knew the time of flight. Searching further, we see that we have enough information to obtain the time of flight from Equation 3-13. Applying Equation 3-13 between the top of the building and the point where the ball strikes the ground, we have

$$\Delta y = -20.0 \text{ m} = 0 + \tfrac{1}{2}(-9.80 \text{ m/s}^2)t^2$$

$$t = 2.02 \text{ s}$$

Substituting this time and the distance in the x direction in Equation 3-9 yields

$$v_{0x} = \frac{\Delta x}{t} = \frac{45.0 \text{ m}}{2.02 \text{ s}} = 22.3 \text{ m/s}$$

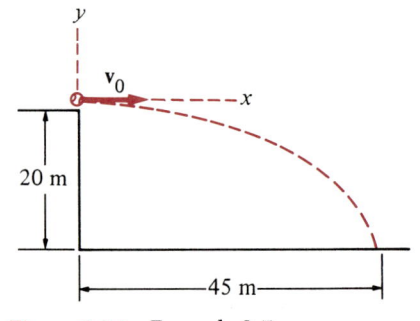

Figure 3-16 Example 3-7.

3-6 Relative Velocities in Two Dimensions

Let us consider the motion of an object undergoing two separate but simultaneous motions. Suppose you have a boat on a river flowing at 0.800 m/s, and assume you can row at a speed of 1.500 m/s with respect to the water. Instead of going upstream, you would like to row across the river. If you head your boat straight across stream from point A toward point B as in Figure 3-17, you see that you will make headway toward the other river bank at the rate of 1.500 m/s but you will also be swept downstream by the current at the rate of 0.800 m/s. We recognize that we have a vector addition problem. The equation for the resultant motion may be written as

$$\mathbf{v}_{B/E} = \mathbf{v}_{B/R} + \mathbf{v}_{R/E} \tag{3-21}$$

which is read as "the velocity of the boat with respect to the earth equals the velocity of the boat with respect to the river plus the velocity of the river with respect to the earth."

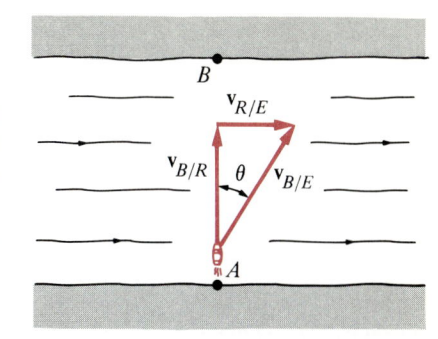

Figure 3-17 The boat heads directly across the river from A toward B with velocity $\mathbf{v}_{B/R}$. The current in the river carries the boat downriver with velocity $\mathbf{v}_{R/E}$. The resulting motion of the boat to an observer on the bank is along the line of resultant $\mathbf{v}_{B/E}$.

Example 3-8 Using the velocities given above, determine the velocity of your boat with respect to the earth if you head directly across the river.

Solution Since the two vectors to be added are perpendicular, they form the sides of a right triangle. The magnitude of the velocity of the boat with respect to the earth can be found from the pythagorean theorem. The angle the velocity makes with the line AB may be found from the definition of the tangent. Thus

$$v_{B/E} = \sqrt{v_{B/R}^2 + v_{R/E}^2} = \sqrt{(1.50 \text{ m/s})^2 + (0.800 \text{ m/s})^2} = 1.70 \text{ m/s}$$

$$\tan \theta = \frac{v_{R/E}}{v_{B/R}} = \frac{0.800 \text{ m/s}}{1.50 \text{ m/s}} \qquad \theta = 28.1°$$

Therefore $v_{B/E} = 1.70$ m/s at 28.1° downstream from AB.

Example 3-9 Suppose you wish to cross the river directly. (That is, the velocity of your boat with respect to the earth must be along the line AB.) Which direction should you head your boat, and what will be the speed of your boat with respect to the earth?

Solution The problem conditions require the resultant to lie along the line straight across the river. Certainly we still wish to apply Equation 3-21. However, we now want the resultant $\mathbf{v}_{B/E}$ to have *no* component either upstream or downstream. Thus we must head our boat such that $\mathbf{v}_{B/R}$ has a component *upstream* exactly equal in magnitude to that of $\mathbf{v}_{R/E}$ carrying the boat downstream. In this manner we may use part of the boat's velocity to cancel the effect of the current (Figure 3-18). Thus we see that we again have a right triangle and can use the pythagorean theorem and one of the trigonometric functions to solve for the magnitude of $\mathbf{v}_{B/E}$ and the direction of $\mathbf{v}_{B/R}$. We must be careful, however. In this problem, $\mathbf{v}_{B/R}$ is the hypoteneuse of the right triangle. Therefore

$$(v_{B/R})^2 = (v_{B/E})^2 + (v_{R/E})^2$$

or $$(v_{B/E})^2 = (v_{B/R})^2 - (v_{R/E})^2 = (1.50 \text{ m/s})^2 - (0.800 \text{ m/s})^2$$

$$v_{B/E} = 1.27 \text{ m/s}$$

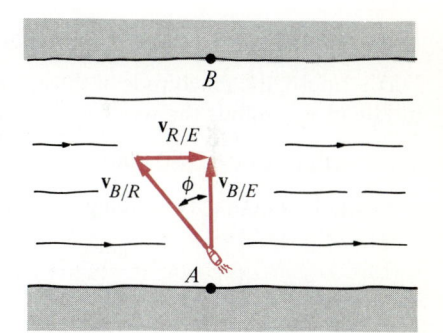

Figure 3-18 To reach B, the boat leaving A must direct its velocity $\mathbf{v}_{B/R}$ upstream at an angle such that when the current velocity $\mathbf{v}_{R/E}$ is added, the resultant velocity $\mathbf{v}_{B/E}$ points directly cross-river.

and
$$\sin\theta = \frac{v_{R/E}}{v_{B/R}} = \frac{0.800 \text{ m/s}}{1.500 \text{ m/s}} \qquad \theta = 32.2°$$

Thus to move directly across the river, you must head your boat at an angle of 32.2° upstream from the line AB. Your speed with respect to the earth will be 1.27 m/s, or less than the speed of the boat through the water, since part of its motion is used to cancel the current.

Example 3-10 A small airplane flies at an average speed of 180 km/h. It heads due west and travels for 3.00 h, at which time the pilot discovers that he is passing over a town which has a displacement 450 km at 30° S of W from his starting point. Calculate (a) the average velocity of the airplane with respect to the earth during that 3-h interval and (b) the average wind velocity during that same period.

Solution (a) We know the change in the airplane's displacement and the interval of time which passed and thus may calculate its resultant velocity. Designating P as airplane, A as air, and E as earth, from the definition of average velocity we have that the resultant velocity $\mathbf{v}_{P/E}$ is

$$\mathbf{v}_{P/E} = \frac{\Delta\mathbf{s}}{\Delta t} = \frac{450 \text{ km at } 30° \text{ S of W}}{3.00 \text{ h}} = 150 \text{ km/h at } 30° \text{ S of W}$$

(b) This resultant velocity is made up of two components, the motion of the plane through the air $\mathbf{v}_{P/A}$ and the motion of the air with respect to the earth $\mathbf{v}_{A/E}$:

$$\mathbf{v}_{P/E} = \mathbf{v}_{P/A} + \mathbf{v}_{A/E}$$

where $\mathbf{v}_{A/E}$ is the average wind velocity we seek. This is shown in Figure 3-19. Since $\mathbf{v}_{P/A}$ was given to be due west, the southerly components of $\mathbf{v}_{A/E}$ and $\mathbf{v}_{P/E}$ are equal. Thus, any southward drift of the plane was supplied by the motion of the air:

$$v_{A/E}\sin\theta = v_{P/E}\sin 30° = (150 \text{ km/h})(\sin 30°) = 75.0 \text{ km/h}$$

Furthermore, the magnitude of the easterly component of $\mathbf{v}_{A/E}$ equals the magnitude of $\mathbf{v}_{P/A}$ minus the westerly component of $\mathbf{v}_{P/E}$, or

$$v_{A/E}\cos\theta = 180 \text{ km/h} - [(150 \text{ km/h})(\cos 30°)] = 50.1 \text{ km/h}$$

We can obtain $\tan\theta$ by dividing the equations for the two components of $v_{A/E}$:

$$\frac{v_{A/E}\sin\theta}{v_{A/E}\cos\theta} = \tan\theta = \frac{75.0 \text{ km/h}}{50.1 \text{ km/h}}$$

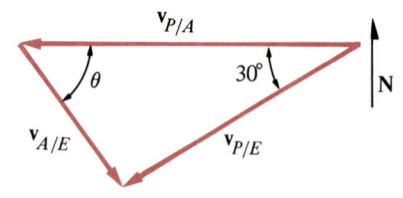

Figure 3-19   Example 3-10.

Therefore, $\theta = 56.3°$ S of E, and the magnitude of $\mathbf{v}_{A/E}$ is

$$v_{A/E} = \sqrt{(75.0 \text{ km/h})^2 + (50.1 \text{ km/h})^2} = 90.2 \text{ km/h}$$

Thus the average wind velocity during the 3-h interval is

$$\mathbf{v}_{A/E} = 90.2 \text{ km/h at } 56.3° \text{ S of E}$$

For most of this and the last chapter we have been concerned with the motion of objects. We have been developing skills to allow us to predict velocities and positions of objects at any given time from a knowledge of initial conditions (initial velocity, initial position, acceleration, etc.). In the next chapter we begin to examine the causes of changes of motion.

Minimum Learning Objectives

After studying this chapter, you should be able to:
1. Define:
   - projectile
   - resolution of vectors
   - resultant
   - scalar quantities
   - trajectory
   - vector components
   - vector quantities
2. Add and/or subtract vectors by both scale drawing (graphical method) and resolution into components (analytical method) to determine their resultants.
3. Resolve a given vector into components along specified axes.
4. Describe a constant-speed motion with changing velocity.
5. Apply the concept of the independence of the horizontal and vertical motion of a projectile to consider each separately.
6. Verify the steps in the mathematical derivation of the formulas for projectile motion, range, and maximum height.
7. Given the initial velocity of a projectile near the surface of the earth, determine:
   a. Its horizontal and/or vertical position at any later time.
   b. Its horizontal and/or vertical speed or its total velocity at any later time.
   c. The time after its release at which it will attain any horizontal or vertical distance given.
   d. The distance and time at which it will intersect a horizontal surface of given height relative to the point of release and the instant of release.

Problems

Throughout this text, the problems are of three levels of difficulty: Standard, somewhat difficult (indicated by a boldface black number preceding the problem), and most difficult (indicated by a colored number).

3-1 The nearest carryout restaurant is four blocks south and five blocks west of your room. Construct a graph showing the successive displacements you follow to get a hamburger. Add the two graphically to get a single resultant. Choose a convenient scale, and use a protractor to get the angle.

3-2 (a) Add the vectors 4.00 m north, 6.50 m north, 3.00 m south, 9.00 m north, 5.00 m south, and 1.50 m south. (b) What vector must be added to the resultant of part a to yield a net displacement of 0 m from the starting point?

3-3 Determine the two components (north and east) of the velocity of an automobile traveling 21.0 m/s at 30° N of E.

3-4 What force must be added to one of 30.0 newtons due east to obtain a resultant of 25.0 newtons at 45° N of E?

3-5 A professor starts from her home and walks 2.00 km south, then turns and walks 4.00 km at 30° N of W. Finally, she walks 3.00 km directly north. Find by resolution and addition of components her resultant displacement from home.

3-6 Graphically add the following displacements by the head-to-tail method: 8.00 km at 45° S of W, 5.00 km at 30° N of W, and 12.0 km at 60° E of N.

3-7 Add and subtract the following vectors graphically: $\mathbf{a} = 6.00$ m in the positive y direction and $\mathbf{b} = 4.00$ m at 30° downward from the negative x direction. That is, show on your graph the resultant $\mathbf{r} = \mathbf{a} + \mathbf{b}$ and $\mathbf{R} = \mathbf{a} - \mathbf{b}$.

3-8 Analytically resolve the vectors $\mathbf{A}$, $\mathbf{B}$, and $\mathbf{C}$ shown in Figure 3-20 into their x and y components.

3-9 A water molecule is formed when two hydrogen atoms are

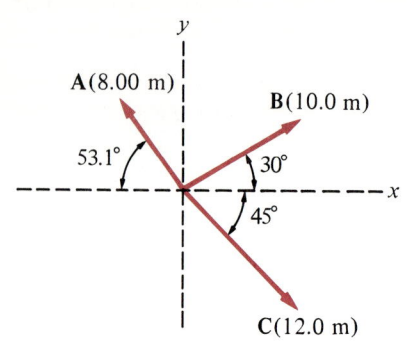

Figure 3-20 Problem 3-8.

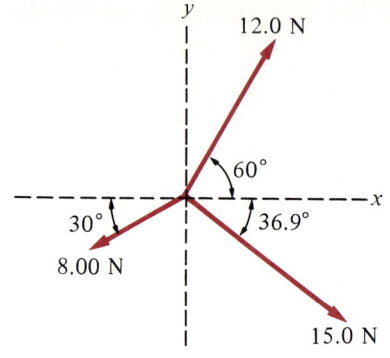

Figure 3-22 Problem 3-13.

bonded to an oxygen atom. The lines along which each hydrogen-oxygen force acts form an angle of 104°. The electrostatic force of attraction in each of the two bonds is about 5.00×10^{-6} N. Determine the magnitude of the resultant force of attraction acting on the oxygen atom.

3-10 (a) What vector displacement **B** must be added to **A** to obtain a resultant vector which is 18.0 m at 30° E of N of an origin if **A** is 10.0 m south from the origin? (b) What vector **D** must be subtracted from **C** to obtain a resultant vector **R** which is 8.00 m at 45° S of W if **C** is 3.00 m at 45° N of W?

3-11 A child stands on the edge of a playground merry-go-round 2.00 m from the center. It rotates one complete revolution in 2.50 s. Calculate (a) the circumference of the circle in which the child moves and (b) the child's average speed.

3-12 The ball in Figure 3-21 is moving along a circular path with the accelerations shown. Given that a_r is radial toward the center of the circle with a magnitude 14.0 m/s² and a_t is tangent to the circular path with a magnitude 10.0 m/s², calculate the resultant acceleration. (Can you determine which way the ball is moving?)

3-13 Calculate the resultant of the forces shown in Figure 3-22.

3-14 A large water beetle travels along a semicircular path of radius 0.600 m in 23.0 s. What is the beetle's (a) average speed, (b) average velocity, and (c) average acceleration as it completes the semicircle?

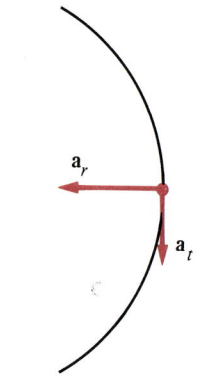

Figure 3-21 Problem 3-12.

3-15 A bald eagle in level flight at a height of 135 m drops the fish it just caught. If the eagle's speed is 25.0 m/s, at what horizontal distance from the drop point will the fish hit the ground?

3-16 A driver of an automobile traveling at 35.0 km/h due west increases speed to 50.0 km/h as she turns with the road to a heading of 45° S of W. The maneuver takes 8.00 s. Calculate the average acceleration during the maneuver.

3-17 Determine the maximum height of an object projected at an initial velocity of 100 m/s at 27° above the horizontal.

3-18 A target pistol is fired horizontally toward a bull's-eye which is at the same height as the pistol and at a distance of 120 m. The muzzle velocity of the bullet is 200 m/s. Calculate (a) the time from firing at which the bullet strikes the target and (b) the distance below the bull's-eye the bullet strikes.

3-19 A baseball is thrown at an unknown angle above the horizontal and has a horizontal component of initial velocity of 25.0 m/s. If it takes exactly 3.00 s to return to its initial height, calculate (a) its horizontal range, (b) its initial vertical component of velocity, and (c) its initial angle of projection.

3-20 A tennis ball thrown at a velocity of 25.0 m/s at 53.1° above the horizontal lands exactly 3.00 s later on top of a building. Calculate (a) the horizontal distance from the point of release to the point where the ball landed and (b) the height of the top of the building above the point of release.

3-21 A ball rolls off a filing cabinet 1.70 m high and lands 0.700 m from the base. Calculate (a) the time of descent and (b) the initial speed of the ball.

3-22 A certain professor can drive a golf ball from the tee with an initial speed of 30.0 m/s. At the sixth hole on the golf course where he plays, the tee and green are separated horizontally by 80.0 m and the green is at the same level as the tee. At what angle should the professor drive the ball on this hole?

3-23 An air rifle is fired at an angle of 30° to the horizontal. The pellet strikes a tree 40.0 m away at the same height as the muzzle. Calculate (a) the initial speed of the pellet, (b) the maximum height, and (c) the time the pellet was in the air.

3-24 Begin with Equation 3-17 and derive an expression for the range of a projectile and its maximum height. Remember when $x = R$, $y = 0$ and when $x = R/2$, $y = h$.

3-25 How long will it take a shell fired from a cliff at an initial velocity of 800 m/s at an angle 30° below the horizontal to reach the ground 150 m below?

3-26 While running, the springbok of southern Africa occasionally bounds high into the air for no apparent reason. If a springbok running at 12.0 m/s leaps, raising its body 2.00 m, (a) what is its total velocity at takeoff and (b) what distance will it travel horizontally during its leap?

3-27 A ski jumper takes off at an angle of 20° above the horizontal with a speed of 30.0 m/s and lands on a downward slope 10.0 m vertically below the point of takeoff. Calculate (a) time of flight and (b) horizontal distance traveled.

3-28 The archerfish of southeast Asia shoots a jet of water at an insect to dislodge the insect from the leaf on which it is sitting. The insect and leaf are on the surface of the water and 2.50 m horizontally from the position of the archerfish. If the jet of water is ejected from the surface at a speed of 5.00 m/s, at what angle to the horizontal must the archerfish shoot in order to hit the insect?

3-29 A football kicked at an angle of 50° to the horizontal is caught 4.00 s later at the same level it is kicked. What is the maximum height of the ball?

3-30 After fielding a sharply hit ball in a softball game, an outfielder throws it back to the catcher. The ball is caught at the same height it was thrown exactly 3.00 s after it left the fielder's hand. Calculate (a) the initial vertical component of velocity and (b) the ball's maximum height.

3-31 At what speed must a softball leave the bat toward the center field fence in order to be a home run if the angle of elevation of the ball is 35° when (a) the fence is 104 m from home plate and its top is the same distance above the ground at which the bat struck the ball and (b) the fence is 90.0 m from home plate and its top is 10.0 m higher than the point where the bat struck the ball?

3-32 A motorboat must head upstream at an angle of 25° to a line directly across a river in order to travel along that line. Its speed with respect to the water is 10.0 m/s. (a) Calculate the speed of the river. (b) How long will it take the boat to cross if the river is 800 m wide?

3-33 An airplane is to be flown due west at a speed of 700 km/h with respect to the ground. The wind is blowing at 150 km/h at an angle of 30° S of E. Determine the necessary heading and speed of the airplane with respect to the air.

3-34 Automobile A travels south at 20.0 m/s, while automobile B travels west at 25.0 m/s. (a) Calculate the velocity of A with respect to B and the velocity of B with respect to A. (b) If they start from the same corner, how far are they separated after 20.0 min?

3-35 A motorboat is headed directly across a 1.50-km-wide river and crosses in 0.300 h. During that time, the boat was swept downstream 0.900 km by the current. Calculate (a) the speed of the river with respect to the earth, (b) the direction the boat would have to head to go directly crossstream, and (c) the time to go directly across.

3-36 An ant is running at a speed of 0.100 m/s on a conveyor belt which is moving at 0.250 m/s. The ant's velocity makes an angle of 45° with the direction in which the belt is moving. (a) How long will it take the ant to cross the 0.800-m-wide belt? (b) How far will the ant have moved along the direction of the belt? (c) What is the ant's speed with respect to the earth while on the belt?

3-37 A motorboat on a lake is traveling south at 12.0 m/s. A pennant at the bow indicates the wind is blowing from a direction 30° S of W. A flag on shore shows the true wind direction is from 20° N of W. What is the magnitude of the wind's actual velocity?

3-38 A subway train is accelerating at 6.00 m/s². At the time when its speed is exactly 30.0 m/s, a passenger drops a book from a height of 1.50 m above a mark on the floor. (a) How much time will it take for the book to reach the floor? (b) Where will it hit?

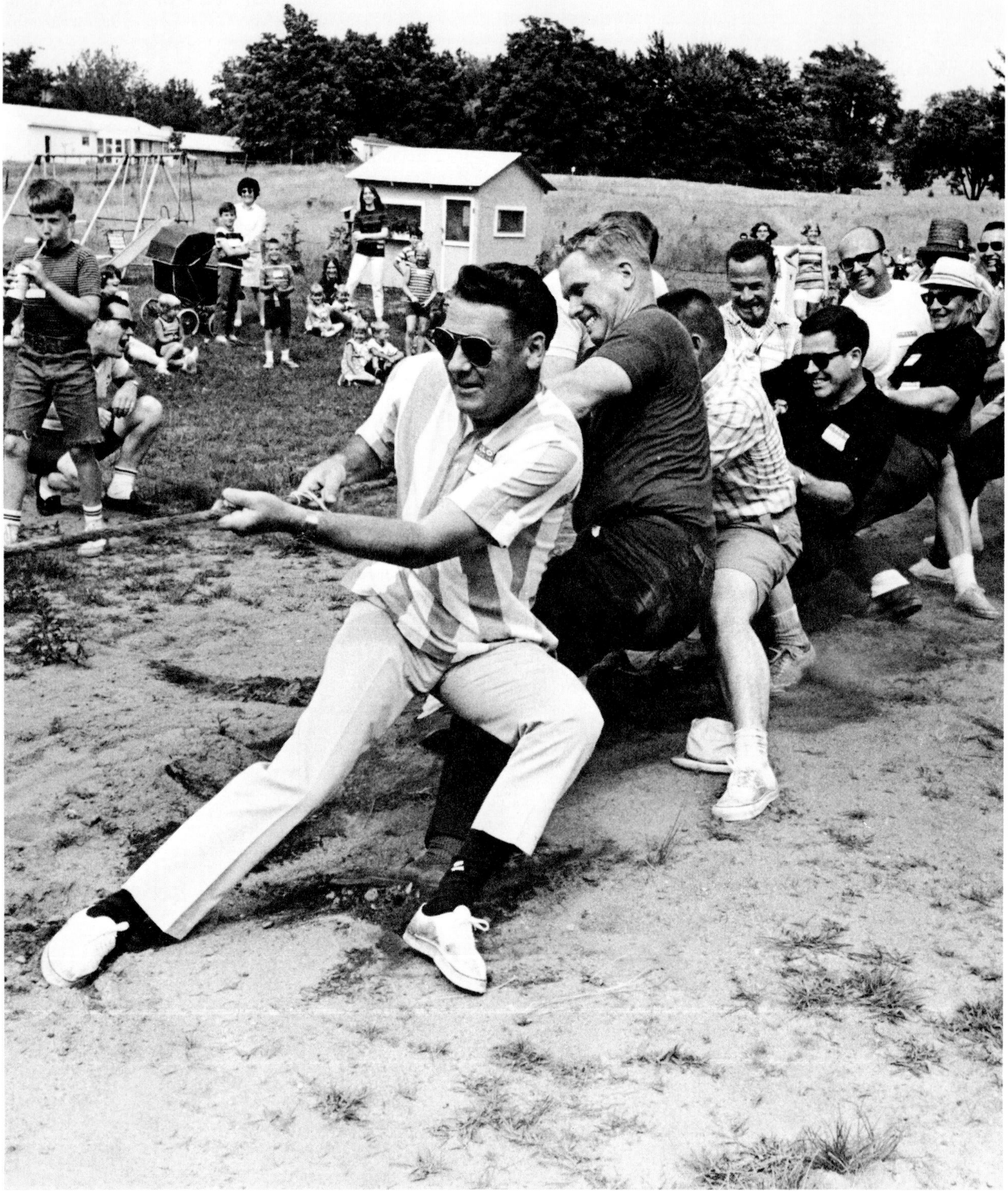

4 Newton's Laws of Motion

4-1 Introduction

After spending the last two chapters learning to describe how things move, we are now ready to consider the question "Why do things move?" The Greeks thought a push or a pull, a *force,* was necessary to produce motion, and in the absence of forces, moving objects would grind to a halt and remain motionless. They assumed that the natural tendency of matter is to be at rest, and that only in the continued presence of a push or pull of some type could this tendency be overcome. At first glance this may seem reasonable.

Galileo (Figure 4-1) pondered the causes of motion and began to modify the Greek viewpoint. He rolled balls down ramps of varying slopes to "dilute" the action of gravity to a point where he could measure the resulting motion. His first conclusion was that falling objects accelerate at a constant rate. He also noted that a ball coming off the ramp did not stop but continued rolling for an indefinite time. On a hard, flat surface, the ball would roll longer than on a rougher or softer surface. Galileo concluded that a force was necessary to stop the ball, and that in the absence of such a stopping force, the motion would persist. The natural state without forces was the persistence of whatever motion, or lack of motion, already existed.

Figure 4-1 Galileo Galilei (1564–1642) is often called the father of modern science. He was the great popularizer of the idea that the earth moved about the sun, which resulted in his trial and conviction by the Inquisition. Under house arrest for the last nine years of his life, he completed the studies of motion he had begun as a young man. In his *Discourses and Mathematical Demonstrations Concerning Two New Sciences* (1638), he established modern kinematics and laid the groundwork for the science of mechanics. (*New York Public Library*)

A sled on ice persists in its motion because the frictional stopping force provided by the ice is too small to perceptibly change the motion. This is not true if the sled reaches a patch of dry concrete and a large frictional force abruptly stops the sled. Galileo asserted that in the absence of all forces, including frictional forces, motion persists at constant speed in a straight line. Newton was born in the year Galileo died, and went on to produce a complete science of **dynamics,** the study of the causes of changes of motion.

4-2 Newton's First Law of Motion

In 1687 Sir Isaac Newton (Figure 4-2) published a three-volume work titled *Philosophiae Naturalis Principia Mathematica* (Mathematical Principles of Natural Philosophy). In it he asserts that experiment is the only correct method of establishing the validity of ideas. In his work, which is known simply as the *Principia,* Newton set forth three laws of motion which are the basis for dynamics. The **first law** establishes the so-called natural state:

> Every body continues in its state of rest, or of uniform motion in a straight line, unless it is compelled to change that state by forces impressed upon it.

4-2 Newton's First Law of Motion

Figure 4-2 Sir Isaac Newton (1642–1727) established modern mechanics, beginning with the unconnected, partially formed ideas which preceded him, and produced the first complete mathematical science. By the time he was 23 years old, he had framed the laws of motion, and the law of gravitation, and developed integral and differential calculus. Despite the fame his work brought him, Newton remained modest, writing, "If I have seen farther, it is by standing on the shoulders of giants." (*New York Public Library*)

This law is sometimes interpreted as stating that each object has an inherent property—**inertia**—which tends to maintain its state of motion (or rest). A measure of the inertia of a body is its **mass.** The larger the mass, the bigger is a body's inertia and the larger is the force required to create a given change in the body's state of motion. For this reason, Newton's first law is often called the **law of inertia.**[1]

Note that forces may indeed act on a body while the body maintains its state of motion. Two people pushing from opposite sides of a swinging door with the same force may have no apparent effect on the door's state of motion. The two forces do not compel a change, and we say they are balanced. We may find the first law easier to understand if we say that *a body will maintain a constant velocity* unless *acted upon by an unbalanced force.*

Consider the student shown at rest studying in Figure 4-3a. The forces acting on the student may be added vectorially to yield the resultant force **F**:

$$\mathbf{F} = \mathbf{F}_1 + \mathbf{F}_2 + \mathbf{F}_3 + \mathbf{F}_4 + \mathbf{F}_5 + \mathbf{F}_6 \qquad (4\text{-}1)$$

Since the student has a constant velocity (of zero), the first law asserts that **F** must be zero, as shown in the graphical addition of Figure 4-3b. The standard

---

[1] One author's interesting recollection of the principle of inertia occurred one early winter day. During a lecture in a windowless auditorium, a fine freezing rain had blanketed the campus. Students were not aware of this as they emerged from the building, and their first step off the dry concrete doorsill found them on a smooth ice surface with a slight downhill run for 2 m leveling off to a long, straight sidewalk. Boosted by this slight headstart, whole clusters of students slid smoothly past their intersection 5 m from the door, making heroic but useless efforts to turn onto the correct sidewalk. Before long a spectator's gallery inside the building was cheering each effort to turn and the equally desperate attempts to stay upright. In the absence of forces, a mass in motion tends to remain in motion at constant speed in a straight line—the principle of inertia.

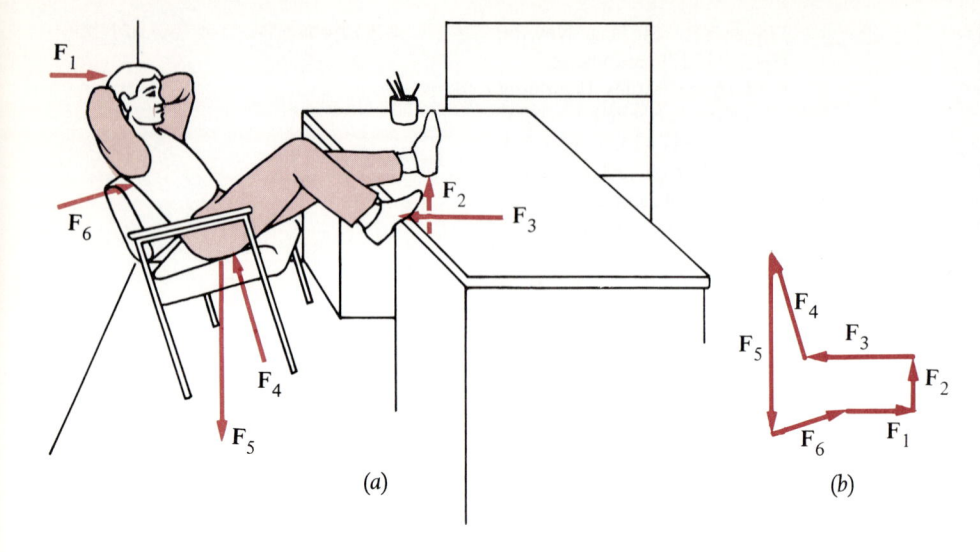

Figure 4-3 (a) The student is in a constant state of motion at rest. Newton's first law asserts that this lack of change is possible only if the vector sum of all the forces acting on the student is zero. (b) The forces acting on the student added as vectors. The head of $\mathbf{F}_6$ indeed ends exactly on the starting point for the tail of $\mathbf{F}_1$. Thus, the sum of the forces is zero, or, in symbols, $\Sigma_i \mathbf{F}_i = 0$.

way to write a situation involving many similar added terms is to use a summation sign Σ (capital sigma), with a dummy index or subscript to specify a general term. Thus Equation 4-1 could be written as

$$\mathbf{F} = \sum_{i=1}^{6} \mathbf{F}_i \qquad (4\text{-}2)$$

where the notation below and above the sigma indicates that the index i should be allowed to take on all values from 1 to 6. Equations 4-1 and 4-2 are identical mathematical statements, and we will use Σ as a summation sign as needed in the text.

We may now state the first law in symbols as

$$\text{If } \sum_i \mathbf{F}_i = 0, \text{ then } \mathbf{v} = \text{constant}$$

(where the general variable i on the Σ indicates that we must sum over *all* the forces acting). In words, the first law says that *in the absence of an unbalanced force, motion does not change*. Human history shows that this is neither a simple nor an obvious idea, so let us examine its implications:

1. It removes forces as the cause of motion: motion will continue without forces; motion is a natural state.

2. Perhaps just as important, it removes every other physical quantity and property from consideration as a cause of *changes* in motion. Length cannot change motion; mass cannot change motion; temperature cannot; color cannot; time cannot—only forces can change the state of a body's motion. The first law states that a body's motion remains constant unless *forces* cause a change.

3. It provides a qualitative definition of the word **force:** forces are quantities which may cause changes in the state of motion of material objects.

4-3 Newton's Second Law of Motion

The foundation for much of the science of mechanics is the rather concise statement of Newton's **second law:**

> The change of motion is proportional to the motive force impressed and is made in the direction of the straight line in which the force is impressed.

Newton defined "motion" as the product of mass and velocity and defined the "change" as the time rate of change. We may write Newton's second law in symbols. Suppose an object of mass m and velocity $\mathbf{v}$ is acted on by a resultant external force $\mathbf{F}$. Newton's second law applied to this case is

$$\mathbf{F} \propto \frac{\Delta(m\mathbf{v})}{\Delta t}$$

where the symbol $\propto$ means "is proportional to" and the right side of the proportionality is the rate of change of $m\mathbf{v}$ with time. We may always change a proportionality to an equation by inserting the appropriate constant. In systems of units which have already defined a unit for force, this is done, and the second law becomes

$$\mathbf{F} = k \frac{\Delta(m\mathbf{v})}{\Delta t}$$

The k is then specified by experimental measurement. In SI units, the decision is made to have the constant be *exactly equal* to 1 and allow the second law to determine the magnitude of the force unit we will use thereafter. In other words, in SI units we write Newton's second law as

$$\mathbf{F} = \frac{\Delta(m\mathbf{v})}{\Delta t} \tag{4-3}$$

In most instances, the mass of an object is constant. In those cases the numerator on the right of Equation 4-3 can be written

$$\Delta(m\mathbf{v}) = (m\mathbf{v})_2 - (m\mathbf{v})_1 = m(\mathbf{v}_2 - \mathbf{v}_1) = m\,\Delta\mathbf{v}$$

Then Equation 4-3 becomes

$$\mathbf{F} = m \frac{\Delta\mathbf{v}}{\Delta t} \tag{4-4}$$

or

$$\mathbf{F} = m\mathbf{a} \tag{4-5}$$

since, by definition, $\mathbf{a} = \Delta\mathbf{v}/\Delta t$. Thus, whenever the mass of an object is constant, Newton's second law states that the resultant force acting on the object equals the product of the body's mass with its acceleration. Both the force and acceleration are vector quantities and must be in the same direction. All vector equations require the right and left sides to be numerically equal in magnitude and to agree in direction. Newton's second law is so often used in this form that Equation 4-5 comes to many people's minds when the name of the law is stated. While this is not entirely wrong (it's certainly an example of the second law), we should try to remember that the second law is more than Equation 4-5, and we will present some problems in which the mass does change. Recall also that $\mathbf{F}$ is the resultant, or net unbalanced, force (sometimes simply called the net force) acting on the object. With this in mind, we might often write Equation 4-5 as

$$\sum \mathbf{F} = m\mathbf{a} \qquad (4\text{-}6)$$

This, of course, means that the vector sum of *all* the forces acting on an object is equal to the product of its mass and its acceleration.

We noted that the first law provided a qualitative definition of the word *force*. Forces are quantities which may cause changes in motion. Now we see that the second law gives a *quantitative* definition of force. Equations 4-3 through 4-6 tell us how to obtain the numbers. We find the magnitude of the resultant force acting on an object by measuring its mass and its acceleration. Then we multiply those quantities together. We see from Equation 4-5 that the unit of force equals the unit of mass times the unit of acceleration. In SI units this means that the unit of force is a kilogram-meter per second squared ($kg \cdot m/s^2$). To honor Sir Isaac Newton, this unit is called a **newton** (symbol N). Thus

$$1 \text{ N} = 1 \text{ kg} \cdot \text{m/s}^2$$

From Equation 4-5 we see that an unbalanced force of 1 N will impart an acceleration of 1 m/s^2 to a mass of 1 kg. Also, a force of 2 N will impart an acceleration of 1 m/s^2 to a mass of 2 kg *or* an acceleration of 2 m/s^2 to a mass of 1 kg, and so on.

Example 4-1 A student of mass 70.0 kg is walking at constant velocity of magnitude 3.00 m/s down the hall when he meets a friend coming out of a room. If he stops in 0.400 s, what is the magnitude of the average stopping force between his feet and the floor?

Solution We must first find the magnitude of the student's acceleration and then apply Equation 4-5. From the definition of the average acceleration,

$$a = \frac{\Delta v}{\Delta t} = \frac{0 - 3.00 \text{ m/s}}{(0.400 - 0) \text{ s}} = -7.50 \text{ m/s}^2$$

The minus sign means that the acceleration is directed opposite to that of the initial velocity. Since we are seeking magnitudes only, we ignore the sign, and writing Newton's second law as a scalar equation,

$$F = ma = (70.0 \text{ kg})(7.50 \text{ m/s}^2) = 525 \text{ N}$$

Example 4-2 Two forces are applied to the block of wood of mass 2.50 kg shown in Figure 4-4. The magnitude of $\mathbf{F}_1$ is 5.00 N, and the magnitude of $\mathbf{F}_2$ is 20.0 N. What is the acceleration of the block?

Solution In this problem we must consider the vector nature of force. However, since only one dimension is involved, we can indicate directions with plus and minus signs. Taking the direction to the right as positive, the forces become

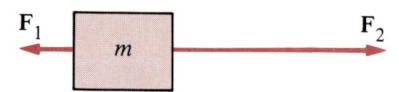

Figure 4-4 Example 4-2.

$F_1 = -5.00$ N and $F_2 = +20.0$ N. Applying Newton's second law in the form of Equation 4-6, we have

$$\sum \mathbf{F} = m\mathbf{a}$$
$$\mathbf{F}_1 + \mathbf{F}_2 = m\mathbf{a}$$
$$\mathbf{a} = \frac{\mathbf{F}_1 + \mathbf{F}_2}{m} = \frac{(-5.00 + 20.0) \text{ N}}{2.50 \text{ kg}} = \frac{15.0 \text{ N}}{2.50 \text{ kg}} = +6.00 \text{ m/s}^2$$

Example 4-3 The forces shown acting on the lead brick in Figure 4-5a have magnitudes 20.0, 30.0, and 25.0 N for $\mathbf{F}_1$, $\mathbf{F}_2$, and $\mathbf{F}_3$, respectively, and act at the angles shown. The magnitude of the brick's acceleration is 4.24 m/s². Calculate (a) the direction of the acceleration and (b) the mass of the brick.

Solution We need to remember that Newton's second law tells us that the direction of the resultant force and the acceleration are identical. Also, we may determine the mass by simply dividing the magnitude of the resultant force by the magnitude of the acceleration. Thus we must calculate the resultant force. Its direction is the answer to part (a), and its magnitude can be used to calculate the answer to part (b).

(a) Figure 4-5b shows the force vectors. The resultant force may be found by adding the separate forces as vectors:

$$\mathbf{F} = \mathbf{F}_1 + \mathbf{F}_2 + \mathbf{F}_3$$

Then
$$F_x = F_{1x} + F_{2x} + F_{3x}$$
$$F_y = F_{1y} + F_{2y} + F_{3y}$$

and
$$F_{1x} = -20.0 \cos 60° = -20.0(0.500) = -10.0 \text{ N}$$
$$F_{1y} = +20.0 \sin 60° = +20.0(0.866) = +17.3 \text{ N}$$
$$F_{2x} = +30.0 \cos 36.9° = +30.0(0.800) = +24.0 \text{ N}$$
$$F_{2y} = +30.0 \sin 36.9° = +30.0(0.600) = +18.0 \text{ N}$$
$$F_{3x} = +25.0 \cos 30° = +25.0(0.866) = +21.7 \text{ N}$$
$$F_{3y} = -25.0 \sin 30° = -25.0(0.500) = -12.5 \text{ N}$$

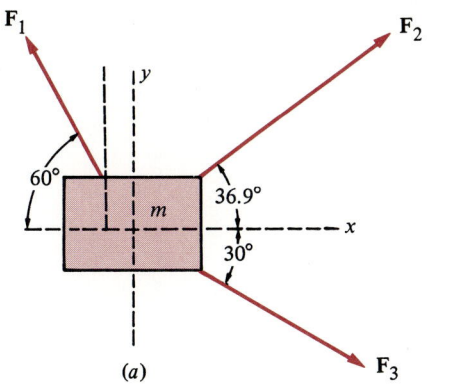

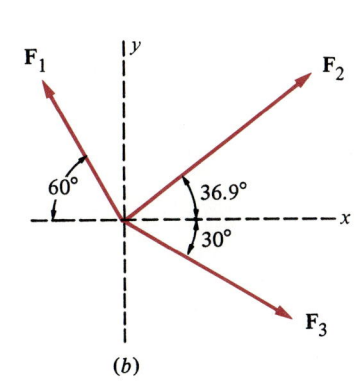

Figure 4-5 Example 4-3.

From which
$$F_x = -10.0 \text{ N} + 24.0 \text{ N} + 21.7 \text{ N} = +35.7 \text{ N}$$
$$F_y = +17.3 \text{ N} + 18.0 \text{ N} - 12.5 \text{ N} = +22.8 \text{ N}$$
$$F = \sqrt{F_x^2 + F_y^2} = \sqrt{(35.7 \text{ N})^2 + (22.8 \text{ N})^2} = 42.4 \text{ N}$$

and
$$\tan \theta = \frac{22.8}{35.7} = 0.639$$

$$\theta = 32.6° \text{ above the positive } x \text{ axis}$$

(b)
$$m = \frac{F}{a} = \frac{42.4 \text{ N}}{4.24 \text{ m/s}^2} = 10.0 \text{ kg}$$

4-4 Weight Distinguished from Mass

In Chapter 2 we dropped a standard (16.0-lb) bowling ball to measure the acceleration due to gravity, g. A balance reveals the mass of the bowling ball to be 7.26 kg, and when released it fell downward with an acceleration of 9.80 m/s². With what force is the ball being pulled downward to have this acceleration? Newton's second law provides an immediate answer:

$$F = ma = (7.26 \text{ kg})(9.80 \text{ m/s}^2) = 71.1 \text{ N downward}$$

This value of 71.1 N of downward force is the **gravitational force** acting on the ball to accelerate it downward and is called the **weight** of the ball. The weight of a body is that force exerted on the body by the pull of gravity. Actually, 16.0 lb is also the weight of the ball but reported in force units of the British engineering system. In SI units forces are given in newtons, and weight is a particular force. From this example, we may infer that

$$71.1 \text{ N} = 16.0 \text{ lb}$$

Thus $\quad 1.00 \text{ N} = 0.225 \text{ lb} \quad$ or $\quad 1.00 \text{ lb} = 4.45 \text{ N}$

The calculations of the weight of a body from a knowledge of its mass and the mass of a body from a knowledge of its weight are both *frequently encountered* and are *quite important*. The basic relationship between force and mass is Newton's second law in the form of Equation 4-5. In the gravitational case this becomes

Force due to gravity = mass times acceleration due to gravity

or
$$\text{Weight} = m\mathbf{g}$$

$$\mathbf{w} = m\mathbf{g} \qquad (4\text{-}7)$$

The weight of an object is a force, a vector quantity. It is proportional to the acceleration due to gravity — the proportionality constant being the object's mass. Equation 4-7 shows that mass and weight are very different quantities. Mass is a scalar quantity, while weight is a vector quantity. Mass is measured in kilograms, while the units of weight are newtons (kg · m/s²). Everyday usage of the measures of these quantities has had the effect of obliterating the distinction. First, we appear to ignore the vector nature of weight. Everybody knows it acts straight down, so we simply do not state the direction of the vector quantity. We simply say that "She weighs 520 N" or "He has a weight of 850 N." The direction is understood and unstated. This perhaps encourages the unwary to accept that no direction is needed for weight. Second, we often seem to interchange the quantities or consider them equivalent in common usage.[1]

On the earth where the acceleration due to gravity has a magnitude of 9.80 m/s², an object of mass equal to 1 kg has a weight of magnitude 9.80 N. If we moved the object to the surface of the moon, we would find that its weight now had a magnitude of 1.62 N. Clearly the amount of material of the object and hence its mass would not have decreased. There is a change in the object's weight because of the different acceleration due to gravity on the moon. If the object were projected very far away from all other bodies to an unoccupied corner of space, its weight could be immeasurably small, while its mass remained 1 kg. We often find it convenient to ignore the vector nature of weight and write Equation 4-7 as

$$w = mg \qquad (4\text{-}8)$$

The downward direction is accepted as being understood because "everybody knows that!"

4-4 Weight Distinguished from Mass

The distinction between weight and mass and the calculation of one from the other should be given great attention until mastered once and for all, right now. Failure to distinguish between weight and mass is one of the most common student errors in problems and on tests.

Example 4-4 A typical college physics textbook weighs 15.0 N while lying on your desk. Calculate (a) its mass and (b) its weight if it were transported to the moon. The acceleration due to gravity at the surface of the moon is $g_M = 1.62$ m/s².

Solution We may apply Equation 4-8 at each of the two locations.
(a) At your desk, $g = 9.80$ m/s². Thus $w = mg$ becomes

$$m = \frac{w}{g} = \frac{15.0 \text{ N}}{9.80 \text{ m/s}^2} = 1.53 \text{ kg}$$

(b) Now the weight on the moon is

$$w = mg_M = (1.53 \text{ kg})(1.62 \text{ m/s}^2) = 2.48 \text{ N}$$

(downward, of course).

4-5 Newton's Third Law of Motion

If you push your hand downward on your desk, you can feel the desk pushing upward on your hand. As you push on a shopping cart handle to accelerate it down the supermarket aisle, you sense the handle pushing back against your hands. Forces never occur singly. They always occur in pairs which are equal in

---

[1] For example, the evening newscaster might say, "Agents confiscated two kilos of heroin in the largest drug bust in the city's history." He might add, almost as an afterthought, "A kilo is one kilogram and is equal to two and two-tenths pounds." The physics student should (perhaps silently) object that, a "*mass* of one kilogram has a *weight* of magnitude about two and two-tenths pounds near the surface of the earth."

4 Newton's Laws of Motion

Figure 4-6 A leftward force is accelerating the boxer's head in that direction. This force is matched by a rightward force on the fist, decelerating it in the leftward direction. It is as likely that the rightward force may break the hand as it is that the leftward force may break the jaw except for the fact that the hand has a great deal more padding. *(UPI/Bettmann Newsphotos)*

magnitude, oppositely directed, and on two *different* bodies. This has become known as Newton's **third law:**

> To every action there is always opposed an equal reaction, or the mutual actions of two bodies upon each other are always equal and directed to contrary parts.

As with the second law, we must be careful not to misinterpret Newton's statement. In present-day language, the third law might be clearer if we say "Whenever one object exerts a force on a second object, the second object exerts a force equal in magnitude in the opposite direction on the first." It must be very clear that the two forces to which the third law refers (often called an action-reaction pair) *always act on different bodies.* If an automobile towing a float in a parade exerts a force of 1000 N to the west on the float, the float exerts a force of 1000 N to the east on the automobile. Neither force in an action-reaction pair could exist without the other (Figure 4-6). Forces occur only in matched sets, equal in magnitude and opposite in direction *at each instant of time,* with neither force coming before the other in time.[1]

---

[1] It is almost unfortunate that the English version (the *Principia* was in Latin) uses the words *action* and *reaction*, since each force in the pair is really the reaction to the other. In English, the temptation is to interpret "action and reaction" as "cause and effect." You put your coins in a candy machine (a cause) and, subsequently, a candy bar is (you hope) dispensed from the machine (an effect). This is certainly *not* the sort of thing Newton's third law addresses as action and reaction.

Newton's third law can be written in symbols as

$$\mathbf{F}_{AB} = -\mathbf{F}_{BA} \quad (4\text{-}9)$$

which may be read as "the force of body A acting on body B is equal to the negative of the force of body B acting on body A." Figure 4-7 illustrates both the first law and the third law. The bowling ball is at rest on a laboratory table. Since its state of motion is constant, we know that its weight $\mathbf{F}_{EB}$ is being balanced by another force. We have shown this as $\mathbf{F}_{TB}$, our designation for the normal (perpendicular) force being applied to the ball by the table. Even though these two forces must be equal in magnitude and oppositely directed to balance each other, they both are acting on the ball. Thus they are *not* an action-reaction pair according to Newton's third law because action-reaction pairs act on *different* bodies. We may clarify this distinction by writing all forces in the form of Equation 4-9. Thus $\mathbf{F}_{EB}$ is the force of the earth on the ball, the ball's weight, and can *only* be part of a pair with $\mathbf{F}_{BE}$, the force the ball exerts on the earth; those forces exist even if the table should vanish.

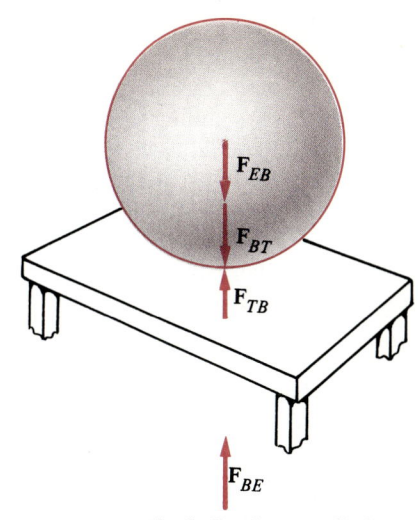

Figure 4-7 The ball is being pulled toward the earth by its weight, and the table is exerting an upward force on the ball, holding it in place. This is *not* an action-reaction pair since both forces are acting on the ball.

4-6 Tension

We often transmit a force by using a cord or rope of some sort. We infer that the pull acts along the length of the cord, with no component to either side since the cord is free to move to the side if pulled in that direction. Forces which tend to pull matter apart along some axis are called tensile forces, or simply **tension**. The force transmitted by a cord is a tension, frequently represented by the symbol T. We may visualize each small segment of the cord being pulled by both neighboring segments, tending to stretch it, and the segment pulling back on each to avoid a change in length. Picture a weight suspended at rest from the ceiling by a length of thread. If the weight is to remain at rest, the thread must exert an upward force equal in magnitude to the weight. If the thread is cut into two parts, an upward force equal in magnitude to the weight must be applied to the part attached to the weight to keep it from falling. Further, a downward force equal in magnitude to the weight must be applied to the part attached to the ceiling to keep the loose thread from snapping upward and to keep its state of motion unchanged. Thus, the tension in the thread is equal in magnitude to the weight.

In Figure 4-8a, we see this illustrated by two masses linked by a length of cord which is held taut because m_2 is being pulled to the right by the hand

The important distinction about a tension in a rope or cord is that this force must act along the line of the rope. If any sideways component of force exists, the rope is free to move to the side in response. This is very different from the forces between rigid bodies. For example, the forces on a hammer do not have to be along the line of your arm, or that of its handle.

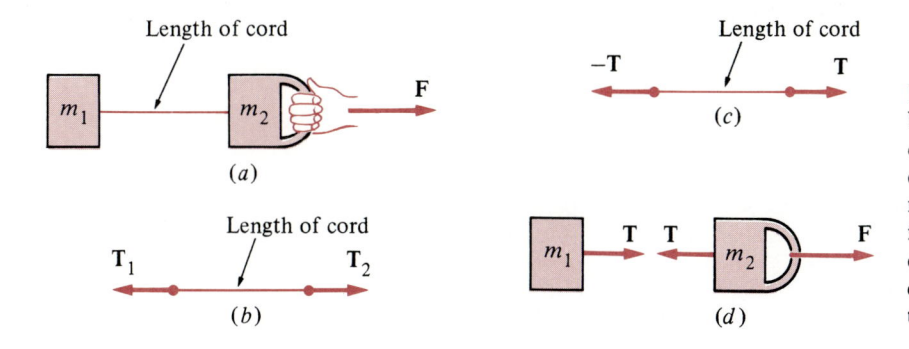

Figure 4-8 (a) Masses m_1 and m_2 are being pulled to the right. (b) The connecting cord has the forces $\mathbf{T}_1$ and $\mathbf{T}_2$ exerted on it by the masses. (c) If the mass of the cord is negligible, the two forces are equal in magnitude and opposite in direction. (d) Each block drawn as an isolated free body with all the forces acting on it shown.

65

gripping the handle on m_2. The masses are both exerting a force on the cord. The mass m_2 is pulling to the right on the cord, and Newton's third law tells us that the cord is pulling to the left on m_2. Also, unless the cord breaks or stretches freely, the cord is pulling to the right on m_1. Finally, m_1 is pulling to the left on the cord. In Figure 4-8b, we show just the length of cord of Figure 4-8a with the forces being exerted on it. Suppose the mass of the cord is very small and can be neglected with respect to m_1 and m_2. Then the magnitude of $\mathbf{T}_1$ is practically equal to that of $\mathbf{T}_2$, since their difference is the net unbalanced force on the cord. By the second law, this unbalanced force must be equal to the mass of the cord times its acceleration. But, if the mass of the cord is negligible and it is accelerating with the system, the unbalanced force on it must be negligible also and $T_1 \approx T_2$. Of course, if the cord is at rest or moving with a constant velocity, then Newton's first law insists that the magnitudes of T_1 and T_2 are precisely equal. When this is so, we might use the symbol T for both forces, as in Figure 4-8c. In this and most cases which we will consider, the cord simply transmits the force.

Example 4-5 In Figure 4-8a, m_1 is 4.00 kg, m_2 is 6.00 kg, and **F** has a magnitude of 20.0 N. Assume the mass of the cord is negligible, and calculate (a) the acceleration of the masses and (b) the tension in the cord.

Solution Newton's second law may be applied to any mass individually even when it is part of a system of masses being considered as a whole. Here we use a technique called **isolating the bodies** to consider each mass separately. Figure 4-8d shows each mass individually with the forces acting on it. Note that we have indicated the tension in the cord acting on each mass as having a magnitude T.

(a) Choosing the positive direction to the right, we now apply Newton's second law to each mass separately and obtain

for m_1: $\qquad T = m_1 a$

for m_2: $\qquad F - T = m_2 a \qquad$ or $\qquad T = F - m_2 a$

Equating these two expressions for T, we get

$$F - m_2 a = m_1 a$$

$$20.0 \text{ N} - (6.00 \text{ kg})(a) = (4.00 \text{ kg})(a)$$

$$a = \frac{20.0 \text{ N}}{10.0 \text{ kg}} = 2.00 \text{ m/s}^2$$

(b) Now using the first equation, we can determine the tension:

$$T = (4.00 \text{ kg})(2.00 \text{ m/s}^2) = 8.00 \text{ N}$$

Example 4-6 A platform is supported by a single cable, as in Figure 4-9a. A block of marble sits on a bathroom scale placed in the center of the platform. The combined mass of the platform and scale is 20.0 kg. The marble has a mass of 10.0 kg. Calculate the reading on the bathroom scale if the tension in the cable is (a) 294 N, (b) 196 N, and (c) 392 N.

Solution We expect the bathroom scale to display a reading equal to the force it is exerting on the block. That force, which we might designate F_s, and the

(a)

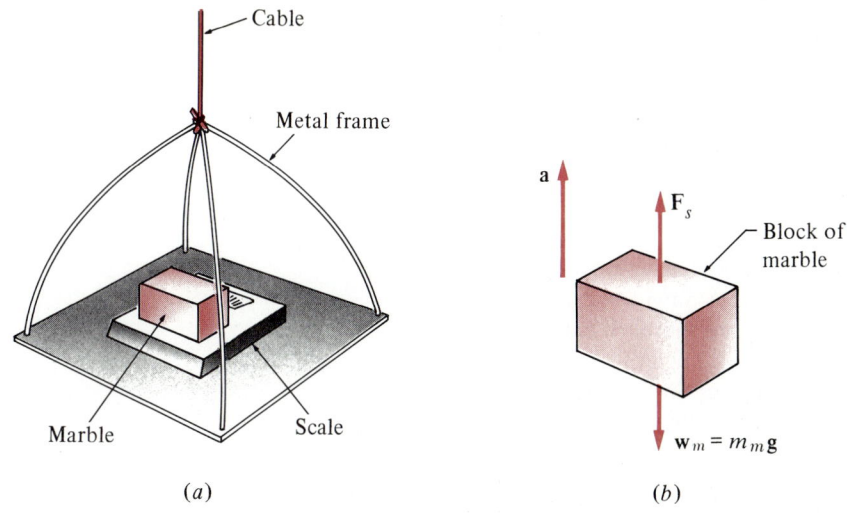

Figure 4-9 Example 4-6.

weight of the marble are the only two forces acting to change the state of motion of the marble. Therefore we may obtain one equation which includes F_s as an unknown by applying Newton's second law to the marble. In Figure 4-9b, we have isolated the body to make this application easier to see.

Choosing the upward direction as positive and assuming the block is accelerating upward, we may use $F_{net} = ma$; here, the net force is $F_s - w_m$, and therefore

$$F_s - w_m = F_s - (m_m g) = m_m a$$

or

$$F_s - 98.0 \text{ N} = (10.0 \text{ kg})(a)$$

Now all we need do to determine F_s is to find the acceleration of the marble for each of the tension values listed in the problem. The acceleration of the marble is the same as that of the platform and the bathroom scale. We know that the entire system has a mass of 30.0 kg, hence a weight of (30.0 kg)(9.80 m/s²) = 294 N.

(a) When the tension in the cable is 294 N, the resultant force on the system is zero and so is the acceleration. From this we have the scale reading as

$$F_s = 98.0 \text{ N} + (10.0 \text{ kg})(0) = 98.0 \text{ N}$$

(b) When the tension in the cable is 196 N, the net force acting on the system is −98.0 N (196 N − 294 N and choosing upward as positive). We determine the acceleration by applying Newton's second law to the whole system:

$$T - w = -98.0 \text{ N} = ma$$

$$a = \frac{-98.0 \text{ N}}{30.0 \text{ kg}} = -3.27 \text{ m/s}^2$$

and the scale reading can be found from

$$F_s - 98.0 \text{ N} = (10.0 \text{ kg})(-3.27 \text{ m/s}^2)$$

$$F_s = 65.3 \text{ N}$$

(c) With the tension in the cable 392 N, we again apply Newton's second law to the whole system and have

$$F = T - w = ma$$

$$a = \frac{T - w}{m} = \frac{(392 - 294) \text{ N}}{30.0 \text{ kg}} = 3.27 \text{ m/s}^2$$

and
$$F = 98.0 \text{ N} + (10.0 \text{ kg})(3.27 \text{ m/s}^2) = 131 \text{ N}$$

Thus we see that in part (a) the scale was simply balancing the weight of the marble. In part (b) the system was accelerating downward ($a = -3.27$ m/s^2). If the scale supported no part of the weight of the marble block, the block would accelerate downward at 9.80 m/s^2 (and right through the scale). Instead, the scale supported some (65.3/98.0) of the weight of the marble, leaving an unbalanced force on the block to cause it to accelerate downward with the rest of the system. In (c) the scale not only had to exert the 98.0 N upward to support the weight of the marble but an additional 32.7 N to accelerate it upward at 3.27 m/s^2. These numbers parallel your experience in a high-speed elevator. When the elevator is not accelerating, its floor interacts with you like any other floor, supporting your weight. When accelerating downward, you feel the floor drop out from under you and the force against your feet lessens. Upward acceleration gives you the opposite feeling, that of being crushed against the floor, as the force exerted between feet and floor is increased so that the upward force can overcome your inertia and additionally accelerate you upward.

4-7 Friction

Any force which always acts to oppose the start or continuance of sliding motion between surfaces is a **frictional force,** or **friction.** If you thrust a book across a desktop, it slides away from your hand at some initial velocity but quickly comes to rest. To change its velocity from the value at which it left the contact with your hand to zero, the book had to have a negative acceleration (a deceleration). What provided this force to decelerate the book to rest? By Newton's laws, the book should continue its motion at constant velocity forever in the absence of forces. Since it does not, we are required to conclude that there is a force acting on the book parallel to the tabletop in the direction opposite its motion, the direction of the deceleration. If we go around to the other side of the desk and repeat the experiment, we find that this force does not now act to increase the velocity but again acts to oppose and stop the motion. Indeed, no matter which direction we slide the book across the desk, the force always acts to oppose the motion and bring the book to rest. Thus, this force is an example of a frictional force.

Experiment shows that there are at least two distinct frictional forces operating. If the book is initially at rest, it requires a greater force to break it free from its original contact with the table and start it moving than it does to keep it moving once it has been started. This is quite generally the case. We call the frictional forces opposed to the beginning of motion **static friction** (variable up to a maximum value) and those opposed to the continuance of motion **kinetic friction** (constant in magnitude). In both cases we find that:

1. The magnitude of the frictional force does *not* depend upon the area of contact between the objects.

2. The magnitude of the force varies greatly with the nature of the two surfaces in contact.
3. For given surfaces, the maximum magnitude of the frictional force is approximately proportional to the magnitude of the force pressing the surfaces into contact. This is called the **normal,** or perpendicular, **force.**

This third point allows us to obtain a mathematical expression for the frictional force. Designating the frictional force by f and the normal force by F_N, we have

$$f \propto F_N$$

We may write this in equation form by the inclusion of an appropriate constant of proportionality. For the case of *static friction,* we write

$$f_s \leq \mu_s F_N \qquad (4\text{-}10)$$

where μ_s is called the **coefficient of static friction** and depends upon the nature of both the surfaces which are in contact. Since both f_s and F_N are in force units, μ_s is unitless. Note that f_s is less than or equal to the right side of Equation 4-10. This recognizes that f_s for the undisturbed book on the table is zero, and that f_s grows as you push harder on the book, eventually to reach its maximum value just before the book breaks free of the initial contact. At the instant before it breaks free, we say **motion impends.**

For *kinetic friction,* we have

$$f_k = \mu_k F_N \qquad (4\text{-}11)$$

where f_k is the force of kinetic friction and μ_k is called the **coefficient of kinetic friction,** which is unitless and depends upon the nature of the two surfaces sliding over each other. Table 4-1 lists values of μ_s and μ_k for various sets of surfaces.

TABLE 4-1 Typical Values of Average Coefficients of Friction*

Surfaces	μ_s	μ_k
Steel on steel (mild)	0.74	0.47
Steel on steel (hard)	0.78	0.42
Aluminum on steel	0.61	0.47
Copper on steel	0.53	0.36
Cast iron on cast iron	1.10	0.15
Copper on cast iron	1.05	0.29
Most metals (light-oil lubrication)	0.15	0.08
Glass on glass	0.94	0.40
Steel on ice	0.10	0.06
Steel on Teflon	0.04	0.04
Wood on leather	0.5	0.4
Waxed wood on dry snow	0.05	0.03
Waxed wood on wet snow	0.15	0.10
Smooth rubber on wet concrete	0.5	0.4
Grooved rubber on wet concrete	0.8	0.7
Rubber on dry concrete	0.9	0.8
Unwaxed wood on dry snow	...	0.08
Bone on bone	...	0.03
Bone on joint (synovial fluid lubrication)	...	0.003

*Actual values of these coefficients may vary widely from these averages. Values depend on moisture and other possible lubricants, condition of surface (rough or smooth), direction of grain of wood, etc.

Example 4-7 A professor wants to slide a box of books weighing 800 N across the classroom floor. If $\mu_s = 0.600$ and $\mu_k = 0.450$ between the floor and the box, calculate (a) the horizontal force required to get the box moving and (b) the horizontal force necessary to keep the box moving at constant speed.

Solution From the free-body diagram shown in Figure 4-10, we recognize that the box has no acceleration in the vertical direction, and thus the normal force pressing the floor to the box is equal in magnitude to the weight of the box of books. Also we use the equal sign in Equation 4-10 since we must raise the horizontal force to a magnitude equal to the maximum static frictional force to cause motion to begin. Thus

(a) $$f_s = \mu_s F_N = (0.600)(800 \text{ N}) = 480 \text{ N}$$
(b) $$f_k = \mu_k F_N = (0.450)(800 \text{ N}) = 360 \text{ N}$$

Note that had we kept the horizontal force at the value required to begin the motion of the box, 480 N, we would have exceeded the force required to maintain the motion against kinetic friction by some 120 N, and this unbalanced force would have accelerated the box horizontally.

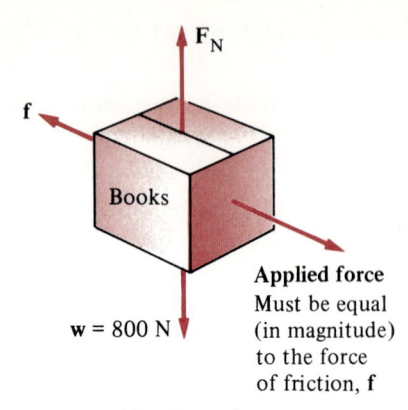

Figure 4-10 Example 4-7.

Example 4-8 A 60.0-kg swimmer emerges from the pool and runs along the poolside at 5.00 m/s. If the coefficient of static friction between wet skin and ceramic tile is 0.350, calculate (a) the maximum force of static friction which can act on the swimmer, (b) the shortest distance in which the swimmer can stop without sliding, and (c) the distance required to stop if $\mu_k = 0.250$ and the swimmer slides to a stop.

Solution Again, Equations 4-10 and 4-11 must be applied to find the forces of friction. Those forces can then be used to obtain the relevant accelerations from which the stopping distances may be calculated.

(a) The normal force pressing the swimmer to the poolside surface is the swimmer's weight mg. Thus

$$f_s = \mu_s F_N = \mu_s mg = (0.350)(60.0 \text{ kg})(9.80 \text{ m/s}^2) = 206 \text{ N}$$

(b) Assuming that the swimmer can keep the static friction at this maximum value and not begin sliding, we may calculate the acceleration from Newton's second law as

$$a = \frac{f_s}{m} = \frac{206 \text{ N}}{60.0 \text{ kg}} = 3.43 \text{ m/s}^2$$

Given the initial velocity and the acceleration, finding the distance required to stop is a kinematics problem, and we use Equation 2-13. Choosing the original direction as positive, we have $a = -3.43$ m/s² since friction is acting to stop the motion.

$$2a \, \Delta s = v^2 - v_0^2$$

$$\Delta s = \frac{v^2 - v_0^2}{2a} = \frac{0 - (5.00 \text{ m/s})^2}{2(-3.43 \text{ m/s}^2)} = 3.64 \text{ m}$$

(c) This analysis is nearly identical to (a) and (b) but uses Equation 4-11 since we have kinetic friction. Thus

$$f_k = \mu_k F_N = (0.250)(60.0 \text{ kg})(9.80 \text{ m/s}^2) = 147 \text{ N}$$

and
$$a = \frac{f_k}{m} = \frac{147 \text{ N}}{60.0 \text{ kg}} = 2.45 \text{ m/s}^2$$

$$\Delta s = \frac{v^2 - v_0^2}{2a} = \frac{0 - (5.00 \text{ m/s})^2}{2(-2.45 \text{ m/s}^2)} = 5.10 \text{ m}$$

Comparing the two distances, we see that the swimmer can stop in a shorter distance under the action of static friction. This is difficult to do, however, since the force at a stationary foot will grow until it exceeds f_s and sliding begins. As one foot begins to slide, it is necessary to shift the weight to the other foot immediately and to keep repeating this stutter-step until motion ceases. This action becomes intuitive with practice, and you have probably seen someone doing it without realizing why it works to shorten the stopping distance.

Example 4-9 The two coefficients of friction between two pieces of material are often determined with a block and a plane surface, as shown in Figure 4-11a. The plane surface is hinged at one end to allow adjustment of the angle the plane makes with the horizontal. The block and plane may be made of the same or of different materials. The block is placed on the plane, and the angle of the plane is increased slowly until the block breaks loose and begins its motion. This angle θ_s is maintained, and the block accelerates down the plane. The angle is quickly reduced to a smaller angle θ_k so that the block will slide down the plane at constant speed. A careful experiment, repeated many times, reveals that $\theta_s = 37°$ and $\theta_k = 23°$ for a particular set of surfaces. Calculate μ_s and μ_k for these surfaces.

Solution Figure 4-11b shows a free-body diagram of the block, where details outside the block are omitted and all forces acting on the block are shown. Subscripts have been omitted, as this figure could represent either the static or the kinetic case. In either case, the block is not accelerating, and we apply Newton's first law (if $\Sigma \mathbf{F} = 0$, then $\mathbf{v} = $ constant, and vice versa) to the block. We must recognize that since the plane is not horizontal, the normal force is *not* simply the weight of the block but only a component of the weight. Choosing

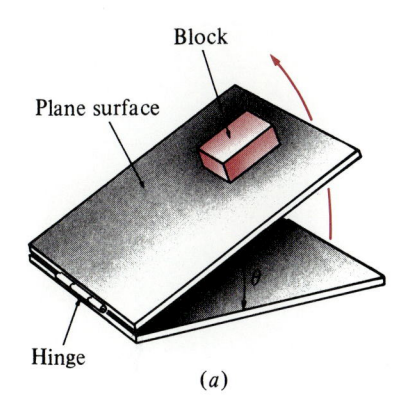

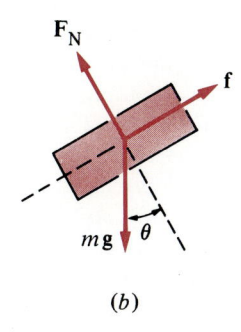

Figure 4-11 Example 4-9. Note that the angle θ the weight vector makes with the perpendicular to the plane in part *b* is the same as the angle of the plane from the horizontal in part *a*.

upward along the plane as the positive x direction and perpendicularly outward from the plane through the block as the positive y direction, we write the first law as two scalar equations:

$\Sigma F_x = 0$:
$$f - mg \sin \theta = 0$$
$$f = mg \sin \theta$$

and

$\Sigma F_y = 0$:
$$F_N - mg \cos \theta = 0$$
$$F_N = mg \cos \theta$$

In both cases (static and kinetic friction),

$$\mu = \frac{f}{F_N} = \frac{mg \sin \theta}{mg \cos \theta} = \tan \theta$$

Thus,
$$\mu_s = \tan \theta_s = \tan 37° = 0.75$$
$$\mu_k = \tan \theta_k = \tan 23° = 0.42$$

While this is a perfectly general result, the values obtained by using a tangent of the angle are only as good as the measurements taken and their reproducibility. It is notoriously difficult to be precise with frictional measurements. For barely perceptible reasons, different parts of the same plane or different faces of the same block may behave differently. Thus, all numbers are averages representing a greater or smaller range of measured values. Values of μ_s or μ_k are generally known to one or two significant figures only. For our work we will ignore this, but engineers producing useful devices from real materials cannot.

The invention of the wheel introduced **rolling friction** as a substitute for sliding friction. Just as kinetic friction is generally lower than static friction, rolling friction is generally much lower than either. A 100.0-N steel block resting horizontally on a steel plane will require a force of 74.0 N to break free and move ($\mu_s = 0.74$). Once moving, it will require a force of 47.0 N to have the block slide at a constant speed across the surface ($\mu_k = 0.47$). If the same block is fashioned into a cylinder, however, only 0.15 N is required to have it roll across the surface at constant speed. Thus, $\mu_R = 0.0015$, where $f_R = \mu_R F_N$ for rolling surfaces.

Finally, we point out the difference between Equations 4-10 and 4-11 and physical laws. These two equations are **empirical** relationships—they derive from experience only, not from some insightful general statement summarizing the behavior of nature. Our knowledge of friction depends upon observation alone with practically no theory to explain it or help us predict whether frictional forces should be large or small in a particular situation. Frictional statements hold true for only certain surfaces and then for only a limited range of forces. Newton's laws, as an example of physical laws, are held to be correct at all levels of *direct* human experience and admit *no* exceptions until we stray far from that range (to near the speed of light or to submicroscopic sizes). Further, there is no way to prove a physical law by using more basic principles and some fancy mathematics—there are no physical principles more basic. Physical laws are held to be correct simply because no violation of them has ever been shown.

Minimum Learning Objectives

After studying this chapter, you should be able to:
1. Define:
   - coefficient of friction
   - dynamics
   - empirical
   - force
   - frictional force
   - gravitational force
   - impending motion
   - inertia
   - kinetic friction
   - mass
   - newton [unit]
   - normal force
   - rolling friction
   - static friction
   - tension
   - weight
2. State in words and by practical example:
   a. The principle of inertia
   b. Newton's second law of motion
   c. The distinction between static and kinetic friction
   d. The meaning of an action-reaction pair of forces
3. Calculate the force required to produce an observed acceleration on an object of given mass.
4. Determine the weight of a given mass at the earth's surface.
5. Determine the mass of a given weight at the earth's surface.
6. Determine the normal component of a given force acting against a surface of given geometry.
7. Given a force, determine its appropriate reaction force.
8. Resolve a situation of multiple forces and objects to determine the net unbalanced force acting on each object.
9. Determine the coefficient of friction of a surface from given forces and masses.
10. Apply the appropriate value of the coefficient of friction to determine the magnitude of the force opposing motion.

Problems

4-1 Determine the magnitude of the force which must be applied to a 500-kg wagon parallel to a hill of angle 20° to keep the wagon from rolling down the hill.

4-2 Two forces of 15.0 N, directly south, and 20.0 N, directly east, act on a mass. What single third force is required to keep the mass in uniform motion?

4-3 An automobile of mass 1.50×10^3 kg accelerates from rest to a speed of 20.0 m/s in 12.0 s. Calculate the magnitude of the net force acting on the auto.

4-4 What is the acceleration of a bowling ball of mass 7.00 kg on which is exerted an average force of 210 N?

4-5 Calculate the magnitude of the force a playful cat applies to a 0.0350-kg ball of yarn which gives the ball an acceleration of 12.0 m/s².

4-6 Two forces $F_1 = 10.0$ N and $F_2 = 25.0$ N act at 90° to each other on a mass of 15.0 kg. Calculate the resultant acceleration.

4-7 Calculate the average force a major league catcher must apply to stop the pitcher's fastball in a distance of 0.300 m. The speed of the ball when it reaches the mitt is 45.0 m/s. Take the mass of the ball to be 0.142 kg.

4-8 A bullet traveling at 250 m/s strikes a large tree and is brought to rest in 3.00×10^{-3} s. Calculate (a) the average acceleration of the bullet and (b) the average force exerted by the tree if the mass of the bullet is 2.00×10^{-2} kg.

4-9 After coal is unloaded at an electric generating station, the engine is to pull the 100 railcars along a level stretch to a point where the train can be loaded with solid waste for transport to a disposal site. The engine is pulling the empty cars through a steady rain. Each car is taking on water at a constant rate of 0.850 kg/s (0.244 gal/s). Neglecting frictional losses, calculate the force the engine must provide to pull the train at a steady speed of 4.00 m/s. (*Hint:* The mass of the train is changing with time, while the velocity is to remain constant.)

4-10 A tennis ball traveling to the east at 45.0 m/s is struck by a racket (held by a player, of course), and the force of impact changes its velocity to 40.0 m/s to the west. If the ball was in contact with the racket for 0.0400 s, calculate (a) the average acceleration of the ball during contact and (b) the average force during contact if the mass of the ball is 0.0570 kg.

4-11 Sand is allowed to flow vertically in a steady stream from a hopper onto a conveyor belt (Figure 4-12). The motor driving the belt can provide a maximum force of 80.0 N. Neglecting friction, calculate (a) the maximum speed of the

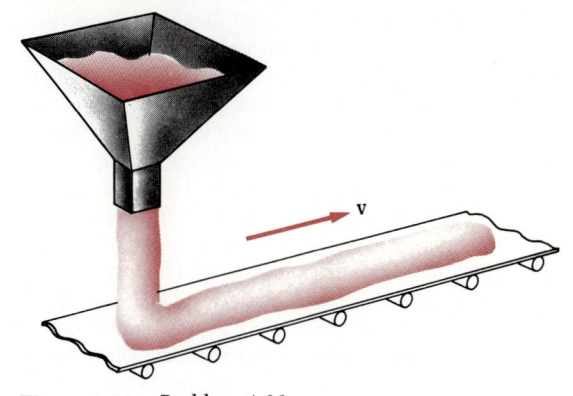

Figure 4-12 Problem 4-11.

73

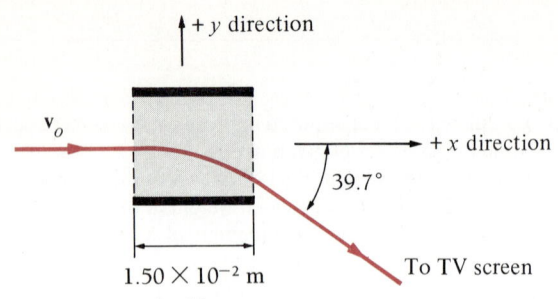

Figure 4-13 Problem 4-12.

belt if sand is dropping on the belt at the rate of 50.0 kg/s and (b) the maximum amount of sand per second which can be dropped if the belt is to move at a constant speed of 4.00 m/s. (*Hint:* The mass is changing with time, while the velocity is constant.)

4-12 The image on a television screen is formed by a beam of very high speed electrons striking a phosphorescent material on the inside of the tube. This beam is moved horizontally and vertically by the application of electric forces perpendicular to the beam's original direction. Figure 4-13 shows the electrons moving in the x direction with a speed of 5.00×10^6 m/s as they enter a region where a force is applied in the negative y direction. The region in which the force exists is 1.50×10^{-2} m long, and as the electrons leave, their velocity is 6.50×10^6 m/s at 39.7° below their original direction. (a) Calculate the constant acceleration the electrons undergo. (b) If the mass of each electron is 9.11×10^{-31} kg, determine the magnitude of the electric force.

4-13 An upward force of 1.20×10^4 N acts on an elevator of mass 2.00×10^3 kg. Calculate the elevator's acceleration.

4-14 A 60.0-kg sprinter starts a 100-m dash, exerting an accelerating force of 400 N. Calculate (a) the sprinter's initial acceleration and (b) the time interval this force must be exerted to bring the sprinter to a speed of 9.00 m/s.

4-15 The record jump recorded in the Calaveras County annual frog jumping contest noted in the *Guinness Book of World Records* was 17 ft 6.75 in (5.35 m) by a frog with a mass of 0.415 kg. Assuming that the jump was at a 45° angle to the horizontal and that the push-off phase of the jump lasted 0.0500 s, (a) what average force was generated in the frog's leg muscles to produce the jump? (b) Using the fact that the value of the maximum force per cross-sectional area generated by muscle tissue is 78.0 N/cm², calculate the minimum diameter of each of the frog's thighs.

4-16 A bullet of mass 0.0300 kg leaves the muzzle of a rifle with a speed of 500 m/s. The rifle barrel is 0.700 m long. Calculate (a) the average acceleration the bullet undergoes and (b) the average force on the bullet.

4-17 An 80.0-kg stuntman with a frontal body area of 0.800 m² jumps off a 5.00-m-high platform and does a bellyflop onto a sandpile, leaving an average depth of impression of 0.100 m in the sand. (a) What is the average force on his body while he is landing in the sand? (b) What fraction of the force per unit area required to produce tissue damage (approximately 5.50×10^5 N/m²) does he sustain? (c) Suppose instead he lands feet first (area = 6.00×10^{-2} m²) and creates a pit 0.500 m deep. What would the average force be?

4-18 A rubber stopper weighing 0.400 N in a flask being heated by a freshman chemistry student suddenly accelerates 2.00×10^{-2} m until it leaves the flask and flips straight upward to the ceiling 2.50 m above. It just barely makes it to the ceiling with zero speed before it falls back down. Calculate (a) the stopper's speed as it left the flask, (b) the stopper's average acceleration while in the flask, and (c) the average net force exerted on the stopper as it was accelerating upward.

4-19 A 4.00×10^{-3} kg bullet is brought to a stop by penetrating 3.00×10^{-2} m into a block of wood. (a) If the muzzle velocity of the gun firing the bullet is 400 m/s, what was the average stopping force? (b) Assuming the force to be approximately constant, how long does the wood take to stop the bullet? (c) The bullet has a cross-sectional area of 1.50×10^{-4} m². The breaking strength of bone is about 1.00×10^8 N/m². Would this bullet be able to shatter a leg bone if it hit a broad surface flat on?

4-20 You are standing on a bathroom scale placed on the floor of an elevator and note the scale reading while you are stopped. Shortly after the door closes, you sense motion and note that the scale reading is now 1.25 times your weight. (a) Determine the magnitude and direction of the elevator's acceleration. (b) You remain in the elevator for several stops and at one point glance down and note that the scale reads 0.600 times your weight. Determine the magnitude and direction of your acceleration during this reading. Can you determine from your acceleration whether the elevator is moving upward or downward?

4-21 A wheel of cheese weighing 125 N is hung from a spring scale attached to the roof of an elevator. What will the scale read when (a) the elevator accelerates upward at 3.00 m/s² and (b) the elevator accelerates downward at 4.00 m/s²?

4-22 Table 4-1 shows that the typical coefficient of static friction between smooth rubber and wet concrete is 0.500 and the coefficient of kinetic friction between smooth rubber and wet concrete is 0.400. Using these values and assuming you wish to bring your automobile to a stop from a speed of 25.0 m/s, calculate the stopping distance if (a) you jam on the brakes and lock the wheels so the tires slide, (b) you apply steady pressure to the brakes so the tires do not slide.

4-23 A boxer wishes to land a right punch on his opponent. His glove is 1.00 m from the opponent's jaw, and the opponent has a reaction time of 0.100 s. (a) How great an acceleration is necessary to have the punch land before the opponent can react and duck? (b) If the boxer's arm and glove have an effective mass of 5.00 kg, how great a force must be exerted by his muscles to deliver the punch? (c) Assume that essentially all this force is delivered by the deltoid muscle of the shoulder (it isn't). The force that may be generated by a muscle is proportional to its cross-sectional area. Muscles generate a force of about 78.0 N/cm² (100 lb/in²). What is the minimal cross-sectional area of the boxer's deltoid muscle necessary to deliver the blow in parts a and b?

4-24 An airbag in a car is designed to provide 0.300 m of stopping distance for the passenger in a collision. (a) If the driver is normally some 0.500 m from the steering wheel, how long does the airbag have to inflate in a 25.0 m/s collision that stops the automobile? (b) How long will the airbag have to stop the motion of the driver? (c) If an area of 0.400 m² of the driver contacts the airbag, what is the average force of the airbag on the driver during the negative acceleration? (Assume an effective upper-body mass of 40.0 kg for the driver.) (d) Given that skin and muscle tissue can split at about 5.50×10^5 N/m², is tissue damage likely to result during the stopping process? (e) Without the airbag, the driver would come to rest during impact with the wheel and dashboard in 0.002 s. Compare the negative acceleration of this situation with that when using an airbag.

4-25 An astronaut on Mars discovers that it requires a force of 1.80 N to impart an acceleration of 3.20 m/s² to a large wrench. Calculate (a) the mass of the wrench and (b) its weight ($g_{Mars} = 3.74$ m/s²).

4-26 Using a tow rope of breaking strength 2.70×10^3 N, what is the maximum acceleration one can impart to a 1.80×10^3 kg automobile?

4-27 Determine your weight in newtons and your mass in kilograms. How much would you weigh on the moon?

4-28 An astronaut takes a scale with her to the surface of Enceladus, one of the moons of Saturn, and discovers she weighs 190 N on that surface. Assume that she weighed 950 N (with all her equipment) on the earth and has not lost or gained any weight during the journey. Calculate (a) the astronaut's mass and (b) the acceleration due to gravity on Enceladus.

4-29 An automobile with driver (total mass 1.20×10^3 kg) starts up a hill of 15° at 25.0 m/s. If the engine is shut off and the gearshift placed in neutral, calculate (a) the normal force acting on the automobile, (b) the net force acting down the hill, (c) the acceleration, and (d) the distance the automobile moves up along the hill before it stops. (Neglect kinetic friction.)

4-30 An Atwood machine consists of two masses connected by a string and hanging over a pulley. Neglecting the mass of the string and the friction of the pulley, calculate the acceleration of the system and the tension in the string if one of the masses is 0.800 kg and the other is 0.950 kg.

4-31 A small ball is hung from the roof of a bus by a long string. Determine the angle the string makes with the vertical when the bus has a horizontal acceleration of 2.50 m/s².

4-32 A rope climber has a mass of 60.0 kg. Calculate the average tension in the rope when the climber is (a) moving upward at a constant velocity, (b) accelerating upward at

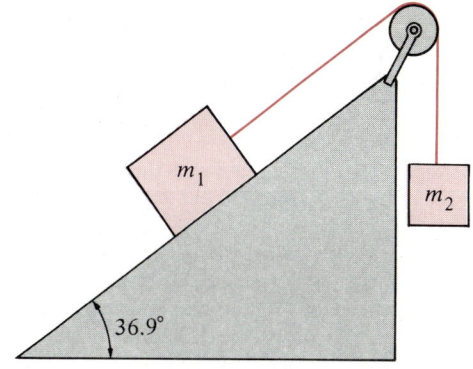

Figure 4-14 Problem 4-34.

1.20 m/s², and (c) accelerating downward at 1.80 m/s².

4-33 Consider a boy swinging a bucket of water in a vertical circle. List *all* the forces that you can imagine are involved in the situation.

4-34 If the pulley in Figure 4-14 introduces no real or apparent friction, the mass of the cord connecting the two blocks is negligible, and μ_k between m_1 and the inclined plane is 0.450, calculate the acceleration of the masses and the tension in the cord ($m_1 = 6.00$ kg, $m_2 = 8.00$ kg).

4-35 Consider a stone on a level place on the earth. Make a drawing showing all the forces acting, and list the forces.

4-36 In Figure 4-15, $m_1 = 6.00$ kg, $m_2 = 9.00$ kg, and $m_3 = 8.00$ kg. Assume that friction and the mass of the cords are negligible. Determine the acceleration of the system and the tension (T_A and T_B) in each of the cords.

4-37 A box of books of mass 22.0 kg is sliding across the floor, and its (negative) acceleration is measured to be 4.00 m/s². Calculate (a) the magnitude of the frictional force and (b) the coefficient of kinetic friction.

4-38 An automobile starts from rest and accelerates smoothly to a speed of 20.0 m/s in 8.00 s. If the automobile has a mass of 1.50×10^3 kg and the tires do not slip, (a) what net

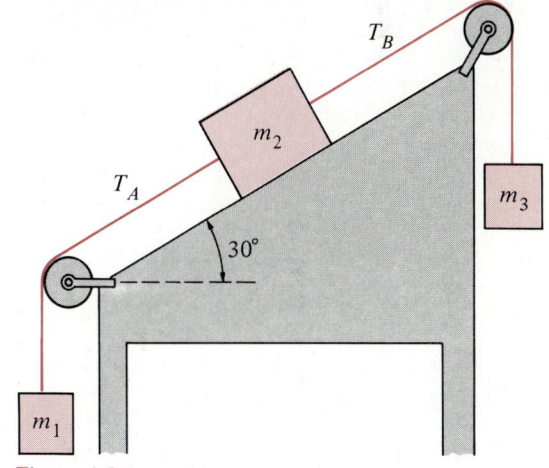

Figure 4-15 Problem 4-36.

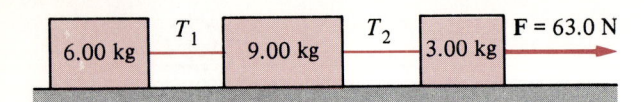

Figure 4-16 Problem 4-40.

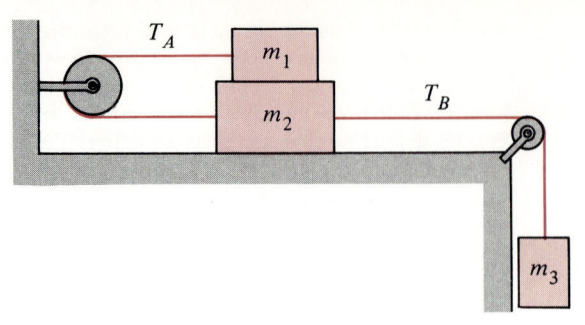

Figure 4-18 Problem 4-45.

force must the tires exert parallel to the road? *(b)* Determine the minimum coefficient of static friction for which this happens.

4-39 A box of sandwiches and soft drinks is placed in the aisle of a tour bus moving at a speed of 20.0 m/s. If the coefficient of static friction between the box and the aisle floor is 0.500, calculate the shortest distance in which the bus can be stopped without making the box slide.

4-40 Determine the acceleration of the system and the tension in each of the cords in Figure 4-16 *(a)* if friction is negligible and *(b)* if the coefficient of kinetic friction between the blocks and the surface is $\mu_k = 0.300$.

4-41 A student pushes her book against a vertical wall with a force of 25.0 N. Determine the minimum coefficient of static friction between the book and wall such that the 1.50-kg book will remain stationary.

4-42 A sled with a child (total mass 55.0 kg) is pulled on a level surface by a rope making an angle of 30° with the horizontal. If the rope has a tension of 100 N, calculate *(a)* the normal force (the force being exerted upward) on the sled, *(b)* the coefficient of kinetic friction if the sled is moving at a constant velocity, and *(c)* the coefficient of kinetic friction if the sled has a forward acceleration of 0.100 m/s².

4-43 The box of books shown in Figure 4-17 weighs 200 N and is being lowered down the ramp. The coefficient of kinetic friction between the box and the ramp is 0.300. *(a)* Calculate the force of kinetic friction acting on the box. *(b)* Determine the force a person must apply to lower the box at constant speed.

4-44 A car leaves 40.0 m of skid marks in attempting to stop before a collision. The marks indicate that the wheels were locked and sliding over the dry concrete surface. Damage indicates that the speed at collision was 30.0 km/h. How fast was the car going when the driver applied the brakes?

4-45 Referring to Figure 4-18, $m_1 = 4.00$ kg, $m_2 = 8.00$ kg, and $m_3 = 12.0$ kg. The coefficient of kinetic friction between m_1 and m_2 is 0.500, and the coefficient of kinetic friction between m_2 and the table is 0.400. If m_3 is released from rest, calculate the acceleration of the system and the tensions T_A and T_B in the two cords.

4-46 Table 4-1 lists values for the coefficient of kinetic friction of waxed and unwaxed wood on dry snow. Let us determine the effect of a skier waxing the skis before use. A hill has a 50.0-m-long slope at 25° to the horizontal. Calculate the final speed of the skier coming down this hill with *(a)* waxed skies and *(b)* unwaxed skies. *(c)* Calculate the final speed for the same skier using waxed skis but on wet snow, instead of the dry "powder" skiers prefer.

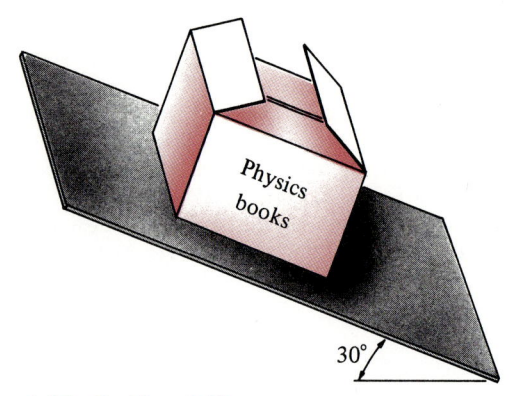

Figure 4-17 Problem 4-43.

76

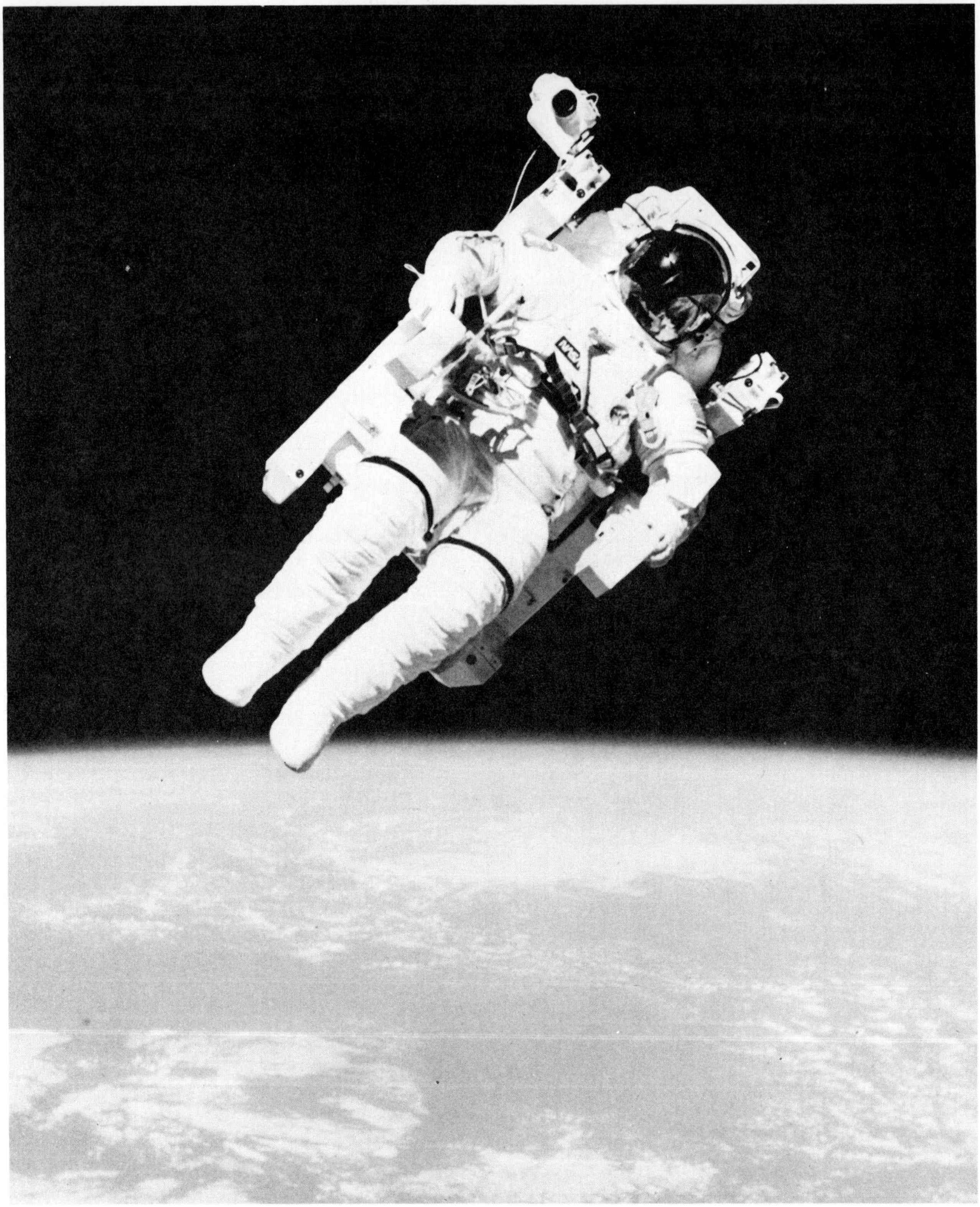

5 Rotational Kinematics and Gravity

5-1 Introduction

In this chapter we will extend our description of motion to rotating systems, which will then allow us to consider Newton's law of gravitation and apply it to the solar system, the most studied rotating system throughout human history.

Objects may move without changing their overall position. A phonograph record on a rotating turntable constantly occupies the same volume of space. Up to now we have considered the motion of an entire object from place to place, which is called **translational motion.** If we move a can of soda on our desk 40 cm to the right, each point on the can (or in it) ends up displaced 40 cm to the right. However, a particular point which is 1.00 cm north of another point on the can remains 1.00 cm north of that point after the motion is completed. During translational motion, any two points on a body or system maintain a constant relative displacement. This is not the case in rotational motion. While two points on a rotating phonograph record maintain a constant separation of distance, the direction from one to the other is constantly changing. During **rotational motion,** the distance between two points on an object or system is fixed but the line joining one point to the other changes orientation.

A spinning top may display a pure rotational motion, and a falling bowling ball may be an example of a purely translational motion. Quite commonly, however, both types of motions are executed at the same time. A yo-yo spins as it moves up and down its string, and the earth rotates on its axis as it moves through space about the sun. All motions of bodies or systems of bodies, regardless of how complex, may be separated into a series of translational and rotational parts occurring simultaneously. We can already describe the translational motion of the examples given. Our first task in this chapter is to develop the ability to describe rotational motions.

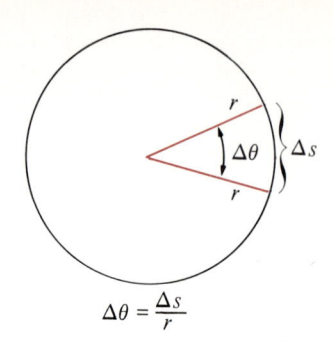

Figure 5-1 The angle $\Delta\theta$ in radians is defined as the arc length Δs divided by the length of the radius r.

5-2 Angular Measure

Angles are used as a measure of the divergence of two intersecting lines. We may imagine the angle as a measure of the amount of turning necessary to bring one of the lines into coincidence with the other. You are familiar with the use of degrees (°) as a measure of angle. Two sides which are mutually perpendicular have an angle of 90° between them. A line turned through a complete circle about one end and returned to its original position will have turned through 360°; thus we say that there are 360° in a circle.

The main objection to using a circle divided into degrees is that it introduces unnecessary complexity into formulas involving angles. For example, **F** = m**a** in a system *only* when units of **F** are chosen such that one unit of *m* times one unit of **a** is exactly one unit of **F**; otherwise, **F** = km**a**, where *k* is a constant necessary to adjust the varying units. (**F** in pounds does *not* equal *m* in grams times **a** in light-years per century per second!) Likewise, angles in degrees would introduce constants in a wide variety of equations, making them needlessly complex. There is, however, a natural system of measure for angles based on the relationships of parts of a circle which avoids making formulas carry conversion constants. This is the system called radian measure. Imagine that we put a length of string on a circle and cut it to the exact length of the circle's radius. If we then lay this string along the circle's circumference, the ends of the string will subtend an arc of definite size as seen from the center of the circle. We call this angle one **radian**. A radian is the angle subtended by one radius of length stretched out along the circle's circumference.

Consider an arbitrary angle in Figure 5-1, where we see two lines extending from the center of a circle intersecting points on its circumference. The points of intersection define an arc length of Δs between the lines, which are separated by an angle of $\Delta\theta$. By definition, the angle $\Delta\theta$ measured in radians is equal to the arc length Δs divided by the radius r of the circle:

$$\frac{\Delta s}{r} = \Delta\theta \quad \text{rad} \quad (5\text{-}1)$$

A *radian* is the angle, as seen from the center of the circle, subtended by its radius when stretched out along its circumference. *Learn this very well* so that the use of radians for angular measure becomes comfortable. Experience indicates that students who skip past this idea pay for it again and again.

Note that as a measure of angle the radian has no dimensions at all; it is simply a length divided by a length, which is unitless. Of course, Δs and r must both be lengths measured in the same system of units. We will often carry the word *radian* (or, more likely, the abbreviation rad) along as if it were a unit to confirm that we are using radian measure rather than some other angular measure. We will also sometimes find it convenient to treat the radian as a unit in dimensional analysis. Nevertheless, an angle measured in radians is unitless.

We can see from Equation 5-1 that $\Delta\theta$ is exactly 1 rad when the arc length subtended is exactly equal to the length of one radius. Furthermore, whenever $\Delta\theta$ represents a complete circle, we see that $\Delta s = 2\pi r$ (the circumference), and

$$\Delta\theta \text{ (circle)} = \frac{\Delta s}{r} = \frac{2\pi r}{r} = 2\pi \text{ rad}$$

Thus 2π rad $= 360°$, or 1 rad $= 360°/2\pi = 57.3°$. Table 5-1 shows these and other equivalencies.

We frequently encounter the word *revolution* (abbreviated r) in connection with rotational or angular motion. Although it is often not considered a true measure of angle, we can certainly make a connection between revolutions and other angular measures. A body which has turned through one complete revolution has swept out $360°$, or 2π rad. Thus

$$1 \text{ r} = 360° = 2\pi \text{ rad}$$

Finally, using Equation 5-1, we may calculate the linear distance traveled by a point on a rotating body. To do this, we need only know the angle through which the body turned and the radius (distance from the point in question to the center of rotation).

TABLE 5-1

Radians	Degrees
2π	360°
$3\pi/2$	270°
π	180°
$2\pi/3$	120°
$\pi/2$	90°
$\pi/3$	60°
1	57.3°
$\pi/4$	45°
$\pi/6$	30°

Any hand calculator beyond the simplest available is set up to handle radian measure directly. Learn how your calculator converts to handle angles entered in radians [usually one button (DRG) or a slide switch (R-D)] and practice using them; e.g., what is $\sin(\pi/4)$ rad?

Example 5-1 The wire on a rotating spool of electric cable is at an average distance of 0.800 m from the center of the spool. A worker pulls out a length of cable Δs, and the spool turns through an angle of 27.0 rad. What length of cable was unwound from the spool?

Solution From Equation 5-1, we see that $\Delta s = r\, \Delta\theta$ and

$$\Delta s = (0.800 \text{ m})(27.0 \text{ rad}) = 21.6 \text{ m}$$

Example 5-2 If a horse on a merry-go-round is 4.00 m from the center and turns through one-third a revolution ($\tfrac{1}{3}$ r), what linear distance does it travel?

Solution We must convert to radians and use Equation 5-1:

$$\tfrac{1}{3} \text{ r} = \tfrac{1}{3}(2\pi) \text{ rad}$$

Now,

$$\Delta s = r\, \Delta\theta = (4.00 \text{ m})(\tfrac{2}{3}\pi \text{ rad}) = 8.38 \text{ m}$$

5-3 Angular Velocity

Consider a complex but quite ordinary motion, as in Figure 5-2. This shows a gymnast executing a motion called a walkover. Since the body is upright at the beginning and at the end of the motion and upside down at the center, it is clear that a rotation has taken place. A flashing stroboscope, recording the gymnast's position every 0.1 s, reveals an overall leftward motion of the entire body accompanied by a simultaneous counterclockwise revolution. Consider a point near the center of the gymnast's body, between the hips at the approximate position of the navel. If we mark this point for each image in the photograph, we obtain a record of its motion, as in Figure 5-3. We see that it records a right-to-

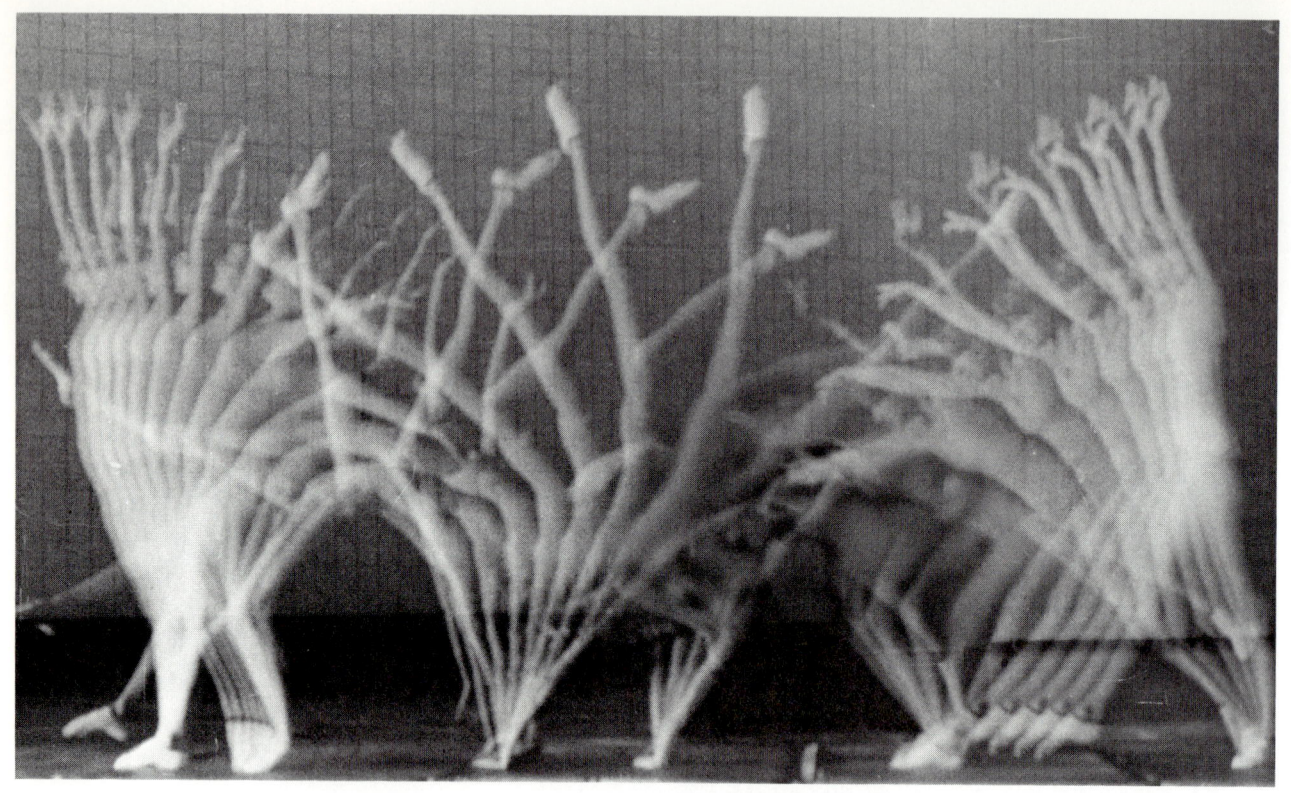

Figure 5-2 A gymnast doing a walkover. (D. Riban)

left motion at increasing then decreasing speed as the maneuver is executed, accompanied by an initial dip, some rise and fall, and then a final rise to the end position. The vertical motion of this point is surprisingly small for this complicated movement (indeed, the smaller it is, the higher the judges' score is likely to be). Taking the same point as the origin for each image, we may draw in a displacement vector to the top of the gymnast's head, as in Figure 5-4a. We see that the motion begins with this vector near vertical, from where it begins an increasingly rapid counterclockwise rotation to end up near vertical and not rotating. The pattern in this sequence is seen better if we draw all the separate displacement vectors as radiating from a common point, as in Figure 5-4b. Of course, this is exactly what they do in reality, as each had its origin at the same given point on the gymnast. Disregarding any additional complexity, as in the separate motions of the limbs, we have separated the motion into its two component parts, a right-left translational motion in Figure 5-3 and a simultaneous rotation about a point internal to the body in Figure 5-4. We may now analyze the rotational motion.

With very few exceptions, we will deal with one rotation at a time about one axis at a time. Thus the sense of the rotation will be either clockwise or counterclockwise about the axis, and we may keep track of this sense of rotation with plus and minus signs. By simply choosing a convenient sense of rotation as positive and the other as negative, the sign of any quantity automatically tells us whether it is clockwise or counterclockwise. For the motion of the gymnast shown, the rotation is counterclockwise, and we will choose this direction (sense of rotation) as positive.

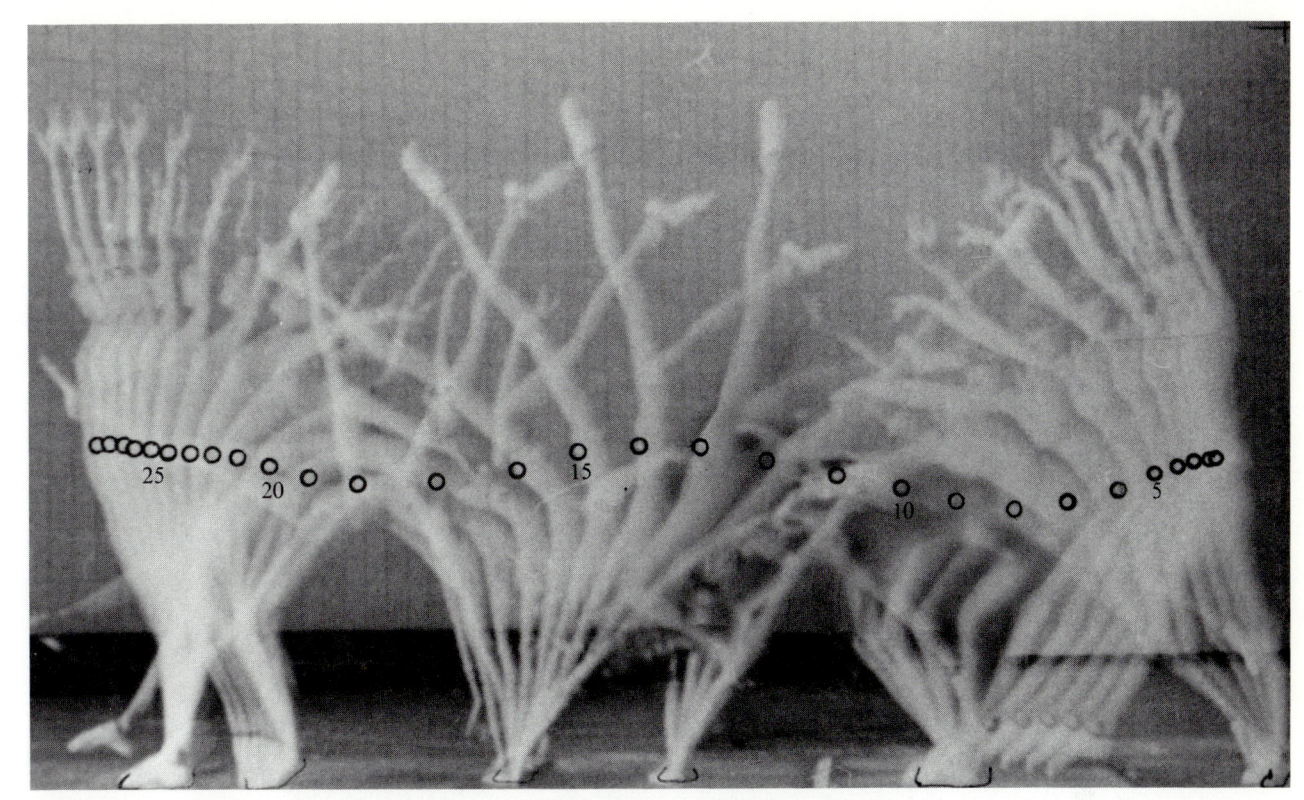

Figure 5-3 A central point on the gymnast is the approximate center of rotation of the body, and its motion is almost purely translational. The high point indicated by circles 1 and 2 is equalled at circles 13 and 14 and at the final circles. The lowest point is at circle 8, where all the gymnast's weight has been transferred to the forward foot.

Angular velocity is defined as the time rate at which an angle is changing. Standard practice uses Greek letters to represent angular quantities, and angular velocity is represented by a lowercase omega (ω). Thus, from the definition, we have

$$\omega = \frac{\Delta\theta}{\Delta t} = \frac{\theta_2 - \theta_1}{t_2 - t_1} \qquad (5\text{-}2)$$

The units of angular velocity are radians per second (rad/s).

As an example, we may measure in Figure 5-4 that the upper body of the gymnast rotates through a quarter of a revolution, $\frac{1}{2}\pi$ rad, between the first and ninth positions shown. Since each of these positions is 0.1 s apart, this represents a time interval of $\Delta t = (0.9 - 0.1)$ s $= 0.800$ s. We calculate the gymnast's average angular velocity for the interval to be

$$\omega = \frac{\Delta\theta}{\Delta t} = \frac{\frac{1}{2}\pi \text{ rad}}{0.800 \text{ s}} = 1.96 \text{ rad/s}$$

This implies that if the gymnast continued turning at this rate, she would make (1.96 rad/s)/(2π rad/turn) = 0.313 turns/s, which is called the **frequency** f of the repeating motion. Conversely, she would require 1/(0.313 turns/s) =

83

(a)

Figure 5-4 (a) A vector indicating the position of the trunk for each image extends from the center of rotation to the top of the gymnast's head. (b) To separate the rotational motion from the translational motion, we redraw the vectors to have them radiate from a stationary center of rotation. The rate of rotation is greatest in the region from 17 to 21.

3.20 s/turn to make each complete rotation. This is called the **period** T of the motion and is the reciprocal of frequency: $T = 1/f$.

As we have done with the definition of linear velocity, we might write Equation 5-2 in more general form as

$$\omega = \frac{\theta - \theta_0}{t - t_0} \tag{5-3}$$

where θ represents a variable angle occurring at a time t and θ_0 is the angle at time t_0 when we started our measurement. The analogy with linear motion continues if we agree to start our clock and measurement at the same time. Then $t_0 = 0$, and we may solve Equation 5-3 as

$$\Delta\theta = \omega t \tag{5-4}$$

Note the analogy to Equation 2-11, substituting change in angular displacement for change in displacement and angular velocity for linear velocity.

(b)

Example 5-3 Calculate the average angular velocity of the second hand of a wall clock.

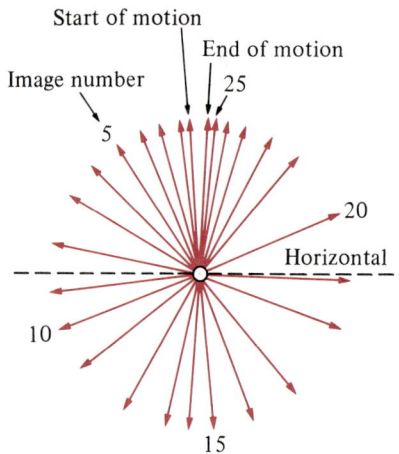

84

Solution The second hand turns through an angle of 2π rad in 60.0 s. From the definitions, we have

$$\omega = \frac{\Delta\theta}{\Delta t} = \frac{2\pi}{60.0 \text{ s}} = 0.105 \text{ rad/s}$$

While we have obtained the angular velocity of the second hand, how would we obtain the linear speed of its tip if we knew its length? In a time Δt the hand sweeps out an angle $\Delta\theta$, while its tip moves along an arc of length Δs. Applying Equation 5-2, we have that the average angular velocity during the interval is

$$\omega = \frac{\Delta\theta}{\Delta t}$$

We may substitute for $\Delta\theta$ by using Equation 5-1 and obtain

$$\omega = \frac{\Delta s}{r \, \Delta t}$$

Now, $\Delta s/\Delta t$ is simply the average speed v_{av} of the tip of the hand, and r is constant as noted. Thus

$$\omega = \frac{v_{av}}{r}$$

or, since the second hand is moving at constant speed,

$$\omega = \frac{v}{r} \tag{5-5}$$

or, finally,

$$v = r\omega \tag{5-6}$$

Thus if we knew the second hand to be 0.120 m long from its center to its tip, we would obtain

$$v = r\omega = (0.120 \text{ m})(0.105 \text{ rad/s}) = 0.0126 \text{ m/s}$$

Equation 5-6 applies to any point moving in a circle at a fixed distance r from the center of the circle. Application of the limiting process of Section 2-3 could show that Equations 5-5 and 5-6 apply to instantaneous values of ω and v as well.

Example 5-4 A housefly chooses a phonograph record as its landing field and lights 0.120 m from the center of the turntable. The record is turning at $33\frac{1}{3}$ r/min (revolutions per minute). At what linear speed will the fly be carried into collision with the near-stationary phonograph pickup arm?

Solution We must first express the angular velocity in radians per second and then apply Equation 5-6. Thus

$$\omega = (33\tfrac{1}{3} \text{ r/min})(2\pi \text{ rad/r})(1 \text{ min}/60 \text{ s}) = 3.49 \text{ rad/s}$$

Therefore $v = r\omega = (0.120 \text{ m})(3.49 \text{ rad/s}) = 0.419 \text{ m/s}$

Example 5-5 The earth has an average radius of 6380 km and completes one revolution in 24.0 h. (*a*) What is the linear speed of a point on the earth's equator due to this motion (in kilometers per hour)? (*b*) What is the linear speed of Philadelphia at 40° N latitude?

Solution (*a*) Note that every point on earth has an angular velocity of $(2\pi \text{ rad})/(24 \text{ h}) = 0.262$ rad/h due to the earth's rotation. The linear speed of each point will depend on its radius of rotation. For the equator,

$$v = r\omega = (6.38 \times 10^3 \text{ km})(2\pi \text{ rad}/24 \text{ h}) = 1.67 \times 10^3 \text{ km/h}$$

(*b*) The main complication here is that Philadelphia at 40° N latitude is turning through a smaller circle daily than a point on the equator. A point at 40° N (or $\frac{40}{360} \times 2\pi = 0.222\pi$ rad $= 0.698$ rad) is closer to the axis of rotation than 6380 km, and we must calculate the effective radius r_{eff}, as in Figure 5-5. We may see that

$$\cos 40° = \cos 0.698 \text{ rad} = \frac{r_{\text{eff}}}{r_E}$$

$$r_{\text{eff}} = r_E \cos 0.698 = (6380 \text{ km})(0.766) = 4.89 \times 10^3 \text{ km}$$

$$v_{\text{eff}} = r_{\text{eff}} \omega = (4.89 \times 10^3 \text{ km})(0.262 \text{ rad/h}) = 1.28 \times 10^3 \text{ km/h}$$

We should note then that angles are not really vectors. It is true that these quantities have both magnitude and direction (sense of rotation). However, vector arithmetic requires certain properties of variables, most notably that they be commutative, that is, **a** + **b** = **b** + **a**. This is not true of rotations, as we see in Figures 5-6 and 5-7. In these figures a gymnast executes the same set of

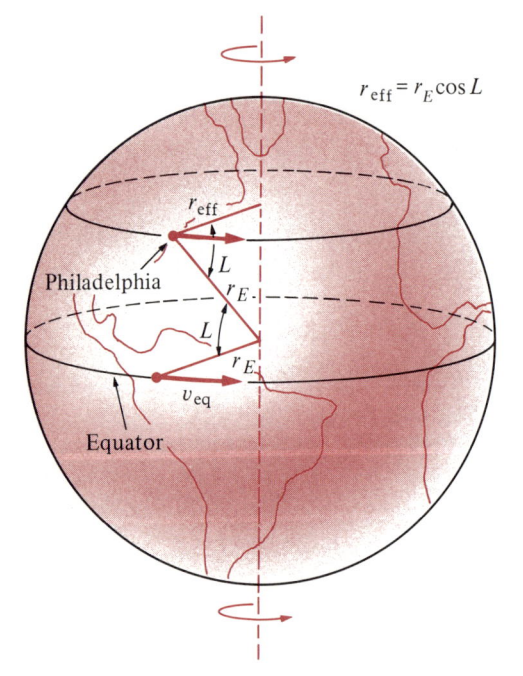

Figure 5-5 Example 5.5.

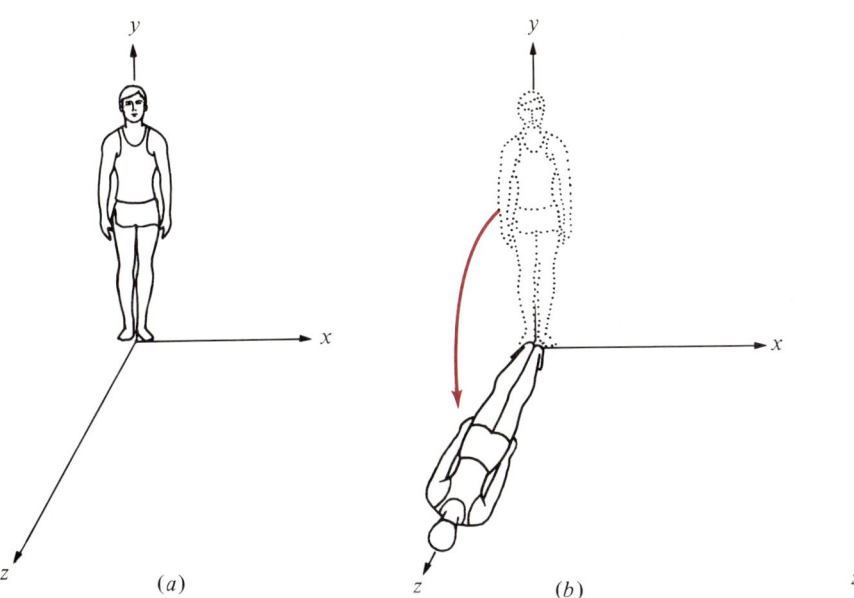

rotations but in two different sequences, leaving him in strikingly different positions. Thus, rotations are *not* vectors since they are not commutative. However, (1) rotations which are confined to a single axis are commutative, and we may treat rotations as vector quantities in this case, and (2) if we concern ourselves with time rates of change using small increments of rotation, the

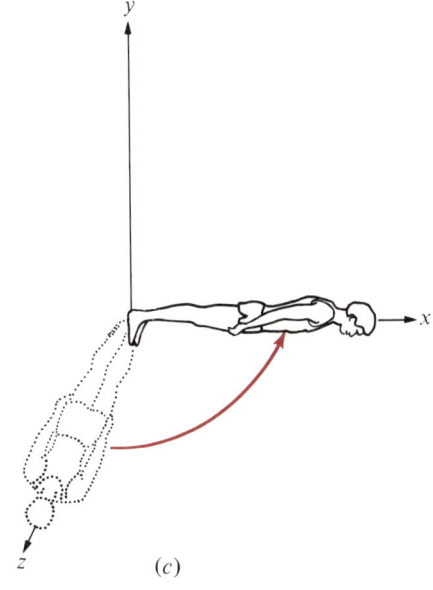

Figure 5-6 The effect of sequential rotations. (*a*) The gymnast begins standing erect facing south. (*b*) He rotates 90° about a horizontal east-west axis. (*c*) He now rotates 90° about a vertical axis, as indicated, winding up facedown on the east-west line with his head toward the east.

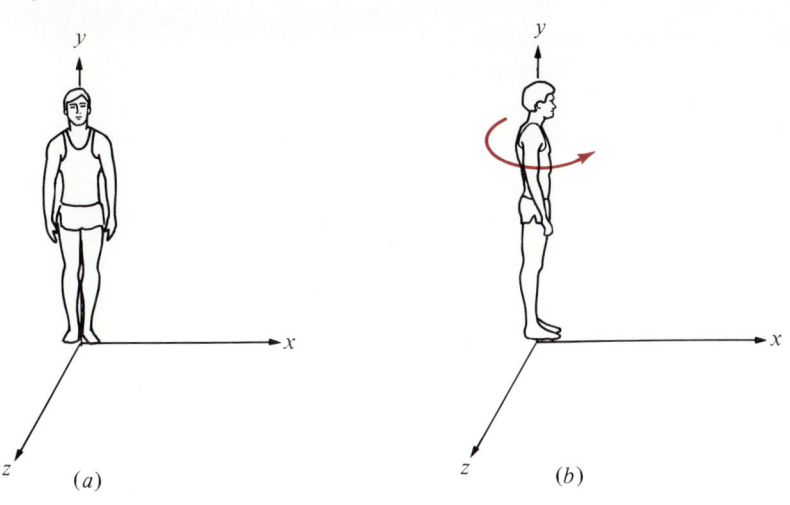

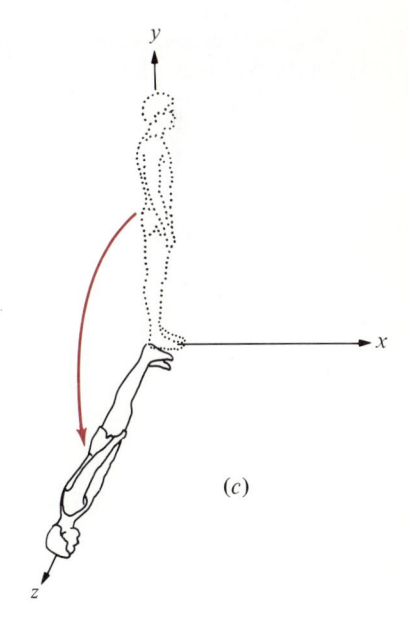

Figure 5-7 Repeating the same rotations in the opposite order. (*a*) The gymnast begins in the starting position. (*b*) He rotates 90° about a vertical axis in the same sense of rotation as before. (*c*) He executes a 90° rotation about a horizontal east-west axis, winding up lying on his side, head pointed south, facing east. Although the rotations were the same as in Figure 5-6, reversing their order has left us with a very different result. Rotations are not commutative.

commutative law is again obeyed. Thus, angular velocities may nearly always be manipulated as vectors. We will continue to treat rotations as vectors since vector arithmetic is the only method we have defined to handle quantities with both magnitude and direction. However, do not try to generalize the vector treatment to large sequential rotations, as in the solution of Rubic's Cube, for example.

5-4 Uniform Rotational Acceleration

The photograph of the gymnast shows several clear angular accelerations. Figure 5-8 reproduces the early portion of this motion. The first seven positions in the photograph show the entire body increasing its turn rate ω about the stationary rear foot. **Angular acceleration** is defined as the time rate of change of angular velocity. In symbols,

$$\alpha = \frac{\Delta \omega}{\Delta t} = \frac{\omega_2 - \omega_1}{t_2 - t_1} \tag{5-7}$$

where the Greek lowercase alpha (α) is the average angular acceleration during the time interval $t_2 - t_1$ and ω_2 and ω_1 are the angular velocities at t_2 and t_1, respectively. Note the similarity of this defining equation to that for linear acceleration.

Measuring from the photograph of Figure 5-8, the angular velocity between positions 3 and 4, ω_{3-4}, is 1.22 rad/s, while between positions 4 and 5, ω_{4-5} is 1.92 rad/s. Since this change in ω occurred over a time of 0.1 s, we may calculate the average angular acceleration for the interval to be

$$\alpha_{3-5} = \frac{\omega_{4-5} - \omega_{3-4}}{0.1 \text{ s}} = \frac{(1.92 - 1.22) \text{ rad/s}}{0.1 \text{ s}} = 7.00 \text{ rad/s}^2$$

We have chosen to measure these motions about the stationary pivot point of the rear foot. However, the body was held essentially rigid during this rotation, and measurement from the central point of the body, as before, would yield a very similar result. While rotations may be measured about any point,

TABLE 5-2

Translational	Rotational
$v = v_0 + at$	$\omega = \omega_0 + \alpha t$
$v_{av} = \dfrac{v + v_0}{2}$	$\omega_{av} = \dfrac{\omega + \omega_0}{2}$
$\Delta s = v_{av} t$	$\Delta \theta = \omega_{av} t$
$\Delta s = v_0 t + \frac{1}{2} a t^2$	$\Delta \theta = \omega_0 t + \frac{1}{2} \alpha t^2$
$2a \Delta s = v^2 - v_0^2$	$2\alpha \Delta \theta = \omega^2 - \omega_0^2$

The complete analogy of form between rotational equations and translational equations frees the learner from having to master large numbers of new equations. While we have derived them from the start, the forms are not new, only the area of application.

the rear toe here is only a good choice for the first several positions. After that, this foot moves in a rapid, high arc over the lower body, and using the toe as the origin would make the motion confusing. (To some extent, this is the same problem which must be faced by inhabitants of a moving earth in trying to analyze the motions of the solar system.) The central point on the body was chosen in Figure 5-4 and used as the center of rotations simply because it is close to being the axis for the entire rotation and displays the most nearly constant translational motion of any point on the body. (As mentioned, the better the execution of the walkover, the less change in translational motion this point will show.)

Confining our discussion to those cases where the angular acceleration is constant, thus $\alpha = \alpha_{av}$, we generalize the definition by rewriting Equation 5-7 as

$$\alpha = \frac{\omega - \omega_0}{t} \tag{5-8}$$

where ω is the angular velocity at a time t and ω_0 is the angular velocity when we agree to start our clock at $t_0 = 0$. Solving for ω, we have

$$\omega = \omega_0 + \alpha t \tag{5-9}$$

Since α is constant, we see that Equation 5-9 is the slope-intercept equation for a straight line of slope α and intercept ω_0. Also the average angular velocity for the first t seconds is

$$\omega_{av} = \frac{\omega + \omega_0}{2} \tag{5-10}$$

We may use Equations 5-9 and 5-10 along with Equation 5-4 to derive two more relationships among the quantities as follows. Substituting Equation 5-10 into 5-4, we have

$$\Delta\theta = \frac{\omega + \omega_0}{2} t \tag{5-11}$$

Now, using Equation 5-9 to eliminate ω yields

$$\Delta\theta = \frac{(\omega_0 + \alpha t) + \omega_0}{2} t$$

$$\Delta\theta = \omega_0 t + \tfrac{1}{2}\alpha t^2 \tag{5-12}$$

If we had eliminated t instead of ω, we would obtain

$$\Delta\theta = \frac{\omega + \omega_0}{2} \cdot \frac{\omega - \omega_0}{\alpha}$$

Finally,
$$2\alpha\,\Delta\theta = \omega^2 - \omega_0^2 \tag{5-13}$$

Now, let us collect some equations from this section in Table 5-2 and compare them to the translational kinematic equations of Section 2-4. We see that the relationships are identical in form. We can obtain the rotational equations from the translational equations simply by substituting θ for s, ω for v, and α for a.

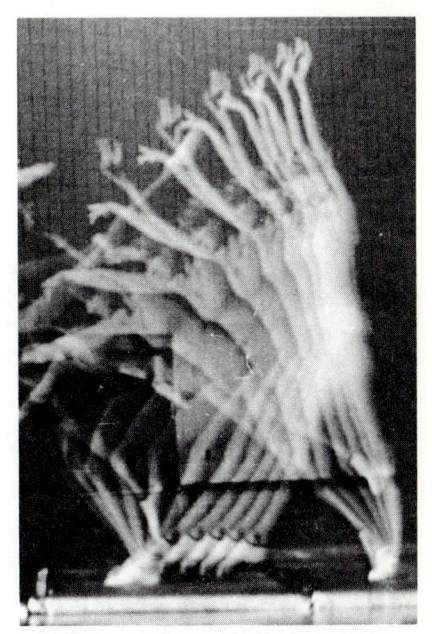

Figure 5-8 Detail of the first several positions of Figure 5-2. The angular separation between positions is increasing—there is an angular acceleration.

Tangential Acceleration Whenever the angular velocity of a body is changing, the linear speed of any point on the body is changing. Thus, each point has a component of translational acceleration due to its changing speed. Consider the propeller tip of a small airplane at the end of a runway preparing to take off. With the engine idling, the propeller tip rotates about the axis with constant angular velocity. The pilot pushes the throttle forward, and the engine rapidly increases its turn rate toward its maximum revolutions per minute. The propeller experiences an angular acceleration to increase its angular velocity. The propeller tip is, at each instant, being accelerated in the direction of its forward motion, or tangent to the circle in which it is moving (Figure 5-9). This **tangential acceleration a_t** is directly proportional to the angular acceleration of the propeller. Rewriting the definition of angular acceleration and substituting from Equation 5-5 gives us

$$\alpha = \frac{\Delta \omega}{\Delta t} = \frac{\Delta(v/r)}{\Delta t} = \frac{1}{r}\frac{\Delta v}{\Delta t}$$

thus

$$\alpha = \frac{a_t}{r} \quad \text{or} \quad a_t = \alpha r \quad (5\text{-}14)$$

This acceleration is often labeled a_t to indicate that it is tangential to the path of motion. Repeating Equations 5-1, 5-6, and 5-14, we have

$$\Delta s = r\,\Delta\theta \qquad v = r\omega \qquad a_t = r\alpha$$

Note that the translational quantities and their rotational analogs are proportional to each other, with the constant of proportionality being the radius of the motion.

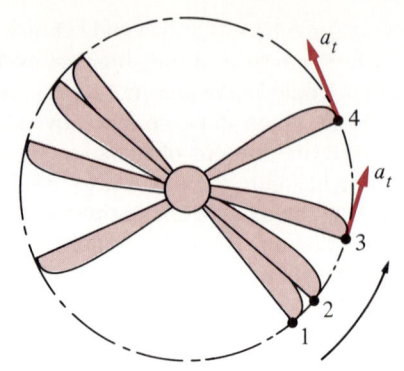

Figure 5-9 A propeller tip as an airplane engine idles at constant revolutions per minute undergoes uniform circular motion, moving in a perfectly circular path at constant speed. As the engine turn rate is increased, the propeller undergoes an angular acceleration and is speeded up in the instantaneous direction of forward motion.

Example 5-6 An antique windup gramophone plays back recordings which are to rotate at 78 r/min, or 8.17 rad/s. When the turntable of such a record player is up to speed, a brake is applied which brings the turntable to rest in 6.00 s. Calculate (a) the angular acceleration and (b) the angle through which the record turned while it came to rest.

Solution (a) Since we know the initial and final angular velocity as well as the time to come to rest, we simply apply the definition of average angular acceleration in Equation 5-8:

$$\alpha = \frac{\omega - \omega_0}{t} = \frac{0 - 8.17 \text{ rad/s}}{6.00 \text{ s}} = -1.36 \text{ rad/s}^2$$

The minus sign indicates that α is opposite the direction of ω and that the rotation is slowing down.

(b) $\Delta\theta = \omega_0 t + \tfrac{1}{2}\alpha t^2$
$= (8.17 \text{ rad/s})(6.00 \text{ s}) + \tfrac{1}{2}(-1.36 \text{ rad/s}^2)(6.00 \text{ s})^2 = 24.5 \text{ rad}$

5-5 Circular Motion and Centripetal Acceleration

Consider again the motion of the airplane propeller as it turns at a constant rate. Since the turn rate is constant, $\alpha = 0$ and thus $a_t = 0$. But, acceleration is defined as the time rate of change of *velocity*. The velocity of the propeller tip is constantly pointing in a different direction and hence is constantly changing. Thus, there must be a linear acceleration of the propeller tip even though the speed is constant, $\alpha = 0$ and $a_t = 0$. Since no component of this acceleration can be tangential to the circle of motion, the acceleration must be purely **radial acceleration**, that is, directed toward or away from the center of rotation at every instant. The case of constant-speed motion in a circle is called **uniform circular motion** and is an important case of motion.

Figure 5-10a illustrates the position of the propeller at two instants as it executes uniform circular motion, showing the velocity vector of the tip for each position. The angle swept out by the propeller blade between the two positions must be the same as the angle between these velocity vectors. Why? The velocity is always tangential to the circle and thus is perpendicular to a radial line. Therefore, $\mathbf{v}_2$ is perpendicular to r_2 and $\mathbf{v}_1$ is perpendicular to r_1. Thus the angle between $\mathbf{v}_2$ and $\mathbf{v}_1$ is the same as the angle between r_2 and r_1 (Figure 5-10b). Further, the magnitudes of $\mathbf{v}_2$ and $\mathbf{v}_1$ are the same (remember the speed is constant), as are the magnitudes of r_2 and r_1. For small angles the arc length Δs (Figure 5-10c) is approximately equal to the cord length Δr, and by similar triangles we have

$$\frac{\Delta v}{v} = \frac{\Delta s}{r}$$

where v is the magnitude of either $\mathbf{v}_2$ or $\mathbf{v}_1$ and r is the magnitude of either r_2 or r_1. Thus, for this equation, we take advantage of the fact that corresponding sides of similar triangles are proportional. Rewriting and dividing both sides by Δt, we have

$$\frac{\Delta v}{\Delta t} = \frac{v}{r}\frac{\Delta s}{\Delta t}$$

The left side is the magnitude of the average acceleration, and $\Delta s/\Delta t$ is the magnitude of the average velocity during the time interval. Thus

$$a = \frac{v}{r}v \quad \text{or} \quad a = \frac{v^2}{r} \tag{5-15}$$

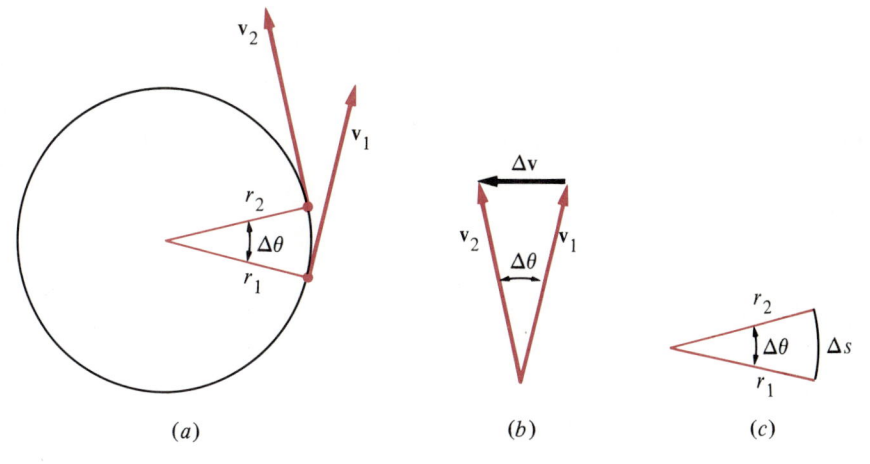

5-5 Circular Motion and Centripetal Acceleration

With two-dimensional motion, it was often convenient to break up vectors into x and y components. For rotational motion, we do the same thing by dividing rotations into radial and tangential components. A purely radial acceleration cannot influence the tangential velocity any more than a vertical acceleration could influence a horizontal velocity.

Figure 5-10 (a) Sequential positions of the propeller tip, marked by the end of r_1 and r_2, have instantaneous velocities $\mathbf{v}_1$ and $\mathbf{v}_2$, respectively. Since the turn rate is constant, the magnitude of $\mathbf{v}_1$ is the same as that of $\mathbf{v}_2$. (b) The change in velocity $\Delta \mathbf{v}$ between $\mathbf{v}_1$ and $\mathbf{v}_2$. If the vector $\Delta \mathbf{v}$ is applied at the midpoint of the motion between $\mathbf{v}_1$ and $\mathbf{v}_2$, it will point directly at the center of rotation. (c) The change in position of the propeller tip between r_1 and r_2 is the arc length Δs.

Remember that this is the net acceleration for circular motion at constant speed, and it is purely radial. Alternatively, from the definition $\mathbf{a} = \Delta\mathbf{v}/\Delta t$, we know that the direction of $\mathbf{a}$ is always in the direction of $\Delta\mathbf{v}$, or $\mathbf{v}_2 - \mathbf{v}_1$. This vector subtraction is shown in Figure 5-10b, where we see that $\Delta\mathbf{v}$ points in the direction of the center of the circle. This will always be the case with uniform circular motion. Of course, we already knew that the acceleration had to be radial, that is, perpendicular to the velocity vector at all times, in this case. If not, there would be a component of the acceleration either along the direction of velocity or opposed to it, in which case the propeller would be either increasing or decreasing in speed. Since speed is constant, the acceleration is purely radial and is directed toward the center of the circle. We call this a **centripetal** (meaning center-seeking) **acceleration** a_c. Since it always points toward the center, it is continually changing direction; thus *centripetal acceleration is never constant* (although it will have constant magnitude for uniform circular motion). If the propeller were increasing or decreasing in speed but remained in circular motion, it would still have a component of acceleration toward the center, and this component could be found by using Equation 5-15. Of course, in such a case it would also have a tangential acceleration which could be found from Equation 5-14.

Any object moving in a circle must be subject to a net force with a component toward the center to change its direction. This force is called the **centripetal force** F_c. If we picture a ball being swung around on the end of a string, F_c is the force the string exerts on the ball to pull it out of a straight-line motion and into a circular pathway. If the string breaks, the ball, no longer subjected to the centripetal force, will move in the straight line defined by its velocity vector at the instant the string broke. Anytime, anywhere an object is moving at constant speed in a circle, a force is required to produce an acceleration of magnitude v^2/r to maintain the circular motion, and the force must always be directed toward the center of the circle. A smaller unbalanced force allows the object to spiral outward, increasing the radius of motion, while a larger force will "reel in" the object toward the center. What provides this force is irrelevant to the preceding kinematic argument, but the force must exist for uniform circular motion to result. We might express the condition for circular motion in equation form as

$$F_c = ma_c$$

Example 5-7 The housefly of Example 5-4 was found to be moving at a constant speed of 0.419 m/s at a radius of 0.120 m on the phonograph record. Calculate (a) the centripetal acceleration of the fly and (b) the minimum coefficient of static friction needed to provide the centripetal force (assume radial motion impends).

Solution (a) We observe that uniform circular motion is occurring and thus may apply Equation 5-15 to find the acceleration:

$$a = \frac{v^2}{r} = \frac{(0.419 \text{ m/s})^2}{0.120 \text{ m}} = 1.46 \text{ m/s}^2$$

(b) The observed motion requires a centripetal force, and its only source can be the force of friction between fly and record. If the surface were made perfectly frictionless, the fly would slide off tangent to its circle of motion. Since radial motion impends, any increased speed would have the required magnitude of the centripetal force exceed the available frictional force. Thus, at the speed considered, the force of static friction is at its maximum, and we use the equality

$$f_s = \mu_s F_N = \mu_s mg$$

since the normal force pressing the surfaces together is equal to the weight of the fly, mg. Now Newton's second law asserts that the unbalanced force, whatever supplies it (here f_s), is equal to the mass to which it is applied times the acceleration of that mass. But we know the acceleration of the fly. It is a mass in circular motion and thus has an acceleration of v^2/r. Thus

$$\mu_s mg = ma = m\frac{v^2}{r} \quad \text{or} \quad \mu_s g = \frac{v^2}{r}$$

and

$$\mu_s = \frac{v^2}{rg} = \frac{(0.419 \text{ m/s})^2}{(0.120 \text{ m})(9.80 \text{ m/s}^2)} = 0.149$$

People moving in a circle often speak of a centrifugal (center-fleeing) force that flings them to the outside of the circle, but this is a fictitious force, as Figure 5-11 shows. In (a), the bus tires are twisted at an angle to the forward direction, providing a force directed toward the center of curvature of the road. This deflects the bus into the curve. In (b), the force at the front tires acts as a centripetal force deflecting the bus into a curved path. The passenger, however, continues moving in the original straight line in the absence of any applied force. At (c), the passenger eventually meets the bus wall, which provides the centripetal force necessary to alter her path. Unaccustomed to having walls come over and exert forces on the body, she describes being "flung to the

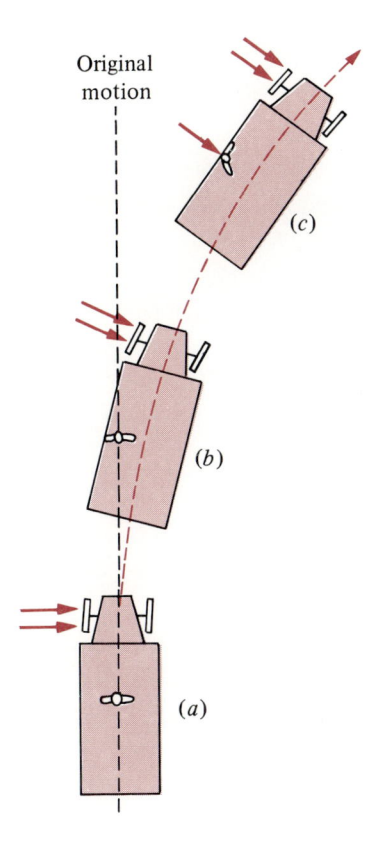

Figure 5-11 Both bus and passenger move in a straight line at constant speed unless acted on by an outside force.

outside wall" in the curve and ascribes this to a fictitious "centrifugal force." In reality, her inertia tending to maintain straight-line motion accounted for her motion, while the wall moved laterally to meet her.

When an automobile is being driven around a curve in a road, a centripetal force must constantly deflect it from straight-line motion into the curved path. If the roadway is level, this centripetal force is provided by friction between the tires and the road (although excessive speed or an icy surface may make turning impossible, and the car will slide off the road). However, we often find that the roadway on a highway is banked, or slanted, inward toward the center of curvature. In such a case, a component of the normal force supplied by the roadway acts toward the center of the curve and provides some or perhaps all of the centripetal force needed. At one particular speed for a given angle of banking, no frictional force is required between the tires and roadway at all. Figure 5-12a shows an automobile on such a roadway, traveling at the proper speed. Figure 5-12b shows an expanded, head-on view of the automobile with the forces acting on it. The normal force $\mathbf{F}_N$ acts perpendicular to the surface of contact, and the car's weight $m\mathbf{g}$ acts vertically downward.

Example 5-8 We wish to design a road which has a radius of curvature of 300 m. Calculate the angle of bank necessary so that a driver need not turn the steering wheel if he maintains a speed of 25.0 m/s through the curve.

Solution Since we are not given the mass of the automobile, we must assume that it is not needed and find a way to eliminate it from our formulas. We choose upward as the positive y direction and toward the center of the circle as the positive x direction. We may write Newton's second law for each of these directions independently. Realizing there is no acceleration in the y direction, we write

$$\sum F_y = 0 \qquad F_N \cos\theta - mg = 0$$

and

$$\sum F_x = ma \qquad F_N \sin\theta = \frac{mv^2}{r}$$

Solving the y equation for F_N, we have $F_N = mg/\cos\theta$, which may be substituted into the x equation to yield

$$\frac{mg}{\cos\theta} \sin\theta = \frac{mv^2}{r}$$

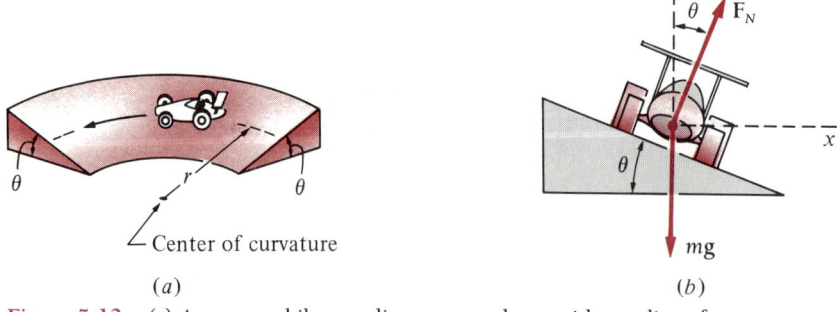

Figure 5-12 (a) An automobile traveling on a roadway with a radius of curvature r and banked at an angle θ. (b) The forces acting in a vertical plane perpendicular to the car's motion.

or
$$\tan \theta = \frac{v^2}{gr} = \frac{(25.0 \text{ m/s})^2}{(9.80 \text{ m/s}^2)(300 \text{ m})} = 0.213$$

$$\theta = 12.0°$$

Note again that at this angle of bank the driver need not turn the steering wheel to have the car proceed through the curve if traveling at the correct speed. Indeed, at this speed, the driver must keep the tires pointed straight ahead, for if the wheels turn to the inside of the curve, the car will go off the road on the inside, while the same is true for turning outward. The big advantage of banked curves is that no radial *frictional* force need act. Thus the automobile may negotiate the banked curve with little difficulty even if the surface is wet or icy provided that it travels at the proper speed. Just as we expected, the angle of bank is independent of the mass of the vehicle. The largest tractor trailer and the smallest motorcycle should negotiate the curve at the same speed. High-speed racing curves for cars (or bobsleds) are different and are curved to provide greater banking as the vehicle climbs the outside wall of the curve. Thus, for any likely speed there is some pathway where the curve can be followed without steering being required.

5-6 Newton's Law of Universal Gravitation

Prior to an understanding of Newton's first law, the law of inertia, the wrong question had been asked about heavenly objects like the moon. "What keeps the moon up, why doesn't it fall?" is the wrong question. Instead we should ask "What keeps the moon from moving away from the earth in a straight line tangent to its orbit?" Instead of a repulsive force acting centrifugally to keep the moon away from the earth in its observed orbit, we need a centripetal force pulling the moon toward the earth and making it bound to us forever. While he was contemplating this in the country, according to Newton, an apple fell from a nearby tree. He immediately wondered if the force pulling the apple to the earth could be extended to provide the required centripetal force to account for the orbit of the moon and, if so, how could he describe this force. His conclusion became the **law of gravitation:**

> Every two mass particles in the universe are attracted to each other by a force which is proportional to the product of the two masses and inversely proportional to the square of the distance between them.

In symbols,

$$F_G = G \frac{m_1 m_2}{r^2} \tag{5-16}$$

where m_1, m_2 = masses of the two particles (often called **point masses**)
r = distance between m_1 and m_2
G = **universal gravitational constant** which makes the proportionality an equation for any two particles

The value of G in SI units is

$$G = 6.67 \times 10^{-11} \text{ N} \cdot \text{m}^2/\text{kg}^2$$

Note the following points about Newton's law of universal gravitation:

1. It states that the force is one of attraction; thus the force is along the line joining the two bodies.
2. Forces act on each body; they are an action-reaction pair.
3. The law states that the force is between mass particles. How do we treat the force between a tennis ball and the earth? Even if the tennis ball can be considered to be a particle, what distance are we talking about? Newton provided an answer. He proved that, to bodies outside a uniform sphere, the spherical body acts as a mass particle with all its mass concentrated at its center.

Thus we may say that r is the distance between the bodies' centers when the bodies may be treated as uniform spheres. Further, when the distance between the centers is very, very large compared to the dimensions of one of the bodies, that body may be treated as a particle regardless of its shape.

Consider a body of mass m_0 on the surface of the earth. After Equation 5-16, we may write the force of gravity acting on it in two ways:

$$F_G = \frac{G m_0 M_E}{r_E^2} \quad \text{or} \quad F = m_0 g$$

where we have replaced m_1 and m_2 in Equation 5-16 by m_0 and M_E (the mass of the earth) and r by r_E (the radius of the earth). Thus we see that

$$g = \frac{GM_E}{r_E^2}$$

Note that this predicts that the value of g depends on the values of the three constants on the right-hand side of the equation, but not on the mass of the object itself! All objects *at the earth's surface* ($r = r_E$) should accelerate downward at the same rate whatever their mass, as indeed they do if unimpeded. This equation was used with the first measured value of G along with accepted values for g and r_E to make the first determination of the mass of the earth. The presently accepted values of the mass and radius of the earth are given in Table 5-3. From these we may calculate g to check for consistency:

$$g = \frac{(6.67 \times 10^{-11} \text{ N} \cdot \text{m}^2/\text{kg}^2)(5.98 \times 10^{24} \text{ kg})}{(6.38 \times 10^6 \text{ m})^2}$$

$$= 9.80 \text{ m/s}^2$$

Of course, the acceleration due to gravity becomes smaller as the height above the earth's surface increases ($r > r_E$).

Example 5-9 Determine the acceleration due to gravity at a distance of 10.0 km above the earth.

Solution First, we determine r.

$$r = r_E + 10.0 \text{ km} = 6.38 \times 10^6 \text{ m} + 1 \times 10^4 \text{ m} = 6.39 \times 10^6 \text{ m}$$

Thus, at 10 km above the earth,

$$g = \frac{GM_E}{r^2} = \frac{(6.67 \times 10^{-11} \text{ N} \cdot \text{m}^2/\text{kg}^2)(5.98 \times 10^{24} \text{ kg})}{(6.39 \times 10^6 \text{ m})^2} = 9.77 \text{ m/s}^2$$

TABLE 5-3 Selected Data for Bodies in the Solar System

Body	Mass, kg	Average Radius, m	Average of Orbital Radius, m	Sidereal Period, s	Average Density, $\times 10^3$ kg/m^3	Average Acceleration due to Gravity on Surface, m/s^2
Sun	1.991×10^{30}	6.960×10^8	...	...	1.410	273.72
Mercury	3.181×10^{23}	2.433×10^6	5.8×10^{10}	7.602×10^6	5.431	3.578
Venus	4.883×10^{24}	6.053×10^6	1.08×10^{11}	1.941×10^7	5.256	8.874
Earth	5.979×10^{24}	6.376×10^6	1.49×10^{11}	3.156×10^7	5.519	9.807
Mars	6.418×10^{23}	3.380×10^6	2.28×10^{11}	5.936×10^7	3.907	3.740
Jupiter	1.901×10^{27}	6.976×10^7	7.78×10^{11}	3.743×10^8	1.337	26.010
Saturn	5.684×10^{26}	5.822×10^7	1.426×10^{12}	9.296×10^8	0.688	11.170
Uranus	8.616×10^{25}	2.347×10^7	2.869×10^{12}	2.651×10^9	1.603	10.490
Neptune	1.001×10^{26}	2.272×10^7	4.495×10^{12}	5.200×10^9	2.272	13.250
Pluto*	1.08×10^{24}	5.700×10^6	5.900×10^{12}	7.837×10^9	1.65	2.21
Moon (Luna)	7.354×10^{22}	1.738×10^6	3.84×10^8	2.361×10^6	3.342	1.620

* Mass, density, and acceleration due to gravity of Pluto are very uncertain: mass = $(1.08 \pm 1.00) \times 10^{24}$ kg; density = $(1.65 \pm 1.57) \times 10^3$ kg/m^3; average acceleration due to gravity = 2.21 ± 2.07 m/s^2.

This is less than a $\frac{1}{3}$ percent change from g on the earth's surface. Therefore, unless we are concerned with more than three-place accuracy, "near" the earth can mean as much as 10 km away, and we may still consider g to be constant.

Determining G required the measurement of the gravitational force between known masses situated at a known distance from each other. This in turn required the measurement of an incredibly tiny force and was first accomplished by Lord Cavendish in 1798, using the apparatus shown in Figure 5-13. A fiber fixed at one end was attached at its other end to a mirror and to a rod

Figure 5-13 The apparatus used by Cavendish to determine the value of G.

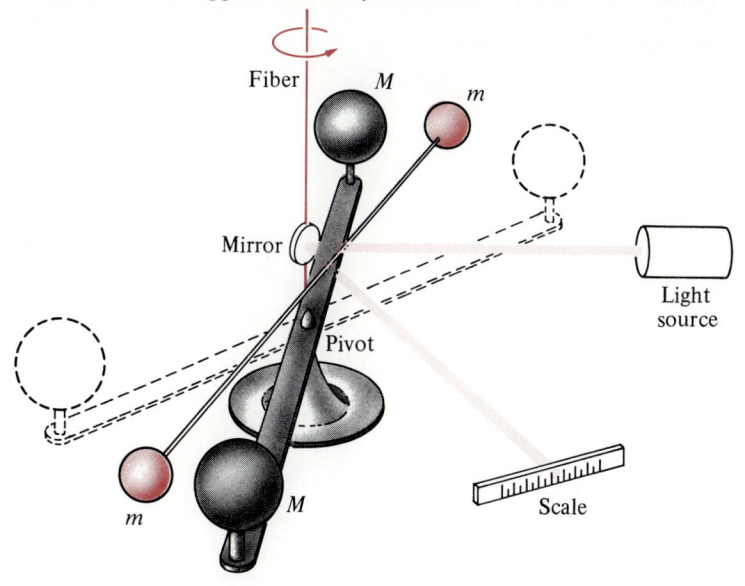

supporting two small masses m. When a rod supporting two much larger masses M was pivoted as shown in Figure 5-13, the gravitational attraction between M and m caused the smaller masses to move toward the larger ones and the rod to turn through a small angle θ. As this happened, the mirror also rotated through the same angle, causing the light beam to move across the calibrated scale. Because θ is directly proportional to the gravitational force exerted by M and m on each other, Cavendish was able to determine F_G and use Equation 5-16 to calculate G.

Newton's law of gravity accounts nicely for the observed motions of the moon (Figure 5-14). If the centripetal force $F_c (= ma_c)$ which keeps the moon in its orbit is due to gravity, we may write that $F_c = F_G$, or

$$\frac{M_M v_M^2}{r_{EM}} = \frac{GM_M M_E}{r_{EM}^2} \tag{5-17}$$

where M_M is the mass of the moon and r_{EM} is the distance between the center of the earth and the center of the moon. The period T_M of the moon in its orbit (originally, "moonth," later "month") has been measured with increasing precision throughout human history. We will manipulate Equation 5-17 to obtain an expression for this period and calculate its value. Since T_M is the time for the moon to complete one full circle about the earth, during that time it travels a distance of $2\pi r_{EM}$. Thus, for one cycle,

$$v_M = \frac{\Delta s}{\Delta t} = \frac{2\pi r_{EM}}{T_M}$$

Substituting this expression for v_M into Equation 5-17 and then dividing both sides by M_M gives us

$$\frac{(2\pi r_{EM}/T_M)^2}{r_{EM}} = \frac{GM_E}{r_{EM}^2}$$

which yields

$$T_M^2 = \frac{4\pi^2 r_{EM}^3}{GM_E} \quad \text{or} \quad T = 2\pi \sqrt{\frac{r_{EM}^3}{GM_E}}$$

Using the earth-moon distance of $r_{EM} = 3.84 \times 10^8$ m along with the other known quantities, we obtain the period of the moon to be

$$T_M = 2\pi \sqrt{\frac{(3.84 \times 10^8 \text{ m})^3}{(6.67 \times 10^{-11} \text{ N} \cdot \text{m}^2/\text{kg}^2)(5.98 \times 10^{24} \text{ kg})}}$$

$$= 2.37 \times 10^6 \text{ s}$$

The moon actually moves 360° around the earth in space in

$$T_M = 27.322 \text{ days} = 2.36 \times 10^6 \text{ s}$$

which is in excellent agreement with the predicted value.

Two comments are necessary concerning this calculation. First, T_M is the **sidereal period** of the moon, or the time it takes to revolve around the earth with respect to the stars. This is not the same as the **synodic period,** or time between full moons, which is about 2 days longer. Since the earth moves about the sun as the moon orbits the earth, it takes more than 360° of revolution between lineups with the sun. Second, the earth is *not* rigidly fixed in space, and the mass of the moon is *not* negligible compared to that of the earth. As a result, the earth-moon system is a bit like a dumbbell with unequal balls on each end,

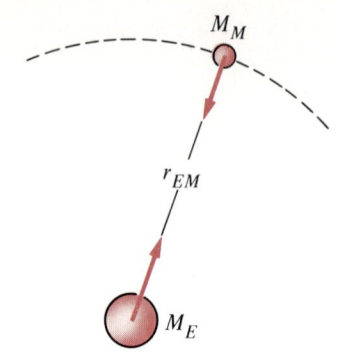

Figure 5-14 The moon moves about the earth in an orbit of radius r_{EM}, and we know that a centripetal force $F_c = mv^2/r_{EM}$ is required to maintain this circular motion. We also know that this force must be directed toward the center of the circle, the earth. The only source for such an earth-directed force acting on the moon is the gravitational attraction of the earth for the moon F_G. Thus, we conclude that $F_c = F_G$.

say a bowling ball and a baseball. Twirling such a system on a frictionless floor might reveal at first glance the baseball orbiting the bowling ball. A closer look shows the bowling ball moving in a much smaller countercircle. This is the same as if you hold the hands of a small child and move her rapidly in a circle around you. You lean back a little and while most of the motion is with the child, your body is tracing out a small circle at the same time (even if your feet remain in essentially the same spot). Likewise, the earth and the moon are properly regarded as orbiting each other. Neglecting this leads to the small difference between T_{calc} and T_{meas} for the period of the moon above. A more careful application of Newton's laws yields exactly the observed period.

The moon is a natural satellite of the earth, but since 1957 we have been placing artificial satellites in various orbits and have even journeyed to the moon. This would not have surprised Newton, who discussed the mechanics of orbiting objects in the *Principia*. To make his point, he suggested the following: Throw a rock horizontally from a mountain top. The rock curves toward the earth and strikes away from and below you. Repeat doing this, each time throwing the rock a bit harder. Eventually, the rock has an enormous initial horizontal velocity, and as its path curves toward the earth, the earth's surface itself is curving away beneath it. The rock falls forever about the earth, never getting any closer to the ever-receding surface. This is called *free-fall*, or *being in orbit*. Newton was there long before Sputnik, or NASA.

Example 5-10   Calculate the speed of an artificial satellite of mass m placed in a circular orbit 180 km above the surface of the earth.

Solution   Recognize again that the centripetal force required for the circular motion will be provided by the force of gravity. Thus $F_c = F_G$, and we write

$$\frac{mv^2}{r} = \frac{GmM_E}{r^2} \quad \text{or} \quad v = \sqrt{\frac{GM_E}{r}}$$

The radius of the circular motion will be $r_E + 180$ km, or

$$r = 6.38 \times 10^6 \text{ m} + 0.180 \times 10^6 \text{ m} = 6.56 \times 10^6 \text{ m}$$

Thus
$$v = \sqrt{\frac{(6.67 \times 10^{-11} \text{ N} \cdot \text{m}^2/\text{kg}^2)(5.98 \times 10^{24} \text{ kg})}{6.56 \times 10^6 \text{ m}}}$$
$$= 7.80 \times 10^3 \text{ m/s}$$

Minimum Learning Objectives

After studying this chapter, you should be able to:

1. Define:
   - angular acceleration
   - angular velocity
   - centripetal acceleration
   - centripetal force
   - frequency
   - gravitation
   - period
   - point mass
   - radial acceleration
   - radian
   - rotational motion
   - sidereal period
   - synodic period
   - tangential acceleration
   - translational motion
   - uniform circular motion
   - universal gravitational constant
2. Convert with ease from degrees to radians, and vice versa.
3. Work problems with angular measures expressed in radians.
4. Understand and use the analogy between translational motion equations and rotational motion equations to obtain the correct form for use from a familiar equation.
5. Determine the speed and/or acceleration of an object moving in uniform circular motion from its period of motion and the radius of its motion.
6. Equate a force law derived from a consideration of the observed motion with one describing the nature of the force involved (as in Equation 5-17) to produce a form allowing for the determination of an unknown quantity.

Problems

5-1 As a bicycle is pushed up a ramp, the front wheel turns exactly three and one half times. (a) Calculate the angle in radians through which the wheel turned. (b) Use 0.350 m as the radius of the wheel, and calculate the length of the ramp.

5-2 Convert (a) 25° to radians, (b) 1.60 rad to degrees, and (c) 840° to revolutions.

5-3 Wire is wound on a rather large spool which is rotated at an angular speed of 75.0 r/min. If the radius of the spool (assumed constant) is 0.900 m, how long will it take to wind a 1.00×10^4 m length of wire?

5-4 Convert (a) 150 revolutions per minute to radians per second, (b) 30° per second to radians per minute, and (c) 45 revolutions per minute to degrees per second.

5-5 An automobile travels at 28.0 m/s and has tires with an outside diameter of 0.720 m. Calculate the angular speed of the wheels.

5-6 A beam of light is shined through a slot on the edge of a spinning wheel which has 500 slots separated by 500 "teeth" of equal width and equally spaced around its perimeter. The light travels to a mirror 800 m away and is reflected back to the wheel. (a) Determine the time required for the light to make the round trip if it moves at 3.00×10^8 m/s. (b) Calculate the angle the wheel must spin through in order that the light arrives just in time to pass through the next slot. (c) What must be the angular speed of the wheel in radians per second and in revolutions per minute?

5-7 The curve of Figure 5-15 represents the angle through which the wheel of an automobile has turned from the time the auto left a stoplight. During which segment(s) of the curve was the automobile (a) increasing forward speed, (b) decreasing forward speed, (c) stopped, (d) moving with constant speed, (e) increasing speed backward, and (f) decreasing speed backward?

5-8 How long must a bicycle wheel accelerate at 0.400 rad/s² in order to increase its angular speed from 3π to 5π rad/s?

5-9 Many of the early phonograph tables turned at 78.0 r/min. (a) How many radians per second were these tables turning? (b) What constant angular acceleration would a turntable have to stop in 20.0 s from its top speed?

TABLE 5-4

Angle from Rest through which Ride Turned $\Delta\theta$, rad	Instantaneous Angular Velocity ω, rad/s
3.00	2.06
5.50	2.79
8.00	3.37
11.0	3.95
16.0	4.77

5-10 A carnival's centrifuge ride starts from rest and accelerates smoothly. The angle through which it has turned and its corresponding angular velocity are measured at the same instant five times while it is accelerating. Table 5-4 shows the data gathered. Determine the acceleration by constructing a graph of ω^2 vs. $\Delta\theta$.

5-11 An electric drill is turning at 1500 r/min when it is turned off. If it requires 5.00 s to coast to a stop, calculate (a) its angular acceleration and (b) the angle through which it turns while stopping.

5-12 An automobile with tires 0.720 m in diameter accelerates from 15.0 to 27.0 m/s in 8.00 s. (a) How far must the outer surface of the tires travel in this time? (b) Calculate the angular acceleration of the wheels.

5-13 A brake is applied to a wheel rotating at 5.00 rad/s. The wheel stops after completing exactly 12.0 revolutions. Calculate (a) the angular acceleration and (b) the speed the wheel would have had if the brake had been disengaged after 1.00 s.

5-14 A grinding wheel in a metal shop requires 40.0 s to reach a top speed of 500 r/min when it starts from rest. Calculate (a) its angular acceleration in radians per second per second, (b) the angle through which it turns coming up to speed, and (c) the time when its speed was 50.0 rad/s.

5-15 A wheel begins at rest and has a constant angular acceleration α. At a time t_1 its angular speed is 45.0 rad/s, and 5.00 s later its speed is 60.0 rad/s. Calculate (a) its angular acceleration α, (b) the value of the time t_1, and (c) the angle through which the wheel turned in the first t_1 seconds.

5-16 A midget auto-racing car starts from rest and accelerates at 1.50 m/s² on a circular track of radius 120 m. Calculate (a) the time to complete one lap, (b) the car's linear speed when it is halfway around, and (c) the angular acceleration of an imaginary line joining the car with the center of the track.

5-17 One of the steam-driven turbines in an electric generating plant has a radius of 2.58 m. It starts from rest and accelerates smoothly at an angular acceleration of 0.500 rad/s² until it reaches an angular speed of 600 r/min. (a) Determine the linear acceleration of a point on its outer edge while it is coming up to speed. (b) Calculate the final linear speed of a point on its outer edge. (c) Determine the time required to reach its final speed.

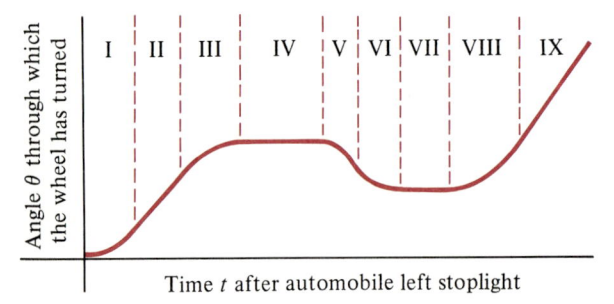

Figure 5-15 Problem 5-7.

5-18 Calculate (a) the angular velocity of the earth in radians per second, (b) the linear velocity of a point on the equator (with respect to the earth's axis), and (c) the centripetal acceleration of a point on the equator.

5-19 At the 1982 World's Fair in Knoxville, Tennessee, there was a Ferris wheel ride which was 42.4 m in diameter. At maximum speed each chair had a tangential velocity of magnitude 3.50 m/s. Calculate (a) the maximum angular velocity and (b) the centripetal acceleration of someone in a chair on the circumference.

5-20 Determine the angular speed of an ultracentrifuge such that the centripetal acceleration of a point 0.0300 m from the axis shall have a value of $1.00 \times 10^5 g$ (that is, 1.00×10^5 times the acceleration due to gravity). Calculate the linear speed of that point.

5-21 A father watching his child on a merry-go-round notes that it takes exactly 11.0 s for the child to complete one revolution. The horses in the child's line are placed on a circle of diameter 9.00 m. Calculate (a) the angular speed of the merry-go-round, (b) the linear speed of the child, (c) the centripetal acceleration of the child, and (d) the centripetal force which the horse must be exerting on the child if the child's mass is 30.0 kg.

5-22 Figure 5-16 shows an arrangement known as a conical pendulum. A small mass m moves in a horizontal circle on the end of a string of length L. The string makes an angle θ with the vertical and sweeps out a cone shape as the mass moves. Carefully analyze the dynamics of the mass to show that for any angle $\theta < \pi/2$ the speed of the mass is

$$v = \left(gL \frac{\sin^2 \theta}{\cos \theta} \right)^{1/2}$$

5-23 A beam of electrons is caused to move in a circular path of radius 2.00 m at a speed of 4.00×10^7 m/s. Calculate (a) the angular speed of an imaginary line joining one of the electrons with the center, (b) the centripetal acceleration of one of the electrons, and (c) the centripetal force on one electron if $m = 9.11 \times 10^{-31}$ kg.

5-24 An airplane is diving at a speed of 250 m/s. The pilot wishes to pull out of the dive while maintaining the same speed. Calculate the minimum radius of curvature of the airplane's path to avoid the total acceleration on the pilot exceeding five times that of gravity.

5-25 At an end-of-semester party, someone accepts a challenge to swing a bucket of water in a vertical circle without spilling any water. If the distance from the center of rotation to the bottom of the bucket is 1.20 m, determine the minimum angular speed required to keep the water in the bucket.

5-26 A playground merry-go-round is pushed to an angular speed of 20.0 r/min. A 35.0-kg child holds on to the rail at the outer edge 2.00 m from the center of rotation. Calculate (a) the linear velocity of the child, (b) the centripetal acceleration of the child, (c) the force the child must exert on the rail and platform to remain on the merry-go-round, and (d) the distance the child must move toward the axis of rotation to reduce the required force to one-quarter its original value.

5-27 The breaking strength, or ultimate strength under tension, of cast iron is 3.70×10^8 N/m². A model of a certain steam turbine may be treated mathematically as a mass of 500 kg rotating in a circle, held by an arm of length 3.00 m and cross-sectional area 0.180 m². Determine the maximum angular velocity the system may have without flying apart.

5-28 (a) How fast would the earth have to rotate so that a spring scale at the equator would show exactly one-half a person's weight at the pole? (b) At this angular velocity, how long would a day be?

5-29 In an amusement park the Ferris wheel rotates at 0.400 rad/s. An 80.0-kg man sits on a bathroom scale as he rides at a distance of 9.00 m from the axis of rotation. What does the scale read when the man is (a) at the top of the ride and (b) at the bottom?

5-30 A 2.50×10^3 kg automobile is to negotiate a curve of radius 200 m at a speed of 30.0 m/s. (a) Determine the centripetal force which must act on the automobile. (b) Calculate the coefficient of static friction required if the curve is not banked. (c) Determine the angle of bank such that the driver need not turn the steering wheel.

5-31 Calculate the gravitational force a 7.26-kg bowling ball exerts on a 0.226-kg baseball when the distance between their centers is 0.300 m.

5-32 Calculate (a) the speed of the earth in its orbit around the sun and (b) the centripetal acceleration, assuming a circular orbit.

5-33 Consider the gravitational forces that the earth and the moon exert on each other. At what distance from the earth are these two forces balanced, that is, exactly equal in magnitude but oppositely directed?

5-34 Calculate the distance above the earth that an artificial satellite must be placed if it is to remain in a position above a point on the earth's equator.

5-35 An astronaut is in a spaceship in a circular orbit 3.19×10^6 m above the earth. His weight (with spacesuit and other equipment) on earth was 1.20×10^3 N. Calculate (a) the astronaut's mass, (b) his weight in orbit (could he detect this on a bathroom scale?), (c) the speed of his spaceship, and (d) the angular speed of the orbit.

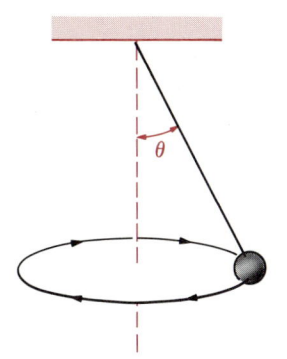

Figure 5-16 Problem 5-22.

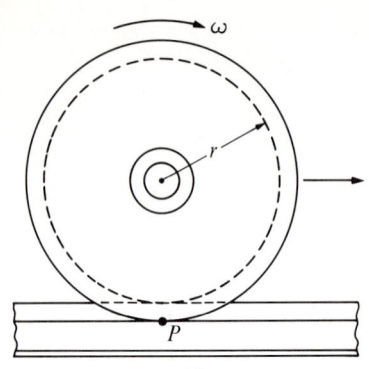

Figure 5-17 Problem 5-37.

5-36 How many times each day does an artificial satellite in an equatorial orbit 500 km above the earth pass over Singapore (or just south of it)?

5-37 The wheel of a railcar shown in Figure 5-17 is rolling with an angular speed of 60.0 rad/s about its axle. It is 0.480 m from the center of rotation to the rail. The flanged rim of the wheel extends beyond the rail an additional 2.50×10^{-2} m. Calculate (a) the linear velocity of the axle and (b) the linear velocity of point P at the edge of the flange directly below the point of contact. (c) Relate this situation to the fact that the moon moves about the earth as the earth revolves in orbit around the sun. Is there any time or relative position such that the moon moves backward with respect to the earth's orbit?

5-38 At what height above the earth's surface will the acceleration due to gravity be decreased (a) by 1 percent compared to that on earth? (b) By 5 percent?

5-39 Consider the sun-Mercury system and assume a circular orbit. Calculate (a) the centripetal force on Mercury, (b) the orbital speed of Mercury, and (c) the period of Mercury (i.e., the length of Mercury's year).

5-40 Io, one of the inner moons of Jupiter, has a period of 1.53×10^5 s and is in a circular orbit of radius 4.20×10^8 m from Jupiter's center. Using this information, calculate the mass of Jupiter.

5-41 The solar system is about 1.00×10^4 parsecs from the gravitational center of the Milky Way galaxy. (The parsec is an astronomical unit of distance: 1 parsec = 3.084×10^{16} m.) The orbital velocity of the solar system about the center of the galaxy is approximately 3.00×10^5 m/s. Using these numbers, (a) calculate the mass of the Milky Way galaxy. (b) What is the ratio of the mass of the Milky Way galaxy to that of the sun?

5-42 Astrologers assert that the positions of the planets influence a child at birth and determine such things as personality, future events, etc. Show that any such influence could not be gravitational by calculating and comparing the gravitational force on a 3.00-kg infant at birth due to (a) Mars at its closest approach of 5.80×10^{10} m, (b) Venus at its closest approach of 3.86×10^{10} m, (c) the moon at 3.84×10^8 m, and (d) a 60.0-kg obstetrician at an average distance of 0.500 m during the birth.

5-43 Halley's comet is in an extremely elliptical orbit, moving past the orbit of Neptune to about 5.35×10^9 km from the sun at its farthest out point (aphelion) and passing inside the orbit of Venus to some 8.94×10^7 km from the sun at perihelion, or closest approach. (a) What is the speed of Halley's comet in its orbit at aphelion and perihelion? (b) What is the greatest distance across the orbit of Halley's comet?

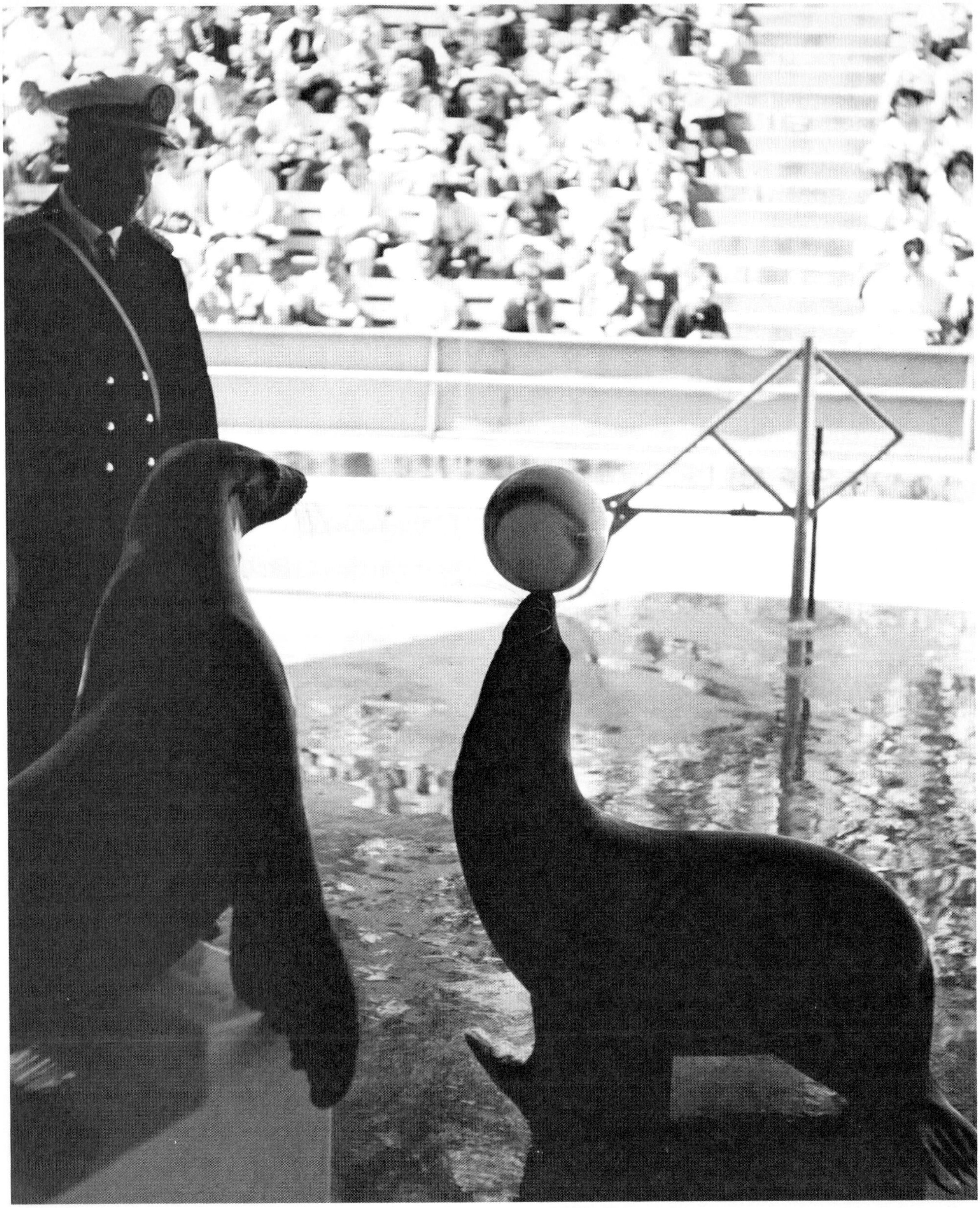

6 Equilibrium and Torques

6-1 Introduction

Equilibrium is a state of balance. As used in physics, **equilibrium** implies that the state of motion of an object is not changing. However, we remember from our study of Newton's laws that an object will maintain uniform motion only if it is *not* acted on by an unbalanced force. Further, we recall that uniform motion means a constant velocity and may also apply to a body at rest. (Zero is a perfectly good constant.) When a state of rest is maintained by an object, we say it is in **static equilibrium.** When a state of constant nonzero velocity is maintained by an object, we say it is in **dynamic equilibrium.**

Both static and dynamic cases are covered by what is called the **first condition of equilibrium:**

> For an object to be in translational equilibrium, the vector sum of all forces acting on the object must be zero.

The same idea may be expressed with great compactness mathematically. If **a** = 0, then

$$\sum_i F_i = 0 \qquad (6\text{-}1)$$

Forces are vectors, and to add them we must take a vector sum in three-dimensional space. We must add in all values of F covered by the index i, which, since no limits are given to i, implies we must add all existing forces. While precise, we rarely do the most general analysis indicated by this expression. If an object is not accelerating at all, it is then surely not accelerating in the x direction, or in the y direction, or in the z direction. Thus, we may apply the first condition of equilibrium to each and every dimension separately. Most frequently, we break the general statement down into three simultaneous scalar equations,

$$\sum_i F_{ix} = 0 \qquad \sum_i F_{iy} = 0 \qquad \sum_i F_{iz} = 0 \qquad (6\text{-}2)$$

where the first of these asserts that "the sum of all components of force acting in the x direction must be zero," and likewise with the y and z directions. Usually we will restrict ourselves to problems in a single plane. When that is the case, we just select a convenient set of x and y axes and ignore the z direction.

Example 6-1 A mirrored sphere weighing 200 N is suspended from the ceiling of a ballroom by two cables, as shown in Figure 6-1a. What is the tension in each of these cables?

Solution This is a problem in static equilibrium since no part of the system is changing its motion. We must first select a point at which to apply the first condition of equilibrium. Note that all points are in equilibrium, and any point could be chosen. However, the point P where the cables attach to the top of the ball is the easiest choice for analysis, since all the forces involved must act through that point. The forces exerted by the cables must act along the cables, thus through P, and the weight of the ball acts straight downward, again through P, as shown in Figure 6-1b. We resolve each force into horizontal (x) and vertical (y) components and apply the first condition of equilibrium to each dimension separately. Choosing upward and rightward as positive, we may write

$$\sum F_x = 0 = T_{2x} + T_{1x}$$
$$= T_2 \cos 30° + (-T_1 \cos 45°)$$
$$0 = 0.866 T_2 - 0.707 T_1 \qquad (1)$$

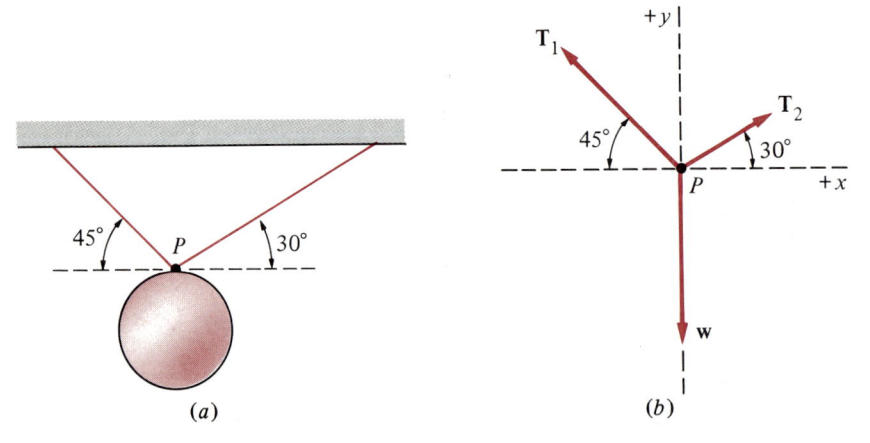

Figure 6-1 Example 6-1.

and

$$\sum F_y = 0 = T_{1y} + T_{2y} - w$$
$$= T_1 \sin 45° + T_2 \sin 30° - 200 \text{ N}$$
$$0 = 0.707T_1 + 0.500T_2 - 200 \text{ N} \quad (2)$$

6-1 Introduction

Now we may solve Equations (1) and (2) simultaneously to obtain

$$T_1 = 179 \text{ N} \quad \text{and} \quad T_2 = 146 \text{ N}$$

An automobile in motion along a smooth horizontal surface provides a particularly common and interesting example of *dynamic equilibrium*. The more the accelerator pedal is depressed, the greater the rate of flow of gasoline to the engine and the greater the force generated propelling the car forward. Yet, for every setting of the gas pedal, there is some final velocity at which the speed of the car levels off. This implies that the forces that retard motion increase with speed, and that for any given forward force there exists a speed at which the retarding force has grown large enough to produce equilibrium. The principal retarding force is that of **air resistance,** which is small at the speeds of normal human motions but increases greatly with increasing speed, as shown in Figure 6-2. You have noticed this if you have held your arm out of a car window at high speed. In general, air resistance is proportional to the frontal area of the moving object and increases with the square of speed. Thus moving objects try not to present more frontal area to the air than is necessary, and for high-speed motion, shapes are streamlined to lower the *effective* frontal area. Even trucks have over-the-cab wind deflectors to create a smooth airflow over the blunt front of their trailers. We will consider airflow in detail in Chapter 13.

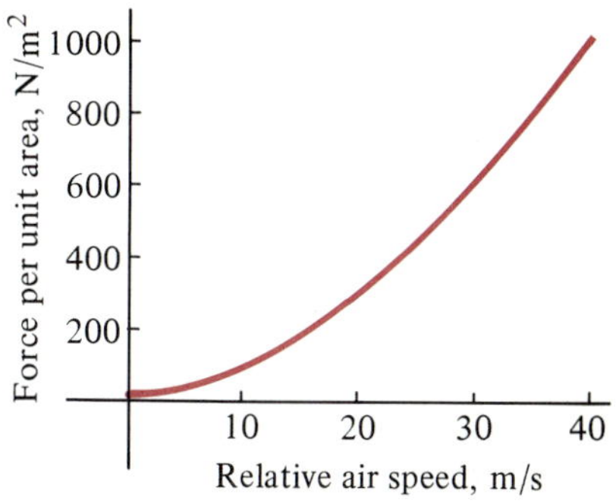

Figure 6-2 The force of air resistance per effective square meter of surface perpendicular to the relative motion of the air vs. the relative speed of the object and the air.

Example 6-2 An average intermediate-sized automobile presents about 2.0 m² of effective frontal area to the direction of motion. (*a*) Determine the forward force which must be provided by its engine to maintain constant speeds of 10 and 30 m/s (neglecting internal friction). (*b*) At a speed of 30 m/s, how much more gasoline must be used driving into a 5.0-m/s headwind than in driving with a 5.0-m/s tailwind?

Solution (*a*) Since $a = 0$, both parts are problems in equilibrium. Thus, the force of the engine is used to balance external forces and maintain the given speeds. Reading from the graph of Figure 6-2, we see that at 10 m/s there is an air resistance of about 65 N/m². Thus for our automobile we have $F = AR = (2.0 \text{ m}^2)(65 \text{ N/m}^2) = 130 \text{ N}$. At 30 m/s the graph shows the resistance to be 600 N/m², yielding a wind resistance for the car of 1200 N. Comparing these numbers, we can see why we save gasoline at lower speeds and why engineers work hard to minimize effective frontal area.

(*b*) This part requires that we read the graph to determine the air resistance for 25 and for 35 m/s. These are the rates that a car must move through the air if traveling 30 m/s with a 5.0-m/s tailwind and headwind, respectively. The values are about 420 and 820 N/m², respectively. Doubling these for the 2.0-m² area on the car yields the two values of force required. The striking point of this is that the difference between the headwind and the tailwind means that the car can encounter almost twice the resistance at the same speed along the highway. Therefore, all else being equal, an automobile will consume almost twice as much fuel with a 5.0-m/s headwind as it will with a 5.0 m/s tailwind in order to maintain a 30-m/s ground speed.

6-2 Torques

It is perfectly possible for the vector sum of the forces on an object to be zero and a change in motion still to occur (Figure 6-3). For example, you can easily support the weight of a horizontal meterstick with a single finger. However, there is only one position on the meterstick where you may rest it on the finger and expect that it could remain motionless: its center. Supported anywhere but below its center, the meterstick will begin to rotate off your finger immediately. *Where* the force is applied counts in trying to balance an object. Two vertical forces equal in magnitude and opposite in direction may act on an object to produce no change in motion, but this is not enough to ensure that no change will result. If the forces are horizontally displaced so that they are not applied to the same point or do not act along the same line, their effect will be to begin a rotation of the object—they produce an angular acceleration as discussed in Chapter 5.

The quantity with which we must concern ourselves here is called **torque**, a measure of the tendency of forces to cause angular accelerations. A good example is an ordinary door. It is much easier to open a door if you push on it at a point nearer the doorknob and away from the hinges. The door moves with much less force if force is applied away from the hinges rather than close to the hinges. If we push or pull right at the hinge, no amount of force will cause the door to rotate. This leads us to believe that besides the magnitude of the force, the distance of that force from the axis of rotation is important in defining

Figure 6-3 The motion of a ball carrier tackled simultaneously from different directions and at different heights will change even if the forces of the two tackles are equal in magnitude and exactly opposite in direction. By itself, the first condition of equilibrium does not ensure that the motion of an object will not change.

torque. Further, the direction of the force is of importance. A given force applied at the edge of the door has the greatest effect if it is perpendicular to the plane of the door. If the force is applied at the same place but inclined 45° to the plane of the door, its effect is not as large. If applied at the edge but parallel to the plane of the door, that is, toward or away from the hinges, there is no effect at all, only a thrust against the hinges.

We now know that the following are important in determining the magnitude of a torque (Figure 6-4): (1) the magnitude of the applied force, (2) the distance of the axis of possible rotation from the point of application of the force, and (3) the direction of the force. Clearly, the mathematical definition of torque will have some complexity and must be followed carefully.

Figure 6-5a shows a hinged door viewed from above. The force **F** is applied at point P, which has a displacement **r** from the axis of rotation of the system, and **F** makes an angle of θ with **r**. The torque produced by **F** acting at P about the axis of rotation is

$$\tau = rF \sin \theta \qquad (6\text{-}3)$$

where the Greek letter τ (tau) is used to indicate a torque and r and F are the magnitudes of the respective vectors.

While Equation 6-3 correctly defines the physical quantity which determines angular accelerations, the torque, the expression looks arbitrary and is difficult to visualize in various physical situations. Scientists and engineers often find it easier to picture if stated in a slightly different way. In Figure 6-5b we have extended the line along which the force is acting, the **line of action** of **F**, and drawn a line perpendicular to the line of action from the axis of rotation. The length of this perpendicular line, l, is exactly r times the sine of θ. Rearranging Equation 6-3, we may write

$$\tau = F(r \sin \theta) \qquad (6\text{-}4)$$

or

$$\tau = Fl \qquad (6\text{-}5)$$

where $l = r \sin \theta$. We call l the **moment arm** of the force and define it as the perpendicular distance between the axis of rotation and the line of action of the force. Thus, we may now define *torque* as the product of the magnitude of the force and its moment arm, Equation 6-5. In other words, to determine the torque about a particular axis, we need only determine the moment arm and multiply it

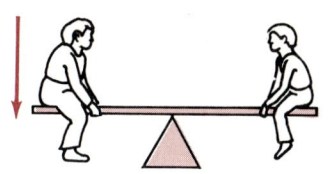

(a) The magnitude of the force counts

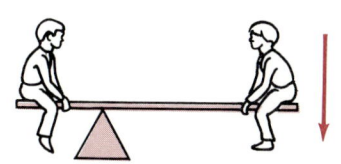

(b) How far from the pivot point the force is applied counts

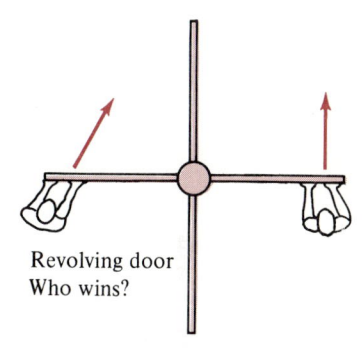

Revolving door
Who wins?

(c) The angle of application of the force at the point of contact counts

Figure 6-4 Torque magnitude is influenced by (a) the magnitude of the applied force, (b) the distance between the point of force application and the pivot, and (c) the direction in which the force acts.

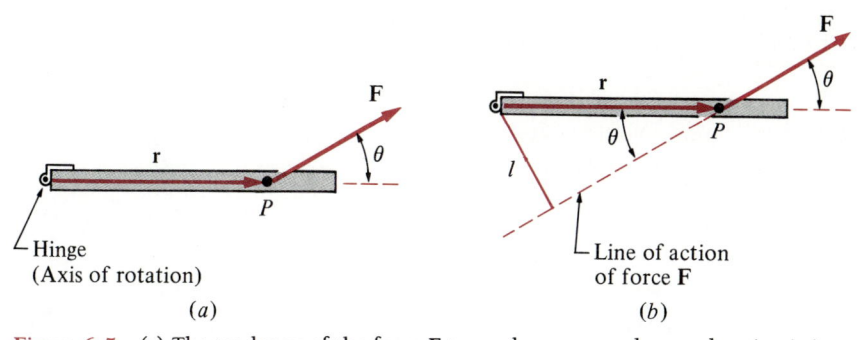

Figure 6-5 (a) The tendency of the force **F** to produce an angular acceleration is its torque, given by $\tau = Fr \sin \theta$. (b) An alternative way to picture torques is by extending the line along which **F** acts.

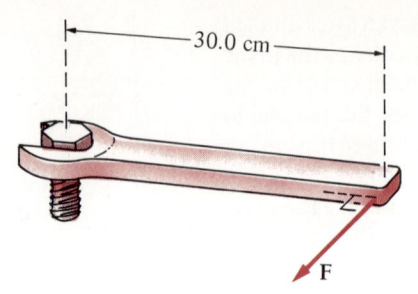

Figure 6-6 Example 6-3.

by the magnitude of the force. The units of torque are force units times distance units, or newton-meters (N·m) in SI units.

Example 6-3 An instruction manual used in assembling an engine cautions you not to overtighten the parts or damage may result. The manual recommends a torque of 54.0 N·m. What force should be applied perpendicular to the handle of a 30.0-cm-long wrench to achieve the desired torque (Figure 6-6)?

Solution Since the force is applied perpendicular to the handle, the moment arm of the torque will be the length of the handle, 0.300 m. Thus, using Equation 6-5, we obtain

$$\tau = Fl \quad \text{or} \quad F = \frac{\tau}{l} = \frac{54.0 \text{ N·m}}{0.300 \text{ m}} = 180 \text{ N}$$

Example 6-4 Calculate the torque that a 16.0-N force imparts to the door of Figure 6-5. The point of application of the force is 0.500 m from the hinge, and the force acts at an angle of 30° to the plane of the door.

Solution This is a straightforward application of any of the defining equations for torque. Using Equation 6-4, we have

$$\tau = F(r \sin \theta) = (16.0 \text{ N})(0.500 \sin 30°) = 4.00 \text{ N·m}$$

This is the magnitude of the torque, and examination of the figure shows that it would tend to set up a counterclockwise rotation about the axis of rotation.

We may use Example 6-4 to interpret the defining equation of torque in yet another way. In Figure 6-7, we have resolved **F** into two components at its point of application to the door, one perpendicular to the plane of the door and one parallel to it. Note that Equation 6-4 can be rewritten as $\tau = (F \sin \theta)r$, where we associate the sine of the angle with the force. Actually, $F \sin \theta$ is the component of **F** which is perpendicular to **r**, $F_\perp$. Thus we may write

$$\tau = F_\perp r \qquad (6-6)$$

This defines torque as the product of the distance between the axis of rotation and the point of application of the force r times the component of the force

Figure 6-7 A third method to visualize torques is to resolve the force vector into two components at the point where the force is applied.

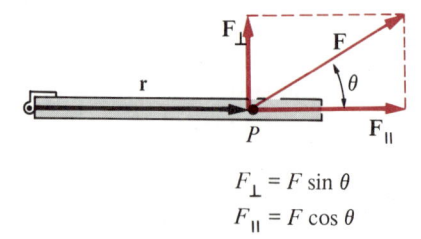

$F_\perp = F \sin \theta$
$F_\parallel = F \cos \theta$

110

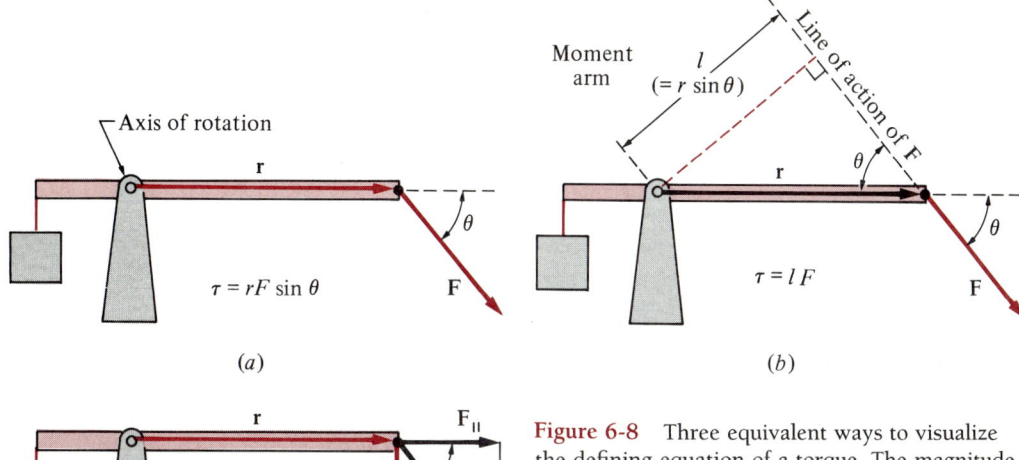

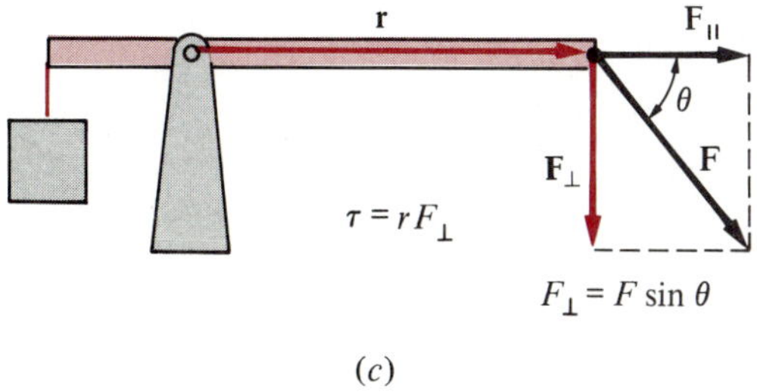

Figure 6-8 Three equivalent ways to visualize the defining equation of a torque. The magnitude of the torque is (a) the distance from the axis of rotation to the point of force application times the magnitude of the force times the sine of the angle between them, (b) the perpendicular distance from the axis of rotation to the line of action of the force times the magnitude of the force, (c) the component of the force perpendicular to r times r.

which is perpendicular to r, $F_\perp$. This form of the equation recognizes that it is the perpendicular component of force on the door in Example 6-4 which will cause all the rotation of the door. The component of the force parallel to the plane of the door can act only to pull against the hinges and has no effect on rotation.

We have now had three interpretations of torque, as illustrated in Figure 6-8. Each provides a consistent method for setting up and determining the torques in physical situations. How you choose to picture torques is a matter of convenience and generally up to you: in terms of the two vectors involved and the angle between them, in terms of the moment arm of a force, or in terms of the component of force perpendicular to the vector between the axis of rotation and the point of application of the force. All these methods yield the identical value for the torque and one that is consistent with the physical effect observed.

Direction of Torques Torques clearly involve direction. If we exert a torque on the bolt of Example 6-3 in one way, we tighten it, but if the torque is in the other sense, or direction, we loosen it. Thus, it is essential to be able to describe the sense of a torque in order to predict its effect. We will do this in an informal way suitable for most problems and in a more precise fashion necessary for later complications.

The vectors **F** and **r** define a plane in which the rotation caused by the torque will take place. The axis of rotation is always a line perpendicular to this plane. Viewed from a particular side, these rotations may be either clockwise or counterclockwise about the axis of rotation. As we did with motion in one dimension, we may handle this by defining a positive direction, or **sense of rotation,** and using algebraic signs to indicate direction. We will follow the common (but not universal) practice of calling torques which tend to set up a counterclockwise change in rotation **positive torques** and those which would set up a clockwise change in rotation **negative torques.** Thus, for example, the torque in Example 6-3 which was used to tighten the engine bolts was a negative torque, while the torque in Example 6-4 was a positive torque. Whenever a problem considers torques in only one plane, we may now assign both a magnitude and a sign to indicate the sense of the change in rotation the torque would tend to impart and add torques as scalars.

Complex situations involving torques about several axes of rotation or the interaction of a torque with other physical variables may require that a torque be treated as a vector. That is, the torque must have not only a defined magnitude from Equation 6-3 but also a single, uniquely defined direction in three-dimensional space assigned to it. How to do this is not immediately obvious, and it will require a convention, that is, an arbitrary rule with which everyone may agree to define uniquely the direction of a torque. The force vector on the wrench in Example 6-3 will assume every possible direction in the plane of rotation at some time during a complete turn. Likewise, the wrench handle, and the **r** vector, will point in continually varying directions within the plane during the rotation. Both these vectors, however, are confined to and define the plane of rotation. The only direction uniquely specified by a defined plane is that direction which is perpendicular to the plane. For example, having a horizontal plane does not, by itself, define the directions of north, south, east, or west, but it does specify uniquely the up-down direction.

We define a **right-hand rule** (RHR) to specify the direction of a torque:

> The direction of a torque is the direction of advance of a right-handed wood screw if rotated in the sense of the torque.

Figure 6-9 shows such a screw being rotated clockwise in the *yz* plane. The screw advances in the negative *x* direction as a result of this rotation; thus, that is the direction of the torque. We may see that a clockwise rotation in the plane of the page would have an assigned direction which was perpendicularly into the page, and a counterclockwise torque would be directly out of the page. If the bolt of Example 6-3 was mounted vertically so that the wrench was rotated in a horizontal plane, the tightening torque was directed downward, in the direction the tightening bolt moved under the action of the torque. Practice with the right-hand rule (RHR) will allow you to apply it rapidly, and we will find it useful in Chapter 9.

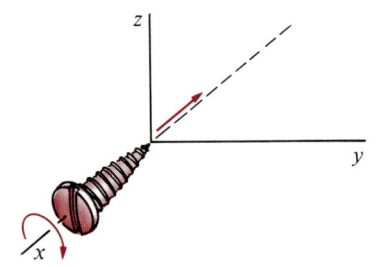

Figure 6-9 A clockwise rotation of the screw in the *yz* plane would cause the screw to advance into the page in the negative *x* direction, which is the direction of the torque defined by the right-hand rule.

6-3 Second Condition of Equilibrium

The state of motion of a body in equilibrium is not changing. This implies that no net force is acting on the body. But, an angular acceleration, that is, one that changes the state of rotation, also changes the motion. A rotating Ferris wheel is in quite a different state of motion than one at rest, even if each remains at the same location in space. Thus, we are led to the **second condition of equilibrium:**

> For an object to be in rotational equilibrium, the sum of all torques acting on it must be zero.

We recognize that if the angular velocity is zero, it will be zero about any axis of rotation. If the Ferris wheel is not rotating, it is also not rotating about a point on the ground that we choose, or about the tip of the Washington Monument, or any other point. Therefore, for a body in static equilibrium, we can sum the torques about *any* axis we choose and set the sum equal to zero. Mathematically, we state the second condition as

$$\sum_i \tau_i = 0 \tag{6-7}$$

This is a powerful physical statement, particularly in combination with the first condition of equilibrium, allowing us to solve a wide variety of problems.

Example 6-5 An artist wishes to construct a mobile which will rotate freely in horizontal planes. It is to be made of two 1.00-m-long rods of negligible mass fastened by wires and supporting three masses, as shown in Figure 6-10. Mass m_1 equals m_2, and either one is twice as large as m_3. At what point along each rod must the wires be fastened for the rods to balance horizontally?

Solution There are two separate problems in rotational equilibrium here. First, the lower horizontal rod must balance by itself, then it must balance the upper rod when it is attached. We define the unknown lengths to the points of attachment of the wires as x_2 and x_1, as shown on the diagram. We are not given the values of the masses, so we must not need this information to solve the problem.

Applying the second condition of equilibrium to the lower rod only, we note that the point where the wire attaches to the rod is a natural choice for the **fulcrum,** or **pivot** point, where the axis of rotation passes through the plane of the page. If we do not choose this as the point about which to calculate the torques, we will have to include a term for the torque caused by the upward pull

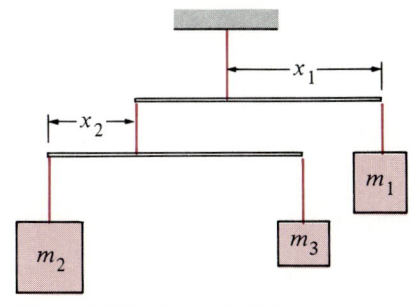

Figure 6-10 Example 6-5.

of the wire at this point. (We could get this, since the upward force must hold up the lower rod and masses, and by the first condition of equilibrium it would have to equal the weights of m_2 and m_3, but let us avoid the first condition for this problem.) Choosing the support point for the fulcrum allows us to ignore the upward force of the wire at that point, *since the moment arm of this force will be zero and it will exert no torque.* Remember this idea, for it is often useful in eliminating one unknown in problems. The moment arm of the weight of m_2 about this axis is x_2, while the moment arm of the weight of m_3 will be $1.00 \text{ m} - x_2$. We write

$$\sum \tau = 0 = (m_2 g)x_2 - (m_3 g)(1.00 \text{ m} - x_2)$$

g divides out of the expression, and we may rearrange to obtain

$$0 = (m_2 + m_3)x_2 - m_3(1.00 \text{ m})$$

but we know that $m_2 = 2m_3$. Substituting this, we obtain

$$0 = (2m_3 + m_3)x_2 - m_3(1.00 \text{ m})$$
$$= 3x_2 - 1.00 \text{ m}$$

or,
$$x_2 = \frac{1.00 \text{ m}}{3} = 0.333 \text{ m}$$

Turning our concern to the upper rod, we note that the force at the left end will be the weight of $m_2 + m_3$, or $\frac{3}{2} m_1 g$. Again, calculating torques about the point of support of the wire, we have

$$0 = \tfrac{3}{2}(m_1 g)(1.00 \text{ m} - x_1) - m_1 g x_1$$

Dividing out $m_1 g$, we obtain

$$x_1 = \tfrac{3}{5} \text{ m} = 0.600 \text{ m}$$

Up to this point, the examples used have had a natural fulcrum, or pivot point, about which rotation could be anticipated. However, once a known force is applied to a given point on a rigid body, the torque is perfectly defined about any point through which we choose to pass an imaginary axis of rotation — fulcrum or not. Since no angular acceleration will occur for a problem in static equilibrium, the choice of a pivot is purely arbitrary and is made to be as mathematically convenient as possible.

6-4 Center of Gravity and Center of Mass

Figure 6-11 shows a meterstick hanging by a string and free to rotate at this point of support with a 200-g mass attached to the 5.00-cm mark near one end. The meterstick is balancing in a horizontal position and is in equilibrium. Clearly the weight of the mass is exerting a counterclockwise torque about the point of support. What is exerting the clockwise torque about that point to make the sum of the torques zero? After a bit of reflection and noticing that the support string is not fastened at the center of the meterstick, you probably concluded correctly that the unbalanced weight of the meterstick was providing the balancing torque. However, this presents us with a problem in calculation

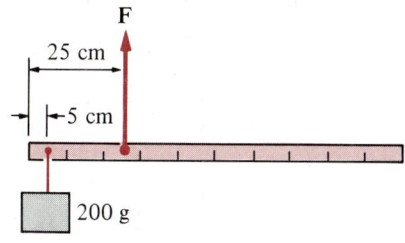

Figure 6-11 A meterstick in rotational equilibrium.

even if we know the weight of the entire meterstick. We note that some of the material of the meterstick is located right at the pivot and will supply no torque, while some of the weight is on the side of the hanging mass and surely provides counterclockwise torque. Furthermore, the segment of the meterstick in the leftmost 5 cm is much farther from the pivot than the 5 cm immediately to the pivot's left and will have a much greater moment arm and contribute much more counterclockwise torque. How are we to treat an object with mass distributed along its entire length and which, like the meterstick, has matter on both sides of the axis of rotation?

The answer to this is quite simple. A rigid body like the meterstick behaves, for the purposes of calculation, as though it were mostly a weightless shell with all its weight concentrated at a single point. The **center of gravity** (CG) of a body or system may be defined as that point through which all the weight of the body may be thought to act for the purposes of calculation. Thus, rather than worry about the separate contribution to the total torque of each segment of the meterstick, we treat it as though all its weight was at the single point, its CG. Locating the CG of a real object can be quite simple. If we suspend an object, say a sheet of paper or a pencil, between index finger and thumb so that it is free to rotate, it will swing until the CG comes to rest directly below the point of support. If this were not so, the weight acting at the CG would have a moment arm about the pivot and rotate the object until it was so. (In fact, this is the operating principle upon which the balance works.) Suspending the object from two different points and marking a vertical line downward from each locates the CG at their intersection.

Example 6-6 The meterstick of Figure 6-11 is supported by a string at the 25.0-cm mark and is balanced by the weight of a 200-g mass attached at the 5.00-cm mark. What is the weight of the meterstick?

Solution We already know that the meterstick alone can be balanced by an upward force at the 50.0-cm mark. Therefore, its CG is directly above this point. Choosing the support point as a natural pivot about which to calculate torques, we see that the CG is 25.0 cm to the right of the pivot. The 0.200-kg mass, of weight mg, or 1.96 N, is 20.0 cm to the left of the pivot. Substituting these values, using w for the unknown weight of the meterstick, we use the second condition of equilibrium and obtain

$$0 = (1.96 \text{ N})(0.200 \text{ m}) - w(0.250 \text{ m})$$

$$w = 1.57 \text{ N}$$

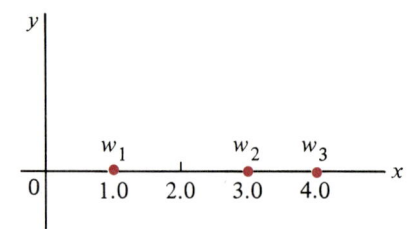

Figure 6-12 Example 6-7.

Example 6-7 Find the x coordinate of the center of gravity of the three weights shown in Figure 6-12 placed at 1.00, 3.00, and 4.00 m from the origin 0. Let $w_1 = 1.00$ N, $w_2 = 2.00$ N, and $w_3 = 2.00$ N.

Solution Assume the CG is at some unknown coordinate l from the origin. According to the definition of the CG, we could place a single body of weight $w = w_1 + w_2 + w_3$ at the CG and obtain, for example, the same torque about 0 as is obtained when the torques due to the individual weights are added. Thus

$$wl = w_1 l_1 + w_2 l_2 + w_3 l_3$$

$$l = \frac{w_1 l_1 + w_2 l_2 + w_3 l_3}{w_1 + w_2 + w_3}$$

$$= \frac{(1.00 \text{ N})(1.00 \text{ m}) + (2.00 \text{ N})(3.00 \text{ m}) + (2.00 \text{ N})(4.00 \text{ m})}{5.00 \text{ N}}$$

$$= 3.00 \text{ m}$$

Example 6-8 A bookshelf made of a uniform wooden board 1.50 m long weighs 20.0 N and is supported by two thin metal rods each 0.0500 m from its end, as shown in Figure 6-13. A book weighing 16.0 N is placed upright on the shelf at a distance of 0.400 m from the right metal rod. Calculate the force each rod must exert on the board to maintain equilibrium.

Solution The known weights of the shelf, w_S, and the book, w_B, are acting downward. Since we are told the shelf is uniform, its CG is at its midpoint, and w_S acts at that point. We have two unknowns, F_R and F_L, the upward forces exerted on the board by the metal rods. Thus, we need any two independent equations to solve for these two unknown forces. We choose to apply the two conditions of equilibrium to obtain these equations. Using the first condition of equilibrium, we sum the forces in the vertical direction, choosing upward as positive, obtaining

$$F_L + F_R + (-w_S) + (-w_B) = 0$$

$$F_L + F_R + (-20.0 \text{ N}) + (-16.0 \text{ N}) = 0$$

$$F_L + F_R = 36.0 \text{ N}$$

We can choose any axis to apply the second condition since the shelf is in static rotational equilibrium about all points. A convenient axis is a line along the right metal rod. This has the advantage of eliminating F_R from the equation because F_R has zero moment arm about this axis. Thus, we obtain

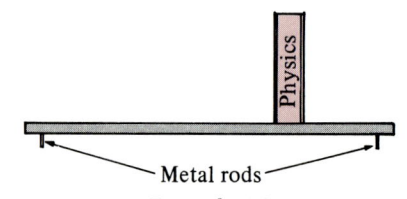

Figure 6-13 Example 6-8.

Again, choosing an axis through a point where a force acts as the pivot (or *center of moments*) about which to calculate torques in a given problem eliminates any torque set up by that force, since it will have zero moment arm about the axis.

$$-F_L l_L + w_s l_s + w_B l_B = 0$$

$$-F_L(1.40 \text{ m}) + (20.0 \text{ N})(0.700 \text{ m}) + (16.0 \text{ N})(0.400 \text{ m}) = 0$$

where the negative sign indicates that F_L alone tends to set up a clockwise torque. This yields

$$F_L = 14.6 \text{ N}$$

Substituting this into our first equation above yields

$$14.6 \text{ N} + F_R = 36.0 \text{ N} \quad \text{or} \quad F_R = 21.4 \text{ N}$$

Instead of using the first condition of equilibrium in this problem, note that we could have just as easily applied the second condition a second time about a different axis to obtain an independent equation. Applying it with an axis of rotation at F_L would yield $F_R(1.40 \text{ m}) - (20.0 \text{ N})(0.700 \text{ m}) - (16.0 \text{ N})(1.00 \text{ m}) = 0$ as our second equation. Solving this yields $F_R = 21.4 \text{ N}$, as expected.

Example 6-9 A 5.00-m-long uniform ladder weighing 300 N is leaning against a smooth vertical wall, as in Figure 6-14. A man weighing 750 N stands on a rung 4.00 m up along the ladder from its base. Determine (a) the upward force the ground exerts on the ladder, (b) the force the wall exerts on the ladder, (c) the frictional force the ground must exert on the ladder, and (d) the minimum value of the coefficient of static friction required to keep the ladder from slipping.

Solution We designate the force of the wall as F_W, of friction as f_s, the normal force at the ground as F_N, and the weights of man and ladder as w_M and w_L. Because the wall is smooth, there is no vertical component of force on the ladder at the wall (no friction), and hence F_W is perpendicularly outward on the ladder. Choosing rightward and upward as positive directions, we apply the first condition of equilibrium in both the horizontal and vertical directions to obtain two independent equations,

$$f_s - F_W = 0 \quad \text{and} \quad F_N - w_L - w_M = 0$$

Figure 6-14 Example 6-9.

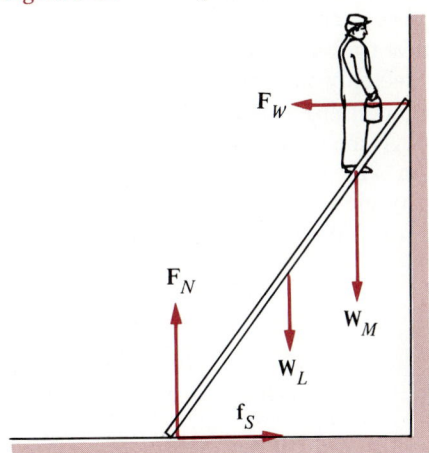

(a) Solving the second equation, we get

$$F_N - 300 \text{ N} - 750 \text{ N} = 0 \quad \text{or} \quad F_N = 1050 \text{ N}$$

(b) Now we can apply the second condition of equilibrium about any axis. A good choice is one through the point where the ladder meets the ground. Both F_N and f_s act through this point, and neither one will contribute a term to the equation if torques are calculated about this point. Moment arms for the other forces become 1.50 m for w_L, 2.40 m for w_M, and 4.00 m for F_W. You should be able to obtain these values. Again choosing counterclockwise torques to be positive, we obtain

$$F_W(4.00 \text{ m}) - (750 \text{ N})(2.40 \text{ m}) - (300 \text{ N})(1.50 \text{ m}) = 0$$

$$F_W = 563 \text{ N}$$

(c) Using this with the equation for the first condition of equilibrium above, we determine that

$$f_s = 563 \text{ N}$$

(d) Since we are asked for the minimum value of μ_s, we may assume motion impends with this value of f_s, and use the equal sign in the definition of static friction, $\mu_s = f_s/F_N$, giving us

$$\mu_s = \frac{f_s}{F_N} = \frac{563 \text{ N}}{1050 \text{ N}} = 0.536$$

While μ_s between ladder and ground could be larger than this, it could not be smaller or the ladder would begin to slide. Note that as the man ascends the ladder farther, the value of the coefficient of friction required to keep the base of the ladder from slipping will increase.

An analogous term to center of gravity is **center of mass** (CM). In a situation where the acceleration due to gravity is constant from position to position, as near the surface of the earth, these two points will be coincident. In the general case, however, these points need not be the same. Torques are sometimes called *moments of force,* or forces times moment arms. The center of mass can be found by using *moments of mass,* or masses times moment arms. The center of mass may be defined as that point which has the same translational motion as though all the mass and all the forces were concentrated there. If we consider a spacecraft in deep space, far from any significant gravitational effect, it is essential to know its center of mass. There is one, and only one, line along which we may push a spacecraft in a given direction without setting up a rotation. If the force vector of the thrust acts on a line through the center of mass, a purely translational acceleration results. If not, the sum of the moments of mass is not zero and a rotational acceleration also occurs. While we should recognize that there is a distinction, we will treat the CM and the CG as equivalent throughout the text.

6-5 Application: Equilibrium and the Human Body

One system in which balance is quite important is in the erect posture of the human being. We must learn to keep our CG above our base of support or fall

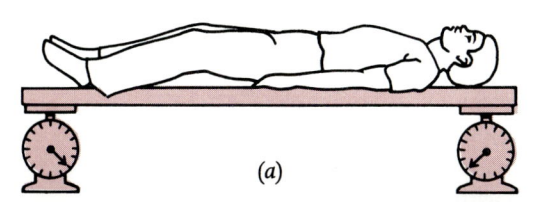

	Pivot location as percentage of height	Center of mass as percentage of height	Mass as percentage of body total	

```
                                    93           7    Head
              Neck 91
                                  72  Upper arm  7
              Shoulder 81         55  Lower arm  4  13 Arms
                                  43  Hand       2
                                  71 ------- 46     Trunk
              Elbow 62
                                  58 ------- 100   Total body
              Hip 52
              Wrist 46            42  Upper leg  21
                                  18  Lower leg  10  34 Legs
                                   2  Foot        3
              Knee 28

              Ankle 4                         (b)
```

Figure 6-15 (a) The center of mass of the body can be determined by placing the individual on a light body-board supported at two known points by scales. Each scale supports a different portion of the body's weight, and the second condition of equilibrium is used to locate the center of mass. (b) Detailed knowledge of the distribution of mass in the body and the location of pivot points is necessary to properly design products to be used by people. These figures represent a typical adult male from studies by NASA and others.

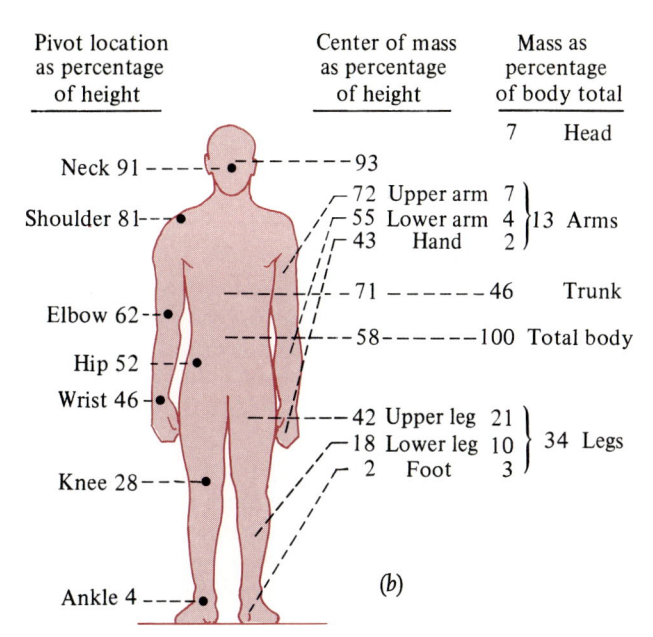

over. If you are standing on one foot, should you lean forward such that the vertical line downward from your CG strikes forward of your toes, a torque exists to cause you to rotate from rest into a forward roll. In such a situation, we have learned to thrust the arms or the free leg rearward to move our CG farther back and over our support base. Mastering the fine muscular corrections of body attitude necessary to keep the CG above the base provided by our feet is a serious learning challenge for the infant, and fun to watch. To be able to predict the effects of forces and torques on the human body, it is necessary to determine the CG of the entire body and its separately movable parts experimentally, as in Figure 6-15.

One of the earliest studies of how human beings move was conducted by Leonardo da Vinci. From his dissections, he concluded that the muscles of the body were mainly designed to pull, not push. Muscles contract powerfully on demand: the maximum force exerted is about 7×10^5 N/m^2 of cross-sectional area (or about 100 lb/in^2). The changes in length associated with muscular contraction are quite small, however. To obtain rapid, sweeping motion from these minor length changes in the muscles, all higher life forms have systems of rigid levers, their skeletons, to magnify the motion. If a point on the lower arm bone a few centimeters from the elbow is raised to be only 2 or 3 cm closer to a fixed point on the upper arm, we find that this action swings the hand, at the end of the lower arm, through a large, rapid arc. The effect is much like using long stilts, where a small motion near the axis of rotation becomes a long arc stride at the far end.

Muscles must occur in opposed sets. If you have one muscle to extend your arm, for example, you will need another to pull it back or it will have to stay extended. (Of course, these opposed muscles would not be used at the same time any more than you use drive and reverse in your car at the same time.) The best example of a set of opposed muscles is that of the biceps, which raises the forearm, and the much smaller triceps, which pulls to lower it (usually assisted by gravity).

Example 6-10 The arm depicted in Figure 6-16a belongs to a 2.00-m-tall, 60.0-kg male of normal build and is shown handing a 16.0-N book to another person. The upper arm bone is inclined 45° to the horizontal. What are the magnitude and direction of the force exerted by the biceps muscle shown, attached to the lower arm bone 5.00 cm from the elbow, to maintain equilibrium?

Solution This is a complex problem with minimal guidance being given in its statement, requiring us to make many assumptions. We are told that this is a male of typical body build and may therefore use the data of Figure 6-15b with some confidence. We note that 6 percent of the typical body's mass is in the lower arms and hands, or 3 percent for each side of the body. Thus, w_A, the weight of the lower arm and hand, is 0.03×60 kg $\times 9.80$ m/s$^2 = 17.6$ N. From Figure 6-15b, we calculate the center of gravity of the lower arm and hand combined to be at 51 percent of the body height when standing erect, while we are told the elbow is at 62 percent of height. Since the man is 2.00 m tall, we calculate l_A, the distance from the elbow to the point of application of w_A, to be 0.11×2.00 m = 22.0 cm. Assuming the book to be centered in the hand, we put w_B at 19 percent of body height from the elbow, or 0.19×2.00 m = 38.0 cm = l_B. This leaves three unknowns from Figure 6-16b: the push P of the upper arm bone against the lower arm bones, the force F exerted by the biceps, and the angle of pull θ_F of the biceps. (Though we note that θ_F is nearly parallel with θ_P.) We have three unknowns and will require three independent equations to solve for them. Using upward and to the right as positive directions, we apply the first condition of equilibrium in the horizontal and vertical directions separately to yield

$$F_x - P_x = 0 \quad \text{or} \quad F \cos \theta_F - P \cos \theta_P = 0 \quad (a)$$

and in the vertical direction, $F_y - P_y - w_A - w_B = 0$, or

$$F \sin \theta_F - P \sin \theta_P - w_A - w_B = 0 \quad (b)$$

For a third equation we apply the second condition of equilibrium about the natural fulcrum of the elbow. This eliminates any term containing P, which acts through this point and thus has zero moment arm about this axis.

$$(F \sin \theta_F) l_F + w_A l_A + w_B l_B = 0 \quad (c)$$

We now have three equations in three unknowns and will rewrite them, inserting known values and omitting extra zeros for clarity:

$$F \cos \theta_F - 0.707 P = 0 \quad (a')$$
$$F \sin \theta_F - 0.707 P - 17.6 \text{ N} - 16 \text{ N} = 0 \quad (b')$$
$$(F \sin \theta_F)(0.05 \text{ m}) - (17.6 \text{ N})(0.22 \text{ m}) - (16 \text{ N})(0.38 \text{ m}) = 0 \quad (c')$$

The last of these, Equation c', yields $F \sin \theta_F = 199$ N. Substituting this into Equation b' yields $P = 234$ N; thus we have determined one of our unknown variables. Using this value of P in Equation a', it becomes $F \cos \theta_F = 165$ N. However, we recall from trigonometry that $\sin^2 \theta + \cos^2 \theta = 1$, and thus we may write

$$F^2 \cos^2 \theta_F + F^2 \sin^2 \theta_F = F^2 \quad (1)$$

or

$$(165 \text{ N})^2 + (199 \text{ N})^2 = F^2$$

and

$$F = 259 \text{ N} \qquad \theta_F = 50.3°$$

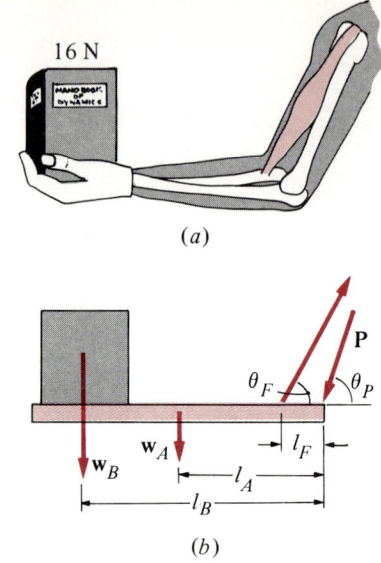

Figure 6-16 Example 6-10.

The first item we note from these figures is the relatively great force (259 N) the biceps must exert to support a small object (16.0 N) in the hand. In this case the muscular force was 16 times the weight of the book. By attaching muscles near the pivot points between bones, nature "buys" a rapid motion through a large arc at the end of the bone. Conversely, the penalty "paid" for this magnified motion is that the force applied near the pivot must be much larger than the force that can be exerted at the free end of the bone.

Minimum Learning Objectives

After studying this chapter, you should be able to:
1. Define:
   - air resistance
   - center of gravity
   - center of mass
   - conditions of equilibrium
   - dynamic equilibrium
   - equilibrium
   - fulcrum
   - line of action
   - moment arm
   - pivot
   - right-hand rule
   - sense of rotation
   - static equilibrium
   - torque
2. State the first condition of equilibrium and the second condition of equilibrium in words and mathematically.
3. Identify the following variables for each force in a problem involving equilibrium:
   a. Moment arms of each force
   b. Normal component of each force
   c. The direction of a given torque vector, using the convention of the right-hand rule
   d. Line of action of each force
   e. (Parallel) radial component of each force from an arbitrary pivot
4. Determine the center of gravity of a series of discrete masses of simple geometry rigidly attached to each other.
5. Apply the first condition of equilibrium to determine the magnitude and direction of a single unknown force acting on a rigid body in equilibrium.
6. Read through and analyze the normal description of diverse physical systems, like a structure or a part of the body, to visualize the pertinent forces and points of their application in order to set up a problem to be solved by using the conditions of equilibrium.
7. Set up and solve by the conditions of equilibrium three equations in three unknown forces, or angles of application of forces, for rigid bodies in equilibrium.

Problems

6-1 Determine the force **F** necessary to keep the brass ring in Figure 6-17 stationary.

6-2 A traffic light weighing 300 N is suspended over Main Street by two cables, each pulling at an angle of 15° with the horizontal. Calculate the tension in the cables.

6-3 Identify and determine the magnitude and direction of the forces acting on a 0.390-kg can of soft drink which is (a) sitting stationary on your desk and (b) sliding at a constant velocity down a 15° ramp.

6-4 A painting with its backing and frame weighs 20.0 N. It is hung from a nail such that the wire makes an angle of 30° below the horizontal to each corner of the frame. Calculate (a) the tension in the wire and (b) the downward force on the nail.

6-5 An automobile stuck in the mud can often be pulled to firmer ground by one person if the rope used is stout enough. In Figure 6-18, the rope is secured to the automobile and pulled very taut before being wrapped around the tree. The owner of the automobile can pull to the side with a force $P = 500$ N, which moves the rope so that it is now 5° away from its original direction, as shown. Determine the tension in the rope and hence the force applied to the automobile.

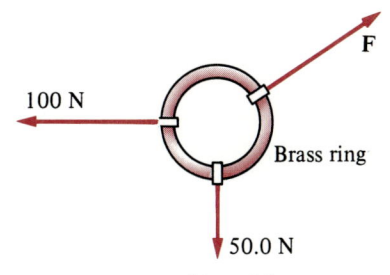

Figure 6-17 Problem 6-1.

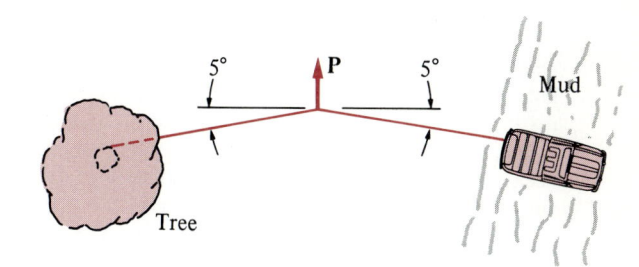

Figure 6-18 Problem 6-5.

6-6 A child weighing 400 N is sitting on a 100-N sled. The coefficient of kinetic friction between the snow and sled runners is 0.0500. Calculate (a) the horizontal force with which a playmate must pull to move the child and sled at a constant speed and (b) the force with which the playmate would have to pull if it is to be applied at an angle of 30° with the horizontal. (Be careful! What is the effect at this pull on the normal force?)

6-7 A football player is pushing at a constant speed a practice sled on which the line coach stands to provide additional weight. The sled and coach together weigh 2500 N. If the player must exert a force of 750 N, determine (a) the force of friction between the sled and the turf and (b) the coefficient of kinetic friction between the sled and the turf.

6-8 Use Figure 6-2 to determine the force exerted on a rather large piece of plywood being carried by a man when the wind is blowing at 15.0 m/s (a) if the effective area is 3.00 m² and (b) if the man turns so the effective area is 0.400 m².

6-9 One of the bicycles in a race is being ridden at a constant speed of 20 m/s. The force propelling the rider forward is 80.0 N. Use Figure 6-2 to determine the effective area the bicycle and rider are presenting to the wind (assume all the resistive forces are coming from the air).

6-10 A small boat equipped with an outboard motor presents an effective area of 0.500 m² to the direction of motion. Use Figure 6-2 to determine the resistive force the water is applying to the boat when it is moving with a speed of (a) 10.0 m/s and (b) 20.0 m/s. (Note that the density of water is about 770 times that of air. Thus the force per unit area in Figure 6-2 must be multiplied by 770 to obtain the correct numbers for water resistance.)

6-11 A block of mass m slides at a constant velocity down a plane surface inclined at an angle θ with the horizontal. Prove that the coefficient of kinetic friction is equal to the tangent of θ.

6-12 Calculate the torque being applied to a screwdriver about an axis along its center. Assume that a single force of 25.0 N acts at a distance of 1.50×10^{-2} m away from but perpendicular to the axis.

6-13 A force of 140 N is applied perpendicular to a baseball bat at 0.660 m from one end and 0.200 m from the other end. Calculate the magnitude of the torques about axes through either end and perpendicular to the bat.

6-14 A rigid rod of negligible weight is 0.660 m long and is to be balanced horizontally on a knife edge 0.400 m from one end. That end has a downward force of 9.00 N applied to it. Calculate the force required at the other end to maintain balance.

6-15 One of the first physical accomplishments of an infant is to hold its head in an upright position. The center of mass of the head is 0.02 m in front of the pivot where the spinal column supports the head. The muscles at the back of the neck exert a tension to keep the head from rotating forward and act vertically downward 0.05 m from the pivot. If the baby's head weighs 14.0 N, what is the tension in the neck muscles?

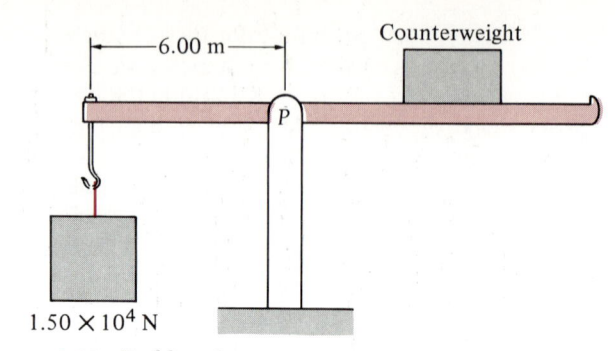

Figure 6-19 Problem 6-16.

6-16 An example of a European crane is shown in Figure 6-19. If the lift arm is 6.00 m, how far from point P must one move the counterweight of 1.00×10^4 N to balance a load of 1.50×10^4 N?

6-17 A wheel weighing 75.0 N is supported at rest on a plane tilted 30° to the horizontal by a force applied parallel to the plane at the axis of the wheel. What is the magnitude of the force necessary to prevent the wheel from rolling down the plane?

6-18 The irregularly shaped body in Figure 6-20 is subjected to the forces shown. Determine the magnitude, direction, and distance along the y axis that an additional force must be applied to keep the body in equilibrium.

6-19 A circular table has four legs spaced at equal intervals around the circumference. The tabletop is uniform and weighs 100 N. Each leg weighs 20.0 N. Determine the smallest weight which when placed on the table will cause it to tip.

6-20 A structural shape called a truss is constructed from two uniform beams of length 3.00 m each. The beams are hinged at the top and stood to form an isosceles triangle with the ground with base angles of 50°. A horizontal tie-rope cable 0.500 m off the ground joins the beams. If the

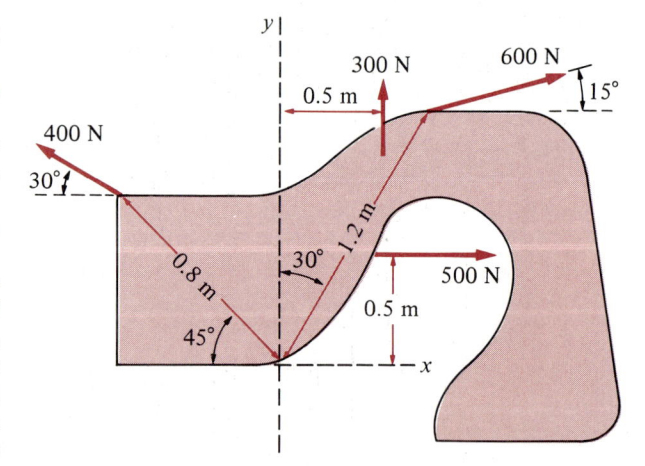

Figure 6-20 Problem 6-18.

122

left beam weighs 600 N and the right beam weighs 400 N, calculate the tension in the cable.

6-21 A gardener is moving some topsoil in a wheelbarrow (Figure 6-21). The total weight of the wheelbarrow and load is 600 N, and the distance from handle to axle is 1.30 m. To maintain an angle of 25° while moving along, an upward vertical force of 250 N must be applied. Calculate (a) the torque the gardener is applying about the point of contact of the wheel with the ground and (b) the distance from the line of action of the center of gravity of the wheelbarrow and load to the point of contact of the wheel with the ground.

6-22 A uniform iron bar 2.80 m long is attached to a wall by a hinge at one end. It is held in a horizontal position perpendicular to the wall by a 300-N vertical force 2.00 m from the wall. Determine the bar's (a) weight and (b) mass.

6-23 Determine the position of the center of gravity of a system of two weights 14.0 and 42.0 N which are separated by a horizontal distance of 1.60 m.

6-24 An I-beam is being hoisted by a cable at one end to an upper floor of a building under construction. The beam is uniform, weighs 5.00×10^3 N, and is 6.00 m long. With one end still on the ground, the tension in the cable is approximately 3.00×10^3 N when the beam is at an angle of 40° with the horizontal. What is the force being exerted on the ground?

6-25 Two painters are above the first floor of a house on a scaffold which weighs 400 N, is 3.00 m wide, and is supported by two ropes attached 0.250 m in from each end. The first painter, of mass 80.0 kg, is standing at the center, while the second painter, of mass 65.0 kg, stands 1.00 m from the left end of the scaffold. Calculate the tension in each rope.

6-26 To recruit students the physics department wishes to put a

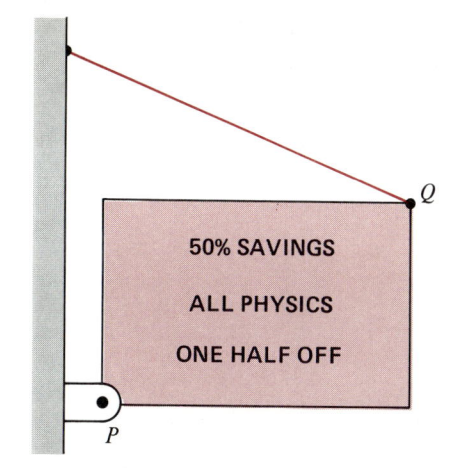

Figure 6-22 Problem 6-26.

street sign over the door of the science building, as shown in Figure 6-22. The sign is to be supported by a hinge pin through its lower left corner at point P and by a cable attached to point Q which makes an angle of 30° with the top of the sign. The sign is a perfect rectangle, measuring 1.20 m high by 1.80 m wide and weighing exactly 400 N. Calculate the tension in the cable and the force on the hinge pin.

6-27 A young girl weighing 400 N and her friend who weighs 500 N wish to make a teeter-totter from a plank 4.00 m long. The plank is uniform and weighs 300 N. If each girl sits 0.250 m from the end of the board and the heavier girl is on the left end, at what distance from the left end should the board be placed across the pipe so that it will balance with the girls on it?

6-28 A bridge over a small stream has a span 20.0 m long, weighs 4.00×10^5 N, and is supported on cement abutments at each end. A car of mass 1.20×10^3 kg is being followed on the bridge by a truck of mass 6.0×10^3 kg. When the center of gravity of the car is 3.00 m from the right end of the bridge, the center of gravity of the truck is 7.00 m from the right end. Determine the force being exerted on each end of the bridge.

6-29 The boom of the crane in Figure 6-23 is uniform, 3.00 m long, and weighs 500 N. It supports a tree stump and associated earth weighing 800 N at its far end. The boom makes an angle of 53.1°, and the tie-rope is at an angle of 36.9° (both with the horizontal). Calculate the tension in the tie-rope.

6-30 A man painting under the eaves of his house wishes to lean out to the right to reach and paint as far as possible without moving the ladder. The ladder in Figure 6-24 is 5.00 m long, and the feet are placed 3.00 m away from the wall. The maximum frictional force the wall can exert on the ladder is 80.0 N. The painter's center of gravity is just at the position of the middle of his belt buckle, and he has a mass of 85.0 kg. The ladder is uniform, 0.500 m wide, and

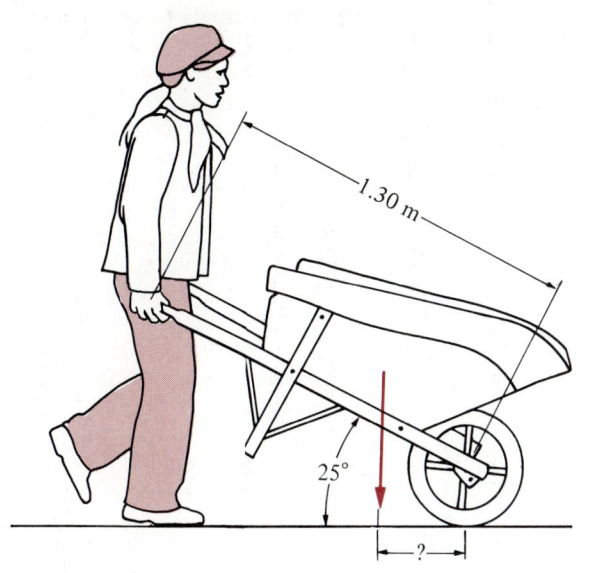

Figure 6-21 Problem 6-21.

123

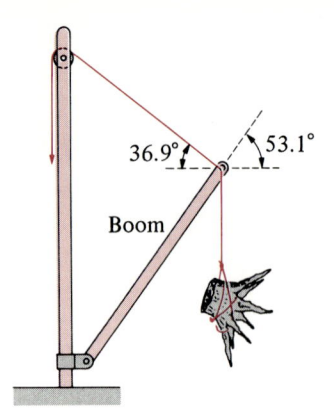

Figure 6-23 Problem 6-29.

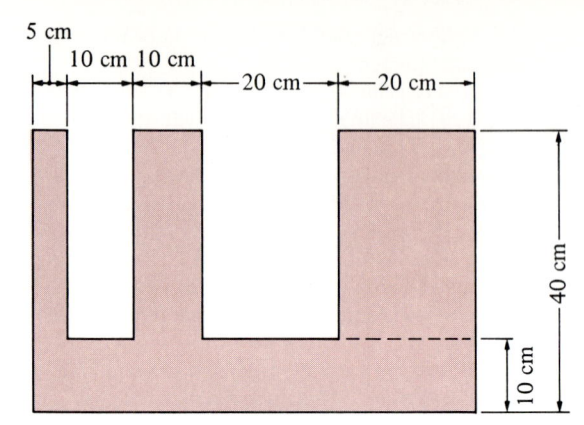

Figure 6-25 Problem 6-34.

weighs 300 N. How far can the man lean out before the ladder slips on the wall? That is, calculate the distance beyond the right edge of the ladder he may place his center of gravity. (You may assume the left leg of the ladder is just about to lift and hence exerts no force on the ground. Why doesn't it matter how far up the ladder he is?)

6-31 A kitchen door is 0.900 m wide, 2.00 m high, and weighs 200 N. It is hung with hinges 0.250 m from both the top and bottom. A child of mass 40.0 kg is hanging from the outer end of the top to swing on the door. Calculate the horizontal and vertical components of force on each hinge. (What did you have to assume?)

6-32 A hydrogen atom consists of an electron "orbiting" a central proton. Determine the average radial distance of the center of mass from the geometric center of a hydrogen atom in its lowest energy state. The mass of the proton is 1.67×10^{-27} kg, and the mass of the electron is 9.11×10^{-31} kg. Assume the average distance from the center of the proton to that of the electron is 5.30×10^{-11} m. If the proton is assumed to be a sphere of radius 1.37×10^{-15} m, is the center of mass inside or outside the proton's "surface"?

6-33 An automobile weighs 1.00×10^4 N, and its center of gravity is 1.20 m forward of the rear axle. Six bags of sand weighing 500 N each are loaded into the trunk 0.800 m behind the rear axle, and the driver and a friend weighing a total of 1500 N sit in the front seat, 1.80 m forward of the rear axle. Now how far is the center of gravity of this system from the rear axle?

6-34 The shape in Figure 6-25 is made of a flat piece of metal of uniform thickness. Determine the position of the center of gravity of the shape.

6-35 A uniform meterstick weighs 2.00 N. A weight of 5.00 N is attached to a point 0.100 m from one end of the stick, and a weight of 9.00 N is attached to the other end. (a) Determine the position of the center of gravity of the system if the meterstick is horizontal. (b) Is the position of the point calculated in part a the same as that for the center of mass?

6-36 A tractor is dragging a large log out of the woods along a level surface. The rope is attached to a hook at the top of one end of the log. To maintain a constant speed, the tension in the rope is 2.00×10^3 N and makes an angle of 20° with the horizontal. (a) If the log weighs 2.50×10^3 N, calculate the magnitude of the normal force and the coefficient of kinetic friction. (b) The log is 0.500 m in diameter and 4.00 m long. Determine the line of action of the normal force. That is, assuming the weight force acts at the geometric center, determine the perpendicular distance from the line of action of the normal force to the geometric center.

6-37 The sauropods were the largest of the dinosaurs; the largest measured was almost 30 m from head to tail. The skeletal structure of the neck of the more modest sauropod shown

Figure 6-24 Problem 6-30.

124

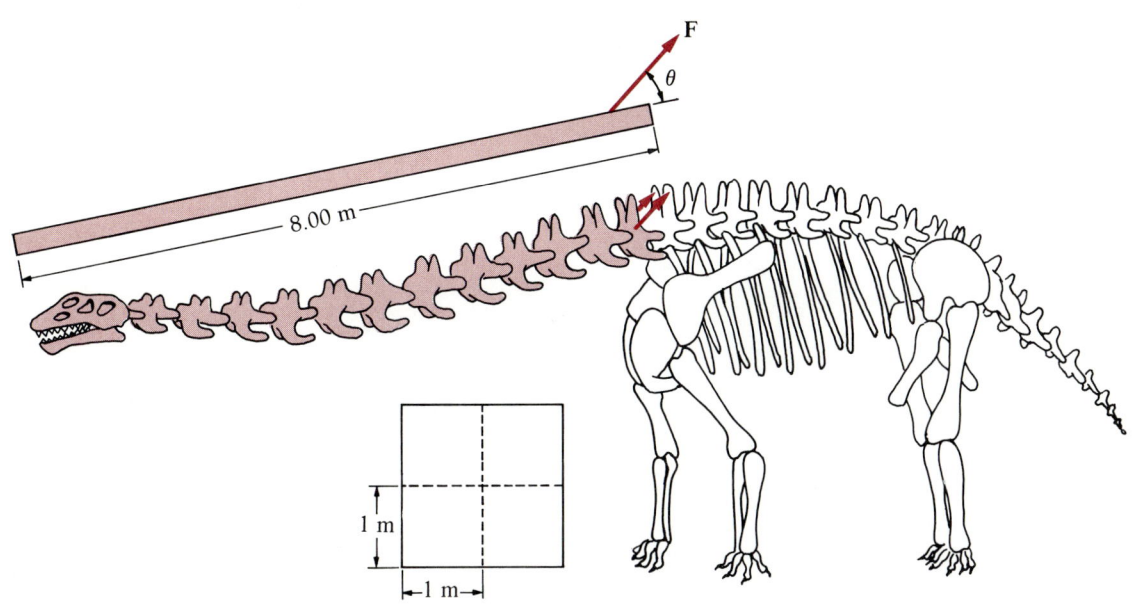

Figure 6-26 Problem 6-37.

in Figure 6-26 extends 8.00 m from the snout to the first vertebra (the 13th) braced from underneath. The mass of the head and neck is estimated to be 1200 kg, with the center of gravity 5.00 m from the snout. What force must the muscles of the 13th vertebra exert to hold the neck as shown if they exert their pull on the 12th vertebra at an angle of 40°?

6-38 A man weighing 900 N balances on the ball of one foot. Figure 6-27 depicts the Achilles tendon attached to the heel bone at one end and the gastrocnemius muscle of the leg at the other. The lower leg bone is pushing on the foot bone. T is the tension in the Achilles tendon, which makes an angle of 30° with the vertical, P is the push of the tibia and is directed at an unknown angle θ_p from the vertical, and F_N is the normal force being applied by the floor and is, of course, 900 N. The perpendicular distance from the line of action of T to the anklebone (the point of application of P) is 0.0500 m. The perpendicular distance from the line of action of F_N to the anklebone is 0.0120 m. Calculate the values of T, P, and θ_p.

6-39 At its simplest approximation the biting-chewing system of the human jaw is a lever system, with the muscles acting some 2.50 cm in front of the pivot in the jaw. The molars, the canines, and the incisors act at 2.50, 4.00, and 4.50 cm in front of the pivot, respectively. (a) If the cross-sectional area of the jaw muscles is 8.00 cm², what is the greatest force they can exert in pulling the lower jaw against the upper jaw? (b) What is the greatest torque the jaw muscles can develop in rotating the jaw about its pivot? (c) What force can be exerted at each of the three types of teeth? (Muscles can exert 78.0 N/cm².)

6-40 A 1.83-m-tall man of mass 80.0 kg is bent over at the hip joint so that his trunk is horizontal, with his arms dangling straight downward. (a) Use the distribution-of-body-mass chart of Figure 6-15b to determine the horizontal distance of the center of mass of the entire upper body from the hip joint while in this position. (b) Muscle tension along the entire back keeps the upper body from sagging floorward. Approximate this with a single force F_m acting 12° from the horizontal at a point midway between the hip and neck joints. Calculate F_m.

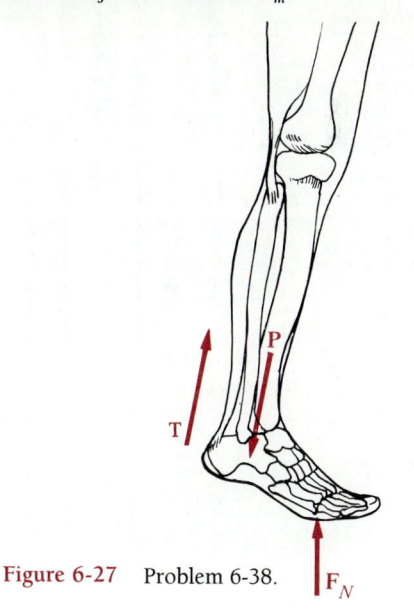

Figure 6-27 Problem 6-38.

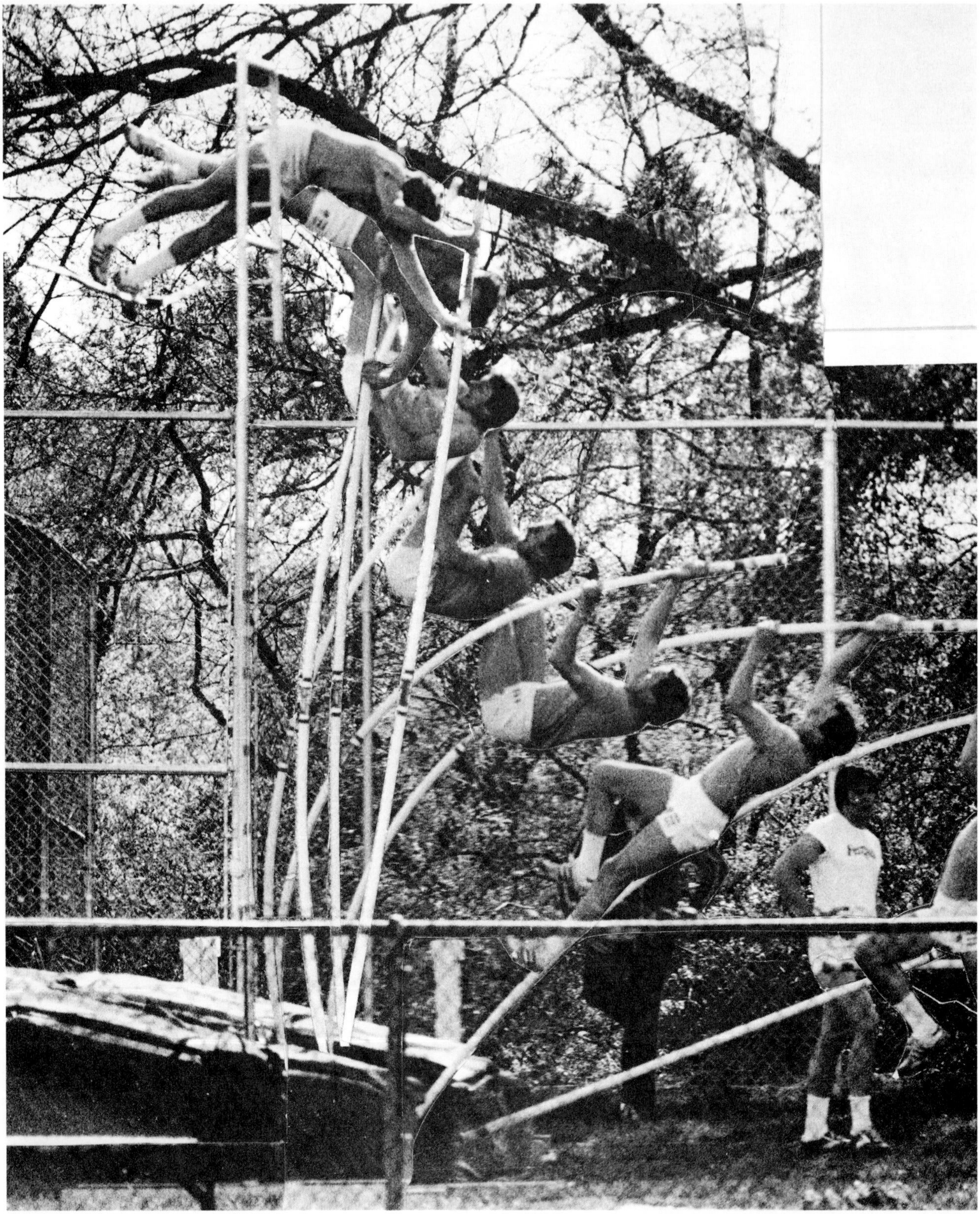

7 Work and Energy

7-1 Introduction

Up to this point we have considered forces in detail, allowing us to predict a wide variety of confirmable results. Yet, the discussion of torques, particularly the idea of the lever in the last chapter, poses a puzzle. Consider the following: Few people can crack a walnut in their hand, yet even children can accomplish this task with a nutcracker. A standard nutcracker consists of two pivoted arms about 20 cm long and operates as a second-class lever. The resistance, the nut, is placed close to the pivot, say 5 cm. A force is now applied to the ends of the handles. At the instant before the nut cracks, the system is in equilibrium and a small additional force breaks the nut. Say the force F applied in that instant was 50 N. The second condition of equilibrium allows us to calculate the resistance force R of the nut as

$$Fl_F - Rl_R = 0$$

or

$$R = \frac{Fl_F}{l_R} = \frac{(50 \text{ N})(0.20 \text{ m})}{0.05 \text{ m}} = 200 \text{ N}$$

where R is the force exerted by the nut on the inside of the nutcracker, F is the force exerted by the hand, and l_F and l_R are the respective moment arms.

By using the nutcracker, we have taken a force easily exerted by the hand and transformed it into a force which could not (usually) be exerted by the hand. Such devices are common enough that we have a name for them. Any device which transforms the size or direction of a force is called a **machine**; the nutcracker is a simple machine.

As another example, if you reached down, grabbed the front bumper of your car and seriously tried to lift it off the ground, you are unlikely to succeed. Yet even the most slightly built of us can use the jack in the car's trunk to accomplish precisely the same task. Or, yet more extreme, suppose you try to stop a speeding car. You stand in the road, prepared to exert a 400-N stopping force on the hood of the onrushing vehicle. You find that the car is prepared to see your 400 N and raise you several thousand more, maybe raising you into the next block in the bargain. Yet, your same 400-N force applied to the brake pedal of the same car brings tons of metal to a screeching halt.

Considering these examples, we ask "If nature allows us liberally to transform the size (and/or direction) of forces, what limits what is possible in a given situation?" If we are clever enough, can we design a machine powered by our own muscles to move us 400 km/h down the road? More immodestly, can we invent a super brake pedal such that when we apply our foot that force alone brings the earth to a halt in its orbit? An old adage says "there's no such thing as a free lunch." What is the hidden cost of using a machine? Surely stupendous accomplishments require stupendous effort on our part, or do they? Can the labors of Hercules be reduced by suitable machines so that they can be performed by the muscles of a typical sixth grader? In the three examples we began with, what didn't nature allow us to transform? What was constant?

Most of these speculations direct us toward a separate physical entity we have not yet defined called energy. The concept of energy is subtle and only emerged in its near modern form about a century ago, some two centuries after Newton's birth. It will take some time and many intermediate steps to develop an adequate statement on the matter, and your picture of energy will evolve throughout the book. Yet, we hope to close this chapter with a sound mental picture of the concept.

7-2 A Sample of Accomplishment

We will investigate what is necessary to accomplish the same task in several different ways. Suppose we have a 1000-N keg of nails which we wish to raise 3.00 m to the second story of a building we are constructing. An individual could simply exert a 1000-N upward force, lift the keg, and carry it up a ladder. However, few individuals are capable of exerting and sustaining a lifting force this size. A more convenient method might be to use a single pulley placed above the spot we want the keg. We tie a rope around the keg, run the rope over the pulley, and pull downward on the free end to lift the keg. An easy way to accomplish this is to have two workers on the second floor with total weight slightly more than 1000 N simply hang on the free end of the rope and use their weight to lift the keg. The effect of the pulley in this case is to transform the direction in which a force must be exerted to lift the keg, but lifting still requires a 1000-N force.

A single individual could raise the keg by using a double-strand pulley, as in Figure 7-1a. If the free end of the rope is pulled downward with a force of

500 N, this sets up a tension in the rope of 500 N. Ignoring friction, this tension is set up in all segments of the rope. The lower, moving pulley wheel is supported by two segments of rope. If the tension in each of these segments is 500 N, we have enough force to raise the keg (again, ignoring friction and the weight of rope and pulleys). Since 500 N is still a large force for a human being to sustain, we might switch to the more elaborate block and tackle of Figure 7-1b. Here, five segments of rope support the lower pulley wheels and keg. If the tension all along the rope is 200 N, each of the segments exerts an upward force this size, and together they may raise the keg.

We have reduced the force required to raise the keg to one-fifth its weight. Actually, we could make it any size we wish, within reason, by choosing an appropriate machine. Suppose we lack the apparatus to rig a pulley system; we might still be able to construct a ramp leading to the second floor and roll the keg up. When the keg is on the ramp, we may resolve its weight into two components — one parallel to the ramp, $F_\parallel$, and one perpendicular to the ramp, $F_\perp$, as in Figure 7-2. Component $F_\parallel$ acts to roll the keg down the ramp, but $F_\perp$ does not — it acts to crush or snap the ramp itself. Leaving $F_\perp$ to the structure of the ramp, we see that to roll the keg up we need only overcome $F_\parallel$. But, $F_\parallel = w \sin \theta$, so by choosing an appropriate value of θ we can make $F_\parallel$ any size we wish. As we allow θ to approach 90°, the ramp rises almost straight up and $F_\parallel$ approaches the weight in magnitude. Or, allowing θ to become smaller and approach zero, $F_\parallel$ declines to become vanishingly small. Using a reasonable ramp between these extremes, we may calculate $F_\parallel$. Say that our ramp is 10.0 m long to rise the 3.00 m to the second floor. Then, $\sin \theta = 3.00 \text{ m}/10.0 \text{ m} = 0.30$, and $\theta = 17.6°$. We calculate $F_\parallel$ as

$$F_\parallel = mg \sin \theta = (1000 \text{ N})(0.300) = 300 \text{ N}$$

Thus, barring friction or any sagging or warping of the ramp (which changes θ in the vicinity of the keg), a force of 300 N could be used to roll the keg up the ramp.

We could continue this exploration for some time; after all, we haven't tried a lever system to raise the keg, or a jack like that in your car, or a hydraulic jack like that at a service station, or a catapult to shoot the nails to the second floor, or a whole host of other possible machines. Yet, we have enough cases to examine this situation more closely for a nonvarying element. Of course, in each of these cases one constant was the task accomplished. A 1000-N keg of nails was raised 3.00 m to the level of the second floor. But we wish to determine in each case what physical quantity was constant. Clearly, the force exerted was not constant, varying from the full weight to one-fifth of this. However, once the commonality in all these cases is seen, it is really quite straightforward. Each

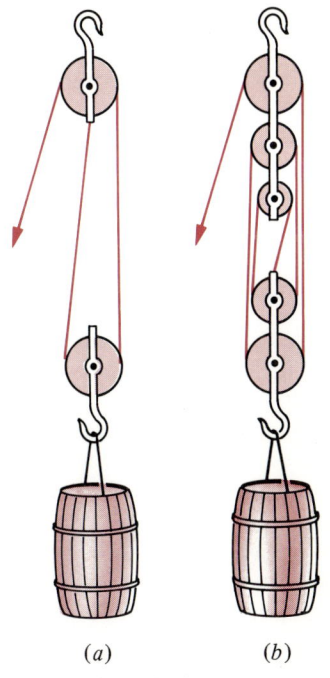

(a)   (b)

Figure 7-1 (a) A double-strand pulley system lifts the keg with the force F applied at the free end of the rope. (b) The block and tackle has five strands of rope supporting the lower, movable block from which the keg is hung.

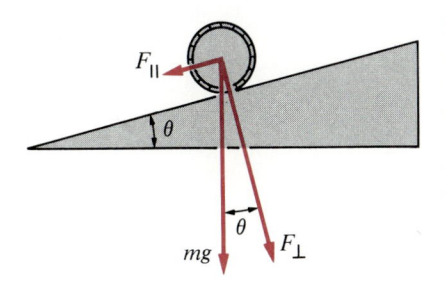

Figure 7-2 The weight of a keg on a ramp resolved into two components: $F_\perp$, which acts perpendicularly into the ramp, and $F_\parallel$, which acts parallel to the surface of the ramp. It is $F_\parallel$ that a worker must overcome to roll the keg up the ramp.

129

time the force which the human being had to exert to lift the keg was reduced, the distance through which this reduced force had to be exerted by the human being was increased. A steep ramp is short but requires a large force to roll the keg up. A shallow-slope ramp is long to reach the same height but requires a much smaller force. However, this force must be exerted along the entire length of the longer ramp. Consider the following cases.

Case 1. Direct Lift   Here, $F = 1000$ N and $d = 3.00$ m. Thus

$$Fd = (1000 \text{ N})(3.00 \text{ m}) = 3000 \text{ N} \cdot \text{m}$$

Case 2. Double-Strand Pulley   (See Figure 7-1a.) The force required was 500 N, but through what distance did human muscles have to exert that pull? Imagine that you grab the rope and pull it 1.00 m toward you to lift the weight. The rope is of fixed length, and if you pull 1.00 m past the top pulley and out of the system, the rope encircling the pulley wheels must be 1.00 m shorter. The rope has two segments which support the keg, and with 1.00 m less rope they will each become 0.500 m shorter, lifting the keg 0.500 m in the process. To raise the keg the full 3.00 m, the two strands supporting the keg must shorten by 3.00 m each, and thus the rope must become 6.00 m shorter. Your hands in pulling the rope must pull 6.00 m of rope past you; your muscular force is exerted through a distance of 6.00 m.

$$Fd = (500 \text{ N})(6.00 \text{ m}) = 3000 \text{ N} \cdot \text{m}$$

Case 3. Block and Tackle   (See Figure 7-1b.) Five strands of rope must each shorten by 3.00 m to raise the keg. Thus, 15.0 m of rope must be pulled away from that encircling the pulleys and past the hand of the worker.

$$Fd = (200 \text{ N})(15.0 \text{ m}) = 3000 \text{ N} \cdot \text{m}$$

Case 4. Inclined Plane (the Ramp)   Here $F = 300$ N and $d = $ length of ramp $= 10.0$ m. Thus

$$Fd = (300 \text{ N})(10.0 \text{ m}) = 3000 \text{ N} \cdot \text{m}$$

In summary, a task requiring 3000 N · m was accomplished by a variety of means. In each case the human being exerted a force through a distance to accomplish the task. There was no obvious restriction on the size of the force that could be used, but, in each case, as force was reduced, the distance through which it had to be exerted was increased such that the product Fd remained 3000 N · m. Experiments lead us to believe that this is a quite general result and that *no* machine or method could allow us to accomplish the same task with Fd less than 3000 N · m (it will most often be more, allowing for friction, etc.). This is our first glimpse of a very fundamental limitation of nature, and we will now proceed more carefully.

7-3   Work

The sled in Figure 7-3 is pulled along at constant speed by a force **F** which makes an angle of θ with the displacement $\Delta \mathbf{s}$. The component of force in the direction of the displacement is $F \cos \theta$. We define the **work** done as the product of the component of force in the direction of the displacement with the magnitude of the displacement. In equation form,

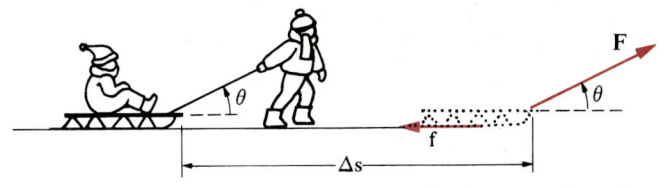

Figure 7-3 The sled is pulled through a displacement Δs under the action of **F**, which is opposed by the frictional force **f**. Also acting, but not shown, are the downward weight of rider and sled and the upward normal force of the surface.

$$W = (F \cos \theta) \Delta s$$

or
$$W = F \Delta s \cos \theta \qquad (7\text{-}1)$$

The units of work are those of the units of force times the unit of length. In SI units, work is in newtons times meters (N · m), as we have indicated. We give this unit the special name of the *joule* (J) in honor of the English physicist James Prescott Joule (1818–1889), who first showed the quantitative equivalence of various forms of energy. One **joule** of work is done by a force of one newton applied to a body which moves one meter in the direction of the force. It should be emphasized that *work does not have a direction; it is a scalar quantity.* A listing of other units in which work and energy are commonly discussed along with their equivalents in joules is presented in Table 7-1.

The common use of the word *work* is varied and relatively imprecise. However, science requires one clear, precise meaning of a term, and when the word *work* is used in a scientific context, it will have the meaning of Equation 7-1 (or its equivalent) only. A force must be exerted on a body *and* the body must move such that a component of force is in the direction of motion for work to be done. Work is a measure of accomplishment, not effort.

Work is done whenever a force is exerted through a distance. The force need not be in the direction of the displacement, but only that component of the force in the direction of the displacement does work.

TABLE 7-1 Common Energy Units

Unit	Abbreviation	Use or Equivalent	Joules
1 electronvolt	eV	Discussion of energy at atomic level	1.60×10^{-19}
1 million electronvolts	MeV	Discussion of energy at nuclear level	1.60×10^{-13}
1 erg	erg	Energy unit in cgs system	1.00×10^{-7}
1 joule	J	SI standard energy unit	1.00
1 foot-pound	ft · lb	Energy unit in British engineering system	1.356
1 calorie	cal	Heat	4.184
1 British thermal unit	Btu	Heat in British engineering system	1.054×10^3
1 Calorie	Cal	Dietary calorie = 1000 cal	4.184×10^3
1 kilocalorie	kcal	Standard chemical energy unit	4.184×10^3
1 kilowatthour	kWh	Electricity bills	3.6×10^6
1 therm	therm	Natural gas bills	1.054×10^8
1 quad	Q	Discussion of national energy use	1.054×10^{18}

Example 7-1 The sled in Figure 7-3 is being pulled along a horizontal surface at a constant speed by a towline which makes an angle of 30° with the horizontal. The tension T in the line is 50.0 N. A frictional force f of 43.3 N acts on the sled. Calculate the work done by **T** and **f** while the sled is pulled a distance of 20.0 m.

Solution Since the tension makes an angle of 30° with the displacement, we may apply the defining equation of work directly:

$$W = T \Delta s \cos 30°$$
$$= (50.0 \text{ N})(20.0 \text{ m})(0.866) = 866 \text{ J}$$

We must be more careful with the frictional force, noting that it always acts to retard the motion and hence is shown directed to the left in the drawing. This is at an angle of 180° to the displacement. Thus

$$W = f \Delta s \cos 180°$$
$$= (43.3 \text{ N})(20.0 \text{ m})(-1.00) = -866 \text{ J}$$

This is an interesting development, and we see that work can be negative as well as positive. Work is not directional and is a scalar quantity, but that does not preclude it from having a negative value. Note that the sum of the work done on the sled by the rope and by friction is zero. Since the weight and the normal force of the ice on the sled do no work (since there is no vertical displacement), the net work done by all forces is zero.

Equation 7-1 may also be written as

$$W = F_\| \Delta s \tag{7-2}$$

where $F_\|$ is the component of the force in the direction of the displacement or parallel to the displacement (hence $F_\|$) and Δs is the magnitude of the displacement. This form can be useful in situations where the force changes its direction while doing work on a body. We will consider instead the situation illustrated by the graph of Figure 7-4, where a force acting on a body in the x direction, the direction of displacement, undergoes two abrupt changes in the magnitude of its x component. The net work done by this force as the body moves from the origin to the end of the third increment is

$$W = F_{1x} \Delta x_1 + F_{2x} \Delta x_2 + F_{3x} \Delta x_3$$

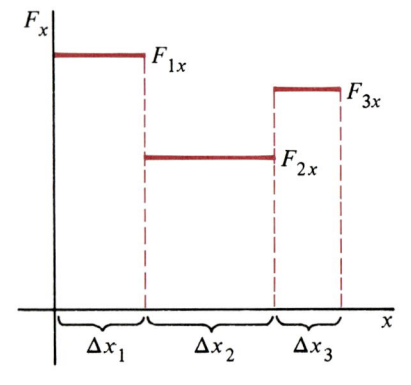

Figure 7-4 A variable force component in the x direction assumes three discrete values in moving an object through three x displacements. The work done by F_{ix} during the entire motion is the area under the F_x vs. Δx curve, which is made up of three rectangles and can be obtained easily.

Figure 7-5 The leftmost spring is at its equilibrium length l. The second spring has one unit of weight added and has stretched downward an amount x under the applied force. The third spring supports two weight units and has stretched twice as far from the equilibrium length as the second spring, or $2x$. The fourth spring bears three weight units and stretches a distance $3x$. (D. Riban)

where F_{1x}, F_{2x}, and F_{3x} are the three parallel components of the force and Δx_1, Δx_2, and Δx_3 are the magnitudes of the displacement that the body undergoes while these components are acting. Note that each of these products is simply the area of a rectangle on the graph, each of height F_{ix} and of width Δx_i. Thus, the sum is just the total area between the F_x vs. x curve and the x axis. This can be stated as

$$\text{Work} = \text{area under a force}_x \text{ vs. } x \text{ curve}$$

This concept of work as the area under a $F_\parallel$ vs. Δs curve is completely general and may be extended to more complicated situations where the force is quite variable.

Sometimes it is relatively easy to find the area under a curve of changing force. For instance, an ordinary spring, like that on a screen door, requires a larger and larger force to stretch it farther and farther from its equilibrium length l (Figure 7-5). A graph of the applied force F vs. this stretch x (Figure 7-6) shows a straight line passing through the origin. Thus, F is directly proportional to x for the region of extension examined. That is, if we double the distance of

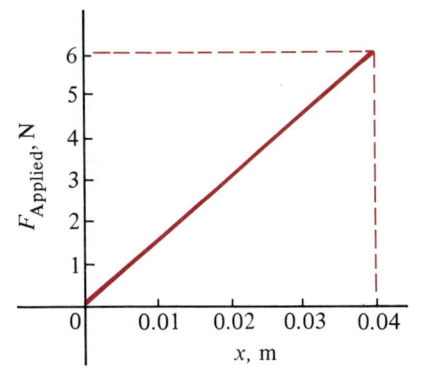

Figure 7-6 The *force applied* to stretch a spring from its equilibrium length (where $x = 0$) vs. the *change in length* x.

The spring is an important example which we will encounter many times in later chapters. Many physical systems behave like the classical coiled spring discussed here; for example, the mathematics equally well describes an atom in a solid displaced from its equilibrium position. Learning the fundamentals of this case at this point will save you from relearning them later.

133

stretch, the force doubles, and so on. This can be written as the proportion

$$F_{app} \propto x$$

or, by including a constant of proportionality, as the equation

$$F_{app} = kx \qquad (7\text{-}3)$$

where the symbol k is called the **spring constant** for the particular spring. The form of this equation is that of a straight line $y = mx + b$, where the intercept b is zero and the slope of the line is k. From the graph of Figure 7-6, we may determine the spring constant for the particular spring considered to be 150 N/m.

Example 7-2 Calculate the work done by (a) the applied force and (b) the restoring force exerted by the spring of Figure 7-6 as it is stretched from its equilibrium length ($x = 0$) to an additional length of 0.0400 m.

Solution (a) We note that the work done by the applied force is the area under the curve and is simply the area enclosed by the triangle formed by the lines $F_{app} = 0$, $x = 0.0400$ m, and $F_{app} = kx$. The base of this triangle along the x axis is 0.0400 m, and the height of the triangle is 6.00 N. Thus

$$\text{Work} = \text{area of triangle} = \tfrac{1}{2}(\text{base})(\text{height})$$

or
$$W = \tfrac{1}{2}(0.0400 \text{ m})(6.00 \text{ N}) = 0.120 \text{ J}$$

This is a positive number because the applied force and the displacement of the end of the spring are in the same direction. This also gives us a hint about the way to approach the second part of the problem.

(b) The restoring force that the spring exerts in attempting to return to its equilibrium length is the reaction force to the applied force. Hence, it is in the opposite direction to the applied force *and* the displacement. The work done *by the spring* as it is stretched should then be equal in magnitude to that done *on the spring* by the applied force, but it should be negative. Thus, $W_{spring} = -0.120$ J.

The equation for the work done by the two forces of Example 7-2 may be written in more general terms. When the spring has been stretched a distance x, the applied force must be equal to kx. Therefore the triangle in the graph has a base of length x and a maximum height of kx. Thus we write

$$W_{app} = \tfrac{1}{2}(x)(kx) = \tfrac{1}{2}kx^2 \qquad (7\text{-}4)$$

and
$$W_{spring} = -\tfrac{1}{2}kx^2 \qquad (7\text{-}5)$$

These equations easily summarize the work done on and by any spring, or other device, which obeys Equation 7-3. Thus, it applies to a stretched suspender, a car antenna pulled to the side at its tip, a plank fastened at one end and deflected laterally at the free end, and many more situations.

Forces may change in direction as well as in magnitude. Suppose a child wishes to start a playground merry-go-round (rotating platform) moving and to get it to go faster and faster. Figure 7-7 shows the child applying a force perpendicular to the radial handholds and running in a circle. The force applied

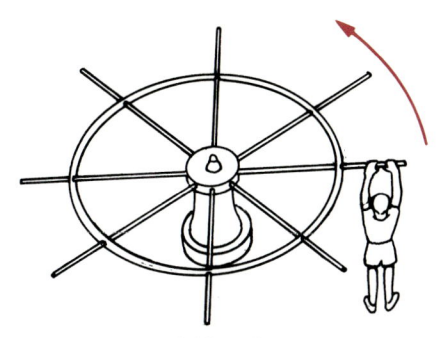

Figure 7-7 A child pushing on a playground merry-go-round applies a force perpendicular to the handhold and tangent to the circumference.

is always perpendicular to a handhold or tangent to the circle. After running a short distance, the child has traveled along an arc of the circle of length Δs. The point of application of the force has also undergone a displacement of Δs. Further, the displacement and the force were in the same direction at every instant. The child does work on the merry-go-round of

$$W_c = F \, \Delta s$$

Now, if the merry-go-round has turned through an angle of $\Delta \theta$ and the handholds are at a radius of r, we may write

$$\Delta s = r \, \Delta \theta \quad \text{and} \quad W = Fr \, \Delta \theta$$

But, Fr is simply the equation for the torque being applied by the child since the force applied was always perpendicular to r. Thus we write

$$W = \tau \, \Delta \theta \tag{7-6}$$

as the work done involving rotational displacement.

Example 7-3 A screwdriver handle has a diameter of 5.08 cm and turns through an angle of 28 rad as a screw is tightened. The force exerted by the hand is tangent to the circumference of the handle and is 50.0 N. Calculate the torque exerted and work done in tightening the screw.

Solution Since the force is tangent to the circumference, it is always perpendicular to the radius r from the center of rotation to the point(s) of application of the force. Thus, $\tau = F_\perp r = Fr$, and we write

$$\tau = (50.0 \text{ N})(0.0254 \text{ m}) = 1.27 \text{ N} \cdot \text{m}$$

Then, using Equation 7-6, we have

$$W = \tau \, \Delta \theta = (1.27 \text{ N} \cdot \text{m})(28 \text{ rad}) = 35.6 \text{ J}$$

Note that while the units of torque are newton-meters, we do not call them joules since torque does not represent work. We see here that a torque is a perpendicular force component times a distance and requires no motion to exist. Work is a parallel force component times the distance moved. The units are similar, the concepts are not.

7-4 Power

Suppose we wish to drag a very large fallen tree from a highway. We could loop a cable around the tree and pull it from the road with a heavy-duty, well-constructed gasoline engine. We could also do the job with a properly geared electric kitchen clock motor. No matter which method was used, the same amount of work would be done on the tree. However, the gasoline engine might take a matter of minutes, while the clock motor might require months or years to perform the task. Of course the amount of time required is important. There is a quantitative difference among various motors and engines which takes time into account. The descriptive term used is *power*. We say the gasoline engine is more powerful than the clock motor. We define **power** as the amount of work done per unit time. In symbols this is

$$P = \frac{W}{\Delta t} \tag{7-7}$$

Power has dimensions of work divided by those of time. In SI units, this is joules per second. We have a special name for this unit as well. We call it a *watt* (W) to honor James Watt (1736–1819), who is credited with early development of the steam engine — an ancestor of the powerful engines of today. One **watt** is the power being developed or delivered when an amount of work equal to one joule is done in one second. Another unit of power in common usage is the **horsepower** (hp). Motors and engines are often rated as 1 hp, 10 hp, $\frac{1}{3}$ hp, and so on. Watt proposed this unit as the rate at which he determined an average horse could do work. Since horses differ, this has been standardized so that one horsepower is exactly equal to 746 watts.

The definition of power can be presented in several different ways. Suppose we substitute Equation 7-2 into Equation 7-7. We obtain

$$P = \frac{W}{\Delta t} = \frac{F_\parallel \Delta s}{\Delta t}$$

but

$$\frac{\Delta s}{\Delta t} = v$$

thus we may write

$$P = F_\parallel v \tag{7-8}$$

For angular motions, we may use Equation (7-6) to substitute for work and obtain

$$P = \frac{W}{\Delta t} = \frac{\tau \Delta \theta}{\Delta t}$$

or

$$P = \tau \omega \tag{7-9}$$

Example 7-4 How much work is done by an automobile engine in accelerating the automobile and driver (total mass 1.50×10^3 kg) from rest to 25.0 m/s (55.9 mi/h or 89.5 km/h) over a distance of 200 m? Assuming the force exerted by the engine to be constant, calculate the average power during the acceleration and the instantaneous power developed at the automobile's final speed.

Solution To determine the work, we must first determine the force, and for this we need the acceleration. The acceleration may be found from the observed motion by using Equation 2-13:

$$2a\,\Delta s = v^2 - v_0^2$$

$$a = \frac{(25.0 \text{ m/s})^2 - (0)^2}{2(200 \text{ m})} = 1.56 \text{ m/s}^2$$

Next, using this acceleration, we may calculate the force from Newton's second law as

$$F = ma = (1.50 \times 10^3 \text{ kg})(1.56 \text{ m/s}^2) = 2.34 \times 10^3 \text{ N}$$

Thus, the work may be calculated to be

$$W = (2.34 \times 10^3 \text{ N})(200 \text{ m}) = 4.68 \times 10^5 \text{ J}$$

To calculate the average power, we need the time required to reach the final speed. This can be found from the motion given, using

$$\Delta s = v_0 t + \tfrac{1}{2}at^2$$

Substituting, we obtain

$$200 \text{ m} = (0)t + \tfrac{1}{2}(1.56 \text{ m/s}^2)t^2$$

$$(0.780 \text{ m/s}^2)t^2 - (200 \text{ m}) = 0$$

of the form

$$t^2 - 256 \text{ s}^2 = 0$$

which factors to

$$(t + 16 \text{ s})(t - 16 \text{ s}) = 0$$

yielding

$$t = 16.0 \text{ s}$$

Thus

$$P_{av} = \frac{W}{t} = \frac{4.68 \times 10^5 \text{ J}}{16.0 \text{ s}}$$

$$= 2.93 \times 10^4 \text{ W} \quad \text{or} \quad 39.2 \text{ hp}$$

The instantaneous power at 25 m/s is

$$P = F_\| v = (2.34 \times 10^3 \text{ N})(25.0 \text{ m/s})$$
$$= 5.85 \times 10^4 \text{ W} \quad \text{or} \quad 78.4 \text{ hp}$$

Example 7-5 The engine of an outboard motor is rated at 10.0 hp. When it is developing maximum power, the boat travels at 20.0 m/s. Calculate the force resisting the motion of the boat.

Solution In this dynamic equilibrium, the force opposing the motion just equals the forward thrust of the engine, which we may calculate directly to be

$$P = 10 \text{ hp}(746 \text{ W/hp}) = 7.46 \times 10^3 \text{ W}$$

$$F = \frac{P}{v} = \frac{7.46 \times 10^3 \text{ W}}{20.0 \text{ m/s}} = 373 \text{ N}$$

7-5 Kinetic Energy

We do work on objects continually throughout the day. Much of this work is necessary to combat frictional forces; however, much of it is not. Consider, for example, your interaction with a bowling ball during a game. When you begin, the ball is motionless in your hands. During your forward movement, you exert a force on the ball through a distance to get it moving forward. The force you exert is not combating friction, it is an unbalanced force accelerating the ball. The work done speeds up the ball in a motion toward the pins. This is quite different from pushing a wheelbarrow or pulling a sled, where you are constantly working against frictional forces while perhaps maintaining constant speed. The ball is changed from a motionless state to one of considerable forward motion. We must do work to accelerate a body; forces must be exerted through distances. The work we do will be positive if we increase its speed and negative if we act to decrease its speed. Another way of saying and picturing this is to say that to speed up a body, we must do work on it. To slow the body down, we must allow it to do work on us.

Consider a resultant force F of constant magnitude acting along the direction of the displacement Δs. This force acts on a body of mass m to change its speed from v_i to v_f. We could express the work of the applied force as

$$W = F \, \Delta s$$

From Newton's second law, we can write $F = ma$ or

$$W = (ma) \, \Delta s$$

From Equation 2-13 we know that

$$2a \, \Delta s = v_f^2 - v_i^2$$

or

$$a \, \Delta s = \frac{v_f^2 - v_i^2}{2}$$

Thus

$$W = m \frac{v_f^2 - v_i^2}{2}$$

$$W = \tfrac{1}{2} m v_f^2 - \tfrac{1}{2} m v_i^2 \qquad (7\text{-}10)$$

A quantity $\tfrac{1}{2} m v^2$ has been changed from its initial value to its final value. We call this quantity the body's energy of motion, or its **kinetic energy**. In symbols,

$$KE = \tfrac{1}{2} m v^2 \qquad (7\text{-}11)$$

It represents the minimum amount of work which must be done to take the mass m from rest to a speed of v, or the amount of work m can do on the environment in slowing down from a speed of v to rest. We rewrite Equation 7-10 to become

$$W_{\text{net}} = KE_f - KE_i \qquad (7\text{-}12)$$

or

$$W_{\text{net}} = \Delta KE \qquad (7\text{-}13)$$

Equation 7-13 is called the **work-energy theorem**. It states that the work done by the resultant force acting on a body is equal to the change in the kinetic energy of the body.

Work is done to change the speed of a mass from rest to a final speed v. The minimum amount in a frictionless situation is $W = \tfrac{1}{2} m v^2$. This work is not "lost"; it is recoverable in a variety of ways since a mass in motion can and will do work in the process of stopping, which is why cannonballs were popular. We designate this work stored in the motion of a mass as its kinetic energy to emphasize its recoverability.

Example 7-6 What is the kinetic energy of the automobile of Example 7-4 after it has reached its final speed?

Solution From Equation 7-11,

$$KE_f = \tfrac{1}{2}(1.50 \times 10^3 \text{ kg})(25.0 \text{ m/s})^2 = 4.68 \times 10^5 \text{ J}$$

This is equal to the work which was done on the automobile by its engine to reach that speed starting from rest. We calculated the work in Example 7-4. Note that their being equal is required by Equation 7-13.

7-6 Potential Energy

Energy is often defined as the ability to do work. A body may be able to do work even though it has no kinetic energy. For example, the keg of nails we described in Section 7-2 could be attached to a rope looped over a single pulley and lowered. If the other end of the rope were attached to another keg of nails, the second could be raised as the first was lowered. The work to raise the second would be supplied by the weight of the first acting through the rope. This could take place very, very slowly so that for all practical purposes we could say the speed of both bodies, hence the kinetic energy involved, was negligible.

Outside forces applied to a body may give the body the ability to do work without changing the body's speed. Even though the net work done on a body is zero and hence its kinetic energy has not changed, it may have had an increase in its ability to do work. In the example, we say that when the first keg is at its upper level, it has **energy of position** due to its height above the ground because it can apply its weight to do work. Another example of an ability to do work without changing speeds is the spring of Example 7-2. Work equal to $\tfrac{1}{2}kx^2$ was done to stretch the spring an amount x from its equilibrium length. The spring stores that work and is able to do that amount of work on an object as it contracts to its original length. A compressed spring, like that on the plunger of a pinball machine, clearly does work when released as it returns to its original length. The stored, or *potential*, energy of the spring is called the **energy of configuration.**

Both these kinds of energy of position and of configuration may simply be there, doing nothing, waiting to be tapped to do work. In these cases the configuration was physical, but it need not be. A bottle of nitroglycerin possesses the ability to exert forces through distances, e.g., push the walls outward. In this case the configuration is chemical, not physical, and much less obvious to inspection. In each of these cases, the object has the potential to do work, and we define the ability of a body or system to do work because of its position or configuration as its **potential energy** (PE).

Your physical intuition will certainly tell you that a bowling ball stored at some height above the floor has the ability to exert forces through distances (to do work) on the way down. If a loose bowling ball is stored on a closet shelf, we would not have to be safety experts to recognize a possible hazard. In rolling off the shelf the ball could fall into a bin of walnuts, cracking their shells. But it clearly takes work to crack a walnut shell — a force must be exerted through a distance. The bowling ball possesses this ability to do work because it is at a greater height in the gravitational field of the earth; it has increased gravita-

---

Work done on a mass, say in pounding a rock into fragments, may be lost in the sense that we may not know how to get that work back. But, as in the case of the work stored in the motion of a mass (kinetic energy), there are situations where the work done on an object goes to some well-defined "place" from which we do know how to get it back, as with a compressed spring or with a mass stored at a height and free to drop. Such stored work is called potential energy to emphasize its recoverability.

tional potential energy. If the ball is released, the force of gravity will act through a downward distance, doing work on it.

The potential energy (or more correctly, the change in potential energy) acquired by a spring as it is extended or compressed a distance x from its equilibrium length is equal to the work which was done to extend or compress it. From Equation 7-4 we write

$$\Delta PE_s = \tfrac{1}{2}kx^2$$

where ΔPE_s is the change in potential energy of the spring and k is the spring constant.

Let us also obtain an expression for the gravitational potential energy of a weight stored at some height. We must determine the work required to raise it to the height in question. Consider a book of mass m raised from rest on the floor to rest on a tabletop of height h. To raise the book from the floor through the displacement $\Delta y = h$, the work done had to be

$$W_{app} = F\, \Delta y = mg(y_f - y_i) = mgh$$

Note that the work which must be done does not depend on where we place our coordinate system but only on the difference between the initial and final position in the vertical direction. In the case of the 7-kg bowling ball stored on a 2.0-m-high closet shelf, we might be tempted to say its potential energy is just $mgh = (7\text{ kg})(9.8\text{ m/s}^2)(2.0\text{ m}) = 137$ J. This would be correct relative to the floor, which we chose as $h = 0$, but what if the ball broke through the floorboards and kept going with no regard whatever for our choice of a zero level? The choice of coordinates is arbitrary, while the work depends only on the difference in height before and after the fall. Further, the work done does not depend on which particular path the object takes. A bowling ball bouncing down the stairs would produce the same change in height as one falling through the floorboards. Remember that no work will be done against gravity as the object moves in the horizontal direction! Why?

The book in our example is at rest on the floor before it is moved and at rest on the table afterward; thus there is certainly no change in kinetic energy. Therefore, there is no *net* work done on the book; otherwise, both sides of Equation 7-13 would not equal zero. Since the applied force does positive work on the book, we know another force or other forces must have done an equal amount of negative work. This other force is the gravitational force, the weight of the book. The weight is downward, while the displacement is upward. Thus the work done by the gravitational force is

$$W_g = mg\, \Delta y \cos 180°$$
$$= mg(y_f - y_i)(-1)$$
$$W_g = -mgh \qquad (7\text{-}14)$$

We again note that the amount of work is independent of the actual path taken by the book.

From Equation 7-14 we may obtain the expression for the change in potential energy. The gravitational force does negative work as the height is increased. It will clearly do positive work if we allow the height to decrease (say we allow the book to drop). Thus the potential energy of the book, its ability to do work because of its position, increases with height. As we can see, the change in gravitational potential energy is exactly the negative of the work done by gravity. This can be written

$$PE_f - PE_i = \Delta PE_g = -W_g \qquad (7\text{-}15)$$

As the book is moved to the table,

$$\Delta PE_g = -(-mgh) = mgh \qquad (7\text{-}16)$$

Example 7-7 *(a)* Calculate the change in potential energy as a 1.50-kg textbook is moved from the floor to the top of a laboratory worktable which is 0.800 m high. *(b)* What would be the change in potential energy if the book were then allowed to drop to the floor?

Solution *(a)*

$$\begin{aligned}\Delta PE &= -W_g = mgh \\ &= (1.50 \text{ kg})(9.80 \text{ m/s}^2)(0.800 \text{ m}) = 11.8 \text{ J}\end{aligned}$$

(b) Recall that as the book falls, gravity does positive work. Thus the change in potential energy as it falls is

$$\Delta PE = -11.8 \text{ J}$$

We could also recognize that when the book falls, its initial y coordinate is greater than its final y coordinate, and thus $y_f - y_i$ is a negative number, $-h$.

7-7 Work and Energy

Although we postulated that the book in Example 7-7 was at rest before and after it was moved, this was not necessary. This condition only allowed us to set the applied force equal in magnitude to the weight. The applied force could have been much larger or smaller. The book could have had a speed (and hence a kinetic energy) before or after it was moved from the floor to the tabletop. Let us consider the more general situation where the book has both an initial and final kinetic energy and the applied force is not necessarily equal in magnitude to the weight. If only the applied force and gravity are acting, we can obtain the net work done as

$$W_{net} = W_{app} + W_g$$

Equation 7-13 becomes

$$W_{app} + W_g = \Delta KE$$

or

$$\begin{aligned}W_{app} &= \Delta KE - W_g \\ W_{app} &= \Delta KE + \Delta PE\end{aligned} \qquad (7\text{-}17)$$

where we have used Equation 7-15 to substitute for $-W_g$.

Equation 7-17 is a modified form of the work-energy principle. It is often more useful than the original form. Note we have dropped the subscript from the change in potential energy, because the change in potential energy need not be due to the gravitational force. It could be due to any force like gravity or that of a spring or any other force which depends only on the object's position, shape, or composition.

Forces which depend only on an object's position, shape, or composition are called **conservative forces**. The gravitational force, the elastic force of a spring, and several others which we may encounter later are conservative forces. Work which is done against these forces is stored as changes in potential energy and can be recovered as useful work. **Nonconservative forces** abound, but the most apparent is that of friction. If we slide a body across a rough surface, we have to do work against friction but there is no obvious way to recover the energy. In the most general case of motion, there are both conservative and nonconservative forces acting, and we write the work-energy theorem as

$$W_{net} = W_{nc} + W_c = \Delta KE \quad (7\text{-}18)$$

or
$$W_{nc} = \Delta KE + \Delta PE \quad (7\text{-}19)$$

where W_{nc} is all the work done by nonconservative forces, and we have substituted ΔPE for $-W_c$, the negative of all the work done by conservative forces.

We are, of course, aware that the change in potential energy could have several different components due to the gravitational force, the elastic force of a spring, and so on; thus $\Delta PE = \Delta PE_g + \Delta PE_s + \ldots$.

Example 7-8 A boy starts from rest on his sled and slides down the hill (Figure 7-8). As he reaches the lower level, the speed of the sled is 10.0 m/s. If the lower level is 15.0 m below the starting point and the boy and sled together have a mass of 60.0 kg, calculate the total work done by friction (the nonconservative force acting) as the sled goes down the hill.

Solution We must apply Equation 7-19 after obtaining expressions for ΔKE and ΔPE:

$$\Delta KE = KE_f - KE_i = \tfrac{1}{2}mv_f^2 - 0$$

since it starts from rest and $v_i = 0$. Also

$$\Delta PE = mg(y_f - y_i)$$

Thus
$$W_f = \tfrac{1}{2}(60.0 \text{ kg})(10.0 \text{ m/s})^2 + (60.0 \text{ kg})(9.80 \text{ m/s}^2)(-15.0 \text{ m})$$
$$= -5.82 \times 10^3 \text{ J}$$

Figure 7-8 Example 7-8.

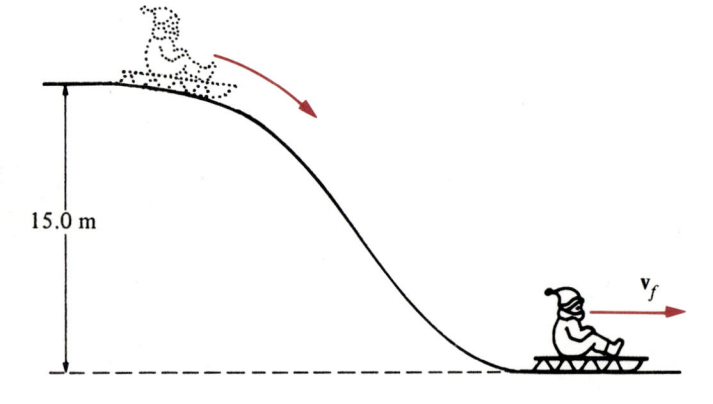

This result begins to show some of the power of the work-energy theorem. We have calculated the work done by the force of friction even though we had no knowledge of its direction or magnitude. Neither had we any knowledge of the shape of the hill.

7-8 Conservation of Mechanical Energy

Many situations arise in which the frictional forces and other nonconservative forces are not acting or are at least do negligible work during the course of some motion. When that occurs, Equation 7-19 becomes

$$\Delta KE + \Delta PE = 0 \tag{7-20}$$

or
$$KE_f - KE_i + PE_f - PE_i = 0$$

and
$$KE_f + PE_f = KE_i + PE_i \tag{7-21}$$

We may state from Equation 7-20 that provided the work done by nonconservative forces can be neglected, then the sum of the changes in the kinetic energy and potential energy is zero. This does not mean that the changes must vanish individually but that only their sums vanish. Equation 7-21 provides no new physics. It is merely Equation 7-20 rewritten. It does, however, provide us with a little more insight. The symbols on the left represent all the final values of energy, while those on the right represent the initial values. We say that this sum, the kinetic energy plus the potential energy, is the **total mechanical energy.** We may thus say Equation 7-21 states that, provided the work done by nonconservative forces can be neglected, the total mechanical energy stays the same. We also say that the mechanical energy is constant, or *conserved*. These two equations 7-20 and 7-21 are statements of the **conservation of mechanical energy.**

To test the conservation of mechanical energy, we choose a system in which nonconservative forces are minimal. One of the best of these is a simple pendulum. The resistance of the air being pushed out of its path is negligible over a single swing of the pendulum. Other than this, there is little to oppose its motion. Figure 7-9 shows a stroboscopic photograph of a pendulum during half

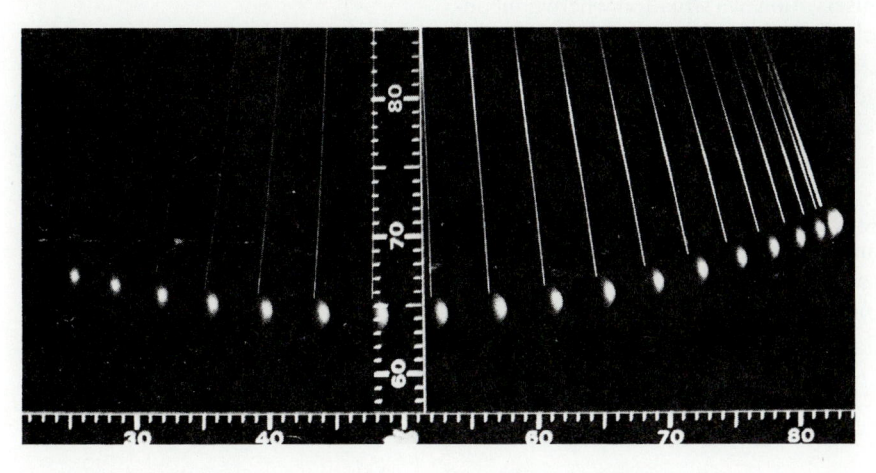

Figure 7-9 Half swing of a pendulum taken by the light of a stroboscope flashing every 0.050 s. *(D. Riban)*

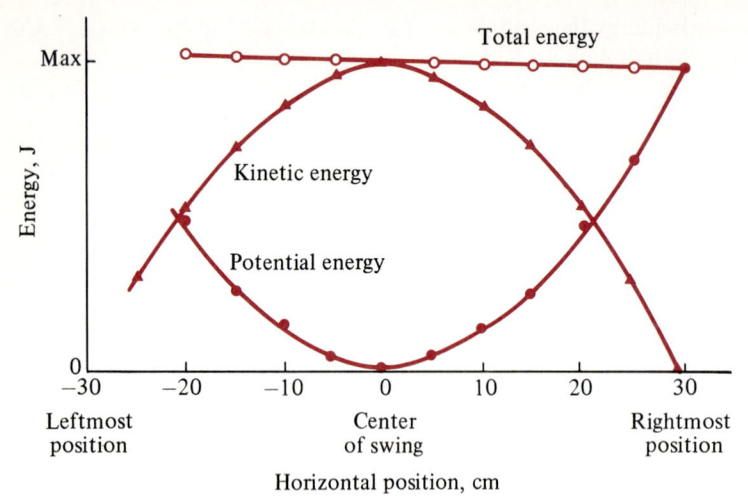

Figure 7-10 Potential energy, kinetic energy, and their sum vs. horizontal position in the photograph of Figure 7-9. Zero height is taken as the lowest point in the half swing, and at this point $h = 0$ and PE = 0. As PE grows in either direction from this point, KE diminishes such that their sum, the total mechanical energy, is nearly constant. The slight downhill slope of total mechanical energy may be attributed to work done against frictional forces.

a complete cycle. The separate images were taken by firing a high-intensity lamp every 0.050 s during the swing. The scale shown in the photograph allows us to measure the distance moved between flashes. For each interval between recorded positions of the pendulum, we may obtain the average height of the center of the bob, h_i, and the distance the center of the bob moved during the interval, Δs_i. The potential energy of each interval is calculated as mgh_i, and the kinetic energy in each interval is

$$\text{KE} = \tfrac{1}{2}mv_i^2 = \tfrac{1}{2}m\left(\frac{\Delta s_i}{\Delta t}\right)^2$$

Figure 7-10 is KE, PE, and their sum, the total mechanical energy, vs. the horizontal position from the photograph. The constant mechanical energy displayed in this graph demonstrates its conservation in a situation where nonconservative forces are minimal. Still, the slight decline of the total energy during the swing allows us to estimate the size of the nonconservative forces acting. These are clearly quite minor compared to the forces transferring energy between increased speed (KE) and increased height (PE) of the bob.

Example 7-9 Assume that the work done by all nonconservative forces acting on the roller coaster car in Figure 7-11 may be neglected. The car has a mass of 1.50×10^3 kg and a horizontal velocity at point A of 25.0 m/s. Calculate (a) the speed of the car when it reaches the top of the loop-the-loop (point B) and (b) its speed at point C.

Figure 7-11 Example 7-9.

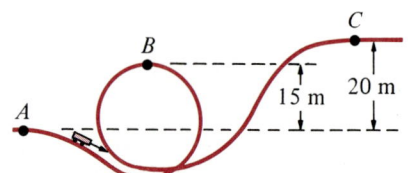

Solution (a) From Equation 7-20, we have

$$\tfrac{1}{2}mv_B^2 - \tfrac{1}{2}mv_A^2 + \Delta \text{PE} = 0$$

144

$$\tfrac{1}{2}(1.50 \times 10^3 \text{ kg})v_B^2 - \tfrac{1}{2}(1.50 \times 10^3 \text{ kg})(25.0 \text{ m/s})^2$$
$$+ (1.50 \times 10^3 \text{ kg})(9.80 \text{ m/s}^2)(15.0 \text{ m}) = 0$$

which yields

$$v_B = 18.2 \text{ m/s}$$

(b) Similarly, applying Equation 7-20 between points A and C yields

$$v_C = 15.3 \text{ m/s}$$

Example 7-10 A small lump of metal of mass 1.50 kg is attached to a spring of spring constant $k = 200$ N/m. The mass is released from rest when the end of the spring is at its normal equilibrium point ($x = 0$), and its weight pulls the spring downward. Calculate (a) x_m, the maximum distance that the mass moves below the point of release, and (b) the speed of the mass when its displacement below its release point is $\tfrac{1}{2}x_m$. Neglect any work done by friction.

Solution (a) Since no nonconservative forces are doing work, we may apply the conservation of energy. At both its point of release and at the maximum stretch of the spring, the mass is at rest; hence $\Delta KE = 0$. Thus the change in energy is the sum of the changes in the gravitational potential energy and that of the spring, or

$$\Delta PE_g + \Delta PE_s = 0$$
$$mg(-x_m) + \tfrac{1}{2}kx_m^2 - 0 = 0$$
$$(1.50 \text{ kg})(9.80 \text{ m/s}^2)(-x_m) + \tfrac{1}{2}(200 \text{ N/m})x_m^2 = 0$$
$$x_m = 0.147 \text{ m}$$

Note that $x_m = 0$ is a perfectly good algebraic solution to the equation. It is also physically correct in that the mass is stopped at $x = 0$ as well as at $x = 0.147$ m.
(b) At $x = \tfrac{1}{2}x_m = 0.0735$ m, the mass is moving. The change in energy from the point of release is

$$\Delta KE + \Delta PE_g + \Delta PE_s = 0$$

or

$$\tfrac{1}{2}mv^2 - 0 + mg(-\tfrac{1}{2}x_m) + \tfrac{1}{2}k(-\tfrac{1}{2}x_m)^2 = 0$$

Omitting units, we have

$$\tfrac{1}{2}(1.50)v^2 + (1.50)(9.80)(-0.0735) + \tfrac{1}{2}(200)(-0.0735)^2 = 0$$

or

$$v = 0.849 \text{ m/s}$$

Note again the power of applying the principle of conservation of mechanical energy. We found the speed without writing the acceleration or force laws (or even knowing them).

7-9 An Application of Mechanical Energy

We will examine a real situation to clarify these ideas about energy at work and their application. Figure 7-12 shows a pole-vaulter from the middle of his run through the peak of his vault. The duration between successive photographs

145

was 0.20 s. The mass of the vaulter is 70 kg, and his height is 1.7 m. The horizontal location of his center of gravity changed by 5.0 m between positions 3 and 6 at a fairly constant rate. Thus, $v = \Delta s/\Delta t = 5.0 \text{ m}/0.60 \text{ s} = 8.3 \text{ m/s}$. Using this speed, his initial kinetic energy before the vault was

$$\text{KE} = \tfrac{1}{2}mv^2 = \tfrac{1}{2}(70 \text{ kg})(8.3 \text{ m/s})^2 = 2.4 \times 10^3 \text{ J}$$

Beginning with position 7, the vaulter jams the pole into the socket at the base of the jump. In successive photographs, we see the pole bend through greater and greater arcs as the kinetic energy is changed into the elastic potential energy of the bent pole. A bent pole is a form of compressed spring, and its potential energy is $\text{PE} = \tfrac{1}{2}k(\Delta s)^2$, where Δs is the displacement of the tip of the pole from its rest position along the arc of a circle. Fastening one end of this pole, in the laboratory, we add weight to the free end and measure its displacement. This indeed produces a graph of form $F = kx$, whose slope yields the value $k = 2.6 \times 10^2$ N/m for our energy equation. Using 3.4 m as the maximum deflection of the end of the pole (position 9), its elastic potential energy is

$$\text{PE} = \tfrac{1}{2}k(\Delta s)^2 = \tfrac{1}{2}(2.6 \times 10^2 \text{ N/m})(3.4 \text{ m})^2 = 1.5 \times 10^3 \text{ J}$$

Comparing the position of the vaulter's center of gravity between position 9 and those immediately after it, we see that the forward motion is comparatively small and that the motion of the center of gravity changes from forward to upward. This introduces a complication. Until the thrust off the ground, the vaulter's strides have been propelling him forward at a nearly constant maximum speed. As he leaves the ground, his last stride is an unusually forceful upward thrust to acquire vertical motion. While most of the upward motion of the vault will come from the stored energy of the bent pole (after all, how high can you jump compared to how high you can vault?), the upward thrust of the jump off the ground never appears in the deformation of the pole. To treat this, note that the bending of the pole from position 9 to 10 changes very little. The motion observed is a rotation of the entire pole forward about its contact point with the ground. Thus, the motion of the vaulter's center of gravity between

Figure 7-12 Photocomposite of a pole-vaulter from his run toward the bar to his peak position in clearing the bar. The vaulter's navel is the approximate position of the normal center of mass of his body (but not the actual CM, allowing for the distorted position of the body). Note that the running speed is quite constant before the pole is dropped into the slot at the base of the bar. (D. Riban)

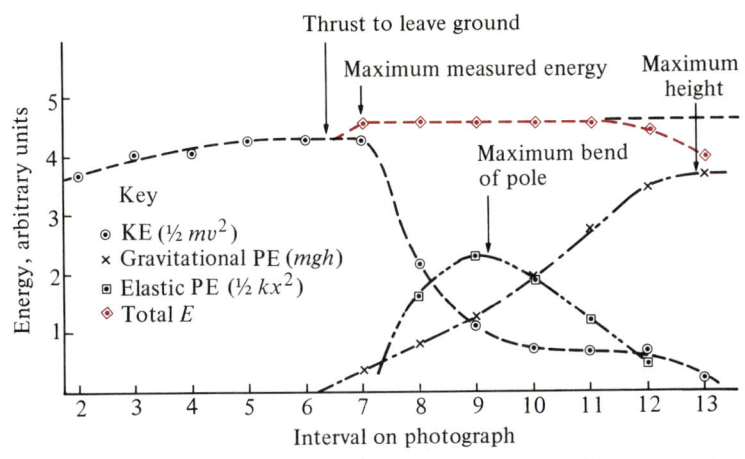

Figure 7-13 Graph produced from measurements taken from Figure 7-12: kinetic energy, gravitational potential energy, elastic potential energy of the bent pole, and the sum of these three (total energy) vs. the position (for PE_G, PE_E) or interval between positions (for KE).

these positions represents his kinetic energy at kickoff with the pole at maximum bending. Measurement from the photographs yields a speed of 5.0 m/s and a kinetic energy of 8.8×10^2 J. This must be added to the elastic potential energy of the pole to obtain the total energy at lift-off of 2.4×10^3 J.

Note that the motion, vertical or horizontal, is quite small near the top of his vault. Figure 7-13 is a position-by-position analysis of the various energies calculated from the composite photograph. Plotted separately is the kinetic energy measured between each set of photographs in the series, the gravitational potential energy at each position, the elastic potential energy of the bent pole for each position, and the total of these energies. That the energy changes from one form to another while the total energy is preserved is an eloquent testimonial to the soundness of the concept of conservation of mechanical energy. Here we have a real experimental situation, far from the controlled environment of the laboratory. Considering the difficulties of measuring the various forms of energy, the consistency of the graphs with the ideas we have been discussing and the relative constancy of total energy are striking. That the total energy declines slightly after the initial impact of the pole is not surprising since we have made no attempt to assess such nonconservative work as heating in the pole and work against air resistance.

7-10 Conservation of Total Energy

To this point we have been considering the more apparent energies of relatively large objects. We classified these as the energy of motion (kinetic) and the energy of position or configuration (potential). Together these make up the macroscopic components which we can discern by observation. We called those mechanical energy. That there are other, perhaps more subtle, forms of energy may not be so obvious. Suppose, for example, we give our book a

TABLE 7-2 Energy Release by Reaction of Various Substances

Material and Amount	Energy Release Common Unit	Energy Release Joules	Energy Equivalent, J/kg
1 cord wood	2.0×10^6 Btu	2.11×10^9	8.63×10^5
1 ton TNT	4.0×10^6 Btu	4.2×10^9	4.6×10^6
1 g carbohydrate	4.1 kcal	1.7×10^4	1.0×10^7
1 gal methanol	6.0×10^4 Btu	6.32×10^7	2.0×10^7
1 g protein	5.6 kcal	2.3×10^4	2.3×10^7
1 ton coal	2.5×10^7 Btu	2.76×10^{10}	2.8×10^7
1 g fat	9.4 kcal	3.9×10^4	3.9×10^7
1 bbl crude oil	5.6×10^6 Btu	5.9×10^9	4.3×10^7
1 gal gasoline	1.3×10^5 Btu	1.32×10^8	4.4×10^7
1 ft³ natural gas	1.03×10^3 Btu	1.09×10^6	5.5×10^7
1 Hiroshima-size nuclear bomb	10-kiloton TNT equivalent	4.2×10^{13}	4.6×10^{10}
1 kg uranium (fission)	200 MeV/fission	8.0×10^{13}	8.0×10^{13}
1 kg hydrogen (fusion to helium)	26.7 MeV/fusion	6.4×10^{14}	6.4×10^{14}
Mass-energy conversion	931 MeV/u	. . .	9.0×10^{16}

horizontal push and then release it. It slides for a short distance across our desk and comes to rest. When we released it, there was some kinetic energy which was gradually lost. Where did it go? It is not available for recovery, as were the components of mechanical energy. However, a very sensitive thermometer would show that the tabletop and the book were both slightly warmer than they were before. We will see (Chapter 15) that the kinetic energy lost is exactly equal to the thermal energy gained by the table and book. This thermal energy was produced by the frictional force doing work. Recall that friction is a nonconservative force. We may rewrite Equation 7-19, dividing the work done by nonconservative forces into components which produce heat (W_{HP}) and those which do not (W_{NH}). Then

$$W_{nc} = W_{HP} + W_{NH} = \Delta KE + \Delta PE$$

If we define the thermal energy produced (ΔTE) as the negative of W_{HP}, we would obtain

$$W_{NH} = \Delta KE + \Delta PE + \Delta TE$$

where W_{NH} is the work done by all nonconservative forces which do not produce heat.

We can continue this process, dividing W_{NH} into the work done by those forces which produce, say, nuclear energy, electric energy, magnetic energy, and so on. As a further continuation, we define the negative of the work done by each type force as a change in that kind of energy. We would then move those terms to the right side of the equation. After this process has been carried to its logical conclusion, there would be no work terms left. Then we have that zero is equal to the sum of the changes of all possible energy terms. Stated another way, we would say that the net change in energy is zero and total energy is conserved.

Table 7-2 lists several energy sources, their common unit of measure, and the energy equivalent in joules per kilogram.

Minimum Learning Objectives

After studying this chapter, you should be able to:
1. Define:
   - conservation of mechanical energy
   - conservation of total energy
   - conservative force
   - energy
   - horsepower [unit]
   - joule [unit]
   - kinetic energy
   - machine
   - nonconservative force
   - potential energy (chemical, elastic, gravitational)
   - power
   - spring constant
   - total mechanical energy
   - watt [unit]
   - work
   - work-energy theorem
2. Determine the component of a given force contributing to the work done.
3. Calculate the work accomplished by a given force acting on a body with a specified displacement.
4. Assess the work done from a force vs. displacement curve.
5. Determine the power of a machine performing work at a given rate.
6. Calculate the kinetic energy of specified objects moving at a given speed.
7. Determine the work done and the power in situations involving rotary motion.
8. Apply the work-energy theorem to problems with and without nonconservative forces acting.
9. Calculate gravitational potential energy of objects.
10. Calculate the elastic potential energy of springlike systems either from a force vs. displacement curve or from the spring constant and displacement.
11. Understand the application of measurement techniques (such as the pendulum and pole-vault examples) to real physical systems to determine various energies.
12. Understand how the extension of the processes of analysis of this chapter can lead to a law of conservation of total energy.

Problems

7-1 Determine the work done by a professor in applying an average force of 300 N to push an automobile 800 m to the service station.

7-2 A book weighing 16.0 N is pushed 1.20 m across a desk by a horizontal force F of 9.00 N. The coefficient of kinetic friction between the book and the desk is $\mu_k = 0.300$. Determine (a) the work done by F, (b) the normal force and the frictional force, (c) the work done by the frictional force, and (d) the net work done on the book.

7-3 A man lifts a large sack of corn weighing 400 N from the floor to his shoulder 1.50 m above the floor (after he stands up). Determine (a) the work done by the man on the corn, (b) the work done by the corn on the man, and (c) the work done by the force of gravity acting on the corn.

7-4 A 750-N bellhop carries a suitcase weighing 200 N up a flight of stairs to the second floor some 3.00 m higher than the original position. Determine (a) the work the bellhop does against gravity to move the suitcase and (b) the net work the bellhop does against gravity.

7-5 One child is pulling another on a sled up a snow-covered hill of angle 15°. The rope is pulled parallel to the hill, and the sled moves at constant speed. The riding child and sled together have a mass of 45.0 kg, and there is a coefficient of kinetic friction $\mu_k = 0.0800$ between the sled runners and the snow. Calculate (a) the normal force the hill applies to the sled, (b) the component of weight acting down the hill, (c) the frictional force, (d) the force being applied by the child pulling, (e) the work done on the sled and rider by the puller as the sled moves 15.0 m up along the hill, (f) the work done by the frictional force, and (g) the net work done on the sled and rider.

7-6 The graph in Figure 7-14 is a plot of the component of force $F_\parallel$ in the direction of motion acting on a moving object. Calculate the work done by the force as the body moves from zero to 20 m.

7-7 The coefficient of kinetic friction between the 6.00-kg block and the surface in Figure 7-15 is $\mu_k = 0.300$. The spring constant is $k = 750$ N/m. Calculate (a) the fric-

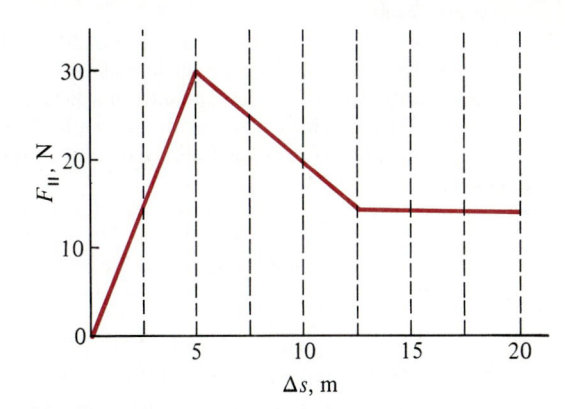

Figure 7-14   Problem 7-6.

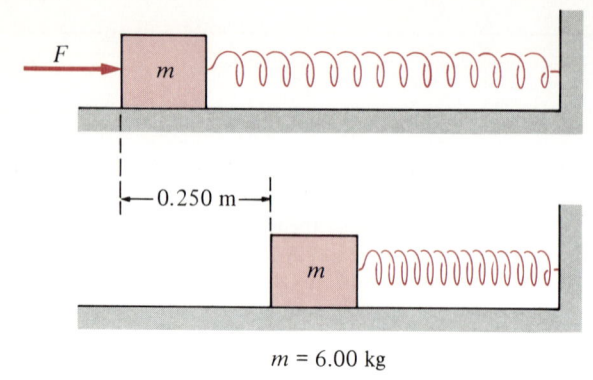

Figure 7-15 Problem 7-7.

tional force, (b) the work done by the frictional force, (c) the work done by the spring, and (d) the work done by F if the block moves very slowly.

7-8 The formula for the energy developed in an earthquake is

$$\log E = 4.40 + 1.50M$$

where $\log$ = logarithm to the base 10
E = energy, J
M = measured magnitude of E on the Richter scale

Prior to an earthquake, motions in different directions move one slab of the earth past another. The rock near the fault, or crack, between these slabs bends as these motions accumulate, as friction holds the two surfaces locked together. At some point the accumulated force exceeds the frictional force, and the rock snaps free along the surface, producing the earthquake.

An earthquake measuring 6.2 on the Richter scale occurs along a fault 400 m long and 100 m deep. Assuming that the rock behaves like a spring as it distorts, what is the spring constant of the rock if the motion along the fault is 2.00 m during the earthquake?

7-9 How much work must an electric motor do in 1 min to be delivering 746 W (1 hp)?

7-10 A sled is towed by a rope pulled at an angle of 30° to the horizontal. The tension in the rope equals 45.0 N. The sled moves at a constant speed on level ground for a distance of 10.0 m in 4.00 s. Calculate (a) the work done by T and (b) the average power expended during the 4.00-s time period.

7-11 A small horse weighing 5.00×10^3 N walks slowly up an inclined trail which rises 1 m for every 6 m of its length. At the end of an hour, the horse has moved 3.6 km ($2\frac{1}{4}$ mi) along the trail. What is the average power developed by the horse in watts and in horsepower?

7-12 The engine of a particular lawnmower develops a maximum of 3.00 hp. The piston is driven 500 times per minute a distance of 0.100 m on the power stroke. Calculate the average force on the piston.

7-13 At what speed would a plow horse have to walk to deliver 1 hp while exerting a force of 500 N to pull the plow?

7-14 An automobile travels 500 m on a level highway at a rate of 25.0 m/s. It presents an effective area of 2.40 m² to the direction of motion. Use Figure 6-2. Determine (a) the force of air resistance on the auto, (b) the work done against the air resistance, and (c) the average power being developed to move against the wind.

7-15 Energy from the sun falls on the earth at the rate of approximately 800 W/m² when the sun is shining perpendicular to a surface. A typical automobile presents about 3.20 m² effective area to the zenith. (a) Determine the amount of energy incident on an automobile during a day in June, assuming the sun shines for 16.0 h at an average angle of 50° to the zenith in your latitude. (Remember that most of the time the sun is at a very large angle with the zenith.) (b) If 12.0 percent of this energy could be converted to electric energy and stored in batteries for later use, how long could it power an automobile which is developing 80.0 hp?

7-16 Calculate the speed of a baseball ($m = 0.142$ kg) thrown by a fastball pitcher if it acquires a kinetic energy of 125 J.

7-17 A child starts from rest and pushes a friend on a sled, exerting a force of 50.0 N for a distance of 40.0 m. Suppose she does it again over the same course but now pushes with only 45.0 N for the first 15.0 m. Determine the average force she must exert over the last 25.0 m to impart the same final speed as she did on the first run.

7-18 In a football game, a running back has twice as much kinetic energy and two-thirds the mass of the linebacker chasing him. If the linebacker can increase his speed by 2.00 m/s, they will both have the same kinetic energy. Determine the original speed of each player.

7-19 An ocean liner rolls to an angle of 25° and stays there for a short time. A table of mass 120 kg in the captain's cabin begins to slide across the deck. The coefficient of friction between the deck and the table is $\mu_k = 0.200$. After the table moves 3.00 m, determine (a) its kinetic energy and (b) the work the captain must do to stop it.

7-20 A sled and child (total mass 48.0 kg) are being pulled by a rope which makes an angle of 25° with the horizontal. The tension in the rope is 68.0 N, and there is a coefficient of kinetic friction of $\mu_k = 0.100$ between the runners and the snow. Calculate (a) the work done by the tension force as the sled moves 8.00 m, (b) the work done by friction, and (c) the change in kinetic energy of the child and sled.

7-21 A freshman is pledging to join a fraternity, and one of his duties is to mow the lawn. He is pushing on the mower handle with a force of 50.0 N at an angle of 10° below the horizontal. The handle makes an angle of 30° with the horizontal. The mower has a mass of 25.0 kg. (a) Determine the work that must be done as the mower moves 10.0 m. (b) If the mower has a speed of 0.800 m/s at the beginning of the 10.0 m, what speed will it have at the end?

7-22 At what height above the earth would a small airplane of mass 600 kg have a potential energy of 1.78×10^6 J?

7-23 Table 7-1 defines one dietary calorie (1 Cal) as equaling 4184 J. A popular fast-food chain sells a double-hamburger sandwich which will give you approximately 530 Cal. If you eat one of these, and immediately plan to use this energy to climb Mount Everest, how high could you climb while expending an equivalent amount of energy? Assume your mass is 60.0 kg and all the energy is expended to lift you vertically.

7-24 The water going over the famous Horseshoe portion of Niagara Falls drops a distance of 57.0 m. If 3.57×10^6 kg goes over the falls every second, how much power is delivered to the pool below?

7-25 A 50.0-kg athlete is running at a rate of 4.00 m/s on a treadmill inclined uphill at an angle of 8°. At what rate is she doing work?

7-26 An hourglass is filled with 1.30 kg of sand. It takes exactly 1 h for all the sand to move from the top to the bottom. The vertical distance moved by the center of mass is 0.100 m. Determine (a) the change in potential energy of the sand and (b) the average power developed during the hour.

7-27 To be on time for class, a physics professor goes from the basement research laboratory to the second floor (taking the steps three at a time) in 9.00 s. The vertical distance from the basement to the second floor is 7.00 m, and the professor has a mass of 82.0 kg. Calculate (a) the change in potential energy and (b) the average power developed, assuming no change in kinetic energy.

7-28 What is the largest power output an average human being can sustain for a reasonable amount of time? One candidate may be among the winners of the annual Empire State Building climb. Competitors race up some 86 floors to the lower observatory of the building. The current record is 12 min 20 s for a climb of 262 m. Assuming an average mass for the winner of 70 kg (158 lb), determine the average power over the duration of the climb in watts and horsepower.

7-29 One of the U.S. automobile companies is testing a two-passenger car to be sold to people who make short trips (commuting, shopping, etc.). The car has a mass of 400 kg and is loaded (including the driver) with another 150 kg. It is to be rated at a maximum of 35.0 hp. (a) Determine whether the test car can be expected to accelerate from 15.0 to 25.0 m/s in 12 s on a level road. (b) Determine the angle of the hill of the greatest slope that the test car can be expected to negotiate at a constant speed of 20.0 m/s.

7-30 The Great Pyramid of Cheops is 125 m high, and the topmost block has a mass of 9000 kg. (a) What is the potential energy of this stone block relative to ground level? (b) If a ramp rising 1.00 m for every 10.0 m of length was used to slide the block to the top, what was the force required to raise this block? Take the coefficient of friction between block and ramp as 0.250. (c) If the average worker could exert a force of 500 N, how many workers would be required to raise the block? (d) Through what distance would the pull of the workers have to act to raise the block? (e) What was the total work done by the workers? (f) What was the work done against friction during the raising of the stone block? (g) What was the useful work done?

7-31 Glaciers can move material uphill in a given locality as long as the net motion of mass is downhill. Consider the basin of Lake Superior. The water surface is 183 m above sea level, while the deepest sounding in the lake is 405 m, or farther below current sea level than its surface is above it. Furthermore, no exit from the lake has any considerable depth. As are all the Great Lakes, Lake Superior is an excavated pit carved by a glacier out of yielding rock layers bounded by more resistant ones. The lake has a width of 260 km and a length of 560 km, with a total surface area of 1.46×10^5 km². The average depth of the lake is 30.8 m. The mass density of the rock that once filled this basin averaged about 2.50×10^3 kg/m³. Disregarding the energy required to break up the rock, (a) what was the energy required just to lift the rock out of the basin and deposit it on the states of the upper Midwest which are about the same height as the present lake surface? (b) How many barrels of oil would this take (assuming 100 percent efficiency of energy conversion)? (c) How many atomic bombs of the size that destroyed Hiroshima (10-kiloton TNT equivalent) would it require with 100 percent efficiency of conversion? (d) How many 20-megaton hydrogen bombs would it take? (e) Do these numbers give you a new respect for the energy of glaciers? If the glacier scooped out Lake Superior in 10,000 years, what was the average power in watts and in horsepower? (f) If the 20-megaton bomb detonates completely in 1/10,000 s, what is its average power in watts and in horsepower?

7-32 A truck of mass 3.00×10^3 kg coasts down a hill 200 m long at an angle of 20°. The speed limit at the bottom of the hill is 10.0 m/s. Determine how much work the brake must do to avoid breaking the speed limit.

7-33 The spring in Figure 7-16 has an unstretched length 0.750 m and a spring constant $k = 1.00 \times 10^3$ N/m. The block of mass $m = 6.00$ kg is released from rest 0.800 m along the plane from the end of the spring and slides down the plane. The coefficient of friction between the block and the plane is 0.350. Calculate (a) the work done by friction, the

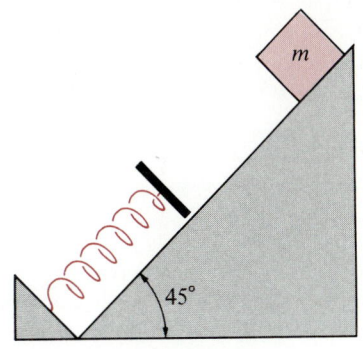

Figure 7-16 Problem 7-33.

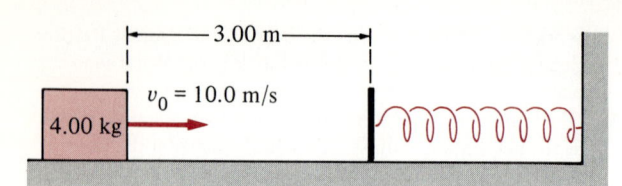

Figure 7-17 Problem 7-34.

change in potential energy, and the change in kinetic energy as the block moves from its starting point until it just reaches the spring, and (b) the maximum compression of the spring.

7-34 A 4.00-kg block in Figure 7-17 is projected across a rough horizontal surface toward a spring 3.00 m away. The initial speed of the block is 10.0 m/s, and the coefficient of friction between the block and the surface is 0.250. Calculate (a) the kinetic energy of the block just as it reaches the spring, (b) the distance the spring is compressed from its equilibrium length (use $k = 700$ N/m), and (c) the distance away from the end of the spring where the block finally comes to rest.

7-35 Many years ago, "Gabby" Hartnett, a major league baseball catcher, was given the opportunity to catch a baseball which had been dropped from the top of the Washington Monument 167 m above the position of his catcher's mitt. Neglect air friction (is this a good assumption?), and calculate (a) the speed of the ball just as it struck the mitt, (b) the work which had to be done to bring the 0.142-kg ball to rest, and (c) the average force Gabby had to apply if he stopped the ball in 0.300 m.

7-36 An automobile of mass 1.00×10^3 kg is parked at the top of a hill of slope 15°. Its parking brake fails, and it rolls 80.0 m down the hill. At the bottom its speed is 8.00 m/s. Calculate (a) the change in kinetic energy of the car, (b) the change in potential energy, and (c) the work done by friction.

7-37 (a) What change in kinetic energy is undergone by a boulder of mass 38.0 kg which is dropped from rest off a cliff 90.0 m above a roadbed? (b) What is the boulder's speed just before it strikes the road? (Neglect air friction.)

7-38 A world-class high jumper of mass 62.0 kg runs at a speed of 9.00 m/s toward the bar. During the jump, he raises his center of mass 1.25 m. Calculate (a) his forward kinetic energy just before he leaves the ground, (b) his potential energy at his highest point, and (c) the percentage of forward kinetic energy converted to potential energy.

7-39 A pole-vaulter with a mass of 50.0 kg is running at a speed of 8.50 m/s just before she plants the pole in the socket. If all her kinetic energy is converted to potential energy, how high can she vault?

7-40 A man of mass 90.0 kg and a girl of mass 35.0 kg step on a spring platform, depressing it 0.200 m from its original position, where it comes to rest. If the man now steps off sideways, what is the maximum height the girl will rise?

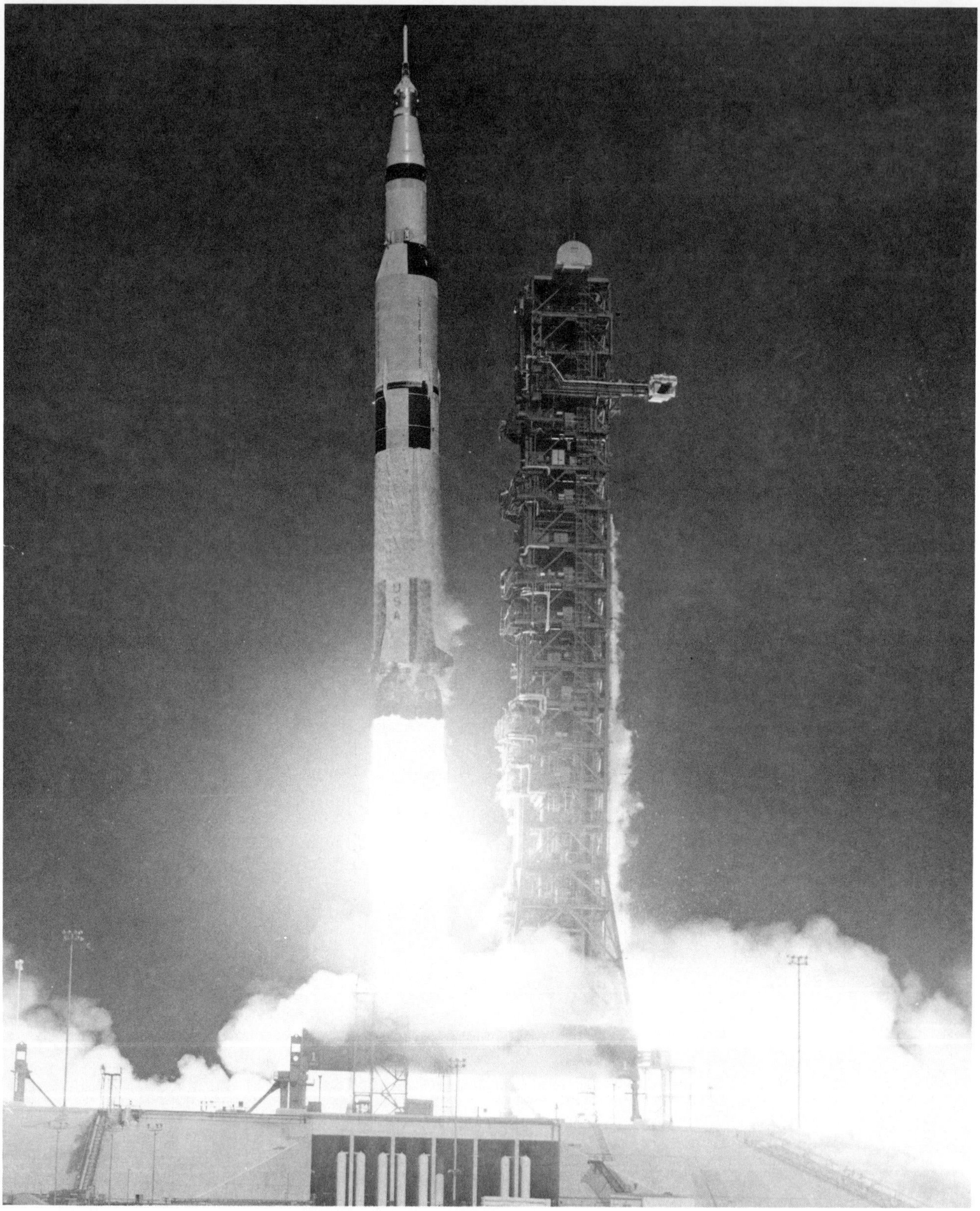
USA

8 Impulse and Linear Momentum

8-1 Introduction

For the first seven chapters, we considered single bodies or connected systems. We often subjected that body or system to outside or applied forces and torques, but we confined our attention to the effects on that body or system. A body subjected to outside forces is not alone. At least two bodies are required for there to be an interaction. Therefore, we consider in this chapter the effects on both interacting bodies. Actually, you have considerable experience on how bodies behave during interactions. Most sports activities, from a game of marbles to surfboarding, require an interaction (Figure 8-1). You have experienced the collisions of baseballs with bats and tennis balls with rackets, and you have thrown objects with your hands. Movies and TV have drenched us all in a slow-motion montage of spectacular automobile collisions. Thus, if you "search the file," you have the data on which this chapter is built. All that will be new is the precise treatment and analysis of data you have been accumulating all your life. The goal of the examination of interactions is to increase our power to predict what will happen to the bodies.

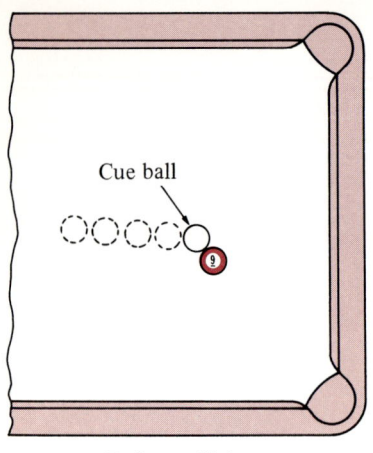

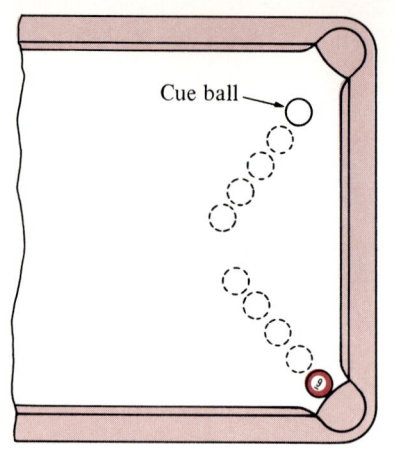

Figure 8-1 In this billiard shot, we would like the cue ball to send the nine ball into the corner pocket. What about the cue ball, however? We do not want it to roll to another pocket or to end up in an impossible position for the next shot. Clearly we must consider the consequences of the collision on both balls to play the game well.

8-2 Impulse

To make some predictions about collisions, we apply our knowledge to search for new relationships. Newton's third law asserts that every interaction between two bodies involves two forces, the force of the first body acting on the second body and the reaction force of the second body acting on the first body, which are equal in magnitude but oppositely directed at every instant in time during the interaction.

The time a force acts is important to its result. Consider a constant resultant force **F** acting on a body of mass m for a time interval Δt. The body will have an acceleration during that time interval, and we may write Newton's second law as

$$\mathbf{F} = m \frac{\Delta \mathbf{v}}{\Delta t} \qquad (8\text{-}1)$$

This is a vector equation, and $\Delta \mathbf{v}/\Delta t$, the acceleration, must agree in direction with **F**. Multiplying both sides of Equation 8-1 by Δt, we obtain

$$\mathbf{F}\,\Delta t = m\,\Delta \mathbf{v} = m(\mathbf{v}_f - \mathbf{v}_i)$$
$$\mathbf{F}\,\Delta t = m\mathbf{v}_f - m\mathbf{v}_i \qquad (8\text{-}2)$$

where $\mathbf{v}_f$ and $\mathbf{v}_i$ represent the final and initial velocities of the body. This result states that if we are able to find or calculate the quantity $\mathbf{F}\,\Delta t$, we may determine $m\mathbf{v}_f$ from a knowledge of $m\mathbf{v}_i$. Or, the quantity $\mathbf{F}\,\Delta t$ gives rise to a change in the quantity $m\mathbf{v}$ from its initial to its final value. These quantities are so important that we give them specific names. We call the product of the force with the time interval the **impulse**. We call the product of the mass with the velocity the **linear momentum**. Equation 8-2 then is stated as *the impulse is equal to the change in linear momentum*. This is the **impulse-momentum theorem** and has a status comparable to the work-energy theorem.

The units of both sides of Equation 8-2 must, of course, be the same. The units of impulse are those of force times time. In SI units, this is newton-seconds (N·s). The units of linear momentum are those of mass times velocity, or kilogram-meters per second (kg·m/s). These units are the same, but it is common practice to keep newton-seconds for impulse and kilogram-meters per second for linear momentum in order to distinguish the two quantities.

We used a constant force in our derivation of Equation 8-2, but we recognize that this is no limitation on the conclusion. If the force is varying, we simply must find its average value over the time interval and write Equation 8-2 as

$$\mathbf{F}_{av} \Delta t = m\mathbf{v}_f - m\mathbf{v}_i \qquad (8\text{-}3)$$

This is the same "impulse equals the change in momentum" result as before. Under certain circumstances, it may be difficult to determine the average value of the force, but the result still applies. Note also that the values of momentum are always instantaneous values at the start and end of the interval Δt.

Example 8-1 A senior in a small foreign automobile (total mass 700 kg) slows down to a speed of 8.00 m/s as she approaches a yellow light and is struck directly from behind by a freshman driving a student bus. The bus exerts an average force of 2.00×10^3 N for an interval of 2.00 s. Calculate (a) the initial linear momentum, (b) the impulse, (c) the final linear momentum, and (d) the final velocity of the senior and her car.

Solution (a) Since this interaction takes place along a single line, we choose the initial direction of the automobile as positive and may work with scalars. From the definition, we have

$$\text{Initial momentum} = mv_i = (700 \text{ kg})(8.00 \text{ m/s})$$
$$= 5.60 \times 10^3 \text{ kg·m/s} \qquad \text{positive direction}$$

(b)
$$\text{Impulse} = F_{av} \Delta t = (2.00 \times 10^3 \text{ N})(2.00 \text{ s})$$
$$= 4.00 \times 10^3 \text{ N·s} \qquad \text{positive direction}$$

(c) From Equation 8-3, we have

$$mv_f = F_{av}\Delta t + mv_i$$
$$= 4.00 \times 10^3 \text{ N·s} + 5.60 \times 10^3 \text{ kg·m/s}$$
$$= 9.60 \times 10^3 \text{ kg·m/s} \qquad \text{positive direction}$$

(d)
$$v_f = \frac{9.60 \times 10^3 \text{ kg·m/s}}{700 \text{ kg}} = 13.7 \text{ m/s} \qquad \text{positive direction}$$

Both impulse and linear momentum are vectors. The direction of impulse is the direction of the applied force, while the directions of the initial and final momenta are the directions of the initial and final velocities, respectively. In every case $\mathbf{F}$ and $\Delta \mathbf{v}$ must agree in direction, but neither $\mathbf{v}_f$ nor $\mathbf{v}_i$ need be in the direction of $\mathbf{F}$, only $\Delta \mathbf{v}$. For the moment, we will consider the one-dimensional case and account for the direction with simply a plus or minus sign.

Forces may vary during an interaction, but we may determine the impulse from a graph of the magnitude of the force vs. time as the area under the curve

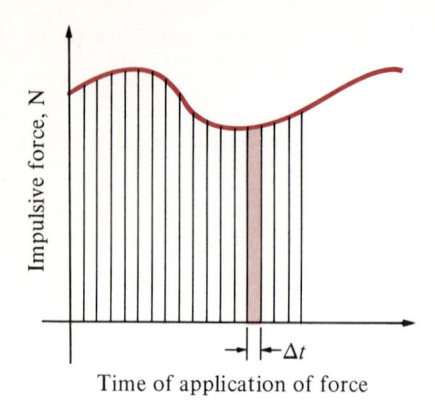

Figure 8-2 A continuously variable force vs. time. The total impulse delivered by the force is the area under the curve and could be determined by a number of methods. Here, we choose to divide the area into thin rectangles of uniform width Δt_i and variable height F_i. The sum of the areas of these rectangles yields the impulse.

(Figure 8-2). In the last chapter, we saw that the total work was the area under a graph of the parallel component of force vs. the magnitude of its displacement. Similarly, impulse is the area under a force vs. time curve. We may, for example, divide the area into very narrow rectangles, find the area of each of these rectangles, and add them all to obtain the total area and hence the impulse.

Example 8-2 A student wishes to launch a paper clip with a plastic ruler (Figure 8-3a). The ruler, as pulled back, acts like a spring, with force proportional to the distance displaced. The ruler is released, and the paper clip is propelled horizontally. The curve in Figure 8-3b shows the force applied to the paper clip vs. time during the release from rest. If the mass of the paper clip is 1.50×10^{-3} kg, calculate (a) the magnitude of the impulse, (b) the magnitude of the final horizontal linear momentum (just as it leaves the ruler), and (c) the final horizontal speed of the paper clip.

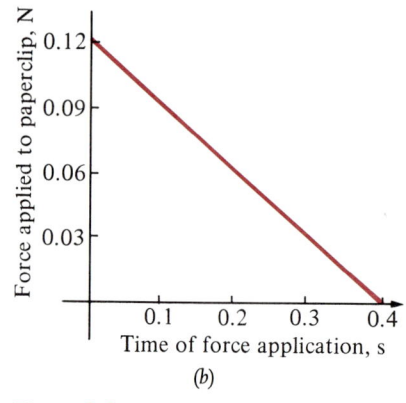

Figure 8-3 Example 8-2.

Solution (*a*) The magnitude of the impulse is the area under the curve, and this area in Figure 8-3*b* is a triangle; thus we may write

$$\text{Impulse} = \tfrac{1}{2} \text{ base times height}$$
$$= \tfrac{1}{2}(0.40 \text{ s})(0.12 \text{ N}) = 0.024 \text{ N} \cdot \text{s}$$

(*b*) The paper clip is released from rest. Thus the initial linear momentum is zero, and the magnitude of the final linear momentum is just equal to the impulse, or

$$mv_f = 0.024 \text{ kg} \cdot \text{m/s}$$

(*c*) The final speed is simply the final linear momentum divided by the mass:

$$v_f = \frac{0.024 \text{ kg} \cdot \text{m/s}}{1.5 \times 10^{-3} \text{ kg}} = 16 \text{ m/s}$$

8-3 Collisions

Probably nothing attests more to the mathematical precision with which collisions occur than to watch a billiard expert repeatedly bring off a seemingly impossible trick shot. Yet, the fact that most of us expend time and effort in practicing our backhand, our golf swing, our soccer kick, or whatever says the same thing statistically. After all, if collisions *did not* work with mathematical precision, then perfecting the tilt of a racket, the angle, the positioning of the foot, or whatever would not optimize the outcome. The first mark of the professional is consistency of effort—doing exactly the same thing the same way each time so that the result is predictable. This is all an eloquent testimony to the reproducibility of collision phenomena.

Let us apply the impulse-momentum theorem, Equation 8-3, to two bodies in collision. Take m_1 and m_2 as the masses of the two bodies, with $\mathbf{v}_{1i}$ and $\mathbf{v}_{2i}$ as their respective velocities before the collision and $\mathbf{v}_{1f}$ and $\mathbf{v}_{2f}$ as their velocities after the collision (Figure 8-4). During the collision, body 1 exerts an average force $\mathbf{F}_{12}$ on body 2, while body 2 exerts an average force $\mathbf{F}_{21}$ on body 1. Application of Equation 8-3 to body 1 yields

$$\mathbf{F}_{21} \Delta t = m_1 \mathbf{v}_{1f} - m_1 \mathbf{v}_{1i} \qquad (8\text{-}4)$$

where Δt is the time the bodies are in contact. Applying Equation 8-3 to body 2 yields

$$\mathbf{F}_{12} \Delta t = m_2 \mathbf{v}_{2f} - m_2 \mathbf{v}_{2i} \qquad (8\text{-}5)$$

where Δt is the same for both forces. (After all, contact begins at the same instant for each body, and ends at the same instant—when body 1 breaks contact with body 2, clearly body 2 has lost contact with body 1.) The forces are an action-reaction pair and are thus related by Newton's third law of motion as

$$\mathbf{F}_{12} = -\mathbf{F}_{21}$$

Therefore the left side of Equation 8-4 is equal to the negative of the left side of Equation 8-5. Similarly, the right side of Equation 8-4 will be equal to the negative of the right side of Equation 8-5, and we write

$$m_1 \mathbf{v}_{1f} - m_1 \mathbf{v}_{1i} = -m_2 \mathbf{v}_{2f} + m_2 \mathbf{v}_{2i}$$

Since we need to describe effects on more than one object throughout this chapter, the careful use of subscripts is necessary. This can look very forbidding in longer equations. Resolve *now* to take the time to understand how these subscripts are assigned on your first reading of the material, and you will save time overall.

Figure 8-4 A one-dimensional collision.

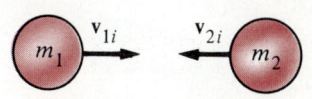

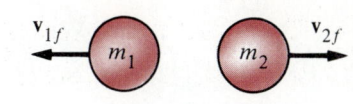

Collecting the final-velocity terms on the left and the initial-velocity terms on the right, we obtain

$$m_1\mathbf{v}_{1f} + m_2\mathbf{v}_{2f} = m_1\mathbf{v}_{1i} + m_2\mathbf{v}_{2i} \qquad (8\text{-}6)$$

This is a very important result! It states that the vector sum of the linear momenta of the two bodies *after* they collide is equal to the vector sum of the linear momenta of the two *before* they collide. Another way to say this is that the vector quantity, resultant linear momentum, is of the same magnitude and direction after the collision as it was before the collision. Thus Equation 8-6 is a statement of the **conservation of linear momentum** for this case. Note that in the derivation of Equation 8-6 there was an implicit condition that $\mathbf{F}_{12}$ and $\mathbf{F}_{21}$ are the only forces (or at least the only important forces) acting during the collision. With this modification, we state the following as *the law of conservation of linear momentum*:

> In a system in which the resultant of all outside forces can be considered to be negligible, the total linear momentum of the system remains constant.

Another way to state this is that the resultant momentum before the collision equals the resultant momentum after the collision. We emphasize that linear momentum is a *vector* quantity. Also note from Equation 8-6 that there is no restriction on the individual terms in the equation. In general, the linear momentum of any individual body may (and most often does) change a great deal. It is only the *resultant* vector afterward which must equal the *resultant* before a collision or interaction.

Example 8-3 An air cart of mass 0.300 kg is initially moving to the right on a linear air track at a speed of 8.00 m/s (Figure 8-5). It encounters a second air cart of mass 0.500 kg which is initially moving to the left at 2.00 m/s. A small bit of putty placed at a strategic location on one of the carts ensures that they will

Figure 8-5 Example 8-3. The air track, a nearly frictionless apparatus ideal for studying collisions, is a tube with a cross section in the shape of a hollow 45° right triangle, with the right angle as the top vertex. Tiny holes are drilled into the upward slanting surfaces, and air pumped into the tube exits from these holes at high velocity. The cart, with flat right angle faces, is placed on the track and is suspended in the air jets, literally floating on air.

stick together after the collision. Calculate (a) the net linear momentum before and after the collision and (b) the speed of the carts after the collision.

Solution (a) We may again take care of the vector nature with plus and minus signs since the interaction takes place in one dimension. Choose the direction to the right as positive and that to the left as negative. If the 0.300-kg cart is m_1 and the 0.500-kg cart is m_2, we have $\mathbf{v}_{1i} = +8.00$ m/s and $\mathbf{v}_{2i} = -2.00$ m/s. Thus

$$m_1\mathbf{v}_{1i} + m_2\mathbf{v}_{2i} = (0.300 \text{ kg})(8.00 \text{ m/s}) + (0.500 \text{ kg})(-2.00 \text{ m/s})$$
$$= +1.40 \text{ kg}\cdot\text{m/s} \quad \text{(to the right)}$$

Equation 8-6 states that this is the same value of the resultant linear momentum after the collision.

(b) After the collision, the carts are stuck together; therefore

$$\mathbf{v}_{1f} = \mathbf{v}_{2f}$$

We call this value $\mathbf{v}_f$ and write

$$1.40 \text{ kg}\cdot\text{m/s} = m_1\mathbf{v}_{1f} + m_2\mathbf{v}_{2f} = m_1\mathbf{v}_f + m_2\mathbf{v}_f = (m_1 + m_2)\mathbf{v}_f$$

or

$$1.40 \text{ kg}\cdot\text{m/s} = (0.300 \text{ kg} + 0.500 \text{ kg})v_f = (0.800 \text{ kg})v_f$$
$$v_f = +1.75 \text{ m/s} \quad \text{(to the right)}$$

> Momentum is a vector quantity, and thus momenta must be added as vectors. We know that we may deal with these directional properties in one dimension by assigning a positive direction and attaching a plus sign to magnitudes of vectors in that direction and a minus sign to magnitudes of vectors in the direction opposite. Be careful to do this properly! One of the most frequent student errors is to report total momentum obtained by adding magnitudes of separate momenta in opposite directions.

8-4 Elastic Collisions in One Dimension

In all collisions for which external forces may be neglected, vector momentum is conserved. Yet, there certainly is a difference between a pitched baseball striking a stationary bat and the same baseball striking a blob of wet clay, of the same mass as the bat, suspended by a rope above home plate. Clearly more is involved in the outcome of a collision than just the conservation of momentum.

In Example 8-3 the bodies are both moving at a faster rate before they collide than they are moving together after the collision. Thus, there is less energy of motion (kinetic energy) after the collision than there was before. Although momentum was conserved, kinetic energy was *not* conserved. Energy lost during a collision is generally converted to heat by internal friction or a permanent deformation (of the putty in our example). Collisions exist ranging from those in which no kinetic energy is lost to those in which all the kinetic energy is lost. We classify the collisions in which no kinetic energy is lost as **elastic collisions.** In an elastic collision the sum of the kinetic energy of the individual bodies undergoing a collison is the same number before and after the collision.

In the one-dimensional elastic collision of Figure 8-6, we can use conservation of linear momentum and conservation of kinetic energy to solve for the two unknown velocities after the collision if we know the velocities before the collision.

Choosing the direction to the right as positive, we apply the conservation of linear momentum to yield

$$m_1 v_{1f} + m_2 v_{2f} = m_1 v_{1i} + m_2 v_{2i} \qquad (8\text{-}7)$$

No generality is lost by choosing to draw all the vector velocities in the

161

Figure 8-6 An elastic collision in one dimension. The figure is drawn as though all motion is to the right. This is not required, but if the sign of v_{1f}, for example, comes out negative in the calculations, we will know this choice was incorrect.

same positive direction, since if some of them happen to be to the left, the numerical values will be negative numbers. The conservation of kinetic energy implies that

$$\tfrac{1}{2}m_1v_{1f}^2 + \tfrac{1}{2}m_2v_{2f}^2 = \tfrac{1}{2}m_1v_{1i}^2 + \tfrac{1}{2}m_2v_{2i}^2 \tag{8-8}$$

Assuming we know the masses and the velocities before the collision, we wish to solve for the velocities after the collision in terms of those known quantities. The solution proceeds most readily from Equations 8-7 and 8-8 as follows.

Multiply Equation 8-8 by 2, and collect the terms common to body 1 on one side of the equation and those terms common to body 2 on the other side. Thus

$$m_1v_{1f}^2 - m_1v_{1i}^2 = m_2v_{2i}^2 - m_2v_{2f}^2$$

or
$$m_1(v_{1f}^2 - v_{1i}^2) = m_2(v_{2i}^2 - v_{2f}^2)$$

This can be factored as

$$m_1(v_{1f} - v_{1i})(v_{1f} + v_{1i}) = m_2(v_{2i} - v_{2f})(v_{2i} + v_{2f}) \tag{8-9}$$

Equation 8-7 can be rearranged as

$$m_1v_{1f} - m_1v_{1i} = m_2v_{2i} - m_2v_{2f}$$

or
$$m_1(v_{1f} - v_{1i}) = m_2(v_{2i} - v_{2f}) \tag{8-10}$$

Note that Equation 8-10 states that the first two factors on the left side of Equation 8-9 are equal to the first two factors on the right side of Equation 8-9. Therefore we can divide these out of Equation 8-9 to obtain

$$v_{1f} + v_{1i} = v_{2i} + v_{2f} \tag{8-11}$$

By solving Equation 8-11 for either v_{2f} or v_{1f} in terms of the other three variables and substituting directly into Equation 8-7, we can (after a little algebra) obtain

$$v_{1f} = \frac{m_1 - m_2}{m_1 + m_2}v_{1i} + \frac{2m_2}{m_1 + m_2}v_{2i} \tag{8-12}$$

and
$$v_{2f} = \frac{2m_1}{m_1 + m_2} v_{1i} + \frac{m_2 - m_1}{m_1 + m_2} v_{2i} \qquad (8\text{-}13)$$

These solutions for the final velocities are perfectly general and apply to all one-dimensional elastic collisions. It is worthwhile to examine them for special cases in which they may be simplified. One such case is when the second mass is at rest before the collision ($v_{2i} = 0$), causing the last term on the right side of both equations to drop out and reducing the equations to

$$v_{1f} = \frac{m_1 - m_2}{m_1 + m_2} v_{1i}$$

and
$$v_{2f} = \frac{2m_1}{m_1 + m_2} v_{1i}$$

Such a use in physics occurs when nuclear physicists employ a beam of particles as projectiles to bombard the stationary atoms of a target. If the bombarding particle is of mass m_1 and the target particle of mass m_2, we distinguish three cases of interest. (In each case, convince yourself that the conclusion is justified from the reduced equations above.)

Case 1. The target particle is much more massive than the projectile, or $m_2 \gg m_1$. The implication is that

$$v_{1f} \approx -v_{1i} \qquad \text{and} \qquad v_{2f} \approx 0$$

Can you see why?

This is just about what you would expect to happen if a table tennis ball (m_1) collided with a bowling ball (m_2) which was initially at rest. The minus sign indicates that m_1 bounces back where it came from. Note also that it has about the same speed after the collision as it had beforehand. That is, it did not lose much kinetic energy.

Case 2. The target particle is much less massive than the projectile, or $m_1 \gg m_2$. Then

$$v_{1f} \approx v_{1i} \qquad \text{and} \qquad v_{2f} \approx 2v_{1i}$$

(Are you convinced that this is true?)

This is the reverse of the first situation. Now the bowling ball (m_1) is moving and collides with a suspended table tennis ball which is initially at rest. Note that the table tennis ball does not have much effect on the bowling ball. Again m_1 does not lose much kinetic energy.

Case 3. The target and projectile are of about the same mass, or $m_1 \approx m_2$. Then

$$v_{1f} \approx 0 \qquad \text{and} \qquad v_{2f} \approx v_{1i}$$

This could be two bowling balls (or tennis balls). Note that the two masses exchanged velocities. This exchange is the result even if both bodies have nonzero initial velocities. To see this, go back to Equations 8-12 and 8-13. By now it should be a lot easier for you to manipulate these equations to this result.

Example 8-4 Assume the two air carts of Example 8-3 are again given the same initial velocities but only after the putty has been removed and the carts fitted with bumpers to ensure that the collision is elastic. Calculate the final velocities v_{1f} and v_{2f} of the two carts after the elastic collision. (Note that $m_1 = 0.300$ kg, $m_2 = 0.500$ kg, $v_{1i} = 8.00$ m/s, and $v_{2i} = -2.00$ m/s.)

Solution Using Equations 8-12 and 8-13, we have

$$v_{1f} = \frac{0.300 \text{ kg} - 0.500 \text{ kg}}{0.300 \text{ kg} + 0.500 \text{ kg}}(8.00 \text{ m/s}) + \frac{2(0.500 \text{ kg})}{0.300 \text{ kg} + 0.500 \text{ kg}}(-2.00 \text{ m/s})$$

$$= -4.50 \text{ m/s}$$

and

$$v_{2f} = \frac{2(0.300 \text{ kg})}{0.300 \text{ kg} + 0.500 \text{ kg}}(8.00 \text{ m/s}) + \frac{0.500 \text{ kg} - 0.300 \text{ kg}}{0.300 \text{ kg} + 0.500 \text{ kg}}(-2.00 \text{ m/s})$$

$$= +5.50 \text{ m/s}$$

8-5 Elastic Collisions in Two Dimensions

Many collisions of interest take place which require more than one dimension to describe. Since many practical collision problems involving only two bodies take place in a single plane, we limit the following discussion to two dimensions. Consider the billiard ball situation of Figure 8-7. We choose the x axis in the direction of the cue ball before the collision. To maintain a consistent notation, we have labeled the cue ball m_1 and the nine ball m_2. The angles their directions of motion after collision make with the original direction of the cue ball are θ_1 and θ_2. Now let us apply the conservation of linear momentum to this collision. We have

$$m_1 \mathbf{v}_{1i} = m_1 \mathbf{v}_{1f} + m_2 \mathbf{v}_{2f}$$

since $\mathbf{v}_{2i} = 0$. In component form, this becomes

$$m_1 v_{1i} = m_1 v_{1f} \cos \theta_1 + m_2 v_{2f} \cos \theta_2 \tag{8-14}$$

Figure 8-7 In a two-dimensional elastic collision, a billiard ball of mass m_1 and velocity $\mathbf{v}_{1i}$ approaches a ball of mass m_2 initially at rest. Following the collision, the balls have velocities $\mathbf{v}_{1f}$ and $\mathbf{v}_{2f}$, respectively, and move away from the dashed line representing the original line of motion of m_1 at angles θ_1 and θ_2, respectively.

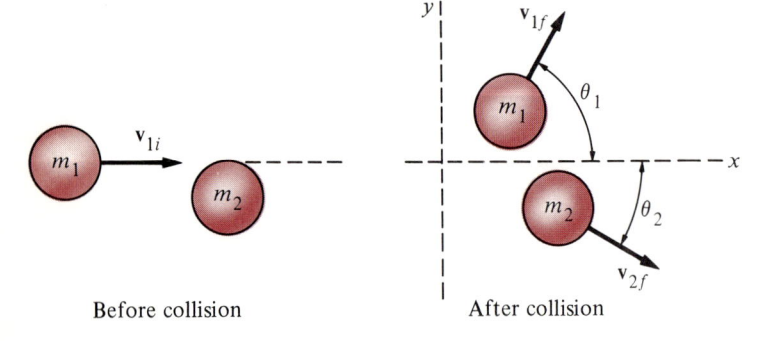

Before collision

After collision

in the x direction and

$$0 = m_1 v_{1f} \sin \theta_1 - m_2 v_{2f} \sin \theta_2 \qquad (8\text{-}15)$$

in the y direction.

Even if we know the value of the masses and the initial velocities, we can see that we still have four unknowns (two magnitudes of velocities after the collision and two angles, or two x components and two y components of the velocities after the collision) and only two equations. We can solve for the unknowns only if we have some more information.

If this is an elastic collision (a pretty good approximation for billiards), we immediately have one more equation, the conservation of kinetic energy:

$$\tfrac{1}{2} m_1 v_{1i}^2 = \tfrac{1}{2} m_1 v_{1f}^2 + \tfrac{1}{2} m_2 v_{2f}^2 \qquad (8\text{-}16)$$

This is most helpful, but we still need one more relationship to solve the problem. This is often obtained by measuring one of the angles after the collision, as in several end-of-chapter problems.

While this is as far as we can go with a general solution to two-dimensional elastic collisions, the billiards example provides a special case. Here, the masses of the balls are equal, and we may consider the situation where the second ball is at rest before collision. Letting $m_1 = m_2 = m$, the conservation of momentum equation becomes

$$m\mathbf{v}_{1i} = m\mathbf{v}_{1f} + m\mathbf{v}_{2f} \qquad (8\text{-}17)$$

The two terms on the right side of the equation must add vectorially to give the term on the left, as in Figure 8-8. Assuming this is an elastic collision, the other equation which must be satisfied is

$$\tfrac{1}{2} m v_{1i}^2 = \tfrac{1}{2} m v_{1f}^2 + \tfrac{1}{2} m v_{2f}^2 \qquad (8\text{-}18)$$

Applying the law of cosines to the triangle of Figure 8-8b, we obtain

$$(mv_{1i})^2 = (mv_{1f})^2 + (mv_{2f})^2 + 2(mv_{1f})(mv_{2f}) \cos(\theta_1 + \theta_2)$$

> Momentum is a vector; hence we may assert that the conservation of momentum holds in the x, y, and z directions separately. This yields independent, simultaneous equations for each dimension. Kinetic energy is a scalar, having only magnitude; thus its conservation will yield only a single equation.

Figure 8-8 Vector addition of momentum vectors for the collision of billiard balls. (a) Figure 8-7 with $m_1 = m_2 = m$. (b) The momentum of the first ball after collision is the vector $m\mathbf{v}_{1f}$ at the angle θ_1. At the tip of this, we begin the vector $m\mathbf{v}_{2f}$, representing the magnitude and direction of the momentum of the second ball after collision. By the law of conservation of momentum, the vector sum of these two vectors must have the same magnitude and direction as the total momentum vector before the collision, $m\mathbf{v}_{1i}$.

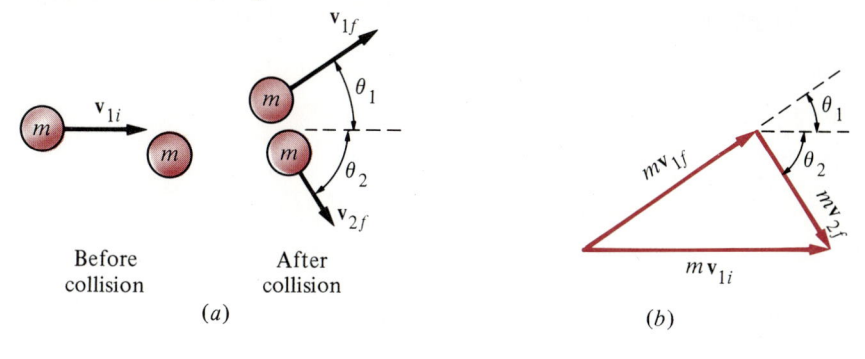

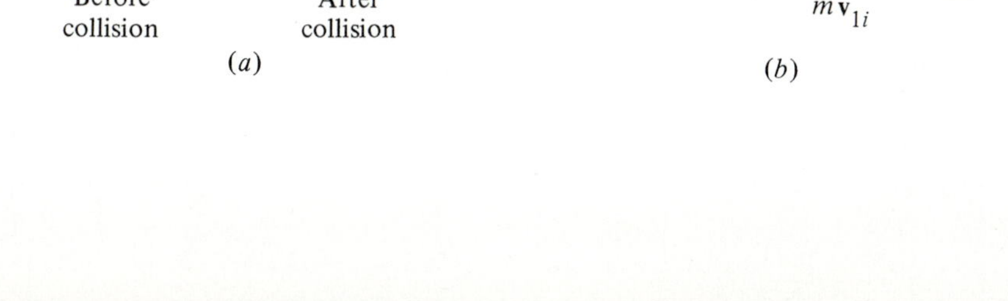

To compare some of these terms to those in Equation 8-18, we multiply this equation by $1/(2m)$, which gives us

$$\tfrac{1}{2}mv_{1i}^2 = \tfrac{1}{2}mv_{1f}^2 + \tfrac{1}{2}mv_{2f}^2 + mv_{1f}v_{2f}\cos(\theta_1 + \theta_2) \quad (8\text{-}19)$$

Note that Equations 8-18 and 8-19 are identical except for the last term of Equation 8-19. These equations must be satisfied simultaneously, and it is evident that this can happen only if the last term on the right side of Equation 8-19 vanishes. That is,

$$mv_{1f}v_{2f}\cos(\theta_1 + \theta_2) = 0$$

There are three separate mathematical conditions which can make this equation equal zero, all of which are physically possible and have a reasonable interpretation:

1. $v_{1f} = 0$ Collision is head on; cue ball stops and target ball goes on
2. $v_{2f} = 0$ No collision; cue ball missed
3. $\cos(\theta_1 + \theta_2) = 0$ Angle of separation after collision is 90°

Thus the cue ball and the initially stationary target ball must separate at an angle of 90° after a glancing collision. This fact can be used to predict the direction the cue ball will go after sinking the target ball and should help you win your next billiard match.

8-6 Inelastic Collisions

An **inelastic collision** is one in which the kinetic energy afterward is *not* equal to the kinetic energy before the interaction.

Example 8-5 Show that the collision of Example 8-3 was inelastic by calculating the change in kinetic energy.

Solution We have

$$\Delta KE = KE_f - KE_i$$

where
$$KE_i = \tfrac{1}{2}m_1 v_{1i}^2 + \tfrac{1}{2}m_2 v_{2i}^2$$

and
$$KE_f = \tfrac{1}{2}(m_1 + m_2)v_f^2$$

since the two carts stuck together after the collision. Substituting,

$$KE_i = \tfrac{1}{2}(0.300 \text{ kg})(8.00 \text{ m/s})^2 + \tfrac{1}{2}(0.500 \text{ kg})(2.00 \text{ m/s})^2 = 10.6 \text{ J}$$

$$KE_f = \tfrac{1}{2}(0.300 \text{ kg} + 0.500 \text{ kg})(1.75 \text{ m/s})^2 = 1.23 \text{ J}$$

thus
$$\Delta KE = -9.4 \text{ J}$$

The change in kinetic energy is negative, and thus kinetic energy is lost during the collision.

Example 8-6 Figure 8-9 shows a ballistic pendulum, a device which can be used to measure the speed of a bullet. The bullet has a mass m_1 and a speed v_{1i} as

8-6 Inelastic Collisions

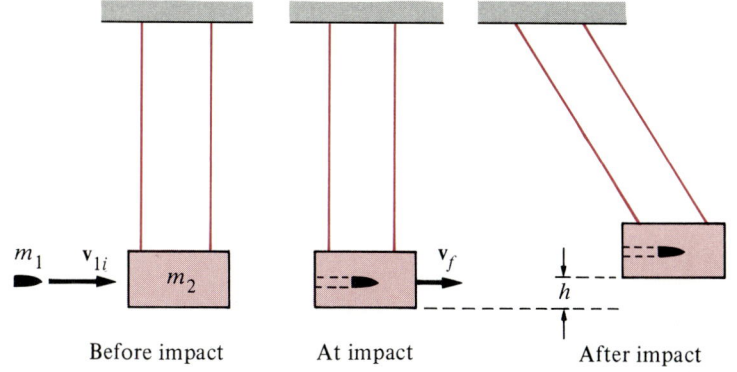

Figure 8-9 Example 8-6. At the instant after the collision, the bullet-block system of mass $m_1 + m_2$ moves off at a speed of v_f. Since the pendulum mounting conserves mechanical energy, the speed v_f determines the height h to which the system rises. From measuring h, v_f may be calculated, and from this and the masses, v_{1i} may be determined.

it approaches the block, which may be made of wood. The block has a mass m_2 and is initially at rest. Upon collision, the bullet embeds itself in the block. This collision takes place in a time interval which is very short when compared with the time the pendulum takes to swing. Thus we may assume that the supporting cords remain essentially vertical during the collision, and hence while the collision takes place, there is no external horizontal force on the bullet-block system. In that case, we know that momentum is conserved during the collision.

Immediately after the collision, the bullet-block system is moving to the right with a speed v_f. The pendulum now swings until it reaches a maximum height h above its original position. During the swing, the only force acting which does work is that of gravity. Therefore we may apply the conservation of mechanical energy to the system *after* the collision. Derive an expression for the initial speed of the bullet in terms of the masses and the final height of the system. Also, if $m_2 = 999 m_1$, show that 99.9 percent of the original kinetic energy is lost during the collision.

Solution We have two separate interactions: the collision (during which linear momentum is conserved) and the swing (during which mechanical energy is conserved).

The collision takes place in one dimension. Therefore we take the initial direction to be positive and apply the conservation of linear momentum, obtaining

$$m_1 v_{1i} = (m_1 + m_2) v_f$$

Now if we apply the conservation of mechanical energy to the swing from the position immediately after the collision to the position where the system comes to rest, we have

$$\Delta KE + \Delta PE = 0$$

or

$$0 - \tfrac{1}{2}(m_1 + m_2) v_f^2 + (m_1 + m_2) g h = 0$$

This equation yields

$$v_f = \sqrt{2gh}$$

Substituting this result in the first equation, we obtain

$$v_{1i} = \frac{m_1 + m_2}{m_1} \sqrt{2gh}$$

To determine the fraction of kinetic energy lost *during the collision*, we must write

$$f = \frac{\Delta KE}{KE_i} = \frac{KE_f - KE_i}{KE_i}$$

Now
$$KE_f = \tfrac{1}{2}(m_1 + m_2)v_f^2$$

which can be written (using the conservation of momentum equation to substitute for v_f) as

$$KE_f = \frac{1}{2}(m_1 + m_2)\left(\frac{m_1 v_{1i}}{m_1 + m_2}\right)^2$$

$$= \frac{1}{2}\left(\frac{m_1^2}{m_1 + m_2}\right)v_{1i}^2$$

Further,
$$KE_i = \tfrac{1}{2}m_1 v_{1i}^2$$

By direct substitution, we obtain

$$f = \frac{m_1}{m_1 + m_2} - 1 = -\frac{m_2}{m_1 + m_2}$$

When $m_2 = 999 m_1$, we have

$$f = -\frac{999}{1000} = -0.999$$

A collision in which the two bodies stick together is said to be **perfectly inelastic** since, for the initial conditions, the maximum amount of kinetic energy will be lost. We note that not all the kinetic energy was lost in the previous examples. This is not surprising. In order to conserve linear momentum we had to have a nonzero component of velocity after the collision. (Can you think of a kind of collision in which *all* the kinetic energy *could* be lost?) Let us rewrite Equation 8-11 and examine it more closely.

$$v_{2f} - v_{1f} = v_{1i} - v_{2i}$$

We see that for a one-dimensional elastic collision, the relative velocity of separation of the two bodies after the collision, on the left, equals the relative velocity of approach of the two before the collision on the right. We also know that the relative velocity of separation for a perfectly inelastic collision is zero (the bodies stick together). Now most collisions are somewhere between these two extremes and may be called partially elastic (or partially inelastic). For any collision, we define the **coefficient of restitution** as

$$\epsilon = \frac{v_{1f} - v_{2f}}{v_{2i} - v_{1i}} \qquad (8\text{-}20)$$

where the Greek lowercase epsilon (ϵ) represents the coefficient of restitution. This typically varies from 1 for an elastic collision to 0 for a perfectly inelastic

8-6 Inelastic Collisions

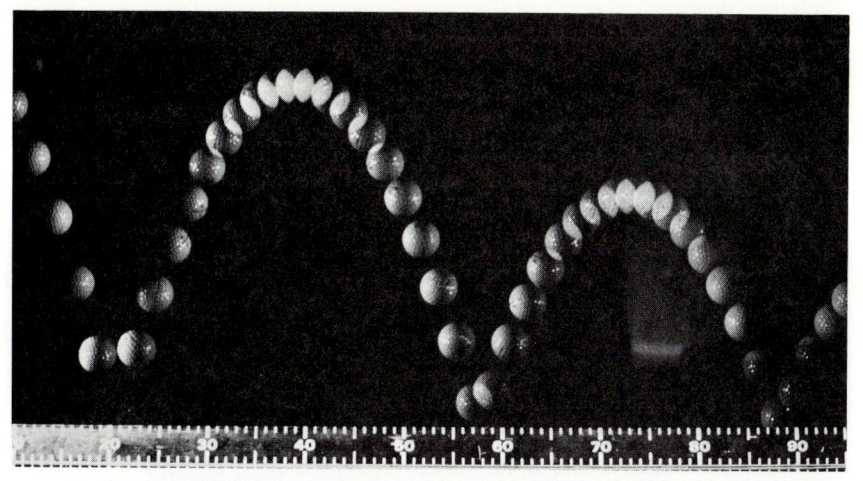

Figure 8-10 A golf ball bouncing on a concrete surface. The coefficient of restitution may be determined from the height of release and the peak height of the first bounce (or from the ratio of heights of any two successive peaks). (C. Riban)

collision. The coefficient of restitution thus is some measure or indicator of the elasticity (or nearness to elasticity) of a collision.

Most games or sports involving a ball must have quite elastic collisions. You can give a fast test of the nature of the collision by dropping the ball from a height onto its intended impact surface. If the ball bounces to its original height of release, the collision was perfectly elastic. Why? No ball achieves this, but many come close (Figure 8-10). Table 8-1 shows some typical values of the coefficient of restitution for a few representative collisions.

When two automobiles collide, there is usually a great deal of kinetic energy lost. A small amount of the lost kinetic energy shows up as sound, heat, and even light (sparks), but most of it does work in permanently deforming fenders, radiators, and other useful components.

TABLE 8-1 Coefficient of Restitution of Various Balls

Ball	Coefficient of Restitution
"Superball"	0.89
Golf ball	0.80
Basketball	0.76
Soccer ball	0.76
Volleyball	0.74
Solid rubber ball	0.73
Tennis ball:	
Well worn	0.71
New	0.67
Baseball	0.55
Softball	0.32

Example 8-7 A 2000-kg limosine (m_2) is traveling at a speed of 20.0 m/s to the north on a through street (Figure 8-11a) when it is struck by a 3000-kg panel

169

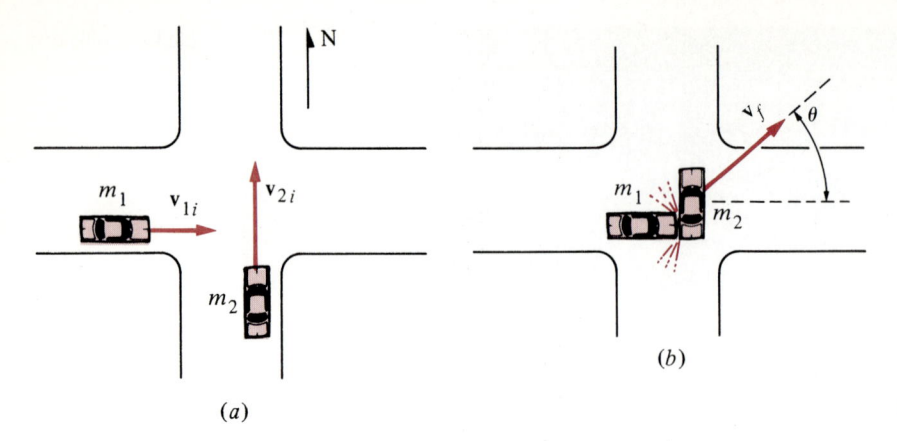

Figure 8-11 Example 8-7.

truck (m_1) traveling to the east at 10.0 m/s. The vehicles lock together (Figure 8-11b). Determine (a) the direction the two leave the point of impact and (b) the kinetic energy lost.

Solution (a) Writing conservation of linear momentum equations in the east and north directions yields

$$m_1 v_{1i} = (m_1 + m_2) v_f \cos \theta$$

and

$$m_2 v_{2i} = (m_1 + m_2) v_f \sin \theta$$

or

$$(3000 \text{ kg})(10.0 \text{ m/s}) = (3000 \text{ kg} + 2000 \text{ kg}) v_f \cos \theta$$

and

$$(2000 \text{ kg})(20.0 \text{ m/s}) = (3000 \text{ kg} + 2000 \text{ kg}) v_f \sin \theta$$

which yield

$$6.00 \text{ m/s} = v_f \cos \theta$$

and

$$8.00 \text{ m/s} = v_f \sin \theta$$

Dividing the second equation by the first gives us

$$\tan \theta = \tfrac{4}{3} = 1.33 \quad \text{or} \quad \theta = 53.1° \text{ N of E}$$

Substituting θ into either equation and solving for v_f, we obtain

$$v_f = 10.0 \text{ m/s}$$

(b)
$$\Delta KE = \tfrac{1}{2}(m_1 + m_2) v_f^2 - \tfrac{1}{2} m_1 v_{1i}^2 - \tfrac{1}{2} m_2 v_{2i}^2$$

Substituting the proper values and solving,

$$\Delta KE = -3.00 \times 10^5 \text{ J}$$

A body moving with a constant velocity will retain its linear momentum provided no *external* forces act on it. This could serve as a statement of Newton's first law. But what if a body explodes and flys apart? If no external forces act during the explosion, linear momentum will still be conserved.

Example 8-8 Consider a bomb of mass m, initially at rest, which explodes into two pieces. One piece of mass $2m/3$ is projected in the positive x direction with a speed v. In what direction and with what speed V would we expect the other fragment to go?

Solution The other fragment has a mass of $m/3$. The linear momentum before the explosion was equal to zero. Thus it will still be zero after the explosion, and

$$0 = \frac{2m}{3} v + \frac{m}{3} V$$

Therefore

$$V = -2v$$

or the smaller fragment goes off after the collision at twice the speed of the larger and in the opposite direction.

SPECIAL TOPIC

The Rocket

The simple example of firing a rifle illustrates that any body which ejects a mass in one direction acquires a momentum in the other direction. In a very real sense, the typical rocket operates very much like the rifle. The ejected mass is in the form of very hot gases (products of fuel combustion) originating in the rocket engine. In Figure 8-12, the large amount of hot gas which has already been ejected in imparting the first few meters per second of speed is evident.

To consider the physics of a typical rocket, we ignore gravity by considering a rocket in outer space. We assume that at a time t the rocket and exhaust gases may be depicted as in Figure 8-13. The structure of the rocket and the remaining unburned fuel at time t have a mass M and a speed V_i. In a very short time interval Δt, there is a small quantity, say Δm, of hot gases ejected. We are concerned with the impulse imparted to the rocket during this time, so we will measure the velocity

Figure 8-12 The launch of *Apollo 11* on July 16, 1969, carrying astronauts Armstrong, Aldrin, and Collins to the first landing on the moon. The Saturn V first-stage rocket engines are expelling gases downward, providing a 33.4-MN thrust to raise the rocket. (NASA)

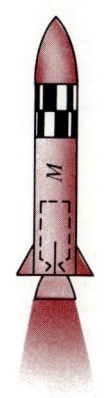

Figure 8-13 Rocket and fuel are initially of mass M_0, but the burning and ejecting of fuel makes the mass variable. The remaining mass M is given an impulse during the interval Δt by ejecting gases of mass Δm at a relative speed of v.

of the ejected gases with respect to the rocket in a coordinate system which is fixed to the moving rocket. Using $-v$ as the velocity of the gas with respect to the rocket gives a component of linear momentum of the ejected gases during the Δt time interval as

$$\text{Change in momentum} = -(\Delta m)v$$

This quantity is equal to the impulse imparted to the ejected gases during Δt, and hence there is an impulse of equal magnitude imparted to the rocket and remaining fuel. If during the interval Δt the rocket increases its speed from V_1 to V_2, the rocket's change in linear momentum can be written as

$$M(V_2 - V_1) = M\,\Delta V$$

where we have ignored the small change in mass due to the gases ejected during Δt. Setting this equal to the impulse imparted to the rocket (the negative of the impulse imparted to the gases), we have

$$M\,\Delta V = (\Delta m)v \quad (8\text{-}21)$$

We may obtain Newton's second law for the force on the rocket by dividing this by the time interval Δt. Thus

$$F = M\frac{\Delta V}{\Delta t} = \frac{\Delta m}{\Delta t} v \quad (8\text{-}22)$$

Here, F is the magnitude of the thrust acting on the rocket. From this equation, we see that the rocket force being exerted is the product of the time rate of ejection of the mass of hot gases, $\Delta m/\Delta t$, and the speed of the ejected gases with respect to the rocket, v.

This thrust accelerates the rocket, and to attain a large final speed, the thrust must be as large as practical. This could be accomplished either by increasing v, the speed of the ejected gases with respect to the rocket, or by increasing $\Delta m/\Delta t$, the rate of ejection of the gases. The magnitude of v may be increased by proper shaping of the inside of the rocket engine or by using a fuel which burns at higher temperatures. Both these tactics have practical upper limits. The first requires that we decrease the cross section of the jet stream of exhaust gases. If we try to make the cross section too small, we increase v but only at the expense of decreasing $\Delta m/\Delta t$. The limit on burning at higher temperatures is determined by the melting point of the structure of the rocket engine's interior. For practical purposes, the best ejection speed we can expect is between 2.00×10^3 and 2.5×10^3 m/s.

Any further increase in the thrust of a rocket requires manipulating the rate of ejection of gases ($\Delta m/\Delta t$). The largest thrust would be obtained if all the fuel were burned at once and ejected in the shortest time interval, but the rocket would then have to withstand a huge acceleration. The payloads of both instruments and/or astronauts have upper limits of acceleration without permanent damage.

Summary of Definitions of Terms

M_0 = initial mass of rocket and fuel at $t = 0$
M_R = final mass of rocket and structure, the payload
M = mass of payload and remaining fuel at time t
Δm = mass of gases ejected in any time interval Δt
ΔM = change in mass of payload and remaining fuel in any time interval Δt ($= -\Delta m$)
v = constant speed of hot gases ejected (measured with respect to the rocket)
V = speed of M at time t

Equation 8-21 can yield some limits on the increase in velocity of the rocket. We write it as

$$M\,\Delta V = -\Delta M\,v$$

which may be rewritten as

$$\frac{\Delta V}{v} = -\frac{\Delta M}{M} \quad \text{or} \quad \frac{\Delta M}{M} = -\frac{1}{v}\Delta V$$

This is a well-studied mathematical form of an equation which is encountered very often in physical situations. Its solution from calculus is

$$V_f = v \ln \frac{M_0}{M_R} \quad (8\text{-}23)$$

where we have assumed $V = 0$ at $t = 0$ and substituted

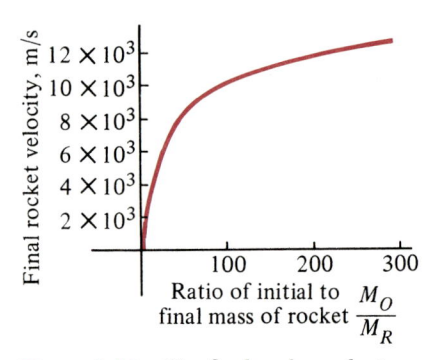

Figure 8-14 The final rocket velocity vs. the ratio of initial to final mass of the rocket. With a smaller and smaller part of the rocket's initial mass as eventual payload, the final velocity which can be attained increases. Figures assume a rocket exhaust velocity of 2.14×10^3 m/s, typical of the space shuttle.

from the definition of terms given in the table. The ultimate velocity V_f of the rocket thus depends on the natural logarithm of the ratio of the initial and final masses of the rocket. To attain a large final velocity, a great deal of the initial mass must be burned away. Equation 8-23 is plotted in Figure 8-14, assuming arbitrarily that $v = 2.14 \times 10^3$ m/s.

The escape velocity to leave the earth is about 11.2×10^3 m/s. From Figure 8-14, we see that a single rocket would have to have an M_0/M_R ratio equal to about 180 to attain this velocity. That is, only about 0.5 percent of the original rocket could be payload and structure to hold the fuel. Actually, when air resistance and the force of gravity (both of which must enter an equation of motion) are taken into account, the proportion of the initial mass available for payload is further reduced to about 0.25 percent. This makes for a very tiny payload or a very, very large rocket. To reduce the initial size of the rocket and also allow for larger payloads, rockets have been designed in multiple stages, leaving detachable parts of their structure behind as their fuel is expended.

Rockets must also be stable. A long, thin thing like a rocket with the force applied at its rearmost end has a powerful tendency to flop over in flight. Fireworks rockets solve this by attaching the rocket to a long stick which tends to prevent rotation during flight. For large rockets this is not practical, and a system is required that senses any tendency of the rocket to fall to the side and adjusts its thrust to the side to correct the motion.

Most of the early problems with large rockets were originally solved by the American physicist Robert H. Goddard (1882–1945) (Figure 8-15). He perfected the shape of the combustion chamber and devised the system to keep it from melting, a system that is still in use. The liquid fuels are circulated in coils around the outer walls of the chamber on their way to injection and burning. He eventually devised gyroscopes to detect small lateral accelerations and correct the rocket's motion by activating vanes to deflect the rocket exhaust. This was the forerunner of the system used in the V-2 rocket of World War II.

In most of the sophisticated rockets of today, the stability problem is solved by mounting the entire engine so that it can be swung in a range of directions to provide the thrust needed to correct the motion.

Perhaps the most sophisticated system yet developed is that of the space shuttle (Figure 8-16). Two solid-fuel boosters generate 11.6 MN each to assist in lift-off and early acceleration. The three main liquid-fueled engines of the shuttle generate 2.1 MN of thrust each. This system is complex because the steerable liquid-fuel engines are necessary to balance and correct

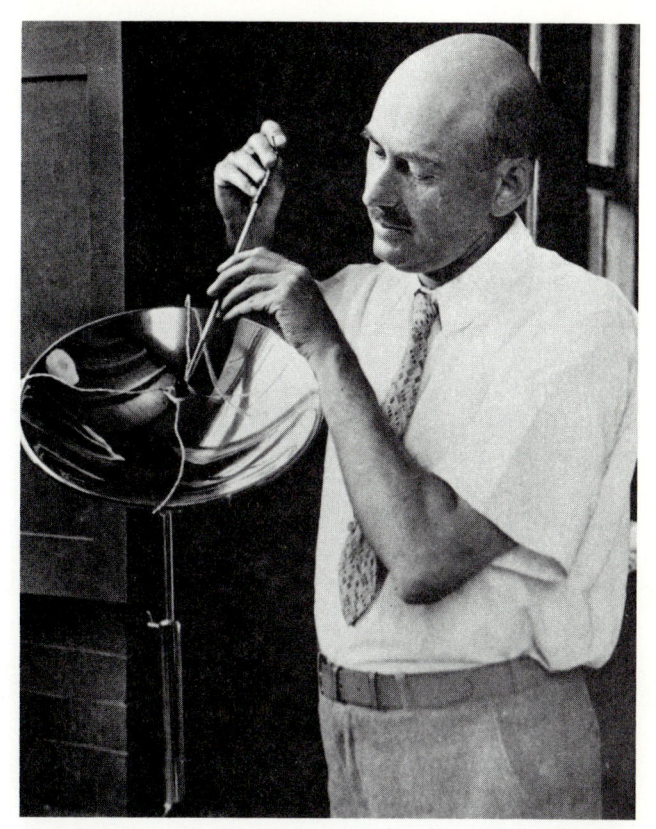

Figure 8-15 Dr. Robert H. Goddard, a pioneer in rocketry. Goddard was a physics professor for most of his life and first published on rocketry in 1919. By 1923 he was testing liquid-fuel engines and went on to stabilized, multistage rockets, working largely alone. He died in 1945 before any serious commitment to space developed and was largely unrecognized for his achievements. Twelve years after his death, the first artificial satellite was orbiting the earth, and 24 years later human beings were on the moon. *(NASA)*

the thrust of the solid-fuel rockets. Within a few seconds, these engines must go from not operating to almost their full thrust. Only when convinced that the liquid-fuel engines are running smoothly can the computers ignite the solid-fuel boosters, since there is no turning back from this point. Once ignited, the boosters will burn until fuel is exhausted. Even if (as was true in the space shuttle Challenger disaster) at ignition smoke is detected where it shouldn't be, it is too late. One minute into Challenger's launch, when a small drop in the right booster pressure indicated a growing opening in the side of the rocket, it was impossible to shut down or correct. Any attempt to detach the fragile shuttle while the forces of the boosters are being managed is likely to initiate an equivalent disaster. As more and more mass is ejected, the constant thrust of the booster rockets produces more and more acceleration. To keep the total acceleration acceptable for human beings ($a < 3g$'s), the liquid-fuel rockets are throttled back to a lower fraction of their maximum thrust.

The rockets lift the 68-Mg orbiter along with its 29.5-Mg payload to an orbit 200 km above the earth. This mass represents about 100 compact cars parked in orbit. This is a far cry from the original American satellite which was approximately the size of a grapefruit.

Figure 8-16 The space shuttle represents a complex wedding of solid-fuel and liquid-fuel engines. The airplanelike orbiter is the only part of the vehicle to enter space; the largest part of the assembly is the liquid-fuel tank supplying the orbiter engines during blast-off. After the single-purpose thoroughbred rockets of the moon program, the shuttle represents a largely reusable workhorse system for moving mass into orbit. (NASA)

Minimum Learning Objectives

After studying this chapter, you should be able to:
1. Define:
   - coefficient of restitution
   - conservation of linear momentum
   - elastic collision
   - impulse
   - impulse-momentum theorem
   - inelastic collision
   - linear momentum
   - perfectly inelastic collision
2. Determine the impulse from a force vs. time curve.
3. Determine the average value of the force from the impulse and the time.
4. Determine the change in momentum from the impulse.
5. Apply the law of conservation of momentum in one dimension.
6. Determine the angle of motion and/or velocity of an object after an elastic collision from conditions before and the motion of the second object afterward.
7. State the conditions under which elastic collisions are ineffective in transferring kinetic energy between colliding bodies.
8. Determine the coefficient of restitution of an object from data obtained from a collision.
9. Determine the final speed of an object with a known coefficient of restitution from the change in motion of the object striking it.
10. Utilize the correct modification of the impulse-momentum theorem in situations where mass of an object is not constant.

Problems

8-1 Calculate (a) the average force required by a bat to impart an impulse of 8.00 N·s to a softball in 6.00×10^{-2} s and (b) the impulse imparted to the ball if the force calculated in part a lasts for 0.200 s.

8-2 Determine the magnitude of the linear momentum of (a) a baseball ($m = 0.142$ kg) traveling at 30.0 m/s and (b) an automobile ($m = 1.40 \times 10^3$ kg) traveling at 20.0 m/s.

8-3 Determine the linear momentum of a student ($m = 55.0$ kg) who is walking west at a speed of 2.40 m/s.

8-4 Figure 4-6 shows a punch landing on its intended target, an 80-kg boxer. If the impulse of the blow is 60.0 N·s, at what speed would the head initially recoil as it absorbs the blow (7 percent of body mass is in the head)?

8-5 An automobile of mass 1.20×10^3 kg (including load) is traveling at 30.0 m/s when the driver sees a sign saying "Speed Limit—15.0 m/s." (a) Calculate the change in linear momentum that the auto and its load must undergo in order to obey the speed limit. (b) If the force of the brakes is applied for 5.00 s to change speeds, what average force is being applied to the car?

8-6 An automobile of mass 800 kg is traveling at a speed of 25.0 m/s. Calculate (a) the impulse required to stop and (b) the magnitude of the braking force if the automobile is to stop in 5.00 s.

8-7 A rifle bullet has a muzzle velocity of 800 m/s. It takes 3.00×10^{-3} s to travel the length of the rifle barrel. The average force exerted on the rifle is 8.00×10^3 N. Calculate (a) the impulse and (b) the mass of the bullet.

8-8 A baseball weighing 1.39 N is thrown by a pitcher with a velocity of 40.0 m/s toward home plate. The batter makes solid contact and returns the ball at 55.0 m/s toward third base at an angle of 135° to the original direction. Calculate (a) the change in linear momentum and (b) the average force exerted on the ball if the contact lasted 5.00×10^{-3} s.

8-9 An 800-kg automobile with driver is traveling due east at a speed of 20.0 m/s when it comes to a sharp 60° turn to the right. While it is turning, the driver increases the speed so that when the road straightens, the auto is now headed at 60° south of east at a speed of 25.0 m/s. (a) Calculate the change in linear momentum. (b) If the turn took place in 8.00 s, determine the average force exerted on the tires by the road during the turn.

8-10 A linear air track such as that in Figure 8-5 reduces friction between the cart and track to such a small value as to make it negligible in almost all cases. The air track with carts can be used to approximate (with a very high degree of accuracy) frictionless motion, frictionless collision, and so on, in one dimension. Consider the situation shown in Figure 8-17a. A long rubber band (could be a number of them looped together) is fixed to two upright rods at right angles to and slightly above a linear air track. The rubber band is stretched to the left by pushing an air cart along the track from the right side of the rubber band. The cart of mass 0.300 kg is released from rest, and the magnitude of the force exerted by the rubber band vs. time is approximated by Figure 8-17b. Determine (a) the magnitude of the impulse, (b) the magnitude of the final linear momentum, and (c) the final speed of the cart.

8-11 Figure 8-18 shows the force exerted on a baseball which is about to leave the ballpark. (a) To a good approximation,

(a)

(b)

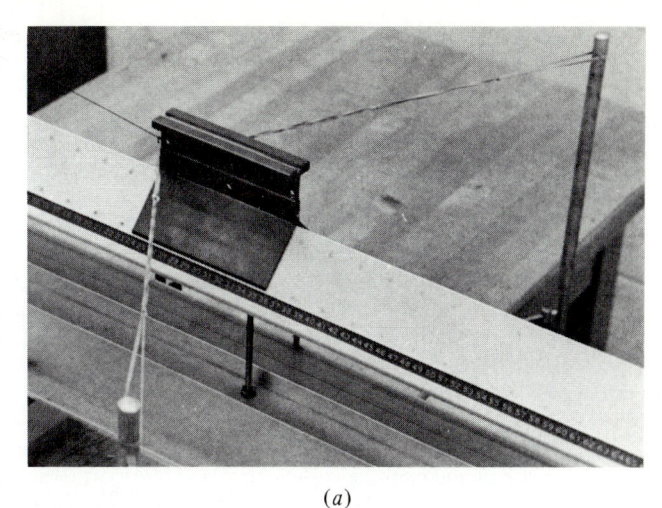

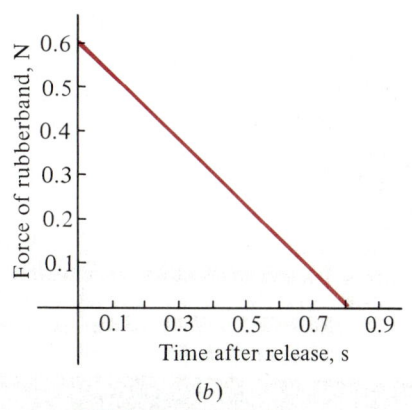

Figure 8-17 Problem 8-10.

determine the impulse experienced by the ball. (b) If the ball has a mass of 0.142 kg and the pitcher has given it a speed of 45.0 m/s, determine the speed at which the ball leaves the bat.

8-12 A general physics student, eager to get out of the room after class, slips and runs into the wall instead of the doorway,

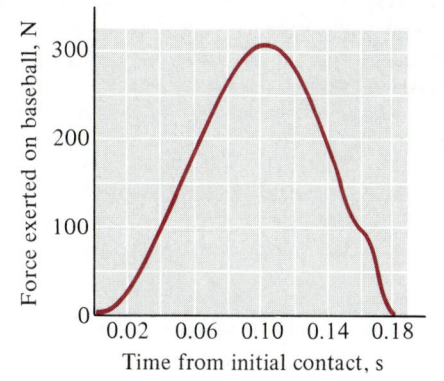

Figure 8-18 Problem 8-11.

175

confirming that experience is consistent with the impulse-momentum theorem. The wall exerts an average force on the student of 800 N for 0.500 s. (a) Calculate the impulse the wall exerts on the student. (b) If the student has a mass of 58 kg and is moving at 5.00 m/s just before the wall is contacted, with what speed does she bounce away from the wall?

8-13 Two balls, each of mass 0.250 kg, are thrown at the wall of the lecture hall by a professor. Each strikes the wall with a speed of 40.0 m/s. Ball A is a "superball" and rebounds with a speed of 38.0 m/s. Ball B is made of soft clay and sticks to the wall. Calculate (a) the change in momentum of each of the balls and (b) the impulsive force in each case if both collisions took place in an average time of 0.120 s.

8-14 A student laboratory helper of mass 80.0 kg runs at a speed of 6.00 m/s and jumps on a stationary laboratory cart. The cart has a mass of 100 kg. (a) Calculate the speed of the cart and student after he jumps aboard. (b) Determine the change in kinetic energy which took place during the collision.

8-15 A 6.00×10^4 kg railroad freight car moving with a speed of 3.00 m/s collides with a second freight car moving toward it at a speed of 5.00 m/s. They couple together and both stop. Calculate the mass of the second freight car.

8-16 A neutron of mass 1.67×10^{-27} kg is moving at a speed of 3.60×10^7 m/s when it collides elastically head-on with a helium nucleus of mass 6.63×10^{-27} kg. (a) Determine the velocity of each of the particles after the collision. (b) How much kinetic energy did the neutron lose during the collision?

8-17 A beryllium nucleus (mass 1.51×10^{-26} kg) is moving with a speed of 3.00×10^6 m/s when it collides elastically with a stationary deuterium nucleus of mass 3.34×10^{-27} kg. After the collision, both nuclei move off in the same direction. Determine the velocity of each nucleus after the collision.

8-18 A 0.250-kg billiard ball moving at 12.0 m/s makes an elastic head-on collision with a 7.00-kg bowling ball which is initially at rest. Calculate the final velocity of (a) the billiard ball and (b) the bowling ball.

8-19 In a billiards game the cue ball of mass 0.200 kg collides with the ten ball (same mass). The cue ball leaves the collision at an angle of 15° from its original direction, and the ten ball leaves at an angle of 75° on the other side of the original direction extended. (The collision is elastic.) If the speed of the cue ball just after the collision is 5.00 m/s, calculate the initial speed of the cue ball and the speed of the ten ball immediately after the collision.

8-20 A large firecracker at rest explodes into two pieces of masses 0.120 and 0.180 kg. The smaller piece moves off to the south at 15.0 m/s. Determine the velocity of the larger piece.

8-21 A truck with load of mass 3.50×10^3 kg is traveling east at 25.0 m/s, following an automobile of mass 1.50×10^3 kg which is traveling east at 15.0 m/s. After the inevitable collision, the vehicles are locked together. (a) Calculate the velocity of the vehicles after the collision. (b) Calculate the change in kinetic energy.

8-22 An experimental golf ball rolls off a table 1.00 m high. It rebounds to a maximum height of 0.870 m. Calculate (a) the speed of the ball just before it hits the floor (this gives you the relative velocity of approach of the ball and floor), (b) the speed of the ball just after the collision (this gives the relative velocity of separation of the ball and floor), and (c) the coefficient of restitution for this collision.

8-23 A ball, mass 0.215 kg, moving at a speed of 35.0 m/s strikes a wall at an angle of 50° to the wall. There is no change in the component of velocity parallel to the wall. The coefficient of restitution for the collision is 0.580. (a) Determine the angle the velocity of the ball makes with the wall after the collision. (b) Determine the magnitude and direction of the change in linear momentum of the ball.

8-24 A bullet of mass 4.00×10^{-2} kg is fired at a speed of 900 m/s into a wooden block of mass 2.00 kg which is at rest on a rough surface. After the bullet embeds itself in the block, the block moves 18.0 m along the surface before coming to rest. (a) Calculate the coefficient of friction between the block and the surface. (b) Calculate the kinetic energy change which took place during the collision.

8-25 A bullet of mass 0.0900 kg is fired into a wooden block of mass 0.550 kg. The bullet embeds itself in the block, and the two move off at a speed of 85.0 m/s after the collision. (a) Calculate the original speed of the bullet. (b) Determine the kinetic energy lost during the collision.

8-26 A wooden croquet ball, moving at a speed of 8.00 m/s, strikes a second ball of identical mass head-on. The second ball was initially at rest, and the coefficient of restitution for the collision is 0.7. Determine (a) the velocity of the two balls after the collision and (b) the change in kinetic energy during the collision. Take $m = 0.420$ kg.

8-27 An automobile of mass 1200 kg moving at a speed of 10.0 m/s is struck from behind by a second automobile of mass 1000 kg traveling at 18.0 m/s. Calculate the final velocity of each car if the coefficient of restitution for the collision is 0.5.

8-28 A bowling ball of mass 7.00 kg is dropped from a height of 1.40 m when a student is handing it to a friend. The ball strikes the floor and bounces to a height of 0.800 m. Determine (a) the change in linear momentum of the bowling ball due to the collision with the floor, (b) the impulse of that collision, (c) the average force if the two are in contact for 0.280 s, and (d) the coefficient of restitution for this collision.

8-29 During a football game a ball carrier weighing 900 N breaks loose and heads directly toward the goal line at a speed of 9.00 m/s. A defensive back weighing 800 N is moving directly across the field at a speed of 11.0 m/s when he tackles and holds on to the ball carrier. Calculate (a) their velocity immediately after impact and (b) the kinetic energy lost in the collision.

8-30 A car of mass 1.80×10^3 kg is traveling west at 30.0 m/s when it strikes a second car of mass 1.20×10^3 kg which was stationary before the collision. After the collision, the heavier car moves off in the direction 30° south of west at a speed of 20.0 m/s. *(a)* Determine the magnitude and direction of the lighter car's velocity after the collision. *(b)* Determine the change in kinetic energy which took place because of the collision.

8-31 A swimmer of mass 45.0 kg jumps horizontally backward at a speed of 8.00 m/s from a rowboat. If the boat and swimmer were initially at rest and the boat has a mass of 150 kg, how fast does the boat move forward?

8-32 A small rocket is to attain a final velocity of 900 m/s. The fuel and geometry are such that the hot ejected gases will have a speed of 400 m/s with respect to the rocket. Calculate the ratio of the initial mass (rocket structure plus fuel) to the final mass (payload).

8-33 Determine the magnitude of the acceleration of a small rocket at the instant when it has a mass (structure plus remaining unburned fuel) of 5.00 kg. Its engine is consuming fuel at the rate of 6.00×10^{-2} kg/s, and the hot gases have a speed of 800 m/s with respect to the rocket.

8-34 A 55-kg basketball player leaps vertically into the air to make a jump shot. In the air she thrusts the 0.800-kg ball away from her at a speed of 15.0 m/s. If the ball moves horizontally away from her, what is her horizontal speed after the shot?

8-35 Sand is poured into a pan on the top of a spring scale at the rate of 0.0250 kg/s from a height of 1.80 m. What will the scale read at exactly 2.00 s after the sand begins to strike the pan? (Ignore the weight of the pan.)

8-36 A movable cannon weighing 1.50×10^4 N (including its wheels) fires a projectile weighing 60.0 N at a velocity of 5.00×10^2 m/s at an angle of 30° above the horizontal. Calculate the recoil velocity of the cannon.

8-37 A sailboat has a mass of 200 kg and is carrying a passenger of mass 60.0 kg at a speed of 8.00 m/s. The passenger runs and jumps off the end of the boat with a speed of 10.0 m/s with respect to the boat. Calculate the velocity of the boat immediately after the (former) passenger jumps if *(a)* she jumps forward or *(b)* she jumps backward.

8-38 Sand is dropped vertically at the rate of 10.0 kg/s onto a conveyor belt moving at 1.50 m/s. Determine the power required to drive the belt.

8-39 A small child is pulling a wagon when a cloudburst occurs. The child was applying a force of 22.0 N horizontally in order to keep the wagon moving at a constant speed before the rain started. If the wagon is taking on water at the rate of 2.50 kg/min, what force will the child have to exert to maintain the original speed?

8-40 A gardener watering a bean patch holds the hose so that the stream leaves the nozzle horizontally with a speed of 16.0 m/s. If the hose delivers 4.00 kg water/min to the garden, what average horizontal force must the gardener exert on the hose?

8-41 A soldier holds a submachine gun which fires bullets of mass 0.100 kg at a muzzle velocity of 180 m/s. If the gun can fire 300 bullets/min, what average force must the soldier exert to hold the gun while firing at a maximum rate?

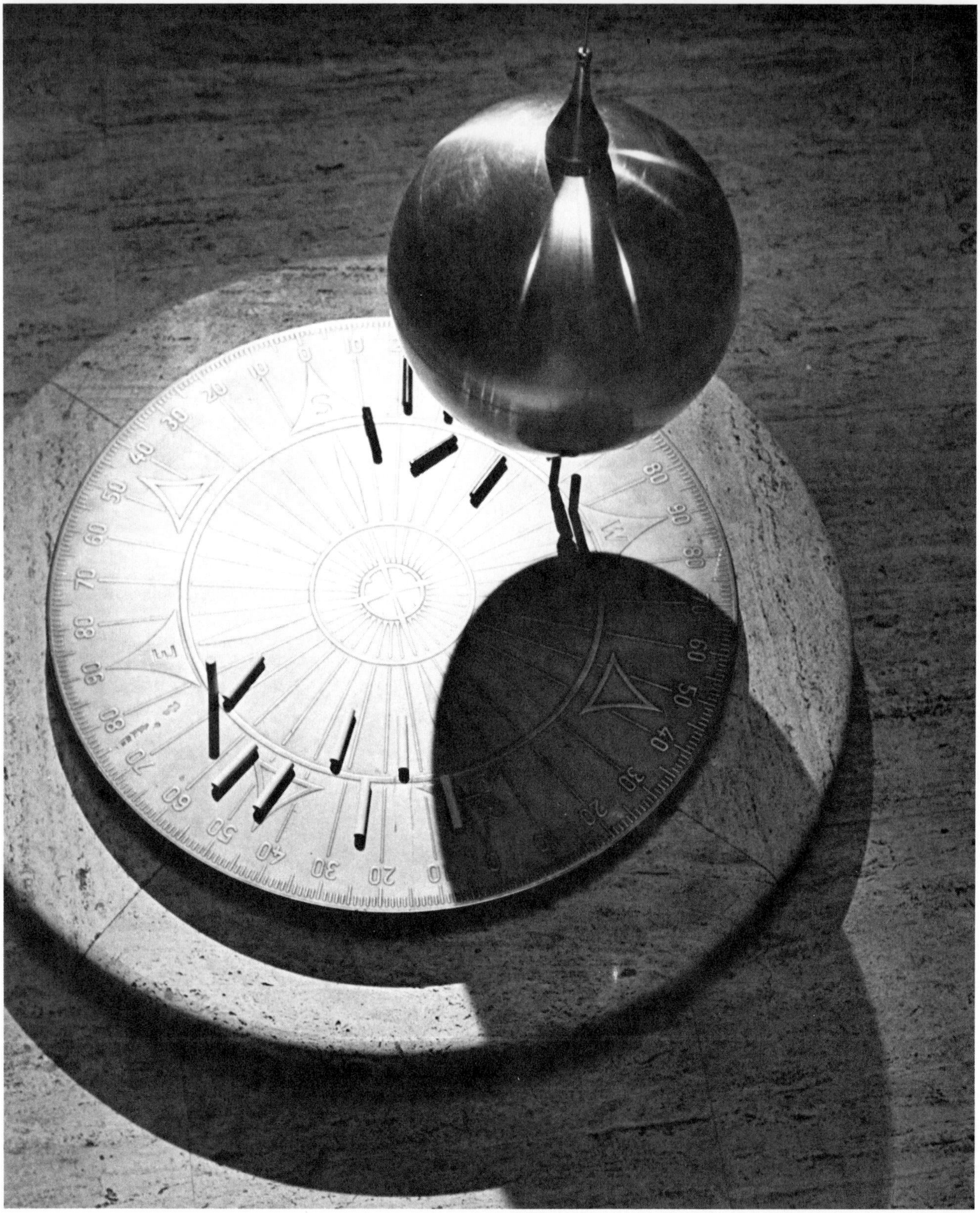

9 Rotational Dynamics

9-1 Introduction

We are now ready to study the causes of changes in rotational motion. The second condition of equilibrium (Chapter 6) states that a body (or system) will remain in rotational equilibrium unless it is acted on by an unbalanced torque. We used this condition to aid in calculating unknown forces acting on bodies in equilibrium; however, when examined, it can be seen that its implications are much broader. It states that no physical quantity except torques *can* cause changes in rotational motion; only torques can cause angular accelerations, just as Newton's first law states that only forces can cause translational accelerations. Torque can be considered an angular analog to force in every way. In fact, as we will see in this chapter, there are angular analogs to all of Newton's laws of motion.

With angular accelerations, we find that there is an additional complication beyond torques to consider, however, since the distribution of mass influences the rotation that results from a given torque (Figure 9-1).

Figure 9-1 The rotating playground platform teaches children the basics of rotational dynamics. They quickly learn that the greatest acceleration results when the running child pushes tangentially to the circumference and that, for a given push, the acceleration is greater with riding children near the center, not at the edge. (D. Riban)

9-2 Torques and Moments of Inertia

Consider a very small body of mass m attached to the end of a string and accelerating in a circle of radius r as a result of an applied force F acting tangent to the circle (Figure 9-2). If Newton's second law is applied to this situation, we have

$$F = ma$$

where the vector notation is dropped and we recognize that the acceleration is tangent to the circle because the applied force is tangent.[1] It was seen in Chapter 5 that the tangential acceleration and the angular acceleration are related by

$$a = \alpha r$$

Furthermore, we may write an equation for the torque that F is imparting to the system about the point P as

$$\tau = rF$$

since F is perpendicular to r. Substituting for F yields

$$\tau = rma$$

Finally, substituting for a gives

$$\tau = rm\alpha r$$

or

$$\tau = mr^2\alpha \qquad (9\text{-}1)$$

Thus, we see that the dynamics of this situation may be described either translationally with Newton's second law or by using Equation 9-1 as a perfectly reasonable alternative.

Figure 9-2 The small mass m is considered a point mass if its dimensions are small compared to its distance from the axis of rotation. A force **F** acts along the tangent line to the circle and imparts a tangential acceleration to m.

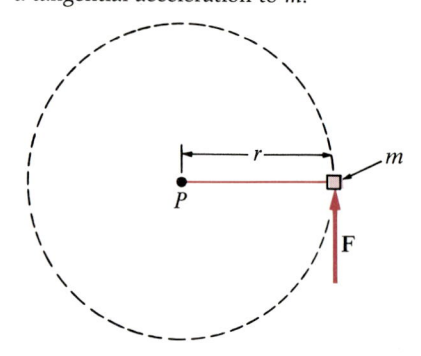

---

[1] Of course the mass has a centripetal acceleration as well, but that is due to the force being applied by the string and is, in any case, perpendicular to the tangential component.

9-2 Torques and Moments of Inertia

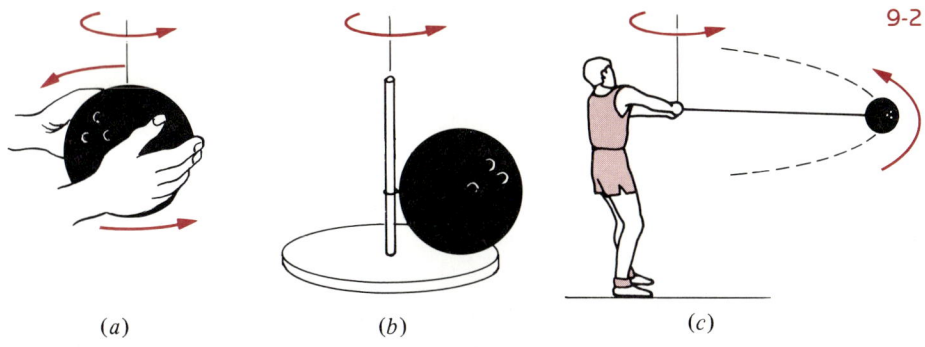

Figure 9-3 (a) A spun bowling ball acquires an angular velocity ω about the vertical axis through its center of mass. (b) The ball secured at its circumference to the vertical shaft of a turntable is given the same angular velocity ω about the shaft in the same period of time. (c) The ball at the end of a rope is whirled about in a circle, acquiring the same ω in the same Δt.

Consider Equation 9-1 further; τ is the angular analog to force, and α is the angular analog to translational acceleration. The quantity mr^2 does not describe any facet of the speed or acceleration, nor has it anything to do with forces and torques. It depends only on the geometry and mass and can be interpreted as *the system's resistance to angular accelerations*. For a point mass m at a distance r away from an axis of rotation, we call mr^2 the **moment of inertia** I.[1] In that case, Equation 9-1 becomes

$$\tau = I\alpha \tag{9-2}$$

The SI unit for moment of inertia is kilogram-meter squared ($kg \cdot m^2$).

Mass cannot be the only consideration in resistance to angular acceleration, and to picture this consider the following three situations. First, imagine a 7.00-kg bowling ball resting on a frictionless desktop. It is not too difficult to rest our hands on each side of the ball, then pull them rapidly in opposite directions to exert a torque on the ball and give it spin, as in Figure 9-3a. Say that the ball acquires a spin such that a particular finger hole passes the north side of the desk once every second. It would have an angular speed of 1 r/s, or 2π rad/s.

Second, imagine we have a small round eyelet screwed into the ball. We place the ball so that this eyelet slides down over the tall central shaft of a phonograph turntable on the desk, as in Figure 9-3b. We hold the top of the central shaft and push the ball, now resting on the turntable, so that it acquires a rotation about this new vertical axis one radius away from its center. Assume a torque is exerted for the same time interval as we used to set the ball spinning. Will a greater or smaller torque have to be exerted to achieve the same final angular velocity? Actually, one rotation per second is 60 r/min, or about twice the normal turning speed of most turntables. You should be able to picture a bowling ball off-center on your turntable rotating at this speed, destroying the bearings of the turntable, bending the central shaft, and possibly pulling the whole assembly off the desk.

---

[1] The term **rotational inertia**, used in some books, is perhaps more descriptive in that the rotational inertia measures resistance to rotational accelerations in the same way that translational inertia, mass, measures resistance to translational accelerations. However, we will use the traditional term *moment of inertia*.

Finally, fasten a 2.00-m-long rope to the eyelet in the bowling ball. Now whip it around in a circle so that the ball makes one complete revolution each second (Figure 9-3c). Would it require a greater torque during the same time interval to achieve this angular speed than in starting the ball spinning on your desktop? Actually, most people could not manage the forces necessary for this task. The ball would be moving at a linear speed of $v = \omega r = (2\pi$ rad/s$)(2$ m$) = 12.6$ m/s. This is about 28.2 mi/h, and at that speed a flying bowling ball is a potentially lethal weapon. The centripetal force the puller would have to exert on the rope to keep the ball moving in a circle is

$$F = \frac{mv^2}{r} = \frac{(7.00 \text{ kg})(12.6 \text{ m/s})^2}{2.00 \text{ m}} = 556 \text{ N}$$

This is well over 100 lb of tension on the rope.

What do these three mental experiments tell us? The mass of the bowling ball did not change, but in each case a greater torque was required to achieve the same final angular velocity. Clearly, the *position of the mass relative to the axis of rotation makes a difference*. Resistance to angular acceleration depends not only on m but also on the geometrical distribution of the mass, the r^2 part of Equation 9-1 for a point mass.

Now as you might suspect, a point mass is a very special and rather simple case, and forces do not always act tangentially to a circle of rotation. So how do we calculate the moment of inertia for a more complex case?

The body shown in Figure 9-4 can be considered as made up of many mass particles, or point masses, labeled $m_1, m_2, \ldots, m_i, \ldots$. Each particle experiences forces and hence possible torques about the axis through P, the axis of rotation. Each particle is kept at a fixed distance from P during the rotation of the rigid body. Let us label these distances as $r_1, r_2, \ldots, r_i, \ldots$, respectively. An arbitrary particle m_i is shown in the figure along with the net perpendicular component of force acting on it. We may apply Newton's second law to each particle. For example, the ith particle obeys

$$F_i = m_i a_i \qquad (9\text{-}3)$$

The angular acceleration of all points on the rigid body will be the same, and we may write

$$F_i = m_i r_i \alpha \qquad (9\text{-}4)$$

where we have recognized that $a_i = r_i \alpha$. Finally, if we multiply both sides by r_i, we have

$$r_i F_i = m_i r_i^2 \alpha \qquad (9\text{-}5)$$

The formula $I = mr^2$ for a point mass may be used *only* when nearly all the mass is at an effective distance of r which is large compared to the size of the object with mass. Normally, blind substitution of the distance between the axis of rotation and the center of mass simply will not work. Consider what such a substitution would imply for the spinning bowling ball on the tabletop.

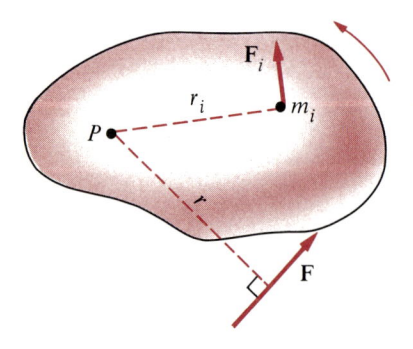

Figure 9-4 The rigid, irregularly shaped body rotates about an axis through P under the torque supplied by the external force **F**. An arbitrary particle m_i is shown along with the perpendicular component F_i of the net force acting on it.

9 Rotational Dynamics

The left side is simply the torque on the ith particle about an axis through P. Thus we have

$$\tau_i = m_i r_i^2 \alpha \tag{9-6}$$

The net torque acting on the body is simply the sum of the torques on the individual particles. Thus

$$\tau = \tau_1 + \tau_2 + \cdots + \tau_i + \cdots$$

Substituting Equation 9-6 yields

$$\tau = m_1 r_1^2 \alpha + m_2 r_2^2 \alpha + \cdots + m_i r_i^2 \alpha + \cdots$$
$$= (m_1 r_1^2 + m_2 r_2^2 + \cdots + m_i r_i^2 + \cdots)\alpha$$

or

$$\tau = \left(\sum_{i=1}^{n} m_i r_i^2\right)\alpha \tag{9-7}$$

Again we obtain a quantity which depends not on the motion but only on the geometry. The quantity in parentheses in Equation 9-7 is the moment of inertia for the extended body, and we write

$$I = \sum_{i=1}^{n} m_i r_i^2 \tag{9-8}$$

The magnitude of the net torque acting on the body of Figure 9-4 is the magnitude of **F** times its **moment arm** r.[1] Thus

$$\tau = rF = I\alpha \tag{9-9}$$

To apply Equation 9-9, we need to know the moment of inertia for the body in question. Note that the moment of inertia depends on the axis about which the body is rotating as well as the body's shape and distribution of mass. Even when these properties — axis, shape, distribution of mass — are known, however, it is very difficult and sometimes practically impossible to use Equation 9-8 to obtain the moment of inertia. An expression for I can often be found by using integral calculus, where the summation becomes an integration. In any case, I can be found experimentally by measuring the angular acceleration acquired by a body to which a known torque has been applied.

Example 9-1 The net force acting on the body of Figure 9-4 has a magnitude of 12.0 N, and its moment arm about P is 0.600 m. The measured angular acceleration that the body acquires is 8.00 rad/s². Calculate (a) the applied torque and (b) the moment of inertia of the body.

---

[1] The force, hence the torque, acting on an individual particle is really being exerted by the other mass particles as well as (perhaps) some portion of the external forces. Forces that the mass particles of the body exert on each other are called internal forces to distinguish them from the external force or forces acting on the body. Fortunately, the internal forces all occur in pairs, equal in magnitude and opposite in direction, according to Newton's third law. Each action-reaction pair has the same moment arm and gives no net torque on the body. Thus the net torque on the body is due only to the external forces. This allows us to apply Equation 9-9 without knowing the nature and/or magnitude of the internal forces.

Solution *(a)* We apply the first part of Equation 9-9 to obtain the torque:

$$\tau = rF = (0.600 \text{ m})(12.0 \text{ N}) = 7.20 \text{ N} \cdot \text{m}$$

(b) We now apply the second part of Equation 9-9 to obtain I:

$$\tau = I\alpha$$

$$I = \frac{\tau}{\alpha} = \frac{7.20 \text{ N} \cdot \text{m}}{8.00 \text{ rad/s}^2} = 0.900 \text{ kg} \cdot \text{m}^2$$

Example 9-2 A physics student in a pizza parlor decides to conduct an experiment to determine the moment of inertia of a classmate sitting on a swivel stool. Using a spring scale and a digital stopwatch, the student makes careful measurements. With the subject's feet extended, a force of 20.0 N is applied at the ankle such that the moment arm of the force about the axis of rotation is 0.850 m. When the force is applied for 2.00 s, the subject and the top of the stool acquire an angular speed of 6.00 rad/s. Calculate *(a)* the applied torque, *(b)* the angular acceleration, and *(c)* the moment of inertia of the classmate and the stool top.

Solution Part *(a)* requires the first part of Equation 9-9:

$$\tau = rF = (0.850 \text{ m})(20.0 \text{ N}) = 17.0 \text{ N} \cdot \text{m}$$

(b) We may use the definition of angular acceleration:

$$\alpha = \frac{\Delta \omega}{\Delta t} = \frac{6.00 \text{ rad/s} - 0}{2.00 \text{ s}} = 3.00 \text{ rad/s}^2$$

(c) The second part of Equation 9-9 will yield the moment of inertia:

$$\tau = I\alpha$$

$$I = \frac{\tau}{\alpha} = \frac{17.0 \text{ N} \cdot \text{m}}{3.00 \text{ rad/s}^2} = 5.67 \text{ kg} \cdot \text{m}^2$$

The moments of inertia of a number of homogeneous objects have been calculated, using calculus, in terms of the mass and some convenient characteristic dimension of the body. Some of these are listed in Table 9-1 and illustrated in Figure 9-5. Note that the position of the axis must be given.

TABLE 9-1 Moments of Inertia

Body	Axis through Center of Mass	Characteristic Dimension	Moment of Inertia
Hoop	Cylindrical axis	Radius r	mr^2
Hoop	Diameter	Radius r	$\frac{1}{2}mr^2$
Solid sphere	Diameter	Radius r	$\frac{2}{5}mr^2$
Solid cylinder	Cylindrical axis	Radius r	$\frac{1}{2}mr^2$
Solid cylinder	Diameter	Radius r, length l	$\frac{1}{4}mr^2 + \frac{1}{12}ml^2$
Thin rod	Perpendicular to rod	Length l	$\frac{1}{12}ml^2$
Circular disk	Diameter	Radius r	$\frac{1}{4}mr^2$
Rectangular plate	Perpendicular to plate	Length a, width b	$\frac{1}{12}m(a^2 + b^2)$

9-2 Torques and Moments of Inertia

Hoop (horizontal) Hoop (vertical) Solid sphere

Solid cylinder (vertical) Solid cylinder (horizontal) Thin rod

Circular disk Rectangular plate

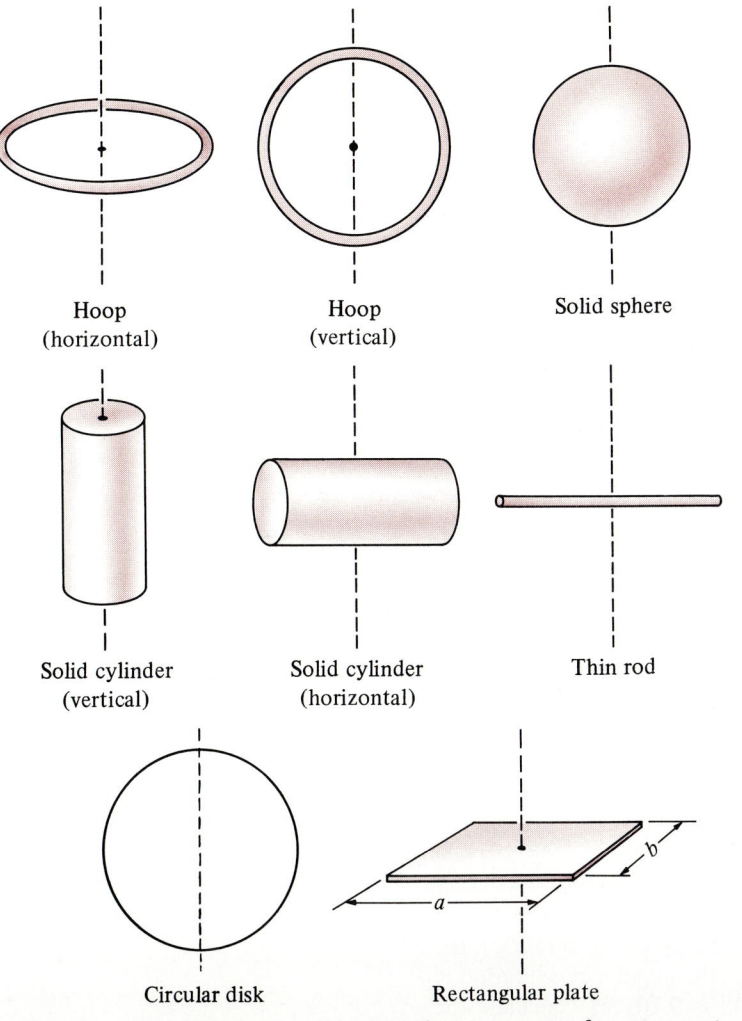

Figure 9-5 The regularly shaped bodies whose moments of inertia are given in Table 9-1.

Example 9-3 Determine the value of the moment of inertia of (a) a solid sphere of mass 6.00 kg and radius 0.100 m about a diameter and (b) a solid cylinder of mass 40.0 kg and radius 0.0500 m about its cylindrical axis.

Solution We need only apply the expressions from Table 9-1. Thus:

(a) $I_{sphere} = \frac{2}{5}mr^2 = \frac{2}{5}(6.00 \text{ kg})(0.100 \text{ m})^2 = 0.0240 \text{ kg} \cdot \text{m}^2$

(b) $I_{cylinder} = \frac{1}{2}mr^2 = \frac{1}{2}(40.0 \text{ kg})(0.0500 \text{ m})^2 = 0.0500 \text{ kg} \cdot \text{m}^2$

While the expressions given in Table 9-1 are not necessarily obvious, they are at least plausible. For instance, choosing the cylindrical axis through the center of a hoop ensures that all mass particles are at the same distance r from the axis. For that situation, Equation 9-8 becomes

185

$$I = r^2 \sum_{i=1}^{n} m_i$$

since r is the same for all particles. Further, the summation of all the mass particles is simply the total mass of the hoop. Thus we can see why $I = mr^2$ is the expression for the moment of inertia of the hoop about its cylindrical axis.

Each expression in Table 9-1 is given about an axis through the center of mass of the body. This is convenient because we often encounter physical situations where the body is rotating about an axis through its center of mass. Frequently, however, the axis of rotation is parallel to the axis of symmetry but some distance away from that axis. The expression for the moment of inertia about such an axis could be very difficult to calculate and even harder to find in a table, because so many variations are possible. Fortunately, a simple relationship exists, the **parallel-axis theorem,** which helps us obtain an expression for the moment of inertia about any axis parallel to the axis through the center of mass:

> The moment of inertia of an object about any axis is equal to the moment of inertia about a parallel axis through the center of mass plus the product of the mass of the object and the square of the distance between the two axes.

In symbols, this can be written

$$I = I_0 + md^2 \qquad (9\text{-}10)$$

where I = moment of inertia about the arbitrary axis
I_0 = moment of inertia about a parallel axis through the center of mass
m = mass of the object
d = distance between the two axes

Example 9-4 Previously we considered a 7.00-kg bowling ball to which a torque was applied, giving it a final angular speed of 1.00 r/s = 2π rad/s. If the required torque is applied in a 1.00-s time interval, the angular acceleration of the ball is 2π rad/s². The radius of a bowling ball is 0.109 m. Calculate the moment of inertia of the ball (a) about an axis through its center, (b) about an axis tangent to the ball, and (c) about an axis 2.00 m away from the ball. (d) Determine the torque required to impart an acceleration of 2π rad/s² in each of the three cases.

Solution (a) We use an expression for the moment of inertia about an axis through the center of mass from Table 9-1.

$$I_0 = \tfrac{2}{5} mr^2 = \tfrac{2}{5}(7.00 \text{ kg})(0.109 \text{ m})^2 = 0.0333 \text{ kg} \cdot \text{m}^2$$

(b) To complete parts (b) and (c), we simply apply the parallel-axis theorem, Equation 9-10:

$$I = I_0 + md^2 = 0.0333 \text{ kg} \cdot \text{m}^2 + (7.00 \text{ kg})(0.109 \text{ m})^2$$

(since the distance from the tangent line to the center is one radius) and

$$I = 0.116 \text{ kg} \cdot \text{m}^2$$

(c) Here, $d = 2.00 \text{ m} + 0.109 \text{ m} = 2.11 \text{ m}$. Thus

$$I = I_0 + md^2 = 0.0333 \text{ kg} \cdot \text{m}^2 + (7.00 \text{ kg})(2.11 \text{ m})^2$$
$$= 31.2 \text{ kg} \cdot \text{m}^2$$

(d) Using $\tau = I\alpha$, we may calculate the torques as

$$\tau_a = (0.0333 \text{ kg} \cdot \text{m}^2)(2\pi \text{ rad/s}^2) = 0.209 \text{ N} \cdot \text{m}$$

$$\tau_b = (0.116 \text{ kg} \cdot \text{m}^2)(2\pi \text{ rad/s}^2) = 0.729 \text{ N} \cdot \text{m}$$

$$\tau_c = (31.2 \text{ kg} \cdot \text{m}^2)(2\pi \text{ rad/s}^2) = 196 \text{ N} \cdot \text{m}$$

The dramatic differences between the torques required for the same angular acceleration certainly justify the conclusion that the position of the mass relative to the axis makes a difference!

9-3 Work and Rotational Kinetic Energy

In Chapter 7 we saw that a torque applied to a body which moved through an angle $\Delta\theta$ about a fixed axis did an amount of **rotational work**

$$W = \tau \, \Delta\theta \tag{7-6}$$

We may substitute for τ in this equation by using Equation 9-9 and obtain

$$W = I\alpha \, \Delta\theta \tag{9-11}$$

Now if ω_i and ω_f are the initial and final angular speeds, we know from Equation 5-13 that

$$2\alpha \, \Delta\theta = \omega_f^2 - \omega_i^2$$

Solving this for $\alpha \, \Delta\theta$ and substituting into Equation 9-11 yields

$$W = I \frac{\omega_f^2 - \omega_i^2}{2}$$

or

$$W = \tfrac{1}{2}I\omega_f^2 - \tfrac{1}{2}I\omega_i^2 \tag{9-12}$$

Equation 9-12 states that when a torque does work on a rotating body to change its angular speed, that work shows up as a change in the quantity $\tfrac{1}{2}I\omega^2$. This is directly analogous to Equation 7-10, which states that when a force does work on a translating body to change its translational speed, that work shows up as a change in the quantity $\tfrac{1}{2}mv^2$. Both changed quantities are defined as kinetic energy. We distinguish between these two forms of kinetic energy by using the subscripts T for translation and R for rotation. Thus Equation 9-12 becomes

$$W = \tau \, \Delta\theta = \tfrac{1}{2}I\omega_f^2 - \tfrac{1}{2}I\omega_i^2$$

or

$$\tau \, \Delta\theta = \Delta KE_R \tag{9-13}$$

since $KE_R = \tfrac{1}{2}I\omega^2$. This is the rotational form of the *work-energy theorem*.

Example 9-5 A 35.0-kg uniform sphere of radius 0.200 m is confined to rotate about an axis tangent to the spherical surface. A constant net torque of magnitude 20.0 N·m is applied about the axis while the sphere increases its angular speed from 10.0 to 20.0 rad/s. Calculate (a) the moment of inertia of the sphere about the given axis, (b) the change in rotational kinetic energy of the sphere, and (c) the angle through which the sphere turned.

Solution (a) To obtain the moment of inertia, we must use an expression from Table 9-1 and then apply the parallel-axis theorem, Equation 9-10. The distance from the center of mass to the parallel axis is exactly one radius. The parallel-axis theorem applied to the situation is then

$$I = \tfrac{2}{5}mr^2 + mr^2$$
$$= \tfrac{7}{5}mr^2$$
$$= \tfrac{7}{5}(35.0 \text{ kg})(0.200 \text{ m})^2 = 1.96 \text{ kg} \cdot \text{m}^2$$

(b) Part b may be solved by applying the definition of rotational kinetic energy:

$$\Delta KE_R = \tfrac{1}{2}I\omega_f^2 - \tfrac{1}{2}I\omega_i^2$$
$$= \tfrac{1}{2}(1.96 \text{ kg} \cdot \text{m}^2)(20.0 \text{ rad/s})^2 - \tfrac{1}{2}(1.96 \text{ kg} \cdot \text{m}^2)(10.0 \text{ rad/s})^2$$
$$= 294 \text{ J}$$

(c) Finally, the rotational form of the work-energy theorem yields the angle we seek:

$$\tau \, \Delta\theta = \Delta KE_R$$

$$\Delta\theta = \frac{\Delta KE_R}{\tau} = \frac{294 \text{ J}}{20.0 \text{ N} \cdot \text{m}} = 14.7 \text{ rad}$$

Just as we saw with the translational case, the work done by a net applied torque in the rotational analog may be either positive or negative. The net applied torque does positive work on the body if the body's rotational kinetic energy increases. The applied torque does negative work if there is a decrease in rotational kinetic energy. Of course the net torque could be the result of two or more torques. These components of the net torque may tend to cause angular acceleration in opposite directions. Thus one torque may be doing positive work while another is doing negative work for the same rotation of a body.

Example 9-6 The pulley in Figure 9-6 is a circular disk of mass 6.00 kg and radius 0.150 m. The tensions in the cord on each side of the pulley are $T_1 =$

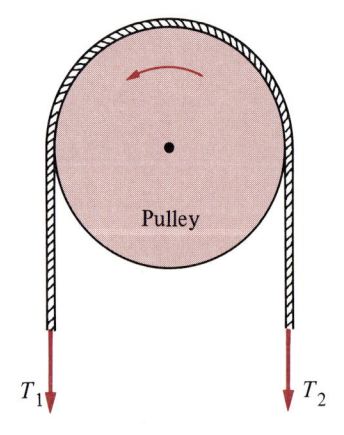

Figure 9-6 Example 9-6.

18.0 N and $T_2 = 15.0$ N. Determine (a) the torques due to each tension, (b) the work done by each torque as the pulley turns through an angle of 20.0 rad, and (c) the final angular speed of the pulley if it starts from rest.

Solution The moment arm for each tension force is the radius of the pulley. Since the two tend to accelerate the pulley in opposite directions, we must choose one direction as positive and the other as negative. We expect the larger force T_1 to prevail and the pulley to acquire an angular acceleration according to the tendency of T_1. Thus, we choose counterclockwise as positive and clockwise as negative.

(a) Part a requires that we apply the definition of torque:

$$\tau_1 = rT_1 \qquad\qquad \tau_2 = -rT_2$$
$$= +(0.150 \text{ m})(18.0 \text{ N}) \qquad = -(0.150 \text{ m})(15.0 \text{ N})$$
$$= +2.70 \text{ N} \cdot \text{m} \qquad\qquad = -2.25 \text{ N} \cdot \text{m}$$

(b) Parts b and c may be solved by using the rotational analogs of work and the work-energy theorem. From $W = \tau \Delta\theta$, we have

$$W_1 = (+2.70 \text{ N} \cdot \text{m})(20.0 \text{ rad}) \qquad W_2 = (-2.25 \text{ N} \cdot \text{m})(20.0 \text{ rad})$$
$$= +54.0 \text{ J} \qquad\qquad\qquad = -45.0 \text{ J}$$

(c) Applying the rotational form of the work-energy theorem yields

$$W = W_1 + W_2 = \Delta KE_R$$
$$+54.0 \text{ J} - 45.0 \text{ J} = \tfrac{1}{2}I\omega_f^2 - \tfrac{1}{2}I\omega_i^2$$

Now $\omega_i = 0$ (it starts from rest), so

$$\tfrac{1}{2}I\omega_f^2 = 9.00 \text{ J}$$

Since I is the moment of inertia of a disk about its center of mass (in this application the disk acts like a very thin cylinder), we have

$$I = I_0 = \tfrac{1}{2}mr^2 = \tfrac{1}{2}(6.00 \text{ kg})(0.150 \text{ m})^2$$

thus

$$\tfrac{1}{2}(0.0675 \text{ kg} \cdot \text{m}^2)\omega_f^2 = 9.00 \text{ J}$$
$$\omega_f = 16.3 \text{ rad/s}$$

This problem could also be solved for the angular acceleration by using Equation 9-9 and the final angular speed determined from kinematic considerations (see Problem 9-24).

9-4 Translational and Rotational Energy Combined

Although only pure rotations about a fixed axis have been considered to this point, many everyday situations include a body or bodies which are both rotating and translating. A common example of combined motion is that of a wheel rolling along, which may provide a great deal of insight, so we will examine it in detail.

A body which is rolling but not slipping may be considered from two

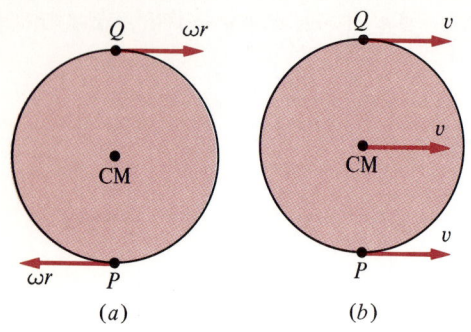

Figure 9-7 The motion of a uniform body rolling to the right with a translational velocity of its center of mass of **v** may be viewed as a combination of two simultaneous motions: (a) a clockwise rotation about the center of mass of angular velocity ω and (b) a rightward translation of each point on the body at velocity **v**. The motion of each point on the body, for example, P or Q, is the vector sum of these separate motions.

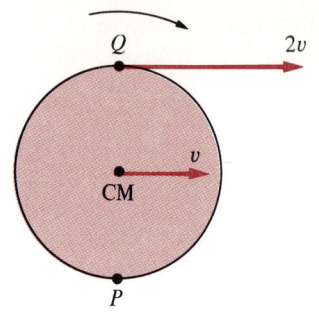

Figure 9-8 An alternative treatment of a rolling body recognizes that the point of contact with the surface P is at rest during the instant of contact.

points of view. The first view assumes the body is rotating about its center of mass while the whole body including the center of mass is translating, as presented in Figure 9-7. The relationship between the translational velocity v and the angular speed ω is $v = \omega r$. The reason for this may be clearer if we consider the second view shown in Figure 9-8. This view notes that since the body is not slipping, the point or line where the body touches the rolling surface is instantaneously stationary while the rest of the body rotates about the instantaneously "fixed" axis through that point. Here we see that point P is fixed, and the line which joins P, Q, and the center of mass has an instantaneous angular speed, say ω. We can then say that the instantaneous translational velocity of the center of mass has a magnitude equal to ωr. Further, the instantaneous translational velocity of point Q is ω times $2r$, or $2\omega r$. If we add the velocities of the two parts of Figure 9-7, we obtain a duplicate of Figure 9-8. Thus either view can be used, and we pick the one which is more convenient for a given problem.

It is not very difficult to show that both views yield the same value of kinetic energy. For the first view, we write the kinetic energy as the sum of the rotational component and the translational component. Thus

$$KE = KE_R + KE_T \tag{9-14}$$

$$KE = \tfrac{1}{2}I_0\omega^2 + \tfrac{1}{2}mv^2 \tag{9-15}$$

For the second view, the kinetic energy is purely rotational, and we have

$$KE = \tfrac{1}{2}I\omega^2 \tag{9-16}$$

Here, however, I is the moment of inertia about an axis through P and is related to I_0 through the parallel-axis theorem. Since the distance between an axis through the center of mass and point P is the radius of the body, we write

$$I = I_0 + mr^2$$

Now if we substitute into Equation 9-16, we have

A rolling object may be treated either as a case of simultaneous translation and rotation about the center of mass *or* as a case of pure rotation about the instantaneous point of support at the surface, but these views cannot be combined. Study the cases until you will not try to mix the treatments together, which cannot work, but is often tried and indicates that the student's physical picture of the situation is confused.

$$KE = \tfrac{1}{2}(I_0 + mr^2)\omega^2$$
$$= \tfrac{1}{2}I_0\omega^2 + \tfrac{1}{2}mr^2\omega^2$$

which is identical with Equation 9-15 when we use $v = r\omega$.

Example 9-7 A billiard ball of mass m and radius r rolls without slipping down a plane inclined at an angle θ with the horizontal. It starts from rest and moves downward a vertical height h from its original position. Apply the principle of conservation of mechanical energy to obtain an expression for the final translational speed of the center of mass of the sphere.

Solution The conservation of mechanical energy can be written as

$$\Delta KE + \Delta PE = 0$$

or

$$\Delta KE_T + \Delta KE_R + \Delta PE = 0$$

We apply this between the top and bottom of the incline by using Equation 9-15:

$$\tfrac{1}{2}mv_f^2 - \tfrac{1}{2}mv_i^2 + \tfrac{1}{2}I_0\omega_f^2 - \tfrac{1}{2}I_0\omega_i^2 + mgh_2 - mgh_1 = 0$$

Since the sphere started from rest, $v_i = \omega_i = 0$. Further, $h_2 - h_1 = -h$. Thus we may simplify as

$$\tfrac{1}{2}mv_f^2 + \tfrac{1}{2}I_0\omega_f^2 - mgh = 0$$

Now substituting for the moment of inertia of a sphere about the center of mass and recognizing that $v_f = \omega_f r$, we obtain

$$\tfrac{1}{2}mv_f^2 + \tfrac{1}{2}(\tfrac{2}{5}mr^2)\omega_f^2 = mgh$$
$$\tfrac{1}{2}mv_f^2 + \tfrac{1}{5}mv_f^2 = mgh$$
$$\tfrac{7}{10}v_f^2 = gh$$
$$v_f = \sqrt{\tfrac{10}{7}gh}$$

If the sphere does not roll down the plane but slides on a surface of negligible friction, the final speed of the center of mass would be $v_f = \sqrt{2gh}$. This is greater than the speed of the rolling sphere. The reason is that part of the kinetic energy acquired by the rolling sphere must go to rotation, and only part goes to translation. When the sphere does not roll, all the acquired kinetic energy is translational.

In the previous example, we stated that the sphere rolls without slipping. It is important to recognize that a frictional force *must* act on the sphere in order that it have an angular acceleration about its center of mass. The necessary torque about the center of mass can only be the result of the frictional force acting parallel to the plane. The only other two forces acting on the sphere (the gravitational force and the normal force) are acting through the center of mass and consequently do not give rise to torques about that point (Figure 9-9).

The fact that friction acts on the sphere introduces a question. If a force of friction is being exerted on the sphere and the sphere moves down the plane, why is the principle of conservation of mechanical energy applied? The answer

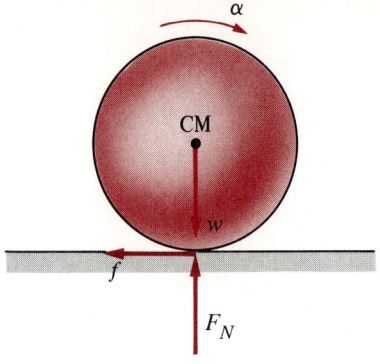

Figure 9-9 The rolling sphere has an angular acceleration about its center of mass. Of the three forces acting on the sphere, only the frictional force f can give rise to the necessary torque. Both the other forces act through the center of mass. On a frictionless incline, the sphere would slide down, not roll.

is that the point of application of the force of friction is fixed. That is, the instantaneous center of rotation does not move as the force of friction acts upon it. Hence the force of friction does no work. Remember that a force does work if it acts on a body and the point of application has a component of motion along (or opposed to) the direction of the force. Thus application of conservation of mechanical energy is appropriate provided the sphere does not slip.

9-5 Angular Momentum

Imagine the playground merry-go-round again. If it is rotating at a speed of ω_i, what happens as a child standing by the side jumps on? As another child jumps off? In each case the angular speed changes—first decreasing, then increasing. We have seen in translational motion that, in the absence of outside forces, momentum is conserved. Similarly, rotating systems tend to preserve their rotations in the absence of outside torques. To this point we have considered only those situations for which the mass and geometry of a rotating system were not changing. When this is true, I is constant by definition, and hence the rotational motion of such a system changes according to

$$\tau = I\alpha$$

where τ = resultant external torque about an axis through the center of mass
I = net moment of inertia about that axis
α = resultant angular acceleration about that axis

In the more general case when the moment of inertia changes with time, we must write

192

TABLE 9-2 Collection of Translational and Rotational Dynamic Analog Quantities and Equations

Translational			Rotational
Mass	m	I	Moment of inertia
Force	$F (= ma)$	$\tau (= I\alpha)$	Torque
Work	$F \Delta x$	$\tau \Delta \theta$	Work
Power	Fv	$\tau\omega$	Power
Linear momentum	$p = mv$	$L = I\omega$	Angular momentum
Kinetic energy	$\tfrac{1}{2}mv^2$	$\tfrac{1}{2}I\omega^2$	Kinetic energy
Work-energy	$W = \Delta KE_T$	$W = \Delta KE_R$	Work-energy
Impulse-momentum	$F \Delta t = \Delta(mv)$	$\tau \Delta t = \Delta(I\omega)$	Impulse-momentum

$$\tau = \frac{\Delta(I\omega)}{\Delta t} \tag{9-17}$$

This reduces to $\tau = I(\Delta\omega/\Delta t)$ for constant I and thus is entirely consistent with $\tau = I\alpha$.

Examining Equation 9-17, we see that if a torque τ acts for a time Δt, it imparts an **angular impulse**

$$\tau \Delta t = \Delta(I\omega)$$
$$\tau \Delta t = (I\omega)_f - (I\omega)_i \tag{9-18}$$

In other words, Equation 9-18 states that the quantity $\tau \Delta t$ gives rise to a change in the quantity $I\omega$ from its initial to its final value. Defining $I\omega$ as the **angular momentum** (symbol L) makes Equation 9-18 the rotational analog of the impulse-momentum theorem.

Note that we have now accumulated a set of angular equations which have exact analogs in the equations of translational dynamics, presented in Table 9-2, simply substituting I for m, τ for F, L for p, α for a, and ω for v. Thus, there are no new equations to learn, only new ideas.

Example 9-8 A child pushes with a force of 100 N tangential to the rim of a playground merry-go-round for 3.00 s. The radius of the merry-go-round is 1.50 m, and its moment of inertia about its axis of spin is 114 kg·m². The initial angular speed of the merry-go-round is 0.500 rad/s. Calculate (a) the applied torque and (b) the final angular speed.

Solution (a) First, we apply the definition of torque:

$$\tau = rF = (1.50 \text{ m})(100 \text{ N}) = 150 \text{ N} \cdot \text{m}$$

(b) Now a straightforward application of Equation 9-18 yields

$$\tau \Delta t = (I\omega)_f - (I\omega)_i$$

or

$$\omega_f = \frac{\tau \Delta t + I\omega_i}{I}$$

$$= \frac{(150 \text{ N} \cdot \text{m})(3.00 \text{ s}) + (114 \text{ kg} \cdot \text{m}^2)(0.500 \text{ rad/s})}{114 \text{ kg} \cdot \text{m}^2}$$

$$= 4.45 \text{ rad/s}$$

Suppose the angular impulse acting on a body or system is zero (or negligible). Then Equation 9-18 indicates that the change in angular momentum is zero also. This is equivalent to stating that the angular momentum is constant or conserved. This can be restated as the **law of conservation of angular momentum:**

> In a system in which the net external torque can be considered to be negligible, the angular momentum of the system remains constant.

The system under consideration may include only one or several bodies. If there is only one body and its moment of inertia is constant, we see that its angular velocity must be constant as well. If the moment of inertia of a single body is changed while the external torque which acts is negligible, then the angular speed of the body must also change. Whenever there are several bodies involved, the law of conservation of angular momentum requires only that the resultant of all the components of angular momentum add to the same number before and after some interaction which imparts a negligible external torque. We use the uppercase L as the symbol for angular momentum, and the law of conservation of angular momentum may be stated in symbols as

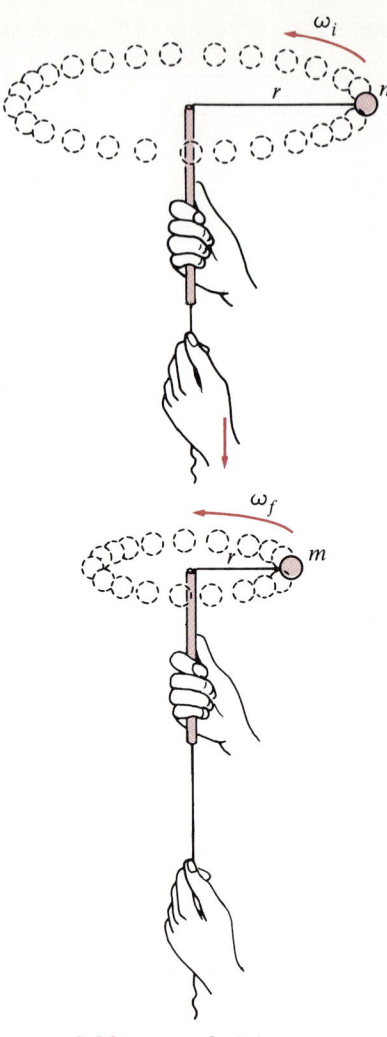

$$\text{If } \tau \, \Delta t = 0, \text{ then } \Delta L = 0 \qquad (9\text{-}19)$$

Figure 9-10 Example 9-9.

Example 9-9 Figure 9-10 shows a small 0.100-kg mass attached to a string which passes through a vertical tube. The mass is initially rotated in a horizontal plane in a circle of radius 0.250 m at an angular speed of 5.00 rad/s. The string is pulled downward so that the radius of the circle is decreased to 0.150 m. Calculate the new angular speed.

Solution Since the force pulling on the mass is radially toward the center of the circle, it imparts no torque about an axis through the center. Thus we apply the conservation of angular momentum:

$$\Delta L = 0$$

becomes

$$I_f \omega_f - I_i \omega_i = 0$$

For a small mass m at a distance r from the center of mass, we have $I = mr^2$. Thus

$$mr_f^2 \omega_f = mr_i^2 \omega_i$$

$$\omega_f = \frac{(0.250 \text{ m})^2 (5.00 \text{ rad/s})}{(0.150 \text{ m})^2}$$

$$= 13.8 \text{ rad/s}$$

That the final speed is not dependent on what mass is used is perhaps not wholly unexpected, since the mass does not change.

There are many uses of the conservation of angular momentum in games and sports. A diver doing a somersault tucks into a ball to reduce the moment of inertia and hence increase angular speed. Ice skaters and ballet dancers, by

Figure 9-11 The ice skater begins her twirl with leg and arms extended and turns at an initial rate of ω_0. Decreasing the distance of the mass of the limbs from her center of rotation will lower her moment of inertia. Since $L (= I\omega)$ is conserved, lowering I requires an increase in ω. Thus, pulling the limbs tightly to the body changes the slow rotational motion to a fast spin about the axis of rotation. *(UPI/ Bettman Newsphotos)*

extending or retracting their arms, may control their speeds of rotation (Figure 9-11). A mechanical example requiring consideration of conservation of angular momentum is the helicopter. If a helicopter had only a single propeller (or rotor), when it increased its rotational speed the main body of the helicopter would acquire rotational motion in a direction opposite to that of the rotor in order to conserve angular momentum. To avoid this difficulty, helicopters are built either with two rotors that turn in opposite directions or with a separate auxiliary tail rotor mounted in a vertical plane to provide an opposing torque keeping the body from rotating.

9-6 Directional Property of Torque and Angular Momentum

In Section 6-2 the right-hand rule (RHR) to specify the direction of torques was introduced. The direction of the torque vector is along the axis of change of rotation (or tendency of change of rotation). This specification of direction can be extended to nearly all rotational quantities. Angular velocities, angular accelerations, angular momentum all have a direction which we specify as being along the axis of rotation and can be found by the RHR.

Many of our previous equations are more generally applicable when they are written by using vector notation. For instance, Equation 9-17 may become

$$\tau = \frac{\Delta(I\omega)}{\Delta t} = \frac{\Delta \mathbf{L}}{\Delta t} \qquad (9\text{-}20)$$

When the axis of rotation is fixed, we know we can then take care of the directions of rotational quantities with plus and minus signs. However, in a problem for which the axis is not fixed but may rotate about a line perpendicular to itself, the directional property of these quantities must be accounted for by using vectors.

A common example of such a situation occurs with a moving motorcycle. If you sit on the motionless cycle with your feet off the ground, the cycle will topple over. Yet, that same cycle in motion is incredibly stable, to the point that it may be steered by an experienced cyclist merely by shifting the body's weight to either side. Understanding this requires treating simultaneous rotations about different (and perpendicular) axes. Consider the wheel of Figure 9-12a, to which we apply the torque shown, about its natural axis of rotation. It will acquire an angular velocity, hence an angular momentum about this axis. By the RHR, the torque, final angular velocity, and final angular momentum **L** will all be to the right and into the page along the axis of rotation shown. Should another torque be applied to the wheel along the same axis, by the RHR it will

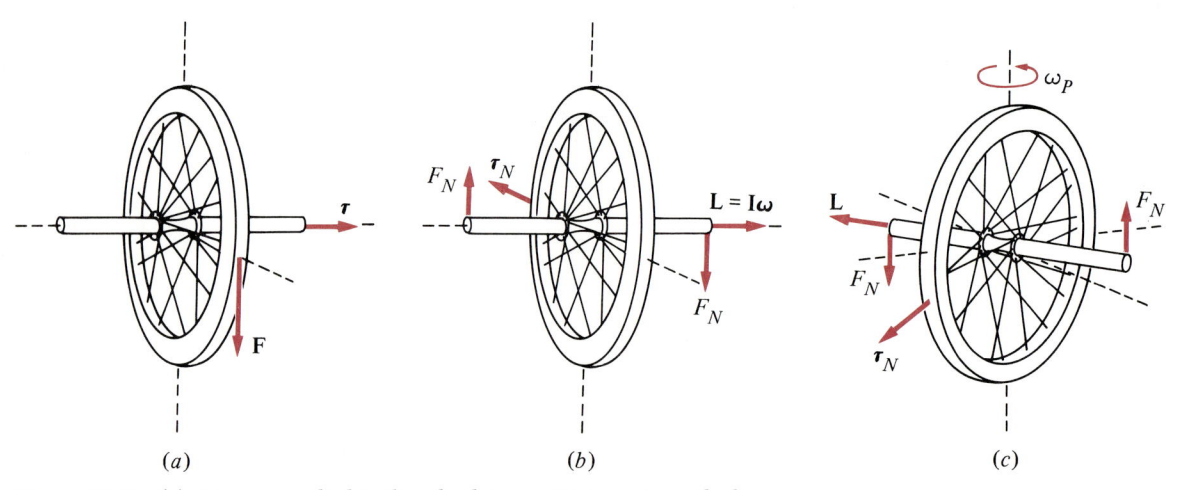

Figure 9-12 (a) A torque applied to the wheel causes it to acquire a clockwise rotation. (b) A new torque τ_N applied to the wheel at right angles to the direction of the angular momentum would appear to cause the wheel to tumble about a horizontal line. (c) Instead, however, the wheel rotates about a vertical line through the center of mass with the spin angular momentum vector following, or appearing to chase, the applied torque vector.

9-6 Directional Property of Torque and Angular Momentum

Figure 9-13 The motorcyclist has shifted his weight off center to the right of the picture. This sets up a clockwise torque (vector toward the motorcycle's rear) as viewed from the front. It might seem as though this torque should cause cyclist and rider to fall to their left. This is exactly what it would do if there were no original angular momentum. Since the front wheel is spinning and has an angular momentum vector to the right, however, the torque causes it to twist about a vertical axis, steering the cycle in the direction of the lean. (*AIP Niels Bohr Library*)

only change the magnitude of **L**, either increasing it or decreasing it. However, should a torque be applied along a different axis, this is not the case. Consider, for simplicity, the new torque τ_N shown in Figure 9-12b acting perpendicular to the natural axis of rotation of the wheel. This torque would tend to rotate the top of the wheel away from us and into the page and the bottom of the wheel out of the page and toward us. By the RHR, τ_N acts from right to left along a horizontal line through the center of mass of the wheel. Now, Equation 9-20 asserts that the direction of change of **L** is in the direction of this new applied torque. Figure 9-12c shows **L** with τ_N applied to induce a change. Since τ_N is perpendicular to **L**, it cannot change its magnitude, only its direction. Thus the action of τ_N is not to topple the wheel sideways out of a vertical rotation plane but to rotate the wheel about a vertical axis more into the plane of the page, or counterclockwise as viewed from directly above.

We now have the reason for the stability of a motorcycle in motion. If the rider shifts her weight to the right in Figure 9-13, her weight has a moment arm to produce a torque similar to τ_N in the previous discussion. This will neither speed up nor slow down the rotation of the wheel, but it will tend to rotate its front edge in the direction of the shift of weight. Thus, rather than topple in the direction of the shift, the cycle tends to steer in that direction to re-right the system. The same situation occurs if the rider shifts her weight to the left. Within wide limits, a moving motorcycle is self-righting due to the vector nature of rotations.[1] Note that the direction of the applied torque of both Figures 9-12 and 9-13 is continually changing as the wheel turns about the vertical axis.

---

[1] Generally, this effect is not adequate for a bicycle with a much smaller moment of inertia for its spinning wheel. For most bicycles, the axis of rotation of the front wheel is moved ahead of the steering axis by a forward-curved fork holding the wheel. This provides the increased stability required. The front edge of the wheel "falls" in the direction of a lean, steering the bicycle in that direction.

In each case the applied torque is always perpendicular to **L**. This changing direction of the orientation of the wheel due to an applied torque is called **precession.** The angular speed of precession ω_p is shown in Figure 9-12c. An applied torque which has a component perpendicular to a body's angular momentum will produce a change in the *direction* of the angular momentum. The angular speed of precession can be shown to be (see Problem 9-43)

$$\omega_p = \frac{\tau}{L} \tag{9-21}$$

The phenomenon of precession shows up in spinning tops, gyroscopes, and even the earth (Problem 9-44). Their axes of rotation change direction when torques are applied not parallel to L.

Minimum Learning Objectives

After studying this chapter, you should be able to:
1. Define:
    - angular impulse
    - angular momentum
    - conservation of angular momentum
    - moment arm
    - moment of inertia
    - parallel-axis theorem
    - precession
    - rotational kinetic energy
    - rotational work
2. Understand why the distribution of the mass of a rotating system as well as its magnitude influences the angular acceleration it will experience for a given torque.
3. Calculate the moment of inertia of a set of point masses about a given axis.
4. Apply the parallel-axis theorem to obtain the moment of inertia of an object of standard geometry (having I_0 listed in Table 9-1) about an axis of rotation other than through its CM.
5. Calculate the angular acceleration that results from a given torque acting on an object with a known moment of inertia, and vice versa.
6. Calculate the change in the angular motion of a specified system from the work done, and vice versa.
7. Perform calculations involving work and kinetic energy on systems undergoing simultaneous changes in both rotational and translational motions.
8. Apply the law of conservation of angular momentum to a system free of external torques to obtain the final motion of the system after internal changes occur, given the nature of the system and its initial motions.
9. Understand directional considerations involving torques and angular momentum and the phenomenon of precession.

Problems

9-1 Calculate the angular acceleration of a wheel with a moment of inertia equal to 4.00 kg·m² about its axis which is being acted on by a torque of 84.0 N·m about the axis.

9-2 (a) Determine an expression for the moment of inertia of the system shown in Figure 9-14 about an axis through the center of mass in terms of the mass of each body and the distance r between them. (b) This could represent a carbon monoxide molecule (CO) with mass of carbon = 2.00×10^{-26} kg, mass of oxygen = 2.67×10^{-26} kg, and $r = 1.13 \times 10^{-10}$ m. Using these values, determine the magnitude of the angular speed of a CO molecule when its rotational kinetic energy is 9.66×10^{-21} J.

9-3 Calculate the moment of inertia of a rectangular plate 0.200 m by 0.500 m about a perpendicular axis through its center. Take the mass of the plate to be 18.0 kg.

9-4 Determine the moment of inertia of a circular disk of mass 0.600 kg about a diameter if the disk's radius is 0.150 m.

9-5 Determine the moment of inertia of a solid cylinder of mass 3.00 kg and radius 0.0500 m about an axis parallel to the cylinder's axis and 0.600 m away from the axis.

9-6 Wire cable on a large spool is wrapped at an average radius of 0.350 m. The moment of inertia of the spool with the cable is 5.60 kg·m². What force applied to the cable will give the spool an angular acceleration of 9.00 rad/s²?

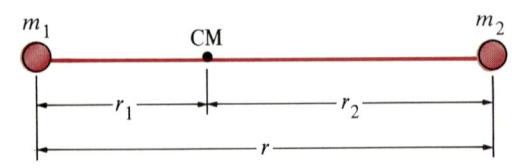

Figure 9-14 Problem 9-2.

9-7 Friction exerts a retarding torque of 45.0 N·m on a certain body about its natural axis of rotation. The body's moment of inertia about that axis is 180 kg·m². Calculate the magnitude of an applied torque necessary to give the body an angular speed of 40.0 rad/s in 60.0 s.

9-8 A force is applied for 3.00 s to a certain body, giving it a net external torque of 10.0 N·m about an axis through its center of mass. The body has a moment of inertia about its center of mass of 1.50 kg·m² and an initial angular speed of 12.0 rad/s about that axis. Calculate the body's angular speed at the end of the 3-s interval.

9-9 A playground merry-go-round is rotating with an angular speed of 6.00 rad/s. If its moment of inertia about its axis of spin is 114 kg·m², (a) what torque must be applied to bring it to rest in 8.00 s? (b) Through what angle will the merry-go-round turn while coming to rest?

9-10 A child applies a force of 15.0 N to a door at the knob (0.700 m from hinge) and perpendicular to the door. The door has a moment of inertia about an axis through the hinges of 1.20 kg·m². Determine (a) the angular acceleration of the door and (b) the time required to slam the door shut 90° from a resting position if the force is sustained.

9-11 A flywheel which may be assumed to have all its mass of 150 kg concentrated at a distance of 1.20 m from the axis of rotation is subjected to a constant net torque for 35.0 s. Calculate (a) the moment of inertia, (b) the torque required to produce an angular acceleration of 0.200 rad/s², and (c) the angle through which the flywheel turns while the torque acts if the wheel starts from rest.

9-12 The human eye may be treated as a sphere of mass 7.25×10^{-3} kg and radius 1.15×10^{-2} m which rotates in a spherical socket about a diameter. You may experiment to determine that you can shift your gaze 90° from one side to the other without moving your head in a time interval of 0.250 s. Assuming the eye is accelerated (angular) uniformly through 45° and then decelerated uniformly (same magnitude acceleration) to the new direction of your gaze, (a) calculate the magnitude of the torque exerted on the eye by the rectus muscles and (b) determine the net force exerted by the rectus muscles, assuming the forces are exerted tangential to the eye. (c) Recalling that the maximum force exerted by a muscle is 7.00×10^5 N/m² of cross-sectional area, what is the minimum cross section of each muscle (compare this to the cross section of a standard no. 5 rubber band, which is approximately 1.00×10^{-6} m²)? (d) Calculate the maximum angular speed of the eye in this maneuver. (e) Calculate the maximum centripetal acceleration of the cornea in units of g.

9-13 A small mass of 0.600 kg is attached to a long string that is wrapped around a cylinder which is free to rotate about its axis of symmetry. The cylinder has a mass of 5.00 kg and a radius of 0.100 m. The small mass is allowed to descend under the action of gravity. Determine the time required for the small mass to descend 0.800 m from rest.

9-14 Show that the linear acceleration of the center of mass of a sphere rolling down a plane inclined at an angle θ to the horizontal is given by the expression $a = \frac{5}{7} g \sin \theta$.

9-15 Repeat Problem 9-14 for a cylinder and show that $a = \frac{2}{3} g \sin \theta$.

9-16 Determine how long it takes a bowling ball 0.218 m in diameter starting from rest to roll 15.0 m down a plane surface inclined at 25° to the horizontal.

9-17 A grinding wheel in a workshop is a uniform disk of radius 0.100 m and rotates at 1.20×10^3 r/min. A tool being sharpened is pressed against the wheel with a force of 8.00 N. If the coefficient of kinetic friction between the wheel and the tool is 0.650, calculate (a) the torque exerted on the grinding wheel and (b) the power dissipated at this angular speed.

9-18 Determine an expression for the moment of inertia of a solid cylinder of mass m and radius r about an axis parallel to the cylindrical axis but (a) along the edge of the cylinder and (b) two-thirds times the radius from the cylindrical axis.

9-19 A 7.00-kg bowling ball of radius 0.109 m starts from rest and rolls down a plane surface inclined at an angle of 30° with the horizontal. After the ball has rolled 12.0 m, what will be its (a) angular speed and (b) total kinetic energy?

9-20 A centrifuge with a moment of inertia of 3.50 kg·m² is to be brought up to 1,000,000 r/min in 500 s. Calculate the power which must be delivered by the electric motor driving the centrifuge.

9-21 Friction between m_1 and the table shown in Figure 9-15 may be neglected. When released from rest, m_2 descends 0.675 m in the first 0.750 s. Calculate the tensions in the cord and the moment of inertia of the pulley. Use $m_1 = 1.60$ kg, $m_2 = 0.800$ kg, and radius of pulley = 0.120 m.

9-22 The pulley in Figure 9-16 has a moment of inertia of 3.00 kg·m² and a radius of 5.00×10^{-2} m. Determine the tensions in the cord and the acceleration of the system. Use $m_1 = 3.00$ kg, $m_2 = 8.00$ kg, $\theta_1 = 30°$, and $\theta_2 = 53.1°$. Neglect friction between the blocks and the sliding surfaces.

9-23 In a softball game, a batter, starting from rest, swings a 0.800-m-long bat of mass 0.600 kg in a half circle in 0.400 s. If the bat can be considered to be a uniform cylinder of radius 3.00×10^{-2} m rotating about one end, determine (a) the torque the batter must exert and (b) the final kinetic energy of the bat.

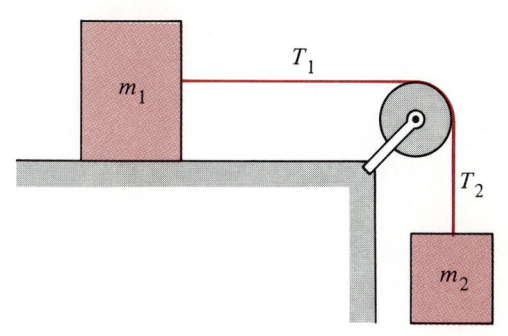

Figure 9-15 Problem 9-21.

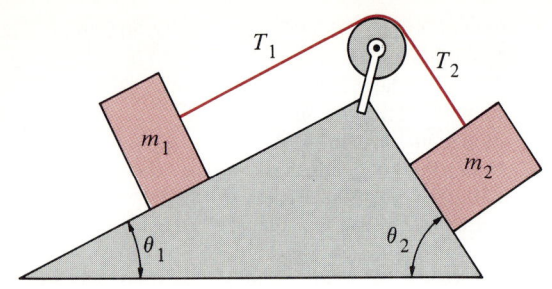

Figure 9-16 Problem 9-22.

9-24 Apply Equation 9-9 to the pulley of Figure 9-6 to obtain the angular acceleration. Then use the appropriate rotational kinematic equation to confirm the answer for the final angular speed obtained in Example 9-6.

9-25 Calculate the rotational kinetic energy of the bowling ball considered in Example 9-4 if it has an angular speed of 2π rad/s and is spinning about (a) an axis through its center, (b) an axis tangent to the ball, and (c) an axis 2.00 m away from the ball. (d) Determine the power delivered in accelerating the ball from rest to 2π rad/s in 1 s for each of the three axes.

9-26 Calculate the (a) rotational and (b) translational kinetic energy of a ball of moment of inertia about a diameter 5.00×10^{-2} kg·m² if the ball is rolling without slipping at 60.0 rad/s. The radius of the ball is 0.120 m.

9-27 An orange and a can of frozen orange juice both have radii of 3.50×10^{-2} m and masses of 0.340 kg. They are released from rest at one end of a wooden plank tilted at an angle of 12° with the horizontal. Both roll without slipping. Calculate (a) the moment of inertia of each body about an axis through its center of mass, (b) the linear speed of each after it has traveled 3.00 m along the plank, and (c) the time for each body to travel the 3.00 m.

9-28 A 0.0568-kg tennis ball is fixed to the end of a 2.20-m string attached to the ceiling. The string is pulled taut, as in Figure 9-17, so that it makes an angle of 50° with the vertical. The

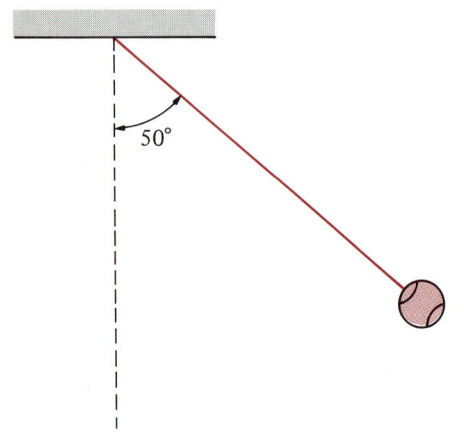

Figure 9-17 Problem 9-28.

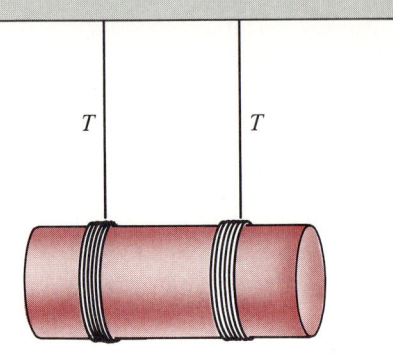

Figure 9-18 Problem 9-30.

ball is released from rest. Calculate the angular speed of the string as the ball passes its lowest point.

9-29 An empty steel drum with a radius of 0.400 m and mass of 60.0 kg is rolling along a level floor so that its center of mass is moving at 2.20 m/s. Treat the drum as a hollow cylinder and calculate its (a) angular speed, (b) rotational kinetic energy, and (c) translational kinetic energy. (d) Determine the work which must be done to stop the drum.

9-30 The solid cylinder in Figure 9-18 is supported by two lines wrapped tightly around the cylinder and attached firmly to a horizontal rod. The cylinder has a mass of 4.00 kg and a radius of 6.00×10^{-2} m. Under the influence of gravity, the cylinder begins to unwind from the position shown. Calculate the tension in the two cords (assume the same in each) and the linear acceleration of the center of mass.

9-31 A hoop and a solid cylinder both of mass 2.00 kg and radius 0.100 m roll without slipping on the roller coaster, like the course shown in Figure 9-19. They start from rest at point A and roll to point B, 3.00 m below point A. Calculate (a) the angular and linear speed of the hoop when it reaches point B, and (b) the angular and linear speed of the cylinder when it reaches point B. (c) Compare the translational, rotational, and total kinetic energies of the two bodies at point B.

9-32 Calculate the angular speed of a wheel with a moment of inertia equal to 9.00×10^{-2} kg·m² when its angular momentum is 3.78 kg·m²/s.

9-33 An athlete runs around a circular track 400 m in circumference in 55.0 s. Calculate the athlete's angular momentum at a constant speed.

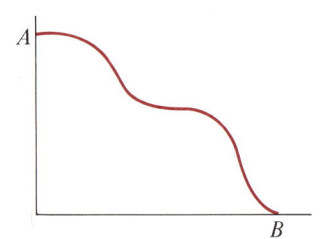

Figure 9-19 Problem 9-31.

9-34 A small eyelet is screwed into one end of a 0.450-kg meterstick. The eyelet is threaded over a horizontal rod such that the meterstick can swing freely in a vertical plane about one end. The stick is held in a horizontal position and released from rest. Calculate (a) the torque due to gravity on the stick about the rod as an axis, (b) the angular acceleration of the stick the instant it is released, (c) the angular momentum of the meterstick at the instant it reaches a vertical orientation, and (d) the kinetic energy of the stick when it reaches a vertical orientation.

9-35 A large solid cylinder of mass 30.0 kg and radius 0.250 m rotates about its cylindrical axis with an angular speed of 5.00 rad/s. For this situation, calculate (a) the moment of inertia, (b) the magnitude of the angular momentum, (c) the kinetic energy, and (d) the magnitude of a constant tangential force applied to a rope wrapped around the cylinder which will stop the cylinder in exactly one revolution.

9-36 A phonograph record can be considered to be a uniform cylinder of mass 8.50×10^{-2} kg and radius 0.150 m. Calculate (a) the angular momentum and kinetic energy of a phonograph record rotating at $33\frac{1}{3}$ r/min and (b) the angular momentum and kinetic energy of a 5.00×10^{-4} kg fly resting on the rotating record at 0.100 m from the axis of rotation.

9-37 The child in Example 9-8 gets the merry-go-round up to its final speed of 4.45 rad/s. Then, from a standing position, the 40.0-kg child jumps onto the moving merry-go-round and holds to its rim at a radius of 1.50 m. Calculate the new angular speed.

9-38 If the moon increased its distance from the earth by 10 percent without changing its spin (1 rotation per revolution) and remained in a stable orbit at 1.10 times its present orbital distance, (a) how would its angular momentum change? (Be careful! Remember that its angular speed may change also.) (b) If this change in angular momentum is obtained by a transfer to or from the earth's intrinsic angular momentum (rotation), how long would the day on earth be?

9-39 Actual measurements of marine fossils which deposit a layer of new shell each day, of varying thickness by season, indicate that the earth's rotation has changed (see Problem 9-38). In the Cretaceous Period 150 million years ago, there were about 400 days/year (396 ± 10). Assuming the earth to be a homogeneous sphere, what percentage of its current distance from the earth was the moon in the Cretaceous Period? Can you think of a mechanism for the transfer of angular momentum from earth to moon?

9-40 The small ball in Figure 9-10 has a mass of 0.100 kg and is rotating with an angular speed of 18.0 rad/s at a radius of 0.250 m. The string is pulled down so that the radius is shortened to 0.150 m. Calculate (a) the new angular speed, (b) the change in kinetic energy, and (c) the work done in shortening the radius.

9-41 A student stands on a demonstration turntable rotating at 4.00 rad/s, holding two 1.50-kg dumbbells at arm's length 0.800 m from the axis of rotation. The moment of inertia of the student about the axis of rotation is 2.00 kg·m². Calculate (a) the angular momentum of the system and (b) the kinetic energy of the system. If the student now pulls in the dumbbells to a new radius of 0.100 m, what is the new (c) angular speed and (d) kinetic energy?

9-42 The stars are divided on the basis of spectral types into the series O, B, A, F, G, K, M, R, N, and S which range from very hot, bright stars to cool, dim ones. Early spectral types (O, B, A), which are single stars, almost invariably are in very rapid rotation, measured in hours, in spite of their usually greater diameters. Later types (F, G, K, . . .) generally have slow rotations measured in days. In general, early spectral types condense from gas and dust clouds very rapidly, ignite thermonuclear reactions promptly, and exhaust their fuel early compared to later spectral-type stars. Why should such stars possess large amounts of angular momentum compared to later spectral types? Consider the solar system: (a) The sun rotates once on its axis in an average of 25 days (mass = 1.99×10^{30} kg; radius = 6.96×10^8 m). What is the angular momentum of the sun's rotation? (b) The mass of Jupiter (1.90×10^{27} kg) is greater than the rest of the planets put together. Jupiter is 7.78×10^{11} m from the sun and orbits it in 3.74×10^8 s. What is the angular momentum of Jupiter in its orbit? (c) Assuming the other planets together have two-thirds the angular momentum of Jupiter (about correct), what percentage of the total angular momentum of the solar system is possessed by the sun's rotation? (d) If the planets of the solar system were eliminated and their angular momentum was transferred to the sun, what would be the period of its rotation?

9-43 Beginning with Equation 9-20, derive Equation 9-21. (Hint: Draw a vector diagram and first show that the magnitude of $\Delta \mathbf{L}$ is equal to the product $L \times \Delta\theta_p$, where $\Delta\theta_p$ is the angle through which $\mathbf{L}$ turns in a time Δt.)

9-44 The north-south axis about which the earth spins does not remain aligned in a fixed direction in space. It precesses about an axis perpendicular to its orbital plane just about once every 26,000 years, so no one lives long enough to notice the effect of the precession (e.g., changing North Stars from time to time). The reason for this precession is that the earth is not a perfect sphere but bulges at the equator because of its spin. Because of the earth's orientation with respect to its orbital plane, the sun and moon exert a net unbalanced torque on the equatorial bulge, causing the precession. The moment of inertia of the earth about its axis of spin is approximately 9.74×10^{37} kg·m², and it rotates once per day. (a) Calculate the earth's angular momentum about its axis of spin. (b) Determine the angular speed of precession of the earth's axis, and calculate the magnitude of the unbalanced torque which causes the precession.

9-45 The bicycle wheel of Figure 9-12 has a moment of inertia about its symmetry axis of 1.25 kg·m² and a mass of 9.00 kg. It is suspended by a rope attached to the end of the axle

0.250 m from its center of mass. Its angular speed about its symmetry axis is 20.0 rad/s. Calculate (a) its angular momentum, (b) the torque applied to the wheel, and (c) the angular speed of precession.

9-46 A competition for high school students called the Physics Olympics was originated at the authors' institution in 1975 as part of our school's centennial. It has since become quite popular nationally. One of the original events was the slow bicycle race. Riders cover a 20.0-m course with a bicycle, maintaining forward motion, and the longest elapsed time wins. Our longest time for an unmodified 10-speed bicycle was 480 s. Assume 1.5 pedal turns per full turn of the wheel, a 35.0-cm radius wheel, and a 17.0-cm-long pedal radius. (a) What was the average angular velocity of the rear wheel during the run? (b) What was the average angular velocity of the pedals? (c) What was the average linear speed of the cyclist's feet turning the pedals? (d) At the bottom of the pedal stroke, what was the average net velocity of the cyclist's foot to a stationary observer? (e) What was the frequency of the pedaling during the run?

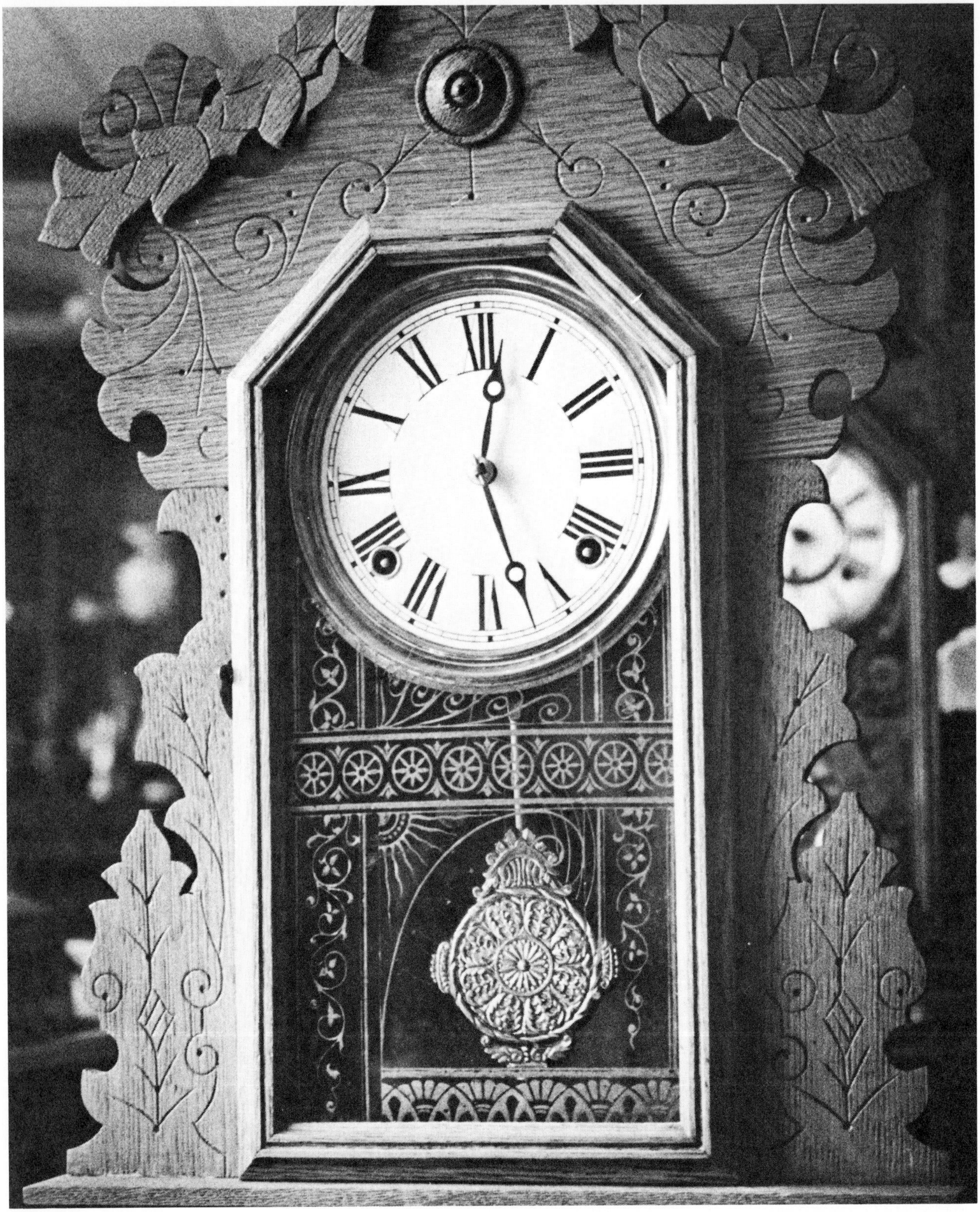

10 Simple Harmonic Motion

10-1 Introduction

Many motions repeat over and over again in regular time intervals. We have already encountered the orbital motion of planets and satellites. In addition, a pendulum, a plucked guitar string, the heart, and the balance wheel of a watch all repeat a motion. Even the atoms of a solid execute repeating motions about an equilibrium position.

Any motion which repeats exactly is characterized by the time required for one complete motion, the **period** τ, and these motions may all be grouped under the generic name of **periodic motion.** One complete performance of a motion requiring exactly one period of time is called a **cycle.** As we have seen, circular and planetary motions are two examples of cyclic motion. For motion which takes place back and forth over a fixed path, as in a pendulum swing or in a violin string, each complete back-and-forth motion is called a **vibration,** or **oscillation.** Therefore, back-and-forth periodic motion is often called **vibratory motion,** or **oscillatory motion.** Such a motion may repeat in very short time intervals, making it more convenient to describe in terms of its *frequency* $f = 1/\tau$, rather than its period.

For the most part, this chapter will be confined to the very important special case of periodic motion called **simple harmonic motion** (SHM). This is motion that may be described by a single sine function (or a single cosine function). The name comes from the fact that we call functions which are expressible in terms of sines and cosines **harmonic functions.** SHM is studied because it is common, and its results may be generalized to all periodic motion by mathematical techniques.

10-2 Producing Simple Harmonic Motion

Consider the apparatus shown in Figure 10-1. If the cart is moved sideways and then released, it will bounce back and forth in a repeating motion. A complete description of the motion would allow us to predict the horizontal displacement of any given point on the mass (say the center of mass) at any instant after the release. At the top of the mass, a felt-tip marker has been rigidly attached, with its tip protruding behind the mass to a roll of paper. The paper moves vertically upward at a constant speed while the mass oscillates sideways, generating a record of displacement vs. time by the felt-tip marker on the paper. Figure 10-2a reproduces the graph generated. The shape of the curve is familiar from trigonometry. Starting, as we did, from the maximum horizontal position, the curve is the **cosine curve.** (This is identical in shape to the **sine curve,** but the sine curve would begin at the equilibrium position of the mass at time zero and proceed upward to its maximum.) Figure 10-2b reproduces the standard cosine curve, or the graph of x vs. θ for the function $x = A \cos \theta$. The similarity of the

Figure 10-1 The air cart attached to springs at both sides is free to oscillate along a horizontal line. Even though the motion of the cart is along a straight line, its displacement through time generates the curve of a trigonometric function. (D. Riban)

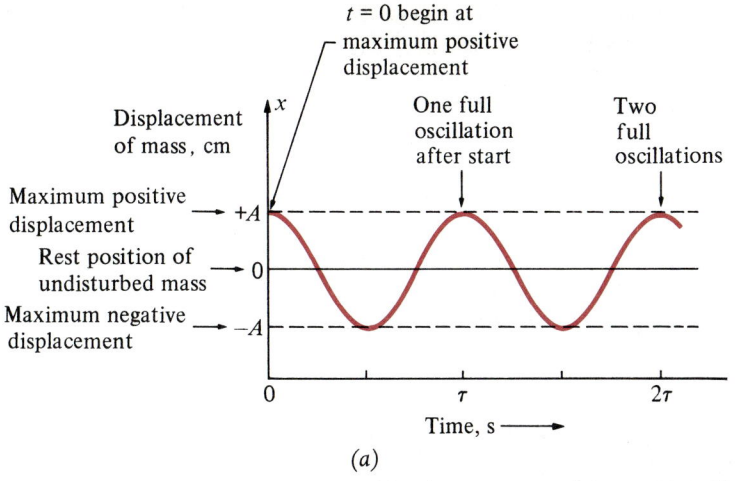

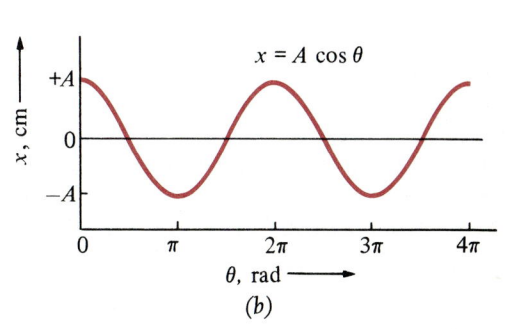

Figure 10-2 (a) The curve generated by the apparatus of Figure 10-1. The mass was released from its farthest rightward position, $x = +A$, at time $t = 0$, and returned to this position in one period at $t = \tau$. (b) The graph of the function $x = A \cos \theta$. The form is identical to that shown in part *a* if it is scaled so that one cycle, or a $\Delta \theta$ of 2π rad, represents the same length as the period τ in (a).

experimental and theoretical curves allows us to expect that the description of this motion will involve the use of a trigonometric function.

Let us examine other moving systems by using the same type of apparatus. Figure 10-3 shows our felt-tip marker attached to the top of a large

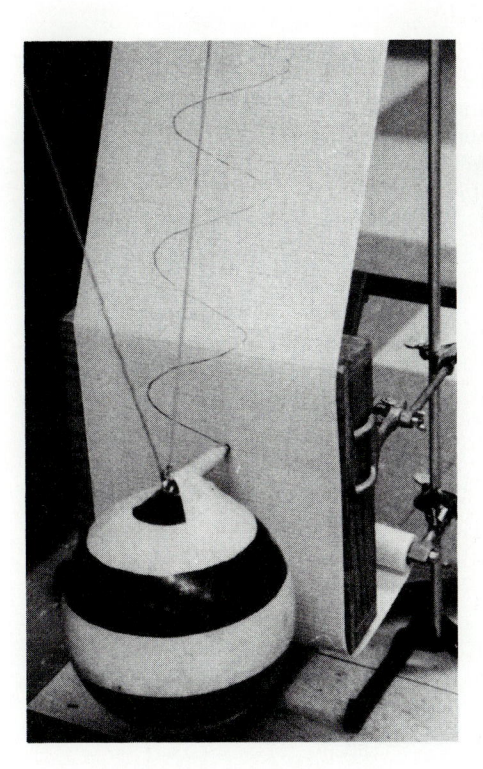

Figure 10-3 A large pendulum using a bowling ball as a bob swinging in a vertical plane. The marker traces the instantaneous x displacement of a point on the ball vs. time and generates a curve similar to that of the oscillating cart of Figure 10-1. (D. Riban)

pendulum bob free to swing horizontally. The paper behind the bob moves from bottom to top in the photograph, and as the bob swings, a displacement vs. time curve for the bob is produced on the paper. The curve generated is similar to those of Figure 10-2. More complex is the motion in the system shown in Figure 10-4. The phonograph turntable is allowed to rotate freely. A peg at the edge of the turntable protrudes upward through a slotted bar that may freely roll to the left or to the right. Thus the motion of the bar records only the left-right, or x, displacement of the peg. The felt-tip marker is attached to the bar and records this displacement on the moving paper behind it.

The curves produced by these two systems are also cosine curves if we choose to begin recording the motion at the instant of maximum positive displacement. We see that several forms of relatively simple repeating motions generate displacement vs. time curves that are sine or cosine curves and are therefore simple harmonic motion.

10-3 A Description of Simple Harmonic Motion

Consider the case of a mass suspended on a spring, as in Figure 10-1. In Chapter 7 we asserted that the force needed to extend (or compress) a spring a distance x from its equilibrium length is proportional to x. That is, to stretch a spring twice as far from rest requires exactly twice as much force, and so on. (Provided, of course, that the force is not great enough to permanently deform the spring.) The externally applied force is in the direction of the displacement x, and thus Equation 7-3 is

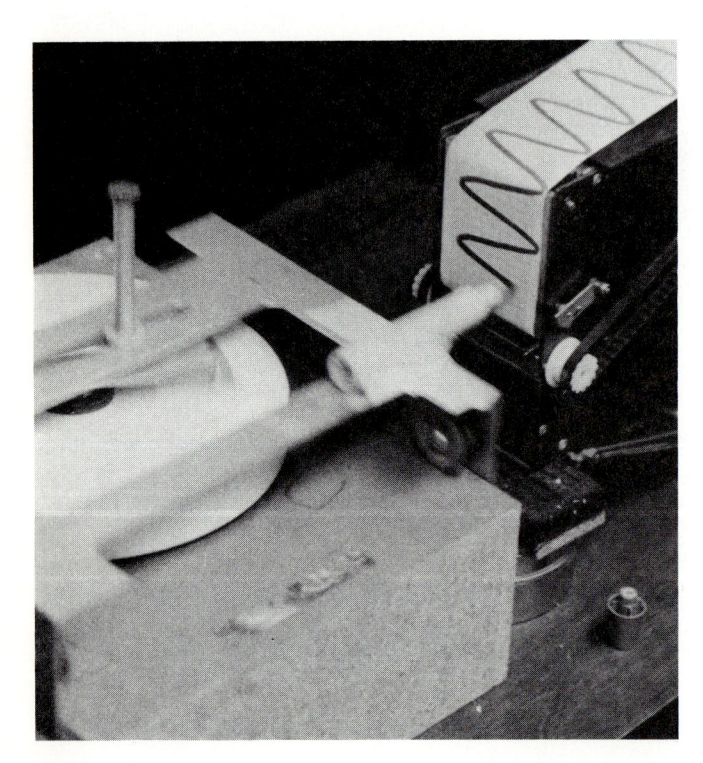

$$F_{app} = kx$$

Figure 10-4 A phonograph turntable in uniform circular motion. As the peg moves in a circle, the motion of the frame shows only the right-left displacement of the peg, and the marker records this. (D. Riban)

Figure 10-5 The motion of a frictionless cart on the end of a spring. (a) The cart is released from a maximum displacement $+A$. (b) As the cart moves toward 0, the force on it declines. (c) At equilibrium, the spring exerts no force on the cart, but the acquired velocity of the cart causes it to (d) continue past equilibrium and begin compressing the spring. At this point, the force of the spring decelerates the cart. (e) At $-A$, the cart has no motion, but a large force from the compressed spring accelerates it back toward equilibrium, reached at (g). It overshoots to its maximum displacement at (i), where the sequence begins again.

where k is the spring constant. Now the stretched spring is exerting an equal and opposite force in reaction to the applied force. This force *due to the spring* is equal in magnitude but opposite in direction to the applied force, or

$$F = -kx \qquad (10\text{-}1)$$

where k is a positive number and the negative sign is shown explicitly to remind us that this *restoring force* and the displacement are in opposite directions.

Picture the motion of a mass-spring system. Imagine a mass m free to slide on a horizontal plane attached to a spring at its left as shown in Figure 10-5. When the mass is moved along the x axis to some displacement A from its equilibrium position 0, the sequential motion illustrated may take place. At the instant of release, the mass is subject to a net force exerted by the spring toward the equilibrium position. The spring contracts, accelerating the mass toward equilibrium and decreasing the displacement of the mass as it moves. As the displacement decreases in magnitude, the force also decreases in accordance with Equation 10-1. At the instant the spring has contracted to its equilibrium length, the displacement and the force are zero. At that point the mass is moving with a relatively large speed in the negative x direction. After all, from the time of release until it reached equilibrium, it was constantly subjected to some force in the negative direction and was gaining speed all the while.

After the mass passes through the equilibrium position, the inertia of the moving mass causes the spring to be compressed. The displacement x is now in the negative direction, and the force exerted by the spring is in the positive direction, causing the speed to decrease. The force increases with increased negative displacement, and the speed decreases until the mass reaches its maximum displacement in the negative direction from the equilibrium position. At this maximum negative displacement, the mass is once again stopped momentarily but is acted on by the maximum force in the positive x direction. Once again it is accelerated toward the equilibrium position with a decreasing force and acceleration as it moves to the right. The mass continues through the equilibrium position at maximum speed and is then subjected to an increasing spring force in the negative direction. It slows until it reaches a maximum positive displacement to complete the first full cycle of periodic motion. In the absence of resistance, the mass will continue to move back and forth in these cycles in regular time intervals.

Whenever Equation 10-1 represents the net force acting on a body of mass m moving in one dimension, we may apply Newton's second law and write

$$ma = -kx$$

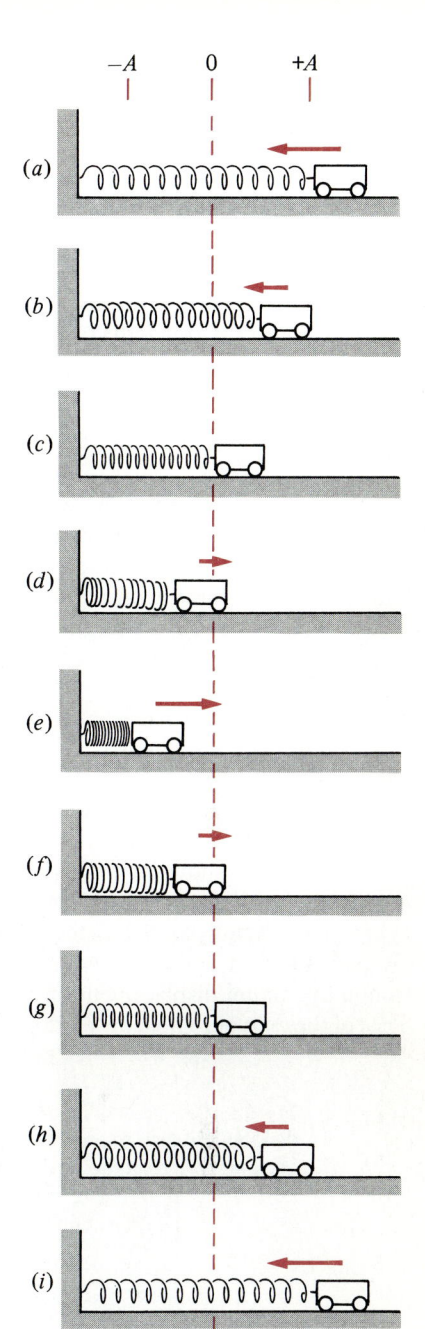

or

$$a = -\frac{k}{m} x \qquad (10\text{-}2)$$

since k and m are both constants for the given situation. This asserts that the acceleration is proportional to the negative of the displacement. Whenever this is the case, the motion that results is defined as *simple harmonic motion* (SHM). In SHM the magnitude of the acceleration must be directly proportional to the magnitude of the displacement, and the acceleration and displacement must be oppositely directed. We will now obtain expressions for displacement, velocity, and acceleration as functions of time for a body undergoing SHM.

10-4 The Reference Circle

The displacement vs. time curve of SHM repeats in a sinusoidal pattern. We expect, therefore, that the mathematical description of this motion will involve a sine (or cosine). Sines and cosines, however, are functions of an angle. It is not immediately obvious in considering the back-and-forth motion of a mass on a spring, which is along a straight line, how to define an appropriate angle to allow us to introduce a trigonometric function. No significant angular relationship is immediately apparent.

In searching for a mechanism to introduce an angular involvement into this motion, we are reminded of the motion of a point on a phonograph turntable as shown in Figure 10-4. Uniform circular motion viewed in one dimension is SHM, yet implicit in this motion are the angular relationships of the circular motion.

Figure 10-6 shows a small cylinder attached to the edge of a disk of radius A which is rotating about its axis of symmetry with a constant angular speed ω. When light shines parallel to the plane of the disk, a shadow is cast on the projection screen. This shadow of the cylinder moves back and forth in SHM from a maximum displacement of $+A$ to a minimum of $-A$. To describe the SHM of the cylinder's shadow on the screen, we use the **reference circle** of the

Figure 10-6 A small cylinder fixed to the edge of a rotating disk. As the disk rotates, the shadow of the cylinder moves back and forth on the screen in simple harmonic motion.

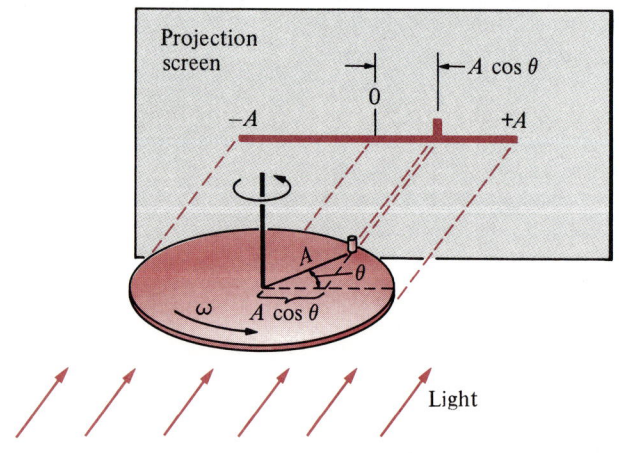

10-4 The Reference Circle

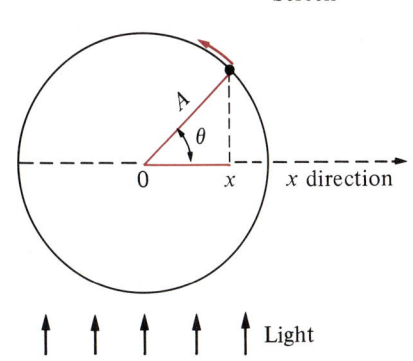

Figure 10-7 A top view of Figure 10-6. The x displacement of the cylinder at any instant is the displacement of the shadow in simple harmonic motion. This, in turn, is determined by where in the cycle of rotation the cylinder happens to be, since $x = A \cos \theta$ in the right triangle shown. Thus, the position of the shadow on the screen is implicitly related to an angle on this reference circle.

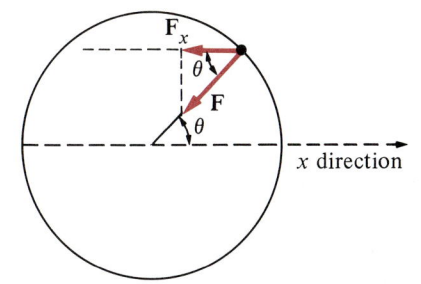

Figure 10-8 The centripetal force F is constant in size and always points toward the center of the circle. Only the x component of this force can influence the x motion of the cylinder. Hence, only F_x is related to the motion of the shadow on the screen. From the right triangle, $F_x = F \cos \theta$ and is always opposed in direction to the displacement.

rotating disk with the real cylinder. (*Usually*, it is a real object in SHM, and the reference circle will be imaginary.)

In Figure 10-7 a reference line, fixed in space as the disk turns, is labeled as the x direction. Parallel rays of light enter the drawing from the bottom and cross the disk to the screen at the top. The x position of the cylinder is determined by dropping a perpendicular to the x axis. The x position of the cylinder then moves on the x axis in exactly the same way as the shadow of the cylinder does on the projection screen.

Applying the definition of the cosine from trigonometry, the x coordinate of the cylinder divided by the radius of the circle is equal to the cosine of θ. Thus we may write

$$\frac{x}{A} = \cos \theta \quad \text{or} \quad x = A \cos \theta$$

Now, if we start timing such that the angle at $t = 0$ is θ_0, then

$$\theta = \omega t + \theta_0$$

and

$$x = A \cos (\omega t + \theta_0) \tag{10-3}$$

You will recall from Chapter 5 that to produce uniform circular motion in a circle of radius A, as in Figure 10-8, a net centripetal force is required. Using $r = A$ and $v = \omega r$, we see that $F = mv^2/r$ becomes

$$F = m\omega^2 A$$

This force may be resolved into components. In particular, we are concerned with the x component. In Figure 10-8 note that the x component is negative and has a magnitude equal to $F \cos \theta$. Thus we may write

$$F_x = -m\omega^2 A \cos \theta$$
$$= -m\omega^2 A \cos (\omega t + \theta_0)$$

Now, since $a_x = F_x/m$, we have

$$a_x = -\omega^2 A \cos(\omega t + \theta_0) \qquad (10\text{-}4)$$

Using Equation 10-3 and substituting, we obtain

$$a_x = -\omega^2 x \qquad (10\text{-}5)$$

Since ω^2 must be a positive number, we may recognize in this expression the defining conditions for SHM in Equation 10-2, $a = -(k/m)x$. Comparing Equation 10-5 with Equation 10-2, we identify ω^2 in the former equation with the ratio of the spring constant to the mass for the case of the oscillating spring. Thus the x component of a body of mass m moving with uniform circular motion at an angular speed ω fulfills the requirement for SHM, provided only that we insist that

$$\omega^2 = \frac{k}{m} \qquad (10\text{-}6)$$

This may be solved for ω to yield

$$\omega = \sqrt{\frac{k}{m}} \qquad (10\text{-}7)$$

Note that any equation like 10-2 may be written in the form of Equation 10-5. Thus we may immediately associate an angular velocity with *any* form of SHM, regardless of the fact that the motion occurs in a straight line. If we know the SHM equation relating the acceleration and displacement, we immediately know an expression for ω.

We are now able to write the equation for the motion of a mass m moving under the action of a spring of spring constant k as

$$x = A \cos\left(\sqrt{\frac{k}{m}}\, t + \theta_0\right) \qquad (10\text{-}8)$$

Example 10-1 If a 100-kg person sits on the right front fender of a car, the spring at that wheel is compressed 5.00 cm from its equilibrium position. Later, the same car while moving hits a bump which compresses the spring 12.0 cm. (*a*) What will be the *period* of the up-down oscillations if 400 kg of the car's mass is supported by the spring? (*b*) Assuming that the car's shock absorbers are very worn and do not reduce the amplitude of the oscillations significantly, what will be the compression/extension of the spring when a second bump is encountered some 2.69 s after the first?

Solution First, it will be necessary to determine the spring constant from the information given. Using this, we may obtain ω and τ.
(*a*) The force applied to the spring was the weight of the 100-kg person, and this produced a deflection of 5.00 cm. Using $F_{app} = kx$, we obtain

$$k = \frac{F_{app}}{x} = \frac{(100 \text{ kg})(9.80 \text{ m/s}^2)}{0.0500 \text{ m}} = 1.96 \times 10^4 \text{ N/m}$$

To obtain ω, we use Equation 10-6, but note that the mass involved here is the mass supported by the spring during its oscillations.

$$\omega^2 = \frac{k}{m} = \frac{1.96 \times 10^4 \text{ N/m}}{400 \text{ kg}} = 49.0 \text{ rad}^2/\text{s}^2$$

$$\omega = 7.00 \text{ rad/s}$$

Since 2π rad constitute one full cycle,

$$\tau = \frac{2\pi \text{ rad/cycle}}{7.00 \text{ rad/s}} = 0.898 \text{ s}$$

Thus, after hitting the bump, the car will vibrate up and down such that it will reach its peak every 0.9 s.

(b) Applying Equation 10-8, we note that the maximum displacement is $A = 12.0$ cm. Since the maximum positive displacement occurred at $t = 0$, the initial angle θ_0 is zero. Thus we may substitute

$$x = A \cos\left(\sqrt{\frac{k}{m}} t + \theta_0\right)$$

$$= (12.0 \text{ cm}) \cos [(7.00 \text{ rad/s})(2.69 \text{ s}) + 0]$$
$$= (12.0 \text{ cm}) \cos (18.8) \text{ rad}$$
$$= (12.0 \text{ cm}) \cos 6\pi = 12.0 \text{ cm}$$

Thus the second bump will occur at exactly that time when the wheel is already compressing upward against the spring at maximum displacement. The result could be unfortunate for the spring, the car, and its occupants. We check the reasonability of this answer by noting that 6π is exactly three complete cycles of vibration, and that the time given to the second bump of 2.69 s is also three times the period we calculated of about 0.9 s.

The reference circle may also be used to obtain an expression for the velocity. Consider Figure 10-9. The body moving in uniform circular motion in a circle of radius A and angular speed ω has a linear speed

$$v = \omega A$$

Now the x component of the velocity is in the negative direction, and we may write

$$v_x = -\omega A \cos (90° - \theta)$$

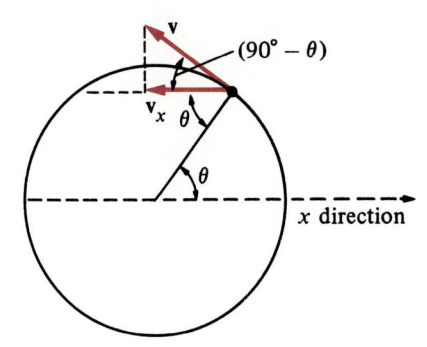

Figure 10-9 Only the x component of the instantaneous velocity of the cylinder is seen in the motion of the shadow on the screen (no longer shown). From the right triangle, $v_x = v \cos (90° - \theta) = v \sin \theta$.

But since $\cos(90° - \theta) = \sin\theta$, we have

$$v_x = -\omega A \sin\theta$$

or, for a mass-spring system,

$$v_x = -\omega A \sin\left(\sqrt{\frac{k}{m}}\,t + \theta_0\right) \qquad (10\text{-}9)$$

Equations 10-5, 10-8, and 10-9 are the basic equations which describe the SHM of a mass-spring system. By substitution for the proper expression for ω, we can and will use them for *all* SHM systems.

Figure 10-10 shows plots which represent the displacement, velocity, and acceleration of a mass undergoing SHM. The curves seem to start (at $t = 0$) at different places. They are similar because each is a sine or cosine curve of the same function of time, namely, $\omega t + \theta_0$. The different starting points depend on whether they are positive or negative functions as well as whether they are sines or cosines. The time-dependent angle $(\omega t + \theta_0)$ is called the **phase angle,** or often simply the **phase** of the motion, with θ_0 as the **initial phase,** the value of the phase when $t = 0$. It is constant, and its value is fixed by the initial conditions of a particular motion. A is also a constant, called the **amplitude,** and is the maximum displacement of the motion. The quantity ω is called the **angular frequency** when it is used this way for SHM.

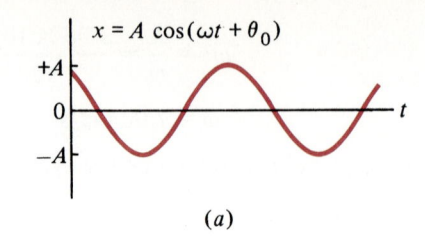

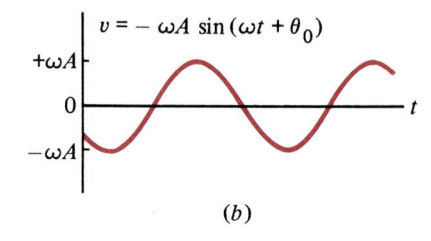

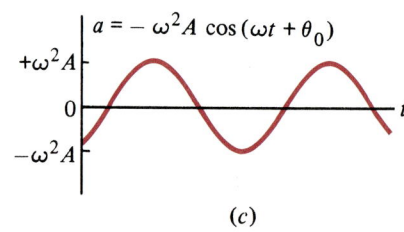

Figure 10-10 (a) Displacement, (b) velocity, and (c) acceleration vs. time curves for an object in simple harmonic motion, starting at the arbitrary position in the cycle of Figures 10-8 and 10-9, where θ_0 (the initial angle) was about 45°, or $\pi/4$ rad.

Example 10-2 A mass $m = 2.00 \times 10^{-2}$ kg is in SHM at the end of a spring ($k = 50.0$ N/m). The clock is started (that is, $t = 0$) when the displacement of the mass is 3.00×10^{-2} m and its velocity is -1.32 m/s. Calculate (a) the angular frequency, (b) the initial phase, and (c) the amplitude of the motion.

Solution (a) This part is determined directly from Equation 10-7:

$$\omega = \sqrt{\frac{k}{m}} = \sqrt{\frac{50.0 \text{ N/m}}{2.00 \times 10^{-2} \text{ kg}}} = \sqrt{2500/\text{s}^2}$$

$$= 50.0 \text{ rad/s}$$

(b) Using ω and substituting the given initial values into

$$x = A\cos(\omega t + \theta_0) \quad \text{and} \quad v = -\omega A \sin(\omega t + \theta_0)$$

yields

$$3.00 \times 10^{-2} \text{ m} = A\cos\theta_0$$

and

$$-1.32 \text{ m/s} = -(50.0 \text{ rad/s})(A\sin\theta_0)$$

Solving for $A\sin\theta_0$ and $A\cos\theta_0$, we have

$$A\sin\theta_0 = 2.64 \times 10^{-2} \text{ m} \quad \text{and} \quad A\cos\theta_0 = 3.00 \times 10^{-2} \text{ m}$$

These may be solved simultaneously by dividing the first equation by the second:

$$\frac{A\sin\theta_0}{A\cos\theta_0} = \frac{2.64 \times 10^{-2} \text{ m}}{3.00 \times 10^{-2} \text{ m}}$$

The amplitude disappears, and since $\sin \theta_0$ divided by $\cos \theta_0$ is equal to $\tan \theta_0$, we have

$$\tan \theta_0 = 0.880 \quad \text{and} \quad \theta_0 = 41.3°$$

(c) We obtain the amplitude by squaring each of the equations and adding them. Thus

$$A^2 \sin^2 \theta_0 = 6.97 \times 10^{-4} \text{ m}^2$$

and

$$A^2 \cos^2 \theta_0 = 9.00 \times 10^{-4} \text{ m}^2$$

Adding these yields

$$A^2 \sin^2 \theta_0 + A^2 \cos^2 \theta_0 = 1.60 \times 10^{-3} \text{ m}^2$$

or

$$A^2(\sin^2 \theta_0 + \cos^2 \theta_0) = 1.60 \times 10^{-3} \text{ m}^2$$

Now we recall that $\sin^2 \theta + \cos^2 = 1$. Therefore

$$A^2 = 1.60 \times 10^{-3} \text{ m}^2 \quad \text{and} \quad A = 4.00 \times 10^{-2} \text{ m}$$

Recall that the period is the time interval required for one complete cycle of the motion. Consider the displacement vs. time graph for SHM in Figure 10-11. The period of the motion (τ) is the time between successive maxima *or* between successive minima. Indeed τ is the time required for one *complete* revolution on the reference circle, no matter where the starting point. We may obtain an expression for the period of this motion from the equation of the curve. A sine or a cosine curve repeats whenever the angle increases by exactly 2π rad; thus we know the phase must increase by 2π when the time increases by τ. At some specific time, say t_1, the displacement, say x_1, can be written as

$$x_1 = A \cos(\omega t_1 + \theta_0)$$

Now the displacement will be x_1 again when the phase angle has increased by 2π:

$$x_1 = A \cos(\omega t_1 + \theta_0 + 2\pi)$$

Figure 10-11 The displacement vs. time curve for a body undergoing simple harmonic motion. Several intervals exactly one period long are marked.

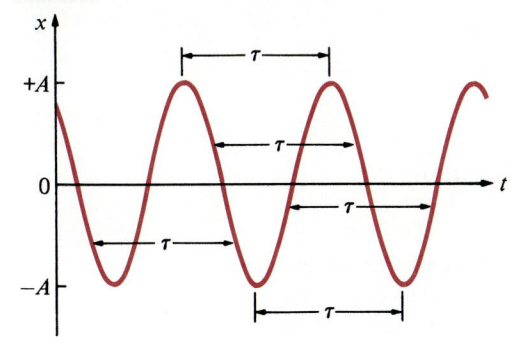

However, this takes place when the time has increased to $t_1 + \tau$, and we could also write

$$x_1 = A \cos[\omega(t_1 + \tau) + \theta_0]$$

The phase angles of these last two equations must be the same. Thus

$$\omega t_1 + \theta_0 + 2\pi = \omega(t_1 + \tau) + \theta_0 = \omega t_1 + \omega\tau + \theta_0$$

which yields

$$2\pi = \omega\tau$$

or

$$\tau = \frac{2\pi}{\omega} \qquad (10\text{-}10)$$

In general, the period is equal to 2π divided by the angular frequency, or, for the mass-spring system,

$$\tau = 2\pi\sqrt{\frac{m}{k}} \qquad (10\text{-}11)$$

Finally we define the **frequency** f of periodic motion as the number of cycles per unit time. Since τ is the time required to complete one cycle, the frequency is equal to the reciprocal of the period:

$$f = \frac{1}{\tau} \qquad (10\text{-}12)$$

or

$$f = \frac{1}{2\pi}\sqrt{\frac{k}{m}} \qquad (10\text{-}13)$$

In SI, frequency units are cycles per second or vibrations per second. To honor Heinrich Hertz, who first produced and measured radio waves, this unit has been given the special name of **hertz** (Hz). Thus, 1 cycle/s = 1 Hz.

Example 10-3   Determine (a) the period and (b) the frequency for simple harmonic motion of the mass-spring system of Example 10-2.

Solution   We apply Equations 10-11 and 10-12 in succession.

(a) $$\tau = 2\pi\sqrt{\frac{m}{k}} = 2\pi\sqrt{\frac{2.00 \times 10^{-2} \text{ kg}}{50.0 \text{ N/m}}}$$
$$= 2\pi\sqrt{(0.0004) \text{ s}^2} = 2\pi(0.02) \text{ s} = 0.126 \text{ s}$$

(b) $$f = \frac{1}{\tau} = \frac{1}{0.126 \text{ s}} = 7.94 \text{ cycles/s} = 7.94 \text{ Hz}$$

10-5 Mechanical Energy in Simple Harmonic Motion

When a mass moves on the end of a spring in SHM, it increases its speed, decreases its speed, stops, reverses, increases its speed again, and so on. As the speed of the mass changes, the kinetic energy associated with the system changes, becoming a maximum, then going to zero, back to a maximum, and repeating. As the mass moves, it does work against the elastic force of the spring. In Chapter 7 we saw that the elastic force of a spring is a conservative force. Hence work done against the spring force is stored as a change in potential energy. If the only force doing work is the elastic force, the change in mechanical energy is zero and the total mechanical energy is constant. Let us examine the kinetic and potential energy of the mass-spring system to see if we may verify this conclusion.

The kinetic energy of a mass m which is moving with a speed v is

$$KE = \tfrac{1}{2}mv^2$$

Using $v = -\omega A \sin(\omega t + \theta_0)$, we have

$$KE = \tfrac{1}{2}m\omega^2 A^2 \sin^2(\omega t + \theta_0)$$

Now since $\omega^2 = k/m$, we see that

$$KE = \tfrac{1}{2}kA^2 \sin^2(\omega t + \theta_0) \tag{10-14}$$

The kinetic energy of the mass-spring system therefore varies with time between values of zero and a maximum of $\tfrac{1}{2}kA^2$.

The potential energy of the system is

$$PE = \tfrac{1}{2}kx^2 \tag{10-15}$$

and from $x = A\cos(\omega t + \theta_0)$, we have

$$PE = \tfrac{1}{2}kA^2 \cos^2(\omega t + \theta_0) \tag{10-16}$$

Finally, the total mechanical energy becomes

$$\begin{aligned} E &= KE + PE \\ &= \tfrac{1}{2}kA^2 \sin^2(\omega t + \theta_0) + \tfrac{1}{2}kA^2 \cos^2(\omega t + \theta_0) \\ &= \tfrac{1}{2}kA^2 [\sin^2(\omega t + \theta_0) + \cos^2(\omega t + \theta_0)] \end{aligned}$$

Once again we recognize that the sum in brackets is equal to unity, and we obtain

$$E = \tfrac{1}{2}kA^2 \tag{10-17}$$

Equation 10-17 not only verifies our conclusion of constant mechanical energy but gives us an expression for it. Consider Figure 10-12, where we have plotted both the kinetic energy and potential energy vs. time. We note that both curves pass through their maximum and minimum values twice during each period.

Figure 10-12 Variation of energy with time for a simple harmonic oscillator, which transforms energy back and forth from kinetic to potential. Their sum, the total mechanical energy (E), is constant if the system is frictionless.

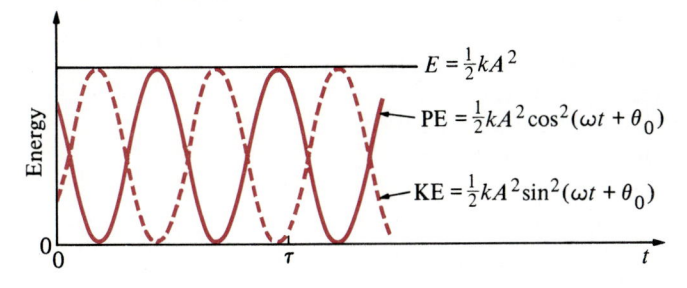

10-5 Mechanical Energy in Simple Harmonic Motion

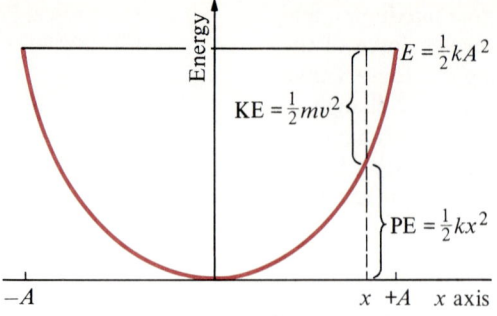

Figure 10-13 Variation of energy with position during an oscillation produces a curve quite different from that of Figure 10-12. For any position x between the extremes of the oscillation from $-A$ to $+A$, the total mechanical energy E is constant and equal to $\frac{1}{2}kA^2$. Part of this total energy is potential energy of magnitude $\frac{1}{2}kx^2$, and the rest is kinetic energy of magnitude $\frac{1}{2}mv^2$.

Further, close examination should convince us that the average values of both the kinetic energy and the potential energy during a cycle are $\frac{1}{2}E$, or $\frac{1}{4}kA^2$.

Figure 10-13 shows the relationships among total energy, potential energy, and kinetic energy as a function of displacement for the mass-spring system. We can see on this figure that as the potential energy becomes larger, the kinetic energy gets smaller. This is demonstrated in the equations also. Furthermore, we may use the equations to obtain an expression for v as a function of x. From $KE + PE = E$, we write

$$\tfrac{1}{2}mv^2 + \tfrac{1}{2}kx^2 = \tfrac{1}{2}kA^2$$

$$\tfrac{1}{2}mv^2 = \tfrac{1}{2}kA^2 - \tfrac{1}{2}kx^2$$

$$v^2 = \frac{k}{m}(A^2 - x^2)$$

$$v = \pm\sqrt{\frac{k}{m}(A^2 - x^2)} \qquad (10\text{-}18)$$

This equation leaves no question that the speed is a maximum when $x = 0$ and zero when $x = \pm A$.

Example 10-4 A mass of 2.00 kg attached to a spring executes simple harmonic motion with an amplitude of 0.120 m. Its kinetic energy is 0.380 J when its displacement is 0.0700 m. Calculate (a) the speed at this position, (b) the spring constant, (c) the total energy, and (d) the frequency.

Solution The speed may be found by using the definition of kinetic energy, while the spring constant can be determined from a manipulation of the energy equations leading to Equation 10-18. The total energy and frequency are then obtained from straightforward applications of Equations 10-17 and 10-13.

(a)
$$KE = \tfrac{1}{2}mv^2$$

$$v = \sqrt{\frac{2KE}{m}} = \sqrt{\frac{2(0.380 \text{ J})}{2.00 \text{ kg}}} = 0.616 \text{ m/s}$$

(b) From conservation of energy, we see that

$$\tfrac{1}{2}mv^2 + \tfrac{1}{2}kx^2 = \tfrac{1}{2}kA^2$$

which may be solved for k to yield

$$k = \frac{mv^2}{A^2 - x^2} = \frac{2(0.380 \text{ J})}{(0.120 \text{ m})^2 - (0.0700 \text{ m})^2}$$

$$= 80.0 \text{ N/m}$$

(c) $\quad E = \tfrac{1}{2}kA^2 = \tfrac{1}{2}(80.0 \text{ N/m})(0.120 \text{ m})^2 = 0.576 \text{ J}$

(d) $\quad f = \dfrac{1}{2\pi}\sqrt{\dfrac{k}{m}} = \dfrac{1}{2\pi}\sqrt{\dfrac{80.0 \text{ N/m}}{2.00 \text{ kg}}}$

$$= 1.01 \text{ Hz}$$

10-6 Some Other Simple Harmonic Motion Systems

In Section 10-3 we saw that any system for which the acceleration and displacement are related by an equation similar to Equation 10-2 is undergoing SHM. Here we consider several examples and along the way obtain relationships for the important quantities, period and frequency.

The first example is that of the **simple pendulum**, a small mass hung on the end of a light string. If the mass is pulled to one side and released, it swings back and forth. Figure 10-3 showed that its displacement vs. time curve was sinusoidal. In Figure 10-14a we see the physical situation at some arbitrary time after the mass has been released. Figure 10-14b shows the mass isolated with the forces acting upon it. We have taken the positive x direction to be away from the equilibrium position (in the direction of increasing angle) and resolved the tension force T into two components. The x component of the net force is then negative, and

$$F_x = -T \sin \phi \qquad (10\text{-}19)$$

Assuming the acceleration in the y direction is negligible,[1] we also have

$$T \cos \phi = mg$$

This can be used to eliminate T in Equation 10-19, yielding

$$F_x = -\frac{mg}{\cos \phi} \sin \phi = -mg \tan \phi$$

Now for small angles, we have

$$\tan \phi \approx \sin \phi = \frac{x}{l}$$

---

[1] This assumption depends on the angle ϕ remaining small enough so that the maximum y displacement remains reasonably small, a so-called small-angle approximation. If the maximum value of ϕ is 15° (approximately $\pi/8$ rad), the deviation of a simple pendulum's motion from SHM is less than 1 percent.

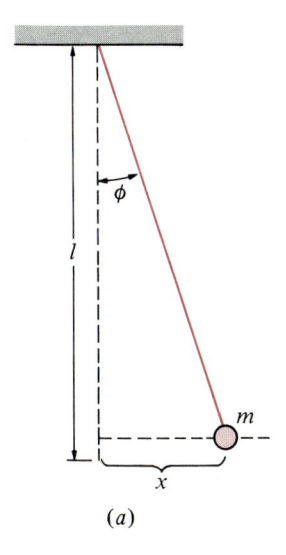

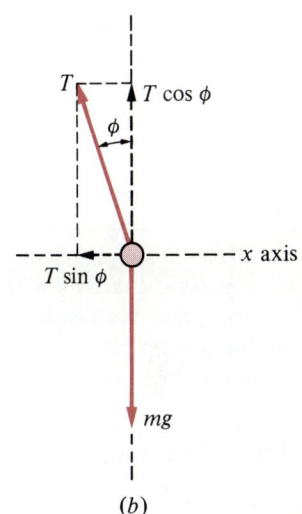

Figure 10-14 (a) A simple pendulum of length l displaced an angle ϕ from vertical. This displaces the bob of mass m horizontally a distance x from the central position. (b) The forces acting on the bob. The weight mg acts vertically downward. The tension T in the string may be resolved into two components. The vertical component, $T \cos \phi$, must balance the weight. The horizontal component, $T \sin \phi$, is unbalanced and accelerates the bob toward the rest position.

and

$$F_x = -mg\frac{x}{l}$$

or

$$F_x = \frac{-mg}{l}x \qquad (10\text{-}20)$$

Equation 10-20 has the form of Equation 10-1, and, within the limits of our assumption, a simple pendulum must execute SHM.

We may divide both sides of Equation 10-20 by the mass to obtain an expression for the acceleration. This yields

$$a = -\frac{g}{l}x$$

Comparing this with Equation 10-5, we see that for a simple pendulum

$$\omega^2 = \frac{g}{l} \quad \text{or} \quad \omega = \sqrt{\frac{g}{l}}$$

Thus the period of a simple pendulum is

$$\tau = \frac{2\pi}{\omega} = 2\pi\sqrt{\frac{l}{g}} \qquad (10\text{-}21)$$

and the frequency is

$$f = \frac{1}{2\pi}\sqrt{\frac{g}{l}} \qquad (10\text{-}22)$$

We see that both the period and frequency are independent of the mass but do depend on the length of the pendulum and the acceleration due to gravity.

Example 10-5 A simple pendulum consists of a small ball suspended so that its center of mass is exactly 0.800 m below the point of suspension. (a) Calculate the period and frequency at a point near the surface of the earth. (b) Calculate the acceleration due to gravity if the pendulum were placed on the planet Venus and its period measured to be 1.89 s.

Solution For both parts of the problem, we simply apply Equation 10-21 and recall that the frequency is the reciprocal of the period.

(a)
$$\tau = 2\pi\sqrt{\frac{l}{g}} = 2\pi\sqrt{\frac{0.800 \text{ m}}{9.80 \text{ m/s}^2}} = 1.80 \text{ s}$$

$$f = \frac{1}{\tau} = \frac{1}{1.80 \text{ s}} = 0.556 \text{ cycles/s}$$

(b)
$$\tau_v = 2\pi\sqrt{\frac{l}{g_v}} \quad \text{yields} \quad \tau_v^2 = \frac{4\pi^2 l}{g_v}$$

or

$$g_v = \frac{4\pi^2 l}{\tau_v^2} = \frac{4\pi^2 (0.800 \text{ m})}{(1.89 \text{ s})^2} = 8.84 \text{ m/s}^2$$

10-6 Some Other Simple Harmonic Motion Systems

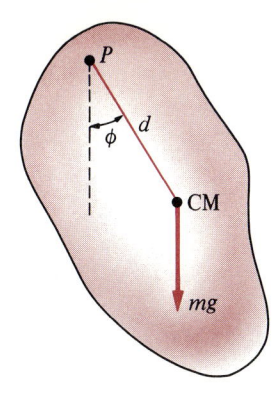

Figure 10-15 A physical pendulum with its center of mass displaced by an angle ϕ from the line extending downward from the pivot point P. The weight of the entire body mg may be thought to act through the CM. If the body is released from this position, a torque rotates the CM to its equilibrium position directly under P.

The second example of a simple harmonic system is that of a **physical pendulum,** a solid body of any mass distribution whatever suspended to swing freely about a horizontal axis, as in the irregularly shaped body shown in Figure 10-15.

The body is suspended from an axis through point P and may swing back and forth in a vertical plane. Figure 10-15 shows the body at some arbitrary time in its motion when the line joining the axis of rotation to the center of mass makes an angle ϕ with the vertical. In the position shown, the weight force gives rise to a torque about the axis of rotation, causing an angular acceleration in the direction of decreasing angle ϕ. The expression for the torque about the axis through P is then

$$\text{Torque} = -mgd \sin \phi$$

We also have

$$\text{Torque} = I\alpha$$

where I is the moment of inertia of the body about an axis through P and α is the angular acceleration. Eliminating torque yields

$$I\alpha = -mgd \sin \phi$$

or

$$\alpha = -\frac{mgd}{I} \sin \phi$$

This equation may not look quite like it, but it is very nearly the same form as that of Equation 10-5 and would be exactly the same form if we could replace $\sin \phi$ with ϕ. Remember that α is related to ϕ in the same way that linear acceleration is related to x. Thus, by again asserting $\sin \phi \approx \phi$, if we restrict the motion to small angles, we have[1]

$$\alpha = -\frac{mgd}{I} \phi \qquad (10\text{-}23)$$

---

[1] If ϕ is measured in radians, the following values show that $\sin \phi$ and ϕ are very nearly equal.

$\phi = 1° = 0.01745$ rad $\sin \phi = 0.01745$
$\phi = 5° = 0.08727$ rad $\sin \phi = 0.08716$
$\phi = 15° = 0.2618$ rad $\sin \phi = 0.2588$

Even at 15°, the difference is only slightly greater than 1 percent.

The period of the physical pendulum for small angular amplitudes is therefore

$$\tau = 2\pi \sqrt{\frac{I}{mgd}} \qquad (10\text{-}24)$$

Example 10-6 Show that the expression for the period of the physical pendulum reduces to that of a simple pendulum when the physical pendulum is a small mass m suspended by a cord of length l.

Solution We need an expression for the moment of inertia of the mass about an axis through the point of suspension. Using l as the distance from the mass to the axis, we have from the definition of moment of inertia that

$$I = ml^2$$

where we have assumed the diameter of the mass is small compared to the length of the cord. Substituting into Equation 10-24 and taking $d = l$ yields

$$\tau = 2\pi \sqrt{\frac{ml^2}{mgl}} = 2\pi \sqrt{\frac{l}{g}}$$

Example 10-7 Determine the frequency of oscillation of a bowling ball with a radius of 0.100 m suspended from an axis which is tangent to the ball.

Solution To obtain an expression for the moment of inertia, we apply the parallel-axis theorem $I = I_0 + md^2$ to a uniform sphere about a tangent axis, yielding

$$I = \tfrac{2}{5}mr^2 + mr^2 = \tfrac{7}{5}mr^2$$

where we have used the fact that the distance d from the tangent axis to the center of mass is equal to the radius. Substituting into Equation 10-24 yields

$$\tau = 2\pi \sqrt{\frac{\tfrac{7}{5}mr^2}{mgr}}$$

$$= 2\pi \sqrt{\frac{7r}{5g}} = 2\pi \sqrt{\frac{7(0.100 \text{ m})}{5(9.80 \text{ m/s}^2)}} = 0.751 \text{ s}$$

The physical pendulum has several very important applications. We have all seen it used as a timekeeper for grandfather clocks and cuckoo clocks. More important to geologists and geophysicists, however, is its application as a portable, accurate method of measuring the acceleration due to gravity. Note that all the quantities appearing in Equation 10-24 may be measured in some other way. Thus, that equation may be solved for g from

$$\tau^2 = 4\pi^2 \frac{I}{mgd} \quad \text{or} \quad g = \frac{4\pi^2 I}{md\tau^2}$$

With I, m, and d known in advance, an instrument can be calibrated to yield g from a measurement of τ.

We mentioned in Section 3.3 that g is somewhat different at the earth's poles than it is at the equator. This difference is attributable in part to the shape of the earth and to the fact that the polar surfaces are closer to the center of mass than the equator. In fact g varies, however slightly, from point to point on the earth's surface due to the different density of material beneath the point where g is measured. For instance, a measurement of g above an underground body of water or a large oil deposit will show a slight decrease, while g above a large deposit of heavy metal may show a slight increase. These slight variations can be measured with great precision by using a physical pendulum. These measurements made over a wide area on the earth's surface may then be plotted on a contour map. When this information is incorporated with the results of other kinds of measurements and a knowledge of the geology of the area, it helps the experienced geologist make excellent estimates of the extent and type of underground deposits.

Finally, any physical system will execute SHM provided the net force and displacement obey the relationship $F = -k_{eff}x$, where k_{eff} is a constant or any combination of constants (perhaps due to two or more springs acting on a mass). There are several end-of-chapter real-challenge problems in which you must determine the **effective spring constant** for a mass being acted on by two or more springs.

10-7 Damped and Driven Harmonic Motion

Most systems in SHM will show a decrease in amplitude as time goes on. They lose energy doing work against friction. When this happens to an oscillating system, we say it is *damped* by friction and the motion is **damped harmonic motion.** Figure 10-16 is a plot of the displacement vs. time for a damped harmonic oscillator.

Figure 10-16 Displacement vs. time for a harmonic oscillator damped by friction. Work is done against friction, and the total mechanical energy ($\frac{1}{2}kA^2$) must decline. Thus amplitude of the oscillations declines in time.

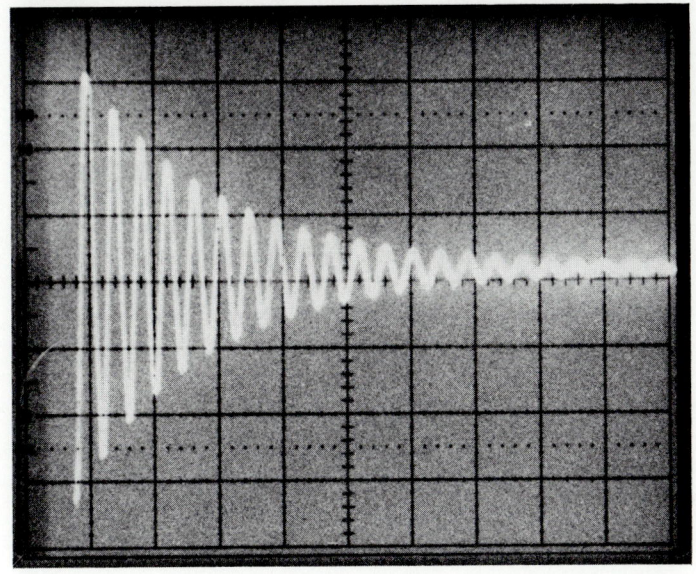

In order to avoid having a damped oscillator run down, the energy lost to friction must be replaced. This is usually done by some external force which simply replaces the lost energy once each cycle. For example, in a typical pendulum clock, a device called the escapement drives the pendulum with a tiny push once each cycle through a series of gears. The motion of the pendulum is said to be **driven, damped harmonic motion.** Note that this driving force, applied once each cycle, is periodic and has the same frequency as the oscillator.

To understand the importance of the timing of the driving force, consider the familiar example of a child on a swing. Normally, to keep the swing going at a reasonable amplitude, energy must be provided during the cycle. Usually either someone gives the swing a small push or the child provides the energy by shifting his or her center of gravity back and forth with respect to the swing each cycle. Now, assume that instead of following the natural frequency of the swing, we push at times determined from a table of random numbers. Statistically, we would expect half the pushes to be applied in the direction opposite to the motion of the swing — pushing forward as the swing is coming back. In such a situation, the energy of the pushes would not add to the energy of the swing and the amplitude would not grow in time.

Next, consider making the pushes periodic but with an arbitrary period compared to the swing. If the swing takes 4.00 s to complete a cycle, we might choose to push every 5.00 s. After an initial push at the peak of the swing, the next push does not occur at the backward peak, but is delayed one-fourth cycle to the bottom of the arc of the swing, etc. Following this for 20.0 s until swing and push again both occur at the peak shows that three out of four pushes in each cycle will add energy to the swinging, but one clearly detracts from the swing. We would say that the frequency at which the swing is being driven by our pushes does not match the natural oscillating frequency of the system, the swing. To obtain the maximum effect on amplitude of our pushes, we must match the driving frequency with the frequency of the oscillator. In other words, for the swing example, we must push with the frequency of the swinging.

A very important phenomenon is illustrated when a damped oscillator is driven by a periodic force of variable frequency. The oscillator moves with the force and will thus oscillate at the frequency of the driving force, whether this frequency is optimal for increasing amplitude or not. Figure 10-17 shows a typical plot of the amplitude of a damped, driven harmonic oscillator vs. driving force frequency. The maximum amplitude occurs at approximately the frequency that the oscillator would have if oscillating freely. That is, the maximum occurs at the point where the driving frequency is about the same as the natural

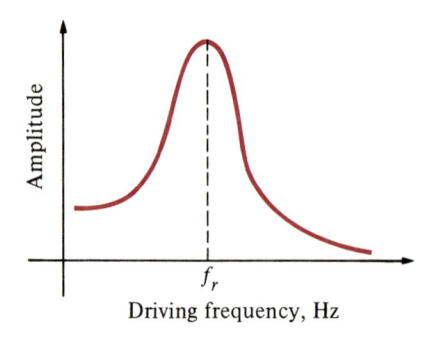

Figure 10-17 The amplitude of a driven, damped harmonic oscillator vs. driving force frequency.

frequency of the oscillator. This phenomenon is known as **resonance,** and this frequency is called the **resonant frequency** f_r.

10-7 Damped and Driven Harmonic Motion

Resonance occurs whenever a periodic impulse or driving force is attempting to drive an oscillator at a natural frequency of the oscillator.

Resonance allows a large and apparent effect to be realized from a quite small and seemingly insignificant input which is properly timed. Very small pushes applied to the child on the swing can lead to a large motion. At resonance, energy input is timed precisely to add to the energy already causing the system to oscillate. As stated earlier, Figure 10-17 is only a typical example of a resonance curve. The maximum amplitude and the exact frequency of resonance f_r depend somewhat on the frictional or damping force. In principle, if there is no damping, the amplitude of the driven oscillator increases to infinity. In practice, something else gives out, and the oscillator breaks. A classic example is the soprano or tenor shattering a crystal goblet by singing a particular note. It is also possible for an automobile which hits a series of bumps at the precise frequency its springs would normally oscillate to buck wildly out of control in a mechanical resonance. Cars have shock absorbers at each spring to damp the motion by converting energy of oscillation to heat, but if these are worn, the likelihood of loss of control increases.

Actually, all physical structures are threatened by resonance. Unless some frictional mechanism exists to dissipate energy as rapidly as it is added to the system at resonance, the structure will eventually fail. The most famed incidence of a resonance destroying a large structure was the collapse of the Tacoma Narrows Bridge in 1941 under the driving force of the wind. The fame rests mostly on the fact that its self-destruction was photographed thoroughly while in progress, rather than such a collapse being unprecedented.

Of course, not all resonances are undesirable. This effect allows us to achieve an unusually large result from a very small input. For example, almost all musical instruments depend upon resonance to produce a sound loud enough to be heard more than a few inches away from its source. Also, a radio wave of incredibly tiny energy can "pump" an electric circuit into resonance and generate a large enough disturbance to be detected, amplified, and put out over a speaker as sound or on a screen as a TV picture.

For most of this century, the only technique for examining the interior of the body without surgery was the x-ray. The ultimate development of x-ray diagnosis was that of the CAT (computerized axial tomography) scanner, which produces a picture of the interior representing one slice through the body. Many such slices can add up to a fairly complete internal examination. The difficulty of x-rays is that they are sensitive only to the density of the materials they are penetrating. Thus, high-density bones show up very clearly, but softer tissues do not. Further, the CAT scanner does use the x-ray, which is an ionizing radiation and presents something of a health risk in itself.

A newer technique for examining the body's interior uses nuclear magnetic resonance (NMR) rather than x-rays (see Figure 10-18). Atoms effectively act like little magnets. When subjected to a magnetic field, they precess or spin about the direction of the field, much as a top precesses or wobbles as it nears the end of its spinning. Using a ring-shaped electromagnet surrounding one plane of the patient's body, the magnetic field may be rapidly varied. When the frequency of this variation is controlled, particular types of atoms are pumped into a resonance, absorbing energy and then reradiating this energy later when

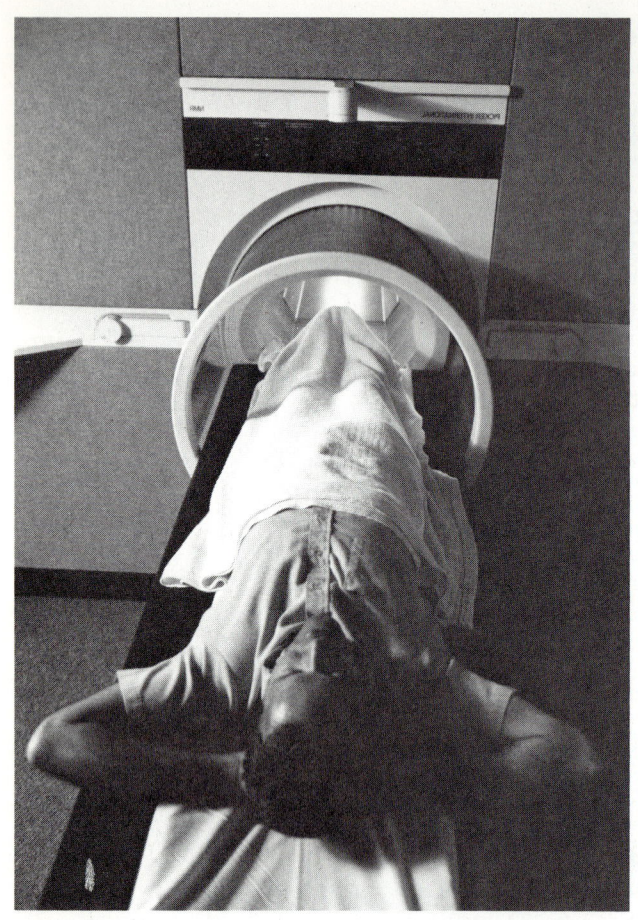

Figure 10-18 NMR in medicine. The patient lies on a sliding, nonmagnetic table which moves through a large ring electromagnet. Hydrogen atoms in the plane of the patient under the ring radiate their resonance energy in all direction. Computer analysis produces a "picture" of an imaginary slice of the patient. (W. McIntyre, Photo Researchers)

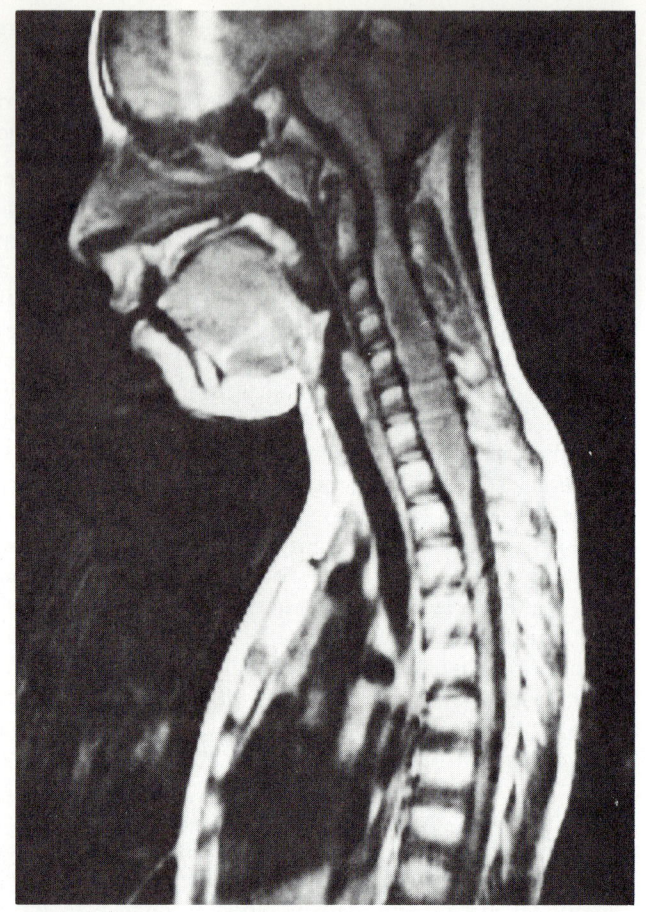

Figure 10-19 In this NMR scan picture (see also color plate VIIIa), black areas represent little or no detected reradiation. This occurs in empty volumes such as the lungs, esophagus, and sinus cavities and also occurs in blood vessels because the blood has moved to a new plane by the time it reradiates. (M. Rotker, Taurus Photos)

the field is turned off. Detectors sample this radiation for the patient, and computers reconstruct an image based on the directions and intensities of reradiated radiations. One advantage of this technique is that it allows selection of what type of atom to excite to resonance by the choice of frequency of the magnetic variation. Most work is done by exciting hydrogen atoms. Water-rich tissues have more hydrogen than fatty tissues and thus show up quite distinctly. Some tumors contrast well with their background tissue because of water differences. Perhaps most interesting is the examination of blood vessels (Figure 10-19). Blood contains a higher percentage of water than body tissues and might be expected to show up well in a NMR scan. However, between the time the plane of the body is excited by the magnet and the time the detectors measure the reradiation, the blood has moved to a new plane of the body. The result is that the interiors of large blood vessels come out blank. The vessel wall is water-rich and shows up well, while fatty deposits on the inside walls of the vessels have less water and become quite distinct from the wall. NMR thus may allow the visual representation of the state of deposit buildup in the circulatory system, area by area, throughout the body.

The phenomenon of resonance is extremely important. It plays crucial roles in acoustics, radio, and television and in many electrical and atomic devices including lasers. We will encounter resonance again and again throughout the text. In the next chapter we will see how wave motion often depends upon resonance. We will also come to recognize that the source of wave motion is most often an oscillator.

Minimum Learning Objectives

After studying this chapter you should be able to:

1. Define:
   - amplitude
   - angular frequency
   - cosine curve
   - cycle
   - damped motion
   - driven motion
   - effective spring constant
   - frequency
   - harmonic function
   - hertz (Hz)
   - initial phase
   - pendulum
   - period
   - periodic motion
   - phase
   - phase angle
   - physical pendulum
   - reference circle
   - resonance
   - simple harmonic motion (SHM)
   - simple pendulum
   - sine curve
   - vibration

2. Recognize the conditions necessary to produce simple harmonic motion.

3. Understand the correspondence between simple harmonic motion and uniform circular motion through the reference circle and thus be confident in the use of angular measures to describe a linear oscillation.

4. Calculate the angular speed (frequency) from the value of the spring constant and the size of the mass in oscillation.

5. Understand the physical meaning of each part of the equation of motion for SHM (Equation 10-8) and apply it to a given situation.

6. Calculate the position and velocity at any time for an object in SHM, given enough information to determine the initial phase, maximum amplitude, and angular frequency.

7. Calculate the period and/or frequency of SHM, given the equation of motion, the angular frequency, or enough information to determine one of these.

8. Calculate the energy involved in SHM from the values of the spring constant and the maximum amplitude.

9. Determine the period of a simple pendulum from its length.

10. Determine the period of a physical pendulum from its specified geometry.

Problems

10-1 A point moves counterclockwise in uniform circular motion at a speed of 8.00 m/s. It takes the moving point 3.00 s for a complete round-trip. The center of the circle is the origin of an xy coordinate system. The clock is started ($t = 0$) when the point is at a position such that the line joining the moving point to the origin makes an angle of $+30°$ with the positive x axis. Write the equation of the x coordinate of the point (same as the x projection of the line) in the form $x = A \cos(\omega t + \theta_0)$, using the proper numerical values for A, ω, and θ_0.

10-2 A mass of 0.100 kg is attached to a spring of constant $k = 40.0$ N/m hung vertically. The mass is pulled 6.00×10^{-2} m below the equilibrium position and released. Write a possible equation for this motion in the form $x = A \cos(\omega t + \theta_0)$, using the proper numerical values for A, ω, and θ_0. Assume $t = 0$ when the mass is first released.

10-3 A mass attached to the end of a spring undergoes simple harmonic motion with an amplitude of 0.120 m at a frequency of 3.00 Hz. The clock is started ($t = 0$) when $x = 0$. Write the equation(s) of motion in the form $x = A \cos(\omega t + \theta_0)$, using the proper numerical values for A, ω, and θ_0.

10-4 The equation for a body of mass 0.100 kg undergoing

227

simple harmonic motion on a spring is $x = 0.300 \cos(1.40t + \pi/6)$. For this motion, determine the (a) amplitude, (b) initial phase, (c) angular frequency, (d) frequency, (e) period, (f) spring constant, (g) total energy, and (h) maximum speed.

10-5 Seen from the plane of their orbits but far away, the moons of Jupiter appear to execute simple harmonic motion about the center of the planet. The moon Ganymede is observed at its greatest rightward position of $+1.07 \times 10^6$ km from the center of Jupiter. At the same time, the moon Europa is at its greatest leftward extension of -6.71×10^5 km. If the periods of these two moons are 7.16 and 3.55 days, respectively, at what position relative to the center of Jupiter will they next appear to meet in their motions? This problem must be done graphically. Thus you must plot the apparent displacement from Jupiter versus time for both moons (on the same graph).

10-6 Viewed from far away in the plane of its orbit, Venus appears to be executing simple harmonic motion about the center of the sun. Venus is 1.08×10^8 km from the sun and moves with a period of 225 days. (a) What is the orbital angular velocity of Venus? (b) What is the orbital linear velocity of Venus? (c) What is the apparent velocity of Venus to the observer as it appears to cross the sun? (d) What is the apparent velocity of Venus as it is at its maximum distance from the sun (maximum elongation)? (e) What is the apparent velocity of Venus when it is one-third, one-half, and two-thirds the distance from the sun to its maximum elongation?

10-7 A mass m undergoes simple harmonic motion with an amplitude of 6.00×10^{-2} m, a period of 0.600 s, and a total mechanical energy of 0.100 J. Determine (a) the spring constant, (b) the angular frequency, and (c) the mass.

10-8 If the tip of a 440-Hz tuning fork moves a total distance of 1.00×10^{-3} m/cycle in SHM, determine its (a) amplitude, (b) period, (c) maximum speed, and (d) maximum acceleration.

10-9 The second hand on a classroom clock has a period of 1.00 min, or 60.0 s. Calculate (a) the frequency of motion of the second hand and (b) the period and frequency of motion of the hour hand.

10-10 A large force applied to and then quickly released from a bathroom spring scale causes it to oscillate with a period of 2.20 s. Determine the frequency of oscillation. Would the period be different if the large force (in the form of a person jumping on it) was applied and stayed there?

10-11 A 50.0-kg dancer oscillates sideways in place to music such that her knees and shoulders have maximum rightward extension when her hips are maximally left, and vice versa. Each extreme position is achieved 2.00 times per second. If we may neglect the motions of the arms and lower legs, the body consists of a set of approximately equal masses in parallel vibration. The head, shoulder, and chest system contains some 32 percent of body mass, as does the lower trunk, hips, and upper thigh system. If the maximum displacement of either of these body regions is 15.0 cm from the normal center line of the body, and the vertical separation of their centers of mass is 30.0 cm, what is the maximum tension in the spinal column connecting these regions at the moment of maximum acceleration?

10-12 What amplitude is required for the simple harmonic motion of a mass of 2.00 kg on a spring of constant 50.0 N/m if the mass is to have a maximum speed of 4.00 m/s?

10-13 Atoms in a solid undergo simple harmonic motion about their equilibrium positions. Typically, the frequency is 1.00×10^{12} Hz, and the amplitude is 1.00×10^{-11} m. For the typical atoms in a solid, determine their maximum (a) speed and (b) acceleration.

10-14 A coffee mug linked to a rubber band undergoes simple harmonic motion with a frequency of 2.20 Hz and an amplitude of 4.00×10^{-2} m. Determine the magnitude of (a) the maximum acceleration and (b) the displacement when the acceleration has a magnitude of 5.00 m/s².

10-15 Two identical masses m are used with two separate springs of constants k_1 and k_2, respectively. The $k_1 m$ system is found to have a period exactly 1.50 times the period of the $k_2 m$ system. Determine the ratio of k_1 to k_2.

10-16 When a particular model of automobile was assembled in the factory, it was determined that the body alone weighed 7.50×10^3 N and that it compressed the springs by 0.120 m from their equilibrium length. Determine (a) the spring constant, (b) the period of vibration with no load, and (c) the frequency of vibration when the automobile was loaded with several passengers weighing a total of 2500 N. The auto has four springs.

10-17 A small mass undergoing simple harmonic motion has a maximum speed of 3.00 m/s and a maximum acceleration of 54.0 m/s². Calculate the (a) angular frequency, (b) period, and (c) amplitude of this motion.

10-18 A mass-spring system is undergoing simple harmonic motion with an amplitude of 0.300 m and an angular frequency of 9.40 rad/s. When the displacement is +0.200 m, calculate (a) the phase, (b) the velocity of the mass, (c) the kinetic energy, and (d) the potential energy. Take the mass as 0.100 kg.

10-19 Assuming the piston in an internal combustion engine moves in simple harmonic motion with an amplitude of 4.00×10^{-2} m, what is its energy when the engine is making 2.00×10^3 r/min? The mass of the piston is 2.50 kg.

10-20 A mass-spring system is undergoing simple harmonic motion with an amplitude A and a total mechanical energy E. The system's energy is doubled by changing the amplitude to A'. What is the ratio of A' to A? (Recognize that the frequency does not change.)

10-21 A mass-spring system undergoing simple harmonic motion is at a point where the displacement is half the amplitude. (a) Determine what fraction of the total mechanical energy is kinetic and what fraction is potential. (b)

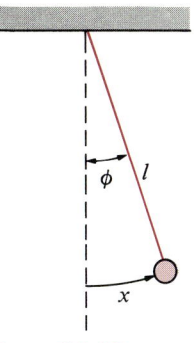

Figure 10-20 Problem 10-26.

Determine the displacement when the kinetic energy and potential energy of the system are equal.

10-22 (a) Determine the length of a simple pendulum which has a period of exactly 2.00 s near the earth's surface. (b) What would its period be on the moon, where the acceleration due to gravity is 1.62 m/s²?

10-23 On a construction job a bucket of sand swings at the end of a 5.00-m rope. For this motion, calculate (a) the period, (b) the frequency, and (c) the length of rope required to double the period.

10-24 A simple pendulum has a length of 0.850 m and undergoes simple harmonic motion with a maximum angle of swing of 10° with the vertical. Calculate its (a) frequency, (b) period, and (c) maximum speed of the mass.

10-25 Calculate the period of a simple pendulum of length 1.50 m if its point of suspension is the overhead of an elevator which is accelerating (a) upward at 2.00 m/s² and (b) downward at 3.50 m/s².

10-26 Show that Equation 10-20 for a simple pendulum may be derived by using the small-angle approximation that $\sin \phi \approx \phi$. Take the x displacement to be along the arc traced out by the mass, as shown in Figure 10-20, and remember that from the definition of radian measure we have $\phi = x/l$.

10-27 A geologist is using a precision physical pendulum which has been adjusted to have a period 3.0000 s at a point where the acceleration due to gravity is exactly 9.8017 m/s². At one of the test points in a field under investigation, the pendulum has a period of 3.0085 s. Determine the acceleration due to gravity at that test point. Use five significant figures throughout.

10-28 Assume your lower leg and foot has a mass of 3.60 kg and a center of mass about 0.180 m below the knee. If the lower leg is allowed to swing freely and pivot at the knee, it undergoes simple harmonic motion with a period of 1.20 s. Calculate the moment of inertia of your lower leg and foot about an axis through the knee.

10-29 The pendulum for a cuckoo clock consists of a thin uniform brass rod rotating about an axis through one end. Table 9-1 shows us that a thin rod has a moment of inertia about a perpendicular axis through its center of mass equal to $\frac{1}{12}ml^2$. Since the rod is uniform, the distance from the axis of rotation to the center of mass is $\frac{1}{2}l$. Determine (a) an expression for the moment of inertia about the axis of rotation by applying the parallel-axis theorem of Chapter 9, (b) an expression for the period in terms of l and g, and (c) the length of the rod if the period is to be 0.867 s.

10-30 A chandelier hangs by a chain and swings in simple harmonic motion with a period of 3.00 s. The distance from the center of mass to the axis of rotation is 0.750 m, and the mass of the chandelier is 12.0 kg. Determine the moment of inertia about (a) the axis of rotation and (b) an axis through the center of mass.

10-31 A physics laboratory exercise uses a ring pendulum, a circular hoop of metal hung from a knife-edge. The moment of inertia of a circular hoop of mass m and radius r about an axis through its center is $I_0 = mr^2$. (a) Using the parallel-axis theorem, derive an expression for the moment of inertia about an axis through its edge. (b) Show that the ring pendulum has a period equal to a simple pendulum of length $l = 2r$.

10-32 Observations of small variations in the wavelength of light emitted by molecules have convinced scientists that molecules are vibrating. A good model for a diatomic molecule like N_2 (nitrogen 2) is to consider it to be two point masses attached to each end of a very stiff spring. If the frequency of vibration of nitrogen is 7.07×10^{13} Hz and nitrogen has a mass of 2.34×10^{-26} kg, calculate the spring constant (see Problem 10-34).

10-33 When a 0.600-kg mass is hung from a certain spring, it may undergo simple harmonic motion with a period of 1.40 s. If the spring is now cut in half and the mass is used with half the spring, what will its new period be?

10-34 Two bodies each of mass m are connected to a spring of constant k, as shown in Figure 10-21. They are pulled apart on a frictionless surface and released. Prove that the frequency of vibration is given by $f = (1/2\pi)\sqrt{2k/m}$.

10-35 Determine the effective spring constant and write an expression for the frequency of simple harmonic motion for the system shown in Figure 10-22.

10-36 Determine the effective spring constant and write an expression for the period of simple harmonic motion for the system shown in Figure 10-23.

10-37 Determine the effective spring constant and write an expression for the frequency of simple harmonic motion for the system shown in Figure 10-24.

10-38 A mass on a frictionless table is supported by two sets of springs, one set in the x direction and another set in the y direction. If the mass is displaced from the equilibrium position in the center of Figure 10-25a and then released or pushed into motion, several different patterns of mo-

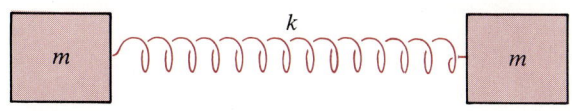

Figure 10-21 Problem 10-34.

229

Figure 10-22 Problem 10-35.

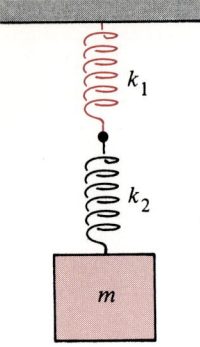

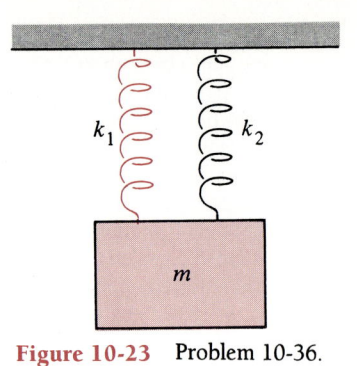

Figure 10-23 Problem 10-36.

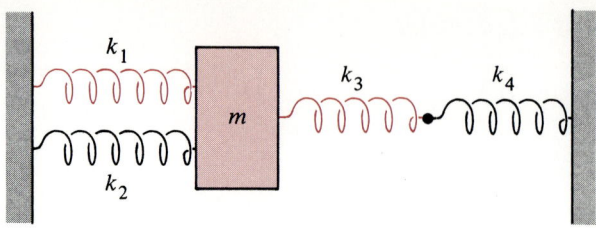

Figure 10-24 Problem 10-37.

tion may result. For example, if the effective spring constant in each direction is the same, the period of motion is the same in both the x and y directions, and $\tau_x = \tau_y$. If also the mass is released so that $A_x = A_y$ and $\theta_{0x} = \theta_{0y}$, the pattern of motion of the mass is shown in Figure 10-25b. Match the following first three statements of conditions with the pattern of motion which *may* result by selecting from Figure 10-25c through e. Can you sketch conditions 4 and 5?

(1) $\tau_x = \tau_y$, $A_x = A_y$, $\theta_{0x} = 0$, $\theta_{0y} = \pi/2$
(2) $\tau_x = \tau_y$, $A_x = A_y$, $\theta_{0x} = 0$, $\theta_{0y} = \pi$
(3) $\tau_x = 2\tau_y$, $A_x = 2A_y$, $\theta_{0x} = 0$, $\theta_{0y} = 0$
(4) $2\tau_x = 3\tau_y$, $A_x = A_y$
(5) $\tau_x = 1/3\tau_y$, $A_x = A_y$

10-39 A child is on a 3.00-m-long swing, and the combined mass of child plus swing is 50.0 kg. Beginning from rest, a push is delivered each swing. The push is of 20.0 N and is exerted through a distance of 10.0 cm (or the equivalent for the earliest pushes). The swing is essentially frictionless. (*a*) What is the period of the motion? (*b*) How many pushes will be required to achieve an arc of swinging of ±15° from the rest position?

10-40 A narrow catwalk 10.0 m long and supported at its ends has an effective mass of 120 kg. It oscillates up and down with an angular frequency of 9.00 rad/s when unoccupied. A worker with a load of combined mass 180 kg walks heavily over the span. At what frequency will the footsteps set the span into resonant vibration?

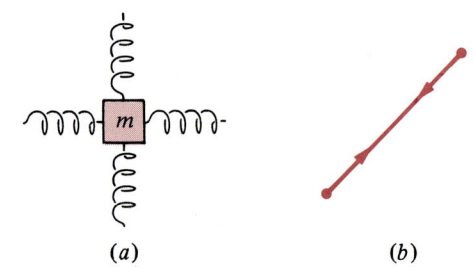

(a) (b) (c) (d) (e)

Figure 10-25 Problem 10-38.

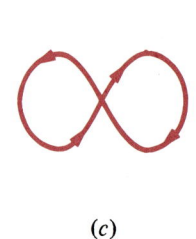

230

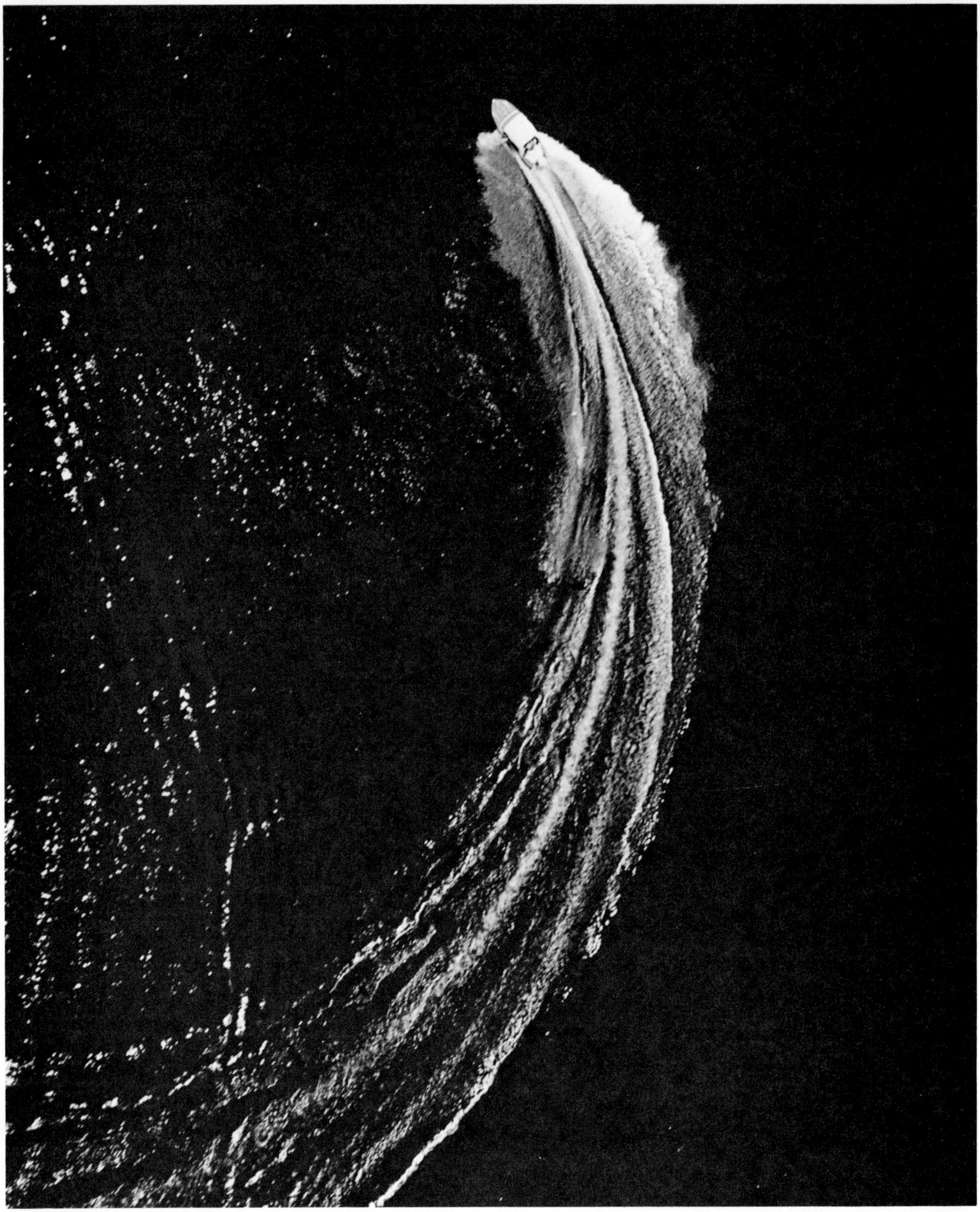

11 Mechanical Waves and Sound

11-1 Introduction

Imagine a motorboat moving on a lake (Figure 11-1). As the boat moves, it generates a traveling disturbance of the water (a **wave**). This disturbance appears to move, or **propagate**, along the surface, encountering other boats and finally the shore. Similarly, a stone thrown into a pond (Figure 11-2) generates waves which move out in a circular pattern in all directions along the surface. In both cases, we observe a universal feature of wave motion: *an effect at one point had its origin at some distant point.* If we watch floating wood chips or leaves, we may note that they bob and sway as the wave passes but essentially remain in their original position after the disturbance has passed. The wave is moving, but it need not carry the water or floating objects along with it. Thus *mechanical wave motion* is the propagation of a disturbance in a medium without the transport of the medium itself. A reasonably sized water wave certainly has the ability to lift a small boat as it passes. Thus we may conclude that the wave possesses energy. The disturbance itself moves, taking energy with it but leaving the material of the medium where it was previously. All material media—solids, liquids, and gases—can carry energy as mechanical wave motion.

Mechanical waves all require a material medium (air, water, string, rope, etc.) in which to propagate. We'll encounter later other waves (light waves, for example) which may propagate in a vacuum.

11 Mechanical Waves and Sound

Figure 11-1 The disturbance generated by the moving boat propagates away from the place where it occurred. *(G. Heilman)*

A model for wave motion assumes that the particles of matter (atoms or molecules) are linked to their nearest neighbors by some kind of elastic force. The disturbance of one particle changes the force between that particle and its nearest neighbors, moving them in response. The motion of this second set of particles in turn stimulates the motion of yet another set in response to the changed force, and so on. In this manner the wave propagates, with the oscillation of each particle affecting the next ones in line. This model produces

Figure 11-2 A localized disturbance of the water produces expanding circular wavefronts moving from the site of the disturbance. If the wavefronts are precisely circular, the speed of the wave propagation is the same in every direction along the water surface. *(C. Riban)*

excellent agreement with experiment, and from it we may note a few implications.

1. A finite time is required for a disturbance to propagate over a distance. Further, if the waves from a stone thrown into a pond are truly circular and centered on the point of disturbance, we suspect that waves propagate in all directions at a constant speed through a uniform medium.
2. The energy of the disturbance is transmitted, while the particles themselves, the medium, remain essentially in place.
3. The actual motion during oscillation of the particles which propagates the wave need not be in any one direction, which allows us to classify types of waves.

Waves are classified according to the relationship between the direction of motion of the particles and the direction of propagation of the energy. A **transverse wave** is one in which the displacement of the particles is transverse (perpendicular) to the direction of propagation of the wave. Figure 11-3a shows a transverse wave moving on a string. While the wave propagates to the right, each segment of string moves up and then down as the wave passes.

A **longitudinal wave** is one in which the displacement of the particles is forward and backward along the direction of propagation of the wave. In Figure 11-3b we see a series of disturbances moving in a spring. The disturbance was created by repeatedly compressing a small portion of the spring at one end and then releasing it. As the wave moves to the right, each coil slides right and then left as the wave passes.

A **torsional wave** is a wave of twist that occurs when the end of a plank of material is rotated back and forth about an axis. In a torsional wave, the displacement of the particles is on an arc surrounding the direction of the propagation of the wave. Such a wave of twist in a long wooden plank is shown in Figure 11-3c.

Waves in a string are transverse, while sound waves are longitudinal. Surface water waves are a combination, with the surface particles moving in both a transverse and longitudinal manner. A floating leaf rises and falls as the wave passes (transverse), but it also sways in the direction of propagation and back again (longitudinal). This complication makes water waves very difficult to analyze. We will work primarily with pure transverse or pure longitudinal waves.

11-2 Traveling Waves

In mechanical wave motion, particles are displaced from their equilibrium position and then return to it. Such disturbances could occur once (a pulse), but most commonly they are repeated periodically in time. The **periodic waves**

Figure 11-3 (a) In the *transverse* wave, each segment of rope moves up and down, perpendicular to the direction of wave motion. (b) In the *longitudinal* wave, each coil moves first with, then against the direction of wave motion. (c) In the *torsional* wave, the displacement is on an arc surrounding the direction of wave propagation.

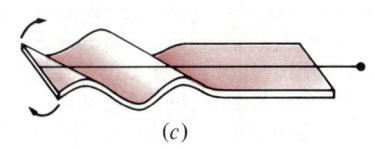

(a) (b) (c)

At time t

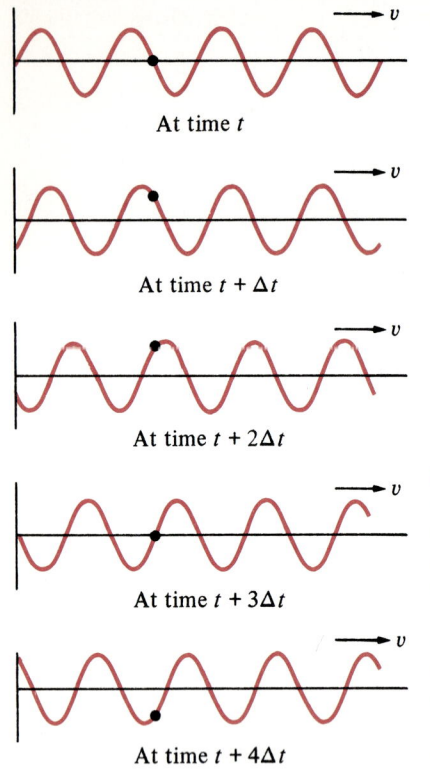

At time $t + \Delta t$

At time $t + 2\Delta t$

At time $t + 3\Delta t$

At time $t + 4\Delta t$

Figure 11-4 A traveling wave in successive positions separated by short time intervals. The wave looks very much like a fixed pattern sliding smoothly to the right, but the pattern is actually being recreated, somewhat displaced, instant by instant by the motions of each segment of the medium. A representative particle of the medium is marked so that its varying displacement may be noted.

resulting from such repeated disturbances are the most important case for study. The pattern repeats in space (and time) and is very often sinusoidal in form.

Most simple repeating waveforms do resemble a repeating sine wave, and measurement reveals that they are indeed **sinusoidal.** We already know that simple harmonic motion is sinusoidal in time. Thus, a mechanical oscillator attached to a medium and moving in simple harmonic motion is creating a sinusoidal disturbance of the medium. This sinusoidal pattern of disturbance then propagates away from this point at a constant velocity. Instant by instant, the entire pattern appears to shift to a position farther away from the source and thus is called a **traveling wave** (Figure 11-4). A mathematical description of this wave must specify the displacement of all particles from their equilibrium position at any time. Consider a wave on a one-dimensional medium like a string. If we assume that the wave is moving in the x direction and the oscillation is in the y direction, an equation should give us y for any x for any time.

Imagine a series, or train, of waves moving from left to right along a clothesline. As the wave train moves, the particles oscillate to maintain the wave's size and shape. The wave train looks very much like a fixed disturbance sliding smoothly along the medium rather than a pattern being recreated instant by instant by multitudes of individual molecular motions. The mathematical strategy to describe the wave depends upon the simplicity of the former view rather than the complex reality of the latter. The easiest way to obtain the displacement equation for a traveling wave is to construct it in a coordinate system which is moving along with the wave. Then we transform our result back to a stationary coordinate system.

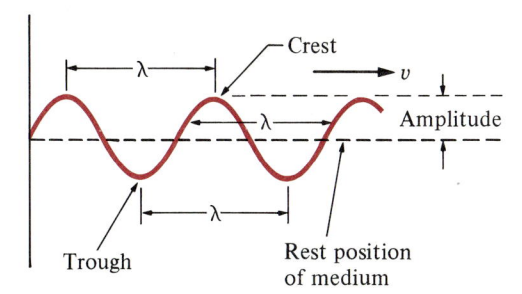

Figure 11-5 A periodic wave repeats its pattern of oscillation in both time and space. The wavelength λ is the minimum distance in which the complete pattern of the disturbance is repeated. The time required to generate one wavelength's worth of wave is the period τ.

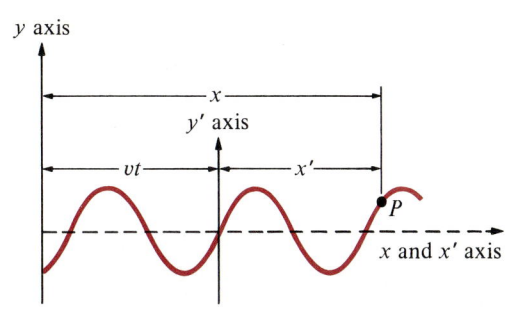

Figure 11-6 Analysis of the traveling wave is simplified if we first define the apparent pattern which seems to be sliding along the medium as though it were frozen. The wave shown is moving along the x axis to the right. We define a moving $x'y'$ coordinate system *fixed to the moving wave*. Thus, point P has coordinates (x',y'), which do not change in time since P is a fixed displacement from the x',y' origin.

The time required for a particle to complete one cycle of its motion is its *period* τ. The number of cycles completed per unit time is the *frequency f* and is the reciprocal of the period. The speed at which the wave moves through the medium is the *speed of propagation v*. In addition to these terms, which should be familiar, we must label other properties which describe how a periodic wave repeats its pattern in space. The minimum distance over which the wave repeats is called the **wavelength,** indicated by a Greek lowercase lambda (λ), as shown in Figure 11-5. We often call the maximum positive displacement of the wave a **crest** and the maximum negative displacement a **trough.** Thus the distance between adjacent crests is one wavelength, as is the distance between adjacent troughs.

In Figure 11-6 a traveling wave is shown moving to the right with a speed v as measured in a stationary coordinate system. The stationary system is the xy system, called the unprimed coordinate system. At a time of our choosing, a second coordinate system is attached to the wave and allowed to move with it at a speed v. Viewed from the second coordinate system, the disturbance is simply a stationary sinusoidal wave in space and is independent of time. The axes of the second system are labeled x' and y', and it is called the primed coordinate system. The x' axis lies along the x axis, so the y' displacements measured perpendicularly from the x' axis are equal to the y displacements, which are measured perpendicularly from the x axis. It is most convenient if we attach the primed system to the wave at a point where it is crossing the axis toward the positive y direction. In this way, the wave as viewed in the primed system is a sine wave. It is also convenient if we start counting time so that $t = 0$ whenever the two coordinate systems coincide ($x = x'$ and $y = y'$ at $t = 0$). This is all shown in Figure 11-6.

Whenever the angle for the sine wave increases by 2π rad, x' must increase by exactly λ, one wavelength. We could also say that as x' increases by any part of a wavelength, the angle must increase by that same part of 2π rad. Thus, if we say that the angle of the sine wave is θ, then

$$\theta = \frac{2\pi x'}{\lambda} \tag{11-1}$$

Examining Equation 11-1 should convince you that θ goes from 0 to 2π as x' goes from 0 to λ. If the maximum y or y' displacement of the wave (remember they are equal) is called y_{max}, we may write

$$y' = y_{max} \sin\left(\frac{2\pi}{\lambda} x'\right) \quad (11\text{-}2)$$

Equation 11-2 gives the particles' y' displacement as a function of x' for the fixed disturbance in the primed coordinate system. It is now necessary to transform this equation into the unprimed coordinate system to take into account that the pattern moves from place to place in time. To do this we note in Figure 11-6 that the arbitrary point P has a horizontal displacement x' in the moving system and x in the stationary system. However, because we started time so that the two coordinate systems coincided at $t = 0$, we have that

$$x = x' + vt$$

or

$$x' = x - vt \quad (11\text{-}3)$$

A straightforward substitution of Equation 11-3 into Equation 11-2 yields

$$y = y_{max} \sin\left[\frac{2\pi}{\lambda}(x - vt)\right] \quad (11\text{-}4)$$

where we have also used the fact that $y' = y$. We again call the angle, the quantity in brackets, the **phase** of the motion.

Equation 11-4 is only one form of the displacement equation for a traveling wave. Several other forms may be obtained by substituting into Equation 11-4 certain relationships which exist among the speed, wavelength, period, and frequency (see Problem 11-7). From the definitions of these terms and Figure 11-4, we see that the wave must travel a distance of exactly one wavelength in a time interval of exactly one period. Thus we have

$$v = \frac{\lambda}{\tau}$$

or, since $\tau = 1/f$,

$$v = \lambda f \quad (11\text{-}5)$$

The relationship among speed of propagation, wavelength, and frequency shown in Equation 11-5 is common to *all* wave motion and is extremely important.

Although we referred to a wave on a string (a transverse wave) as we derived Equation 11-4, that equation could equally well apply to a longitudinal wave. We simply must identify y as Δx, the displacement of the particles from their equilibrium positions back and forth along the x axis. Figure 11-7a shows the equilibrium position of each coil of an undisturbed spring. Figure 11-7b represents a longitudinal wave moving to the right along the spring. Figure 11-7c plots the displacement from equilibrium, Δx, of each of the coils in Figure 11-7b. The plot is clearly sinusoidal in shape.

Sound is the most commonly encountered longitudinal wave. Sound generally results from the motion of matter against the air, as in the buzzing of a mosquito's wings. As the wing moves forward against air molecules, these are compressed and recoil against the zone of molecules farther from the wing. The

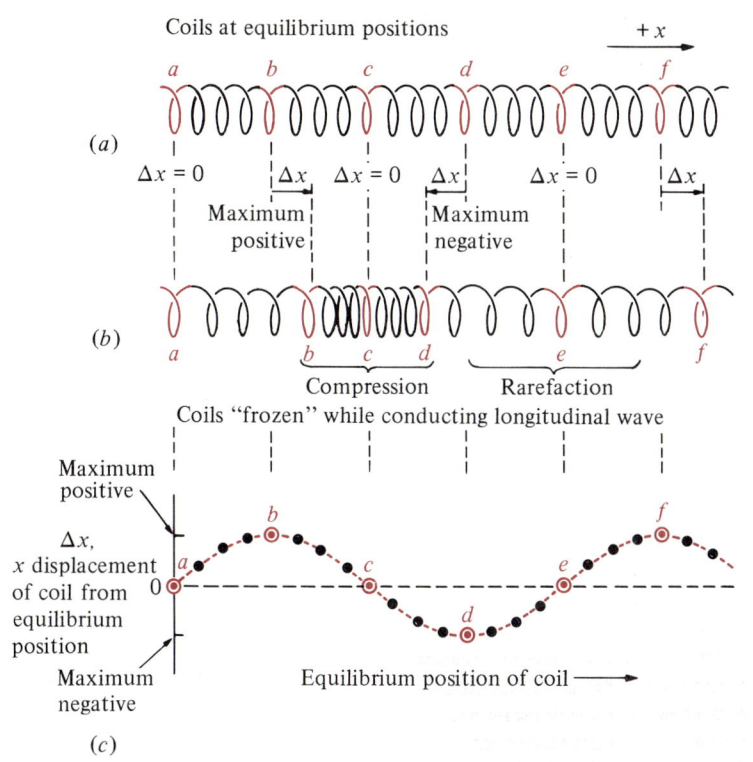

Figure 11-7 (a) A length of spring at rest. Each fourth coil is in color to allow it to be followed as it moves in response to a wave. (b) The coils at some later time at an instant when a periodic wave is moving from left to right across the spring. The zone from *b* to *d* is a zone of *compression*. From *d* to *f*, coil spacing is greater than at equilibrium in a zone of *rarefaction*. Comparison of parts *a* and *b* shows that only coils *a, c,* and *e* are momentarily at their equilibrium positions in part (b). Each other coil has moved a distance Δx from its position in part *a*. (c) The displacement of each coil from its equilibrium position vs. the position of the undisturbed coil is a sinusoidal curve.

shock of this forward thrust thus generates a zone of *compression* which races through layer after layer of air molecules forward from the wing. As the wing is pulled back, the zone in front of it has additional volume into which molecules may expand. As they do, the density, or packing, of the molecules per unit volume is lessened. The more-compressed molecules of the next zone away from the wing react to this by moving into this lower-density zone, and a wave of lesser packing, a **rarefaction,** races through the various layers of molecules following the wave of compression.

We may associate this packing of molecules into volume with air pressure. Thus, sound is a variation in air pressure. While normal sea level atmospheric pressure is 1.013×10^5 N/m², the voice of a person near you in conversation might add a rapid pressure variation of $\pm 2.9 \times 10^{-12}$ N/m². The microphone and the ear both respond to this pressure variation if it is in an appropriate frequency range. Thus, the *y* of Equation 11-4 or the Δx of the coil displacements for a wave on the coil spring of Figure 11-7 may be interpreted

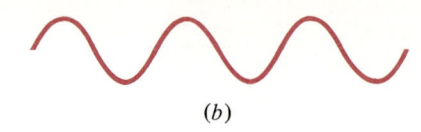

Figure 11-8 (a) A sound wave moving along a tube. The dots represent air molecules. (b) The variation in air pressure along the tube is a sinusoidal wave.

for a sound wave. If we imagine the air to be made of a series of very thin layers of molecules across the thickness of which the sound is propagating, y represents the instantaneous displacement of this layer from its equilibrium position with or against the direction of propagation. Figure 11-8a represents a sound wave moving along a tube, with dots representing individual molecules of air. In the absence of sound, the dots would be evenly spaced in the tube. Instead, we see them crowding together, representing the pressure increase of compressions, and the lowered-packing, lowered-pressure zones of rarefactions following each other down the tube. If we lined the tube with a succession of small devices to measure the pressure, the variation with position along the tube would be as shown in Figure 11-8b.

Example 11-1 The maximum sound from a fire whistle has a frequency of 900 Hz. If sound propagates in air at a speed of 340 m/s, calculate (a) the period of the sound and (b) the wavelength.

Solution (a) The period is found as the reciprocal of the frequency:

$$\tau = \frac{1}{f} = \frac{1}{900/s} = 1.11 \times 10^{-3} \text{ s}$$

(b) For the wavelength, we apply Equation 11-5 directly:

$$v = \lambda f$$

$$\lambda = \frac{v}{f} = v\tau = (340 \text{ m/s})(1.11 \times 10^{-3} \text{ s}) = 0.377 \text{ m}$$

Example 11-2 A traveling wave moves in one dimension with an amplitude of 0.100 m, a frequency of 800 Hz, and a wavelength of 0.0500 m. Write the equation for the particle displacement in the form of Equation 11-4, placing numerical values where they are appropriate.

Solution This is done primarily by inspection, as y_{max} and λ are given. However, we must determine v from f and λ:

$$v = \lambda f = (0.0500 \text{ m})(800/s) = 40.0 \text{ m/s}$$

Thus

$$y = y_{max} \sin\left[\frac{2\pi}{\lambda}(x - vt)\right]$$

becomes

$$y = 0.100 \sin\left[\frac{2\pi}{0.0500}(x - 40t)\right]$$

$$= 0.100 \sin[40\pi(x - 40t)]$$

11-3 Speed of Wave Propagation

We may obtain the speed of propagation of wave motion on a string by making use of the impulse-momentum theorem. The string shown in Figure 11-9a has a mass per unit length ρ_L (called **linear mass density**) and is under a tension force T. Suppose at time $t = 0$ an external force F is applied perpendicular to the string at the left end. After a very short time interval Δt, the string takes on the shape shown in Figure 11-9b. The pulse will have moved along the string at the speed of propagation of wave motion v and will have progressed a distance equal to $v\,\Delta t$. If the left end of the string has moved upward at an average speed V, then it will be a distance $V\,\Delta t$ away from its original position. By similar triangles, we may write

$$\frac{F}{T} = \frac{V\,\Delta t}{v\,\Delta t}$$

or

$$F = T\frac{V}{v}$$

Thus the average impulse applied in a direction perpendicular to the string is equal to

$$\text{Impulse} = T\frac{V}{v}\Delta t$$

The average change in momentum perpendicular to the string during Δt is the mass of that part of the string which moved times its average change in velocity in that direction. The mass which moved is the linear mass density (ρ_L) times the length of the string that moved $(v\,\Delta t)$; thus the average change in momentum perpendicular to the string is

$$\text{Change in linear momentum} = \Delta(mV) = \rho_L v\,\Delta t\,V$$

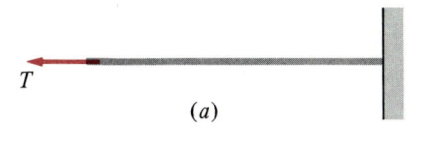

Figure 11-9 (a) A string under uniform tension T acting along its length. (b) The end of the string disturbed by an additional force F acting perpendicular to its length.

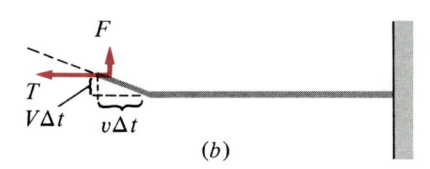

Applying the impulse-momentum theorem, we have

$$\text{Impulse} = \text{change in linear momentum}$$

$$T \frac{V}{v} \Delta t = \rho_L v \, \Delta t \, V$$

Dividing both sides by $V \Delta t$ and solving yields

$$v^2 = \frac{T}{\rho_L}$$

or

$$v = \sqrt{\frac{T}{\rho_L}} \qquad (11\text{-}6)$$

Equation 11-6 shows that the speed of propagation of wave motion in a string depends only on the tension in the string and its linear density.

A similar derivation shows that the speed of longitudinal wave propagation in a fluid is given by

$$v = \sqrt{\frac{B}{\rho_V}} \qquad (11\text{-}7)$$

where ρ_V is the *volume mass density*, the mass per unit volume, and B is a constant for the fluid, called the *bulk modulus*, which describes its behavior on compression and will be discussed in Chapter 12.

Example 11-3 A transverse wave of frequency 120 Hz is applied to a string which has a linear mass density of 6.18×10^{-3} kg/m and is under a tension of 8.00 N. Determine (a) the speed of propagation and (b) the wavelength of the wave.

Solution We must first determine the speed by applying Equation 11-6. Then the wavelength is found by using Equation 11-5.

(a) $$v = \sqrt{\frac{T}{\rho_L}} = \sqrt{\frac{8.00 \text{ N}}{6.18 \times 10^{-3} \text{ kg/m}}} = 36.0 \text{ m/s}$$

(b) $$\lambda = \frac{v}{f} = \frac{36.0 \text{ m/s}}{120 \text{ /s}} = 0.300 \text{ m}$$

Example 11-4 Determine the value of the bulk modulus for water from the experimental evidence that sound travels 60.4 m in 4.00×10^{-2} s in water of density 1.00×10^3 kg/m³.

Solution We must first determine the speed of sound in the water and then apply Equation 11-7.

$$v = \frac{\text{distance}}{\text{time}} = \frac{60.4 \text{ m}}{4.00 \times 10^{-2} \text{ s}} = 1.51 \times 10^3 \text{ m/s}$$

Now from $v = \sqrt{B/\rho}$, we have

$$v^2 = \frac{B}{\rho}$$

or

$$B = v^2\rho = (1.51 \times 10^3 \text{ m/s})^2 \, (1.00 \times 10^3 \text{ kg/m}^3)$$
$$= 2.28 \times 10^9 \text{ N/m}^2$$

11-4 Reflections of Waves and Superposition

When a wave comes to the end of the medium in which it is traveling and to the boundary of a new medium, the energy must go somewhere. There are several possibilities: It is either reflected back to the original medium, transmitted through the new medium, absorbed by the new medium, or some combination of these. In most instances, the combination is the alternative that actually happens. However, two important special cases occur where virtually all the energy is reflected.

The end of the medium may be **fixed**, or **closed**, that is, not free to vibrate in the manner required by wave propagation. Such a fixed end occurs, for example, when a clothesline is attached to a hook firmly anchored in a wall. Figure 11-10 shows a **pulse** (easier to follow than a continuous wave train) approaching such a fixed end. As the wave reaches the wall, the line exerts an upward force on the immovable hook. The hook exerts an equal in magnitude but oppositely directed reaction force on the end of the line. This reaction force generates a pulse on the line which then travels along the line away from the hook. We say that the reflection of a wave from a closed end is **inverted in phase** from the original wave, a crest becoming a trough, and so on.

By contrast, the end of the medium may be **free**, or **open**, that is, capable of oscillating in the manner required by wave propagation. Figure 11-11 simu-

Figure 11-10 Reflection of a pulse from a *fixed end* not free to vibrate in the manner required for wave propagation. Note that the reflected pulse is inverted in phase and direction.

Figure 11-11 Reflection of a pulse from a *free end* free to vibrate transversely in the manner of the incoming wave. The reflected wave retains its original phase but is inverted in direction.

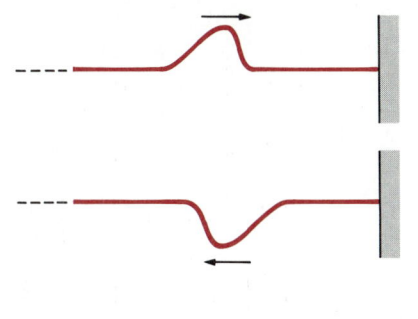

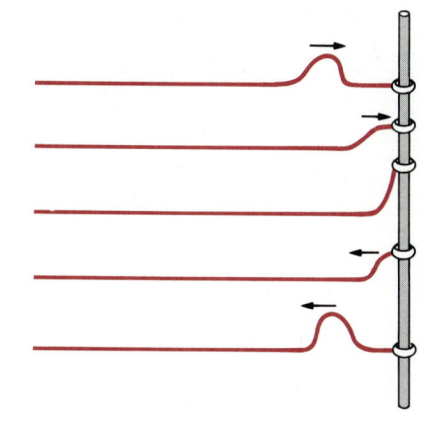

lates this by attaching the line to a ring which may slide frictionlessly along a vertical rod while maintaining a horizontal tension in the line. As the lead portion of the pulse encounters the ring, it exerts an upward force which begins to accelerate the ring. By the second position, the ring has arrived at its new equilibrium position for the peak of the pulse, but with considerable velocity. The ring overshoots this position, while the force of the pulse in the line now acts downward. The reaction force of the ring exerts an upward pull on the line, forming an upward pulse which will travel to the left along the line. We say that on reflection from a free end, a wave does not change phase.

Most ends to a medium are neither completely free nor completely fixed. Instead, the medium connects to a new medium at a joining point (called an interface). A wave will generally have different speeds of propagation in the two media usually due to their having different densities and elastic properties. Figure 11-12a shows two such media with a pulse in the first, of density ρ_L, traveling toward the second, of density ρ'_L. If ρ'_L is greater than ρ_L, medium 2 will present more inertia to resist wave propagation than does medium 1. Figure 11-12b shows that the reflected pulse is inverted by 180° in phase as though it had encountered a fixed end, like the anchored hook. The transmitted portion of the pulse in medium 2 propagates with a lower speed and is less far from the interface at any instant than the reflected pulse. Note, too, that the transmitted pulse is shortened; in other words, a periodic wave would have a shorter wavelength in material 2. However, the timing does not change at the interface. Hence, if the lead end of the pulse is 0.1 s ahead of its trailing edge in material 1, the same will be true in material 2. For a periodic wave, we would say that the frequency is unchanged.

Figure 11-12c shows the situation where ρ_L is greater than ρ'_L and medium 2 has a higher speed of propagation. As the pulse encounters the interface, medium 2 presents less resistance to oscillation (inertia) than medium 1; it accelerates and overshoots much like the ring on the vertical rod of Figure 11-11. The reflection back into medium 1 is erect as from a partially free end. The transmitted pulse is longer and moving faster than the original pulse, and if it were a periodic wave, its wavelength would be increased and its frequency would be unchanged. Note that in all cases the transmitted pulse is erect.

Several applications of wave transmission require the avoidance of reflections in order to transmit most of the energy. With mechanical waves, the way to do this is to avoid abrupt changes in density. For example, two strings of very different density (a tug-of-war rope and a clothesline) could be connected by a segment whose density changes almost continuously from the density of the first medium to that of the second. This continuous transition avoids the abrupt changes which give rise to reflections.[1]

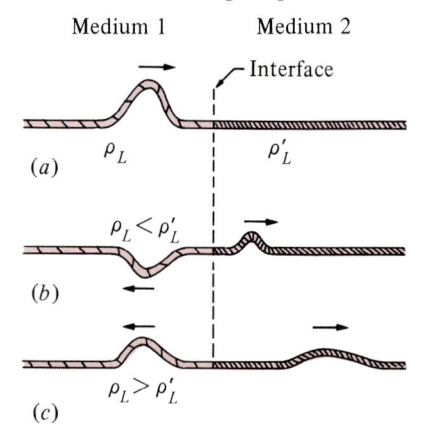

Often, two or more waves pass through the same point in a medium at the same time. When that happens, the displacement of the medium is found from the **principle of superposition:**

> The actual displacement at any point due to the presence of two or more disturbances is equal to the algebraic sum of the displacements due to the individual disturbances.

This assumes that the individual displacements are small, so that the resultant displacement never exceeds the ability of the system to oscillate.

---

[1] Nonmechanical examples abound also. Electric and electronic components are carefully matched at the interfaces between any two in order to reduce reflections and transmit maximum signals. The general problem of controlling wave reflections is common to all wave systems.

11 Mechanical Waves and Sound

Is the sidewall of a bathtub an open or a closed end for water waves? The key distinction as to whether an end is open or closed is not one of appearance. It is whether the particles of the medium at that end can (open) or cannot (closed) vibrate in the manner required for propagation of the wave. Water molecules at the tub wall certainly can slosh up and down, making it an open end for the transverse component of the wave but not for the longitudinal component.

Figure 11-12 (a) A pulse in a string of linear mass density ρ_L approaching the interface to a medium with a different linear mass density ρ'_L. (b) If $\rho'_L > \rho_L$, the interface acts as a fixed end and the pulse reflected back into the original string is inverted in phase. The pulse entering the new medium is unchanged in phase, but its speed and length are determined by the properties of the medium. (c) If $\rho'_L < \rho_L$, the reflected pulse is not inverted in phase. The transmitted pulse moves away from the interface at a higher velocity and is longer although representing the same duration of disturbance as the original pulse.

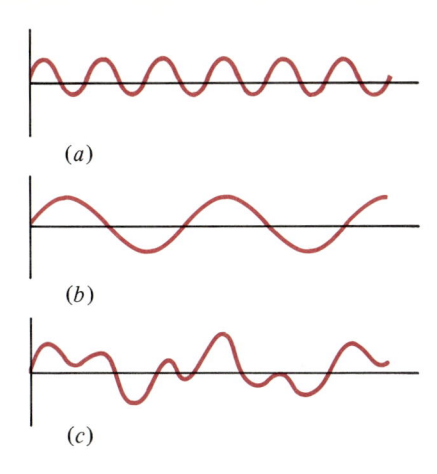

Figure 11-13 (*a* and *b*) Two waves traveling on the same medium at the same time. (*c*) By the principle of superposition, each position here is the vector sum of the waves shown in (*a*) and (*b*). Since this pattern is generated by sinusoidal disturbances, it repeats regularly in time and space. Indeed, the rightmost part begins the repetition of the pattern begun at the left.

The term **interference** is most often used to describe the physical effects of the superposition of two or more waves. When two crests or two troughs are superimposed, the resultant crest or trough has a greater displacement than either of the two individually (twice as great if the two amplitudes are equal). This is *constructive interference*. When a crest of one wave is superimposed on a trough of another, the resultant displacement is smaller in magnitude than that of the larger amplitude wave. This is *destructive interference*.

Figure 11-13*a* and *b* shows two waves traveling separately on the same medium. These waves may be added to form the complex (not sinusoidal) wave shown in Figure 11-13*c*. A vertical line running across the three drawings of Figure 11-13 at any location will allow you to estimate the displacement of the wave in (*a*) and (*b*). Vectorial addition of these displacements will be seen to yield the total displacement for that point as shown in Figure 11-13*c*. The principle of superposition is a great aid in determining how waves add together. It is also used in analyzing a complex wave to determine its constituent parts. We use the principle of superposition frequently, and the following three sections might be considered to be applications of that principle.

11-5 Standing Waves on a String

The waves considered to this point have been moving freely along a medium. Waves from a rock thrown into a pond visibly move outward from the source of the disturbance. This is why we called them traveling waves. If you have experience with a small boat, you may be familiar with traveling waves that strike the boat with a particular frequency and rock it repeatedly. A different pattern is possible, referred to as *chop,* where the surface at various positions rises and falls in wavelike action but without any apparent horizontal movement of the pattern from place to place. This situation can be produced in an anchored clothesline. As your hand pumps the free end of the line in a quick up-and-down motion, a wave disturbance will be seen to race away from the hand along the clothesline toward the fixed end. If you stop after a short burst of wiggling, this may be all that you will notice. However, if you continue pumping new waves into the free end after the original waves have had time to reach the other end and reflect, another pattern may emerge. Often, the string will be seen to oscillate up and down at certain points while other points remain largely motionless. No motion of an overall pattern from right to left, or vice versa, will be apparent. This is a **standing wave.**

Standing waves may be produced whenever a train of waves crosses its own reflection. Since one train of waves is moving right to left while another is proceeding left to right, it is not surprising that the overall appearance of the medium shows no horizontal movement of an overall disturbance pattern. Indeed, certain positions exist where the incoming wave, or **incident wave**, and the reflected wave will always differ in phase by 180° and, by the principle of superposition, always add to essentially zero disturbance at that point. This produces a point called a **node** which remains nearly at rest at all times. At other places, halfway between nodes, the two sets of waves will alternately agree and disagree in phase — from maximum positive disturbance to maximum negative disturbance and repeat. Such positions will oscillate strongly and are called **antinodes**. Because this is a very important case with many applications, we will develop its mathematics carefully.

Figure 11-14 is a sequence showing a transverse wave in a string moving to the right (black line). After reflection, one can see the reflected wave (color line) progressing to the left. The incident wave and the reflected wave are passing through the same points in the medium, and hence the principle of superposition applies. (We will continue to use the transverse wave in a string as an example since it is easier to visualize as well as to illustrate. We must keep in mind, however, that the equations could equally well apply to a longitudinal wave.) Once again we wish to obtain a relationship which describes the transverse displacement y in terms of the horizontal displacement x and time t. If the incoming wave is y_1 and the reflected wave is y_2, we have

$$y_1 = y_{max} \sin\left[\frac{2\pi}{\lambda}(x - vt)\right]$$

and

$$y_2 = y_{max} \sin\left[\frac{2\pi}{\lambda}(x + vt)\right]$$

We have used the fact that the only change to the derivation of Equation 11-4 which would be necessary if the wave were moving to the left instead of to the right is to replace the minus sign between x and vt with a plus sign. From the principle of superposition, we may write the total transverse displacement as

$$y = y_1 + y_2 \quad (11\text{-}8)$$

Now y_1 and y_2 may be rewritten (using $f = v/\lambda$) and expanded as the sine of the sum and differences of two angles.[1]

$$y_1 = y_{max} \sin\left(\frac{2\pi}{\lambda}x - 2\pi ft\right)$$

$$y_1 = y_{max} \sin\frac{2\pi x}{\lambda} \cos 2\pi ft - y_{max} \cos\frac{2\pi x}{\lambda} \sin 2\pi ft \quad (11\text{-}9)$$

and

$$y_2 = y_{max} \sin\left(\frac{2\pi x}{\lambda} + 2\pi ft\right)$$

$$y_2 = y_{max} \sin\frac{2\pi x}{\lambda} \cos 2\pi ft + y_{max} \cos\frac{2\pi x}{\lambda} \sin 2\pi ft \quad (11\text{-}10)$$

---

[1] Applying the identity $\sin(\theta_1 \pm \theta_2) = \sin\theta_1 \cos\theta_2 \pm \cos\theta_1 \sin\theta_2$.

11 Mechanical Waves and Sound

Figure 11-14 (a) A moving wave proceeds rightward to a fixed end. (b) A short time later, the original wave has been reflected and proceeds leftward (colored line) across the incoming wave (black line).

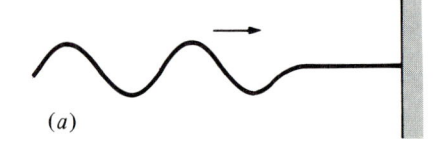

(a)

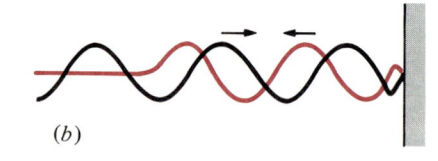

(b)

Whereupon adding Equations 11-9 and 11-10 to obtain y yields

$$y = 2y_{max} \sin \frac{2\pi x}{\lambda} \cos 2\pi ft \qquad (11\text{-}11)$$

Equation 11-11 is often written as

$$y = (2y_{max} \cos 2\pi ft) \sin \frac{2\pi x}{\lambda} \qquad (11\text{-}12)$$

Any wave which obeys Equation 11-12 is called a *standing wave*. This equation has two separate parts. The expression in parentheses is a function of time, while the separate sine term is not. This *time-independent* sine term defines a maximum size for the disturbance at any point along the medium regardless of time. We say that it defines the envelope which will contain the disturbance on the medium at all times. As time progresses, the disturbance may vary but only within the defined envelope. Note that as the position along the medium x axis takes on different values, $\sin(2\pi x/\lambda)$ grows and shrinks as a sine wave. In particular, whenever the expression $2x/\lambda$ takes on a whole-number value n, the value of $\sin(2\pi x/\lambda)$ becomes $\sin n\pi = 0$. Thus, the time-independent portion of the expression requires that the disturbance at certain positions be *nodes*. Within the envelope determined by the sine expression, the time-dependent cosine expression in parentheses in Equation 11-12 determines the instantaneous displacement for any time.

If we stretch a rubber band between forefinger and thumb, then displace the center of its outer strand and let it snap back, we will witness a very complex action. The disturbances we created at the center propagate to both right and left along the band, where they encounter either thumb or finger, reflect, and come back. The separate waves cross each other to reflect at the other end, cross again, etc. This action will repeat until the original energy of the disturbance is lost to the fingers or the air. Since the pattern of disturbance on the band changes very rapidly, we are not capable of seeing the pattern of instantaneous displacement of portions of the band. Instead, we see all displacement of the band blurred together through time, the extremes of which are the envelope of the standing wave.

Figure 11-15a shows such an extended time view of a vibrating string. The positions of nodes are clearly defined. The positions of maximum disturbance midway between nodes are the antinodes. For comparison, Figure 11-15b is a photographic record of the instantaneous displacement pattern of the same string undergoing the same vibration. This photograph was recorded with a high-speed light flash lasting some thirty-millionths of a second and is definitely *not* what the eye would be able to detect. These two photographs really examine the individual parts of Equation 11-12 separately. Figure 11-15b shows the string for a particular value of t and records a sharply defined position for each small segment of the string. It is not clear from this photograph that this is a standing wave and not a traveling wave. That is, it is not clear that the pattern shown is the superposition of waves and will not just move left or right with time. Figure 11-15a was taken in about a tenth of a second of time. Individual segments are not precisely defined in their y displacements since they varied over their entire possible range during the time interval of the photograph. Instead we see the time-independent envelope of the standing wave, which is what the eye usually perceives for a string in rapid oscillation.

The position of nodes for a standing wave may be precisely defined. Consider the rubber band example. Since the band is held against thumb and forefinger, these must be fixed ends. The band cannot vibrate where it is fixed; thus the ends of the band must be nodes. Likewise, the endpoints of the vibrating string in Figure 11-15 are fastened rigidly, cannot vibrate, and must be nodes. There may or may not be additional nodes between these fastened endpoints. If not, the string is vibrating in its simplest possible manner, and we have what is called its *fundamental* mode of vibration. Clearly, the string in Figure 11-15 has many nodes. We can derive the position of all the nodes, using some properties of the sine function. Consider once again the string of length l stretched along the x axis shown in Figure 11-16. At any given instant, the configuration of the string tracing out each of those envelopes must obey Equation 11-12. The frequency and wavelength must be different for each

Figure 11-15 (a) A standing wave as the eye might see it. The blur of the dropped ball in the background indicates the time during which the photograph was taken. (b) High-speed photograph of the same string undergoing the same vibration. The crispness of the falling ball in the background indicates that we are observing the *instantaneous* condition of the string. (D. Riban)

11-5 Standing Waves on a String

Fundamental mode of vibration (first harmonic) (a) — String length = half wavelength

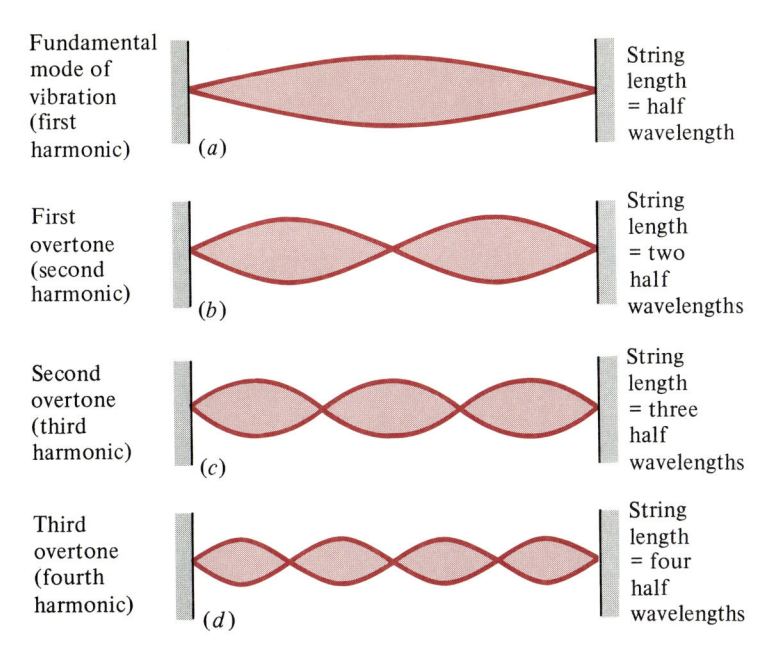

First overtone (second harmonic) (b) — String length = two half wavelengths

Second overtone (third harmonic) (c) — String length = three half wavelengths

Third overtone (fourth harmonic) (d) — String length = four half wavelengths

Figure 11-16 When a string under tension is displaced and released, only waves of certain wavelengths form standing waves on the precisely defined medium; the rest rapidly die out. Each end of the string is fixed and cannot vibrate and thus must be a node for any standing-wave pattern that emerges. The simplest pattern is when the nodes are a half wavelength apart with an antinode between, as shown in (a). The second simplest pattern has one node between the nodes at the end of the string and is shown in (b), and so on.

pattern, but the configuration must be a sine wave obeying $\sin(2\pi x/\lambda)$. We know that this is zero when $x = 0$ and the first end is a node. How can it be zero at the other end? We know there must be a node at the other end since it is fastened and cannot vibrate. Therefore we insist that $\sin(2\pi x/\lambda)$ must be zero when $x = l$, or

$$\sin \frac{2\pi l}{\lambda} = 0$$

Now the sine of an angle is zero whenever the angle is any integral multiple of π radians. Thus our condition is satisfied if

$$\frac{2\pi l}{\lambda} = n\pi \quad \text{where } n = 1, 2, 3, \ldots$$

Solving, we obtain

$$\lambda = \frac{2l}{n} \quad n = 1, 2, 3, \ldots \quad (11\text{-}13)$$

The condition required for standing waves in a string of length l fixed at both ends is that the wavelength be one of the values

$$2l, \frac{2l}{2}, \frac{2l}{3}, \frac{2l}{4}, \ldots$$

These first four are shown in Figure 11-16.

Once the string is selected (the mass per unit length is known) and the tension determined, Equation 11-6 fixes the speed of the wave motion in the string. From this point on, the frequency of vibration of a standing wave depends on the wavelength. Solving Equation 11-5 for the frequency yields

$$f = \frac{v}{\lambda}$$

Now substituting for λ from Equation 11-13 gives us

$$f_n = \frac{nv}{2l} \qquad n = 1, 2, 3, \ldots \qquad (11\text{-}14)$$

Thus the frequency of vibration of a standing wave in a string fixed at both ends may take on the values

$$\frac{v}{2l}, \frac{2v}{2l}, \frac{3v}{2l}, \ldots \quad \text{and no others}$$

These are the natural frequencies of vibration for the fixed string. The smallest natural frequency for any system which has more than one is called the **fundamental frequency**. The accepted convention is to call all natural frequencies other than the fundamental, **overtones**. Thus, for the fixed string, the fundamental frequency is $f_1 = v/2l$. All other natural frequencies (including those which are not integral multiples of the fundamental) are overtones. For example, $2v/2l$ is the first overtone, $3v/2l$ the second, and so on.

Figure 11-17a shows a frequency analysis for the sound produced when striking the note A on the piano. The accepted frequency for concert A is 440.0 Hz, which is the fundamental vibration for the string. The analysis reveals this is the most intense (loudest) vibration present. However, vibrations of 880.0 Hz, 1320.0 Hz, etc., are also present with varying intensity. These are the first, second, . . . overtones of the string, or $2f_1$, $3f_1$, . . . in frequency. Comparing this analysis with the possible modes of vibration for a standing wave in Figure 11-16 shows the largest amount of energy (hence the loudest

Figure 11-17 (a) A sound spectrum for the sound produced from a piano striking the note A (440 Hz). In addition to the fundamental frequency, an entire series of overtones are present in lesser intensities. (b) A similar diagram for the same note played on a violin displays an entirely different pattern of overtone intensities.

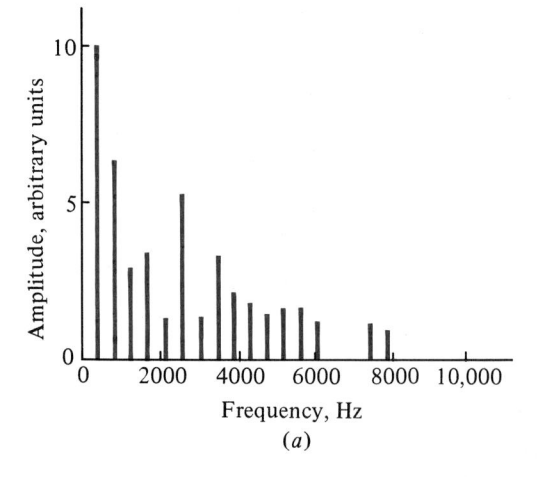

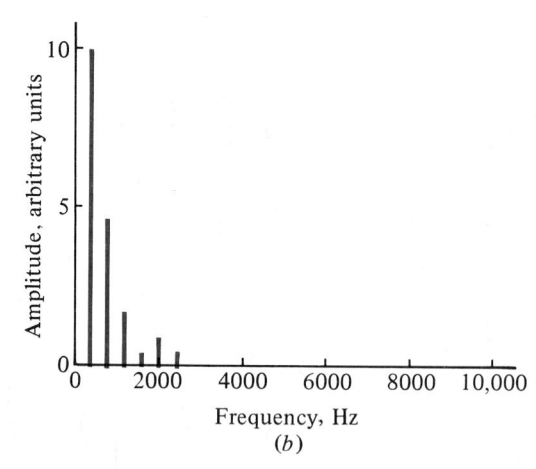

sound) associated with the fundamental. Lesser energies are involved in the other modes of vibration, but these overtones will be present whenever A is sounded on the piano.

Whenever an overtone is an integral multiple of the fundamental frequency, it is called a **harmonic.** The number of the harmonic of a certain frequency is equal to the integer which must be multiplied by the fundamental to obtain the frequency. Hence $f_2 = 2(v/2l) = 2f_1$, and $f_3 = 3(v/2l) = 3f_1$. The first overtone thus is also the second harmonic, the second overtone is the third harmonic, and so on. Note that the fundamental frequency is always the first harmonic. In the possible natural frequencies of the fixed string, all harmonics are present. This is not true of all systems. In many systems only certain harmonics may be present in the overtones. In some systems, none of the overtones are harmonics.

Why is it that a piano, a violin, and a horn, for example, can play the same note yet sound quite distinct to the ear? The series of frequencies present in a vibration leads us to an answer. Figure 11-17b shows the frequency analysis of a violin playing the note A. Observe that the relative strengths of overtones is quite different from the same note played on the piano. In part, this difference accounts for the musical quality of a sound. Both instruments produce the same fundamental, but the mixture of overtones accompanying it is quite different and distinct for each. This makes the same note produced on different instruments sound quite unique to the ear.

Example 11-5 The string of Example 11-3 supported a speed of propagation of 36.0 m/s and was conducting waves with a wavelength of 0.300 m. Suppose it is 0.450 m long and is fixed at both ends. (a) At what harmonic and overtone is the string vibrating? (b) How many antinodes are there?

Solution (a) Since we know both λ (0.300 m) and l, we may apply Equation 11-13 to find n. This is the number of the harmonic. Also, since it is a string fixed at both ends, we know that the number of the overtone is the harmonic number minus 1.

$$\lambda = \frac{2l}{n}$$

or

$$n = \frac{2l}{\lambda} = \frac{2(0.450 \text{ m})}{0.300 \text{ m}} = 3$$

Thus it is vibrating in the third harmonic, which is the second overtone. (b) Close examination of Figure 11-16 shows that the number of antinodes is equal to the harmonic number; thus the number of antinodes is equal to n, which is 3. In general, antinodes occur every half wavelength along the string, as do nodes.

11-6 Standing Longitudinal Waves

Imagine that we play a constant frequency, say 1000 Hz, through a speaker aimed at a nearby reflecting surface, like a flat, hard wall. Under certain conditions, if we move a microphone along the axis of the speaker from the speaker to

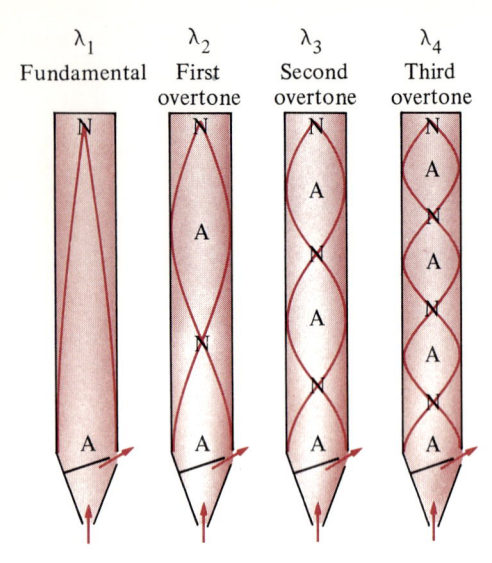

| λ_1 | λ_2 | λ_3 | λ_4 |
| Fundamental | First overtone | Second overtone | Third overtone |

Figure 11-18 Standing waves in a closed-end organ pipe. (The transverse-type standing waves in colored lines are to aid visualization only, as the waves in air are longitudinal.)

the wall, it will detect a variation in the loudness of the sound. This could be revealed by the built-in meter of a recorder, or by recording the signal detected by the microphone as it moves along this line and listening to the playback. Even though the speaker output was kept constant, the measured intensity would be found to vary. Loud zones, louder than the sound from the speaker alone if playing in an open field, would alternate with quiet zones where the intensity drops to near zero. We would be sampling a standing sound wave in air as waves from the speaker cross their reflected waves from the wall.

A simpler case to study is that of sound waves confined to the length of a tube, as used in pipe organs and many other musical instruments. At the end of the tube, there is an abrupt change in the medium and sound is reflected. If the end is closed, it acts as a fixed end, and the reflected sound changes phase by 180° with the incident sound. There is always a node at a closed end. If the end of the pipe is open to the atmosphere, it still represents a change of medium and some energy is reflected. The reflected wave has about the same phase as the incident wave,[1] and an antinode results. The organ pipe shown in Figure 11-18 has its upper end closed. The bottom end of the pipe is marked with an A for antinode. (The triangular chamber below is a separate compressed air chamber opening at the side of the tube to direct an airstream across the tube end and excite the air column to vibration.) Similarly, the position near the closed end is marked with an N for node. Both these conditions must be true for any standing wave in the pipe. However, between these positions, there may or may not be other nodes and antinodes. The fundamental frequency and the first three overtones are depicted in the four views. In the illustration of the fundamental, we see that the length of the pipe between the opening and the end is the distance between a node and an adjacent antinode. This is one-fourth a wavelength. Thus

$$l = \frac{\lambda_1}{4}$$

---

[1] The exact point of reflection depends on the ratio of the diameter of the pipe to the wavelength of the sound. If this ratio is very small, as is true for most musical instruments, the open end is very nearly a displacement antinode, and the reflected wave is in phase with the incident wave.

For the first overtone, we see that

$$l = \frac{3\lambda_2}{4}$$

Continuing with the second and third overtone, we have

$$l = \frac{5\lambda_3}{4} = \frac{7\lambda_4}{4}$$

The pattern developed is that

$$l = \frac{(2n-1)\lambda_n}{4}$$

or

$$\lambda_n = \frac{4l}{2n-1} \qquad n = 1, 2, 3, \ldots$$

The natural frequencies of the closed pipe are then

$$f_n = \frac{v}{\lambda_n} = \frac{(2n-1)v}{4l} \qquad n = 1, 2, 3, \ldots \qquad (11\text{-}15)$$

where v is the speed of sound in air. We note that although all the overtones are harmonics, only the odd harmonics are possible.

Figure 11-19 shows the fundamental and the first three overtones in an organ pipe open at both ends. For the open pipe, there must be antinodes at both ends. For the fundamental, we see that the length of the pipe is equal to the distance between two adjacent antinodes, or one-half wavelength. Thus

$$l = \frac{\lambda_1}{2}$$

For the first overtone, we see that

$$l = \frac{2\lambda_2}{2} = \lambda_2$$

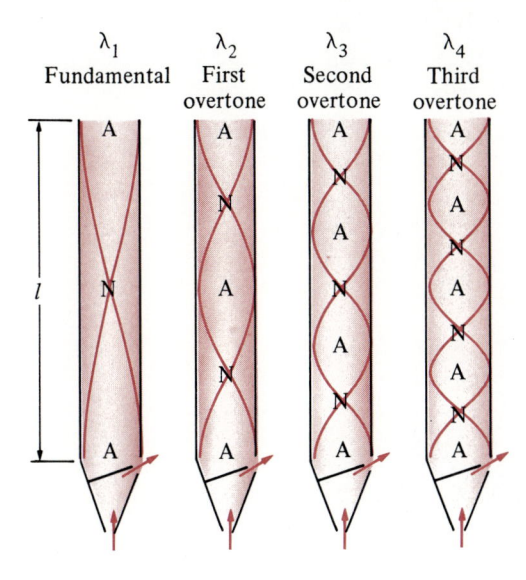

Figure 11-19 Standing waves in an open-end organ pipe. (Again, the transverse standing waves in colored lines are not a physical representation.)

11-6 Standing Longitudinal Waves

253

and for the second and third overtones,

$$l = \frac{3\lambda_3}{2} = \frac{4\lambda_4}{2}$$

In general,

$$l = \frac{n\lambda_n}{2}$$

or

$$\lambda_n = \frac{2l}{n} \qquad n = 1, 2, 3, \ldots$$

Thus the natural frequencies of the open pipe are

$$f_n = \frac{v}{\lambda_n} = \frac{nv}{2l} \qquad n = 1, 2, 3, \ldots \qquad (11\text{-}16)$$

For the open pipe, we see that all overtones are harmonics and all harmonics are possible.

All musical instruments have many natural frequencies which *may* be excited. A musical note specifies the fundamental frequency to be sounded by the instrument. When sounded, the amplitudes of each of the various natural frequencies accompanying the fundamental may be relatively large or vanishingly small. The specific overtones which are excited and their amplitudes determine tone quality. Most often the fundamental determines what we perceive as pitch.

Example 11-6 If the speed of sound in air[1] is 345 m/s, what length of organ pipe is required for a fundamental frequency of 45.0 Hz if the pipe is (*a*) closed or (*b*) open? (*c*) What would be the wavelength in each case?

Solution Since the fundamental is the first harmonic, we apply Equation 11-15 with $n = 1$ for part *a* and Equation 11-16 with $n = 1$ for part *b*.
(*a*) For the closed pipe,

$$f_n = \frac{(2n-1)v}{4l_c}$$

$$f_1 = \frac{v}{4l_c}$$

or

$$l_c = \frac{v}{4f} = \frac{345 \text{ m/s}}{4(45.0/\text{s})} = 1.92 \text{ m}$$

(*b*) For the open pipe,

$$f_n = \frac{nv}{2l_o}$$

$$f_1 = \frac{v}{2l_o}$$

---

[1] For examples and end-of-chapter problems, we assign reasonable values for the speed of sound in a range from 330 to 350 m/s. This value varies with air temperature and is near the lower end of the range at freezing and near the upper end on a hot summer day.

or

$$l_0 = \frac{v}{2f} = \frac{345 \text{ m/s}}{2(45.0/\text{s})} = 3.83 \text{ m}$$

(c) We first recognize that the wavelength is the same for both cases and then use the relationship among speed, frequency, and wavelength.

$$v = \lambda f$$

$$\lambda = \frac{v}{f} = \frac{345 \text{ m/s}}{45.0/\text{s}} = 7.67 \text{ m}$$

11-7 Beats

When two waves of equal amplitude but slightly different frequency are superimposed, alternating constructive and destructive interference occurs in a phenomenon known as **beats** (Figure 11-20). Consider two tuning forks, one vibrating at 400 Hz and the other at 401 Hz. Assume that they are started in oscillation and are in phase. Initially, both forks generate a crest at the same time, followed by a trough, etc. If the forks are equidistant from the ear and producing waves of the same intensity, the waves will add to produce a louder tone than either fork separately. However, consider the situation 0.5 s later. Fork B has to accomplish one more vibration per second than fork A. In 0.5 s it will be half a vibration ahead of fork A. Thus, while A is giving off a crest at $t = 0.500$ s (if the forks were at crest at $t = 0.00$ s), fork B will have advanced in the oscillation cycle half a wave beyond crest to a trough. By the principle of superposition, the intensity at the ear now drops to zero. As t approaches 1.00 s, the forks come back into phase and again produce a louder sound together than either separately. A moment's reflection should convince you that if the initial frequencies were 400 and 402 Hz, the forks would agree and disagree twice per

Figure 11-20 (a) Displacement vs. time for two waves of slightly different frequency passing through the same point in space. (b) The resultant of the two waves due to superposition shows a regular fluctuation in amplitude from zero to twice the amplitude of either wave, illustrating the phenomenon of *beats*.

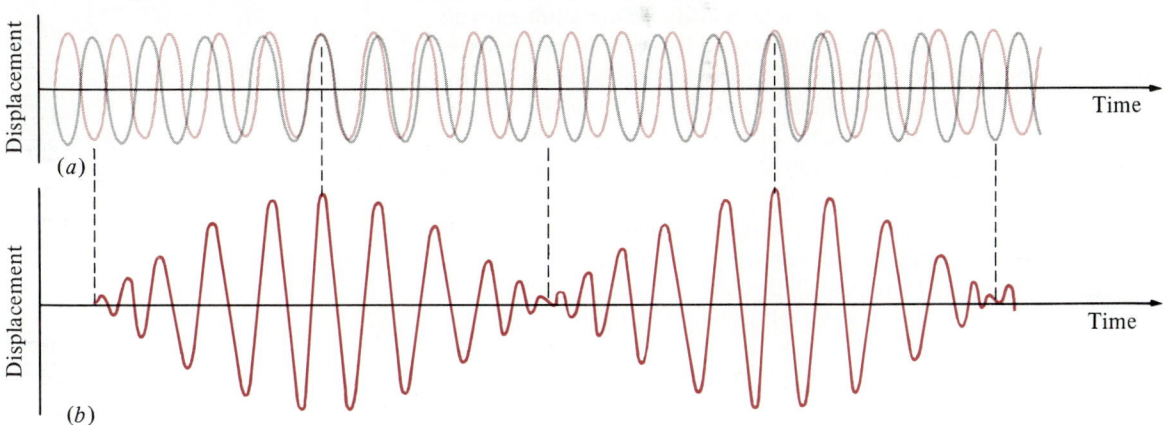

second to produce two beats per second. In general, the frequency of the beats f_b is just

$$f_b = f_1 - f_2 \qquad (11\text{-}17)$$

If the two components are sound waves, the variation in amplitude gives rise to a variation in loudness. As the frequencies of the two components get closer together, the beat frequency gets smaller, making a longer and longer time between beats. This is used by piano tuners and guitarists to tune their instruments. Two strings may be tuned to the same frequency by increasing the tension on the lower (or decreasing tension on the higher) while plucking both. This process is continued until the time between beats gets so long that it is no longer discernible.

Example 11-7 When two tuning forks are struck at the same time, three beats per second are heard. One of the forks is known to vibrate at 440 Hz. Determine the possible frequencies of the other.

Solution This is an application of Equation 11-17. However, we do not know whether the unknown is f_1 or f_2. Indeed, either is possible. Thus, using $f_b = f_1 - f_2$, either

$$3 = 440 - f_2 \quad \text{or} \quad 3 = f_1 - 440$$

and the unknown frequency is either 437 Hz or 443 Hz.

11-8 Sound Intensity Level

A physics text does not often point out that observation of physical phenomena, since it is done by human beings, often has a subjective nature. A good general definition of sound might be a sensation which can be detected by the sense of hearing. Physicists or acoustical engineers, however, would insist on being more specific. They would note that sound is a longitudinal wave and that one pure sound may differ from another in only two properties, frequency and intensity. Thus the experts would say a sound is completely specified when its frequency and intensity are known.

We are familiar with the term frequency as the number of vibrations per second. It has been established that the average healthy young adult ear can detect sounds with frequencies between about 20 and 20,000 Hz. **Intensity** is defined as the energy per unit time per unit area that the sound wave carries across an imaginary area placed perpendicular to the sound. Intensity is measured in joules per second per square meter ($J/s/m^2$) or in watts per square meter (W/m^2). Perception of loudness is not altogether objective and varies with frequency. Thus again we employ the average young adult ear, which can just barely hear sounds of frequency 1000 Hz at an intensity of $I_0 = 1.00 \times 10^{-12}$ W/m^2. Further, the healthy young ear will hear the 1000-Hz sound increase in loudness as intensity increases all the way up to about 1.00 W/m^2.* Above this intensity, the sound is generally felt rather than heard and causes

---

* There is quite a large range between the intensities of the softest and loudest sound to be heard. The loudest is 1,000,000,000,000 (1 trillion) times the intensity of the softest one.

pain to the ear. We call the smallest intensity for which a sound is heard the **threshold of hearing**. The intensity above which sound is felt is called the **threshold of pain** (or **threshold of feeling**).

Although our perception of the loudness of a sound increases with increasing intensity, it is not linear. If we perceive a second sound to be twice as loud as a first sound, we find the intensity of the second is about 10 times that of the first. Further, a second sound perceived to be three times as loud as the first has about 100 times the intensity. In an attempt to compromise between the objective (intensity) and subjective (loudness) as well as to create a reasonable range of numbers, a kind of measure was devised called **sound intensity level** (SIL). The SIL of a sound of intensity I is defined as 10 times the common logarithm (base 10) of the ratio of the intensity I to the reference intensity I_0. In equation form,

$$\text{SIL} = 10 \log \frac{I}{I_0} \qquad (11\text{-}18)$$

SIL has no real units, but we give the name **decibel** (dB) to sound levels reported in this way. The reference intensity is chosen as the threshold of hearing at 1000 Hz of 10^{-12} W/m². Figure 11-21 shows SIL vs. frequency. Curves of equal perceived loudness are shown, including the thresholds of hearing and pain.

Figure 11-21 Human auditory response for a normal ear, from the lowest perceived sound to the loudest. The human ear is most sensitive near frequencies of 4000 Hz, but the plateau of the curves near 1000 Hz is used to standardize sound intensity levels. At 40 Hz the ear can just barely perceive a sound with an intensity level of 58 dB, while the same ear perceives sounds of 1000 or 6000 Hz with a SIL of 0 dB.

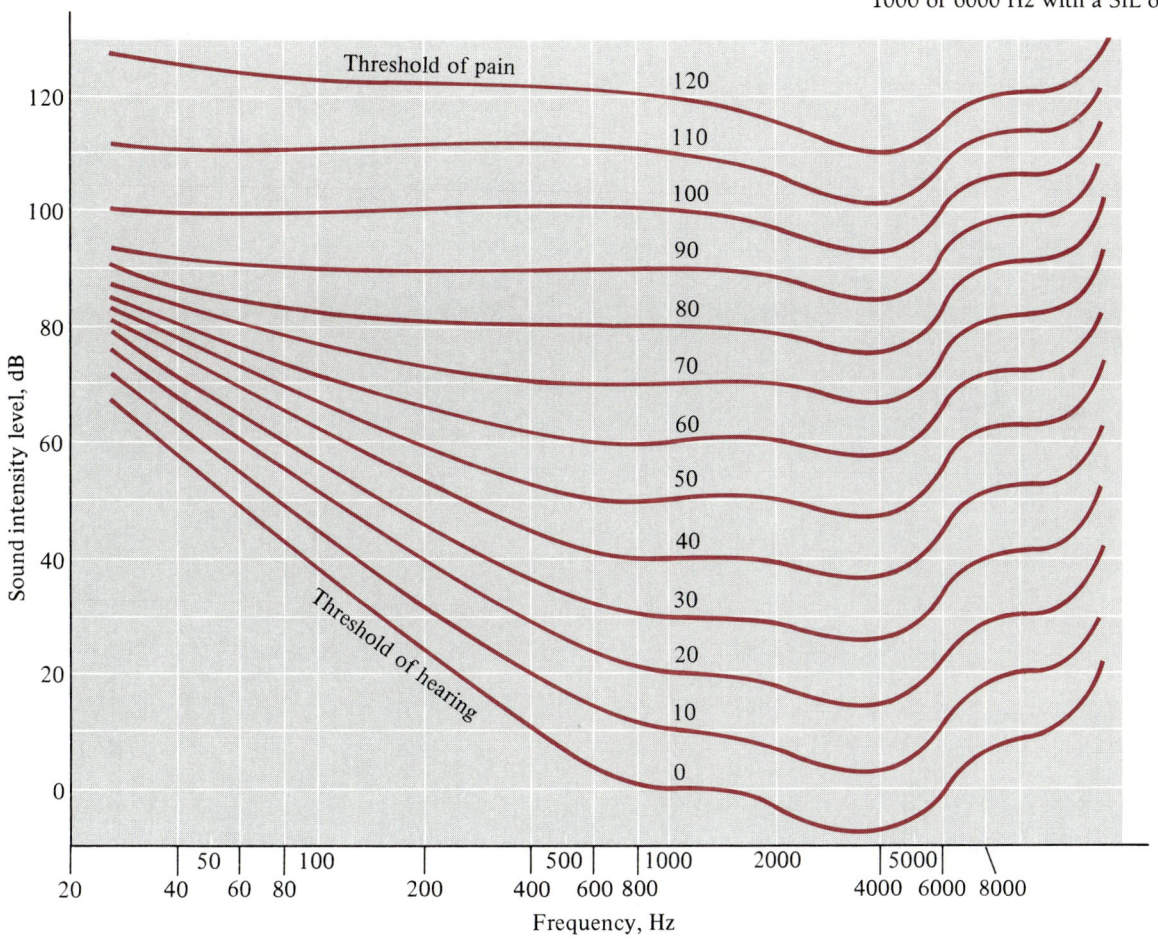

Example 11-8 At a certain distance away from a trumpet being played, the SIL is measured to be 85 dB. Calculate (a) the intensity of this sound and (b) the SIL of three such trumpets played simultaneously.

Solution (a) We must simply solve Equation 11-18 for I, given SIL and I_0.

$$SIL = 10 \log \frac{I}{I_0}$$

$$85 = 10 \log \frac{I}{10^{-12}}$$

$$\log \frac{I}{10^{-12}} = 8.5$$

$$\frac{I}{10^{-12}} = 10^{8.5}$$

$$I = 10^{-3.5} = 3.16 \times 10^{-4} \text{ W/m}^2$$

(b) To solve this part, we must recognize that it is intensities which add. Thus the intensity of three identical trumpets playing equals three times the intensity of one of them. Using the intensity, we can determine SIL.

$$I_3 = 3I_1 = 9.48 \times 10^{-4} \text{ W/m}^2$$

$$SIL = 10 \log \frac{9.48 \times 10^{-4}}{10^{-12}} = 89.8 \text{ dB}$$

Example 11-9 An automobile driver in the city is talking to a passenger. The windows are all open. If the SIL of conversation is 65 dB and the SIL due to traffic and street noise is 70 dB, determine (a) the sound intensity and (b) the SIL in the automobile.

Solution (a) We solve Equation 11-18 for the individual intensities and add them directly.

$$65 = 10 \log \frac{I_1}{I_0}$$

$$\frac{I_1}{10^{-12}} = 10^{6.5}$$

$$I_1 = 10^{-5.5} = 3.16 \times 10^{-6} \text{ W/m}^2$$

$$70 = 10 \log \frac{I_2}{I_0}$$

$$I_2 = 10^{-5} \text{ W/m}^2$$

Thus

$$I = I_1 + I_2 = 1.32 \times 10^{-5} \text{ W/m}^2$$

(b) We may directly apply Equation 11-18.

$$SIL = 10 \log \frac{1.32 \times 10^{-5}}{10^{-12}} = 71.2 \text{ dB}$$

11-9 The Doppler Effect

The proper frequency of a sound wave (or any other wave) is determined by the source. The number of cycles per second depends on how often the source vibrates, and once produced cannot be changed by interaction with the medium or an observer. Under certain circumstances, however, the frequency as *measured* by an observer may be different than that being produced by the source. If we listen to an automobile horn as the automobile passes us at high speed, we hear an apparent drop in frequency. That is, the pitch (which we interpret as frequency) of the sound from the horn as the automobile approaches us is higher than the pitch as the automobile recedes. This phenomenon of an apparent change in frequency due to the motion of the source and/or observer is called the **Doppler effect,** after Christian Doppler, who first called attention to the effect in 1842. We may determine its cause if we consider two sources of waves, one stationary and one moving.

In Figure 11-22 we see at s a stationary source of waves. The circles, which represent adjacent crests of the emitted waves, are concentric as we might expect since the wave speed v is the same in all directions and the source is stationary. The distance between adjacent crests in Figure 11-22 is exactly one wavelength. The largest circle represents the crest of a wave produced the longest time ago. An observer O, if stationary, would have exactly the same number of crests go past per unit time as left the source per unit time. Thus the observed frequency f_o would be the same as the source frequency f_s.

Consider Figure 11-23, where the source of waves is moving to the right with a speed v_s and the observer is moving to the right with a speed v_o. We will derive a relationship between the source frequency f_s and the observed frequency f_o. The crests all form circles, but each is centered on the position at which the source was when it sent out that particular crest. The crests move out from their respective centers with a speed v. *All speeds are measured with respect to the medium,* and *the direction from the source to the observer is positive.* For the derivation, consider a time interval Δt during which the source moves from the position s' to the position s. The distance the source moves during Δt equals $v_s \Delta t$, and the distance the first crest moves equals $v \Delta t$. Thus the distance from s to point P equals $(v - v_s)\Delta t$. During the time Δt, the number of waves emitted is $f_s \Delta t$. Therefore the apparent wavelength in front of the source is

$$\lambda_a = \frac{(v - v_s)\Delta t}{f_s \Delta t} = \frac{v - v_s}{f_s}$$

The speed of the waves with respect to the observer is $v - v_o$. Hence the frequency at which the observer has waves go past is

$$f_o = \frac{v - v_o}{\lambda_a} = \frac{v - v_o}{(v - v_s)/f_s}$$

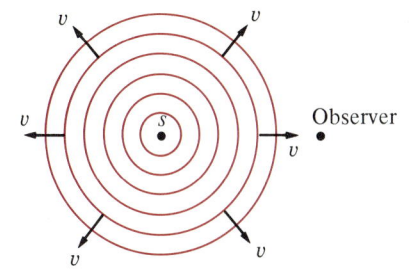

$$f_o = \frac{v - v_o}{v - v_s} f_s \qquad (11\text{-}19)$$

Figure 11-22 The stationary source at s generates sound waves of fixed frequency which propagate as ever-expanding circles centered on the source. Each circle represents a crest of the sound wave.

Figure 11-23 A moving source also generates expanding circular waves, each centering on the source at the instant the wave was generated. Because the source is moving, the circles are not concentric, but concentrate in the direction of forward motion and spread farther apart in the rearward direction. This alters the frequency measured by an observer from the transmitted frequency in the phenomenon known as the Doppler shift.

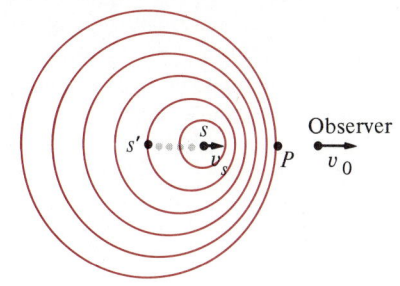

Equation 11-19 holds true for all possibilities provided it is applied using the conditions of its derivation. The direction from source to observer is positive. If the source is moving toward the observer, v_s is a positive number. If the source is moving away from the observer, v_s is a negative number. If the observer is moving away from the source (in the direction from source to observer), v_o is a positive number; otherwise it is negative. If the motion of the source and/or observer is not along the line joining the two, v_s and v_o must be taken as the components of the source and observer velocities along this line. In any case, v_s and v_o are always to be measured with respect to the medium, and v is always a positive number.

Example 11-10 What frequency is observed by someone driving toward a fire station at 20.0 m/s when the whistle is blowing? The frequency of the fire whistle is 900 Hz, and the speed of the sound is 340 m/s.

Solution This is a direct application of Equation 11-19, recognizing that v_o is a negative number (that is, $v_o = -20.0$ m/s).

$$f_o = \frac{v - v_o}{v - v_s} f_s = \frac{340 - (-20)}{340 - 0} (900) = 953 \text{ Hz}$$

Example 11-11 An automobile with a horn of frequency 600 Hz is traveling toward an intersection when the driver sounds the horn. A pedestrian standing at the intersection measures the frequency to be 640 Hz. How fast is the automobile traveling? Take the speed of sound to be 340 m/s.

Solution This is another direct application of Equation 11-19, but this time we recognize that v_o is zero and we must solve for v_s.

$$f_o = \frac{340 \text{ m/s} - 0}{340 \text{ m/s} - v_s} f_s$$

$$(640/\text{s})(340 \text{ m/s} - v_s) = (340 \text{ m/s})(600/\text{s})$$

$$2.176 \times 10^5 - 640 \, v_s = 2.040 \times 10^5$$

$$640 \, v_s = 1.36 \times 10^4$$

$$v_s = 21.3 \text{ m/s}$$

Electromagnetic waves — light, radio, and radar waves — all exhibit the Doppler effect. However, the relationship between source frequency and observed frequency is different from Equation 11-19. Instruments applying the Doppler effect are used to measure the speed of various vehicles. The instrument emits a radar signal which when bounced from a moving vehicle is doppler-shifted as though the moving vehicle were a source. The speed of the vehicle is determined by the amount of the Doppler shift.

When light from distant stars and galaxies is compared to light of the same type of source on the earth, a Doppler effect is seen. The wavelengths from sources on increasingly distant galaxies are increasingly longer. They are shifted toward the red end of the spectrum. This red shift is taken as evidence that the universe is expanding.

Minimum Learning Objectives

After studying this chapter, you should be able to:
1. Define:
   - antinode
   - beats
   - crest
   - decibel
   - Doppler effect
   - fixed (closed) end
   - free (open) end
   - fundamental
   - harmonic
   - incident wave
   - intensity
   - interference
   - inverted in phase
   - linear mass density
   - longitudinal wave
   - node
   - overtone
   - periodic wave
   - phase
   - propagate
   - pulse
   - rarefaction
   - sinusoidal
   - sound intensity level
   - standing wave
   - superposition
   - threshold of hearing
   - threshold of pain
   - torsional wave
   - transverse wave
   - traveling wave
   - trough
   - wave
   - wavelength
2. Develop a mental model of mechanical wave motion.
3. Distinguish between transverse and longitudinal waves.
4. Convert frequency to period, and vice versa.
5. Calculate any one of the three variables wavelength, frequency, and wave velocity, given the values of the other two.
6. Calculate the wave velocity on a string, given the tension and the linear mass density.
7. Calculate the disturbance at any point on a medium, given the separate wave disturbances at that point at that time.
8. Describe wave behavior at the ends of the medium, fixed and free, as well as at the boundary to a medium of a given but different density.
9. Follow and understand the derivation for the y displacement of a standing wave at any point in space or time.
10. Calculate the positions of nodes for a standing wave, given the speed of wave propagation and the geometry of the medium.
11. Calculate the frequencies of the fundamental and overtones for standing waves on strings or in air columns.
12. Calculate the beat frequency, given two wave frequencies.
13. Calculate the sound intensity level in decibels, given the sound intensity in watts per square meter, or vice versa.
14. Calculate the Doppler shift of frequency, given the velocities of the source and observer and the frequency of the source.

Problems

11-1 Radio waves are transverse waves for purposes of classical electromagnetic theory. The upper and lower limits of the broadcast FM band are 88.0×10^6 to 108×10^6 Hz. To what wavelengths do these limits correspond?

11-2 A tuning fork produces a frequency of 512 Hz. (*a*) Calculate the wavelength of this sound in air where the speed of sound is 345 m/s. (*b*) Determine the velocity of this sound in water if the wavelength in water is 2.93 m.

11-3 Bats navigate and hunt insects by using ultrasonic pulses (longitudinal waves of frequency greater than normal human hearing can detect). They emit waves of frequency 3.85×10^4 Hz. Calculate the wavelength of this sound when the speed of sound in air is 345 m/s.

11-4 Simple depth indicators for boats operate by sending out a sound pulse from a transducer under their hull and timing the echo from the bottom. Suppose a pleasure boat is operating with such an echo sounder. (*a*) The density of seawater is 1024 kg/m³, and its bulk modulus is 2.20×10^9 N/m². What is the speed of sound in ocean water? (*b*) How long does it take a pulse of sound to travel from the transducer 1.0 m under the water to the bottom, 18.0 m below the surface, and back? (*c*) Suppose the boat now moved up a freshwater river (density = 1000 km/m³; bulk modulus = 2.28×10^9 N/m²). If the echo sounder indicated 10.0 m of water under the hull, what is the real depth of the water? (*d*) Why is this source of error unlikely to get a boat operator into trouble even if overlooked?

11-5 An oscilloscope makes a 10.0-cm sweep of the screen and is set to sweep 100 times per second. (*a*) If a pattern showing waves which are 2.8 cm long is recorded on the screen, what is the frequency of the wave being displayed? (*b*) If exactly three complete waves are displayed on the screen, what is the frequency of the source of waves?

11-6 A traveling transverse wave is represented by the equation $y = 1.50 \sin(\pi x - 50t)$, where x and y are in centimeters when t is in seconds. For this wave, determine the (*a*) amplitude, (*b*) wavelength, (*c*) frequency, (*d*) period, and (*e*) the speed of propagation.

11-7 Show that each of the following is a form of Equation 11-4:

261

$$y = y_0 \sin\left[2\pi\left(\frac{x}{\lambda} - ft\right)\right]$$

$$y = y_0 \sin\left[2\pi\left(\frac{x}{\lambda} - \frac{t}{\tau}\right)\right]$$

$$y = y_0 \cos\left[\frac{2\pi}{\lambda}(x - vt) - \frac{\pi}{2}\right]$$

11-8 A traveling transverse wave has an amplitude of 0.100 m, a frequency of 200 Hz, and a velocity of propagation of 600 m/s. For this wave, write the equation for displacement in the form of Equation 11-4, inserting numerical values wherever you can.

11-9 A mariner in a canoe is crossing a bay but is having difficulty coping with the waves, which have an amplitude of 0.750 m and a period of 2.86 s. While atop a crest, he notes the bow wave of a large motorboat proceeding toward him from a direction perpendicular to the surface waves. The bow wave is a single crest with an amplitude of 1.50 m and is some 20.0 m away, proceeding toward him at a rate of 2.00 m/s. What will be the total displacement of the water under the canoe at the instant the bow crest passes?

11-10 Calculate the diameter of a steel piano wire in which the velocity of wave propagation is 300 m/s when it is under a tension of 250 N. Steel has a volume mass density of 7.89×10^3 kg/m^3.

11-11 Two strings are made of the same material and are under the same tension T. The second string has twice the diameter of the first. Determine the ratio of the velocities of propagation of transverse waves in the two strings.

11-12 A long line of cars with 10.0 m between successive brakelights is proceeding along an expressway at 45 km/h. A dog runs onto the expressway, causing a driver to hit his brakes to stop. Each successive driver reacts to the brake lights of the car before him or her with a typical reaction time of 0.15 s. The result, as seen from a helicopter, is a wave of suddenly illuminated brake lights proceeding backward from the point of the original incident through the mass of cars. This is accompanied by a compression of cars in the vicinity of the original action which also propagates rearward through the mass of cars. What is the velocity of rearward propagation of the wave of brakelights as measured from the helicopter?

11-13 Determine the speed of propagation of longitudinal waves in (a) glass of density 2.50×10^3 kg/m^3 and bulk modulus 3.10×10^{10} N/m^2 and (b) lead of density 11.3×10^3 kg/m^3 and bulk modulus 5.20×10^{10} N/m^2.

11-14 Seismic waves generated by an earthquake are of three types: a longitudinal pressure wave, or P wave, a transverse shear wave, or S wave, and surface waves, which may be responsible for much of the structural damage near the location of the earthquake. The velocities of P and S waves are related to the modulus of rigidity of the rocks, μ. In general, $V_S = \sqrt{\mu/\rho}$ and $V_P = \sqrt{(B + \frac{4}{3}\mu)/\rho}$, where ρ is the density and B is the bulk modulus. (a) At a depth of 33 km, at the start of the mantle of the earth, $V_S = 4.36$ km/s and $V_P = 7.76$ km/s. At this level, the density increases from 2.87 times that of water in the lower crust to about 3.32 times the density of water in the upper mantle. Calculate the modulus of rigidity and bulk modulus for the rocks of the upper mantle. (b) The upper crustal rocks within a dozen or so kilometers of our feet have an average density of 2.65 times that of water. If the same values of μ and B hold as for the rocks of the mantle in part a, what are V_P and V_S for these upper rocks? (c) Assume your values of V_S and V_P are correct for the upper rocks of the earth. If a P wave arrived at your institution's seismograph, followed in 3.5 s by an S wave from the same earthquake, how far away did the earthquake occur? (d) The earth has a radius of 6371 km. At a depth of 2898 km below the surface, S waves abruptly stop and P waves drop in speed from 13.6 to 8.1 km/s. From the model for mechanical wave motion, what could account for this behavior?

11-15 Two waves on the same string obey the equations $y_1 = 2.00 \sin [\pi(x - 20 t)]$ and $y_2 = 2.00 \sin [\pi(x - 20 t + \frac{1}{4})]$, where x and y are in centimeters and t is in seconds. Determine the points on x where at $t = 0$ there will be (a) total cancellation and (b) maximum reinforcement. $\left(\text{Hint: } \sin A + \sin B = 2\cos\dfrac{A - B}{2} \sin\dfrac{A + B}{2}\right)$

11-16 Sound from an explosion in air strikes the surface of water. The sound transmitted through the water strikes the steel hull of a submarine. The sound transmitted through the hull enters the air inside the submarine. The speeds of sound in air, water, and steel are, respectively, 340 m/s, 1.53×10^3 m/s, and 5.90×10^3 m/s. If a crest (compression) from the explosion is followed: (a) What is the phase of the reflected wave from the water surface? (b) What is the phase of the transmitted wave in the water? (c) What is the phase of the reflected wave from the outside of the submarine hull? (d) What is the phase of the wave transmitted through the steel? (e) What is the phase of the wave reflected from the boundary between the steel hull and the air inside the submarine? (f) What is the phase of the wave arriving at the ears of the people in the submarine? (g) What is the frequency of the wave in each of the media if the original frequency was 17.0 Hz? (h) What was the original wavelength of the waves in air? (i) What are the wavelengths of the waves in each of the other media?

11-17 A light wave in air enters a pane of glass. The speed of the wave in the glass is lower than in air. It then encounters the back surface of the glass and emerges into the air. The wavelength of the light wave in air is 5.00×10^{-7} m. (a) Is the portion of light reflected from the front of the glass pane inverted in phase or not? (b) Is the portion of the light moving through the glass and reflected off the back surface of the pane inverted in phase or not? (c) If the light travels only two-thirds as fast in glass as in air, what is its wavelength in the glass?

11-18 A suspension bridge has cables which are 400 m long between supports across its main span supporting a total mass of 2.00×10^6 kg. Assume that this mass is distributed evenly along the cable and that the tension in the cable is 6.50×10^7 N. (a) What is the fundamental frequency of vibration of the bridge? (b) What frequency will set the bridge into vibration by halves with a node at the center?

11-19 By plotting the resultant displacement according to the principle of superposition, one can show that two waves of the same frequency and amplitude traveling in opposite directions give rise to standing waves. (a) Plot the two waves y_1 and y_2 at $t = 0$, 0.0500, and 0.100 s in the interval from $x = 0$ to $x = 1.00$ cm.

$$y_1 = 0.200 \sin [2\pi(x - 5t)]$$

$$y_2 = 0.200 \sin \left[2\pi(x + 5t) + \frac{2\pi}{3}\right]$$

where y and x are in centimeters when t is in seconds. (b) What is the wavelength of the standing wave?

11-20 The wave velocity of vibrations of the A string on a violin is 496 m/s in its fundamental mode of 440 Hz. Calculate (a) the fundamental wavelength for the A string, (b) the length of the A string, and (c) the wavelength of the second overtone.

11-21 A 2.00-m string is fixed at both ends. Calculate (a) the longest wavelength of a natural vibration of the string and (b) the frequency of vibration if the velocity of propagation is 450 m/s.

11-22 A clothesline 4.50 m long has a mass of 0.610 kg. One end is fastened to a house, and the other end runs over a pulley. (a) If a child hanging from the free end and climbing the rope has a weight of 350 N (78.6 lb), what is the velocity of a disturbance traveling in the clothesline? (b) What will be the natural (fundamental) frequency of oscillations in the clothesline?

11-23 A steel piano string fixed at both ends is to vibrate in its fundamental mode when struck at middle C (approximately 262 Hz on the equally tempered scale). If the string is 1.20 m long and has a mass of 6.00×10^{-3} kg, (a) what must be the tension in the string? (b) What is the wavelength of the highest harmonic which may be heard by someone whose upper limit is 1.50×10^4 Hz?

11-24 (a) If a guitar string is plucked precisely at its midpoint, what harmonics will not be present in the sound produced? (b) If it is plucked one-third of the way from one end, which harmonics will be missing? (c) If it is plucked at the one-quarter point along its length, which harmonics are absent?

11-25 The strings on a guitar have an effective length of 0.630 m. When in tune, the highest string has a fundamental frequency of 330 Hz. When the second string is shortened (by pressing it above the fifth fret) to an effective length of 0.470 m, it should have that same frequency. Recognizing that the tension and mass per unit length in the second string will not change when its effective length is restored to the original, determine the fundamental frequency of the second string. (*Hint:* First determine the speed of wave propagation in the second string.)

11-26 Orchestras like to play with more "brilliance" than their competitors. Thus, since many instruments are tuned by ear, musical scales have increased in frequencies through time. Between Handel's *Messiah* and "The Stars and Stripes Forever" of John Phillip Sousa, the frequency for the same note had increased by 10 percent. This creates a problem for any instrument designed and built for a specific scale. For example, consider the flute as an open-end pipe. The lowest note the flute can play is middle C, with a frequency of 261.63 Hz on the currently accepted international scale (A = 440 Hz). How long would a flute have to be (a) to play Handel's middle C of 250 Hz, (b) to play Sousa's middle C of 274 Hz, and (c) to play the international scale middle C of today (262 Hz)? (Use 345 m/s as the velocity of sound.)

11-27 Tufo is a native of the Staten Islands and has been making sets of pipes to play music for years. The pipes are a set of hollow plant stems of varying lengths, each playing a different note when blown across one open end. By accident one pipe has a hole punched in it near its midpoint. Tufo discovers that if he holds his finger over the hole, the pipe plays its original note, but if he removes his finger to open the hole, the note played is much higher in pitch. He gets the brilliant idea, discovered in several societies, that he can use a single pipe to play his entire musical scale if he pierces it with many finger holes. Like many simple societies, Tufo's uses the five-note (pentatonic) scale. The lowest pitched note is close to our high C at 523 Hz and proceeds to frequencies of 588, 697, 785, 882, and 1046 Hz to complete one octave (or doubling of frequency). (a) How long a pipe must Tufo have to begin with? (b) At what distances from the blown end of the pipe should he pierce it for the finger holes to produce the five additional notes? (Use 345 m/s as the velocity of sound.)

11-28 The first three natural frequencies of a certain organ pipe are 68.0, 204, and 340 Hz. (a) Is the pipe open or closed? (b) How long is the pipe? (c) What is the wavelength of the fundamental mode of vibration? (Use 345 m/s as the velocity of sound.)

11-29 Determine the frequency of an organ pipe 4.50 m long if it is (a) closed at one end or (b) open at both ends. Take the speed of sound in air to be 340 m/s.

11-30 Organ pipe A is closed at one end and is the same length (0.700 m) as pipe B, which is open at both ends. (a) If pipe A vibrates in its second overtone, and pipe B vibrates in its first overtone, determine the ratio of the frequency of A to that of B. (b) How long is pipe C, which is open at both ends, if its fundamental frequency is equal to the frequency of the first harmonic of pipe A?

11-31 In flying a two-engine, propeller-driven airplane, we find that a strong shuddering vibration is felt through the body of the airplane every 2.0 s. What is the likely source of the vibration, and how may it be eliminated?

11-32 A singer performing on a stage can generate a sound intensity level (SIL) of 70 dB in the tenth row of an auditorium. How many singers would it take to generate 95 dB at the same distance away?

11-33 The operator of a jackhammer during its use is exposed to SILs of 105 to 115 dB. Calculate the intensity of the sound in watts per square meter for these two limits.

11-34 Whenever sound waves can move freely in all three dimensions away from their source, the sound intensity varies as the inverse square of the distance ($I \propto 1/d^2$). Thus a listener three times as far away would be exposed to one-ninth the sound intensity. Using this relationship, calculate the following: *(a)* The SIL at a rock concert measured 20 m from the loudspeakers is found to be 110 dB. How close would you have to move to the speakers to arrive at the threshold of pain? *(b)* The deep-pitched foghorn of a large ocean liner can just be heard at a distance of 16 km (10 mi). What is the SIL for a passenger standing on deck 100 m from the horn as it sounds? (Why do not all exposed persons on the ship collapse in pain?) *(c)* The sound of the great eruption of Krakatoa (1883) in Indonesia was reported by Chilean coast guard ships over 3000 mi away as "distant cannonfire." If this sound was at the threshold of hearing, what would be the SIL at a distance of 1 mi from the eruption (assuming the inverse square law of sound intensity was valid—it wasn't)?

11-35 A frequency meter is used to measure the pitch of an automobile engine during a race. As the car approaches the position along the track where the meter is, the meter records a frequency of 580 Hz. After the car passes and is receding from the meter, the measured frequency of the engine noise is 460 Hz. *(a)* What is the frequency of the engine noise being generated? *(b)* If the car has an eight-cylinder engine, at what revolutions per minute is the engine operating? *(c)* What is the speed of the car along the track? ($v_{sound} = 340$ m/s)

11-36 The martian *mauk* takes its cell creature *eek* to an amusement park to share the latest craze. Patrons are shot nearly vertically from a cannon to a great height and fall back into an air pad on a net near the cannon. Note that *eek* always screams at a pitch of 800 Hz. What is the frequency of the sound *mauk* will hear in each of the following cases? (All are different.) *(a)* Mauk is on the ground and *eek* is receding upward at 40 m/s as it screams. *(b)* Mauk is on the ground and *eek* is falling 40 m/s as it screams. *(c)* Mauk is shot from the cannon and ascending 40 m/s as *eek*, on the ground, screams. *(d)* Mauk is falling 40 m/s toward the ground as *eek*, on the ground, screams. *(e)* Mauk and *eek* are shot from adjacent cannons. Mauk is ascending 40 m/s toward *eek*, who is falling 40 m/s as it screams. *(f)* Mauk is falling 40 m/s away from *eek*, who is rising 40 m/s as it screams. *(g)* Mauk is rising 40 m/s toward *eek*, who is also rising 40 m/s as it screams. ($v_{sound} = 340$ m/s.)

11-37 A sonar detector on a ship sends out high-frequency sound pulses. The operator listens for echoes and compares the pitch of a large echo (possibly a submarine) with the echoes from irregularities (bubbles, kelp, small fish) in the ocean. If the pitch of the echo from the supposed submarine is higher, the operator says "up Doppler" and assumes that the submarine has a component of motion toward the ship. If the source frequency is known to be 9.00×10^3 Hz and the ship is moving due west at 10.0 m/s, determine the speed and direction of the submarine when it is detected due west of the ship with a "down Doppler" frequency of 8.94×10^3 Hz. The speed of sound in seawater is 1.53×10^3 m/s. (*Hint:* First, determine the frequency of the sound after it is reflected from the assumed stationary irregularities.)

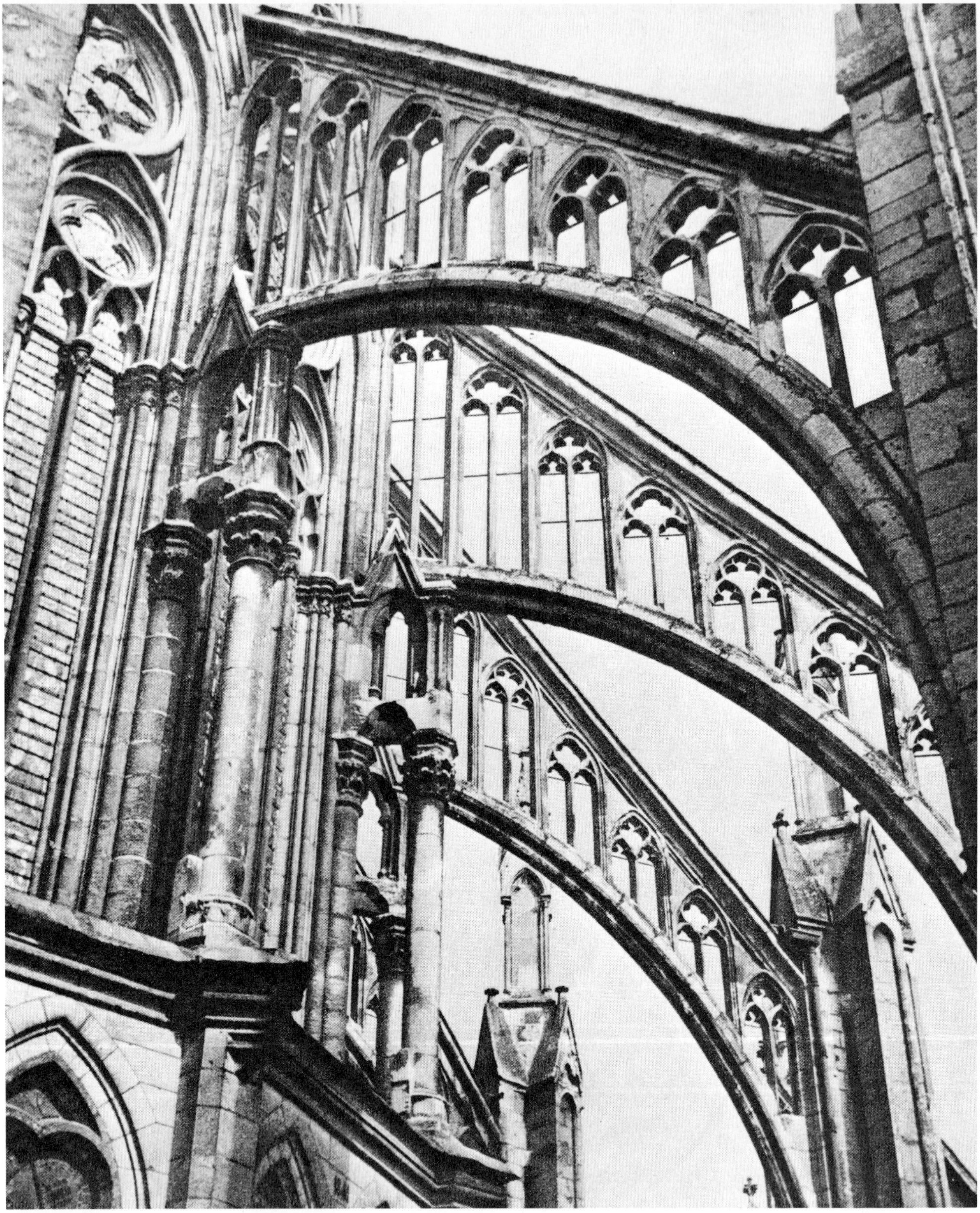

12 Some Properties of Materials

12-1 Introduction

Consider the way an ordinary paper clip is used. We exert forces or torques to spread the clip, which deform it, or change its shape. We insert papers and release the clip. As the clip snaps back, it applies forces to both sides of the papers, holding them together. A rubber band behaves in a similar way. Pulling on both ends of a rubber band causes it to extend its length. Provided the applied forces are not too large, the rubber band returns to its original shape and size when it is released. If the forces and/or torques are very large, the paper clip and the rubber band may be deformed so far that they will not return to their original shapes, and, of course, they may break. Handkerchiefs may be squeezed together, paper may be folded, and even wooden boards and metal plates covering a ditch may bend noticeably when someone walks on them.

Up to now, we have considered the effects of forces on objects only in terms of the accelerations or changes in rotation they may produce. Now it is time to consider the effects of the forces on the matter of the object itself.

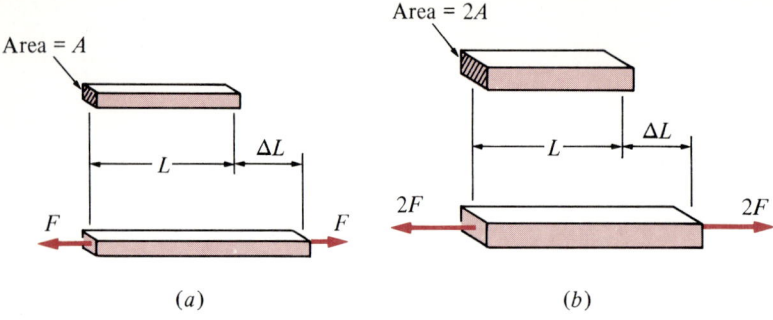

Figure 12-1 (a) A section of rubber band of length L and cross-sectional area A is stretched a distance ΔL by a tension of magnitude F. (b) If we replace the original piece of rubber band with one that is identical in length but has twice the cross-sectional area, 2A, we must use twice the original tension to stretch it the same distance ΔL.

12-2 Stress and Strain

A moderate force applied to a piece of rubber band causes the rubber to stretch a small distance. A second piece of rubber band of the same material and length but of twice the thickness requires two times the force to stretch the same distance (Figure 12-1). By doubling the thickness of the band, we have doubled the cross-sectional area of the rubber being stretched. A single piece of rubber with three times the cross section requires three times the force for the same elongation, and so on. You may conclude that the important quantity in determining the stretch for a given length of band is the ratio of the magnitude of the applied force to the cross-sectional area. Indeed it is! We define a new quantity called **stress** as the force per unit area of cross section. The SI unit of stress is newtons per square meter (N/m²). In equation form, we may write

$$\text{Stress} = \frac{F}{A} \tag{12-1}$$

While we chose the rubber band material because of its familiarity, the concept of stress may be applied to any kind of material.

Example 12-1 Determine the stress placed on a piece of steel wire 1.50×10^{-3} m in diameter which is used to support a mass of 3.00 kg.

Solution We need only apply the definition of stress, Equation 12-1. First, however, we must determine the force applied to the wire and the cross-sectional area of the wire. The force applied to the wire is simply the weight of the 3.00-kg mass. Thus

$$F = mg = (3.00 \text{ kg})(9.80 \text{ m/s}^2) = 29.4 \text{ N}$$

The area is that of a circle of diameter 1.50×10^{-3} m:

$$A = \frac{\pi d^2}{4} = \frac{\pi (1.50 \times 10^{-3} \text{ m})^2}{4} = 1.77 \times 10^{-6} \text{ m}^2$$

Therefore

$$\text{Stress} = \frac{29.4 \text{ N}}{1.77 \times 10^{-6} \text{ m}^2} = 1.66 \times 10^7 \text{ N/m}^2$$

This seems to be a very large number (it is), but we will see that it is not an unusually large stress.

The rubber band limits our consideration to a force of **tension,** the attempt to pull the atoms of a material apart. Picture instead a metal cylinder, say an empty soft drink can. In a moment of exuberance, you might grab each end of the can and try to pull it apart with tension. The distortion to tensile stress is much less obvious than with the rubber band. However, the soft drink can offers other strategies for distortion. For example, we may apply the palm of each hand to either end of the can and attempt to crush it. This is a force of **compression,** the attempt to force the atoms closer together. Alternatively, we may grab each end and twist them in opposite rotational senses, applying torsion and producing **torsional stress.** Or, we could hold each end rigidly before us, keeping the axis of the can perpendicular to our line of sight, and, without allowing the can to rotate, move one end directly away from us and the other toward us. This attempt to rip the can apart by sliding one layer of atoms over another produces **shear stress.**

For the moment, we will consider tension and compression. Figure 12-2 shows a rod subjected to both of these types of stress. When the rod is not under a load, its length is L. Under a tension of magnitude F its length is increased by an increment ΔL_T, while under compression its length is decreased by ΔL_C. These two incremental changes in length need not be the same magnitude. In either case, we may deduce that the magnitude of ΔL depends on the magnitude of L. Consider Figure 12-3. We see that if under tension the bar increases by ΔL, then each half elongates by $\frac{1}{2}\Delta L$ (or each third increases by $\frac{1}{3}\Delta L$). Further, a bar of half the length under the same tension elongates by half as much, and so on. The conclusion is that the change in length is proportional to the original length. A bar 1.00 km long stretching 1.00 m is much different from a bar 1.00 m long stretching an additional 1.00 m. To account for these differences, we define **strain** as the ratio of the change in length to the original length. Note that strain is unitless, a length divided by a length (just as long as we measure both in the same system).

$$\text{Strain} = \frac{\Delta L}{L} \qquad (12\text{-}2)$$

Figure 12-3 (a) Two metal rods, identical except that one has a length of L and the other $\frac{1}{2}L$, before being subjected to tensile stress. (b) Under tension, the rods elongate, such that the longer rod acquires a length of $L + \Delta L$. If the identical tension is applied to the shorter rod, its new length is $\frac{1}{2}(L + \Delta L)$.

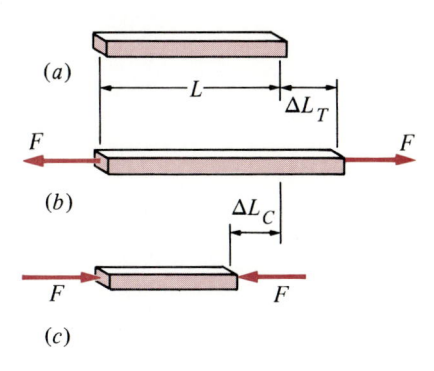

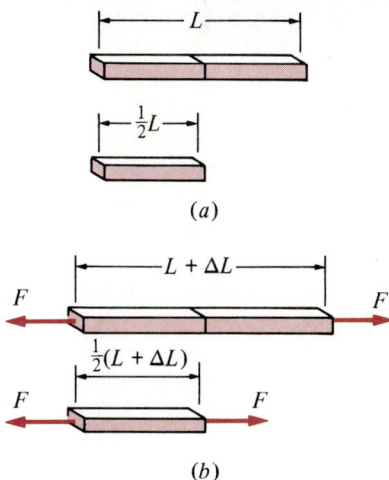

Figure 12-2 (a) A metal rod of length L in the absence of external forces. (b) Under tension, the rod elongates by an amount ΔL_T. (c) Under a compressional force, the rod shrinks from its original length by an amount ΔL_C.

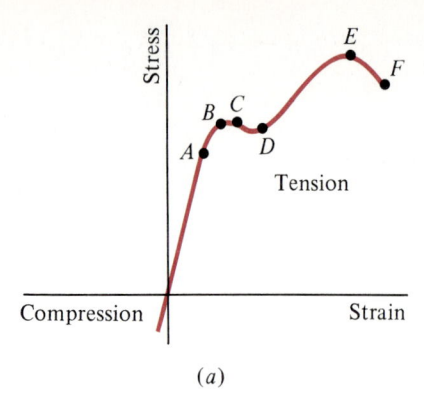

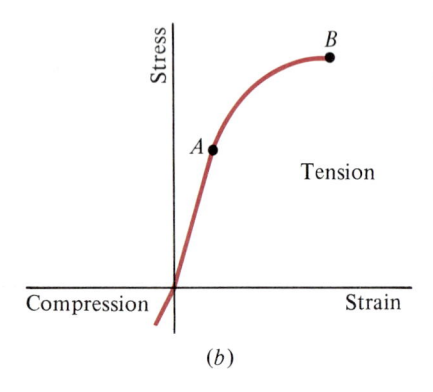

Figure 12-4 Typical stress vs. strain curves for (a) a ductile material, such as copper, rolled aluminum, or mild steel, and (b) a brittle material, such as high carbon steel, cast iron, or bone. Most use of materials must take place in the proportional zone below point A so that structures are not deformed or destroyed in use.

Strain may be recognized as the fractional change in length. If the strain is 0.15, the material has changed length by 15 percent, and so on. Whenever a stress is applied to an object, strain results. A straightforward method to investigate stress as a function of strain is to set up an experiment with a rod of known length and cross section, place it under various measured tensions and compressions, and measure the change in length. We then calculate the stress corresponding to a given strain and construct a graph of stress vs. strain. Such experiments have been done for many materials and detailed graphs constructed. Two typical examples are shown in Figure 12-4. Part *a* shows the stress vs. strain curve for a ductile material, one that can be elongated significantly or hammered thin without rupture. Many metals are ductile. Figure 12-4b shows the stress vs. strain curve for a brittle material, which has a greater tendency to rupture than to deform permanently.

Let us compare the features of these curves. For the ductile material under tension, we see that as the stress increases from moderate values, the stress vs. strain curve is a straight line up to point A, called the **proportional limit.** That is, up to this point, stress is proportional to strain. If the stress is increased beyond point A, the strain begins to increase more rapidly. At point B, the material is said to have reached its **elastic limit.** If the tension is released at this point, the material will return to its original length. If the stress is relieved after the material has passed the elastic limit, it will recover but with a permanent deformation, as with an excessively bent paper clip. A further increase in stress causes the material to reach point C, the **yield point**. At the yield point, **plastic flow** begins. In the plastic-flow region (from C to D), there is a large increase in strain with no increase (and perhaps a decrease) in stress. At point D, the material begins to change structure microscopically, becoming more nearly brittle. This is called **strain hardening**. Further increases in stress beyond D finds renewed resistance until point E is reached. At point E the material reaches its **ultimate strength** (under tension), a point at the largest stress the material can sustain. Beyond point E, the strain increases with a decrease in stress until the **breaking point** F is reached where the material ruptures.

The brittle material's stress vs. strain curve shown in Figure 12-4b does not display quite so many features. A brittle material also has a proportional limit, point A. Up to this point, the stress is again directly proportional to the strain. If the stress is increased beyond point A, the strain increases more rapidly until the breaking point B is reached, which is also (very nearly) the ultimate strength point, and at B the material ruptures.

Under moderate compressive stresses, both ductile and brittle materials would show that stress is proportional to strain. Further, the slope of the proportional portion of the curve under compression is not necessarily the same as the slope under tension. We rarely wish to fracture or permanently deform materials. Thus, the most important portion of the stress vs. strain curve (and, of course, the easiest to analyze) is that portion showing a straight-line relationship where stress is proportional to strain. We examine this more closely in the next section.

12-3 Elastic Moduli

For the straight-line part of the stress vs. strain curve, stress is proportional to strain, and we can write

$$\text{Stress} \propto \text{strain}$$

or

$$\frac{F}{A} \propto \frac{\Delta L}{L} \tag{12-3}$$

The direct proportion between stress and strain was first formalized by Robert Hooke and announced in 1678. It subsequently became known as **Hooke's law**.

The proportionality constant which makes Equation 12-3 an equality is called the **elastic modulus** of the material. In particular, for tension or compression it is called **Young's modulus** (symbol Y), and we have

$$\frac{F}{A} = Y \frac{\Delta L}{L} \tag{12-4}$$

From Equation 12-4 we see that

$$Y = \frac{F/A}{\Delta L/L}$$

Since a larger value of Y means that a larger force per unit area is required for a given elongation, we say that the elastic modulus is a measure of the difficulty of deforming a certain material.

Example 12-2 Young's modulus for the wire of Example 12-1 is 2.00×10^{11} N/m². (a) Determine the strain on the wire for the stress of 1.66×10^7 N/m². (b) If the wire is exactly 3.00 m long under no load, what is its increase in length?

Solution This is a direct application of Equation 12-4. Thus:

(a)
$$\frac{\Delta L}{L} = \frac{F/A}{Y} = \frac{1.66 \times 10^7 \text{ N/m}^2}{2.00 \times 10^{11} \text{ N/m}^2}$$
$$= 8.30 \times 10^{-5}$$

(b)
$$\Delta L = (8.3 \times 10^{-5})(3.00 \text{ m})$$
$$= 2.49 \times 10^{-4} \text{ m} \quad \text{or} \quad 0.249 \text{ mm}$$

While this elongation is not very large, there are many situations in which it can be significant.

Recognize that Young's modulus is equal to the slope of the stress vs. strain curve before it reaches the proportional limit. This slope is not necessarily the same under tension as it was under compression. It is, however, an experimental fact that for homogeneous materials, such as metal wires pulled through a die, these slopes are about the same. For inhomogeneous or composite materials such as bone, wood, or concrete, this is not the case. Hence, for inhomogeneous materials, Young's modulus for tension is different from Young's modulus for compression. A number of representative values are shown in Table 12-1.

TABLE 12-1 Approximate or Average Values of Young's Modulus for Tension and Compression, Ultimate Tensile Strength, and Ultimate Compressive Strength for Selected Materials

Material	Young's Modulus for Tension, 10^{10} N/m²	Young's Modulus for Compression, 10^{10} N/m²	Ultimate Tensile Strength, 10^7 N/m²	Ultimate Compressive Strength, 10^7 N/m²
Aluminum	7.1	7.0	20	35
Blood vessel	0.00002			
Bone	1.6	0.9	12	17
Brass	9	9	31	
Brick	2	2	0.3	5
Concrete	3	3	0.2	4
Glass	7	7	7	35
Granite	5	...	1	20
Hair (human)	1	...	10	
Hardwood (from deciduous trees)	1.5	1	6	7
Lead	1.6	1.6		
Marble	6	6	...	9
Muscle	...	...	0.06	
Nylon	0.008	...	7	
Polystyrene	0.3	0.3	5	10
Rubber	0.0001			
Silk	0.6			
Spider's thread	0.3			
Steel	21	21	55	41
Tendon	0.1		7	

Example 12-3 (a) Using the value for the ultimate compressive strength for bone from Table 12-1, determine the maximum force the human upper leg bone, the femur (approximate cross-sectional area = 5.50×10^{-4} m²), can sustain. (b) Suppose an 85.0-kg professor drops from rest down to a concrete pad, landing erect, heels first with his knees locked. The concrete will stop his motion essentially instantly, but the flesh of the heel and pads in the joints will compress by perhaps $\Delta y = 0.400$ cm after initial contact to cushion the impact somewhat. From what height can the professor jump before breaking the femurs due to exceeding the compressive strength of bone as in part *a*?

Solution (a) The maximum force is simply the ultimate compressive strength times the cross-sectional area. Thus

$$F_{max} = (1.70 \times 10^8 \text{ N/m}^2)(5.50 \times 10^{-4} \text{ m}^2)$$
$$= 9.35 \times 10^4 \text{ N}$$

(b) Assuming the force exerted on the professor by the concrete is essentially constant, we need to find an expression for the constant acceleration of his center of mass as he stops. We may find the expression for his speed on contact by using kinematics or conservation of energy considerations.

From $2a \, \Delta y = v^2 - v_0^2$ applied between the point of contact and final stop, we have

$$2a(0.00400 \text{ m}) = 0 - v_0^2$$
$$a = -(125 \text{ m}^{-1}) v_0^2$$

(Negative because the direction of the acceleration is opposite to that of the vertical displacement.) Now applying conservation of mechanical energy between the point of drop and the concrete pad yields

$$\Delta KE + \Delta PE = 0$$
$$\tfrac{1}{2} m v_f^2 - 0 + mg(h_f - h_i) = 0$$

but $h_f - h_i = -h$, and we obtain

$$v_f^2 = 2gh$$

Here, v_f, the speed at the end of his drop, is the speed at the start of impact (or v_0 from the first part of the problem). Therefore

$$2gh = v_0^2$$

and in magnitude

$$a = (125 \text{ m}^{-1})(2gh)$$

Thus, using Newton's second law,

$$F = ma = (85.0 \text{ kg})(250 \text{ m}^{-1})(9.80 \text{ m/s}^2)h$$
$$= 2.08 \times 10^5 h$$

Now the maximum force which may be sustained by both femurs is

$$F_{max} = 2(9.35 \times 10^4) = 1.87 \times 10^5 \text{ N}$$

thus

$$h_{max} = \frac{1.87 \times 10^5 \text{ N}}{2.08 \times 10^5 \text{ N/m}} = 0.899 \text{ m!!}$$

Thus if the professor steps off a lecture table 0.900 m high and locks his knees, he could very well fracture both femurs from the compressive forces alone. However, one rarely locks one's knees while landing from a height. Most often the knees are bent, and the center of mass is stopped over a much larger distance than 0.400 cm. Thus the acceleration and hence the force are dramatically reduced.

Furthermore, the force is not constant. It is an impulsive force rising rapidly from a magnitude of zero at the instant of impact to a very large value before it quickly returns to the normal force equal in magnitude to the professor's weight. The average force is exceeded by nearly a factor of 2. This means that the maximum height the professor may drop with knees locked is more like 0.500 m.

Finally, if someone attempted this maneuver, any breaks would occur at the weak link in the chain of bones absorbing the shock. For most people, the weak link is not that part of the femur under compression or even the lower leg bone, the tibia, with its smaller cross-sectional area. The weak link is that portion of the femur which extends sideways to the hip joint. Even this is not subject to compression under the drop being considered but most probably fractures due to a combination of tensile and shear stresses.

As previously mentioned, there are other types of deformations. If you lean against a flagpole embedded in the ground, your horizontal force is neither clearly trying to push the atoms of the pole closer to each other nor trying to pull them apart. In many physical situations, the force applied to an object has the effect of attempting to slide one layer of atoms past another layer.

Consider Figure 12-5, which could represent a metal block or a stack of magazines. The tangential force F is applied to the top layer as shown. The area across which this force is producing deformation is the area of each surface or layer which the force attempts to displace by sliding it over another layer. We define that area as the surface with the greatest displacement in response to the force, and it is the area of the top face of the block in the diagram. As before, stress = F/A, so

$$\text{Shear stress} = \frac{F}{A}$$

The strain will be the lean Δs divided by that perpendicular length over which this lean was accomplished, or

$$\text{Shear strain} = \frac{\Delta s}{L}$$

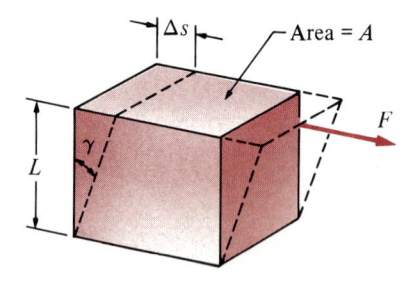

Figure 12-5 A block of material is deformed by *shear stress*. The bottom of the block is firmly anchored, and a tangential force F is applied at its top. This force attempts to slide the area A of the top face of the block across the lower layers. The block leans in the direction of the force with an angle of shear γ.

If we define the **angle of shear** at the undisturbed surface as γ in the diagram, then shear strain = $\tan \gamma$. The **shear modulus** is defined for this situation as the ratio of the shear stress to the shear strain. Using the symbol S for shear modulus, we have

$$S = \frac{F/A}{\Delta s/L} \tag{12-5}$$

Note that shear strain is dimensionless, and S has dimensions of force per unit area. In SI units, S is in newtons per square meter (N/m²).

Example 12-4 A shearing force of 4.00×10^4 N is applied to a cube of brass ($S = 3.5 \times 10^{10}$ N/m²) measuring 0.150 m on each edge. Determine (a) the shear strain, (b) the maximum displacement, and (c) the angle of shear.

Solution (a) After the shear stress is determined, the shear strain is found by applying Equation 12-5. The area of one face of the cube is $(0.150 \text{ m})^2 = 2.25 \times 10^{-2}$ m², and the shear stress is

$$\frac{F}{A} = \frac{4.00 \times 10^4 \text{ N}}{2.25 \times 10^{-2} \text{ m}^2} = 1.78 \times 10^6 \text{ N/m}^2 = S\frac{\Delta s}{L}$$

Thus the shear strain is

$$\frac{\Delta s}{L} = \frac{1.78 \times 10^6 \text{ N/m}^2}{S} = \frac{1.78 \times 10^6 \text{ N/m}^2}{3.5 \times 10^{10} \text{ N/m}^2}$$

$$= 5.08 \times 10^{-5}$$

(b) Using this result and the definition of shear strain, the maximum displacement may be calculated as

$$\Delta s = (5.08 \times 10^{-5})(0.150 \text{ m}) = 7.62 \times 10^{-6} \text{ m}$$

(c) The angle of shear is equal to the angle whose tangent is the shear strain.

$$\gamma = \tan^{-1} (5.08 \times 10^{-5}) = 2.91 \times 10^{-3} \text{ deg}$$

This should give you some further idea of the magnitudes of the quantities involved in deformations. The force applied was quite large (more than 4 tons), yet the angle of shear is very tiny.

A final mechanical deformation we wish to consider is that of a change in volume. Such a change of an object can occur when a uniform force is exerted perpendicular to all surfaces of the object. The magnitude of the force per unit area perpendicular to a surface is called **pressure.** The symbol P is used for pressure, and

$$P = \frac{F}{A} \tag{12-6}$$

Pressure has dimensions of force per unit area. In SI units, pressure is in newtons per square meters (N/m²); however, when it refers to pressure, this unit is given a special name, **pascal** (Pa), in honor of Blaise Pascal, who did early experimental work in fluid pressures. From the definition,

$$1 \text{ Pa} = 1 \text{ N/m}^2$$

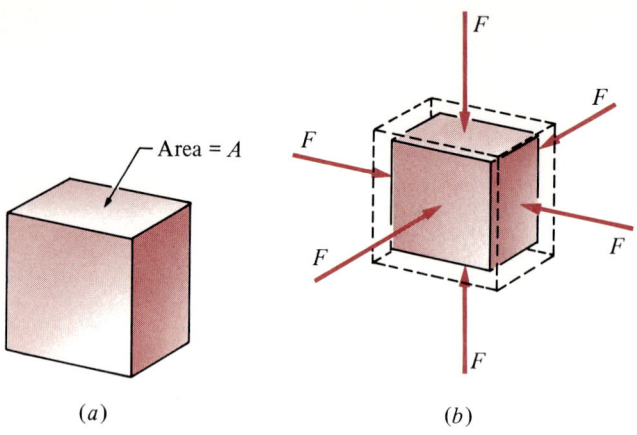

Figure 12-6 (a) A cube of volume V with each face having an area of A. (b) Uniform forces F applied to each face compress the block by an amount ΔV.

Now consider the cube in Figure 12-6a. Each face has a cross-sectional area A. Each face is subjected to a total force F perpendicular to the surface in Figure 12-6b, and the cube decreases in volume by an amount ΔV. Volume stress is defined as the pressure (F/A), and volume strain is defined as the ratio of the change in volume to the original volume. Thus

$$\text{Volume stress} = P$$

$$\text{Volume strain} = \frac{\Delta V}{V}$$

The **bulk modulus** is defined for the kind of situation occurring in Figure 12-6 as the ratio of the volume stress to the volume strain. Using B as the symbol for bulk modulus, we have

$$B = \frac{P}{\Delta V/V} = P\frac{V}{\Delta V} \tag{12-7}$$

where we are using magnitudes only since the fractional change in volume is negative if the pressure is increased. Also, since the volume strain is dimensionless, the bulk modulus has the unit of pressure, the pascal.

Example 12-5 **Density** is defined as the mass per unit volume, or $\rho = m/V$, where the lowercase Greek rho (ρ) is used for density. If the density of copper is given as 8.93×10^3 kg/m³, determine the change in density of a lump of copper when it is subjected to uniform external pressure of 4.00×10^9 Pa. The bulk modulus for copper is 1.31×10^{11} Pa.

Solution Assuming that the mass of the copper does not change, we write that the relative change in density is equal to the relative change in volume as a good approximation. Therefore

$$\frac{\Delta \rho}{\rho} = \frac{\Delta V}{V}$$

Equation 12-7 may be solved for $\Delta V/V$ to yield

$$\frac{\Delta V}{V} = \frac{P}{B}$$

Thus we have

$$\frac{\Delta \rho}{\rho} = \frac{P}{B}$$

or

$$\Delta \rho = \frac{P}{B}\rho$$

$$= \frac{4.00 \times 10^9 \text{ N/m}^2}{1.31 \times 10^{11} \text{ N/m}^2}(8.93 \times 10^3 \text{ kg/m}^3)$$

$$= 273 \text{ kg/m}^3$$

Note that even under this tremendous pressure, almost 40,000 times atmospheric pressure, the lump of copper decreases in volume and hence increases in density by only about 3 percent.

The greater the bulk modulus, the greater the pressure required to achieve a given shrinkage in volume of the material. A quantity often used in volume deformation is called **compressibility**, defined as the reciprocal of the bulk modulus. Using the Greek lowercase kappa (κ) for compressibility and applying Equation 12-7, we have

$$\kappa = \frac{1}{B} = \frac{1}{P}\frac{\Delta V}{V} \tag{12-8}$$

From this definition, we say that the compressibility of a substance equals the fractional decrease in volume per unit increase in pressure. Compressibilities for some common materials are given in Table 12-2.

TABLE 12-2 Approximate or Average Values of Shear Modulus, Bulk Modulus, Compressibility, and Ultimate Shear Strength for Some Selected Materials

Material	Shear Modulus S, 10^{10} N/m²	Bulk Modulus B, 10^{10} N/m²	Compressibility κ, 10^{-11} m²/N	Ultimate Shear Strength, 10^7 N/m²
Aluminum	2.6	7.8	1.3	13
Bone	1.0	0.8	13	
Brass	3.5	6.8	1.5	20
Brick	...	...	...	1.5
Concrete (reinforced)	1	1.5	6.7	0.2
Glass	2.3	3.1	3.2	
Granite	...	4.5	2.2	1.4
Hardwood	1	...	...	0.8
Lead	0.54	5.2	1.9	1.2
Marble	...	7	1.4	0.9
Steel	8.2	12.4	0.8	41
Acetone	...	0.07	140	
Glycerine	...	0.5	21	
Mercury	...	2.7	3.7	
Water	...	0.2	48	

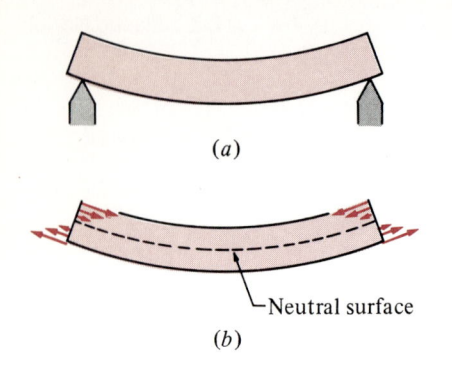

Figure 12-7 (a) A wooden beam supported at its ends and sagging under its own weight. (b) Examining the forces in successive vertical layers of the beam, we find that somewhere inside the beam is a neutral surface which is under neither tension nor compression. Layers lower than this surface experience increasing tension to the bottom of the beam. Layers above this surface experience increasing compression as the upper ends of the beam rotate inward, shortening the length of the layers.

12-4 Applications and Extensions

Each of the major types of deformations has been treated and explained separately. However, they often occur in combinations. Some portions of a beam, block, or board are under tensile stress, while other parts are under compressive or shear stress. Figure 12-7a shows a wooden beam supported near each end and sagging under its own weight. The bottom of the beam is subjected to tension forces trying to stretch the wood and pull it apart. The top is subjected to compression forces at the same time. We would also expect shear forces (not shown) to come into play, at least near the supports.

Building materials, particularly concrete and brick, often have much larger ultimate compressive strength than they have ultimate tensile strength. If a structure is to be built using such a material, it is important for the designer (for both safety and economy) to build into the design compressive loads where possible rather than tension loads.

If we examine Equation 12-4, we see another extension of elastic moduli. Solving that equation for the applied force, we have

$$F = \frac{YA}{L} \Delta L$$

Now if we replace the combination of constants YA/L by a single constant k and replace the variable ΔL by x, we have

$$F = kx \qquad (12\text{-}9)$$

This states that the elongation of an object under tension (or its decrease under compression) is directly proportional to the applied tension (or compression) force. Equation 12-9 is always applicable to bodies which obey Hooke's law. This was first introduced in Chapter 7 (Equation 7-3) and derived for a spring under tension. As we see, however, it has much wider application as a direct result of Hooke's law.

12-5 Strength of Materials

The values of elastic moduli, ultimate strengths, and so on, are presented as approximate or average values, because these so-called constants vary greatly with very small variations in the composition of the materials involved. Various grades of steel are a good example. For instance, tempered steel with a 1.20 percent carbon content has an ultimate tensile strength of 124×10^7 N/m², while tempered steel with a 0.49 percent carbon content has an ultimate strength of 67×10^7 N/m². Further, any discontinuity, inhomogeneity, or impurity could change the constants.

Thus only approximate or average values may be tabulated. We also note that the constants are obtained experimentally and are not the result of theory. In this sense, a study of the properties of materials is similar to a study of friction. The relationships with which one must deal are empirical.

All this is not to say that a study of the strength of materials is fruitless. Mechanical and architectural engineers and architects have honed the study to a fine edge. They have learned and published numerous volumes on the strength of materials. More is being learned all the time. In addition, we know that we must take into account the uncertainty of the strength of materials due to small variations in composition, discontinuities, and the like. In building, the structure is always designed to handle many times the greatest anticipated stress by specifying what is called the **safety factor.** Some common safety factors are listed in Table 12-3.

For example, consider a timber bridge which must withstand a varying load. Table 12-3 shows that we should use wood of such thickness and length that it will accept at least 10 times the expected maximum load.

The safety factors of Table 12-3 are large for a reason besides variations in composition. Consider the way you might break a wire paper clip or a stiff wire coat hanger. Usually you bend the wire back and forth many times until a crack develops and the wire breaks. This illustrates a phenomenon called **fatigue.** When we apply a stress and then remove it many times in succession, the ultimate strength of every material becomes smaller and smaller. This fatigue phenomenon, apparently due to microscopic changes in the material structure, must also be accounted for in safety factors and structural design.

Another consideration in design is the fact that supporting pillars, posts, and even animal legs very rarely crush under compressive forces. The most common failure is by **buckling.** Compressive forces are almost never exactly along the axis of even a perfect cylinder. Thus, large opposing torques may develop about an axis or axes perpendicular to the pillar's surface. The pillar may then bend and perhaps buckle, fracturing at the bend. Resistance to buckling (called *bending strength*) may be increased by the proper shaping of structural members. I beams and H beams are much better able to resist bending than a square beam or cylindrical rod made of the same material and having the same material cross section. Likewise, a hollow cylinder resists bending better than a solid cylinder of the same length and mass (hence the same material cross-sectional area). Most bones are designed according to this last criterion — they are

TABLE 12-3 Selected Safety Factors Used in Building Design for the Conditions Shown

Material	Safety Factor		
	For Constant Load	For Varying Load	For Impact Load
Structural steel	4	6	10
Timber	8	10	15
Brick, stone, unreinforced concrete	15	25	40

hollow cylinders. Hence, they resist bending and buckling better than if they were solid and had the same material cross-sectional area.

Finally, the relationships used in design calculations have the implicit assumption that the forces producing stress are uniformly distributed over the cross-sectional area in question. This is not often true, and therefore a small part of any given structural member may have to support much more than another part of that same member.

When we consider all the uncertainties involved in the strength of a structure, we should not be surprised at the magnitude of the safety factors used in design. In point of fact, it is perhaps surprising that so few structures fail and collapse.

SPECIAL TOPIC
Strength and Scaling Laws

While considering the structure and properties of materials, let us look at some natural structural designs which seem to have stood the test of time. Figure 12-8 shows two familiar animals side by side for comparison. While similar, there are structural differences between an elephant and a mouse. Ignoring the specialized trunk feature of the elephant, we can see that it has very thick, stumplike legs compared to the extremely thin limbs of the mouse. Why is this? We know a mouse runs rather well, can jump several times its height, and is generally quite nimble. Would we have a better elephant if it were built in the same proportions—say, one that could jump to several times its height? Is the elephant a bad design? For answers, we must consider what we mean by the laws of *scaling*, those laws which describe how the size and proportion of objects are related to the strength of materials.

Certain properties of physical bodies are directly related to the dimensions of the body. For example, the weight of an aluminum cylinder is directly proportional to its volume, while the ultimate tensile or compressive strength of the cylinder is directly proportional to its cross-sectional area. Likewise, the weight of a leg (the elephant's or mouse's) is proportional to its volume,

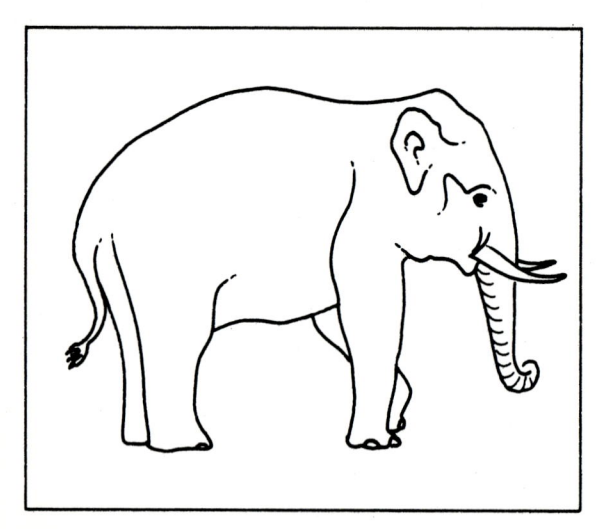

Figure 12-8   There are considerable structural differences between these two familiar animals. Note especially that the elephant's legs are so much thicker than those of the mouse.

while its strength under tension or compression is proportional to the cross section of the bone. For an animal as large as an elephant to be supported by its legs, the leg must be disproportionally thicker than that of a small animal.

We know the volume of an object has units of length cubed. The volume of the box in Figure 12-9a, for example, is the product of the length of the box times the width of the box times its height. That is,

$$\text{Volume} = \text{length} \times \text{width} \times \text{height}$$

or

$$V = lwh$$

Suppose each dimension is changed by the same proportion. This is done by multiplying each dimension by the same constant **scaling factor.** If we choose the Greek lowercase lambda (λ) to represent this factor, we have that the new length is λl, the new width is λw, and the new height is λh. If the scaling factor λ is greater than 1, we are *scaling up*. If λ is less than 1, we are *scaling down*. The λ chosen for Figure 12-9b is approximately 1.5. In any event, the scaled volume is

$$\begin{aligned}\text{Scaled volume} &= (\lambda l)(\lambda w)(\lambda h) \\ &= \lambda^3 lwh \\ &= \lambda^3 (\text{original volume})\end{aligned}$$

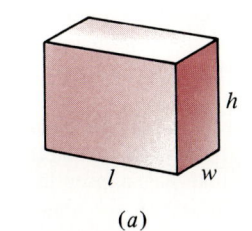

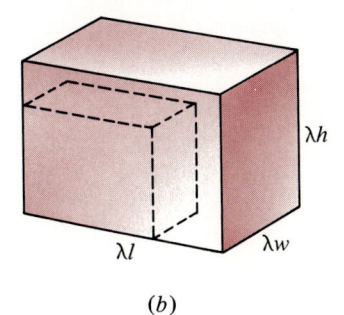

Figure 12-9 (a) A box of length l, height h, and width w has a volume $V = lwh$. (b) The box in (a) scaled up by a *scale factor* λ. The new volume of the box is $(\lambda l)(\lambda h)(\lambda w) = \lambda^3 lwh$.

We see that if an object is scaled such that each dimension is multiplied by λ, its volume is multiplied by λ^3.

Similarly, consider the area of the top of the box in Figure 12-9. The original area is

$$\text{Original area} = lw$$

The scaled area is

$$\begin{aligned}\text{Scaled area} &= (\lambda l)(\lambda w) = \lambda^2 lw \\ &= \lambda^2 (\text{original area})\end{aligned}$$

Thus, as dimensions are scaled by the factor λ, areas are multiplied by λ^2.

As an illustration, consider a 6-ft-tall man weighing 900 N. Let us scale him up to 12 ft and increase every dimension proportionately by using $\lambda = 2$. His volume and hence his weight should be multiplied by λ^3. Thus

$$\begin{aligned}\text{Weight of 12-ft man} &= (2)^3 \text{ (weight of 6-ft man)} \\ W_{12} &= 8(900 \text{ N}) = 7.20 \times 10^3 \text{ N}\end{aligned}$$

Now suppose we make the perfectly proportioned man 600 ft tall. Thus $\lambda = 100$, and his weight must be

$$\begin{aligned}W_{100} &= (100)^3 \text{ (900 N)} \\ &= 9.00 \times 10^8 \text{ N}\end{aligned}$$

Recognizing that all new cross-sectional areas are scaled as λ^2, consider the bones supporting the 600-ft man. In particular each lower leg bone, called the tibia, must support half his weight as he stands with both feet firmly planted. The cross section of the tibia of a 6-ft man is approximately 3.50×10^{-4} m². For the 600-ft man, that area is scaled by $(100)^2$, or the cross-sectional area of the tibia of the 600-ft man is 3.50 m².

Table 12-1 indicates that the ultimate strength per unit area under compression for bone is 1.70×10^8 N/m². Therefore, the 600-ft man's tibia will crush if it is subjected to a compressive force of

$$(1.70 \times 10^8 \text{ N/m}^2)(3.50 \text{ m}^2) = 5.95 \times 10^8 \text{ N}$$

Thus, as long as our 600-ft giant stands on both feet, each tibia must support only 4.50×10^8 N. However, if he lifts one foot to walk (he surely could not run or jump), the other tibia will be crushed because his weight cannot be supported by a single tibia. Of course, there are many other problems with which a 600-ft man would have to deal. Since the ultimate strength under tension of tendons and ligaments depends only on the square of the scale factor, it is likely he would tear tendons, rip muscles loose, and worse if he tried to move. Further, since the strength of muscles depends on their cross section, he probably would not have the strength to move anyway (see Problems 12-20 and 12-21). While there are certain specialized designs in

281

animals (trunks, horns, tails, long necks) which we might wish to change, it is highly unlikely that we can build a better elephant by using mouse blueprints. Contemplate for a moment an ordinary housefly scaled up to 6 ft. Could its legs hold it off the floor? Could its wings support it in flight?[1]

The mouse and the elephant are testimony that animals are not scaled up in perfect proportion. If they were, we would expect the volume and thus the weight to scale up proportional to the cube of some dimensions, perhaps the height or length. A dog which is twice as long as a cat would weigh eight times as much as the cat. Stated another way, we would expect perfectly scaled up animals to increase their length or height proportionally to the weight to the one-third power. Thus an elephant weighing 5.00×10^4 N and a small mouse weighing 0.500 N would have their lengths related as

$$\frac{L_e}{L_m} = \left(\frac{5.00 \times 10^4}{0.5}\right)^{1/3}$$
$$= (1 \times 10^5)^{1/3}$$

or

$$L_e = 46.4 \, L_m$$

If the mouse is 10.0 cm long, we would expect the elephant to be 4.64 m, a very long elephant. Actually, animals scale up rather more slowly with weight. Assume that the length L of an animal is proportional to its weight W to an unknown power u as

$$L \propto W^u$$

Measurement shows that for cattle $u = 0.24$ and for primates $u = 0.28$.

Designers of large buildings, bridges, and so on must be aware of the laws of scaling. Even the growth of a biological cell is probably limited by scaling laws. The volume and hence the oxygen requirement of a cell scale as λ^3. The surface area of the cell which must pass the oxygen scales only as λ^2. Thus, as the cell increases in size, it will eventually become oxygen-starved unless it divides to increase its surface-to-volume ratio.

Minimum Learning Objectives

After studying this chapter, you should be able to:
1. Define:
   - buckling
   - bulk modulus
   - compression
   - density
   - elastic limit
   - elastic modulus
   - fatigue
   - Hooke's law
   - pascal (Pa) [unit]
   - plastic flow
   - pressure
   - proportional limit
   - safety factor
   - scaling factor
   - shear modulus
   - shear stress
   - strain
   - strain hardening
   - stress
   - tension
   - Young's modulus
2. Calculate the deformation of a specified object from the force applied, the materials used, the geometry of the situation, and tabular values of moduli and ultimate strength.
3. Understand how properties of materials limit the types and designs of structures.
4. Calculate how a change of scale requires changes in the relative proportions of the parts of an object.
5. Select from a list of materials appropriate ones for particular structural purposes based on the ability of the material to withstand the types of stress to which it will be subjected.
6. Understand the empirical nature of statements involving the strength of materials and that "pure" stress-strain calculations idealize the situations found in real structures, necessitating the use of considerable safety factors in the design of stuctures.

Problems

12-1 An unstretched piece of rubber band is 3.00 mm wide and 1.20 mm thick. It is placed under a tension force of 2.00 N. Calculate (a) the cross-sectional area of the rubber band and (b) the stress.

---

[1] The lift force in flight is proportional to the area of the wings (λ^2), while the fly's weight is proportional to its volume (λ^3).

12-2 A 9.00-m-long nylon climbing rope stretches 0.630 m under the weight of the climber. Calculate the strain on the rope.

12-3 A tug-of-war at the spring picnic is expected to put a rope under tension of 1.60×10^3 N. If the material the rope is made of can take a maximum stress of 1.00×10^7 N/m², calculate the minimum diameter the rope must have.

12-4 A steel wire 4.00 m long and 1.50 mm in diameter is placed under a tension of 500 N. It stretches 5.40×10^{-3} m. Calculate (a) the strain, (b) the stress, and (c) Young's modulus for steel from these data.

12-5 A pendulum consists of a 2.50-cm-diameter steel ball of mass 0.0640 kg attached to the end of a 2.00-m-long string. The string has a diameter of 1.00 mm and a Young's modulus of 3.50×10^7 N/m². The ball is released from rest when the string makes an angle of 45° with the vertical. Calculate (a) the ball's speed at its lowest point, (b) the force exerted on the ball by the string at its lowest point, and (c) the increase in length of the string.

12-6 A marble column is 20.0 cm in diameter and 3.00 m long. (a) What maximum load can it support? (b) Assuming it obeys Hooke's law, by how much will the column be compressed?

12-7 A 3.00-m-long wire 3.57×10^{-3} m in diameter is stretched 9.00×10^{-5} m by a tension force of 20.0 N. Calculate (a) the tensile stress, (b) the tensile strain, and (c) Young's modulus for tension.

12-8 The professor steps off the demonstration table and free-falls 1.00 m to the floor. If the cross section of each tibia is 3.50×10^{-4} m² and he lands on both feet, calculate the average force exerted on his legs which would cause the bones to decrease in length by 3.00×10^{-2} percent.

12-9 A 900-kg automobile is being towed by a 3.00-m-long towrope. The diameter of the rope is 1.20 cm, and its Young's modulus is 2.50×10^8 N/m². What is the minimum elongation of the rope when the car is being accelerated at 0.750 m/s²?

12-10 A 7.00-kg bowling ball is swung in a vertical circle at the end of a 1.50-m-long nylon rope. The rope has a diameter of 9.00×10^{-3} m and a Young's modulus of 8.00×10^7 N/m². (a) Calculate the angular speed necessary such that the rope elongates by 5.00 cm when the bowling ball is at its lowest point. (b) At this angular speed, what is the elongation of the rope when the bowling ball is at its highest point?

12-11 The professor brings a piece of fishing line to the laboratory and announces it has a diameter of 7.00×10^{-4} m. He then tells the laboratory group that they are to determine Young's modulus for the line and leaves. The group measures the length of the line to be 0.850 m. They then clamp one end and hang a 4.00-kg mass on the other end. Measurement shows that the line stretches exactly one additional centimeter. What value for Young's modulus should they report to the professor when he returns?

12-12 How much energy is required to stretch a wire of length L, cross section A, and Young's modulus Y by ΔL? (Hint: Derive an expression for the work done, recalling that the force required increases linearly with increasing elongation.)

12-13 During a town picnic in the 1970s, an attempt was made to set a record for the number of people engaged in a tug-of-war. Unhappily, the rope broke, although real tragedy was avoided and only some broken limbs resulted. The rope in question was a 2-in-diameter nylon line. (a) Assuming no flaw in the rope, what force was required to snap the rope? (b) Assuming that an average person (many were children) exerted a force of 300 N, how many people were engaged in the pull? (Be careful with this. How many people would be needed if the rope were tied to a large tree?) (c) Is it reasonable that there was no flaw in the rope? (d) Assuming that the stress was concentrated in the central 50.0 m of rope, by how much did this section elongate before snapping? (e) Assume an average value of the force and calculate the energy stored in this central 50.0 m of rope at the time of snapping. (f) A stick of TNT stores 9.55×10^5 J of chemical potential energy. Rate the energy stored by the rope at snapping in equivalent sticks of TNT. (g) The mass of the rope was about 1.00 kg/m. What was the initial acceleration of each 25.0-m segment away from the point of the break? (h) If this acceleration lasted 0.0100 s, what would be the speed of the 25.0-m rope segment? (i) Compare the kinetic energy of this rope segment with that of a 7.00-kg bowling ball falling from a building. From what height would the ball have to fall to acquire equivalent kinetic energy? (Would you like to catch either?)

12-14 A steel rivet 1.00 cm long and 3.00 mm in diameter is subjected to a shearing force of 5.00×10^4 N. Calculate (a) the shear strain and (b) the distance one end moves with respect to the other.

12-15 A gelatin cube 8.00 cm on each edge is subjected to a shearing force of 0.300 N. The angle of shear measures 21°. Calculate (a) the shear stress, (b) the shear strain, and (c) the shear modulus for gelatin.

12-16 A steel bolt with a diameter of 1.00 cm has threads cut 1.00 mm deep at one end with a pitch of 6.00 threads/cm. The nut attached to the bolt contacts four complete turns of the threads at any one time. What force must be applied to the end of a 25.0-cm-long wrench tightening the nut to shear the threads completely off the bolt?

12-17 A glass sphere has a radius of 10.0 cm at standard atmospheric pressure (1.01×10^5 Pa). Calculate the change in its radius if (a) it is taken to the moon (pressure essentially zero) and (b) it is placed at the bottom of the ocean, where the pressure is 8.00×10^7 Pa.

12-18 Calculate the change in the density of water when the pressure is increased from 1.01×10^5 Pa at sea level to 1.58×10^7 Pa 1 mi beneath the surface. ($B = 2.20 \times 10^9$ N/m²)

12-19 Deep-sea diving machines are essentially hollow steel spheres, with external tanks of oil, steel pellets, etc., added. The basic sphere is 1.70 m in radius at the surface. (a) What is its area and the volume at the surface? (b) What is the volume of the sphere in the Mariana Trench, deepest spot in the oceans, some 11,034 m deep, if the pressure on the sphere is 1.07×10^8 Pa (1.55×10^4 lb/in²)? (c) If an air-filled balloon were tied to the outside of the sphere and had a volume of 0.500 m³ at the surface, what would its volume be in the trench?

12-20 In the Special Topic, we encountered a perfectly proportioned 600-ft man scaled up from a 6.00-ft man by using a scaling factor of $\lambda = 100$. The giant must weigh 9.00×10^8 N if the 6.00-ft man weighs 900 N. We know that the gastrocnemius (calf muscle) of the 6-ft man has a cross-sectional area of 1.10×10^{-2} m² and that muscle can exert a force of up to 7.00×10^5 N/m². Determine (a) the maximum force which can be supported by each calf muscle of the 6.00-ft man (How does this compare to his weight?), (b) the cross section of each calf muscle of the giant, and (c) the maximum force each of the giant's calf muscles can support. (How does this compare to his weight?)

12-21 The maximum girth of the biceps muscle of the 6.00-ft man of the Special Topic is about 0.200 m, yielding a cross section of 3.18×10^{-3} m². Using the fact that muscle fiber can exert a maximum force of 7.00×10^5 N/m², determine (a) the maximum force the six-footer's biceps muscle can exert, (b) the cross section of the biceps of the 600-ft giant, (c) the maximum force the giant's biceps can exert, (d) the ratio of the maximum force the six-footer's biceps can exert to the weight of the six-footer, and (e) the ratio of the maximum force the giant's biceps can exert to the giant's weight. (No wonder the giant cannot do chin-ups!)

12-22 The span for a floor between supports is to be increased from 3.00 to 6.50 m. The strength of a beam is given by $S = kwh^3/l$, where w = width, h = height, l = length, and k is about 4.0×10^8 N/m³ for building lumber. (a) What size beams should replace the 2×8 floor beams (where 8.00 in is the vertical dimension) used previously to provide the same support? (Assume a constant width of 2.00 in.) (b) Actually, the formula presented is for a static load. If you were to walk across the floor whose support you calculated in (a), you would find it perceptibly springy to your footsteps. Also, in general, long, thin planks or boards are quite susceptible to twisting or torsional stress. Because of these reasons, assume that we increase the width of the new beams to 4.00 in and design in a safety factor of 4 to cover any uncertainty. What is the new height of the beams?

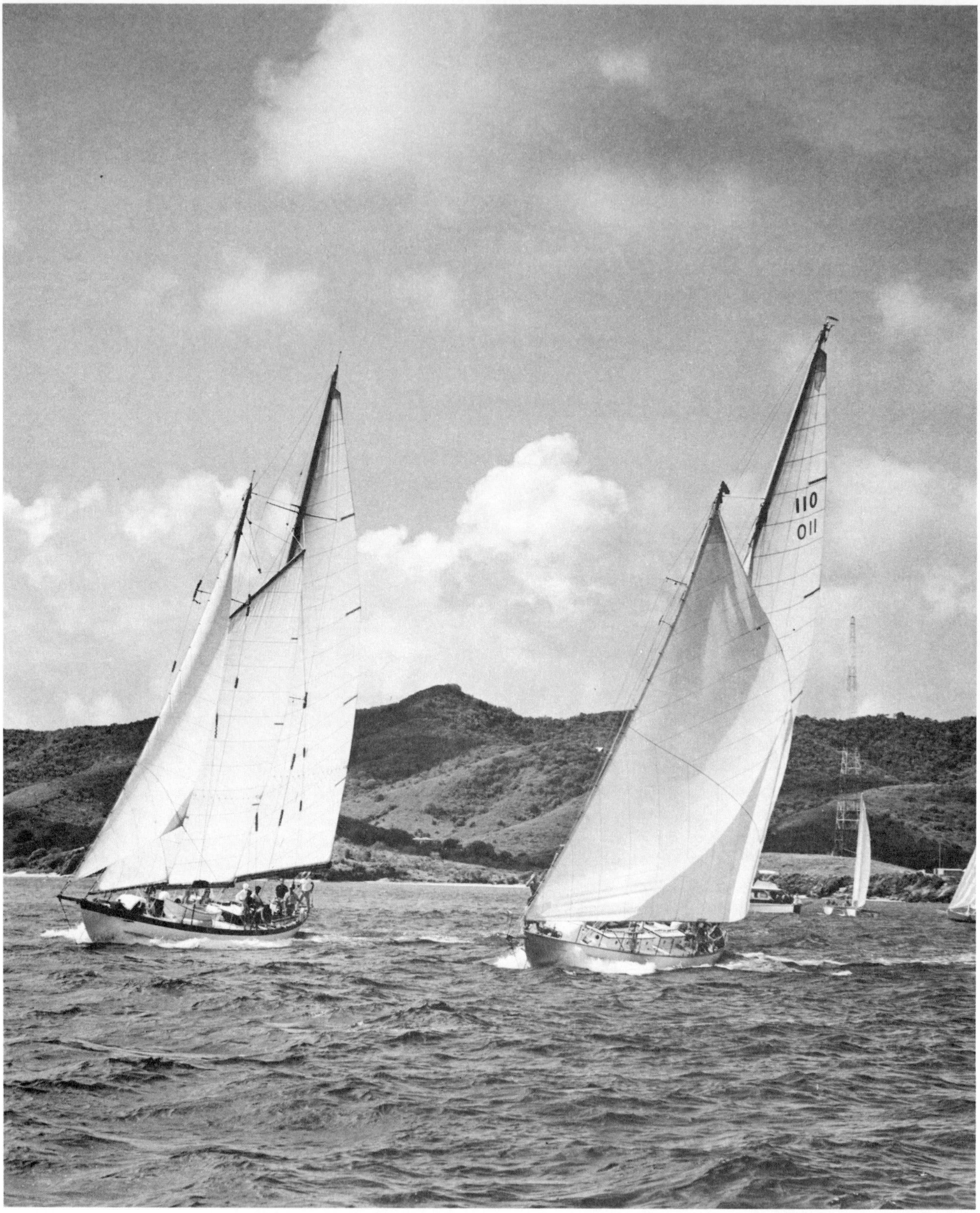
110
011

13 Mechanics of Fluids

13-1 Introduction

Forces may be transmitted through nonrigid materials. For example, if you slam a door in a tightly sealed house, windows several rooms away from the door rattle and drapes shudder. Air pumped into the tires of a car effectively lifts the car from a rigid contact with the road surface. The shock wave in air from an explosion can knock nearby buildings flat. The same shock waves from an explosion in water are even more effective, such that a near-miss of a bomb on an armored warship is generally more effective in sinking than a direct hit would be.[1]

So far, we have looked at different objects and materials that, more or less, retained their sizes and shapes under the action of forces and torques. This common property is used to classify objects — bowling balls, books, bricks, and so on — as **solids.** All other substances flow when unbalanced forces act on them. They are therefore called **fluids.** This definition of fluid encompasses a broad spectrum of substances — air, water, butter at room temperature, molasses, even glass.

---

[1] A near-miss crushes in the side of the ship, opening seams and allowing water to enter. A bomb exploding inside the ship has less effect on the outer walls, which are solidly braced by relatively incompressible water. Instead, decks and interior walls sustain most of the damage.

Fluids may be further divided into two general categories: **liquids** and **gases.** The distinction between them is that they differ to a great degree in the compressibilities. Gases may be compressed with ease, while liquids (as we saw in Chapter 12) are for most purposes practically incompressible. In this chapter, we will consider primarily liquids which flow rather easily. Gases will be covered in the next chapter.

13-2 Density

You have little trouble walking upstream against a 5 km/h current of air. Fighting the same flow waist-deep in a stream of water requires considerable effort, while opposing a similar current in a mudflow is beyond your capacity. The distinction is largely one of **density.**

You will recall that the density ρ of a homogeneous substance is defined as the mass of the material per unit volume:

$$\rho = \frac{m}{V} \tag{13-1}$$

where m is the mass of the object and V is its volume. In SI units, density is given in kilograms per cubic meter (kg/m^3). The densities of a selected number of substances are listed in Table 13-1. These densities are given for homogeneous materials, but we may also work with average densities for inhomogeneous materials.

TABLE 13-1 Approximate Value of Mass Density for Some Selected Homogeneous Materials at 20°C (unless Otherwise Noted)

Material	Density, 10^3 kg/m^3
Solids	
Aluminum	2.70
Balsa wood	0.12–0.20
Bone	1.70–2.00
Brass	8.67
Brick	1.40–2.20
Carbon (diamond)	3.52
Carbon (graphite)	2.25
Cement	2.70–3.00
Cherry wood (black)	0.534
Fir wood	0.415
Glass (common)	2.40–2.80
Gold	19.30
Ice (0°C)	0.917
Iron	7.89
Lead	11.30
Oak wood (white)	0.710
Platinum	21.40
Silicon	2.42
Tungsten	19.30
Uranium	18.70
Walnut wood	0.562
Liquids	
Acetone	0.791
Carbon tetrachloride	1.594
Ethyl alcohol	0.810
Glycerin	1.260
Mercury	13.60
Water (pure)	1.000
Water (sea)	1.030

Example 13-1 The mass of the first Mr. Universe winner was 92.6 kg. If his average density was 1.04×10^3 kg/m^3, determine his volume.

Solution This is a straightforward application of Equation 13-1:

$$\rho = \frac{m}{V}$$

or

$$V = \frac{m}{\rho} = \frac{92.6 \text{ kg}}{1.04 \times 10^3 \text{ kg/m}^3} = 8.90 \times 10^{-2} \text{ m}^3$$

Liquids flow under the action of unbalanced forces, so that if an amount of liquid is stationary, or static, we know the net force acting on that amount is zero. For this reason, the study of liquids at rest is called **hydrostatics.** We will take up the study of liquids in motion in Section 13-6 under the name *hydrodynamics.* Just as we discovered magnitudes and directions of unknown forces from a study of solids in equilibrium, we may determine the value of unknown quantities from an analysis of fluids at rest. However, we quickly discover that working with forces is not convenient, and the most practical approach is to deal with pressures.

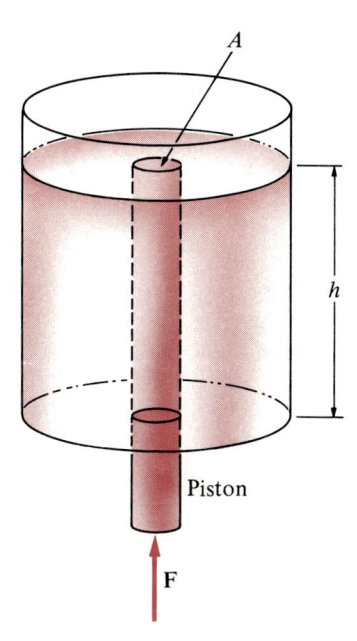

Figure 13-1 An upward force of F must be supplied to the piston to keep it in place. The size of F is determined by the depth of the liquid and its density. The piston must support the dashed-in column of liquid above it of weight mg.

13-3 Pressure and Pascal's Principle

In Chapter 12 we introduced the concept of pressure as the magnitude of the force per unit area. Now we know that a fluid, by definition, cannot support shearing forces without flowing. Thus we deduce that any forces acting on a fluid at rest must be in a direction perpendicular to the fluid surface or surfaces.

How can we determine the hydrostatic pressure at some point in a liquid? That is, we wish to determine the magnitude of the force per unit area exerted by the liquid. Consider the glass container in Figure 13-1. It is filled to a height h with a liquid of density ρ. A small hole of cross section A in the bottom of the container holds a leak-tight movable piston. The magnitude of the additional force F that must be applied to keep the piston from moving is equal to the weight w of the column of liquid above the piston. Thus we have

$$F = w = mg$$

where m is the mass of the column of liquid above the piston. Now from Equation 13-1, we have

$$m = \rho V$$

where V is the volume of the column. Finally, since $V = Ah$, we can write

$$F = \rho V g = \rho A h g$$

or

$$P = \frac{F}{A} = \rho g h \qquad (13\text{-}2)$$

Thus the pressure at the bottom of the container due to the liquid is equal to the product of the density of the liquid with the acceleration due to gravity and the height of the liquid. Further, by changing the vertical position of the top of the

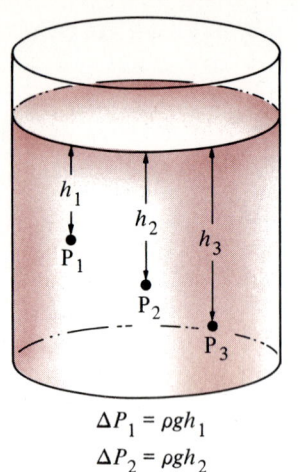

Figure 13-2 The pressure due to a liquid at a point below the liquid surface varies with the depth in the liquid h. This pressure difference is $\Delta P_n = \rho g h_n$.

$$\Delta P_1 = \rho g h_1$$
$$\Delta P_2 = \rho g h_2$$
$$\Delta P_3 = \rho g h_3$$

piston and hence the height of liquid above it, we find that Equation 13-2 is true for any h in the liquid. This is shown in Figure 13-2.

Pressure has no direction. The force due to the fluid exerted on any surface is perpendicular to the surface area regardless of the orientation of the surface. A very thin disk of area A placed at a depth h in a liquid, as in Figure 13-3, has a downward force of magnitude $F_d = \rho g h A$ acting on its upper surface and an upward force of magnitude $F_u = \rho g h A$ acting on its lower surface due to the liquid.

Note that we say that Equation 13-2 gives the pressure *due to the liquid*. Suppose an additional pressure is applied externally to the liquid. For example, a typical container at the earth's surface has the pressure of the atmosphere acting on its upper surface. After all, if the upper surface is open to the atmosphere, it must support the weight of the column of air above it. Standard atmospheric pressure is defined as the average value at sea level of about $P_A = 1.013 \times 10^5$ Pa. Thus at any point at a depth h in the container, the net or

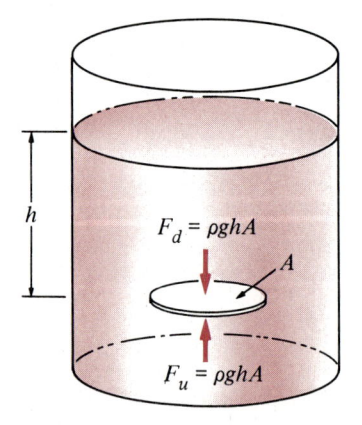

Figure 13-3 A very thin disk of area A oriented horizontally at a depth h in a liquid of density ρ. The forces due to the liquid acting on the disk on the bottom and top are equal in magnitude.

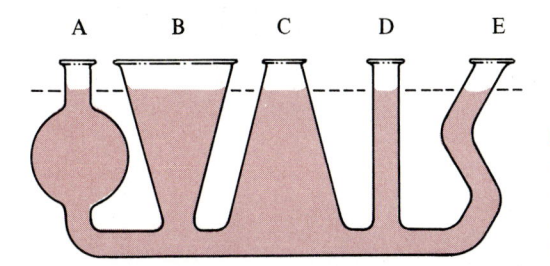

Figure 13-4 Liquid poured into interconnecting containers will settle to the same level in all of them, resulting in the hydrostatic paradox.

absolute pressure is equal to the sum of the pressure due to the liquid and atmospheric pressure:

$$P_{abs} = P_A + \rho g h \qquad (13\text{-}3)$$

A good understanding of Equation 13-3 cleared up a phenomenon which was known as the *hydrostatic paradox*. When liquid is poured into one of a number of interconnected containers of different shapes, as in Figure 13-4, the liquid stands at the same level in each container. It was considered a paradox that containers with greater volume did not have greater pressures at the bottom and thus force the liquid to move to the smaller containers. However, Equation 13-3 explains that the only causes of pressure changes from point to point in a liquid are changes in the vertical location of the points in question. Hence, at the same depth below the surface, the pressure is the same in all the vessels independent of the amount of liquid above. This leads to the familiar statement that "water seeks its own level."

Extending the reasoning which leads to Equation 13-3, Blaise Pascal formulated and then stated what subsequently became known as **Pascal's principle:**

> When a change in pressure is applied to an enclosed fluid, the change is transmitted undiminished to every point in the fluid and to the walls of the container.

Pascal reasoned that if the pressure of an enclosed liquid is changed by any means, whether by the atmosphere or by inserting a piston in the container and pushing (or pulling), as in Figure 13-5, the pressure at any depth must undergo precisely the same change.

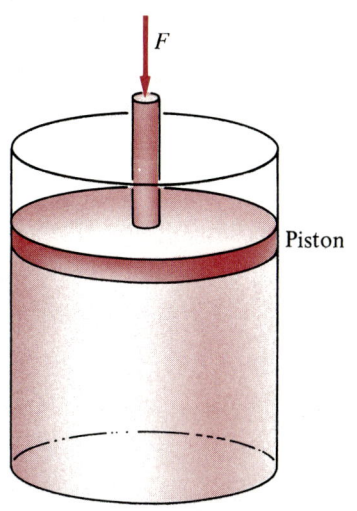

Figure 13-5 The increase (or decrease) in pressure on the confined liquid due to the force exerted on the piston is transmitted to every point in the liquid and to the walls of the container.

The implications of this principle are enormous. Applications of it in the form of the **hydraulic press** (Figure 13-6) are involved in many everyday devices. A relatively small force F_1 applied to the confined liquid through a small piston of cross section A_1 gives rise to an increase in pressure $\Delta P = F_1/A_1$ on the liquid. By Pascal's principle, this increase is transmitted to every point in the enclosed liquid and to the walls of the container. In particular, this increase in pressure is now exerted on the larger piston of cross section A_2. This gives rise to an increased force $F_2 = \Delta P A_2$ on the larger piston. Since we know that the pressure increase is the same for both pistons,

$$\frac{F_1}{A_1} = \frac{F_2}{A_2}$$

and we see that

$$F_2 = \frac{A_2}{A_1} F_1 \qquad (13\text{-}4)$$

Figure 13-6 A hydraulic press, which works on Pascal's principle.

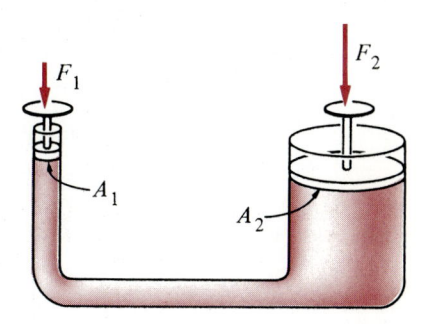

Thus the hydraulic press in its simplest form is merely a force multiplier, much as other machines we have studied. The multiplying factor is equal to the ratio of the areas of the two pistons involved. If the ratio of those areas is large enough, even a small child could provide the force necessary to lift a large truck. In fact, hydraulic lifts in automobile service stations and hydraulic jacks operate on this principle, as do many other everyday devices and commercial and industrial machines: some dentist's chairs, most barber chairs, the brake system for automobiles and trucks, sheet metal cutters, industrial stamping machines, etc.

Example 13-2 A 2.00×10^3 kg vehicle is to be lifted for servicing by a hydraulic lift pressure of 2.50×10^5 Pa. Calculate the radius of the lift piston.

Solution We must first determine the weight of the vehicle, which is the minimum force we must exert to lift the vehicle. This force along with the pressure will yield the lift piston's cross section, from which we may calculate its radius.

$$w = mg = (2.00 \times 10^3 \text{ kg})(9.80 \text{ m/s}^2) = 1.96 \times 10^4 \text{ N}$$

Now since $P = F/A = w/A$, we have

$$A = \frac{w}{P} = \frac{1.96 \times 10^4 \text{ N}}{2.50 \times 10^5 \text{ N/m}^2} = 7.84 \times 10^{-2} \text{ m}^2$$

The cross section of a cylinder is πr^2, where r is the radius, and thus

$$\pi r^2 = 7.84 \times 10^{-2} \text{ m}^2$$

$$r = \sqrt{\frac{7.84 \times 10^{-2} \text{ m}^2}{\pi}} = 0.158 \text{ m}$$

13-4 Measurement of Pressure

We live in the atmosphere and come to think of atmospheric pressure as being normal. Indeed we define a tire as flat if the pressure inside it is only 1 atmosphere. Most measuring devices for pressure (called gauges) are calibrated to read zero when the air in the object being measured is at the same pressure as that of the atmosphere. **Gauge pressure** is therefore equal to the absolute pressure minus the atmospheric pressure. In symbols,

$$P_G = P_{abs} - P_A \tag{13-5}$$

Gauges vary in form, but nearly all depend on principles we have already learned. The most fundamental gauge is an **open-tube manometer,** as shown in Figure 13-7. A J or U tube partially filled with liquid is connected to the vessel whose pressure we wish to measure. The liquid settles to different heights on different sides of the tube under the action of P, the pressure in the vessel, and P_A, atmospheric pressure in the open tube. Point 1 on the left is the top surface of the liquid in the tube, at which the pressure is P. Point 2 is at the same vertical

13-4 Measurement of Pressure

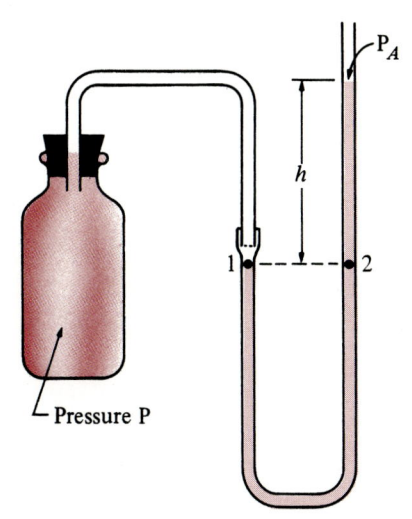

Figure 13-7 The pressure inside a flask of confined gas of density ρ being measured by an open-tube manometer. At point 1, the surface of the liquid is at the pressure of the gas. The top of the right-hand column is at atmospheric pressure P_A. Point 2 is at the same height as point 1 but a depth h below the liquid surface. The difference between P and P_A, the gauge pressure, is $\rho g h$.

location in the liquid and hence at the same pressure as point 1 (by Equation 13-2). However, the pressure at point 2 is just the sum of P_A and the pressure due to a column of liquid of height h, thus allowing us to use Equation 13-3.

Example 13-3 Determine the gauge pressure and the absolute pressure of the vessel shown in Figure 13-7 if $h = 28.0$ cm and the liquid is mercury.

Solution The gauge pressure is $\rho g h$, and the absolute pressure is the gauge pressure plus the pressure of the atmosphere:

$$P_G = \rho g h = (1.36 \times 10^4 \text{ kg/m}^3)(9.80 \text{ m/s}^2)(0.280 \text{ m})$$
$$= 3.73 \times 10^4 \text{ Pa}$$

$$P_{abs} = P_G + P_A = 3.73 \times 10^4 \text{ Pa} + 1.013 \times 10^5 \text{ Pa}$$
$$= 1.39 \times 10^5 \text{ Pa}$$

The same principle is used for measuring atmospheric pressure. A long glass tube is filled with liquid (usually mercury) and inverted in a container of the liquid which is open to the atmosphere (Figure 13-8). When the bottom of the inverted tube is opened, the liquid descends to a fixed level, leaving a vacuum above it in the tube. The pressure of the remaining liquid column is balanced by the pressure of the atmosphere; hence

$$P_A = \rho g h$$

Mercury is the liquid most often used for this measurement, since it has the largest density and thus requires a shorter tube than any other liquid. Devices used to measure atmospheric pressure are called **barometers.**

Various older unit systems are still used to report pressure. Standard atmospheric pressure is 1.013×10^5 Pa. A pressure of exactly 1.00×10^5 Pa is called one **bar.** Thus atmospheric pressure is approximately equal to a bar. Television meteorologists still report atmospheric pressure in units of inches of mercury, supported as in Figure 13-8. Even blood pressure measurements (see Special Topic) are reported in millimeters of mercury, called **torr.** Standard atmospheric pressure is 760 torr, or 760 mmHg, or 29.92 in Hg.

Figure 13-8 In the mercurial barometer, a tube sealed at one end is filled with mercury and inverted into a dish of mercury. The mercury in the tube descends to a height h, where the pressure of the mercury column equals the pressure of the atmosphere on the liquid surface in the dish.

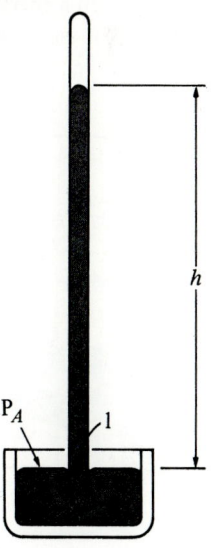

13-5 Buoyancy and Archimedes' Principle

Bodies immersed in water appear to weigh less. You may find it somewhat difficult to lift a classmate while you are both in the lecture hall, but you can do so with ease if you are both shoulder-deep in water. Further, when released under water, a wooden board or block will rise to the surface and float partially submerged. Any body submerged or partially submerged in a fluid will experience an upward force (called a **buoyant force**) due to the fluid and appear to weigh less than it really does. The reason for this is that the force on a submerged body is due to the pressure, which varies with depth. Therefore the pressure (and hence the force) on the lower portion of the body (at greater depth) is larger than the pressure (and force) on the upper portion.

Consider Figure 13-9. The rectangular block of area A and thickness d is immersed in a liquid of density ρ. The downward force on its upper surface is equal to the pressure at that surface, $\rho g h$, times the area A. The upward force on the block's lower surface is equal to the pressure at that surface $\rho g(h + d)$ times its area A; thus the net upward force, that is, the buoyant force, F_B on the block due to the fluid is

$$F_B = F_u - F_d = \rho g(h + d)A - \rho g h A = \rho g d A$$

but dA is the volume V of the block. Thus

$$F_B = \rho g V \qquad (13\text{-}6)$$

Equation 13-6 is a statement of a very important general principle first formulated by Archimedes (287–212 B.C.). **Archimedes' principle** is:

> A body immersed in a fluid is buoyed up by a force equal to the weight of the fluid which the body has displaced.

The density of the liquid is ρ, and the volume of liquid displaced in this case is certainly equal to the volume of the block. Thus ρV is the mass of liquid displaced, while $\rho g V$ (the buoyant force) is the weight of the liquid displaced.

Example 13-4 An irregularly shaped piece of metal is attached to a strong thread and hung from a spring balance in order to determine its weight. While suspended in air, the metal weighs 38.0 N. When the spring is placed over a

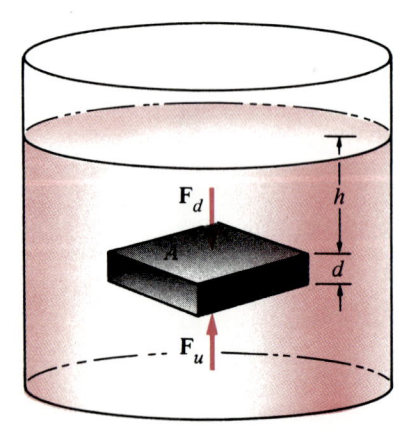

Figure 13-9 A rectangular body submerged in a liquid of density ρ has acting on it a net upward force of magnitude $\rho g d A = \rho g V$.

bucket of water so that the metal hangs completely submerged in water, the scale reads 23.9 N. Calculate (a) the volume and (b) the average density of the metal.

13-5 Buoyancy and Archimedes' Principle

Solution (a) We recognize that the difference between the weight of the object when it is suspended in air and when it is suspended in water is (neglecting the buoyant force of the air) equal to the weight of water displaced by the metal. This may be expressed in equation form and solved for the volume.

$$W_A - W_W = \rho_w V g$$

$$38.0 \text{ N} - 23.9 \text{ N} = (1000 \text{ kg/m}^3)(V)(9.80 \text{ m/s}^2)$$

$$V = 1.44 \times 10^{-3} \text{ m}^3$$

(b) Once we know the volume of the object, we can determine its density from its weight when suspended in air.

$$\rho = \frac{m}{V} = \frac{W_A}{gV}$$

$$= \frac{38.0 \text{ N}}{(9.80 \text{ m/s}^2)(1.44 \times 10^{-3} \text{ m}^3)}$$

$$= 2.69 \times 10^3 \text{ kg/m}^3$$

The net force on an object floating at rest in some liquid must be equal to zero. The volume of the liquid displaced equals the volume of the submerged portion of the object, yet has weight equal to the entire floating object. Thus, in order that it float with even a tiny part exposed, any object must have a density less than the liquid in which it floats.

Example 13-5 A long oak post 10.0 cm by 10.0 cm has a density of 710 kg/m³. If it is floated in water, what fraction of the post will float above the surface of the water?

Solution Since the post is floating, the net force is zero. The weight of the post (downward) must be exactly equal in magnitude to the buoyant force (upward). Thus we must formulate suitable expressions for the magnitude of these two forces and set them equal.

The weight of the post is the product of its mass and the acceleration due to gravity:

$$w = mg = \rho_p V_p g$$

where ρ_p is the density of the post and V_p is the volume of the post. Now if the post floats with a fraction of its volume fV_p above the water, then it is displacing a volume of water equal to V_p minus fV_p:

$$V_{disp} = V_p - fV_p = (1 - f)V_p$$

The weight of the water displaced and therefore the buoyant force is

$$F_B = \rho_w V_{disp} g = \rho_w (1 - f) V_p g$$

thus we have

295

$$\rho_p V_p g = \rho_w(1-f)V_p g$$
$$\rho_p = (1-f)\rho_w$$
$$f = 1 - \frac{\rho_p}{\rho_w} = 1 - \frac{710 \text{ kg/m}^3}{1000 \text{ kg/m}^3} = 0.290$$

13-6 Hydrodynamics and Continuity

In principle, there are several possible ways to describe the motion of a large amount of fluid. We could consider the fluid divided into a very large number of droplets, or mass particles. Each particle then would be described by Newton's laws of motion. While this method might be possible in principle, we would find that the large number of independent equations required would make the description too complex to be practical. We are fortunate in that many real examples of fluid motion may be described by using a model which considers the fluid to move as a whole.

Consider a pipe like that in Figure 13-10, with a changing diameter and direction, in which a liquid is flowing from the lower left at point 1 to the upper right at point 2. In a time interval, say Δt, liquid flowing past point 1 at velocity v_1 will have moved a distance $\Delta x_1 = v_1 \Delta t$. The volume of liquid which has flowed past point 1 in that time interval is $A_1 \Delta x_1 = A_1 v_1 \Delta t$. The mass of that volume of liquid is $\rho A_1 v_1 \Delta t$. Similarly, the mass of liquid which flows past point 2 during a Δt interval must be $\rho A_2 v_2 \Delta t$, where we have assumed that the liquid is incompressible and thus the density ρ is the same at both points. Now if the pipe remains full and rigid, the same mass of liquid must flow past point 2 during the Δt interval as flows past point 1.[1] Thus we may set these two increments of mass equal to each other and write

$$\rho A_1 v_1 \Delta t = \rho A_2 v_2 \Delta t$$

If we cancel the common factors, we have

$$A_1 v_1 = A_2 v_2 \qquad (13\text{-}7)$$

Equation 13-7 is the result of conservation of mass and expresses the continuity of flow for incompressible fluids. It is often called the **continuity equation.** Note that $A_1 v_1$ and $A_2 v_2$ are, respectively, the volumes of liquid per unit time flowing past points 1 and 2. Thus the cross section of a pipe times the average speed of flow is said to be the **volume flow rate.** The symbol Q is often used to represent the volume flow rate; thus $Q = Av$.

Example 13-6 The large artery into which the heart pumps oxygenated blood is called the aorta. In a normal resting adult, the heart pumps blood into the aorta at an average volume flow rate of about 9.00×10^{-5} m³/s. Calculate the average speed of the blood in an aorta of inside diameter 1.40 cm.

---

[1] If more liquid flows past point 1 than point 2, then the pipe would have to expand or leak liquid somewhere. On the other hand, if more liquid flows past point 2 than point 1, then the pipe would eventually become empty or it would have to have some other source of liquid.

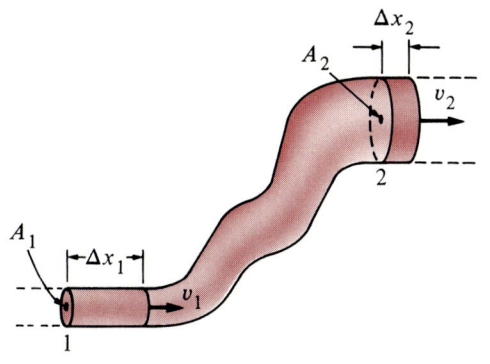

Figure 13-10 A pipe carrying a moving fluid varies in diameter, shape, and elevation between points 1 and 2.

Solution Since the volume flow rate is simply the product of the cross-sectional area with the speed, we must first find the area and then divide that value into the volume flow rate.

$$A = \pi r^2 = \pi(7.00 \times 10^{-3} \text{ m})^2 = 1.54 \times 10^{-4} \text{ m}^2$$

$$v = \frac{Q}{A} = \frac{9.00 \times 10^{-5} \text{ m}^3/\text{s}}{1.54 \times 10^{-4} \text{ m}^2} = 0.584 \text{ m/s}$$

We can solve Equation 13-7 for v_2 and obtain

$$v_2 = \frac{A_1}{A_2} v_1$$

Thus, if the cross-sectional area of the pipe is smaller at point 2 than at point 1, the liquid moves faster at point 2. In other words, the liquid moves faster at narrow places and slower at wider places. This has significant implications, as we will see in the following sections.

13-7 Work-Energy and Bernoulli's Equation

The work-energy principle may be applied to fluid flowing in a closed pipe, assuming such flow to be frictionless. In Figure 13-11, liquid is shown flowing from the lower left to the upper right part of the diagram. Assume the liquid flows for a time interval Δt during which the pipe is full and rigid. First, we know that the conditions for the continuity equation are satisfied and $A_1 v_1 = A_2 v_2$. Now consider the flow for a time interval Δt. The net result can be represented by assuming that a quantity Δm of liquid is transferred from one part of the pipe to another and its speed changed from v_1 to v_2. This change is accomplished by the net work done on the liquid by the forces due to pressure. We wish to apply the form of the work-energy principle which states that the net work done by the nonconservative forces (those due to pressure in this case) is equal to the sum of the changes in kinetic and potential energy.

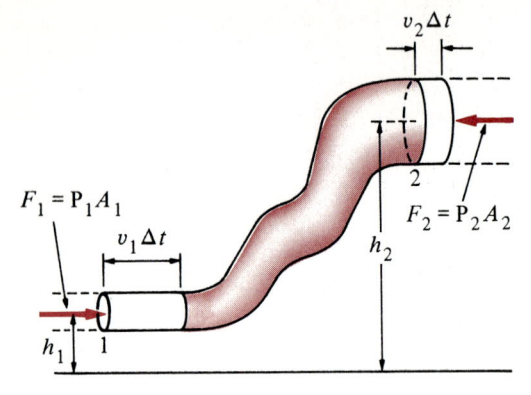

Figure 13-11 Defining the same situation as Figure 13-10 in terms of heights, speeds, and forces allows the application of the work-energy theorem and leads to Bernoulli's equation.

The force on the liquid at point 1 is $P_1 A_1$. Thus the work done by this force as the fluid moves a distance $v_1 \, \Delta t$ in the same direction is

$$W_1 = P_1 A_1 v_1 \, \Delta t$$

Similarly, the work done by the force due to the pressure at point 2 during the Δt interval is

$$W_2 = -P_2 A_2 v_2 \, \Delta t$$

It is negative because the liquid moves in a direction opposite to that of the applied force. Since the mass is $\rho A v \, \Delta t$, the change in kinetic energy of the fluid during Δt is

$$\Delta \mathrm{KE} = \tfrac{1}{2}(\rho A_2 v_2 \, \Delta t) v_2^2 - \tfrac{1}{2}(\rho A_1 v_1 \, \Delta t) v_1^2$$

Finally, the change in potential energy of the fluid during Δt is

$$\Delta \mathrm{PE} = (\rho A_2 v_2 \, \Delta t) g h_2 - (\rho A_1 v_1 \, \Delta t) g h_1$$

The modified work-energy equation becomes

$$W = W_1 + W_2 = \Delta \mathrm{KE} + \Delta \mathrm{PE}$$

$$P_1 A_1 v_1 \, \Delta t - P_2 A_2 v_2 \, \Delta t = \tfrac{1}{2} \rho A_2 v_2 \, \Delta t \, v_2^2 \\ - \tfrac{1}{2} \rho A_1 v_1 \, \Delta t \, v_1^2 + \rho A_2 v_2 \, \Delta t \, g h_2 - \rho A_1 v_1 \, \Delta t \, g h_1$$

Note that each term contains a Δt and either an $A_1 v_1$ or an $A_2 v_2$. Since we know from the continuity equation that $A_1 v_1 = A_2 v_2$, we may divide each term in the modified work-energy equation by either $A_1 v_1 \, \Delta t$ or $A_2 v_2 \, \Delta t$. This then gives us

$$P_1 - P_2 = \tfrac{1}{2} \rho v_2^2 - \tfrac{1}{2} \rho v_1^2 + \rho g h_2 - \rho g h_1$$

which may be written as

$$P_1 + \tfrac{1}{2} \rho v_1^2 + \rho g h_1 = P_2 + \tfrac{1}{2} \rho v_2^2 + \rho g h_2 \qquad (13\text{-}8)$$

Equation 13-8 is known as **Bernoulli's equation.** It is named for Daniel Bernoulli, who first formulated it in 1738. Note that in the equation, all the terms on the left apply to position 1 and all the terms on the right apply to position 2. Thus we may write Bernoulli's equation as

$$P + \tfrac{1}{2} \rho v^2 + \rho g h = \text{constant} \qquad (13\text{-}9)$$

as long as we realize that each term applies to the same point.

It is interesting to note that Equation 13-3 is a special case of Equation 13-8 when the speed of the flow is zero. Setting $v_1 = v_2 = 0$ in Equation 13-8 yields

$$P_1 + \rho g h_1 = P_2 + \rho g h_2$$

or

$$P_1 = P_2 + \rho g (h_2 - h_1)$$

Example 13-7 Water flows into the basement of a two-story house through a 2.50-cm-diameter pipe. When the only water running in the house is the cold-water faucet in the bathroom some 3.00 m above the basement pipe, the speed of flow in the basement pipe is 0.160 m/s and the pressure is 1.80×10^5 Pa. The bathroom faucet is being fed by a tube 0.700 cm in diameter. Determine (a) the speed of the water in the tube feeding the faucet and (b) the pressure in the tube.

Solution Part a can be solved by using the continuity equation to find v_2. Then a straightforward application of Bernoulli's equation provides the answer to part b.

(a)
$$A_1 v_1 = A_2 v_2$$

$$v_2 = \frac{A_1}{A_2} v_1 = \frac{\pi r_1^2}{\pi r_2^2} v_1$$

$$= \frac{r_1^2}{r_2^2} v_1 = \frac{(1.25 \text{ cm})^2}{(0.350 \text{ cm})^2} (0.160 \text{ m/s}) = 2.04 \text{ m/s}$$

(b) $P_2 = P_1 + \frac{1}{2}\rho v_1^2 - \frac{1}{2}\rho v_2^2 + \rho g(h_1 - h_2)$
$= 1.8 \times 10^5 \text{ Pa} + \frac{1}{2}(1000 \text{ kg/m}^3)(0.160 \text{ m/s})^2$
$\quad - \frac{1}{2}(1000 \text{ kg/m}^3)(2.04 \text{ m/s})^2 + (1000 \text{ kg/m}^3)(9.80 \text{ m/s}^2)(-3.00 \text{ m})$
$= 1.49 \times 10^5 \text{ Pa}$

Note that this is absolute pressure and must be greater than atmospheric pressure or water will not flow out of the faucet.

Bernoulli's equation answers several questions, from "How do airplanes fly?" through "Why do curve balls curve?" We consider some of these next.

13-8 Applications of Fluid Flow Equations

Let us consider several special cases for a better understanding of the implications of Bernoulli's equation. Figure 13-12 shows a very large tank filled with liquid of density ρ and open to the atmosphere. There is a small hole in the side of the tank at some distance h below the surface of the liquid. We will obtain an expression for the speed of efflux of liquid through the hole. First, we note the pressure is atmospheric both at the top surface of the tank and at the flowing

Figure 13-12 If we assume that the flow is essentially frictionless so that Bernoulli's equation applies, the velocity of efflux from a small hole in a large open tank depends only on the distance of the hole below the surface.

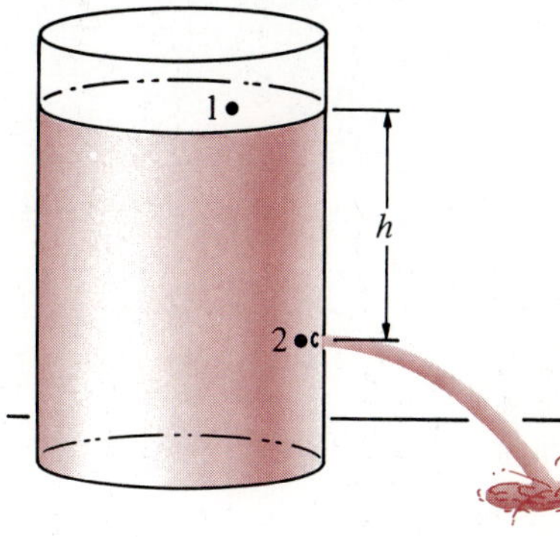

stream just outside the hole. Thus, when applied to these two points, Equation 13-8 becomes

$$P_A + \tfrac{1}{2}\rho v_1^2 + \rho g h_1 = P_A + \tfrac{1}{2}\rho v_2^2 + \rho g h_2$$

Subtracting P_A from both sides and rearranging yields

$$\tfrac{1}{2}\rho v_2^2 = \tfrac{1}{2}\rho v_1^2 + \rho g(h_1 - h_2)$$

Now assuming that A_1 is much, much greater than A_2, the continuity equation requires that v_1 be much, much less than v_2. We assume that $\tfrac{1}{2}\rho v_1^2$ may be neglected when compared to the other terms in the equation. (See Problem 13-22, which explores whether this is a reasonable assumption.) We therefore set $v_1 = 0$ and solve the equation for v_2, obtaining

$$\begin{aligned} v_2 &= \sqrt{2g(h_1 - h_2)} \\ &= \sqrt{2gh} \end{aligned} \qquad (13\text{-}10)$$

This is the same as the speed that one would predict by employing either kinematical considerations or the conservation of energy for a body falling through a height h in a region of constant gravitational acceleration g.

Another special case is that of liquid flowing in a horizontal pipe. Since h is the same value for any two points along such a pipe, the potential energy term is the same for both points. For this situation, such as that shown in Figure 13-13, Equation 13-8 becomes

$$P_1 + \tfrac{1}{2}\rho v_1^2 = P_2 + \tfrac{1}{2}\rho v_2^2 \qquad (13\text{-}11)$$

From the continuity equation, we know that the speed of flow is much larger at the constriction of point 2 than at point 1. Further, we see from Equation 13-11 that if v_2 is much larger than v_1, then P_2 must be much *smaller* than P_1 in order that the equality hold. Using a measurement of this pressure difference and the cross-sectional areas at points 1 and 2, one can solve for the velocities and the volume flow rate.

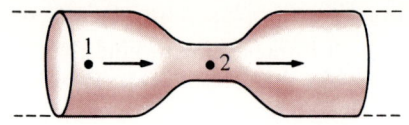

Figure 13-13 Barring a significant density change in the fluid, in order for the same mass to pass point 2 as point 1 in a given time, the speed of fluid flow past point 2 in the constriction must be much greater than that past point 1.

Example 13-8 In a water pipe, the difference in pressure between the main pipe and the constricted section of a Venturi meter like that shown in Figure 13-14 is 1.25×10^5 Pa. The areas of the main pipe and the constriction are 3.00×10^{-2} m² and 7.50×10^{-3} m², respectively. Determine (a) the speed of the water at the constriction and (b) the volume flow rate of water in the pipe.

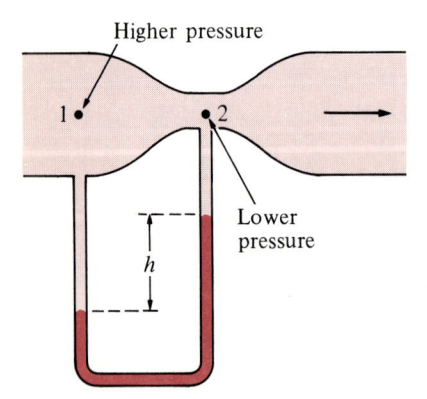

Figure 13-14 Bernoulli's principle asserts that if the height change of the fluid is negligible, any increase in the speed of flow must be accompanied by a drop in pressure. The decreased pressure at point 2 requires the additional pressure of the liquid column $\rho g h$ to balance the pressure at point 1.

Solution (a) We must apply both the continuity equation (Equation 13-7) and Bernoulli's equation for a horizontal pipe (Equation 13-11) and solve them simultaneously to obtain the speed at the constriction.

$$A_1 v_1 = A_2 v_2$$

$$(3.00 \times 10^{-2} \text{ m}^2)\, v_1 = (7.50 \times 10^{-3} \text{ m}^2) v_2$$

$$v_1 = 0.250 v_2$$

and

$$P_1 + \tfrac{1}{2}\rho v_1^2 = P_2 + \rho v_2^2$$

becomes (if we suppress the units for clarification)

$$P_1 - P_2 + 500 v_1^2 = 500 v_2^2$$

Substituting for v_1, we have

$$1.25 \times 10^5 + 500(0.25 v_2)^2 = 500 v_2^2$$

$$469 v_2^2 = 1.25 \times 10^5$$

$$v_2 = 16.3 \text{ m/s}$$

(b) We use this speed with the given cross section of the constriction to obtain the volume flow rate.

$$Q = Av = A_2 v_2 = (7.50 \times 10^{-3} \text{ m}^2)(16.3 \text{ m/s}) = 0.122 \text{ m}^3/\text{s}$$

13-8 Applications of Fluid Flow Equations

This important special case of the Bernoulli equation, called **Bernoulli's principle**, where height change is negligible, may be stated as:

As the velocity of fluid flow increases, the pressure exerted by the fluid decreases.

The action of the Bernoulli principle may be seen easily in the following experiment. Hold two sheets of paper parallel to each other, one in each hand, and blow sharply between them. Your intuition might indicate that the air jet would push the sheets outward, away from the center. Actually, the opposite happens. Blowing between the sheets causes the sheets to pull toward each other. (Try it.) The pressure exerted by the moving air between the sheets is lowered, and the atmospheric pressure acting on the outside of the sheets pushes them together.

You have probably encountered this effect on the highway when passed by a large truck. The airflow through the constriction between car and truck is increased in velocity to allow the air to pass. As a result, you feel the car pulled toward the truck. Actually, the car is pushed toward the truck by the greater air pressure on the side away from the truck. This effect is even more pronounced with boats passing near each other in the water.

In an atomizer spray, air is forced from a compressible bulb through a restriction immediately above a tube extending down into a liquid (Figure 13-15). The fast-moving air jet lowers the pressure at the top of the tube and allows the atmospheric pressure on the surface of the liquid to force the liquid up the tube. It enters the air jet, where it is broken up into a fine mist of droplets by the airflow and carried away. This principle is used in perfume bottles and in

Figure 13-15 The atomizer works by Bernoulli's principle.

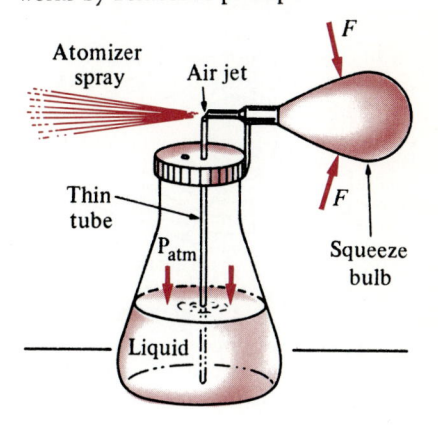

301

garden insecticide sprayers. A similar principle is used to vaporize gasoline droplets in an automobile carburetor.

Perhaps the most widespread application of the Bernoulli principle is in flying. The shape of the wing on an airplane is designed to take advantage of this principle of fluid flow. The cross-sectional view of a wing shown in Figure 13-16 is such that the air flowing over the top of the wing must travel a longer distance than the air flowing under the wing. Thus the air traveling over the top of the wing flows faster than that passing underneath. The air pressure therefore is lower above the wing than beneath it, and a net upward force (lift) results. It is this lift which raises the plane from the ground and allows it to fly.

In a completely similar manner, it is the principle of fluid flow which usually allows a sailboat to move through the water. (In fact, the airplane wing may be thought of as a horizontal sail providing lift rather than forward propulsion.) The air pressure on the convex, or outward-bulged, part of the sail is reduced since the air flowing past it must move a greater distance than the air moving across the chord behind the sail. Thus, air pressure behind the sail is greater and pushes it and the boat in the forward direction. Generally, sailboats move much less well with the wind behind rather than across the sails.

In many sports, a spinning ball in flight may follow a curved path. Figure 13-17 is a top view of a table tennis ball spinning clockwise and traveling to the right. Because the ball is not perfectly smooth and air is not frictionless, layers of air near the ball are pulled around with the ball's spin. The velocity of these layers due to the ball's spin add vectorially with the relative motion of the air past the ball due to its flight. Thus the net velocity of the air near the ball is greater on the side where the spin is in the direction of the airstream (the bottom in the diagram), and the pressure is less on that side. The ball feels a net force due to this velocity-induced pressure differential toward that side and curves in that direction.

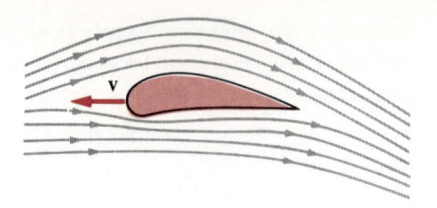

Figure 13-16 Cross section of an airplane wing moving horizontally to the left. Lines indicate the flow of the air around the wing.

13-9 Viscosity

The last section was the first place we alluded to frictional forces associated with fluids. Yet, if you have ever tried to run through waist-deep water, you know that the water's resistance to flow dominated the motion. We can imagine

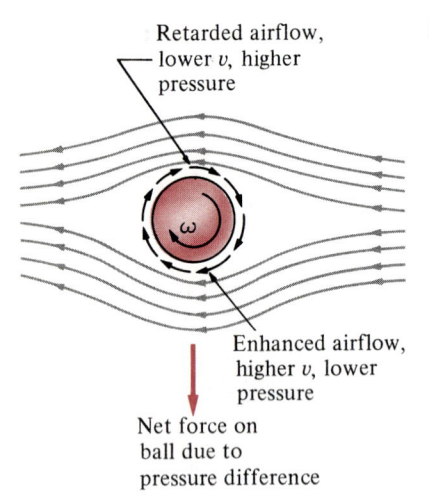

Figure 13-17 A top view of a table tennis ball spinning with an angular speed ω and moving to the right. The lines represent the relative motion of the air to the left.

fluids made up of very thin layers stacked parallel to the surface across which flow occurs. In this model, the layer against the surface does not move, while each successive layer is moving slightly faster than the previous one. Between layers, a frictional force acts by action-reaction to slow the faster layer and speed up the slower layer. In some liquids, like water and alcohol, the frictional forces are rather small, the liquid flows easily, and we say they have low **viscosity**. In others, like glycerin, molasses, and tar, the frictional forces are much larger, and we say they are highly **viscous**. The **coefficient of viscosity** is a measure of the frictional forces in liquids, hence, their resistance to flow.

When fluid is flowing past a wall, that part of the fluid touching the wall is not moving at all. The average speed of individual droplets increases in moving away from the wall until it reaches a maximum, as shown in Figure 13-18. The fluid is moving parallel to the wall in the x direction, while the distance away from the wall is measured in the y direction. Thus the slope of the graphed line is equal to the change in v_x with respect to a change in y, or

$$\text{Slope} = \frac{\Delta v_x}{\Delta y}$$

This quantity is called the **velocity gradient**. If there is a large velocity gradient, there is a large frictional force f_x. Also, the greater the wall area A_{wall}, the larger the frictional force. Thus we can write the proportionality

$$f_x \propto \frac{\Delta v_x}{\Delta y} A_{\text{wall}}$$

which becomes an equation by inserting the proper proportionality constant. Thus

$$f_x = \eta \frac{\Delta v_x}{\Delta y} A_{\text{wall}} \tag{13-12}$$

where the lowercase Greek letter η (eta) is the coefficient of viscosity for the fluid. Equation 13-12 can be used to determine both the magnitude and units of η. In the SI system, η is obtained in newton-seconds per square meter (N·s/m²). We give this combination the special name **poiseuille** (approximate pronunciation: pwah-zoy′) in honor of Jean Poiseuille, a nineteenth-century French physician and anatomist who studied fluid flow in an attempt to understand blood flow in the vessels of small animals. Thus we measure the coefficient of viscosity in poiseuilles (Pl), where

$$1 \text{ Pl} = 1 \text{ N·s/m}^2$$

Table 13-2 give the coefficient of viscosity for some selected fluids.

Poiseuille studied the flow rate of viscous fluid through a pipe. Figure 13-19 shows the velocity profile of a viscous liquid flowing steadily in a pipe. At the wall, the fluid is stationary, but its velocity is greater and greater at farther distances from the wall, reaching a maximum value v_m at the center of the pipe (a distance R from the wall). The velocity profile actually forms a parabola of revolution about the pipe's central axis. Assume that the pipe is of length l and of uniform radius R between points 1 and 2, which have pressures P_1 and P_2, respectively. The volume flow rate of the liquid is given by

$$Q = A v_{av}$$

where it may be shown that $v_{av} = \frac{1}{2} v_m$ and $A = \pi R^2$. Thus

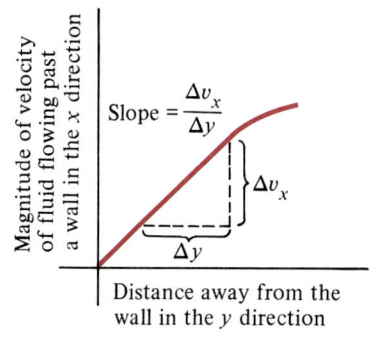

Figure 13-18 The speed of a fluid flowing past a wall vs. distance from the wall. The fluid against the wall ($y = 0$) is not moving ($v_x = 0$). As the distance from the wall increases, so does the velocity of the fluid.

TABLE 13-2 Coefficient of Viscosity of Selected Fluids at Atmospheric Pressure and 20°C (unless Otherwise Noted)

Fluid	Coefficient of Viscosity, Pl
Gases	
Air	1.82×10^{-5}
Ammonia	9.74×10^{-6}
Carbon dioxide	1.47×10^{-5}
Mercury (200°C)	4.50×10^{-5}
Oxygen	2.03×10^{-5}
Liquids	
Blood (37°C)	2.08×10^{-3}
Ethyl alcohol	1.20×10^{-3}
Glycerin	1.41
Oil (heavy machine)	0.661
Oil (olive)	0.138
Water (40°C)	0.653×10^{-3}

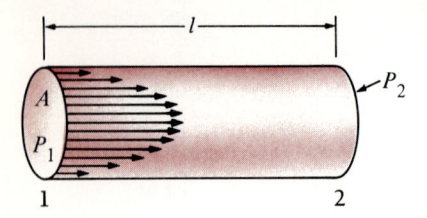

Figure 13-19 Viscous fluid flowing in a pipe. The arrows represent fluid velocity as a function of distance from the wall. The velocity is greatest at the pipe's center. The variation in velocity along any plane containing the tube axis is a parabola centered on the axis, with its maximum on the axis and dropping to zero at the wall.

$$Q = \frac{\pi R^2 v_m}{2} \tag{13-13}$$

Another expression which includes the parameters of the flow may be obtained from the fact that the flow is steady-state. That is, at a given point the flow is time-independent under the forces due to pressure and the viscous, or frictional, forces. Now the forces on the liquid between points 1 and 2 are $P_1 A$ to the right, $P_2 A$ to the left, and f_x (the viscous force) to the left. For equilibrium we have $\Sigma F_x = 0$. Choosing the direction to the right as positive and inserting the forces, we obtain

$$P_1 A - P_2 A - f_x = 0$$

or

$$f_x = (P_1 - P_2)A$$

Substituting for f_x from Equation 13-12 yields

$$f_x = \eta \frac{\Delta v_x}{\Delta y} A_{\text{wall}} = (P_1 - P_2)A \tag{13-14}$$

Using more advanced mathematics, one can show that for a parabolic velocity profile,

$$\frac{\Delta v_x}{\Delta y} = \frac{2v_m}{R}$$

in a pipe of radius R. Making this substitution in Equation 13-14 and using the fact that $A_{\text{wall}} = 2\pi R l$, we have

$$\eta \frac{2v_m}{R} 2\pi R l = (P_1 - P_2)A$$

or

$$4\pi \eta l v_m = (P_1 - P_2)A$$

Now if we use the cross-sectional area A of the pipe, which is πR^2, and solve for v_m, we obtain

$$v_m = \frac{(P_1 - P_2)R^2}{4\eta l}$$

Finally, substituting this into Equation 13-13 yields

$$Q = \frac{(P_1 - P_2)R^2}{4\eta l} \frac{\pi R^2}{2}$$

$$Q = \frac{(P_1 - P_2)\pi R^4}{8\eta l} \tag{13-15}$$

This is called **Poiseuille's law** because he was the first to formulate it, and it is of considerable practical importance in predicting fluid flow through tubes.

Example 13-9 In an average person, an artery in the upper thigh has an interior radius of 1.80 mm. If blood flows through this artery at the volume flow rate of 2.25×10^{-7} m³/s, calculate (a) the average and maximum speed of the blood, (b) the pressure difference in an 8.00-cm length of the artery, and (c) the power being delivered to move this blood.

Solution (a) The average speed is found from the volume flow rate and the cross-sectional area, while the maximum speed is simply twice the average speed.

$$Q = A v_{av} \quad \text{or} \quad v_{av} = \frac{Q}{A}$$

but

$$A = \pi R^2 = \pi (1.80 \times 10^{-3} \text{ m})^2 = 1.02 \times 10^{-5} \text{ m}^2$$

Thus

$$v_{av} = \frac{2.25 \times 10^{-7} \text{ m}^3/\text{s}}{1.02 \times 10^{-5} \text{ m}^2} = 2.21 \times 10^{-2} \text{ m/s}$$

and

$$v_m = 2v_{av} = 2(2.21 \times 10^{-2} \text{ m/s}) = 4.42 \times 10^{-2} \text{ m/s}$$

(b) This can be solved by applying Poiseuille's law:

$$Q = \frac{(P_1 - P_2)\pi R^4}{8\eta l}$$

or

$$P_1 - P_2 = \frac{8Q\eta l}{\pi R^4}$$

$$= \frac{8(2.25 \times 10^{-7} \text{ m}^3/\text{s})(2.08 \times 10^{-3} \text{ Pl})(0.0800 \text{ m})}{\pi (1.8 \times 10^{-3} \text{ m})^4}$$

$$= 9.08 \text{ N/m}^2$$

This is a very small pressure difference, equivalent to less than 0.07 torr.
(c) Recall that average power is the product of the resultant force with the average velocity. Since average power $= F_R v_{av}$ and $F_R = (P_1 - P_2)A$, we have

$$\begin{aligned} \text{Power}_{av} &= (P_1 - P_2)A v_{av} \\ &= (P_1 - P_2)Q \\ &= (9.08 \text{ N/m}^2)(2.25 \times 10^{-7} \text{ m}^3/\text{s}) \\ &= 2.04 \times 10^{-6} \text{ W} \end{aligned}$$

13-10 Stokes' Law

Bodies moving through a viscous fluid experience a retarding force due to the viscosity. In 1845, Sir George Stokes formulated **Stokes' law**, which states that the force on a spherical body of radius R moving with a speed v through a fluid with a coefficient of viscosity η is

$$F_D = 6\eta\pi R v \qquad (13\text{-}16)$$

Consider a small sphere falling in a viscous fluid, as shown in Figure 13-20. The forces acting on a sphere are its weight, the buoyant force, and the Stokes' law drag force. As the sphere is released in the viscous fluid, it accelerates. But as its speed increases, so does the drag force in accordance with Equation 13-16. This force eventually reaches a value such that the sphere is in equilibrium, falling at a constant speed called its **terminal speed.** At equilibrium, the vector sum of the forces acting on the sphere is zero. Choosing upward as positive and converting the vector equation $\Sigma F = 0$ to a scalar equation, we have

$$F_D + F_B - mg = 0 \qquad (13\text{-}17)$$

Equation 13-16 gives an expression for F_D, and the buoyant force is

$$F_B = \rho_F V g = \rho_F (\tfrac{4}{3}\pi R^3) g$$

where ρ_F is the density of the fluid and R is the radius of the sphere. We may rewrite the weight of the sphere as

$$mg = \rho_S V g = \rho_S (\tfrac{4}{3}\pi R^3) g$$

where ρ_S is the density of the sphere. Substituting into Equation 13-17 yields

$$6\pi\eta R v_T + \rho_F(\tfrac{4}{3}\pi R^3)g - \rho_S(\tfrac{4}{3}\pi R^3)g = 0$$

which may be solved for the terminal speed as

$$v_T = \frac{2R^2 g}{9\eta}(\rho_S - \rho_F) \qquad (13\text{-}18)$$

This equation may also be used to obtain the coefficient of viscosity of a fluid by measuring the terminal velocity of a sphere of known radius and density.

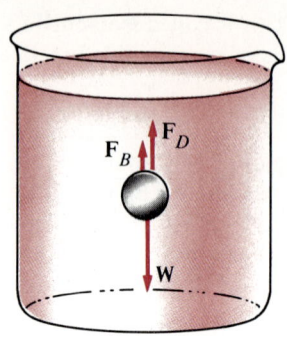

Figure 13-20 The weight force, the buoyant force, and the Stokes' law drag force acting on a sphere falling through a viscous medium.

Example 13-10 Calculate the terminal speed of 1.000-mm-diameter steel ball bearings falling in glycerin. $\rho_{\text{steel}} = 7.86 \times 10^3$ kg/m^3.

Solution This is a direct application of Equation 13-18:

$$v_T = \frac{2(1 \times 10^{-3} \text{ m})^2 (9.80 \text{ m/s}^2)}{9(1.41 \text{ Pl})} (7.86 \times 10^3 \text{ kg/m}^3 - 1.26 \times 10^3 \text{ kg/m}^3)$$

$$= 1.02 \times 10^{-2} \text{ m/s}$$

This is only 1.02 cm/s, which is slow enough to be measured with an ordinary stopwatch.

13-11 Turbulence

To this point in the chapter, we have acted on the assumption that the fluid flow is smooth. During this type of flow, called **streamline flow** (or **laminar flow**), each particle of fluid which passes through a particular point follows the same path as every other particle which passes through that point. This unchanging path along which the particles flow is called a **streamline.** There is also another type of flow, called **turbulent flow,** in which the motion is very irregular and the patterns of flow are constantly changing. Particles move in little **eddies,** or whirlpools. Friction is much greater in turbulent flow. Both types of flow are

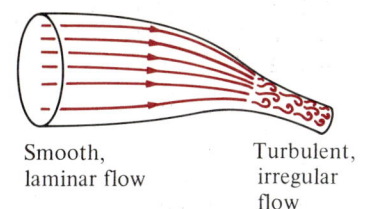

Figure 13-21 A section of pipe in which a fluid is flowing narrows to a constriction, requiring increased flow velocity. Regular streamlines indicate the initial zone of *laminar* flow. These break up into the swirling, chaotic motion of *turbulent* flow as the fluid proceeds into the constriction.

Smooth, laminar flow

Turbulent, irregular flow

13-11 Turbulence

illustrated in Figure 13-21. This difference can best be observed in the smoke rising from a cigarette in still air, as in Figure 13-22.

Whether fluid flow is streamline or turbulent depends on several physical quantities associated with the flow. In 1883, Osborne Reynolds formulated an expression yielding a number which is used to distinguish between the two types of flow. This became known as the **Reynolds number** (N_{Re}) and for flow in a pipe is given by

$$N_{Re} = \frac{\rho v_{av} D}{\eta} \tag{13-19}$$

where ρ = density of the fluid
v_{av} = its average speed
η = coefficient of viscosity
D = diameter of the pipe

Figure 13-22 Directly above the cigarette, the smoke streams in the sheetlike layers of laminar flow. Higher, the smoke breaks into the swirls and curls of turbulent flow. (D. Riban)

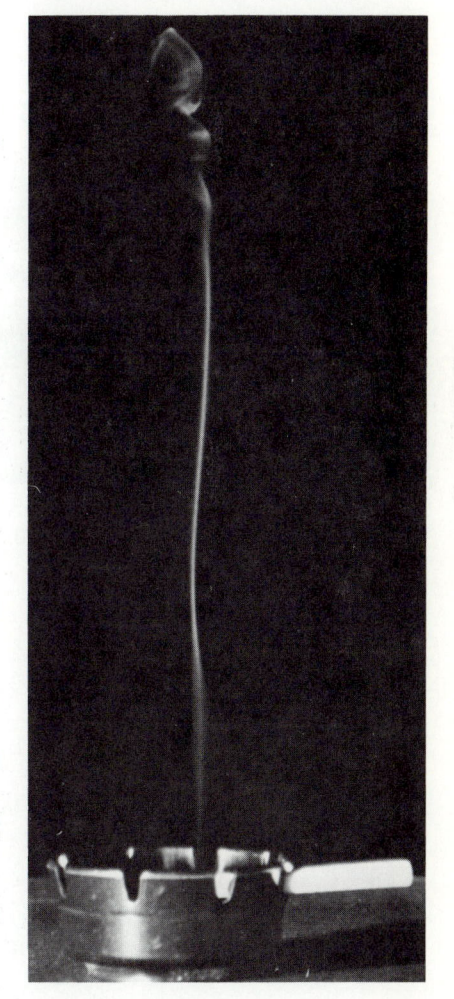

All the dimensions in Equation 13-19 cancel, and N_{Re} is unitless. Investigation shows that flow is streamline if N_{Re} is less than about 2000 and turbulent when N_{Re} is greater than 3000. If N_{Re} lies between 2000 and 3000, the flow is not stable. Local conditions in the pipe (discontinuities and so on) may cause the flow to change from one type to the other.

In Chapter 6, we saw that the retarding force on a vehicle moving through air is proportional to the square of the vehicle's speed. Actually this describes the drag force due to the turbulent flow of the air. Whereas Stokes' law represents the drag force for streamline flow, the turbulent flow drag force for motion of a body in any fluid is given by

$$F_d = \tfrac{1}{2}\rho A_{eff} v^2 \tag{13-20}$$

where ρ = density of the fluid
A_{eff} = effective area presented by the moving body to the direction of motion
v = body's speed

For most relatively large bodies moving in air or water, the flow of the fluid around the body is turbulent. Thus Equation 13-20 applies, and the drag force increases with the square of the body's speed. In practical situations—automobiles, airplanes, boats, even submarines—we attempt to reduce the drag by proper shaping (called streamlining) to reduce the effective area.

The ratio of the effective area as given in Equation 13-20 to the actual frontal area of an object is called the **drag coefficient.** Considerable effort is expended in wind-tunnel tests with models to reduce this for all types of vehicles. Particularly noteworthy have been efforts made with automobiles to meet federally mandated fuel efficiency standards. Most older cars had relatively flat frontal areas exposed to the air, and protuberances which created turbulence. Newer models are inevitably engineered with low drag coefficient as well as general styling in mind.

SPECIAL TOPIC
Measuring Blood Pressure

The measurement of human blood pressure is very important to medical diagnosis. When the heart contracts, a high-pressure pulse of blood is ejected into the arteries. The pressure rises momentarily then drops to a minimum before the next heartbeat. Direct measurement of blood pressure would require surgery. Instead, a **sphygmomanometer** (Figure 13-23) is used for an indirect measurement.

To obtain someone's blood pressure readings, the inflatable cuff is wrapped around the upper arm and fastened snugly. The upper arm is normally at about the same elevation as the heart. Thus, the pressure in the upper-arm artery is approximately the same as that in the heart. Using the hand-held rubber bulb with a release valve, air is pumped into the cuff, causing it to expand and squeeze the arm. The person making the measurement listens with a stethoscope to the blood flow in the upper-arm artery as the external pressure is increased. When the pressure in the cuff exceeds the maximum pressure in the artery, the artery is collapsed so that no blood gets through and no sound can be heard. Air is then slowly released from the cuff, decreasing the pressure on the arm. As the pressure decreases, nothing is heard until it drops just below the maximum value on each cycle of the heart. At this maximum internal pressure, the artery can reopen momentarily and a small amount of blood can squirt through. This makes a detectable sound. At this point, the gauge pressure in the cuff is read from the attached manometer and corresponds to the maximum blood pressure. In a healthy young adult, this value may be near 120 torr (mmHg) and is known as the **systolic pressure.** As the pressure in the cuff is further decreased, the sounds continue since the artery is still collapsed for part of the heart cycle. As soon as the pressure is lower than the minimum arterial pressure, the artery remains open and the periodic sounds disappear. Again the pressure is read on the manometer and corresponds to the minimum blood pressure. For a young healthy adult, this may be near 80 torr and is called the **diastolic pressure.** Blood pressures are reported systolic and then diastolic as 120 over 80, sometimes written as 120/80.

Blood pressure is affected by viscosity of the blood and working of valves in the blood vessels as well as age, sex, and certain medical conditions. High blood pressure, called **hypertension,** is statistically linked to several abnormal medical conditions. Most medical professionals consider systolic pressures in excess of 160 torr (170 torr for a person over 45 years of age) and diastolic pressures in excess of 95 torr (105 torr for a person over 45) to be serious enough to warrant further tests.

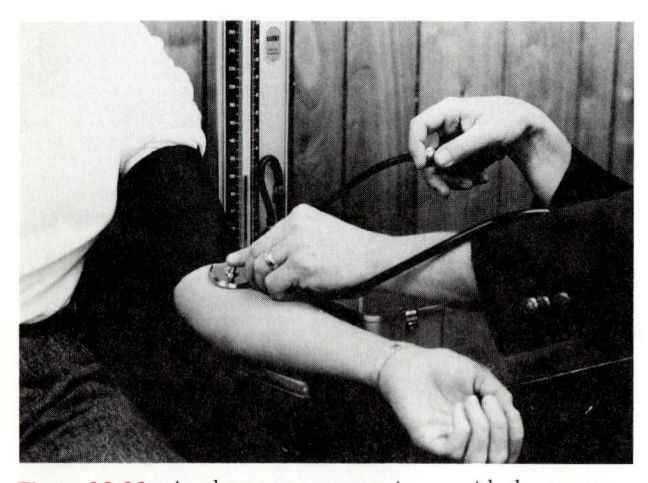

Figure 13-23 A sphygmomanometer in use with the pressure cuff attached to the upper arm and inflated.

Minimum Learning Objectives

After studying this chapter, you should be able to:
1. Define:

absolute pressure	barometer	buoyant force	gauge pressure
Archimedes' principle	Bernoulli's equation	continuity equation	hydraulic press
bar [unit]	Bernoulli's principle	drag coefficient	hydrodynamics
		eddy	hydrostatics
		fluid	laminar flow

308

manometer
pascal (Pa) [unit]
Pascal's principle
poiseuille (Pl) [unit]
Poiseuille's law
Reynolds number
sphygmomanometer
Stokes' law
streamline
terminal speed
torr [unit]
turbulent flow
velocity gradient
viscosity, coefficient of
volume flow rate

2. Calculate pressure as a function of depth in a fluid of known density.
3. Calculate the force exerted on a given area by the pressure in a confined liquid.
4. Calculate the buoyant force on an object immersed in a fluid.
5. Determine the density of an object of given volume from its weights when immersed in fluids of different known densities.
6. Understand the application of the work-energy equation to a fluid system yielding the Bernoulli equation.
7. Calculate pressure changes for fluids flowing through closed systems of varying geometry.
8. Understand the generation of lateral forces on objects moving relative to a fluid via the Bernoulli principle.
9. Calculate the flow rate of a viscous fluid through a pipe of known diameter and length, given the pressure difference between ends of the pipe.
10. Determine the terminal velocity of a regularly shaped object falling through a viscous fluid.
11. Understand the difference between laminar and turbulent flow and why the latter generates increased drag on an object.

Problems

13-1 Calculate the mass of a solid platinum cylinder 0.100 m long and 4.00×10^{-2} m in diameter.

13-2 If an iceberg has a mass density of 917 kg/m³, what fraction of the iceberg is below the surface of the seawater?

13-3 A bowling ball with a hidden cavity is made of a material of density 1.30×10^3 kg/m³. If the outside diameter of the ball is 0.218 m and the ball weighs 48.0 N, what is the volume of the cavity?

13-4 A submarine changes its density to dive or to surface. Since its mass is fixed, its volume is varied. Tanks open to the sea at the bottom are either flooded, to become part of the sea and not a part of the volume of the craft, or pumped full of compressed air, to drive out the water and increase the effective volume of the vessel. A typical World War II submarine has a mass of 1200 tonnes (1 tonne = 1000 kg). What volume would the floodable tanks have to be to allow it to float with 0.100 of its volume above water, or to attain a density of 1050 kg/m³ to sink?

13-5 Air is moving at a steady speed of 10.0 m/s. The surface of a skyscraper in this airflow is 20.0 m wide and 600 m high. If the column of air encountering the side of the skyscraper is stopped in its forward motion and then forced to deflect sideways around this obstacle at zero forward velocity, what is the force exerted on the building by the wind? Take the density of air to be 1.29 kg/m³.

13-6 A U-tube open to the atmosphere on both sides has mercury on both sides to a height of 100 mm above the bottom. Water is to be poured into one side of the U-tube. How high must the top of the water be if the water-mercury interface is (a) 50 mm above the bottom or (b) exactly at the bottom (i.e., mercury to a height of 200 mm on the other side)?

13-7 A ginger ale bottle 8.50×10^{-2} m in diameter is fitted with a rubber stopper through which is inserted a very long glass tube oriented vertically. If the bottom of the bottle can withstand a maximum force of 100 N before cracking, to what height above the bottom of the bottle may the tube be filled with liquid without destroying the bottle if the liquid is (a) water or (b) carbon tetrachloride?

13-8 Calculate the gauge pressure and the absolute pressure at the bottom of the bottle of Problem 13-7 when the bottle cracks.

13-9 In hydraulic disc brakes, a force applied to the brake pedal causes pads to pinch a disc rotating with the wheels. If 400 N of force is applied to the brake pedal, this pushes against a piston (Figure 13-24). If the pistons at-

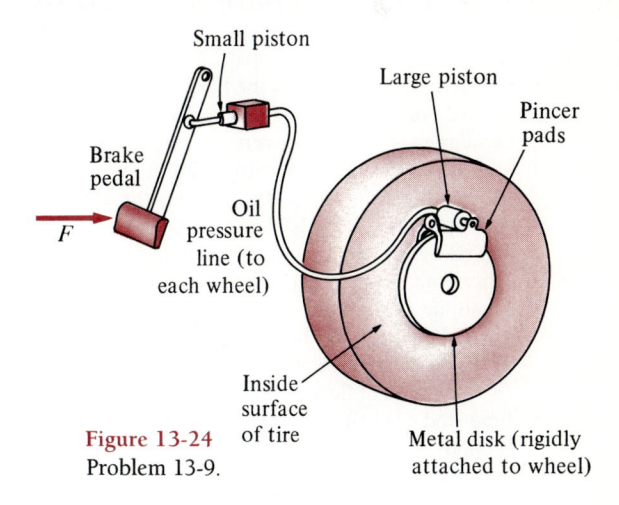

Figure 13-24 Problem 13-9.

tached to the pads have an area 10 times as large (with one at each of four wheels), what is the total force stopping the wheels from rotating?

13-10 Calculate the height of the liquid in a glycerin-filled barometer when the atmospheric pressure is 1.013×10^5 N/m².

13-11 Suppose the earth's atmosphere had a constant density of 1.29 kg/m³ (instead of decreasing with height). (a) What would be the height of the atmosphere which would give rise to the same pressure at the earth's surface (1.013×10^5 N/m²)? (b) What height of water of constant density would yield the same pressure?

13-12 Determine the density of a wooden block if it floats half immersed in water between a layer of water and a layer of oil (density = 628 kg/m³).

13-13 One way to determine the purity of a sample of gold is to weigh it in both air and water and use the measured values to determine its density as we did with the metal of Example 13-4. Indeed, the problem Archimedes was considering when he discovered his principle was purportedly an assignment by the king to determine whether the crown was pure gold. Suppose the crown weighed 18.0 N when suspended in air. How much would it weigh when suspended completely submerged in pure water if it is 100 percent gold?

13-14 Calculate the area of the smallest rectangular oak plank (density 710 kg/m³) 0.100 m thick in water that will just support a professor weighing 820 N.

13-15 A 1-kg laboratory mass weighs 9.80 N when suspended in air and 7.92 N when suspended under water. Determine (a) its density and (b) its weight when suspended completely submerged in glycerin.

13-16 During a sneeze, a volume of 100 cm³ of air is ejected through the nasal passages in a time of 0.100 s. What is the speed of the airflow at the nostrils if they have an area of 2.00 cm²?

13-17 Water enters the science building through a pipe 7.50×10^{-2} m in diameter at a pressure of 3.50×10^5 Pa. The pipe immediately narrows to 5.00×10^{-2} m. If, at a particular time, water is entering at a speed of 2.00 m/s, determine (a) the speed and (b) the pressure in the narrow part of the pipe.

13-18 Oxygenated blood leaves the heart via the aorta at a rate of 70.0 cm³/s. This blood eventually moves through fine capillaries, where it delivers its oxygen and enters the system of veins to return to the heart. If a typical capillary has a radius of 5.00×10^{-4} cm and the average blood flow speed is 4.00×10^{-4} m/s, how many capillaries are served by the heart at any one time?

13-19 A horizontal pipe 5.00 cm in diameter is carrying water at a volume flow rate of 4.00×10^{-3} m³/s. At some point along the pipe, it has been separated and a length of pipe of inside diameter 1.50 cm has been inserted. Calculate (a) the speed of flow in both the 5.00- and 1.50-cm sections and (b) the pressure difference between the two sections.

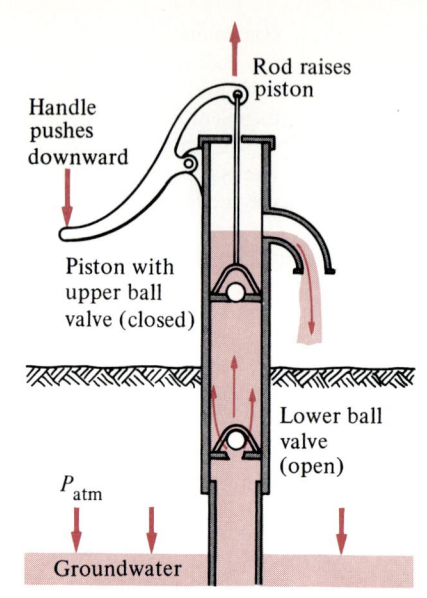

Figure 13-25 Problem 13-20.

13-20 The hand water pump (Figure 13-25) is increasingly less common. As the handle was pumped, a ball valve let air escape from the chamber of the pump. On the upstroke of the piston, the air could not return from above because of the tightly seated ball valve. Instead, fluid from below raised to fill the chamber through the lower, fixed ball valve. Eventually, water from the underground water level would rise to replace the air evacuated from the chamber and be ejected during each upstroke of the piston at the spigot. What was the deepest the water level could be below the chamber and still have the pump deliver water?

13-21 How high will the water level rise in a cylindrical bucket 0.300 m in diameter being filled from a faucet at 1.50×10^{-4} m³/s if there is a circular hole of area 8.00×10^{-5} m² in the bottom of the bucket? (Hint: The level reaches a steady-state maximum when water is flowing out as fast as it flows in.)

13-22 A cylindrical oil tank 30.0 m in diameter is filled with oil of density 650 kg/m³. A hunter accidentally fires a bullet which strikes the tank, creating a small circular hole of radius 7.50×10^{-3} m at a distance of 6.00 m below the top surface of oil. (a) Assuming the velocity of the top surface may be neglected, calculate the speed of efflux of oil through the bullet hole. (b) Using the speed calculated in part a, apply the continuity equation to calculate the approximate speed of the top surface.

13-23 A large cylindrical tank is open to the atmosphere and is filled with water to a height H above the bottom of the tank. A small hole is drilled in the tank at a distance h below the water surface (see Figure 13-26). (a) Derive an expression in terms of H and h for the distance R away from the tank where the water initially strikes the

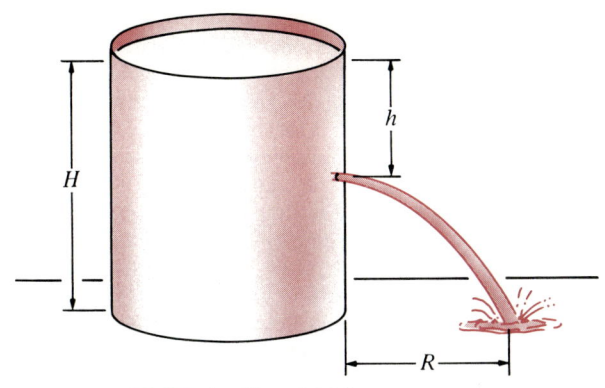

Figure 13-26 Problem 13-23.

ground. (b) Apply the expression to determine R when $H = 8.00$ m and $h = 2.00$, 4.00, and 6.00 m.

13-24 What is the maximum height that a stream from a firehose can reach if the absolute pressure in the city mains is 4.00×10^5 Pa?

13-25 The pressure of groundwater found 60.0 m below the surface and under a cap of impermeable rock is measured to be 7.50×10^5 Pa by a drilling team. If their 10.0-cm-inside-diameter borehole is left open and unobstructed, the water spouts as an artesian well. What height above the surface does the spout reach?

13-26 A water pistol fires a stream of water horizontally from 1.20 m above the floor. It strikes the floor 3.00 m away from the muzzle. Calculate the maximum pressure in the pistol's water chamber.

13-27 Figure 13-27 illustrates a Pitot tube. This device applies Bernoulli's equation to measure fluid speeds. A small cylinder is placed along the axis of a larger cylinder and a gentle cone shape used to connect their ends. The tubes are connected to opposite arms of a manometer. Measurement is made by orienting the tube so the fluid flows toward the cone end. At the opening of the smaller cylinder, the speed of the fluid is zero, while at the holes in the larger cylinder, the speed is v. The potential energy terms of Bernoulli's equation may be neglected. (a) Show that the speed of fluid flow may be written as $v = \sqrt{2(P_A - P_B)/\rho}$, where $P_A - P_B$ is the pressure difference between the smaller cylinder opening and the holes in the larger cylinder and ρ is the fluid density. (b) The pressure difference $P_A - P_B$ is equal to $\rho'gh$, where ρ' is the density of the manometer fluid. Thus we may write $v = \sqrt{2\rho'gh/\rho}$. If a Pitot tube is used with a mercury-filled manometer to measure the airspeed of a small airplane, determine the pressure difference and the speed for a mercury height difference of 23.5 cm. Use 1.20 kg/m³ as the density of air at the flight level.

13-28 The tube shown in Figure 13-28 is being used to transfer water from container A to container B at a lower level. In this application, the tube is called a **siphon**. If the inside diameter of the siphon tube shown is 1.50×10^{-2} m and the elevation of point 1 is 0.930 m above point 2, calculate (a) the pressure at point 1, (b) the speed of the water at points 1 and 2, and (c) the volume flow rate. (Note that the tube discharges to the atmosphere. Thus the pressure at point 2 is atmospheric.)

13-29 A Venturi meter (Figure 13-14) has a pipe diameter of 0.120 m and a constriction diameter of 0.0500 m. If the mercury height difference in the manometer is 0.500 m when water is flowing in the pipe, calculate (a) the pressure difference, (b) the speed of water in the main pipe, and (c) the volume rate of flow of water.

13-30 If liquid flowing through a pipe encounters a constriction, the speed of flow must increase and the absolute pressure must decrease. If the constriction is small enough, the absolute pressure may drop below atmospheric. A tube connected to such a constriction can serve as a vacuum pump when the liquid flows. Such an arrangement is called an aspirator. If water is flowing through a pipe 2.50×10^{-2} m in diameter at an absolute pressure of 3.00×10^5 Pa at a volume flow rate of 2.00×10^{-3} m³/s,

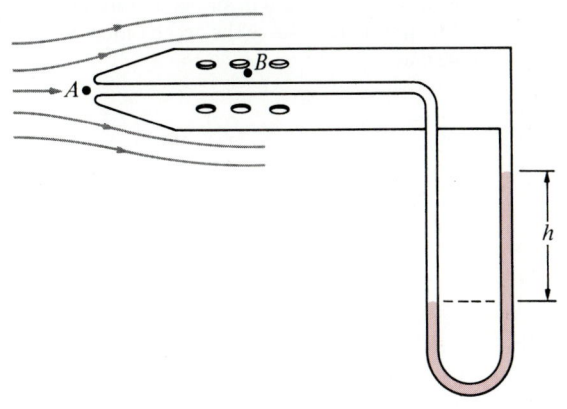

Figure 13-27 Problem 13-27.

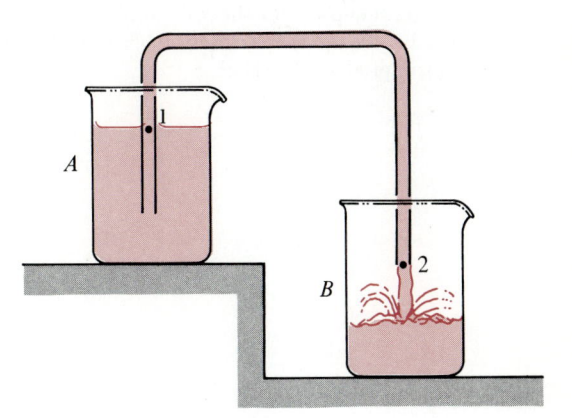

Figure 13-28 Problem 13-28.

what must be the diameter of the constriction if the pressure at the constriction is to be 1.00×10^4 Pa?

13-31 A good rule of thumb used in aircraft design is that economic flight requires about 1.00×10^3 N/m² of lift on the wing area. What is the minimum speed of flow over the upper surface of the wing when the speed of air past the lower surface is 90.0 m/s? Take the density of air to be 1.29 kg/m³.

13-32 Assume air is blowing horizontally past a sail on a boat in the sideways direction such that its speed is 9.00 m/s over the front surface and 8.00 m/s across the chord behind the sail. Use the density of air as 1.29 kg/m³, and determine the propelling force the wind applies to a sail of area 6.00 m².

13-33 A Boeing 747 airplane has a mass of 3.33×10^5 kg and a wing area of 600 m². When the plane is in level flight and the pressure on the upper wing surface is 4.00×10^4 Pa, what is the pressure on the lower wing surface?

13-34 A small plane has a 1.50×10^4 N takeoff weight and a 16.8-m² wing area. (a) What average force must be developed per square meter of wing to lift the plane? (b) If the takeoff speed is 56.0 m/s, what is the wind velocity over the upper surface of the wing? (c) A Boeing 747 has 310 m² of lifting surface, each square meter of which must lift 600 N. If the cruising speed in air at a quarter the normal surface density is 200 m/s, what must the speed of airflow over the upper surface be?

13-35 In golf, a slice (for a right-handed golfer) is a hit ball that begins its motion in the straight-out direction but then strongly curves off to the right of its intended direction. The slice is caused by a frustratingly large number of possible faults during the swing of the club. (a) According to Bernoulli's principle, in which direction must the ball be rotating as seen from above in order to slice? (b) One cause of the slice can be that the clubhead is moving sideways as well as forward when it strikes the ball. Which motion of the clubhead—right to left or left to right, as seen from behind the swinging golfer—would produce a slice? (c) Another cause for a slice may be that the clubface is not perpendicular to the straight-ahead direction at impact with the ball. Would an open clubface (angled to the right) or a closed clubface (angled to left of forward) produce the slice?

13-36 According to its own commercials, the outstanding physical property of a particular ketchup is its high viscosity. Describe the experimental setup you would use to measure the viscosity of this ketchup in poiseuilles.

13-37 Water flowing from a faucet of diameter 1.50×10^{-2} m at a speed of 1.00 m/s contracts into a smaller and smaller diameter stream as it falls. (a) Why is this so? (b) Calculate its diameter after it has fallen 0.300 m.

13-38 There is a pressure drop of 9.80×10^3 Pa along a 100-m-long horizontal pipe 6.00×10^{-2} m in diameter carrying heating oil (coefficient of viscosity 6.80×10^{-2} Pl). Calculate (a) the volume flow rate, (b) the average speed of flow, and (c) the power expended by the pump on the oil to keep it flowing at a steady rate.

13-39 Fluids are to flow through a pipe 5.00×10^{-2} m in diameter. What is the maximum velocity for streamline flow if the fluid is (a) air, (b) water, (c) glycerin, and (d) mercury? (Density of air is 1.29 kg/m³.)

13-40 An air bubble rising in a viscous liquid will quickly reach terminal speed. Calculate the terminal speed of a bubble 3.00×10^{-3} m in diameter if it is rising in (a) water and (b) glycerin.

13-41 Determine the terminal speed of a 90.0-kg parachutist who is free-falling prior to opening the chute. Assume turbulent flow. Take the effective area of the parachutist to be 0.380 m² and the density of air to be 1.29 kg/m³.

13-42 The parachute limits the parachutist's terminal speed by presenting a larger effective area to increase the drag force. Assume turbulent flow and (a) determine the effective chute area required to limit the terminal speed of a 900-N parachutist to 2.50 m/s. (b) What would be the terminal speed of a 500-N parachutist who used the same chute?

13-43 Determine the maximum speed for streamline flow of water through a 7.50×10^{-2} m firehose.

13-44 A brass ball bearing of radius 1.00×10^{-3} m is released in a container of water. Calculate (a) its terminal speed and (b) its speed when it is still accelerating at 2.50 m/s². (Assume streamline flow.)

13-45 Determine the distance a marble 1.50×10^{-2} m in diameter would have to fall in air before turbulent flow would set in around it. Assume the acceleration of the marble is constant.

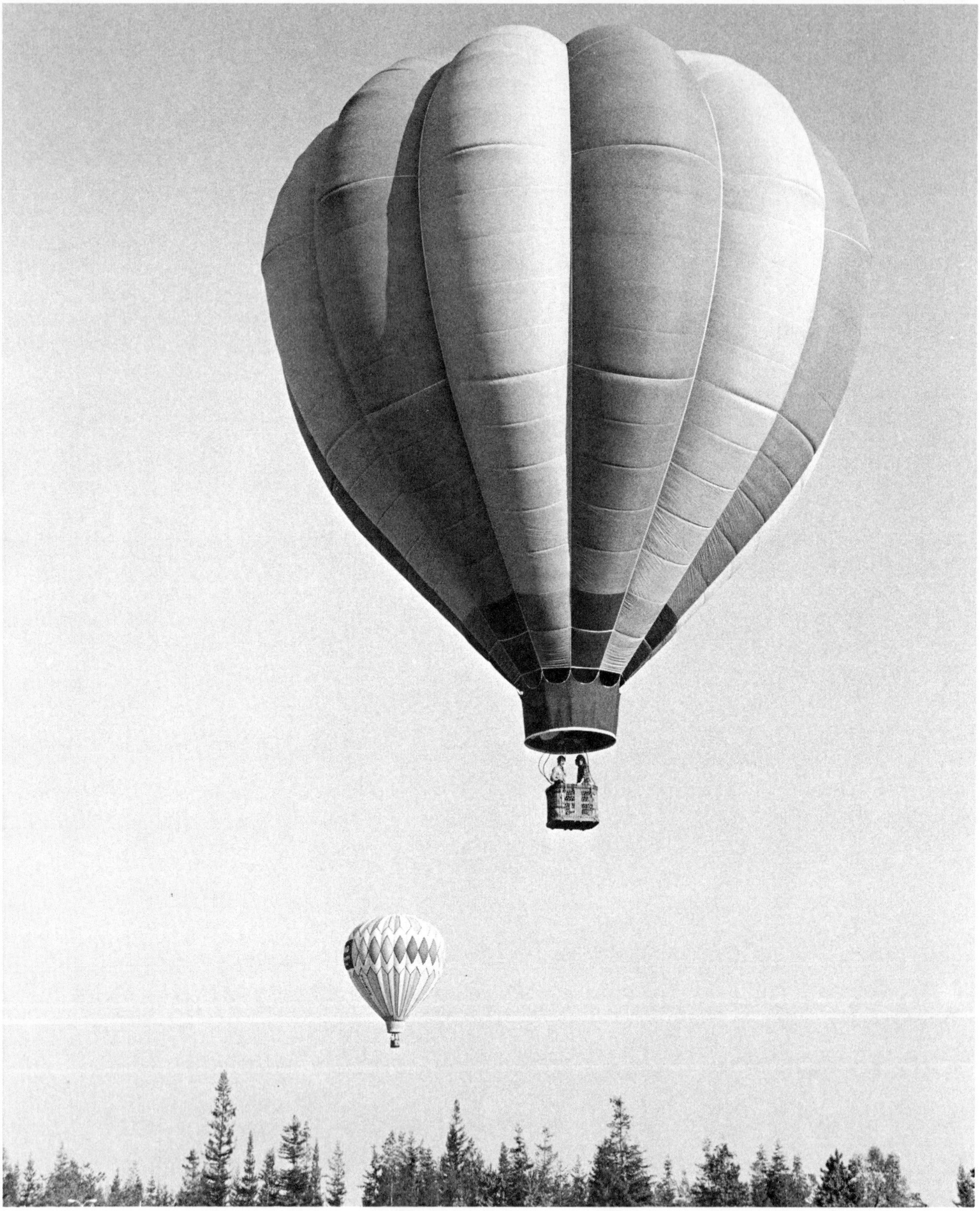

14 Temperature, Gases, and Kinetic Theory

14-1 Introduction

Kinetic energy "lost" to friction is not recoverable in any immediately obvious way, but a sensitive thermometer detects an increase in temperature on the surfaces where friction acted. Here we begin to examine the connection between mechanical energy and heat.

Words such as *hot* or *cold* are relative in their use, but there is considerable agreement on comparisons between objects. Everyone will agree that a barbeque grill being used to cook hot dogs is hotter than the skin on our hands. Likewise, we will agree that the ice cooling our lemonade is colder than our hands. The occurrence that allows us to judge hotter and colder when objects touch is that whenever two objects are in contact, energy is always transferred from the hotter object to the colder object. The hotter object transfers what we call **thermal energy** to the colder object. Thermal energy transferred from one object to another is called **heat.**

When two objects, one hotter and one colder, are placed in contact, the energy transfer proceeds only for a finite time. When it finally stops, we note that neither object is now hotter nor colder than the other, and we say they are in **thermal equilibrium** and are at the same temperature. Whenever two objects *are* at the same temperature, there will be no net transfer of heat between them if they are placed in contact. It is an experimental fact that two objects or systems which are separately in thermal equilibrium with a third are in thermal equilibrium with each other. This statement is often called the **zeroth** law of thermodynamics.

14-2 Temperature

To measure temperature, we need to find a physical property which will change in a predictable and reproducible manner as a body gets hotter and colder. Several such properties are available. Galileo used the change in volume of a confined gas as a measure of temperature. This is still a precise means of measurement but is not very convenient, because a gas **thermometer** must be recalibrated each time it is used. Furthermore, it is difficult to imagine medical patients calmly accepting a softball-size gas bulb in their mouths for measurement. The common liquid-in-glass thermometer was the result of work to produce an instrument which would remain usable for an indefinite period once manufactured.

A standard thermometer consists of a capillary tube expanded at one end to form a bulb and partially filled with a visible liquid such as mercury or alcohol (Figure 14-1). The air is pumped out of the space above the liquid and then the top of the tube is sealed. Temperature measurement is based on the fact that most liquids expand with increasing temperature more rapidly than do solids. As the temperature increases, the volume of the liquid increases more rapidly than that of the tube. As a result, the level of the liquid rises in the tube. Conversely, when cooled, the liquid appears to drop in the tube since it reduces in volume more rapidly than the tube and bulb do.

The changes in the length of a liquid column in such a tube are reproducible. To calibrate such a device, it is brought into thermal equilibrium with masses of material at reference temperatures. For example, it could be placed in contact with a block of ice. At equilibrium, a mark at the top of the liquid level would then represent the temperature of ice. Choosing two such reference temperatures would allow us to divide the space between the marks by an arbitrary number of divisions to establish a **temperature scale.**

Other properties of matter vary reproducibly with temperature, making them suitable for thermometers. The electric resistance of certain materials varies with temperature (thermistor). The voltage difference between two ends of a conducting wire held at different temperatures increases as the temperature difference gets larger (thermocouple). Certain liquid crystals have a variety of temperature-dependent color effects. For that matter, at very high temperatures, most materials exhibit color changes with increasing temperature (the heating element on electric stoves becomes red at very high temperatures—optical pyrometer). All these properties and others can be and are used in the construction of thermometers.

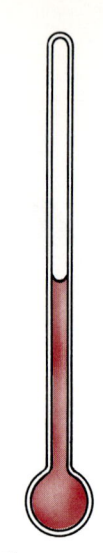

Figure 14-1 The most common type of thermometer uses a liquid sealed in glass. The length of the liquid column varies reproducibly with temperature.

14-3 Temperature Scales

In 1714, Gabriel Fahrenheit introduced the first precision mercury-in-glass thermometer and the first widely accepted temperature scale. He arbitrarily selected two widely spaced, reproducible temperatures and marked the top of

14-3 Temperature Scales

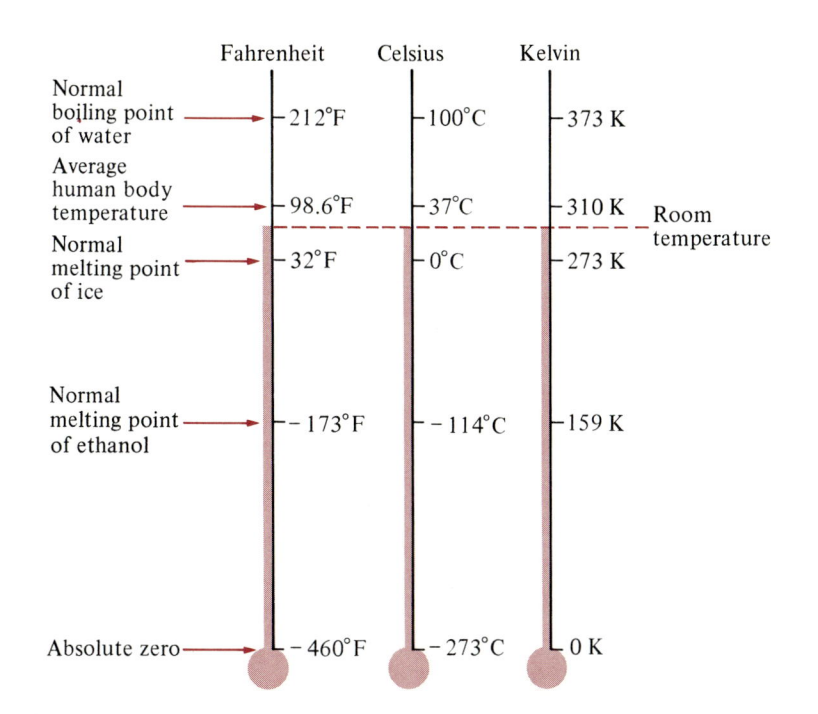

Figure 14-2 The three temperature scales differ in the values they assign specific temperatures. The Fahrenheit and Celsius scales are arbitrary and allow comparison of temperature changes only, since a reading of "2" does not represent twice as much of anything as a reading of "1" does.

his mercury column for each reference. His choice for a low temperature was the coldest ice-salt solution he could produce in the laboratory, which he called zero degrees. At the other extreme, he selected human body temperature as 96 degrees. He found that an ice-water mixture had a temperature of 32 on this scale and that water normally boiled at about 212 degrees. This scale became known as the **Fahrenheit temperature scale,** and its readings are reported with a degree mark and an uppercase F. Thus water freezes at 32°F and boils at 212°F.

Fahrenheit's choice of a scale was unfortunate. Outside temperatures routinely go well below zero on his scale in cold climates, and the average human body core temperature is 98.6°F. Furthermore, neither reference point was easily reproducible in practice. Perhaps most seriously, this scale produced arbitrary numbers for the most frequently used references, the freezing and boiling points of water. In 1742 Anders Celsius announced a new scale, now called the **Celsius scale,** based on these references. On this scale, the normal melting point of ice is 0 and the normal boiling point of water is 100. Thus, water freezes at 0°C and boils at 100°C.

Both the Fahrenheit and the Celsius scales are arbitrary in their choice of reference points. A third scale, the **Kelvin scale** (discussed more fully in Section 14-7), is called an **absolute temperature** scale because it defines its zero as the coldest possible temperature. A degree on the Kelvin scale is the same magnitude as the degree on the Celsius scale. The normal melting point of ice has the value of 273.15 K (read "273.15 **kelvins**" rather than "273.15 degrees kelvin"). Figure 14-2 depicts several common points on all three scales.

If we know a temperature on one of the three scales, we can convert to either of the other scales. Using t_F, t_C, and T to represent temperatures on the Fahrenheit, Celsius, and absolute scale, respectively, we have to three significant figures that

$$T = t_c + 273 \tag{14-1}$$

317

Further, since there are 180 Fahrenheit degrees but only 100 Celsius degrees between the normal melting point of ice (32°F, 0°C) and the normal boiling point of water (212°F, 100°C), each Fahrenheit degree is equal to five-ninths of a Celsius degree. Thus we may find the Celsius reading by simply subtracting 32 from the Fahrenheit reading and multiplying by $\frac{5}{9}$. In equation form,

$$t_C = \tfrac{5}{9}(t_F - 32) \tag{14-2}$$

Equation 14-2 may be solved for the Fahrenheit temperature to yield

$$t_F = \tfrac{9}{5}t_C + 32 \tag{14-3}$$

Thus we may obtain the Fahrenheit reading from the Celsius reading by simply multiplying the Celsius reading by $\frac{9}{5}$ and adding 32 to the result of the multiplication.

Example 14-1 Show that Equation 14-3 is the proper conversion relationship at both the normal melting point of ice and the normal boiling point of water.

Solution Substitute $t_C = 0$ and then $t_C = 100$ to verify that $t_F = 32$ and 212, respectively.

Normal melting point: $t_F = (\tfrac{9}{5})(0) + 32 = 32°\text{F}$

Normal boiling point: $t_F = (\tfrac{9}{5})(100) + 32$
$= 180 + 32$
$= 212°\text{F}$

Example 14-2 Determine the temperature where the Fahrenheit scale and the Kelvin scale have the same reading.

Solution We must first combine Equations 14-1 and 14-3 to eliminate t_C. We then set $t_F = T$ in the resulting equation.

$$t_F = \tfrac{9}{5}t_C + 32.0$$

But

$$t_C = T - 273$$

Substituting yields

$$t_F = \tfrac{9}{5}(T - 273) + 32.0$$

The scales will have the same reading when $t_F = T$. Thus

$$T = \tfrac{9}{5}(T - 273) + 32.0$$
$$= 1.80T - (1.80)(273) + 32.0$$
$$0.800T = (1.80)(273) - 32.0$$
$$T = 574 \text{ K}$$

Thus, when $T = 574$ K, it is also true that $t_F = 574°\text{F}$.

---

14 Temperature, Gases and Kinetic Theory

14-4 Thermal Expansion

Most objects or substances expand when their temperatures are increased provided the pressure is constant. A property frequently used is the change in length (or area or volume) of a substance as the temperature changes. If a thin metal rod of length L is heated such that its temperature is increased by an amount ΔT, experiment shows that it increases in length by an amount ΔL. The magnitude of ΔL is directly proportional to both ΔT and L, and

$$\Delta L \propto L \, \Delta T$$

also depends on the properties of the material being heated. Using a proportionality constant that is specific for each material, we write

$$\Delta L = \alpha L \, \Delta T \tag{14-4}$$

where the lowercase Greek alpha (α) is the proportionality constant called the **linear coefficient of thermal expansion,** defined from Equation 14-4 as the change in length per unit length per degree change in temperature. Table 14-1 shows values of α for some selected materials.

TABLE 14-1 Linear Coefficient of Thermal Expansion for Selected Solids at 20°C

Material	α, 10^{-6} K^{-1}
Aluminum	25
Brass	18
Brick and concrete	~10
Glass	~10
Iron and steel	12
Rubber	~80

Example 14-3 Steel rails 8.00 m long are laid end to end in the winter when the temperature is $-10°C$. How much space should be left between them to allow for expansion in the summer when the temperatures could reach $50°C$?

Solution We simply find the increase in L (that is, ΔL) due to the increase in temperature by applying Equation 14-4.

$$\Delta L = \alpha L \, \Delta T = \alpha L (T_2 - T_1)$$

From Table 14-1, we see that the linear coefficient of thermal expansion for steel is $\alpha = 12 \times 10^{-6}$ K^{-1}. Thus

$$\Delta L = (12 \times 10^{-6} \text{ K}^{-1})(8.00 \text{ m})[50 - (-10)] \text{ K}$$
$$= 5.76 \times 10^{-3} \text{ m}$$

This is more than half of 1 cm, and this gap left for expansion accounts for the "clickety-click" sound of the train as it moves along the rail.

Experiment shows that the same form of relationship as Equation 14-4 holds for thermal expansion of area and volume. That is,

$$\Delta A = \gamma A \, \Delta T \tag{14-5}$$

where A is the area and the lowercase Greek gamma (γ) is used for the **area coefficient of thermal expansion.** Finally,

$$\Delta V = \beta V \, \Delta T \tag{14-6}$$

where V is the volume and the lowercase Greek beta (β) is used for the **volume coefficient of thermal expansion.** Table 14-2 lists values of β for some selected liquids. Equation 14-6 states that when a substance of volume V is heated such that its temperature is increased by an amount ΔT, there will be an increase in volume ΔV. This volume increase is proportional to the product $V \, \Delta T$, and the proportionality constant β is the change in volume per unit

TABLE 14-2 Volume Coefficient of Thermal Expansion for Selected Liquids at 20°C

Material	β, 10^{-6} K^{-1}
Acetone	~1500
Ethanol	~1100
Glycerin	~500
Mercury	~180
Water	~210

volume per degree change in temperature. It can be shown (see Problem 14-17) that α, γ, and β are related as

$$\gamma = 2\alpha \qquad \beta = 3\alpha \qquad (14\text{-}7)$$

The values of these coefficients are not constant over the whole temperature scale, and the values in Tables 14-1 and 14-2 are given near room temperature. Further, their magnitudes depend on the external pressure. Thus all values are given for standard atmospheric pressure. Finally since the units of length (or area or volume) cancel, these coefficients have units of reciprocal temperature. In SI, we choose the absolute scale, and the coefficients have units of K^{-1} (read "per kelvin").

Example 14-4 Suppose the rails of Example 14-3 had been butted against each other at $-10°C$. What would be the stress each would have to withstand at $50°C$ in order that they *not* increase in length by buckling?

Solution Recall from Chapter 12 that for most materials, stress (F/A) is proportional to strain ($\Delta L/L$) and that the proportionality constant was Young's modulus Y, or $F/A = Y \Delta L/L$. We must equate the increase in length due to thermal expansion to a decrease in length due to stress so that the length of the rails remains unchanged. We have

$$\Delta L_T = \alpha L \, \Delta T \quad \text{and} \quad \Delta L_S = \frac{F}{A}\frac{L}{Y}$$

Thus

$$\frac{F}{A}\frac{L}{Y} = \alpha L \, \Delta T$$

$$\frac{F}{A} = \alpha Y \, \Delta T$$

$$= (12 \times 10^{-6} \, K^{-1})(21 \times 10^{10} \, N/m^2)(60 \, K)$$
$$= 1.51 \times 10^8 \, N/m^2$$

Almost 1500 times atmospheric pressure, or over 10 tons of force for every square inch of contact. Thus, we had better leave a gap between them.

14-5 Gas Laws

While working in Oxford, England, to improve an early version of an air pump, Robert Boyle (1627–1691), an Irish physicist, was able to demonstrate that the volume of a confined quantity of gas varied inversely as the absolute pressure provided the temperature was held constant. This is now known as **Boyle's law** and may be written as

$$V \propto \frac{1}{P}$$

or in equation form as

$$V = \frac{C_1}{P}$$

or
$$PV = C_1 \qquad (14\text{-}8)$$

where C_1 is the proportionality constant. If the absolute pressure on a confined gas is changed from P_1 to P_2 at constant temperature, the volume changes from V_1 to V_2. Equation 14-8 indicates these quantities can be related as

$$P_2 V_2 = P_1 V_1 \qquad (14\text{-}9)$$

since both sides of Equation 14-9 must be equal to the same constant.

Example 14-5 The chamber in a bicycle pump contains air at standard atmospheric pressure. The valve is closed so no air escapes, and a downward push of the handle decreases the volume of the air by 30 percent. What is the new pressure of air in the chamber (assume no temperature change)?

Solution We apply Boyle's law in the form of Equation 14-9. We do not have a numerical value for the initial and final volumes, but we do know that the final volume is 70 percent of the initial volume. That is, $V_2 = 0.70 V_1$. Since P_1 is standard atmospheric pressure, $P_1 = 1.013 \times 10^5$ Pa. (Recall that $1.0 \text{ N/m}^2 = 1.0$ Pa.) Thus

$$P_2 V_2 = P_1 V_1$$

becomes

$$P_2 (0.70 V_1) = (1.013 \times 10^5 \text{ Pa}) V_1$$
$$P_2 = 1.45 \times 10^5 \text{ Pa}$$

Or, subtracting the pressure of the atmosphere (1.01×10^5 Pa), this could be a gauge pressure of about 4.5×10^4 Pa, or approximately 6.5 lb/in².

Nearly a whole century after Boyle's death, Jacques Charles (1747–1823), a French physicist, discovered the relationship which became known as **Charles' law**. The volume of a confined gas is directly proportional to its absolute temperature provided the pressure is held constant. In symbols, this becomes

$$V \propto T$$

or

$$V = C_2 T \qquad (14\text{-}10)$$

where C_2 is another proportionality constant. Equation 14-10 may also be expressed as

$$\frac{V_2}{T_2} = \frac{V_1}{T_1}$$

where V_1 and T_1 are the initial values of volume and absolute temperature, and V_2 and T_2 are the values of the same quantities after a change has occurred in a confined gas at constant pressure.

Gases obey Equations 14-8 and 14-10 under a fairly wide range of conditions. These laws may be combined as

$$\frac{PV}{T} = C_3 \qquad (14\text{-}11)$$

or

$$\frac{P_2 V_2}{T_2} = \frac{P_1 V_1}{T_1} \qquad (14\text{-}12)$$

where C_3 is yet another constant. You may convince yourself that Equation 14-8 results from Equation 14-11 at constant temperature. Also, Equation 14-10 results from Equation 14-11 when the pressure is held constant.

Example 14-6 The gauge pressure of a certain amount of gas occupying a volume of 0.0500 m³ at a temperature of 27°C is 9.80×10^4 Pa. Calculate its new pressure if the volume is decreased to 0.0200 m³ and the temperature increased to 127°C.

Solution We may find the final pressure by direct substitution into Equation 14-12. First, however, we must change the pressure and temperatures to absolute values. The pressures and temperatures appearing in the gas laws must be absolute pressures and absolute temperatures.

$$P_1 = P_G + P_A = 9.80 \times 10^4 \text{ Pa} + 1.01 \times 10^5 \text{ Pa}$$
$$= 1.99 \times 10^5 \text{ Pa}$$

$$T_1 = t_{C1} + 273 = 27 + 273 = 300 \text{ K}$$

$$T_2 = t_{C2} + 273 = 127 + 273 = 400 \text{ K}$$

Thus

$$\frac{P_2 V_2}{T_2} = \frac{P_1 V_1}{T_1}$$

becomes

$$P_2 = \frac{P_1 V_1 T_2}{T_1 V_2} = \frac{(1.99 \times 10^5 \text{ Pa})(0.0500 \text{ m}^3)(400 \text{ K})}{(300 \text{ K})(0.0200 \text{ m}^3)}$$

$$= 6.63 \times 10^5 \text{ Pa}$$

Again we note that this is absolute pressure. The gauge pressure may be found by subtracting standard atmospheric pressure (1.01×10^5 Pa). Thus the final gauge pressure is 5.62×10^5 Pa.

Example 14-7 A weather balloon used to determine upper-atmosphere wind velocities is filled with helium. At the earth's surface, a typical balloon has a volume of 0.180 m³ at standard atmospheric pressure and a temperature of 20°C. Determine the volume of the balloon after it has risen to a height of about 20 km, where the pressure is 1.00×10^4 Pa and the temperature is -30°C.

Solution We apply Equation 14-12 once again. Further, we must convert to absolute temperature value. (The pressures are already absolute.)

$$T_1 = t_{C1} + 273 = 20 + 273 = 293 \text{ K}$$

$$T_2 = t_{C2} + 273 = -30 + 273 = 243 \text{ K}$$

Thus, from

$$\frac{P_2 V_2}{T_2} = \frac{P_1 V_1}{T_1}$$

we have

$$V_2 = \frac{P_1 V_1 T_2}{T_1 P_2} = \frac{(1.01 \times 10^5 \text{ Pa})(0.180 \text{ m}^3)(243 \text{ K})}{(293 \text{ K})(1.00 \times 10^4 \text{ Pa})}$$

$$= 1.51 \text{ m}^3$$

This is more than eight times the original volume.

14-6 Ideal Gases

Equation 14-11 asserts that the pressure of a gas times its volume divided by the absolute temperature is a constant, $PV/T = C_3$. This is a form of the most important of gas laws, called the **ideal gas law.** The constant C_3 depends on the amount of gas and may be written

$$C_3 = nR$$

where n is the amount of gas present and R is a constant for all gases.

The amount of gas is measured in a unit called a **mole** (mol) which contains a fixed number, called an **Avogadro's number** (N_A), of particles.

$$N_A = 6.02 \times 10^{23} \text{ molecules/mol}$$

The mole, or gram molecular mass, is the numerical value of a substance's atomic or molecular mass expressed in grams. For example, on the periodic table of elements, which is duplicated in Appendix C, we may find that carbon atoms have an atomic mass of 12. If we took 12 g of carbon, we would have taken 6.02×10^{23} atoms of carbon. For a molecular substance, the smallest parts of which are composed of more than one atom joined together, we must take the molecular mass in grams to have N_A particles. Thus, carbon dioxide, CO_2, has one carbon atom joined to two oxygen atoms as its smallest part, its molecule. With 12 units of mass for the carbon atom and 16 units for each of the two oxygen atoms, we obtain $12 + 16 + 16 = 44$ as the molecular mass of carbon dioxide. If we take 44 g of CO_2, we will have N_A molecules of CO_2, 1 mol. Further, 1 mol of any gas occupies the same volume at the same temperature and pressure. At *standard temperature and pressure* (STP) of 1.0 atm and 0°C, a mole of any gas occupies 2.24×10^{-2} m³ (or 22.4 L). Thus, we would find that at STP, 2.24×10^{-2} m³ of CO_2 gas would have a mass of 44 g.

The constant R is called the **universal gas constant** and has the value

$$R = 8.314 \text{ J/(mol} \cdot \text{K)}$$

The ideal gas law is most often written in the form

$$PV = nRT \qquad (14\text{-}13)$$

An **ideal gas** is one that has no tendency to liquefy no matter how low the temperature becomes. *Real* gases tend to deviate from this behavior at very high pressures or low temperatures (see Special Topic II). We may often need to remind ourselves that the values of pressure and temperature for which Equation 14-13 applies are the *absolute* values. Gauge pressures must be converted to absolute pressures (units of Pa, or N/m² in SI). Also, any temperatures in degrees Celsius or degrees Fahrenheit must be converted to their values in kelvins before Equation 14-13 is used. It's easy to accept the fact that the absolute values are required if we think of some real example of a relationship among the three variables. For instance, we certainly do not expect the product of the pressure and volume of air inside an automobile tire to vanish whenever the temperature drops to freezing (0°C).

14-7 Absolute Zero

At constant volume, Equation 14-13 becomes

$$P = \frac{nR}{V} T = C_4 T \qquad (14\text{-}14)$$

where C_4 is the constant nR/V. We see that the pressure varies directly with the absolute temperature. Thus, if we have some convenient way to maintain a constant volume of a confined gas and to measure its pressure, Equation 14-14 provides the relationship and the property needed for a thermometer.

Any rigid gas-tight container will do nicely as a constant-volume confinement, allowing us to measure pressure with an open-tube mercury manometer. Such a constant-volume gas thermometer is shown in Figure 14-3. The can and the attached tubing to the zero marked on the scale form the fixed volume. If pressure readings are taken with a constant-volume gas thermometer for different temperatures, the absolute pressure may be plotted vs. the Celsius temperature to obtain a graph, as in Figure 14-4.

The fact that pressure drops with temperature in a straight line can lead us

Figure 14-3 The constant-volume gas thermometer measures pressure changes to determine temperature. As its temperature increases, the gas exerts downward pressure on the mercury surface. Raising the mercury reservoir increases the external pressure and returns the gas level to the zero mark. The new pressure of the gas at this point is atmospheric pressure plus the added pressure of the mercury column, $\rho g h$, (or, minus $\rho g h$ if the mercury reservoir would have to be dropped below the zero mark to sustain a constant volume of gas).

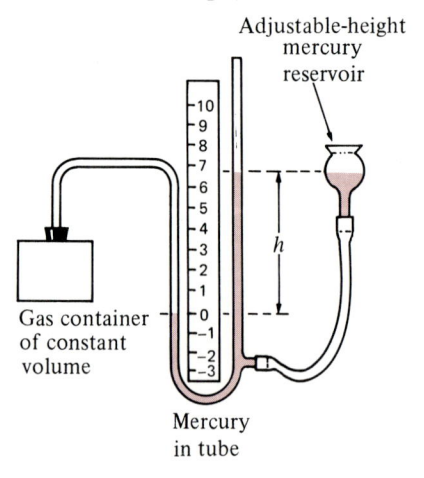

Figure 14-4 A calibration graph of measured pressure vs. temperature for a constant-volume gas thermometer. The variation of pressure with temperature is a straight line over a considerable range.

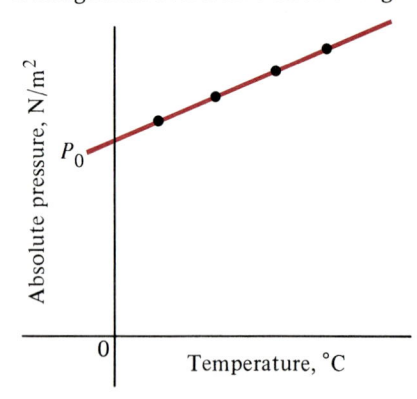

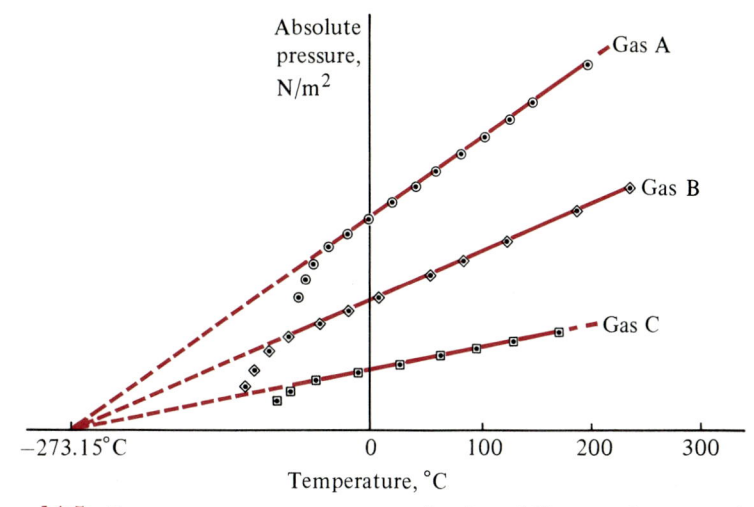

Figure 14-5 Pressure vs. temperature curves for three different real gases each reveal the same straight-line behavior until very low temperatures are reached.

to ask the question "At what temperature will the gas pressure drop to zero?" Figure 14-5 depicts typical pressure vs. temperature plots for an arbitrary amount of three real gases. At higher temperatures, it can be seen that the data points for each gas fall on the straight line for that gas. As the temperature is lowered in each case, the points begin to deviate from the straight line as the gas starts to liquefy. If the straight-line portions are extrapolated to lower temperatures, they meet at $P = 0$ at a temperature of $-273.15°C$. This point is called **absolute zero**, is the same for all gases used, and represents the zero point on the Kelvin temperature scale. A large bonus obtained from using the constant-volume gas thermometer is that it yields a reference point which is independent of the substance used to measure it.

The Kelvin scale is used in most scientific work. To define its temperature scale, a second reference point, in addition to absolute zero, is required. There is a single combination of pressure and temperature where all three **phases**—solid, liquid, and gas—may coexist. This point is called the **triple point.** The triple point of H_2O where ice, water, and steam all exist in thermal equilibrium occurs at a pressure of 4.58 mmHg (about 610 Pa or less than 1 percent of standard atmospheric pressure). The temperature at the triple point of water is defined as 273.16 K (0.01°C).

A thermometer must be a practical measuring instrument. In the laboratory, temperatures must be measured with precision over a very wide range. This requires a number of different kinds of thermometers since any particular type has a limited temperature range. For instance, mercury freezes below $-39°C$ and can hardly be poured into a capillary tube. Even gases liquefy at certain low temperatures. Helium is the best gas for low temperatures since it has the lowest boiling point. However, it, too, is useful only above about 1 K. Devices which may depend on other properties such as color, resistance, or voltage are often needed to cover all possible desired temperature readings. Since we might go from device to device in traversing the whole range, we need a number of temperature reference points to calibrate the different devices.

TABLE 14-3 Defined Fixed Reference Points of the International Practical Temperature Scale*

State and Material	T, K	t_c, °C
Triple point of hydrogen	13.81	−259.34
Boiling point of hydrogen	20.28	−252.87
Boiling point of neon	27.402	−246.048
Triple point of oxygen	54.361	−218.789
Boiling point of oxygen	90.188	−182.962
Triple point of water	273.16	0.01
Boiling point of water	373.15	100.0
Freezing point of zinc	692.73	419.58
Freezing point of silver	1235.08	961.93
Freezing point of gold	1337.58	1064.43

* With the exception of the triple points, the assigned values are at a pressure of 1 atm.

Table 14-3 is a list of such reference points. Figure 14-6 shows typical temperature ranges of some common types of thermometers.

14-8 Kinetic Theory of Gases

When a solid melts, the liquid that is formed occupies about the same volume as the original solid. However, when the liquid is boiled to form a vapor, its volume expands enormously. At atmospheric pressure, the steam from boiling water

Figure 14-6 Various types of thermometers must be used over different temperature ranges. For example, optical pyrometers measure the peak color (wavelength) emitted by incandescent materials and may be used with molten metals or the interiors of furnaces. The familiar liquid-in-glass thermometers have particularly limited ranges.

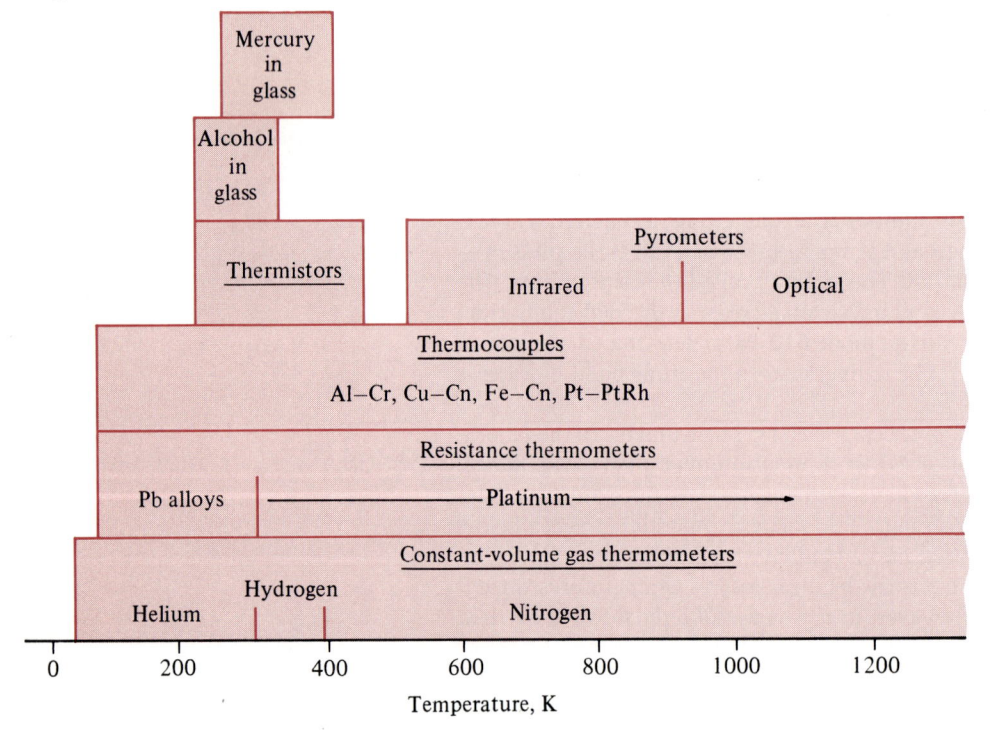

occupies about a thousand times as much volume as the liquid water from which it came. In 1808, John Dalton proposed the modern atomic theory of matter which asserted that normal matter was made up of incredibly small constituent units, called atoms, or their simple combinations, called molecules. Interpretation of the expansion of matter upon becoming a gas requires that we examine the behavior of these constituent atoms or molecules. Are we to believe that a molecule, which was apparently of the same physical volume as a solid and as a liquid, suddenly increases greatly in size upon becoming a gas? This is unlikely.

An alternative interpretation known as the **kinetic theory of matter** holds that all matter is in motion. In a solid, the motion of the individual atom (or molecule) is that of oscillation about a fixed position in the structure of the solid. In a liquid, the molecules may physically move from place to place but are bound to their neighboring molecules by forces of attraction. In a gas, this theory asserts, molecules have acquired enough energy of motion to dominate the attractions of fellow molecules and may move freely in space. This view holds that most of the volume of a gas is empty space with molecules moving rapidly through it, colliding with each other and the confining walls.

Suppose we confine a quantity of gas in a cylinder with a leak-tight piston inserted as in Figure 14-7 and place a weight on the piston. Observation shows that the weighted piston will quickly come to rest as the gas is compressed. However, it will not compress the gas all the way to the bottom of the cylinder. The gas will exert a force on the bottom of the piston due to its pressure, and the piston will remain at some equilibrium position. The force exerted by the gas appears to be steady, but the kinetic model of a gas does not support the idea of a steady force. Instead, the force must be interpreted as the average effect of billions and billions of impulses each second on the piston due to collisions of the tremendous number of gas molecules moving around in the cylinder. We will explore the implications of this model by applying Newton's laws, making the following assumptions:

1. A small volume of confined gas contains an extremely large number of molecules moving in random directions and colliding with the walls of the container.
2. Each molecule can be considered to be a point mass occupying essentially no volume.
3. The only forces exerted on molecules occur when they collide with the wall or each other, and these collisions are perfectly elastic.
4. Newton's laws of motion may be applied to the molecules.

For simplicity, consider a quantity of gas confined to a box, as in Figure 14-8. The box is oriented such that its edges are along the x, y, and z axes, and the length of these sides are l_x, l_y, and l_z, respectively. The molecules are moving about in random fashion, and each molecule has the same mass m. To determine the force and hence the pressure exerted by the gas, let us find the force due to one molecule colliding with the wall and sum over the total number of molecules.

Consider a single collision of a single molecule with one wall, as depicted in Figure 14-9. Our representative molecule m is shown both before and after the collision. The vectors $\mathbf{v}_o$ and $\mathbf{v}_f$, representing the original and final velocities, are of the same magnitude since the collision is elastic. Because the wall is

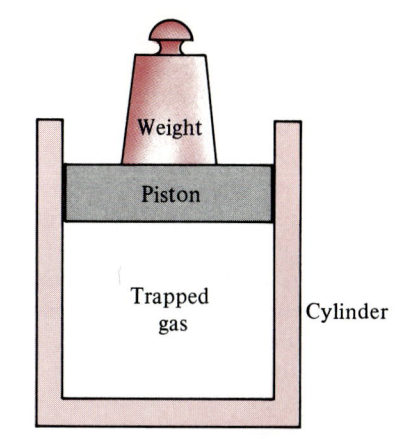

Figure 14-7 The weighted piston is held in equilibrium by the pressure exerted by the trapped gas.

Figure 14-8 A rectangular box of dimensions l_x by l_y by l_z contains N molecules in random, rapid motion. We are interested in the force exerted by the repeated collisions of the molecules on the shaded face of the box of area A.

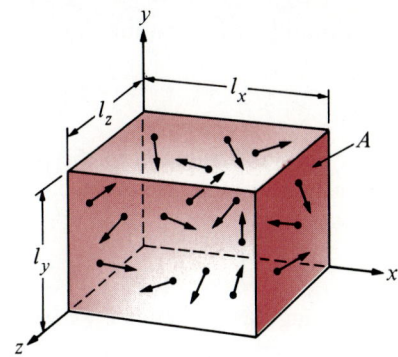

327

parallel to the y and z axes and the collision is elastic, no component of force is exerted on the molecule in the y or z directions. Thus only the x component of the velocity changes during the collision. This molecule will continue to move about and, in general, will bounce off all the interior walls over and over again. We wish to find the pressure due to this molecule at the wall shown. To do this, we need to determine the average force exerted on this wall due to repeated collisions of the molecule. This force divided by the area of the wall will yield the pressure due to this molecule's motion.

Newton's third law tells us that the force the molecule exerts on the wall, $\mathbf{f}_w$, is equal in magnitude but opposite in direction to the force the wall exerts on the molecule, $\mathbf{f}_m$. In symbols, $\mathbf{f}_w = -\mathbf{f}_m$. The average force exerted on our representative molecule due to repeated collisions with the wall is, according to Newton's second law, equal to the time rate of change of momentum of the molecule caused by collisions with that wall. However, the time rate of change of momentum is equal to the product of the change of momentum per collision with the number of collisions per unit time. This can be written in equation form as

$$f_m = \frac{\text{change in momentum}}{\text{collision}} \times \frac{\text{no. of collisions}}{\text{time}} \qquad (14\text{-}15)$$

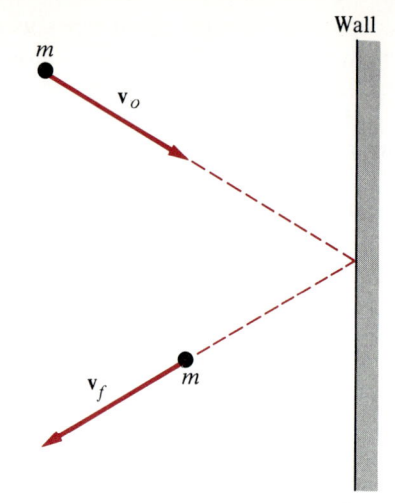

Figure 14-9 A representative molecule of mass m, moving randomly, approaches the wall of area A at a velocity of $\mathbf{v}_o$. If the collision with the wall is to be elastic, the velocity after collision $\mathbf{v}_f$ must have the same magnitude as $\mathbf{v}_o$.

For our representative collision, the y and z components of momentum do not change. We need only consider the x component of the change in momentum. We write

$$\Delta(mv) = \Delta(mv_x) = mv_{xf} - mv_{xo}$$

But $v_{xf} = -v_{xo}$, and therefore

$$\Delta(mv_x) = -2mv_x \qquad (14\text{-}16)$$

where we have dropped the o from the subscript for simplification. The number of collisions per unit time with the wall in question is simply the reciprocal of the time required for the molecule to make a round trip in the container, returning to the original wall. This is equal to the distance the molecule must travel in the x (and minus x) direction divided by its x component of speed. The time for one round trip Δt can be written $\Delta t = 2l_x/v_x$. Finally the number of collisions per unit time the molecule makes with the wall is

$$\frac{\text{Collisions}}{\text{Time}} = \frac{1}{\Delta t} = \frac{v_x}{2l_x}$$

Substituting these quantities into Equation 14-15, we have

$$f_m = (-2mv_x)\frac{v_x}{2l_x} = -\frac{mv_x^2}{l_x} = -f_w$$

Finally, the contribution to the pressure at the wall in question due to our representative molecule is equal to f_w divided by the area A of the wall:

$$P_1 = \frac{mv_x^2}{Al_x}$$

Using this expression, we would like to determine the pressure P at the wall due to a very large number N of molecules confined in this box. Assume the molecules have speeds $v_1, v_2, v_3, \ldots, v_n$. We can write

$$P = \frac{mv_{1x}^2}{Al_x} + \frac{mv_{2x}^2}{Al_x} + \cdots + \frac{mv_{nx}^2}{Al_x}$$

or

$$P = \frac{m}{Al_x} \sum_{i=1}^{N} v_{ix}^2$$

Using the fact that $Al_x = V$ (since $A = l_y \times l_z$) and defining an average square x component of velocity $\langle v_x^2 \rangle$ such that

$$N\langle v_x^2 \rangle = \sum_{i=1}^{N} v_{ix}^2$$

we obtain

$$P = \frac{m}{V} N\langle v_x^2 \rangle \tag{14-17}$$

Note that the average value of the velocities is zero since there must be just as many molecules moving to the right as to the left. However, the average value of the squares of the velocities is certainly not zero. Further, since all components of velocity are equally probable in random motion,

$$\langle v_x^2 \rangle = \langle v_y^2 \rangle = \langle v_z^2 \rangle$$

and

$$\langle v^2 \rangle = \langle v_x^2 \rangle + \langle v_y^2 \rangle + \langle v_z^2 \rangle$$

or

$$\langle v^2 \rangle = 3 \langle v_x^2 \rangle$$

Substituting, we have

$$P = \frac{Nm}{V} \frac{\langle v^2 \rangle}{3}$$

or

$$PV = \frac{Nm\langle v^2 \rangle}{3} \tag{14-18}$$

Note that Equation 14-18 has exactly the same form as the ideal gas law, $PV = nRT$, provided we may equate the right side of this equation with nRT.

$$\frac{Nm\langle v^2 \rangle}{3} = nRT \tag{14-19}$$

We may solve Equation 14-19 for the average kinetic energy of a single molecule. Thus

$$\frac{\langle KE \rangle}{\text{Molecule}} = \tfrac{1}{2} m \langle v^2 \rangle = \frac{3}{2} \frac{nRT}{N}$$

Recalling that N is the number of molecules confined in our box and that it is simply the number of moles of gas confined (n) times the number of molecules per mole (N_A), or $N = nN_A$, we have

$$\tfrac{1}{2}m\langle v^2\rangle = \frac{3}{2}\frac{nRT}{nN_A} = \frac{3}{2}\frac{R}{N_A}T \qquad (14\text{-}20)$$

R and N_A are well-known universal constants, and for their ratio we use the symbol k and call it the **Boltzmann constant**. It has the value

$$k = \frac{R}{N_A} = \frac{8.314 \text{ J/(mol·K)}}{6.024 \times 10^{23} \text{ molecules/mol}}$$
$$= 1.38 \times 10^{-23} \text{ J/(molecule·K)}$$

Equation 14-20 becomes

$$\tfrac{1}{2}m\langle v^2\rangle = \tfrac{3}{2}kT \qquad (14\text{-}21)$$

Note what this equation is implying. We have vaguely defined temperature as a measure of the "hotness" of a system. Equation 14-21 may be rearranged to solve for the absolute temperature, yielding T equal to a constant $(\tfrac{3}{2}k)^{-1}$ times the average molecular kinetic energy. Thus, according to kinetic theory, *absolute temperature is a direct measure of the average, random, translational kinetic energy of the molecules of a system.* **This defines temperature.** When you place your hand in cold water, heat flows from your hand to the water. Since you are at a higher temperature, your molecules are moving with greater average kinetic energy. In the innumerable collisions with less-energetic water molecules that will occur, the tendency will be for your molecules to lose kinetic energy to increase the kinetic energy of the water molecules. "Heat flows from hotter to colder" implies that in collisions, molecules with greater kinetic energy tend to be slowed down and slower molecules speeded up.

Furthermore, we have said that energy lost to friction shows up as increased temperature. Why? Consider a mass of water flowing over a waterfall. The potential energy of the water is converted to kinetic energy as the water falls. This is an external, ordered kinetic energy since all the water gains speed in the same direction. What happens as the water hits the rocks below? The motion is not lost, but **randomized.** Molecules are diverted in every direction, and instead of sharing an external, translational kinetic energy, we now have increased random, internal kinetic energy—increased temperature.

We may solve Equation 14-21 for $\langle v^2\rangle$ as $\langle v^2\rangle = 3kT/m$. Now we define v_{rms} as the **root-mean-square velocity,** which is the square root of the average square velocity:

$$v_{\text{rms}} = \sqrt{\langle v^2\rangle}$$

$$v_{\text{rms}} = \sqrt{\frac{3kT}{m}} \qquad (14\text{-}22)$$

In general, v_{rms} would not be quite the same as v_{av} for the molecules, since the squaring procedure will emphasize larger values of v.

Example 14-8 (a) Determine the average kinetic energy of molecules of a confined gas at a temperature of 300 K. (b) Calculate the root-mean-square speed of the molecules if the gas is hydrogen with a mass of 3.35×10^{-27} kg/molecule.

Solution (a) Equation 14-21 is interpreted as stating that the average kinetic energy of any gas molecule depends only on the absolute temperature and may be written as

$$\text{KE}_{\text{av}} = \tfrac{1}{2}m\langle v^2\rangle = \tfrac{3}{2}kT$$

Thus

$$\tfrac{1}{2}m\langle v^2\rangle = \tfrac{3}{2}[1.38 \times 10^{-23} \text{ J/(molecule}\cdot\text{K)}](300\text{ K})$$
$$= 6.21 + 10^{-21} \text{ J/molecule}$$

(b) The solution to this part is obtained by applying Equation 14-22.

$$v_{\text{rms}} = \sqrt{\frac{3kT}{m}}$$
$$= \sqrt{\frac{3[1.38 \times 10^{-23}\text{ J/(molecule}\cdot\text{K)}](300\text{ K})}{3.35 \times 10^{-27}\text{ kg/molecule}}}$$
$$= 1.93 \times 10^3 \text{ m/s}$$

SPECIAL TOPIC I
How Hot Are You?

That human body temperature is 98.6°F is an agreed-upon fiction if taken literally. If you were to measure the temperatures of a very large group of people and plot the number of persons with any given temperature vs. that temperature, you would obtain a bell-shaped curve having a mean temperature between 98 and 99° and a range of about ±1.5°. The mean would be lowest in the early morning and highest in the afternoon. Furthermore, the temperatures obtained would vary with the method of measurement. Normal measurements are taken orally, rectally, or under the arm, and all methods show variations upon repetition, with the oral method being the least reliable.

We say that the human body maintains its temperature. Actually, this is not true, as anyone who suffers from cold feet in the winter can tell you. The body attempts to maintain a constant temperature within a limited volume containing critical organs — the head and the abdominal and thoracic cavities — the so-called core of the body. The temperature of the rest of the body is allowed to vary widely. The primary reason for maintaining a constant core temperature is to allow chemical reaction to occur at a steady rate. As the temperature drops, the respiratory needs of tissues decline. (This fact is used in transplant surgery when organs are cooled to limit their respiratory needs.) If a naked individual is placed in a large chamber in which the temperature may be kept at any desired value, we may see this effect. In a very cold environment, say more than halfway from body temperature to the freezing point of water, the core temperature will drop less than a degree, measured rectally. The average skin temperature, however, will fall over a third of the way to the environmental temperature, and the feet will hover only a few degrees above the air temperature. The lower limit of cold tolerance of human tissue is not clear. On occasion, limbs have even recovered from being frozen.

The response of the body to cold makes great sense thermally. About half the body's volume lies within 5 cm of the surface. Trying to maintain the temperature of this peripheral zone would be futile. Instead, mechanisms to limit the heat loss from this zone are used. Immediately under the skin is a fat layer through which heat passes very slowly. This is used to limit heat loss. As the temperature against the bare skin drops, small blood vessels along the skin constrict to confine the blood under this fat layer, with maximum constriction occurring for a skin temperature of 80.0°F. Only enough blood is let through to keep external cells alive. Normal blood flow in the fingers nearly equals the finger's volume each minute. When the cold response is triggered, the volume of blood allowed to the finger can drop to about 1 percent of this. Within reason, the living tissue of the fingers will tolerate this reduced supply of oxygen and nutrients, since as the temperature drops so does the need of the cells. You have probably observed the aftereffects of this restriction of blood flow without realizing it. When coming in from the cold, the skin becomes very red and flushed looking. This is caused by the opening of small vessels to the skin to restore the circulation now that the need to conserve heat is over. The increase of blood supply beyond normal levels gives the reddish look.

Other methods limit heat loss from the core. Normally, blood flowing to the extremities is at about the same temperature as blood returning to the heart. Even with the fat layer of the skin and reduced flow, this warm blood would lose considerable heat in the arm and hand, for example. Measurements of temperatures inside the blood vessels show that blood to the hand is cooled almost 30°F before supplying the arm. Incoming and outgoing vessels are located close together in the body, and the bulk of this heat is lost by the warm blood to the colder returning blood before it reaches the core. Thus, much of the heat of the body's core is retained. Also, the blood reaching the extremities is very much cooler and will lose less heat to the outside. How hot you are, therefore, depends on the location of the measurement in your body. Your system takes 98.6°F blood at the heart and delivers it at 72°F to the hand. Meanwhile, the returning blood has been warmed to about 90°F before it enters the core. This all takes place in the incredibly efficient heat exchange of your body.

SPECIAL TOPIC II
States of a Real Gas

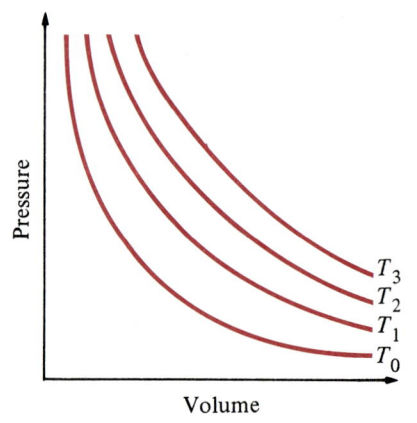

The behavior of a gas is frequently depicted in terms of graphs. Perhaps the most common of these is the pressure vs. volume curve of Figure 14-10. For a given number of moles of enclosed gas kept at a constant temperature, all combinations of pressure and volume over a reasonable range are connected by a smooth line. By the ideal gas law ($PV = nRT$), if the right-hand quantities are fixed, the product of pressure times volume will be a constant. Thus, as one doubles, the other will be reduced to half, etc. Figure 14-10 displays this behavior. If the enclosed gas were kept at a different, higher temperature, we would find that pressure and volume varied much as before but along a new (hyperbolic) curve farther from the origin than the first. Increasing the temperature in steps and then, for each new temperature, generating the curve representing possible PV combinations yields a family of curves, as in Figure 14-11.

The major difficulty with the curves generated is that they represent only ideal gases. A real material (steam, for example) would follow the curves reasonably well at high temperatures and/or low pressures but would deviate at very low temperatures and very high pressures. Consider steam above the normal boiling point of water. At low pressures, the molecules are substantially independent of each other since the

Figure 14-10 The pressure of an enclosed ideal gas as a function of volume for some constant temperature T_0. Each point on the curve identifies a possible state of the gas.

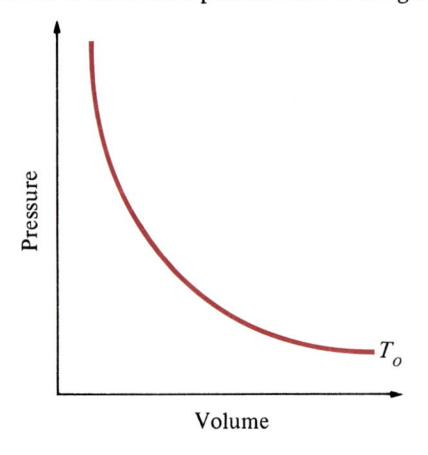

Figure 14-11 Adding energy to the ideal gas system of Figure 14-10 can increase the temperature from T_0 and produce a new curve of possible states of P and V for each temperature, where $T_3 > T_2 > T_1 > T_0$.

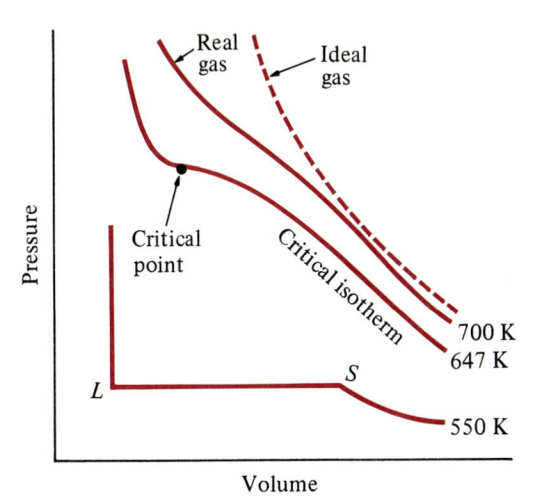

Figure 14-12 The behavior of a real gas.

average distance between them is very large relative to their size. As we force the molecules closer together by increasing the pressure, the normal attractions they have for each other become more significant, and volume decreases more rapidly with increasing pressure than the ideal gas law would predict. If the energy (temperature) is high enough, we may simply note this high-pressure deviation from prediction, and that is all (Figure 14-12, upper curve). Should the energy be much lower, high pressure may force the molecules to clump together and begin to form a liquid even though the temperature is well above the boiling point at atmospheric pressure (Figure 14-12, lower curve). The form of this curve is completely different. Following it from the right end, as the pressure increases, the volume contracts along the familiar curve. Then, at point S, we reach the saturation point for the vapor, where liquid will begin to form. Liquid and vapor coexist at this pressure, but a very minor change in pressure fully condenses the vapor into a liquid until point L is reached. At this point, the steam is fully condensed into liquid water, which is not readily compressible. Thus, the curve ascends very steeply with only minor volume change for increasing pressure. Between the upper and lower curves of Figure 14-12, there is one PV curve called the **critical isotherm,** for temperature 374.1°C (647.25 K). Above this curve, no amount of pressure will liquefy the gas. Below this curve (temperature), the steam is properly called a **vapor** since it can be liquefied by an increase in pressure only. The saturation points and points of complete liquefaction of the steam of all curves below the critical isotherm can be connected in a separate curve called the **saturation curve,** as shown in Figure 14-13. The single point at

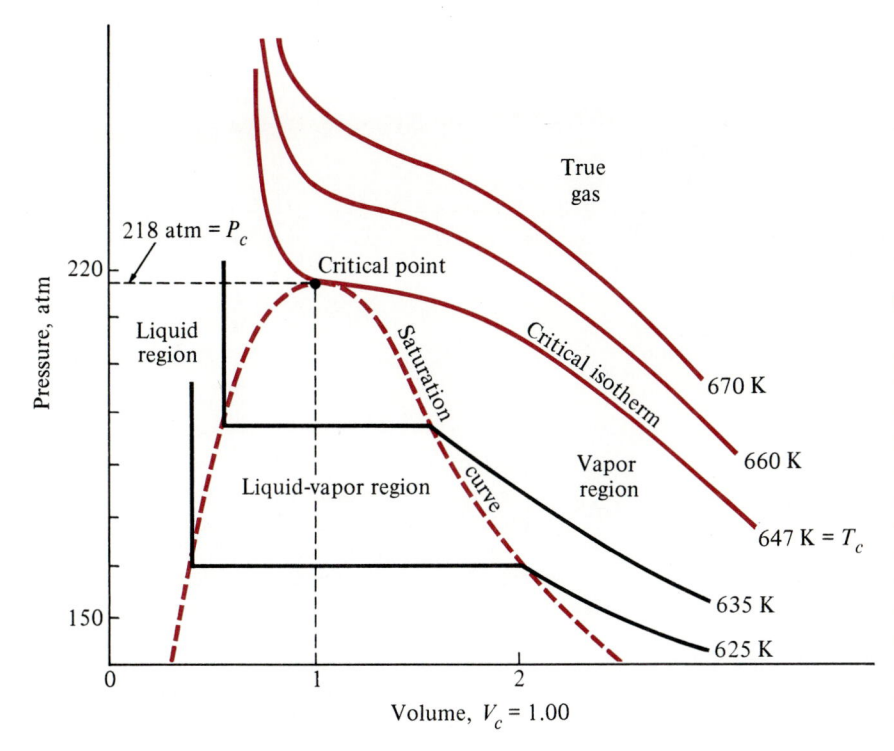

Figure 14-13 Connecting the saturation points and liquefaction points for all isotherms gives a saturation curve. Within the saturation curve is a liquid-vapor region where steam and liquid water may coexist. Below the critical isotherm but not compressed enough to reach saturation is the region where the steam is a vapor and may be condensed to water by increasing pressure only. Above the critical isotherm, the steam is a true gas and will not liquefy at any pressure. The critical isotherm touches the saturation curve at a single point, the critical point of water, which occurs at 218.3 atm and 647.25 K.

333

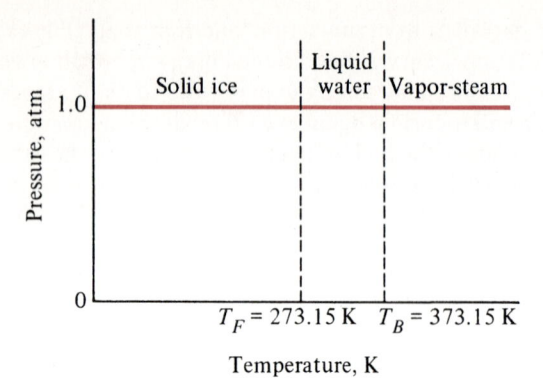

Figure 14-14 A pressure vs. temperature plot for the states of water limited to the range of most human experience (at about 1.0 atm) identifies the normal boiling point and freezing point of water.

which the critical isotherm touches the saturation curve is called the **critical point** for the material.

We normally experience common materials only at conditions with which we are personally comfortable. For example, we defined the freezing point and melting point of water to determine two points on the Celsius scale. These had to be defined for a particular pressure to be meaningful. Figure 14-14 shows the pressure vs. temperature behavior with which we are familiar. Clearly, the phase of water might well depend upon pressure. Would 50°C water remain a liquid exposed to the near vacuum of space? Would water boil at 100°C under 100 atm of pressure? If you answered "yes" to either of these, you would have trouble understanding freeze drying of foods or pressure cookers. The complete pressure vs. temperature curve for water is shown in Figure 14-15. We note that the boiling and freezing points we define for water are an accident of our normal atmospheric pressure. Indeed, below a pressure of 0.00602 atm, liquid water cannot exist at any temperature. Below this point, solid water changes directly to a vapor in the process of **sublimation** without passing through the liquid state at all. At the right of the liquid-to-gas curve, we find the line ending abruptly at the critical point, the same one as in Figure 14-13. Above this temperature of 647.25 K, water cannot be liquefied at any pressure. Of particular interest is the intersection of the lines on the curve, the *triple point* for water. This is the one and only combination of pressure and temperature where solid, liquid, and vapor forms of water can coexist in equilibrium, 273.16 K and 0.00602 atm of pressure (610 N/m²). The precision of the conditions required to be at the triple point of a substance makes it ideal as the reference value upon which to build a temperature scale. One part of the curve for water shown is relatively unusual among materials. The melting curve between the solid and liquid states has a negative slope; that is, increasing pressure can cause solid ice to melt to form liquid water with no change in temperature. For

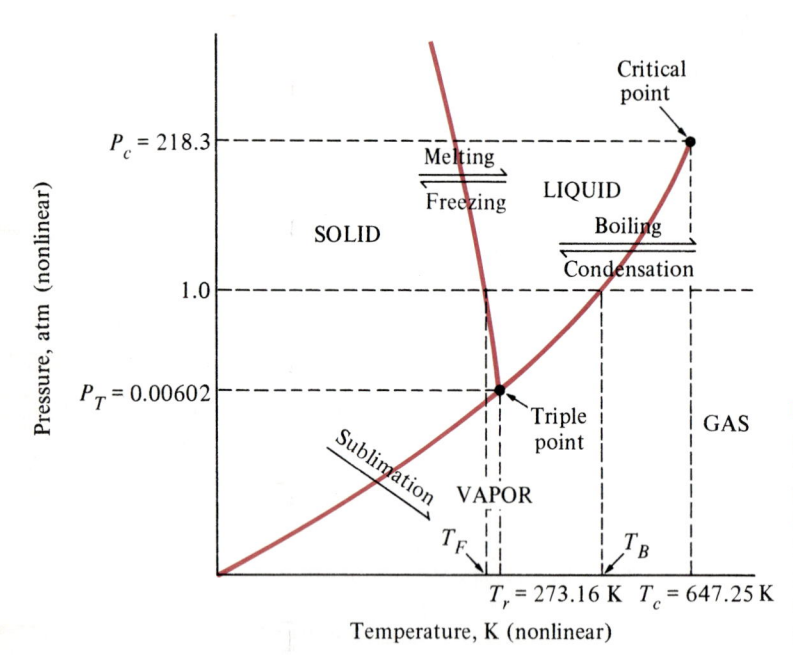

Figure 14-15 A fuller view of the pressure vs. temperature behavior of water displays much more detail, which must be graphed on nonlinear scales to be seen clearly. The critical temperature T_C and the triple point temperature T_T are basic immutable properties of water not dependent on other conditions or systems of measurement.

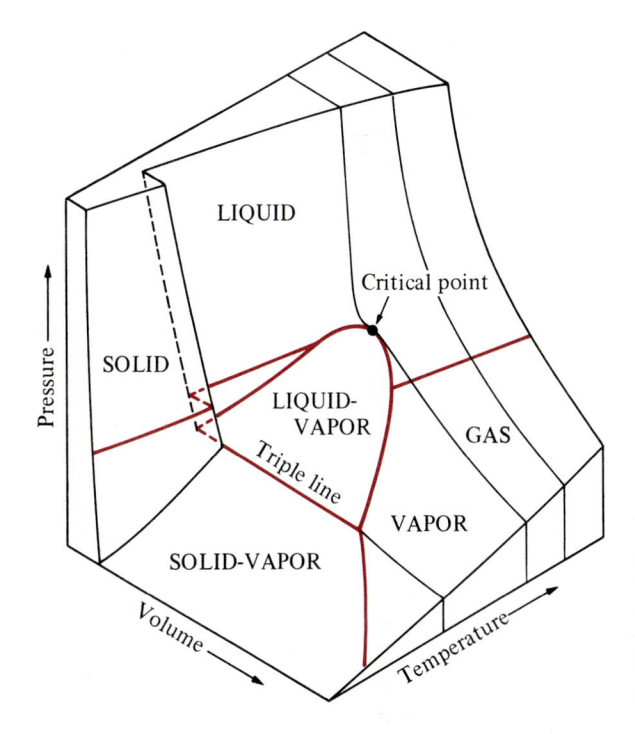

Figure 14-16 A section of a *PVT* surface for water. Viewed perpendicular to the pressure-volume plane, one may imagine Figure 14-13. If viewed from the pressure-temperature plane, one may imagine Figure 14-15. Note that in three dimensions, the triple point is really a line.

most materials, increasing pressure would tend to do the opposite, since the solid form is more densely packed than the liquid and high pressure favors the solid phase. This is not true with water, which achieves its maximum density as a liquid at 4°C and expands as it warms or cools from that temperature. This ensures that lakes and oceans freeze from the top down and icebergs float, very handy behavior for the inhabitants of earth.

Pressure, volume, and temperature may be regarded as the *x*, *y*, and *z* variables to produce a three-dimensional phase model for a material. For a given amount of a given material, the phase is perfectly determined for any set of values of *P*, *V*, and *T*, as shown in Figure 14-16. For example, a slice of this model parallel to the *PV* plane would produce one of the *P* vs. *V* curves we have seen. Such a phase model is difficult to illustrate on a flat page and is much easier to see as a three-dimensional model.

Minimum Learning Objectives

After studying this chapter, you should be able to:

1. Define:

   absolute temperature
   absolute zero
   Avogadro's number
   Boltzmann constant
   Boyle's law
   Celsius scale
   Charles' law
   coefficients of thermal expansion
   Fahrenheit scale
   heat
   ideal gas
   ideal gas law
   kelvin (K) [unit]
   Kelvin scale
   kinetic theory of gases
   mole (mol) [unit]
   phase (of matter)
   root-mean-square temperature
   temperature scale
   thermal energy
   thermal equilibrium
   thermometer
   triple point
   universal gas constant
   vapor

2. Describe how to construct a thermometer.
3. Convert readings between commonly used temperature scales.
4. Derive a conversion formula between two arbitrary temperature scales, given two common readings on each scale.
5. Distinguish between temperature and heat.
6. Calculate length (and/or volume) changes, given the temperature change and the appropriate coefficient of expansion.
7. Calculate the pressure, volume, and/or temperature changes of an enclosed gas, given the initial set of values and two of the three final values.
8. Calculate the pressure, volume, temperature, or amount (in moles, or on occasion, in grams) of a gas for which three of these variables are given.

335

9. Understand and appreciate the difficulty of defining an invariable standard for the calibration of thermometers.
10. Convert freely between moles, grams, and numbers of molecules for simple molecular substances.
11. Calculate the root-mean-square (rms) velocity of molecules for simple molecular substances.
12. Calculate the ratio of rms velocities for molecules of different masses at the same temperature.

Problems

14-1 To five significant figures and in degrees Fahrenheit, what is (a) the boiling point of oxygen and (b) the freezing point of zinc? (See Table 14-3.)

14-2 At what temperature do the Fahrenheit and Celsius scales give the same numerical reading?

14-3 The highest and lowest temperatures ever recorded on earth are $+136.4$ and $-126.9°F$, respectively. Convert these to degrees Celsius and to kelvins.

14-4 On Belcare in the Zebulon star system, the thermometer scale sets zero as the normal boiling point of water. The normal melting point of ice is set at $+250°B$ (for degrees Belcare). Construct an equation which can be used to (a) convert t_c (degrees Celsius) to t_B (degrees Belcare) and (b) convert t_B (degrees Belcare) to t_F (degrees Fahrenheit).

14-5 The atmosphere of the planet Arbela is mainly composed of two gases which liquefy at -253 and $-112°C$, respectively. The Arbelon temperature scale (degrees Arbela) arbitrarily sets $0°A$ as the boiling point of the first and $10°A$ as the boiling point of the second. Write an expression to convert (a) degrees Arbela to kelvins, (b) degrees Arbela to degrees Celsius, and (c) degrees Celsius to degrees Arbela. (d) What temperature (degrees Arbela) is the normal boiling point of water?

14-6 A brick wall is 100.0 m long at $30.0°C$. How long will it be at $-30°C$? (Use four significant figures.)

14-7 A steel piston is machined at room temperature ($20°C$) to be 15.00 cm wide and fit precisely in a cylinder. How wide will the piston be at an operating temperature of $1200°F$? (Use four significant figures.)

14-8 The steel rails of the Siberian railroad are 5300 km long in the summer at a temperature of $25°C$. How long are the rails in the winter at $-50°C$? If each rail is 10.0 m long, how large an expansion joint must there be between rails in the winter to allow for this length change?

14-9 A steel bridge spanning a small river is 75.0 m long at $20°C$. During the year, its temperature may be as low as $-30°C$ and as high as $+40°C$. By how much may its length change during the year?

14-10 At $20°C$ a steel sphere is exactly 2.5000 cm in diameter and exactly 0.0400 cm larger than the inside diameter of an aluminum ring. Calculate the temperature at which the sphere will be able to pass through the ring.

14-11 A 100-cm³ glass test tube is filled to the brim with water at $0°C$. How much water will overflow if the test tube and its contents are heated to $100°C$?

14-12 A steel wire 3.00 m long is stretched to span an outside gateway to a college during the summer when the temperature is $30°C$. Calculate the additional stress on the wire on a cold day when the temperature is $-20°C$.

14-13 A brass cylinder is to be used to plug a circular hole in a brass plate by shrink fitting. The plate is heated to $900°C$ and the brass cylinder at $20°C$ is placed in the hole. The plate is then allowed to cool, and as it does, it compresses the cylinder to hold it in place. If the hole is 2.5000 cm in diameter at $20°C$, what is the cylinder's maximum possible diameter at $20°C$? (Use five significant figures.)

14-14 What is the original length of an aluminum rod which contracts by exactly 2.50×10^{-3} m when it is removed from boiling water and plunged into an ice water bath?

14-15 A steel ruler is calibrated to be accurate at $20°C$. This ruler is used to measure the length of an aluminum rod at $0°C$ and yields a reading of 30.000 cm. Calculate the reading you would expect if both the rod and ruler were at $20°C$. (Use five significant figures.)

14-16 A structural steel beam 10.0 m long is placed in position when the outside temperature is $35°C$. If its ends are welded rigidly in place, what tension will be exerted on the beam on a day when the temperature drops to $-20°C$? (Assume the beam's cross-sectional area is 2.00×10^{-2} m².)

14-17 The temperature of a rectangular plate of sides L_1 and L_2 in length is increased by an amount ΔT. The sides expand by ΔL_1 and ΔL_2, respectively. The linear and area coefficients of thermal expansion of the plate are α and γ. Consider Figure 14-17. Ignore the small corner contribution (ΔL_1 times ΔL_2), and apply Equations 14-4 and 14-5 to show that $\gamma = 2\alpha$.

Figure 14-17 Problem 14-17.

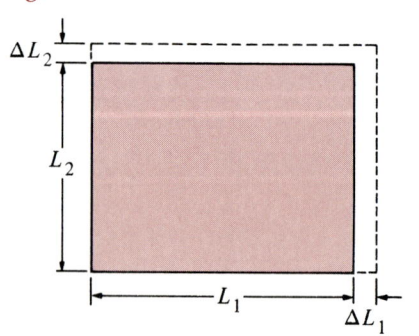

14-18 A cylindrical drinking glass 12.0 cm high is inverted over a swimming pool and forced downward to a depth of 1.50 m. Assume constant temperature, and calculate how high water rises in the glass.

14-19 Determine the compression ratio (ratio of maximum volume to minimum volume) for a diesel engine which operates between the limits of 45°C at 1 atm and 600°C at 50 atm.

14-20 A constant-pressure gas thermometer has a volume of 3.50 L at 20°C. After heating, its new volume is 5.00 L. (a) Calculate its new temperature. (b) What is its volume when the temperature is 50°C?

14-21 A constant-pressure gas thermometer, as used by Galileo, has a volume of 1000.0 cm³ of air in a glass bulb leading to a long, thin tube with an inside diameter of 5.00 mm. Inside the tube, a bead of liquid, some 8.00 cm from the glass bulb, seals the trapped gas off from the other end of the tube, which is open to the air. As the air in the bulb warms, it expands and moves the liquid bead toward the open end of the tube. Disregarding expansion of the glass, if the air is at 17°C, by how much would the liquid bead be moved down the stem if the temperature increased to 20°C? (Use five significant figures.)

14-22 The absolute pressure in an automobile tire is 3.10×10^5 Pa (about 30 lb/in² gauge) when the temperature is 20°C. The temperature increases to 75°C due to friction acting on the tire during a trip on the highway. Calculate the final (a) absolute pressure and (b) gauge pressure in pounds per square inch.

14-23 A scuba diver at a depth of 14.0 m releases an air bubble 4.00 cm in diameter. The temperature is 15°C at that depth and 25°C at the surface of the water. Determine the diameter of the bubble when it reaches the surface.

14-24 The interior of a tire may be approximated as a cylindrical shell. A particular tire has an inside diameter of 40.0 cm, an outside diameter of 1.00 m, and a thickness of 30.0 cm. If such a tire is inflated to a gauge pressure of 30.0 lb/in² at 20°C, how many moles of gas are contained inside the tire?

14-25 If 2.00 mol of gas occupies a volume of 12.0 L at a temperature of 50°C and a pressure of 4.42 atm, determine the universal gas constant.

14-26 A quantity of gas which may be treated as ideal has a volume of 8.00×10^{-3} m³ at 20°C and a pressure of 1.01×10^5 Pa. (a) Determine the number of moles of gas present. (b) Calculate its volume if the temperature is increased to 100°C at constant pressure. (c) Calculate its pressure if the temperature is increased to 100°C at constant volume.

14-27 A laboratory cylinder contains a volume of 40.0 L of helium gas at a pressure of 1.40×10^6 Pa and a temperature of 70°C. A valve is opened and the gas released to a final pressure of 1.00×10^5 Pa and a temperature of 25°C. Calculate (a) the number of moles of helium originally in the cylinder and (b) the final volume occupied by the gas.

14-28 A low-pressure constant-volume gas thermometer has a pressure of 8.00 torr (mmHg) at -10°C. (a) What is the temperature if the thermometer is heated until the pressure is 12.0 torr? (b) What pressure should you expect if the thermometer is plunged under boiling water?

14-29 The *Hindenburg* zeppelin of Germany was the largest airship ever built, with a length of 245 m and a volume of 2.00×10^5 m³. Less well known is that the United States built several large airships. The largest of these was the *USS Akron* (length, 239 m), launched in 1931 and designated as an aircraft carrier. The *Akron* could launch and recover its five fighter planes while it was in flight. One option available to the American airship was denied the *Hindenburg*. The United States did (and does) control the world supply of helium gas, and we could use this in place of the inflammable hydrogen in the *Hindenburg*. Assume both these airships had the same volume. What would be the relative lifting power of the two filled with these different gases? (Remember air is essentially N_2 with a molecular mass of 28, while hydrogen, atomic mass = 1, exists as H_2 and helium exists as independent atoms with an atomic mass of 4.00. Recall also that a mole of any gas at standard temperature and pressure occupies 22.4 L of volume.)

14-30 According to the Avogadro hypothesis, a fixed volume of any gas (or combination of gases) contains a fixed number of molecules at standard temperature and pressure (0°C, 1.00 atm). Thus, 22.4 L of any gas at 0°C and 1.00 atm contains 6.02×10^{23} molecules. The most common gas of the air is nitrogen gas, N_2, with a molecular mass of 28.0. As we raise the humidity, we add more and more water vapor molecules to the air. (a) Should this make the air more dense or less dense? (b) At standard conditions, 1.00 m³ of dry air has a mass of 1.293 kg. What would the mass of 1.00 m³ of air be at 22°C and 1.00 atm? (c) Saturated air (with 100% relative humidity) contains 19.22 g water vapor/m³ at 22°C and 1.00 atm. What is the density of 1 m³ of such air? (d) What is the percentage difference in density between dry air at 0°C and 1.00 atm and saturated air at 22°C and 1.00 atm?

14-31 How many molecules per cubic centimeter are there in a cylinder held at a pressure of 1.00×10^{-10} Pa (at a temperature of 20°C)?

14-32 Gas is confined in a tank at a pressure of exactly 70 atm at a temperature of 20°C. The pressure valve is opened, and a quantity of gas is released as the temperature is raised to 45°C. The pressure in the tank is now 40 atm. What fraction of the gas was released?

14-33 A hot air balloon holds a volume of air equal to 800 m³ and has a total mass of 400 kg including its passengers. Assume the pressure inside the air bag is the same as the outside pressure and that the heater maintains a uniform temperature inside. (a) Calculate the temperature to which the air in the balloon must be heated in order for lift-off to occur when the outside temperature and pressure are 20°C and 1.00×10^5 Pa. (b) Calculate the inside temperature required to maintain an altitude of 2.00×10^3 m where the temperature and pressure are 5°C and 8.00×10^4 Pa. (Take the average molar mass of air to be 29.0 g/mol = 0.0290 kg/mol.)

14-34 (a) Determine the number of moles of gas in a volume of 1.00×10^4 cm³ at a temperature of 100°C and pressure 3.00×10^3 Pa. (b) If the gas in part *a* is nitrogen, what is its mass?

14-35 The equation for the complete combustion of octane, a major component of gasoline, is

$$2C_8H_{18} + 25O_2 \rightarrow 16CO_2 + 18H_2O$$

Since all reactants enter the cylinder of the automobile as gases, the equation implies that 27 volumes of reactants yield 34 volumes of products. Disregard heat loss to the cylinder walls. (a) If an appropriately proportioned mixture of octane vapor and oxygen gas is taken into the cylinder at 0°C and 1 atm of pressure and then compressed to one-tenth its original volume, what will the pressure be if the temperature remains constant? (b) If this mixture is ignited and increases the temperature to 760°C before any change in volume, what will the pressure become? (c) The number of moles of gas increases during combustion, as indicated by the chemical equation. What change does this have on the pressure? (d) If the head of the piston has a diameter of 15.0 cm, what is the total force tending to push the piston away from the gases at this point? (e) Gasoline combustion actually takes place not with pure oxygen gas but with air, which is only 21 percent oxygen, with the balance of the gases being largely inert during the reaction. In which of the steps above would this fact change the calculations?

14-36 Twenty particles are moving around in a box. The speeds of the particles are distributed as follows: three particles have speeds of 1.00 m/s, four have speeds of 2.00 m/s, seven have 3.00 m/s, three have 4.00 m/s, two have 5.00 m/s, and one has 6.00 m/s. Determine (a) the average speed and (b) the rms speed.

14-37 (a) What is the root-mean-square velocity of oxygen gas molecules (O_2) at room temperature (20.0°C)? (Molecular mass ~ 32.0 u.) (b) What would be the root-mean-square velocity of helium atoms at the same temperature ($m_{He} \approx 4.00$ u)? (c) Of carbon tetrachloride molecules (CCl_4) at the same temperature ($m_{CCl_4} \approx 154$ u)?

14-38 If an object attains a speed of 11.2 km/s (the escape velocity) directed away from the earth, it will escape the earth's gravitational pull. (a) The temperature of the earth's surface is about 300 K. What is the root-mean-square velocity of an oxygen atom ($M_{mass} = 16$) released in the earth's atmosphere? A hydrogen atom ($M_{mass} = 1$)? (b) Root-mean-square velocities represent an average value over a great number of molecules. Sizable numbers of molecules are traveling faster and similar numbers slower than this value. If the root-mean-square velocity is of the same order of magnitude as the escape velocity, the fastest of the molecules will constantly trickle away into space. If their numbers are large enough, a planet may lose one or all types of molecules of its atmosphere to space. The general rule of thumb for retaining an atmosphere for about 10^9 years is that the root-mean-square velocity must be less than 0.2 times the escape velocity. Based on this criterion, is the earth in danger of losing oxygen atoms released in its atmosphere? Hydrogen atoms? (c) About 20 percent of the atmosphere of Jupiter is hydrogen by weight (mostly in the forms of methane and ammonia). Give at least two reasons why Jupiter would not be likely to lose a hydrogen atom liberated in its atmosphere.

14-39 At what temperature is the rms speed of nitrogen and oxygen equal to the escape velocity (a) of the moon (2.38×10^3 m/s) and (b) of the earth (11.2×10^3 m/s)? Do your answers hint at why the earth has an atmosphere and the moon does not? (See problem 14-38.)

14-40 For electrons in an electron gas at 20°C, calculate (a) the average kinetic energy, (b) the root-mean-square speed ($m_e = 9.11 \times 10^{-31}$ kg), and (c) the root-mean-square speed if the particles are neutrons, so-called thermal neutrons ($m_n = 1.67 \times 10^{-27}$ kg).

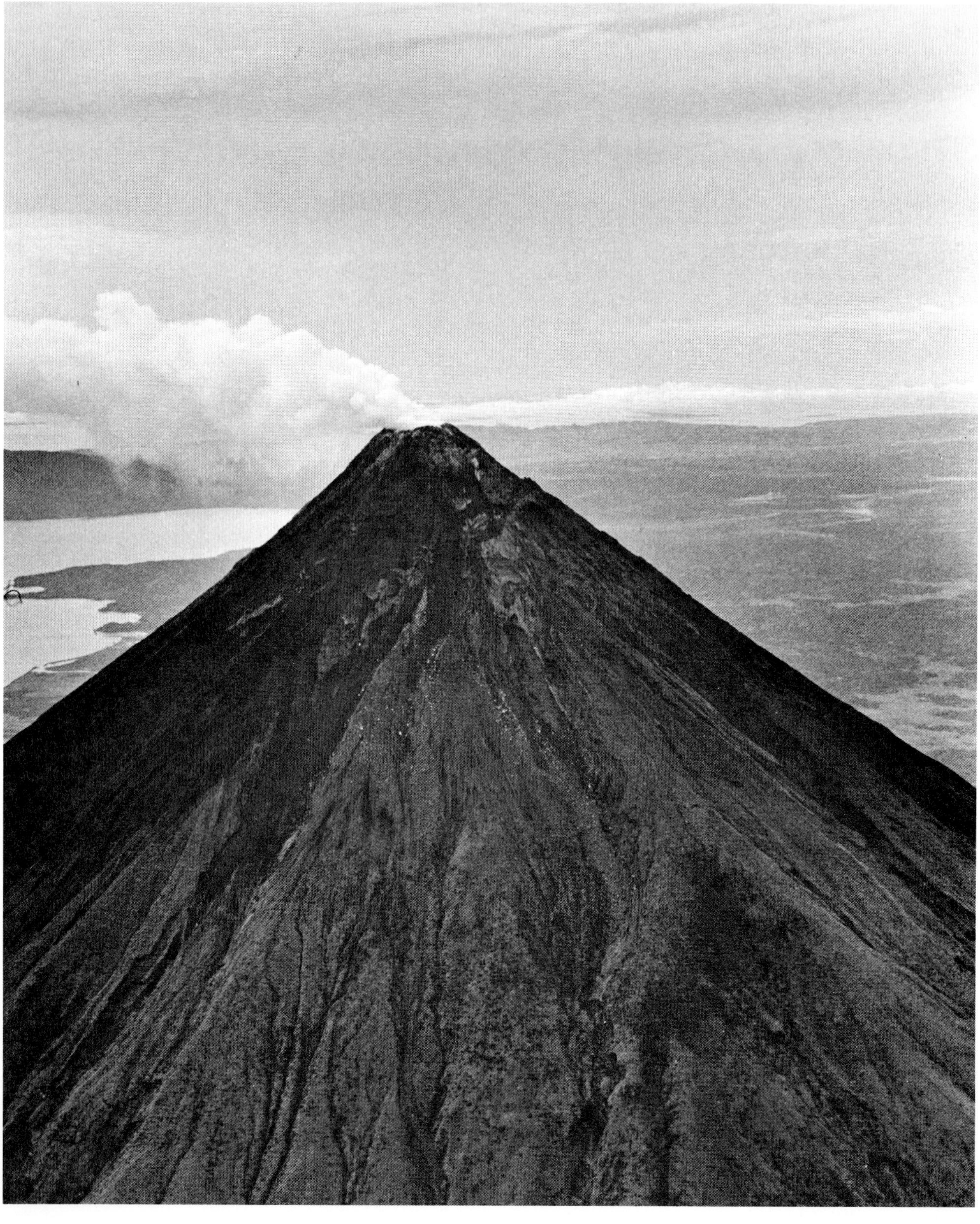

15 Heat and Heat Transfer

15-1 Introduction

Throughout the eighteenth century and into the nineteenth, it was generally believed that heat was a fluid called **caloric** which was transferred from hotter bodies to colder bodies. Certain properties of caloric were postulated to explain the known behavior of heat and of matter which was absorbing or expelling heat. While supervising the boring of cannon at Munich in 1798, Count Rumford (Benjamin Thompson) noted that sharp drill bits seemed to generate less heat than dull drill bits. Further, the water used to cool the cannon would continue to boil as long as the drilling continued. Even if heat was a fluid that could be transferred by friction, eventually the drill bit and cannon would have to run out of caloric and the water would cease to boil. Since boiling did not stop, he inferred that heat was not a fluid at all. He concluded that mechanical work by the bits on the cannon generated heat and that heat was some form of motion.

Very early, a unit of heat called the **calorie** (cal) was defined as the amount of heat necessary to raise the temperature of one gram of water by one degree Celsius. The word *calorie* is derived from the caloric theory and is definitely not an SI unit. Also, it is not the Calorie associated with the energy value of food. That **Calorie** (Cal) is 1000 cal, or 1 kcal. Thus a slice of cheese has about 100 Cal, or 100,000 cal. (It is interesting to speculate what dieters might think of that.) Any reference we make to the food Calorie will always have that word capitalized, while the calorie as defined above will always begin with the lowercase letter. [An analogous definition in the British engineering system is that of the **British thermal unit** (Btu). One Btu is the amount of heat necessary to raise the temperature of one pound of water by one degree Fahrenheit. Thus 1.00 Btu = 252 cal.] If mechanical energy and heat are two forms of the same quantity, shouldn't we be able to change one form into the other? The answer was provided by James Prescott Joule, for whom the SI unit of energy is named (Figure 15-1).

Joule took on the mission of determining the amount of mechanical energy which would be equivalent to 1 cal. The apparatus he used in his most frequently cited experiment to make this determination is shown in Figure 15-2. The mass m descends a measured distance at a constant speed. As it does, the string attached to it turns the paddle wheel apparatus, stirring the water. Since the descending mass moves at constant speed, there is no change in its kinetic energy. Thus its loss in potential energy should show up as a gain in random thermal energy of the water, and an increase in temperature should show on the thermometer. Knowing the mass of water available, we should be able from this measured temperature increase to calculate the number of calories absorbed by the water. Setting this equal to the magnitude of the mechanical energy lost by the descending mass, we may obtain an equivalence between heat energy and mechanical energy. Experiments like this and many quite different have been conducted again and again. The present accepted equivalence, called the **mechanical equivalent of heat**, is

$$1.000 \text{ cal} = 4.184 \text{ J}$$

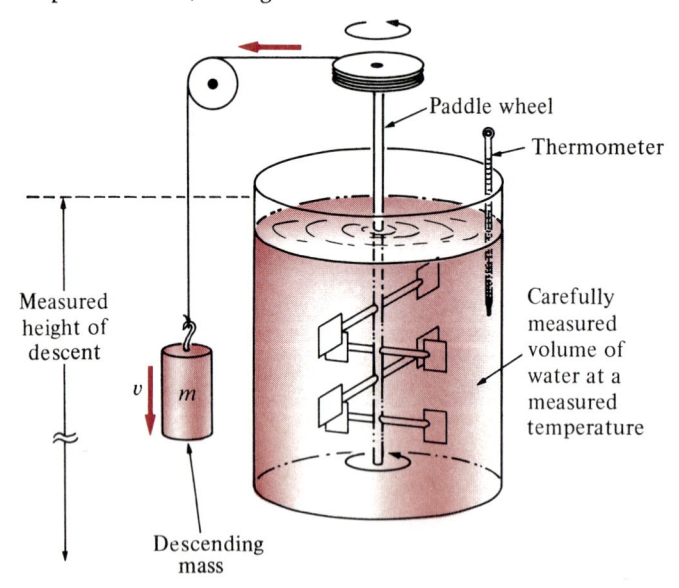

Figure 15-1 James Prescott Joule (1818–1889) was an innovative experimenter whose lifelong obsession was to measure the amount of mechanical energy necessary to produce a given amount of heat—the mechanical equivalent of heat. He measured this equivalence in such a variety of ways that it is rare even today to find a method he did not try. Joule's work was initially ignored, but his tenacity eventually established his ideas, and he became one of the first individuals with a clear appreciation of the concept of conservation of energy. (*Culver Pictures*)

Figure 15-2 Joule's most noted apparatus for measuring the mechanical equivalent of heat used a descending mass. As the mass lowered slowly, its potential energy was used to turn paddle wheels, stirring a known mass of water.

Example 15-1 One of the fast-food chain stores which you might frequent serves a superburger that a national magazine listed as delivering a total of 543 Cal (kcal). Assuming 20 percent of the energy absorbed may be converted to gravitational potential energy by the muscles, how high could an 80.0-kg student climb up a mountain on the fuel from such a sandwich?

Solution We must convert the food energy to joules by using the mechanical equivalent of heat. Setting 20 percent of this result equal to $mg\,\Delta h$ will then allow us to solve for the 80.0-kg student's increase in height, which would yield an equivalent energy.

$$\text{Energy} = 543 \text{ Cal} \times \frac{1000 \text{ cal}}{1 \text{ Cal}} \times \frac{4.184 \text{ J}}{\text{cal}} = 2.27 \times 10^6 \text{ J}$$

Considering the 20% efficiency of muscles, we have

$$\Delta E = (2.27 \times 10^6 \text{ J})(0.20) = 4.54 \times 10^5 \text{ J}$$

$$mg\,\Delta h = 4.54 \times 10^5 \text{ J}$$

$$\Delta h = \frac{4.54 \times 10^5 \text{ J}}{(80.0 \text{ kg})(9.80 \text{ m/s}^2)} = 5.80 \times 10^2 \text{ m}$$

This is the height of a 150-story building, quite a penance for a dieter to pay for eating a single hamburger.

15-2 Specific Heat Capacity

Normally, we expect that when we add heat to a body, its temperature will increase. The amount of heat necessary to raise the temperature of a given body by one degree is the heat capacity for that body.[1] The usefulness of this concept is limited by the fact that each and every object, in general, has a different heat capacity. As you might expect, careful measurement would show that the heat capacity increases in proportion to mass. Thus, the heat capacity divided by mass has the same value for objects made of the same material. This heat capacity per unit mass depends only on what the material is. We call this the specific heat capacity, or simply the **specific heat**. The specific heat is defined as the amount of heat required to raise a unit mass a unit temperature interval.

$$c = \frac{\Delta Q}{m\,\Delta T} \qquad (15\text{-}1)$$

where c is the specific heat of a body of mass m and ΔQ is the heat required to increase the temperature by an amount ΔT. Specific heat has units of energy per mass per degree temperature change. In SI, the units of c are J/(kg·°C), which is the same as J/(kg·K) since kelvin intervals are the same magnitude as Celsius intervals. (You might even note that the definition of the calorie is simply a statement of the specific heat of water in calories per gram per degree Celsius.) Table 15-1 lists the value of the specific heat of some selected materials.

---

[1] Don't be misled by the name. Heat capacity is *not* the amount of heat a body can hold.

TABLE 15-1 Specific Heat of Selected Materials at 25°C (except as noted)

Material	Specific Heat, J/(kg·K)
Solids	
Aluminum	902
Brick	840
Concrete	880
Copper	384
Germanium	322
Glass (crown)	674
Glass (flint)	494
Gold	129
Granite	804
Human body (av)	3500
Ice (av −50 to 0°C)	2090
Iron (or steel)	449
Lead	128
Silicon	712
Silver	235
Sulfur	736
Wood	1760
Liquids	
Benzene	1720
Ethanol	2430
Mercury	140
Water	4180
Gases	
Air (av)	1000
Carbon dioxide	833
Nitrogen	1040
Oxygen	913
Steam (100°C)	2010

Example 15-2 (a) How much heat must be added to a 4.00×10^{-3} kg steel ball bearing in order to increase its temperature from 20 to 50°C? (b) By how much will the temperature of the ball increase if it is made of gold rather than of steel?

Solution Both parts of the example require that we apply Equation 15-1. We obtain the proper specific heat values from Table 15-1.
(a) We must solve for ΔQ by using Equation 15-1.

$$\Delta Q = cm \, \Delta T$$
$$= [4.49 \times 10^2 \text{ J/(kg} \cdot \text{K)}](4.00 \times 10^{-3} \text{ kg})(30 \text{ K})$$
$$= 53.9 \text{ J}$$

(b) We must solve for ΔT and use the ΔQ we have just calculated along with the specific heat of gold.

$$\Delta T = \frac{\Delta Q}{cm} = \frac{53.9 \text{ J}}{[1.29 \times 10^2 \text{ J/(kg} \cdot \text{K)}](4.00 \times 10^{-3} \text{ kg})}$$
$$= 104 \text{ K} \quad \text{(or 104°C)}$$

Is it possible that Archimedes could have discovered whether the crown was pure gold from a knowledge of specific heats?

We have hinted in several places that perhaps adding heat to a substance does not always ensure that its temperature will increase. Let us examine a reason why this may be so.

15-3 Latent Heat of Phase Change

The normal **melting point** of ice was defined as that temperature at which ice and water coexist in thermal equilibrium under standard atmospheric pressure. We did not state that there needed to be any specific ratio of the amount of ice to water, because any ratio will do. Chilled drinks are served over ice so that the temperature of the mixture will drop to 0°C. Furthermore, the mixture stays at 0°C until the last of the ice melts.

Consider a block of ice at -50°C to which we begin adding heat uniformly at a constant rate. The resulting temperature may be plotted vs. time, yielding the graph of Figure 15-3.[1]

The important parts of the curve of Figure 15-3, for present considerations, are the horizontal portions. In the two regions designated II and IV on the graph, heat was added for a time with *no* accompanying increase in temperature. In region II, the heat energy is being used to break the bonds which hold the molecules to their nearest neighbors, melting the ice. In region IV, the heat

---

[1] This graph is not 100 percent correct, since, in practice, the lines representing increasing temperature are not perfectly straight but have slopes varying slightly. For example, the slope of the first part of the curve should be a little steeper near -50°C and a little less steep near 0°C because the specific heat of ice is smaller near -50°C than it is near 0°C. Most specific heats vary somewhat with temperature, but since this variation is small, we will take the specific heat to be constant.

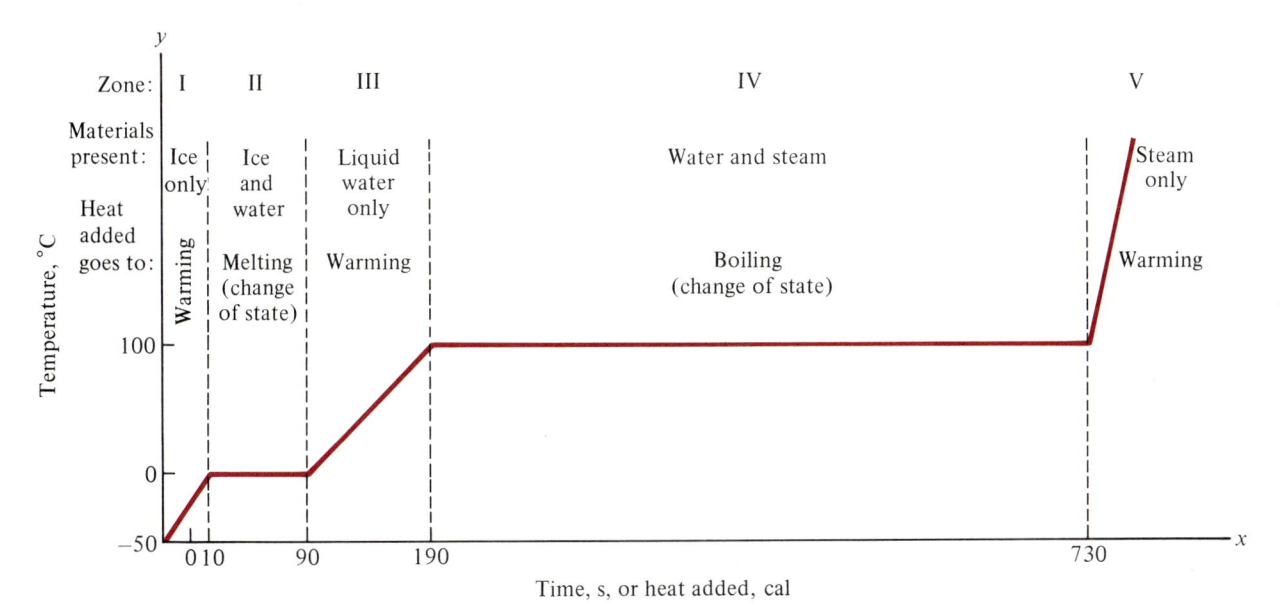

Figure 15-3 The heating curve for water plots the Celsius temperature of a fixed quantity of water vs. time as heat is added at a constant rate. This is the same as plotting temperature vs. heat added to the material.

energy added is absorbed by those molecules which escape altogether from the liquid phase, boiling the water away. In each of the two regions, a certain amount of heat is required to **change the phase** of every gram of substance. The heat required to change the phase of a substance does not result in a temperature increase; it is said to be hidden or **latent**. Specifically, the latent **heat of fusion** L_f is the amount of heat per unit mass required to melt a substance. To change solid ice at 0°C to liquid water at 0°C, each gram of ice requires 80 cal. Thus, for water, $L_f = 80$ cal/g. The latent **heat of vaporization** L_v is the amount of heat per unit mass required to vaporize a liquid at its **boiling point**. To change liquid water at 100°C to gaseous steam at 100°C, each gram of water requires 540 cal. Thus, for water, $L_v = 540$ cal/g. The units given for these latent heats are historic or traditional. In SI units, we have that for water

$$L_f = 3.35 \times 10^5 \text{ J/kg} \quad \text{and} \quad L_v = 2.26 \times 10^6 \text{ J/kg}$$

To determine the amount of heat ΔQ required for a mass m of substance to change phase, we must multiply the value of the mass by the latent heat required for that change. Thus, in general,

$$\Delta Q = mL \qquad (15\text{-}2)$$

If the substance is changing from the solid to the liquid phase or from the liquid to the gaseous phase, it absorbs ΔQ joules of heat. If it is changing in the other direction, the substance must liberate ΔQ joules of heat. For example, to convert 5.00 kg of ice to water at 0°C, the ice must absorb

$$\Delta Q = mL_f = (5.00 \text{ kg})(3.35 \times 10^5 \text{ J/kg}) = 1.68 \times 10^6 \text{ J}$$

On the other hand, to condense 5.00 kg of steam to water at 100°C, the steam must liberate

$$\Delta Q = mL_v = (5.00 \text{ kg})(2.26 \times 10^6 \text{ J/kg}) = 1.13 \times 10^7 \text{ J}$$

The graph of Figure 15-3 is dominated by region IV, representing the vaporization of the water. If a kilogram of ice at $-50°C$ is converted to steam at $+200°C$, we find that the total heat delivered is 3.32×10^6 J and that nearly 70 percent of this heat is used to convert the water at $100°C$ to steam at $100°C$. Whenever water boils or steam condenses, 2.26×10^6 J of heat must flow for every kilogram which changes phase. This implies that per unit mass, steam at $100°C$ is a great deal more dangerous to people than water at $100°C$. Although contact with either can be quite unpleasant, the heat absorbed by a person's hand placed in boiling water is due to only the temperature difference, but if the hand is in contact with steam at $100°C$, it must also absorb the latent heat as the steam condenses. To put this number in perspective, recognize that the heat liberated by a kilogram of steam just condensing to water is about one-twelfth the heat liberated by the complete combustion of a kilogram of coal.

An important benefit of the latent heat of vaporization is evaporative cooling. Since every kilogram of water which evaporates must absorb at least 2.26×10^6 J, moisture on the human body (as in perspiration) will take away a great deal of heat as it evaporates. This is extremely important to the process of human body temperature control.

Example 15-3 A 2.00×10^{-2} kg ice cube at $0°C$ is dropped into a vacuum bottle originally holding 0.400 kg of water at $35°C$. Determine the final temperature after thermal equilibrium is attained. Assume that any loss or gain of heat by the vacuum bottle is negligible.

Solution The mass of ice gains heat as it melts to become water at $0°C$. This amount of water (previously ice) must then gain heat as it is warmed to the final temperature. It gains this heat from the original 0.400 kg of water which loses heat as it cools from $35°C$ to the final temperature. Since we are to ignore heat flow to or from the vacuum bottle, the net heat flow must be zero. That is, if ΔQ_1 is the heat necessary to melt the ice, ΔQ_2 is the heat to warm the melted ice to its final temperature, and ΔQ_3 is the heat flow from the original 0.400 kg of water, then

$$\Delta Q_1 + \Delta Q_2 + \Delta Q_3 = 0$$

But, the heat to melt the ice is just

$$\Delta Q_1 = m_i L_f = (2.00 \times 10^{-2} \text{ kg})(3.35 \times 10^5 \text{ J/kg}) = 6.70 \times 10^3 \text{ J}$$

and the heat to warm the resulting ice water, initially at $0°C$, to a final temperature T is just

$$\Delta Q_2 = cm_i \, \Delta T_2 = [4.180 \times 10^3 \text{ J/(kg} \cdot °\text{C)}](2.00 \times 10^{-2} \text{ kg})(T - 0)°C$$
$$= 83.7T \text{ J}$$

where T, the final temperature of the mixture, is unknown. Note that we used the specific heat of water since the ice is now melted. Finally, the heat surrendered by the warm water as it cooled to the final temperature T is

$$\Delta Q_3 = cm_w \, \Delta T_3 = [4.180 \times 10^3 \text{ J/(kg} \cdot °\text{C)}](0.400 \text{ kg})(T - 35)°C$$
$$= 1.67 \times 10^3 T - 5.86 \times 10^4 \text{ J}$$

Thus, substituting into our original equation, we have

$$6.70 \times 10^3 + 83.7T + 1.67 \times 10^3 T - 5.86 \times 10^4 = 0$$
$$1.75 \times 10^3 T = 5.19 \times 10^4$$
$$T = 29.6°C$$

The previous example is an illustration of **calorimetry**, the measurement of the thermal-energy transfer between objects or systems. Various special devices called **calorimeters** are designed to be used to make such heat-transfer measurements. A typical laboratory calorimeter is shown in Figure 15-4. A metal cup is supported inside a larger vessel by a plastic or fiber disk. The air space between the cup and the outer vessel along with the chosen nonconducting properties of the disk ensures that the cup is fairly well insulated from its environment. When water is placed in the cup and heat is added, part of the heat goes to the water and part to the cup. When a hot body is added to the cup, the whole contents will come to thermal equilibrium at some final temperature. It is assumed by what is called the method of mixtures that the heat given up by the hot body in reaching the final temperature is equal to that absorbed by the cup and its original contents. Thus the net flow of thermal energy is zero.

Example 15-4 A 3.00×10^{-2} kg ice cube at a temperature of $-25°C$ is added to a 0.200-kg copper calorimeter cup containing 0.500 kg of water at 50°C. The mixture comes to thermal equilibrium at a final temperature of 42°C. Calculate the value of the latent heat of fusion of ice these experimental data yield.

Solution We assume the method of mixtures holds and that the net heat transfer is zero. The cup and original contents lose heat as they cool from 50 to 42°C. Meanwhile, the ice cube gains heat as its temperature is increased from -25 to 0°C. It gains more heat as it melts at 0°C. Finally, the melted ice (now water) gains heat as its temperature is increased from 0 to 42°C. We must write expressions for each of these heat transfers and add them, setting the result to zero. Taking ΔQ_1 and ΔQ_2 as the heat flows from the calorimeter cup and original water, respectively, we have

$$\Delta Q_1 = c_{copper} m_{cup} \Delta T_{cup}$$
$$= [384 \text{ J/(kg} \cdot °C)](0.200 \text{ kg})(42 - 50)°C$$
$$= -614 \text{ J}$$

$$\Delta Q_2 = c_{water} m_{water} \Delta T_{water}$$
$$= [4180 \text{ J/(kg} \cdot °C)](0.500 \text{ kg})(42 - 50)°C$$
$$= -1.67 \times 10^4 \text{ J}$$

Now if we take ΔQ_3 as the heat gained by the ice as it warms to 0°C, ΔQ_4 as the heat used to melt it, and ΔQ_5 as the heat the melted ice gains as it warms to 42°C, we have

$$\Delta Q_3 = c_{ice} m_{ice} \Delta T_{ice}$$
$$= [2090 \text{ J/(kg} \cdot °C)](0.0300 \text{ kg})[0 - (-25)]°C$$
$$= 1.57 \times 10^3 \text{ J}$$

$$\Delta Q_4 = m_i L_f = (0.0300 \text{ kg})(L_f \text{ J/kg})$$
$$= 0.0300 L_f \quad \text{J}$$

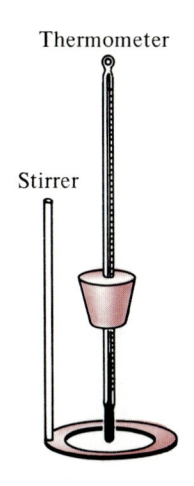

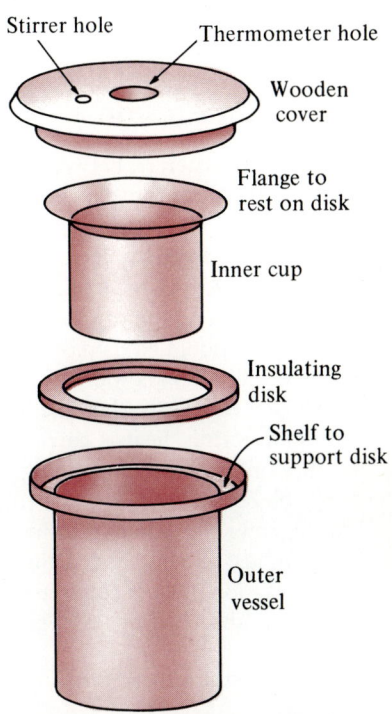

Figure 15-4 Exploded view of a calorimeter. Heat exchange occurs in the inner cup, which is isolated by the rest of the apparatus to minimize heat gain or loss. The inner cup and its contents approximate an isolated thermal system for which $\Delta Q_{tot} = 0$. Thus, any part of the system gaining heat does so with heat gained from other members of the system.

$$\Delta Q_5 = c_{\text{water}} m_{\text{water}} \Delta T_{\text{ice water}}$$
$$= [4180 \text{ J/(kg} \cdot {}^\circ\text{C})](0.0300 \text{ kg})(42-0){}^\circ\text{C}$$
$$= 5.27 \times 10^3 \text{ J}$$

Thus

$$\Delta Q_1 + \Delta Q_2 + \Delta Q_3 + \Delta Q_4 + \Delta Q_5 = 0$$

becomes

$$(-614) + (-1.67 \times 10^4) + (1.57 \times 10^3) + 0.03L_f + 5.27 \times 10^3 = 0$$
$$-1.05 \times 10^4 + 0.03L_f = 0$$
$$L_f = 3.50 \times 10^5 \text{ J/kg}$$

Can you think of a reason for the discrepancy between this experimental value and the accepted value of $L_f = 3.35 \times 10^5$ J/kg?

15-4 Conduction

Heat may be transferred between bodies by several methods. As an analogy, suppose you wish to transfer a book to the last person in the row from your place up front. Three options are apparent: (1) Pass the book to the person behind you, with instructions to continue passing it until it reaches the back — *conduction;* (2) give the book to a passing student walking up the aisle to deliver at the rear — *convection;* or (3) throw the book to the rear seat — *radiation.*[1]

In conduction, heat is passed from molecule to molecule through a substance, much as the book was passed from person to person down the row. Note that the heat is transferred but that the molecules themselves retain their places.

In convection, heat is carried from its source to its destination by the actual motion of the matter doing the carrying, just as the person (matter) carrying the book physically moved down the aisle to deliver the book to the end of the row. Note that at both ends the heat is picked up or delivered by conduction, just as you handed the person the book.

In radiation, heat is converted to an electromagnetic wave which leaves the source, is absorbed by the receptor of the energy, and requires nothing in between.

We will study these forms of heat transfer in sequence, beginning with conduction.

When you hold a cup filled with hot coffee, heat is transferred to your hand. Likewise, when you hold a cold glass of lemonade, your hand loses heat. In both cases, heat is transferred through the wall of a container of finite thickness (the cup or the glass) from a higher temperature on one side of the container to a lower temperature on the other side. **Conduction** is heat transfer by the passing of energy from layer of molecules to layer of molecules through a substance. The equation which describes a heat transfer due to conduction is derived empirically but may almost be obtained from a little concentrated thought.

---

[1] This analogy is used to give you some feel for the mechanism of heat transfer, but you should recall that heat is not a material body or any kind of substance. **Heat** is transferred thermal energy.

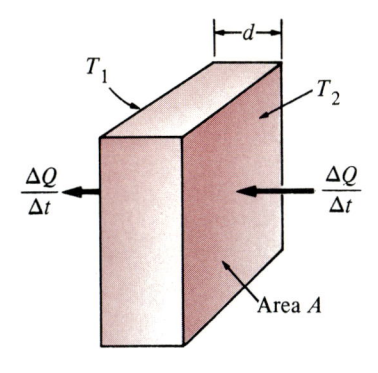

Figure 15-5 Heat transfer by conduction through a block of material is directly proportional to the duration of transfer, the area, and the temperature difference and is inversely proportional to the thickness of the material.

Consider Figure 15-5. The left side of the solid block of material of thickness d and cross section A is held at a fixed temperature T_1, while the right side is held at T_2. If T_2 is greater than T_1, we expect an amount of heat ΔQ to be conducted from right to left through the block. The question is, "How much heat flows?" Well, what might we expect? Surely ΔQ depends on the time; if the conditions shown exist for 20 s, twice as much heat will flow than if they exist for only 10 s. Thus we might write $\Delta Q \propto \Delta t$, where Δt is the time interval. Also we might expect that ΔQ is proportional to the area. If the area were twice its present magnitude, it is reasonable to suspect that twice as much heat would flow. Thus we write $\Delta Q \propto A$. Also, the larger the temperature difference, the more heat we expect to flow. After all, if $T_2 - T_1$ is zero, we would not expect *any* net heat flow. Thus $\Delta Q \propto T_2 - T_1$.

The dependency on thickness is a little harder to see. Consider the situation when the thickness is increased. There would then be a smaller temperature difference between adjacent molecules in the block, and we might expect less heat to flow. In which case, we might write $\Delta Q \propto 1/d$. Combining all these proportionalities, we obtain

$$\Delta Q \propto \frac{A(T_2 - T_1)\Delta t}{d}$$

Finally, the heat flow must depend on the type of material. We use this fact to define the proportionality constant which makes the above proportion an equation, and we have

$$\Delta Q = \frac{kA(T_2 - T_1)\Delta t}{d} \qquad (15\text{-}3)$$

where k is called the **thermal conductivity** of the substance in question and is large for a **good conductor**, like metals, and small for poor conductors, or **good insulators**, like gases. In SI units, the thermal conductivity is expressed in W/(m·K) or in W/(m·°C). The expression for conductive heat flow is often written

$$\frac{\Delta Q}{\Delta t} = \frac{kA(T_2 - T_1)}{d} \qquad (15\text{-}4)$$

where $\Delta Q/\Delta t$ is the rate of flow of energy expressed in watts (J/s). Table 15-2 gives the thermal conductivity for a variety of materials.

TABLE 15-2 Thermal Conductivity of Some Selected Materials near 0°C

Material	Thermal Conductivity k W/(m·°C)	Btu·in/(h·ft²·°F)
Air (20°C)	0.023	0.16
Aluminum	230	1600
Brick (dry)	0.40	2.8
Concrete	0.8	5.6
Copper	400	2800
Fiberglass	0.046	0.32
Germanium	67	470
Glass wool	0.036	0.25
Gold	318	2200
Ice	2.2	15
Iron (or steel)	84	580
Plasterboard	0.16	1.1
Silicon	170	1200
Silver	430	3000
Water (20°C)	0.59	4.1
Window glass	1.0	6.9
Wood (across grain)	0.14	0.97

Example 15-5 (a) How much heat is lost through the outside wall of a dormitory room overnight (8 h) when the average outside temperature is −15°C and the average inside temperature is 22°C? The wall is 2.50 m by 3.00 m, is 0.150 m thick, and has an average thermal conductivity of 0.850 W/(m·K). (b) How much does it cost to replace the heat lost by electric resistive heating at $0.08/kWh, where 1 kWh = 3.60 × 10⁶ J?

Solution (a) This requires a straightforward application of Equation 15-3.

$$\Delta Q = \frac{kA(T_2 - T_1)\Delta t}{d}$$

Since $A = (2.50 \text{ m})(3.00 \text{ m}) = 7.50 \text{ m}^2$ and $t = (8 \text{ h})(3600 \text{ s/h}) = 2.88 \times 10^4$ s, we have

$$\Delta Q = \frac{[0.850 \text{ W/(m·°C)}](7.50 \text{ m}^2)(37°C)(2.88 \times 10^4 \text{ s})}{0.150 \text{ m}}$$

$$= 4.53 \times 10^7 \text{ J}$$

(b) We must determine the number of kilowatthours used and multiply by $0.08.

$$\Delta Q = 4.53 \times 10^7 \text{ J} \times \frac{1 \text{ kWh}}{3.60 \times 10^6 \text{ J}} = 12.6 \text{ kWh}$$

$$\text{Cost} = (\$0.08/\text{kWh})(12.6 \text{ kWh}) = \$1.01$$

(We had only one small wall exposed for 8 h. Consider how this applies to a whole house for a month.)

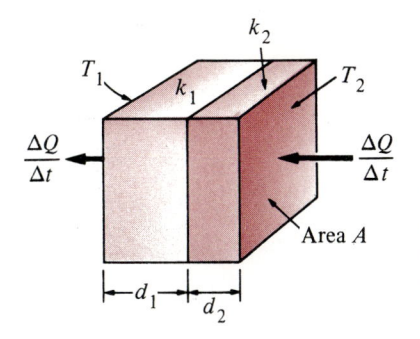

Figure 15-6 Two blocks made of different material and having different thicknesses. The temperature along the surface of contact is some value T intermediate between T_1 and T_2.

Most walls are made of several layers of different materials. These materials may have very different thermal conductivities. This is particularly true when we add some insulating material to decrease heat loss. Even this situation may be solved by using Equation 15-4. Figure 15-6 depicts two such layers. We recognize that once a steady state is reached, the same amount of heat per unit time must flow out of the cold side as flows into the warm side. Further, if we assume that the interface where the two layers meet is at some unknown intermediate temperature T, we may write

$$\frac{\Delta Q}{\Delta t} = \frac{k_2 A (T_2 - T)}{d_2} \quad \text{and} \quad \frac{\Delta Q}{\Delta t} = \frac{k_1 A (T - T_1)}{d_1}$$

We may eliminate the unknown temperature (see Problem 15-28) to obtain

$$\frac{\Delta Q}{\Delta t} = \frac{A(T_2 - T_1)}{d_1/k_1 + d_2/k_2} \qquad (15\text{-}5)$$

If a third layer of material of thickness d_3 and thermal conductivity k_3 is added, the heat flow rate becomes

$$\frac{\Delta Q}{\Delta t} = \frac{A(T_2 - T_1)}{d_1/k_1 + d_2/k_2 + d_3/k_3}$$

Thus we need only add the ratio of the thickness of the new material to its thermal conductivity in the denominator to determine the new heat flow rate due to conduction.

Notice that adding another layer decreases the heat flow rate, just as you might expect. We say that each new layer provides additional insulation. The ratio d/k for each material is a unique measure of its ability to provide additional insulation. Insulating materials are rated and sold by this number. It is perhaps unfortunate that this ratio is most commonly given in the British engineering units of ft$^2 \cdot$ °F $\cdot$ h/Btu, called **R-value,** where R stands for thermal resistance. You may have heard that in extremely cold climates, it is recommended that houses have ceiling insulations of R = 31 and wall insulations of R = 16 or more. Table 15-3 shows typical R-values for some common building and insulation materials. We may convert between British engineering and SI units, knowing that 1 ft$^2 \cdot$ °F $\cdot$ h/Btu = 0.1763 m$^2 \cdot$ °C/W.

TABLE 15-3 Approximate R-Value in British Engineering Units and d/k in SI Units for Some Selected Materials

Material	R-Value, ft²·°F·h/Btu	d/k, m²·°C/W
Brick, common, 4″	0.8	0.14
Carpet, 1″:		
On fibrous underlay	2.1	0.37
On foam rubber	1.2	0.21
Cellulose fiber, 1″	3.7	0.65
Concrete (cinder) block, 8″	1.7	0.30
Fiberglass, 1″	4.0	0.71
Glass, single pane, 1/8″	0.02	0.004
Glass, thermopane	0.44	0.077
Gypsum or plasterboard, 3/8″	0.3	0.05
Mineral fiber batt:		
2 3/4″	7.0	1.2
3 1/2″	11.0	1.9
6 1/2″	19.0	3.3
Mineral (or glass fiber) wool, 1″	3.7	0.65
Polystyrene, expanded, extruded, 1″	5.2	0.91
Styrofoam, 3/4″	4.2	0.74
Still air, 1″	0.17	0.63
Wood siding, 1 1/2″	1.9	0.33
Wood subfloor, 3/4″	0.9	0.16

Example 15-6 The ceiling in a house in Alaska is insulated to have an R-value of 31. On a cold day, the attic above the ceiling has a temperature of 14°F (−10°C), while the inside of the house is maintained at 68°F (20°C). Determine the rate of heat loss through the ceiling if the ceiling has an area of 646 ft² (60.0 m²).

Solution This may be solved by using either British engineering or SI units, depending on whether we want our answer in Btu/h or in watts. Let us stay with SI by first converting our R-value to SI units. Then, since $R = d/k$, we apply Equation 15-4. In SI,

$$d/k = (31)(0.1763 \text{ m}^2\cdot°\text{C/W}) = 5.47 \text{ m}^2\cdot°\text{C/W}$$

$$\frac{\Delta Q}{\Delta t} = \frac{kA(T_2 - T_1)}{d} = \frac{A(T_2 - T_1)}{d/k}$$

$$= \frac{(60.0 \text{ m}^2)[20 - (-10)]°\text{C}}{5.47 \text{ m}^2\cdot°\text{C/W}} = 329 \text{ W}$$

This is about 20 percent of the heat rate loss of the dormitory wall in Example 15-5, even though the ceiling had six times the area. This illustrates the importance of insulation in cold climates.

15-5 Convection

Convection is the process by which heat is transferred from one point to another by the actual transport of warm fluid. We categorize convection as either forced or natural. One example of *forced convection* is the typical hot-air furnace shown in Figure 15-7, where heated air is forced to circulate by a blower. The modern hot-water heating system is another example of forced convection. Heated water is pumped to radiators or fin-covered pipes. The radiator (or hot fins) in turn heats the air near it by conduction. The heat is then transferred around the room by *natural convection,* shown in Figure 15-8. Much of the movement of air near the surface of the earth is due to natural convection. For example, as air near the earth's surface is heated, it becomes less dense and rises. The cooler air moving in to take its place is the origin of certain surface winds.

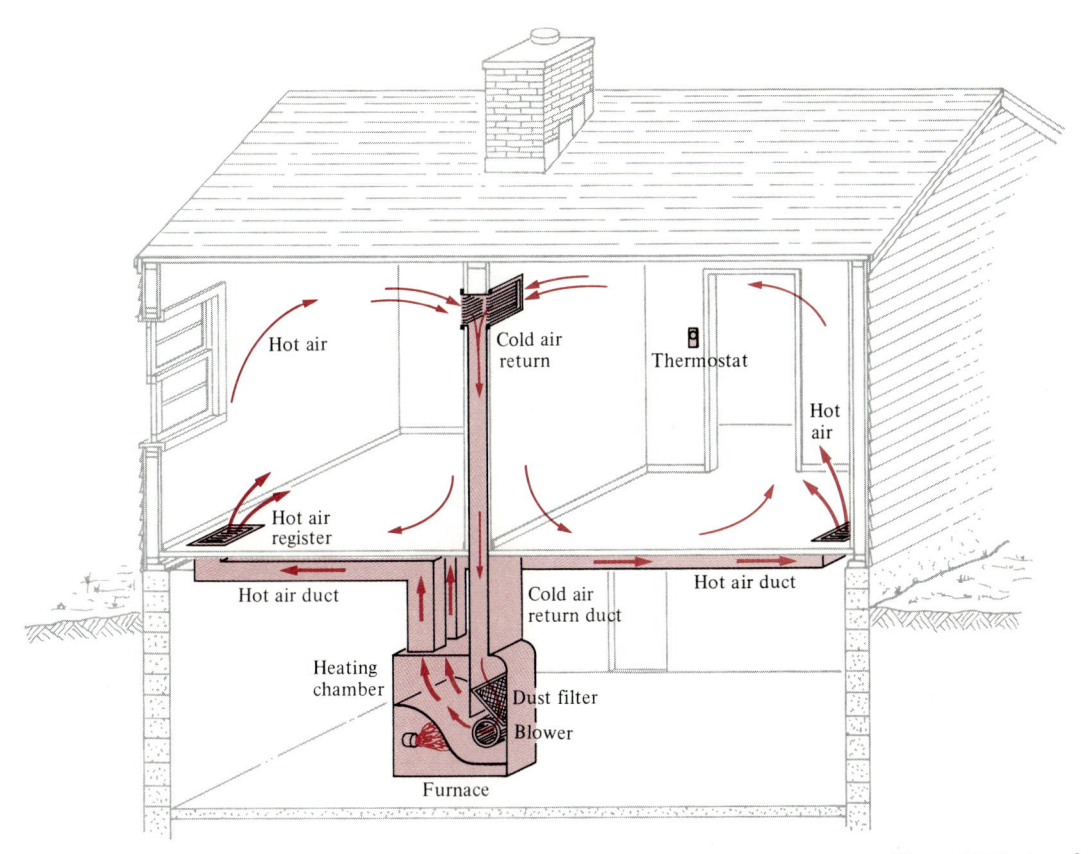

The equation for heat transfer from a moving fluid to a stationary surface is

$$\frac{\Delta Q}{\Delta t} = hA(T_2 - T_1) \tag{15-6}$$

where $\frac{\Delta Q}{\Delta t}$ = heat transferred to or from the surface per unit time
h = convection coefficient
A = area of the stationary surface
T_2 = temperature of the fluid
T_1 = temperature of the surface

Figure 15-7 In a forced-air heating system, a blower forces hot air from the furnace through ducts leading to registers in the floor near the outside wall of each room. The hot air rises, losing heat to the cooler walls and air already in the room. The air's motion, as well as its lower density, causes it to rise along the outer wall and deflect along the ceiling, losing heat. The cooler air descends along the inside wall of the room and returns along the floor to the register, completing a convective loop. Openings toward the center of the house lead to a cold-air duct which returns air to the furnace for rewarming and recirculating.

353

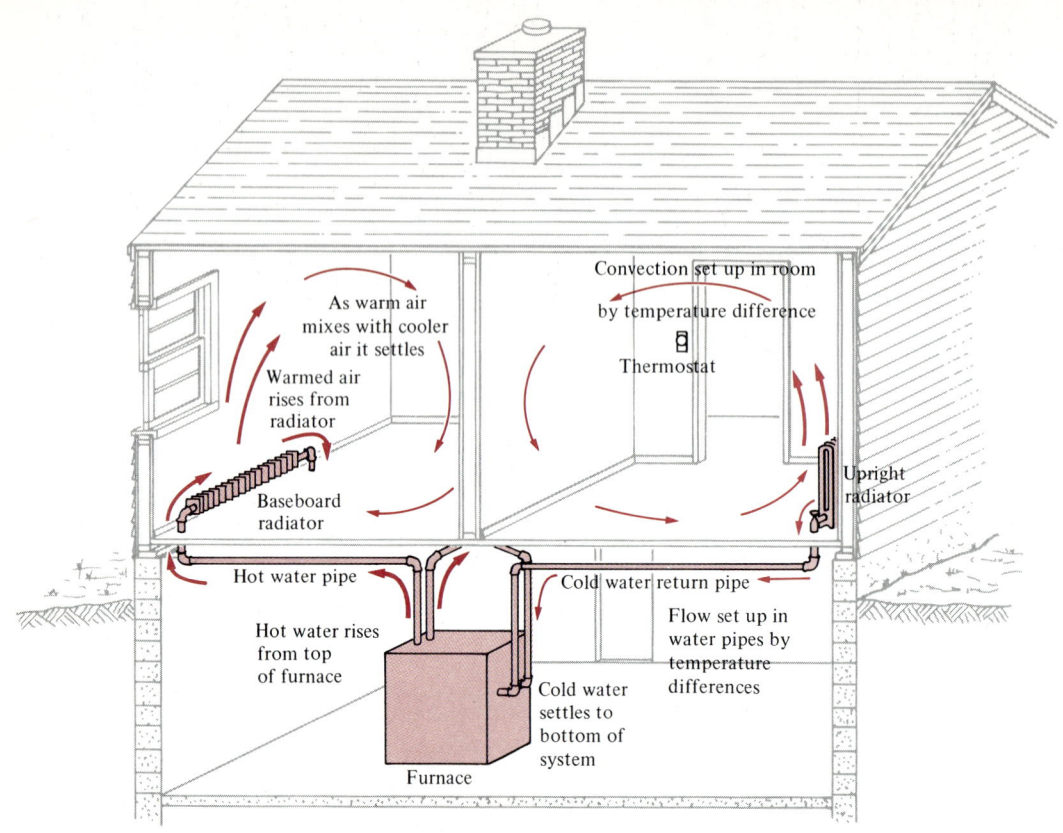

The **convection coefficient** h depends on many variables, such as the shape of the surface (flat or curved), its orientation (horizontal or vertical), type of fluid (gas or liquid), the fluid's characteristics (density, specific heat, thermal conductivity, etc.), the velocity of the fluid, and other factors including possible evaporation, condensation, etc. It may even depend on the magnitude of the temperature difference. The value of the convection coefficient is most often determined experimentally by careful measurement in well-controlled experiments. Many such experiments have been carried out, and handbooks are available which include tabulated values of h. Graphs and nomograms can be used for determining convection coefficients in home heating and cooling situations.

Figure 15-8 A hot-water heating system encloses the combustion chamber of the furnace with a water jacket, where water is heated. Hot, less dense, water rises to the top of the water jacket and is pumped into hot water pipes. The hot water goes to a radiator, where it warms the air by conduction. In the rooms, natural convection of air is established to maintain a warm, livable temperature.

Example 15-7 The inside surface of an outside wall of a house is maintained at a constant temperature of 5°C on a day when the temperature of the outside air is -10°C. How much heat is lost by natural convection from the 8.00 m by 4.00 m wall in a 24-h day? The average convection coefficient is $h = 3.49$ J/(s·m²·°C).

Solution We apply Equation 15-6 and multiply the result by the number of seconds in 24 h.

$$A = (8.00 \text{ m})(4.00 \text{ m}) = 32.0 \text{ m}^2$$

$$\frac{\Delta Q}{\Delta t} = hA\,\Delta T$$

$$= [3.49 \text{ J}/(\text{s} \cdot \text{m}^2 \cdot °\text{C})](32.0 \text{ m}^2)[5 - (-10)]°\text{C}$$
$$= 1.68 \times 10^3 \text{ J/s}$$

where
$$\Delta t = (24 \text{ h})(3600 \text{ s/h}) = 8.64 \times 10^4 \text{ s}$$
Thus
$$\Delta Q = (1.68 \times 10^3 \text{ J/s})(8.64 \times 10^4 \text{ s}) = 1.45 \times 10^8 \text{ J}$$

15-6 Radiation

As you stand in front of a roaring fire in a fireplace with a glass enclosure, you feel warmth from the fire. You also know that you can stand outside on a clear day and be warmed by the sun. The mechanism by which heat is transferred in each case is not conduction (you are not in contact with the fire or the sun) or convection (fluid does not pass from the fire or the sun to your body). **Radiation** is the process by which all bodies emit heat in the form of electromagnetic waves.

Electromagnetic waves, first mentioned in Chapter 11, include waves of visible light, radio, radar, and more. Electromagnetic waves are categorized in part by their wavelength. Figure 15-9 gives a rough idea of the major divisions by wavelength in what is called the electromagnetic spectrum.

Consider sunlight falling on a surface. If all the light is absorbed and *none* is reflected, the surface appears to be flat black. Conversely, if all the sunlight falling on a surface is randomly reflected and none is absorbed, the surface appears to be perfectly white. We define a perfect **blackbody** as a perfect absorber and emitter of electromagnetic radiation. Figure 15-10 shows the results of measuring the emitted intensity (power per unit area) from a blackbody as a function of the wavelength of the emission for several temperatures. You might notice that most of the thermal energy is emitted in the infrared region nearest to the visible region of electromagnetic radiation. Joseph Stefan made the measurement of the *total* power emitted by a blackbody in 1879 and showed that it was proportional to the fourth power of the absolute temperature. Five years later, Ludwig Boltzmann derived the same relationship from theory. The relationship is known as the **Stefan-Boltzmann law** and is written

$$\left.\frac{\Delta Q}{\Delta t}\right|_{\text{emitted}} = \sigma A T^4 \qquad (15\text{-}7)$$

where $\frac{\Delta Q}{\Delta t}$ = radiated, or emitted, power

A = area of the blackbody

T = body's absolute temperature

The quantity σ (Greek lowercase sigma) is known as the **Stefan-Boltzmann constant** and in SI units has the value

$$\sigma = 5.67 \times 10^{-8} \text{ W/(m}^2 \cdot \text{K}^4)$$

Although all bodies are not perfect emitters, it requires only a slight modification to Equation 15-7 to apply the relationship to all bodies. Thus for any body we write

$$\left.\frac{\Delta Q}{\Delta t}\right|_{\text{emitted}} = \epsilon \sigma A T^4 \qquad (15\text{-}8)$$

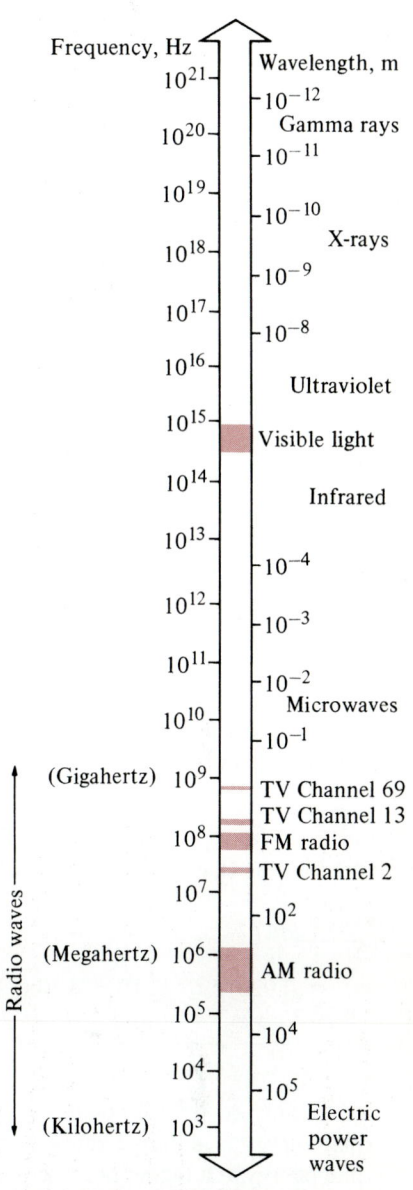

Figure 15-9 The electromagnetic spectrum varies continuously from very long, low-frequency waves (bottom) to extremely short, high-frequency waves (top). Bodies at ordinary temperatures emit (radiate) most of their power in the infrared region. It is common practice, therefore, to call radiation from the infrared portion of this spectrum **thermal radiation.**

355

where ϵ is a quantity called the **emissivity** of the body. The emissivity is a measure of how near the body is to being a perfect emitter. Its value lies between 0 and 1 and depends on the material and state of the surface. Shiny surfaces have smaller values of ϵ than rough surfaces. A perfect reflector has $\epsilon = 0$, while a perfect absorber has $\epsilon = 1$.

Equation 15-8 is known as **Stefan's law** and applies to all objects and bodies including the human body. ($\epsilon = 0.98$ for the human body at thermal wavelengths and is remarkably independent of skin color.) All bodies are both emitting *and* absorbing radiant energy all the time according to Equation 15-8. A body's net gain of thermal power due to radiation is equal to the difference between the radiant energy absorbed and the radiant energy emitted. For example, a person whose skin temperature is T_1 sitting in a room where all the walls, floor, and ceiling are at a temperature T_2 would gain thermal power at a rate

$$\left.\frac{\Delta Q}{\Delta t}\right|_{gain} = \epsilon \sigma A T_2^4 - \epsilon \sigma A T_1^4$$

$$= \epsilon \sigma A (T_2^4 - T_1^4) \qquad (15\text{-}9)$$

Of course, if the temperature of the surroundings T_2 is lower than the person's skin temperature T_1, Equation 15-9 would yield a negative number, indicating a net loss.

Note in Equation 15-9 that the temperature difference between bodies, ΔT, may *not* be used. $T_2^4 - T_1^4$ is *not* ΔT^4, and this constitutes the most common student error in applying the Stefan-Boltzmann relationship in heat transfer. The radiation emitted by the emitter depends on its absolute temperature, not on the temperature of a possible absorber.

Example 15-8 Determine the electric power which must be supplied to the filament of a light bulb operating at 3.00×10^3 K. The total surface area of the filament is 8.00×10^{-6} m², and its emissivity is 0.920.

Solution We apply Stefan's law, Equation 15-8, to determine the power radiated by the filament. This must be supplied electrically.

$$\frac{\Delta Q}{\Delta t} = \epsilon \sigma A T^4$$

$$= (0.920)[5.67 \times 10^{-8} \text{ W/(m}^2 \cdot \text{K}^4)](8.00 \times 10^{-6} \text{ m}^2)(3.00 \times 10^3 \text{ K})^4$$

$$= 33.8 \text{ W}$$

Not only do separate bodies which are at different temperatures radiate thermal energy at different rates, but one part of a single body may radiate thermal energy at a higher or lower rate than another part because of nonuniform temperature distribution. Because most of the thermal radiation is in the infrared region, the temperature distribution on the surface of an object can be determined by measuring the intensity of the infrared radiation from that object. This may be done by either of two methods: (1) A photograph may be taken with a camera using infrared-sensitive film. (2) A special infrared camera similar to a television camera may take pictures and display them on a matching television screen. Higher-temperature areas give off more radiation and show up lighter on the picture or screen. If they are properly calibrated, these thermal pictures, called **thermograms,** can be used to measure absolute temperature levels of objects or parts of objects. More commonly, thermograms are used qualitatively. Figure 15-11 is a thermogram of a building. White areas show the so-called hot spots, where most of the heat is lost by radiation. Thermograms are also an important diagnostic tool for medicine (Figure 15-12). The variation

Figure 15-10 The intensity of emitted electromagnetic radiation vs. wavelength for a perfect blackbody at three temperatures. The temperatures are for the range of a pool of molten metal (2000 K) to a star somewhat cooler than the sun (4000 K). By a change of scale, they could also approximate the relative curves of a block of dry ice (200 K), the human body (300 K), and a hot griddle (400 K), although none of these is a perfect blackbody.

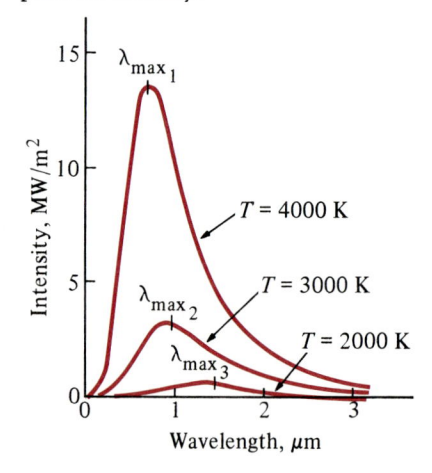

in skin temperature over the human body will show as lighter or darker areas on a thermogram. Many abnormalities may be indicated by a thermogram of the human body. For instance, any condition which restricts blood flow may show as a darker area, and cancerous tumors which are growing rapidly or any area where metabolic activity is higher than average will show as lighter spots.

15-7 Evaporation and Humidity

On the human body, some of the water molecules of perspiration gain enough kinetic energy in collision with others to break away and leave the skin. The molecules which leave are the most energetic and take away a great deal of heat. The amount of energy taken away per unit mass of water evaporated is equivalent to that required to increase the temperature of the water to the boiling point and then have the water absorb the latent heat of vaporization. Thus the net heat absorbed per unit mass is

$$\frac{\Delta Q}{\Delta m} = c\,\Delta T + L_v \qquad (15\text{-}10)$$

Figure 15-11 A thermogram of a building showing areas of heat loss may be used to make alterations designed to conserve energy. (*Daedalus*)

Assuming the surface temperature of the skin is about 30°C (303 K), we see that

$$\frac{\Delta Q}{\Delta m} = [4.18 \times 10^3 \text{ J/(kg·K)}](70 \text{ K}) + 2.26 \times 10^6 \text{ J/kg}$$

$$= 2.55 \times 10^6 \text{ J/kg}$$

This is quite a lot of energy being taken away for each kilogram of water evaporated.

We see the truth of the statement that "evaporation is a cooling process." The water which evaporates must have absorbed at least the latent heat of vaporization from the pot or glass or pond or skin or whatever body it escaped. Thus evaporation reduces the average thermal energy of the molecules left behind. One example is that of canteens often carried by hikers, campers, or soldiers. Normally, water is kept in the metal canteen, which is covered by a cloth jacket (usually made of canvas). The canteen is filled with water and the cloth jacket soaked before a journey. As the water on the jacket evaporates, at least a portion of the latent heat is drawn from the water inside the canteen.

Figure 15-12 Thermograms in medical diagnoses. Lighter areas may indicate higher than normal metabolic activity, whereas darker areas may be associated with restricted blood flow. Thermography has the advantage over x-rays in that it does not expose the patient to any additional radiation. (*A. Tsiaras/Science Source*)

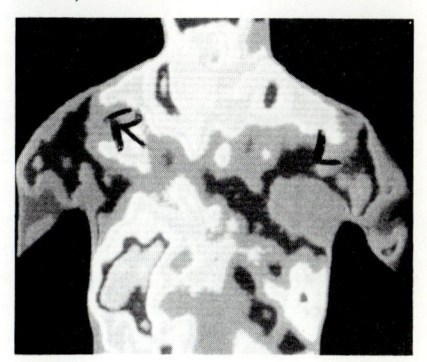

Example 15-9 At the beginning of a planned hike, a U.S. Army recruit fills his canteen with 1.00 kg of water at 30°C. He then soaks the canvas canteen jacket in water from the same source and snaps it firmly into place, covering the canteen. The canvas absorbs 0.150 kg of water. If the water on the jacket evaporates during the first part of the hike by drawing 10 percent of the necessary energy from the water in the canteen, what is the temperature of the water in the canteen just as the jacket gets dry? (Ignore the effect of the metal canteen.)

Solution We first must determine the amount of heat energy absorbed by the evaporating water. Then, realizing that one-tenth of this is supplied by cooling the water inside the canteen, we determine the change in temperature of that water as it delivers that heat to the jacket. From the change in temperature, we may determine the final temperature of the contents of the canteen.

From the results of applying Equation 15-10 to 30°C water (above), we have

$$\Delta Q = 2.55 \times 10^6 \, \Delta m$$
$$= (2.55 \times 10^6 \text{ J/kg})(0.150 \text{ kg})$$
$$= 3.83 \times 10^5 \text{ J}$$

$$\Delta Q_{contents} = (0.1)(\Delta Q) = 3.83 \times 10^4 \text{ J}$$

This cools the contents of the canteen, since

$$\Delta Q_{contents} = cm \, \Delta T$$
$$3.83 \times 10^4 \text{ J} = [4180 \text{ J/(kg} \cdot \text{°C)}](1.00 \text{ kg})(\Delta T)$$
$$\Delta T = 9.16\text{°C}$$

Thus the final temperature of the contents of the canteen is

$$T = 30\text{°C} - 9.16\text{°C} = 20.8\text{°C}$$

People who have used canteens such as those described know that sometimes they do not work so well. Furthermore, we are nearly all aware that there are times when our human cooling system does not seem to work very well. This is especially true during what is often described as a muggy summer day. We might guess that our perspiration is not evaporating quite so rapidly. That is true, but why not? The answer has to do with the ability of the air to accept additional moisture and a term called *relative humidity*.

When a liquid such as water is put in an open bowl and the bowl is placed in a closed container, some of the water will begin to evaporate. In general, the water vapor will not continue to form indefinitely. On occasion, some of the water vapor molecules which are moving about above the liquid phase will strike the liquid surface and be captured. In due time, there will be just as many molecules returning to the liquid phase as there are leaving it. At this point, we say that the air in the container is **saturated** with water vapor and that the air is holding as much water as is normally possible. The term **humidity** is used to describe the water vapor content of the atmosphere. Formally, the **relative humidity** (RH) is the ratio of the density of water vapor actually present in the air to the density of water vapor possible when the air is saturated. The relative humidity is most often multiplied by 100 and reported as a percentage. In equation form, this may be written as

$$RH = \frac{\rho}{\rho_s} \times 100\% \qquad (15\text{-}11)$$

where RH = relative humidity
 ρ = actual vapor density
 ρ_s = vapor density at saturation

Note that any system of units may be used for the vapor densities so long as they are the same for both. RH is a dimensionless quantity. The higher the relative humidity, the less additional water vapor may join the air. In fact, when the relative humidity is 100%, there are on the average just as many water molecules condensing and returning their heat to a surface as there are evaporating and taking heat away, and evaporation ceases. It is not surprising that hot weather causes a great deal more discomfort when the humidity is high, and the statement that "it's not the heat; it's the humidity" has a large element of truth.

TABLE 15-4 Density of Water Vapor and Partial Pressure of Water Vapor in Saturated Air for a Limited Range of Temperatures

Temperature t_c, °C	Density of Water Vapor at Saturation ρ_s, g/m³	Partial Pressure due to Water Vapor P_s, mmHg
−20	0.89	0.783
−16	1.28	1.142
−12	1.81	1.644
−8	2.54	2.340
−4	3.52	3.291
0	4.84	4.58
4	6.33	6.10
8	8.22	8.04
12	10.57	10.52
16	13.51	13.64
20	17.12	17.55
24	21.55	22.40
28	26.93	28.38
32	33.45	35.70

Relative humidity is very strongly dependent on temperature, as we see in Table 15-4. Just as high humidity stops us from losing enough heat in hot weather, the opposite problem also occurs in cold weather. If the humidity is very low, we may lose heat to evaporation much faster than we would like. Thus we might prefer to maintain a rather low relative humidity when the weather is very warm and a rather high humidity during colder weather. Of course, we cannot control the outside weather (at least not at present), but we can and often do influence humidity inside houses and other living spaces. Air conditioners and dehumidifiers remove water vapor from the air, while humidifiers inject water vapor into the air.

The importance of humidifying inside air in the wintertime can be illustrated rather easily. Suppose the outside air on a moderately cold day is −4°C (approximately 25°F) and is saturated with water vapor. That is, the outside air has a relative humidity of 100%. We see from Table 15-4 that this air holds 3.52 g/m³ of water. If that air is brought inside and heated to 20°C (68°F), it can now hold 17.12 g/m³. The air expands to about 1.09 times its original volume according to Charles' law. Thus the heated air holds only (3.52/1.09)g/m³ = 3.23 g/m³. Its relative humidity is now RH = (3.23/17.12) × 100 = 18.9%. At a relative humidity of 18.9%, moisture would evaporate from our skin rather rapidly, and we may get a chill. It is generally conceded that the optimum value of relative humidity for both health and comfort is between 40 and 50%. This allows the body's cooling system to work as necessary while maintaining enough moisture to prevent drying of the skin and mucous membranes.

As a final point on relative humidity, we note that at a given temperature, the density of water vapor is directly proportional to that portion of the atmospheric pressure, the **partial pressure,** due to the water vapor (see Problem 15-38). Thus we often determine the relative humidity from

$$\text{RH} = \frac{P}{P_s} \times 100\% \tag{15-12}$$

where P is the actual partial pressure of the water vapor and P_s is the partial pressure of water vapor at saturation.

Example 15-10 The inside of a large two-story house has a volume of 500 m³. On a cold day it is heated to 24°C with air that has been drawn from the outside at −8°C. If the outside air is saturated with water vapor, determine (a) the relative humidity in the house when the humidifier is not working, (b) the amount of water which must be added to make the relative humidity equal to 45%, and (c) the partial pressure of the water vapor before the additional water is added.

Solution (a) We first determine the density of water vapor in the air after it has been heated. We find the new volume of a cubic meter of air which has been heated from −8 to 24°C by applying Charles' law. We assume the pressure is constant. Thus

$$\frac{V_2}{T_2} = \frac{V_1}{T_1}$$

$$V_2 = V_1 \frac{T_2}{T_1} = 1.00 \text{ m}^3 \times \frac{297 \text{ K}}{265 \text{ K}} = 1.12 \text{ m}^3$$

Therefore, the density of water vapor in the house is equal to the number of grams of water vapor held by a cubic meter at −8°C divided by the new volume after the air has been heated to 24°C:

$$\rho = \frac{2.54 \text{ g}}{1.12 \text{ m}^3} = 2.27 \text{ g/m}^3$$

Thus

$$\text{RH} = \frac{\rho}{\rho_s} \times 100 = \frac{2.27}{21.55} \times 100 = 10.5\%$$

(b) To obtain a relative humidity of 45%, the air must have 45% of the saturated water vapor density at 24°C.

$$\rho_{45} = (0.45)(21.55 \text{ g/m}^3) = 9.70 \text{ g/m}^3$$

Thus we must add the difference between the water vapor density at 45% saturation and the actual water vapor density.

$$\Delta \rho = 9.70 \text{ g/m}^3 - 2.27 \text{ g/m}^3 = 7.43 \text{ g/m}^3$$

The mass of water which must be added is the product of the necessary increase in density and the total volume in the house.

$$m = (\Delta \rho)V = (7.43 \text{ g/m}^3)(500 \text{ m}^3) = 3.72 \times 10^3 \text{ g} \quad \text{or } 3.72 \text{ kg}$$

This is 3.72 L or over 1 gal of water. If the air is changed several times per hour, as is often recommended, you can see that the humidifier will have to put gallons and gallons of water vapor into the air every day.
(c) From Equation 15-12, we have that

$$P = (\text{RH})(P_s) = (0.105)(22.40 \text{ mmHg}) = 2.35 \text{ mmHg}$$

15-7 Evaporation and Humidity

Example 15-11 A certain lake is 2.00 km long by 1.00 km wide. Both the lake water and the air are at 20°C at 8 A.M. when a constant wind begins carrying the air across the short axis of the lake at 4.00 m/s. Before encountering the lake, the air has a constant relative humidity of 20%. After crossing the lake, the bottom 20.0 m of air has risen in relative humidity to 25%. The wind lasts for 8 h. Assuming no other heating or cooling occurs, (a) what mass of water is evaporated from the lake during the 8-h period? (b) If the lake averages 10.0 m deep, what percentage of its water is this? (c) Assuming the water temperature remains uniform throughout its depth, what is the final temperature of the lake after the 8-h period? (d) What is the average temperature of the 20-m-high layer of air coming off the lake?

Solution (a) We must first obtain the moisture added to a single cubic meter of air. From Table 15-4, at 20°C, ρ_s for air = 17.12 g/m³. Twenty percent of this is 3.424 g/m³, and 25% of this is 4.28 g/m³. Thus, the air in crossing the lake gains

$$(4.28 - 3.424) \text{ g/m}^3 = 0.856 \text{ g/m}^3$$

of water vapor from the lake. The 20.0-m-tall layer of air affected is 2.00 km wide and moving at 4.00 m/s. Thus the volume of air leaving the lake per unit time is

$$\frac{\Delta V}{\Delta t} = vwh = (4.00 \text{ m/s})(2000 \text{ m})(20 \text{ m}) = 1.60 \times 10^5 \text{ m}^3/\text{s}$$

which has gained a mass of water amounting to

$$\frac{\Delta m}{\Delta t} = \frac{\Delta V}{\Delta t} \Delta \rho = (1.60 \times 10^5 \text{ m}^3/\text{s})(0.856 \text{ g/m}^3)$$
$$= 1.37 \times 10^5 \text{ g/s} = 1.37 \times 10^2 \text{ kg/s}$$

or in 8 h,

$$\Delta m = \frac{\Delta m}{\Delta t} \Delta t = (1.37 \times 10^2 \text{ kg/s})(8 \text{ h})(3600 \text{ s/h})$$
$$= 3.95 \times 10^6 \text{ kg}$$

Since 10^3 kg is a metric ton, this implies that the lake loses about 4000 tons of water in the 8 h.

(b) The total volume of the lake is

$$V_{\text{lake}} = lwh = (2000 \text{ m})(1000 \text{ m})(10.0 \text{ m}) = 2.00 \times 10^7 \text{ m}^3$$

Since water has a density of 10^3 kg/m³, the total mass of the lake water is

$$m = V\rho = (2.00 \times 10^7 \text{ m}^3)(10^3 \text{ kg/m}^3) = 2.00 \times 10^{10} \text{ kg}$$

The water lost as a percentage is

$$\frac{\Delta m}{m_{\text{tot}}} \times 100 = \frac{3.95 \times 10^6 \text{ kg}}{2.00 \times 10^{10} \text{ kg}} \times 100 = 1.98 \times 10^{-2}\% = 0.0198\%$$

This is about one five-thousandth of the lake's volume. Expressed differently, with no input of water, the lake could lose moisture at this rate for 5000 days before completely drying up.

(c) From Equation 15-10, we may calculate the heat lost by evaporation of the water:

$$\frac{\Delta Q}{\Delta m} = c\,\Delta T + L_v$$
$$= [4.18 \times 10^3 \text{ J/(kg·K)}](80 \text{ K}) + (2.26 \times 10^6 \text{ J/kg})$$
$$= 2.59 \times 10^6 \text{ J/kg}$$

Thus the total heat lost is

$$\Delta Q = \Delta m \frac{\Delta Q}{\Delta m} = (3.95 \times 10^6 \text{ kg})(2.59 \times 10^6 \text{ J/kg}) = 1.02 \times 10^{13} \text{ J}$$

and the drop in lake temperature is

$$\Delta T = \frac{\Delta Q}{cm} = \frac{1.02 \times 10^{13} \text{ J}}{[4180 \text{ J/(kg·°C)}](2.00 \times 10^{10} \text{ kg})}$$
$$= 1.22 \times 10^{-1}\text{°C}$$

Thus, even with the large volume of water evaporated away, the lake temperature drops only 0.122°C in the 8 h.

(d) The air heated constitutes a band 2 km wide by 20 m high by (4.00 m/s × 8 h × 3600 s/h) long:

$$V_{air} = (2000 \text{ m})(20 \text{ m})(4.00 \times 8 \times 3600 \text{ m}) = 4.61 \times 10^9 \text{ m}^3$$

The heat added per cubic meter will be

$$\frac{\Delta Q}{V} = \frac{1.02 \times 10^{13} \text{ J}}{4.61 \times 10^9 \text{ m}^3} = 2.21 \times 10^3 \text{ J/m}^3$$

Now since the density of relatively dry air at 20°C is approximately 1.21 kg/m³, the air absorbed an amount of heat per unit mass of

$$\frac{\Delta Q}{\Delta m} = \frac{2.21 \times 10^3 \text{ J/m}^3}{1.21 \text{ kg/m}^3} = 1.83 \times 10^3 \text{ J/kg}$$

We may calculate the increase in air temperature from

$$\Delta T = \frac{\Delta Q}{c\,\Delta m} = \frac{1.83 \times 10^3 \text{ J/kg}}{1000 \text{ J/(kg·°C)}} = 1.83\text{°C}$$

Thus the air should leave the lake at 21.83°C.

SPECIAL TOPIC
Keeping Your Cool

In Chapter 14, we considered the ability of the body to retain heat to avoid a hypothermic reaction. However, it is also true that elevated temperatures in the core of the body are not easily tolerated. Thus, the problem of the body often is to shed excess heat rather than to retain it. During periods of high activity, the body generates internal heat from muscular actions at a considerable rate. A first guess might be that the rate of generating heat might be proportional to body bulk, that is, volume or mass. This is not the case. Late in the last century, it was determined experimentally that the rate of internal heat production, or **metabolism,** of a very wide range of mammals at rest varied as the surface areas of their body. Studies of hundreds of people of diverse body size disclosed the same relationship: metabolism divided by body area tends to be a constant. We may define **basal metabolic rate** as the energy generated by the body while resting (lying down, but awake) divided by body area. For people, this tends to average about 44.7 W/m². While there are small differences by age, sex, general health, etc., more than 90 percent of the population come within 15 percent of this value. The hardest part of conducting such measurements tends to be in obtaining an accurate value for body area. The generally accepted formula for people is

$$A = 0.202 m^{0.425} h^{0.725}$$

where A = body area, m^2
m = body mass, kg
h = individual's height, m

Thus, a 50.0-kg individual of height 1.45 m would have a body surface area of 1.40 m^2 and would generate energy when resting at a rate of

$$\left.\frac{\Delta Q}{\Delta t}\right|_{\text{muscles}} = (1.40 \text{ m}^2)(44.7 \text{ W/m}^2) = 62.6 \text{ W } (\pm 15\%)$$

The human metabolic rate varies with activity. While sleeping, the rate drops to a low of about 41 W/m^2. Escalating activities, like sitting, standing, walking, and running increase the rates to about 58, 70, 160, and 696 W/m^2, respectively. While moderate work requires only slightly more energy than walking, the random muscle twitches of shivering raise the metabolic rate to 290 W/m^2, the same as midlevel exercising, like bicycling. At our most active, we generate energy internally at a rate more than 20 times as great as at our least active times. The body must lose heat at the rate it gains heat from its environment and from the production of internal heat, or, if gains exceed losses, the body temperature will rise.

Consider an extreme example of body heating. Imagine a 50.0-kg 1.60-m tall female picnicking on a reservoir beach in the desert. From her dimensions, we may calculate that she has a body surface area of 1.50 m^2. Imagine her running along the beach in a bathing suit, and let us calculate the heat lost and gained by her body. The air temperature is a hot 40°C (104°F), and she is running at a speed of 4.00 m/s. The sun is up at an angle of 60° above the horizon on a windless, cloudless day.

In general, conduction is a minor source of human heat gain or loss unless areas of bare skin come in direct contact with a good conductor. Grabbing a car's door handle on a winter morning with your bare hand, or the metal spoon in a cup of hot coffee can convince you rapidly that conduction is not minor in such cases. However, in still air, the heat exchange of exposed skin by conduction is minor. In general, the body warms or cools the layer of air immediately against the skin until the layer is at body temperature. The fine hairs on the body then hold this layer against the skin and effectively use it to insulate the body from the external air. Clothes do much the same thing but are even more effective in trapping air layers. The value of clothes in keeping us warm varies directly with their ability to trap and maintain cells of air near the skin. The twisted fibers of wool are good at this, and fine-bristled feathers, like down, are excellent. However, air is such a poor conductor of heat that unless a new layer of air is recycled against the skin continuously, loss to it tends to be small.

When air is continually exchanged at the skin, we are considering convection. In our example, the subject is running through the air. Essentially all her skin is exposed and is continually swept free of its air lining.[1] Thus, we may calculate the heat gained by her body from the warmer environmental air by using Equation 15-6.

$$\left.\frac{\Delta Q}{\Delta t}\right|_{\text{conv}} = hA(T_{\text{air}} - T_{\text{skin}})$$

Here, $A = 1.50$ m^2, $T_{\text{air}} = 40°$C, and $T_{\text{skin}} = 37°$C. The convection coefficient h depends upon wind velocity. As the air speed rises from zero, its value increases rapidly at first and then more slowly. For wind speeds from 1.00 m/s increasing in increments of 1.00 m/s, its value is 15, 21, 24.5, 28, ... W/(m$^2 \cdot$K). Thus, for our running speed, 4.00 m/s, which is the wind speed here, we have $h = 28.0$ W/(m$^2 \cdot$K) and

$$\left.\frac{\Delta Q}{\Delta t}\right|_{\text{conv}} = [28.0 \text{ W/(m}^2 \cdot \text{K)}](1.50 \text{ m}^2)(3 \text{ K}) = 126 \text{ W}$$

While this is a substantial rate of heat gain, it should be pointed out that the most important consideration with convection is usually heat loss in the winter. Skin areas exposed to cold winds lose heat at a very high rate and can be threatened by freezing. This is usually described in terms of a **windchill index.** The windchill index gives the air temperature *in still air* at which convection heat losses are the same as at some given combination of temperature and wind velocity. For example, from Figure 15-13, we can see (dashed lines) that $-10°$F air with a 20 mi/h wind cools naked skin at the same rate as still air of $-52°$F. For comparison, if our runner were exposed under these conditions, the rate of heat loss would be

---

[1] A standing, naked body exposes about 80 percent of its surface for convection or radiation. Curling the body into a ball exposes less. Running in a bathing suit exposes over 95 percent, so we use the entire body surface area in the calculation.

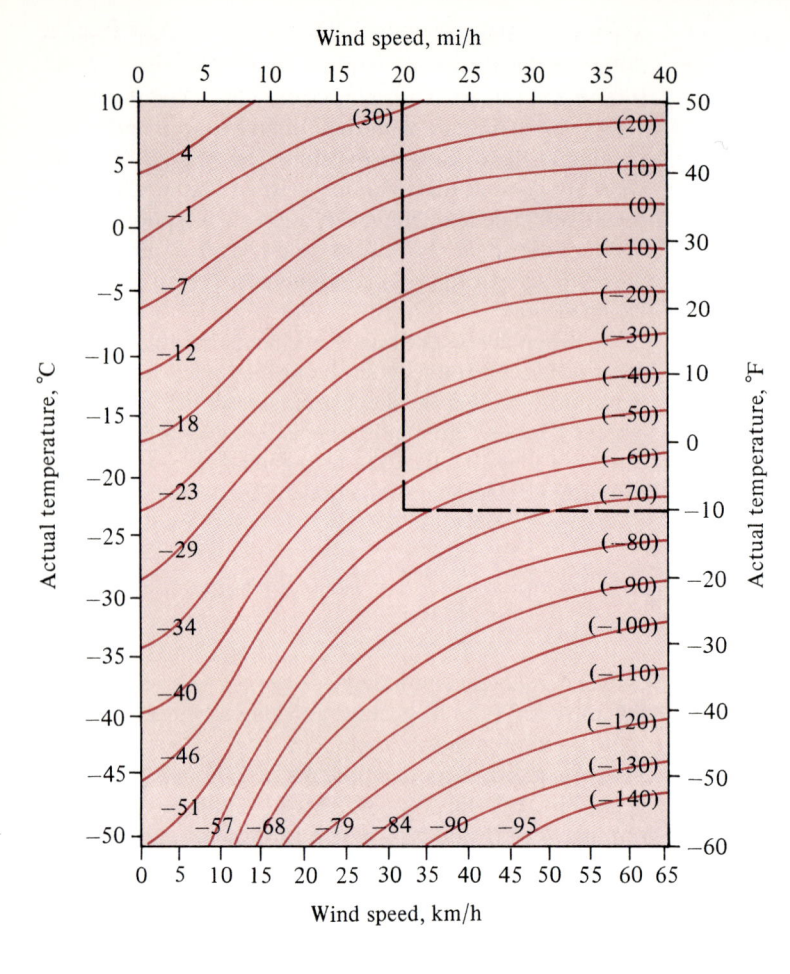

Figure 15-13 The windchill temperature for bare skin as a set of curves on a graph of temperature vs. wind velocity. The colored lines represent the windchill temperature and are labeled in °C (left side and bottom) and °F (right side and top enclosed in parentheses). The windchill temperature for a given set of conditions is found at the intersection of a line representing the actual temperature with a line representing the wind speed. In most cases, we must interpolate (estimate between known values). We can see on the graph that the intersection of the 20 mi/h dashed line with the −10°F dashed line falls between the curves representing windchill temperatures of (−50) and (−60)°F.

$$\left.\frac{\Delta Q}{\Delta t}\right|_{conv} = [37 \text{ W/(m}^2 \cdot \text{K)}](1.50 \text{ m}^2)(-23°\text{C} - 37°\text{C})$$
$$= -3330 \text{ W}$$

or a higher cooling rate than several room air conditioners running simultaneously (which is why few people run around in bathing suits in such conditions).

Half our bather's body is exposed to sunlight at any given time. Sunlight striking a surface perpendicularly delivers energy at a rate of 1390 W/m² at the top of the earth's atmosphere. In cloudless, dry air with few particles, about 900 W/m² makes it to the surface. Our bather is running vertically and the sun is 60° from perpendicular to the flat area of her body, so the energy delivered will be approximately

$$\left.\frac{\Delta Q}{\Delta t}\right|_{sun} = \tfrac{1}{2}(1.50 \text{ m}^2)(900 \text{ W/m}^2)(\cos 60°) = 338 \text{ W}$$

This is *not* the same as the energy absorbed by the body. While the human body is nearly a perfect blackbody for emission or absorption of radiation at normal environmental temperatures, it is *not* so for the radiations of sunlight. Light skin will reflect up to 40 percent of the sun's energy and dark skin 20 percent. Thus, our bather has an emissivity for solar radiations of about 0.60 and absorbs[1]

$$(338 \text{ W})(0.60) = 203 \text{ W}$$

Heat is also gained from the environment by radiation and may be calculated from Equation 15-9.

---

[1] Light-colored clothes also have the effect of reflecting a substantial portion of this solar radiation. They also reduce heat gain by conduction and convection by trapping cooled air layers next to the body. All these are excellent reasons why dwellers of the Sahara wear light woolen robes and did not invent the bikini.

$$\left.\frac{\Delta Q}{\Delta t}\right|_{rad} = \epsilon \sigma A(T_{env}^4 - T_{skin}^4)$$
$$= (1.0)[5.67 \times 10^{-8} \text{ W/(m}^2 \cdot \text{K}^4)]$$
$$\times (1.50 \text{ m}^2)(313^4 - 310^4) \text{ K}^4$$
$$= 30.8 \text{ W}$$

Thus far, our bather has shown heat gain from convection, solar radiation, and environmental radiation. To this we must add the heat generated by the activity of the muscles used in vigorous running. This amounts to $(696 \text{ W/m}^2)(1.5 \text{ m}^2) = 1044 \text{ W}$. Summing all these sources, the total rate of heat gain by the body is

$$\left.\frac{\Delta Q}{\Delta t}\right|_{tot} = \left.\frac{\Delta Q}{\Delta t}\right|_{conv} + \left.\frac{\Delta Q}{\Delta t}\right|_{sun}$$
$$+ \left.\frac{\Delta Q}{\Delta t}\right|_{rad\text{-}envir} + \left.\frac{\Delta Q}{\Delta t}\right|_{muscles}$$
$$= 126 \text{ W} + 203 \text{ W} + 30.8 \text{ W} + 1044 \text{ W}$$
$$= 1404 \text{ W}$$

This is greater than the rate of heat generation of a typical four-slice toaster in action.

To maintain the body core temperature, the body must lose heat at the same rate it is gaining heat. Only one mechanism is left, evaporation. People normally lose heat at a rate of 20 W at all times just due to the moisture lost in breathing and water loss from cells near the surface. In addition, we are equipped with a full, heavy-duty cooling system triggered when the body needs to eliminate heat. Sweat glands cover the body surface with a film of water for evaporation. Most animals do not have such an extensive system as we do. The same conditions in which a person can perform useful work may leave a dog panting in the shade. The dog pants to evaporate moisture from its mouth and tongue to cool itself. People, without a heavy body covering of hair, use most of the entire body surface for the same purpose. The average human surface contains over 2 million sweat glands to cover the surface with perspiration as necessary. To evaporate a kilogram of water from the skin, the water must first absorb energy equivalent to that required to raise it to the boiling point and then enough energy to vaporize it. Of course, evaporation is not a boiling process—little bubbles do not form on the skin and the water temperature never approaches 100°C. Evaporation occurs molecule by molecule, as particular molecules acquire enough energy to join the vapor state and are lost. According to Section 15-7, some 2.55×10^6 J is lost for each kilogram of evaporated perspiration. To dissipate heat at the rate of 1404 W, our running bather will have to perspire away water at a rate of

$$\left.\frac{\Delta Q}{\Delta t}\right|_{tot} = \frac{\Delta Q}{\Delta m}\frac{\Delta m}{\Delta t}$$
$$\frac{\Delta m}{\Delta t} = \frac{(\Delta Q/\Delta t)_{tot}}{\Delta Q/\Delta m} = \frac{1404 \text{ W}}{2.55 \times 10^6 \text{ J/kg}}$$
$$= 5.51 \times 10^{-4} \text{ kg/s}$$
$$= 1.98 \text{ kg/h}$$

This is just about 2 L (or a little more than 2 qt) of water per hour and well within human capability. Heat-adapted individuals can perspire water at a rate of 4 kg/h over short periods. Of course, any perspiration that drips off the body rather than evaporating produces little cooling. The extraordinary capabilities of the human cooling system have been recognized for a long time. An experimenter in 1775 had men stay in an oven at 250°F for 15 minutes without ill effects—a time and temperature chosen because they are sufficient to cook a beefsteak. That's some cooling system you carry around with you!

Minimum Learning Objectives

After studying this chapter, you should be able to:

1. Define:

blackbody	convection	latent heat	specific heat (capacity)
boiling point	convection coefficient	mechanical equivalent of heat	Stefan-Boltzmann constant
Btu [unit]	electromagnetic wave	melting point	Stefan-Boltzmann law
caloric	evaporation	partial pressure	thermal conductivity
calorie (cal) [unit]	heat	phase change	thermal energy
Calorie (Cal) [unit]	heat of fusion	R-value	thermogram
calorimeter	heat of vaporization	radiation	
conduction	humidity	relative humidity	
	infrared	saturated	

2. Convert between SI and commonly used measures of energy delivered and/or properties of materials reported involving heat.
3. Formulate and execute the calculations of calorimetry, where heat is exchanged between two or more thermally confined substances, including situations involving the change of phase of one or more of the substances.
4. Recognize situations which will involve heat transfer and determine its type: conduction, convection, radiation, and/or evaporation.
5. Calculate the heat transferred by conduction, convection, radiation, and/or evaporation, given the appropriate variables and constants (or the means to determine them).
6. Use information commonly supplied on energy billings and with or for building materials to determine heat exchanges and their cost.

Problems

15-1 The professor's office is lighted by four 160-W light fixtures. If 95 percent of the energy is converted into heat, how many calories of heat are produced in the office in an hour?

15-2 A 90.0-kg hiker carries a 20.0-kg backpack. He hikes along a gentle uphill slope such that his average metabolic rate is 350 W, while his average vertical component of motion is 0.180 m/s. Determine (a) how long a time interval will pass before he uses up the fuel from a 1.20×10^3 Cal meal and (b) how high he climbs from the starting point.

15-3 In an experiment similar to Joule's (illustrated in Figure 15-2), a 2.00-kg mass is allowed to fall 4.00 m at a constant speed, turning the paddles in the water. (a) Assuming negligible losses from the system, how many joules of energy should be delivered to the water? (b) If there is 0.400 kg of water present, by how much should the water temperature rise?

15-4 A small ball of lead of mass 25.0 g is dropped from the top of the science building and lands on a sidewalk 20.0 m below. Calculate (a) the kinetic energy just before impact and (b) the rise in temperature of the lead if 50 percent of the kinetic energy is converted to heat (the rest goes to permanent deformation, sound, etc.).

15-5 Upper Yosemite Falls is the highest free-leaping falls in the United States, with an uninterrupted drop of 436 m. (a) Assuming no gain or loss of heat, how much warmer should water at the bottom of the falls be than at the top? (b) What percentage of the water would have to evaporate during the fall to cause enough heat loss to leave the water temperature unchanged? (Assume an initial water temperature of 10°C.)

15-6 An electric power plant generates 1000 MW of electric energy with an efficiency of 40 percent; that is, for every 4.00 W produced in electricity, 6.00 W of waste heat is generated and must be disposed of. The heat is deposited in a stream which is 30.0 m wide and averages 3.00 m deep. If the average velocity of the water is 1.50 m/s, how much warmer is the water downstream from the power plant's heat exchangers than the water reaching the plant upstream? (Neglect the energy given up by the water to radiation and evaporation.)

15-7 (a) How much heat (in joules) is required to raise the temperature of 1 m³ of water by 1°C? (b) How much water at 100°C could be converted to steam at 100°C by this quantity of heat?

15-8 An ethyl-alcohol thermometer consists of 90.0 mm³ of ethanol at 25°C inside a glass bulb with walls 1.00 mm thick having 105 mm² of contact area with the alcohol. (a) How much heat would the alcohol have to absorb to increase in temperature by 1.00°C? (b) If the thermal conductivity of the glass is 0.400 W/(m·°C), about how long after being immersed in a 26°C reservoir will it take for sufficient heat to be transferred through the glass to warm the alcohol? (*Note:* Use an average ΔT of 0.500°C.) (c) The volume coefficient of thermal expansion of the glass is 30.0×10^{-6}/K. By how much will the volume of the glass bulb increase with the 1.00°C temperature rise? (d) The volume coefficient of thermal expansion for ethanol is 1.10×10^{-3}/K. By how much will the volume of the ethanol increase with the 1.00°C temperature increase? (e) What volume of ethanol will be forced to leave the glass bulb following the temperature increase? (f) If it is desired to have the markings for each degree Celsius 0.500 cm apart on the stem of the thermometer, what radius of capillary tube must be used?

15-9 Hurricanes are notable for (among other things) their high winds and the great amounts of rainfall they can deliver in a short time over a large area. Assume that a hurricane deposits 10.0 cm of rain (4 in) over 1.00×10^4 km² of the Gulf Coast. The rain temperature averages 25°C. All this rain originated from the air over the area in question, extending to a maximum altitude of 20.0 km. Assume that the density of the air is constant throughout this volume at the sea level value. If all the energy left with the air by the condensing moisture went into increased kinetic energy of the air *in one particular direction*, what would the speed of this air mass be if it started at a speed of zero before the rain began?

15-10 The volume of all the earth's oceans is 1.32×10^9 km³ (or 1.32×10^{18} m³). The volume of all the earth's ice caps and glaciers is 2.90×10^{16} m³. Assume that the current average ocean temperature is 17°C. If all the glaciers and ice caps melted by absorbing heat from the oceans, what would the resulting average water temperature be? (Of course, if this happened, sea level would rise by nearly 100 m, so if you live anywhere near an ocean, you might not care about thermal effects; you would be too busy swimming.)

15-11 Show that the heat necessary to convert 1.00 kg of ice originally at a temperature of $-50°C$ to steam at a final temperature of $200°C$ is equal to 3.32×10^6 J. That is, calculate the heat required to (a) warm the ice to $0°C$, (b) melt the $0°C$ ice to water at $0°C$, (c) warm the water to $100°C$, (d) vaporize the water to steam at $100°C$, and (e) warm the steam to $200°C$.

15-12 The design of a passive solar house requires that heat be stored while the sun is shining to be released during the night. In a particular design, 1.00×10^{10} J is stored in a material which is kept in large cans and heated from 20 to $45°C$ by the sun during the day. What total volume of cans is required if the material is (a) water or (b) Glauber's salt, which has a density of 1.46×10^3 kg/m^3, a melting point of $32°C$, specific heat of 1.93×10^3 J/(kg·K) in the solid phase and 2.85×10^3 J/(kg·K) in the liquid phase and latent heat of fusion 2.43×10^5 J/kg?

15-13 For health reasons, it is generally not desirable to seal a house too thoroughly. Most houses completely exchange the inside air in 20 to 30 min, which means that the outside air in winter must be continuously warmed to the inside temperature. This presents a problem for a solar heated house at night when the sun is not up. To store heat, heavy masonry solar walls inside the house may be exposed to direct heating by the sun during the day to release this heat at night. Another approach is to store solar heat in the phase change of a salt that melts near room temperature. Suppose such a salt has a density of 2500 kg/m^3 and a heat of fusion of 2.00×10^5 J/kg. The solar house has a volume of 400 m^3 and exchanges the internal air at $20°C$ for external air at $0°C$ twice an hour. How great a mass, and volume, of the salt must be available to store the heat required for warming the new air for 14 h in the winter until the next sunrise? ($\rho_{air} = 1.29$ kg/m^3)

15-14 Snow of average density 100 kg/m^3 accumulates to a depth of 0.200 m on a flat dormitory roof during a winter storm. If the roof measures 40.0 m by 60.0 m and the snow initially at $-15°C$ is melted entirely by heat conducted through the roof from the inside, (a) how much heat must be supplied by the furnace of the dormitory? (b) The dormitory has a natural gas–fired furnace. How much does it cost to melt the snow if natural gas costs \$5.00/1000 ft^3 and 1000 ft^3 supplies 1.09×10^9 J?

15-15 In the early part of the twentieth century, much of home refrigeration (where it existed) used an icebox. A large block of ice was purchased periodically from an iceman who placed the block in the top compartment of the icebox. Perishable food was then stored in the bottom compartment. The top compartment had an external area of approximately 1.50 m^2. Its walls were 5.00×10^{-2} m thick and filled with sawdust, so the walls had an effective thermal conductivity of 0.0800 W/(m·°C). If the interior of the box was at $8°C$ and the exterior at $25°C$, how many kilograms of ice would melt in 1 day?

15-16 Lake Superior is 580 km long and averages 150 km wide. During the week, an icy wind blows south from Canada, converting the lake's surface from water at the freezing temperature to an ice layer averaging 10.0 cm thick. (a) How much heat was lost by the water to the air during the week? (b) The energy released in the explosion of 1000 tons of TNT is called a kiloton (1 kiloton = 4.18×10^{12} J). The nuclear bombs which destroyed Hiroshima and Nagasaki were approximately 10-kiloton devices. How many Hiroshima-size nuclear bombs would have to be detonated to equal the energy lost by Lake Superior during the week? (c) Consider the weather implications of the answer to part b for residents just south of the Great Lakes compared to equally north residents of the Great Plains.

15-17 Chicken soup (essentially water, thermally) is ladled from a pot at boiling into a ceramic bowl. (a) Some 600 cm^3 of soup is placed in the 200-g bowl [specific heat 700 J/(kg·K)] at $20°C$. Assuming no external heat loss, what will the temperature of the soup be when the bowl is warmed to its temperature? (b) The soup is placed on the table at $90°C$ and is allowed to sit for 2.00 min. During that time its temperature drops to $80°C$. Assuming 80 percent of the heat loss is due to evaporation from the surface of the soup, what mass of soup evaporated away? (Neglect the heat lost by the bowl.) (c) When the soup is at $80°C$, two chilled dumplings [$c = 1100$ J/(kg·K)] of mass 100 g each at a temperature of $10°C$ are added to the bowl. Assuming no further heat gain or loss to the outside (and neglecting heat transfer of the bowl), what will the final temperature of the soup-dumpling mixture be?

15-18 Ethanol at $10°C$ is mixed with benzene at $40°C$. The final temperature of the solution is $22°C$. (a) What fraction of the total mass is benzene? (b) If there is 2.50 kg of ethanol, how much additional benzene at $40°C$ would have to be added so that the final temperature would be $30°C$?

15-19 (a) Calculate the final temperature when 1.00 kg of water at $100°C$ is mixed with 1.00 kg of ice at $0°C$. (b) What is the ratio of water at $30°C$ to ice at $0°C$ which will produce water at $4°C$?

15-20 A mass of 25.0 g of ice at $-30°C$ is placed in a calorimeter cup of mass 0.400 kg which contains 0.100 kg of water at $40°C$. The mixture comes to an equilibrium temperature of $19.8°C$. Calculate the specific heat of the calorimeter cup.

15-21 Exactly 1.00 kg of boiling water at $100°C$ is poured into a copper pot initially at $20°C$. The two come to an equilibrium temperature of $80°C$. Calculate (a) the mass of water originally at $20°C$ which would have absorbed the same amount of heat as the pot and reach the same final temperature and (b) the mass of the copper pot.

15-22 In the interest of science, a 50.0-kg physics student agrees to be confined in a very large calorimeter with air supplied through a small inlet tube. The calorimeter is made of 300 kg of aluminum. If the temperature of the calorimeter and its contents increases by 0.100°C/min, (a) how much heat is generated by the student per hour? (b) What fraction of the student's mass in white bread at 1.09×10^7 J/kg (2.60 Cal/g) would the student have to eat per day to supply this heat? (Assume the specific heat capacity of the student is the same as that of water.)

15-23 A dormitory for 200 students is 45.0 m long by 12.0 m wide by 7.00 m high. The outside walls are of 4.00-in brick over 8.00-in-thick concrete blocks. Some 25 percent of the total outside wall area is single glazed windows. The roof is plasterboard, fiberglass, concrete, and tar with a combined R-value of 25. (Neglect the volume of interior walls and furnishings.) The outside temperature is $-10°C$. (a) If the inside air is replaced by outside air once every 20 min, how much heat is necessary to warm the air to 25°C each hour? (b) Assuming a uniform inside temperature of 25°C, how much heat is lost to conduction through the side walls (with their windows) and the roof each hour? (c) If the dormitory is heated with coal containing 2.50×10^7 Btu/ton and costing \$32/ton and the heat delivery is 85 percent efficient (15 percent of the heat content of the coal is lost to use between burning and delivery), what is the cost per day of heating the dormitory? (d) If the heat came from heating oil containing 5.4×10^6 Btu/bbl (1 bbl = 42 gal) costing \$1.30/gal and was delivered to the dormitory at 85 percent efficiency, what would be the cost? (e) If the heat came from natural gas containing 1.00×10^6 Btu/mcf (1 mcf = 1000 ft³ of gas at 1 atm and 60°F) costing \$6.00/mcf and delivered with 85 percent efficiency, what would be the cost? (f) What would be the energy cost of heating the building directly with electric resistance heaters (100 percent efficiency) if electricity costs \$0.09/kWh?

15-24 An aluminum rod 0.400 m long and 5.00×10^{-3} m in diameter is used to stir a sugar-water solution which is held at a temperature of 108°C. If the other end of the rod is at room temperature of 25°C, determine how much heat flows along the rod in 5.00 min.

15-25 The botany professor's vegetable garden is a plot 10.0 m by 12.0 m and is covered with snow to a depth of 0.150 m during spring break. The snow has a thermal conductivity of 0.110 W/(m·°C), and the upper surface of the snow is at $-15°C$ while the ground is at 0°C. Calculate the amount of heat lost by the garden to conduction through the snow in a 3-h period.

15-26 A heavy aluminum pot 0.500 cm thick conducts heat at the rate of 125 J/(s·cm²). What is the temperature of the outside surface if the temperature of the inside of the pot is 100°C? The pot's surface area is 900 cm².

15-27 In the physics laboratory, there are 12 windows with a total area of 27.0 m². The glass is 3.17×10^{-3} m thick and has a thermal conductivity of 1.00 W/(m·°C). (a) How much heat is lost in 24 h by conduction through these windows on an average day when the temperature difference between the inside and outside of the windows is 4.00°C? (b) How much does it cost to heat the laboratory on this day if it is heated by burning natural gas at \$5.00/mcf and 1 mcf supplies 1.09×10^9 J? (Ignore convection and radiation losses.)

15-28 Show that the result of adding a second layer of insulation of thermal conductivity k_2 and thickness d_2 is given by Equation 15-5. That is, eliminate the intermediate temperature T between the previous equations to show that Equation 15-5 is obtained.

15-29 (a) How much heat is lost in a 24-h period by conduction through a picture window 1.50 m high by 2.50 m wide if the glass is 0.600 cm thick and the inside temperature of the house is 25°C warmer than the outside? (b) What would the heat loss be for glass twice as thick? (c) What would the heat loss be if the window consisted of two panes of glass separated by a still-air layer 1.00 cm wide?

15-30 A child suffering from high fever has a skin temperature of 40°C. He is lowered into a bathtub of 20°C water such that 1.00 m² of his surface is in contact with the water. (a) Calculate the heat loss through the skin, which acts like a 3.00-mm-thick layer having a thermal conductivity $k = 0.360$ W/(m·°C). (b) Calculate the same rate through the first 1.00 cm of water if the skin temperature drops quickly to 33°C and stays there. (c) In such a situation, does it make sense to agitate the water to ensure its continuous flow over the body surface to promote heat transfer?

15-31 When heat loss from exposed skin reaches 1250 W/m², the U.S. Weather Bureau reports the condition as bitter cold. (a) This condition is reached in still air at a temperature of about $-70°C$ ($-94°F$). Calculate the thermal conductivity of the body if the average thickness from the body's core at 37°C and the skin is 5.00 cm. (b) Bitter cold conditions can be reached at a hotter temperature with a wind blowing. For example, a 1.00 m/s wind of air at $-25°C$ will chill exposed flesh at this rate. At 10.0 m/s, the air temperature can be only 0°C and produce the same result. Assume that the exposed skin temperature in each case is 20°C. Calculate the convection coefficient for each wind speed given.

15-32 When the air is still, the convection coefficient for heat loss by bare skin is about 5.00 W/(m²·°C). For the wind blowing at 4.00 m/s over bare skin, the convection coefficient is about 28.0 W/(m²·°C). Assume the skin temperature is 32°C and calculate at what temperature in still air the heat loss by convection would be the same as the heat loss by convection when the wind is blowing 4.00 m/s and is 0°C.

15-33 The natural convection coefficient for a cylindrical pipe of outside diameter d and temperature difference Δt between the outside of the pipe and the surrounding fluid is given by

$$h = 1.32 \times 10^{-8} \left(\frac{\Delta t}{d}\right)^{1/4}$$

where h is in W/(m²·°C) when Δt is measured in degrees Celsius and d is measured in meters. In one prominent fraternity house, a hot-water pipe with an outside diameter of 1.25 cm runs 12.0 m through an unheated basement. During cold weather, the average temperature in the basement is 10°C and the average temperature of the pipe is 50°C. Calculate (a) the convection coefficient for this situation and (b) the heat transferred from the pipe by natural convection in 24 h.

15-34 Calculate the surface temperature of the sun if it is determined that its output of thermal radiation is 6.75×10^7 W/m².

15-35 Suppose the emissivity of the pipe's surface of Problem 15-33 is 0.600. Calculate (a) the rate of heat loss due to radiation, (b) the rate of heat gain due to absorption of radiation, (c) the net rate of heat loss, and (d) the net heat loss to radiation in 24 h.

15-36 As seen from the sun, the earth would look like a flat disk of radius 6.38×10^6 m. Solar energy is delivered to this disk at an intensity of 1390 W/m². (a) What is the total solar power incident on the earth? (b) Imagine the earth to be a smooth, perfectly absorbing rock sphere with no atmosphere. As the earth rotates, all its surface is exposed to solar radiation. What would be the average solar power falling on each square meter of the earth's surface? (c) To attain thermal equilibrium, the surface would have to radiate away exactly the same power it received. At what surface temperature would the earth's radiation balance the solar input? (d) Is your answer to part c a reasonable estimate for the average surface temperature of the earth? Why or why not? The radiation-balance method used in this problem would work very well for Mercury or the moon. What is different about the Earth?

15-37 Assume the temperature of the surface of the sun is about 5800 K. (a) What is the power radiated by each square meter of the sun's surface? (b) The sun's radius is 6.96×10^8 m. What is the power output of the entire sun? ($A_{\text{sphere}} = 4\pi R^2$.) (c) This radiated energy spreads out uniformly in space in all directions and is only lost when it encounters some absorber in space. Thus, the total radiated power crossing an imaginary sphere the size of the earth's orbit equals the power radiated at the sun's surface. How much power would be absorbed by a perfect absorbing surface of 1 m² pointed directly at the sun at the distance of the earth from the sun ($R_{\text{orbit}} = 1.50 \times 10^{11}$ m) in space?

15-38 If a sample of gas such as air consists of a mixture of different kinds of gases with n_1 mol of type 1, n_2 mol of type 2, and so on, the ideal gas law becomes

$$PV = (n_1 + n_2 + \cdots)RT$$

Partial pressures are defined by the relationships

$$P_1V = n_1RT \qquad P_2V = n_2RT$$

and so on. Thus we have

$$P = P_1 + P_2 + \cdots$$

This equation is known as **Dalton's law of partial pressures:** The total pressure of a mixture of gases is equal to the sum of the individual partial pressures, each obtained as if it were the only gas in the volume V. Equation 15-12 is obtained by using the fact that the density of water vapor is directly proportional to the partial pressure of the water vapor. Prove this proportion by using the ideal gas law and the definition of a mole.

15-39 At the Physics Club picnic, lemonade is kept in a large metal container. The system absorbs net heat from its surroundings at a rate of 35.0 W. The container is covered by a wet beach towel to keep it cool. If exactly half the water which evaporates from the towel gains its energy from the lemonade and container, (a) how much water must evaporate per second to keep the lemonade cold and (b) how much water must evaporate during the 5.00-h picnic? ($T_{\text{lemonade}} = 5°C$)

15-40 (a) The inside of a house is maintained at 20°C and 40% relative humidity. What is the maximum temperature of the outside of a cold drink glass on which condensation begins? (b) On a summer evening, the temperature is 28°C and the relative humidity 80%. Will condensation form on water pipes in which the water is at a temperature of 20°C?

15-41 The volume of air in a sorority house is 800 m³. With windows open on a hot day in the summer, the temperature of the air climbs to 28°C and the relative humidity in early evening is 75%. After the windows are closed, how much water must be removed by the air conditioning system so that the relative humidity is 45% when the temperature gets down to 20°C?

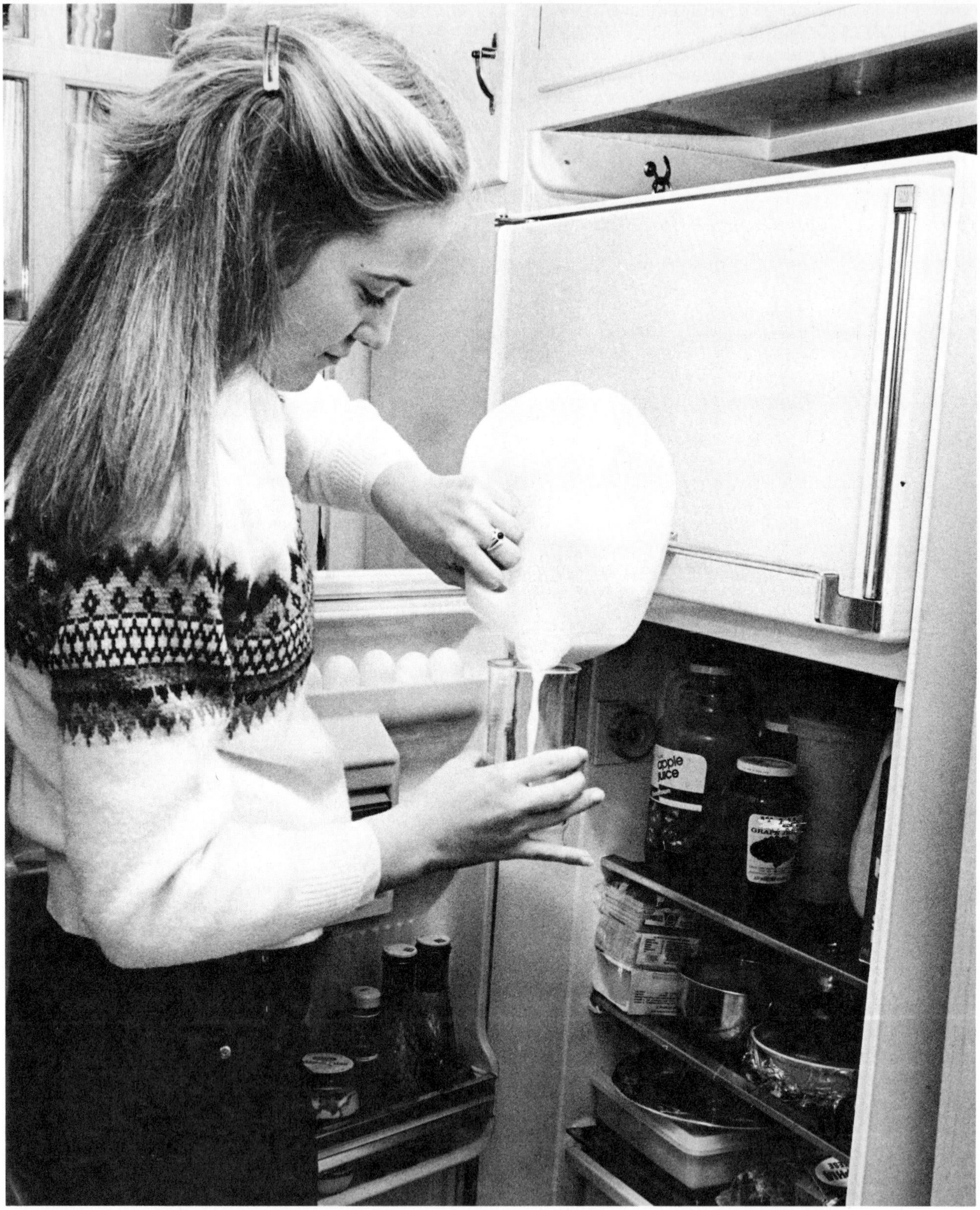
apple
juice

16 Thermodynamics

16-1 Introduction

In the previous chapter, we saw several examples of the transformation of mechanical energy into thermal energy: friction, the waterfall, and so on. We are also aware that the opposite transformation takes place. If a covered pot of water is placed on a stove and the burner turned to high heat, eventually the lid will begin to chatter. The heat is conducted through the bottom of the pot, boiling the water and eventually increasing the pressure of the steam generated so much that the force it exerts on the lid is large enough to lift it. As soon as the lid is lifted a crack, the steam pressure is momentarily relieved and the upward force on the lid reduced. The lid then drops back, closing off the escape route and allowing the pressure to build once more. Repeated very rapidly, this cycle yields the chattering noise of the lid. The result of all this is that heat has been transformed into mechanical energy.

This is a very important point, since where there is mechanical energy, work can be done. Some clever person might design a mechanism to use some of the mechanical energy to do useful work—the pulley and gear arrangement shown in Figure 16-1, for example. Such an arrangement is a **heat engine** since it converts heat into useful mechanical work (however inefficient). To be very

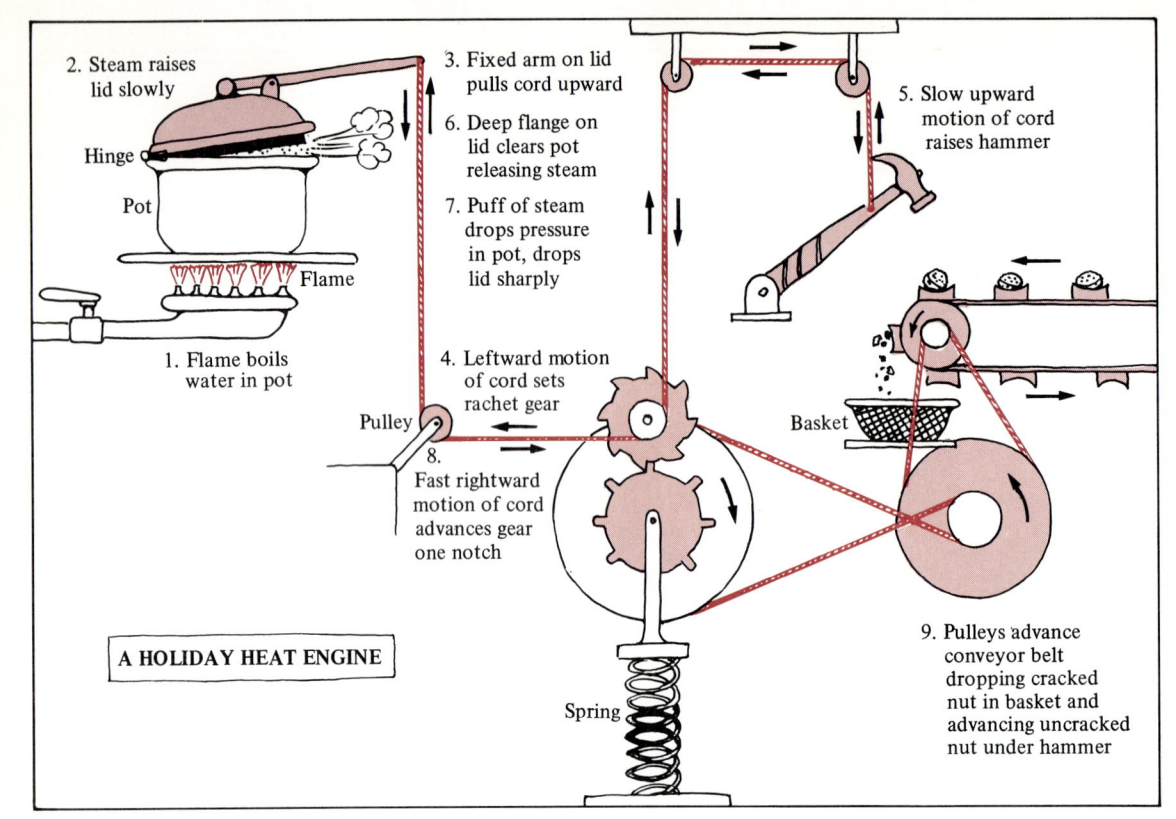

Figure 16-1 An automatic nutcracking system. To crack a nut, you must exert a force through a distance; that is, you must do work. While this system is fanciful, your intuition should tell you that it could be made to operate as intended. Heat can be a source of work, and the device shown is a heat engine.

efficient, an engine must convert much of the energy supplied to it into useful work while losing a minimum of energy to the environment. We can see that our pot would surely lose energy to radiation and convection as well as all the energy of the steam which escapes when the lid jiggles.

Heat engines, their efficiencies, and several other topics are dealt with in the study of **thermodynamics,** a word that stems from the Greek words *therme,* meaning "heat," and *dynamic,* meaning "action." Thermodynamics is in part the study of the transformation of heat to mechanical energy.

As we examine the heat engine idea in more depth, we will determine what physical processes are taking place and what mechanism could increase power and efficiency. We will deal with such questions as "What portion or portions of the mechanism are important?" "What principles apply?" "How can these principles be stated quantitatively?"

Answering the first question allows us to define a **system** as a definite quantity of matter enclosed by real or imagined boundaries. In dealing with any thermodynamic process, we must first identify the system. For the simple case of water boiling in a covered pot, the water and steam could constitute the system. Once the water begins to boil, we assume the pot and lid are absorbing about as much heat per unit time as they are giving up and are hence in an average equilibrium state. The water, on the other hand, is absorbing more heat than it is giving up. Some of the heat goes to the change in phase, and a small part goes into working on the lid. This now leads us to an answer to the second question. The work-energy principle applies. What remains is to determine how we should state the principle.

16-2 The First Law of Thermodynamics

The energy possessed by a system is its internal energy. Heat and work are the only two ways a system may gain or lose energy. Any system which absorbs heat while it does work will, in general, have its internal energy change. In the ideal case, the increase in internal energy ΔU equals the heat added minus the work done:

$$\Delta U = \Delta Q - W \tag{16-1}$$

where ΔQ is the heat added to the system and W is the work done by the system. Equation 16-1 is a mathematical statement of the **first law of thermodynamics.**

It is customary to use the sign convention indicated when applying Equation 16-1. An increase in the internal energy of the system is a positive number, while a decrease is a negative number. Heat added to the system is a positive number, while heat released by the system is a negative number. Finally, work done *by* the system on its surroundings is a positive number, while work done *on* the system is a negative number. This convention, while not universal, is used in most textbooks. It arose in this fashion primarily because thermodynamics developed in large part in conjunction with the development of heat engines. In these applications, one is most interested in having the system perform useful work *on* its surroundings.

Example 16-1 Determine the amount of thermal energy which must be added to a system if the system is to expand by doing 800 J of work while its internal energy decreases by 300 J.

Solution This is a straightforward application of Equation 16-1, recognizing that the change in internal energy is a negative number and the work done is a positive number.

$$\Delta U = \Delta Q - W$$

or

$$\Delta Q = \Delta U + W = (-300 \text{ J}) + (+800 \text{ J}) = +500 \text{ J}$$

The first law of thermodynamics is a statement of the conservation of energy or of the work-energy principle. It also leads to something more. Suppose a system such as a confined gas is changed from one state to another by changing its pressure, volume, and temperature. The *state* of the gas is completely specified if we know the quantity (number of moles) and the pressure, volume, and temperature. Once the gas is confined, its quantity is constant. We may thus give its original state as P_1, V_1, and T_1 and its final state as P_2, V_2, and T_2.

We may calculate the difference between the heat added and the work done during the change. This difference, $\Delta Q - W$, is the change in internal energy of the system as it goes from one state to the other. We may make this change in many different ways. For instance, we may add a lot of heat to the system while the system does a great deal of work to reach the final state. Or we may add only a moderate amount of heat while the system does a much smaller amount of work in moving to the final state. If we carry this experiment out again and again, varying the amount of heat and work for each trial but always starting at state 1 and ending at state 2, we discover that the difference between the heat added to the system and the work done by the system is always the same. Further, this difference is independent of how the system gets from state 1 to state 2. Thus this difference, the change in internal energy between states 1 and 2, is independent of the path traveled by the system but depends only on the endpoints. Indeed, the value of the internal energy of any system does not depend on the history of that system but only on its state (pressure, volume, and temperature in the case of a gas). In fact, Equation 16-1 defines internal energy from a macroscopic point of view.

Whenever a change or several successive changes are made to a system and the system is returned to its original state, the net change in internal energy is zero. Under those circumstances, Equation 16-1 becomes

$$\Delta U = 0 = \Delta Q - W$$

or

$$\Delta Q = W$$

Thus a system taken about a closed path and returned to its original state will have done a net amount of work on its surroundings equal to the net heat added to the system. This is precisely the way a heat engine works. A working fluid, perhaps steam, air, or combustion products as a system, is carried through a cycle, a sequence of operations which do useful work and return the engine to its original state. Whenever a thermodynamic system undergoes a change of state, a **thermodynamic process** is said to have taken place. We will explore some specialized thermodynamic processes for an ideal gas in Section 16-4.

> A heat engine, like an air conditioner or an automobile engine, uses a working fluid, usually a gas. The gas is taken from one state of P, V, and T through others under varying circumstances and returned to its original state to begin again. The net effect of this cycle is the production of motion from heat (for the automobile engine) or removal of heat from an enclosure (for the air conditioner). Since we are trying to understand the science of such devices, not build one, we focus in this chapter on the cycle of changes on the working fluid—the system—not on the engineering details of the devices.

16-3 Molar Heat Capacity

Assume we have n mol of an ideal gas confined to a cylinder fitted with a leak-tight movable piston, as in Figure 16-2. The system of gas obeys the equation $PV = nRT$. If the gas exerts a force F on the piston and expands, moving the piston a small distance Δx, the work done by the gas is

$$W = F\,\Delta x$$

However, since $F = PA$ from the definition of pressure, we have

$$W = PA\,\Delta x$$

Further, $A\,\Delta x$ is nothing more than the change in volume ΔV of the confined gas. Thus

$$W = P\,\Delta V \qquad (16\text{-}2)$$

This is true if the displacement is small (so the pressure change can be considered to be negligible) or even if the displacement is very large provided the pressure is held constant. Of course, if $\Delta x = 0$, then $\Delta V = 0$ and $W = 0$ as well.

Since we wish to determine any changes in our system as we add heat, we must be concerned with specific heat capacities. However, when we are dealing with gases, it is most advantageous to use heat capacity per mole, called **molar heat capacity** and defined as

$$C = \frac{1}{n}\frac{\Delta Q}{\Delta T} \qquad (16\text{-}3)$$

where C is the molar heat capacity and ΔQ is the heat which must be added to n mol to cause a temperature rise of ΔT. This is not really an SI quantity, but it should be reported in J/(mol·°C) or J/(mol·K). A problem with gases which is not generally true for liquids and solids is that while heat is added, a gas may change volume dramatically. But, an expanding gas does work. Since this work may come from the heat being added, it must take more heat to raise the temperature of an expanding gas by 1°C than that of the same quantity of gas held at a constant volume. Under ideal conditions, only two values of molar heat capacity are useful: the molar heat capacity when the gas is held at **constant volume** C_V and the molar heat capacity when the gas is held at **constant pressure** C_P.

C_V may be determined by adding a known quantity of heat ΔQ to our n mol of gas, measuring the temperature increase while keeping the piston clamped in place ($\Delta V = 0$), and substituting into Equation 16-3. C_P may be determined by maintaining a constant amount of weight (therefore a constant force) on the top of the piston as the heat is added. Since the force and the area of the piston are constant, the pressure of the gas will remain constant. Measurement of ΔT and substitution into Equation 16-3 yields C_P. Note that C_P must be larger than C_V, since the amount of heat added during the constant-pressure operation must be greater than that added during the constant-volume process. After all, during the constant-pressure process, only part of the heat added goes into a temperature increase while the rest is used as the gas does work, raising the piston and weight. During the constant-volume process, all the heat added goes into a temperature increase.

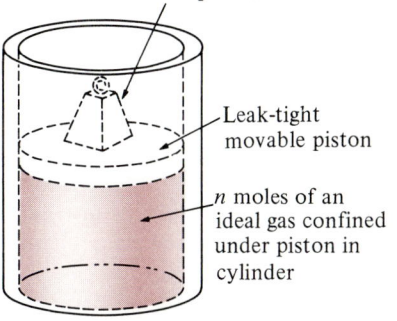

Figure 16-2 Through most of this chapter, we work with a system comprising n mol of an ideal gas enclosed in a cylinder by a sliding, leak-tight weighted piston.

16-4 Reversible Thermodynamic Processes

To simplify calculations and also to obtain ideal theoretical results as a limiting case for real processes, we will consider some **reversible thermodynamic processes,** defined as those that take place in such a way that the system is always in thermal equilibrium. In practice, this can be approximated by having all changes take place so slowly that the system is never more than an infinitesimal step away from equilibrium. All real processes are irreversible to a greater or lesser extent.

We can investigate reversible thermodynamic processes and their consequences graphically as well as mathematically, since such processes can be represented by curves on P, V, and/or T graphs. The processes we will consider are those which take place when some quantity is not allowed to change. Each process has a name loosely derived from the variable or quantity that is held constant:

1. Isovolumetric process ($\Delta V = 0$, volume constant)
2. Isobaric process ($\Delta P = 0$, pressure constant)
3. Isothermal process ($\Delta T = 0$, temperature constant)
4. Adiabatic process ($\Delta Q = 0$, no heat flow)

Case 1. *Isovolumetric Process* Suppose a confined gas undergoes an **isovolumetric process.** Such a process may be depicted by either Figure 16-3a or 16-3b. In Figure 16-3a, the flame adds heat to the system; the ice in Figure 16-3b is taking heat away from the system. In either case, the thumbscrews shown ensure that the volume of our ideal gas remains constant, but the temperature and hence the pressure may change. Since the work done by an ideal gas is $W = P\,\Delta V$, and for our process $\Delta V = 0$, we have $W = 0$. Therefore, for an isovolumetric process, the first law becomes

$$\Delta U = \Delta Q - 0$$

or simply

$$\Delta U = \Delta Q$$

For constant volume, Equation 16-3 yields

$$\Delta Q = nC_V(T_2 - T_1) \qquad (16\text{-}4)$$

Thus, we have

$$\Delta U = nC_V(T_2 - T_1) \qquad (16\text{-}5)$$

This is a very important result. First, it states that the change in internal energy (and by implication the internal energy itself) of an ideal gas depends only on the absolute temperature of the gas. Second, although here we obtained the expression for ΔU in an isovolumetric process, Equation 16-5 is *always* the proper expression for ΔU as the system changes temperaure from T_1 to T_2. Remember that when the first law was introduced, we saw that during a change of state ΔU depends only on the endpoints and *not* on the path between those points.

On a PV diagram as in Figure 16-4, an isovolumetric process is represented by a straight line parallel to the pressure axis. In fact, any straight line parallel to the pressure axis represents *some* isovolumetric process. (The name *isovolumetric,* being rather cumbersome, is often shortened to **isometric.**) The straight lines representing isometric processes are called **isomets.** Note that lines of constant pressure, called **isobars,** are also plotted on the graph of Figure 16-4.

Case 2. *Isobaric Process* Now let us look at the confined gas as it undergoes an **isobaric process,** that is, one in which the pressure does not change. In Figure 16-5a and b, a constant downward force F is applied to a peg attached to the piston. Since the piston has a constant area, the pressure it exerts, hence the pressure of the gas, is held constant. In Figure 16-5a, the system is presumably absorbing heat, whereas in Figure 16-5b, it could be releasing heat. In either case, the maintaining of a constant force on a constant area ensures that the pressure will not change. However, both the temperature and volume may change. For an ideal gas in an isobaric process, we may write the first law of thermodynamics as

$$\Delta U = \Delta Q - W$$
$$nC_V(T_2 - T_1) = \Delta Q - P\,\Delta V$$

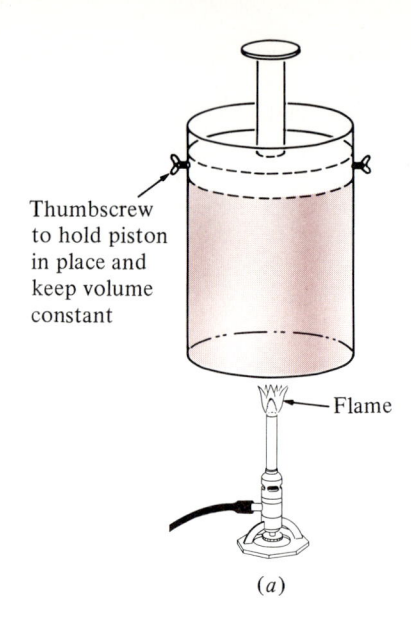

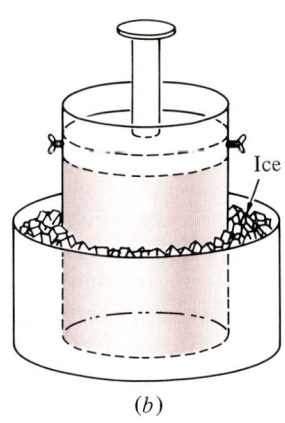

Figure 16-3 The system of confined ideal gas. (*a*) If we heat the system, the added heat must increase the internal energy of the gas, which shows up as a temperature increase of the gas. (*b*) If the system is cooled, lost heat must show up as a drop in internal energy, since no work can be done on the gas (by compressing it) to compensate for the lost heat.

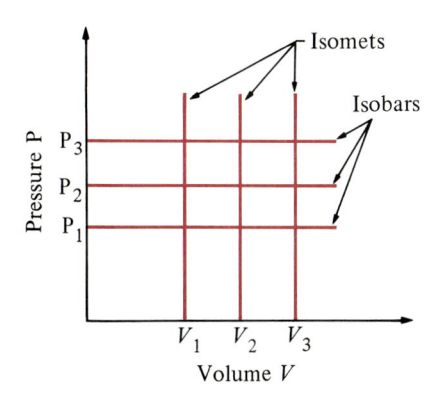

Figure 16-4 On a PV graph for the n mol of gas in the system we are using, each point is a separate state of the system. Thus, a line on such a graph connects states of the system.

For constant pressure, Equation 16-3 yields

$$\Delta Q = nC_P(T_2 - T_1) \tag{16-6}$$

Further, since we are dealing with an ideal gas,

$$P \Delta V = P(V_2 - V_1) = nR(T_2 - T_1)$$

or, at constant pressure,

$$W = nR(T_2 - T_1) \tag{16-7}$$

Thus the first law becomes

$$nC_V(T_2 - T_1) = nC_P(T_2 - T_1) - nR(T_2 - T_1)$$

This may be solved to give us

$$C_P - C_V = R \tag{16-8}$$

Note that Equation 16-8, derived for an ideal gas, tells us that the difference between C_P and C_V is the universal gas constant R [8.31 J/(mol·K)]. Table 16-1 lists the measured molar heat capacities for various real gases. Note that $C_P - C_V$ differs very little from R, with a percentage difference of less than 8 percent even for the worst cases, the real polyatomic gases.

TABLE 16-1 Molar Heat Capacities of Various Gases at Low Pressure and at 25°C

Gas	C_P, J/(mol·K)	C_V, J/(mol·K)	$C_P - C_V$, J/(mol·K)
Monatomic			
He	20.77	12.46	8.31
Ne	20.77	12.46	8.31
A	20.77	12.46	8.31
K	20.77	12.46	8.31
Xe	20.77	12.46	8.31
Diatomic			
H_2	28.72	20.40	8.32
N_2	29.05	20.73	8.32
O_2	29.39	21.08	8.31
CO	29.13	20.82	8.31
Polyatomic			
CO_2	36.91	28.42	8.49
SO_2	40.32	31.35	8.99
H_2S	34.57	25.92	8.65

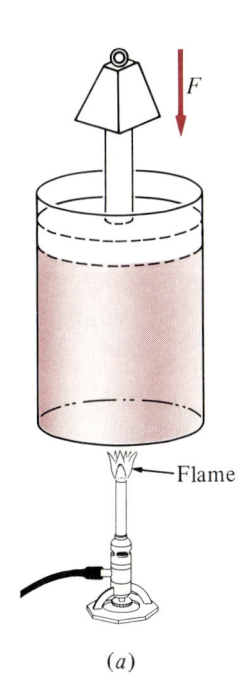

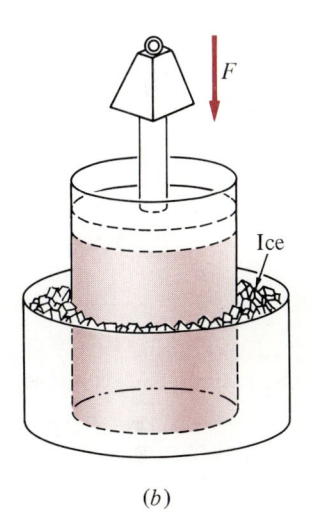

Figure 16-5 A system at constant pressure, making any transition between states an *isobaric process*. (a) Some of the heat added increases the internal energy of the gas, and some does work by lifting the weighted piston. (b) When the system is cooled, the heat drawn away comes from both the internal energy of the gas and the work done on the gas by the compressing force F.

377

Example 16-2 A cylinder contains 2.00 mol of helium gas at 27°C. The gas is kept at a constant pressure of 1.50×10^5 N/m² and heated to 77°C. Determine (a) the amount of work done by the gas, (b) the change in its internal energy, (c) the net heat added to it, and (d) its change in volume.

Solution We may treat helium as an ideal gas and utilize the data from Table 16-1.

(a) This can be solved by using Equation 16-7.

$$W = nR(T_2 - T_1)$$
$$= (2.00 \text{ mol})[8.314 \text{ J/(mol} \cdot {}°\text{C})](77 - 27)°\text{C}$$
$$= 831 \text{ J}$$

(b) We apply Equation 16-5.

$$\Delta U = nC_V(T_2 - T_1)$$
$$= (2.00 \text{ mol})[12.46 \text{ J/(mol} \cdot {}°\text{C})](77 - 27)°\text{C}$$
$$= 1.25 \times 10^3 \text{ J}$$

(c) To determine the net heat added, we may either apply Equation 16-6 or solve for ΔQ in the first law. We will do both to convince ourselves of consistency.

$$\Delta Q = nC_P(T_2 - T_1)$$
$$= (2.00 \text{ mol})[20.77 \text{ J/(mol} \cdot {}°\text{C})](77 - 27)°\text{C}$$
$$= 2.08 \times 10^3 \text{ J}$$

Alternately,

$$\Delta U = \Delta Q - W$$

or

$$\Delta Q = \Delta U + W$$
$$= 1.25 \times 10^3 \text{ J} + 0.831 \times 10^3 \text{ J}$$
$$= 2.08 \times 10^3 \text{ J}$$

(d) Finally, we realize that $W = P \Delta V$, and use the given value of pressure along with the solution to part a to solve for ΔV.

$$W = P \Delta V$$

or

$$\Delta V = \frac{W}{P}$$

$$= \frac{831 \text{ J}}{1.50 \times 10^5 \text{ N/m}^2} = 5.54 \times 10^{-3} \text{ m}^3$$

Case 3. Isothermal Process We now wish to look at the confined gas in an **isothermal process.** We might carry out such a process by surrounding the cylinder with a large amount of material at constant temperature which can readily absorb or give up heat without changing temperature. Such an arrangement is often called a constant-temperature reservoir, or simply a **heat reservoir.** The heat reservoir will absorb heat from or deliver heat to the cylinder as required to maintain a constant temperature as the gas compresses or expands during the isothermal process. Since the temperature remains constant, the internal energy of an ideal gas does not change during an isothermal process. Thus the first law of thermodynamics becomes

$$\Delta U = 0 = \Delta Q - W$$

or

$$\Delta Q = W = P \, \Delta V \qquad (16\text{-}9)$$

In general, both the pressure and volume change during an isothermal process and Equation 16-9 may be applied only to small increments, since P is not constant.

Since we are dealing with an ideal gas as our system, we have that

$$PV = nRT = \text{constant}$$

for an isothermal process. Figure 16-6 is a PV diagram illustrating several isothermal processes. Note the different curves, called **isotherms,** for different constant temperatures. Figure 16-7 shows a segment of a particular isotherm. The system might be thought to move along the isothermal path between the points A and B. The shaded area shows the increment of work done as the gas goes through a very small expansion ΔV while its pressure does not change very much. From this we conclude that the work done as the system moves isothermally from A to B may be found by adding up all such little areas and obtaining the area under the curve. Thus

$$W_{AB} = \Sigma P \, \Delta V \qquad \text{(sum from } A \text{ to } B\text{)}$$

but for an ideal gas, we have $PV = nRT$, or

$$P = \frac{nRT}{V}$$

Thus

$$W_{AB} = nRT \, \Sigma \, \frac{\Delta V}{V} \qquad \text{(from } V_A \text{ to } V_B\text{)}$$

The mathematical process for analyzing such a sum, carried out by varying a system in infinitesimal steps, is the central concern of the integral calculus. As we proceed from V_A to V_B in finer and finer steps of ΔV, it can be shown that the sum $\Sigma(\Delta V/V)$ approaches closer and closer to (reaches the limit of) the expression $\ln(V_B/V_A)$. Or

$$\lim_{\Delta V \to 0} \Sigma \, \frac{\Delta V}{V} \to \ln \frac{V_B}{V_A}$$

and

$$W_{AB} = nRT \ln \frac{V_B}{V_A} \qquad (16\text{-}10)$$

for an isothermal process. Since the points A and B were arbitrary and it really did not matter whether the process was an expansion or compression, Equation 16-10 may be applied to any isothermal process carried out on an ideal gas. Equation 16-9 can therefore be written as

$$\Delta Q = W = nRT \ln \frac{V_2}{V_1} \qquad (16\text{-}11)$$

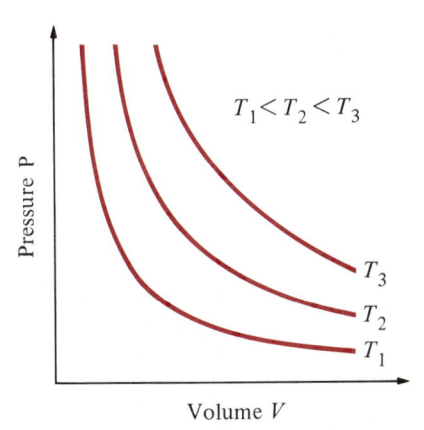

Figure 16-6 Isotherms produced by confined ideal gas kept at constant temperature. Each curve connects all possible states of the gas at the fixed temperature.

Figure 16-7 The work done by (increasing V) or on (decreasing V) an ideal gas is $P \, \Delta V$. This work is always given by the area contained under the graph bounded by the curve, the V axis, and the vertical lines at V_A and V_B.

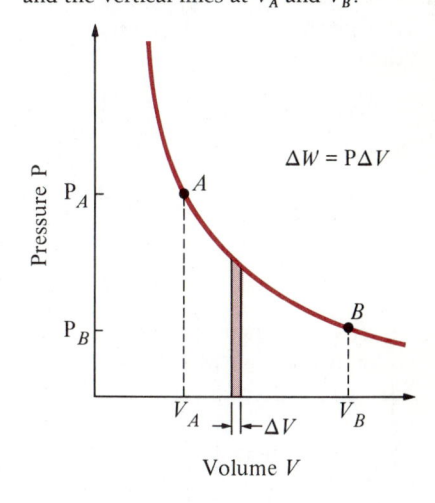

379

Note that if V_2 is greater than V_1, Equation 16-11 describes an isothermal expansion, and the work done by the system is positive. Of course, if V_1 is greater than V_2, we have an isothermal compression. Further, V_2/V_1 is a fraction less than 1 whose logarithm is therefore negative. Thus work must be done *on* the system, or the work done by the system is a negative number.

Example 16-3 A tire on the professor's automobile blows out. It is later estimated that the air inside expanded isothermally from 0.0125 to 0.0400 m³ at a temperature of 290 K to a final pressure of 1.00×10^5 N/m². Show how the professor calculates the work done by the air during expansion, assuming the air may be treated as an ideal gas.

Solution Assuming our professor is versed in physics, she knows she should apply Equation 16-11.

$$W = nRT \ln \frac{V_2}{V_1}$$

Now assuming the ideal gas law and that the process is isothermal, we have

$$P_1 V_1 = P_2 V_2 = nRT$$

Thus we may substitute the product of the final pressure and volume for nRT.

$$W = P_2 V_2 \ln \frac{V_2}{V_1}$$
$$= (1.00 \times 10^5 \text{ N/m}^2)(0.0400 \text{ m}^3) \ln \frac{0.0400 \text{ m}^3}{0.0125 \text{ m}^3}$$
$$= 4.65 \times 10^3 \text{ J}$$

Case 4. Adiabatic Process Finally, let us consider an **adiabatic process** applied to the confined gas. Since this requires that no net heat be transferred to or from the system, we must surround the cylinder with a perfect insulator while expansion or compression takes place. Since $\Delta Q = 0$ during an adiabatic process, we write the first law of thermodynamics as $\Delta U = 0 - W$ or

$$W = -\Delta U$$

The prefix *iso-* is derived from the Greek *isos*, meaning "equal" or "alike." The word *adiabatic* comes from the Greek *adiabatos*, meaning "impassable." During an adiabatic process, no heat is to pass into or out of the system.

If during the process the temperature changes from T_1 to T_2, we must as usual write

$$\Delta U = nC_V(T_2 - T_1)$$

Thus

$$W = -nC_V(T_2 - T_1)$$

or

$$W = nC_V(T_1 - T_2) \qquad (16\text{-}12)$$

We see from Equation 16-12 that during an adiabatic expansion when positive work must be done by the system, the temperature of the system must decrease. This can be interpreted as meaning that internal energy must be converted to the work necessary for an adiabatic expansion. The other side of the coin is that the internal energy and hence the temperature of an ideal gas must increase when the system undergoes an adiabatic compression.

Suppose our confined gas undergoes a small adiabatic process. We could write the first law as

$$nC_V \Delta T = -P \Delta V \tag{16-13}$$

We can also show that for small possible changes in all three variables, the ideal gas equation $PV = nRT$ becomes

$$P \Delta V + V \Delta P = nR \Delta T \tag{16-14}$$

where ΔV, ΔP, and ΔT are the small changes in volume, pressure, and temperature that our gas has undergone.

Let us solve Equation 16-13 for ΔT and substitute into Equation 16-14. We have

$$P \Delta V + V \Delta P = nR \frac{-P \Delta V}{nC_V}$$

Canceling n and substituting for R from Equation 16-8 yields

$$P \Delta V + V \Delta P = (C_P - C_V) \frac{-P \Delta V}{C_V} = -\frac{C_P}{C_V} P \Delta V + P \Delta V$$

or

$$\frac{\Delta P}{P} + \frac{C_P}{C_V} \frac{\Delta V}{V} = 0$$

For large changes in pressure and volume, this requires the same summation (integration) process which yielded Equation 16-10. Thus if we designate C_P/C_V as γ, we have

$$\ln P + \gamma \ln V = \text{constant}$$

Using the rules for logarithms, this becomes

$$PV^\gamma = \text{constant} \tag{16-15}$$

Obtaining Equation 16-15 was rather involved, but it has a number of important facets. It is the equation of the adiabatic curves (also called **adiabats**) which we plot on a PV diagram to show the progress of an ideal gas undergoing an adiabatic process. Consider Figure 16-8, on which several of these adiabats are shown. Notice there is a similarity with isotherms, which are shown in dashed lines for comparison. We see that although both sets of curves approach the axes asymptotically, the adiabats drop more steeply with increasing volume because the exponent of V is $\gamma = C_P/C_V$ and $C_P > C_V$. In practice, most rapid processes are nearly adiabatic since the system simply does not have the time for heat transfer to or from the environment to occur. Values of γ may be calculated from the data in Table 16-1 and are summarized in Table 16-2.

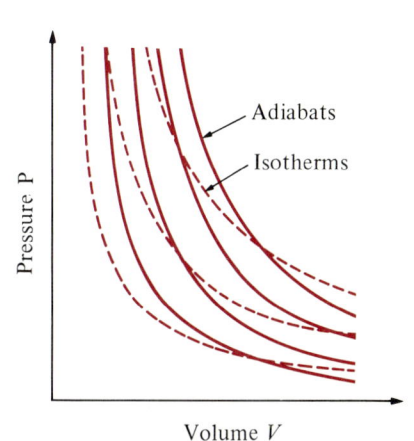

Figure 16-8 Several adiabats for different initial internal energies. The dashed isotherms show that temperature drops along each adiabat as it moves to higher volume.

TABLE 16-2

Gas	$\gamma = C_P/C_V$
Monatomic	
He	1.67
Ne	1.67
A	1.67
K	1.67
Xe	1.67
Diatomic	
H_2	1.41
N_2	1.40
O_2	1.39
CO	1.40
Polyatomic	
CO_2	1.30
SO_2	1.29
H_2S	1.33

Example 16-4 During an adiabatic compression, the volume of a quantity of nitrogen (N_2) is decreased to one-fourth its original value. Calculate the ratio of the final pressure to the initial pressure of the nitrogen.

Solution We wish to apply Equation 16-15 but in a more convenient form.

$$PV^\gamma = \text{constant}$$

implies that $P_1 V_1^\gamma = P_2 V_2^\gamma$. Solving this for the ratio of P_2 to P_1 yields

TABLE 16-3 Summary of Thermodynamic Equations for an Ideal Gas

$$\Delta U = nC_V \Delta T \quad \text{(always)} \quad (16\text{-}5)$$
$$C_P - C_V = R \quad (16\text{-}8)$$

Isothermal process: Constant temperature

$$\cancel{\Delta U}^{\,0} = \Delta Q - W \quad \text{since} \quad \Delta U = nC_V \cancel{\Delta T}^{\,0}$$
$$\Delta Q = W = nRT \ln \frac{V_2}{V_1} \quad (16\text{-}11)$$

Adiabatic process: No heat transfer

$$\Delta U = \cancel{\Delta Q}^{\,0} - W$$
$$= -W = nC_V \Delta T \quad (16\text{-}12)$$
$$PV^\gamma = \text{const} \quad \gamma = \frac{C_P}{C_V} \quad (16\text{-}15)$$

Isobaric process: Constant pressure

$$\Delta U = \Delta Q - W$$
$$W = P\,\Delta V = nR\,\Delta T \quad (16\text{-}7)$$
$$\Delta Q = nC_P\,\Delta T \quad (16\text{-}6)$$

Isovolumetric process: Constant volume

$$\Delta U = \Delta Q - \cancel{W}^{\,0} \quad \text{since} \quad W = P\cancel{\Delta V}^{\,0}$$
$$\Delta U = \Delta Q = nC_V\,\Delta T \quad (16\text{-}5)$$

$$\frac{P_2}{P_1} = \frac{V_1^\gamma}{V_2^\gamma} = \left(\frac{V_1}{V_2}\right)^\gamma$$

In Table 16-2 we see that for nitrogen $\gamma = 1.40$. Thus

$$\frac{P_2}{P_1} = \left(\frac{V_1}{V_2}\right)^{1.40}$$

and since $V_2 = V_1/4$, we have that $V_1/V_2 = 4$. Therefore

$$\frac{P_2}{P_1} = 4^{1.40} = 6.96$$

Some of the results and conditions associated with the more important thermodynamic processes are summarized in Table 16-3.

16-5 Heat Engines and Refrigerators

As we mentioned in Section 16-1, a heat engine operates by converting heat to mechanical work. During this process, the system is taken around a closed path which can be illustrated on a *PV* diagram. The heat engine may be thought of as *accepting* an amount of heat Q_H from a high-temperature reservoir each cycle, converting part of that heat to useful work W, and expelling the remainder Q_L to a low-temperaure reservoir (Figure 16-9). The net heat absorbed by the engine is $\Delta Q = Q_H - Q_L$, where Q_L is the magnitude of the heat *expelled* and the minus sign is shown explicitly. This is just equal to the work done by the engine

Figure 16-9 The energy flow of a heat engine. An amount of heat Q_H is picked up from a high-temperature reservoir. Some of this heat is converted by the engine into useful work W; the balance, Q_L, is deposited in a low-temperature reservoir.

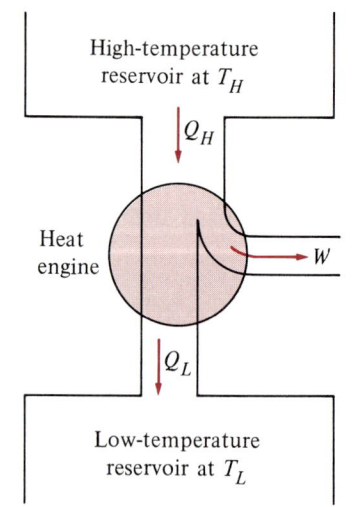

during the cycle since the engine returns to its original state and the net change in internal energy is zero. That is, for each cycle of a heat engine,

$$\Delta U = 0 = \Delta Q - W$$

$$W = \Delta Q = Q_H - Q_L$$

We define the **efficiency** ϵ of a heat engine (or any engine) as the ratio of the useful output energy to the input energy. For a heat engine, the useful output is the work and the input is the heat absorbed from the high-temperature reservoir:

$$\epsilon = \frac{W}{Q_H} \qquad (16\text{-}16)$$

We often write

$$\epsilon = \frac{Q_H - Q_L}{Q_H} \qquad (16\text{-}17)$$

or sometimes

$$\epsilon = 1 - \frac{Q_L}{Q_H} \qquad (16\text{-}18)$$

Example 16-5 An automobile engine operates with an efficiency of 32 percent. It does 8.00×10^3 J of work during each cycle. Calculate (a) the amount of heat absorbed per cycle at the high temperature of the burning gases and (b) the amount of heat the engine delivers as waste to the low-temperature reservoir (the outside air) during each cycle.

Solution (a) We apply Equation 16-16 and solve for Q_H.

$$\epsilon = \frac{W}{Q_H}$$

or

$$Q_H = \frac{W}{\epsilon} = \frac{8.00 \times 10^3 \text{ J}}{0.32} = 2.50 \times 10^4 \text{ J}$$

(b) This requires that we recognize that the work done per cycle is the difference between the heat absorbed and the heat rejected by the engine. Solving $W = Q_H - Q_L$ for Q_L gives

$$Q_L = Q_H - W = 2.50 \times 10^4 \text{ J} - 8.00 \times 10^3 \text{ J}$$
$$= 1.70 \times 10^4 \text{ J}$$

Thirty-two percent is a very good efficiency for an automobile engine, and yet the waste heat is more than twice the useful work.

The efficiency is a number between 0 and 1 but is sometimes multiplied by 100 and reported as a percentage. On a PV diagram, a general heat engine cycle may be represented as any irregular closed curve, such as that shown in Figure 16-10. Assume the system starts at point 1 and moves clockwise around the cycle. As the system moves from 1 to 2, the volume increases and work is done

Figure 16-10 The work done as our system starts at point 1 and moves to point 2 is W_1, the sum of the colored and gray shaded areas. As the system returns from 2 to 1, the work done by the environment on it is W_2, the gray shaded area. The net work done by the system is then $W_1 - W_2$, the colored area enclosed by the curve representing the full cycle.

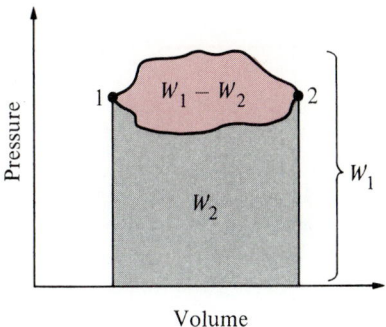

by the system on its surroundings. This work, W_1, is the area under the curve running from 1 to 2, that is, the sum of the gray and colored shaded areas. As the system moves from 2 back to 1 along the lower part of the cycle, the volume decreases and work is done on the system by its surroundings. This work, W_2, is the gray shaded area. The net work done by the system in one cycle is $W_1 - W_2$, the work done by the system as it expands from 1 to 2 minus the work done on the system as it is compressed from 2 to 1. Therefore, the net work done per cycle is the colored area enclosed by the curve. It should be emphasized that every point on the PV plane is an equilibrium point. Thus, only reversible processes may be plotted on a PV diagram. For any irreversible process, only the initial and final points may be plotted.

Any reversible cycle may be run either forward or backward. For example, the system shown in Figure 16-10 may be moved along the bottom part of the curve from 1 to 2 and back along the top part of the curve from 2 to 1. If the cycle were reversed in this way, the system would be a **refrigerator** rather than an engine. A net amount of work W would have to be done to remove an amount of heat Q_L from the lower-temperature reservoir. These two quantities would combine to form the amount of heat Q_H deposited at the higher-temperature reservoir. A refrigerator is represented schematically in Figure 16-11. The work done on the system combines with the heat removed from the lower temperature and is expelled at the higher temperature. Figure 16-11 is simply Figure 16-9 with the direction of the arrows reversed.

For refrigerators, a quantity called the **coefficient of performance** (COP) is defined to describe the cooling ability of the refrigerator for the energy used. The COP is the ratio of the amount of heat removed from the lower-temperature reservoir to the amount of work which must be done. This is the ratio of the useful return to what it costs in energy. Thus,

$$\text{COP} = \frac{Q_L}{W} \qquad (16\text{-}19)$$

or

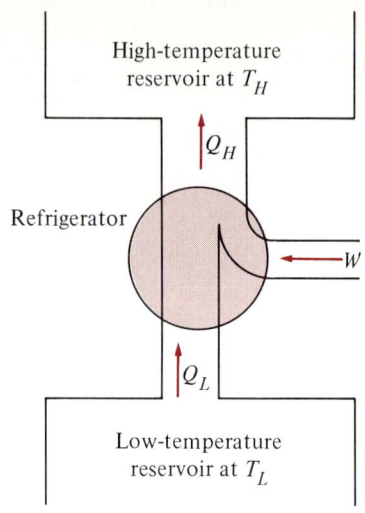

$$\text{COP} = \frac{Q_L}{Q_H - Q_L} \qquad (16\text{-}20)$$

Figure 16-11 In a refrigerator, work W is done on the system as heat Q_L is absorbed from the low-temperature reservoir. This heat, plus the energy representing the work input, $W + Q_L$, is then deposited in the high-temperature reservoir as Q_H, such that $Q_H = W + Q_L$.

Example 16-6 Determine the coefficient of performance for a system like that of Example 16-5 if it were run backward as a refrigerator.

Solution This is a straightforward application of Equation 16-19.

$$\text{COP} = \frac{Q_L}{W} = \frac{1.70 \times 10^4 \text{ J}}{8.00 \times 10^3 \text{ J}} = 2.13$$

16-6 A Practical Cycle: The Otto Cycle

The **Otto cycle** closely approximates the common gasoline engine used in automobiles, and is of more practical interest. It is named after Nicholas Otto, a German engineer who manufactured gasoline (internal combustion) engines in the latter part of the nineteenth century. The most widely used internal combustion engine is called a four-stroke cycle engine because four separate move-

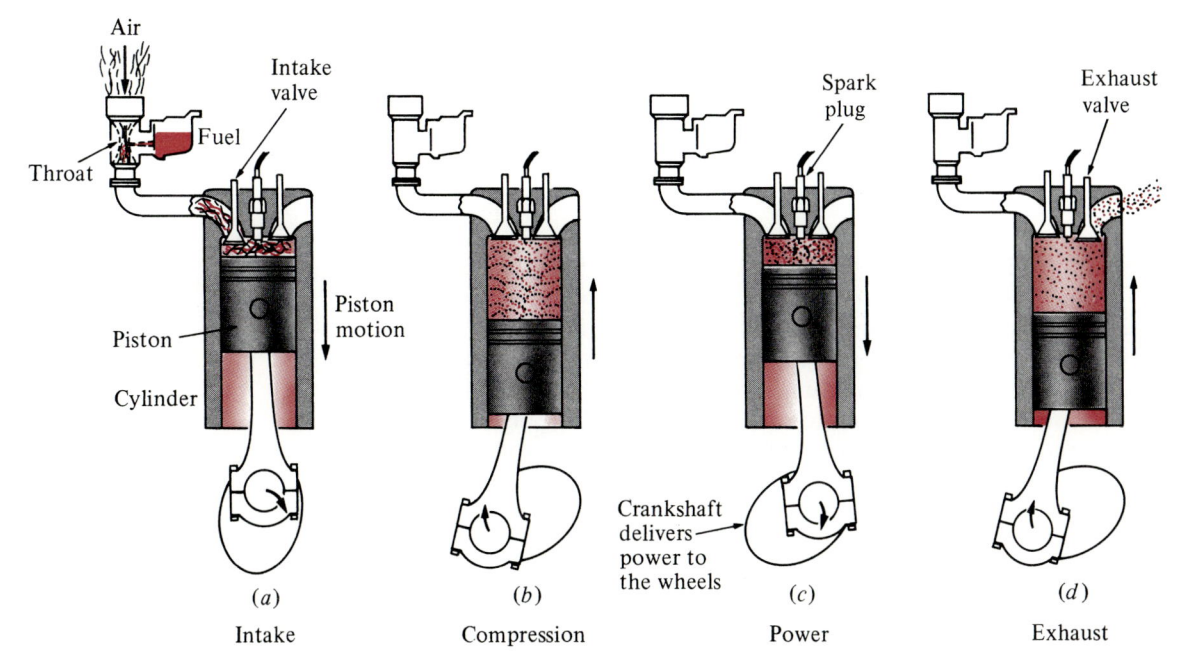

Figure 16-12 The four-stroke cycle engine produces useful work from heat. Air rushing through the throat of the carburetor vaporizes a jet of fuel, and the air-fuel mixture is the working fluid of the engine.

ments or **strokes** of a piston in a cylinder take place during each cycle. Figure 16-12 shows the start of each stroke for a one-cylinder engine as it proceeds through the cycle. Briefly, the cycle is as follows:

In the **intake stroke** (Figure 16-12a), the piston moves downward, the intake valve is open, and a mixture of air and fuel moves into the space behind the piston through the intake valve.

In the **compression stroke** (Figure 16-12b), the piston moves back upward, rapidly compressing the air-fuel mixture since the valves are now closed. Near the end of the stroke, the spark plug fires, igniting the mixture.

In the **power stroke** (Figure 16-12c), the piston moves back down under the action of the large pressure increase caused by the increase in temperature of the burning mixture. Work is done during this stroke by the expanding gases. The valves remain closed.

In the **exhaust stroke** (Figure 16-12d), the piston moves back up, forcing the combustion products out through the now-open exhaust valve. The exhaust valve then closes, and the cycle is ready to begin again.

A diagram of the Otto cycle, the idealized version of the internal combustion engine cycle, appears in Figure 16-13. The system passes through the cycle in the order ABCDEFA. At A, the exhaust valve has just closed and the piston is up. The intake stroke occurs from A to B as the fuel-air mixture is drawn in under constant pressure. The compression stroke takes place from B to C and is shown as an adiabat. This approximation is a good one since the compression takes place very rapidly and there is no time for heat to be absorbed or expelled by the system (gas-air mixture). The spark plug firing and the rapid rise in temperature is shown as C to D. The power stroke shown from D to E is also

Figure 16-13 The Otto cycle. The shaded area represents the useful work done during the cycle, while the area between line CB and the V axis represents the waste heat. To increase the useful work, compression could be carried past C closer to the P axis, line DE could be raised by generating more heat from the burning fuel, etc.

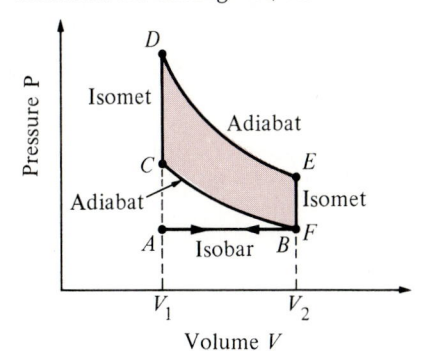

385

adiabatic because of the speed with which it takes place. At the end of the power stroke, shown by the isomet from E to F, the exhaust valve opens and the pressure drops to atmospheric at constant volume. The exhaust stroke then decreases the volume at constant pressure from F to A back to the beginning.

The Otto cycle is often drawn without showing the A to B and F to A portions. This is fine, since those two isobaric processes taken together do not contribute to the net work done during the cycle. As usual, the net work per cycle is the area enclosed by BCDEB. Heat is added to the system during the isometric process C to D. This heat added could be used along with the work done to calculate the efficiency of the engine. Derivations show that the efficiency of an Otto cycle may be expressed in terms of V_2/V_1, the ratio of the largest to smallest volume of the system, and γ, the ratio of molar heat capacities for the fuel-air mixture. If V_2/V_1 (called the **compression ratio**) is 10, the Otto cycle efficiency is approximately 60 percent. While the Otto cycle is highly idealized, ignoring energy losses to friction, heating engine parts, and so on, it does yield an upper limit that one might expect from a common internal combustion engine. Another very special engine cycle is discussed in the next section.

16-7 An Ideal Cycle: The Carnot Cycle

In 1824, a French engineer named Sadi Carnot announced the design of an idealized reversible cycle which represents, in principle, *the most efficient heat engine possible*. This cycle, known as the **Carnot cycle**, is shown in Figure 16-14. As you can see, the area enclosed, hence the work done per cycle, is bounded by two adiabats and two isotherms. Heat is absorbed by the system at a heat reservoir of high temperature T_H and expelled at the low temperature T_L. The system moves clockwise on the cycle. Work is done *by* the system as it expands along the top isotherm from A to B and along the adiabat from B to C on the right. Work must be done *on* the system to compress it along the bottom isotherm from C to D and along the adiabat from D to A on the left.

We can obtain an expression for the efficiency of a Carnot cycle in terms of the absolute temperature. First, we note that for a system in a Carnot cycle, heat is absorbed only during the isothermal process at the higher temperature and expelled only during the isothermal process at the lower temperature. If we label these Q_H and Q_L, the efficiency may be written by using Equation 16-16 as

$$\epsilon = \frac{W}{Q_H}$$

where the first law tells us that the net work done per cycle may be written as

$$W = Q_H - Q_L$$

since the system is returned to its original state at the end of each cycle. Thus Equation 16-17 applies to the Carnot cycle, and we write

$$\epsilon = \frac{Q_H - Q_L}{Q_H}$$

for a Carnot cycle.

From Equation 16-11, we know that the heat added at T_H is

Figure 16-14 The Carnot cycle. The cycle is bounded by two isotherms at the extreme temperatures and connected by two adiabats. The useful work done per cycle is the shaded area.

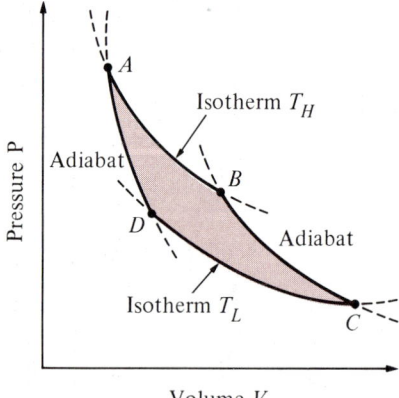

$$Q_H = nRT_H \ln \frac{V_B}{V_A}$$

while the heat expelled at T_L is

$$Q_L = nRT_L \ln \frac{V_C}{V_D}$$

Thus we can write that efficiency of a Carnot cycle is

$$\epsilon = \frac{nRT_H \ln(V_B/V_A) - nRT_L \ln(V_C/V_D)}{nRT_H \ln(V_B/V_A)}$$

It can be shown that $V_B/V_A = V_C/V_D$.[1] Thus the efficiency of a Carnot cycle becomes

$$\epsilon = \frac{T_H - T_L}{T_H} \qquad (16\text{-}21)$$

or

$$\epsilon = 1 - \frac{T_L}{T_H} \qquad (16\text{-}22)$$

Therefore, the efficiency of a Carnot cycle depends only on the absolute temperatures of the high- and low-temperature reservoirs. The higher T_H or the lower T_L, the greater the efficiency.

In a practical heat engine such as those used to generate electricity, the high temperature is provided by the fuel (gas, oil, coal, or nuclear). The low temperature is usually the ambient temperature of some part of our environment (the atmosphere, a river, lake, or pond). Most often, we have little control over the low temperature. Thus, to obtain optimum efficiency, we require the maximum practical high-temperature reservoir. In practice, we find limitations on this maximum. In addition to the natural combustion temperatures of various fossil fuels, the materials used for combustion chambers are limited in the highest temperature they can sustain in operation without failure.

Example 16-7 Modern fossil-fuel-fired plants have combustion chambers made of materials which can be operated at 500°C. The reactors of nuclear power plants must be made of materials which will sustain operation at 300°C. If the waste heat of two such plants is delivered to cooling ponds at 25°C, calculate the Carnot efficiency of each type of plant.

Solution We must first convert to absolute temperatures. Then we may apply Equation 16-21 or 16-22 to each situation.

Coal:
$$T_H = 500 + 273 = 773 \text{ K}$$
$$T_L = 25 + 273 = 298 \text{ K}$$

$$\epsilon = \frac{T_H - T_L}{T_H} = \frac{773 \text{ K} - 298 \text{ K}}{773 \text{ K}} = 0.614 \qquad \text{or } 61.4\%$$

---

[1] This proof requires the use of the ideal gas law and Equation 16-15 along with some algebraic manipulation. You might give it a try.

Nuclear:
$$T_H = 300 + 273 = 573 \text{ K}$$
$$T_L = 298 \text{ K}$$
$$\epsilon = 1 - \frac{298 \text{ K}}{573 \text{ K}} = 0.48 \quad \text{or } 48\%$$

The Carnot efficiency is the ideal thermal efficiency and may be misleading when applied to a practical situation. In practice, both mechanical and thermal losses reduce the overall efficiency from that of the ideal. Friction in the bearings and other moving parts accounts for most of the mechanical losses. Frictional losses can be reduced but never completely eliminated. More important for a heat engine or refrigerator are the thermal losses, such as the hot gases leaving the smokestack of a fossil-fuel-fired system or the warm air which rushes inside when the refrigerator door is opened. The maximum overall efficiency of modern power plants (either fossil or nuclear) is reduced to about 35 to 40 percent. An interesting consequence is that nearly two-thirds of the heat generated is released to the environment. This waste heat gives rise to what has been labeled **thermal pollution** but is a requirement for operation.

A Carnot cycle may be postulated to operate backward as a Carnot refrigerator since the processes are reversible. The coefficient of performance for a Carnot refrigerator can be shown to be

$$\text{COP} = \frac{T_L}{T_H - T_L} \quad (16\text{-}23)$$

16-8 The Second Law of Thermodynamics

Would you expect the water in a swimming pool on the earth's equator to suddenly give up heat to the atmosphere and freeze solid in a fraction of a second? Could the water in a glass sitting on the lecture table do that? If you accidentally grasped a red-hot element on an electric stove, what would you say the chances were that your hand would get colder while the element got hotter? If your reaction to the above is typical, you are saying "That's absurd! Things simply don't happen in that way." Of course, that reaction is entirely correct; things do not happen that way, but why not? These processes do not violate the first law of thermodynamics (since we can account for all the energy). Yet we know from experience that these and a whole host of other processes we might name just do not take place in nature. Processes always proceed naturally in one direction. When a hot body and a colder body are placed in contact, the hot body loses heat and the colder body gains heat, which leads to an attempt to formulate the second law of thermodynamics.

The second law of thermodynamics is not easily stated in compact form, and there are many different correct ways to say it. It places limits on what processes are possible. Several early attempts stated specific limitations, but as the science progressed, the statements became more general. Historically, the laws of thermodynamics began with the statement of the second law by Rudolf Clausius in 1850: "*It is impossible for a self-acting machine, unaided by an external agency, to convey heat from one body to another at a higher temperature.*" In 1851, Lord Kelvin announced his statement of the second law, which we paraphrase as: *It is impossible to construct a heat engine which can convert its heat* input *completely* into work. These two statements can be shown to be equivalent in that if either is violated, the other is violated also.

Scientists believe it is necessary to express discovered laws and relationships in a quantitative way. It is not obvious that either of these statements of the second law may be so stated, but when Clausius discovered that the ratio of heat to absolute temperature had special properties which related to the second law, a quantitative formulation was born. Naming this ratio **entropy** (from the Greek "turning into"), Clausius proceeded to a quantitative law. If a system absorbs a quantity of heat ΔQ at an absolute temperature T, its change in entropy ΔS is defined as

$$\Delta S = \frac{\Delta Q}{T} \tag{16-24}$$

The SI units for entropy are joules per kelvin (J/K). It is found that the change in entropy of an isolated system is never negative. We will see that this leads to a final statement of the second law.

Example 16-8 Determine the change in entropy of 2.00 kg of ice at 0°C which is melted to form 2.00 kg of water at 0°C.

Solution We determine the heat absorbed by using the latent heat of fusion of ice and then apply Equation 16-24.

$$\Delta Q = mL_f = (2.00 \text{ kg})(3.35 \times 10^5 \text{ J/kg}) = 6.70 \times 10^5 \text{ J}$$

Remembering that we must use absolute temperature, $T = 273$ K, we then have

$$\Delta S = \frac{\Delta Q}{T} = \frac{6.70 \times 10^5 \text{ J}}{273 \text{ K}} = 2.45 \times 10^3 \text{ J/K}$$

The ice of Example 16-8 absorbed heat. Thus ΔQ and hence ΔS are positive. If water freezes to become ice, the change in entropy is negative. However, water could not do this in isolation. Its surroundings would have to absorb heat and therefore have a positive change in entropy. Consider the net change in entropy of the system of Example 16-9.

Example 16-9 We place 2.00 kg of water at 100°C and 2.00 kg of water at 0°C together in a perfectly insulated calorimeter cup which absorbs no heat. That is, assume the system is 4.00 kg of water and is completely isolated. If the system reaches a final temperature of 50°C, what is the net change in entropy?

Solution We must find the change in entropy of each half of the system separately and add these algebraically to find the net change. To apply Equation 16-24, we will use the average temperature for each half (75°C for the hot water and 25°C for the cold water). This is an excellent approximation for the conditions we have chosen, but for an exact answer and a method which works for any condition, one must use integral calculus.
First, we calculate the heat loss of the hot water:

$$\begin{aligned}\Delta Q_{hot} &= cm\,\Delta T_{hot} \\ &= [4.18 \times 10^3 \text{ J/(kg} \cdot °\text{C)}](2.00 \text{ kg})(50 - 100)°\text{C} \\ &= -4.18 \times 10^5 \text{ J}\end{aligned}$$

16-8 The Second Law of Thermodynamics

Using this heat loss, the entropy change of the hot water is

$$\Delta S_{hot} = \frac{\Delta Q_{hot}}{T_{av}} = \frac{-4.18 \times 10^5 \text{ J}}{(273 + 75) \text{ K}} = -1.20 \times 10^3 \text{ J/K}$$

Now, the heat gained by the cold water is

$$\Delta Q_{cold} = cm \, \Delta T_{cold}$$
$$= [4.18 \times 10^3 \text{ J/(kg} \cdot {}^\circ\text{C)}](2.00 \text{ kg})(50 - 0)^\circ\text{C}$$
$$= +4.18 \times 10^5 \text{ J}$$

giving the cold water an entropy change of

$$\Delta S_{cold} = \frac{\Delta Q_{cold}}{T_{av}} = \frac{+4.18 \times 10^5 \text{ J}}{(273 + 25) \text{ K}} = +1.40 \times 10^3 \text{ J/K}$$

Finally, the total entropy change of the isolated system is

$$\Delta S = \Delta S_{hot} + \Delta S_{cold}$$
$$= -1.20 \times 10^3 \text{ J/K} + 1.40 \times 10^3 \text{ J/K} = +200 \text{ J/K}$$

Note that Example 16-9 yields a positive net change in entropy for the isolated system. As was stated, it is found that the net change in entropy of an isolated system is never negative. In fact, we most often state the **second law of thermodynamics** as follows:

> The only possible processes which an isolated system can undergo are those during which the net entropy either increases or remains constant.

This can be shown to be equivalent to the statements of Clausius and Kelvin. Further, it is equivalent to the many different correct statements of the second law. In symbols,

$$\Delta S \geq 0 \quad \text{(for an isolated system)}$$

is a good statement of the second law.

Let us further probe the question of why certain processes may proceed in one direction only. What is the problem? No energy is lost to the system in Example 16-9 during the process. Something must be lost, however, or the process could be reversed, heating two of the kilograms of water to 100°C while the other two kilograms cool to 0°C. We repeat that this does not happen, and the answer to "what is lost?" is *opportunity*. When the two halves of the system were at a 100°C temperature difference, we could have run a small heat engine by using the hot water as the high-temperature reservoir and the cold water as a low-temperature reservoir. We could then have extracted a certain amount of work from the system. After the system has reached the equilibrium temperature of 50°C, this opportunity is no longer available. Thus you will often see *entropy* defined as being a measure of the internal energy of a system which is unavailable to do work.

There is also a statistical approach to the definition of entropy which we briefly consider. A coin has two possible states when flipped, heads or tails. If 100 coins are flipped, it is not impossible for all 100 to land heads up, merely improbable. Dice have six possible states, one to six. The probability of rolling 100 dice and producing 100 sixes is even less. Molecules have many, many possible states, and a tiny sample contains billions of molecules; thus, the likelihood of molecules conspiring to all lapse into some improbable state simultaneously is infinitesimal. For example, the likelihood of all the molecules in cold lemonade striking the faster-moving molecules of your hand in such a manner as to lose kinetic energy in the collisions—burning your hand and freezing the lemonade in the process—is vanishingly small and is never observed. Most of the time, "cold" molecules absorb energy from "hotter" molecules during a collision. Thus we see that the net energy moves from the hotter hand to the cooler lemonade. As a system proceeds toward a more probable state, its entropy increases and the system becomes more random or less ordered. Ordered systems, such as 100 sixes, are less probable than randomized systems.

As another example, consider the combustion of gasoline. The molecules of gasoline are highly ordered structures, whereas the combustion products of gasoline dispersed in the air are much less so. Only a highly organized mechanism driven by external energy can make gasoline from carbon dioxide and water. Spilled on the ground, however, gasoline will eventually degrade to CO_2 and H_2O naturally. Natural processes in isolated systems proceed toward higher degrees of randomness, or **disorder.** Another way of saying this is that their entropy tends to increase.

What we are saying in these examples is that entropy is a measure of the disorder of a system. Ice at 0°C is a more ordered system than water at 0°C. Thus water, which has much more random thermal motion of molecules, has a larger entropy than ice. Gases, which have much greater thermal disorder, have larger entropies than liquids, and so on. Thus, another way to state the second law of thermodynamics is "Microscopic disorder of a system and its surroundings does not spontaneously decrease."

Before closing this chapter, we should state that many scientists list a *third law of thermodynamics:*[1]

> It is impossible to reach the absolute zero of temperature in a finite number of steps of physical processes.

This is made plausible by noting that

$$\epsilon = 1 - \frac{T_L}{T_H} \tag{16-22}$$

The efficiency of a Carnot cycle would be 100 percent if the low-temperature reservoir were at absolute zero. But we saw that the efficiency may also be written as

$$\epsilon = 1 - \frac{Q_L}{Q_H} \tag{16-18}$$

This will be 100 percent only if Q_L, the heat deposited to the low-temperature reservoir, is zero. This would violate the Kelvin statement (and hence the Clausius statement and all other statements) of the second law. Therefore, the attainment of absolute zero, which is then used as a low-temperature reservoir for a Carnot cycle, violates the second law of thermodynamics and is impossible.

---

[1] Some scientists insist that what is called the third law is really only an extension of the second law. In any case, scientists nearly all accept the statement that absolute zero is unattainable.

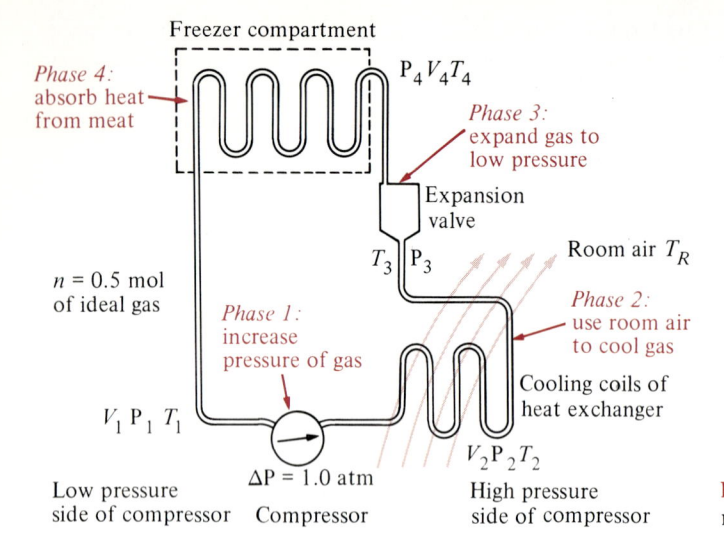

Figure 16-15 Schematic diagram of a refrigerator.

16-9 Application: The Refrigerator

Many of the relationships of the preceding sections were developed after heat engines became so widely used that there was a real need to understand the principles by which they worked. The relationships may seem fairly abstract to you at first reading, so we will devote this entire section to working through the operating cycle of a common heat engine, the refrigerator.

At first, consideration might lead you to question whether the refrigerator violates a form of the Clausius statement of the second law of thermodynamics: heat flows naturally from hotter objects to colder ones. It does not, but to understand this, we will examine the complete cycle of operations consisting of four phases. Phase 1, a gas is compressed, raising it to a high temperature; phase 2, it is kept pressurized but cooled to nearly room temperature; phase 3, it is expanded into a partial vacuum, dropping its temperature well below the freezing point of water; and phase 4, it passes near to the matter to be cooled, absorbing enough heat to return to its initial condition (Figure 16-15). After this, the cycle repeats.

We will begin with a cool slab of beef of mass 1.00 kg, recently purchased unfrozen and at 5°C. Our objective is to freeze this meat, which is placed in the freezer compartment of our refrigerator. To calculate what is happening, we will use an ideal refrigerator, that is, a refrigerator using a confined ideal gas. Real refrigerators use nonideal gases, which actually work much better for this purpose. However, the expressions we have derived are valid only for ideal gases, and the treatment of nonideal gases is more complex. Arbitrarily, we have chosen the gas neon as a coolant. Examination of Table 16-4 gives its important properties. Neon behaves very much like an ideal gas unless very high pressures or low temperatures are achieved.

Phase 1: Compression

We start with 0.500 mol of gas confined in tubes near the meat and at the same temperature of 5°C. The gas is initially at a pressure of 0.75 atm (7.60×10^4 N/m²). The gas is fed to a compressor which increases its pressure by

---

Real refrigerators use about 5 mol of refrigerant, usually Freon 12, or dichlorodifluoromethane, CCl_2F_2. Most of this is in a liquid state since Freon liquefies at room temperature when under increased pressure. Ideal gases are not allowed any inclination to liquefy—that is why they are ideal. Because of its change of state during the cycle, Freon is a much better refrigerant than an ideal gas.

1.00 atm over its inlet pressure. Since this process occurs very quickly, there is little time for heat exchange in the compressor and it is essentially adiabatic. We may calculate the outlet pressure to be

$$P_2 = P_1 + 1.00 \text{ atm}$$
$$= 0.75 \text{ atm} + 1.00 \text{ atm} = 1.75 \text{ atm} \quad (= 1.77 \times 10^5 \text{ N/m}^2)$$

We may calculate the original volume of the gas from the ideal gas law:

$$V_1 = \frac{nRT_1}{P_1} = \frac{(0.500 \text{ mol})[8.31 \text{ J/(mol} \cdot \text{K)}](278 \text{ K})}{7.60 \times 10^4 \text{ N/m}^2}$$
$$= 0.0152 \text{ m}^3 = 15.2 \text{ L}$$

TABLE 16-4 Properties of Neon Gas

$C_P = 20.77$ J/(mol·K)
$C_V = 12.46$ J/(mol·K)
$\gamma = 1.67$
Mol. mass = 20.18 g/mol
Freezing pt. = $-248.67°$C
Boiling pt. = $-246.05°$C
Specific heat = 1000 J/(kg·K)

Alternatively, we could have obtained the original gas volume from the knowledge that 1 mol of a gas occupies 22.4 L of volume at 0°C and 1.00 atm. We have 0.500 mol, which would thus occupy 11.2 L at standard conditions. Converting to our initial conditions of 0.75 atm and 5°C, we obtain

$$\frac{P_0 V_0}{T_0} = \frac{P_1 V_1}{T_1}$$

$$V_1 = V_0 \frac{P_0}{P_1} \frac{T_1}{T_0}$$

$$= 11.2 \text{ L} \times \frac{1.00 \text{ atm}}{0.75 \text{ atm}} \times \frac{278 \text{ K}}{273 \text{ K}}$$

$$= 15.2 \text{ L*}$$

Following compression, both temperature and volume are unknowns. We may solve for the volume by using Equation 16-15 for an adiabatic process:

$$P_1 V_1^\gamma = P_2 V_2^\gamma$$

$$V_2 = \sqrt[\gamma]{V_1^\gamma \frac{P_1}{P_2}}$$

$$= \sqrt[1.67]{(15.2 \text{ L})^{1.67} \times \frac{0.75 \text{ atm}}{1.75 \text{ atm}}}$$

$$= 9.15 \text{ L}$$

Since work was done on the gas and heat transfer did not occur, there will also be a change in internal energy, hence, in temperature. From the ideal gas law, we calculate the new temperature.

$$\frac{P_1 V_1}{T_1} = \frac{P_2 V_2}{T_2}$$

$$T_2 = T_1 \frac{P_2}{P_1} \frac{V_2}{V_1}$$

$$= 278 \text{ K} \times \frac{1.75 \text{ atm}}{0.75 \text{ atm}} \times \frac{9.15 \text{ L}}{15.2 \text{ L}}$$

$$= 390 \text{ K} \quad (= 117°\text{C})$$

---

* If you are a critical reader, you might note that 15.2 L is a considerable volume. It is roughly equivalent to four 1-gal milk jugs in the refrigerator. You might question whether the volume of the fine tubes in the refrigerator adds up to this much. Actually, the 0.500 mol we are using represents the gas in the entire system. Refrigerators operate in a continuous loop, but we are imagining that all our gas is in one phase of operation at once.

Thus the gas is now very hot compared to room temperature since it lost no heat as it was compressed. Since $\Delta Q = 0$, $\Delta U = -W$ and we may calculate the work done on the gas:

$$\Delta U = nC_V \Delta T = (0.5 \text{ mol})[12.46 \text{ J/(mol} \cdot \text{K)}](390 - 278) \text{ K} = 698 \text{ J}$$

This is a positive change in the internal energy, and thus this amount of work was done on the gas by the compressor, or $W = -698$ J. (The work done *by* the gas is negative.)

Phase 2: Cooling the High-Pressure Gas

The hot, high-pressure gas is circulated through fine tubes, usually underneath or behind the refrigerator. Room air is circulated across these tubes to cool the gas. Since the gas is confined to the volume of the tubes, we will treat this as an *isometric* process, $\Delta V = 0$; hence, $W = 0$ and $\Delta U = \Delta Q$. The gas is cooled from T_2 to room temperature of 20°C, or 293 K.

$$\Delta U = \Delta Q = nC_V \Delta T = (0.5 \text{ mol})[12.46 \text{ J/(mol} \cdot \text{K)}](293 - 390) \text{ K} = -604 \text{ J}$$

This is lost heat and represents lost internal energy or decreased temperature. It differs in amount from the work done because the temperature of the gas is not returned all the way to its starting point at 5°C. Furthermore, since $\Delta V = 0$, the volume remained at 9.15 L. We may calculate the new pressure:

$$P_3 = P_2 \frac{T_3}{T_2} = 1.75 \text{ atm} \times \frac{293 \text{ K}}{390 \text{ K}} = 1.31 \text{ atm}$$

Thus, we now have a cool, high-pressure gas.

Phase 3: Expansion

The gas is now sprayed through a fine valve into a partial vacuum created by the intake side of the compressor evacuating the tubing. Assume the new pressure to be maintained at 0.75 atm by the compressor. The initial expansion is quite rapid, hence, *adiabatic*, but only small amounts of gas are expanded at a time through the valve. We will ignore this and assume that all 0.500 mol is expanded to the 0.75-atm environment. To calculate the new volume of the gas, we use Equation 16-15:

$$P_3 V_3^\gamma = P_4 V_4^\gamma$$

$$V_4 = \sqrt[\gamma]{V_3^\gamma \frac{P_3}{P_4}}$$

$$= \sqrt[1.67]{(9.15 \text{ L})^{1.67} \times \frac{1.31 \text{ atm}}{0.75 \text{ atm}}} = 12.8 \text{ L}$$

We may calculate the temperature T_4 at this new volume from the ideal gas law:

$$T_4 = T_3 \frac{V_4}{V_3} \frac{P_4}{P_3}$$

$$= 293 \text{ K} \times \frac{12.8 \text{ L}}{9.15 \text{ L}} \times \frac{0.75 \text{ atm}}{1.31 \text{ atm}} = 235 \text{ K}$$

The gas is now some 38°C *below* the freezing point of water.

Phase 4: Absorbing Heat from the Freezer

16-9 Application: The Refrigerator

The cooled gas is now circulated near the meat to be frozen. The specific heat of our gas is about 1000 J/(kg·K), and we have

$$(0.500 \text{ mol})(20.18 \text{ g/mol}) = 10.1 \text{ g} = 0.0101 \text{ kg of gas}$$

Since the mass of the meat is 100 times the mass of the gas, we should not expect a great amount of cooling of the meat to occur. The specific heat of the meat is about 4000 J/(kg·K).

Assuming the net heat flow is zero, we have

$$0 = \Delta Q_{gas} + \Delta Q_{meat}$$
$$= m_{gas} C_{gas} \Delta T_{gas} + m_{meat} C_{meat} \Delta T_{meat}$$
$$= (0.0101 \text{ kg})[1000 \text{ J/(kg·K)}](T_f - 235 \text{ K})$$
$$+ (1.00 \text{ kg})[4000 \text{ J/(kg·K)}](T_f - 278 \text{ K})$$
$$= 10.1 T_f - 2374 + 4000 T_f - 1{,}112{,}000$$

$$T_f = 277.9 \text{ K}$$

Here, the fourth significant figure was retained to show the amount of the cooling for one cycle of the gas. Treating the meat as essentially impure water thermally, calculation shows that it will need to lose some 360,000 J to freeze and be cooled to a final temperature of $-10°C$ (near where most freezers keep frozen foods). Evaluating the first term of the last calculation, we note that our gas can absorb about 434 J from the meat for each cycle. Thus, the number of cycles of our gas through the refrigerator to freeze and cool the meat will be

$$N = \frac{360{,}000 \text{ J}}{434 \text{ J/cycle}} = 830 \text{ cycles}$$

This is a very rough number, but, recall that a real refrigerator uses a much greater mass of a nonideal gas and thus requires fewer cycles. Note also that the gas at the meat is now at the starting conditions of 278 K and 0.75 atm and ready to repeat the cycle.

In summary we might note that each transfer of heat—hot gas to room air, meat to cold gas—was from a hotter substance to a colder one. The net effect of the cycle is to take heat from the meat and release it to the room air along with the heat representing the work done by the compressor. The heat lost by the meat in one cycle was 434 J. The work done by the compressor was 698 J. We may calculate the coefficient of performance for our refrigerator as

$$\text{COP} = \frac{Q_L}{W} = \frac{434 \text{ J}}{698 \text{ J}} = 0.622$$

We could also verify that for each heat transfer, the change in entropy was positive. For example, the meat lost heat and the cold gas gained heat. Using the mean temperature of each during the exchange for the calculation, we obtain

$$\Delta S_{tot} = \Delta S_{meat} + \Delta S_{gas}$$
$$= \frac{-434 \text{ J}}{278 \text{ K}} + \frac{434 \text{ J}}{257 \text{ K}}$$
$$= -1.56 \text{ J/K} + 1.69 \text{ J/K} = +0.13 \text{ J/K}$$

Finally, 830 cycles of the gas is required to freeze and cool the meat. Each cycle requires some 698 J of work from the compressor, for a total of 579,000 J of work to accomplish the task. Since 1 W = 1 J/s, we will need 579,000 Ws of work, or

$$(579 \text{ kWs})(1 \text{ h}/3600 \text{ s}) = 0.161 \text{ kWh}$$

The work done by the compressor comes from the electricity provided it. A typical charge for electric energy is \$0.080/kWh. Thus, freezing the meat will cost

$$(0.161 \text{ kWH})(\$0.080/\text{kWh}) = \$0.013$$

This adds less than a cent per pound to the cost of the meat. While our refrigerator had a fair coefficient of performance, a real refrigerator does even better and also keeps the cost of energy under a cent a pound. Society has found that this is a small price to pay for the reduced spoilage and prevention of disease provided by refrigeration.

Minimum Learning Objectives

After studying this chapter, you should be able to:

1. Define:
   - adiabat
   - adiabatic process
   - C_P
   - C_V
   - Carnot cycle
   - coefficient of performance
   - compression ratio
   - compression stroke
   - disorder
   - efficiency
   - engine stroke
   - entropy
   - exhaust stroke
   - first law of thermodynamics
   - heat engine
   - heat reservoir
   - intake stroke
   - isobar
   - isobaric process
   - isolated system
   - isomet
   - isometric process
   - isotherm
   - isothermal process
   - isovolumetric process
   - molar heat capacity
   - Otto cycle
   - power stroke
   - reversible thermodynamic process
   - second law of thermodynamics
   - system
   - thermal pollution
   - thermodynamic process
   - thermodynamics

2. Write the appropriate form of the first law of thermodynamics for isobaric, isothermal, isometric, and adiabatic processes for an ideal gas.
3. Distinguish the molar heat capacity at constant pressure and at constant volume. Indicate which is larger and why, and identify the magnitude of their difference.
4. Calculate the work done in isobaric, isothermal, and adiabatic processes, given the appropriate changes of state in an ideal gas.
5. Calculate the change in the internal energy of a system, given the work done on or by it and the heat transferred to or from it.
6. Calculate the work done in a heat engine cycle operating with an ideal gas and employing defined processes between specified states.
7. Calculate the efficiency of any engine, given the input energy and the useful output energy.
8. Calculate the coefficient of performance of a refrigerator, given the amount of heat removed and the amount of work done.
9. Calculate the entropy change of a system, given the heat transfers occurring and the temperatures at which they occur.

Problems

16-1 A system is changed from state A to state B, increasing its internal energy by 4.00×10^4 J while 6.00×10^4 J of heat is added. The system is then returned to state A while 3.00×10^4 J of heat is taken away. Calculate (a) the work done by the system as its state was changed from A to B, (b) the work done by the system returning to A, and (c) the net work done in the round-trip.

16-2 How much heat in calories is added to a system if the system's internal energy is increased by 4.00×10^3 J, and 2.50×10^3 J of work is done (a) by the system and (b) on the system?

16-3 If 1.50×10^3 cal of heat is added to a system, how much work is done by the system if its internal energy (a) increases by 4.00×10^3 J and (b) decreases by 4.00×10^3 J?

16-4 A new tire of internal volume 3.24×10^{-2} m³ is mounted on the wheel of a car. Air is compressed adiabatically from 20°C and 1.00 atm to a gauge pressure of 30.0 lb/in²

($P_{abs} = 3.08 \times 10^5$ N/m²) to fill the tire. (a) How much work is done on the air? (Use the average value of the pressure while the tire is being filled.) (b) What is the temperature of the air inside the inflated tire before the rubber is warmed? (c) By how much has the internal energy of the air been increased? (Air is essentially nitrogen gas, N_2, a diatomic molecule.)

16-5 Six moles of an ideal monatomic gas initially at a pressure of 1.50×10^5 Pa and volume of 0.300 m³ absorbs 1.60×10^4 J of heat in a reversible process. (a) How much work is done by the gas if the process is isometric? Isobaric? (b) What is the final temperature if the process is isometric? Isobaric?

16-6 When 2 mol of hydrogen gas (H_2) and 1 mol of oxygen gas (O_2) react, 2 mol of water vapor (H_2O) is formed and 2.42×10^5 J of heat is liberated for *each* mole of vapor formed. Assume that this reaction is conducted in the adiabatic enclosure of a calorimeter initially at 140°C and at an absolute pressure of 4.00×10^5 N/m². (Note that the number of moles of vapor changes almost instantly during the reaction and before pressure, volume, and temperature adjustments occur.) (a) If the chamber has a constant volume, what is the new pressure on the walls? What is the new temperature? (b) If, instead, the chamber is isobaric and may change volume, what amount of work will be done? What will the change in volume be? For this water vapor (at very high pressure and temperature), use $C_V = 42.0$ J/mol·K and $C_P = 50.3$ J/mol·K.

16-7 An airliner with a pressurized 450-m³ cabin kept at 20°C and 1.00 atm of pressure is flying at high altitude with an outside pressure of 0.420×10^5 N/m² and a temperature of −50°C. (a) If a crack develops in the cabin wall, how much work is the internal air capable of doing in expanding to the outside? (b) How high into the air would this amount of work propel an automobile (mass = 1000 kg) initially at ground level?

16-8 A quantity of gas initially at a pressure of 2.00×10^5 Pa and a volume of 0.600 m³ has 3.50×10^3 J of heat added to it as it expands isobarically to a volume of 0.800 m³. Calculate (a) the work done by the system during this expansion and (b) the change in internal energy of the system.

16-9 A World War II submarine had a mass of about 1.20×10^6 kg. The volume of the sub could be divided into two parts: (i) the pressure hull of the boat itself, with crew quarters, machinery, etc., having a volume of 1.10×10^3 m³, and (ii) diving tanks outside the hull of volume 2.00×10^2 m³, which could be filled with air or water. When submerging, air was allowed out of the tanks to be replaced by seawater, lowering the effective volume of the boat. To ascend, stored compressed air from inside the hull was forced into the top of the tanks, forcing the seawater out through valves at the bottom and increasing the boat's effective volume. The density of seawater is 1.025×10^3 kg/m³ at 15°C. Neglect the mass of the air involved. (a) With the tanks fully filled with air, what percentage of the sub would be out of water? (b) What is the density of the submarine when the tanks are filled with water? (c) What percentage of the tanks needs to be water-filled to match the density of seawater and allow the sub to remain at a given depth (neutral buoyancy)? (d) At a depth of 70.0 m, what is the absolute pressure of the air in the tanks above the seawater? (Do not forget atmospheric pressure.)

16-10 The stretch in the rubber walls of a balloon keeps the air inside at about 0.101×10^5 N/m² (0.10 atm) above the atmospheric pressure. The balloon is inflated with 37°C air from the lungs and goes from a volume of 10.0 to 1800 cm³. How much work is done?

16-11 How much work is done against the atmosphere in expansion when 0.500 kg of water evaporates at 30°C?

16-12 For the system of Problem 16-5, calculate (a) the initial temperature and (b) the final volume, assuming the heat is added in an isothermal process.

16-13 Two moles of a diatomic ideal gas initially at a pressure of 1 atm and a volume of 4.00×10^{-2} m³ is compressed adiabatically to a volume of 8.00×10^{-3} m³. Calculate (a) the final pressure, (b) the initial and final temperatures, (c) the work done *by* the system, and (d) the change in internal energy of the system.

16-14 Some volcanos erupt violently; others are spectacular but not particularly violent. In fact, their eruptions are tourist attractions (e.g., Kilauea on Hawaii). The difference is partly one of dissolved gases in the molten rock. We will explore the implications of dissolved gases. At some combination of temperature and pressure reduction, gas bubbles will begin to form in molten rock underground. Assume that we have a 1.00-cm³ bubble of gas in molten rock of temperature 1400°C and pressure of 3.04×10^7 N/m² (300 atm). As the molten rock rises, the bubble is confined in a largely adiabatic enclosure [$C_V = 30.0$ J/(mol·K)]. (a) What is the internal energy of the gas bubble at original depth? (b) What will be the internal energy of the gas bubble at 100°C and 1.00 atm at the surface? (c) How much work will be done by the gas during the transition? (d) Assume that this work is done against a 1.00-kg mass of rock initially at rest. What final upward velocity will the rock have? (1.00 m/s = 2.24 mi/h; what is the rock's speed in miles per hour?) (e) Most ejected rock is in the form of a fine powder. What mass would the rock have to have while receiving this energy to attain the escape velocity of the earth of 11.2×10^3 m/s?

16-15 Since rock is a poor conductor of heat, the earth's interior is nearly an adiabatic enclosure. This approximation allows us to calculate the rate at which temperature should increase with depth as being $\Delta T/\Delta y = g\beta T/C_P$, where g = acceleration due to gravity, β = volume coefficient of thermal expansion of rock (~2×10^{-5}/K), C_P = heat capacity at constant pressure, about 0.2 times that of water [or 838 J/(kg·°C)], and T = temperature at a given depth. Assume that at a depth of 200 km the conditions are adiabatic and $T = 1000$°C. (a) What would the

397

adiabatic lapse rate $\Delta T/\Delta y$ be? (b) The earth is 6376 km in radius. What temperature would this predict for the earth's center?

16-16 An ideal gas is allowed to expand adiabatically to three times its original volume. Calculate the ratio of its final pressure to its initial pressure if the gas is (a) monatomic and (b) diatomic.

16-17 One intercity travel scheme uses tubes like subway tubes but with trains tightly fitted to the walls. Air would be pumped out of a tube in front of an accelerating train to behind a train accelerating in the other tube in the other direction. Thus, air pressure would propel (partially) the trains. To brake the train to a stop, the tunnel in front of it would be sealed and the speeding train would be gradually decelerated by the increasing air pressure on its front surface. Assume a frictionless train of mass 6.00×10^3 kg moving at 200 m/s (447 mi/h) down the tube is to be decelerated to 10.0 m/s. The cross-sectional area of the train is 8.00 m², and a section of the tunnel ahead some 2.00 km long is sealed. Assuming adiabatic compression, how far will the train move along the tunnel until the desired speed is reached? Is this scheme reasonable?

16-18 A normal explosive is a material that rapidly changes from a solid or liquid to a gas with the evolution of heat. A 20.0-g bullet is placed into a rifle barrel 1.00 m from its exit point. Some 5.00 cm³ of explosive propellant behind the bullet will produce 0.333 mol of gas with the evolution of 1.26×10^6 J of energy as heat. The expanding gases push against the back of the bullet and accelerate it down the barrel at a great rate. The process occurs so rapidly that it is adiabatic. (a) The instant after explosion, the gas is confined to the original volume of 5.00 cm³. Disregarding the heat evolved, what would the pressure of this gas be at the temperature of its surroundings of 15°C? (1.0 L = 10^3 cm³.) (b) Additionally, before any expansion takes place, the heat from the reaction is added to the gas. To what temperature are the gases raised? What is the final pressure? [Assume $C_V = 30$ J/(mol·K).] (c) The cross-sectional area of the barrel and/or bullet is 1.00 cm². What is the force on the bullet at the moment of pushing it down the barrel? (d) In the absence of frictional forces, what would the acceleration of the bullet be? (e) In reality, there are very large frictional forces involved. The bullet is made of a soft metal, somewhat larger than the barrel. As a result, it is crushed into curved grooves in the metal (riflings) and rapidly spun for greater stability of its trajectory. When the bullet has moved to the exit of the barrel, what is the volume occupied by the gases? What is the pressure at this volume? (The process is adiabatic, and assume $\gamma = 1.28$.)

16-19 We will calculate the rate at which dry air cools as it rises. This is called the dry-adiabatic lapse rate. It is adiabatic since the surrounding air is rising also and thus in thermal equilibrium at all times. Imagine a 1.000-m cube of air initially at 20°C and 1.00 atm of pressure. (a) Pressure drops as we ascend at a rate where the pressure for any height h in meters is $P = P_0 e^{-ch}$, where $c = 1.150 \times 10^{-4}$/m. Calculate the new pressure when our cube of air has risen 100.0 m. (You will need to carry four significant figures in this problem.) (b) As the pressure drops, the volume of the air is expanded. Calculate the new volume of our quantity of air for an adiabatic expansion. (c) Since the process was adiabatic, no heat was lost or gained, but work was done ($W = P \Delta V$). Thus, we expect the temperature has changed. Calculate the new temperature of our cube of air 100.0 m higher than its beginning. (d) The dry-adiabatic lapse rate is usually given as a drop of 1°C/100 m of rise. Note that this is the *dry* rate since affairs could be quite different for humid air. Suppose our air was initially saturated with moisture. How much water vapor would 1.00 m³ contain from Table 15-4? How much could it hold after rising 100 m and undergoing a 1°C temperature change? (e) Energywise, what happens when water vapor is condensed into liquid water in the air? (Energy is transferred from what to what?) About how much energy is involved for our cubic meter of air? Will the *wet*-adiabatic lapse rate be higher or lower than the *dry*-adiabatic lapse rate? (f) Put the preceding answers into a coherent explanation of the following: "Wherever a prevailing moist wind encounters a mountain range, the windward side of the mountains is quite moist, with large amounts of precipitation. The leeward side of the mountains is dry, and if the terrain on this side drops to its original elevation, a hot, dry desert results, hotter than the original air temperature on the windward side."

16-20 As an extension of Problem 16-19, consider the following applied case. Air moving eastward from the Pacific Ocean crosses the low, central valley of California. It then rises over the gentle western slopes of the Sierra Nevada mountain range (running north-south at the eastern edge of most of the state). The air is elevated almost 4.00 km to the Sierra Nevada crest before dropping into the valleys, the lowest of which is Death Valley nearly 100 m below sea level. Assume a dry-adiabatic lapse rate of 1°C/100 m and a wet-adiabatic lapse rate of 0.65°C/100 m. Give the precipitation regime (moist, dry, very dry) and the temperature in degrees Celsius for the following, assuming that saturated Pacific Ocean air at 20°C enters the state: (a) the central valley (elevation approximately sea level); (b) the western Sierra slope (average elevation 2.0 km); (c) the Sierra crest (elevation 4.0 km); (d) the eastern Sierra slope (average elevation 2.0 km); (e) the eastern valleys (elevation approximately sea level). Recall the numbers in this problem the next time someone puzzles that one can stand in a hot, dry desert and look up at the snow-covered ski slopes of a mountain range.

16-21 A heat engine absorbs 5.00×10^4 J during each cycle at the high-temperature reservoir and exhausts 3.00×10^4 J at the low temperature. Calculate (a) the work done per cycle and (b) the thermal efficiency of the engine.

16-22 A heat engine with a thermal efficiency of 25 percent absorbs 2.00×10^4 J during each cycle at the high-temperature reservoir. Calculate (a) the work output per cycle and (b) the amount of heat exhausted per cycle.

16-23 The work output per cycle for a heat engine is 3.00×10^4 J, and its thermal efficiency is 30 percent. Calculate (a) the amount of heat absorbed at the high temperature during each cycle and (b) the amount of heat exhausted per cycle.

16-24 A steam engine has a work output of 2.50×10^4 J each cycle while it loses 5.00×10^3 J to friction, sound, etc. If it exhausts 5.00×10^4 J each cycle, determine (a) the amount of heat absorbed per cycle and (b) the overall efficiency.

16-25 A heat pump is supplied with 4.00×10^4 J of work during each cycle while it exhausts 1.00×10^5 J to the high-temperature reservoir. Calculate (a) the heat absorbed from the low-temperature reservoir and (b) the COP of the heat pump.

16-26 A refrigerator with a coefficient of performance equal to 4.00 removes 2.00×10^3 J from its interior during each cycle. Calculate (a) the amount of work input required during each cycle and (b) the amount of heat exhausted each cycle to the atmosphere.

16-27 Home air conditioners are rated in terms of an energy efficiency ratio (EER). This is the number of Btu/h an air conditioner will remove per watt of electrical power delivered. Since 1.00 Btu/h = 0.293 W, the number is really a unitless measure. A recent national catalog listed air conditioners with an EER range of 8.8 to 5.1. What is the range of the coefficient of performance for this line of air conditioners?

16-28 It requires 3.50×10^3 J of work each cycle for a heat pump to absorb 1.40×10^4 J of heat from the low-temperature reservoir. Calculate (a) the COP of the heat pump and (b) the amount of heat exhausted to the high-temperature reservoir during each cycle.

16-29 A home refrigerator operates between 25 and $-15°C$ with 30 percent of the coefficient of performance of an ideal refrigerator. How much heat per second is removed from the cold reservoir if the refrigerator uses 0.300 hp (224 W)?

16-30 A refrigerator is supposed to absorb 6.00×10^4 J of heat per cycle from the low-temperature reservoir while exhausting 8.00×10^4 J to the high-temperature reservoir. (a) How much work must be done per cycle? (b) What is the refrigerator's COP?

16-31 The theoretical efficiency of an Otto cycle (ideal internal combustion) engine is given by $\epsilon = 1 - 1/(V_2/V_1)^{\gamma-1}$, where V_1 and V_2 are the minimum and maximum volumes of the gases in the cylinder and V_2/V_1 is the compression ratio. (a) Assume an ideal gas of diatomic molecules and calculate the theoretical efficiency of an Otto cycle engine with a compression ratio of 10. (b) What compression ratio is required if the engine is to have a theoretical efficiency of 0.75?

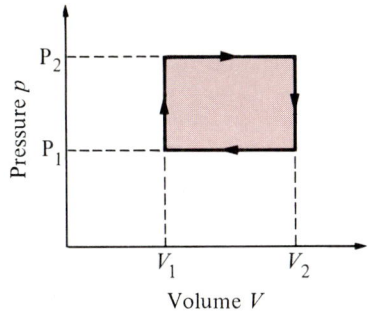

Figure 16-16 Problem 16-32.

16-32 Refer to the PV diagram shown in Figure 16-16. Determine (a) the work per cycle for such an engine if $P_1 = 1.00 \times 10^5$ N/m², $P_2 = 1.60 \times 10^5$ N/m², $V_1 = 0.200$ m³, and $V_2 = 0.600$ m³, (b) the heat added per cycle, and (c) the heat engine efficiency. (Assume a monatomic ideal gas.)

16-33 A heat engine used to run a turbogenerator operates between 700°C and 100°C and works at 60 percent of its Carnot efficiency. Determine how much heat the engine absorbs per cycle if it must do 6.00×10^4 J of work per cycle.

16-34 Calculate the changes during a cycle of an internal combustion engine operating with diatomic ideal gases.

Step 1: An air-fuel mixture at 20.0°C and 1.00 atm of pressure is compressed adiabatically to one-eighth its original volume. The cylinder diameter is 15.0 cm, and the piston motion is 7.00 cm.

Step 2: A spark ignites a chemical reaction which raises the temperature to 1200°C in a small fraction of a second (isovolumetric). (a) Calculate the new pressure at this temperature. (b) Calculate the change in the internal energy of the gas.

Step 3: The gas pushes the piston downward to its original position in an adiabatic expansion. Calculate (c) the new temperature and pressure, (d) the work done by the gas, and (e) the final internal energy of the enclosed gas.

Step 4: The chamber is opened to the air at 20.0°C and 1.00 atm pressure, and the gas is allowed to expand out of the cylinder. (f) Calculate the percentage of the previously enclosed gas expanding out of the cylinder, presuming no heat exchange with the outside air.

Step 5: The piston is pushed upward against the 1.00-atm gas remaining to remove most of it from the chamber. (g) Calculate the work done on the gas.

Step 6: The cycle now begins again at step 1. (h) Calculate the total work done throughout the cycle.

16-35 A Carnot engine operates between 400°C and 100°C. (a) Calculate its efficiency. (b) If it is reversed to operate as a refrigerator between these same temperatures, calculate its COP.

399

16-36 A human being is a very sophisticated heat engine. Let us treat a human being in the most simple-minded application of the efficiency formula for a heat engine. (a) Food Calories are "burned" in the cells at 37°C (body temperature) and waste heat "exhausted" to the environment at the earth's average temperature of about 12°C (54°F). What is the thermal efficiency of the human body? (b) Of course, human beings are really very sophisticated energy converters lacking "little furnaces" in the cells. By actual measurement, a bicycle racer who trains for 8.00 h/day requires an increase in food energy of 3700 Cal beyond his normal, unexercising diet. During the hours of racing, he works at an average rate of 0.13 hp, or 97.0 W. From this data, what is the actual efficiency of energy conversion in the human body? (1.00 Cal = 4184 J.)

16-37 The maximum density of water is achieved at 4°C. This implies that both colder water and hotter water are more buoyant and will float on the 4°C layer (as does ice). Also implied is that the deepest layers of all large bodies of water are near 4°C. Actually, most of the ocean is at this temperature except for the upper few hundred meters. A heat engine could be constructed by using a warm surface current at 30°C as a high-temperature reservoir and the deep, cold water as the low-temperature reservoir. What would the maximum thermodynamic efficiency of such a heat engine be?

16-38 A Carnot engine has an efficiency of 25 percent with a low-temperature reservoir at 100°C. Determine (a) the temperature of the high-temperature reservoir, (b) the value that the high temperature must be in order to increase the efficiency to 75 percent if the low-temperature reservoir does not change, and (c) the value to which the low-temperature reservoir must be decreased to obtain a 75 percent efficiency if the high temperature is kept as in part a.

16-39 A Carnot ice maker (refrigerator) works between reservoirs at temperatures of 23°C and 0°C. If it works long enough to make 50.0 kg of ice, determine (a) the net amount of heat absorbed from the water, (b) the coefficient of performance, (c) the amount of heat exhausted to the room, and (d) the energy (as work) supplied to the refrigerator.

16-40 A Carnot refrigerator takes 4.00×10^3 J of heat per cycle from the freezer compartment at -10°C and ejects heat to the kitchen at a temperature of 23°C. Calculate (a) the coefficient of performance, (b) the work supplied to the engine each cycle, and (c) the heat per cycle ejected to the room.

16-41 A Carnot engine operates from a high-temperature reservoir at 800 K and absorbs 1.60×10^3 J of heat at that temperature during each cycle. If it rejects 6.00×10^2 J to the low-temperature reservoir, calculate (a) the thermal efficiency of the engine and (b) the temperature of the low-temperature reservoir.

16-42 Which process has the greater change in entropy, 1.00 kg of steam condensing to water at 100°C or 5.00 kg of ice melting to water at 0°C?

16-43 Work amounting to 1.50×10^4 J is done on a system of gas as it undergoes an isothermal process at 100°C. The change of entropy of the system is $+50.0$ J/K. (a) How much heat is added to the system? (b) What is the change in internal energy of the system?

16-44 If 25.0 kg of water at 60°C is mixed with 40.0 kg of water at 10°C, calculate (a) the final temperature of the mixture, and (b) the change in entropy of the system. (Use the individual average temperature for each portion of water when calculating entropy changes.)

16-45 A heat engine absorbs 5.00×10^3 J at a high temperature of 250°C and exhausts 3.00×10^3 J at a low temperature of 20°C. Determine the change in entropy of (a) the high-temperature reservoir, (b) the low-temperature reservoir, and (c) the whole system.

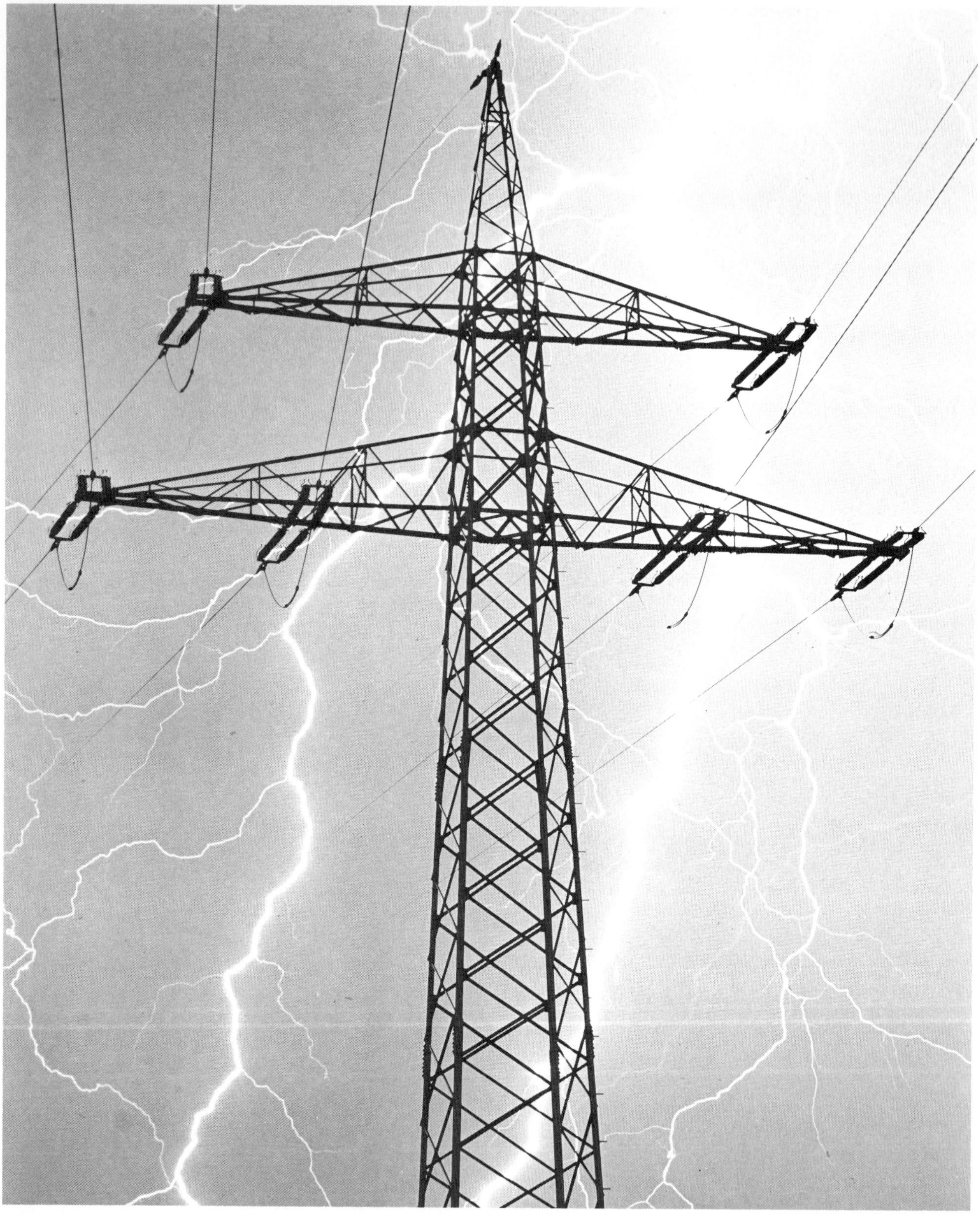

17 Electrostatic Forces

17-1 Introduction

The results and practical uses of electrical phenomena are familiar to all of us, and you could probably name several dozen examples in a few mintues. In addition to examples which might be considered manufactured, there are natural occurrences. Lightning is probably the most notable, but we are all aware of others: the shock when you touch the metal doorknob on a dry winter day, the crackling sound when you comb your hair, even the static cling we hear about in TV commercials. The expression **static electricity** is a part of our everyday vocabulary. (**Static** means "stationary," or "not in motion.") Electricity is important to our society and warrants very close examination. In fact, the next seven chapters of this text are devoted to topics which are directly or indirectly related to electrical phenomena.

17-2 Electric Charge

Greek tradition credits Thales of Miletus with the discovery of electricity. He noted that when he rubbed the semiprecious stone amber with wool, the amber acquired the ability to pick up small scraps of wool. The Greek word for amber is *elektron,* from which the study of this attraction was called **electricity.** Actually, the attraction effect is easy to produce, and the tools needed are available to you almost constantly. The plastic shaft of a ballpoint pen is an adequate substitute for amber, as are most combs. If you run the shaft of the pen across your hair several times, you will find that it acquires the ability to attract and lift dust specks off your desktop.

One of the best combinations for electrical effects is to rub hard black rubber (called *ebonite,* and found in many combs) with a fur (dry human hair is good; cat and rabbit fur are better). The combination of a rubber rod and fur has become fairly standard for laboratory and classroom demonstrations. To display an electric attraction, some light, yet visible, chunk of matter is desirable. Early experimenters used bits of cork or the dried inner stem, the pith, of large grasses. The easiest material for this purpose today is a small chunk of Styrofoam broken from a coffee cup.

If a small ball of light material is suspended from a cotton or silk thread, it displays an attraction for a rubber rod that has been rubbed with fur. The ball no longer hangs vertically but is pulled sideways toward the rod, as in Figure 17-1. Some materials do not display this power of attraction after being rubbed with fur; for example, no metals do. For those which do, the magnitude of the attraction depends on the materials. Investigators needed to describe these experiments and introduced special terms. For example, the rubber rod is said

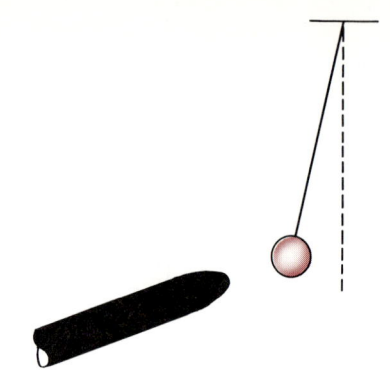

Figure 17-1 An electrically neutral object is attracted toward a charged rod. While such attractive forces are common, the object must be very light and freely suspended to show a displacement.

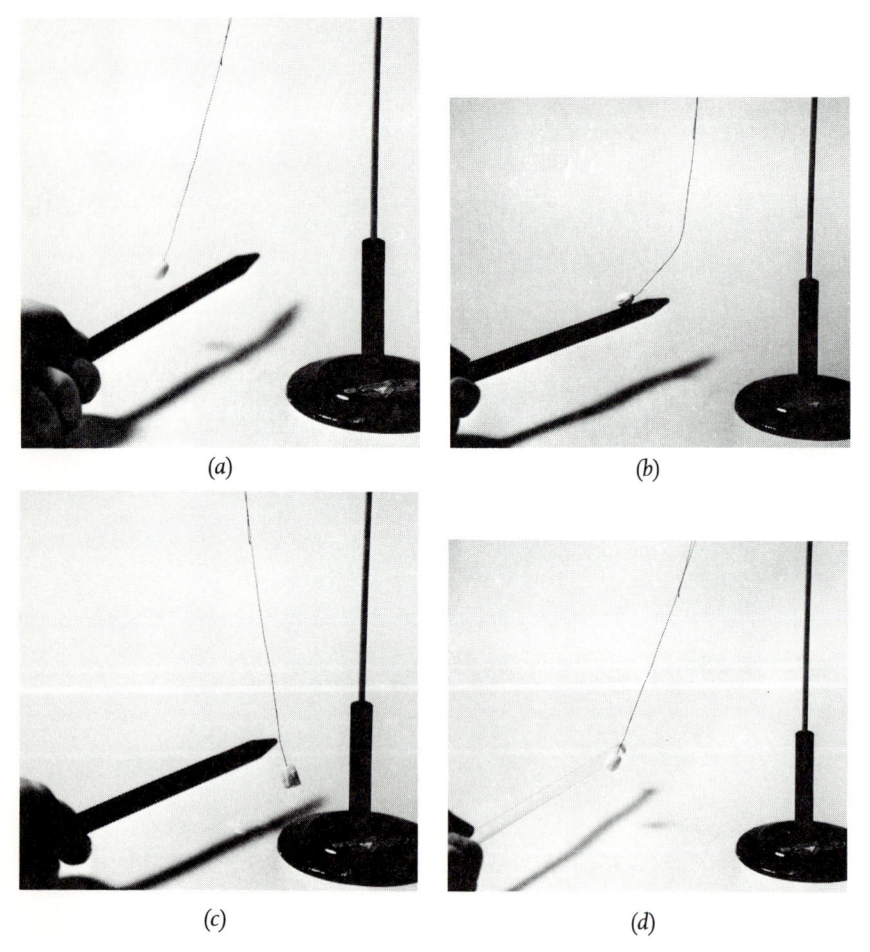

Figure 17-2 (*a*) A neutral pith ball is attracted to a rubber rod rubbed with fur. (*b*) If the ball is allowed to touch the rod, it clings momentarily and then jumps away. (*c*) Now the ball is repelled from the rod. (*d*) The same ball is now attracted to a glass rod which has been rubbed with silk. (*D. Riban*)

to be **charged** when it is rubbed with fur. Two charged rubber rods suspended from threads near each other will show a weak repulsion for one another. The effect is easier to observe if a small ball of pith, cork, or Styrofoam is used (Figure 17-2). At first, the ball is attracted to the charged rod. However, if the ball is allowed to touch the rod, it is then repelled (after clinging for a short time). The conclusion is that something is transferred to the ball when it touches the rod. This was the same "something" that was placed on the rod when it was rubbed with fur. This something is called **electric charge.** Further experiments show that the pith ball which is being repelled by the rubber rod is attracted to a glass rod which has been rubbed with silk. The process works in reverse. If the uncharged ball is touched by the charged glass rod, it will be repelled (after a short clinging period) by the glass. Now, however, the ball is attracted to the charged rubber rod.

If the thread holding the ball is replaced by a fine metal wire connected to the earth, the repulsion for the charged rod is quickly lost. This suggests that some materials, such as metals, allow charge to move away through them, while others such as cotton or silk do not. Materials which allow charge to move, more or less freely, from place to place are called **conductors** of electric charge. Materials which seem to confine charge to wherever it is placed are called **insulators.**

The results of these miniexperiments can be reproduced again and again. How do we interpret them? What do we postulate for cause and effect? The interpretation that is consistent with these and many other observations is that there are two kinds of charge. Benjamin Franklin is generally credited with having arbitrarily named the charge appearing on a glass rod that has been rubbed with silk as positive (+) and that appearing on hard rubber which has been rubbed with fur as negative (−). The experiment with the two different rods leads to the conclusion that *like charges repel each other* and *unlike charges attract.* We may picture charged bodies as shown in Figure 17-3a and b, with plus and minus signs to indicate the charge. The ultimate interpretation of all these experiments rests on atomic theory.

As we saw in Chapter 14, scientists agree that all matter is made up of atoms, the smallest constituents which retain all the chemical properties of the

17-2 Electric Charge

Assigning a name to electric charge, of course, does not explain it. Charge simply exists, no matter what we choose to call it. Its existence can no more be explained than that of mass or of space. Like mass, charge is a fundamental property we find in matter, with implications which it is the province of science to explore if not to explain.

Figure 17-3 (a) A negatively charged pith ball is repelled by the charge on the rubber rod. (b) The same ball is attracted by the charge on a glass rod rubbed with silk.

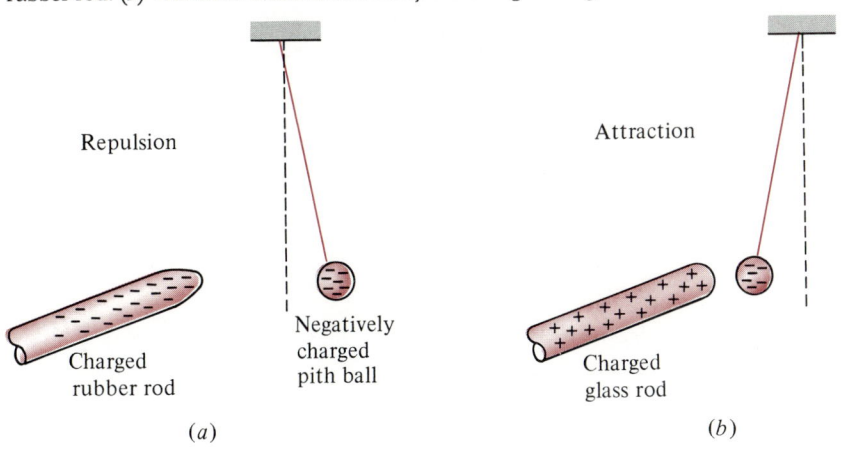

405

element. You will recall that there is a tremendous number of atoms in even a very small quantity of matter. For example, just the graphite in a typical lead pencil contains more than 10^{22} carbon atoms. In a simple model, atoms are built up of three somewhat more fundamental particles: **electrons, protons,** and **neutrons.** The neutrons and protons in any given atom are bound tightly together in a small volume called the **nucleus.** The electrons are pictured as orbiting the nucleus somewhat like the planets orbit the sun (Figure 17-4). Electrons and protons carry charges, while neutrons are uncharged, or neutral. In accordance with convention, the charge on the proton is positive, while the electron carries a negative charge. Further, the charge on the proton is equal in magnitude to the charge on the electron. A "normal" atom has the same number of electrons as it has protons and is thus electrically neutral. Electrons are much less massive than protons and account for less than 0.02 percent of the typical atom's mass. This atomic model may be used to help us interpret electrostatic phenomena.

It is postulated that, in general, whenever a body is charged, it is electrons that have been moved (does this seem reasonable?). When a body acquires a negative charge, it has somehow obtained an excess of electrons over its neutral state. However, a body which acquires a net positive charge is said to have lost some electrons from its neutral state and is left with fewer electrons than protons. The act of rubbing two insulators together tranfers electrons from one to the other.

When hard rubber is stroked with fur, the rubber acquires some of the fur's electrons, giving the rubber a net negative charge while the fur is left with a residual net positive charge. Similarly, a glass rod stroked with silk gives up some electrons to the silk, becoming positively charged as the silk becomes charged negatively. When any two insulators are rubbed together, most often there will be a net transfer of electrons from one to the other. Which of the two will lose electrons and which will gain depends on the substance. The list in Table 17-1 is called the triboelectric series (from the Greek *tribo*, meaning "to rub"). In general, when two substances on the list are rubbed together, the lower-numbered one becomes positively charged.

The differences between electric conductors and insulators are also explained by the atomic model. In neutral insulators, almost every electron is bound to a single nucleus and is not free to wander about. Some of the electrons in metals, however, are only bound to the substance as a whole and not to a specific nucleus. These so-called free electrons may move around throughout the metal, being attracted toward positive charges and repelled from other negative charges. They may move rather freely so long as they remain within the metal. When a body has equal amounts of positive and negative charges, the charges are said to be **balanced,** and the body is neutral or has no net charge. Moreover, when we say a body has a charge, we really mean that the body has a net unbalanced charge. To more fully explore electrical effects, one must be able to detect the presence of unbalanced charges, even in small amounts.

17-3 The Electroscope

A sensitive device for detecting the presence of unbalanced electric charge is the **electroscope** (Figure 17-5), an electrically isolated metal system. One end of the system is accessible to the experimenter, and the other end is usually sealed and has some delicate arrangement for the detection of unbalanced charge. The

Realize that this atomic model is a very oversimplified picture. It will, however, serve us very well until we deal with atomic theory in more detail in Chapter 30.

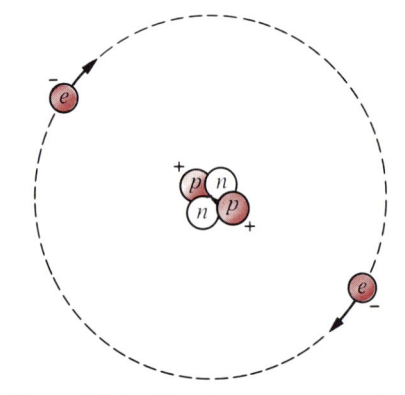

Figure 17-4 The planetary model of the atom. The vast majority of the volume of the atom is outside the nucleus and is occupied only by negatively charged electrons of negligible mass.

TABLE 17-1 The Triboelectric Series

1.	Asbestos
2.	Glass
3.	Mica
4.	Wool
5.	Cat's fur
6.	Silk
7.	Cotton
8.	Resin, sealing wax
9.	Ebonite
10.	Sulfur

earliest such systems used a very thin sheet of gold leaf attached at one end to a metal plate. Gold was used because it could be beaten into a thinner sheet than any other material. If a charged rod is brought near the knob of an electroscope, the bottom of the gold leaf deflects or stands away from the plate to which it is attached, indicating that there are unbalanced charges at the bottom of the electroscope and the gold leaf is being repelled from the plate. If the charged rod is withdrawn, the leaf collapses to its rest position once again. What is happening? The interpretation is that if a negative rod, for example, is brought near the knob, the protons in the knob are attracted toward the rod and the electrons in the knob are repelled. While the protons (along with the rest of the nucleus) are more or less fixed, some of the electrons are free to move. However, the metal system is isolated, not touching anything except insulators. Thus the electrons which move away from the negatively charged rod are forced to the bottom of the metal and onto the plate and the gold leaf. These are thus both now negatively charged and repel each other. The leaf does most of the moving and stands out away from the plate. When the negative rod is removed from the vicinity of the knob, the electrons quickly redistribute and the leaf hangs limp. Can you describe what might be a good interpretation of the leaf being deflected when a positively charged rod is brought near the knob?

If the knob is actually touched by the charged rod, the gold leaf is deflected and remains so when the rod is withdrawn. We say the electroscope is now charged. Apparently, charge is physically transferred from the rod to the electroscope by contact. Charging by direct contact with another charged object is called **charging by conduction.** Belief that charge is actually transferred is strengthened by noting that on occasion there will be a crackling sound as the rod approaches the knob. Following this, the electroscope will be charged just as if it had been touched by the rod. The rod partially discharges to the electroscope with a small spark that jumps across the gap between these objects. Charge really does move. In both these cases (touching and spark jumping gap), the charge on the electroscope is the same as the charge on the rod. Can you think of a way to test this statement experimentally?

To restore a charged electroscope to neutrality, you have only to touch the knob momentarily with your finger. While you probably feel nothing, the fact that the electroscope discharges suggests that human beings are conductors of electricity. This provides us with another method of charging an electroscope. Consider Figure 17-6. If a charged rubber rod is brought near the knob, but not allowed to touch it, the leaf will deflect. While the rod is held in this position, the knob is touched with the other hand. As this is done, the leaf collapses. Now, after the hand is removed, the rod is withdrawn. As the rod is taken away, the leaf once again deflects, indicating that the electroscope is charged. Interestingly, while the rubber rod is negatively charged, the charge that resides on the electroscope will be positive. This process is called **charging by induction.** Can you explain what is happening at each stage of the illustrations of Figure 17-6?

Figure 17-5 The electroscope detects the presence of small, unbalanced electric charges. The metal knob is connected to the metal plate but electrically insulated from the case. In the absence of an unbalanced charge, the foil leaf hangs against the plate. Charges placed on the knob spread freely to the plate and leaf. Then, because plate and leaf have the same type of charge, they repel each other and the foil leaf moves away from the plate.

This is a good place to develop your mental model of what is happening in electrostatic situations. Recall that the electrons of a conductor are free to rearrange within the conductor in response to a force, while electrons in an insulator are largely fixed in place.

17-4 Coulomb's Law

The phenomenon of charges exerting forces on one another is firmly established. The nature of the force requires further investigation. How does it depend on the amount(s) of charge? How does it vary with the distance between

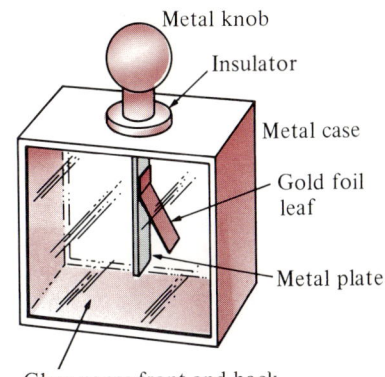

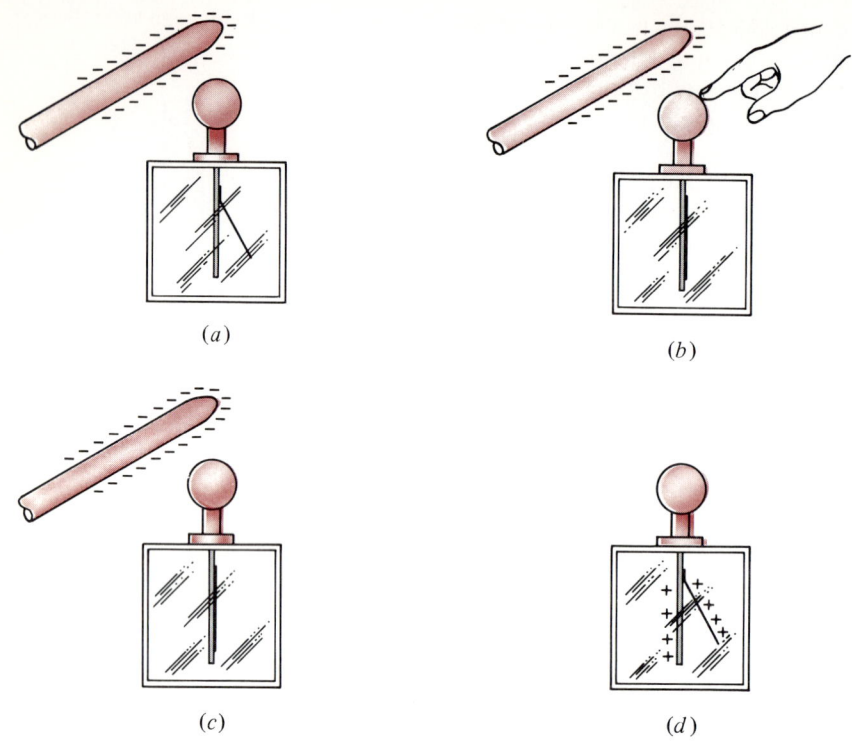

Figure 17-6 Charging by induction. (a) The charged rod is brought near but not touching the electroscope knob, and the leaf of the electroscope deflects. (b) With the rod still in place and the leaf deflecting, the knob is momentarily touched with a finger; the leaf collapses. (c) The finger is removed with the rod still in position; the leaf remains collapsed. (d) The rod is removed; the leaf again deflects, revealing the presence of a net charge.

charges? Are there any other quantities upon which the forces between charges depend? The answers to these questions were provided by Charles Augustin de Coulomb in 1784, when he announced the results of a series of beautiful (conceptually clear) experiments designed to investigate the force between charges.

Coulomb took advantage of the fact that when two small identical metal balls are insulated from their surroundings, charged, and allowed to touch, they share the net charge equally. He verified this and then measured the force each exerted on the other for various distances of separation. He varied the charges by allowing one or the other to touch other identical but uncharged metal balls. In this way, the charge could be made half as large or one-quarter as large or one-eighth as large as the original charge by successively touching and discharging the identical metal ball. Varying both the charge and the distance allowed Coulomb to determine the force relationship.

Coulomb discovered that the magnitude of the force between two charges is directly proportional to the product of the magnitude of the two charges and inversely proportional to the square of the distance between them:

$$F \propto \frac{q_1 q_2}{r^2}$$

where F = magnitude of the force on either charge due to the presence of the other charge
q_1, q_2 = magnitudes of the two charges
r = distance between the centers of the charges

All that remains to convert the proportionality into an equation is to determine a proportionality constant, say k_e. We have then what has become known as **Coulomb's law**:

$$F = k_e \frac{q_1 q_2}{r^2} \tag{17-1}$$

The value of k_e depends on the system of units chosen. You may remember from our discussion of units of the quantities involved in Newton's second law that there are at least two possible choices at this juncture. We could choose a value for k_e and use Equation 17-1 to define the basic unit of charge.[1]

Using the SI system of units, we choose a value for the basic unit of charge and use Equation 17-1 to determine the magnitude of the constant k_e by measuring the force between two known charges at a known distance of separation and solving for k_e.

The unit of charge is named the **coulomb** to honor the scientist. As we will see in Chapter 21, the coulomb (C) of charge is a derived unit in the SI system and is defined operationally in terms of electric current, the time rate of flow of charge in an electric circuit. In the meantime, we can gain some appreciation for the magnitude of the coulomb. The smallest charge measured and verified is that on a single electron and is equal in magnitude to approximately 1.60×10^{-19} C. This is such a fundamental number that it is common practice to assign a specific symbol, usually e, to represent it. Thus $e = 1.60 \times 10^{-19}$ C, and the charge on the proton is $+e$, while that on the electron is $-e$. One coulomb becomes a charge equivalent to that on about 6.25×10^{18} electrons.

A coulomb is really quite a lot of charge. If we could isolate two separate, 1-C charges and hold them exactly 1 m apart, we would find the force that each exerts on the other to be equal in magnitude to approximately 9.00×10^9 N. When we substitute these values into Equation 17-1, we evaluate the constant k_e as follows:

$$F = k_e \frac{q_1 q_2}{r^2}$$

$$k_e = \frac{Fr^2}{q_1 q_2} = \frac{(9.00 \times 10^9 \text{ N})(1.00 \text{ m})^2}{(1.00 \text{ C})(1.00 \text{ C})} = 9.00 \times 10^9 \text{ N} \cdot \text{m}^2/\text{C}^2$$

Thus the constant of proportionality in Coulomb's law, Equation 17-1, is 9.00×10^9 N·m²/C² in SI units. We also note that the force a 1.00-C charge exerts on another 1.00-C charge at a distance of 1.00 m is slightly more than 1 million tons.

Charge is a basic property of matter, and it is necessary to define a fifth fundamental unit (along with meters, kilograms, seconds, and kelvins). The difficulty is that the mobility of charges makes it impossible to keep a standard coulomb of charge on a shelf somewhere for comparison like we do with a standard kilogram. For this reason, the precise experimental definition of the coulomb is deferred to Chapter 21, where we will see that it is more convenient to make the coulomb a derived unit.

---

[1] For many years this choice was made by using the centimeter-gram-second (cgs) system of units. In the cgs system, the unit of force, defined by Newton's second law, is called the dyne and is equal to 1.00×10^{-5} N. The constant k_e is defined to be equal to one, exactly, and the basic unit of charge, called an **electrostatic unit** (esu), is determined from Equation 17-1. We insist that when two charges, each of magnitude one esu, are separated by exactly one centimeter, each will exert a force of magnitude exactly one dyne on the other. Thus we produce the new unit, esu, and determine its magnitude since each of the other quantities is already well defined.

Example 17-1 Determine the magnitude of the electrostatic force exerted by the proton on the electron in an ordinary hydrogen atom when their distance of separation is 5.30×10^{-11} m. (This is approximately equal to their average distance of separation in the model presented.)

Solution We must apply Coulomb's law, knowing that the proton and electron have charges of the same magnitude equal to 1.60×10^{-19} C.

$$F = k_e \frac{q_1 q_2}{r^2}$$

$$= \frac{(9.00 \times 10^9 \text{ N} \cdot \text{m}^2/\text{C}^2)(1.60 \times 10^{-19} \text{ C})(1.60 \times 10^{-19} \text{ C})}{(5.30 \times 10^{-11} \text{ m})^2}$$

$$= 8.20 \times 10^{-8} \text{ N}$$

Note that a force of this magnitude can impart an acceleration larger than 9.00×10^{22} m/s² to an electron.

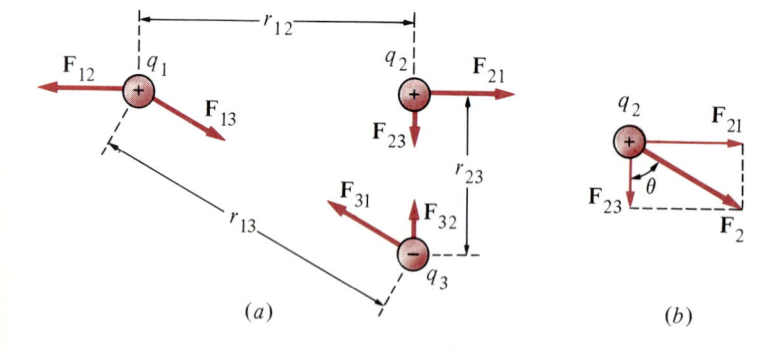

Figure 17-7 Two small, positively charged objects with charges q_1 and q_2 separated by a distance r. The repulsive forces on the objects are equal in magnitude and opposite in direction. Both forces act along the extension of the line joining the centers of the two objects.

The force one charge exerts on another *has* direction, and we must not ignore it. For charges which are essentially points, the force is along the line on which the two charges lie, and it is repulsive if they are of the same sign, attractive if they are of opposite sign. Figure 17-7 shows two charges whose magnitudes and separation are given. The force exerted on each is depicted, also. The two forces shown are equal in magnitude but oppositely directed, as they form an action-reaction pair by Newton's third law. The subscript notation, used before, can be read as follows: $\mathbf{F}_{21}$ is the force exerted on charge 2 due to charge 1.

If a third charge were placed in the vicinity of the two shown in Figure 17-7, it would exert an additional force on each of the others. In turn, it would feel a separate force due to the presence of each of the two. This is shown in Figure 17-8a. A negative charge q_3 has been placed at distances of r_{13} from q_1 and r_{23} from q_2. The vectors representing the individual forces on each of the charges are shown. If we knew all the distances and the magnitude of q_3, we could determine the resultant force acting on each of the charges. Figure 17-8b shows the individual forces acting on q_2 and the resultant ($\mathbf{F}_2$) of those forces. The two forces must be added vectorially to find the resultant.

Example 17-2 Determine the resultant force on charge q_2 shown in Figure 17-8. Take $q_3 = -2.00 \times 10^{-8}$ C, $r_{13} = 2.50$ m, and $r_{23} = 1.50$ m. Calculate the magnitude of $\mathbf{F}_2$ and the angle θ shown in Figure 17-8b.

Figure 17-8 (a) An object with charge q_3 is introduced to the system of Figure 17-7. This produces two new sets of forces, attractive in this case since q_3 is negative. The total electric force on any object is the vector sum of the individual forces on that object. (b) Vector resolution of the forces on object 2.

410

Solution First, we apply Coulomb's law to find the magnitudes of F_{21} and F_{23}.

$$F_{21} = \frac{k_e q_1 q_2}{r_{12}^2}$$
$$= \frac{(9.00 \times 10^9 \text{ N} \cdot \text{m}^2/\text{C}^2)(4.00 \times 10^{-8} \text{ C})(3.00 \times 10^{-8} \text{ C})}{(2.00 \text{ m})^2}$$
$$= 2.70 \times 10^{-6} \text{ N}$$

and

$$F_{23} = \frac{k_e q_2 q_3}{r_{23}^2}$$
$$= \frac{(9.00 \times 10^9 \text{ N} \cdot \text{m}^2/\text{C}^2)(3.00 \times 10^{-8} \text{ C})(2.00 \times 10^{-8} \text{ C})}{(1.50 \text{ m})^2}$$
$$= 2.40 \times 10^{-6} \text{ N}$$

We note the three charges are at the three vertices of a 3-4-5 right triangle. Thus $\mathbf{F}_{21}$ and $\mathbf{F}_{23}$ are perpendicular to one another. We may therefore find the magnitude of $\mathbf{F}_2$ by applying the pythagorean theorem, and the angle θ from the trigonometric definition of the tangent function.

$$F_2 = \sqrt{F_{21}^2 + F_{23}^2} = \sqrt{(2.70 \times 10^{-6} \text{ N})^2 + (2.40 \times 10^{-6} \text{ N})^2}$$
$$= 3.61 \times 10^{-6} \text{ N}$$

and

$$\theta = \tan^{-1} \frac{F_{21}}{F_{23}} = \tan^{-1} \frac{2.70 \times 10^{-6} \text{ N}}{2.40 \times 10^{-6} \text{ N}} = \tan^{-1} 1.125$$

Thus

$$\theta = 48.4°$$

Figure 17-9 Example 17-3.

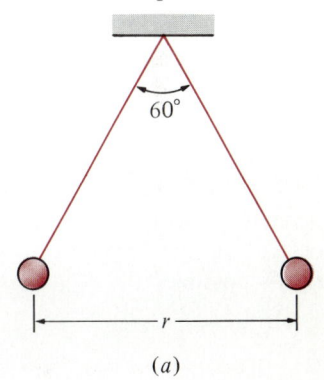

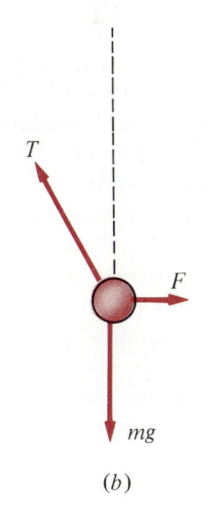

Example 17-3 Two small metal balls each of mass 0.100 kg are given equal positive charges and suspended from the same point at the ends of silk threads of length 0.500 m. As shown in Figure 17-9a, the repulsive force causes the masses to hang in equilibrium such that the angle between the threads is 60°. Calculate the magnitude of the charge on each ball.

Solution This may be solved as a straightforward equilibrium problem, setting the resultant force acting on either ball equal to zero. In Figure 17-9b, we see the ball on the right isolated with all the forces acting on it: T is the tension in the thread, mg is the weight of the ball, and F is the coulomb repulsive force. Each thread stands at an angle of 30° with the vertical, and the distance between the balls is also $r = 0.500$ m since the point of suspension and the two balls are at the vertices of an equilateral triangle. We choose a coordinate system with the positive y axis vertically upward and the positive x axis to the right. Thus the first condition of equilibrium allows us to sum the forces in the x and y directions separately, and we have

$$\Sigma F_x = 0: \quad F - T \sin 30° = 0$$
$$\Sigma F_y = 0: \quad T \cos 30° - mg = 0$$

Solving the first equation for T yields

$$T = \frac{F}{\sin 30°}$$

This can be substituted into the second equation to obtain

$$\frac{F}{\sin 30°} \cos 30° - mg = 0$$

or

$$F = mg \tan 30°$$

Coulomb's law gives an equation for F as

$$F = \frac{k_e q^2}{r^2}$$

Thus

$$\frac{k_e q^2}{r^2} = mg \tan 30°$$

or

$$q = \sqrt{\frac{r^2 mg \tan 30°}{k_e}}$$

$$= \sqrt{\frac{(0.500 \text{ m})^2 (0.100 \text{ kg})(9.80 \text{ m/s}^2)(0.577)}{9.00 \times 10^9 \text{ N} \cdot \text{m}^2/\text{C}^2}}$$

$$= 3.96 \times 10^{-6} \text{ C}$$

Examination of Example 17-2 for three charges shows that when the number of independently positioned, interacting charges becomes quite large, the calculation of forces using Coulomb's law will become very, very tedious, if indeed still practically possible.

17-5 The Electric Field

A given electric charge at a specified position in space far from any other charges establishes the electrostatic force on any other charge placed in some nearby position. This is similar to the situation with gravitational force. The presence of the earth establishes the magnitude and direction of the gravitational force on any small mass introduced at some point in its vicinity. We say that the earth modifies the space around itself, establishing a unique value for the "force per unit mass" for each point. We describe the total effect by saying that the earth has established a gravitational field about itself, a volume in which $\mathbf{F}/m$ is defined at each point.

It is very useful to interpret electrostatic forces in terms of a similar electric field. Suppose a standard charge, or **test charge**, say q_0, is placed at a point some distance r from a known charge q, as in Figure 17-10. Coulomb's law is used to determine the magnitude of the electrostatic force acting on q_0 as

$$F = \frac{k_e q q_0}{r^2}$$

Figure 17-10 The standard charge q_0 placed a distance r from the known charge q experiences a force $\mathbf{F}$. The electric field at any point is the force acting per unit positive charge at that point in space, $\mathbf{E} = \mathbf{F}/q_0$.

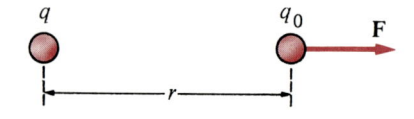

17-5 The Electric Field

If q_0 is removed and a charge half as large is put in its place, the magnitude of the force exerted will now be one-half as large. If the charge is doubled, the force will double, and so on. The ratio of the magnitude of the force exerted on any charge to the magnitude of that charge is the same number for any and all charges placed at the original position. It seems that this number, the force per charge, should be associated with the position or point in space rather than with the magnitude of the standard charge. In fact, this idea is universally accepted as being correct. The charge q is said to modify the space around it! At every point in the space surrounding q, there exists a unique value of electrostatic force per unit charge. Thus we define a quantity called the **electric field intensity** at a point as the electrostatic force per unit positive charge placed at that point:

$$\mathbf{E} = \frac{\mathbf{F}}{q_0} \tag{17-2}$$

where $\mathbf{E}$ is the electric field intensity at a specific point in space and $\mathbf{F}$ is the electrostatic force exerted on a positive charge q_0 placed at that point. The SI units of electric field intensity are newtons per coulomb (N/C).

One important consequence of the definition of electric field intensity is that we may easily determine the electrostatic force which will be exerted on a known charge q to be placed at a point where the field is known. That is, we have

$$\mathbf{F} = q\mathbf{E} \tag{17-3}$$

where $\mathbf{F}$ is the force exerted on the charge q when it is at a point in space where the value of the field is equal to $\mathbf{E}$.

> The direction of the electric field at a point is always the same as the direction of the force that would be exerted on a positive test charge placed at that point. Of course, the force exerted on a negative test charge placed at that same point would be exerted in the opposite direction.

Example 17-4 Calculate the force on an electron placed at a point where the electric field intensity is 2.00×10^4 N/C toward the west.

Solution This is an application of Equation 17-3, but we must be wary. The electron possesses a negative charge. Hence the force acting on the electron is toward the east when $\mathbf{E}$ is toward the west. When we use $-e$ as the charge on the electron, Equation 17-3 becomes

$$\mathbf{F} = -e\mathbf{E}$$

Dropping the vector notation to calculate magnitude,

$$F = eE = (1.60 \times 10^{-19} \text{ C})(2.00 \times 10^4 \text{ N/C})$$

and

$$F = 3.20 \times 10^{-15} \text{ N toward the east}$$

We may also use Coulomb's law to derive an expression for the electric field intensity in the vicinity of any point charge. If we place our positive test charge q_0 at a distance r from a charge q, Coulomb's law states that the magnitude of the force exerted on q_0 is

$$F = \frac{k_e q q_0}{r^2}$$

Hence the magnitude of the electric field intensity at a distance r away from q is

$$E = \frac{F}{q_0} = \frac{k_e q q_0}{r^2 q_0}$$

$$E = \frac{k_e q}{r^2} \tag{17-4}$$

The direction of **E** due to a point charge is away from the charge if the charge is positive and toward it if it is negative (Figure 17-11).

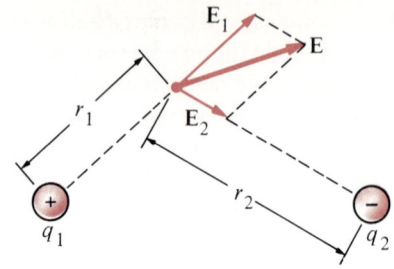

Figure 17-11 The net electric field intensity at an arbitrary point in the vicinity of two point charges is equal to the resultant of the fields due to the individual charges. The magnitudes of E_1 and E_2 are found with Equation 17-4, and E is calculated by vector algebra. Because it can be used to find the magnitude and direction of **E** for all points in the space surrounding these charges, this method may be used to "map" the **E** field throughout the space surrounding any distribution of point charges.

Example 17-5 Determine the electric field intensity at the site of the electron due to the proton in a hydrogen atom when they are separated by 5.30×10^{-11} m.

Solution This is a direct application of Equation 17-4. The magnitude is

$$E = \frac{k_e q}{r^2} = \frac{k_e e}{r^2} = \frac{(9.00 \times 10^9 \text{ N} \cdot \text{m}^2/\text{C}^2)(1.60 \times 10^{-19} \text{ C})}{(5.30 \times 10^{-11} \text{ m})^2}$$

$$= 5.13 \times 10^{11} \text{ N/C}$$

The electric field due to the proton is directed away from it since the proton is positively charged.

Experimentally, to determine **E** at a point in some region of space, we might place our positive test charge q_0 at that point and measure the electrostatic force exerted upon it. However, if we think of **E** as a kind of modification of the space due to the proximity of charges, will not q_0 also modify the space? Will not the introduction of another charge alter the very quantity we are attempting to measure? Actually, if the charges causing the original field are fixed in place, the field caused by q_0 will not alter their positions nor will it change the quantity it is being used to measure. For our purposes, we need worry only if the modification due to q_0 causes the original charges to move or redistribute. In that case, we would have to use a very tiny magnitude test charge—the smaller the better—in order to have a minimum effect on the measurement.

If a conducting object, say a metal sphere, is placed in an electric field, electrons in the metal will experience a force and may move. Indeed, they must move so long as any component of the net force lies along the conductor. Thus, electrons would migrate in the conductor in the direction opposite to the applied electric field (Figure 17-12). Such migration is brief for an isolated conductor. As electrons crowd toward one side of the metal sphere, the electric field caused by the charge separation will rapidly cancel the effects of the external field within the metal, and electron migration will cease.

If we change the external electric field (by placing a test charge in the vicinity, for example), the electrons will redistribute until the net field inside the metal is again zero. Thus we must use tiny test charges if we wish to measure the electric field intensity near a metal or due to charges distributed on a metal. Furthermore, in any electrostatic situation, there will be no net electric field inside a conductor. If there were, the free charges would move under the action of the field, and the situation would not be static. Thus, the electrostatic field will terminate at the surface of a conductor.

Also, there can be no net charge inside a conductor in the static case. Any small volume inside a conductor which contains a net unbalanced charge would

Figure 17-12 (a) An electrically neutral metal sphere in a region with no electric field remains everywhere neutral. (b) In the presence of an external electric field, charges in the metal migrate, leaving the side in the direction of the field positive and the opposite side negative.

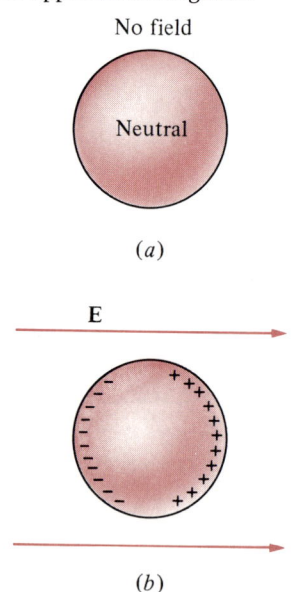

have an associated electric field, causing other free electrons to accelerate, voiding the static hypotheses. Thus the interior of conductors in the electrostatic case must be neutral, and unbalanced charge may reside only on the exterior of the conductor.

Finally, we may conclude that any electrostatic field which terminates at a conducting surface will be perpendicular to that surface at the point where it terminates. If it were not perpendicular, it would have a parallel component at the conductor's surface, causing electrons to redistribute until the parallel component was canceled.

17-6   Electric Field Lines of Force

The electric field is a vector field; that is, at each point within a volume of space, it has both magnitude and direction. Furthermore, both the magnitude and the direction may vary from point to point. This is difficult to picture. To visualize an electric field, we use the construct or model showing *lines of force,* a technique first introduced by Michael Faraday as a graphic representation of the electric field. A **line of force** is an imaginary line drawn with arrowheads fixed to the line such that the direction at any point in space is the same as the direction of the electric field intensity at that point.

Figure 17-13a shows an isolated positive point charge with the lines of force in the plane of the paper. Note that the lines are straight and are directed away from the charge. We might expect this, since the force on any positive test charge placed in the vicinity would be one of repulsion. We should remind ourselves that even though the only lines of force drawn are those in the plane of the page, they really extend outward in all directions, including toward you and toward the front and back covers of the text. The lines of force in the plane of the page due to an isolated negative charge are shown in Figure 17-13b. We see that they are similar to those of the positive charge but are directed toward the negative charge. Should we expect this?

In the general case, where there are several charges in close proximity, the lines of force follow curved paths. Figure 17-14a shows a positive and negative charge of equal magnitude along with the lines of force drawn in the plane of the page. In Figure 17-14b we see the lines of force in the vicinity of two equal positive charges. In both cases, we see the curvature of the lines of force. Several points worth noting concerning lines of force in an electric field are:

1. They are imaginary and are used only to help us "map" the electric field.
2. The electric field at any point is directed along a line tangent to the line of force at that point.
3. The lines of force are closer together where the field is stronger, and farther apart where it is weaker. In actuality, a line of force could be drawn through every point in space. However, we choose to draw only a small number of lines and space them so that the number crossing a unit area placed perpendicular to the direction of the field is proportional to the magnitude of the electric field intensity at each point.
4. Electrostatic lines of force may originate and terminate only on charges. They begin on positive charges and terminate on negative charges.
5. Lines of force can never cross each other. The electric field exerts a force on a

(a)

(b)

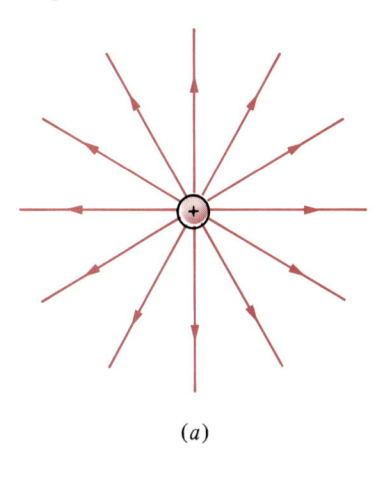

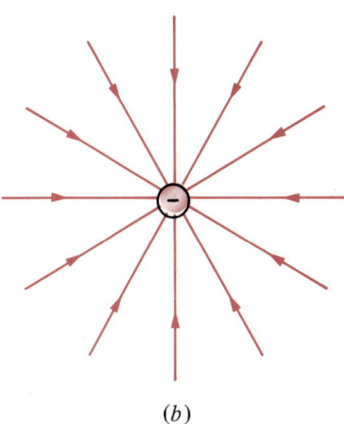

Figure 17-13   (a) The lines of force around a positive charge are directed radially outward, indicating that a positive charge placed in this vicinity will experience an outward repulsive force due to the charge. (b) The lines of force surrounding a negative charge are directed inward, indicating an attractive force for any positive test charge placed nearby. In both cases, the strength of the field is proportional to the concentration of force lines in the area of a positive test charge.

(a)

(b)

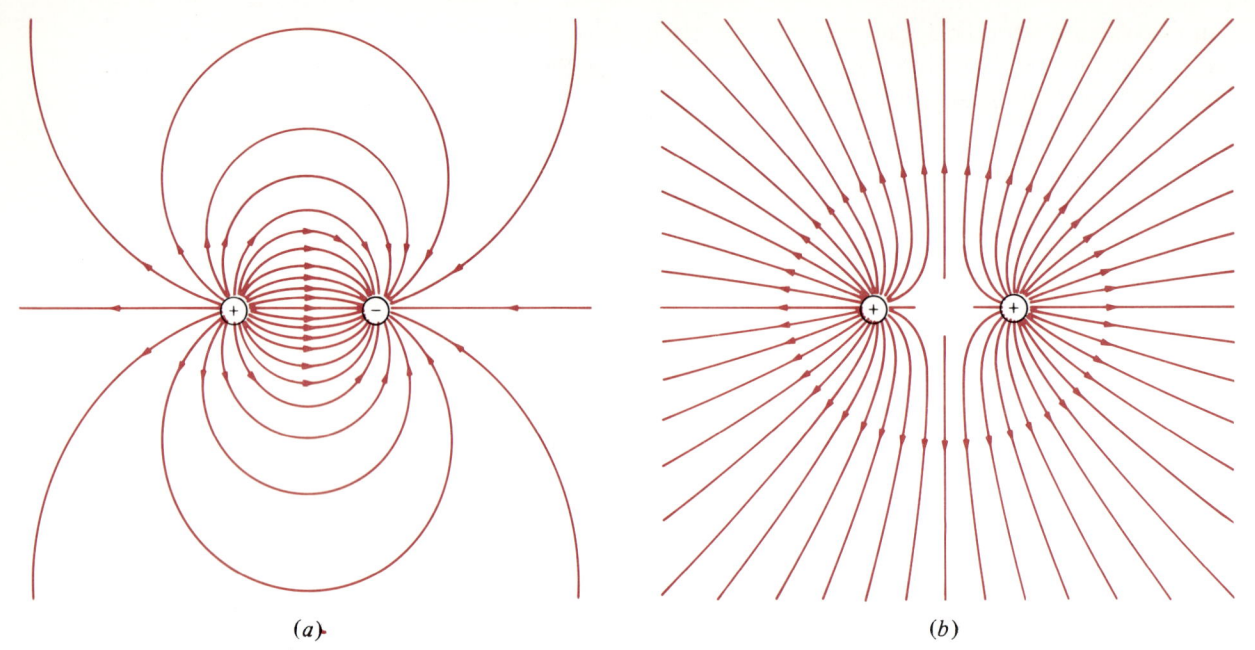

Figure 17-14 Electric field surrounding (a) equal charges of opposite sign and (b) equal positive charges.

test charge in a single direction at every point in space. Thus, the electric field must have a single value everywhere.

The electric field is defined for any distribution of charges, and we may, in principle, determine the nature of the distribution of charges from a knowledge of the field. For example, if we use a test charge to map the electric field in a region of space of unknown charge distribution and obtain lines of force resembling those of Figure 17-14a, we would probably conclude that the field was due to opposite charges separated by some distance.

17-7 Electric Dipoles

An important distribution of charge is one in which equal-magnitude positive and negative charges are separated by some distance. Such an arrangement is called an **electric dipole.** The dipole may be due to only two charges, as shown in Figure 17-14a, or it may be due to the separation of the centers of positive and negative charge, like that on the sphere in Figure 17-12b. The dipole may be produced by charge rearrangements due to an external field (like the metal sphere), or it may be naturally occurring and permanent. Perhaps the most common permanent dipole is that of the ordinary water molecule (Figure 17-15a). The hydrogen atoms are bonded to the oxygen atom such that both are off to one side of the oxygen atom. This, plus the fact that the electrons spend more time in the vicinity of the oxygen atom, gives rise to a permanent separation between the average position of positive charge and negative charge. For purposes of its electric dipole effect, the water molecule could be replaced by

416

17-7 Electric Dipoles

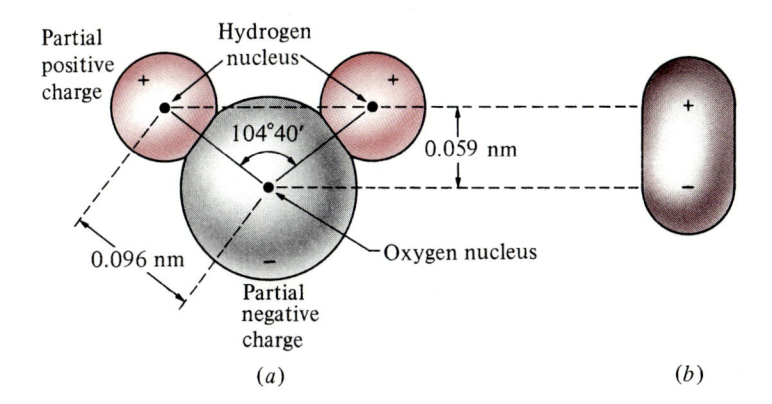

Figure 17-15 (a) The water molecule: two hydrogen atoms attached at an angle of 104°40′ to an oxygen atom. (b) The molecule represents a permanent dipole with unequal sharing of electrons, which leaves the hydrogens partially positive and the oxygen partially negative.

two opposite charges separated by some distance d, as shown in Figure 17-15b. We define a direction for the dipole as that from the negative charge to the positive. The strength of the dipole, the **dipole moment,** is defined as the product of the magnitude of either of the charges with the distance between the two. The water molecule has a dipole moment of about 6.00×10^{-30} C·m. The permanent electric field about a neutral water molecule set up by the dipole allows the molecule to exert forces at a distance from itself and accounts for many of the unique properties of water.

Except under certain conditions, most substances do not show any macroscopic dipole properties. Most molecules have no permanent electric dipole, so this is not too surprising. For molecules with permanent dipoles, random thermal motion and random orientation usually assure that the dipole fields cancel even a short distance from the substance. Figure 17-16 represents an object of substance with no permanent dipole moments, first in a region where there is no electric field and then in a region where a uniform electric field exists. The representative molecules in Figure 17-16a are depicted as circles, with no charges shown. The molecules are electrically neutral, and the average position of the centers of both positive and negative charges are at the same point. In Figure 17-16b, the charge distribution has been distorted by the electric field. The centers of positive and negative charge are shown to be displaced a small distance from each other by the field, and the molecules are now said to have **induced dipole moments.**

Figure 17-17 shows an object with molecules which have permanent dipole moments before and after the object has been placed in a region of

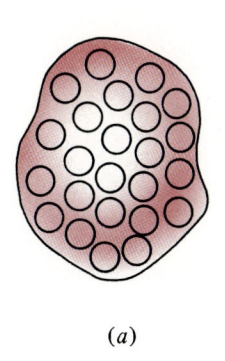

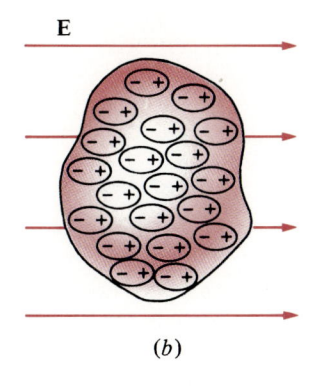

Figure 17-16 (a) A neutral piece of matter made up of electrically neutral molecules in the absence of an external electric field. (b) In an electric field, the molecules acquire an induced dipole moment. Positive charges migrate slightly in the direction of the field, and/or negative charges migrate in the direction against the field.

uniform electric field. Before the object is immersed in the field (Figure 17-17a), the dipoles are oriented in all different directions so that the individual fields effectively cancel at a very short distance away. In Figure 17-17b, the torque provided by the electric field has caused the dipoles to twist so that they line up in the direction of the field. The dipole alignment in 17-16b and 17-17b gives rise to an **induced field** inside the object due to the newly formed or newly aligned dipoles. This induced field is opposite in direction to the external field. Its effect, when added vectorially to the external field, is to reduce the net electric field inside the object. This property of dipoles causing a decrease in an impressed, external field is crucial in the fabrication of large capacitors (Chapter 18). Permanent dipoles, especially those of water molecules, provide the mechanism by which microwave ovens cook food. The standing microwave provides rapid oscillation of the electric field, which adds kinetic energy by continually reorienting the water molecules in the food. This kinetic energy is rapidly converted to increased temperature (and hence cooking) of the rest of the food by molecular collisions. Note that unless a material has some molecules possessing a permanent dipole, it will not experience this heating effect due to microwaves.

Dipole effects account for most of the attraction of uncharged insulators in a nonuniform electric field, such as a neutral pith ball near a charged glass rod as shown in Figure 17-18. Note that the alignment of the dipoles in the pith ball causes the side closest to the rod to have an effective negative charge and hence be attracted. The side farther away is repelled. The lines of force are getting farther apart with increased distance from the rod since the field is weaker farther away. Hence, the force of repulsion on the positive side is smaller in magnitude than the force of attraction on the closer negative side. The net force on the neutral pith ball is, therefore, one of attraction. A similar diagram could be shown if the rod were negatively charged, yielding attraction in that instance, also. Figure 17-19 shows the effect of an electric field due to a charged rod on a stream of water. This effect is also due to the permanent dipole moments of the water molecules.

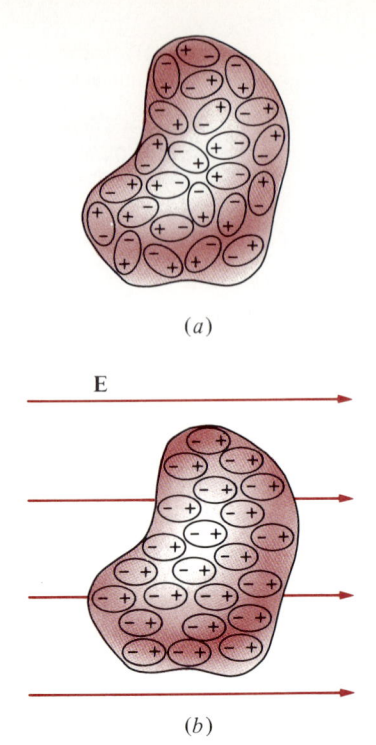

Figure 17-17 (a) Molecules with a permanent dipole moment may have no net external electrical effect in the absence of an external field because the dipoles are aligned randomly. (b) With an external field, dipoles align with the external field and may display external electrical effects.

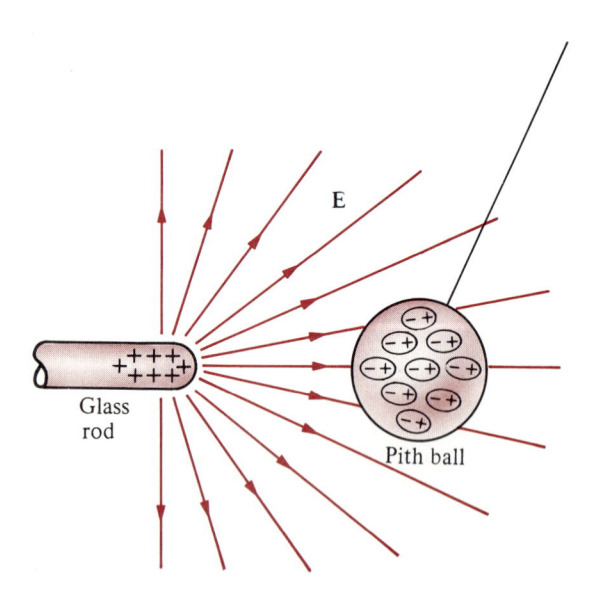

Figure 17-18 In the spreading electric field of the charged rod, the molecules of the pith ball are polarized and become induced dipoles.

418

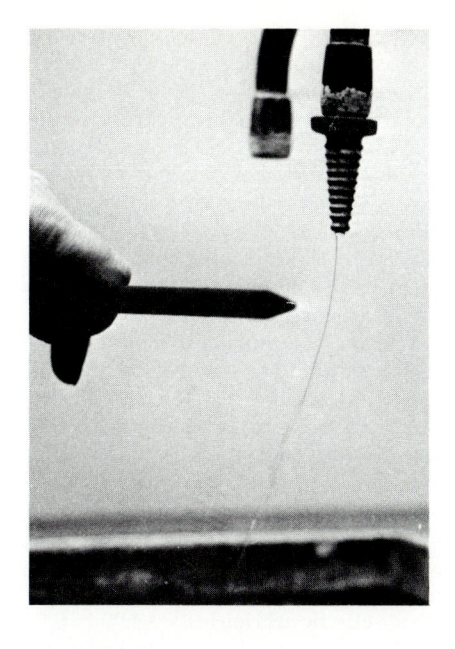

Figure 17-19 A negatively charged rubber rod attracting a stream of water. Because they are permanent dipoles, the water molecules align in the field of the rod such that their positive hydrogen sides are toward the rod and their negative oxygen sides are away from it. Since the positive side of the molecule is closer to the rod than the negative side, the attractive force is slightly greater than the repulsive force for the negative side, and the net force is one of attraction. (*D. Riban*)

Minimum Learning Objectives

After studying this chapter, you should be able to:
1. Define:
   - charge (electric)
   - conduction, charging by
   - conductor
   - coulomb (C) [unit of charge]
   - Coulomb's law
   - dipole
   - dipole moment
   - electric field intensity
   - electricity
   - electron
   - electroscope
   - field
   - induced dipole moment
   - induction, charging by
   - insulator
   - lines of force
   - negative charge
   - net charge
   - neutron
   - nucleus
   - positive charge
   - proton
   - static (electricity)
   - test charge

2. Specify the conditions necessary for producing static electric charges by friction.
3. Interpret specific static-electrical phenomena in terms of the shifts of charged particles constituting the matter involved, e.g., charging and discharging electroscopes.
4. Calculate the magnitude and direction of (*a*) the forces between point charges of given magnitude and geometry, (*b*) the force on objects of given charge in a uniform electric field, and (*c*) the electric field established at a given location by one or more point charges of given magnitude and position.
5. Understand the information being conveyed by a field diagram showing lines of force.
6. Understand the form of an electric dipole and how its dipole moment allows it to interact with an external electric field.

Problems

17-1 The centers of two protons (assumed spheres) in a helium nucleus are thought to be separated by about 1.40×10^{-15} m. Calculate the repulsive force between protons at this separation. (You might wish to speculate why the nucleus does not fly apart all by itself.)

17-2 At what distance away from a proton would the magnitude of the electrostatic force on an electron equal the electron's weight near the earth's surface? (The mass of the electron is 9.11×10^{-31} kg.)

17-3 In a print-copying process, dark pigment grains are attracted onto the paper by the charges on a plate behind the sheet. If each grain carries an excess charge of $-10e$ and the plate is effectively a point charge of 5×10^{-6} C placed 0.200 cm from the grains, what is the magnitude of the force that pulls the dark grains toward the paper?

17-4 (a) Calculate the distance between two small bodies which are each charged to 2.00×10^{-6} C such that the force they exert on each other has a magnitude of 1.00 N. (b) How many electrons are there in a charge of -2.00×10^{-6} C?

17-5 Suppose the moon had a net negative charge equal to $-q$ and the earth had a net positive charge equal to $+10q$. What value of q would yield the same magnitude force that we now attribute to gravity?

17-6 The net charge on a sodium ion is $+e$, while that on a chloride ion is $-e$. The distance between the nuclei of these ions in a salt crystal is 2.82×10^{-10} m. (a) Assuming each charge is effectively concentrated at the ion's nucleus, what force pulls these ions together? (b) How large a force would be required to pull off all the ions along one edge of a salt cube 1.00 cm on each edge?

17-7 Suppose the following were possible. We take two 1-qt canning jars and collect 1.00 kg of pure electrons ($m = m_e = 9.11 \times 10^{-31}$ kg; $q = -e = -1.60 \times 10^{-19}$ C) in each jar. The first jar is securely fastened at the south pole of the earth, while the second is held by a rope attached to a stake at the north pole ($R_e = 6.38 \times 10^6$ m). Gravity pulls the north pole electrons downward (neglect the weight of the jar), while the repulsion of the other electrons, a full earth diameter away, tends to lift them away from the earth and into space. (a) When we cut the rope holding the electrons at the north pole, which wins, gravity or electric repulsion? (b) What is the net force on the jar of electrons? (c) What would the initial acceleration of the jar be? (d) Just for fun (and neglecting relativity), how long would it take to reach the speed of light ($c = 3.00 \times 10^8$ m/s), accelerating at this rate? (e) What does this imply about the relative size of electric and gravitational forces? (f) Can you formulate a scientific response to the question "Why is it unlikely that 1.00 kg of pure electrons will ever be collected in one place at one time?"

17-8 Three charges, $q_1 = +2.00$ µC, $q_2 = -5.00$ µC, and $q_3 = +5.00$ µC, are placed in a straight line along the y axis at 1.00, 2.00, and 3.00 m, respectively, from the origin. (a) Calculate the magnitude and direction of the electrostatic force on q_2. (b) What charge must be placed at the origin in order that q_2 be in equilibrium?

17-9 Two charges, one of $+3.00 \times 10^{-4}$ C and the other of -8.00×10^{-5} C, are placed along a line 0.400 m apart. Where along the line must a third charge be placed such that the resultant electrostatic force acting on it is zero? (Did you guess that the magnitude and the sign of the third charge are not relevant?)

17-10 Two small balls of the same mass m are given identical charges of 8.00×10^{-7} C and are attached to fine silk threads 1.00 m long and hung from a common point on the ceiling. If the threads separate at an angle of 22°, what is the mass of each ball?

17-11 Two point charges, one of $+0.300$ µC and the other of -0.500 µC, are placed on adjacent corners of a square 0.800 m on a side. A third charge of $+0.600$ µC is placed at the corner diagonally opposite to the negative charge. Calculate the magnitude and direction of the force acting on the third charge. For the direction, determine the angle the force makes with the edge of the square parallel to the line joining the first two charges.

17-12 A charge of $+2.00 \times 10^{-8}$ C is placed at the origin of coordinates. A charge of -4.00×10^{-8} C is 3.00 m from the origin along the positive x axis, and a charge of $+5.00 \times 10^{-8}$ C is 5.00 m along the negative y axis. Calculate the magnitude and direction of the net force acting on the $+2.00 \times 10^{-8}$ C charge.

17-13 Determine the magnitude of an upward electric field which will balance the weight of a proton near the earth's surface.

17-14 A tiny drop of oil of mass 2.50×10^{-15} kg has one excess electron. Calculate the electric field intensity which is necessary to hold the drop suspended against the force of gravity near the earth.

17-15 In a region where the electric field intensity is 4.00×10^4 N/C, determine the acceleration which will be imparted to (a) an electron, (b) a proton, and (c) a small sphere of mass 1.00×10^{-3} kg and charged to 5.00×10^{-8} C.

17-16 A proton in an electric field is accelerated from rest to a speed of 1.50×10^6 m/s in a distance of 0.400 m. Calculate (a) its acceleration, (b) the magnitude of the electric field intensity, and (c) the work done by the electric field.

17-17 In Figure 17-20, an electron is projected from the origin of coordinates in the positive x direction with a speed v_0. It moves in a region of uniform electric field intensity which is in the negative y direction. Derive an equation for the y coordinate of the electron as a function of x.

17-18 The force on a -2.00×10^{-5} C charge placed at the origin is 3.00 N at a direction of 30° above the negative x axis. Determine the magnitude and direction of the electric field intensity at that point.

Figure 17-20   Problem 17-17.

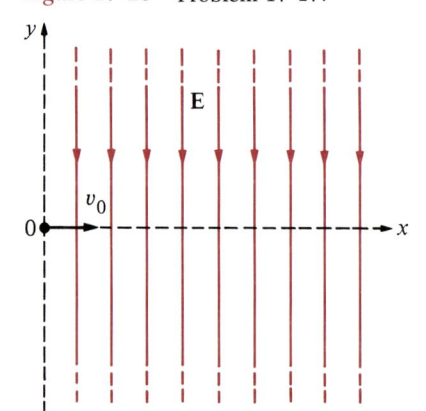

17-19 Determine the magnitude of the charge on a small sphere which gives rise to an electric field intensity of 4.00×10^3 N/C at a distance of 0.300 m away from the sphere.

17-20 Two small bodies are oppositely charged, one of $+2.00 \times 10^{-8}$ C and the other of -2.00×10^{-8} C. Determine the magnitude of the electric field intensity which will provide the force necessary to keep them separated by a distance of 0.500 m.

17-21 The maximum electric field intensity which can be sustained in dry air without breakdown is about 8.00×10^5 N/C. Calculate the radius required of a metal sphere that can be charged to 1.00 C in dry air.

17-22 Two point charges, one of $-5.00\,\mu C$ and the other of $8.00\,\mu C$, are placed on a straight line 3.00 m apart. Calculate the electric field intensity *(a)* midway between the charges and *(b)* at 1.00 m from the negative charge and 4.00 m from the positive charge along the line.

17-23 A charge of $+50.0\,\mu C$ is placed at the origin of coordinates. A second charge of $-40.0\,\mu C$ is placed 3.00 m away along the positive x axis. Determine the magnitude and sign of a third charge which must be placed 6.00 m along the negative x axis so that the electric field at $x = 2.00$ m is zero.

17-24 Two point charges of $+4.00\,\mu C$ each are placed at adjacent corners of a 2.00-m square, while two other point charges of $-4.00\,\mu C$ each are placed on the other corners. Calculate the electric field intensity at *(a)* the center of the square, and *(b)* the point on the side midway between a positive and negative charge.

17-25 Three charges each of 2.00×10^{-8} C are placed at three of the four corners of a 0.500-m square. Calculate the electric field intensity at *(a)* the center of the square and *(b)* the fourth corner.

17-26 Sketch the electrostatic lines of force for the charge distribution shown in Figure 17-21*a*.

17-27 Sketch the electrostatic lines of force for the charge distribution shown in Figure 17-21*b*.

17-28 Sketch the electrostatic lines of force for the charge distribution shown in Figure 17-21*c*.

17-29 An ammonia molecule has a net positive charge of $+10e$ and a net negative charge of $-10e$. The time average positions of the centers of positive and negative charge coincide unless the molecule is placed in an electric field. Suppose the molecule is placed in such a field of intensity $E = 2.00 \times 10^5$ N/C. *(a)* What will the distance of separation of the average positions of positive and negative charge be so the average attractive force holding the molecule together is equal to the force due to the field which pulls it apart? *(b)* The measure of the strength of a dipole is called the dipole moment and is equal to the product of magnitude of one of the charges with the distance of separation. Calculate the induced dipole moment for the ammonium molecule when placed in a field of $E = 2.00 \times 10^5$ N/C.

17-30 Derive an expression for the magnitude of the electric field intensity at a large distance r from but along the line of a dipole of charges $+q$ and $-q$ separated by a distance a.

17-31 A water molecule may be modeled as an electric dipole with two charges of opposite sign and equal in magnitude to that on the electron but separated by 3.75×10^{-11} m. Calculate the maximum torque exerted on a water molecule placed in an electric field of $\mathbf{E} = 2.40 \times 10^3$ N/C.

17-32 One of the strongest common dipole moments is that between the carbon and double-bonded oxygen of the organic acid (carboxyl) group

$$-C\begin{matrix}\diagup\!\!\!\diagup\, O\\ \diagdown\, OH\end{matrix}$$

If the nuclear separation of carbon and oxygen is 1.20×10^{-10} m and the dipole moment is 7.00×10^{-30} C·m, what is the effective charge separated in the dipole?

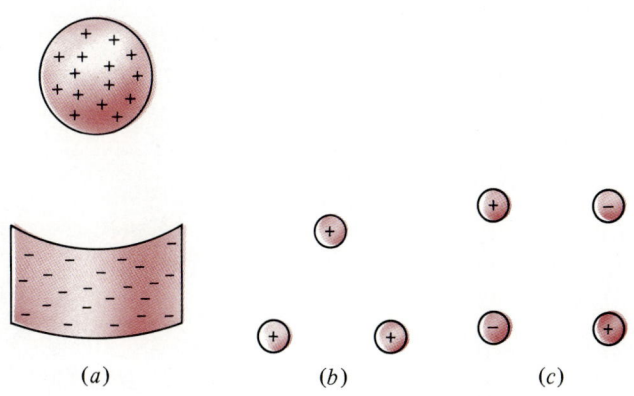

Figure 17-21 *(a)* Problem 17-26; *(b)* Problem 17-27; *(c)* Problem 17-28.

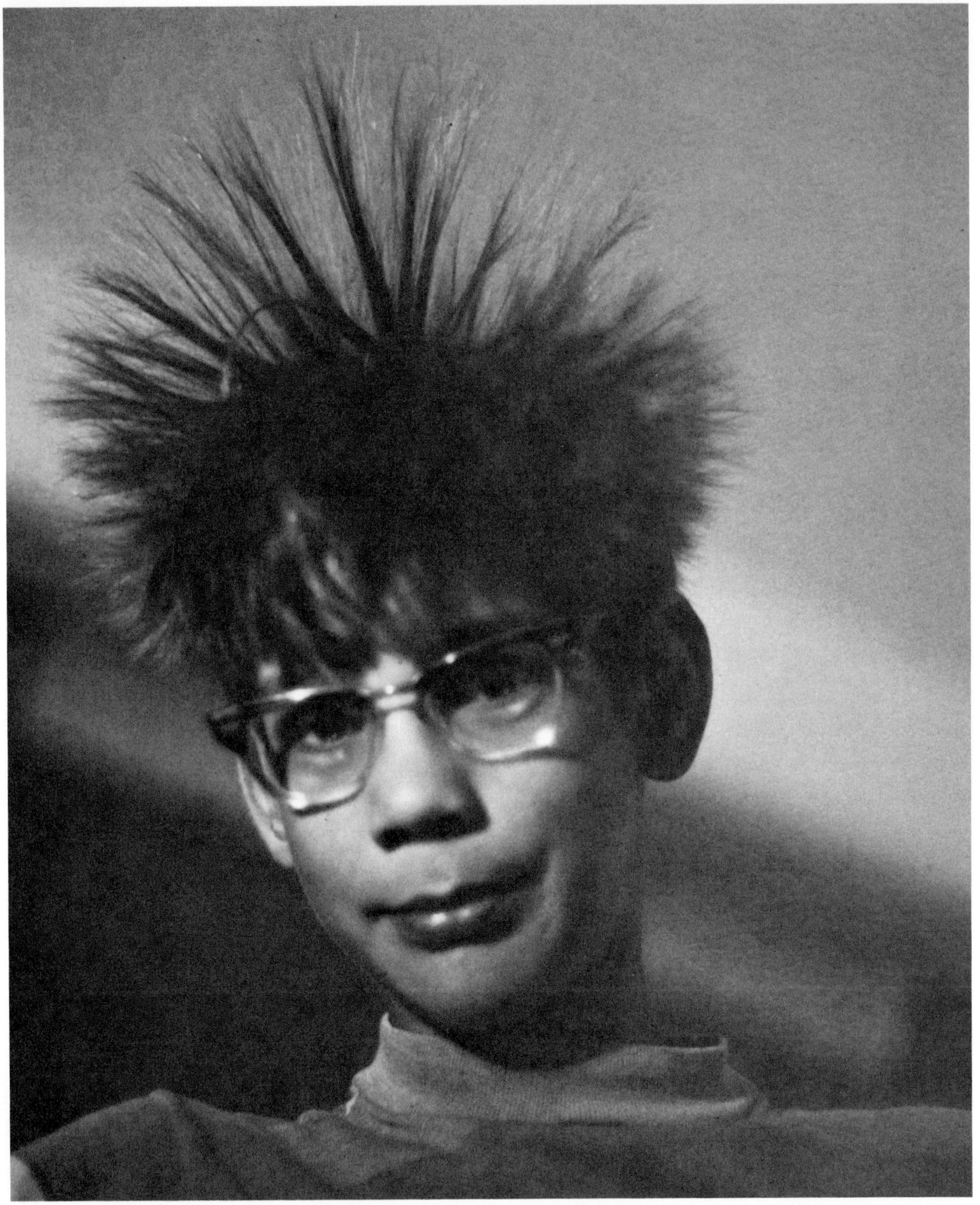

18 Electrostatic Energy and Capacitance

18-1 Introduction

Electrostatic forces obey the same laws of mechanics as other forces. If a charged mass is placed in a region of space where there is an electric field, an electrostatic force is exerted upon it. If the mass is released, it will be accelerated by the electric force, and its subsequent motion can be determined by Newton's laws. Thus we may use the techniques developed in Chapter 4 to determine the acceleration of the mass (Figure 18-1) and then calculate its velocity and displacement at any time after its release. Recall, however, that there is another approach we could take to predict the future behavior of the mass: we could apply the work-energy principle, since the electric force certainly does work on the mass and the accelerating mass has had a change in kinetic energy. First, we determine the displacement between the original position and any new position the ball acquires after its release. A knowledge of the electric force, say $\mathbf{F}_e$, then allows us to calculate the work done by the electric field from

$$W = F_e \, \Delta s \cos \theta$$

where F_e = magnitude of the electric force
Δs = magnitude of the displacement
θ = angle between the two

The work-energy principle allows us to simply set the work equal to the change in kinetic energy. Using that, we can calculate velocities, and so on.

Thus, either Newton's laws *or* the work-energy principle may be used to describe the mechanics of charged bodies being acted upon by electric forces (or any other forces, for that matter). It is not obvious which approach is preferable. However, in many instances in dealing with electric forces, we may not know or may not care about the details of the path of motion or the force law involved. In such instances, we opt for the work-energy or a conservation of energy approach. This is indeed the case for electric fields. The application of the work-energy principle is preferable in the solution of most electrical problems. In this and succeeding chapters we pursue this method.

18-2 Potential Energy and Potential

In early discussions of energy, we distinguished between conservative and nonconservative forces. Conservative forces depend only on position, shape, or configuration, and the work done by a conservative force in moving a particle between two points is independent of the path taken. The electrostatic force which one charge exerts on another is a conservative force, with all that that entails. We may, for example, determine the change in electric potential energy when two charged objects are moved with respect to one another. From its definition, the change in potential energy is equal to the negative of the work done by a conservative force. Consider the two charges q and q_0 in Figure 18-2. Their separation has been increased from the distance r_a to a distance r_b. The work done by the Coulomb force during the displacement has a magnitude

$$W = F_{av}\, \Delta r$$

and is positive since the force and displacement are in the same direction. The magnitude of the average force each charge exerts on the other is

$$F_{av} = \frac{k_e q q_0}{r^2_{av}}$$

If we choose to make the approximation that $r^2_{av} = r_a r_b$ and recognize that $\Delta r = r_b - r_a$, we have

$$F_{av} = \frac{k_e q q_0}{r_a r_b}$$

and

$$W = \frac{k_e q q_0}{r_a r_b}(r_b - r_a)$$

Thus

$$W = k_e q q_0 \left(\frac{r_b}{r_a r_b} - \frac{r_a}{r_a r_b}\right)$$

$$= k_e q q_0 \left(\frac{1}{r_a} - \frac{1}{r_b}\right)$$

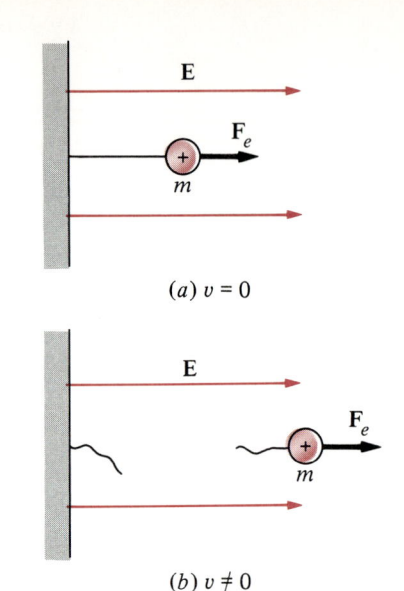

Figure 18-1 (*a*) A charged ball is held in an electric field by a nonconducting string. (*b*) When the string is broken, the ball will accelerate in the field.

Figure 18-2 Charge q_0 is repelled by the fixed charge q from separation r_a to separation r_b, a distance of Δr. A force has been exerted through a distance; hence work has been done, and the potential energy may have been changed.

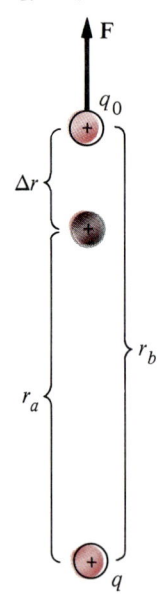

Finally,
$$\Delta PE = -W$$
yields
$$\Delta PE = k_e q q_0 \left(\frac{1}{r_b} - \frac{1}{r_a} \right) \tag{18-1}*$$

Equation 18-1 represents the change in potential energy which takes place when the separation of two charges is changed from r_a to r_b. For our derivation, we chose the charges to be of the same sign and r_b to be greater than r_a. Thus Equation 18-1 yields a negative number, and the change in potential energy is negative. It works just as well if we choose to move the charges closer together, making r_b less than r_a. In this case, the work done by the Coulomb force is negative (because the force and displacement are in opposite directions) and the change in potential energy is positive.

Example 18-1 Determine the change in electric energy when a proton is moved toward a uranium nucleus of charge $+1.47 \times 10^{-17}$ C. The original separation of the particles is $r_a = 6.00 \times 10^{-11}$ m, and their final separation is $r_b = 2.00 \times 10^{-11}$ m.

Solution We recall that the charge on a proton is $+e = +1.60 \times 10^{-19}$ C and then simply apply Equation 18-1.

$$\Delta PE = k_e q q_0 \left(\frac{1}{r_b} - \frac{1}{r_a} \right)$$
$$= (9.00 \times 10^9 \text{ N} \cdot \text{m}^2/\text{C}^2)(1.47 \times 10^{-17} \text{ C})(1.60 \times 10^{-19} \text{ C})$$
$$\times \left(\frac{1}{2.00 \times 10^{-11} \text{ m}} - \frac{1}{6.00 \times 10^{-11} \text{ m}} \right)$$
$$= 7.06 \times 10^{-16} \text{ N} \cdot \text{m} = 7.06 \times 10^{-16} \text{ J}$$

We see from the result of Example 18-1 that Equation 18-1 works quite well when the charges are of the same sign. We may even make use of the signs of the charges as *algebraic* signs in the equation. If we include the charge signs, Equation 18-1 is still a correct representation. If the charge signs are the same (either both positive or both negative), their product is positive. If the charges are of opposite sign, their product is negative. Also, they attract each other, and their potential energy increases with increasing distance just as Equation 18-1 would predict. Thus, using the signs of the charges as algebraic signs will always give us consistent results in calculating potential energy changes.

Another very important quantity arises from an examination of the equation. Note that if q_0 is doubled in magnitude, the change in potential energy will be doubled. If q_0 is tripled, ΔPE is tripled, and so on. We see that the ratio of ΔPE to q_0 is independent of the magnitude of q_0. It appears that we have another quantity which may be assigned to the space surrounding the charge q, as was

---

* Although we use the approximation $r^2_{av} = r_a r_b$ in its derivation, Equation 18-1 is an exact result and may be obtained from a calculus-based derivation.

the electric field. The quantity is **electric potential**. Electric potential difference is defined as the change in electric potential energy per charge as a test charge is moved between two points. Using q_0 as the symbol for the test charge and employing Equation 18-1, we have

$$V_{ab} = \frac{\Delta PE}{q_0} \tag{18-2}$$

$$V_{ab} = \frac{k_e q q_0}{q_0}\left(\frac{1}{r_b} - \frac{1}{r_a}\right)$$

or

$$V_{ab} = k_e q \left(\frac{1}{r_b} - \frac{1}{r_a}\right) \tag{18-3}$$

The symbol V_{ab} is used to represent the change in potential between the points at distances r_a and r_b from the charge q. In the SI system of units, potential is measured in joules per coulomb (J/C) and is given the name **volt** (symbol V) in honor of Alessandro Volta (1745–1827), an Italian scientist credited with inventing the electric battery. If we can determine the potential difference between two points, we can calculate the change in potential energy for any charge which is to be moved between those points.

> The change in potential between points and the unit in which it is measured, the volt, are both represented by a capital V. Do not confuse the variable for the unit, or vice versa. Usually, context makes the intended one quite clear. Because change in potential is a variable quantity, it is represented in this text by an italic capital V; the abbreviation for volt, because it is not a variable quantity, is represented by a roman (nonitalic) capital V.

Example 18-2 A very small ball is charged to -3.00×10^{-6} C. (a) Calculate the potential difference between an initial position 0.200 m away from the charge and a final position 0.800 m away. (b) What change in potential energy takes place if another ball which has been charged to $+6.00 \times 10^{-8}$ C is moved between the two positions?

Solution (a) We apply Equation 18-3, but we must be careful with signs.

$$V_{ab} = k_e q \left(\frac{1}{r_b} - \frac{1}{r_a}\right)$$

$$= (9.00 \times 10^9 \text{ N} \cdot \text{m}^2/\text{C}^2)(-3.00 \times 10^{-6} \text{ C})\left(\frac{1}{0.800 \text{ m}} - \frac{1}{0.200 \text{ m}}\right)$$

$$= +1.01 \times 10^5 \text{ V} \quad (\text{J/C})$$

(b) This is nearly as straightforward. We apply Equation 18-2, identifying q_0 as $+6.00 \times 10^{-8}$ C and using the results of part a.

$$\Delta PE = q_0 V_{ab} = (+6.00 \times 10^{-8} \text{ C})(1.01 \times 10^5 \text{ J/C})$$
$$= 6.06 \times 10^{-3} \text{ J}$$

Since this is positive, it represents an increase in potential energy and is the amount of work which would have to be done on the sphere to move it from one position to the other.

A lack of feeling for what a volt "is" can separate a mechanical problem chugger from the individual with a good intuitive grasp of electrical phenomena. Your automobile battery is labeled "12 volts," but what exactly does this mean? If the terminals of your battery indeed measured 12.0 V, that would imply that if 1 C of electrons (6.25×10^{18} of them) flowed from the negative terminal out through the path supplied and back to the positive terminal, somewhere during that trip the electrons would have performed 12.0 J of work. No less work and no more work than 12.0 J could be accomplished by each coulomb of charge. Conversely, if we force a coulomb of charge backward, that is, against the electric field of the battery, pulling electrons from the positive terminal and forcing them into the negative terminal (this procedure is called *charging the battery*), we must do 12.0 J of work for each coulomb so transferred.

The voltage between given points tends to be the constant element in a wide variety of electrical situations. Flashlight batteries (cells) come labeled "1.5 volts," transistor radio batteries "9.0 volts," and so on. It is essential for your ability to think of electrical phenomena to appreciate what these mean. When you are about to insert a two-pronged plug into a 120-V wall socket, visualize what you already know about the situation electrically. Every coulomb of charge that comes out of the wiring in the wall through one prong of the plug, flows through the appliance, and reenters the wall via the other prong of the plug will have done 120 J of work somewhere between leaving the wall and returning to it. This work could be in making a bulb light, a toaster heat, a motor turn, or a speaker produce sound. It could also be used in just heating the wires. Wherever and however the work is done by the moving charges, it amounts to 120 J of work for each and every coulomb transferred. How many coulombs flow in a given time, if any, depends in part on other properties of the circuit, but the work done by each is known if the voltage is determined. This intuitive view is quite useful in studying electricity.

18-2 Potential Energy and Potential

Example 18-3 Johnny is bored in science class one day, so he unbends a paper clip and inserts the free ends into the 120-V outlet at his desk. In the flash that follows, the circuit breakers trip, shutting off the electricity but not before a 1.00-cm length of the 0.600-mm-diameter paper clip is melted completely away. Assuming 20 percent of the energy of the electric surge went into melting this section of the paper clip, how many coulombs of charge flowed during the flash? (What happens to Johnny?)

Solution First, we will need to determine the mass of the paper clip melted, then the heat required to melt that amount. Since this is 20 percent of the energy released, multiplying by 5 will give us the total energy of the flash. Substituting into $V = W/q$ will give us the charge transferred. We must first obtain the volume of the melted metal. The mass is then equal to that volume times its density.

$$\begin{aligned} \text{Vol} &= (\pi r^2)l = \pi(0.300 \text{ mm})^2(1.00 \text{ cm}) \\ &= \pi(3.00 \times 10^{-4} \text{ m})^2(1.00 \times 10^{-2} \text{ m}) \\ &= 2.83 \times 10^{-9} \text{ m}^3 \end{aligned}$$

Paper clips are made of steel, the properties of which depend upon the particular mixture. However, steel is predominantly iron, so we will use the constants for iron.

$$\rho_{\text{iron}} = 7.86 \times 10^3 \text{ kg/m}^3$$

$$\begin{aligned} m = \rho(\text{vol}) &= (7.86 \times 10^3 \text{ kg/m}^3)(2.83 \times 10^{-9} \text{ m}^3) \\ &= 2.22 \times 10^{-5} \text{ kg} \end{aligned}$$

Room temperature is 20°C; iron melts at 1535°C, has a specific heat of 449 J/(kg·°C), and a heat of fusion of 2.89×10^5 J/kg.

$$Q = mc\,\Delta T + mL_f$$
$$= (2.22 \times 10^{-5}\text{ kg})[449\text{ J/(kg}\cdot{}°\text{C})](1535 - 20)°\text{C}$$
$$+ (2.22 \times 10^{-5}\text{ kg})(2.89 \times 10^5\text{ J/kg})$$
$$= 15.1\text{ J} + 6.42\text{ J} = 21.5\text{ J}$$

This is the energy to melt the section of paper clip and represents (21.5×5) J = 108 J released in the flash.

$$V = \frac{W}{q}$$

$$q = \frac{W}{V} = \frac{21.5\text{ J} \times 5}{120\text{ J/C}} = 0.896\text{ C}$$

18-3 "Absolute" Potential

In every application of work-energy or conservation of energy, we are concerned only with *changes* in potential energy. Thus the choice of a zero position is arbitrary, and we choose one which is most convenient for solving a given problem. Later, we will see that the most convenient choice for **zero potential** is usually the biggest source of "free" charges available. (This is most often the earth itself; thus we frequently call it *ground potential*.)

When dealing with a point charge, we find it convenient to choose our zero level to simplify the mathematics. This can be seen if we rewrite Equation 18-3 as

$$V_{ab} = \frac{k_e q}{r_b} - \frac{k_e q}{r_a}$$

But remember that V_{ab} is really the change in potential when the separation is changed from a to b. Or, we can say that it is the final potential minus the initial potential. Thus we might wish to write

$$V_{ab} = V_b - V_a \qquad (18\text{-}4)$$

With this identification, we may define the potential at a point which is an r_b distance away from q as

$$V_b = \frac{k_e q}{r_b}$$

and the potential at a point which is an r_a distance away from q as

$$V_a = \frac{k_e q}{r_a}$$

Since these are arbitrary distances, we have, by implication, defined the absolute potential at a point for any distance r away from a point charge q

$$V = \frac{k_e q}{r} \qquad (18\text{-}5)$$

18 Electrostatic Energy and Capacitance

The magnitude of a change does not depend upon the choice of a zero level for a variable. The change in height from the basement to the second story is three floors, regardless of whether we choose ground level or the rooftop as the zero level for height.

We call this expression the **absolute potential**, but it is really the potential with respect to a zero level whose choice is hidden in the development. Consider Equation 18-5. As r gets larger and larger, V will get smaller and smaller, so that as r approaches infinity, V approaches zero. Thus, the chosen zero level is at infinity. In other words, the zero of absolute potential due to a point charge is an infinite distance away from the charge. This choice simplifies the mathematical expression. Thus the absolute potential or the potential at a point given by Equation 18-5 is really the change in potential energy per unit charge which takes place when a charge is moved from an infinite distance away to the point in question. We note that the potential at a point may be positive or negative (or zero). Further, if the point in question is near several charges, the net potential is equal to the algebraic sum of the potential due to each of the charges. Both these statements are the direct result of potential being a scalar quantity.

Example 18-4 Two charges $q_1 = +2.00 \times 10^{-6}$ C and $q_2 = -5.00 \times 10^{-6}$ C are separated by 0.400 m. Determine the potential at a point midway between them.

Solution We simply apply Equation 18-5 at a point $r = 0.200$ m distance from each charge and add them algebraically. We have

$$V_1 = \frac{k_e q_1}{r_1} = \frac{(9.00 \times 10^9 \text{ N} \cdot \text{m}^2/\text{C}^2)(+2.00 \times 10^{-6} \text{ C})}{0.200 \text{ m}}$$
$$= +9.00 \times 10^4 \text{ V}$$

and

$$V_2 = \frac{k_e q_2}{r_2} = \frac{(9.00 \times 10^9 \text{ N} \cdot \text{m}^2/\text{C}^2)(-5.00 \times 10^{-6} \text{ C})}{0.200 \text{ m}}$$
$$= -2.25 \times 10^5 \text{ V}$$

Therefore,

$$V = V_1 + V_2 = +9.00 \times 10^4 \text{ V} + (-2.25 \times 10^5 \text{ V})$$
$$= -1.35 \times 10^5 \text{ V}$$

The negative potential means that a test charge loses potential energy as it is moved from infinity to the point in question.

In general, we see that when we wish to calculate the potential at a point due to any number n of charges, we simply write the algebraic sum as

$$V = V_1 + V_2 + \cdots + V_n = \sum_{i=1}^{N} V_i$$

where $V_i = k_e q_i / r_i$. Therefore, we have

$$V = k_e \sum_{i=1}^{N} \frac{q_i}{r_i} \qquad (18\text{-}6)$$

Suppose we allow a test charge (perhaps a proton) to move freely under the action of an electric field generated by point charges. The change in potential energy plus the change in kinetic energy must sum to zero. If the test charge moves from a point labeled a to a point labeled b, we would have

or
$$\Delta KE + \Delta PE = 0$$
or
$$KE_b - KE_a + q_0 V_{ab} = 0$$

$$\tfrac{1}{2}mv_b^2 - \tfrac{1}{2}mv_a^2 = -q_0 V_{ab}$$
$$\tfrac{1}{2}mv_b^2 - \tfrac{1}{2}mv_a^2 = q_0 V_{ba} \qquad (18\text{-}7)$$

Can you justify substituting V_{ba} for $-V_{ab}$?

Example 18-5 Determine the final speed of a proton (mass = 1.67×10^{-27} kg) which is allowed to move freely from rest at a very large distance away to the midpoint between the two charges described in Example 18-4.

Solution We may employ Equation 18-7 and solve for v_b, realizing that $v_a = 0$ since the proton starts from rest. Also, for our problem, V_{ab} is simply the potential at the point in question, the solution of Example 18-4.

$$\tfrac{1}{2}mv_b^2 - 0 = q_0 V_{ba} = -q_0 V_{ab}$$

$$v_b^2 = \frac{-2q_0 V_{ab}}{m}$$

$$v_b = \sqrt{\frac{-2q_0 V_{ab}}{m}}$$

$$= \sqrt{\frac{-2(1.6 \times 10^{-19}\ \text{C})(-1.35 \times 10^5\ \text{J/C})}{1.67 \times 10^{-27}\ \text{kg}}}$$

$$= 5.09 \times 10^6\ \text{m/s}$$

Equation 18-7 shows that any charged particle allowed to accelerate through a potential difference will undergo a change in kinetic energy. For instance, a proton accelerating through a potential difference of 1.00 V has a change in kinetic energy given by

$$\Delta KE = qV_{ba} = (+1.60 \times 10^{-19}\ \text{C})(1.00\ \text{J/C}) = 1.60 \times 10^{-19}\ \text{J}$$

This is a very small number reported in the proper SI unit. Thus, for many purposes, it is convenient to use a unit developed to report energies associated with atoms, molecules, and elementary particles. That unit is called the **electronvolt** (eV), defined as the energy acquired by an electron when it moves through a potential difference of one volt. From its definition, we have that

$$1\ \text{eV} = 1.60 \times 10^{-19}\ \text{J}$$

Electronvolts, kiloelectronvolts (keV), and megaelectronvolts (MeV) have become standard in atomic and particle physics even though the electronvolt is not an SI unit.

18-4 Potential Difference and Electric Field

A relationship exists linking the potential difference between two points and the electric field intensity between the points. This is most easily illustrated in a uniform electric field. Consider the test charge q_0 moved a distance Δs parallel to an electric field, as shown in Figure 18-3. The force exerted on the charge by the electric field has a magnitude equal to $q_0 E$ and is directed to the right. Since the motion is also to the right, the work done by the field is

$$W = F\,\Delta s = q_0 E \Delta s$$

The change in potential energy is equal to the negative of the work done by the conservative field. Thus

$$\Delta PE = -W = -q_0 E\,\Delta s$$

Dividing by the magnitude of the test charge yields

$$\frac{\Delta PE}{q_0} = -E\,\Delta s$$

The left side of the equation is the potential difference from Equation 18-2, which we may write as ΔV. Then we have

$$\Delta V = -E\,\Delta s$$

Solving for E yields

$$E = -\frac{\Delta V}{\Delta s} \tag{18-8}$$

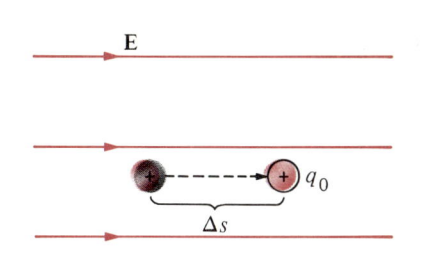

Figure 18-3 Charge q_0 in a region of uniform electric field moves a distance Δs due to the force produced by the field.

Thus the electric field intensity is equal to the negative of the rate of change of voltage with displacement. The negative sign simply means that the electric field intensity points in the direction opposite to that of increasing voltage. The rate of change of voltage with displacement is called the **potential gradient.** Therefore, we often say that the electric field is equal to the negative of the potential gradient. In many instances, especially in dealing with electric circuits, it is customary to drop the delta (Δ) and even all subscripts from the change in potential and simply write V for the **voltage.** Once you get used to the symbol, it usually does not cause any confusion, and it does simplify notation. Equation 18-8 may also be used for large distances in an electric field if we realize that we calculate the average value of E. Thus, if we have a voltage of V (remember potential difference) over a displacement of magnitude d, we would write

$$E_{av} = -\frac{V}{d} \tag{18-9}$$

Note that the units of the left side of the equation are newtons per coulomb (N/C), while those of the right side are volts per meter (V/m). You should be able to prove that these are the same.

Example 18-6 The potential difference across the human heart just before a beat has a typical value of 2.50×10^{-3} V, or 2.50 millivolts (mV). The magnitude of the displacement over which this voltage occurs is about 0.150 m. The area near the lower left rib is positive, and that nearest the sternum (breastbone)

431

is negative. Determine the average value of the electric field intensity through the heart at this time.

Solution We apply Equation 18-9, but we recognize that the sternum is at a lower potential than the rib. Thus, as we move from the sternum to the rib, we find that the potential change is positive.

$$E_{av} = -\frac{V}{d} = -\frac{2.50 \times 10^{-3} \text{ V}}{0.150 \text{ m}} = -1.67 \times 10^{-2} \text{ V/m}$$

The minus sign indicates that the direction of E_{av} is toward the sternum, while the voltage is increasing as we move away from the sternum.

Since the magnitude of the electric field intensity equals the rate of change of voltage with displacement, the electric field is zero in any region in which the voltage is constant. Conversely, in any region in which the electric field intensity is zero, the voltage remains constant. This is graphically illustrated by considering a charged metal sphere of radius r_0, which is shown in Figure 18-4 along with the electric field intensity and the potential associated with it. First note that the magnitudes of E and V are plotted only from $r = 0$ outward, since they are the same in all directions. Next, note that outside the sphere, both quantities are represented by the same equations as those for a point charge, because outside a uniform spherical distribution of charge, both E and V are the same as though all the charge were concentrated at the center of the sphere.[1] The proof of this statement is similar to that used by Newton to prove that the gravitational force outside a uniform spherical distribution of mass is the same as if all the mass were concentrated at its center. A complete proof requires the use of calculus. Finally, note in Figure 18-4 that inside the sphere, the electric field intensity drops immediately to zero while V retains the value it had at the surface. This is the result of the sphere's being metal. As we discussed in Section 17-5, there is no electric field nor net charge on the interior of a conductor in the electrostatic case. Since there is no electric field, the potential cannot change and still remain consistent with Equation 18-8.

We may construct imaginary lines or surfaces connecting all points with

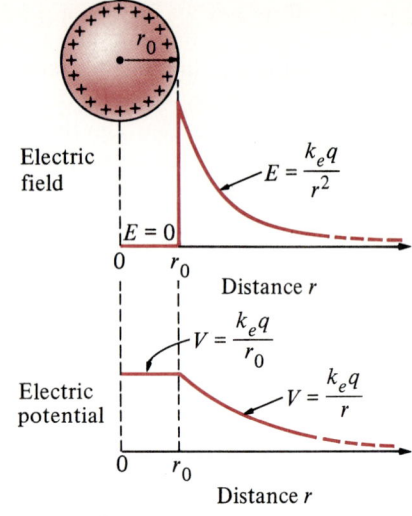

Figure 18-4 Electric field and electric potential vs. distance from a positively charged metal sphere. The electric field within the sphere must be zero if charges are to remain static. This implies that the potential within the sphere is the same as at the surface.

---

[1] A useful relationship is derived from this fact. If a sphere of radius r is given a uniformly distributed surface charge q, we may write the magnitude of E at a point very near the surface as

$$E = \frac{k_e q}{r^2}$$

By multiplying and dividing by 4π, we obtain

$$E = 4\pi k_e \frac{q}{4\pi r^2}$$

The area of the sphere is $A = 4\pi r^2$. Thus we have

$$E = 4\pi k_e \frac{q}{A}$$

This is sometimes written as $E = 4\pi k_e \sigma$, where the lowercase Greek sigma (σ) is the **surface charge density,** or the charge per unit area on the sphere. Thus the magnitude of the electric field intensity at a point very near the surface is proportional to the charge per unit area on the surface. This conclusion actually applies to any surface charge and will be used in Section 18-5.

the same potential. We call these constructs **equipotential lines** or **equipotential surfaces** (or sometimes, just **equipotentials**). An equipotential surface is one on which all points have the same value of potential. Since there is no net electric field intensity between two points which are at the same potential, charge could be moved along such a surface, with no work done by or against the field. The equipotential surfaces surrounding a point charge are a family of concentric spheres which form circles where they intersect a plane. In Figure 18-5, we see both the equipotential surface (really equipotential lines) and the lines of force. We note from Figure 18-5 that the equipotential surfaces are circles and the lines of force representing the electric field are radial. Thus the lines of force are perpendicular to the equipotential surfaces, as we might expect. Why? We may use equipotentials to map the electric field. For any charge distribution, we simply choose a zero level and make many voltage measurements in the space surrounding the charges. Connecting all points of the same voltage gives us the equipotential surfaces. The lines of force are then drawn to be perpendicular to the equipotentials. Figure 18-6 shows two more common charge distributions along with the equipotentials and lines of force. Electric field lines must be perpendicular to the surface of any metal in an electrostatic situation, or charges would accelerate on the metal surface. Therefore, no net electric work is required to move a charge along the surface, and it is an equipotential. In fact, the whole interior volume of a metal in any electrostatic situation is an equipotential volume.

Michael Faraday conducted an experiment by using an ice pail (metal bucket) and a charged metal ball which he touched to the inside of the pail. He determined that *all* the charge of the metal ball was transferred to the pail even if the pail was already charged! When excess charge is placed on a conductor, it immediately is distributed to the exterior surface, with no net charge left within

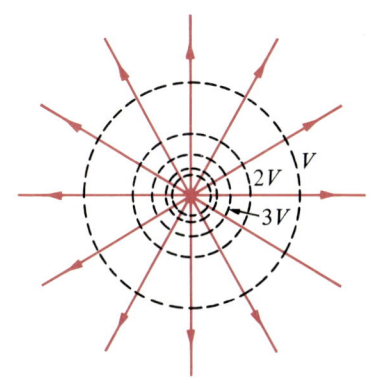

Figure 18-5 Outwardly directed electric field of an isolated positive charge, and concentric equipotential lines. A charge could be moved along an equipotential line, doing no work, since the motion would always be perpendicular to the direction of the electric force.

(a)

(b)

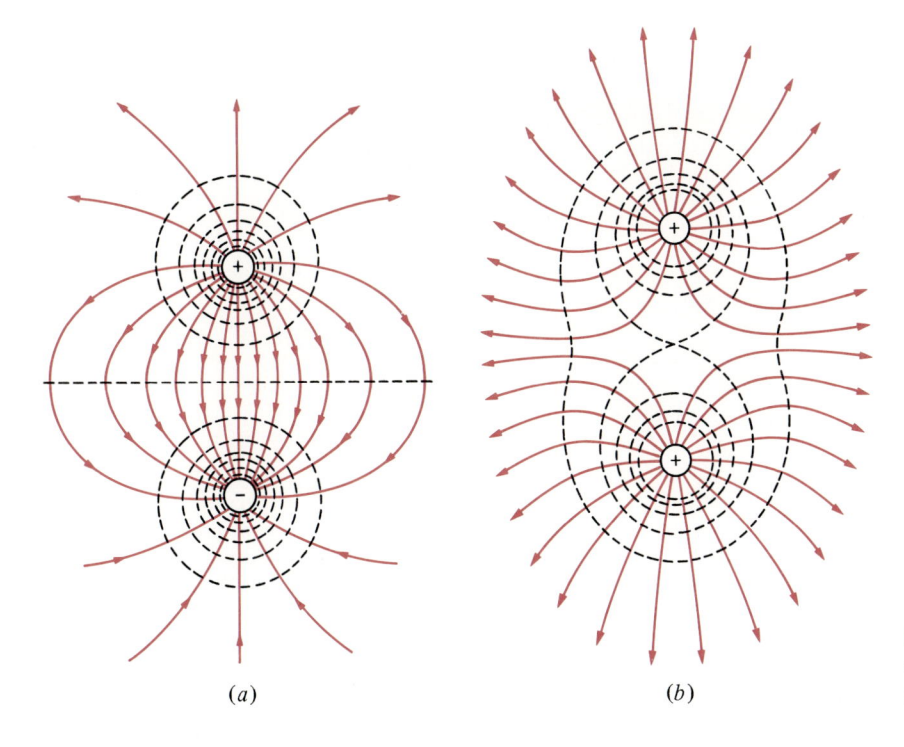

Figure 18-6 Electric field lines and equipotential lines for nearby charges of (a) opposite and (b) like signs.

433

the metal. If we introduce charge in small amounts to the interior surface of a hollow conductor, we may build a larger and larger charge on the external surface. This is depicted in Figure 18-7. Before the metal ball touches the interior of the hollow conductor shown in Figure 18-7a, both are charged. Afterward (Figure 18-7b), the ball has no net charge, as it has all been transferred to the hollow conductor and immediately distributed on the exterior surface. The ball can be removed, recharged, inserted, and touched to the interior again and again. Recall that the potential due to the charged sphere is proportional to the magnitude of the charge on the sphere, as shown in Figure 18-4. Thus we can, *in principle,* increase the potential of the hollow sphere to as large a value as we wish by doing work to increase the charge on it.

In 1930, Robert Van de Graaff applied this idea to develop what is now called the **Van de Graaff generator.** His goal was to obtain very high voltages which could be used to accelerate charged particles (protons, electrons, ions) to very large values of kinetic energy. These particles could then be used as projectiles to probe the nuclear atom. A schematic diagram of a Van de Graaff generator is shown in Figure 18-8. The generator employs a belt made of some insulating material which passes over two rollers or pulleys. The belt is driven by a small electric motor. One of the rollers is at ground potential, and the other is inside a large hollow conducting dome which is supported and insulated from ground. An electric charge is placed on the belt at the lower roller and carried by the belt to the inside of the dome. There it flows to the interior of the dome and is distributed to the outside of the conductor. Charge may be carried continuously and the dome charged to a very high voltage in a short time (Figure 18-9). We stated that, *in principle,* the potential could be increased to as large a value as we wish. *In practice,* the amount of charge which can be held is limited by several

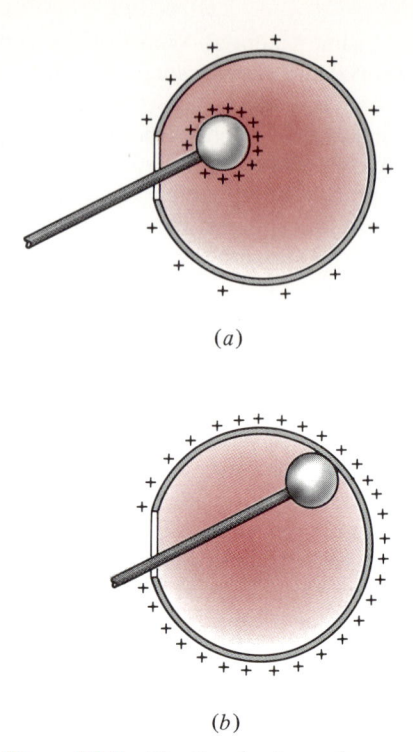

Figure 18-7 The Faraday ice-pail experiment. (*a*) A small, charged metal ball is brought to the inside surface of a large conducting object and touched momentarily. (*b*) The charge on the ball is completely transferred to the large conductor.

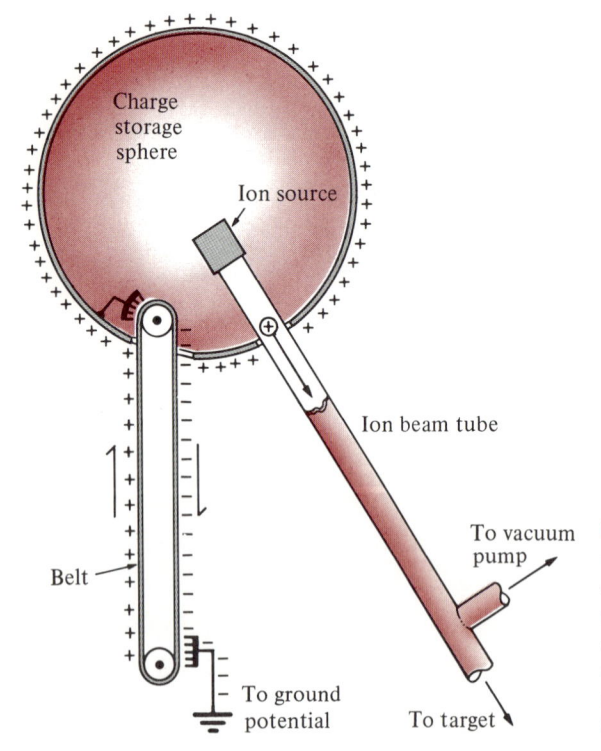

Figure 18-8 A Van de Graaff accelerator. Charges placed on the belt are introduced to the zero electric field inside the surface of the dome, allowing a large charge to accumulate. An ion source inside the dome introduces charged particles to an evacuated beam tube. When such a particle drifts far enough down the beam tube to have left the electric field–free region of the dome, it is accelerated down the beam tube and away from the dome.

factors. No matter how well insulated, the supporting structure for the dome may begin to leak charge when the potential difference between dome and ground gets very large. Another limiting factor is that the large electric field intensity causes the medium in which the dome is immersed to break down and arc. However, with proper modifications to minimize these problems, single Van de Graaff generators can be designed to run as high as 10 million V.

18-5 Capacitors

The Van de Graaff generator is designed to store a large quantity of charge and therefore generate very high voltages. The charge is not used as such but only accumulated to create large voltages. There are many applications for a different kind of device, a **capacitor**, which stores energy in the form of separated charges and releases it to be used whenever circumstances require. A capacitor consists of two conducting surfaces separated by a space that does not conduct.

The most common type of capacitor is the **parallel-plate capacitor**, which comprises two conductors separated by a distance very small with respect to their dimensions. The two conductors are usually given charges equal in magnitude but opposite in sign. This may be accomplished by connecting them to the opposite terminals of a battery. The battery causes one plate to become positively charged while the other acquires an equal-in-magnitude negative charge. Figure 18-10 depicts a charged parallel-plate capacitor. Because the two plates are oppositely charged and very close, rather large quantities of charge may be transferred from one plate to the other with relatively small voltage. We define the **capacitance** C of a capacitor as

$$C = \frac{q}{V} \qquad (18\text{-}10)$$

Figure 18-9 The electrically isolated boy has both hands on the dome of a Van de Graaff generator and thus shares the charge. There is little danger of electric shock because he is insulated from ground potential. *(T. Pix)*

where q is the magnitude of the charge on either plate and V is the voltage (potential difference) between the plates. The SI unit of capacitance is coulomb per volt (C/V), which has been given the name **farad** (F) to honor Michael Faraday. Thus 1 F = 1 C/V. Note that the *net* charge on a capacitor is virtually always zero. When we refer to the charge on a capacitor, the reference is nearly always to the magnitude of charge on either conductor. We may also see that although the definition of capacitance is given by Equation 18-10, the capacitance of a given device does not depend on either the charge q or the voltage V. We have encountered such situations before. Recall that we could write Newton's second law (under certain circumstances) as $m = F/a$, and yet we are well aware that m does not depend on either F or a.

An expression for the capacitance of a parallel-plate capacitor may be obtained as follows. Consider the capacitor of Figure 18-11. Each plate has an area A separated by a distance d. One plate has a charge equal to $+q$, and the other plate has a charge equal to $-q$. The potential difference between the plates is related to the electric field intensity by Equation 18-9. We may write

$$V = Ed$$

where we have dropped the minus sign since we are interested only in the magnitude. Further, the field between parallel plates is constant, and thus E has

Figure 18-10 The parallel-plate capacitor consists of two parallel conducting surfaces which can be charged with opposing sign charges, giving rise to the electric field in the gap between the plates.

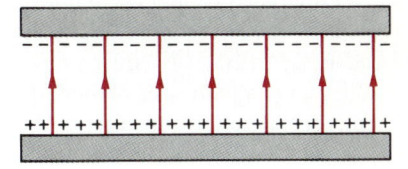

435

been substituted for E_{av}. Now, since the electric field intensity is proportional to the charge per unit area (see the footnote in Section 18-4), we have

$$E = 4\pi k_e \frac{q}{A} \quad \text{and} \quad V = 4\pi k_e \frac{q}{A} d$$

Substituting this into Equation 18-10 yields

$$C = \frac{q}{V} = \frac{q}{4\pi k_e q d/A}$$

$$C = \frac{A}{4\pi k_e d} \quad (18\text{-}11)$$

Thus the capacitance of a parallel-plate capacitor depends on the cross-sectional area of the plates and the distance between them.

Equation 18-11 and many other equations of electricity and magnetism take on a much simpler form if k_e is expressed in terms of a different constant, ϵ_0, by the relationship

$$k_e = \frac{1}{4\pi\epsilon_0} \quad (18\text{-}12)$$

or

$$\epsilon_0 = \frac{1}{4\pi k_e} \quad (18\text{-}13)$$

The lowercase Greek epsilon (ϵ) with a subscript zero is called the electric permittivity of free space and has the value

$$\epsilon_0 = 8.85 \times 10^{-12} \text{ C}^2/(\text{N}\cdot\text{m}^2)$$

With this definition, Equation 18-11 becomes

$$C = \frac{\epsilon_0 A}{d} \quad (18\text{-}14)$$

for the capacitance of a parallel-plate capacitor.

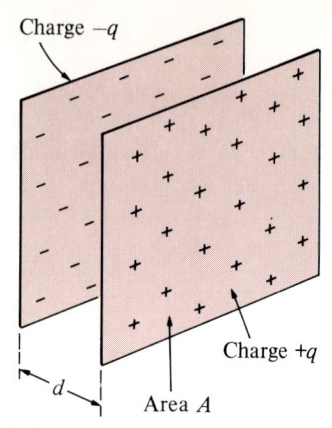

Figure 18-11 Conducting parallel plates each of area A are separated by a distance d and charged to $-q$ and $+q$. The capacitance is the amount of separated charge which can be stored per volt of potential difference applied between the plates.

The permittivity of a material may be associated with the difficulty of establishing a given magnitude electric field in the material. This is "easiest" in empty space; thus ϵ_0 is lower than the permittivity in material substances (see Section 18-7).

Example 18-7 Determine the area A of the plates of a parallel-plate capacitor of capacitance $C = 1.00$ F if the plates are separated by 1.00×10^{-3} m.

Solution We simply solve Equation 18-14 for A and substitute the other known values.

$$A = \frac{Cd}{\epsilon_0} = \frac{(1.00 \text{ F})(1.00 \times 10^{-3} \text{ m})}{8.85 \times 10^{-12} \text{ C}^2/(\text{N}\cdot\text{m}^2)} = 1.13 \times 10^8 \text{ m}^2$$

This is nearly 44 mi² for a capacitance of "only" 1 F.

Obviously 1 F is a very large capacitance. Most capacitors we use are less than 1.00×10^{-6} F, and the units microfarad (μF) (1 μF = 10^{-6} F) and picofarad (1 pF = 10^{-12} F) are common. However, even a 0.0100-μF capacitor could require a plate area of more than 1 m². To keep the volume of a capacitor small, they are often constructed of two metal foils as plates, separated by some thin insulating material, and rolled into a cylinder. The packaged capacitor then

(a) (b)

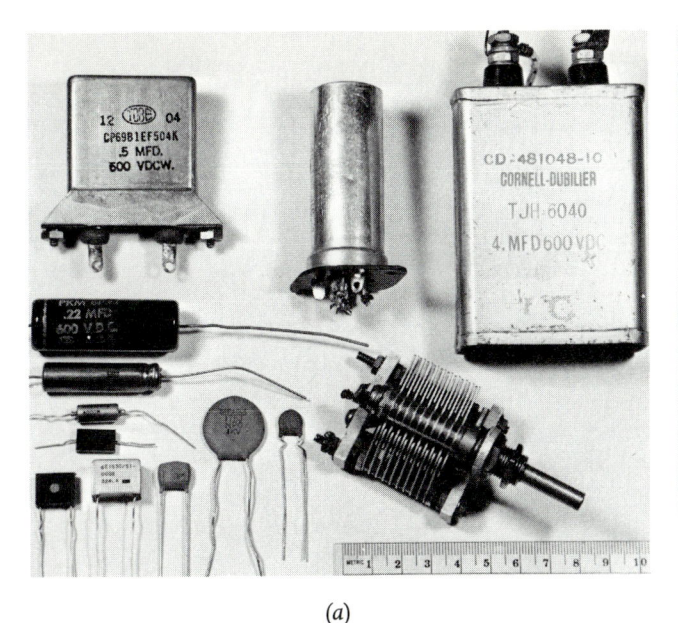

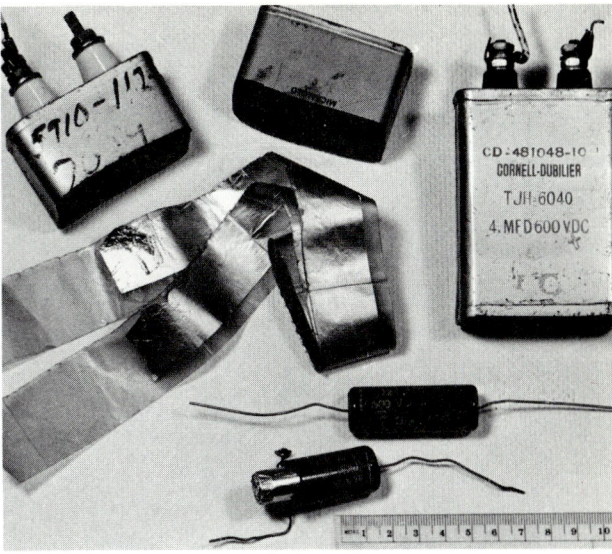

Figure 18-12 A variety of capacitors. (*a*) The small tubular, disk-shaped, and rectangular capacitors are found in most electronic circuits. The variable capacitor at lower right has sets of interleaving plates. (*b*) A large capacitor cut open to reveal the layers of metal foil separated by oil-saturated paper. At the bottom, the plastic coating of a cylindrical capacitor is partly removed to show a similar winding of foil layers and insulators. (D. Riban)

looks like those of Figure 18-12*a*. In Figure 18-12*b*, a large capacitor and a cylindrical capacitor have been taken apart to show their construction.

There are many other types and configurations of capacitors, depending on their intended uses. Variable capacitors are often used to tune radio receivers by changing the area of plates opposite each other. Other capacitors, when properly connected, may be used for smoothing a varying voltage in an electric circuit or to isolate a direct voltage from an alternating voltage, as we will see. Capacitors can even help increase the efficiency of transmission of alternating voltage in power lines.

18-6 Capacitors and Energy

We have said that the advantage of a capacitor is its ability to store charge. Some might say that the capacitor really stores energy in the form of the charge separation and that energy is available as the capacitor is discharged. It is certainly true that the work which must be done in charging the capacitor is stored as potential energy. We most often say that the energy is stored in the electric field. We may determine just how much energy is stored in the electric field associated with a charged capacitor by calculating the change in potential energy which takes place as the capacitor is being charged. Consider the two separate plates of a capacitor of capacitance C initially uncharged. A stream of small increments of charge is moved from one plate to the other until a net charge q has been transferred. As the capacitor is being charged, a voltage is being developed between the plates. The definition of potential difference tells us how to calculate the change in potential energy which takes place when a quantity of charge is moved through a potential difference. The change in potential energy is equal to the product of the magnitude of the charge with the magnitude of the potential difference. In our case, the potential difference (voltage) is changing. It is initially zero as both plates are uncharged. Using Equation 18-10, we see that the final voltage after a net charge q has been transferred may be written as

$$V = \frac{q}{C}$$

Since the voltage increases linearly from zero, its average value is half its maximum value. We may calculate the change in potential energy and hence the energy stored as though all this charge was transferred at the average value of voltage, $\frac{1}{2}q/C$.

$$\Delta PE = q \frac{\frac{1}{2}q}{C}$$

$$w = \frac{\frac{1}{2}q^2}{C} \tag{18-15}$$

where w is the energy stored in the electric field and is equal to the change in potential energy which has taken place. Utilizing Equation 18-10, we may write the energy as

$$w = \tfrac{1}{2}CV^2 \tag{18-16}$$

or

$$w = \tfrac{1}{2}qV \tag{18-17}$$

Example 18-8 An electronic flash mechanism used with a camera is activated when a capacitor is discharged through a filament wire. (a) Determine the minimum value of the capacitor which must be used, for a filament which requires 5.40×10^{-3} J to fire properly if the charging batteries total 12.0 V. (b) Calculate the maximum charge on the capacitor.

Solution (a) Since we know both the energy to be stored and the voltage, we may apply Equation 18-16 to solve part a.

$$w = \tfrac{1}{2}CV^2$$

or

$$C = \frac{2w}{V^2} = \frac{2(5.40 \times 10^{-3} \text{ J})}{(12.0 \text{ V})^2} = 7.50 \times 10^{-5} \text{ F} = 75.0 \text{ }\mu\text{F}$$

(b) We use the definition of capacitance along with the given voltage and calculated capacitance.

$$C = \frac{q}{V}$$

or

$$q = CV = (7.50 \times 10^{-5} \text{ F})(12.0 \text{ V}) = 9.00 \times 10^{-4} \text{ C}$$

---

18 Electrostatic Energy and Capacitance

Since we say that energy is stored in the electric field, we might determine the amount of energy stored in the parallel-plate capacitor as a function of the electric field intensity. We may substitute into Equation 18-16 as follows:

$$w = \tfrac{1}{2}CV^2 = \frac{1}{2}\left(\epsilon_0 \frac{A}{d}\right)(Ed)^2$$

where we have used $C = \epsilon_0 A/d$ and $V = Ed$ for the parallel-plate capacitor. Thus

$$w = \tfrac{1}{2}\epsilon_0 E^2 A d \qquad (18\text{-}18)$$

The quantity Ad is equal to the volume of the capacitor. We often obtain the energy per unit volume as

$$u = \frac{w}{Ad}$$

or

$$u = \tfrac{1}{2}\epsilon_0 E^2 \qquad (18\text{-}19)$$

where u is the energy per unit volume stored in an electric field of intensity E. Although this was derived for the special case of a parallel-plate capacitor, it is applicable to any electric field in free space.

18-7 Dielectric Coefficient

We have treated devices, thus far, as though they existed in a vacuum, but actually most at least have air surrounding them. Most capacitors use some nonconducting material called a **dielectric** between their plates. The dielectric may be paper, some type of plastic, an organic material, or even a metal oxide. In addition to allowing the capacitor to be used at higher voltages without breakdown or arcing between plates, the dielectric increases the capacitance of the capacitor. This is easily shown. If we charge a capacitor which has a vacuum between the plates, we obtain a particular voltage, say V_0, for a given charge q. If we now insert a dielectric between the plates, the voltage drops to a new value V. When the dielectric is removed, however, the voltage returns to V_0, as illustrated in Figure 18-13. The charge on the plates remains unchanged with the insertion of the dielectric. Thus, since the voltage has changed, the capacitance must change also. With a vacuum between the plates, we have the capacitance

$$C_0 = \frac{q}{V_0}$$

After the insertion of the dielectric, the capacitance becomes

$$C = \frac{q}{V}$$

The voltage, hence the capacitance, is different for different dielectrics. Therefore, we define a **dielectric coefficient** for each dielectric material as

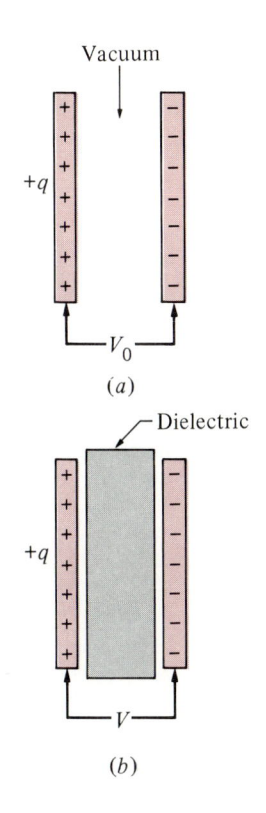

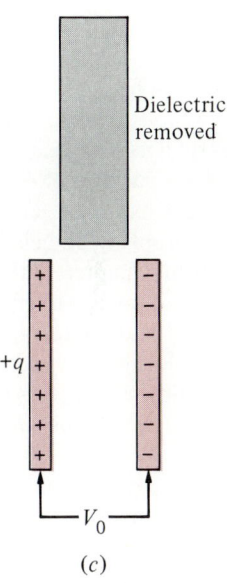

$$K = \frac{C}{C_0} \qquad (18\text{-}20)$$

Figure 18-13 (a) With a vacuum between the plates, a charge of $\pm q$ is produced at voltage V_0. (b) If a dielectric is introduced and the charge is kept at $\pm q$, the voltage between the plates *drops* to V. (c) If the dielectric is removed, the voltage returns to V_0.

439

TABLE 18-1 Approximate Dielectric Constants and Breakdown Strengths at 1 atm and 25°C

Material	Dielectric Constant K*
Vacuum	1.0000
Air	1.005 (0.8)
Beeswax	2.9
Butyl rubber	2.4
Glass:	
Corning 0010	6.3
Pyrex 1710	6.0 (13)
Silica	3.8 (8)
Lead telluride	400
Manganese dioxide	10,000
Neoprene rubber	6.6 (12)
Paper	1.7–4.0 (5–20)
Polyethylene	2.3 (20)
Polystyrene	2.5 (25)
Polyvinyl chloride	4.5
Teflon	2.1 (60)
Titanium dioxide	100 (6)

* The number in parentheses is the maximum electric field which may exist in the substance without electric breakdown occurring and is called the **dielectric strength** of the material. The units of dielectric strength are usually given as kilovolts per millimeter (kV/mm) and are the same as 10^6 V/m. All values are approximate, as both K and the dielectric strength may change drastically with changes in temperature and pressure.

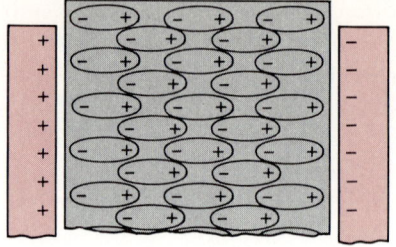

Figure 18-14 A microscopic view showing the neutral molecules of a dielectric becoming induced dipoles in the electric field of a capacitor. The induced negative and positive surfaces of the dielectric produce an electric field partially canceling the field of the capacitor, thus reducing the voltage required to sustain it.

where K is the dielectric coefficient for the material which causes the capacitance to change from C_0 to C when the material is inserted. Being equal to the ratio of two capacitances, K is dimensionless. Also, C is never less than C_0, so K is always equal to 1 or greater. Some typical values of K are listed in Table 18-1. Note that for a vacuum, K equals 1, and for air, K is very nearly equal to 1. We most often use exactly 1 for air.

What property of dielectrics causes an increase in the capacitance? You will recall from Chapter 17 that a nonconducting material placed in an electric field will have dipoles (either induced or permanent) line up along the direction of the electric field intensity. Figure 18-14 illustrates this point by depicting a few of the molecules in a dielectric situated between oppositely charged plates. We see that although there is no net charge on the dielectric, the lining up of the dipoles gives rise to a net *induced* charge on the edges next to the capacitor plates.

The decrease in voltage may be attributed to the decrease in the net electric field intensity vector. For instance, within the dielectric of Figure 18-14, the component of electric field intensity due to the charge on the plates is from left to right, while the component due to the induced charge is from right to left. These two components may be considered to add vectorially, giving a smaller net electric field and hence a smaller voltage difference between the plates.

Another way of viewing this situation is to attribute the decrease in voltage to the effective reduction of charge at the inner surface of the two capacitor plates. Using the dielectric of Figure 18-14 as an example once again, we see that the real charge on the plate and the induced charge on the contacting dielectric are of opposite signs. Thus the surface of contact has a smaller effective charge than it would have without the dielectric in place. Hence the electric field intensity and therefore the voltage difference between plates are smaller.

Both views lead to the same results. A decrease in voltage and hence an increase in capacitance are obtained when a dielectric is placed between capacitor plates.

Equation 18-20 may be written as $C = KC_0$. Therefore, we can write the equation for the parallel-plate capacitor with a dielectric as

$$C = \frac{K\epsilon_0 A}{d} \qquad (18\text{-}21)$$

It is common practice to define the **permittivity** of the dielectric as ϵ, where

$$\epsilon = K\epsilon_0 \qquad (18\text{-}22)$$

With this definition, Equation 18-21 becomes

$$C = \frac{\epsilon A}{d} \qquad (18\text{-}23)$$

Note that the equations which describe the various quantities when a dielectric is in place are the more general and include the same quantities for a vacuum as a special case. After all, $K = 1$ and $\epsilon = \epsilon_0$ for a vacuum.

18-8 Application: Painting Pictures with Electricity

Certain molecules have the ability to absorb energy from rays (like light or x-rays) or from the impact of high-speed particles and to release this energy as light. Materials made of such molecules are called phosphors, and common examples are the material lining the inside of fluorescent lights or some glow-in-the-dark trinkets. A pane of glass lined with a phosphor becomes a screen, and the glow caused by incident rays may be seen through the glass. The most important example is the television picture tube screen. In this section, we will examine how a television picture is produced by electrostatic deflection.[1]

A thin wire, called a **filament,** may be heated in a vacuum by passing electricity through it. When the wire is hot, the thermal agitation of its atoms causes violent collisions between the atoms and passing electrons. Some of these collisions physically eject the electrons from the surface of the wire into the space surrounding it. Left alone, these electrons would drift to the walls of the container or back to the filament. If, however, the entire filament assembly is surrounded with a metal can having a positive potential relative to the filament, ejected electrons from the wire will be accelerated to the can (Figure

---

[1] Actually, most of the deflection of the beam in today's television receivers is done by magnetic fields rather than electrostatic fields. A modest magnetic field can give rise to very large forces and hence large deflections. This allows us to shorten the picture tube (CRT) and hence the television receiver set even for very large pictures. Early sets, however, did use electrostatic deflection exclusively, and electrostatic deflection is still used in most oscilloscopes (Chapter 23).

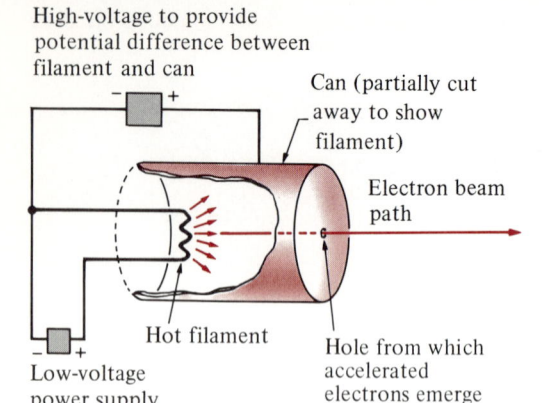

Figure 18-15 In the electron gun of a TV picture tube, electrons are ejected from the hot wire filament into the surrounding evacuated space, where some emerge through a hole in the can and constitute an electron beam.

18-15). Their speed when reaching it depends upon the potential difference, as in Equation 18-7:

$$\tfrac{1}{2}mv^2 - \tfrac{1}{2}mv_0^2 = -qV$$

Assuming the original speed of the electron is zero, the second term on the left disappears. Here, q is the charge on the electron ($-e = -1.602 \times 10^{-19}$ C), and we will assume $V = 1000$ V. We may then solve for the speed as

$$v = \sqrt{\frac{2eV}{m}} = \sqrt{\frac{2(1.602 \times 10^{-19}\ \text{C})(10^3\ \text{V})}{9.11 \times 10^{-31}\ \text{kg}}}$$
$$= \sqrt{3.51 \times 10^{14}\ \text{m}^2/\text{s}^2} = 1.87 \times 10^7\ \text{m/s}$$

This is a very respectable speed, being about 6 percent the speed of light. When electrons strike the can, they are simply absorbed along with their energy. If there is a hole in the end of the can, however, electrons accelerated toward this hole pass through rather than hit the can and emerge as a fine beam at this great speed. The electrons emerging would be decelerated and attracted back toward the can except that the region into which they emerge is kept at the same potential as the can.

We now have a beam of high-speed electrons emerging from an assembly called an **electron gun** at the back of the picture tube directed toward the screen. If undisturbed, these electrons will hit the center of the screen to form a bright dot at the phosphor layer. While undoubtedly something, a bright dot is not a picture. We need to spread the electrons all over the screen and control where they hit and with what intensity.

An actual picture tube face has a spread of about 48° as seen from the neck of the tube. To deflect the electron beam to the top of the screen, we pass the beam between a set of deflection plates similar to a parallel-plate capacitor (Figure 18-16). When a voltage is applied to these plates, the region between them becomes one of uniform electric field. Assume that our plates are 4.00 cm long and separated by 2.00 cm. We may calculate the time that an electron in the beam will take to cross the plates as

$$\Delta t = \frac{d}{v} = \frac{0.0400\ \text{m}}{1.87 \times 10^7\ \text{m/s}} = 2.14 \times 10^{-9}\ \text{s}$$

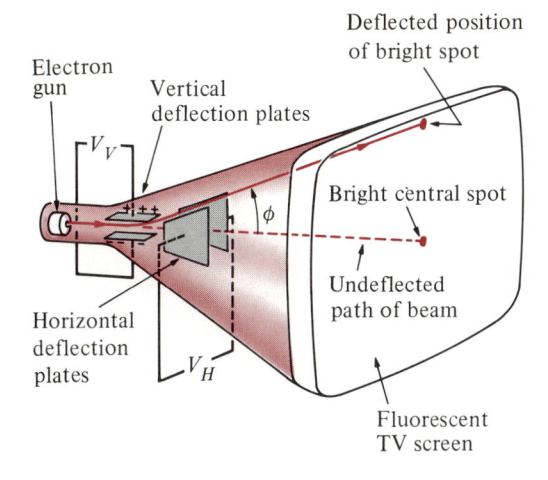

18-8 Application: Painting Pictures with Electricity

Figure 18-16 The electron beam may be deflected to any part of a television screen electrically or magnetically. Here, horizontal and vertical deflection plates, essentially capacitors, are used.

In that short span of time, we must accelerate the beam away from its forward direction toward the top of the screen. The velocity that the electrons must acquire may be calculated simply (Figure 18-17). Since we need the final beam direction to be 24° upward from the horizontal, we need a vertical speed of

$$v_y = v_x \tan 24° = (1.87 \times 10^7 \text{ m/s})(0.445) = 8.32 \times 10^6 \text{ m/s}$$

Given this result, we may calculate the electric field needed between the plates.

$$F_y \Delta t = m v_y \quad \text{(impulse-momentum equation, Chapter 8)}$$

but

$$E_y = \frac{F_y}{q} = \frac{F_y}{e}$$

thus

$$E_y e \Delta t = m v_y$$

$$E_y = \frac{m v_y}{e \Delta t} \quad \text{(note that } v \text{ is } v_y \text{ here)}$$

$$= \frac{(9.11 \times 10^{-31} \text{ kg})(8.32 \times 10^6 \text{ m/s})}{(1.60 \times 10^{-19} \text{ C})(2.14 \times 10^{-9} \text{ s})}$$

$$= 2.21 \times 10^4 \text{ N/C} = 2.21 \times 10^4 \text{ V/m}$$

But, since our plates are 2.00 cm apart, we will need an applied voltage of $(2.21 \times 10^4 \text{ V/m})(0.0200 \text{ m}) = 443$ V. Applying this voltage to the plates will deflect the bright spot to the upper center of the screen.

By using a second set of plates to provide horizontal deflection for the beam, we may move it sideways. In particular, varying the voltage between the horizontal deflection plates smoothly and continuously from, say, +500 V through zero to −500 V would have the effect of sliding or scanning the beam in a streak across the top of the picture tube. Figure 18-18a shows the voltage variation needed on the plates for this horizontal scanning. Following the first scan, we can reduce the voltage between the vertical plates to drop the beam down toward the center and scan a second line (Figure 18-18b). If we synchronize these two actions, we can hold a vertical deflection voltage on the plates

Figure 18-17 Vector diagram for calculating the required deflecting voltage on the vertical deflection plates. Electrons enter the field of the plates with a velocity purely in the x direction, acquired at the electron gun. While between the plates, they must acquire an additional velocity v_y such that their total velocity will be at the correct angle θ to hit the screen where desired.

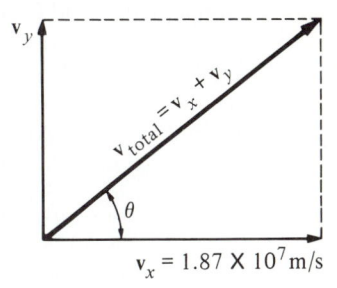

$v_x = 1.87 \times 10^7$ m/s

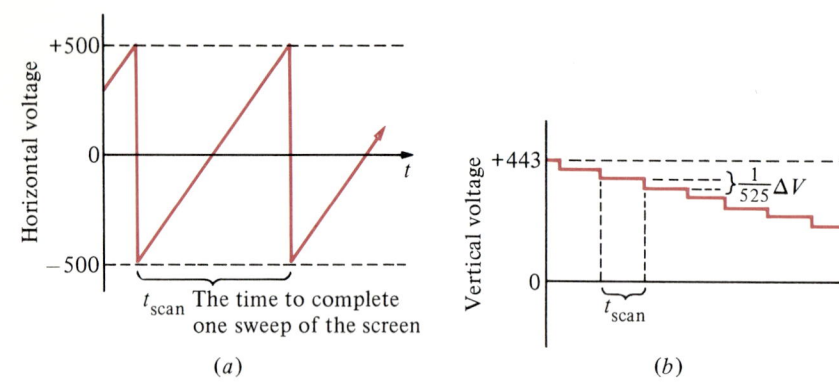

Figure 18-18 (a) Horizontal deflection voltage is varied in a sawtooth wave to slide the electron beam across the screen smoothly and then return it instantaneously for the next scan. (b) Vertical deflection voltage is a stairstep function, holding a voltage long enough for one line scan, then dropping it to lower the beam for the next scan. (c) The result is a smoothly scanning electron beam painting 525 equally spaced horizontal lines on the screen from top to bottom, then returning to the top for the next picture.

long enough to scan one line, then drop it just as we jump the horizontal voltage to return the beam to the left side of the screen for another scan. The result is to "paint" the TV screen with the electron beam, line by line (Figure 18-18c).

The number of lines used to cover the screen is a detail determined by practical considerations, including the number of TV stations in a given area and the state of electronic technology when that number was decided. The United States and Japan use 525 lines from the top to the bottom of the tube to make a picture (one scan). Many European nations use 625 lines per picture scan, yielding a sharper, more detailed picture.[1]

We may calculate the maximum allowable time to scan one line, t_{scan}, in the U.S. system. It requires 525 lines to paint a full picture on the screen with the electron beam. The human eye remembers a picture for about $\frac{1}{16}$ s before it is ready to process a new image. If the TV screen were not repainted within this $\frac{1}{16}$ s, the eye could notice smearing, streaking, or flickering of the image, much as occurs in some old-time movies. Thus, a picture should be completed in a maximum of $\frac{1}{16}$ s, and each line can only take $\frac{1}{525}$ of this time.

$$\text{Max } t_{scan} = \tfrac{1}{525} \times \tfrac{1}{16} \text{ s} = 0.000120 \text{ s}$$

Actually, between pictures, additional information is transmitted (representing the black band between pictures when the vertical hold is off). In practice, the typical 60-Hz electricity makes it convenient to repaint the entire picture 30 times each second. The real picture production rate used is 15,750 lines per second, yielding $t_{scan} = 0.0000635$ s.

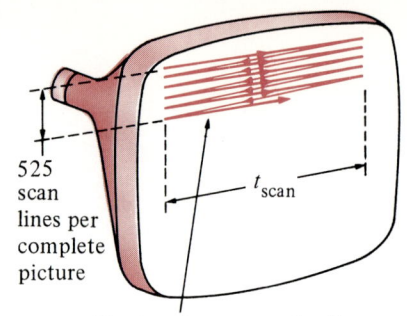

---

[1] An international commission met in September 1985 to determine a method to deliver high-definition television at 1125 lines per scan (as a worldwide standard of the future) without disrupting the present system. When this is eventually instituted, it should make video picture quality comparable to 35-mm photography in both sharpness and color fidelity.

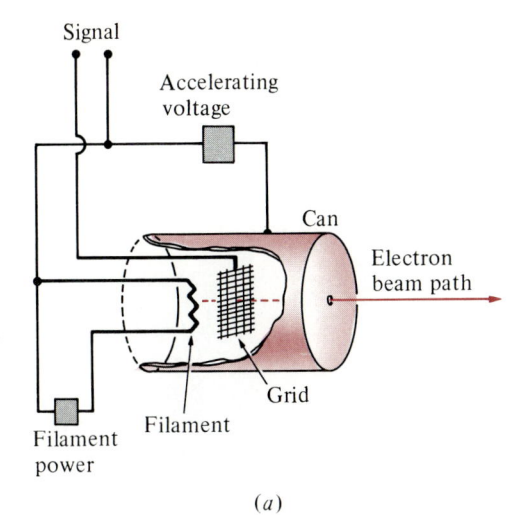

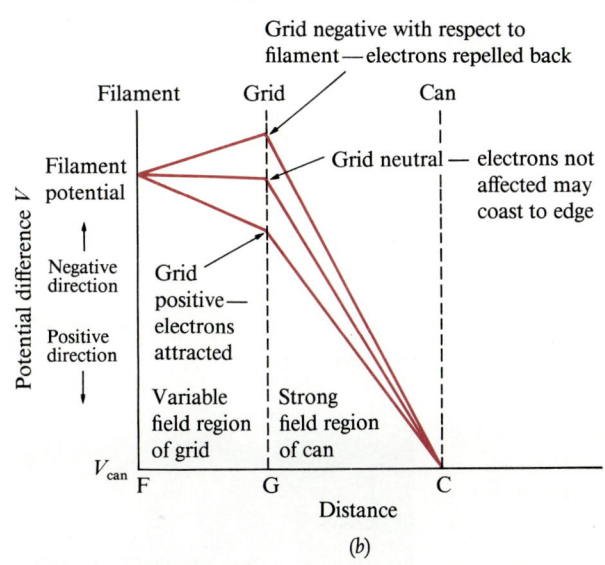

18-8 Application: Painting Pictures with Electricity

Figure 18-19 (a) A grid placed between the filament and can of the electron gun. (b) When the grid is made negative with respect to the filament, the electrons are repelled back to the filament and no beam is produced, resulting in a dark portion of the line during scanning. Reversing the grid to a small positive potential encourages the electron stream and results in a bright spot on the screen.

Of course, we do not have a picture yet in this story, only a bright picture tube made up by a rapidly scanning electron beam on the face of the tube. To produce a picture, we need to make the screen brighter in some places and darker in others. This is quite simple. Back in the electron gun that produced the beam, we include a fine-wire **grid** between the hot filament and the can (Figure 18-19). If this grid is at the same potential as the filament, it will have little effect on the electrons released. If the grid is made temporarily negative with respect to the filament, it will tend to repel electrons back toward the filament and thus greatly reduce or shut off the flow of electrons from the electron gun.

This voltage fluctuation is obtained from the **signal** broadcast by the

445

Figure 18-20 (a) A TV screen showing a well defined picture which appears continuous. (b) A 4× magnification of a part of the picture shows a granular look, with uneven edges to objects. (c) A 16× magnification shows the individual dots of phosphors which emit light at various intensities to produce the picture. (D. Riban)

television station. A very tiny variation in voltage between this grid and the filament can change the electron flow from a torrent to none at all; we say that the electron beam **amplifies** the slight variation fed to the grid. If we connect the grid to the antenna receiving the signal from the TV station, we have the variation in brightness required for each position of the screen (Figure 18-20). Color pictures require three electron guns, making three separate pictures on three separate rows of dots of phosphors that will glow in different colors when struck by the beam. Close examination of the screen reveals these dotlike phosphors to the eye.

Minimum Learning Objectives
After studying this chapter, you should be able to:
1. Define:

    absolute potential
    capacitance
    capacitor
    dielectric
    dielectric coefficient
    dielectric strength
    electronvolt (eV) [unit]
    equipotential
    farad (F) [unit]
    parallel-plate capacitor
    permittivity
    potential (electric)
    potential gradient
    surface charge density
    Van de Graaff generator
    volt (V) [unit]
    voltage
    zero potential

2. Apply the work-energy theorem to free charges in an electric field to determine the change in motion.
3. Determine the potential difference between given positions relative to one or more known concentrations of change.
4. Calculate the capacitance of a simple capacitor, given its geometry.
5. Determine the charge stored by a capacitor of known capacitance, given the voltage applied.
6. Determine the energy stored by a given capacitor at a stated voltage.

Problems

18-1 A 12.0-V automobile storage battery will deliver 2.40×10^4 C of charge, without recharging, before being electrically depleted. (a) What total energy is stored by the battery? (b) To what height could the energy stored in a fully charged battery alone lift a 3.00×10^3 kg automobile?

18-2 A 1.50-V flashlight D cell can deliver some 1.20×10^3 C before being exhausted. (a) What is the energy stored by the cell? (b) If used in a flashlight drawing 4.00 C/s, how long will the cell last in continuous operation? (c) If used in a calculator drawing 0.100 C/s, how many hours of operation will it last?

18-3 Two identical charges of $+6.00 \times 10^{-8}$ C are separated by 4.00 m. Calculate (a) the potential difference between the point halfway along the line joining them and a point on that line 0.500 m from either charge, (b) the change in potential energy a -2.00×10^{-6} C charge would undergo if moved between the two points, and (c) the work done by the electrostatic force as the charge is moved.

18-4 Measurement indicates that the potential difference between two points at distances $r_a = 1.40$ m and $r_b = 3.60$ m away from an unknown charge is 157 V. Calculate the magnitude of the unknown charge.

18-5 A light sphere containing 10.0 C of charge is at a potential of 4.00×10^4 V with respect to a wall. The sphere is attracted to the wall, and you wish it to arrive at the wall with a negligible impact (zero velocity). How much work must you do in restraining the sphere as it moves to the wall?

18-6 An energy of 10.4 eV is required to eject the outermost electron (ionize) from an atom of mercury. In the fluorescent light, electrons streaming across the tube collide with atoms of mercury vapor in the tube. The light produced by the tube depends upon the subsequent action of the ionized mercury. If a fluorescent fixture is operating at 120 V with a charge of 0.200 C flowng each second, how many ionizations of mercury atoms can occur each second if 5.00 percent of the energy available goes to this purpose?

18-7 A test charge of -1.50×10^{-8} C is separated from a second charge of 3.00×10^{-6} C by a distance of 5.00 m. (a) Where must the test charge be moved such that the change in potential energy is -5.40×10^{-5} J? (b) What is the potential difference between those two points with respect to the second charge?

18-8 Point charges of -2.00 and $+4.00$ μC are placed along the positive x axis at $x = +1.00$ m and $x = +3.00$ m, respectively. A third charge of $+5.00$ μC is then moved along the y axis in the negative direction from $y = +4.00$ m to the origin. Calculate the change in potential energy as the $+5.00$-μC charge is moved.

18-9 Suppose the earth had a charge of $+1.80 \times 10^{13}$ C and the moon had a charge of -6.00×10^{11} C. Calculate the change in electrostatic potential energy as the moon moves from apogee ($r = 4.05 \times 10^8$ m) to perigee ($r = 3.63 \times 10^8$ m).

18-10 Calculate the change in electric potential energy when an electron in a hydrogen atom moves from an excited state at $r = 2.12 \times 10^{-10}$ m to its lowest energy state at $r = 5.30 \times 10^{-11}$ m.

18-11 Calculate (a) the absolute potential at a point 1.33×10^{-11} m distant from a helium nucleus of charge $+3.20 \times 10^{-19}$ C, (b) the work done by the field as an electron is moved from a very large distance to that point, and (c) the work done by the field as a second electron is moved to the same distance on the opposite side of the nucleus after the first electron is in place.

18-12 A charge of $+3.00 \times 10^{-9}$ C is placed on the x axis of a coordinate system at $x = -2.00$ m, and a second charge of $+2.00 \times 10^{-9}$ C is at $x = +4.00$ m. Determine the magnitude and sign of the charge which must be placed at $y = +0.500$ m so that the absolute potential at the origin is zero.

18-13 Calculate (a) the distance away from a point charge of $+4.00 \times 10^{-8}$ C where the absolute potential is 1.20×10^3 V, and (b) the distance away from that point that a second charge of -2.00×10^{-7} must be placed such that the absolute potential at the point of part a is zero.

18-14 We wish to convince ourselves that the choice of zero level is unimportant to the result when we are dealing with potential differences. We should be able to select any arbitrary distance away from a point charge for which we define the potential as zero and still obtain Equation 18-3 as the potential difference at a distance r_b with respect to a distance r_a. Using r_0 as the zero point, determine the "absolute" potential at distances r_b and r_a from a point charge and subtract to find their difference.

18-15 The potential at point P is 800 V, and the potential at point Q is 200 V. An electron is projected from P toward Q with a speed of 1.90×10^7 m/s. Calculate (a) the electron's initial kinetic energy, (b) the kinetic energy of the electron when it reaches Q, and (c) the potential necessary at Q so that the electron would just stop as it reaches there.

18-16 A proton traveling at a speed of 3.40×10^5 m/s moves from a position with a potential of zero to a position at a potential with respect to the same reference of 300 V. Calculate (a) the initial kinetic energy, (b) the change in potential energy, and (c) the proton's final speed.

18-17 Through what potential difference must a proton be accelerated from rest so that its final speed will be the same as that of an electron which has been accelerated from rest through a potential difference of 500 V?

18-18 An electron, a proton, and a helium nucleus ($m = 6.63 \times 10^{-27}$, $q = +3.20 \times 10^{-19}$ C) are each accelerated from rest through a potential difference of 800 V. (a) Calculate their final speeds. (b) Which particle has the greatest final kinetic energy?

18-19 In the Bohr model of the hydrogen atom, the electron may exist only in certain allowed orbits. The smallest allowed orbit has a radius 5.30×10^{-11} m. The next two are at radii of 2.12×10^{-10} and 4.77×10^{-10} m. When an electron drops from a larger-radius orbit to a smaller-radius orbit, the potential energy loss (ΔPE is a negative number) yields a gain in kinetic energy and some radiant energy (light). If the increase in kinetic energy is equal to the amount radiated, how much light energy is radiated when the electron drops from (a) the third orbit to the second and (b) the second to the smallest?

18-20 A flashlight D cell consists of a zinc cylinder with a central carbon electrode. The space between electrodes is packed with a conducting paste containing manganese dioxide to react with any hydrogen gas produced. (Most batteries produce hydrogen gas which can "coat" an electrode and stop the production of electricity unless taken care of.) While producing current, the zinc metal ionizes, releasing two electrons per atom, and dissolves into the paste. The atomic mass of zinc is 65.4 g/mol. (a) If 10.0 g of zinc is present in a fresh battery, how many coulombs of charge could the battery deliver during its useful life if all the zinc was ionized and dissolved? (b) If this battery was placed in a flashlight drawing 4.00 C/s, how long should it keep the flashlight lit? (c) From your experience with flashlights, is this a reasonable number? What do you conclude?

18-21 A 120-V coffee maker brings 0.750 kg of water initially at 20°C to a boil in 6.00 min after being turned on. Assuming no loss of heat, at what rate is charge passing through the coffee maker?

18-22 The largest operating particle accelerator on earth will accelerate protons to an energy of 500 GeV (gigaelectronvolts, or $\times 10^9$ eV). What would be the energy in joules of an accelerated proton from this machine?

18-23 The specific heat of tungsten is 133 J/(kg·°C), and its melting point is 3410°C. A filament in a television picture tube contains 0.120 g of tungsten. In operation, a 6.30-V potential difference across the filament causes 7.70 C of charge to flow through it each second. In practice, this combination allows the filament to reach a comfortable operating temperature well below the melting point of tungsten, where energy losses equal energy input. Assuming no losses, how long would this combination of voltage and charge flow require to raise the tungsten from room temperature (20.0°C) to its melting point?

18-24 Two metal plates A and B are separated by 0.600 cm and attached to an electric circuit such that a potential difference is maintained between them. An electron is ejected from A with a speed of 2.00×10^7 m/s and arrives at B with a speed of 8.00×10^6 m/s. Calculate (a) the potential difference between the plates and (b) the magnitude and direction of the electric field intensity in the space between the plates.

18-25 When a helium nucleus (charge $+3.20 \times 10^{-19}$ C) is moved from point P to point Q, the electrostatic field does 1.28×10^{-16} J of work. Calculate (a) the potential difference between the two points and (b) the average value of the electric field intensity between the points, knowing that they are separated by a distance of 0.0800 m.

18-26 In Section 18-4, we asserted that at any point outside a uniform spherical distribution of charge, both the electric field intensity and the potential are the same as though all the charges were concentrated at the center of the sphere. (a) Show that very near a sphere of radius r the magnitude of the electric field intensity and the absolute potential are related by $E = V/r$. (b) Determine the minimum value of r for a sphere which is to have an absolute potential of 2.00×10^6 V in dry air ($E_{max} = 8.00 \times 10^5$ V/m). (c) If the sphere is placed in a pressurized tank and the pressure of the gases surrounding it is raised, the electric field required for breakdown is increased. Calculate the maximum absolute potential for a sphere of radius 0.300 m in a pressurized tank where the breakdown field is 2.50×10^7 V/m.

18-27 Electrons in a color TV set are accelerated across a potential difference of 6000 V in the electron gun. The beam is to be deflected through an angle of 30° by the uniform electric field of capacitor plates 3.00 cm long and separated by 3.00 cm. What voltage must be applied to the plates?

18-28 A positive particle in the vacuum tube of a Van de Graaff accelerator is repelled by the accumulated positive charge on the conducting sphere of the machine. Such a repulsive force is first experienced when the positive particle drifts outside the radius of the sphere. If a sphere of radius 40.0 cm has a repulsive potential of 50,000 V, what is the charge stored on the sphere? (Hint: The charge on the sphere from a distance acts as though concentrated at its center.)

18-29 (a) Calculate the area of the plates of a 1.00-μF parallel-plate capacitor if the plates are separated by 2.00×10^{-4} m in air. (b) What would the capacitance be if the separation was changed to 8.00×10^{-4} m?

18-30 You are to construct a 0.500-μF capacitor by using metal foil as plates and waxed paper 3.00×10^{-4} m thick with a dielectric constant 2.00 between the plates. (a) Calculate the area of each foil you will need. (b) What thickness should the waxed paper be if the capacitance is to be 2.00 μF, using the same-size foils?

18-31 A parallel-plate capacitor has plates separated by 2.00×10^{-4} m, and each has an area 1.50 m². Calculate the maximum value to which the capacitor can be charged without breakdown if the space between the plates is filled with (a) air, (b) Teflon, and (c) neoprene rubber.

18-32 A capacitor, formed of two parallel plates each of area 0.500 m² and separated by a distance of 2.00×10^{-4} m, is given an unknown charge. When the space between the plates is filled with air, the potential difference between them is 250 V. (a) Calculate the magnitude of the charge on each plate. (b) If the space is now filled with polyvinyl chloride, what is the new voltage?

18-33 A capacitor is formed of two parallel plates each of area 1.20 m² and initially separated by 1.50×10^{-4} m with air between the plates. (a) What will be the potential difference between the plates when they are charged to $+7.00 \times 10^{-8}$ and -7.00×10^{-8} C? (b) Suppose that without losing charge the plates are separated to a distance of 6.00×10^{-4} m. What is the new potential difference? (c) Calculate the electric field intensity in each case.

18-34 A parallel-plate capacitor is to be constructed to have an electric field intensity between the plates of 2.00×10^4 V/m whenever they are charged to 3.60×10^{-7} C. (a) Calculate the cross-sectional area of each plate. (b) If the distance between the plates is 4.00×10^{-2} m, what is the potential difference?

18-35 A capacitor used for a flashgun has a capacitance of 50.0 μF. When fired, the capacitor must supply 0.0500 J. Assuming it discharges completely, calculate (a) the original voltage and (b) the charge on the capacitor.

18-36 A capacitor has metal foil plates of area 0.600 m² separated by beeswax of thickness 2.00×10^{-3} m. It is used in an electric circuit with a voltage of 240 V. For this element, calculate (a) the capacitance, (b) the electric energy stored, and (c) the energy density.

18-37 A defibrillation unit passes electric energy through a patient's heart to "clamp" the heart stopped. When the heart is released as the current stops, it often resumes normal beating rather than the arrhythmic jerking called fibrillation. A particular model defibrillation unit works by completely discharging a 16.0-μF capacitor through the patient. The maximum electric energy available noted on the dial of the machine is 400 W·s (J). When the capacitor is fully charged, determine (a) the voltage and (b) the charge stored.

18-38 A 3.00-μF capacitor with air between the plates is charged to 7.50×10^{-5} C and disconnected from the battery. A dielectric of $K = 5.00$ is now inserted into the space between the plates. Calculate (a) the potential difference with air as the dielectric, (b) the energy with air, (c) the potential difference after the dielectric is inserted, and (d) the energy with the dielectric in place. Can you account for an energy difference?

18-39 A 20.0-μF capacitor with Pyrex glass as a dielectric is connected to a 45.0-V battery. After it has been fully charged, the battery is disconnected and the dielectric is removed so that there is air between the plates. Calculate (a) the original charge, (b) the energy stored with Pyrex glass as the dielectric, (c) the charge after the Pyrex glass is removed, and (d) the energy stored with air as the dielectric. (e) What work was done to account for the change in energy?

18-40 A parallel-plate capacitor of capacitance 1.50×10^{-8} F is charged by a 120-V source, and the source is removed. The plates of the capacitor are then pulled apart until the capacitance has been reduced to 4.00×10^{-9} F. Calculate (a) the energy stored in the capacitor after the source is removed, (b) the potential difference between the plates after they have been forced apart, and (c) the final energy stored in the capacitor.

18-41 A parallel-plate capacitor is fully charged to 3.60×10^{-6} C by a 50.0-V voltage source, and then the source is disconnected. The plates of the capacitor are moved toward each other, cutting the separation to one-fifth its original value. Calculate (a) the original capacitance, (b) the original energy stored, (c) the capacitance after the plates are moved, (d) the final voltage, and (e) the final energy stored.

18-42 A parallel-plate capacitor with air between the plates has a capacitance of 5.00 μF and is connected across a 120-V battery. Later, glass with a dielectric constant of 6.00 is inserted between the plates without disconnecting the battery. Calculate (a) the initial charge on the plates, (b) the initial energy stored in the capacitor, (c) the final capacitance, (d) the final charge on the plates, and (e) the final energy stored in the capacitor.

18-43 The dielectric strength of air depends upon the pressure and the humidity. (This is why static electric effects are best observed in the dry, cold air of winter.) The potential which can be reached by a Van de Graaff generator is limited by the dielectric strength of the air. Most large generators operate in a large pressure vessel under high pressure. One such machine reaches 2.00 MV in a vessel at a pressure of 20 atm with a clearance of 30.0 cm between the charged sphere and the pressure vessel. What is the minimum dielectric strength of the air under such conditions?

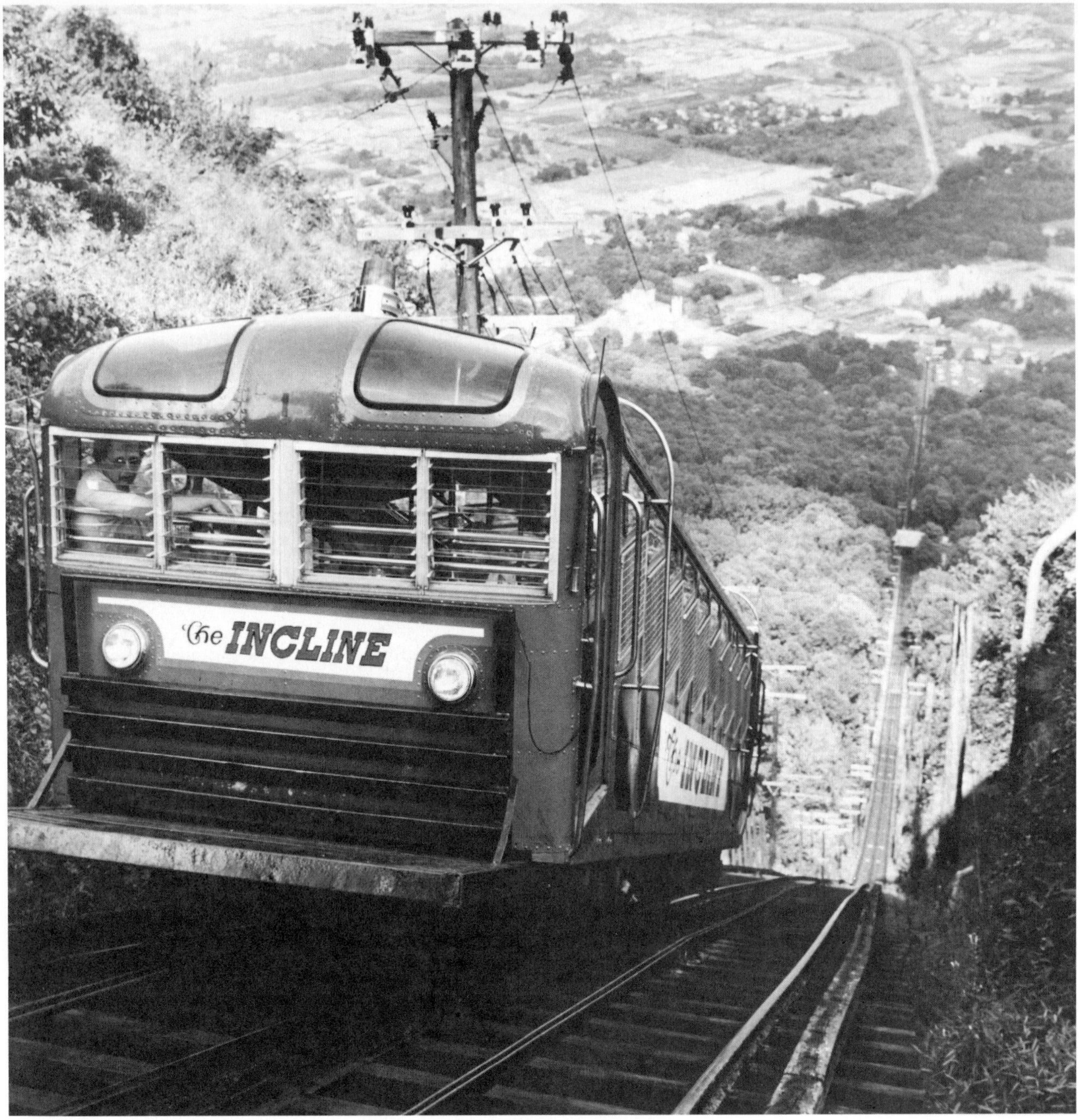
The INCLINE

19 Electric Current, Resistance, EMF

19-1 Introduction

Most applications of electricity with which you are familiar involve moving charges. The electric lights we use daily function as a result of charges flowing through small lengths of wire or through confined gases. Kitchen appliances, stereo systems, television and radio receivers all require moving charges. In most applications, the flow of charge takes place in some conducting material. Such a flow is called current electricity, or simply **current.**

Recall that some of the charges in a conductor are bound not to a single nucleus but only to the substance as a whole. This makes them free to wander about. If a conductor is placed in a region where there is an electric field, a force will be exerted on the free charges, which may then accelerate. Suppose we provide an electric field within a conducting wire. Figure 19-1a shows a leaf electroscope which has been charged to a very high positive voltage. One end of a piece of wire is then attached firmly to a copper water pipe passing into the earth (**ground**), which we take as our zero reference voltage. Now we attach the other end of the wire to the metal knob of the electroscope. This establishes a large potential difference (voltage) between the two ends of the wire and hence a potential gradient (electric field) within the wire. The charges begin to flow at

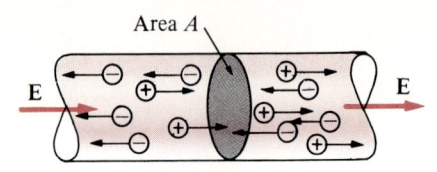

Figure 19-2 Submicroscopic view inside a wire of cross section A. Under the influence of the electric field, positive charges experience a force to the right and negative charges a force to the left.

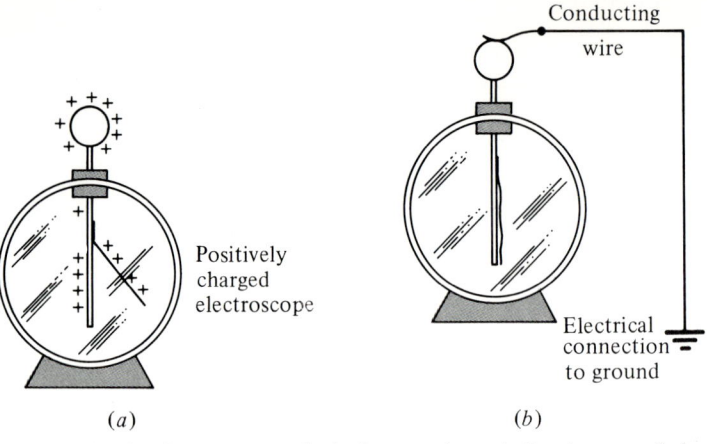

Figure 19-1 (a) Like charges cause the leaf to stand out, indicating an unbalanced charge. (b) When the electroscope is grounded, the leaf collapses, indicating that the unbalanced charge has been neutralized. Either positive charges have moved out of the system or negative charges have flowed in. Either way, a flow of charge, or current, was produced.

once, and the leaf of the electroscope collapses, as shown in Figure 19-1b. Charges flow until the electroscope and the earth are at the same potential. Zero potential difference means zero electric field, hence, no further new flow of charge.

Consider the small time interval during which charge is flowing. Figure 19-2 shows an expanded view of a very small portion of the wire during this interval. The electric field within the segment of wire shown is directed from left to right. A small number of charges are shown moving under the action of the field. The field exerts a force directed to the right on the positive charges and to the left on the negative charges. In general, both types of charges may flow. In metals, like the wire, the free charges are electrons (i.e., negative charges), while the protons are part of the structure and are not mobile. In the fluid (electrolyte) used in automobile batteries, both positive and negative charges (ions) are free to flow. Likewise, positive and negative ions as well as electrons are the charge carriers for the gas in fluorescent lights.

19-2 Electric Current

We wish to deal with the amount of charge flowing past a cross-sectional area, such as the shaded area of Figure 19-2, in a given time interval. A positive charge passing through the area from left to right has the effect of increasing the positive charge on the right side of the area, while a negative charge moving through the area from right to left has the exact same effect. Thus, a positive charge moving in one direction is equivalent to a negative charge moving in the other.

Keeping this in mind, we define *electric* **current** as the time rate of flow of positive charge past a section of conductor, or

$$I = \frac{\Delta q}{\Delta t} \qquad (19\text{-}1)$$

Figure 19-3 The motion of an electron in a wire conducting a current.

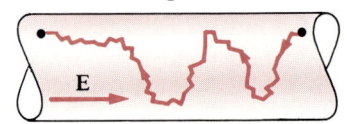

where we have used the symbol *I* for current. The SI unit of current is the **ampere** (A), named in honor of the French scientist Andre Ampere (1775–1836), a pioneer in the development of electricity and magnetism. We most often abbreviate the word *ampere* and report the current in amps. By Equation 19-1, a current of one ampere exists when a charge of one coulomb flows past a section of the conductor in one second, or 1 A = 1 C/s.*

We must also assign a direction to the current, but the direction chosen will in a sense be arbitrary. It has become conventional to define the direction as if all the charges were positive. Thus, **conventional current direction** is the direction that positive charges flow. This convention of positive flow makes the current direction the same as that of the electric field. In the case of the electroscope of Figure 19-1, the conventional current is directed from the electroscope through the wire to the earth. However, we also remember that in metallic conductors it is really the negative charges (electrons) which move, and they move in the direction opposite to that of conventional current.

When the two ends of a conductor are connected across a potential difference, the electric field is established throughout the entire conductor at the speed of light in that conducting medium. An electric force is exerted very rapidly on all the free charges in the conductor. The free charges, which were originally in random (thermal) motion, are accelerated in the direction of the electric field (or opposite to that if they are negative charges). In general, a charge will accelerate for a very brief interval until it collides with an atom or ion and its direction is changed. It is again accelerated by the electric field until it has another collision. It continues in this way, accelerating, colliding, stopping, changing direction, and, in general, following a very tortuous path. Figure 19-3 shows the path of an electron moving through a wire in which there is an electric field. While the electron may be moving at very high speeds between collisions, its overall motion from right to left is at a very slow average rate. This average rate is called the **drift velocity** v_d. In metallic conductors, the drift velocity rarely exceeds several millimeters per second and is most often smaller than this.

We may express the current in a wire in terms of the drift velocity. Consider the free charges moving to the right at the average speed v_d in the segment of wire shown in Figure 19-4. If we wish to find *I*, we must know the quantity of charge Δq which passes through the shaded area *A* in a time interval Δt. During the interval Δt, each charge drifts a distance along the wire equal to v_d times Δt. Thus, any charge which is closer to *A* than $v_d \Delta t$ at the beginning of the interval will pass through *A* during the interval Δt. Also, any charge farther from *A* than $v_d \Delta t$ will not travel far enough during the interval Δt to reach *A*. Thus all the free charges which are inside a cylinder of length $v_d \Delta t$ at the beginning of the time interval will pass through *A* during the interval. The net magnitude of the charge is the product of the number of charge carriers per unit volume with the charge per carrier with the volume of the cylinder. Thus we have

$$\Delta q = nqAv_d \Delta t$$

Charges flow in a conductor and establish a current only if a potential difference exists between parts of the conductor. This is roughly equivalent to saying that marbles at rest will roll on a tray only if an unbalanced sideways force is applied (by, perhaps, tilting the tray).

Figure 19-4 Idealized electrons flowing smoothly to the right at the drift velocity cross a section of the wire of area *A*. In a time period Δt, each electron will travel a distance $v_d \Delta t$. Thus, during Δt only those electrons upstream of *A* by a distance $v_d \Delta t$ or less will cross *A*.

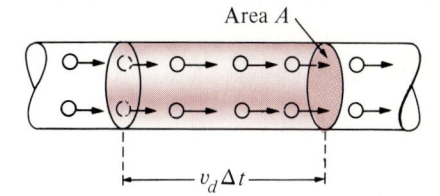

---

* While this is true, it might be a little misleading. The ampere is not a derived unit in SI. As we pointed out in Chapter 17, the coulomb is derived from the ampere because, as we will see in Chapter 21, the ampere can be measured with greater precision. We usually say that one coulomb is the amount of charge which flows past a section of conductor in one second if a current of one ampere exists in the conductor, or 1 C = 1 A · s.

where n represents the number of charge carriers per unit volume and q is the magnitude of the charge on each carrier. The volume of the cylinder is equal to the product of its area A with its length $v_d \Delta t$. Dividing by Δt yields

$$\frac{\Delta q}{\Delta t} = nqAv_d$$

and

$$I = nqAv_d \tag{19-2}$$

If there are several different kinds of charge carriers, each with its own charge and drift velocity, we write

$$I = n_1 q_1 A v_{d1} + n_2 q_2 A v_{d2} + \cdots$$

or

$$I = A(n_1 q_1 v_{d1} + n_2 q_2 v_{d2} + \cdots)$$

$$= A \left(\sum_{i=1}^{n} n_i q_i v_{di} \right) \tag{19-3}$$

Example 19-1 A copper wire of cross-sectional area 7.10×10^{-6} m² carries a current of 20.0 A. Assume there are 8.46×10^{28} free electrons per cubic meter in copper, and calculate (a) the drift velocity of the electrons and (b) the time interval required for 5.00 C of charge to pass a given point in the wire.

Solution (a) This part may be solved by using Equation 19-2 since every quantity with the exception of drift velocity is known.

$$I = nqAv_d$$

or

$$v_d = \frac{I}{nqA}$$

$$= \frac{20.0 \text{ C/s}}{(8.46 \times 10^{28} \text{ el/m}^3)(1.60 \times 10^{-19} \text{ C/el})(7.10 \times 10^{-6} \text{ m}^2)}$$

$$= 2.08 \times 10^{-4} \text{ m/s}$$

While this is a very small speed, it *is* typical. The current would have to be five times as large in this wire in order that the drift velocity be even 1 mm/s.
(b) The definition of current will allow us to solve this. From $I = \Delta q/\Delta t$, we have

$$\Delta t = \frac{\Delta q}{I} = \frac{5.00 \text{ C}}{20.0 \text{ C/s}} = 0.250 \text{ s}$$

This amounts to more than 10^{20} electrons passing each point in the wire during every second. Even though the drift velocity has a very small magnitude, the large charge can be passed because of the enormous numbers of free electrons which are available to move.

19-3 Resistance and Ohm's Law

In moving and colliding with fixed atoms, electrons give up energy to the metal which then appears primarily as thermal energy. Different materials absorb energy from the electrons at different rates. To maintain a constant current, energy must be provided constantly as it is being lost in heating the conductor. The source of this electric energy is the potential difference of a battery or electric generator of some kind.

To produce a continuous current in a conductor, we must connect a constant source of voltage (potential difference) between the two ends of the conductor. Measurement subsequently shows that it requires a larger voltage to produce a bigger current. Measurements of current plotted vs. the corresponding voltage for a given conductor yield a graph like that of Figure 19-5. The current in the conductor is directly proportional to the voltage across the conductor. If the voltage is doubled, the current is doubled, and so on; thus I is proportional to V, and we may write a proportionality as

$$I \propto V$$

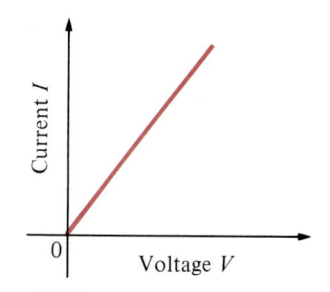

Figure 19-5 Current through a conductor vs. the voltage applied between the ends of the conductor.

This proportionality was discovered by the German educator Georg Simon Ohm (1787–1854) and is now known as **Ohm's law.** It may be made into an equation by inserting a proportionality constant. Usually, the proportionality constant is made a multiplier of the independent variable (V, in this case), but here it is more convenient to write it as a divisor, as

$$I = \frac{1}{R} V \qquad (19\text{-}4)$$

The reciprocal of R is therefore the constant chosen to make the proportionality an equation; R represents the electric **resistance** of the conductor. We see that the larger the value of R, the smaller will be the current I for a given voltage. Thus R is a measure of the tendency of the material to oppose a current. Equation 19-4 may be written as

$$V = IR \qquad (19\text{-}5)$$

or

$$R = \frac{V}{I} \qquad (19\text{-}6)$$

Equation 19-5 is the most common of the three expressions and is most often what is meant when someone refers to Ohm's law. Equation 19-6 defines the resistance of a conductor (and nonconductors as well) as equal to the voltage across its ends which is required to cause a unit current in it. Simply stated, the resistance is equal to the ratio of the voltage across an element to the current through it.

In SI units, electric resistance is in volts per ampere. This has been given the name **ohm** to honor G. S. Ohm. That is,

$$1 \text{ ohm} = \frac{1 \text{ volt}}{1 \text{ ampere}}$$

We use the uppercase Greek omega (Ω) to represent ohms, so $1\ \Omega = 1$ V/A.

For solid metallic conductors, the current increases directly with increasing voltage, yielding a straight line passing through the origin, as in Figure 19-5.

The slope of the straight line is equal to the reciprocal of the resistance. Such materials or electric elements which obey Ohm's law are called **ohmic circuit elements.** For these elements, the effect (current through the element) is linearly related to the cause (voltage across the element). Hence such elements are linear circuit elements. We will encounter other very useful elements which do not obey Ohm's law. That is, a plot of current vs. voltage for these **nonlinear circuit elements** does not yield a straight line. However, we may still define the element's resistance for corresponding values of I and V by using Equation 19-6.

Example 19-2   A copper wire carries a current of 200 A and has a resistance per unit length of 8.00×10^{-4} Ω/m. Calculate (*a*) the resistance of 5.00 m of this wire and (*b*) the voltage across the 5.00-m length.

Solution   (*a*) We multiply the resistance per unit length with the length in which we are interested.

$$R = (5.00 \text{ m})(8.00 \times 10^{-4} \text{ Ω/m}) = 4.00 \times 10^{-3} \text{ Ω}$$

(*b*) This is a straightforward application of Ohm's law.

$$V = IR = (200 \text{ A})(4.00 \times 10^{-3} \text{ Ω}) = 0.800 \text{ V}$$

Thus, across each 5.00 m in the current direction there is a voltage decrease of 0.800 V. This decrease is often called a **potential** (or **voltage**) **drop.**

Under normal circumstances, all conductors have some resistance. For restricting current, however, we often need a large resistance concentrated in a small volume. Thus, high-resistance conductors called **resistors** are made for this purpose. For most circuit applications, they are assumed to have the only nonnegligible resistance in the circuit (Figure 19-6).

Figure 19-6   Some common resistors. Top center: a resistor in a metal heat sink with cooling fins for dissipating thermal energy. Top right: a variable resistor whose value is changed by rotating the protruding shaft. The stripes on the cylindrical units are a code indicating resistance. The wires allow the devices to be connected in a circuit. (D. Riban)

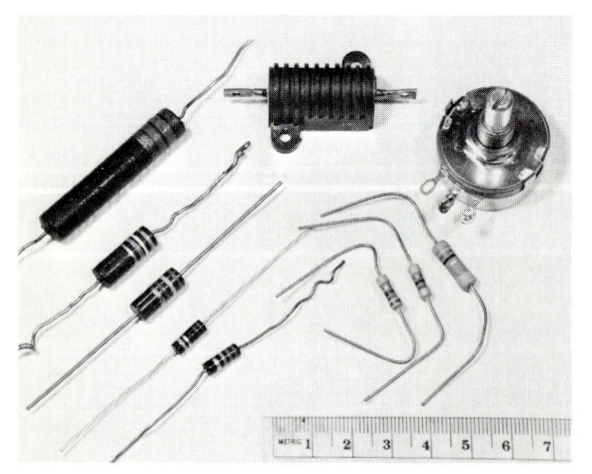

Figure 19-7   Common sources of emf. All the batteries shown work by converting internal chemical energy to electric energy for external use. (D. Riban)

19-4 Voltage Sources; EMFs

To maintain a constant current in a conductor, energy must be continuously supplied to the conductor. From Ohm's law we see that this energy must take the form of a voltage, or potential difference, across the conductor. This can be provided if the ends of the conductor are connected to the two terminals of a storage battery or any other device which converts some other form of energy to electric energy. Such a device which separates positive and negative charge, thereby creating a voltage, is called a source of **electromotive force (emf)**. The emf of such a device is the steady-state voltage across its terminals when no current is being drawn.[1] We often use the symbol $\mathscr{E}$ to represent the emf of a battery, several of which are shown in Figure 19-7. Each of these works by converting chemical energy into electric energy, but there are other possibilities, including those which provide electric energy by converting light, heat, mechanical energy, nuclear energy, and so on.

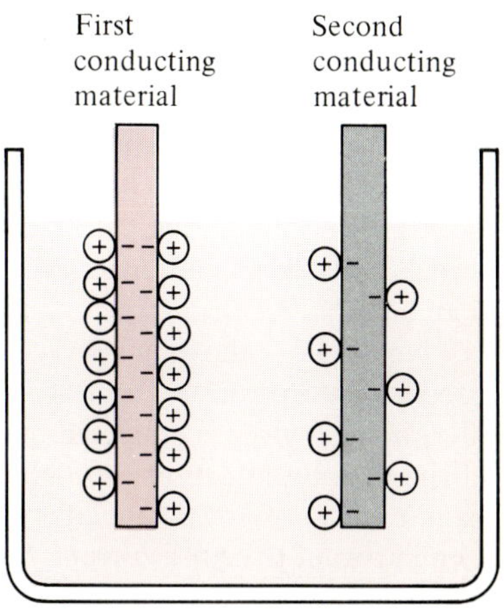

The chemical battery is the most convenient portable emf available. Figure 19-8 shows the simplest kind of chemical **cell**. It consists of two dissimilar metal rods each partially submerged in a dilute acidic solution called an **electrolyte**, which dissolves some of the atoms from each of the rods. The atoms enter the solution as positive ions, leaving electrons behind on the rod. An equilibrium is reached with the rods being negative and the solution positive. Because the rods are of dissimilar metals, one dissolves more than the other and thus becomes more negative. The rod which dissolves more is at a lower potential because it holds more electrons. Hence there is a potential difference between the rods. The magnitude of this potential difference is the emf of the cell. Two or more cells electrically connected constitute a **battery**, although common usage does not honor this distinction.

The part of the rod sticking out of solution and to which an electrical connection can be made is called a **terminal**. In a simple cell, the rod which dissolves less is the positive terminal, and the other is the negative terminal. When the terminals of the cell are connected to the opposite ends of a conductor, electrons flow from the negative to the positive terminal (outside the solution). This disturbs the equilibrium, making the negative rod slightly less negative, so more atoms must be dissolved to reestablish equilibrium. At the same time, the positive rod becomes slightly less positive and absorbs more positive ions to reestablish equilibrium. The result of the cell's providing current is that outside the solution electrons flow from the negative to the positive rod, while within the solution positive ions flow from the negative to the positive rod. When the positive rod is completely covered with ions from the negative rod, the two will be at the same potential. Thus, after some period of use, the positive rod must be cleaned. This is accomplished by sending current through the cell in the opposite direction (using a larger potential difference or emf), forcing the negative rod atoms to leave the positive rod and return to the negative rod. This process is called **charging** the cell or the battery.

The magnitude of the emf of a cell depends on the material used in the rods. For example, the two materials in a common automobile battery are

Figure 19-8 The electrochemical cell consists of two dissimilar metal rods in a conducting solution (electrolyte). In the diagram, the atoms of the left rod have a greater tendency to ionize. This rod thus becomes more negative than the right rod, and if the terminals are connected outside the solution, electrons flow from left to right through the wire connecting the terminals.

The tendency of ions from the negative rod to get deposited on the positive rod as the cell is used has practical importance. It is the basis of electroplating a thin and uniform layer of one metal on another and has many applications.

---

[1] The term is historical and standard, but the word *force* in emf could be a little misleading. We are really dealing with the energy per unit charge provided to a charge as it is moved between the battery terminals. We simply call a source of voltage an emf and downplay what the letters stand for.

metallic lead and lead oxide (in the form of plates rather than rods) in a weak sulfuric acid solution. The emf at equilibrium between lead and lead oxide in weak sulfuric acid is about 2.1 V. Six such sets are required in a nominal 12-V battery. The typical dry cell found in flashlights uses a carbon rod in the center of a cylindrical zinc can. The solution between the two is a rather thick, moist paste (the dry cell is thus not really dry) containing zinc chloride and ammonium chloride laced with manganese dioxide. The emf of a dry cell is between 1.5 and 1.6 V. We ordinarily call them $1\frac{1}{2}$-V cells.

Figure 19-9 shows a simple electric circuit in operation. A wire is connected from the positive terminal of a dry cell to the metal sleeve on a small light bulb. Another wire is connected between the negative terminal of the dry cell and the metal-end cap of the light bulb. The circuit must be **closed**, which means it must present a complete conducting pathway between the terminals of the cell in order to have a current. We see there is a current since the filament of the bulb is glowing. Current will continue only if the circuit is closed. If the wire is disconnected or broken at any point, no charge will flow. Such a circuit is said to be **open**.

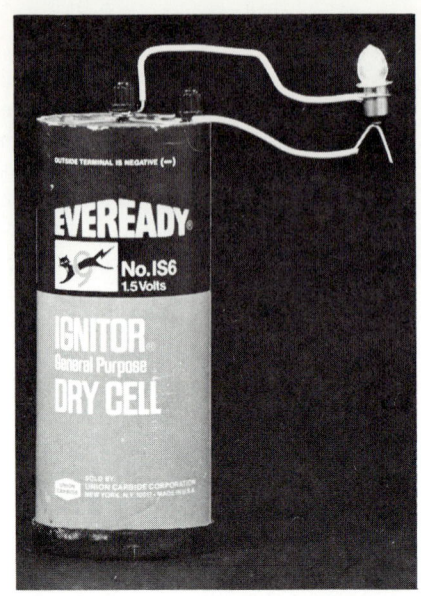

We normally use **schematic** diagrams to illustrate electric circuits. The elements and wires involved are all represented by symbols, some of which are shown in Table 19-1. Variable resistors, capacitors, and sources of emf are rather common, and this variability is most often shown by an arrow drawn diagonally across the symbol. Thus ⌿⊢ represents a variable capacitor and ⌿⌾ a variable resistor.

Note that each symbol includes a short straight line at either end to indicate that it may be part of a circuit. Using the symbols from Table 19-1, we may draw the schematic representing the circuit of Figure 19-9. Such a schematic is shown in Figure 19-10, with a switch added to turn the light on and off. Note that the light filament is shown as a resistor because for purposes of the electric circuit, that is precisely what it is. It has a potential difference across it whenever it carries a current. Electrically, it is of no consequence that some of the energy given up by moving electrons to the atoms of the filament is converted to light rather than to heat.

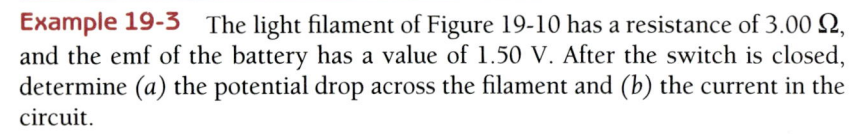

Figure 19-9 The terminals of a dry cell connected to the two electric contacts of a light bulb constitute a simple electric circuit, that is, a complete and unbroken pathway for the flow of charge from one terminal of an emf to the other. (D. Riban)

Example 19-3 The light filament of Figure 19-10 has a resistance of 3.00 Ω, and the emf of the battery has a value of 1.50 V. After the switch is closed, determine (a) the potential drop across the filament and (b) the current in the circuit.

Solution (a) We must recognize that the voltage across the resistor (filament) must be the same as the voltage across the battery, which is equal to its emf.

$$V = \mathscr{E} = 1.50 \text{ V}$$

Note that this assumes negligible resistance in the wires and contacts. Zero resistance for the wires assures that there is no potential drop across the wires. They must obey Ohm's law, which means that the voltage, a product of the current with the resistance, must be zero if the resistance is zero. For most circuit applications, we will assume our connecting wires have no voltage across them.

TABLE 19-1 Some Electric Circuit Elements with Their Symbols

Element	Symbol
Resistor	⌇⌇⌇
EMF	⊣⊢
Wire (R = 0)	────
Capacitor	⊣⊢
Switch	⌁⌁

(b) This part requires that we apply Ohm's law to the resistor.

$$V = IR$$

or

$$I = \frac{V}{R} = \frac{1.50 \text{ V}}{3.00 \text{ }\Omega} = 0.500 \text{ A}$$

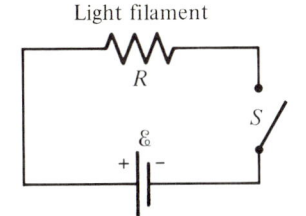

Figure 19-10 Schematic diagram for the circuit of Figure 19-9, with a switch added.

19-5 Terminal Potential Difference

In the solution of Example 19-3 we assumed that all the resistance of the circuit resided in the filament of the light bulb. We assumed that the battery as well as the wires and all contacts had zero or at least negligible resistance compared to that of the filament. In many instances this is a good assumption, but we must be aware that it is not always true. If the concentrated resistors are of small value, it is possible that we may not be able to neglect the resistance of the wires. Further, the contacts — the points in a circuit where wires connect other elements — must be good and solid. Otherwise, we may introduce a nonnegligible contact resistance. Finally, we must recognize that every battery will have a resistance associated with it. We call this the **internal resistance** of the battery. In general, the internal resistance will be quite small when the battery is new, and we may neglect it with impunity. However, as the battery ages, particularly in heavy use, the internal structure acts increasingly to restrict the current; hence internal resistance increases and often may no longer be neglected.

Consider a battery of internal resistance r and emf $\mathcal{E}$ delivering a conventional current I to a circuit, as in Figure 19-11. The points A and B are the battery terminals, and both the emf and the internal resistance are inside the battery. As the resistance has a current, it will have a potential decrease across it of magnitude

$$V_r = Ir$$

Note that B must be at a lower potential than the positive terminal of the emf (because conventional current direction is always from higher to lower potential). Thus the voltage of B with respect to A will be

$$V_{AB} = \mathcal{E} - Ir \qquad (19\text{-}7)$$

This is called the terminal potential difference, or **terminal voltage,** for a battery delivering current. We see that when the battery is delivering current, the terminal voltage will always be less than the emf. We now see why the emf is defined as the voltage across the terminals when no current is being drawn. The

The absence of good, firm electric contacts between circuit elements may increase the resistance enormously at the junction. At the extreme, it could leave a tiny air gap, requiring a spark to jump across or at least a space for oxidation and corrosion. Poor battery contact at the terminals is one of the most common causes of automobiles failing to start, for example.

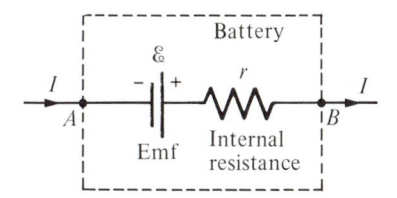

Figure 19-11 Electrically, the battery consists of both an internal emf ($\mathcal{E}$) and an internal resistance r between its terminals.

459

terminal voltage depends on the magnitude of the current, while the emf is fixed, in principle, and depends on the material of the plates or rods. We also note that the terminal voltage is very nearly equal to the emf if the internal resistance of the battery is very small.

Suppose the battery is being charged. That is, suppose the current direction through the battery is opposite to that which would be delivered by the emf. What then is the terminal voltage? Consider Figure 19-12. Once again we see that the emf and the internal resistance are inside the battery between the terminals A and B. This time, however, the current direction is from B to A through the battery. Since the conventional current direction is always from higher to lower potential, we see that B must be at a higher potential than the positive emf terminal. Therefore, the voltage of B with respect to A for a charging battery is

$$V_{AB} = \mathcal{E} + Ir \tag{19-8}$$

Thus the terminal voltage for a charging battery is always greater than its emf; in other words, it takes a source of voltage greater than a battery's emf to charge the battery. A source of voltage which is only slightly greater than a battery's emf can be used to charge the battery, but the current may be quite small and the charging process may take a very long time.

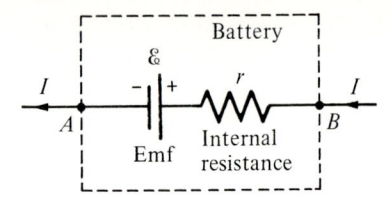

Figure 19-12 When a battery is being charged, a current is passed from B to A, that is, in the direction opposite that of the current generated by the battery's emf. This external emf must be larger than the emf of the battery by an amount determined by the magnitude of the internal resistance and the reverse current desired.

Example 19-4 When a battery of known emf equal to 12.0 V is delivering a current of 2.00 A to a circuit, its terminal voltage is 10.5 V. Calculate (a) the battery's internal resistance and (b) the charging voltage which must be applied to charge the battery at 5.00 A.

Solution (a) We can find the internal resistance by solving Equation 19-7 for r and substituting the other known quantities.

$$V_{AB} = \mathcal{E} - Ir$$

or

$$r = \frac{\mathcal{E} - V_{AB}}{I} = \frac{12.0 \text{ V} - 10.5 \text{ V}}{2.00 \text{ A}} = 0.750 \text{ }\Omega$$

(b) We apply Equation 19-8, using the solution to part a.

$$V_{AB} = \mathcal{E} + Ir = 12.0 \text{ V} + (5.00 \text{ A})(0.750 \text{ }\Omega) = 15.8 \text{ V}$$

19-6 Resistivity

Careful measurement of the resistance of different wires of some single pure material shows that the resistance is larger for longer wires and smaller for wires of larger cross-sectional area. If the length is doubled, measurement shows the resistance is doubled; if the length is tripled, the resistance is tripled. Thus, the resistance of a wire of uniform cross section is directly proportional to the wire's length l. Further, measurement shows that if the cross section of a wire of constant length is doubled, the resistance of the wire is reduced to one-half its original value, and so on. Thus, the resistance of a wire of constant length is inversely proportional to the cross-sectional area A of the wire. These two empirical facts may be combined as

$$R \propto \frac{l}{A}$$

For a given single pure material, the resistance of a wire depends only on the wire's length and cross section. It is also found that the resistance varies from material to material. We may express this as an equation by inserting a proportionality constant, as

$$R = \rho \frac{l}{A} \tag{19-9}$$

where ρ (Greek rho), the proportionality constant, is chosen both to keep proper units and to take into account the dependence of resistance on the type of material. The quantity ρ is called the **resistivity** of the material involved and has SI units of ohms times meters ($\Omega \cdot m$).

Resistivity varies somewhat with temperature, but the variation from material to material is much more dramatic. Table 19-2 gives the value of resistivity for some selected materials. Note that conductors have resistivities in the vicinity of 10^{-6} to 10^{-8} $\Omega \cdot m$, whereas the resistivity of insulators falls in the range from 10^{8} to 10^{17} $\Omega \cdot m$. Thus, the least-resistant insulator listed has over a million, billion times the resistivity of the poorest conductor listed. Sandwiched between these two extremes is a group called the **semiconductors**, materials which are very important to modern electronics. Within this broad classification, boron, for example, has a resistivity more than 1 billion (10^9) times that of carbon, and yet both are designated as semiconductors.

TABLE 19-2 The Resistivity* for Some Selected Materials at 20°C

Material	Resistivity ρ, $\Omega \cdot m$
Conductors	
Silver	1.59×10^{-8}
Copper	1.67×10^{-8}
Gold	2.35×10^{-8}
Aluminum	2.65×10^{-8}
Iron	9.71×10^{-8}
Mercury	9.84×10^{-7}
Semiconductors	
Carbon	1.40×10^{-5}
Germanium	4.60×10^{-1}
Silicon	2.00×10^{3}
Boron	1.80×10^{4}
Insulators	
Wood	10^{8} to 10^{12}
Glass	10^{10} to 10^{14}
Hard rubber	10^{13} to 10^{16}
Fused quartz	5.00×10^{17}

* All values are approximate since they apply to pure substances.

Example 19-5 The physics professor has a large spool of copper wire of cross section 5.00×10^{-7} m². (a) What length would you tell the professor to use if a piece of the wire with a total resistance of 0.450 Ω is wanted? (b) What current would you predict the wire would carry when its ends are connected to the terminals of a 2.75-V battery?

Solution (a) We use the resistivity of copper from Table 19-2 (1.67×10^{-8} $\Omega \cdot m$) and apply Equation 19-9.

$$R = \rho \frac{l}{A}$$

or

$$l = \frac{RA}{\rho} = \frac{(0.450\ \Omega)(5.00 \times 10^{-7}\ m^2)}{1.67 \times 10^{-8}\ \Omega \cdot m} = 13.5\ m$$

(b) From Ohm's law,

$$V = IR$$

or

$$I = \frac{V}{R} = \frac{2.75\ V}{0.450\ \Omega} = 6.11\ A$$

461

Resistivity varies slightly with changes in temperature. Thus the resistance of a conductor varies with temperature also. Experiment shows that this change is proportional to both the temperature change and the original (reference) resistivity. Thus we may write

$$\Delta\rho = \alpha\rho_0 \Delta T \qquad (19\text{-}10)$$

where $\Delta\rho$ = change in resistivity
ρ_0 = reference resistivity
ΔT = change in temperature

The lowercase Greek alpha (α) is used both to make the proportionality an equation and to make the equation's units proper. The quantity α is called the **thermal coefficient of resistivity** and has SI units of $°C^{-1}$ or K^{-1}. The thermal coefficients of resistivity for some selected materials are listed in Table 19-3.

If we choose to write $\Delta\rho = \rho - \rho_0$, where ρ is the new value of resistivity, Equation 19-10 becomes

$$\rho = \rho_0 (1 + \alpha \Delta T) \qquad (19\text{-}11)$$

Substituting into Equation 19-9 yields

$$R = R_0(1 + \alpha \Delta T) \qquad (19\text{-}12)$$

where $R_0 = \rho_0 l/A$.

TABLE 19-3 The Thermal Coefficient of Resistivity for Some Selected Materials (Temperature Range 0 to 100°C)

Material	Thermal Coefficient α, $°C^{-1}$
Conductors	
Mercury	0.00089
Gold	0.004
Silver	0.0041
Aluminum	0.0043
Iron	0.0065
Copper	0.0068
Semiconductors	
Carbon	−0.0005
Germanium	−0.05
Silicon	−0.07

Example 19-6 A copper resistor has a resistance of 150 Ω at 20°C. It is used in a circuit where it is not able properly to dissipate the heat it acquires as the charges flow through it. Its temperature thus rises to 90°C. Calculate its resistance at the higher temperature.

Solution Since we know the change in temperature (+70°C), we determine the value of α for copper from Table 19-3 (0.0068°C^{-1}) and apply Equation 19-12. Therefore

$$R = R_0(1 + \alpha \Delta T)$$

becomes

$$R = (150 \; \Omega)\left[1 + \left(\frac{0.0068}{°C}\right)(70°C)\right] = 221 \; \Omega$$

Table 19-3 indicates that the thermal coefficients of resistivity for conductors are positive numbers and those for semiconductors are negative numbers. This simply means that the resistance of a conductor increases with increasing temperature, while the resistance of a semiconductor decreases with increasing temperature. An interesting phenomena associated with resistivity and temperature is that of **superconductivity.** While the resistivity of conductors seems to decrease rather smoothly as the temperature is decreased, there exists for many materials a *critical* (or *transition*) *temperature* at which the resistivity abruptly drops to some vanishingly small value (essentially zero). This property is called superconductivity, and materials which are below their transition temperature are called superconductors. In superconductors forming a closed circuit, currents once started can, in principle, continue indefinitely without a battery. Though the transition temperature for all presently known superconductors is within several degrees of absolute zero, the phenomenon still opens the way to many applications of zero resistance circuits.

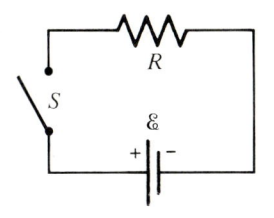

Figure 19-13 A simple circuit containing a pure emf and an external resistor.

19-7 Energy and Power in Electric Circuits

We recall that the electrons moving in a circuit are continuously giving up energy to the atoms of the circuit and then gaining energy from the potential gradient due to the source. We should be able to obtain an expression for the rate at which the rest of the circuit gains energy and the rate at which the source loses it. Conservation of energy dictates that these rates must be equal. Consider a circuit with essentially all its resistance in the resistor R, and a pure emf (Figure 19-13). After the switch is closed, a conventional current I is established. Its direction is from the positive terminal of the battery, through the resistor, and back to the battery's negative terminal. There is a voltage across the resistor which according to Ohm's law is $V = IR$. In a time interval Δt, a quantity of charge Δq flows through the resistor, and we have that $I = \Delta q/\Delta t$, or $\Delta q = I\Delta t$. The amount of energy, say ΔU, which is lost by the charge as it moves through the resistor is equal to the magnitude of the charge times the voltage (the change in potential energy per unit charge):

$$\Delta U = \Delta q\, V$$

or

$$\Delta U = (I\,\Delta t)(IR)$$

Thus the rate of loss of energy by the moving charge is

$$\frac{\Delta U}{\Delta t} = I^2 R$$

This is really just the power being expended by the current (and consequently being absorbed by the material of the resistor). Thus

$$P = I^2 R \qquad (19\text{-}13)$$

Equation 19-13 is known as **Joule's law,** because Joule is credited with discovering that the rate at which heat is produced in a metallic conductor is proportional to the square of the current in the conductor. This phenomenon is now called **Joule heating.**

Using Ohm's law in conjunction with Equation 19-13, we may write the equation for the power dissipated as

$$P = I^2 R = \frac{V^2}{R} = IV \qquad (19\text{-}14)$$

Example 19-7 Commercial resistors are most often rated according to maximum power as well as resistance. For example, a resistor rated at 2 W can absorb and give up 2 J of heat/s without failure. For an 800-Ω resistor rated at 2.00 W, calculate (a) the maximum voltage which can be placed across the resistor and (b) the current the resistor carries at maximum voltage.

Solution (a) This part may be solved by using the second form of power loss from Equation 19-14.

$$P = \frac{V^2}{R}$$

or

$$V^2 = RP$$
$$V = \sqrt{RP} = \sqrt{(800\ \Omega)(2.00\ W)} = 40.0\ V$$

(b) We apply Ohm's law.

$$V = IR$$

or

$$I = \frac{V}{R} = \frac{40.0\ V}{800\ \Omega} = 5.00 \times 10^{-2}\ A$$

By the same kind of reasoning which we used to develop Equation 19-13, an expression for the power being delivered by the emf of Figure 19-13 may be obtained. As the charge flows through the emf, the energy gained by an amount of charge $I\ \Delta t$ is simply

$$\Delta U = \mathcal{E} I\ \Delta t$$

since $\mathcal{E}$ is the voltage of the emf. Thus the power $\Delta U/\Delta t$ being delivered by the emf is

$$P = \mathcal{E} I \qquad (19\text{-}15)$$

We used a pure emf only to simplify the derivation. Suppose we had used a real battery with emf $\mathcal{E}$ and an internal resistance r. The emf would still be delivering power as given by Equation 19-15. However, part of it would be dissipated in the internal resistance of the battery. The power delivered by the battery would be equal to the terminal potential difference times the current, Thus

$$P_D = (\mathcal{E} - Ir)I$$
or
$$P_D = \mathcal{E} I - I^2 r \qquad (19\text{-}16)$$

where P_D is the power being delivered by the battery to the external circuit. When a battery is being charged, the power delivered by the charging source is

$$P_C = \mathcal{E} I + I^2 r \qquad (19\text{-}17)$$

Figure 19-14 Example 19-8.

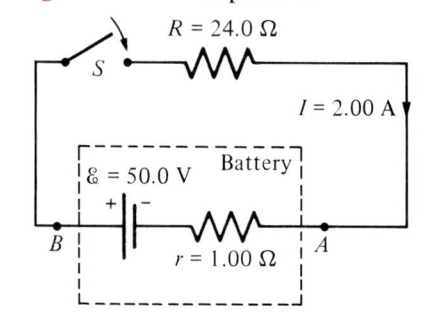

Example 19-8 A battery of emf 50.0 V and internal resistance 1.00 Ω delivers a current of 2.00 A to a 24.0-Ω resistor (shown in Figure 19-14). For this circuit, calculate (a) the terminal voltage of the battery, (b) the power delivered by the battery, (c) the power dissipated in the resistor R, (d) the power dissipated in the internal resistance r, and (e) the power delivered by the emf.

Solution (a) Equation 19-7 will give us the terminal voltage.

$$V_{AB} = \mathcal{E} - Ir = 50.0\ V - (2.00\ A)(1.00\ \Omega) = 48.0\ V$$

(b) The power delivered by the battery is equal to its terminal voltage times the current.

$$P_D = V_{AB}I = (48.0 \text{ V})(2.00 \text{ A}) = 96.0 \text{ W}$$

(c) The power dissipated in the resistors can be found by applying Joule's law.

$$P = I^2R = (2.00 \text{ A})^2(24.0 \text{ Ω}) = 96.0 \text{ W}$$

The power dissipated by the external resistor is equal to the power delivered by the battery to the external circuit, just as we should expect.

(d) $\qquad P = I^2r = (2.00 \text{ A})^2(1.00 \text{ Ω}) = 4.00 \text{ W}$

(e) Finally, the power delivered by the emf is found from Equation 19-15.

$$P = \mathscr{E}I = (50.0 \text{ V})(2.00 \text{ A}) = 100 \text{ W}$$

The power delivered by the emf (100 W) is thus dissipated in the external resistor (96.0 W) and the internal resistance (4.00 W).

Minimum Learning Objectives

After studying this chapter, you should be able to:

1. Define:
   - ampere, amp (A) [unit]
   - battery
   - cell (chemical)
   - charging (a battery)
   - circuit
   - closed circuit
   - conventional current direction
   - current
   - drift velocity
   - electrolyte
   - emf (electromotive force)
   - free charges
   - ground (potential)
   - internal resistance
   - Joule heating
   - Joule's law
   - nonlinear circuit element
   - ohm (Ω) [unit]
   - ohmic circuit element
   - Ohm's law
   - open circuit
   - potential (voltage) drop
   - resistance
   - resistivity
   - resistor
   - schematic
   - semiconductor
   - superconductivity
   - terminal
   - terminal voltage
   - thermal coefficient of resistivity

2. Calculate current, given the charge passing a certain point in a given time.

3. Determine the drift velocity of charge carriers from the current, charge density, and geometry of a specified conductor.

4. Calculate the resistance of a device, given the current and the potential difference across the device.

5. Understand the generation of an emf from the differing tendency of various conducting materials to lose electrons.

6. Interpret the schematic diagram of a simple circuit in terms of components of that circuit and their electrical significance to the circuit.

7. Calculate a battery's internal resistance, given its emf and the terminal voltage it supplies to a known external circuit.

8. Calculate the potential needed to charge a given battery at a known rate.

9. Determine the resistance of a simple conductor from its geometry and the resistivity of its material.

10. Calculate the variation in resistance of a given wire with temperature.

11. Determine the Joule heating in a conductor carrying a given current at a specified voltage.

12. Determine the power supplied by a battery of known terminal voltage delivering a given current to a specified external resistance.

Problems

19-1 A lightning bolt transfers 72,000 C of charge in 0.300 s. What is the average current during the transfer of charge?

19-2 Electrolysis is used to produce aluminum from an ore called bauxite which is primarily an oxide of aluminum (Al_2O_3). It requires three electrons to neutralize every aluminum atom which is then deposited as a metal. The atomic mass of aluminum is 27.0 u, and its density is 2.70×10^3 kg/m³. (a) What net charge is required to plate out the aluminum necessary for a small solar water heater, a block of about 0.200 m³ volume? *(b)* What electrolysis current is required if this much aluminum is to be deposited during an 8.00-h shift?

19-3 A flashlight uses 60.0 cm³ of metal (including that in the batteries) as part of the conductive path for electric charge. Assume an average concentration of 4.52×10^{28} free electrons per cubic meter for the metals. When switched on, a current of 3.00 A flows. *(a)* If the distance through the batteries, through the bulb, and back through the metal case of the flashlight to the rear of the batteries is 42.0 cm, what is the average drift velocity of a free electron in the metal? *(b)* How long would it take the average free electron to complete the trip and return to its starting point?

19-4 Assuming that pure copper has 8.46×10^{28} free electrons per cubic meter, calculate the magnitude of the drift velocity of electrons in a copper wire 2.00×10^{-3} m in diameter which is carrying a current of 15.0 A.

19-5 The manufacturer of an automobile battery which is rated at 100 A·h is claiming that the product of the current (in amperes) being delivered by the battery with the time (in hours) over which that current is delivered will be 100 A·h before the battery must be recharged. *(a)* How long can this battery crank a starter motor which draws 450 A before it runs down? *(b)* How much charge leaves the battery on the way to the starter motor in 1.00 min of cranking?

19-6 The fact that the negative plates (or cathodes) of a battery dissolve into solution while the positive plates (anodes) accumulate material from solution can be used to plate metals onto conducting surfaces. This is done by forcing electrons into the conductor to be plated using a greater, external emf. The excess electrons attract positive metal ions and neutralize them to form metal atoms attached to the electrode. Electroplating produces a thinner, more uniform coating of metal less likely to peel or flake than the coating obtained by dipping in molten metal. *(a)* A metal object is to be silver-plated in a solution containing silver ions lacking one electron each. If 5.00 g of silver is to be deposited on the object, what current would be required to complete the plating in 5.00 min? *(b)* A steel bumper is to be plated with chromium atoms from chromium ions in solution each lacking two electrons. How long must a 100-A current pass through the bumper to plate 400 g of chromium?

19-7 A demonstration Van de Graaff generator is charged to 1.50×10^{-6} C. A lead wire is then attached to the sphere which discharges it at an average current of 2.50×10^{-4} A. Calculate *(a)* the time required to discharge the Van de Graaff and *(b)* the current necessary to discharge the Van de Graaff in 5.00×10^{-5} s.

19-8 In the Bohr model of the hydrogen atom, the electron in its lowest energy state is traveling in a circle of radius 5.30×10^{-11} m at a speed of 2.18×10^6 m/s. Calculate *(a)* the frequency of revolution and *(b)* the current due to the electron's motion.

19-9 A 1.50×10^{-3} m diameter metal wire with 8.40×10^{27} free electrons per cubic meter delivers a charge of 250 C in a ½-h time period. Calculate *(a)* the average current in the wire, *(b)* the average drift velocity, and *(c)* the average distance traveled by an electron in the ½-h period.

19-10 An electric space heater carries a current of 6.00 A at a voltage of 120 V. Calculate *(a)* its resistance and *(b)* the current it will carry if the receptacle voltage drops to 105 V.

19-11 A wire carries a current of 5.00 A and has a resistance per unit length of 3.00×10^{-2} Ω/m. What is the voltage across *(a)* 1.00 m of the wire and *(b)* 20.0 m of the wire?

19-12 The electric resistance of the human body depends on the condition of the skin (among other things). Calculate the current one might expect if one completes the circuit between a 120-V electric outlet and ground when the hands are *(a)* dry ($R = 2.00 \times 10^5$ Ω) and *(b)* wet ($R = 3000$ Ω).

19-13 A piece of wire has electric resistance per unit length of 0.500 Ω/m. *(a)* Determine the length necessary to yield a current of 2.60 A when the wire is connected to a 1.50-V battery. *(b)* What current would this length of wire carry if the battery is 12.0 V?

19-14 When an electric toaster is connected across an emf, a current of 6.00 A exists in the toaster coils, which have a resistance of 20.0 Ω. If the number of electrons per unit volume in the metal is 6.90×10^{27} per cubic meter, calculate *(a)* the magnitude of the emf, *(b)* the number of electrons which flow per second, and *(c)* the cross-sectional area of one wire of the coil if the drift velocity is 2.50×10^{-4} m/s.

19-15 A hand-held calculator is powered by a 9.00-V battery which is rated at 1.30 A·h. This means that the battery may deliver 1.30 A continuously for 1 h, or 0.65 A for 2 h, etc. For purposes of circuit analysis, the calculator may be considered to be a 31.0-Ω resistor in series with the battery. How long will the battery last if the fresh battery is placed in the calculator and the calculator left turned on?

19-16 A battery that has an emf of 8.40 V and an internal resistance of 0.150 Ω is connected to a 2.65-Ω resistor. For this circuit, calculate *(a)* the current, *(b)* the voltage across the battery terminals, and *(c)* the voltage across the 2.65-Ω resistor.

19-17 When there is a current of 5.00 A in a battery circuit, measurement shows a terminal potential difference of 13.8 V. When the same battery is being charged by a current of 1.50 A, its terminal potential difference is 18.3 V. Calculate *(a)* the internal resistance of the battery and *(b)* its emf.

19-18 A potentiometer is an instrument designed to measure voltages without drawing any current from the source. It is very useful for measuring emf's of batteries, since the terminal potential difference of a battery and its emf are the same when no current is being delivered. Suppose we use a potentiometer and discover that the emf of a certain battery is 12.6 V. When this battery is then used in a circuit, we find that its terminal potential difference is 11.9 V when it delivers a current of 3.00 A. Calculate the internal resistance of our battery.

19-19 A battery with an internal resistance of 0.350 Ω requires a terminal potential difference of 23.6 V to charge it at a current of 4.00 A. Calculate (a) its emf and (b) the terminal voltage if the battery is used to deliver 15.0 A.

19-20 The emf of a battery is 45.0 V, and the current it delivers if an ammeter (which measures current) of negligible resistance is connected between its terminals is 35.0 A. Determine the terminal potential difference when a 12.0-Ω resistor is connected to the battery.

19-21 A panel voltmeter of resistance 50.0 Ω used to measure the terminal voltage of a large battery shows a reading of 150 V. A second panel meter of resistance 300 Ω shows a terminal voltage of 160 V with the battery. Determine (a) the internal resistance of the battery, (b) its emf, and (c) the terminal voltage required to charge the battery at 12.0 A.

19-22 A battery of emf equal to 12.0 V has an internal resistance of 0.0800 Ω. Calculate (a) its terminal potential difference when it is used to charge a battery of 9.00 V emf and internal resistance 0.220 Ω at a current of 10.0 A and (b) the terminal voltage of the 9.00-V battery.

19-23 Calculate the resistance of a rod 7.50 cm long and 0.500 cm in diameter if the rod is made of (a) silver, (b) carbon, (c) boron, or (d) fused quartz.

19-24 A 2.00-m length of copper wire 0.160 mm in diameter is connected in a circuit with a 3.00-V flashlight battery. Calculate the current in the wire.

19-25 Ordinary household copper electric wire is designated as 12-gauge and has a diameter of 2.05 mm. Calculate (a) the resistance of a piece of 12-gauge copper wire strung from the basement to a receptacle in the attic some 15.0 m away from the fuse box, (b) the resistance of a piece of 18-gauge copper (1.02-mm diameter) of the same length, and (c) the diameter of aluminum wire which will dissipate heat at the same rate as a 12-gauge copper wire.

19-26 Determine the thermal coefficient of resistivity of a coil of wire with measured resistance 18.0 Ω at 20°C and 23.5 Ω when submerged in boiling water.

19-27 Nichrome, an alloy of nickel and chromium, has properties which make it attractive to use for heating elements in toasters, toaster ovens, etc. Take the resistivity of nichrome to be 1.00×10^{-6} Ω·m and the thermal coefficient of resistivity to be 4.50×10^{-4} °C^{-1}. A toaster using nichrome wire has a resistance of 60.0 Ω at 20°C. When connected to a voltage of 120 V and allowed to heat for a long time, measurement shows it is drawing a current of 1.50 A. Calculate (a) the operating resistance of the heating element, (b) its final temperature, (c) the Joule heating at final temperature, and (d) the energy used in toasting a slice of bread if it takes 1.00 min to do so.

19-28 A coil of copper wire used as a resistor is to operate at 90°C and have a resistance of 240 Ω. (a) What should its resistance be at 20°C? (b) How long a piece of copper wire 0.250 mm in diameter should be used?

19-29 The typical incandescent light bulb emits light when a thin tungsten wire called a filament is raised to a very high temperature by the Joule heating of the current in it. Much of this energy is then radiated as light rather than as heat. At 20°C the resistivity of tungsten is 5.60×10^{-8} Ω·m and its thermal coefficient of resistivity is 4.50×10^{-3} °C^{-1}. Suppose the temperature of the filament of a 100-W bulb operating on a voltage of 120 V is 3000°C. Determine (a) the resistance of the filament at 3000°C, (b) the resistivity of tungsten at 3000°C, (c) the cross-sectional area of the filament if its length is 0.0800 m, and (d) the resistance of the filament at 20°C.

19-30 Calculate (a) the resistance of a 3.00-cm-diameter disk spacer of 1.40-mm-thick hard rubber with resistivity 2.50×10^{15} Ω·m, (b) the leakage current in the spacer when there is 900 V across it, and (c) the power being dissipated.

19-31 A desk lamp has a flexible cord 1.00 m long made up of two-stranded copper wires with a cross section of 8.17×10^{-7} m² each. When turned on, 0.800 A passes through the lamp. (a) What is the drift velocity of free electrons in the lamp cord? (b) How long would it take for an average free electron to emerge from the wall socket, flow through the lamp, and return to the other prong in the wall plug (assuming the flow was always in one direction)? (c) If the outlet is a standard 120-V outlet, what is the resistance of the lamp? (d) If the outlet is a standard 120-V outlet, what is the wattage of the light bulb?

19-32 A battery with an emf of 120 V and an internal resistance of 2.10 Ω is used to deliver 8.00 A to a circuit. Calculate (a) its terminal potential difference, (b) the power being delivered by the battery to the external circuit, and (c) the power being dissipated in the battery's internal resistance.

19-33 An ammeter connected across the terminals of a battery reads 24.0 A. If the resistance of the ammeter and wires is 0.300 Ω, calculate (a) the terminal potential difference of the battery, (b) the internal resistance of the battery if its nominal emf is 12.0 V, (c) the power delivered by the battery to the external circuit, and (d) the power dissipated inside the battery.

19-34 The bulb in a portable electric lantern used for camping dissipates 8.00 W when used with a 6.00-V lantern battery. Calculate (a) the current in the bulb filament, (b) the resistance of the filament, and (c) how long a new battery could keep the lamp operating in continuous use if the battery has a rating of 5.00 A·h.

19-35 A medical defibrillator is designed to deliver an electric shock to stop an irregularly beating (fibrillating) heart in the hope that it can then be restarted with a normal beat. A capacitor charged to 6000 V is discharged across the patient's chest, dropping the capacitor to 100 V in 10 ms. If the body resistance between electrodes is 1000 Ω, what is the energy delivered to the body during the discharge? Use the average of the initial and the final power being delivered.

19-36 A typical lightning bolt discharges in about 0.400 s with a current of about 25,000 A through an effective potential difference of 3.00×10^6 V. (*a*) What is the average power of a typical lightning bolt? (*b*) How much energy is dissipated in a single bolt? (*c*) If all the energy of a bolt is dissipated in a copper rod 2.54 cm in radius and 8.00 m long leading into the ground, how much will the bar's temperature be raised by a single discharge? (*d*) About 150 people per year are killed directly by lightning bolts. Suppose 10 percent of a bolt's energy is dissipated uniformly in a human body of mass 50.0 kg. By how much would the body temperature rise (the body is essentially water)?

19-37 A four-slice electric toaster is rated at 1200 W and designed for use with household 120-V circuits. (*a*) What is the resistance of the toaster? (*b*) When turned on, what current will flow? (*c*) At what rate is heat generated in the toaster? (*d*) If all the heat generated were concentrated into 500 g of water at room temperature, at what rate would the water temperature be rising? (*e*) The nichrome heating wires in the toaster total 2.00 m long if pulled straight. Effectively all the energy converted in the device is converted in the nichrome wire. What is the electric field in the wire during operation?

19-38 Household circuits operate at 120 V. The common circuits are protected from overheating due to excessive currents and consequent Joule heating by 15-A fuses. The fuse melts, breaking the circuit and stopping the flow of electricity at the stated current. (*a*) What is the maximum wattage light bulb or appliance that can be used on the circuit without blowing (melting) the fuse? (*b*) If the circuit is drawing the maximum current, what is the rate of Joule heating in each meter of the 18-gauge (1.02-mm diameter) wire making up the circuit? (*c*) If the specific heat capacity of copper is 384 J/(kg·°C), at what rate are the wires rising in temperature when the circuit is first turned on?

19-39 Given that a battery of emf $\mathscr{E}$ and internal resistance r is connected to another resistor R, (*a*) write an equation for the current in the circuit as a function of $\mathscr{E}$, r, and R. (*b*) Apply Joule's law to obtain an equation for the power delivered to R in terms of $\mathscr{E}$, r, and R. (*c*) The power delivered to R is a maximum when $R = r$. Convince yourself that this may be true by using values for $\mathscr{E}$ and r. For instance, take $\mathscr{E} = 10.0$ V, $r = 1.00$ Ω, and calculate the power delivered to R when $R = 1.10$, 1.00, and 0.900 Ω. This is an example of **impedance matching**, which states that energy is transferred most efficiently between systems when they have the same impedance. Impedance is taken to mean a resistance to energy flow, whether it be electric, mechanical, or whatever.

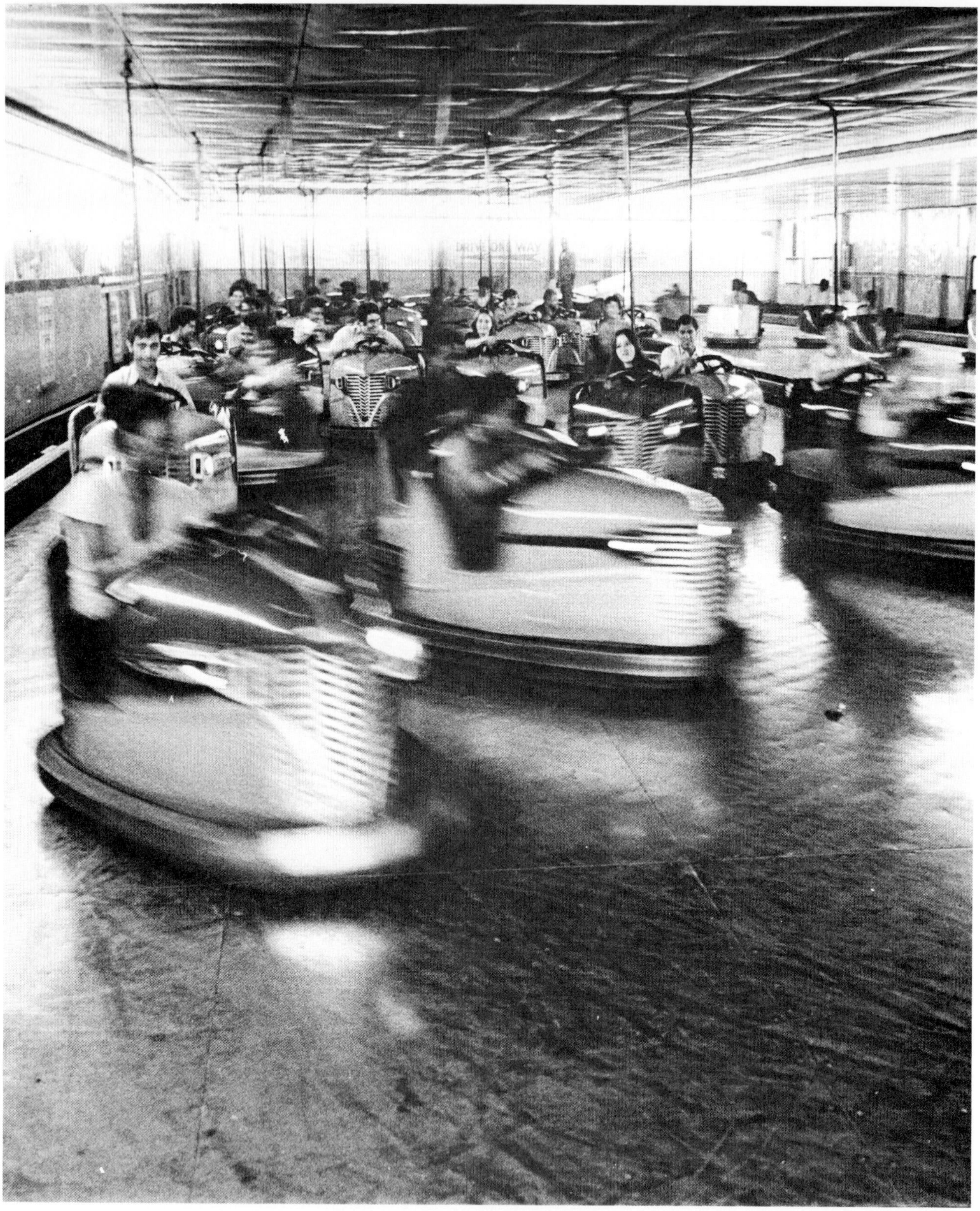

20 Direct-Current Circuits

20-1 Introduction

In electric circuits, the motion of charge is classified as either **direct current** (dc) or **alternating current** (ac). The distinction is that in direct current, the flow of charge is in one and only one direction in any branch of the circuit. The current may vary in magnitude, but it is always in the same direction. In an ac circuit, charge flows through the wires in one direction for a while and is then reversed and flows in the other direction. Figure 20-1 shows several graphs of current vs. time for both direct and alternating currents. Note that in both Figure 20-1b and c the current varies in magnitude but never goes negative (in the other direction). Thus, these are both direct currents. The majority of dc circuits which we encounter, however, have current vs. time graphs that resemble Figure 20-1a. In this chapter, we will concentrate on dc circuits, though many of the concepts will carry over to ac circuits.

We saw our first recognizable dc circuit in Figure 19-9, where the light bulb, wires, and cell form the complete electric pathway for a simple dc circuit. As we have already seen, electrochemical cells provide a fixed potential difference, promoting the flow of charges in one direction (direct current) when connected in a complete circuit. Most portable pieces of equipment (calculators, radio and television receivers, tape recorders, flashlights) depend on chemical batteries and thus use direct current.

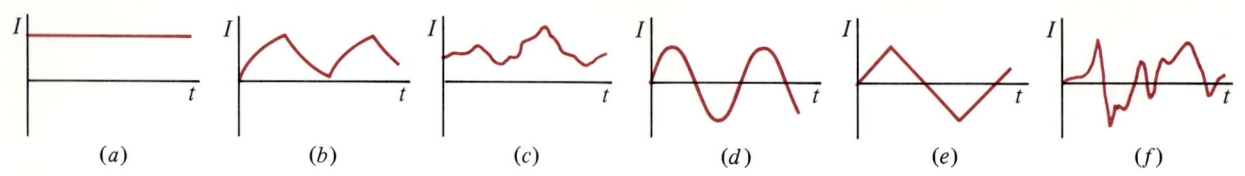

Figure 20-1 A direct current may be (a) steady or (b and c) varying in time. If the current becomes negative, that is, reverses its direction, it is alternating current, as in the (d) sinusoidal, (e) triangular, and (f) irregular ac graphs.

20-2 Batteries in Series and Parallel

In general, most circuits are more complex than that shown in Figure 19-9. For instance, even a flashlight may require more than one cell. There are two ways two cells can be connected to form a single source in an electric circuit. Figure 20-2 shows both possibilities. The resistor shown for each possibility not only completes the circuit but also forms the **load** for each circuit, which is that element or combination of elements on which the electric energy is expended. In Figure 20-2a, the negative terminal of $\mathcal{E}_1$ is connected to the positive terminal of $\mathcal{E}_2$. The external circuit then is connected between the positive terminal of $\mathcal{E}_1$ and the negative terminal of $\mathcal{E}_2$. The same current I exists in both $\mathcal{E}_1$ and $\mathcal{E}_2$. The battery of Figure 20-2a is said to have its cells connected in **series**, one right after the other.

By contrast, Figure 20-2b shows the positive terminals of the two cells connected together and the negative terminals also connected together. Each cell delivers it own current which then combines to become the total current in the circuit. That is, $I_T = I_1 + I_2$. The external circuit is then connected between the common connection of the positive terminals and the common connection of the negative terminals. The battery of Figure 20-2b is said to have its cells connected in **parallel**.

What leads us to choose to connect batteries or cells in series or parallel, and why do we choose one arrangement over the other? We obtain larger voltages by a series connection of batteries. The net emf $\mathcal{E}_s$ of a series arrangement of a number of batteries is

$$\mathcal{E}_s = \mathcal{E}_1 + \mathcal{E}_2 + \mathcal{E}_3 + \cdots$$

and hence the voltage is

$$V_s = V_1 + V_2 + V_3 + \cdots$$

Large-emf batteries are constructed of several smaller-emf batteries or cells. For example, six lead–lead oxide cells are needed to fabricate a nominal 12-V automobile battery, with the six cells connected in series. The 9-V batteries for calculators are often formed of six 1.5-V cells in series.

The reason for connecting batteries in parallel is that the currents of the individual batteries add to produce a larger possible current.[1] For elements which require a very large current, it is often necessary to connect several

Figure 20-2 (a) Cells connected in series. The net emf is the sum of the separate emf's, and the net current is the same as the current through either cell. (b) Cells connected in parallel. The net current is the sum of the separate currents, while the net emf is the same as the separate emf of any cell if all are equivalent.

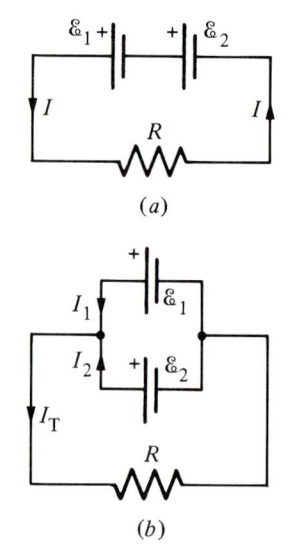

---

[1] The maximum current which can be delivered by a battery depends on the surface area of the plates or rods, because only those atoms in contact with the electrolytic solution will dissolve, and the maximum current depends on the maximum rate at which they may dissolve. Placing several batteries in parallel is equivalent to increasing the area of the plates of a single battery.

batteries in parallel. There is one word of caution. Batteries may be connected in parallel *only* if they all have the same emf, because if one battery has a smaller emf than the others, it will be charged at the expense of the rest. The addition of the smaller-emf battery will reduce the current available to the load rather than increase it. In essence, the net emf of a set of equivalent batteries in parallel is equal to that of one of the batteries.

20-3 Resistors in Series and Parallel

Circuits often require that resistors be connected in combinations of series and parallel. For example, miniature Christmas tree lights appear in a series-connected string, as in Figure 20-3a. The light bulbs appear in the figure as identical resistors because, electrically, that is what they are. Charges flowing through the first resistor must then go through the second, third, etc. If one bulb burns out, the circuit is broken, and all the bulbs go out. This was a problem with very early series strands of lights. Newer, miniature bulbs provide an alternative pathway at the base of the bulb to maintain the circuit when the filament burns out. In Figure 20-3b we see the circuit which represents a string of individual, screw-in Christmas tree bulbs of an earlier generation. Note they are all connected in parallel in that each bulb is independently connected to the same potential difference. Thus one burned-out filament does not influence the others.

Calculations involving series and parallel combinations of resistors may be handled rather easily if the resistors can be reduced to a simple series circuit with one **equivalent emf** and one **equivalent resistance**. First, let us consider the three resistors in a simple series circuit with a pure emf shown in Figure 20-4a. When the switch S is closed, a current I is established in the whole circuit. If we designate the voltage across the resistors as V_1, V_2, and V_3, Ohm's law tells us that

$$V_1 = IR_1 \qquad V_2 = IR_2 \qquad V_3 = IR_3 \qquad (20\text{-}1)$$

Now we also know that we can write

$$V = V_1 + V_2 + V_3 \qquad (20\text{-}2)$$

This must be true since the net loss in potential energy per unit charge as the charge moves through the resistors must be the same as the net gain as it moves through the battery. Potential, just as potential energy, is single-valued. Thus, the voltage between the switch and the positive terminal of the battery should

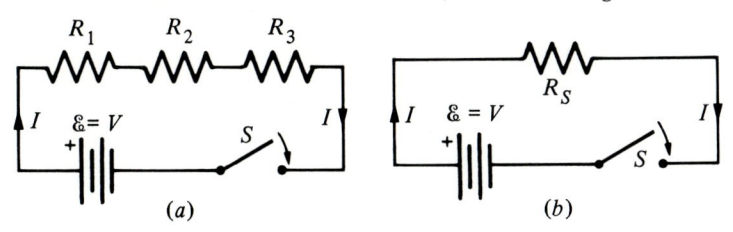

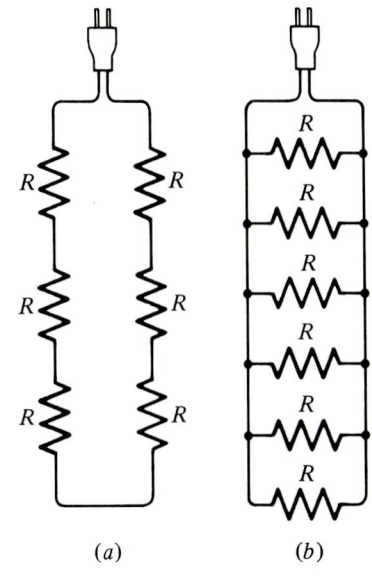

Figure 20-3 (a) Resistors connected in series. The equivalent resistance is the sum of the separate resistances. (b) Resistors connected in parallel. Each resistance experiences and operates at the full voltage of the source.

Figure 20-4 (a) A simple series circuit. When the switch is closed, the current is I. (b) The circuit of Figure 20-4a with the three resistors replaced by a single equivalent resistance such that when the switch is closed, the current is again I.

be the same net value whether we measure across the battery or measure individual values across the resistors and sum them.

We would like to be able to substitute a single equivalent resistor, say R_s, which with the given emf yields exactly the same current. Our circuit would then be that of Figure 20-4b with the current and the emf unchanged. From Ohm's law, we have

$$V = IR_s \qquad (20\text{-}3)$$

Substituting Equations 20-1 and 20-3 into Equation 20-2 yields

$$IR_s = IR_1 + IR_2 + IR_3$$

Finally, dividing this equation by I yields

$$R_s = R_1 + R_2 + R_3 \qquad (20\text{-}4)$$

Thus the equivalent resistance of resistors in series is equal to the sum of the individual resistances.

Example 20-1 Two light bulbs, one of resistance 150 Ω and the other of resistance 250 Ω, are connected in series with a 120-V emf. Calculate (a) the equivalent resistance of the bulbs, (b) the current in the circuit, and (c) the power being dissipated by the 250-Ω bulb.

Solution (a) To solve, we simply apply Equation 20-4, recognizing that we have only two resistors.

$$R_s = R_1 + R_2 = 150\ \Omega + 250\ \Omega = 400\ \Omega$$

(b) We apply Ohm's law.

$$I = \frac{V}{R_s} = \frac{120\ V}{400\ \Omega} = 0.300\ A$$

(c) Using Joule's law, we have

$$P_2 = I^2 R_2 = (0.300\ A)^2 (250\ \Omega) = 22.5\ W$$

It is not very much more difficult to deal with resistors in parallel. Consider Figure 20-5a, where the three resistors are connected in parallel and have a common voltage V across them equal to that of the emf. The current must divide into three separate currents, one through each resistor. These three currents are not necessarily equal to each other, but their sum must equal I*:

$$I = I_1 + I_2 + I_3 \qquad (20\text{-}5)$$

---

* After all, whatever charge per unit time flows into a junction must equal the charge per unit time which flows out of the junction—no more, no less. If more charge per unit time flowed in than flowed out, there would have to be a buildup of charge at the junction or charge would have to disappear. The first is never observed, and the second would violate an established principle, the **conservation of charge.** If less charge per unit time flowed into the junction than flowed out, charge would either have to be depleted at the junction (not observed) or appear from thin air (violation of conservation of charge).

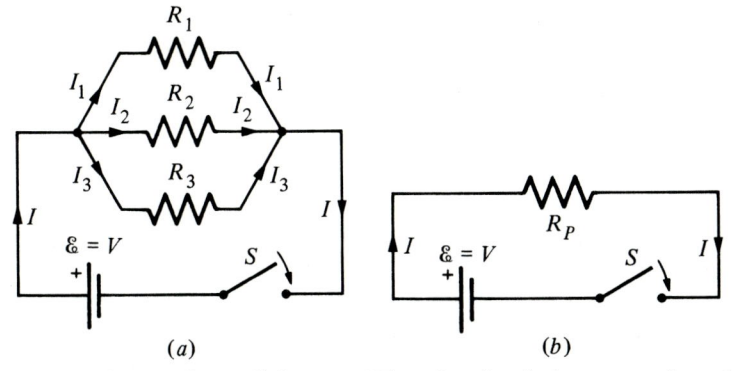

Figure 20-5 (a) A simple parallel circuit. When S is closed, the current through S is the sum of the separate currents through the three resistors. (b) The circuit of Figure 20-5a with the three resistors replaced by a single equivalent resistance such that when the switch is closed, the current through the switch is the same as in part a.

We apply Ohm's law to each resistor and obtain

$$V = I_1 R_1 \qquad V = I_2 R_2 \qquad V = I_3 R_3 \qquad (20\text{-}6)$$

Once again we would like to substitute a single equivalent resistor, say R_p, which with the given emf yields exactly the same net current. With R_p substituted, our circuit is now that of Figure 20-5b. Applying Ohm's law to this circuit yields

$$V = IR_p \qquad (20\text{-}7)$$

Equations 20-6 and 20-7 may be solved for the currents to yield

$$I = \frac{V}{R_p} \qquad I_1 = \frac{V}{R_1} \qquad I_2 = \frac{V}{R_2} \qquad I_3 = \frac{V}{R_3} \qquad (20\text{-}8)$$

Substituting Equation 20-8 into Equation 20-5 yields

$$\frac{V}{R_p} = \frac{V}{R_1} + \frac{V}{R_2} + \frac{V}{R_3}$$

Finally, dividing by V gives us

$$\frac{1}{R_p} = \frac{1}{R_1} + \frac{1}{R_2} + \frac{1}{R_3} \qquad (20\text{-}9)$$

Figure 20-6 Example 20-2.

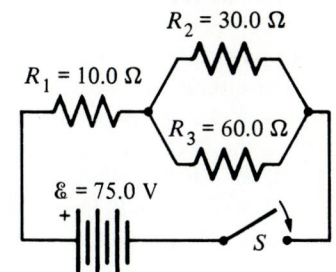

Example 20-2 Reduce the circuit in Figure 20-6 to a simple equivalent circuit. Determine (a) the effective resistance of the circuit, (b) the current being delivered by the battery, (c) the voltage across the $R_2 R_3$ parallel combination, (d) the current in R_2, and (e) the current in R_3.

Solution (a) To reduce this circuit, we first recognize that the $R_2 R_3$ combination may be replaced with an equivalent R_p found by using Equation 20-9. R_1 and R_p are then in series and may be replaced with an equivalent R_s which is equal to R_1 plus R_p. This solves part a, as R_s is then the effective resistance of the circuit. The circuit diagram may be redrawn as Figure 20-7a and finally as

Figure 20-7b, where

$$\frac{1}{R_p} = \frac{1}{R_2} + \frac{1}{R_3} = \frac{1}{30.0\,\Omega} + \frac{1}{60.0\,\Omega} = \frac{3}{60.0\,\Omega}$$

$$R_p = 20.0\,\Omega$$

and

$$R_s = R_1 + R_p = 10.0\,\Omega + 20.0\,\Omega = 30.0\,\Omega$$

(b) This part may be solved by using Ohm's law with the given emf and the effective circuit resistance.

$$I = \frac{V}{R_s} = \frac{75.0\,\text{V}}{30.0\,\Omega} = 2.50\,\text{A}$$

(c) We apply Ohm's law by using the current obtained for part b and the value of R_p from part a to solve for the voltage across R_p.

$$V_p = IR_p = (2.50\,\text{A})(20.0\,\Omega) = 50.0\,\text{V}$$

(d) Ohm's law serves us again to find the current in R_2 by using the voltage across R_p since it is across both R_2 and R_3.

$$I_2 = \frac{V_p}{R_2} = \frac{50.0\,\text{V}}{30.0\,\Omega} = 1.67\,\text{A}$$

(e) Finally, the current in R_3 can be found either by subtracting the current in R_2 from the circuit current or by applying Ohm's law to R_3.

$$I_3 = I - I_2 = 2.50\,\text{A} - 1.67\,\text{A} = 0.83\,\text{A}$$

or

$$I_3 = \frac{V_p}{R_3} = \frac{50.0\,\text{V}}{60.0\,\Omega} = 0.83\,\text{A}$$

We see that solving this circuit does not require complex mathematics or even difficult calculations. The primary requirement is a knowledge of Ohm's law.

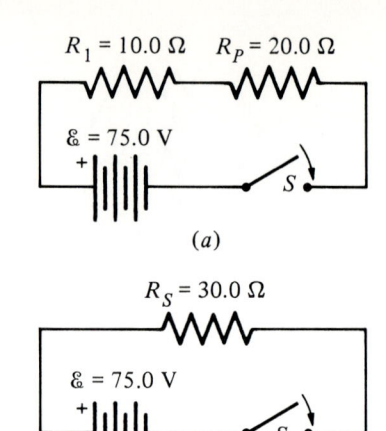

Figure 20-7 Circuit of Figure 20-6 redrawn: (a) substituting R_p for the parallel resistors; (b) substituting R_s for R_1 and R_p in series.

In attempting to solve for unknowns in a circuit, you must analyze it carefully to determine which elements are in a series configuration and which are in parallel. This determination can sometimes be difficult, especially in the laboratory or in any real circuits where the wires are not straight lines and things do not fit together in nice rectangles. It often helps to draw a circuit diagram if none is provided. It even may help to redraw a circuit diagram of a circuit which is difficult to analyze. For instance, the circuit represented by Figure 20-8a may be redrawn as Figure 20-8b. It may not be altogether obvious that the two circuits are the same, but you should be able to convince yourself of their equivalence by following the paths of conventional current in each and comparing the two. Starting at $\mathscr{E}_1$, conventional current I enters junction A, where it divides into I_1, which goes through the R_1R_2 branch, and I_4, which goes through R_4. At junction B, I_1 again divides into I_5, through the R_5R_6 branch, and I_3, through R_3. At junction C, I_5, I_3, and I_4 recombine to form I, which passes through $\mathscr{E}_2$ and R_7 as it returns to $\mathscr{E}_1$.

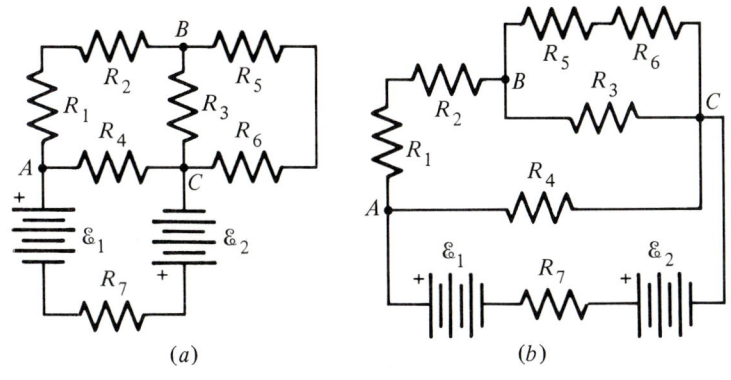

Figure 20-8 (a) A complex circuit. (b) The same circuit redrawn to emphasize parallel and series relationships.

Drawing the circuit as in Figure 20-8b has not required us to do anything new. It may simply help in the analysis of the circuit. Turning our attention to that analysis, we see that it might be easier to accept that $R_5 R_6$ are in series and may be replaced with an equivalent, say R_{s1}, which is simply the sum of R_5 and R_6. Also we see that R_3 is in parallel with R_{s1}. These may then be replaced with the parallel equivalent, say R_{p1}, where

$$\frac{1}{R_{p1}} = \frac{1}{R_3} + \frac{1}{R_{s1}}$$

Continuing, we see that R_1 and R_2 are in series with R_{p1}, and this trio may then be replaced with a series equivalent, say R_{s2}, which is the sum of R_1, R_2, and R_{p1}. Then, since R_{s2} is in parallel with R_4, we may replace them with R_{p2}, where

$$\frac{1}{R_{p2}} = \frac{1}{R_{s2}} + \frac{1}{R_4}$$

We then find that R_{p2} and R_7 are in series and can be replaced by their sum, say R_{s3}. Finally, $\mathscr{E}_1$ and $\mathscr{E}_2$ are in series and may be replaced by their sum, say $\mathscr{E}_s$.

Thus the circuit of Figure 20-8a will have been reduced to a simple equivalent circuit with an emf $\mathscr{E}_s$ and a resistor R_{s3} in series. We could then find the current delivered by the batteries and the current in any element by applying Ohm's law one step at a time. Many people find that redrawing circuits helps immeasurably in circuit analysis. Further, even a person with excellent qualifications and a great deal of experience in electric and electronic circuits may find it impossible to analyze a circuit without a diagram.

Figure 20-9 A relatively simple circuit for which simple substitution of equivalent resistances is not possible, requiring the more powerful circuit analysis technique of Kirchhoff's rules.

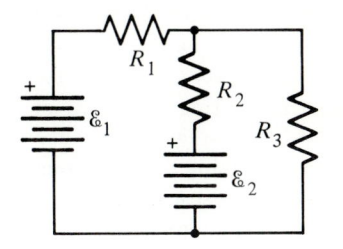

20-4 Kirchhoff's Rules of Electric Circuits

Successively combining series and parallel combinations of resistors to reduce a circuit to a single equivalent resistor in series with a single emf is an excellent method of solving circuit problems. Unfortunately, the method is not always applicable. Figure 20-9 shows a very simple circuit which can neither be reduced any further nor solved by direct application of Ohm's law. The presence of batteries in parallel branches along with a resistor in at least one branch precludes this possibility. Even if there were no resistor in series with either battery, the batteries could not be combined unless they had identical emf's. In general, a whole class of practical circuits exists which cannot be reduced by using combinations of series and parallel resistors and/or emf's. Thus another method must be found.

Fortunately, a general method exists which, in principle, is applicable to *all* circuits, no matter how complex. The general method utilizes **Kirchhoff's rules**, two rules named after the German physicist Gustav Kirchhoff (1824–1887), who is credited with being the first person formally to state them. Kirchhoff's first rule, often called the **point rule**, is:

The algebraic sum of the currents into any point is zero.

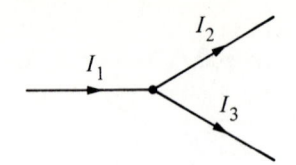

Figure 20-10 Kirchhoff's point rule: The algebraic sum of the currents into any point is zero.

It applies to any point in a circuit but is most useful at **junctions**, where three or more wires are joined. In mathematical notation, the point rule becomes

$$\sum_i I_i = 0 \quad \text{(at a point)} \quad (20\text{-}10)$$

Figure 20-10 shows a junction point of a circuit with one current directed into the point and two others directed out of the point. Currents directed into a point are defined as positive, and those directed out of the point are defined as negative. For the junction point of Figure 20-10, Equation 20-10 translates to

$$I_1 - I_2 - I_3 = 0$$

We have really already used this rule, which is simply a manifestation of the law of conservation of charge. It requires that "what goes in, must come out."

Kirchhoff's second rule, often called the **loop rule**, is:

The algebraic sum of the voltages in traversing any closed loop is zero.

It applies to any closed path, on or off a circuit, but is most useful when applied to a loop of the circuit. In mathematical notation, the loop rule is

$$\sum_i V_i = 0 \quad \text{(around any closed loop)} \quad (20\text{-}11)$$

For example, suppose the currents are as indicated in Figure 20-11 and we make voltage measurements around the closed loop *ABCDA*. Equation 20-11 applied to this loop translates to

$$V_{AB} + V_{BC} + V_{CD} + V_{DA} = 0$$

or

$$+\mathscr{E}_1 - I_1 R_1 + (-\mathscr{E}_2 + I_2 R_2) - I_1 R_4 = 0 \quad (20\text{-}12)$$

Figure 20-11 Complex circuit to illustrate Kirchhoff's loop rule: The algebraic sum of the voltages in traversing any closed loop is zero.

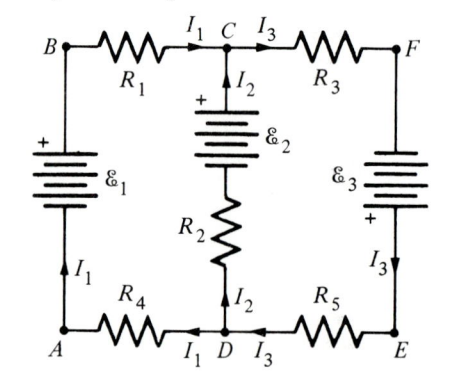

How do we know what signs to use with each of the terms of Equation 20-12? How do we know the directions of the currents in the first place? How do we know the proper direction in which to traverse the loop in making our voltage assignments? The sign has to do with what we mean by traversing the loop and making voltage measurements. To obtain the terms in Equation 20-12, we began at point *A* and measured the voltage across $\mathscr{E}_1$. We agree that we are always measuring the voltage of the second point with respect to the first. Thus, for our first voltage, we are measuring the positive terminal with respect to the negative terminal, and we must obtain $+\mathscr{E}_1$. Next, we measure the voltage across the resistor R_1. Again, we report the voltage of the second point (*C* in Figure 20-11) with respect to the first (*B*). The voltage certainly has a magnitude equal to I_1 times R_1 according to Ohm's law. Further, if I_1 has the direction indicated, point C must be at a lower voltage than point B because conventional current is always directed from higher voltage to lower voltage. Thus the voltage across R_1 to be used in Equation 20-12 is $-I_1R_1$. Note that the voltage across $\mathscr{E}_2$ is then negative because we are going from plus to minus, and the voltage across R_2 is positive because we are going against the current from lower voltage to higher. Finally, you should expect that the voltage across R_4 is negative because we are going with the current from a higher to lower voltage. All this can be summarized in the following easy-to-remember sign conventions.

Sign Convention for Kirchhoff's Rules

1. If we traverse an emf from its minus to its plus terminal, the voltage is $+\mathscr{E}$.
2. If we traverse an emf from its plus to its minus terminal, the voltage is $-\mathscr{E}$.
3. If we traverse a resistor in the same direction as the current in it, the voltage is $-IR$.
4. It we traverse a resistor in a direction opposite to that of the current in it, the voltage is $+IR$.

The response to how we know the direction of the currents originally is "We often do not know!" When we do not, *the directions for currents are assigned arbitrarily*. If we happen to choose the wrong direction for a particular current, the final solution will yield a negative value for that current. No harm is done, provided that whatever direction is chosen is kept throughout the solution. The correct directions for all the currents will be apparent from their signs in the complete solution.

Finally, we do not know in which direction to traverse the loop either. One direction is as good as the other. Suppose we had traversed the same loop of Figure 20-11 in the direction *ADCBA* instead. The resulting equation from the loop rule is

$$+I_1R_4 + (-I_2R_2 + \mathscr{E}_2) + I_1R_1 - \mathscr{E}_1 = 0 \qquad (20\text{-}13)$$

(Can you verify this equation?) Equation 20-13 is the original equation multiplied by a minus sign, which does not change the equality. Thus, we may traverse a loop in either direction. You should also be able to show that it does not make any difference where you start in traversing a loop, provided you get back to that same point.

In applying Kirchhoff's rules to a circuit problem, we will usually know the values of the emf's and the values of the resistors as well as their placements in the circuit. Most often we will wish to obtain the currents. To do this, we must know how many different currents are involved and then obtain enough independent equations to solve for them. For example, in Figure 20-11 there are three separate currents, as labeled, because there are three branches. Thus we need three independent equations relating those currents, which we intend to solve simultaneously. We obtain these equations by applying Kirchhoff's rules to the circuit. Note that despite the fact that there are two junctions, we obtain only one equation from the point rule. To illustrate, applying the point rule at junction C yields

$$I_1 + I_2 - I_3 = 0 \qquad (20\text{-}14)$$

whereas applying the point rule at junction D yields

$$-I_1 - I_2 + I_3 = 0 \qquad (20\text{-}15)$$

Equation 20-15 is simply the negative of Equation 20-14. Thus there is only one independent equation. Because charge flow throughout the circuit is continuous, it is always true that the number of independent equations obtainable from the point rule is one less than the number of junctions.

Since the point rule gives only one equation for the circuit of Figure 20-11, a complete solution requires that we obtain two independent equations from the loop rule. Note, however, that there are three possible closed circuit loops: *ABCDA*, *CFEDC*, or *ABCFEDA*. Once again we will see that there are only two independent equations available. You should be able to verify that the loop law applied to *CFEDC* can yield

$$-I_3 R_3 + \mathscr{E}_3 - I_3 R_5 - I_2 R_2 + \mathscr{E}_2 = 0 \qquad (20\text{-}16)$$

Application of the loop law to *ABCFEDA* yields

$$+\mathscr{E}_1 - I_1 R_1 - I_3 R_3 + \mathscr{E}_3 - I_3 R_5 - I_1 R_4 = 0 \qquad (20\text{-}17)$$

This is *not* independent of the other two loop equations. You can show that Equation 20-17 could also be obtained by adding Equations 20-12 and 20-16.

The generalization for loops is similar to that of the point rule. The number of independent equations obtainable from the loop rule is always less than the number of possible loops (for a circuit of more than one loop). In complex circuits, however, many more possible loops than possible independent equations may be available. We handle this by first applying the point rule to $n - 1$ junctions, where n is the total number of junctions. Then we apply the loop rule to enough loops so that we have the same number of independent equations as there are currents. We must be very careful to make sure *each* element of the circuit is included in *at least one* of the loop equations we have chosen.

For the circuit of Figure 20-11, any two of the three equations form an independent pair, which is really all we need to solve the problem. Thus, for example, if the emf's and resistors are known, we could obtain all the currents in the circuit of Figure 20-11 by the simultaneous solution of

$$I_1 + I_2 - I_3 = 0 \qquad (20\text{-}15)$$

$$+\mathscr{E}_1 - I_1 R_1 - \mathscr{E}_2 + I_2 R_2 - I_1 R_4 = 0 \qquad (20\text{-}12)$$

and

$$-I_3 R_3 + \mathscr{E}_3 - I_3 R_5 - I_2 R_2 + \mathscr{E}_2 = 0 \qquad (20\text{-}16)$$

Now that we have seen Kirchhoff's rules in all their detail, let us condense a bit and provide a prescription for solution of a multiple-loop circuit problem.

Prescription for Applying Kirchhoff's Rules

1. Draw a good circuit diagram including all elements of the circuit, and label all known quantities.
2. Assign current direction in each branch. Choose a different symbol for each separate current.

3. Apply Kirchhoff's point rule to $n - 1$ junctions, obtaining $n - 1$ independent equations.
4. Select enough closed loops from the circuit so that, including the equations from step 3, you will have as many independent equations as there are currents (branches). Be sure that every element in the circuit is included in at least one loop.
5. Apply Kirchhoff's loop rule to the closed loops selected in step 4, obtaining the rest of the needed independent equations.
6. Rewrite and organize all equations in a convenient form.
7. Solve the equations of step 6 as simultaneous equations.

Example 20-3 Apply Kirchhoff's rules to the circuit of Figure 20-12 to obtain all the currents. Assume all quantities are known to three significant figures.

Solution In this example, we will select the currents so that we are certain to have at least one wrong direction, thereby obtaining a negative value for at least one of the currents, just to show it.

Steps 1 and 2. You can see in Figure 20-12 that we have assigned all three current directions toward the top junction. This cannot be correct because all charge would be entering the junction and none leaving it. Thus, at least one of the currents and perhaps two of them are really in the other direction. We will discover which in the final solution.

Step 3. We apply the point rule to only one of the two junctions. Choosing the top junction, we have

$$\sum_i I_i = 0$$

or

$$I_1 + I_2 + I_3 = 0$$

Step 4. There are three possible loops. We must select two of them to obtain two equations. These, along with the equation from step 3, will gives us three equations with three unknowns. It really does not matter which two of the loops we choose since any two will include all elements at least once. Neither does it

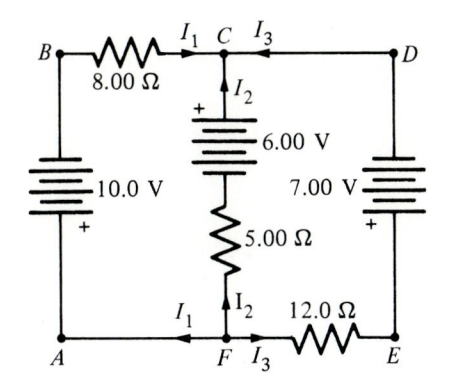

Figure 20-12 Example 20-3.

matter where we start to traverse the loops nor in which direction we go. Let us choose loops ABCFA and ABCDEFA.

Step 5. Applying the loop rule to ABCFA yields

$$\sum_i V_i = 0$$

or

$$-10\text{ V} - I_1(8\text{ }\Omega) - 6\text{ V} + I_2(5\text{ }\Omega) = 0$$

For loop ABCDEFA, we have

$$-10\text{ V} - I_1(8\text{ }\Omega) + 7\text{ V} + I_3(12\text{ }\Omega) = 0$$

We now have the three necessary equations.

Step 6. Rewriting the three equations and omitting units for the time being,

$$I_1 + I_2 + I_3 = 0 \quad (A)$$

$$-8I_1 + 5I_2 = 16 \quad (B)$$

$$-8I_1 + 12I_3 = 3 \quad (C)$$

Step 7. Any method which can be used to solve simultaneous equations will work. You may add and/or subtract equations to eliminate unknowns, use determinants, or whatever. For our example, we choose the method of substituting for one unknown in terms of others. Solving Equations B for I_2 and C for I_3 yields

$$I_2 = \frac{16 + 8I_1}{5} \quad (D)$$

$$I_3 = \frac{3 + 8I_1}{12} \quad (E)$$

When we substitute Equations D and E into Equation A, we have

$$I_1 + \frac{16 + 8I_1}{5} + \frac{3 + 8I_1}{12} = 0$$

Clearing fractions by multiplying each term by 60 yields

$$60I_1 + 192 + 96I_1 + 15 + 40I_1 = 0$$

$$196I_1 = -207$$

$$I_1 = \frac{-207}{196}\text{ A} = -1.06\text{ A}$$

The negative sign means we have chosen I_1 in the wrong direction. It is really directed from B to A through the battery rather than as shown in Figure 20-12 from A to B. Nonetheless, when we use I_1 to obtain I_2 and I_3, we must maintain the minus sign. Rearranging Equations B and C and substituting for I_1 gives us

$$5I_2 = 16 + 8\left(\frac{-207}{196}\right)$$

$$I_2 = \tfrac{370}{245}\text{ A} = 1.51\text{ A}$$

and

$$12I_3 = 3 + 8\left(\frac{-207}{196}\right)$$

$$I_3 = -\tfrac{267}{588}\text{ A} = -0.45\text{ A}$$

Thus I_3 was also chosen in the wrong direction. Finally note that $I_1 + I_2 + I_3 = -1.06 + 1.51 - 0.45 = 0$, and our solution checks.

Example 20-4 Apply Kirchhoff's rules to solve for the unknowns in Figure 20-13. Assume all quantities are known to three significant figures.

Solution Once again we will apply the prescription as listed. This time, however, we see that our three unknowns are not all the currents but a current, an emf, and a resistor. Note that steps 1 and 2 have been completed.
Step 3. The point rule applied to junction C yields

$$I_1 + I_2 - I_3 = 0$$

$$I_1 = I_3 - I_2 = 0.305 \text{ A} - 0.271 \text{ A} = 0.034 \text{ A}$$

Steps 4, 5, and 6. We choose the two smaller loops, and when the loop rule is applied to *ABCFA*, we obtain

$$-I_1(5 \text{ }\Omega) + \mathscr{E} - 4 \text{ V} + I_2(8 \text{ }\Omega) = 0$$

$$\mathscr{E} = 4 \text{ V} - (8 \text{ }\Omega)(0.271 \text{ A}) + (5 \text{ }\Omega)(0.034 \text{ A}) = 2.00 \text{ V}$$

When the loop rule is applied to FCDEF, we have

$$-I_2(8 \text{ }\Omega) + 4 \text{ V} - I_3 R = 0$$

$$(0.305 \text{ A})R = 4 \text{ V} - (8 \text{ }\Omega)(0.271 \text{ A})$$

$$R = 6.01 \text{ }\Omega$$

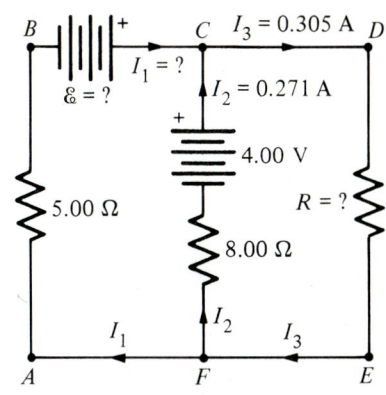

Figure 20-13 Example 20-4.

20-5 Measuring Direct Currents and Voltages

The application of Kirchhoff's rules in an electric circuit allows us to calculate the values of some quantities if we know the values of the others. Some of these knowns must be obtained by measurement, using instruments called **meters** which have been devised for just this purpose. **Voltmeters, ammeters,** and **ohmmeters** are used to measure potential differences, currents, and resistances, respectively. Many portable laboratory meters are constructed to be converted at the flip of a switch to measure two and perhaps all three of these quantities. Such instruments are most often called **multimeters.**

Meters may be either analog devices or digital devices (Figure 20-14). An analog device normally has a thin pointer needle or line which moves along a graduated scale on the instrument. At the position of the pointer, the measured quantity may be read directly on the scale. Digital meters represent the measured quantity by displaying numbers or digits. Both types of meter are in common use, although digital meters seem to be replacing analog meters because of their easy readability.

Whether we are dealing with analog or digital meters, there are many common features of measurement which are quite important. First, the meter must be attached properly to the circuit. For example, a voltmeter must be attached in parallel with the element whose voltage you wish to measure or across the two points between which you wish to measure the potential differ-

20 Direct-Current Circuits

Figure 20-14 Multimeters measure several ranges of different electrical variables. The digital meter (left) gives the reading directly. The analog meter (right) has the reading proportional to the travel of a pointer along a ruled scale. (D. Riban)

ence. In Figure 20-15, the voltmeter is attached to measure the potential difference across resistor R_1, between points A and B.

The circuit symbol for a voltmeter is ─(V)─. The voltmeter is connected between two points which could be at very different potentials. We have shown the connections made between the voltmeter and the points A and B as arrowheads to remind us that the voltmeter connections are often temporary. The voltmeter is rarely considered to be part of the circuit; it is an outside observer looking in.

Also shown in Figure 20-15 is an ammeter, indicated as ─(A)─. To measure the current at a point, the circuit must be physically disconnected at that point and the ammeter connected between the open ends so that all the current passes through it. Thus, an ammeter is part of the circuit, and if removed without making other connections, the current in its branch goes to zero. Ammeters must be connected in series.

When connected, meters become circuit elements having some finite resistance. This has the effect of changing a circuit while a measurement is taking place. Consider Figure 20-15, where the voltmeter is considered as a resistor in parallel with resistor R_1. The effective resistance between A and B is different after the meter is connected than beforehand. Thus, the addition of the meter changes the equivalent resistance of the entire circuit and, more importantly, the voltage between A and B. Hence, the very act of attempting to measure changes the quantity we seek to measure. Ideally, since connected in parallel, a voltmeter should have an infinite resistance to leave the voltage between A and B in Figure 20-15 unaltered. In general, this is not possible, but we recognize that in practice, voltmeters must have a very large resistance with respect to the element whose voltage is being measured.

Likewise, the ammeter is a resistor in series with the elements in whatever branch of the circuit we wish to measure the current. It, too, changes the resistance and hence the current in that branch. This new current is then shown on the scale or displayed on the digits. Since it is connected in series, the

Voltmeters can be used with operating circuits by simply touching the two leads to two points in the circuit. The meter is not part of the circuit. By contrast, to use an ammeter we must cut or disconnect the circuit at a point and connect the meter between the broken ends. The ammeter must be part of the circuit, and its removal stops the current in that branch.

Figure 20-15 A circuit with an ammeter and a voltmeter. The voltmeter is measuring the voltage across R_1. Note that the ammeter is part of the circuit.

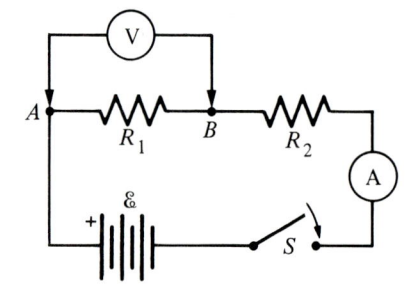

484

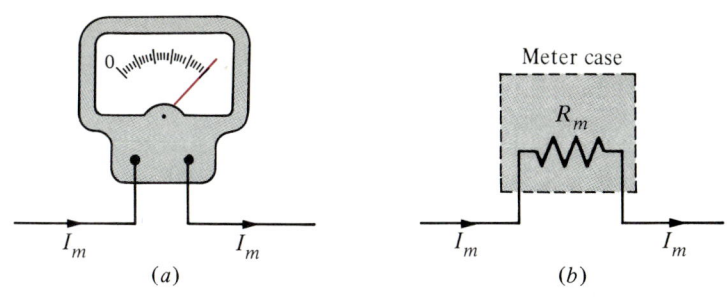

Figure 20-16 (a) An electric meter movement has two external contacts for connection to an electric circuit, between which a current I_m must flow for a measurement. (b) The meter interior presents a continuous pathway for the current and an internal resistance R_m due to the meter movement.

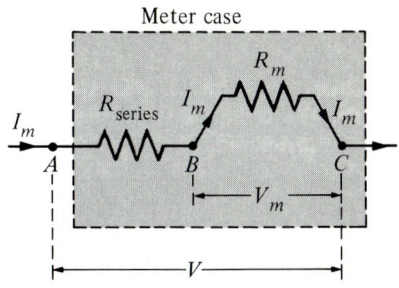

Figure 20-17 To make a voltmeter of the meter movement, a large, known resistance R_{series} is connected in series to restrict the flow to a range which can be measured by the movement. To measure various ranges of voltage, an external switch on the meter may connect internal resistors of various known values in series with the movement.

resistance of an ammeter should be very small, ideally zero, to leave the current undisturbed. (See Problems 20-22, 20-23, and 20-26.)

The heart of any meter is some device which will respond to a very small current. For the moment, we may treat that part of the meter which responds — the meter movement — as an element with some resistance and which yields the largest marked reading on the scale, a *full-scale reading*, with a given current. A knowledge of Ohm's law allows us to construct either a voltmeter or an ammeter from any meter movement. In Figure 20-16a, we see an analog meter movement with resistance R_m showing a full-scale reading with a current I_m, represented in circuit symbols in Figure 20-16b. The movement yields a full-scale reading every time a current of magnitude I_m is directed through it. From Ohm's law, we know that the voltage V_m needed across the movement's terminals to produce this current is $V_m = I_m R_m$. However, we can modify and use this movement to measure a larger voltage than V_m or a larger current than I_m. We may change the meter to make it read full-scale as a voltmeter or an ammeter for different maximum voltages or currents.

Suppose we wish the voltmeter to read, say, V at full-scale, where $V > V_m$, but we still want only V_m across the movement. To do this, we include a resistor in series with the movement inside the meter case, as shown in Figure 20-17. The external terminals of the meter at points A and C have a voltage of V across them. Points B and C represent the terminals of the meter movement. The voltage drop from A to B plus the voltage drop from B to C must equal V. By choosing an appropriate value of R_{series}, any set fraction of the potential drop can occur across the resistor, leaving only V_m as the maximum drop across the meter movement.

We can use the same movement to make an ammeter which will read maximum for a current $I > I_m$. To do this, we provide a path to divert, or shunt, part of the current around the movement, and allow only I_m through the meter movement, as shown in Figure 20-18. The additional path provided is called a **shunt**, or a shunt resistor. By choosing the value of this resistor, we choose what fraction of the total current passes through the meter movement. For either the voltmeter or the ammeter, in all cases, for a full-scale reading on the meter movement, I_m is the current in the movement and V_m is the voltage across it.

The meter movements for all meters are essentially alike in function. What the meter measures (voltage, current, resistance) depends on other components inside the meter case and how these are connected to the movement and the external leads.

Figure 20-18 To make an ammeter of the meter movement, a resistor of low resistance, R_{shunt} is connected in parallel. This allows a substantial, known portion of the current to bypass the movement, leaving I_m small enough to be measured by the meter. To operate in various ranges, a knob can select from various values for R_{shunt}.

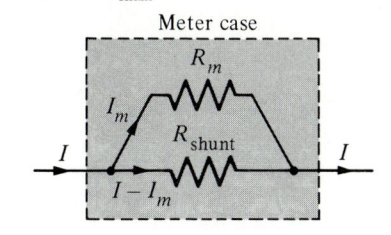

Example 20-5 Determine the value of a series resistor necessary to construct a voltmeter with a full-scale reading of 10.0 V by using a meter movement

485

whose resistance is $R_m = 50.0\ \Omega$ and which deflects full-scale with $I_m = 2.00 \times 10^{-3}$ A.

Solution We apply Ohm's law to the circuit of Figure 20-17, remembering that our maximum voltage V must appear between A and C when there is a current of I_m. Thus

$$V = I_m R_{series} + I_m R_m$$

$$R_{series} = \frac{V - I_m R_m}{I_m} = \frac{10.0\ \text{V} - (2.00 \times 10^{-3}\ \text{A})(50.0\ \Omega)}{2.00 \times 10^{-3}\ \text{A}}$$

$$= 4.95 \times 10^3\ \Omega$$

Example 20-6 Use the same meter movement as in Example 20-5 to construct an ammeter with a full-scale reading of 4.00 A. Find the value of the shunt resistor R_{shunt}.

Solution This time, we apply Ohm's law to the elements of Figure 20-18. We must remember that since R_m and R_{shunt} are in parallel, the same voltage is across each. Further, we remember that $I_{shunt} = I - I_m$.

$$V_m = I_m R_m = I_{shunt} R_{shunt} = (I - I_m) R_{shunt}$$

Thus

$$R_{shunt} = \frac{I_m R_m}{I - I_m} = \frac{(2.00 \times 10^{-3}\ \text{A})(50.0\ \Omega)}{4.00\ \text{A} - 2.00 \times 10^{-3}\ \text{A}} = 2.50 \times 10^{-2}\ \Omega$$

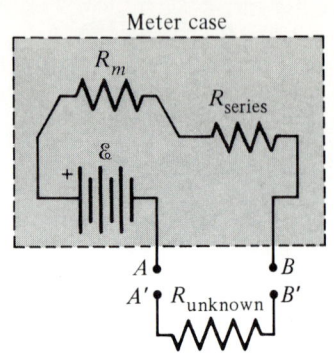

Figure 20-19 To produce an ohmmeter from a meter movement, a resistor and a known emf are placed in series with the movement between the external contacts A and B. In effect, the meter compares the reading when only the known resistance R_{series} is in the circuit with the reading when an additional, external resistance $R_{unknown}$ is placed between the external contacts.

Multimeters allow the user to select different series or parallel resistors to measure different voltage or current ranges by turning a knob or pushing a switch. The meter scale changes to correspond to the readings for the series or parallel resistor selected.

An ohmmeter requires a known emf which is part of the meter system (inside the case). A common version is shown in Figure 20-19. The series resistor is selected so that the meter movement reads full-scale when the leads A and B are connected together, having zero resistance between them. The movement reads zero when the leads are apart (open) and have an infinite resistance between them. Thus, the ohmmeter reads backward for this version, full-scale for no resistance and zero scale for infinite resistance. An unknown resistor is connected A to A' and B to B', and the meter reads between these two extremes. If properly calibrated, the resistance of the unknown may be read from the backward scale. A number of different, internal series resistors may be selected to measure different resistance ranges. Ohmmeters usually are not very accurate, but they do provide a rapid method of determining resistance.

20-6 Capacitors in Parallel and Series

Many electric circuits use more than one capacitor. Further, we often wish to alter the value of the capacitance or use a capacitor of a different value than the ones we have on hand. Thus, we are often required to obtain the **equivalent**

capacitance of several capacitors in series or parallel or both. Consider the circuit shown in Figure 20-20. Capacitors C_1, C_2, and C_3 are connected in parallel with a battery of voltage V. When the switch is closed, a charge q flows from the battery and is distributed on one side of the capacitors as q_1, q_2, and q_3, with an equal charge flowing from the other side of the capacitors back to the other battery terminal. We note that the voltage across each capacitor is the same as that of the battery:

$$V_1 = V_2 = V_3 = V$$

If we wish to replace the parallel combination with a single equivalent capacitor C_p, our definition of capacitance tells us we must have

$$C_p = \frac{q}{V}$$

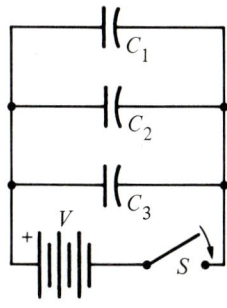

Figure 20-20 Three capacitors connected in parallel with a battery. The equivalent capacitance is the sum of the individual capacitances.

We also know that

$$C_1 = \frac{q_1}{V} \qquad C_2 = \frac{q_2}{V} \qquad C_3 = \frac{q_3}{V}$$

Since

$$q = q_1 + q_2 + q_3$$

we may solve the above for the charge in each case and substitute:

$$C_p V = C_1 V + C_2 V + C_3 V$$

Finally, dividing by V gives us

$$C_p = C_1 + C_2 + C_3 \qquad (20\text{-}18)$$

Thus, the equivalent capacitance of capacitors connected in parallel is simply the sum of their individual capacitances.

Figure 20-21 shows three capacitors connected in series with a battery of voltage V. Once again, when the switch is closed, conventionally a charge $+q$ flows to the negative terminal. This time, however, the charge from the battery flows only onto the left plate of C_1, and the returning charge comes from the right plate of C_3. Charge must then be distributed among the capacitors. A charge $+q$ must go from the right plate of C_2 to the left plate of C_3. Further, a charge of $+q$ must go from C_1 to C_2. In series, each capacitor has the same charge q. Thus, from the definition of capacitance, we have

$$C_1 = \frac{q}{V_1} \qquad C_2 = \frac{q}{V_2} \qquad C_3 = \frac{q}{V_3}$$

and

$$C_s = \frac{q}{V}$$

where C_s is the equivalent capacitance of the series combination. According to Kirchhoff's loop rule, the sum of the voltage across the capacitors must equal the battery voltage:

$$V = V_1 + V_2 + V_3$$

Solving for the voltage in the definition of capacitance and substituting yields

Figure 20-21 Three capacitors connected in series with a battery. The equivalent capacitance is smaller than the capacitance of the smallest capacitor.

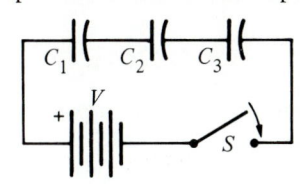

$$\frac{q}{C_s} = \frac{q}{C_1} + \frac{q}{C_2} + \frac{q}{C_3}$$

Finally, after dividing by q, we have

$$\frac{1}{C_s} = \frac{1}{C_1} + \frac{1}{C_2} + \frac{1}{C_3} \qquad (20\text{-}19)$$

Thus the reciprocal of the equivalent capacitance of several capacitors connected in series is equal to the sum of the reciprocals of the individual capacitances.

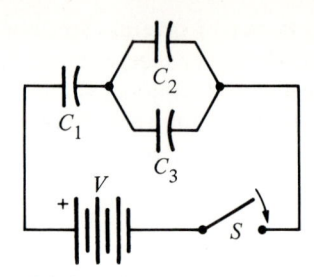

Figure 20-22 Example 20-7.

Example 20-7 For the circuit in Figure 20-22, determine (a) the equivalent capacitance, (b) the charge on C_1, (c) the voltage across the parallel combination, and (d) the charge on C_2. Take $C_1 = 6.00 \times 10^{-6}$ F, $C_2 = 1.00 \times 10^{-6}$ F, $C_3 = 2.00 \times 10^{-6}$ F, and $V = 40.0$ V.

Solution (a) We first note that C_2 and C_3 are in parallel, and that parallel combination is in series with C_1. Thus we add C_2 and C_3 to obtain their parallel equivalent C_p and apply Equation 20-19 to that value along with C_1 to find the circuit capacitance.

$$C_p = C_2 + C_3 = 1.00 \times 10^{-6} \text{ F} + 2.00 \times 10^{-6} \text{ F} = 3.00 \times 10^{-6} \text{ F}$$

and

$$\frac{1}{C_s} = \frac{1}{6.00 \times 10^{-6} \text{ F}} + \frac{1}{3.00 \times 10^{-6} \text{ F}}$$

$$C_s = 2.00 \times 10^{-6} \text{ F}$$

(b) The charge on C_1 is simply the charge delivered by the battery, which is found from the net voltage and the equivalent capacitance.

$$q_1 = q = C_s V = (2.00 \times 10^{-6} \text{ F})(40.0 \text{ V}) = 8.00 \times 10^{-5} \text{ C}$$

(c) We can calculate the voltage across C_1 and subtract it from V to find the voltage across the parallel combination.

$$V_1 = \frac{q_1}{C_1} = \frac{8.00 \times 10^{-5} \text{ C}}{6.00 \times 10^{-6} \text{ F}} = 13.3 \text{ V}$$

$$V_p = V - V_1 = 40.0 \text{ V} - 13.3 \text{ V} = 26.7 \text{ V}$$

(d) Since the answer to part c is also the voltage across C_2, we apply the definition of capacitance again to determine the charge on C_2.

$$q_2 = C_2 V_2 = C_2 V_p = (1.00 \times 10^{-6} \text{ F})(26.7 \text{ V}) = 2.67 \times 10^{-5} \text{ C}$$

20-7 Resistors and Capacitors in DC Circuits

Capacitors often play a part in dc circuits, as noted in several applications mentioned in Chapter 18. For instance, the charging portion of the flash mechanism of Example 18-8 consists of a capacitor, a resistor, and an emf all in series.

After the mechanism has flashed, the circuit is connected as shown in Figure 20-23, and the battery generates current to charge the capacitor. The current decreases in magnitude and finally goes to zero when the capacitor is fully charged. Beginning at the lower right and applying Kirchhoff's loop rule clockwise to the circuit yields

$$V - V_R - V_C = 0$$

where V_R is the voltage across the resistor, equal to IR, and V_C is the voltage across the capacitor, equal to q/C. Substituting these expressions and rearranging the equation gives us

$$IR + \frac{q}{C} = V \qquad (20\text{-}20)$$

Both I and q are changing with time but are related by the definition of current:

$$I = \frac{\Delta q}{\Delta t}$$

This may be substituted into Equation 20-20 and the equation further rearranged as

$$I = \frac{\Delta q}{\Delta t} = -\frac{q}{RC} + \frac{V}{R} \qquad (20\text{-}21)$$

Equation 20-21 involves both the time rate of change of charge ($\Delta q/\Delta t$) and the instantaneous value of the charge (q) in the same equation. As such, it may not be solved by the methods of algebra, but application of calculus yields the expression for the charge on the capacitor at any instant after the voltage is applied as

$$q = CV(1 - e^{-t/RC}) \qquad (20\text{-}22)$$

The quantity t is the time after the charging process began and CV is the maximum value of charge on the capacitor. The quantity RC is called the **time constant** of the circuit and is equal to the time required for the capacitor to reach to within e^{-1} of its maximum charge. Figure 20-24 shows q vs. time for a charging capacitor. As with many other exponential curves, it changes very rapidly at first and then slows as it approaches asymptotically its final value. Because the time constant is equal to the product of the resistance and capacitance, the charging process takes longer when these quantities are larger.

The current in a charging capacitor circuit may be found by substituting the expression for q of Equation 20-22 into Equation 20-21:

$$I = \frac{V}{R} e^{-t/RC} \qquad (20\text{-}23)$$

and the current follows a decreasing exponential curve with the same time constant, RC (Figure 20-25). Once again we see that the current decreases rapidly at first and then slows its rate of change.

In many applications, after the capacitor is charged, a change in the circuit causes it to discharge through a resistor. The camera flash mechanism, for instance, is switched to discharge through a flashtube (resistor). A simplified version of the discharging circuit, which bypasses and ignores the emf, is shown in Figure 20-26. When the photographer pushes the button to cause the flash,

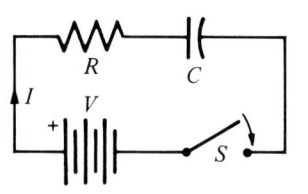

Figure 20-23 Circuit with a resistor in series with a capacitor and a source of voltage, sometimes called an *RC* circuit.

Figure 20-24 The charge stored on a capacitor vs. the time from connecting the capacitor to a voltage source.

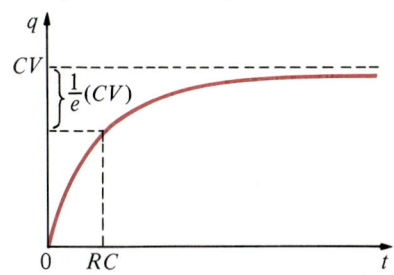

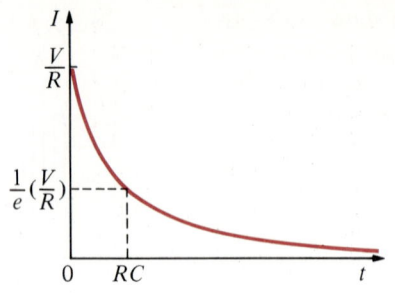

Figure 20-25 The charging current vs. time for the circuit of Figure 20-23. At $t = 0$, the current has a maximum value of V/R which drops as charge is stored on the capacitor. In time RC, the current has dropped to $1/e$ of its maximum.

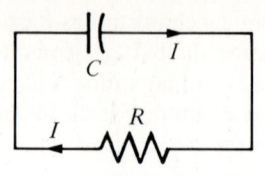

Figure 20-26 With the voltage source electrically removed from the circuit, the capacitor discharges through the resistor.

the charged capacitor begins to discharge, creating the current I in the circuit. Applying Kirchhoff's rule clockwise, beginning at the lower right corner, yields

$$-V_R - V_C = 0$$

This becomes

$$IR + \frac{q}{C} = 0$$

and finally

$$\frac{\Delta q}{\Delta t} = -\frac{q}{RC} \qquad (20\text{-}24)$$

Equation 20-24 has the solution for charge remaining on the capacitor of

$$q = q_0 e^{-t/RC} \qquad (20\text{-}25)$$

where q_0 is the original charge on the capacitor. Once again, we see that RC is the time constant. Now, however, we have a decreasing exponential curve, as shown in Figure 20-27. The curve is of the same form as Figure 20-25 of the charging current vs. time. In practical applications, the values of R and C are selected to adjust the time constant. For instance, the discharging time constant for the flash mechanism should represent a very short time interval.

We may find the discharging current by substituting the expression for q of Equation 20-25 into Equation 20-24. This yields

$$I = -\frac{q_0}{RC} e^{-t/RC} \qquad (20\text{-}26)$$

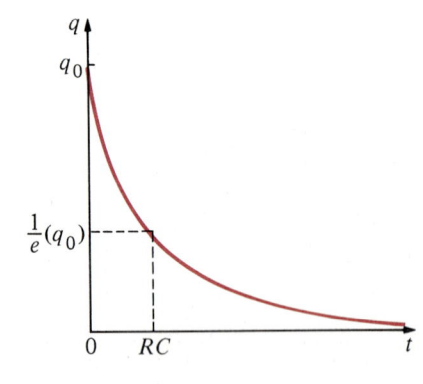

Figure 20-27 Charge remaining on the capacitor vs. time since beginning of discharge. The original charge on the capacitor was q_0, and the charge drops to $1/e$ of this value in a time equal to RC, the time constant of the circuit.

490

The minus sign simply means that the current is really in the other direction, opposite to that of the charging capacitor circuit. The magnitude of the current vs. time is shown in Figure 20-28.

Example 20-8 An artificial heart pacemaker takes advantage of the charging and discharging of a capacitor in RC circuits. One model uses a charging resistor of $2.00 \times 10^6 \, \Omega$. In order that the pacemaker cause the heart to beat 72 times each minute, the maximum time constant of the charging portion of the curve should be 0.800 s. Assume the capacitor must charge for only 0.800 s, and calculate (a) the value of the capacitor, (b) the maximum time available for discharging, and (c) the maximum value of the discharging resistance if the capacitor is to discharge to $1/e$ of its maximum value.

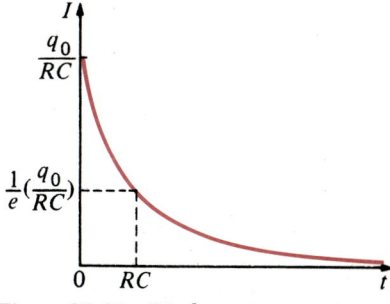

Figure 20-28 Discharge current vs. time since beginning of discharge from a maximum stored charge of q_0. The initial, maximum current is q_0/RC and drops to 36.8 percent (or $1/e$) of this in time RC.

Solution (a) Since we know that the time constant of an R_cC circuit is equal to the product of resistance and capacitance, we find the solution by simply setting R_cC equal to 0.800 s and solving for C.

$$R_c C = 0.800 \text{ s}$$

$$C = \frac{0.800 \text{ s}}{2.00 \times 10^6 \, \Omega} = 4.00 \times 10^{-7} \text{ F} = 0.400 \, \mu\text{F}$$

(b) The solution to this part is found by taking the difference between the time of one beat and the charging time constant.

$$\text{Time per beat} = \frac{60 \text{ s}}{\text{min}} \times \frac{1 \text{ min}}{72 \text{ beats}} = 0.833 \text{ s/beat}$$

$$\Delta t = 0.833 \text{ s} - 0.800 \text{ s} = 0.033 \text{ s}$$

(c) Using the answers to parts a and b, we may solve for the answer to this part. Sine $\Delta t = R_D C$,

$$R_D = \frac{\Delta t}{C} = \frac{0.033 \text{ s}}{4.00 \times 10^{-7} \text{ F}} = 8.25 \times 10^4 \, \Omega$$

Minimum Learning Objectives

After studying this chapter, you should be able to:
1. Define:
   - ac
   - alternating current
   - ammeter
   - conservation of charge
   - dc
   - direct current
   - equivalent capacitance
   - equivalent emf
   - equivalent resistance
   - exponential curve
   - junction
   - Kirchhoff's rules
   - load
   - loop rule
   - meter (device)
   - multimeter
   - ohmmeter
   - parallel circuit
   - point rule
   - series circuit
   - shunt
   - time constant
   - voltmeter
2. Calculate the equivalent emf of batteries or cells connected in series or parallel.
3. Calculate the equivalent resistance of resistors connected in series and/or parallel.
4. Solve for unknown currents, resistances, etc., in a circuit consisting of an emf and some combination of resistances by repeated application of Ohm's law to parts of the circuit.
5. Correctly apply Kirchhoff's rules to a complex circuit and solve the resulting equations for resistances, currents, and/or emf's.
6. Identify the correct usage of meters in electric circuits.
7. Calculate the resistances needed to use a meter movement with specified electrical characteristics to measure a given maximum voltage or current.

8. Identify the proper connection of a resistance with a meter movement to produce a voltmeter and/or ammeter.
9. Calculate the equivalent capacitance of capacitors connected in series or in parallel.
10. Calculate the time required for a capacitor in a circuit with a resistor to achieve any given amount of its maximum stored charge during charging with a given emf or during discharging.

Problems

20-1 The positive terminals of a 12.0-V and an 8.00-V battery are connected together. Their negative terminals are then connected to the opposite ends of a 100-Ω resistor. (*a*) Determine the value of the current in the circuit. (*b*) The 8.00-V battery is disconnected and then reconnected, with its negative terminal connected to the positive terminal of the 12.0-V battery and its positive terminal connected to the free end of the resistor. What is the current now? (*c*) Calculate the power being delivered by the 12.0-V battery in both configurations. (*d*) Calculate the power being dissipated in the 100-Ω resistor in both configurations.

20-2 A room air conditioner rated at 1.08×10^3 W operates on a 120-V house circuit. During the three summer months, it operates for a total of 750 h. Calculate (*a*) the current in the air conditioner, (*b*) the equivalent resistance of the air conditioner, and (*c*) the cost of using this during the summer when electric energy sells for $0.08/kWh.

20-3 For the circuit shown in Figure 20-29, calculate (*a*) the equivalent resistance, (*b*) the power delivered by the battery, and (*c*) the voltage across and the current in the 30.0-Ω resistor.

20-4 A 12.0-V battery is to be used to produce a current of 1.00 A. You have available a bin of 4.00-Ω resistors to limit the current from the battery. However, each resistor is rated at 0.25 W. That is, if the power dissipated in the resistor exceeds 0.25 W, the resistor may be expected to burn out from the heating that results. Describe how you could connect an assemblage of the resistors to produce the required current without burning out any of the resistors.

20-5 A string of 15 light bulbs in a set of Christmas tree lights is connected in series to a 120-V emf. Each bulb dissipates

Figure 20-29 Problem 20-3.

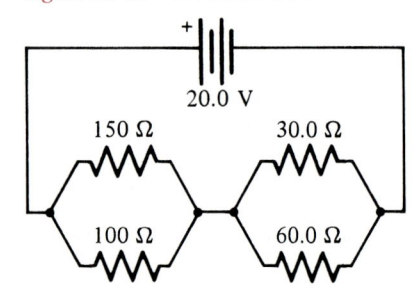

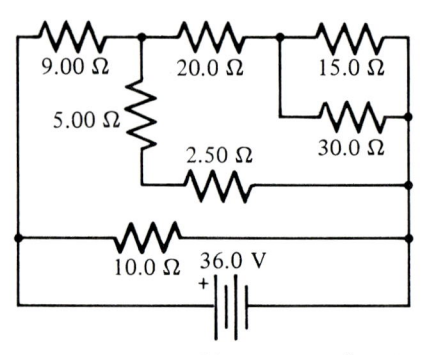

Figure 20-30 Problems 20-6 and 20-8.

4.00 W of power. (*a*) What is the current through the circuit? (*b*) What is the equivalent resistance of the string? (*c*) What is the resistance of each bulb? (*d*) What is the voltage drop across an individual bulb?

20-6 For the circuit shown in Figure 20-30, calculate (*a*) the equivalent resistance, (*b*) the current and power delivered by the battery, and (*c*) the voltage across the 9.00-Ω resistor.

20-7 Three identical bulbs are designed to operate independently with a 120-V emf and are rated at 75.0 W for that use. Instead, the three bulbs are connected in series and placed across the emf. (*a*) What is the equivalent resistance? (*b*) What will be the current through the circuit? (*c*) What power is dissipated in each bulb? (*d*) The bulbs are reconnected so that two bulbs in parallel are placed in series with the third. Draw this circuit arrangement. (*e*) What is the equivalent resistance of the new circuit? (*f*) What is the current through the series bulb? (*g*) What is the current through each of the parallel bulbs? (*h*) What is the power dissipated in the circuit? (*i*) What percentage of this power is dissipated in the series bulb?

20-8 Suppose the 2.50-Ω resistor in the circuit of Figure 20-30 can safely dissipate a maximum of 25.0 W. What is the maximum emf of the battery which can safely be used with the circuit?

20-9 In the circuit of Figure 20-8, assume that each resistor has a value of 12.0 Ω and each cell an emf of 2.00 V. For each of the seven numbered resistors, complete a table giving the current through that resistor, the voltage difference between the ends of that resistor, and the power dissipated in that resistor.

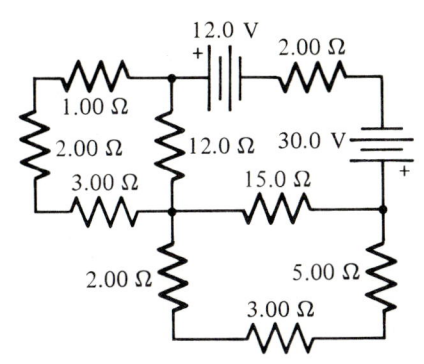

Figure 20-31 Problem 20-14.

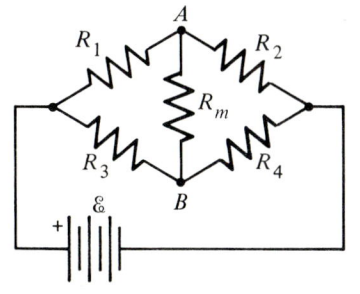

Figure 20-32 Problems 20-15 and 20-21.

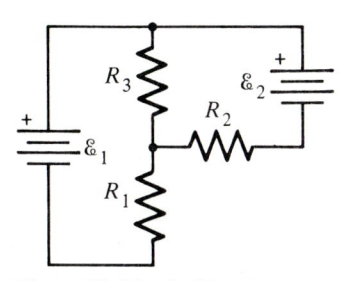

Figure 20-33 Problem 20-17.

Figure 20-34 (a) Problem 20-18, (b) Problem 20-19, and (c) Problem 20-20.

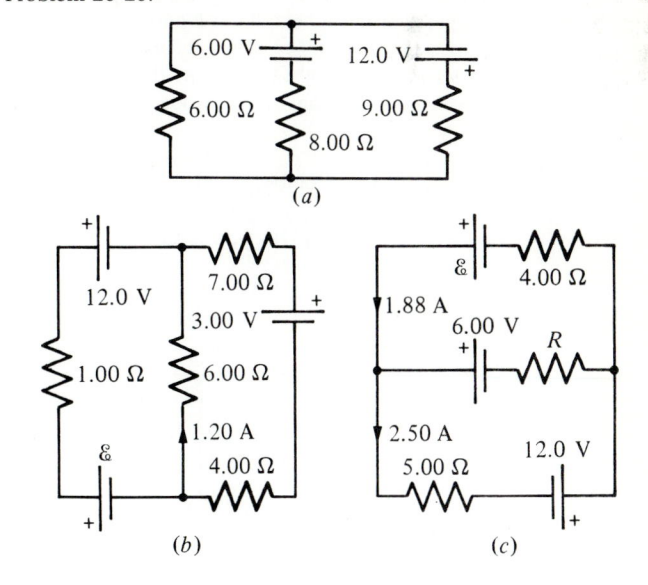

20-10 Show that when two resistors R_1 and R_2 are placed in parallel, the combination has an effective resistance closer and closer to R_1 as R_2 gets larger and larger with respect to R_1. This leads us to want to use a voltmeter with a very large internal resistance so that when it is placed in parallel with one of the resistors of the circuit, the effective resistance between the two points where it is connected will not change appreciably.

20-11 A car has eight 15.0-W parking lights and four 5.00-W instrument lights, all connected in parallel to the battery. The car is driven one dark morning, and the parking lights are left on. The 12-V battery is initially fully charged and rated at 200 A·h. (a) If the battery requires two-thirds of its full charge to ensure the engine starts, how many hours can the car be left with the lights on and still be started? (b) How would this change if the two 150-W headlights were also on?

20-12 Show that the effective resistance of any two resistors connected in parallel is smaller than either of the two.

20-13 A 40.0- and a 30.0-Ω resistor are connected in series. What resistor must be connected (a) in parallel with the 30.0-Ω resistor so that the total combination will have an effective resistance of 60.0 Ω, (b) in parallel with the 40.0-Ω resistor so that the total combination has an effective resistance of 60.0 Ω, and (c) in parallel with both together so that the total combination is 60.0 Ω?

20-14 For the circuit shown in Figure 20-31, calculate (a) the effective resistance, (b) the net current being delivered by the batteries, (c) the current in the 12.0-Ω resistor, and (d) the power dissipated in the 12.0-Ω resistor.

20-15 The circuit diagram in Figure 20-32 is that of a Wheatstone bridge. It is commonly used to measure unknown resistances or to detect small changes in resistances of sensitive circuit elements. The element designated R_m represents the internal resistance of a meter which is used to detect and measure the presence of current in that branch. When one or some combination of the four numbered resistors is adjusted so that the voltages at points A and B are the same, the bridge is said to be balanced, and there is no current in the branch containing the meter. By applying only Ohm's law, show that when the bridge is balanced, the resistances are related by $R_1 = R_2(R_3/R_4)$.

20-16 In the circuit of Figure 20-9, the three resistances are 40.0, 10.0 and 20.0 Ω and the emf's are each 9.00 V. Solve for the current through each resistor.

20-17 (a) In terms of the emf's and resistances, use Kirchhoff's rules for electric circuits to derive an expression for the current in R_2 of Figure 20-33. (b) Show that this will be zero whenever $\mathscr{E}_2/\mathscr{E}_1 = R_3/(R_1 + R_3)$. This is the basis of the device known as the potentiometer, which is used to measure emf's without drawing current from a battery.

20-18 Apply Kirchhoff's rules to the circuit of Figure 20-34a to determine the currents in all the branches.

20-19 Apply Kirchhoff's rules to the circuit of Figure 20-34b to determine the unknown currents and emf.

20-20 Apply Kirchhoff's rules to the circuit of Figure 20-34c to determine the unknown current, resistance, and emf.

20-21 Apply Kirchhoff's rules to the circuit of Figure 20-32 to obtain an equation for the current in the meter (when the bridge is unbalanced) in terms of the emf and the resistances. Then show that the condition for balance of Problem 20-15 arises naturally whenever the meter current is zero.

20-22 A simple series circuit contains a battery of emf 1.50 V and two resistors $R_1 = 4.70 \times 10^4$ Ω and $R_2 = 6.80 \times 10^4$ Ω. (a) Calculate the voltage across each resistor. (b) A voltmeter of internal resistance 2.00×10^4 Ω is used to read voltages by being placed in parallel first with R_1, then with R_2, and finally with the series combination of R_1 and R_2. Calculate the voltage displayed on the voltmeter in each of the three measurements.

20-23 A 3.00-V battery is used to power a circuit containing two small resistors in parallel. If $R_1 = 0.300$ Ω and $R_2 = 0.150$ Ω, (a) calculate the current delivered by the battery and the current in each resistor. (b) An ammeter of internal resistance 0.200 Ω is used to measure the current in each branch of the circuit by placing it in series first with R_1, then with R_2, and finally with the battery. Calculate the current reading which should be displayed on the ammeter in each of the three measurements.

20-24 A meter movement with an internal resistance of 800 Ω gives a full-scale deflection for a current of 100 μA. Draw the circuit, and give the value of resistance necessary to make an ammeter which reads 5.00 A full-scale.

20-25 A meter movement which reads full-scale for a current of 0.200 mA has an internal resistance of 18.0 Ω. Calculate (a) the series resistor necessary to make this movement a voltmeter which reads 25.0 V full-scale and (b) the parallel (shunt) resistance necessary to make this movement an ammeter which reads 8.00 A full-scale.

20-26 The voltmeter of the schematic in Figure 20-35 has an internal resistance of 100 Ω, while the ammeter has an internal resistance of 2.00 Ω. Calculate the error in both the current and the voltage readings due to the presence of the meters. This can be done by determining the voltage across the resistor and the current before the meters are inserted and then subtracting the calculated values of those quantities after putting the meters in the circuit.

20-27 A 45.0-Ω meter movement has been converted to an ammeter which reads 20.0 A full-scale with the addition of a 0.0750-Ω parallel (shunt) resistor. Calculate (a) the maximum current the meter movement can accept and (b) the resistance necessary to convert the meter movement to a voltmeter which reads 100 V full-scale.

20-28 A meter movement has a resistance of 50.0 Ω and deflects full-scale with a current of 300 μA. Design a series-parallel circuit including the meter such that a current of 0.150 A gives full-scale deflection and the voltage across the circuit is 12.0 V.

20-29 In the separate circuits of Figures 20-20 and 20-21, assume that the three capacitors have values of 10.0, 6.00, and 2.00 μF, respectively. In each case the voltage of the source is 9.00 V. Calculate the charge stored by each capacitor and the voltage drop across each capacitor for (a) the circuit of Figure 20-20 and (b) the circuit of Figure 20-21.

20-30 Two capacitors $C_1 = 5.00$ μF and $C_2 = 8.00$ μF are placed in parallel with each other and then charged by a 12.0-V battery. Calculate (a) the equivalent capacitance, (b) the net charge delivered by the battery, (c) the charge on each capacitor, and (d) the energy stored in each capacitor.

20-31 For the circuit shown in Figure 20-36a, calculate (a) the equivalent capacitance, (b) the charge on the 4.00-μF capacitor, (c) the charge on the 6.00-μF capacitor, and (d) the energy stored in the 6.00-μF capacitor.

20-32 For the circuit shown in Figure 20-36b calculate (a) the equivalent capacitance, (b) the charge on the 8.00-μF capacitor, (c) the charge on the 15.0-μF capacitor, (d) the charge on the 4.00-μF capacitor, and (e) the energy stored in the 12.0 μF-capacitor.

20-33 (a) What capacitor can you add and where should it be

Figure 20-36 (a) Problem 20-31, (b) Problem 20-32, and (c) Problem 20-33.

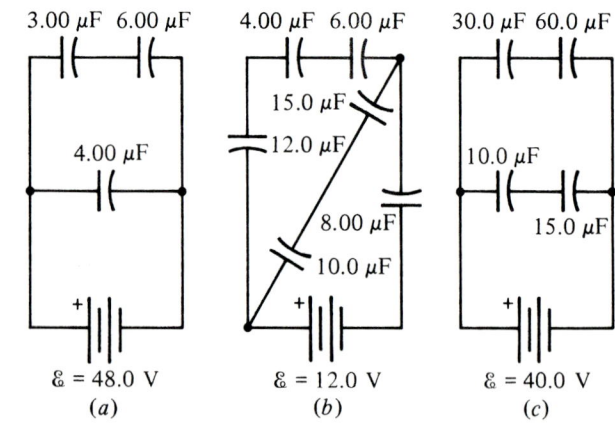

Figure 20-35 Problem 20-26.

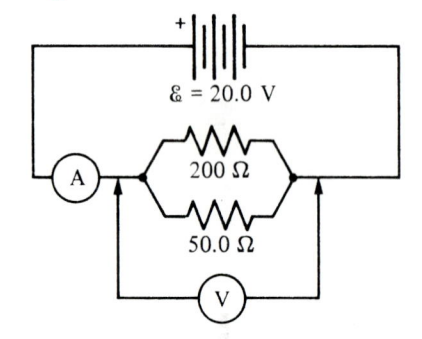

inserted in the circuit of Figure 20-36c to make the equivalent capacitance of the circuit 12.0 μF? (There are two possibilities.) (b) Calculate the charge and energy stored in the added capacitor.

20-34 A 2.00-μF capacitor is charged by a 300-V emf, and a 6.00-μF capacitor is charged by a 150-V emf. The capacitors are disconnected from their sources, and each plate of each capacitor is connected by a conducting wire to a plate of opposite charge on the other capacitor. Calculate (a) the net charge on the two plates which are connected together, (b) the magnitude of the charge on each capacitor, and (c) the voltage across each capacitor.

20-35 A 10.0-μF capacitor is charged to 2.50×10^{-3} C by an unknown voltage source, and a 20.0-μF capacitor is fully charged by an emf of 400 V. The capacitors are disconnected from their sources, and each plate of each capacitor is connected by a conducting wire to a plate of the same charge on the other capacitor. After they are connected to each other, calculate (a) the magnitude of the charge on each capacitor, (b) the voltage across each capacitor, and (c) the energy stored in each capacitor.

20-36 An 80.0-μF capacitor is placed in series with a 45.0-V battery and a resistor. It takes 6.00 s for the capacitor to charge from 0.0 to 22.5 V. Calculate (a) the time constant of the circuit and (b) the value of the resistor.

20-37 It is common practice to use a very large resistor (called a bleeder resistor) in parallel with large capacitors to protect people who work with electric circuits. The idea is that if the capacitor is charged when the circuit is turned off, it will discharge through the resistor rather than through the person. Consider a color television receiver with a 4.00-μF capacitor which is charged by a 2.50×10^4 V potential difference. If its bleeder resistor is 4.00×10^6 Ω, calculate (a) its time constant and (b) the time required to discharge by 99 percent of its original value.

20-38 A fuel-level meter on an automobile contains a 1.00-MΩ resistor. If it takes 4.00 s after the engine is started for the gauge to reveal that the car has a full tank, what is the capacitance in the circuit? (Assume that the gauge reads full when the capacitor is 99 percent charged.)

20-39 Many flashing warning lights used near construction projects are nothing more than a series circuit of a battery, resistor, and capacitor with a neon lamp in parallel with the capacitor. The capacitor is charged by the battery through the resistor. When the voltage across the capacitor exceeds the breakdown voltage of the neon lamp, the lamp becomes a conductor and the capacitor discharges through the lamp. Suppose such a circuit is to be constructed by using a 2.00×10^6 Ω resistor, a 90.0-V battery, and a neon lamp which breaks down when its voltage exceeds 85.5 V. Calculate the capacitance of the capacitor that must be selected in order that the light flash at regular intervals of 1.60 s. (This should be three times the RC charging time constant.)

20-40 A heart defibrillation unit consists of a 20.0-μF capacitor charged by a 6000-V emf. When the unit is used, good electric contact is made with the patient's skin and the electric resistance of the patient is about 1.00×10^3 Ω. For this circuit, calculate (a) the time constant and (b) the time required for the capacitor to discharge 95 percent of its maximum charge through the patient.

20-41 In a brochure, a photographic strobe light is described as delivering a $\frac{1}{2000}$-s burst of light rated at 200 W·s. In testing, a battery pack made up of two 1.50-V AA cells in series recharges the light in 2.00 s when fresh. This time remains quite constant for about 100 flashes then rapidly rises to a charge time of 12.0 s at the 150th flash. (a) What is a more conventional unit to describe the output than the watt-second? (b) What is the charge stored by the light when at full charge? (c) What is the average current through the lamp during a burst of light? (d) Assuming that 95.0 percent of the maximum charge is regarded as fully charged, what is the time constant of the recharging circuit? (e) What is the capacitance of the capacitor in the circuit? (f) What is the resistance of the circuit? (g) To what value has the external voltage of each AA cell dropped by the 150th flash? (h) Assuming the internal resistance of each cell was originally negligible, to what value has it grown by the 150th flash? (i) Assuming that the drop-off after the 100th flash marks the point where 60 percent of the battery energy has been depleted, what is the total energy delivered by the AA cell during its working life? (j) If the two cells cost $0.48, what is the cost per kilowatthour? Compare this with residential electricity at $0.08/kWh.

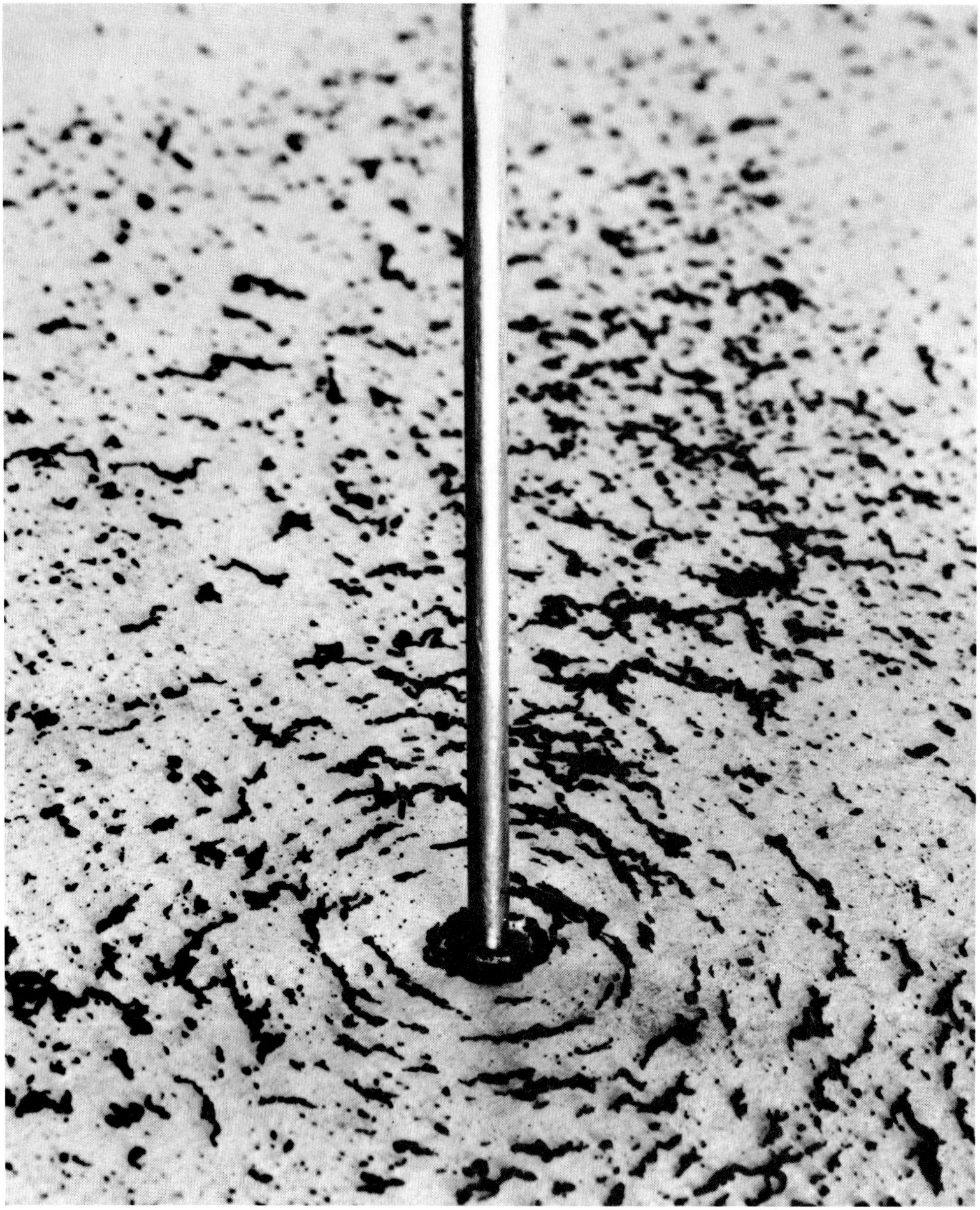

21 Magnetic Phenomena

21-1 Introduction

The common magnet and certain magnetic phenomena are familiar to nearly everyone. Children play with small horseshoe magnets that attract pins and paper clips. A sufficiently strong magnet can stick to an iron or steel cabinet, supporting its own weight and more. Most compasses have a magnetic needle which points toward the northern part of the earth, indicating that the earth, too, has magnetic properties. Magnets are used in galvanometers, electric generators, and audio speakers. In short, because of their many practical applications, magnets can be found virtually everywhere.

The phenomenon of magnetism has been known for over 2500 years. Natural magnets called **lodestones,** a type of iron ore, were found near the ancient city of Magnesia in Asia Minor. These pieces of rock attract each other as well as pieces of soft iron. Often this attraction is observable at a fairly large distance of separation. We say that the magnet alters the space in its vicinity so that another magnet or piece of soft iron placed there will experience a force. Classically, then, we have all the makings of the concept of a field to which we may assign magnetic lines of force.

One can easily visualize these lines of force by observing the pattern acquired by iron filings sprinkled in an area of strong magnetic field, as shown in Figure 21-1. Note that the field appears to be most concentrated near two certain areas of the magnet. These areas are known as **poles.** When suspended from its center and allowed to turn freely, a bar magnet always lines up such that one end is toward the north and the other end toward the south. These ends are then called, respectively, the **north-seeking pole** and the **south-seeking pole,** or simply the north and south poles of the magnet. Simple experiments with two bar magnets show that *like poles repel* each other and *unlike poles attract* each other. That is, a north pole repels another north pole but attracts and is attracted by a south pole.

The lines along which the iron filings align in Figure 21-1 are called the **magnetic lines of force,** or **magnetic field lines.** These are perhaps more readily seen in Figure 21-2. If a small compass were placed in the vicinity of the bar magnet of Figure 21-2, its needle or arrow would align along the magnetic lines of force, pointing from the north pole to the south pole. This is in the *direction* of the magnetic field, which is taken as being from the north pole to the south pole outside the magnet. The field is stronger as one moves nearer to one of the poles, as indicated by the increasing density of field lines.

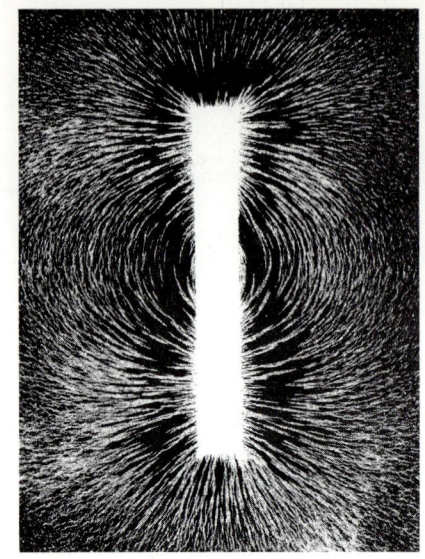

Figure 21-1 Patterns of iron filings around a permanent magnet reveal the presence of forces in the area that we call the magnetic field.

21-2 Magnetic Force on Moving Charges

The magnitude of the force between magnetic poles is proportional to the product of the strength of the poles and inversely proportional to the square of the distance between them. The magnetic force may be defined with an equation similar to Coulomb's law for electrostatics. Indeed, Coulomb himself proposed such an equation. However, this approach is not productive as it has been impossible to produce isolated magnetic poles. Certainly, with macroscopic phenomena, poles always come in pairs: one north pole and one south pole.

For this reason and several others, we use an alternative definition for magnetic effects. A charged particle moving in a region where there is a magnetic field experiences a force due to that field. Figure 21-3a shows a demonstration tube which displays a beam of electrons traveling along the positive x direction. In Figure 21-3b, a magnet has been placed such that its north pole and hence the direction of the magnetic field are aimed in the negative z direction. Note that the beam is deflected downward in the negative y direction. In Figure 21-3c, we see the magnet reversed so that the field is along the positive z direction. Now the beam is deflected upward in the positive y direction. Apparently the force exerted on the moving charges is directed perpendicular to the velocity of the charges and also perpendicular to the magnetic field.

By moving the magnet closer and farther away and observing the change in deflection of the beam, we determine that the magnitude of the force is proportional to the strength of the field in the vicinity of the moving charges. The magnitude of the force is also proportional to the magnitude of the velocity of the moving charges. This is shown by varying their speed while keeping the magnet in a fixed position. Further, the force varies from a maximum value when the beam is at an angle of 90° to the field to zero force when the angle is 0° (or 180°). Finally, by using beams of charges which differ in magnitude as well as sign, it is found that the magnitude of the force is proportional to the magnitude of the charge and is oppositely directed for charges of opposite signs.

Figure 21-2 Lines of force representing the magnetic field around a bar magnet. The lines describe the magnitude and direction of the magnetic force that would be exerted on any idealized, north-seeking pole placed nearby.

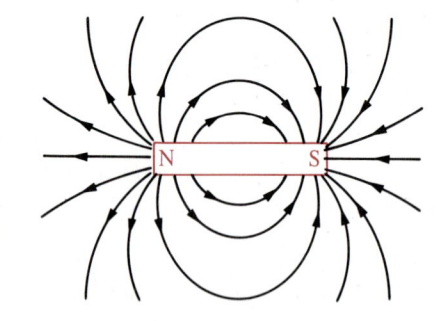

498

These conditions are all fulfilled for the proportionality

$$F \propto qvB \sin \theta$$

where F is the magnitude of the force, q is the magnitude of the charge, v is its velocity, B represents the magnetic field and θ is the smaller angle between **v** and **B**. We may make the constant of proportionality be exactly equal to one and use the equation to define the units of B. Thus

$$F = qvB \sin \theta \qquad (21\text{-}1)$$

and

$$B = \frac{F}{qv \sin \theta} \qquad (21\text{-}2)$$

The quantity B is called the **magnetic induction** (or sometimes, magnetic flux density).[1] The SI units of B from Equation 21-2 are newtons per coulomb times meters per second [N·s/(C·m)] or newtons per ampere times meters [N/(A·m)]. This is called the **tesla** (T) to honor Nikola Tesla (1856–1943), who did much of the definitive work in alternating current and transmission of electric power. Thus

$$1 \text{ T} = 1 \text{ N/(A·m)}$$

We see by Equation 21-1 that if $\theta = 0$ or $180°$, there will be no force exerted on the moving charged particle by the magnetic field. Thus, measurements on moving charged particles in a magnetic field can be used to completely specify **B** in both magnitude and direction. The magnitude is determined from Equation 21-2, while the direction of **B** is that direction in which the magnetic field exerts no force on the moving charges. Figure 21-4 shows the relationship of the vectors representing **F**, **B**, and **v** for positive charges if **v** and **B** are perpendicular to each other. For a negative charge, the force would be opposite, along the negative y direction.

We can remember and determine the relationship among the three vectors by employing a convention known as the **right-hand screw rule** which was developed for this purpose. If the vector representing the velocity of the charge is thought to rotate through the smaller angle ($<180°$) in the plane of **v** and **B** while attached to an ordinary wood screw, the screw will advance in the direction of **F**, as shown in Figure 21-5a. Figure 21-5b shows another right-hand rule. Using your *right* hand, point your fingers in the **v** direction, and orient your hand so that as you bend your wrist, your fingers and the palm of your hand turn to line up along **B**. Your extended thumb points in the direction of **F** if the charge is positive and in the direction opposite to that of **F** if the charge is negative. These rules will require some practice on your part.

(a)

(b)

(c)

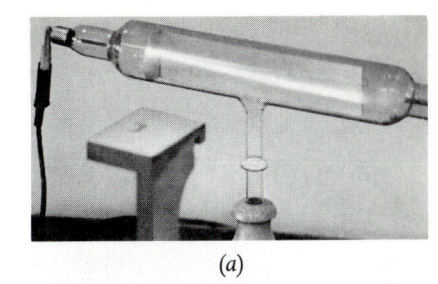

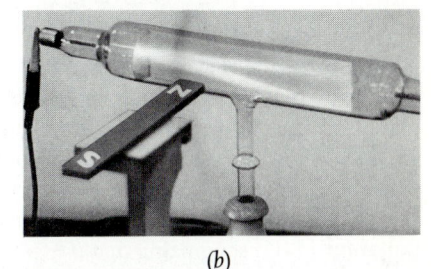

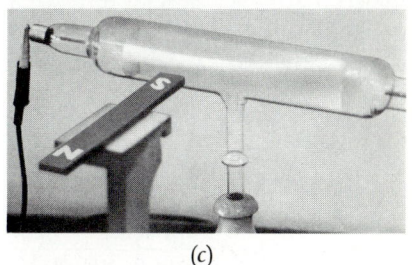

Figure 21-3 Moving charged particles experience a force in a magnetic field but are not attracted or repelled by the magnet. (a) An electron beam tube aligned so that the beam moves along the x axis. (b) A magnet placed parallel to the z axis with its north pole nearest the tube deflects the beam downward in the xy plane. (c) The magnet rotated so that its south pole is nearest the tube deflects the beam upward in the xy plane.

---

[1] You will recall from our study of electricity that **E** is the electric field intensity. Although there is a parallel between **B** and **E**, **B** is not called the magnetic field intensity because, at the time magnetism was first being studied, that name had already been given to the force per unit pole of Coulomb's law of poles, a concept no longer in use.

499

Example 21-1 A ball bearing of mass 5.00×10^{-4} kg is given a charge of 1.00×10^{-5} C and projected directly east at a speed of 2.00×10^3 m/s in a uniform magnetic field. (a) What direction of **B** would ensure that the force due to the magnetic field is up? (b) If **B** is perpendicular to **v**, what magnitude of **B** is necessary to balance the weight of the ball bearing?

Solution (a) We must apply one of the right-hand rules to obtain the direction. Since the force is to be straight upward and is always perpendicular to **B** (as well as to **v**), the **B** vector must lie in the horizontal plane. Therefore, if we use the right-hand screw rule, the velocity vector would have to be thought to rotate in the horizontal plane. For a right-hand threaded screw to advance vertically upward, **v** must be rotated from east toward the north. Thus, the only requirements on the **B** vector to ensure that the magnetic force on the charge be vertically upward is that **B** must lie in the horizontal plane and have a northward component.

(b) This requires application of the first condition of equilibrium along with Equation 21-1, recognizing that $\theta = 90°$ and hence $\sin \theta = 1$. From the answer to part a, we see that **B** for this part must be directed due north. Furthermore, the body is to be in equilibrium in the vertical direction. Thus the sum of the forces in the vertical direction must be zero. Choosing upward as positive, we have

$$qvB \sin \theta - mg = 0$$

and

$$B = \frac{mg}{qv \sin \theta} = \frac{(5.00 \times 10^{-4} \text{ kg})(9.80 \text{ m/s}^2)}{(1.00 \times 10^{-5} \text{ C})(2.00 \times 10^3 \text{ m/s})(1)}$$

$$= 0.245 \text{ T}$$

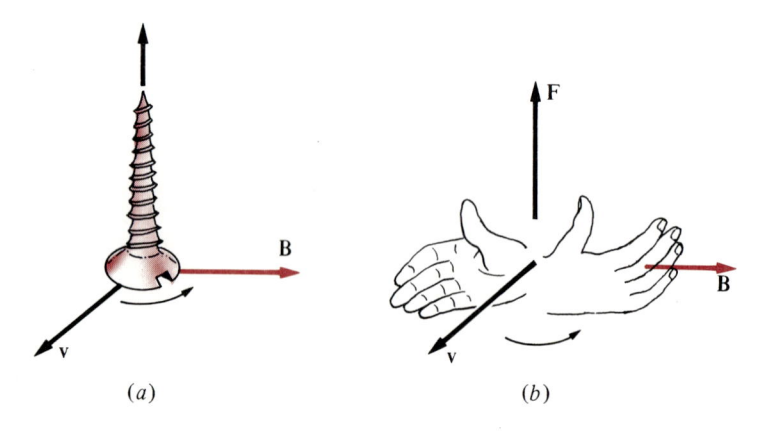

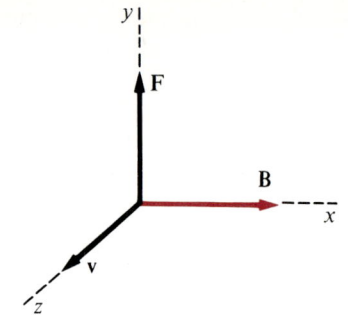

Figure 21-4 The magnetic force **F** on a charged particle is always perpendicular to the direction of the magnetic field *and* to the particle's velocity. Thus no part of **F** is ever parallel or opposed to the motion, and it can neither speed up nor slow down the particle.

The magnetic force on a moving charged particle is *always* perpendicular to the velocity of the particle and is thus centripetal in nature. Therefore the acceleration of the particle due to this force *never* changes the speed but only the direction in which the particle is moving. Figure 21-6 shows a particle of mass m, charge q, and a speed v moving in the plane of the page and entering a

Figure 21-5 The right-hand screw rule. (a) If we rotate **v** into **B** through the smallest angle possible, the direction of **F** on a positive particle will be the same as the motion of a right-handed screw rotated through this angle. (b) If we align the fingers of the right hand with **v** so that they curl to the direction of **B** by the smallest angle possible as we curl the fingers, the right thumb points in the direction of the force on a positive particle.

region of uniform magnetic field directed perpendicularly out of the plane of the page. The magnetic field lines are represented by small circles with dots in their centers, thought of as representing the points of arrows directed outward. To represent a magnetic field directed into the page, we would use an x in the small circles ⊗ as the tail feathers of the arrow directed away from the reader.

The magnitude of the initial magnetic force on the particle is found from Equation 21-1 and is $F = qvB$. Just as the particle enters the region of the field, we see by the right-hand rule that the magnetic force is directed toward the bottom of the figure. This acts to change the direction of the particle's motion. However, even as the particle changes direction, its velocity remains perpendicular to **B**. Thus the magnetic force is constant in magnitude, but its direction changes so it stays perpendicular to v. This centripetal force gives rise to a centripetal acceleration which, recalling from Chapter 5, is $a_c = v^2/r$. The quantity r is the radius of the circle and is often called the radius of curvature of the path for moving charged particles. The magnetic force is constant in magnitude and is directed toward the center of the circle. Thus, applying Newton's second law to this situation yields

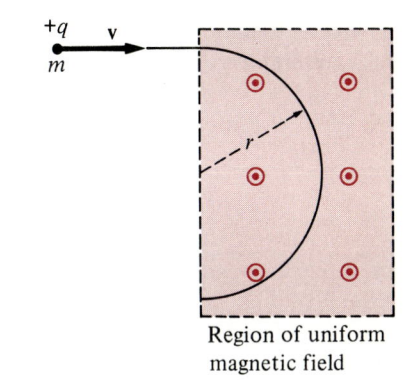

$$F = ma_c$$

or

$$qvB = \frac{mv^2}{r}$$

This may be solved for the radius to yield

$$r = \frac{mv}{qB} \tag{21-3}$$

If the particle's initial velocity is not perpendicular to **B** but has components both perpendicular and parallel to **B**, only the perpendicular component will be affected. In that case, the particle's parallel component remains constant, while the perpendicular component changes direction. The particle thus moves in a helical path with axis in the direction of the magnetic field.

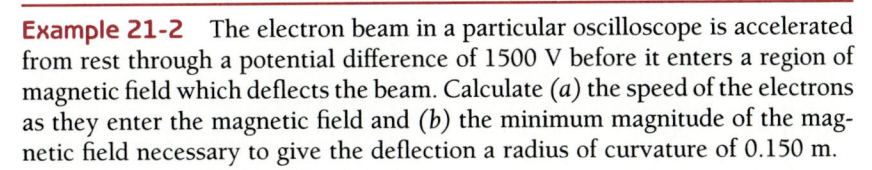

Figure 21-6 A particle of mass m and charge $+q$ moving with velocity v enters a region of uniform magnetic field perpendicular to v. The path of the particle in the field is a circle of radius r.

Example 21-2 The electron beam in a particular oscilloscope is accelerated from rest through a potential difference of 1500 V before it enters a region of magnetic field which deflects the beam. Calculate (a) the speed of the electrons as they enter the magnetic field and (b) the minimum magnitude of the magnetic field necessary to give the deflection a radius of curvature of 0.150 m.

Solution (a) We may find the speed by applying the conservation of mechanical energy. The change in kinetic energy is set equal to the negative of the product of the magnitude of the charge with the potential difference.

$$\Delta KE = -\Delta PE$$

$$\tfrac{1}{2}mv^2 - 0 = -(-e)(V)$$

$$v = \sqrt{\frac{2eV}{m}} = \sqrt{\frac{2(1.60 \times 10^{-19} \text{ C})(1500 \text{ J/C})}{9.11 \times 10^{-31} \text{ kg}}}$$

$$= 2.30 \times 10^7 \text{ m/s}$$

(b) Now we assume **B** is perpendicular to **v** in order to yield the maximum force for the minimum value of **B**. Hence we may apply Equation 21-3 to solve for B.

$$r = \frac{mv}{qB}$$

or

$$B = \frac{mv}{qr} = \frac{(9.11 \times 10^{-31} \text{ kg})(2.30 \times 10^7 \text{ m/s})}{(1.60 \times 10^{-19} \text{ C})(0.150 \text{ m})}$$
$$= 8.73 \times 10^{-4} \text{ T}$$

There are a number of other very useful applications of Equation 21-3. Whenever any four of the quantities in that equation are known or can be measured, we may solve for the fifth. A **mass spectrograph** (Figure 21-7), an instrument used to determine atomic masses, relies on these principles. Positive ions produced at point P in Figure 21-7 are incident on a metal plate in which there is a thin vertical slit s_1. Ions passing through s_1 are accelerated by a potential difference toward a second plate with a slit s_2. Ions passing through s_2 form a narrow beam and then enter the **velocity selector.** Here they pass between two metal plates with opposite charges which create an electric field E directed from the positive to negative plate. Perpendicular to the electric field is a magnetic field **B_1** directed downward into the figure. An ion of charge q moving through the space from left to right with a speed v has two forces exerted upon it: the electric force of magnitude qE directed from the positive to the negative plate and the magnetic force qvB_1 directed oppositely. For those ions of the proper velocity, these forces are equal in magnitude and balance each other, so the ions may move undeviated from their original path and pass through slit s_3. This condition is

$$qE = qvB_1$$

or

$$v = \frac{E}{B_1} = \frac{V}{dB_1} \qquad (21-4)$$

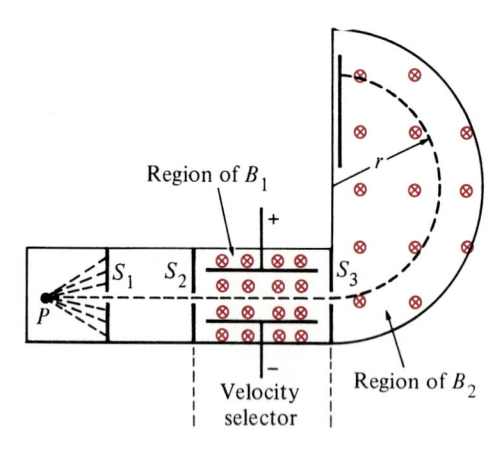

Figure 21-7 The mass spectrograph separates particles having tiny differences in mass. Positive ions from the ion source are sorted by the velocity selector, with only ions of a particular speed passing s_3. More-massive particles will move in a circle of larger radius and less-massive particles in a smaller arc.

where we have used the fact that the magnitudes of the electric field and the potential difference between the plates are related by $V = Ed$ (from Chapter 18), where d is the distance between the plates. Ions with greater speeds experience a larger magnetic force and are deviated toward the positive plate, while ions with smaller speeds are deviated toward the negative plate. Thus only ions with a specific velocity pass through s_3 into the next region of the apparatus.

This next region also has a magnetic field $\mathbf{B}_2$ directed downward into the page but no electric field. Here the ions trace out a circular path of radius r given by Equation 21-3 as

$$r = \frac{mv}{qB_2} \tag{21-5}$$

Substituting the expression for the velocity from Equation 21-4 into Equation 21-5 yields

$$r = \frac{mV}{qdB_1B_2} \tag{21-6}$$

After traveling one semicircle, the ions strike a photographic plate and expose it, just as would light. The distance from s_3 to the point of impact on the photographic plate is one diameter D of the circle, equal to twice the radius of curvature. Thus

$$D = 2r = \frac{2mV}{qdB_1B_2} \tag{21-7}$$

Equation 21-7 may be solved for the mass of the ions to yield

$$m = \frac{qdB_1B_2D}{2V} \tag{21-8}$$

Example 21-3 A mass spectrograph is to be used with positively charged carbon 12 ions of mass $m = 1.99 \times 10^{-26}$ kg and charge $q = +1.60 \times 10^{-19}$ C. The ions are to have a speed of $v = 4.50 \times 10^5$ m/s. The magnetic field in the velocity selector is $B_1 = 0.300$ T, and the plate separation is $d = 0.0200$ m. Determine (a) the voltage across the plates for which the ions will pass undeviated into slit s_3, and (b) the radius of curvature of these ions if $B_2 = 0.200$ T.

Solution (a) This is a straightforward application of Equation 21-4, solving for V.

$$v = \frac{V}{dB_1}$$

$$V = vdB_1 = (4.50 \times 10^5 \text{ m/s})(0.0200 \text{ m})(0.300 \text{ T}) = 2.70 \times 10^3 \text{ V}$$

(b) Here, we may use either Equation 21-5 or Equation 21-6 along with the answer to part a.

$$r = \frac{mv}{qB_2} = \frac{(1.99 \times 10^{-26} \text{ kg})(4.50 \times 10^5 \text{ m/s})}{(1.60 \times 10^{-19} \text{ C})(0.200 \text{ T})} = 0.280 \text{ m}$$

21-2 Magnetic Force on Moving Charges

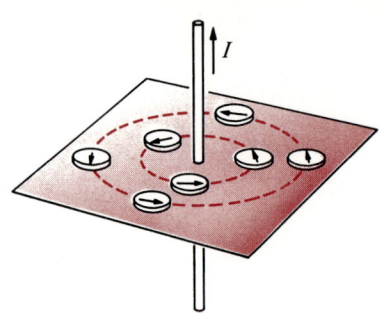

Figure 21-8 The alignment of magnetic compasses in the region surrounding a current-carrying wire reveals the presence of a magnetic field in the region.

21-3 Magnetic Fields Produced by Moving Charges

At this point, you might ask "What causes magnetic fields?" However, this is not the kind of question which prods the experimental physicist to action, because it is not immediately obvious what kind of experiment could answer it or even that an experimental answer exists. Consider instead a more subtle question: "Since a charge moving near a magnet is acted on by a force, should not we expect that a magnet placed in the vicinity of a moving charge will also experience a force due to that moving charge?" Now here is a question an experimental physicist enjoys! It can be answered by as simple an experiment as placing a compass in the vicinity of moving charge, for instance, a current-carrying conductor, and seeing if it reacts to the current. The result of such an experiment is that the compass needle *is* indeed affected: it aligns along a circle surrounding the wire (Figure 21-8). The magnetic field near the current is better revealed by sprinkling iron filings on a sheet of paper through which a current-carrying conductor passes, as shown in Figure 21-9.

The result is that not only have we answered the second question, but we have also started to answer the first. *Currents cause magnetic fields.* Further, we have seen that the direction of the magnetic field due to a current is around the current. We may specify uniquely the direction of the magnetic field due to a current by yet another right-hand rule, as shown in Figure 21-10. If you point the thumb of your right hand along the direction of conventional current in the wire and pretend to grasp the wire, your fingers curl around the wire in the same direction as the magnetic field.

The magnitude of B at a point near a current-carrying wire varies directly with the magnitude of the current and inversely with the perpendicular distance from the wire to the point in question, as experimentally determined by Jean Biot and Felix Savart in 1820. For the wire shown in Figure 21-11, the proportionality may be written as

$$B \propto \frac{I}{a}$$

where I is the magnitude of the current in the wire and B is determined at any point P on a circle of radius a surrounding the wire. This proportionality may be converted to an equation by inserting a proper constant. In anticipation of several other definitions and to simplify other equations, physicists chose to

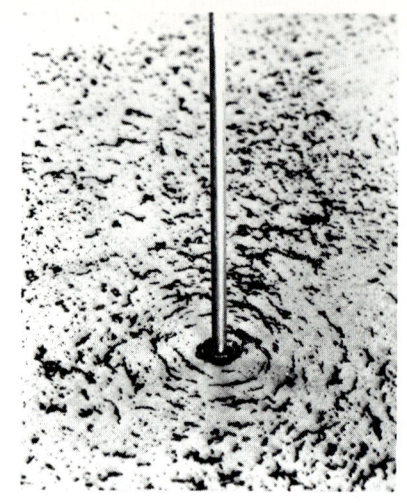

Figure 21-9 The pattern of iron filings around a current-carrying wire reveals the nature of the magnetic field. The field is *not* toward, away from, or parallel to the wire. Instead, magnetic field lines loop around the wire. *(Fundamental Photographs)*

Figure 21-10 If a wire is grasped by the right hand such that the thumb points in the direction of the conventional (positive) current, the fingers curl about the wire in the direction of the magnetic field.

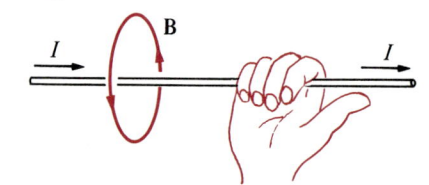

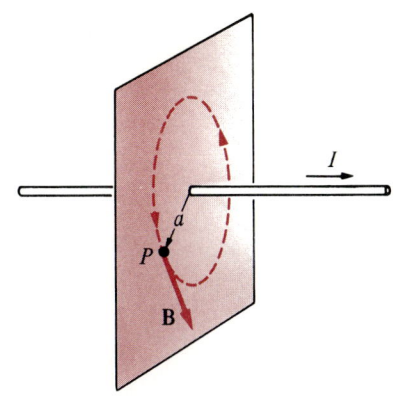

Figure 21-11 The magnetic field at P is of intensity |B| and has a direction perpendicular to the line joining the wire with P.

write the constant as $\mu_0/2\pi$. The equation for the magnetic field near a long, straight current is given by the Biot-Savart law as

$$B = \frac{\mu_0 I}{2\pi a} \quad (21\text{-}9)$$

The Greek lowercase mu with the subscript zero (μ_0) is known as the **magnetic permeability** of free space. In SI units, it has the value

$$\mu_0 = 4\pi \times 10^{-7} \text{ T} \cdot \text{m/A}$$

so that B is in teslas when I is in amperes and a is in meters.

Example 21-4 Determine the magnitude of the magnetic field 0.0200 m from a long, straight wire carrying a current of 25.0 A.

Solution This is a straightforward application of Equation 21-9.

$$B = \frac{\mu_0 I}{2\pi a} = \frac{(4\pi \times 10^{-7} \text{ T} \cdot \text{m/A})(25.0 \text{ A})}{2\pi(0.0200 \text{ m})} = 2.50 \times 10^{-4} \text{ T}$$

Current configurations other than straight wires are used to generate magnetic fields, as shown in Figure 21-12. Note that the lines of force for a circular loop of wire (Figure 21-12a) look very much like what we would expect from a very short bar magnet. For a circular loop of wire with a radius r carrying a current I, the magnetic field at the exact center is

$$B = \frac{\mu_0 I}{2r} \quad (21\text{-}10)$$

Figure 21-12b represents an arrangement called a **solenoid**, a very long wire wound in a tight helical geometry. It is equivalent to many circular loops stacked side by side and carrying a common current. The field due to the loops add inside the cylinder formed and can give a fairly large value of B. Whenever

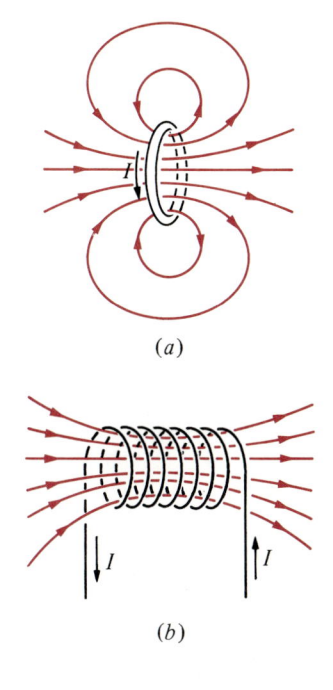

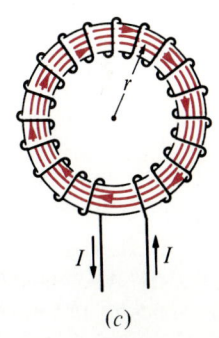

Figure 21-12 (a) The field produced by a loop of current-carrying wire is similar to that of a short bar magnet. (b) Coiling the wire into a solenoid produces a shape with a relatively uniform, concentrated field in its interior. (c) Extending the solenoid and bending it to form a ring produces the toroid and closes the interior magnetic field on itself.

the length L of the cylinder is large compared to the radius, the field inside the solenoid is

$$B = \frac{\mu_0 NI}{L} \quad (21\text{-}11)$$

where N is the total number of loops or turns. We sometimes use $n = N/L$ as the number of turns per unit length and write the equation for the solenoid as

$$B = \mu_0 nI \quad (21\text{-}12)$$

Near the ends, where the field decreases in magnitude as it begins to spread out in space, there are fringe effects. However, the field inside a long solenoid is very uniform over its cross section. This is quite useful in applications which require a uniform field over a large volume.

The arrangement in Figure 21-12c is known as a **toroid**. It may be thought of as a solenoid twisted into a circle or doughnut shape. As you can see, the field lines form closed circles inside the space. The equation for the field at a point P inside a toroid is

$$B = \frac{\mu_0 NI}{2\pi r} \quad (21\text{-}13)$$

where r is the radius of the field line at the point in question and N is the total number of turns. Although the field of the toroid varies inversely with the radius, the radius varies only from the inside to the outside of the doughnut. Whenever the mean radius of the doughnut is large compared to the cross section of the toroid, we often interpret $N/2\pi r$ as the number of turns per unit length. In that case, Equation 21-13 takes the same form as that for a solenoid. The toroidal winding is used to determine magnetic properties of materials. Perhaps the most important use at present, however, is in the Tokamak (Chapter 33), an apparatus designed to control nuclear fusion.

Example 21-5 A solenoid 0.600 m long has two layers of loops of 500 turns each and carries a current of 8.00 A. Calculate the magnitude of **B** near the center.

Solution We must first determine the total number of turns and then apply Equation 21-11. Since there are two layers of windings of 500 turns each, we have

$$N = (2)(500) = 1000 \text{ turns}$$

Therefore,

$$B = \frac{\mu_0 NI}{L} = \frac{(4\pi \times 10^{-7} \text{ T} \cdot \text{m/A})(1000)(8.00 \text{ A})}{0.600 \text{ m}}$$

$$= 1.68 \times 10^{-2} \text{ T}$$

Example 21-6 A long, hollow plastic cylinder is bent in the shape of a doughnut with an inner radius of 0.250 m and an outer radius of 0.300 m. It is wound with a closely spaced toroidal winding of 800 turns and carries a current of 4.50 A. Calculate the magnitudes of (a) the maximum field at the inner radius and (b) the minimum field at the outer radius.

Solution This is a straightforward application of Equation 21-13, which may give us some idea of how large the variation within a toroid might be.

(a) $$B_i = \frac{\mu_0 NI}{2\pi r_i} = \frac{(4\pi \times 10^{-7}\ \text{T}\cdot\text{m/A})(800)(4.50\ \text{A})}{2\pi(0.250\ \text{m})}$$
$$= 2.88 \times 10^{-3}\ \text{T}$$

(b) $$B_o = \frac{\mu_0 NI}{2\pi r_o} = \frac{(4\pi \times 10^{-7}\ \text{T}\cdot\text{m/A})(800)(4.50\ \text{A})}{2\pi(0.300\ \text{m})}$$
$$= 2.40 \times 10^{-3}\ \text{T}$$

21-4 Magnetic Force on Currents

We have now seen that currents produce magnetic fields and that moving charges experience a magnetic force when placed in a magnetic field. Since currents are moving charges, shouldn't we expect that a current placed in a magnetic field will experience a magnetic force? Further, shouldn't we expect that the magnetic field produced by one current will exert a force on a second current placed in the vicinity of the first? Both expectations are correct.

That a magnetic field exerts a force on a current-carrying wire is not too difficult to imagine. Charges moving in a wire will certainly experience a magnetic force if crossing a magnetic field. If the charges are to remain in the wire, the material of the wire must exert a force on the charges that is equal in magnitude but opposite in direction to the magnetic force. By Newton's third law, the reaction to this force exerted *by* the wire is a force exerted *on* the wire that is equal in magnitude and direction to the magnetic force. Thus the force acting on the moving charges due to the magnetic field is transmitted to the current-carrying wire.

Figure 21-13 shows a wire segment of length l carrying a current I through a region of uniform magnetic field. The charges are confined to the wire and move with a drift velocity v_d in a direction making an angle θ with the magnetic field lines. Thus the force acting on an individual charge is, from Equation 21-1,

$$f = qv_d B \sin \theta$$

The net force F acting on the moving charge in this length and hence on the wire as a whole is equal to the total number of moving charges times f. The number of moving charges, say N, is simply the number of free charges per unit volume n times the volume of this segment of wire. If the wire has a cross-sectional area A, we have

$$N = nAl$$

Therefore

$$F = Nf$$

becomes

$$F = nAlqv_d B \sin \theta$$

Figure 21-13 A wire segment of length l in a uniform magnetic field at an angle of θ to the field. The wire carries current I.

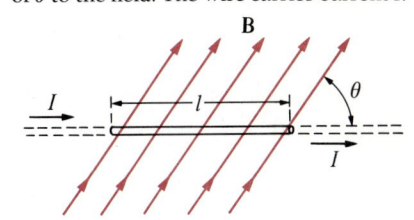

507

Rearranging the right side of this equation,

$$F = nqAv_d lB \sin\theta$$

Now we have from Equation 19-2 that

$$I = nqAv_d$$

Therefore

$$F = IlB \sin\theta \qquad (21\text{-}14)$$

Thus we obtained the magnitude of the force exerted on a current-carrying conductor by a magnetic field. The direction of the force is found by the right-hand rule since the conventional current is in the direction of the movement of positive charges. You should be able to convince yourself that the force acting on the wire segment in Figure 21-13 is out of the page toward you.

Let us now consider the force exerted on a current-carrying wire due to the magnetic field of a second current-carrying wire. Figure 21-14 shows segments of two long, parallel wires separated by a distance a and carrying currents I_1 and I_2. Several of the field lines due to I_1 are shown. Equation 21-9 gives us the field due to I_1 at a distance a. Thus, along its whole length, the wire carrying I_2 is immersed in a magnetic field of magnitude

$$B = \frac{\mu_0 I_1}{2\pi a}$$

and directed at 90° to the direction of I_2. The force on I_2 is given by Equation 21-14 as

$$F_2 = I_2 lB \sin\theta$$

Since $\theta = 90°$, $\sin\theta = 1$. Thus substituting for B and $\sin\theta$, we have

$$F_2 = I_2 l \frac{\mu_0 I_1}{2\pi a} \times 1$$

Rearranging to obtain the force per unit length yields

$$\frac{F_2}{l} = \frac{\mu_0 I_1 I_2}{2\pi a} \qquad (21\text{-}15)$$

You should be able to show that the direction of the force is toward the other wire; that is, the force is attractive. If the currents in parallel wires are in opposite directions, the force is repulsive. You should also be able to show that the magnitude of the force per unit length acting on I_1 due to the magnetic field of I_2 is

$$\frac{F_1}{l} = \frac{\mu_0 I_1 I_2}{2\pi a} \qquad (21\text{-}16)$$

which is identical to the magnitude of the force per unit length acting on I_2. Furthermore, it is also attractive, that is, toward I_2. In fact, these may be thought of as forming an action-reaction pair.

You will recall that we stated that the fundamental SI unit of electricity is the ampere. The interaction of currents through their magnetic fields provides a way to define the ampere operationally:

> One ampere is that unvarying current which if present in each of two parallel conductors of infinite length and one meter apart in free space

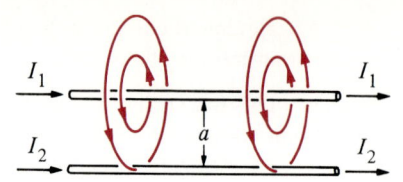

Figure 21-14 Parallel wires separated by distance a carrying currents of I_1 and I_2. Charge movement in one wire occurs in the magnetic field produced by charge movement in the other wire. Thus, the wires exert a force on each other. Can you apply the right-hand rule to determine the directions of the forces?

produces a force per unit length between the wires due to their magnetic fields of exactly 2×10^{-7} N/m.

The operational measurement actually used at the U.S. National Bureau of Standards substitutes coils of wire for the long, straight wires. The planes of the coils are parallel, yielding a constant distance between wires, with the coils connected such that they carry a common current. This current is adjusted to a value which yields the proper force per unit length as measured by a very sensitive balance system. The value of the proper current is the ampere. The coulomb is then a derived unit and is the net charge which passes through a conductor when a steady current of one ampere exists in the conductor for one second.

The definition of the ampere is an operational definition. That is, it defines the unit in terms of the operations that must be performed to produce a current of the magnitude. Such definitions are important in science since they allow the unit to be independently reproduced whenever and wherever necessary.

Example 21-7 Figure 21-15 shows a square loop of wire 0.0500 m on a side situated in a uniform field of $B = 0.100$ T such that the plane of the square is parallel to the field. The wire is part of a circuit carrying a current of 6.00 A in the direction shown. Determine the magnitude and direction of the force acting on each of the wire segments labeled 1 to 4.

Solution We apply the right-hand rule to each of the four segments to determine the direction of the force. Note that the magnetic field is perpendicular to segments 1 and 3 but parallel to segments 2 and 4. The magnitudes of the forces can then be found by utilizing Equation 21-14.

The right-hand rule applied to segment 1 shows that the force is downward into the page. The forces on segments 2 and 4 are zero. Finally, the force on segment 3 is upward out of the page.

The magnitudes of the forces acting on segments 1 and 3 are equal since I, l, B, and θ are the same for both. Thus

$$F = IlB \sin \theta = (6.00 \text{ A})(0.0500 \text{ m})(0.100 \text{ T})(\sin 90°)$$
$$= 0.0300 \text{ N}$$

for both segments 1 and 3.

21-5 Magnetic Properties of Materials

We have seen that currents cause magnetic fields. What causes permanent magnets? Furthermore, why are some materials attracted to permanent magnets while others are not?

Early on, the materials of permanent magnets were thought to contain many miniscule particles which were actually permanent magnets themselves. Such a body could be magnetized by causing a portion of the tiny permanent magnets to align in the same direction. As evidence mounted that currents cause magnetic fields, Ampere suggested that the miniscule permanent magnets were actually tiny current loops within the material. Recall from Figure 21-12a that the field outside a current loop looks very much like the field due to a bar magnet. Ampere's suggestion does not change the original idea much since a body is still magnetized by aligning the small permanent magnets. In Ampere's version, however, it is small current loops within the matter which are being aligned with their planes parallel and their currents all in the same direction.

Figure 21-15 Example 21-7.

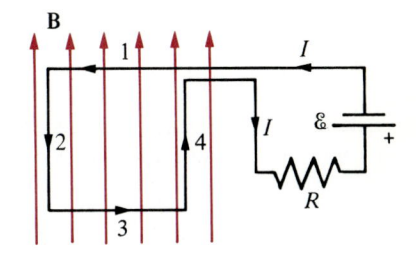

Various experimental evidence supported, or at least did not oppose, the theory. For example, if a magnet is broken, each piece has a north and a south pole. The bar magnet of Figure 21-16a becomes the magnets shown in Figure 21-16b and c, if repeatedly broken. By extrapolation, if we continue this process, each tiny piece down to the atomic level might become a permanent magnet with a north pole and a south pole. The most popular atomic model (Chapter 17) has the electrons circulating in orbits about the nucleus, forming tiny current loops, just as Ampere had speculated. Beyond this, refinements of measurement of atomic spectra led researchers to suggest that electrons and nuclei each have additional associated "spin." The **electron spin** and **nuclear spin** were originally visualized as rotation about an axis through the particle or nucleus similar to the daily rotation of the earth as it revolves about the sun.[1] Spins also give rise to magnetic fields similar to current loops. After all, a spinning sphere of charge could be replaced by a large number of current loops. As a result, we must assume that each atom has the possibility of an associated magnetic field made up of contributions from electron orbital and spin currents and perhaps the nuclear spin.

Most materials do not exhibit magnetic properties unless placed in a strong magnetic field. In Chapter 18, we saw that a material placed between the plates of a charged capacitor changes the magnitude of the electric field, and we used this to define the dielectric coefficients of material. Similarly, a material placed in an external magnetic field of magnitude B_0 changes the magnitude of the field to a new value B. The ratio B/B_0 is called the relative magnetic permeability of the material. In equation form,

$$K_m = \frac{B}{B_0} \qquad (21\text{-}17)$$

Most materials exhibit one of three types of behavior when placed in a magnetic field: diamagnetism, paramagnetism, and ferromagnetism.

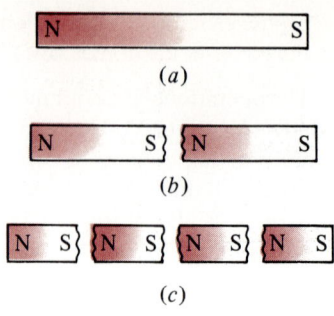

Figure 21-16 (a) A bar magnet possesses clearly distinct north and south poles separated at ends of the magnet. (b) If the bar is snapped in two, each part has both north and south poles. (c) Resnapping the pieces likewise leaves both north and south poles on each fragment, even sections representing the center of the original bar where no poles had been evident.

Diamagnetism ($B < B_0$)

In diamagnetic materials, K_m is very slightly less than 1. It is as though, when placed in the magnetic field, tiny magnets in the material align in a direction opposite that of **B**$_0$ and thus cause the net field to be smaller than **B**$_0$. Actually, diamagnetism is thought to be the result of the magnetic force acting on electrons in the material. It is believed that the applied field **B**$_0$ causes those electrons whose internal magnetic field is in the same direction as **B**$_0$ to decrease in speed, while those whose internal magnetic field is opposed to **B**$_0$ increase in speed. This yields a slightly smaller internal field in the same direction and a slightly larger internal field in the direction opposite to the applied field. The result is a slight decrease in the net field. This is all derivable, using Equation 21-1 along with the right-hand rule. (See Problem 21-37.) The diamagnetic effect is believed to be present in all materials, but it is so small that it can only be detected if the material does not display either paramagnetism or ferromagnetism.

---

[1] We must remind you that this is only a model we use to help us paint mental pictures so the physics will be easier to understand. We must not confuse such a model with what is actually taking place. Reality, as far as the scientist is concerned, is the measured quantity. We will deal more with atomic models in Chapter 29.

Paramagnetism (B > B₀) For paramagnetic materials, K_m is slightly greater than 1. Electrons in atoms normally form pairs, with their separate spins always directed oppositely. Effectively, the spins of such paired electrons cancel to produce no net external magnetic effect. It is believed that the atoms of paramagnetic materials have unpaired electrons which make them tiny magnets. These atoms are partially aligned in the same direction as the applied field, thereby causing a slightly larger net field. The effect is relatively small, though most often larger than and therefore able to mask the diamagnetic effect.

Ferromagnetism (B ≫ B₀) For ferromagnetic materials, K_m is typically much, much larger than 1 (usually more than 100 and often greater than several thousand). Ferromagnetic materials include the elements iron, cobalt, nickel, gadolinium, and dysprosium as well as a number of alloys of these metals. All five elements are unusual in that they have several electrons which are not paired with others of opposite spin but have their spins aligned in the same direction. This, by itself, is not what accounts for the huge magnetic effect. It is that many adjacent atoms of ferromagnetic materials are "locked" together, with their spin orientations aligned by an action called **exchange coupling.** Within a large piece of iron, these locked-together atoms form many small regions called **domains,** within which all the atomic magnets are pointing in the same direction (Figure 21-17). While these domains are very tiny, most often less than 0.1 mm on a side, they still contain large numbers of atoms, with 10^{15} or more in each domain.

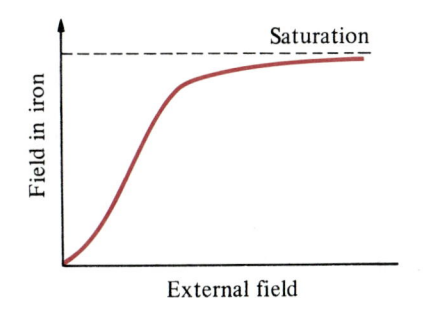

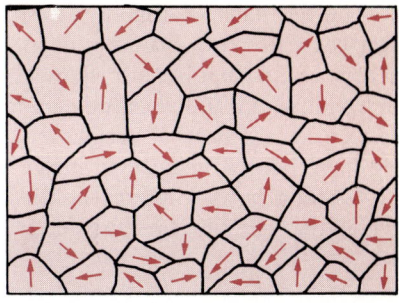

Figure 21-17 Ferromagnetic materials are subdivided into numerous microscopic magnetic domains.

As Figure 21-17 shows, the various domains in a piece of ferromagnetic material do not necessarily have their directions of magnetization aligned with each other. When this iron is placed in an external field of magnetic induction B_0, however, some of the domains may turn to align with the external field, while others which are already in the direction of the field grow in size as their neighboring domains shrink. The iron becomes magnetized and is attracted to the source of the external field.

A piece of iron placed inside a current-carrying coil of wire also becomes magnetized. The field due to the magnetized iron may become very large, usually much larger than the applied field.

Figure 21-18 shows how the field inside a piece of iron increases as the external field is increased. This magnetization is the principle behind the **electromagnet.** In addition to being able to create very large magnetic inductions, another advantage of an electromagnet is that the field may be switched on and off with the current in the coil. This alignment in the presence of an applied external field is also what causes unmagnetized iron to be attracted to a permanent magnet.

Figure 21-18 In the presence of an applied external field, the internal magnetic field in an iron ferromagnet rises quickly from zero as the magnetic domains align. The exact curve depends upon the nature of the iron used, but with all domains aligned, the field in the iron exceeds 1 T.

When the iron is removed from a field or the field is decreased to zero, the iron generally does not lose all its acquired magnetism. It, too, becomes a permanent magnet because many of the domains remain aligned. However, a sharp blow or a little heating may misalign the domains once again, thereby demagnetizing the iron. Various techniques such as annealing while magnetizing have been developed to make magnets more nearly permanent. Figure 21-19 shows how the internal field in a piece of iron varies as the external field is cycled through increases and decreases.

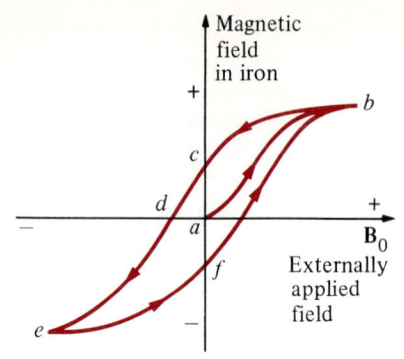

Figure 21-20 When a closed loop of wire moves in a magnetic field [(a), (b), (c)], the interaction of the field and loop produces a current in the wire. If the loop does not move but is (d) expanded or (e) decreased in area, a current appears in the loop during the change. If the magnetic field in the region of the loop changes while the loop is stationary [(f), (g), (h)], a current is induced in the loop.

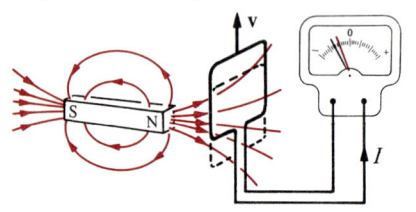

Figure 21-19 As the external field increases, the magnetism increases from (a) zero to (b) near saturation. If B_0 is reduced to zero, some (c) residual magnetism is retained by the iron. To remove this, B_0 may be reversed to point d. If the magnitude of the reversed B_0 is further increased, (e) saturation in the other direction may be approached. Returning B_0 to zero leaves (f) residual magnetism. Repeating the cycle produces a closed curve called a **hysteresis loop**.

Loop moving upward away from position of maximum flux (dashed); flux decreasing; current negative

(a)

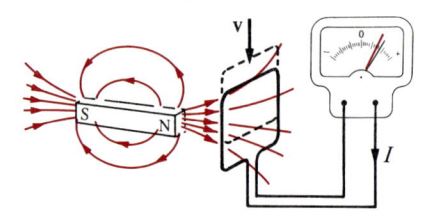

Loop moving downward toward position of maximum flux; flux increasing; current positive

(b)

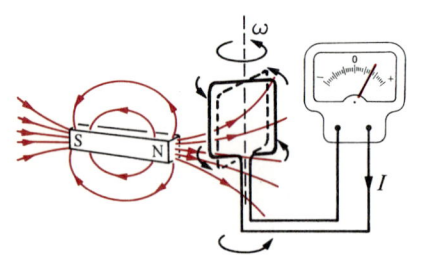

Loop rotating toward position of maximum flux (dashed); flux increasing; current positive

(c)

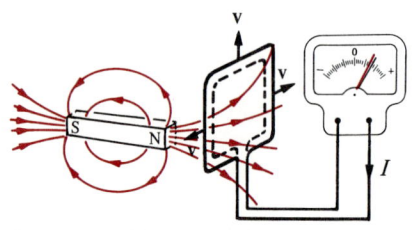

Loop expanding in size; flux increasing; current positive

(d)

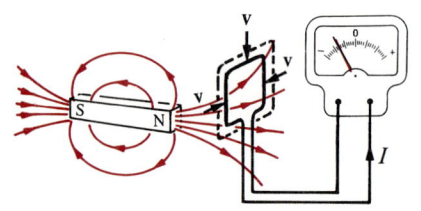

Loop contracting in size; flux decreasing; current negative

(e)

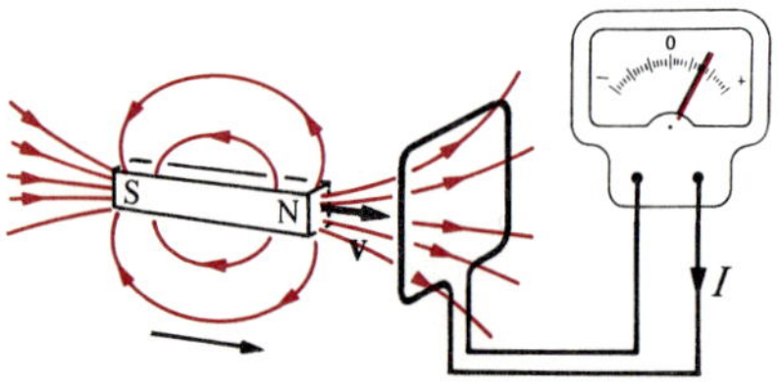

Magnet moving toward loop; flux increasing; current positive

(f)

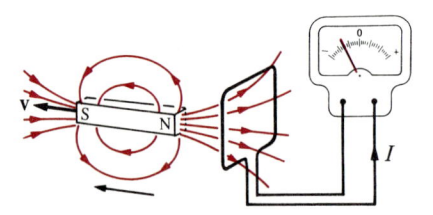

Magnet moving away from loop; flux decreasing; current negative

(g)

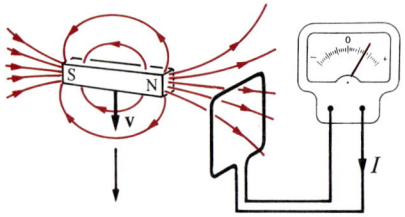

Magnet moving past loop; current positive as flux increases (shown), negative as flux decreases

(h)

21-6 Induced Voltages

Figure 21-20 shows several situations involving the interactions between a magnetic field, motion, and current. The opposite ends of a piece of wire are connected to the terminals of a meter, making a complete circuit, but having no source of voltage. Whenever a loop formed by the wire is moved in a magnetic field, twisted, expanded, or contracted in size, a current is established in the wire. Further, whenever the magnet moves toward or away from or past the loop, a current exists in the circuit during the motion. In each of these cases, some potential difference must have been **induced** in the wire to produce a current.

Let us explore a fairly simple situation to determine the mechanism by which charges flow. Figure 21-21 shows a segment of conducting wire moving with a constant velocity in the $+x$ direction into a region of uniform magnetic field. For simplicity, we assume the direction of the velocity is perpendicular both to the segment of wire and to the magnetic field. Prior to entering the field, there is no magnetic force acting on the free charges in the wire. As the wire enters and moves through the field, however, positive charges experience a magnetic force of magnitude $F = qvB$ in the positive z direction, making the magnetic force on free electrons in the negative z direction. The free charges move toward the ends until the separation of opposite charge causes an internal electrostatic field large enough to assure that the electrostatic force and the magnetic force on each free charge in the wire balance each other. The end of the segment in the positive z direction becomes positively charged, while the end in the negative z direction becomes negatively charged. There is thus an electrostatic potential difference between the ends of the wire. If the ends are connected by another piece of wire which is stationary with respect to the magnetic field, charge will flow due to the difference in potential. The segment of wire moving through the field thus is a source of voltage called a **motional emf.**

We can obtain an expression for the induced motional emf for a segment of wire of length l moving with a constant speed v perpendicular to a magnetic field of induction B. The magnitude of the force exerted on a charge q in the wire is

$$F = qvB$$

The change in energy of q as it moves a length l in the wire is equal in magnitude to the work done on q by the force. Thus

$$W = Fl = qvBl$$

The potential difference between the ends is equal to the change in energy per unit charge. Thus

$$V = \frac{W}{q} = vBl$$

But this is just the magnitude of the motional emf, and we write

$$\mathcal{E} = Blv \tag{21-18}$$

The direction of $\mathcal{E}$ may be found by applying any one of the right-hand rules.

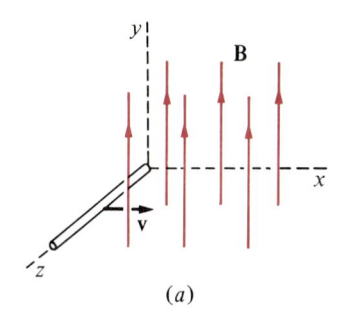

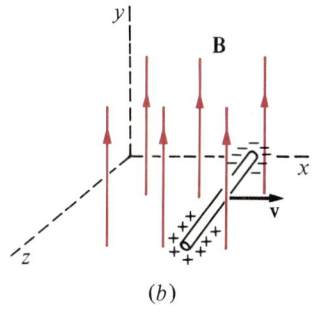

Figure 21-21 (a) A wire moving perpendicular to its length into a uniform magnetic field perpendicular to its motion. Each charged particle in the wire will experience a force. (b) By the right-hand rule, positive charges will experience a force in the $+z$ direction and negative charges in the $-z$ direction.

Example 21-8 Determine the minimum speed with which a 5.00×10^{-2} m segment of wire must move perpendicular to its length and to a magnetic field of $B = 0.200$ T in order that a voltage of 0.150 V be induced across its length.

Solution We apply Equation 21-18 after having solved for v. Thus

$$\mathscr{E} = Blv$$

or

$$v = \frac{\mathscr{E}}{Bl} = \frac{0.150 \text{ V}}{(0.200 \text{ T})(5.00 \times 10^{-2} \text{ m})} = 15.0 \text{ m/s}$$

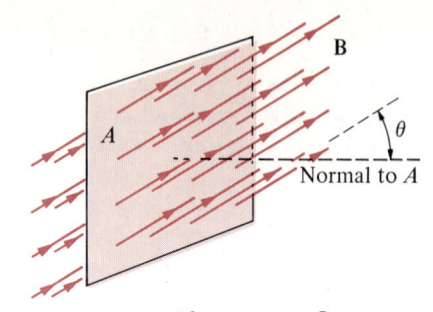

Figure 21-22 The magnetic flux, or amount of the magnetic field penetrating a given area A, depends upon the magnitude of the field, the area, and θ, the angle between B and the perpendicular to the surface of A.

A different and more general way to view the induced voltages indicated in Figure 21-20 involves a quantity called **magnetic flux** and a physical law which has become known as Faraday's law. In Figure 21-22, an imaginary plane surface of cross section A is situated in a uniform magnetic field **B** such that a line drawn perpendicular, or normal, to the surface makes an angle θ with the field lines. The magnetic flux over A is defined as

$$\Phi = BA \cos \theta$$

where the Greek uppercase phi (Φ) is the symbol for magnetic flux. In SI units, magnetic flux is measured in $\text{T} \cdot \text{m}^2$, also known as webers (Wb). Michael Faraday (1791–1867) (Figure 21-23) discovered that the magnetically induced emf in a circuit is proportional to the time rate at which the magnetic lines of force change near the charges which comprise the induced current:

$$\mathscr{E} \propto \frac{\Delta \Phi}{\Delta t} \quad (21\text{-}19)$$

or the induced emf is proportional to the time rate of change of flux. If the emf is induced in a loop of wire or any closed circuit, it gives rise to an induced current.

Measurement shows (and the conservation of energy dictates) that the magnetic field of the induced current always opposes the change which causes it. Heinrich Lenz (1804–1865) first put this opposition effect into words, and it thus became known as **Lenz's law.** We incorporate Lenz's law by using -1 as the constant which makes Proportionality 21-19 an equation. Thus **Faraday's law** becomes

$$\mathscr{E} = -\frac{\Delta \Phi}{\Delta t} \quad (21\text{-}20)$$

The minus sign is due to Lenz and reminds us that the induced current in a circuit gives rise to a magnetic field which opposes the change that caused the current. If we are dealing with a coil of wire with N turns through which the flux is changing, an emf is induced in each turn. Thus in this situation, the net emf is

$$\mathscr{E} = -N \frac{\Delta \Phi}{\Delta t} \quad (21\text{-}21)$$

Figure 21-24 shows a straight conducting wire forced to slide on a U-shaped conductor such that the area intersected by a magnetic field may be

Figure 21-23 Michael Faraday (1791–1867) was an experimental genius. Although he left school early to become a bookbinder's apprentice, science fascinated him, and a gift of tickets to a science lecture at the Royal Institute changed his life. He took a pay cut to become assistant to its director, Sir Humphrey Davy. There, he framed the laws of electrolysis, produced the first electric motor, and devised the concepts of *lines of force* and *fields* to visualize electrical and magnetic phenomena. *(Culver Pictures)*

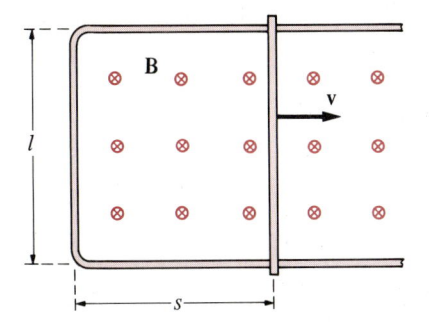

Figure 21-24 As the wire segment slides along its support wires with speed v, the area of the closed loop increases. Since the loop is in a uniform magnetic field, the flux through the loop also increases, inducing a current in the loop.

Lenz's law is not really abstract or confusing. Assume the change in flux through a wire loop is caused by withdrawing the north pole of a magnet from its center along the axis of the loop. Lenz's law asserts that the induced current in the loop will have a direction to set up a magnetic field which will act toward attracting the magnet back. Similarly, the induced field of the current in the loop would have acted to repel the north pole when it was brought toward it in the first place.

increased or decreased. At the time shown, the area of the loop is $A = ls$. Since the angle between the magnetic field and the normal to the area is $0°$, we have

$$\Phi = BA \cos 0° = BA = Bls$$

The flux Φ is changing with time because s is changing with time. In fact,

$$\frac{\Delta s}{\Delta t} = v$$

Thus

$$\frac{\Delta \Phi}{\Delta t} = Bl \frac{\Delta s}{\Delta t} = Blv \qquad (21\text{-}22)$$

According to Equation 21-18, this is exactly equal in magnitude to the motional emf of a wire of length l moving at a speed v in a direction perpendicular both to its length and to the magnetic field. From the right-hand rule, we would deduce that positive charge experiences a magnetic force toward the top of the moving conductor when it moves to the right and toward the bottom when it moves to the left. This indicates that the induced current in the loop is counterclockwise when the conductor moves to the right and clockwise when it moves to the left. Lenz's law insists that the induced current must give rise to a magnetic field through the loop which is out of the page (opposing the increasing flux) when the conductor moves to the right. Also, the induced current must give rise to a magnetic field into the page (opposing the decreasing flux) when the conductor moves to the left. Careful consideration convinces us that the two points of view, motional emf's and Faraday's law, are entirely equivalent for this case. Indeed, it turns out that any motional emf is derivable from Faraday's law.

We should not get the idea that Faraday's law and motional emf's are simply two different ways of looking at the same phenomenon. Faraday's law is much more generally applicable. It may be used to determine the induced emf in situations where the flux change is not due to motion of a magnet or part of a circuit. Figure 21-25 shows a circuit containing a solenoid around which a second loop of wire has been wound and connected to the terminals of an ammeter to form a separate circuit. There is no electrical connection between the two circuits. However, when the switch is closed, the magnetic field of the solenoid changes from zero to its final value in a short but finite time. This changing field causes the flux through the second loop to change with time. According to Faraday's law, we should expect an induced emf in the second loop. An induced emf is found experimentally in spite of the fact that no parts of the circuits are moving. We see by this illustration that Faraday's law encompasses a wider spectrum of physical phenomena.

Figure 21-25 When the switch is closed, current will surge through the solenoid. As this occurs, the ammeter in the physically separate circuit containing a loop surrounding the solenoid will record an induced current in this separate circuit, while the current through the solenoid grows to its steady value.

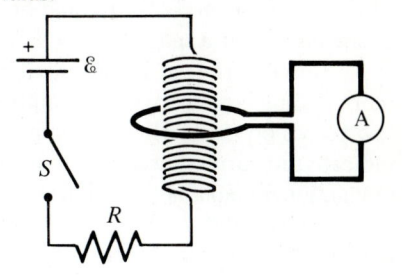

515

Example 21-9 The solenoid in Figure 21-25 has 300 turns and is 0.600 m long. When the switch is closed, the current in the solenoid circuit increases from zero to 4.00 A in 5.00×10^{-4} s. The cross section of the solenoid is 1.60×10^{-3} m². Calculate the magnitude of the induced emf in the second loop during the 5.00×10^{-4} s time interval.

Solution We wish to find the time rate of change of flux through the second loop because its magnitude is equal to the induced emf. This entails finding the change in magnetic flux and dividing it by the time interval over which that change takes place. Since the second loop completely surrounds the solenoid, the flux through the second loop at any time is simply the flux of the solenoid. (We say that there is a flux linkage between the circuits.) Thus we apply Equation 21-11 to find the final flux through the solenoid and divide it by the time interval. Remember that the initial magnetic field and hence the flux are zero.

$$B_f = \frac{\mu_0 N I_f}{L} = \frac{(4\pi \times 10^{-7} \text{ T} \cdot \text{m/A})(300)(4.00 \text{ A})}{0.600 \text{ m}}$$

$$= 2.51 \times 10^{-3} \text{ T}$$

$$\frac{\Delta \Phi}{\Delta t} = \frac{\Delta(BA)}{\Delta t} = \frac{\Delta B \, A}{\Delta t}$$

$$= \frac{B_f - B_i}{\Delta t} A$$

$$= \frac{2.51 \times 10^{-3} \text{ T} - 0}{5.00 \times 10^{-4} \text{ s}} \times 1.60 \times 10^{-3} \text{ m}^2$$

$$= 8.04 \times 10^{-3} \text{ V}$$

SPECIAL TOPIC
Magnetic Field of the Earth

The magnetic field of the earth presents an interesting puzzle. Ancient Greek navigators used the lodestone to indicate north. The first detailed study of magnetism (*de Magnete*) was published by Sir William Gilbert in 1600 A.D. He had a large, spherical lodestone cut to shape to simulate the earth and measured the direction of the field across its surface. This approximated the behavior of the compass on the earth, and he concluded that the entire earth was a magnet.

Actually, the earth's magnetic behavior is much more complex than that of a simple magnet. There are few places on earth where a compass needle will indicate the correct direction of north (Figure 21-26). All navigational maps have the difference recorded somewhere on them (Figure 21-27). Furthermore, the deviation between magnetic north and true north shifts in time. For example, in 1580, a compass needle at London pointed some $11\frac{1}{2}°$ east of true north. By 1819, the same reading had varied some 36° to a point $24\frac{1}{4}°$ west of true north. Today, this value, the **magnetic declination**, at London is some 8° west. Much shorter variations in the field also occur (Figure 21-28). The average magnetic induction at the earth's surface is about 6.3×10^{-5} T near the magnetic poles and 2.1×10^{-5} T at the equator.

In addition to the horizontal component of the magnetic field measured by a compass needle, the magnetic force has a vertical component, or **dip**. A magnetic needle mounted to rotate freely in a vertical plane will point straight downward or upward near the magnetic poles and horizontally near the magnetic equator. These locations do not correspond to the geo-

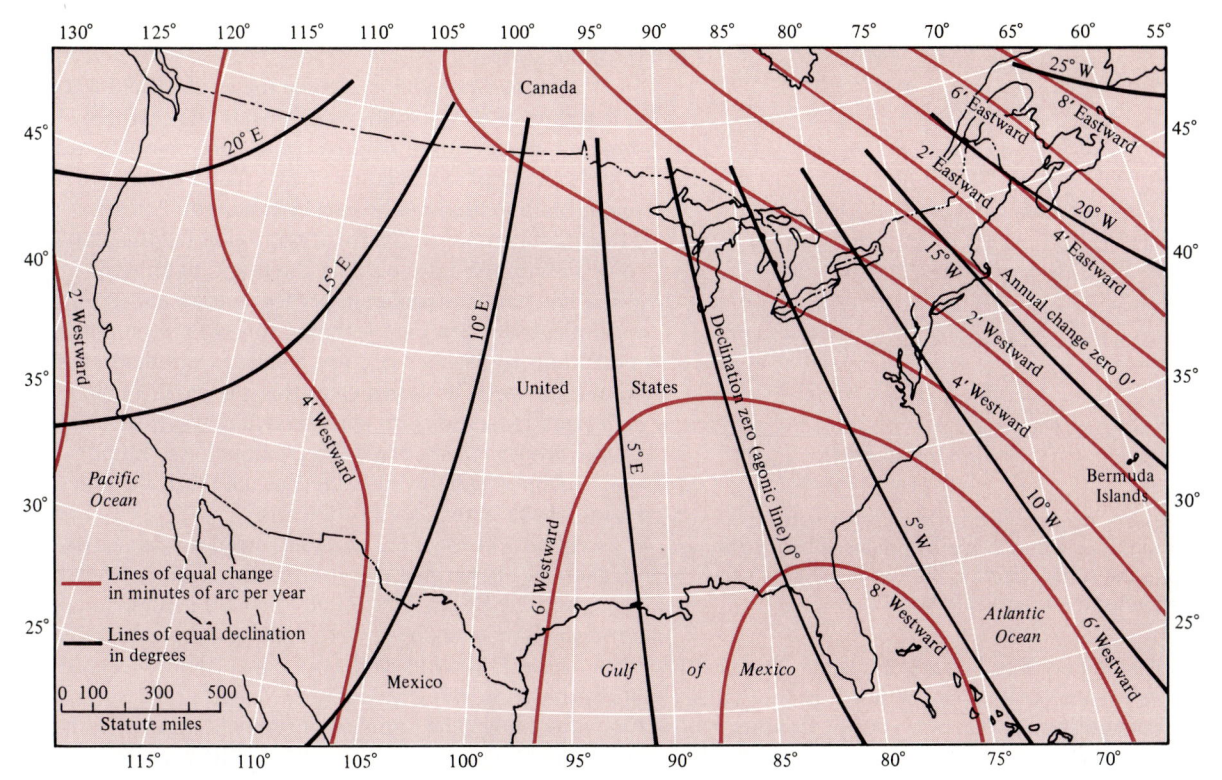

Figure 21-26 Magnetic declination for the United States. On the agonic line of declination = 0, the magnetic compass indicates true north. Off this line, the deviation gets progressively larger. At Seattle, for example, the compass points 24° east of true north, a very serious difference for a navigator.

Figure 21-27 Typical marginal diagram appearing on any precision navigational map. Shown are the differences between magnetic north, astronomical north (the star), and geographic north.

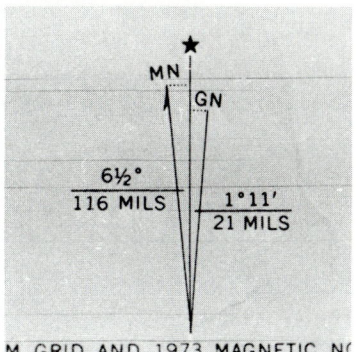

Figure 21-28 The earth's magnetic field is distorted by interaction with streams of charged particles from the sun. The field lines are swept back and elongated away from the sun. Toward the sun, the region dominated by the earth's field is blunted and ends at the magnetopause, where external magnetic effects dominate.

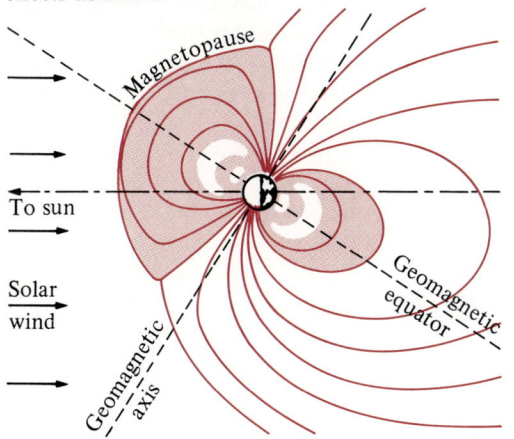

517

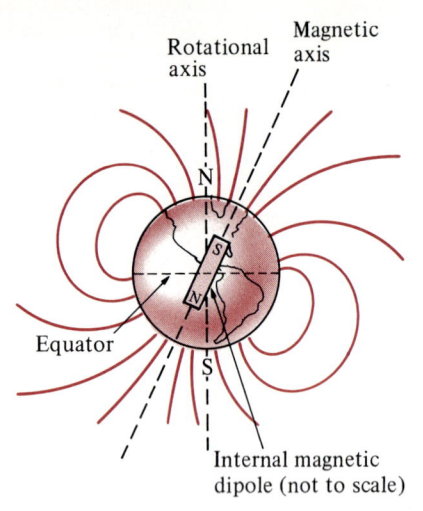

Figure 21-29 The magnetic field at the earth's surface is fairly similar to the field that would be produced by a giant bar magnet deep in the earth's interior, but this is *not* an acceptable explanation for the magnetic field.

graphical poles and equator but are displaced. The north magnetic pole is some $14\frac{1}{2}°$ from the north geographical pole, while the south magnetic pole is $23\frac{1}{2}°$ from its respective geographical pole. This means that the two magnetic poles are not antipodal, or directly opposite each other on the earth.

The origin of the earth's magnetic field is quite perplexing. After Gilbert, some scientists pictured the earth as having a permanent-magnet iron core some 600 km across in its center. Given such an internal bar magnet, the angle of dip observed on the surface would be about correct (Figure 21-29). While it is quite likely that the earth possesses a great deal of iron in its core, the bar magnet idea is not correct. Any permanent magnet can have its magnetism destroyed by heating, as thermal agitation destroys the alignment of magnetic domains. The temperature beyond which such magnetic alignment is impossible is called the **Curie point** for the material in question. For iron, the Curie point is 770°C and increases only very slowly with increasing pressure. However, temperature increases with increasing depth inside the earth at a rate of about 30°C/km, so iron would lose its magnetism within a few tens of kilometers from the surface.

A second possible cause for the earth's field is an actual flow of charge somewhere inside the earth. Such a dynamic model of the field becomes very complex quite rapidly. A simple calculation shows that a current of several billion amperes would be required (Problem 21-39). Of course, we know of no reason for this current to be flowing. What would be the origin of an emf? What mechanism would supply the energy losses to, for example, Joule heating? Actually, this overly simplistic model fails to pass almost any detailed examination and would not work. It seems clear that the earth's field is generated by some complex, nonsymmetric flow of charge whose causes are unknown. Such a dynamo mechanism is usually suspected in the fluid, metallic region of the outer core of the earth.

Whatever mechanism is advanced for the earth's magnetism, its hardest test may be the really long-term behavior of the field. First of all, the field is now much stronger than it has been in the geological past. Magnetism "frozen" in rocks as they cooled provide a record of the direction and strength of the earth's field at the time of cooling. The strength of the field appears to have grown for the past several hundred million years, until today it is twice the strength of a hundred million years ago. Actually, the field is decreasing in strength in historical times from a peak reached some 2000 years ago of half again as large as at present.

An even more difficult test for any proposed mechanism for the earth's field will be to explain the fact that the entire field reverses direction periodically! The vast majority of rock samples indicates that magnetic north was, at the time of their cooling, either fairly much where it is now or in the completely opposite direction. The entire magnetic field of the earth "flips" in direction with a frequency of several times per million years. These **geomagnetic reversals** constitute a severe test for any theory of the earth's interior. (They also constitute a splendid new mechanism for a type of dating of rock strata.) Only the future will tell whether we can determine the mechanism for generating such a complex and changeable field.

Minimum Learning Objectives

After studying this chapter, you should be able to:
1. Define:

B (magnetic field)
diamagnetism
domain
electromagnet
electron spin
Faraday's law
ferromagnetism
hysteresis loop
induced current
induced field
induced voltage
Lenz's law

lodestone
magnetic flux
magnetic induction
magnetic line of force
magnetic permeability
motional emf
nuclear spin
paramagnetism

permanent magnet
pole
right-hand screw rule
solenoid
tesla (T) [unit]
toroid
velocity selector

2. Calculate the force on a moving charge in a magnetic field.
3. Calculate the magnetic induction for the motion of a known charge.
4. Describe correctly the direction of the force on moving electric charges of given sign in a magnetic field.
5. Calculate the mass of a known charge moving in a given magnetic field, given the radius of its orbit and speed.
6. Calculate the field induced near a long, straight wire by the current in the wire.
7. Calculate the central field of a solenoid, given the current, length, and number of loops.
8. Calculate the force on a current-carrying wire from a parallel current-carrying wire at a given distance.
9. Articulate the microscopic causes of macroscopic magnetic effects of matter.
10. Calculate the induced emf in a given length of wire moving at a known speed across a known field.
11. Calculate the magnetic flux through a given area in a known magnetic field, given the orientation of the field to the area.
12. Calculate the magnitude and direction of an induced emf in a loop from the rate of change of the magnetic flux through the loop.

Problems

21-1 Determine the magnitude and direction of the magnetic field in which a beam of electrons of speed 2.00×10^7 m/s projected to the east experiences a force toward the north. Also, the same beam directed toward the south is deflected toward the east and moves in a circular path of radius of curvature 2.15 m.

21-2 Electrons in a color television receiver are accelerated to speeds of 9.00×10^7 m/s. They are then deflected by the magnetic field of the deflection yoke. If the magnetic force on the electron is in a direction perpendicular to the velocity and has a magnitude of 4.32×10^{-13} N, calculate (a) the magnitude of the magnetic field and (b) the acceleration if you assume the electron's mass is 9.11×10^{-31} kg. (c) Actually, we will see in Chapter 27 that an electron traveling at this speed has an effective mass of 9.56×10^{-31} kg. What is the correct acceleration?

21-3 Protons are accelerated in a small Van de Graaff generator by a 2.00×10^6 V potential difference and directed along an evacuated cylindrical tube. A magnet provides a magnetic field perpendicular to the tube to bend the protons along a path of radius of curvature 0.250 m to the final beam tube. Calculate (a) the speed of the protons and (b) the magnitude of the magnetic field of the bending magnet.

21-4 Part of the low-energy component of cosmic rays at sea level is made up of electrons. At a place in the northern hemisphere where the earth's magnetic field has a magnitude of 6.00×10^{-5} T and is pointed into the earth at an angle of 20° with the vertical, calculate (a) the magnetic force on an electron moving straight down with a speed 2.50×10^6 m/s, (b) the acceleration of the electron, and (c) the ratio of the magnetic force to the electron's weight.

21-5 An alpha particle in the cosmic rays moves along a helical path of radius 3.32×10^3 m in a portion of the Van Allen belt of "trapped" charged particles surrounding the earth where the magnetic field has a magnitude 5.00×10^{-6} T. The component of velocity of the alpha particle which is parallel to the magnetic field is 6.00×10^5 m/s. Take the mass of the alpha particle to be 6.64×10^{-27} kg and its charge to be $+3.20 \times 10^{-19}$ C. Calculate (a) the component of velocity perpendicular to the magnetic field and (b) the magnitude of the net velocity of the particle.

21-6 Calculate the magnitude and determine the direction of the magnetic force acting on a proton moving directly east near the earth's surface. Take the earth's magnetic field at that point to be downward at an angle 30° from the vertical toward the north and to have a magnitude 5.00×10^{-5} T. The proton's speed is 8.10×10^6 m/s.

21-7 An electron is moving directly south at a speed of 4.00×10^6 m/s near the earth's north magnetic pole where the magnetic field is straight downward and has a magnitude of 7.20×10^{-5} T. Determine (a) the magnitude of the magnetic force, (b) its direction, and (c) the acceleration of the electron.

21-8 A proton enters a region where there exists a magnetic field of induction 0.150 T directed perpendicular to the path of the proton. The proton traces out a semicircular path and leaves the region of the field. Calculate the radius of curvature and the amount of time the proton takes to complete the semicircle if its initial speed is (a) 5.00×10^5 m/s and (b) 2.00×10^6 m/s.

21-9 The blood flow rate in an artery may be measured continuously during an operation by a magnetic cuff attached around the artery. Blood contains both positive and negative ions flowing past the cuff at some speed v. The cuff generates a magnetic field perpendicular to the blood flow, deflecting positive ions toward one wall of the artery and negative ions toward the other. The resulting charge separation produces a voltage measured by a microvoltmeter. (a) A cuff on an artery of diameter 10.0 mm is energized with a field of 6.00×10^{-3} T. If a

519

voltage of 7.80 μV is measured across the sides of the artery, what is the velocity of blood flow? (*Hint:* This is very similar to towing a wire segment of length *d* through a magnetic field.) (b) What volume of blood flows through the artery each minute?

21-10 Two metal plates are placed parallel at a separation of 1.50 cm and oppositely charged to a potential difference of 500 V. The electric field intensity between the plates is directed horizontally toward the east. A magnetic field of induction 0.0200 T directed straight downward is established between the plates, and the region between the plates is evacuated. An electron with a speed of 2.00×10^6 m/s is projected between the plates in the northerly direction (perpendicular to both fields). Calculate the magnitude and direction of the (a) electric force acting on the electron, (b) the magnetic force, (c) the net force, and (d) the magnetic field necessary so that the net force on the electron is zero.

21-11 A mass spectrometer has a potential difference of 3.00×10^3 V between the plates of the velocity selector. The plates are separated by 2.00 cm, and the magnetic field in that region is 0.250 T. After passing through the selector, a beam of singly charged positive ions is bent in the second magnetic field of magnetic induction 0.500 T. Calculate the mass of the ions deposited on a circle of diameter (a) 22.0 cm and (b) 26.0 cm. (c) What was the magnitude of the velocity of the ions?

21-12 Determine the magnitude of the magnetic induction due to a very long wire carrying a current of 4.00 A if the wire is (a) straight and the measurement is to be made at a distance of 2.00 cm away and (b) wrapped tightly to form a solenoid of 500 turns per meter and the measurement is inside the solenoid.

21-13 An electron in a hydrogen atom revolves in a circular orbit of radius 5.30×10^{-11} m at a speed of 2.19×10^6 m/s. Calculate (a) the current this loop represents and (b) the magnetic field produced at the center of the loop.

21-14 A loop of wire has 100 tightly packed turns, a radius of 8.00 cm, and carries a current of 12.0 A. Calculate (a) the magnetic field at the center of the loop and (b) the current required to produce a magnetic field of 8.00×10^{-5} T.

21-15 A proton moves at a speed of 4.00×10^6 m/s 2.00 cm away from but parallel to a long, straight wire which is carrying a current of 5.00 A. Determine the magnitude and direction of the force on the proton if it moves (a) in the same direction as the current and (b) in the direction opposite to the current. (c) What would be the direction of the force if the proton moved perpendicularly toward and then away from the wire?

21-16 Two air-core solenoids are wound on the same axis, one slightly larger than the other and completely surrounding it. The interior solenoid has 20 turns per centimeter and carries a current of 2.50 A. The exterior solenoid has 12 turns per centimeter and carries a current of 1.50 A in the opposite sense to that of the interior solenoid. (a) Calculate the magnetic field on the common axis of the solenoids. (b) What current is necessary in the exterior solenoid so that the magnetic field on the axis is zero?

21-17 Determine the magnitude of the magnetic field at the average radius of the toroid of Example 21-6 when the doughnut is filled with iron of relative magnetic permeability $K_m = 200$.

21-18 A long solenoid has 12.0 turns per centimeter of length and carries a current of 20.0 A. (a) Assuming no magnetic material in the solenoid or in the vicinity, determine the magnetic induction within the solenoid. (b) In what direction is the force acting on the wire of the solenoid?

21-19 The starter motor of a particular automobile draws a current of 300 A when in use. (a) What is the strength of the magnetic field produced by this current 2.00 cm away from the wire leading to the motor from the battery? (b) How does this compare with the magnetic field of the earth? (c) A magnetic compass was placed at this distance directly above the north-south-running wire carrying positive current southward. Describe the behavior of the compass (i) before the starter motor is engaged, (ii) as the motor is first engaged, and (iii) after the motor has been turning for several seconds drawing a steady current.

21-20 The starter solenoid of an automobile operates at 12.0 V and has a resistance of 1.00 Ω. When the key is turned to the start position, a current through the solenoid creates a magnetic field to force a starting gear into position. What will the field at the center of the solenoid be if it has 200 turns, is 5.00 cm in diameter, and is 8.00 cm long?

21-21 Most speaker systems produce sound by feeding a variable electric signal to a coil in a magnetic field. The resulting attractions and repulsions move a speaker cone and generate sound. In a particular speaker, a large, fixed permanent magnet produces a field of 0.200 T in the space occupied by a 40-turn coil of radius 3.00 cm. What instantaneous current in the coil would produce a force attracting the coil and its attached speaker cone rearward at an initial acceleration of 20.0 m/s² if the mass to be moved is 0.0700 kg?

21-22 A long, straight wire carries a current of 3.00 A in a uniform external magnetic field of $B = 0.0450$ T directed at an angle of 60° with the direction of the conventional current. Calculate the magnetic force per unit length on the wire.

21-23 Two long, parallel wires separated by 4.00 cm carry currents of 12.0 and 20.0 A. (a) Calculate the force per unit length on each wire. (b) What common current in the two wires would cause a force per unit length on each of 1.00 N/m? (c) If the currents are in opposite directions, will the forces be repulsive or attractive?

21-24 A straight copper wire of length 1.50 m and cross-sectional area 2.00×10^{-6} m² carries a current of 6.00 A from north to south. Calculate the magnitude and direction of the magnetic field so that the magnetic force will balance the gravitational force. Take the mass density of copper to be 8.90×10^3 kg/m³.

21-25 Two lines from a dc generator are run 30.0 m north and south between supports and are separated by 50.0 cm. A positive current of 100 A runs out one line and returns by the other. (a) What is the force per unit length on one wire caused by the current in the other wire? (b) Is this force attractive or repulsive? (c) The vertical component of the earth's magnetic field at the wires is 3.50×10^{-5} T acting downward. What is the deflecting force per unit length on the wire due to the flow of charges across a magnetic field? (d) How far apart would the two wires have to be spaced so that these two separate effects would exactly cancel each other? (e) In this case, which line running southward from the generator, the western line or the eastern one, should carry the outward-bound positive current.

21-26 A rectangular loop of wire 0.150 m by 0.300 m carries a current of 8.00 A and is oriented so that the plane of the rectangle is horizontal. It is in a region where there is a magnetic field of induction 5.00×10^{-2} T directed parallel to the short side of the rectangle. Determine the magnitude and direction of the force on each of the four sides of the rectangle.

21-27 The starter motor of an automobile engine may draw as much as 400 A while engaged. If the cables from the battery to the starter motor are 0.800 m long and separated by 1.50 cm, calculate the magnetic force each exerts on the other during starting.

21-28 Soft iron may be easily magnetized, but its field drops to nearly zero once the external field is removed. In some materials, it is desirable to leave a residual magnetism after the field is removed. One of these materials is magnetic recording tape. A recording head has a coil electromagnet which magnetizes domains on the tape and leaves them magnetized. Later, on playback, the varying field of the magnetized tape generates a varying emf in the coil to reproduce the record. (a) A tape traveling past the recording head moves at a speed of 12.0 cm/s. The highest frequency at two domains per cycle which can be recorded on the tape is 11,000 Hz. What is the minimum size of a magnetic domain from these data? (b) A computer may use the recording tape to store data as a sequence of 1s and 0s (binary code). Each stored piece of data is called a **bit,** and eight bits make one sequence called a **byte** which may specify a letter, numeral, or operation. If a personal computer has an internal storage of data, or memory, of 48 kilobytes, how long a strand of recording tape would it take to record the contents of this internal memory?

21-29 A straight segment of wire 10.0 cm long is confined to move in the *xy* plane. It maintains an angle of 30° with the +*y* axis and moves at a speed of 40.0 m/s in the +*x* direction. (a) Calculate the magnitude of the magnetic field directed perpendicular to the *xy* plane such that the motional emf induced between the ends of the wire is 0.700 V. (b) If the wire slows so that the motional emf drops to 0.100 V, what is the wire's new speed?

21-30 Twenty turns of conducting wire are wound in the shape of a rectangle 10.0 cm by 15.0 cm. The plane of the coil is originally parallel to the earth's magnetic field ($B = 6.00 \times 10^{-5}$ T) and is rotated to a new orientation perpendicular to the field in 0.200 s. Calculate (a) the average emf induced in the coil, (b) the strength of the magnetic field necessary so that the average induced emf would be 1.00 V for the same maneuver, and (c) the average current in the coil in each case if the coil resistance is 0.0350 Ω.

21-31 A magnetic microphone uses a thin metal leaf suspended between the poles of a magnet. Sound waves move the leaf back and forth through the magnetic field, generating an emf between the ends of the leaf roughly proportional to the intensity of the sound and having the same frequency. A leaf 1.50 cm long in a magnetic field of 0.100 T is displaced with an amplitude of 0.100 cm by a steady sound with a frequency of 400 Hz. (a) What is the maximum velocity of the leaf as it oscillates in simple harmonic motion in response to the sound? (b) What is the maximum emf induced across the leaf?

21-32 A rectangular loop of wire 20.0 cm by 40.0 cm moves at 15.0 m/s into a region with a uniform magnetic field $B = 0.450$ T, as shown in Figure 21-30. The field is perpendicular to the plane of the loop. Calculate (a) the motional emf being induced in the loop as it is moving into the field region and (b) the current in the loop if its resistance is 3.00×10^{-3} Ω.

21-33 A thin copper rod 0.600 m long rotates about an axis through one end perpendicular to a uniform magnetic field of 0.200 T. Calculate the angular velocity required to give an induced emf across the rod of 3.00 V.

21-34 The magnetic field through the rectangular loop of wire shown in Figure 21-31 is directed perpendicularly out of the plane of the page and is decreasing at the rate of 0.0500 T/s. Calculate (a) the emf being induced in the loop and (b) the current if the resistance of the loop is 2.00×10^{-3} Ω.

21-35 A rectangular 10-turn loop of wire ($N = 10$) of length $l = 20.0$ cm and width $w = 12.0$ cm is rotating at an angu-

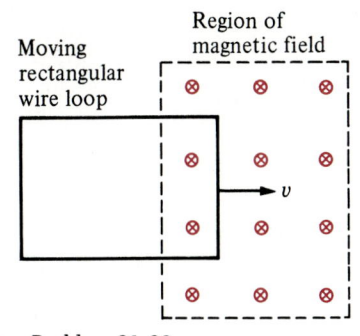

Figure 21-30 Problem 21-32.

521

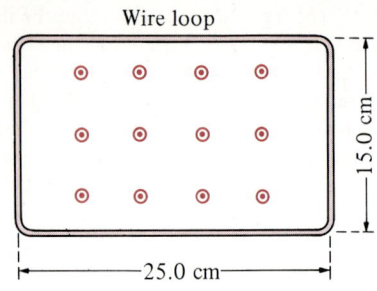

Figure 21-31 Problem 21-34.

lar velocity $\omega = 8.00$ rad/s about an axis along one of the lengths. The loop is situated in a region of uniform magnetic field $B = 0.0600$ T. (a) Determine an *expression* for the emf in the loop as a function of N, ω, B, $A (= lw)$, and time. (b) Calculate the emf in the loop when the normal to the plane of the loop makes an angle of 45° to the magnetic field.

21-36 The current in an 800-turn air-core solenoid which is 0.500 m long and has a radius of 2.50 cm is increasing at the rate of 0.150 A/s. A flat coil of 10 turns is wrapped around the outside of the solenoid and completes a separate circuit with a 2.00-Ω resistor and an ammeter of negligible resistance. Calculate (a) the rate of change of magnetic field through the solenoid, (b) the rate of change of flux through the solenoid, (c) the emf induced in the flat coil, and (d) the reading of the ammeter.

21-37 Assume that a popular model of electron motion (Figure 21-32) shows single-orbital electrons in two separate atoms moving in circles in the plane of the page. The electron moving counterclockwise (a) comprises a current loop which gives rise to a magnetic field inside the loop directed into the page as shown, labeled B_{int}. The electron orbiting clockwise (b) gives rise to an internal field directed out of the page. If an external field B_{ext} directed into the page is imposed in this region, show that if the radii of these orbits do not change, the magnetic force on these electrons will cause a weakening of the internal field into the page and a strengthening of the internal field out of the page, thereby reducing the effect of the external field inside the orbits. This model may help explain the diamagnetic effect associated with all materials.

21-38 Two solenoids are interwound on the same air-filled cylindrical shell of radius 2.40 cm and length 40.0 cm. One solenoid has 800 turns, and the other has 4800 turns. A steady current of 5.00 A flows in the 800-turn solenoid until a switch is opened. The current is reduced to zero in a time interval of 5.00×10^{-5} s. Determine the average emf induced in the 4800-turn coil.

21-39 The magnetic field at a distance x out along the axis of a loop of radius R carrying a current I is given by $B = \mu_0 I R^2 / 2x^3$. The maximum value of the earth's field at the surface $(7.0 \times 10^{-5}$ T) occurs at the south magnetic pole 6370 km from the earth's center. Imagine that this field is due to a belt of charges circulating about the magnetic axis at the boundary to the earth's inner core 1250 km from its center. (a) Determine the current necessary to produce the observed magnetic field at the south magnetic pole. (b) If the circulating charges are electrons, in which direction must they be moving?

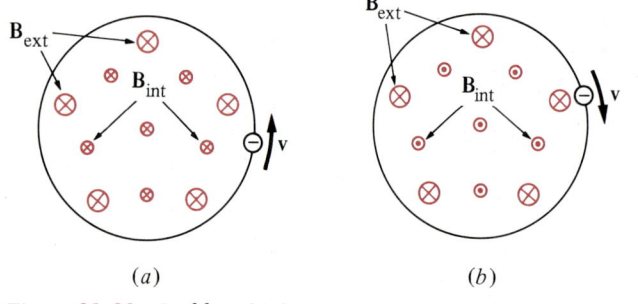

Figure 21-32 Problem 21-37.

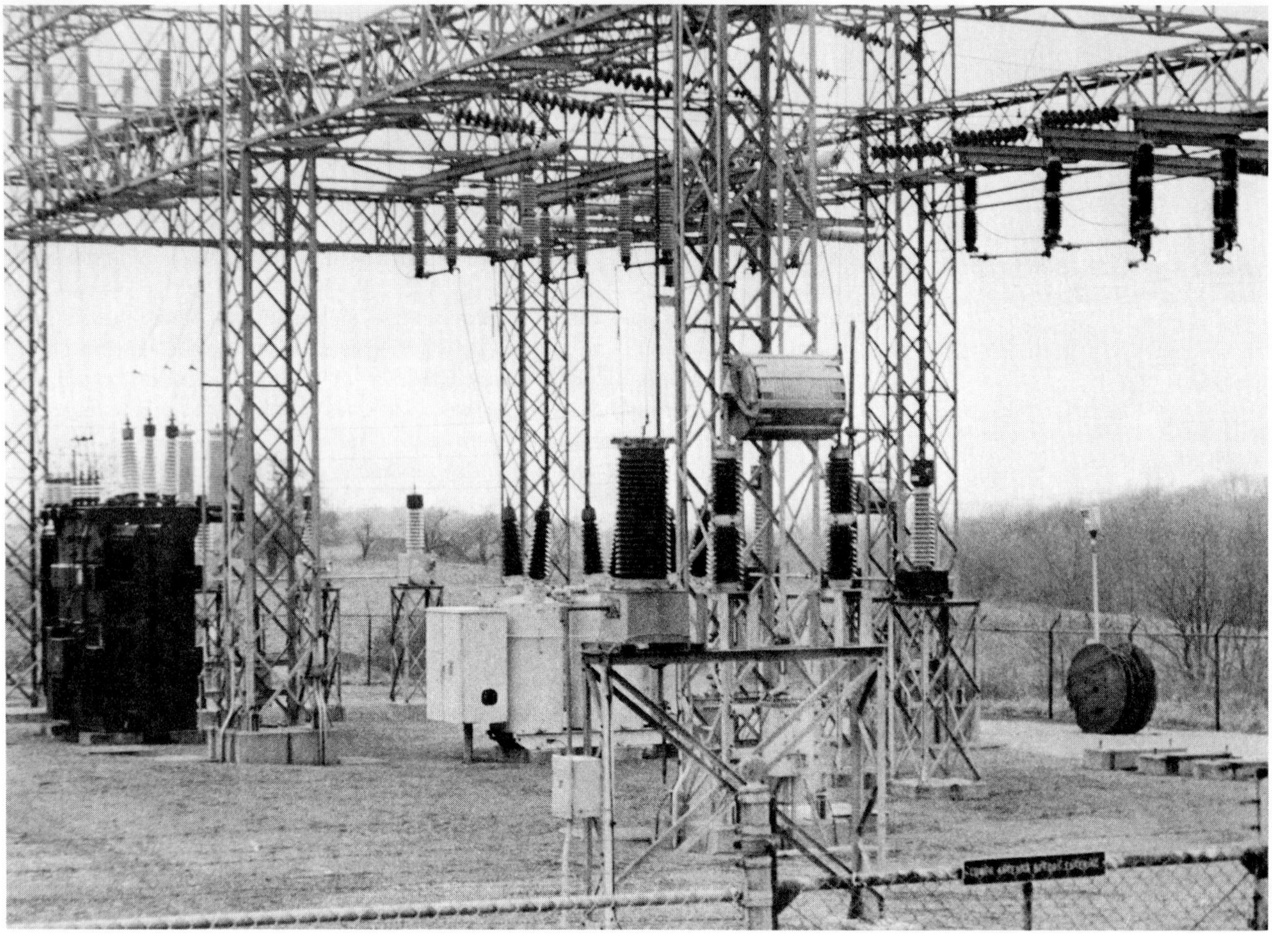

22 Inductance, Motors, and Generators

22-1 Introduction

In Section 21-6 we saw that when a current is changing in a solenoid, an emf is induced in a second loop wrapped around the solenoid, a result of a change in flux linkage between the circuits described by Faraday's law. Likewise, a changing current in the second loop should give rise to an induced emf in the solenoid. In fact, if any two coils of wire *share* a possible flux linkage, a changing current in either one may induce an emf in the other. We say that two such coils possess a common property appropriately called **mutual inductance.**

We may go even further. When the current is changing in an isolated solenoid (or any circuit), it gives rise to a changing field and hence to a changing flux through the solenoid itself. Faraday's law applies to this changing flux as well. The changing flux gives rise to an induced emf and hence to a current in the solenoid which opposes the change. Since this will happen when the solenoid is isolated, that is, all by itself, we call the property of such an arrangement **self-inductance.**

Both mutual inductance and, particularly, self-inductance are important properties in electric circuits. In many circuits, an inductance may be inserted intentionally to take advantage of its properties. In other circuits, the properties of unintentional inductances may hinder the operation and must be minimized. Ultimately, studying inductance leads us to the electric generator, by which practically all our electric power is produced.

22-2 Mutual Inductance and Self-Inductance

We saw that the magnetic field produced by any current carrier is proportional to the magnitude of the current in the conductor. Thus we conclude that the flux set up by the field through some area is also proportional to the current. This provides us with a relationship that we can use to define the properties of mutual inductance and self-inductance. Figure 22-1 shows a coil of wire wrapped around a solenoid. When the switch in the solenoid circuit is closed, a current I_1 begins to grow in the solenoid. If the solenoid has n_1 turns per unit length, its field is

$$B = \mu_0 n_1 I_1$$

The flux through the solenoid and therefore also through the coil is

$$\Phi = BA = \mu_0 n_1 I_1 A$$

where A is the cross section of the solenoid. If there are N_2 turns in the coil, we have by Faraday's law that the emf induced in the coil is

$$\mathcal{E}_2 = -N_2 \frac{\Delta \Phi}{\Delta t} = -\mu_0 n_1 N_2 A \frac{\Delta I_1}{\Delta t}$$

All the factors multiplying the time rate of change of the current are geometrical and constant. For this case, we define the mutual inductance of the coil with respect to the solenoid as

$$M_{21} = \mu_0 n_1 N_2 A$$

using the symbol M_{21} to represent the mutual inductance of coil 2 with respect to coil 1. In general, for any two coils or circuits for which there is a flux linkage, we may write

$$\mathcal{E}_2 = -M_{21} \frac{\Delta I_1}{\Delta t} \qquad (22\text{-}1)$$

It is also seen that a changing current in coil 2 induces an emf in coil 1 given by

$$\mathcal{E}_1 = -M_{12} \frac{\Delta I_2}{\Delta t} \qquad (22\text{-}2)$$

It can be shown (see Problem 22-10) that $M_{12} = M_{21}$, and therefore we can drop the subscripts from M and write

$$\mathcal{E}_2 = -M \frac{\Delta I_1}{\Delta t} \qquad (22\text{-}3)$$

and

$$\mathcal{E}_1 = -M \frac{\Delta I_2}{\Delta t} \qquad (22\text{-}4)$$

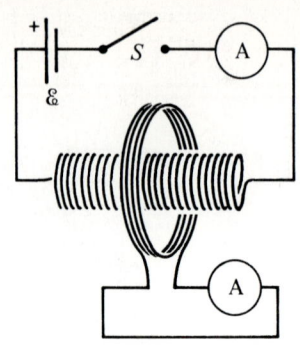

Figure 22-1 Mutual inductance. A large coil in a circuit containing an ammeter but no emf is wrapped around a solenoid that is in a separate circuit containing an emf, switch, and ammeter. There is no electric contact between the circuits, but when the switch is closed, both ammeters will indicate that a current is present.

Operationally, we use either Equation 22-3 or 22-4 to define the mutual inductance of two coils:

$$M = -\frac{\mathscr{E}_2}{\Delta I_1/\Delta t} \quad \text{or} \quad M = -\frac{\mathscr{E}_1}{\Delta I_2/\Delta t}$$

The SI units of mutual inductance are volts per ampere per second [V/(A/s)]. This is given the name **henry** (H) to honor Joseph Henry (1797–1878), an American physicist who did much of the early work on electromagnetic induction. We have that

$$1\text{ H} = 1\,\frac{\text{V}}{\text{A/s}} = 1\,\frac{\text{V}\cdot\text{s}}{\text{A}}$$

22-2 Mutual Inductance and Self-Inductance

Example 22-1 When the current in a coil is changing at the rate of 80.0 A/s, there is an induced emf of 20.0 V in a second coil close-by. What is the magnitude of the mutual inductance between the two coils?

Solution We first solve Equation 22-3 for M and then make a direct substitution.

$$\mathscr{E}_2 = -M\frac{\Delta I_1}{\Delta t}$$

or in magnitude,

$$M = \frac{\mathscr{E}_2}{\Delta I_1/\Delta t} = \frac{20.0\text{ V}}{80.0\text{ A/s}} = 0.250\text{ H}$$

The negative sign in the solution of Example 22-1 is ignored there. Recall that it really is an indication of the direction of the emf induced in the second coil. That emf will be such that it will set up a current in the second coil which will generate a magnetic field to oppose the direction of the changing field due to the first coil in accordance with Lenz's law.

Self-inductance is perhaps even more important. Consider the solenoid circuit in Figure 22-2. When the current through the solenoid changes, the flux through it also changes, and there will be a self-induced emf in the solenoid. If there are n closely wound turns per meter and the switch is closed, a current I in the solenoid gives rise to a magnetic field

$$B = \mu_0 n I$$

The flux through the solenoid is then

$$\Phi = BA = \mu_0 n I A$$

where A is the cross section of the solenoid. If the solenoid has N turns, Faraday's law dictates that whenever the current is changing, there will be an induced emf in the solenoid equal to

$$\mathscr{E} = -N\frac{\Delta\Phi}{\Delta t} = -N\mu_0 n A \frac{\Delta I}{\Delta t}$$

We may define the self-inductance of the solenoid as

$$L = N\mu_0 n A \tag{22-5}$$

which we see again has only geometrically related quantities. This yields for the solenoid or, in fact, for any circuit in which the induced emf is due to a changing current,

Figure 22-2 Self-inductance. When the switch is closed, a current will surge through the solenoid. The changing current implies a changing flux through the solenoid, which induces an emf across its windings.

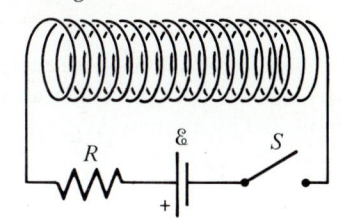

$$\mathcal{E} = -L\frac{\Delta I}{\Delta t} \qquad (22\text{-}6)$$

The self-inductance L is defined operationally as

$$L = -\frac{\mathcal{E}}{\Delta I/\Delta t} \qquad (22\text{-}7)$$

Equation 22-7 may be expressed in words as the self-inductance of a coil in a circuit is equal to the self-induced emf per unit rate of change of current. As we can see from Equation 22-7, self-inductance is also measured in henrys (H).

Every circuit possesses self-inductance since the current in every circuit must follow a closed path and thus gives rise to a magnetic field whose flux lines penetrate that path. For most circuits, this unintentional inductance effect is pretty small and may be neglected. However, we often need and use elements which have a significant self-inductance. These may be fabricated by winding coils, solenoids, and so on, which then provide a concentrated self-inductance in a small volume. Because elements with self-inductance are so commonly used in electric circuits, many people drop the *self* and refer to the property as **inductance** and to the element as an **inductor**. The inductance of a solenoid or coil may be increased several powers of 10 by filling the space of the solenoid with iron and providing a closed iron path for the magnetic field lines. The ferromagnetic effect increases the field attainable and thus the flux and inductance to very large values. Figure 22-3 shows several typical inductors in practical use.

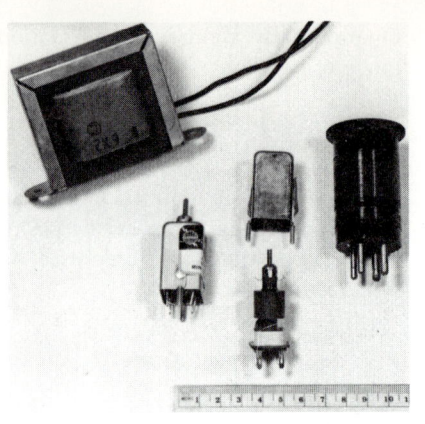

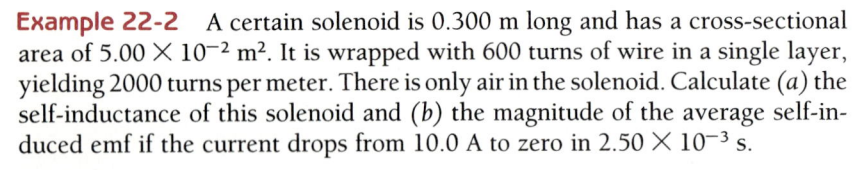

Figure 22-3 Typical circuit inductors. The inductor at the center is displayed intact and removed from its case to show the coil wound on a shaft. The choke coil at upper left is perhaps the most common type of inductor. Much smaller inductors are used in electronics and are frequently encased to protect their coils.

Example 22-2 A certain solenoid is 0.300 m long and has a cross-sectional area of 5.00×10^{-2} m². It is wrapped with 600 turns of wire in a single layer, yielding 2000 turns per meter. There is only air in the solenoid. Calculate (a) the self-inductance of this solenoid and (b) the magnitude of the average self-induced emf if the current drops from 10.0 A to zero in 2.50×10^{-3} s.

Solution   (a) This may be solved with a straightforward application of Equation 22-5.

$$\begin{aligned}L &= N\mu_0 nA \\ &= (600)\,(4\pi \times 10^{-7}\text{ T}\cdot\text{m/A})\,(2000/\text{m})\,(5.00 \times 10^{-2}\text{ m}^2) \\ &= 7.54 \times 10^{-2}\text{ H}\end{aligned}$$

(b) This part requires the use of Equation 22-6.

$$\mathcal{E} = -L\frac{\Delta I}{\Delta t}$$

$$|\mathcal{E}| = 7.54 \times 10^{-2}\text{ H}\left(\frac{10.0\text{ A} - 0}{2.50 \times 10^{-3}\text{ s}}\right) = 302\text{ V}$$

Just as capacitors store energy in an electric field, inductors store energy in a magnetic field. Consider an inductor in which a magnetic field is being established by an increasing current. We know that the induced emf is

$$\mathcal{E} = -L\frac{\Delta I}{\Delta t}$$

This opposes the change which is causing it. Therefore, the source must be supplying a voltage to the inductor of

$$V_L = L \frac{\Delta I}{\Delta t}$$

and an average power

$$P = I_{av} V_L = I_{av} L \frac{\Delta I}{\Delta t}$$

The energy stored in the inductor is equal to the work done as the current increases from zero to its final value. This is

$$W = P \, \Delta t = L I_{av} \, \Delta I$$

The average value of the current as it increases at a constant rate is simply one-half its final value. Also, the change in current is equal to the final current minus zero, or

$$I_{av} = \frac{I}{2} \quad \text{and} \quad \Delta I = I - 0 = I$$

Thus

$$v = L(\tfrac{1}{2}I)(I)$$

or

$$v = \tfrac{1}{2} L I^2 \qquad (22\text{-}8)$$

where v (lowercase upsilon) is the **magnetic energy** stored in the solenoid.

We would like also to determine the energy stored as a function of B by using the example of the solenoid. Equation 22-5 states that for the solenoid,

$$L = N \mu_0 n A$$

and from Equation 21-12, I and B are related by

$$B = \mu_0 n I \qquad (21\text{-}12)$$

Thus Equation 22-8 becomes

$$v = \tfrac{1}{2}(N \mu_0 n A) \left(\frac{B}{\mu_0 n} \right)^2$$

We substitute the number of turns per unit length times the total length of the solenoid for the total number of turns. That is, $N = nl$, where l is the total length of the solenoid. Rearranging the above equation and canceling yields

$$v = \tfrac{1}{2}(Al) \frac{B^2}{\mu_0} \qquad (22\text{-}9)$$

The quantity Al is the volume enclosed by the solenoid. We may therefore obtain an expression for the magnetic energy per unit volume (the energy density) by dividing by Al. Thus

$$u_m = \frac{v}{Al} = \frac{1}{2} \frac{B^2}{\mu_0} \qquad (22\text{-}10)$$

Even though this expression for the energy density was obtained for the special case of the magnetic field of a solenoid, it is applicable to any magnetic field in free space.

22-2 Mutual Inductance and Self-Inductance

Example 22-3 Calculate the (a) energy per unit volume and (b) the total energy stored in the magnetic field of a solenoid. The solenoid is 0.400 m long, has a cross-sectional area of 3.20×10^{-3} m², has 2.00×10^3 turns per meter, and carries a current of 6.00 A.

Solution (a) This requires an application of Equation 22-10. First, however, we must calculate the magnetic field of the solenoid.

$$B = \mu_0 n I$$
$$= (4\pi \times 10^{-7} \text{ T} \cdot \text{m/A})(2.00 \times 10^3/\text{m})(6.00 \text{ A})$$
$$= 1.51 \times 10^{-2} \text{ T}$$

$$u_m = \frac{1}{2}\frac{B^2}{\mu_0} = \frac{1}{2}\left[\frac{(1.51 \times 10^{-2} \text{ T})^2}{4\pi \times 10^{-7} \text{ T} \cdot \text{m/A}}\right] = 90.7 \text{ J/m}^3$$

(b) The answer to part a multiplied by the volume solves this part.

$$v = (Al)u_m = (3.20 \times 10^{-3} \text{ m}^2)(0.400 \text{ m})(90.7 \text{ J/m}^3) = 0.116 \text{ J}$$

22-3 Inductors in Electric Circuits

There are many applications in which an inductor is used to provide an "inertia" in a circuit. Recognize that the action of the inductor is to oppose change. Try to increase the current; the inductor's reaction is to attempt to decrease it. Try to decrease the current, and the inductor will attempt to increase it. Further, the action of the inductor is more pronounced at higher rates of change of current.

While every circuit has at least a small amount of inductance, we assume that any stray inductance is part of the intentional inductors. Similarly, since inductors are coils of wire, each has some resistance, but we assume that any stray resistance of the inductors is combined with that of the intentional resistors. Thus, we deal theoretically with "pure" inductors, "pure" resistors, and so on. Figure 22-4 is a schematic of an inductor and resistor in series with a battery and switch. (Note that the circuit symbol for an inductor looks somewhat like a coil.) When the switch is closed, there is a current in the direction indicated. Beginning at the switch and going clockwise to apply Kirchhoff's loop rule to the circuit yields

$$V - V_R - V_L = 0$$

where V_R and V_L are the voltages across the resistor and inductor, respectively. Substituting IR and $L\,\Delta I/\Delta t$ for these voltages, we obtain

$$V - IR - L\frac{\Delta I}{\Delta t} = 0$$

Solving for $\Delta I/\Delta t$ gives us

$$\frac{\Delta I}{\Delta t} = -\frac{I}{L/R} + \frac{V}{L} \qquad (22\text{-}11)$$

This equation is identical in form to that of Equation 20-21. The solution to

Figure 22-4 An *LR* circuit containing both an inductor and a resistor in series with an emf and a switch.

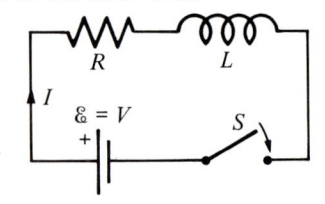

Equation 22-11 also has the same form as Equation 20-22, and the instantaneous current is

$$I = \frac{V}{R}(1 - e^{-[t/(L/R)]}) \qquad (22\text{-}12)$$

The quantity t is the time, beginning at the closing of the switch, and L/R is the time constant of the exponential process. Figure 22-5 shows the graphical relationship of current vs. time. This is nothing more than a plot of Equation 22-12 and is typical of this type of exponential process.

Once there is a current in the inductor, a magnetic flux will have been established through its turns. The inductor will act to oppose the disestablishment of this flux if, for example, the circuit is "shorted" (the battery terminals are connected by a wire and the battery is removed). Figure 22-6 shows a schematic of such a circuit at the instant of shorting. If we apply Kirchhoff's loop rule to this circuit going clockwise from the lower left, we obtain

$$-V_R - V_L = 0$$

This becomes

$$IR + L\frac{\Delta I}{\Delta t} = 0$$

or

$$\frac{\Delta I}{\Delta t} = -\frac{I}{L/R} \qquad (22\text{-}13)$$

Once again, this equation has the same form as Equation 20-24, and thus its solution for the current is of the form of Equation 20-25.

$$I = I_0 e^{-[t/(L/R)]} \qquad (22\text{-}14)$$

Equation 22-14 is, of course, an exponential decay process with a time constant equal to L/R. The current vs. time is shown in Figure 22-7. For both circuits, the inductor acts to spread the time required for the current to reach its final value, which would be almost zero time in its absence. Even if the switch of Figure 22-4 is simply opened after the current has been established, the tendency of L to keep I going will cause a spark across the switch.

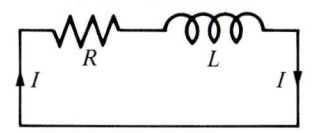

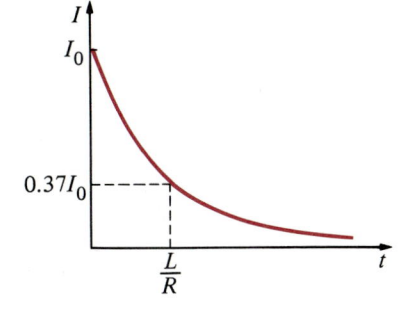

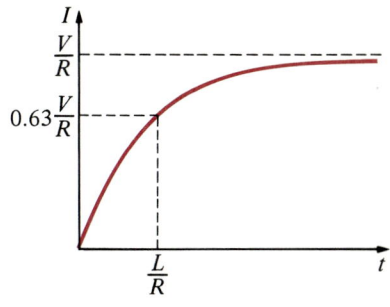

Figure 22-5 The current through an LR circuit vs. time. As with RC circuits, the variation is best described in terms of a time constant, in this case L/R. In this time interval, the current rises from zero to $1 - 1/e$ of its maximum value (about 63 percent of $I_{\max}$).

Figure 22-6 The circuit of Figure 22-4 at the instant the battery is electrically removed. The current does not cease at this instant because the energy stored in the magnetic field of the inductor sustains the current as the field collapses.

Figure 22-7 The current in the circuit of Figure 22-6 vs. time from the shorting of the battery. During the interval equal to the time constant L/R, the current drops to $1/e$ (or about 37 percent) of I_0.

531

Example 22-4 A circuit element of inductance 20.0 H and resistance 5.00 Ω is placed in a series circuit with an emf of 10.0 V and a switch. Calculate (a) the final steady-state current of the circuit after the switch is closed and (b) the current at $t = 1.50$ s.

Solution (a) Both parts may be solved by applying Equation 22-12. For part a, we use a very large value for t, essentially $t \to \infty$.

$$I = \frac{V}{R}(1 - e^{-[t/(L/R)]})$$

As $t \to \infty$, this becomes

$$I_{ss} = \frac{V}{R} = \frac{10.0 \text{ V}}{5.00 \text{ Ω}} = 2.00 \text{ A}$$

(b) This is a straightforward application using the values given.

$$I_{1.5} = (2.00 \text{ A})\left[1 - \exp\left(-\frac{1.50 \text{ s}}{20.0 \text{ H}/5.00 \text{ Ω}}\right)\right]$$
$$= (2.00 \text{ A})(1 - e^{-0.375}) = 0.625 \text{ A}$$

An interesting and useful circuit is one which contains an inductor, a capacitor, and a resistor in series, an *LCR* **series circuit.** Such a circuit (sometimes called a tank circuit) creates electric oscillations which are analogous to the mechanical oscillations of a damped harmonic oscillator that we saw in Chapter 10.

Figure 22-8 shows an idealized circuit in a sequence we may use to investigate electric oscillations. The circuit contains a charged capacitor, an inductor, and a switch but no resistance.

The process described in Figure 22-8 will repeat indefinitely if there are no energy losses in the circuit. The repeating processes are called **electric oscillations.** The energy of the circuit is originally stored in the electric field of the capacitor as $\frac{1}{2}q_m^2/C$, where q_m is the initial (maximum) charge on the capacitor. This energy is transferred to the magnetic field of the inductor, reaching a maximum value of $\frac{1}{2}LI_m^2$, where I_m is the maximum current. As the oscillations continue, the energy is shared by the electric field and magnetic field, each growing and reaching a maximum as the other drops to zero. At some intermediate time, the energy of the circuit is part electric and part magnetic. Let us apply Kirchhoff's loop rule to circuit 2 shown in Figure 22-8 when there is a charge q remaining in the capacitor and a current I in the circuit. Beginning at the lower right and traversing the circuit clockwise yields

$$-V_C - V_L = 0$$

Upon substitution, this becomes

$$-\frac{q}{C} - L\frac{\Delta I}{\Delta t} = 0$$

or

$$\frac{\Delta I}{\Delta t} = -\frac{q}{LC} \qquad (22\text{-}15)$$

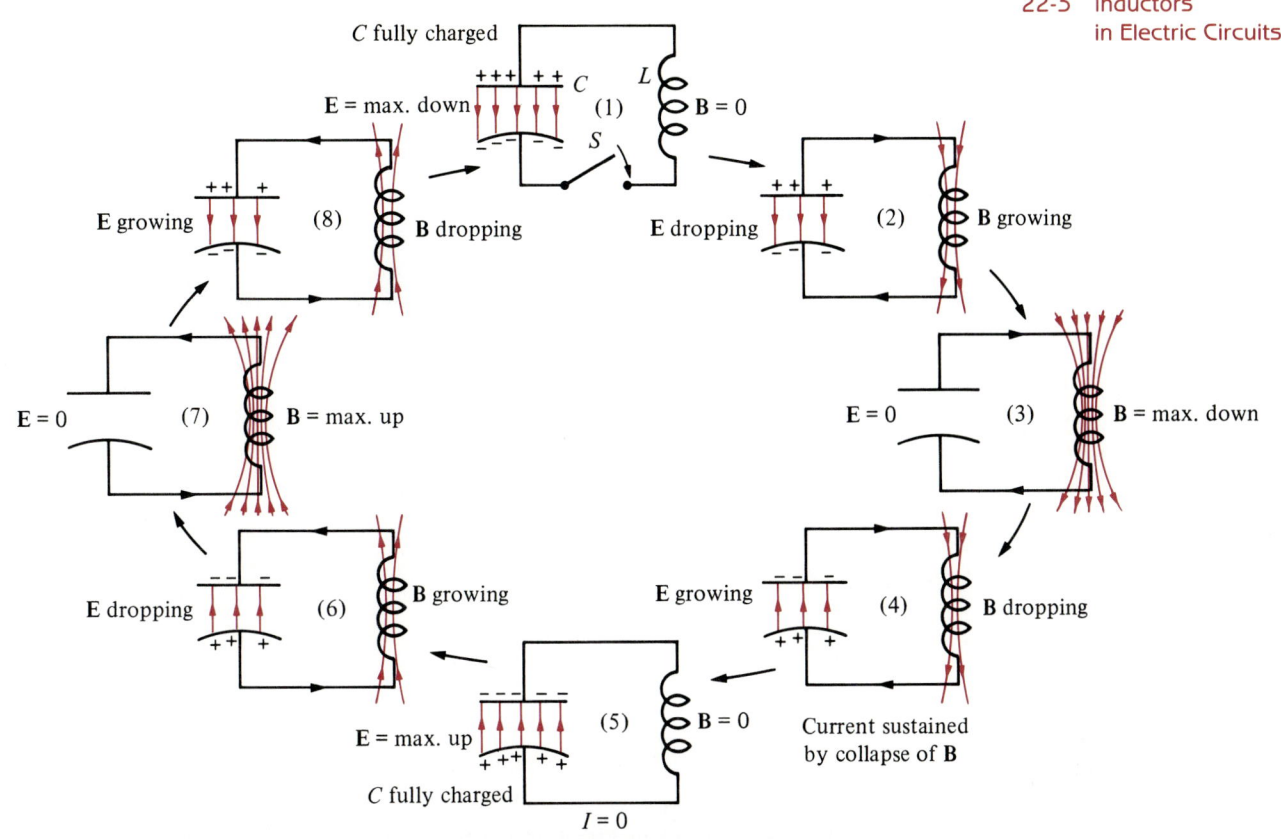

Figure 22-8 Electric oscillation in an *LC* circuit. (1) Energy exists in the separated charge of a charged capacitor at the instant the switch is closed. (2) The discharging capacitor sets up a current through the inductor, producing a growing magnetic field. (3) As the current reaches a maximum, the magnetic field is maximum. (4) As the current begins to drop, the collapsing magnetic field of the inductor sustains the current. (5) The continuing magnetic field collapse sustains the current past the point of zero charge separation on the capacitor and builds up a charge opposite the original charge. The complete disappearance of the magnetic field now allows the capacitor to discharge in the opposite direction (6, 7, 8, 1).

Equation 22-15 is analogous to Equation 10-2, which may be written

$$a = \frac{\Delta v}{\Delta t} = -\frac{k}{m} x \tag{10-2}$$

Because v has the same relationship to x ($v = \Delta x/\Delta t$) as I has to q ($I = \Delta q/\Delta t$), the analogy is complete, and the solution to Equation 22-15 should have the same form as that of Equation 10-2. We recall that one of the forms of the solution to Equation 10-2 is

$$x = A \cos(\omega t + \theta_0)$$

where $\omega = \sqrt{k/m}$ and A is the maximum value of x. Thus we expect an acceptable form of the solution to Equation 22-15 to be

$$q = q_m \cos(\omega t + \theta_0) \tag{22-16}$$

where
$$\omega = \frac{1}{\sqrt{LC}} \quad (22\text{-}17)$$

Equation 22-17 gives the angular frequency of oscillation. Recall from Equations 10-7 and 10-13 that the frequency f is given by
$$f = \frac{\omega}{2\pi} = \frac{1}{2\pi\sqrt{LC}}$$

This is the natural frequency of the circuit and plays an important part when the circuit is driven at different frequencies (Chapter 23).

Equation 22-16 represents a cosine function if q is plotted vs. time. You will recall, however, that this equation was from an ideal circuit for which we postulated no resistance. However, all circuits do have resistance, and the plot of q vs. t for an LCR circuit (Figure 22-9) shows an amplitude which decreases in time, a damped oscillator. The magnitude of R determines how much energy is lost per cycle and hence how rapidly the amplitude decreases. Note that the current in this circuit is not confined to one direction; thus we have in Figure 22-8 an ac circuit.

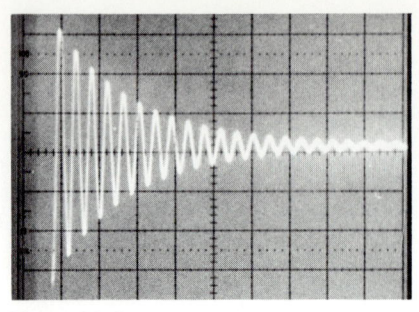

Figure 22-9 Oscilloscope trace of current vs. time in an LCR circuit. The resistance ensures that the energy in the circuit will decrease in time from the sustained oscillation of a pure LC circuit. The curve is that of a damped harmonic oscillator.

22-4 Torque on a Current Loop in a Magnetic Field

In Example 21-7 we saw that the net magnetic force acting on a current loop whose plane is parallel to a magnetic field is zero. Let us consider a somewhat more general situation. In Figure 22-10a, we see a single current-carrying loop of length l and width w in a uniform magnetic field. The field makes an angle θ with the plane of the loop and hence with the ends (of width w). The other sides are perpendicular to the magnetic field **B**. By the right-hand rule, the forces on opposite ends of the loop cancel. Figure 22-10b shows the loop as viewed parallel to the plane along the sides of length l. The forces on these sides both have magnitudes equal to IlB, but they are not along the same line. These forces exert unbalanced torques on the loop, which could cause it to begin to rotate provided it is free to do so. Choosing an axis parallel to the sides and through the center of the loop, we see that the moment arm is the same magnitude for each force and is equal to $(w/2)\cos\theta$. The torque due to either one of the forces has a magnitude

$$\tau_1 = F\frac{w}{2}\cos\theta = IlB\frac{w}{2}\cos\theta$$

Since both torques are in the same direction, the total torque acting on the loop is

$$\tau = 2IlB\frac{w}{2}\cos\theta = IlwB\cos\theta$$

But, lw is just the area A of the loop, and we may write

$$\tau = IAB\cos\theta \quad (22\text{-}18)$$

Note that there should be no net force acting on the leads carrying the current to

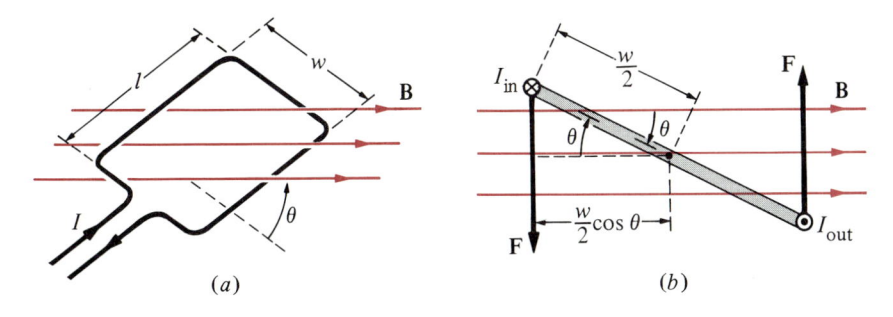

Figure 22-10 (a) A current-carrying coil of length l and width w in a magnetic field perpendicular to the coil length at the instant when the angle between the field and the plane of the coil is θ. (b) The identical situation seen in the plane of the coil looking perpendicular to w and B.

the loop. Since the current is in opposite directions in the two leads, if they are twisted together, the net current in the two leads is zero.

If the current loop is really a coil of N turns, the torque of Equation 22-18 will be exerted on each turn. Thus the total magnetic torque will be

$$\tau_N = NIAB \cos \theta \qquad (22\text{-}19)$$

Finally, we used a rectangular loop, but that was not necessary. Equation 22-19 is correct for any shaped loop provided that A is the area of the loop.

In Chapter 20, we noted that most analog meters utilize a moving coil meter to indicate the presence and magnitude of a current. Figure 22-11 depicts a typical meter movement of what is called a d'Arsonval galvanometer, an instrument that responds to very small currents. The movement consists of a coil of very fine wire mounted on a rod and free to rotate about the axis of the rod. This coil is located between the opposite poles of a permanent magnet. The pole faces are curved, and a stationary, cylindrical iron core is mounted inside

> The radial magnetic field of the galvanometer keeps the field constant at the coil within its limits of rotation regardless of the orientation of the coil. This ensures that the torque on the coil does not depend on its orientation but is proportional to the current in it.

Figure 22-11 (a) The movement of a d'Arsonval galvanometer. (b) A top view shows how a central iron cylinder and curved pole pieces produce radial lines of force so that the field remains uniform through the coil as it rotates.

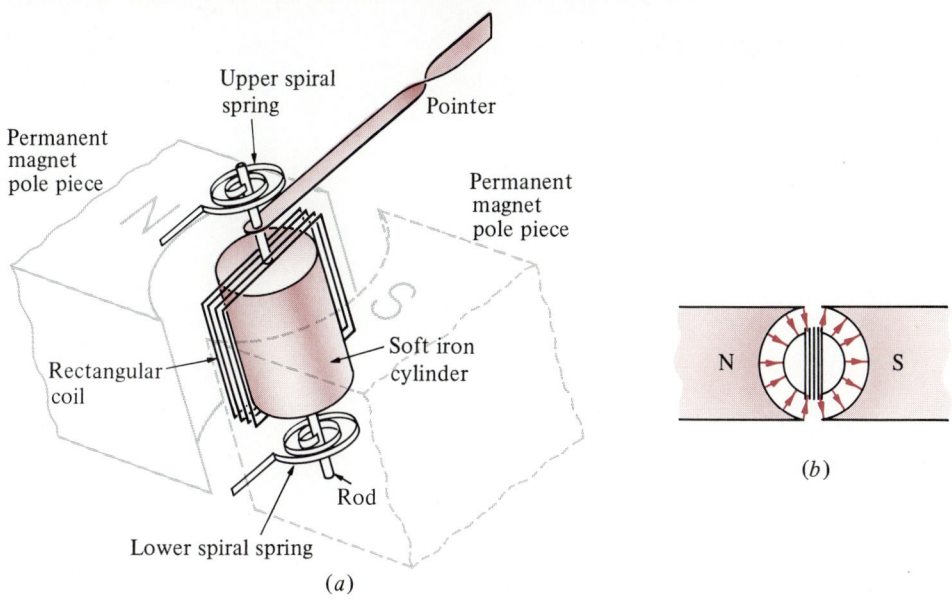

535

the coil in order to concentrate the magnetic field and shape it to be as nearly radial as possible. The coil is connected to the current source through the spiral springs shown. If there is no current, the springs hold the coil and hence the attached pointer in an equilibrium position corresponding to zero.

When charge flows, a torque given by Equation 22-19 is exerted on the coil. If the field is nearly radial, the plane of the coil will always be parallel to the field. Thus $\theta = 0$, $\cos \theta = 1$, and the torque on the coil is

$$\tau_N = NIAB \tag{22-20}$$

This torque starts the coil rotating, which winds the spiral springs. This winding produces an elastic restoring torque which is proportional to the angular deflection, and we may write

$$\tau_R = -K_R \beta \tag{22-21}$$

where τ_R = torque due to the spiral springs
beta (β) = angle of deflection
K_R = effective spring constant

The restoring torque τ_R opposes the deflection of the coil and increases in magnitude until the sum of the torques on the coil becomes zero.

$$\tau_N + \tau_R = 0$$

$$NIAB + (-K_R \beta) = 0$$

$$\beta = \frac{NIAB}{K_R} \tag{22-22}$$

Note that the angle through which the coil rotates is proportional to the current.

Example 22-5 A 50-turn galvanometer coil has an area of 2.00×10^{-4} m² and is placed in a field of 0.600 T. Determine the magnitude of the effective spring constant of the spiral springs if a current of 0.0800 A is to cause a deflection of 60°.

Solution We must apply Equation 22-22 and solve it for K_R.

$$\beta = \frac{NIAB}{K_R}$$

or

$$K_R = \frac{NIAB}{\beta} = \frac{(50)(2.00 \times 10^{-4} \text{ m}^2)(0.0800 \text{ A})(0.600 \text{ T})}{60°}$$

$$= 8.00 \times 10^{-6} \text{ N} \cdot \text{m/deg}$$

22-5 Direct-Current Motors

The torque exerted on a current loop by a magnetic field may be used to provide mechanical energy. In fact, we may think of a dc electric **motor** as a meter coil continuously turning about its axis, converting electric energy to mechanical

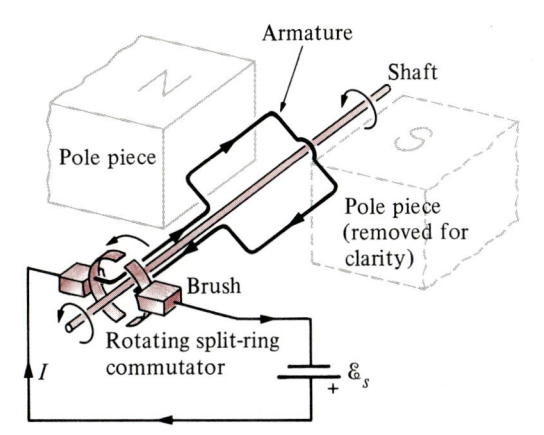

Figure 22-12 A simple one-turn dc motor.

energy. Figure 22-12 illustrates a simple one-turn dc motor. The coil of wire, called the **armature**, is most often wound on an iron core to direct and enhance the magnetic flux. The current is supplied to the armature by some dc source $\mathcal{E}_s$ through sliding contacts called **brushes**. A split ring, called a **commutator**, is attached to the armature and turns with it. The purpose of the commutator is to reverse the current in the armature so that the magnetic torque provided by the field of the permanent magnets is always in the same direction. The commutator is split and oriented to ensure that each brush breaks contact with one of the sides and makes contact with the other just as the net magnetic torque on the armature goes to zero and is about to reverse directions. The armature is mechanically connected to a shaft (usually fixed to the iron core) which turns with the armature to provide the mechanical energy to run a portable electric razor, or whatever.

While the simple arrangement shown in Figure 22-12 works, its operation is not very smooth since the torque is continuously changing. For smoother operation, an armature is constructed of many partial windings, with the plane of each partial winding oriented at a slightly different angle around the shaft. The partial windings are then connected to different segments of a multisegmented commutator (Figure 22-13).

The torque produced by a single winding of N turns is given by Equation 22-19 as

$$\tau_N = NIAB \cos \theta$$

The average area exposed to flux as the armature turns at a constant angular speed is

$$(A \cos \theta)_{av} = \frac{2}{\pi} A = 0.637 A$$

because the moment arm of each winding varies from zero to $w/2$ with the cosine of the angle between the plane of the winding and the magnetic field. Thus the average torque produced by a practical dc motor is

$$\tau = 0.637 NIAB \quad (22\text{-}23)$$

When the motor begins to turn, the windings are rotating in a magnetic field. According to Faraday's law (and Lenz's law), there will be an emf induced in the windings which opposes the changing flux in them. Thus the motor

Figure 22-13 Armature of a dc automobile starter motor, showing the multiple windings and the slotted iron core. The top end of the armature contains the split-ring connections.

produces an emf, often called a **back emf,** which opposes the external power source. This back emf depends on the angular speed ω of the armature and can be shown to be

$$\mathcal{E}_b = 0.637 NAB\omega$$

The current in the armature is limited by both the back emf and the resistance R_c of the armature coils as

$$I = \frac{\mathcal{E}_s - \mathcal{E}_b}{R_c} \qquad (22\text{-}24)$$

When the shaft of the motor is not engaged to provide power, the motor spins freely. In that case, it may reach an angular speed such that the back emf is almost equal to the source voltage, and the motor draws very little current. However, when the shaft is engaged so that the motor is doing mechanical work, more current is drawn from the source. The back emf decreases, and the current increases in response to a mechanical load. In general, the power drawn from the source is

$$P = \mathcal{E}_s I = \mathcal{E}_b I + I^2 R_c \qquad (22\text{-}25)$$

This is equal to the mechanical power delivered by the motor plus the Joule heating loss in the coils.

Example 22-6 For a motor being supplied by a 120-V source, with an armature coil resistance of 1.50 Ω and carrying a current of 2.00 A, calculate (a) the back emf and (b) the mechanical power delivered by the motor.

Solution (a) We simply solve Equation 22-24 for $\mathcal{E}_b$ and substitute the other known quantities.

$$I = \frac{\mathcal{E}_s - \mathcal{E}_b}{R_c}$$

$$\mathcal{E}_b = \mathcal{E}_s - IR_c = 120 \text{ V} - (2.00 \text{ A})(1.50 \text{ }\Omega) = 117 \text{ V}$$

(b) Here we must recognize that the mechanical power is equivalent to the product of the back emf with the current.

$$P_m = \mathcal{E}_b I = (117 \text{ V})(2.00 \text{ A}) = 234 \text{ W}$$

22-6 Electric Generators

Because a coil of wire turning in a magnetic field produces an induced emf, the principle of the motor may be reversed to produce a **generator** which converts mechanical energy into electricity. Consider the coil of Figure 22-14 which is being mechanically turned at an angular speed ω in the magnetic field of the permanent magnets shown. The coil has sides of length l and width w with an emf being induced in each side. The speed of the sides is $v = \omega R = \omega(\frac{1}{2}w)$. Thus there is a motional emf (Section 21-6) induced in each side equal to

$$\mathcal{E}_1 = Blv \cos \theta$$

22-6 Electric Generators

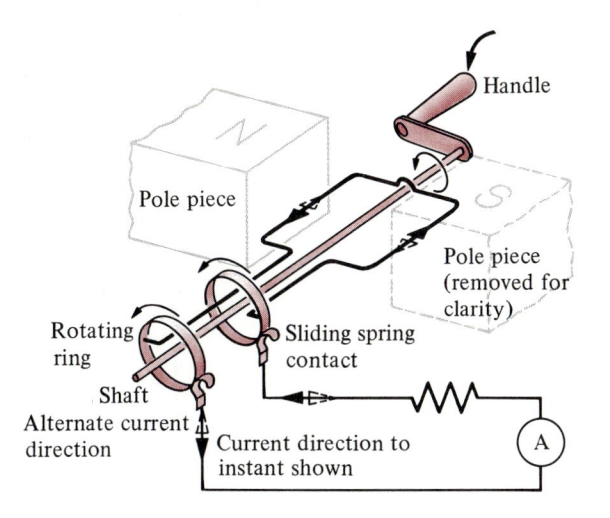

Figure 22-14 A simple one-turn electric generator. The coil is turned in the fixed magnetic field of the pole pieces. The flux through the coil varies and produces an emf. The external circuit is connected to the ring contacts. The flux through the loop alternately increases and decreases as the loop turns, generating an alternating current in the loop. (The dashed arrowheads represent the alternate direction of the current.)

where θ is the angle between **v** and **B**. The net induced emf is then twice this since the emf's from the two sides are the same magnitude and in the same direction. Thus we can write the net emf as

$$\mathscr{E} = 2Bl\omega(\tfrac{1}{2}w)\cos\theta$$

Now rearranging and recognizing that $lw = A$, the area of the coil, yields

$$\mathscr{E} = BA\omega\cos\theta$$

If the coil has N turns, the emf is N times as large. Further, we must recognize that the angle θ is changing with time. In fact, $\theta = \omega t = 2\pi f t$. Thus, in general, we write

$$\mathscr{E} = NBA\omega\cos 2\pi f t \qquad (22\text{-}26)$$

If a current is drawn from the coil through the load resistors, that current produces what is known as a **back torque** opposing the movement of the coil. Thus if a generator is to maintain a constant speed, work must be done in rotating the coil against the back torque. It can be shown that the rate of work, that is, the power, is exactly equal to the electric power delivered to the load plus the Joule heating loss in the coils. The emf generated in this example alternates in direction. Thus the current in the load is alternating, and this is an ac generator.*

We can obtain a dc output if we replace the slip rings with a split-ring commutator which reverses the connections to the load at every half turn just as the emf is about to reverse. The direct current so generated would fluctuate quite a bit and look somewhat like that plotted in Figure 22-15. The negative halves of the sinusoidal variation have been flipped over by the split-ring commutator to provide a direct current I_L to the load resistor. The current from a dc generator can be made very smooth by using a large number of partial

Figure 22-15 Current vs. time for a simple dc generator. The same small angle of rotation produces differing changes in flux depending on the orientation of the coils to the magnetic field, hence a varying emf.

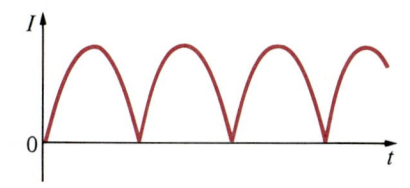

---

* Common ac generators are often called simply **alternators.** This is the terminology for the mechanism installed in most automobiles. A separate circuit must then convert the output of the alternator to direct current in order to charge a battery.

539

windings oriented at different angles with a multisegmented commutator. This is similar to the dc motor. In fact, we often say that, in principle, a dc generator is simply a dc motor used to convert mechanical energy to electric energy, rather than the other way around. While this is technically true, generators are designed primarily to have the smallest possible coil resistance in order that the terminal voltage be as large as possible. Motors, on the other hand, are designed to give a large back emf which requires a very large number of turns and increases the coil resistance. For these and several other reasons, a motor being operated as a generator or a generator being operated as a motor loses efficiency.

Example 22-7 A single rectangular loop of wire is 0.150 m long by 0.100 m wide. (a) At what constant angular speed must it rotate in a magnetic field of 0.200 T in order to have an induced emf of amplitude 50.0 V? (b) What will be the frequency of the alternating voltage?

Solution (a) We obtain an equation for the amplitude of the voltage from Equation 22-26 in terms of ω and solve for it.

$$\mathscr{E}_{max} = NBA\omega$$

or

$$\omega = \frac{\mathscr{E}_{max}}{NBA} = \frac{50.0 \text{ V}}{(1)(0.200 \text{ T})(0.150 \text{ m})(0.100 \text{ m})}$$
$$= 1.67 \times 10^4 \text{ rad/s}$$

(b) Here we must recall from Section 10-4 that $\omega = 2\pi f$, or

$$f = \frac{\omega}{2\pi} = \frac{1.67 \times 10^4 /\text{s}}{2\pi} = 2.65 \times 10^3 \text{ Hz}$$

In most commercial generators, the magnetic field does not come from a permanent magnet but is supplied by a direct current through field-generating coils that rotate about a stationary armature. As the magnetic field rotates, the flux through the armature windings changes with time, generating an induced emf. In the United States, this output is alternating current with a frequency of 60 Hz. This is generally what is available from receptacles in homes, businesses, and laboratories.

22-7 Alternating-Current Motors

Most electricity in the United States comes in the form of 60 Hz ac. For this reason, we must have motors which can convert alternating current to mechanical energy. In some ways, ac motors should be easier to design and more reliable than dc motors, because we need not worry about split-ring commutators and sliding contacts (brushes) since the current does the reversing by itself.

One of the most rugged motor designs, called a **series-wound motor,** includes field coils to generate the magnetic field plus armature windings. The armature does the turning and is connected to the mechanical load. A series-wound motor operates on direct current as well as on alternating current and for this reason is sometimes called a universal motor. It operates just as well on alternating current as on direct current because the field windings and the armature windings are in series. Thus the currents in both reverse at the same instant, and the torque remains in the same direction as before reversal. An auxiliary stationary starting winding is often used with this motor, both to start it and to ensure that it turns in the right direction. The auxiliary winding helps generate a magnetic field whose direction rotates about the armature axis, pulling the armature around with it. When the armature reaches operating speed, the starting winding may be switched out of the circuit.

Another important ac motor design is the **induction motor,** used in many large appliances and most commercial applications. In this design, the alternating current is supplied only to stationary field coils, and the armature windings are not connected to a current source. A current is induced in the armature by the time-varying flux due to the changing field of the field windings.

The ac motor design used for a typical analog (nondigital) electric clock is called a **synchronous motor.** The clock-motor armature rotates in synchronization with the changing magnetic field of the field windings. This is then geared down to turn the second, minute, and hour hands. To maintain proper time, this type of clock must be supplied with exactly 60 Hz ac. At 59 Hz it would lose 1 min every hour. Actually, the frequency of commercial electricity is monitored quite closely. At least once every day, power plants adjust the frequency to catch up or slow down any synchronous motors. The requirement is that the daily counter show exactly 5,184,000 cycles in every 24 h.

22-8 Transformers

You may wonder why commercial and domestic power are alternating current rather than direct current. Actually, the first lighting systems put into operation by Thomas Edison ran on direct current. The history of the emergence of alternating current and the conversion of the already-existent dc systems to ac ones makes for some fascinating reading. People were working at the forefront of understanding, and misinformation and misunderstandings abounded. In any case, alternating current surfaced as the victor in the competition because of several advantages. First, as we have seen, it is easier to generate alternating current than direct current. More important, however, is the fact that alternating current can easily be transformed to very high voltages for efficient transmission (with small energy loss) and then to low voltages for distribution to and use by consumers. The existence of the transformer probably had a larger effect on the development of the electric industry than any one other machine or principle.

A transformer schematic is shown in Figure 22-16. Two coils of insulated wire, called the **primary coil** and the **secondary coil,** are wrapped around a closed iron core. Typically, the primary is supplied by a source of alternating current which produces a time-varying current. This current induces a time-varying magnetic flux in the iron core and hence through the secondary. By Faraday's law, this changing flux induces an emf in the secondary which is then transmitted to the load. If there is negligible resistance in the primary, then the

Figure 22-16 The **transformer** converts ac power from one voltage to another with almost perfect efficiency. The ratio of the emf's across the coils is proportional to the ratio of their respective numbers of turns.

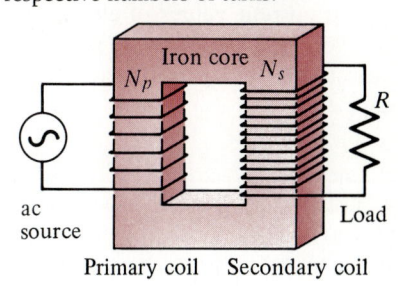

impressed alternating voltage is equal to the back emf in the primary. By Faraday's law,

$$\mathscr{E}_p = -N_p \frac{\Delta\Phi}{\Delta t}$$

where $\mathscr{E}_p$ and N_p are the induced emf and the number of turns in the primary coil and $\Delta\Phi/\Delta t$ is the time rate of change of magnetic flux in the core. The changing flux also gives rise to an induced emf in the secondary of

$$\mathscr{E}_s = -N_s \frac{\Delta\Phi}{\Delta t}$$

We assume that the rate of change of flux through both coils is the same. This is an excellent approximation if the core is iron or some other ferromagnetic material, as the flux lines are confined to the core. Eliminating the changing flux between the two equations, we obtain

$$\frac{\mathscr{E}_p}{\mathscr{E}_s} = \frac{N_p}{N_s} \quad \text{or} \quad \mathscr{E}_s = \frac{N_s}{N_p}\mathscr{E}_p \qquad (22\text{-}27)$$

If the secondary has more turns than the primary, the voltage of the secondary will be greater than that of the primary. The voltage would have been stepped up, and such a transformer is called a **step-up transformer.** If the number of secondary windings is smaller than the number of primary windings, the voltage is stepped down in a **step-down transformer.**

Although it may seem so, we do not get something for nothing in the step-up transformer or incur any net losses in the step-down process. Assuming conservation of energy, the increase in voltage in a step-up transformer must come at the expense of a decrease in current. Assuming no energy losses at all, the power delivered to the primary must be equal to the power from the secondary which is being delivered to the load. Equating expressions for power in the primary and secondary, we have

$$P_p = P_s$$
$$\mathscr{E}_p I_p = \mathscr{E}_s I_s$$

or

$$\frac{\mathscr{E}_p}{\mathscr{E}_s} = \frac{I_s}{I_p} = \frac{N_p}{N_s} \qquad (22\text{-}28)$$

Equation 22-28 shows that the currents in the two coils are inversely proportional to the voltages, provided that there is no energy loss in the transfer. This inverse relationship is central to the efficient transmission of electric power. In practice, transformers approach 100 percent efficiency in converting the voltage and current values of electric power, so there is little loss in their use. However, Joule heating (I^2R) is the prime source of energy loss in long-distance transmission. By raising the voltage, current is reduced and along with it, energy loss.

Example 22-8 Power to be transmitted 25.0 km over cables with resistance of 0.100 Ω/km is to be used in an evaporator requiring 100 A at 220 V. Calculate the Joule heating losses and the output voltage required if (a) a generator is

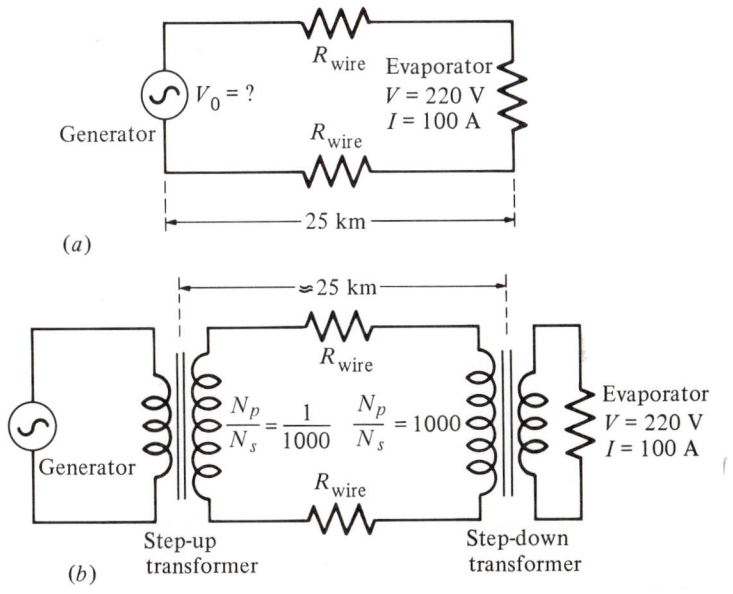

Figure 22-17 Example 22-8. (a) The generator powers the evaporator directly. (b) Step-up and step-down transformers are used.

connected to the evaporator with two 25.0-km pieces of cable and (b) a step-up transformer of $N_s/N_p = 1000$ is used at the generator and a step-down transformer of $N_p/N_s = 1000$ is used at the evaporator end.

Solution (a) We must determine the losses from I^2R after we calculate the total resistance of the cable. We must then calculate the voltage drop across the cables in order to determine the magnitude of the required output voltage. Figure 22-17a shows the circuit as postulated for this part. With two cables 25.0 km long at $R/l = 0.100 \,\Omega/\text{km}$, we have that the net resistance of the cables is

$$R_c = 2(25.0 \text{ km})(0.100 \,\Omega/\text{km}) = 5.00 \,\Omega$$

The Joule heating in the cables is then

$$P = I^2R_c = (100 \text{ A})^2(5.00 \,\Omega) = 5.00 \times 10^4 \text{ W}$$

The output voltage must equal the voltage at the evaporator plus the IR_c drop in the cables. Thus

$$V_0 = 220 \text{ V} + IR_c = 220 \text{ V} + (100 \text{ A})(5.00 \,\Omega) = 720 \text{ V}$$

(b) Here we must find the value of the voltage at which transmission takes place. From this we can determine the current and then the I^2R loss. The voltage drop and the required output voltage may be calculated in the same way as was done in part a. We see in Figure 22-17b the schematic which includes the two transformers. Note that the symbol used to represent a transformer is shown both at the generator end and at the evaporator end. Assuming the output voltage at the generator is 220 V, the first transformer steps up to obtain the transmission voltage. According to Equation 22-27, we have

$$\mathcal{E}_s = \frac{N_s}{N_p} \mathcal{E}_p = (1000)(220 \text{ V}) = 2.20 \times 10^5 \text{ V}$$

From Equation 22-28, we then find the transmission current as

$$I_s = \frac{\mathcal{E}_p}{\mathcal{E}_s} I_p = \frac{220 \text{ V}}{2.20 \times 10^5 \text{ V}} \times 100 \text{ A} = 0.100 \text{ A}$$

Thus the Joule heating in the cables is

$$P = I^2 R_c = (0.100 \text{ A})^2 (5.00 \text{ }\Omega) = 0.0500 \text{ W}$$

The voltage drop is IR_c, which amounts to 0.500 V and can certainly be neglected with respect to 2.20×10^5 V. Thus the output voltage of the generator is approximately the same as the voltage required at the evaporator. We note that the power delivered to the evaporator is

$$P_e = (220 \text{ V})(100 \text{ A}) = 2.20 \times 10^4 \text{ W}$$

The loss in the transmission cables (0.0500 W) is surely negligible with respect to the power delivered if we use the transformers. However, we see in part *a* that if the transformers are not used, we lose more than twice as much energy to Joule heating as we deliver to the evaporator.

At a commercial electric generating station, power is typically produced at 12,000 V ac and may be stepped up to 60,000 or even 230,000 V for long-distance transmissions. The voltage can then be stepped down to perhaps 25,000 V at a substation near the end-use point and then to 120 or 240 V on utility poles outside the building. Figure 22-18a shows a common substation which you might see near your town or city, and in Figure 22-18b you can see a utility pole transformer. Values of currents and voltages mentioned here and used in Example 22-8 are effective values since ac values are continuously changing. The effective value for alternating current is the magnitude of direct current which produces Joule heating in a resistor at the same rate. We will see how effective values are related to peak values of alternating currents and voltages in Chapter 23.

22 Inductance, Motors, and Generators

Figure 22-18 (*a*) An electric utility substation for stepping down high voltages used in long-distance transmission to lower values for local distribution. (*b*) A utility pole transformer performs the final step-down of voltage to 240/120 V for the lines entering individual homes.

(*a*)

(*b*)

Transformers are also used where a number of different voltages are required while the source is a single value. For instance, a television receiver typically is plugged into a receptacle with a voltage of approximately 120 V. However, a much higher voltage (perhaps 25,000 V) is required to accelerate electrons, and a much lower voltage (~6 to 12 V) is needed to "boil" electrons from the filament of the electron gun in the picture tube. Individual circuits in the set may function on voltages from 300 V down to a few tenths of a volt. These voltages may all be obtained by proper use of transformers and ancillary circuits.

An excellent application of the transformer is to provide maximum power transfer with minimum reflection of energy at an interface between two parts of a circuit (an ac power wave is a wave, after all). This is part of a technique called **impedance matching**. (See Problem 22-38 and Chapter 23.)

Minimum Learning Objectives

After studying this chapter, you should be able to:

1. Define:
   - alternator
   - armature
   - back emf
   - back torque
   - brushes
   - commutator
   - electric oscillation
   - generator
   - henry (H) [unit]
   - induction motor
   - inductor
   - magnetic energy
   - motor
   - mutual inductance
   - primary coil
   - secondary coil
   - self-inductance
   - series-wound motor
   - short circuit (-ed)
   - synchronous motor
   - transformer

2. Calculate the mutual inductance of coaxial solenoids, given the geometry of each and the induced emf in one if the rate of change of current through the other is given.
3. Calculate the self-inductance of a solenoid from its geometry.
4. Calculate the energy stored in the magnetic field of a solenoid, given the inductance and the current in the solenoid *or* the geometry of the solenoid and its axial magnetic field.
5. Calculate the current through an inductor of known inductance in a circuit of given R and V at any time after completing the circuit *or* at any time after shorting the circuit once a steady-state current has been reached.
6. Calculate the frequency of oscillation of an *LCR* circuit, given L and C.
7. Calculate the torque on a loop of given geometry in a known magnetic field while it is carrying a specified current.
8. Calculate the torque developed by a simple dc motor, given the area and number of loops in its armature, the current supplied it, and the magnetic field in which it operates.
9. Calculate the back emf of the same motor, given its rate of turn.
10. Calculate the electric power drawn by the same motor, given also its coil resistance.
11. Calculate the emf produced by a generator of given geometry in a known magnetic field operating at a specified turn rate.
12. Distinguish among general types of electric motors, and state some of their properties.
13. Calculate the current and voltage outputs of a transformer, given these same inputs and the ratio of the windings of the coils.

Problems

22-1 Two coils of wire are wrapped around the same axis. Measurement shows that when the current in one coil changes at a constant rate of 200 A/s, an emf of 40.0 V is induced in the other coil. Calculate (a) the mutual inductance between the coils and (b) the rate at which the current must change in one of the coils in order that a 5.00×10^3 V emf be induced in the other.

22-2 A solenoid of 20 turns per centimeter and cross-sectional area 8.00×10^{-4} m² is 0.750 m long and has a second coil of 120 turns surrounding it at its center. If the current in the solenoid is increased linearly from zero to 5.00 A in 4.00×10^{-2} s, calculate (a) the mutual inductance between the two coils and (b) the emf induced in the 120-turn coil.

545

22-3 A coil of wire connected in an electric circuit has an induced emf of 12.0 V appear across it when the current is changing at the rate of 0.500 A/s. Calculate (a) the self-inductance of the coil and (b) the average voltage induced in the coil when the current is decreased from 3.50 A to zero in 2.00 ms.

22-4 Three inductors of 5.00, 10.0, and 15.0 mH are connected together. Draw the circuit and calculate the equivalent inductance in each of the following cases. (The chapter gives no formula for calculating equivalent inductance, but we have done resistors, capacitors, and emf's in the past, so reason out what the relationship must be by example.) (a) All three are in series. (b) The first two are in series with each other and in parallel with the third. (c) All three are in parallel.

22-5 An air-core solenoid is 0.400 m long and has 840 total turns. It is connected in a circuit, and the current is increased from zero at the rate of 0.150 A/s. Calculate (a) the self-inductance of the solenoid, (b) the emf induced in the solenoid, (c) the energy stored in the inductor when the current in the circuit is 15.0 A, and (d) the energy per unit volume stored in the magnetic field if the solenoid has a radius of 2.00 cm.

22-6 The product of the number of turns of a coil of wire with the magnetic flux is often called the flux linkage for the coil. (a) Calculate the flux linkage for a self-inductor of 80 turns with an inductance of 0.100 H carrying a current of 10.0 A. (b) What energy is stored in this inductor?

22-7 The current in an inductor varies with time, as shown in Figure 22-19. If the inductor has an inductance of 5.00 H and a resistance of 10.0 Ω, show on a graph the voltage across the inductor as a function of time.

22-8 Determine the magnetic energy per unit volume at the center of the orbit of an electron in its lowest orbit in a hydrogen atom. The radius of the orbit is 5.30×10^{-11} m, and the electron's angular speed is 6.60×10^{15} r/s.

22-9 A major problem with transplanting an artificial heart into a human being is that of power supply. Heart pacemakers require very little power, so the supply is part of the device and is implanted entirely in the body. An entire artificial heart would be a very different problem. The skin rejects breaks or openings for electric wires, tubes, etc., and tries to grow under and around them rather than fuse to them, leaving channels for eventually fatal infection. One actively researched option is to implant power coils under the skin and induce a current in it from outside the skin. Assume that the largest mutual inductance which can be engineered conveniently between coils above and below the skin is 0.200 H. The power requirement of the artificial heart is 40.0 W. (a) If the maximum voltage desired inside the body would be 15.0 V, what would be the maximum rate of change of current through the outside coil? (b) If the maximum ac voltage is 15.0 V, the *effective voltage* would be $0.707V_{max}$. Similarly, the *effective current* for alternating current is $0.707I_{max}$. What effective current would be needed inside the body to supply the artificial heart's power requirement? (c) The rate of change of a sinusoidal current varying as $I = I_{max} \sin 2\pi ft$ is $\Delta I/\Delta t = 2\pi f I_{max} \cos 2\pi ft$ for any small time interval Δt. To produce the internal power requirement of the circuit without exceeding the internal maximum voltage, what must be the maximum voltage of the external circuit at a frequency of 30.0 Hz?

22-10 In Section 22-2 it was asserted that the mutual inductance between two coils did not depend on how it was measured (that is, $M_{12} = M_{21}$). The most general proof of this requires some advanced mathematics. However, for certain simple geometries, the proof is somewhat less difficult. Consider two solenoids of the same length L wrapped on the same hollow cylinder but insulated from each other. The first solenoid has N_1 turns, while the second has N_2 turns. (a) Use Equations 22-1 and 22-2 along with Faraday's law to show that

$$M_{21} = \frac{N_2 \Phi_{21}}{I_1} \quad \text{and} \quad M_{12} = \frac{N_1 \Phi_{12}}{I_2}$$

(b) Determine the mutual inductance of these solenoids by first assuming that a changing current I_1 in the first coil induces an emf in the second. Then determine the mutual inductance by assuming that a changing current I_2 in the second solenoid induces an emf in the first. Are M_{12} and M_{21} equal?

22-11 For a series circuit containing a 0.800-H inductor, a 0.500-Ω resistor, and a 1.50-V emf in which the switch is closed at $t = 0$, determine for $t = 1.00$ s (a) the current in the circuit, (b) the rate of change of current, (c) the rate at which energy is being stored in the magnetic field, (d) the rate at which energy is being lost to Joule heating, and (e) the rate at which energy is being delivered by the emf.

22-12 A 12.0-H inductor, a 60.0-Ω resistor, and a 24.0-V battery are connected in a series circuit with a switch which is closed at time $t = 0$. Calculate (a) the steady-state current, (b) the time constant for the circuit, and (c) the magnitude of the current in the circuit when $t = 0.100$ s.

22-13 After the switch is closed in a series circuit containing a 20.0-H inductor, a 5.00-Ω resistor, and a 40.0-V emf, determine (a) the steady-state current, (b) the energy stored in the inductor at the steady state, and (c) the

Figure 22-19 Problem 22-7.

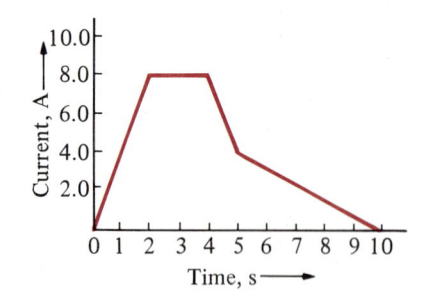

time required for the current to reach 95 percent of its steady-state value.

22-14 Beginning with Equation 22-11, show that the inductive time constant L/R could also be considered as the time required for the current in an inductor-resistor-battery series circuit to reach its final steady-state value if it continued to increase at the constant rate it had at $t = 0$. (*Hint:* Remember that $I = 0$ whenever $t = 0$.)

22-15 The switch is closed in a series circuit containing a 4.00-H inductor, a 10.0-Ω resistor, and an 18.0-V emf. Calculate (*a*) the steady-state current, (*b*) the initial rate of increase of current (remember that $I = 0$ whenever $t = 0$), (*c*) the time which would be required to reach the steady-state value if the current continued to increase at its initial rate, and (*d*) the time constant of the circuit.

22-16 A 3.00-μF capacitor is charged to 200 V and placed in series with a 0.600-H inductor. Calculate (*a*) the frequency of oscillation of the circuit, (*b*) the maximum energy stored in the capacitor, and (*c*) the maximum current in the inductor.

22-17 A 10.0-μF capacitor and a 0.300-H inductor are placed in a series circuit after the capacitor has been charged. Determine (*a*) the natural frequency of oscillation of the circuit and (*b*) the value of a capacitor which must be placed in parallel with the first capacitor in order that the natural frequency be reduced to one-third its original value.

22-18 The *LC* oscillating circuit of Figure 22-8 has a mechanical analog in a child on a swing. There, energy is shuffled between the kinetic energy of the swinging child and gravitational potential energy as the swing rises. In the circuit, the shuffle is between electric energy stored in the capacitor and magnetic energy stored in an established field. Like the swing, in the absence of an energy input, the oscillation dies out in time. Also like the swing, a precisely timed, small energy input can cause the oscillation to grow in time to a large value. For the circuit, the energy input can be a signal of a carefully controlled frequency matching that of the circuit provided by a distant broadcasting station. Normally, a circuit is tuned to a particular frequency by a variable capacitor (like that shown in Figure 18-12). (*a*) If the maximum capacitance of that device is 2.20×10^{-10} F, what size inductor would be needed to produce a circuit which could be tuned to the entire AM radio band with frequencies of 540 to 1600 KHz? (*b*) With this size inductor, what would have to be the minimum capacitance of the variable capacitor?

22-19 A 500-turn rectangular coil of wire measures 6.00 cm by 12.0 cm and is situated in a magnetic field of 0.200 T. Calculate (*a*) the maximum torque on the coil when it carries a current of 0.0300 mA and (*b*) the angle between the plane of the coil and the magnetic field when the torque is 1.50×10^{-5} N·m.

22-20 A 10-turn circular wire loop of diameter 0.150 m is placed in a magnetic field and carries a current of 2.00 A. (*a*) What is the minimum magnetic field which will balance an external torque of 0.120 N·m acting on the coil? (*b*) What net torque would act on this coil if the angle between the plane of the coil and field were 30°?

22-21 A 200-turn galvanometer coil has a cross-sectional area of 2.50 cm². If the magnetic field at the core is 8.00×10^{-2} T and the effective spring constant is 5.00×10^{-7} N·m/deg, calculate (*a*) the angular deflection of the coil for a current of 6.00 mA in it and (*b*) the current required for a 15° deflection.

22-22 We wish to design a galvanometer which will yield a 60° coil deflection when it carries a current of 50.0 mA. We have hair springs available which can give us an effective spring constant of 4.00×10^{-6} N·m/deg. If the magnetic field is 0.100 T, determine (*a*) the number of turns necessary if the coil is 1.20 cm by 1.80 cm and (*b*) the increase in the width of the coil if the number of turns is to be reduced by 100.

22-23 A dc motor is to operate at 12.0 A and deliver mechanical power of 360 W. If the resistance of the armature coil is 0.200 Ω, calculate (*a*) the back emf at operation, (*b*) the source voltage, and (*c*) the starting current.

22-24 The armature coil of a practical dc motor has 80 rectangular turns, each measuring 18.0 cm by 8.00 cm. The magnetic field is 0.0650 T, and the current in the armature is 25.0 A. Calculate (*a*) the average torque produced, (*b*) the angular speed (in revolutions per second) at which the back emf is 20.0 V, (*c*) the source voltage if the armature resistance is 0.160 Ω, and (*d*) the mechanical power delivered by the motor.

22-25 You wish to design a practical dc motor which delivers a mechanical power of 900 W at a current of 20.0 A and provides a torque of 15.0 N·m. Your source has an emf of 50.0 V. Determine (*a*) the back emf, (*b*) the armature resistance, (*c*) the magnetic field if the armature coil is to have 400 turns and an area of 0.0400 m², and (*d*) the angular speed at which the motor will run.

22-26 An anemometer measures wind speed. It consists of three arms with a hollow hemisphere (cup) on each to catch the wind and rotate at nearly the speed of the wind. As the wind cups turn, they rotate a shaft to which is connected the coil of a simple dc generator. The output of the generator is a measure of wind speed. The effective length of the arms of the anemometer is 20.0 cm. The area of the rotating coil is 4.00 cm², and its output is rectified and then "smoothed" by a capacitor to an effective dc current of 0.707 of the maximum current. The maximum reading of the meter to measure wind speed occurs at a current of 234 mA in a circuit with a resistance of 4.00 Ω and is to represent a wind of 100 km/h. If the coil is in a magnetic field of 0.0441 T, determine the number of turns it must have.

22-27 Whenever the output of an automobile generator (alternator) drops below the emf of the battery, the battery takes over the operation of the electric system and is being drained. Since alternator emf depends upon the rate

547

of turn of the engine, long idling periods as in heavy traffic (especially with lights and/or wipers on) can lead to depletion of the battery. To avoid this, the idle rate of the engine can be increased to allow the electric system to be powered by the alternator even at idle. Assume that the alternator is similar to the simplest-design dc generator (it isn't), each turn having an area of 9.95×10^{-4} m². (a) For every rotation of the engine, the alternator turns twice. At an idle rate of 1800 r/min, how many turns would the armature of the alternator need if rotating in a magnetic field of 0.200 T to maintain a maximum emf of 15.0 V? (b) If the drive pulley on the alternator is 10.0 cm in diameter, what is the torque provided to the pulley by the engine at idle if 4.00 A is required to operate the automobile? (Assume perfect efficiency of generation and transmission of power.)

22-28 The armature of an ac generator has 80 turns, and each turn is 18.0 cm by 7.00 cm. The generator is to produce a sinusoidal voltage of amplitude 17.0 V and frequency 40.0 Hz. Calculate (a) the angular speed at which it must turn and (b) the magnetic field required.

22-29 An ac generator is to turn at 3.60×10^3 r/min. Its armature coils are to have a cross-sectional area of 0.0300 m² and turn in a magnetic field of 0.0800 T. (a) How many turns are required if the generator is to have a maximum emf of 170 V? (b) What is the frequency of the output?

22-30 A precision coil has 120 turns, each having an area of 12.0 cm², mounted on a nonmagnetic shaft and sealed into a wooden box. The coil is turned by a wooden shaft from a distant power supply and is rotated at 20.0 Hz. The coil box is reoriented in an open space until a maximum emf is established at some particular facing. If three separate, isolated sites agree in orientation for maximum emf, the values recorded are averaged and the magnetic induction of the earth's field is calculated. The emf generated is alternating current and varies between maximum positive and negative values. The meter records the average absolute value of the emf, which is 0.707 of its maximum (this is true of most ac meters). If the averaged, measured emf at this site is 0.600 mV, what is the value of the earth's magnetic field at the site?

22-31 A windmill has blades with a length of 2.00 m. Air moving at 8.00 m/s is slowed to 5.00 m/s after passing the blades. The density of air is 1.29 kg/m³. The blades of the windmill are connected to a generator regulated to produce an effective emf of 120 V. What effective current may be expected from the windmill's generator with this wind, assuming perfect efficiency of the generator?

22-32 The primary coil of a transformer is connected to a 120-V line and draws a current of 2.00 A. The primary-to-secondary turns ratio is 20 : 1. Calculate (a) the voltage across the secondary coil and (b) the current in the secondary.

22-33 A transformer for a toy electric train steps down the voltage from 120 to 8.00 V. The primary coil has 270 turns and is rated at 40.0 W. Calculate (a) the number of turns in the secondary and (b) the current in the primary and secondary.

22-34 A 12,000-V power line passing through a residential area is tapped and the voltage stepped down to 240 V by an iron-core transformer with 3400 turns in the primary. It is to deliver a maximum of 100 A to the residence. Calculate (a) the number of turns in the secondary, (b) the maximum power delivered to the residence, and (c) the current in the transformer's primary.

22-35 A pole transformer attached to a 12,000-V line provides eight houses with 240-V service, each rated at 100 A maximum. (a) If the secondary coil of the transformer has 20 turns, how many must the primary have? (b) What maximum current must the high-voltage line provide the transformer? (c) Each turn of the secondary has an average effective radius of 15.0 cm. What must the diameter of the copper wire of the secondary be to limit Joule heating in the transformer to 500 W at maximum power? (d) Why can heat generated in the primary coil be largely ignored?

22-36 A hydroelectric plant at a small dam delivers power at peak demand hours at 5.00×10^3 V and 400 A. It is to be delivered to a substation 75.0 km away over transmission lines with a resistance of 0.150 Ω/km. Calculate (a) the Joule heating loss if this is transmitted at 5.00×10^3 V and (b) the voltage to which this must be stepped up if the power loss is to be reduced to 1 percent of that in part a.

22-37 A transformer is to provide a current source for a small electric arc welder. We require that the current in the secondary be five times that in the primary. Calculate (a) the ratio of the number of turns in the secondary to the number in the primary, (b) the voltage in the secondary if the primary voltage is 240 V, and (c) the minimum power delivered to the primary if the secondary current is 150 A.

22-38 Problem 19-39 defines impedance matching as stating that energy is transferred most efficiently between systems when they have the same impedance. This is true for electric impedances as well as mechanical impedances. This is very important in many applications when energy is to be transferred from one circuit to another. A common example is the transfer of energy from an audio amplifier to the speaker coil. The amplifier usually has a high impedance, while the speaker has a very low impedance. This mismatch means that a direct connection between the two would not be very efficient. Fortunately, a transformer can be used to correct such mismatches. Take N_p and N_s as the number of turns in the primary and secondary, respectively. Show that impedances presented by the primary and secondaries (Z_p and Z_s) are related to the number of turns of each by the equation (note that $Z = V/I$)

$$\frac{Z_p}{Z_s} = \left(\frac{N_p}{N_s}\right)^2$$

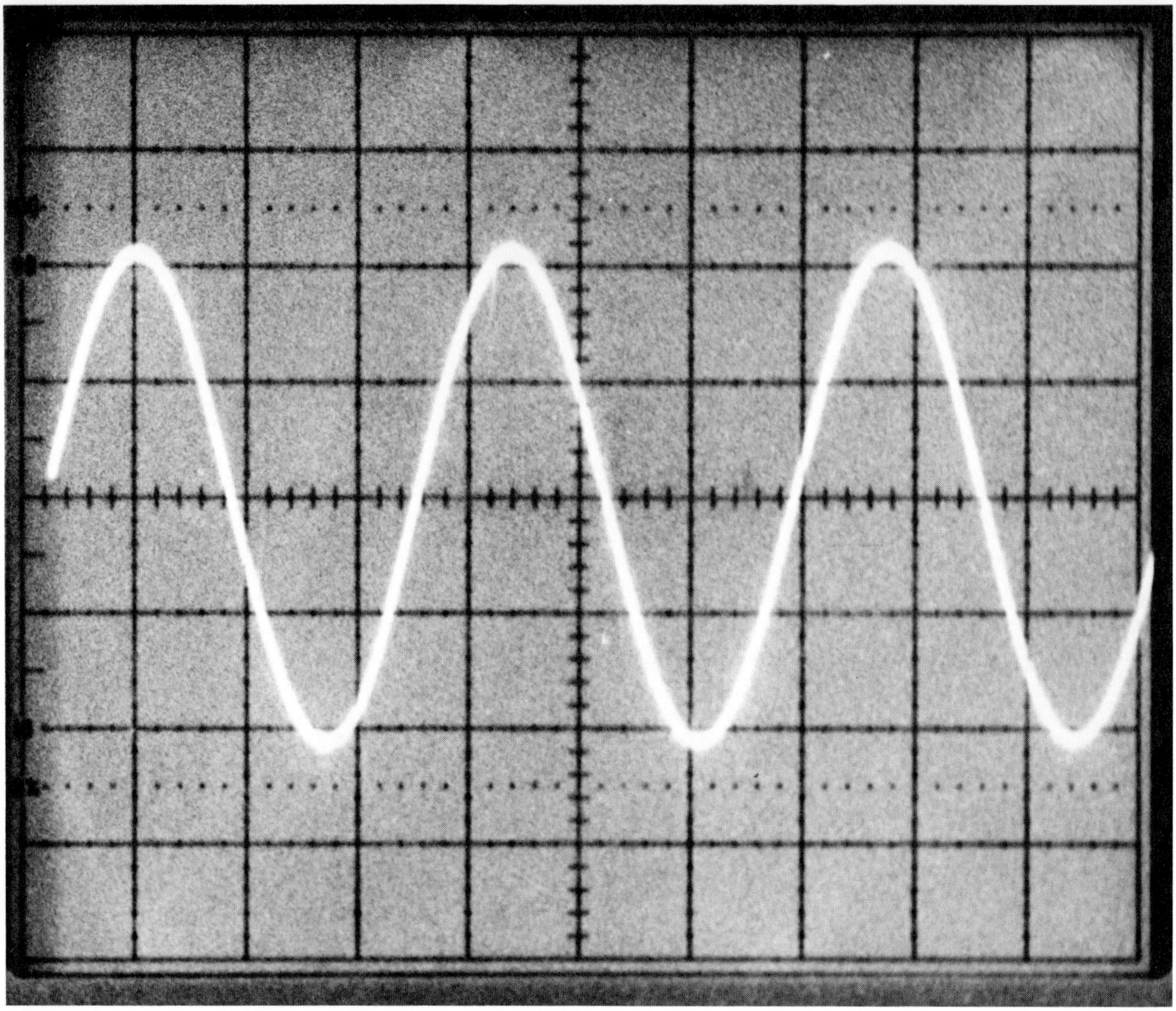

23 Alternating Current and Electrical Safety

23-1 Introduction

In Chapter 20, we began to distinguish between direct current (dc) and alternating current (ac) and explored circuits in which the current was not only direct but also constant in magnitude in every branch. Situations with steady currents are not only easier to analyze but are representative of a wide range of practical applications. In most ac circuits, the current changes magnitude and direction in a periodic manner with fixed frequency. In Chapter 22, we saw that as a loop of wire rotates in a magnetic field, the area it presents to the field varies as the cosine of the angle between the plane of the loop and the field. Thus, an ac generator rotating with constant angular speed gives rise to a sinusoidally varying emf. Therefore, the most common repeating pattern of alternating current is one that varies sinusoidally with time. Figure 23-1 shows this, as well as several other possible periodic voltage vs. time curves, each of which has important applications. The sinusoidal curve of Figure 23-1a is by far the most common and the one we will consider in this chapter. Recall, however, that any periodic wave may be described by a series of sine and cosine functions.

Once the switch is closed in a dc circuit, currents quickly attain a steady-state value, which they then maintain. In ac circuits, the current and voltage are constantly changing. This suggests that certain circuit elements have a very different relationship to an ac circuit than they do to a dc one. For example, a capacitor in a dc circuit acts much like a break in the wire. Once the capacitor is fully charged, no more charge will flow across the branch containing the capacitor. In the ac circuit, the capacitor acts more like an intact wire, provided the time variation of current is more rapid than the time required for the capacitor to acquire full charge. Thus, capacitors stop direct current (after a very short time) but freely pass alternating current.

Perhaps the most serious complication of the ac circuit is the consideration of **phase.** Just because the ac source achieves its maximum voltage at a particular point in time does not imply that all devices in the circuit are also at maximum voltage at that instant. The voltage of some devices **leads** the voltage variation of the source, while that of others **lags** behind it in time. For example, an inductor generates its own emf with a magnitude which is proportional to the rate of change of current through the inductor. Current with sinusoidal variation is always changing in time, and the rate of change of current is also constantly changing — the current is not increasing or decreasing at the same rate from one instant to the next. The implication is that the maximum voltage across an inductor never occurs when source voltage is a maximum. In other words, the inductor voltage is **out of phase** with the source throughout the cycle.

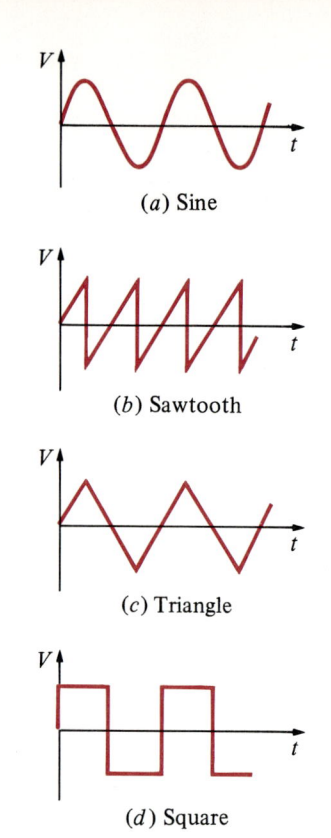

(a) Sine

(b) Sawtooth

(c) Triangle

(d) Square

Figure 23-1 Frequently encountered ac waveforms. The sinusoidal variation of (a) is by far the most common.

23-2 Phasor Diagrams

Given the complications, we will proceed very carefully into the development of ac circuit relationships. To completely specify a periodic alternating voltage or alternating current, we must specify the maximum variation from zero (the amplitude) and the number of cycles per unit time (the frequency). Consider a voltage source which is an alternator, or a sinusoidal wave generator. Equation 22-26 gives us the emf output of the simplest alternator as

$$\mathcal{E} = NBA\omega \cos 2\pi ft \qquad (22\text{-}26)$$

This is the instantaneous voltage and can be written as

$$V = V_0 \cos 2\pi ft \qquad (23\text{-}1)$$

where V_0 is the **voltage amplitude,** or maximum value of the voltage, and is equal to $NBA\omega$ for an alternator turning at a frequency f. (Remember that $\omega = 2\pi f$.)

Recall from studying simple harmonic motion in Chapter 10 that a sinusoidal variation may be written as a sine or cosine function, depending on where the function starts: the **initial phase angle.** Since we always have the option of when to start our clock, the choice is up to us. We consider our alternating voltage source to be represented by the equation

$$V = V_0 \sin (2\pi ft + \phi_0) \qquad (23\text{-}2)$$

The initial phase angle ϕ_0 is intentionally left unspecified. Its value may be fixed or, perhaps, determined by each situation.

In practice we will see that the most important phase relationship is not the initial phase angle but the **phase-angle difference** between the circuit voltage and the circuit current. Because of this, we often find it convenient to choose whatever initial phase angle in an ac circuit yields a current of the form

$$I = I_0 \sin 2\pi ft$$

where I_0 is the maximum value of I and is called the **current amplitude.** This simply means that we start our clock when the current is passing through zero, going from negative to positive. Comparing this expression for the current with Equation 23-2 for the voltage, we see that this choice forces (mathematically) the phase-angle difference between current and voltage to be ϕ_0. Thus, we need only determine the initial phase angle of the voltage, and we will have found the value of the phase-angle difference. The importance of this phase-angle difference will become apparent in Section 23-4.

Just as we did with simple harmonic motion, we may use a reference circle with rotating vectors to aid us. When used with alternating voltages, this construct is called a **phasor diagram.** The phasors are the rotating vectors representing the voltages or currents. Figure 23-2 is the phasor representation of the voltage of Equation 23-2. The maximum voltage V_0 is called the voltage amplitude. The phasor of magnitude V_0 begins at an angle ϕ_0 with the x axis and rotates counterclockwise with an angular speed $\omega = 2\pi f$. At any time t, the angle that the phasor makes with the positive x axis is equal to $2\pi ft + \phi_0$. The projection of the phasor on the y axis at time t is equal to $V_0 \sin(2\pi ft + \phi_0)$, which from Equation 23-2 is the time-varying voltage V. Phasor diagrams are most useful when there are inductors and/or capacitors in a circuit being analyzed, and we will use them frequently.

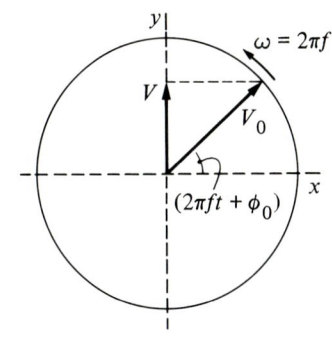

Figure 23-2 AC voltage varies in a continuous, sinusoidal cycle, as shown in this phasor diagram. We picture the V_0 vector rotating at constant angular speed ω in a circle such that its projection on the y axis is V, the instantaneous voltage experienced in the circuit. This process begins at $t = 0$ with V_0 at some arbitrary phase angle ϕ_0.

23-3 Simple AC Circuits

Alternating Current and Resistance

One of the simplest ac circuits possible, a resistor connected in series with an alternating voltage source, is shown in Figure 23-3. The source voltage is across the resistor and is given by $V = V_0 \sin(2\pi ft + \phi_0)$ from Equation 23-2. Since a resistor obeys Ohm's law, we determine the current from

$$I = \frac{V}{R} = \frac{V_0}{R} \sin(2\pi ft + \phi_0)$$

or

$$I = I_0 \sin(2\pi ft + \phi_0) \qquad (23\text{-}3)$$

where

$$I_0 = \frac{V_0}{R} \qquad (23\text{-}4)$$

Equation 23-4 describes the instantaneous relationship between I_0 and V_0; thus the current in the resistor and the voltage across it have the same phase. That is, when the potential difference across the resistor is greatest, the current through the resistor is greatest. Phasors of current and voltage drawn from the same origin would be along the same line and rotate together. The graphs in Figure 23-4 represent voltage vs. time and current vs. time for the resistor of Figure

Figure 23-3 A simple ac circuit containing an ac source and a resistor.

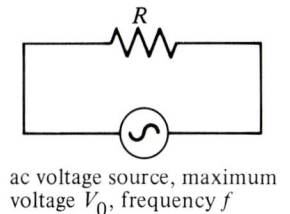

ac voltage source, maximum voltage V_0, frequency f

553

23-3. The relative heights of the curves of voltage and current depend on the value of the resistor as well as on the vertical scales selected for the graphs. Nevertheless, we can see that the two are always exactly in phase.

Example 23-1 Suppose the circuit of Figure 23-3 has a resistance of 25.0 Ω, a voltage amplitude of 100 V, and is being driven at a frequency of 60.0 Hz. Using the proper numerical values, write the equation for the circuit current in the form of Equation 23-3 if the current equals 2.00 A and is increasing at time $t = 0$.

Solution We must determine the values of I_0 and ϕ_0 for our circuit and substitute these along with the given frequency into Equation 23-3. We see from Equation 23-4 that

$$I_0 = \frac{V_0}{R} = \frac{100 \text{ V}}{25.0 \text{ Ω}} = 4.00 \text{ A}$$

At time $t = 0$, Equation 23-3 has the form

$$I = I_0 \sin \phi_0$$

Substituting the values for I and I_0, we have

$$\sin \phi_0 = \frac{I}{I_0} = \frac{2.00 \text{ A}}{4.00 \text{ A}} = 0.500$$

$$\phi_0 = 30° = \frac{\pi}{6} \text{ rad}$$

We choose 30° ($\pi/6$) rather than 150° ($5\pi/6$) because I is increasing at $t = 0$. Thus the proper equation is

$$I = 4.00 \sin\left(120 \pi t + \frac{\pi}{6}\right)$$

Alternating Current and Capacitance

Another simple ac circuit is one which includes only a capacitor and an ac source, as in Figure 23-5. Again, the voltage of the source across the capacitor is $V = V_0 \sin(2\pi f t + \phi_0)$. Since the capacitor voltage and the source voltage must be equal, we have

$$V_C = \frac{q}{C} = V_0 \sin(2\pi f t + \phi_0)$$

We also know that $I = \Delta q/\Delta t$. We can write $\Delta q = C \Delta V$, and thus

$$I = C \frac{\Delta V_C}{\Delta t} = C \frac{\Delta V}{\Delta t}$$

This can be shown to be

$$I = 2\pi f C V_0 \cos(2\pi f t + \phi_0)$$

or

$$I = I_0 \cos(2\pi f t + \phi_0) \qquad (23\text{-}5)$$

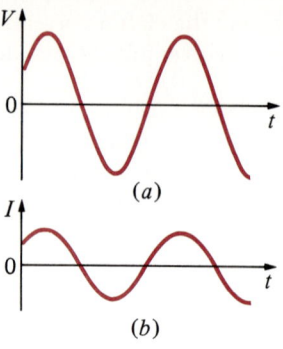

Figure 23-4 Simultaneous graphs of V vs. t and I vs. t across the resistor in Figure 23-3. Since I_0 is attained at the same instant as V_0, current and voltage at the resistor are said to be in phase with each other.

Figure 23-5 A simple ac circuit containing an ac source and a capacitor; V and I are not in phase in this circuit.

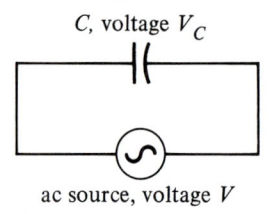

Note that the current is proportional to the rate of change of voltage, $\Delta V/\Delta t$. This is just the slope of a graph of V vs. t (Figure 23-1a) at any point in time. This slope is zero when V reaches its maximum, V_0. The slope is then negative until V reaches $-V_0$. Thus, current must be zero when $V = V_0$ at the maximum and be negative until $I = 0$ when $V = -V_0$. At the instant when $V = 0$ between these, the slope has the largest magnitude as does the current. This is the relationship described in Equation 23-5 and shown in Figure 23-6b.

where

$$I_0 = 2\pi f C V_0 \qquad (23\text{-}6)$$

A cosine function such as Equation 23-5 is 90° ($\pi/2$ rad) out of phase with a sine function of the same angle such as Equation 23-2. Further, the cosine function is 90° "ahead" of the sine function. That is, the cosine function reaches its peak while the sine function is passing through zero and will not reach its peak for 90°. Figure 23-6a shows the response of the system to the varying

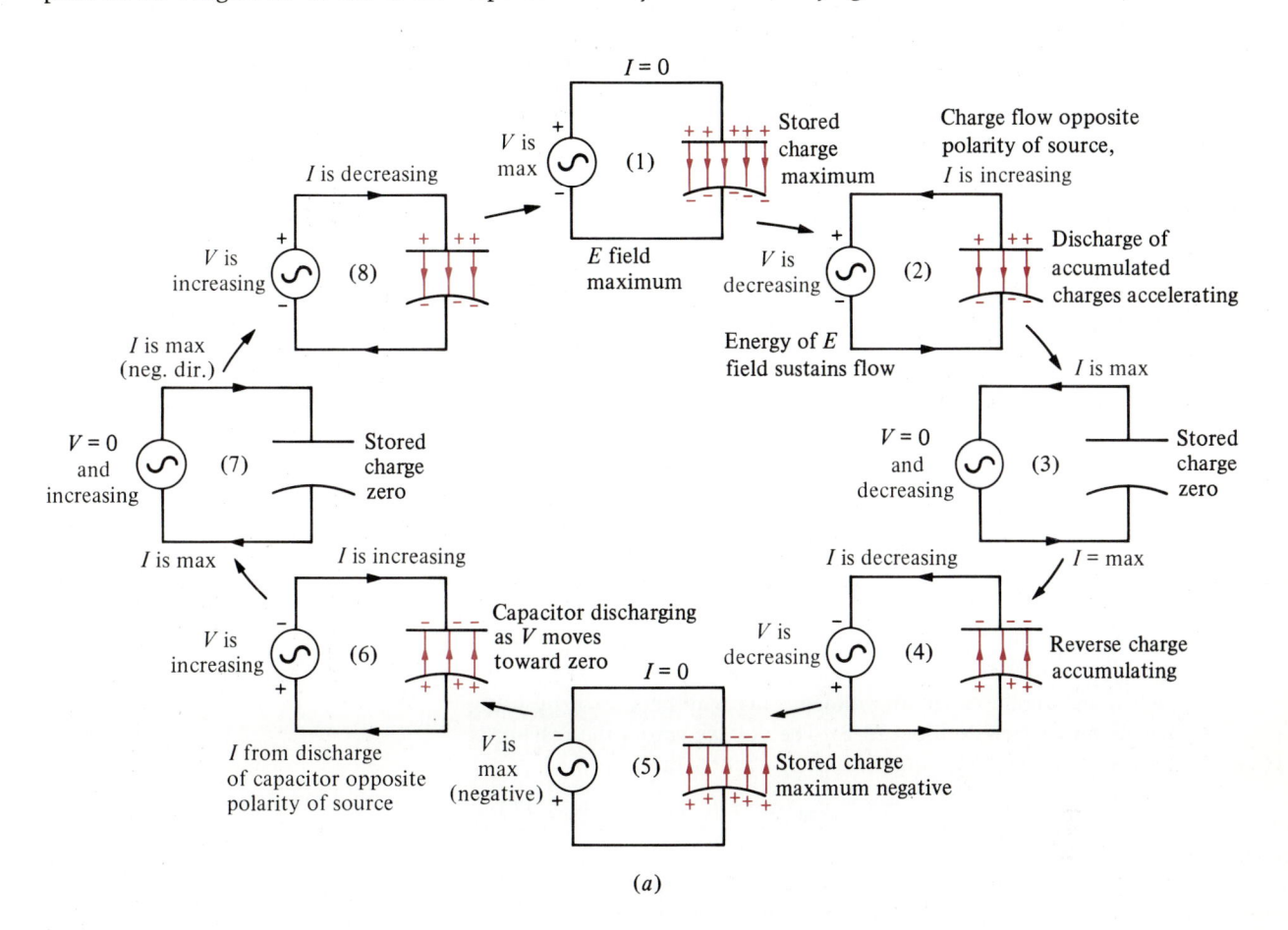

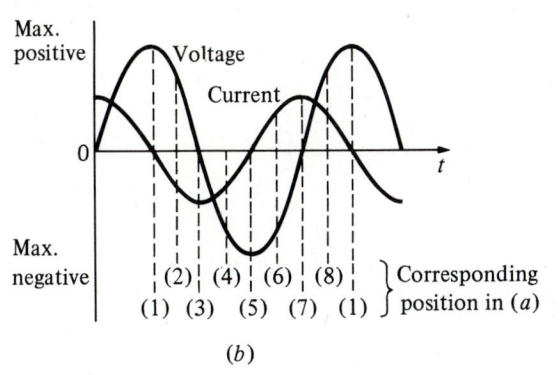

Figure 23-6 (a) The full charge-discharge cycle of the circuit of Figure 23-5. (1) The source is at maximum positive voltage, and the capacitor has its maximum stored charge. (2) The source voltage decreases, and the capacitor begins to discharge. (3) The source momentarily has reached zero potential difference in its cycle, and the last stored charges leave the capacitor as the current reaches its greatest magnitude. (4) The growing negative source polarity maintains the direction of the current and is charging the capacitor. (5) By the end of the half cycle, the source voltage is at its negative maximum value and the capacitor has its maximum negative charge. Any change in source voltage toward zero will begin the discharge of the capacitor and cause a current. The return cycle (6)–(8) repeats (2)–(4) with reversed polarity. (b) Variation of current and voltage across the capacitor.

source voltage, while Figure 23-6b shows Equations 23-2 and 23-5 plotted on the same axes with positions 1 to 8 of Figure 23-6a identified. Note that the voltage does not reach its positive peak until the current reaches zero on the way to a negative peak, and so on. We say that the capacitor current leads the capacitor voltage by 90° or that the capacitor voltage lags the capacitor current by 90°.

A phasor diagram of the current and voltage for this circuit is shown in Figure 23-7. In order that the projection of V_0 on the y axis be $V = V_0 \sin(2\pi ft + \phi_0)$, V_0 must make an angle of $2\pi ft + \phi_0$ with the positive x axis. Since I_0 and V_0 are 90° out of phase, on the diagram I_0 must make an angle $2\pi ft + \phi_0$ with the y axis. Therefore, the projection of I_0 on the y axis is $I = I_0 \cos(2\pi ft + \phi_0)$. Thus the y projections of V_0 and I_0 at any instant in the cycle give the instantaneous values of V and I at the capacitor.

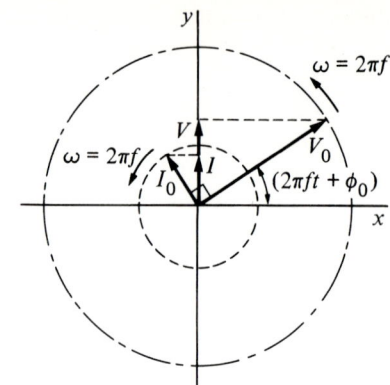

Figure 23-7 A phasor diagram for the circuit of Figure 23-5. I_0 is 90° ahead of V_0 as the vectors rotate at angular speed ω. The current and voltage in the circuit at any instant during the cycle are the y traces of I_0 and V_0, respectively. The variation in I and V as the I_0 and V_0 vectors rotate in time is the graph of Figure 23-6b. Can you identify the position drawn with the instant it represents in the cycle in Figure 23-6b?

Alternating Current with Resistance and Capacitance (RC Circuit)

The voltage across a resistor and the resistor current are always in phase. Thus if a circuit contains both a resistor and a capacitor, an *RC* circuit, the capacitor voltage will lag the resistor voltage by 90°. We must be careful, however, because in that case, the circuit voltage and circuit current are not necessarily in phase. Figure 23-8 shows a circuit which includes a capacitor, a resistor, and an alternating voltage source. We choose the initial phase angle so that the current in the circuit is $I = I_0 \sin 2\pi ft$ and the circuit voltage is $V = V_0 \sin(2\pi ft + \phi_0)$. Since voltage across the capacitor lags the current by 90°, it must be a negative cosine function and may be written as

$$V_C = -\frac{I_0}{2\pi fC} \cos 2\pi ft = -V_{C0} \cos 2\pi ft \qquad (23\text{-}7)$$

where V_{C0} is the maximum capacitor voltage. This may be shown by using trigonometric identities (see Problem 23-8). The voltage across the resistor is equal to the product of the resistance with the current. Thus

$$V_R = RI_0 \sin 2\pi ft = V_{R0} \sin 2\pi ft \qquad (23\text{-}8)$$

where V_{R0} is the maximum resistor voltage.

We may apply Kirchhoff's loop rule to this circuit at any time to obtain a relationship among the voltages. Choosing a time when the current is clockwise and beginning at the lower left traversing the circuit clockwise, we obtain

$$V - V_R - V_C = 0$$

or

$$V = V_R + V_C$$

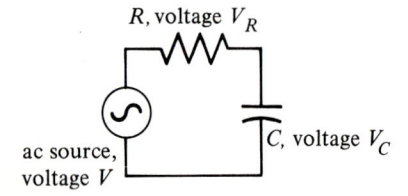

Figure 23-8 A circuit with an ac source, a resistor, and a capacitor: an alternating-current *RC* circuit.

This states that the *instantaneous* voltage of the source equals the sum of the *instantaneous* voltages across the resistor and capacitor. Substituting into this equation yields

$$V_0 \sin(2\pi ft + \phi_0) = V_{R0} \sin 2\pi ft - V_{C0} \cos 2\pi ft \quad (23\text{-}9)$$

The phasor diagram of Figure 23-9 shows the relationship among the voltages. Note that the V_0 phasor is the vector resultant of the other two and that this is necessary if their y projections are to satisfy Equation 23-9. We see that ϕ_0 is negative. Angles measured counterclockwise from the x axis are positive, and negative angles are measured clockwise. This simply means that at time $t = 0$, the voltage of the source has a negative projection on the y axis. Stated another way, we chose to start our time when the voltage source had a negative value in order to obtain the desired form for I. You may recognize at this point that the magnitudes of V_0, V_{R0}, and V_{C0} are related by the pythagorean theorem as

$$V_0^2 = V_{R0}^2 + V_{C0}^2$$

Also, the tangent of ϕ_0 and then ϕ_0 itself can be found from

$$\tan \phi_0 = \frac{-V_{C0}}{V_{R0}}$$

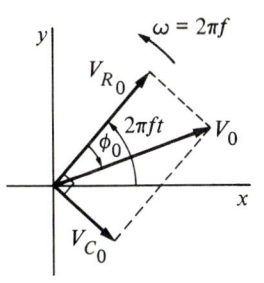

Figure 23-9 A phasor diagram for the voltage relationships of Figure 23-8. Since current and voltage are in phase across the resistor, the orientation of V_{R0} is the orientation of I_0. Because V across the capacitor lags I by 90°, V_{C0} is 90° behind V_{R0}. V_0 is the vector sum of V_{R0} and V_{C0} and lags V_{R0} by ϕ_0. The values of R and C fix the relative lengths of V_{R0} and V_{C0}, whose resultant is V_0. The entire array rotates at angular speed ω, allowing us to determine from the y projections of the vectors the instantaneous voltage across the capacitor V_C, across the resistor V_R, or across the source V.

Example 23-2 An RC circuit contains a capacitor of 3.00 µF, a 45.0-Ω resistor, and is being driven at a frequency of 800 Hz. If $I_0 = 1.50$ A, determine (a) the maximum capacitor voltage, (b) the maximum resistor voltage, (c) the voltage amplitude of the circuit, and (d) the phase difference between the voltage and the current.

Solution For parts a and b, the maximum voltages are determined from Equations 23-7 and 23-8.

(a) $$V_{C0} = \frac{I_0}{2\pi fC} = \frac{1.50 \text{ A}}{2\pi(800 \text{ Hz})(3.00 \times 10^{-6} \text{ F})} = 99.5 \text{ V}$$

(b) $$V_{R0} = I_0 R = (1.50 \text{ A})(45.0 \text{ Ω}) = 67.5 \text{ V}$$

(c) Using the pythagorean theorem, from the phasor diagram of Figure 23-9 we see that

$$V_0^2 = V_{R0}^2 + V_{C0}^2$$
$$V_0 = \sqrt{V_{R0}^2 + V_{C0}^2} = \sqrt{(67.5 \text{ V})^2 + (99.5 \text{ V})^2} = 120 \text{ V}$$

(d) From the definition of the tangent,

$$\tan \phi_0 = \frac{-V_{C0}}{V_{R0}} = \frac{-99.5 \text{ V}}{67.5 \text{ V}} = -1.47$$
$$\phi_0 = -55.8°$$

This is really the phase angle between the resistor voltage and the circuit voltage. However, because the resistor voltage is in phase with the current, it is also the phase-angle difference between the circuit current and the circuit voltage.

Alternating Current and Inductance

The voltage across an inductor also is not in phase with the current. Recall that the energy stored in the magnetic field of an inductor acts to oppose any change. In particular, should the source voltage fall to zero, the collapse of the field will attempt to sustain the current which generated the field. For an inductor L, the voltage is

$$V_L = L \frac{\Delta I}{\Delta t}$$

Thus the voltage is proportional to the rate of change of current or, in other words, to the slope of the current vs. time curve. Figure 23-10 shows a simple circuit containing an inductor and an ac source. For this circuit, the inductor voltage is equal to the source voltage, and we have

$$L \frac{\Delta I}{\Delta t} = V_0 \sin(2\pi f t + \phi_0)$$

or

$$\frac{\Delta I}{\Delta t} = \frac{V_0}{L} \sin(2\pi f t + \phi_0) \qquad (23\text{-}10)$$

This has the solution

$$I = -\frac{V_0}{2\pi f L} \cos(2\pi f t + \phi_0)$$
$$= -I_0 \cos(2\pi f t + \phi_0) \qquad (23\text{-}11)$$

where

$$I_0 = \frac{V_0}{2\pi f L} \qquad (23\text{-}12)$$

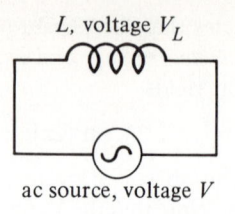

Figure 23-10 A simple circuit containing an inductor and an ac source. The action of the inductor is to oppose a change in the current, which ensures that current and voltage will be out of phase across the inductor.

We may convince ourselves that Equation 23-11 is consistent with Equation 23-10. Figure 23-11a shows the cyclic action of the inductor in sustaining the current after $V = 0$. Figure 23-11b shows the circuit voltage and the circuit current plotted on the same axes with an arbitrary value of ϕ_0. Note that the current passes through zero and has its maximum positive slope at the same instant that the voltage has its maximum positive value. Also, when the slope of the current curve is zero, the voltage is passing through zero, and so on. We say that the inductor voltage leads the inductor current by 90° or that the inductor current lags the inductor voltage by 90°. Figure 23-12 is a phasor diagram of the voltage and current for this circuit. You should be able to convince yourself that the y projections of these phasors yield the proper functions for V and I. Even the negative sign of I emerges naturally.

Figure 23-12 A phasor diagram for the circuit of Figure 23-10. Here I_0 lags V_0 by 90° as both rotate with angular speed ω. The instantaneous value of V and I is given by the y trace of V_0 and I_0. What instant in the cycle of Figure 23-11b is represented by the phasors shown?

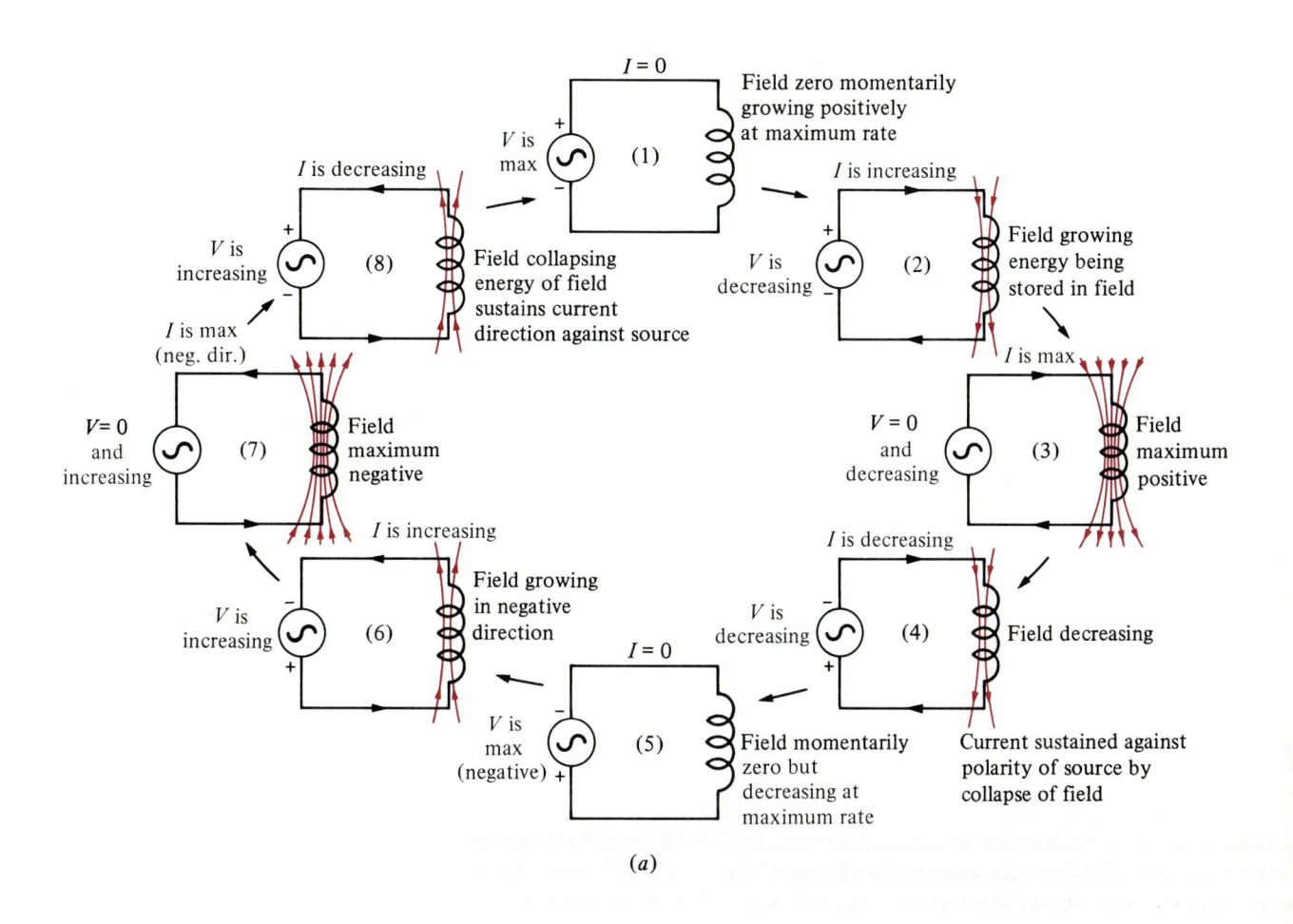

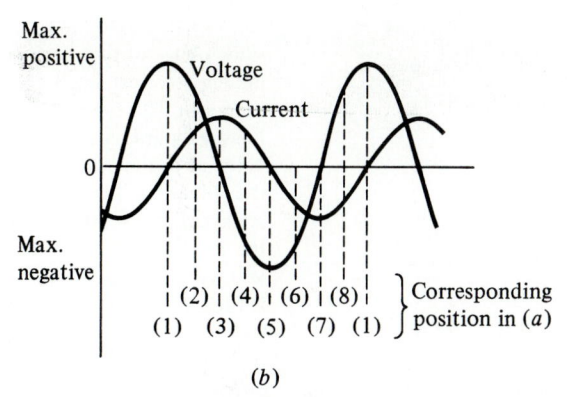

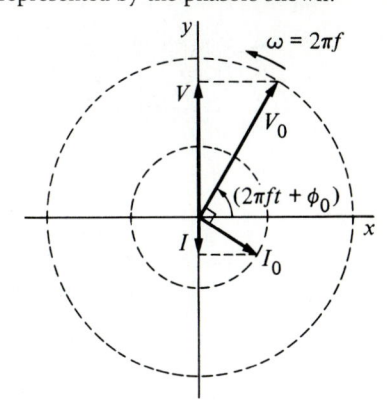

Figure 23-11 The full cycle of an inductor and an ac source. (1) The instant the source is connected at maximum positive voltage to the inductor. (2) As the current through the inductor grows, the magnetic field established by the inductor also grows. (3) The maximum field has been established, and the source has dropped to zero voltage. However, the energy stored in the field prevents the current from falling to zero. Even though (4) the source voltage now has reversed polarity, the energy stored in the inductor sustains a positive current until the field reaches zero at (5). The back cycle of (5)–(8) repeats with reversed polarity. (b) Voltage and current variation throughout the cycle.

559

Alternating Current with Resistance and Inductance (RL Circuit)

Once again, we recognize that if there were a resistor in a circuit containing an inductor, the resistor voltage would lag the inductor voltage by 90° since the resistor's voltage and current are always in phase. Consider Figure 23-13,

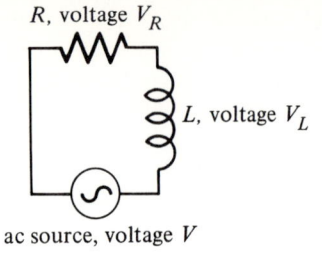

Figure 23-13 An alternating-current LR circuit. The instantaneous voltage across the resistor is V_R, across the inductor is V_L, and of the source is V.

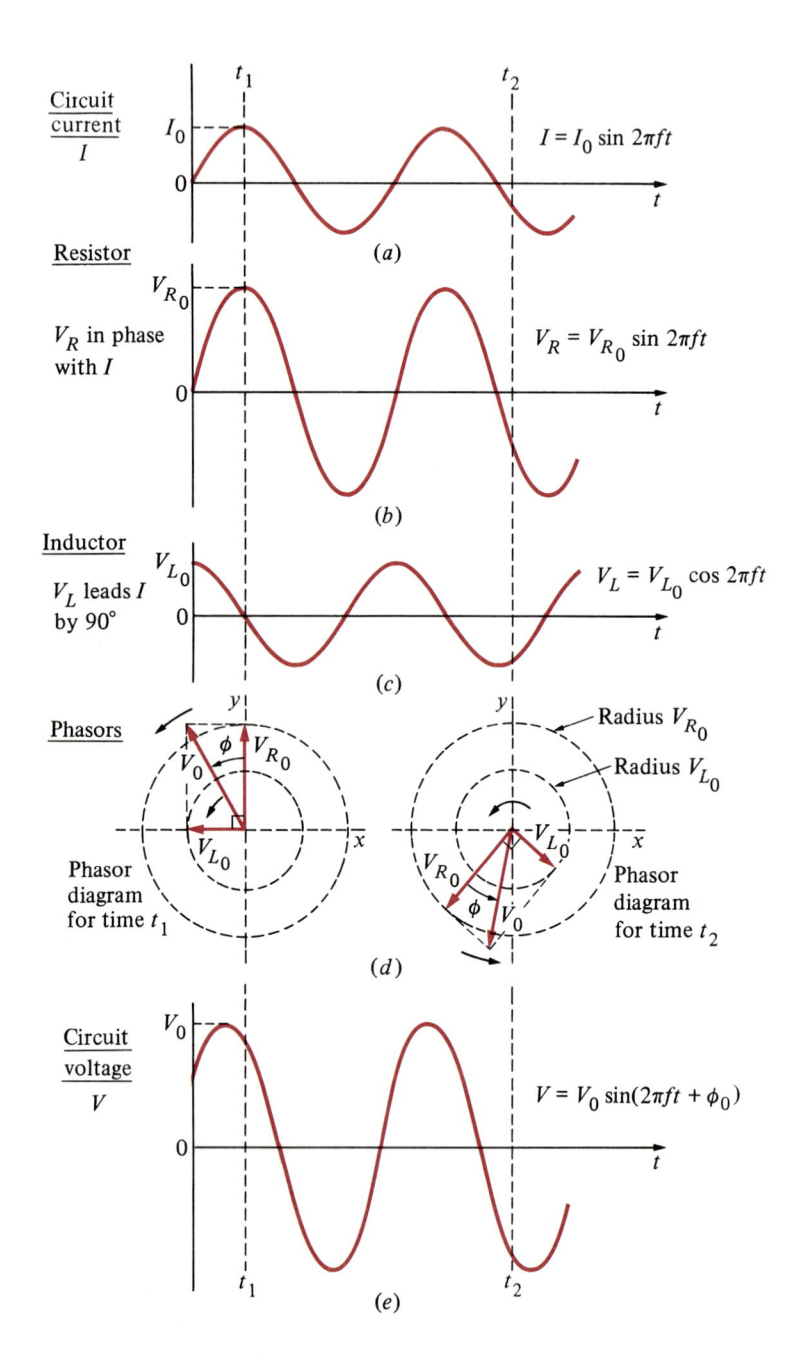

Figure 23-14 Current and voltages for the circuit of Figure 23-13. (a) The initial phase is chosen so that the current is a sine wave increasing from zero at time zero. (b) The resistor voltage has the same phase as the current. (c) The inductor voltage leads the current (and therefore the resistor voltage) by 90°. (d) Phasor diagrams representing circuit voltages at the two times shown. Since the current and resistor voltage are always in phase, the orientation of V_{R0} is the same as that of I_0. The inductor voltage V_{L0} leads the current by 90° and is thus 90° ahead of V_{R0}. The magnitudes of R and L determine the magnitudes of V_{R0} and V_{L0} (at a given frequency). The maximum source voltage V_0 is the resultant of V_{R0} and V_{L0}. All three vectors retain their relative magnitudes and orientations as they rotate through the cycle with angular speed ω ($=2\pi f$). (e) The source voltage, a superposition of the V_R and V_L curves (b and c). Note that the source voltage leads the resistor voltage by the phase-angle difference ϕ_0.

which includes a resistor, an inductor, and an alternating voltage source. It is again useful to choose the initial phase so that the circuit current is $I = I_0 \sin 2\pi ft$ and the circuit voltage is $V = V_0 \sin(2\pi ft + \phi_0)$. We are anticipating that the current and voltage may not be in phase and forcing ϕ_0 to have the same value as the phase difference between them. Since the inductor voltage leads the current by 90°, it must be a positive cosine function, and, in fact,

$$V_L = 2\pi fLI_0 \cos 2\pi ft = V_{L0} \cos 2\pi ft \qquad (23\text{-}13)$$

where V_{L0} is the maximum inductor voltage. This can be shown in a number of ways, including trigonometrically (see Problem 23-15). As usual, the voltage across the resistor is equal to the product of the resistance with the current and is given by Equation 23-8.

We apply Kirchhoff's loop rule to the circuit of Figure 23-13. Choosing a time when the current is clockwise and beginning at the lower right to traverse the circuit clockwise, we have

$$V - V_R - V_L = 0$$

or

$$V = V_R + V_L$$

Substituting, this becomes

$$V_0 \sin(2\pi ft + \phi_0) = V_{R0} \sin 2\pi ft + V_{L0} \cos 2\pi ft \qquad (23\text{-}14)$$

The relationship among these voltages is shown in the phasor diagram of Figure 23-14d. Once again, we see that the V_0 phasor is the vector resultant of the other two phasors, a condition that is required if their projections are to satisfy Equation 23-14. The voltages are also related by the pythagorean theorem, and the phase difference between V and I may be determined from the definition of the tangent.

Example 23-3 A circuit like that of Figure 23-13 contains an alternating voltage source of amplitude $V_0 = 50.0$ V, a resistor of $R = 400\ \Omega$, and an unknown inductor L. When the frequency of the source is 300 Hz, the current is 0.100 A. Calculate (a) the maximum resistor voltage, (b) the maximum inductor voltage, (c) the phase difference between the circuit current and the circuit voltage, and (d) the inductance L.

Solution (a) This is solved by using

$$V_{R0} = I_0 R = (0.100\ \text{A})(400\ \Omega) = 40.0\ \text{V}$$

(b) This requires that we apply the pythagorean theorem.

$$V_0^2 = V_{R0}^2 + V_{L0}^2$$
$$V_{L0} = \sqrt{V_0^2 - V_{R0}^2} = \sqrt{(50.0\ \text{V})^2 - (40.0\ \text{V})^2} = 30.0\ \text{V}$$

(c) We obtain the phase difference from the definition of the tangent.

$$\tan \phi_0 = \frac{V_{L0}}{V_{R0}} = \frac{30.0\ \text{V}}{40.0\ \text{V}} = 0.750$$

$$\phi_0 = 36.9°$$

(d) By straightforward application,
$$V_{L0} = 2\pi f L I_0$$
$$L = \frac{V_{L0}}{2\pi f I_0} = \frac{30.0 \text{ V}}{2\pi (300 \text{ Hz})(0.100 \text{ A})} = 0.159 \text{ H}$$

You should emphasize to yourself that the maximum voltages do *not* add arithmetically or even algebraically but must be dealt with by using *vectors*.

23-4 RCL Alternating-Current Circuits

We are aware that every circuit contains some resistance, some capacitance, and some inductance. The stray inductance and/or capacitance of a given circuit may be small, but the interactions of these elements are frequency-dependent. Thus, while the stray values may be negligible for many dc effects, it is not clear that we can ignore them when dealing with ac effects.

In the ac circuits we have worked with so far, the maximum voltages across the resistor, capacitor, and inductor were, respectively,

$$V_{R0} = RI_0 \qquad V_{C0} = \frac{I_0}{2\pi f C} \qquad V_{L0} = 2\pi f L I_0$$

where I_0 is the current amplitude. Usually we consider the voltage to be the independent variable and the current to be the dependent variable. Therefore, we solve these expressions for the current and its dependence on the other quantities. In a resistor, we obtain that the maximum current is

$$I_0 = \frac{V_{R0}}{R} \tag{23-15}$$

For a capacitor, the maximum current is given by

$$I_0 = V_{C0} 2\pi f C \tag{23-16}$$

Finally, the maximum current in an inductor is

$$I_0 = \frac{V_{L0}}{2\pi f L} \tag{23-17}$$

In all three equations, a voltage across an element gives rise to a current which is limited by some quantity or property. In the case of the resistor, the limiting property is the resistance. The larger the resistance, the smaller the resistor current for a given voltage. For the other two elements, the limiting factor depends on the frequency as well as on the properties of the elements.

It is useful to put Equations 23-16 and 23-17 into the same form as Equation 23-15 by defining a new quantity called **reactance**. We define **capacitive reactance** as

$$X_C = \frac{1}{2\pi f C} \tag{23-18}$$

and **inductive reactance** as

$$X_L = 2\pi f L \tag{23-19}$$

Equations 23-16 and 23-17 then become

$$I_0 = \frac{V_{C0}}{X_C} \quad (23\text{-}20)$$

and

$$I_0 = \frac{V_{L0}}{X_L} \quad (23\text{-}21)$$

From these equations, we see that the units of both capacitive and inductive reactance are ohms (Ω), the same unit as that of resistance. Indeed, reactance has the same effect on sinusoidal alternating voltage that resistance has on any voltage, that of limiting the current. The biggest difference is that reactance is frequency-dependent. Capacitive reactance varies inversely with frequency (that is, it gets smaller as the frequency is increased), and inductive reactance varies directly with frequency (it gets larger as the frequency is increased).

Figure 23-15 shows an *RCL* **series circuit** with a resistor, a capacitor, an inductor, and an alternating voltage source. Once more, we will start our clock such that the circuit current is given by $I = I_0 \sin 2\pi ft$. Over a very wide frequency range, this will be the same current for all elements everywhere in the circuit. We recall that V_R is in phase with I, V_L leads I (hence, V_R) by 90°, and V_C lags I (hence, V_R) by 90°. The voltage equations become

$$V_R = V_{R0} \sin 2\pi ft$$
$$V_C = -V_{C0} \cos 2\pi ft$$
$$V_L = V_{L0} \cos 2\pi ft$$

and we choose to write the source voltage as

$$V = V_0 \sin(2\pi ft + \phi_0)$$

Applying Kirchhoff's loop law gives the instantaneous relationship among these voltages as

$$V = V_R + V_C + V_L$$

This is shown in the phasor diagram of Figure 23-16a. The relative magnitudes of the vectors are arbitrary and reflect only the values of the quantities involved and the frequency of the source. We see that the V_0 phasor must be the vector resultant of the other three and that the y axis projections of these phasors satisfy the voltage equations. The vector addition is not very difficult since V_{L0} and V_{C0} are along the same line in opposite directions. The resultant of adding them is a vector of magnitude $V_{L0} - V_{C0}$ in the V_{L0} direction.[1] We then apply the pythagorean theorem to find the magnitude of V_0.

This might be even easier to see if we use a simpler version of the phasor diagram, one in which the directions of the positive x axis and the resistor voltage are made to coincide. The inductor voltage then lies along the positive y axis, and the capacitor voltage lies along the negative y axis. Figure 23-16b shows the voltages of our *RCL* series circuit plotted on the simpler form. We see that the relationships among the vectors is the same as that of Figure 23-16a, but they all have been rotated clockwise by an angle $2\pi ft$. It is obvious that the

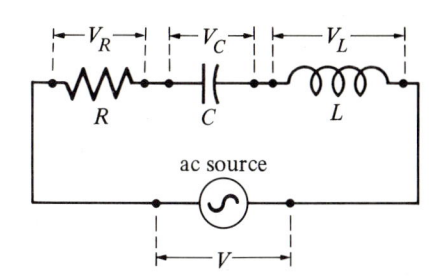

Figure 23-15 An alternating-current *RCL* circuit containing a resistance R, capacitance C, and inductance L in series with an ac source. The voltages measured at any instant across the devices are V_R, V_C, V_L, and V, respectively.

In dc circuits, the opposition to current was named resistance, and $I = (1/R)V$ defined how this limits the current in the circuit. In ac circuits, both inductors and capacitors place limits on currents, though not by expending energy to Joule heating. Reactance is an ac analog to resistance since $I = (1/X)V$. But, unlike R, X varies with the ac frequency.

---

[1] This works even if the magnitude of V_{L0} is less than that of V_{C0}. In that case, the quantity $V_{L0} - V_{C0}$ is a negative number which is treated as a vector of that magnitude in the negative V_{L0} direction.

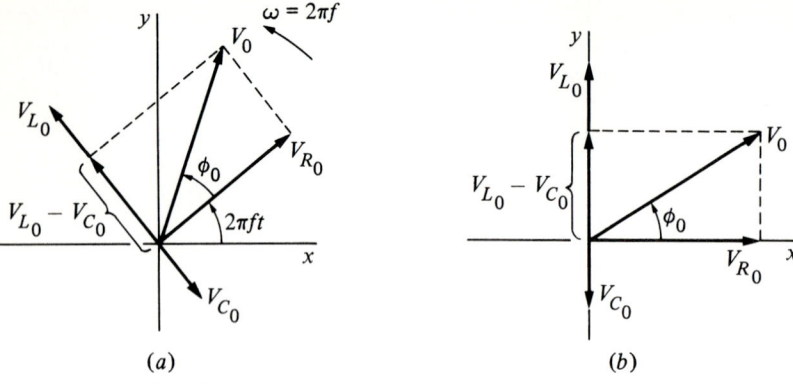

Figure 23-16 (a) A phasor diagram for the circuit of Figure 23-15. Again, V_{R0} is in the direction of the current amplitude phasor I_0. The phase angle ϕ_0 between the current I_0 and voltage V_0 may be positive, negative, or zero, depending upon the relative magnitudes of V_{L0} and V_{C0}. Since V_{L0} leads current by 90° and V_{C0} lags current by 90°, they are always opposed to each other. Their difference determines the phase difference between voltage and current. (b) Figure 23-16a redrawn so that V_{R0}, hence I_0, coincide with the x axis.

vector sum of V_{L0} and V_{C0} is perpendicular to V_{R0}, and it is perhaps easier to visualize the relationship for ϕ_0.

In any case, we apply the pythagorean theorem to obtain

$$V_0 = \sqrt{V_{R0}^2 + (V_{L0} - V_{C0})^2} \qquad (23\text{-}22)$$

Also, the definition of the tangent yields

$$\tan \phi_0 = \frac{V_{L0} - V_{C0}}{V_{R0}}$$

$$\phi_0 = \tan^{-1} \frac{V_{L0} - V_{C0}}{V_{R0}} \qquad (23\text{-}23)$$

We should point out that these equations work for an ac series circuit whether all three circuit elements are involved or not. This limits the number of equations about which we must concern ourselves.

Each of the quantities under the radical of Equation 23-22 can be written as a product of the current amplitude with either a resistance or a reactance. Equation 23-22 can then be written as

$$V_0 = \sqrt{(I_0 R)^2 + (I_0 X_L - I_0 X_C)^2}$$
$$= \sqrt{I_0^2 R^2 + I_0^2 (X_L - X_C)^2}$$
$$= I_0 \sqrt{R^2 + (X_L - X_C)^2} \qquad (23\text{-}24)$$

This is also in the same form as Equation 23-15. We call the radical part of the term the circuit **impedance** and use an uppercase Z as its symbol. Thus

$$Z = \sqrt{R^2 + (X_L - X_C)^2} \qquad (23\text{-}25)$$

for a series circuit, and we have

$$V_0 = I_0 Z \quad \text{or} \quad Z = \frac{V_0}{I_0}$$

These are analogous in form to Ohm's law for a dc circuit, with Z replacing R. Evidently, the impedance must have the units of ohms, the same as those of reactance and resistance. We can also rewrite Equation 23-23 as

$$\phi_0 = \tan^{-1} \frac{I_0 X_L - I_0 X_C}{I_0 R}$$

$$\phi_0 = \tan^{-1} \frac{X_L - X_C}{R} \quad (23\text{-}26)$$

We see that we may even draw a phasor diagram for the impedance, as in Figure 23-17. This is possible because each quantity in Figure 23-17 when multiplied by I_0 becomes the corresponding voltage of Figure 23-16b.

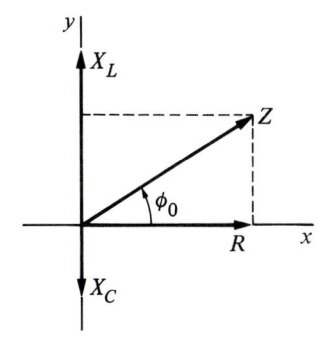

Figure 23-17 Phasor diagram for the impedance of the circuit of Figure 23-15. This is identical in form to Figure 23-16b since when each vector shown here is multiplied by the magnitude of I_0, the vectors of Figure 23-16b result.

Example 23-4 Suppose the RCL series circuit of Figure 23-15 has elements with the following values: $R = 75.0\ \Omega$, $C = 1.50\ \mu F$, and $L = 0.100$ H. The alternating voltage has an amplitude of 40.0 V and a frequency of 500 Hz. Calculate (a) the capacitive and inductive reactances, (b) the impedance of the circuit, (c) the current amplitude, (d) the maximum voltage across the resistor, capacitor, and inductor, and (e) the phase difference between the circuit current and circuit voltage.

Solution (a) This requires that we apply Equations 23-18 and 23-19.

$$X_C = \frac{1}{2\pi f C} = \frac{1}{2\pi (500\ \text{Hz})(1.50 \times 10^{-6}\ \text{F})} = 212\ \Omega$$

$$X_L = 2\pi f L = 2\pi (500\ \text{Hz})(0.100\ \text{H}) = 314\ \Omega$$

(b) To find the impedance, we use Equation 23-25.

$$Z = \sqrt{R^2 + (X_L - X_C)^2} = \sqrt{(75.0\ \Omega)^2 + (314\ \Omega - 212\ \Omega)^2} = 127\ \Omega$$

(c) The impedance may now be used to find the current amplitude directly.

$$I_0 = \frac{V_0}{Z} = \frac{40.0\ \text{V}}{127\ \Omega} = 0.315\ \text{A}$$

(d) The current amplitude times the resistance and reactances yields the voltages.

$$V_{R0} = I_0 R = (0.315\ \text{A})(75.0\ \Omega) = 23.6\ \text{V}$$

$$V_{C0} = I_0 X_C = (0.315\ \text{A})(212\ \Omega) = 66.8\ \text{V}$$

$$V_{L0} = I_0 X_L = (0.315\ \text{A})(314\ \Omega) = 98.9\ \text{V}$$

Is it reasonable that the maximum capacitor and inductor voltages be greater than V_0?

(e) Either Equation 23-23 or 23-26 may be applied to obtain the phase difference.

$$\tan \phi_0 = \frac{X_L - X_C}{R} = \frac{314\ \Omega - 212\ \Omega}{75.0\ \Omega} = 1.36$$

$$\phi_0 = 53.7°$$

565

Resonance in AC Circuits An important phenomenon with this type of circuit is that of **resonance.** We noted that the capacitive reactance decreases from a very large value toward zero as the frequency of an ac circuit increases. Further, the inductive reactance increases from zero to a very large value with increasing frequency. We therefore expect that there is one frequency at which the two reactances are equal in magnitude. At this frequency, which we call the **resonant frequency** f_r, the impedance has a minimum value equal to R and the current is a maximum for a constant-amplitude voltage.

Figure 23-18 is a plot of current amplitude vs. frequency in an RCL series circuit in which the voltage amplitude is held constant as the frequency is varied. The resonant frequency is that frequency for which the current amplitude is a maximum. We may determine an expression for the resonant frequency by simply setting the reactance expressions at that frequency equal to each other:

$$X_L = X_C$$

$$2\pi f_r L = \frac{1}{2\pi f_r C}$$

or

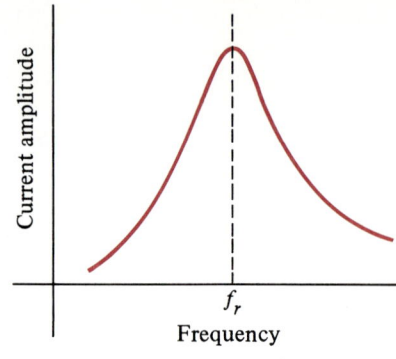

$$f_r = \frac{1}{2\pi \sqrt{LC}} \qquad (23\text{-}27)$$

Figure 23-18 Current vs. frequency for an RCL circuit. Since X_L increases with frequency while X_C decreases, there is a frequency f_r at which $X_L = X_C$ and the current becomes a maximum for the applied voltage. This condition is called resonance, and f_r is the resonant frequency of the circuit.

This is the same expression as the one for the natural frequency of oscillation of an LC circuit which we saw in Chapter 22.

An RCL series circuit is completely analogous to the driven, damped harmonic oscillator discussed in Chapter 10. Each electric element has its mechanical analog. The alternating voltage source is analogous to the periodic driving force in the mechanical system. The inductor acts as an inertia, just like a mechanical mass, in that it resists change. The resistor dissipates energy from the system and is analogous to the mechanical damping force. Finally, the capacitor is the mechanical spring's counterpart, first storing energy and then returning it to the system.

We even saw that in the mechanical system, a very small amplitude force driving at resonant frequency could cause the oscillations to grow quite large. Amplitudes of vibration much larger than the amplitude of the driving force could be attained, provided the damping force was small enough. Mechanical energy is shuffled back and forth between the kinetic energy of the vibrating mass and the potential energy of the spring. In the RCL series circuit, it is possible for a small voltage-amplitude source at resonance to cause very large maximum voltages in the inductor and capacitor if the resistance is small. Energy is shuffled back and forth between the magnetic field of the inductor and the electric field of the capacitor. The analogy between the mechanical and electric systems is nearly perfect, including comparable terms in all the descriptive equations. In fact, many mechanical systems are modeled by using electric circuits in order to predict behavior and test for possible defects, especially unwanted resonances.

Example 23-5 For the circuit described in Example 23-4, determine (a) the resonant frequency, (b) the current amplitude at resonance, (c) the maximum voltage across the inductor at resonance, and (d) the net voltage across the capacitor and inductor together at resonance.

Solution (a) The resonant frequency is calculated by setting the inductive reactance equal to the capacitive reactance. The result is Equation 23-27.

$$f_r = \frac{1}{2\pi\sqrt{LC}} = \frac{1}{2\pi\sqrt{(0.100 \text{ H})(1.50 \times 10^{-6} \text{ F})}} = 411 \text{ Hz}$$

(b) We must recognize that the circuit impedance at resonance is simply R, since $X_L - X_C = 0$. We then apply

$$I_0 = \frac{V_0}{R} = \frac{40.0 \text{ V}}{75.0 \text{ }\Omega} = 0.533 \text{ A}$$

(c) We must determine the inductive reactance at resonance and multiply by the current amplitude.

$$X_L = 2\pi f_r L = 2\pi(411 \text{ Hz})(0.100 \text{ H}) = 258 \text{ }\Omega$$
$$V_{L0} = I_0 X_L = (0.533 \text{ A})(258 \text{ }\Omega) = 138 \text{ V}$$

(d) The net voltage is I_0 times $(X_L - X_C)$, but we already know that $X_L - X_C = 0$ at resonance. Thus, the net voltage across the inductor and capacitor together is zero. Another way to look at this is to recognize that the voltage curves for the inductor and capacitor have equal amplitudes at resonance. Since they are 180° out of phase with each other, they must add to zero.

23-5 Power and RMS Values in AC Circuits

Since both the current and voltage being delivered by an ac source vary with time, the power being delivered may also vary. We may write an expression for the power being delivered at any instant of time t as follows:[1]

$$P = IV = (I_0 \sin 2\pi ft)[V_0 \sin(2\pi ft + \phi_0)]$$

With the use of trigonometric identities and some mathematical manipulation (see Problem 23-31), this power delivered can be shown to be

$$P = \frac{I_0 V_0}{2} \cos \phi_0 - \frac{I_0 V_0}{2} \cos(4\pi ft + \phi_0) \qquad (23\text{-}28)$$

When this is averaged over one or more cycles, the second term on the right vanishes. The first term is not a function of time and is therefore unaffected. Thus the time-average power delivered by an alternating voltage source is

$$P_{av} = \frac{I_0 V_0}{2} \cos \phi_0 \qquad (23\text{-}29)$$

Consider the implications of $\cos \phi_0$ appearing in this equation. It means that the average power in an ac circuit depends on the phase difference between

---

[1] There is no loss of generality by assuming that the current and voltage have the same form as we assumed for several different circuits already. However, there are as many possible forms as there are ways of writing the functions. As you might expect, all forms yield the same time variation—the only difference is where they start at $t = 0$.

the circuit current and the circuit voltage. When $\phi_0 = 90°$ ($\pi/2$ rad), as with a pure inductor circuit, or $-90°$, as with a pure capacitor, the average power delivered is zero. Figure 23-19 shows a graph of Equation 23-28 when $\phi_0 = \pi/2$. We see that the power has twice the frequency of the voltage and the current. Further, it includes the same amount of area above the time axis per cycle as below. Thus it averages to zero.

For a circuit in which $\phi_0 = 0$, the current and voltage are in phase. Either the circuit is purely resistive or the situation is like an *RCL* circuit at resonance. In either case, the average power delivered is

$$P_{av} = \tfrac{1}{2}I_0 V_0 = \tfrac{1}{2}I_0^2 R \tag{23-30}$$

This would be just like Joule's law if the average current had a value

$$I_{av} = \frac{I_0}{\sqrt{2}}$$

We could then write

$$P_{av} = (I^2)_{av} R = \frac{I_0^2}{2} R = \tfrac{1}{2}I_0^2 R$$

Such an average current is defined and is called the **effective current**, or the **root-mean-square (rms) current**. We use the symbol I_{rms} and

$$I_{rms} = \sqrt{(I^2)_{av}}$$

The root-mean-square current is the square root of the average of the square of the alternating current.

It can be shown mathematically for sinusoidal functions (see Problem 23-32) that

$$(I^2)_{av} = \frac{I_0^2}{2} \quad \text{or} \quad (V^2)_{av} = \frac{V_0^2}{2}$$

Thus

$$I_{rms} = \frac{I_0}{\sqrt{2}} = 0.707 I_0 \quad \text{and} \quad V_{rms} = \frac{V_0}{\sqrt{2}} = 0.707 V_0$$

The root-mean-square current is called the effective current because the effective value of alternating current is that which causes Joule heating in a resistor at the identical rate as a direct current of the same magnitude.

The rms values are really the most important of the various measures of current and voltage. We will see that even ac meters are calibrated to read rms values. For that reason and to simplify notation, it is common practice to drop the rms subscript. It is to be understood that any quantity to which we refer is the rms value unless we specifically state otherwise. Thus the rms resistor voltage becomes V_R, and the rms values of the source, inductor, and capacitor voltages become V, V_L, and V_C, respectively. In the most general case of ac circuits, the average power delivered and dissipated is given by Equation 23-29 and may be written as

$$P_{av} = IV \cos \phi_0 \tag{23-31}$$

The quantities I and V are the rms circuit values. We have simply dropped the subscript. The term $\cos \phi_0$ is called the **power factor** for the circuit and may

23 Alternating Current and Electrical Safety

A purely inductive circuit expends zero power. Physically, this is saying that half the time the inductor is storing energy and half the time it is returning it to the voltage source. The same is true of a purely capacitive circuit, although the instant-by-instant details would vary. Of course, any normal circuit will have some resistance, which amends either case somewhat to a net power loss.

Figure 23-19 Voltage, current, and power vs. time for an ac circuit. Power delivered is a function of the phase angle ϕ_0 between current and voltage. Shown is the case when $\phi_0 = \pi/2$, or the voltage leads the current by 90°. Power is the product of the instantaneous voltage with the instantaneous current and varies sinusoidally with a frequency of $2f$.

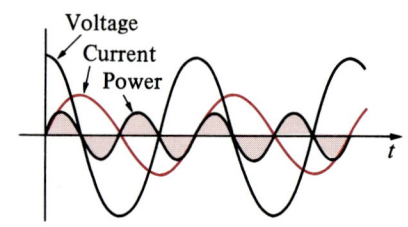

vary from 0 to 1. A low power factor means the voltage and current are way out of phase. When this happens, a large current must be supplied for a given voltage in order to deliver a large power to a load. This can cause unwanted Joule heating losses in transmission lines. Thus the power factor of ac machines is measured in manufacture and increased if necessary by the proper addition of a capacitor.

All rms values in sinusoidal ac circuits are equal to the same constant ($1/\sqrt{2} = 0.707$) times the maximum value. Thus, for a capacitor in an ac circuit, the rms voltage is given by

$$V_C = \frac{V_{C0}}{\sqrt{2}} = 0.707 V_{C0}$$

For an inductor, the rms voltage is

$$V_L = \frac{V_{L0}}{\sqrt{2}} = 0.707 V_{L0}$$

Note that we have again dropped the rms part of the subscripts. Because of this constant relationship, we may draw phasor diagrams by using rms voltages. The magnitude of each phasor is reduced from our originals by a factor of $1/\sqrt{2}$. Otherwise, all relationships among them are the same, including phase differences.

23-5 Power and RMS Values in AC Circuits

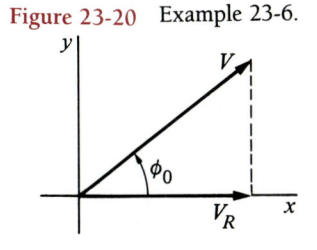

Example 23-6 A milling machine for a metal shop is built to operate on 120 V at 60 Hz. During manufacture, its power factor is measured to be 0.800 at a power of 480 W. The voltage leads the current. Calculate (a) the rms current in the circuit, (b) the rms resistor voltage, (c) the capacitance of the capacitor which when placed in series results in a power factor of 1.00, and (d) the power drawn by the milling machine after the capacitor is in place.

Solution (a) Since the voltage leads the current, the net reactance ($X_L - X_C$) must be positive, and the circuit is more inductive than capacitive. Adding the proper capacitor in series can put the voltage in phase with the current to make $\phi_0 = 0$ and $\cos \phi_0 = 1.00$. The rms current can be found by using Equation 23-31, because all quantities in that equation except I are given. For the circuit,

$$IV \cos \phi_0 = P$$

or

$$I = \frac{P}{V \cos \phi_0} = \frac{480}{(120 \text{ V})(0.800)} = 5.00 \text{ A}$$

(b) A phasor diagram will be helpful. We see in Figure 23-20 that the rms resistor voltage is equal to the rms source voltage times the power factor. It is really the projection of the source voltage on the current phasor, and the relationship we need is

$$V_R = V \cos \phi_0 = (120 \text{ V})(0.800) = 96.0 \text{ V}$$

(c) We can use either the pythagorean theorem or the phase angle to calculate the required capacitor voltage. From that, we can determine the reactance of the needed capacitor and then its capacitance. Once again, we see in Figure 23-20

Figure 23-20 Example 23-6.

that V can be brought into phase with V_R by adding an rms capacitor voltage equal to the y component of V. Thus, in magnitude,

$$V_C = V \sin \phi_0$$

But $\sin \phi_0 = 0.600$, and

$$V_C = (120 \text{ V})(0.600) = 72.0 \text{ V}$$

$$X_C = \frac{V_C}{I} = \frac{72.0 \text{ V}}{5.00 \text{ A}} = 14.4 \text{ }\Omega$$

From $X_C = 1/(2\pi f C)$, we have

$$C = \frac{1}{2\pi f X_C} = \frac{1}{2\pi (60 \text{ Hz})(14.4 \text{ }\Omega)} = 1.84 \times 10^{-4} \text{ F} = 184 \text{ }\mu\text{F}$$

(d) This requires that we find the new rms current and multiply it by the rms voltage. This recognizes that the current and voltage are now in phase, and thus $\cos \phi_P = 1.00$. We could simply square the rms voltage and divide by R, but it might be instructive to see how the current changes. In either approach, we must first find R. Since the resistance will not change, we can use the data we have before the capacitor is added to the circuit to find R. We have

$$R = \frac{V_R}{I} = \frac{96.0 \text{ V}}{5.00 \text{ A}} = 19.2 \text{ }\Omega$$

After the capacitor is added, the new current is in phase with the source voltage. Therefore, the new current is equal to the source voltage divided by the resistance:

$$I = \frac{V}{R} = \frac{120 \text{ V}}{19.2 \text{ }\Omega} = 6.25 \text{ A}$$

The new power is then

$$P = IV = (6.25 \text{ A})(120 \text{ V}) = 750 \text{ W}$$

We see that the power delivered by the source is increased by more than 50 percent. Remember also that this all goes to the resistor, or load, since the added capacitor dissipates no net power.

23-6 AC Meters and Signal Detection

A typical sinusoidal alternating signal, whether it be current or voltage, has a time-average value of zero. Unless the frequency of the signal is very, very small, an ordinary meter cannot respond fast enough to follow the variation. To measure a sinusoidal variation, most ac meters have a built-in circuit which "flips over" one-half the cycle, producing variable direct current. The circuit most often uses rectifiers (Figure 23-21), symbol ———▶︎——— , which present a vanishingly small resistance to conventional current coming from one direction and an enormous, essentially infinite, resistance in the other direction (Chapter 30). This means that a rectifier will pass conventional current practically undiminished in one direction but stop it altogether in the other. The symbols are oriented so that they pass conventional current in the direction of

Figure 23-21 Internal circuit of an ac meter with external contacts at A and B. The rectifiers allow passage of positive current only in the direction indicated by their arrows.

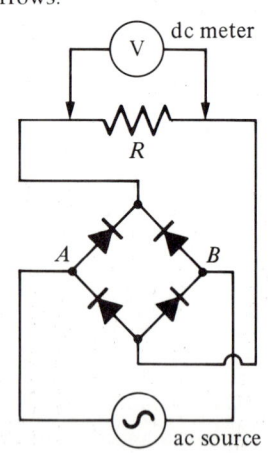

the arrowhead. In Figure 23-21, whether *A* or *B* is positive with respect to the other point, conventional current is directed through the resistor from left to right only. Figure 23-22 shows the voltage as presented to the voltmeter. The colored dashed line represents the voltage as it leaves the source. It is negative for half of each cycle, whereas the voltage across the resistor, though fluctuating, is always positive.

The exact shape of an electric signal and its maximum and minimum values may be very important whether or not the signal is periodic and regular. For instance, technicians attempting to repair a malfunctioning television receiver often must view the signal shape (variation in time) and magnitude (voltage) at various points in the circuits to help isolate the problem. Likewise, a physician may be aided immensely in diagnosing an apparent heart problem by seeing the electrical activity of the heart in action. There are many situations in which a "view" of the exact signal is nearly indispensable.

Most electric signals may best be viewed on an oscilloscope, which uses an electron beam to make visible dots or lines on a phosphor-coated glass plate or screen. The low-inertia, easily deflected electron beam of the cathode-ray tube (CRT) is used to paint a voltage vs. time graph on the screen with a speed no mechanical drawing system can match.

Figure 23-23 shows the face of a typical oscilloscope with the beam sweeping across the screen at a high-enough rate to display a straight line. The horizontal sweep is controlled by a linearly increasing voltage impressed on the horizontal deflection plates which repeats over and over. This sweep is called the sawtooth wave (Figure 23-24), and its frequency may be adjusted by turning one of the oscilloscope knobs. The frequency of the sawtooth on most oscilloscopes is controlled by a knob marked "seconds per division."

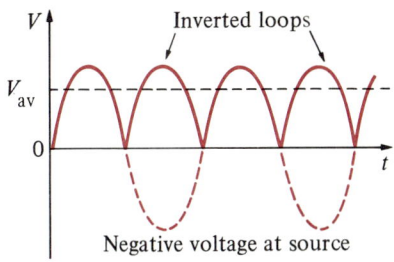

Figure 23-22 Voltage vs. time for the ac-meter circuit of Figure 23-21. Source voltage is sinusoidal alternating current with both positive and negative loops. The rectifier network converts this to the sequence of positive loops shown. To present a steady value to the meter, the variable dc loops usually are smoothed to the flat value V_{av}.

Figure 23-23 The oscilloscope records voltage across its external contacts vs. time. When the external voltage is not varying, as here, a straight line is swept across the screen in the *x* direction by a steadily increasing internal voltage.

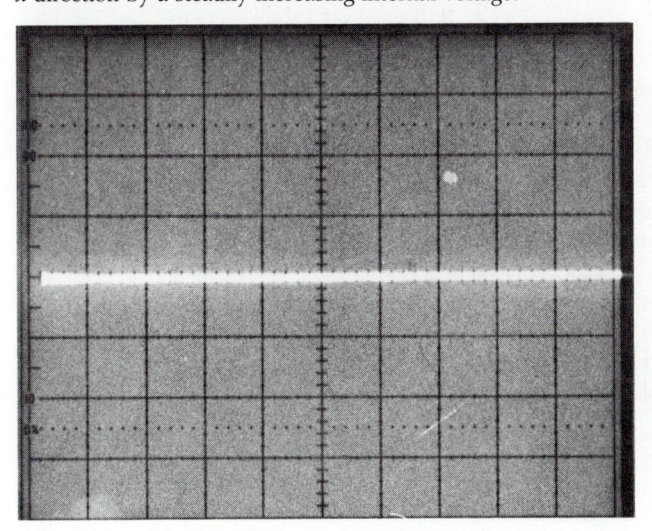

Figure 23-24 A sawtooth wave similar to the horizontal voltage variation of a sweeping oscilloscope. Each cycle represents the time for one sweep of the oscilloscope screen.

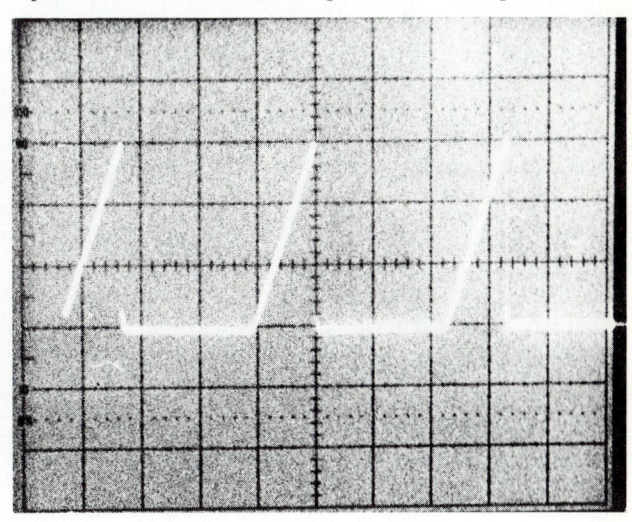

571

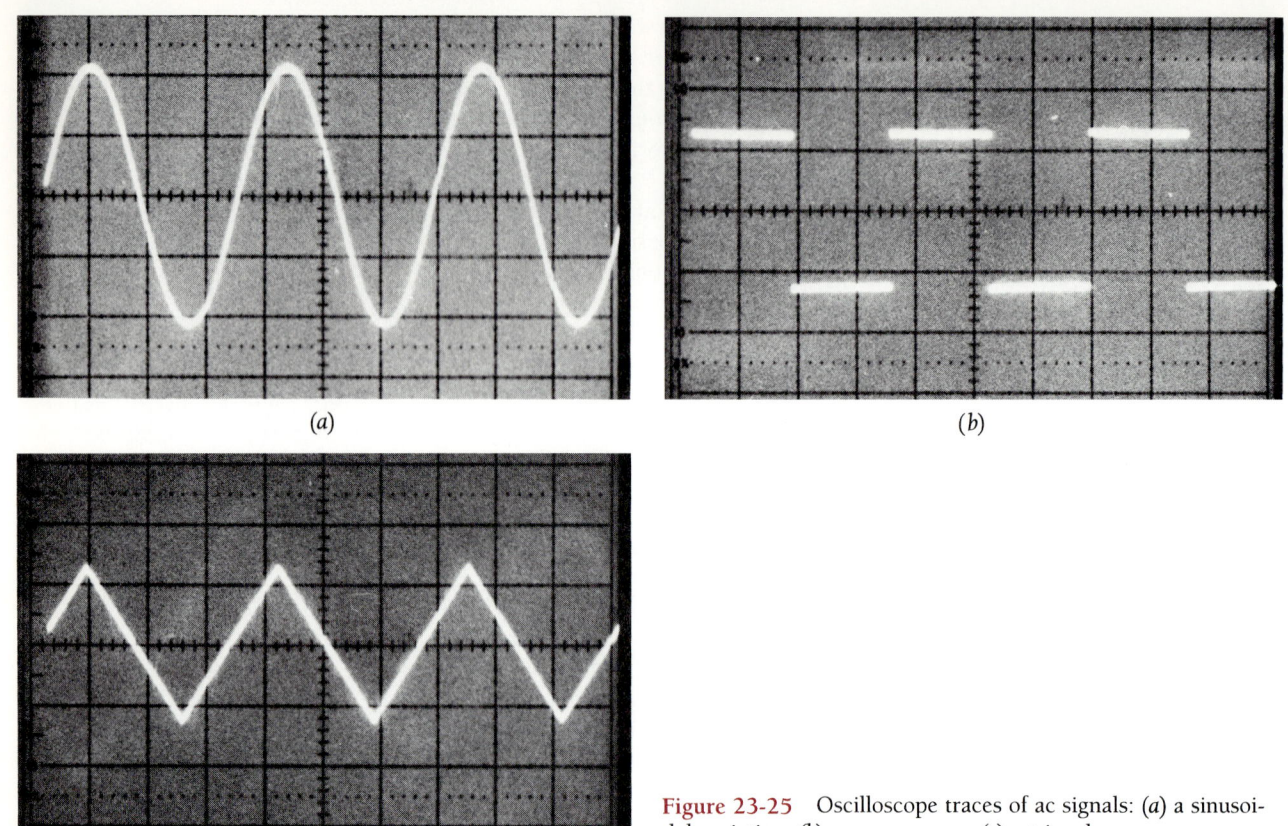

Figure 23-25 Oscilloscope traces of ac signals: (*a*) a sinusoidal variation, (*b*) a square wave, (*c*) a triangle wave.

Figure 23-25 shows oscilloscope traces of a sine wave, a square wave, and a triangle wave. Nearly any phenomenon or physical change that can be converted into a voltage can be displayed and measured on an oscilloscope. Examples are the strength and rate of a heartbeat, temperature, pressure, sound, speed, acceleration, and many, many others. All that is required is the ability to convert the phenomenon to a voltage and to have a working knowledge of the oscilloscope.

Many of the biological signals of the human body have been measured and recorded, but the most common signals monitored are those associated with the heart and the brain. Small electrodes placed in good contact with the skin can pick up signals related to the activity in the human body. The signals can be viewed on an oscilloscope, but unless the oscilloscope is a special one which can also record or store, the signals are not available for later analysis. We often amplify the signals and apply them to a pen in an instrument called a chart recorder, which gives us a permanent record of the signal.

The permanent record of the electrical activity associated with the heart is called an **electrocardiogram** (EKG), and that associated with the brain is an **electroencephalogram** (EEG).[1] Figure 23-26*a* shows an actual EKG trace on a storage oscilloscope. Figure 23-26*b* shows the EKG of the same heart on a piece

---

[1] For many years, the word *electrocardiogram* was abbreviated ECG, and many still use that designation. However, it was found that when spoken either directly or over a loudspeaker or radio, it could be confused with EEG. Thus, many medical professionals have adopted EKG as the abbreviation for electrocardiogram, and it is used routinely in most hospitals.

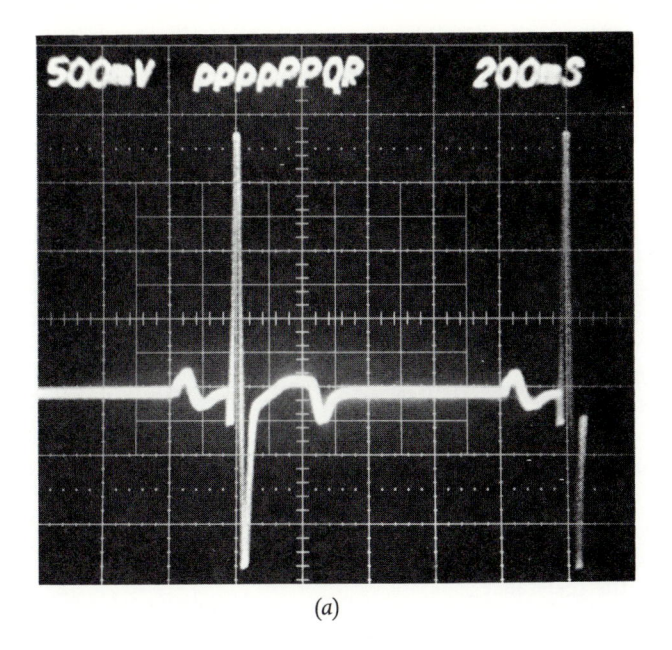

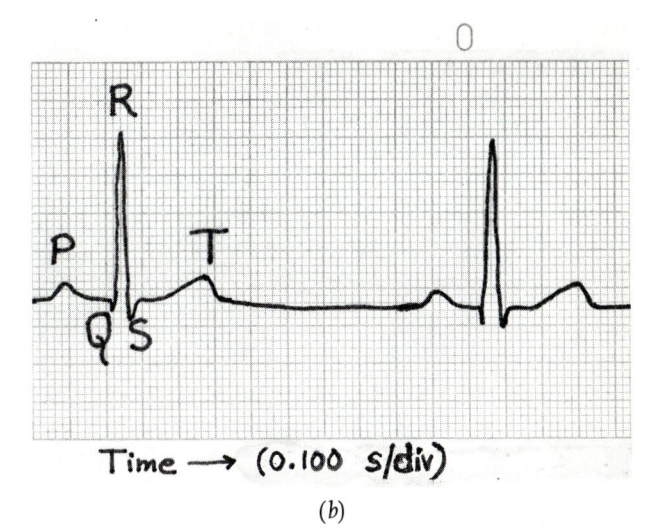

Figure 23-26 (a) Oscilloscope trace of human heartbeat in an electrocardiogram. Voltage variations measured across the human chest are fed to the vertical amplifier of the oscilloscope. (b) A permanent record of the EKG produced on a strip chart recorder.

of chart recorder paper. The human heart can be considered to be an electric dipole with its lower left part positively charged and its upper right negatively charged. We all have a specialized group of nerve cells called the sinoatrial (SA) node which serves as the natural pacemaker. The heart muscle contracts because of a depolarization and then a repolarization of its dipole charge, which is initiated by the SA node.[1] The trace shown in Figure 23-26b has specific peaks and dips called waves and labeled P, Q, R, S, and T as standard notation. The P wave corresponds to the initial electric pulse from the SA node, which triggers a heart cycle. Between points P and Q, the depolarization pulse spreads over the upper heart chambers, which are called the atria. During the QRS period, the pulse spreads over the lower heart chambers (ventricles). Finally, the T wave corresponds to ventricular repolarization, and the cycle begins again. As noted, time is along the x axis.

The EKG is an excellent tool for analyzing heart problems. For example, an unusually deep S wave may indicate a partial clogging of the pulmonary artery or its branches as blood moves toward the lungs. Also, any reflections of the depolarization wave may be a sign of dead tissue in the heart as a result of a heart attack. Information such as this when coupled with other signs and symptoms can allow a cardiologist to pinpoint any trouble and make a good diagnosis.

23-7 Electrical Safety

Current-carrying conductors may become very hot and burn whatever touches them. Even if wires are insulated, it is possible for the Joule heating due to a large current to raise the temperature high enough to cause a fire. Further, we

---

[1] The cycle of depolarization and then repolarization of any nerve membrane creates a short voltage pulse called the action potential. When this pulse propagates along the nerve fibers and reaches a muscle cell, it results in a mechanical contraction of that cell.

are all aware that the energy associated with electricity is capable of doing work on something other than that for which it is intended. Most of us have experienced electric shock as our body (or part of it) completed a circuit between some high voltage and ground. Currents can cause a variety of sensations in living beings, ranging from a mild tickle all the way to instant death. Such hazards make it a priority that electric circuits be designed with safety in mind and that electricity be afforded a healthy respect.

In general, biological effects depend on current, but the current in a closed path depends on the voltage as well as the resistance the path presents. Two important points to remember in considering electrical safety, therefore, are that (1) a current requires a closed path, and thus we should avoid making the human body part of that path, and (2) we should attempt to ensure that the resistance of any unintentional closed path we might create have as large a value as possible.

The resistance associated with the human body can vary a great deal depending on certain physical considerations. Fluids inside the body have many charged ions in solution and are fairly good conductors. Also, the construction of nerve fibers makes them very good conductors. From head to toe, the body's resistance is from 500 to 1000 Ω. Between the ears, the resistance may be below 100 Ω.

These small values might seem inconsistent to those of you who have used an ohmmeter. It is very common for someone who first uses an ohmmeter to measure body resistance by grasping the leads simultaneously. The result may be a resistance of several thousand to a million ohms. This apparent anomaly is because most of our electric resistance is in the skin. However, even that varies with the condition of the skin and the total area of contact. Dry skin may display a contact resistance as high as a million ohms, while if damp with perspiration or wet, this may drop to less than ten-thousand ohms. Immersed in a bathtub, skin resistance might be as low as several hundred ohms. If the skin is punctured or broken by a blister, its contact resistance can be vanishingly small, and the current then is limited only by the body's internal resistance. Given all this, we should be interested in the magnitude of the possible currents we might expect from everyday voltage sources and how they might affect humans.

The voltage of a dry cell is about 1.5 V. If someone touches both terminals of one with dry fingers and contact resistance of 1×10^6 Ω, the current would be

$$I = \frac{1.5 \text{ V}}{1.0 \times 10^6 \text{ } \Omega} = 1.5 \times 10^{-6} \text{ A}$$

At the other end of the spectrum, consider an electric stove with a 240-V line. If the input voltage is contacted by broken skin and the internal resistance is 1000 Ω, we would expect a current of

$$I = \frac{240 \text{ V}}{1000 \text{ } \Omega} = 0.240 \text{ A}$$

Between these extremes, contact with a standard electric outlet by dry skin yields a current of 1.20×10^{-4} A, while the terminals of an automobile battery on wet hands gives rise to a current of 2.00×10^{-3} A.

Table 23-1 shows the physiological effects of currents. Currents below 5 mA may be felt but are generally considered to be harmless. Internal electric pulses cause muscle contractions, and this can happen even if the electricity

TABLE 23-1 Physiological Effects of Current on Healthy Human Beings

Current	Possible Effects
0.3–1.0 mA	Tingle, threshold of sensation
1.0–5.0 mA	Threshold of pain
10–25 mA	Muscle contraction; cannot let go
50–100 mA	Labored breathing
100 mA–2.5 A	Ventricular fibrillation, respiratory arrest
Above 2.5 A	Burns, shock, heart stoppage

comes from an external source. For example, the muscles which control the hand can clamp closed, making it impossible to let go of a wire. Above 100 mA (0.100 A), a current through the heart muscle can cause the ventricle walls to fibrillate, or undergo irregular rippling contractions. This results in a lack of synchronization between filling and pumping actions, and the heart pumps very little blood. Death can result in a minute or two because of lack of oxygen to the brain and to the heart muscle itself unless fibrillation is stopped quickly. Above 2.5 A, the heart is stopped forcibly but may start of its own accord once the current is removed. Medical defibrillation units take advantage of the fact that heart action is stopped by large currents. In defibrillation, a large pulse of current is forced through the heart to completely stop it, with the hope that normal beating will be resumed when the current stops.

Beyond protecting people, we must worry about property and equipment. Numerous fires every day have some electric origin. Further, electric appliances and equipment can be permanently damaged or destroyed by large currents. Many strategems are employed to help protect people and property. Electric wires are covered with insulating material. Accessibility to electric wires is limited by distance and barriers. Receptacles and switches are designed to make human contact with current-carrying wires difficult. Electrical codes require that wires be of sufficient thickness so as not to heat enough to cause a fire under normal use. Every house and building has a main electric service box. The electricity for the structure is directed through **fuses** which are designed to be the weak link and melt if too much current is drawn. The main electric service is then further divided into circuits, each of which has its own fuse or **circuit breaker** that causes the circuit to go dead if too much current is drawn. In addition, many appliances and instruments have their own separate fuses or circuit breakers for further protection. Finally, in most new construction and nearly all laboratories, electric receptacles have a second ground connection called the safety ground. The safety ground is designed to provide an electric path to earth which is separate from the neutral return wire. It is very important in power tools and appliances or equipment in which the high-voltage wire may become shorted to the case. When such a short takes place, the case will safely conduct the current to ground if it is properly connected.

Any or all of these safety designs can be bypassed. Every year, over 2000 people are accidentally electrocuted in the United States. People must be taught good safety practices in the use of electricity. In addition to knowing that we should not place our wet fingers into a light socket while standing on a wet basement floor in our bare feet, we must learn not to use plug-in radios and hair driers while taking a bath. Further, the importance of the safety ground can hardly be overstated, particularly in situations in which the skin is bypassed. For example, when a blood pressure catheter is inserted into a blood vessel near the heart, the patient suddenly presents only 500 to 1000 Ω along a path to ground. In that situation, 120 V can be fatal. It is imperative that this kind of equipment be maintained and grounded properly.

Minimum Learning Objectives

After studying this chapter, you should be able to:
1. Define:

   capacitive reactance
   circuit breaker
   current amplitude
   fuse
   impedance
   inductive reactance
   initial phase angle
   instantaneous voltage
   lag
   lead
   out of phase
   phase
   phase-angle difference
   phasor diagram
   power factor
   RCL circuit
   reactance
   resonance
   resonant frequency
   rms current
   rms voltage
   root-mean-square
   voltage amplitude

2. Draw a phasor diagram for voltage-current relationships in an ac circuit with a source and a resistance.
3. Draw voltage phasors for each component in an alternating-current RCL circuit.
4. Calculate the instantaneous voltage at each component in an RC or RL circuit from the voltage at the source and the phase angle.
5. Calculate the voltages across components in an RC circuit, given values of R and C as well as the frequency and current amplitude.
6. Calculate the voltages across components in an LR circuit, given values of L and R as well as the frequency and current amplitude.
7. Apply Kirchhoff's rules to an RCL circuit.
8. Calculate capacitive and inductive reactances from C or L and the source frequency.
9. Calculate the impedance of an RCL circuit from the values of the components and the frequency.
10. Calculate voltages across each element of an RCL circuit, given the values of the components and the source voltage and frequency.
11. Calculate the resonant frequency of an RCL circuit.
12. Calculate the root-mean-square values of current and/or voltage from maximum values.
13. Calculate the power dissipated in each component of an RCL circuit.
14. Interpret an oscilloscope screen pattern.
15. Understand electric shock hazards and how to avoid them.
16. Understand the use of fuses and circuit breakers in electric circuits.

Problems

23-1 A sinusoidal generator delivers an alternating voltage which varies with time according to the equation $V = 150 \sin(10\pi t + \pi/6)$, where V is in volts and t is in seconds. For this voltage, determine the (a) amplitude, (b) frequency, (c) the time to complete one cycle (the period), and (d) the voltage when $t = 0$.

23-2 The output of an alternating voltage source is given by $170 \sin(120\pi t - \pi/3)$, where V is in volts and t is in seconds. For this voltage, (a) determine the frequency, (b) determine the voltage when $t = 0$ s and when $t = \frac{1}{180}$ s, and (c) draw phasor diagrams for $t = 0$ and $t = \frac{1}{180}$ s.

23-3 Voltage is delivered to a load resistor by a sinusoidal generator at an amplitude of 40.0 V and a frequency of 25.0 Hz. The voltage has a value of -20.0 V at $t = 0$. (a) Using the proper numerical values, write an equation for the voltage as a function of time. (b) If the load has a resistance of 16.0 Ω, write an equation for the current in the load as a function of time. (Use another possible value of the initial phase. What other information would you need to specify uniquely the initial phase angle?)

23-4 A transmission line delivers 50 MW of power at an effective voltage of 230 kV from a dam to a distant city. (a) What is the effective current in the line? (b) What is the maximum current? (c) If 1 percent of the power is lost to Joule heating, what is the resistance of the line? (d) When stepped down for use at an effective voltage of 120 V, what effective current is produced?

23-5 A sinusoidal voltage across a capacitor has an amplitude of 12.0 V at a frequency of 60.0 Hz. Determine (a) the capacitance if the current amplitude is 2.00 A and (b) the frequency when the current amplitude is 5.50 A.

23-6 Using the various devices discussed in this and previous chapters, draw the following circuit: a 120-V ac input is used to charge a capacitor to 12.0 V and hold the charge until a switch discharges it through a xenon strobe lamp (a resistance).

23-7 An alternating voltage of amplitude 30.0 V and variable frequency is connected in series with a 50.0-μF capacitor. Calculate (a) the current amplitude when the frequency is 500 Hz and (b) the frequency if the current amplitude is 2.00 A.

23-8 Beginning with Equation 23-5, that is, $I = I_0 \cos(2\pi ft + \phi_0)$, as representing the current in an ac circuit with a capacitor and a resistor, where the voltage across the capacitor is given by $V_C = V_{C0} \sin(2\pi ft + \phi_0)$, show that if ϕ_0 is chosen to give the current the form $I = I_0 \sin 2\pi ft$, then the capacitor voltage is given by $V = -V_{C0} \cos 2\pi ft$. Use the trigonometric identities for the sine and cosine of the sum and difference of two angles, i.e., $\sin(A \pm B) = \sin A \cos B \pm \cos A \sin B$ and $\cos(A \pm B) = \cos A \cos B \mp \sin A \sin B$.

23-9 A resistor in series with a sinusoidal voltage of amplitude 170 V and frequency 60.0 Hz has a current amplitude of 5.00 A. A capacitor is placed in series with the resistor, and the current is reduced to an amplitude of 3.50 A. Calculate (a) the resistance, (b) the maximum voltage across the resistor after the capacitor is added, (c) the maximum voltage across the capacitor, and (d) the capacitance.

23-10 A 12.0-μF capacitor and a 40.0-Ω resistor are placed in series with a sinusoidal voltage of amplitude 25.0 V and variable frequency. Calculate (a) the maximum voltage across the capacitor when the frequency is 500 Hz and the current is 0.520 A, (b) the phase difference between the voltage and the current, and (c) the maximum voltage across the resistor. A phasor diagram will aid in the solution.

23-11 A 0.200-μF capacitor and a 700-Ω resistor are placed in series with an alternating voltage source of amplitude 15.0 V and variable frequency. Determine (a) the maximum voltage across the capacitor when the current amplitude is 14.0 mA, (b) the frequency at this current amplitude, and (c) the phase difference between the voltage and current.

23-12 A 0.180-H inductor is connected to an alternating voltage source of frequency 50.0 Hz. Calculate (a) the voltage of the source if the maximum current in the inductor is 2.16 A and (b) the current in the inductor if the source voltage is 40.0 V.

23-13 The maximum current in a 0.400-H inductor is 0.800 A at a certain initial frequency. When the frequency is increased by 30 percent, it requires a maximum voltage of 65.0 V to maintain the same current. Calculate (a) the initial frequency and (b) the initial voltage.

23-14 The maximum voltage across a 1.20-H inductor in an ac circuit is 4.20 V when the frequency is 80.0 Hz. Calculate (a) the maximum current, (b) the maximum time rate of change of current, and (c) the frequency for the same voltage which causes a current of 0.0100 A in the inductor.

23-15 Begin with Equation 23-11, that is, $I = -I_0 \cos(2\pi ft + \phi_0)$, as representing the current in an ac circuit with a resistor and an inductor, where the voltage across the inductor is given by $V_L = V_{L0} \sin(2\pi ft + \phi_0)$. Show that if ϕ_0 is chosen to give the current the form $I = I_0 \sin 2\pi ft$, then the inductor voltage is given by $V_L = V_{L0} \cos 2\pi ft$. Use the trigonometric identities given in Problem 23-8.

23-16 A 0.150-H inductor and a 600-Ω resistor are placed in series with an alternating voltage source of amplitude 240 V. The maximum voltage across the resistor is found to be 189 V. Calculate (a) the voltage across the inductor, (b) the current in the circuit, (c) the frequency of the source, and (d) the phase difference between the circuit current and circuit voltage.

23-17 A 100-Ω resistor is used with a 120-V 60-Hz ac source. The power dissipated in the resistor must be limited to 0.250 W. (a) What inductance could be inserted in series with the resistor to keep the power to this level? (b) What resistance in series with the resistor would restrict the power to the same level?

23-18 Measurement shows that the current lags the voltage by 71° in a circuit containing an inductor, a 40.0-Ω resistor, and an alternating voltage source of amplitude 8.00 V and frequency 100 Hz. Calculate (a) the maximum voltage across the resistor, (b) the maximum voltage across the inductor, (c) the maximum current in the circuit, and (d) the value of the inductor.

23-19 The maximum current in a resistor is 0.300 A when it is in series with an alternating voltage source of amplitude 120 V and frequency 60 Hz. Calculate (a) the value of the resistance, (b) the value of the inductor necessary to reduce the current to 0.210 A, and (c) the phase difference between the circuit current and circuit voltage with the inductor in place.

23-20 A 0.065-H inductor is placed in series with an unknown resistor and an alternating voltage source of amplitude 50.0 V and variable frequency. When the frequency is adjusted to 2.50×10^3 Hz, the current amplitude is 38.5 mA, and when the frequency is 1.25×10^3 Hz, the current amplitude is 52.7 mA. Calculate the value of the resistance twice (once for each data set).

23-21 A series circuit with a 15.0-Ω resistor, a capacitor C, and an inductor L is being driven by an alternating voltage source at a frequency of 100 Hz. The rms (effective) voltages are measured across each of the three elements, and the results are $V_R = 30.0$ V, $V_C = 90.0$ V, and $V_L = 50.0$ V. Calculate (a) the current in the circuit, (b) the capacitive and inductive reactances, and (c) the resonant frequency of the circuit.

23-22 Draw a phasor diagram for an RCL circuit with the following characteristics: (i) I is at its maximum positive value of 1.00 A; (ii) the circuit is at resonance; (iii) the source voltage is 12.0 V at 200 Hz; (iv) the value of the inductance is 0.100 H; (v) label your diagram with actual values of V_0, V_R, V_L, V_C, and ω. (a) What are the values of R and C? (b) What is the power dissipated by the circuit?

23-23 An *RCL* series circuit contains a 180-Ω resistor, a 3.00-μF capacitor, and an 80.0-mH inductor with a 400-V amplitude source operating at 200 Hz. Calculate the (*a*) capacitive and inductive reactances, (*b*) the impedance, (*c*) the current amplitude, and (*d*) the phase difference between the current and voltage.

23-24 A speaker has a measured dc resistance of 0.0100 Ω, yet it has the value "8 Ω at 1000 Hz" stamped on it. (*a*) What is the inductance of the speaker coil? (*b*) What is the speaker's impedance at 60 Hz and at 10,000 Hz? (*c*) From these numbers, suggest an explanation as to why high-fidelity systems use several speakers of differing sizes for the reproduction of sound.

23-25 An 80.0-Ω resistor and a 0.150-H inductor are in series with a 120-V amplitude ac source of variable frequency. Calculate (*a*) the impedance of the circuit at 60.0 Hz, (*b*) the phase difference between the current and voltage at 60.0 Hz, and (*c*) the capacitance which must be added in series in order that the circuit be in resonance at 300 Hz.

23-26 An ac circuit contains a 200-Ω resistor and a 1.20-μF capacitor in series with an ac source of amplitude 25.0 V. Calculate (*a*) the impedance at 800 Hz, (*b*) the current at 800 Hz, (*c*) the inductor which when placed in series will give the circuit a resonant frequency of 500 Hz, and (*d*) the inductive reactance at resonance.

23-27 An *RCL* series circuit contains a 1.00×10^5 Ω resistor, a 400-pF capacitor, and a 5.00-H inductor with a 200-V amplitude ac source. At a frequency of 5.00×10^3 Hz, calculate (*a*) the inductive and capacitive reactances, (*b*) the impedance, (*c*) the current, (*d*) the phase difference between the current and voltage, and (*e*) the voltage across the inductor-capacitor combination.

23-28 A 2.00-Ω resistor, a 1.00-μF capacitor, and an 80.0-mH inductor are to be used in an *RCL* resonance circuit with an ac source of constant 10.0 V amplitude and variable frequency. (*a*) On the same graph, plot the capacitive reactance, the inductive reactance, and the impedance vs. frequency on a 100- to 1000-Hz interval by calculating these quantities at each 100 Hz in the interval *and* at the resonant frequency f_r. (*b*) Calculate the current at the resonant frequency and at 99 and 101 percent of the resonant frequency.

23-29 An electric drill with an armature resistance of 8.00 Ω and an inductance of 12.0 mH is used in a 120-V 60-Hz circuit. The power developed by the drill can be increased by the use of a capacitor. (*a*) Calculate the current power factor of the drill. (*b*) What is the phase angle between current and voltage before the capacitor is installed? (*c*) What is the initial impedance of the drill? (*d*) What size capacitor should be installed for maximum power at the drill? (*e*) What is the ratio of power developed after inserting the capacitor to that before?

23-30 A normal 120-W light bulb is almost a pure resistance designed to draw 1.00 A of effective current at an effective voltage of 120 V. A 0.400-H inductor is inserted in series in one of the lines to the bulb, and the line is plugged into a standard socket. (*a*) What power will be delivered to the inductance? (*b*) What do you expect will happen to the brightness of the bulb (before performing any calculations)? (*c*) What is the effective current with the inductor in place? (*d*) What is the power dissipated in the bulb? (*e*) The inductor is removed, and a 5.30-μF capacitor is installed in series with a line to the bulb. What is the effective current at the bulb?

23-31 Use the trigonometric identities given in Problem 23-8 to show that the instantaneous power in an ac circuit,

$$P = IV = (I_0 \sin 2\pi ft)[V_0 \sin(2\pi ft + \phi_0)]$$

becomes

$$P = \frac{I_0 V_0}{2} \cos \phi_0 - \frac{I_0 V_0}{2} \cos (4\pi ft + \phi_0)$$

23-32 Show that if $I = I_0 \sin 2\pi ft$, then $(I^2)_{av} = I_0^2/2$. [*Hint:* Realize that $\cos 4\pi ft = \cos (2\pi ft + 2\pi ft)$, and apply the identities from Problem 23-8 as necessary to solve for $\sin^2 2\pi ft$. Finally, recognize that the average of the cosine over a complete cycle (or any number of complete cycles) is zero.]

23-33 A waffle iron has a dc resistance of 9.00 Ω measured with an ohmmeter. When used with a 120-V dc source, it generates heat twice as fast as when used with a 120-V 60-Hz ac source. (*a*) What is the inductance of the heating coils of the waffle iron? (*b*) What is the power factor of the iron used with alternating current? (*c*) What device could you connect in series in one of the input wires to make the iron heat as rapidly with the alternating current as it did with the direct current? (*d*) What would be the value required of this device? (*e*) Once attached, what would be the effect of this device on heating if the waffle iron is used in a dc circuit again?

23-34 Electric impedances are added in series and parallel similar to the way resistors in series and parallel are added. (Would you know how to go about proving this if you were asked to do so?) Calculate (*a*) the impedance of the circuit of Figure 23-27, (*b*) the effective current in the 30.0-Ω resistor, and (*c*) the effective voltage across the inductor.

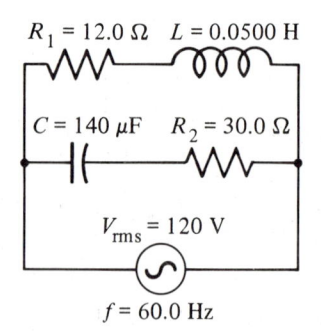

Figure 23-27 Problem 23-34.

$R_1 = 300\ \Omega$ $L = 0.0300$ H

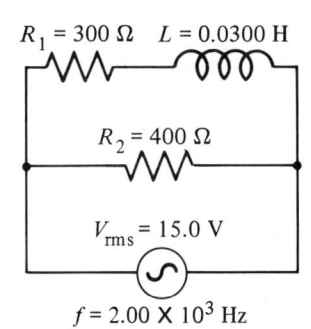

$R_2 = 400\ \Omega$

$V_{rms} = 15.0$ V

$f = 2.00 \times 10^3$ Hz

Figure 23-28   Problem 23-35.

23-35 For the circuit shown in Figure 23-28, determine (a) the impedance, (b) the current being delivered by the source at the frequency shown, (c) the effective voltage across the inductor-resistor combination, (d) the effective current in the inductor, and (e) the effective voltage across the inductor.

23-36 The three oscilloscopes of Figure 23-25 had the following horizontal and vertical settings: (a) 2 ms/division, 5 V/division; (b) 50 μs/division, 0.2 V/division; (c) 10 μs/division, 2 mV/division. Determine the frequency and maximum voltage of each of the three signals.

23-37 Suppose that for the sine wave displayed in Figure 23-25, the oscilloscope's horizontal and vertical controls had been set at 50 ms/division and 2.0 V/division, respectively. (a) What are the period and frequency of the wave? (b) What is the voltage amplitude of the wave? (c) Repeat parts a and b for the square wave and the triangle wave if the horizontal and vertical controls were set at 20 ms/division and 0.50 V/division, respectively, for both.

23-38 Suppose you are told that the sine wave in Figure 23-25a has a frequency of 1.25×10^4 Hz and a voltage amplitude of 5.00 V. (a) What is the sine wave's period? (b) What is the setting of the oscilloscope's horizontal control? (c) What is the setting of the vertical control?

23-39 The user of an electric drill without a safety ground is grasping the metal case of the drill, making excellent electric contact. A partially conductive pathway of metal powder and fragments built up from the use of the drill has collected inside the drill from the hot-wire contact to the case. (a) If the body resistance of the person is 10,000 Ω, what must the resistance of the pathway inside the drill drop to before the person will feel a trickle current through his or her body when using the drill? (b) Suppose a direct short occurred where the hot wire inside the drill touched the case. Will the person notice the current and be able to let go of the drill? (c) Recalculate the internal resistance for the user to sense a current, and give the physiological consequences of a direct, internal short for the following situations: (i) The user is wearing leather gloves working in a dry, cool garage while standing and wearing sneakers. The resistance between drill and ground through the user is 1.50 MΩ. (ii) The user is in a hot, damp basement sweating profusely and holding a copper water pipe with a bare hand while drilling it with the other to install a pressure line to a dehumidifier. Resistance between drill and ground through the person is 1500 Ω.

23-40 The motor of a bench lathe used in a woodshop has worn wiring and develops a 400-Ω resistance path between the hot 120-V wire of the power cord and the steel housing of the lathe. An operator who is electrically grounded lays a wet hand on the housing, establishing a resistance path of 1000 Ω to ground. (a) Calculate the shock current. (b) Is it dangerous? What are the possible physiological effects?

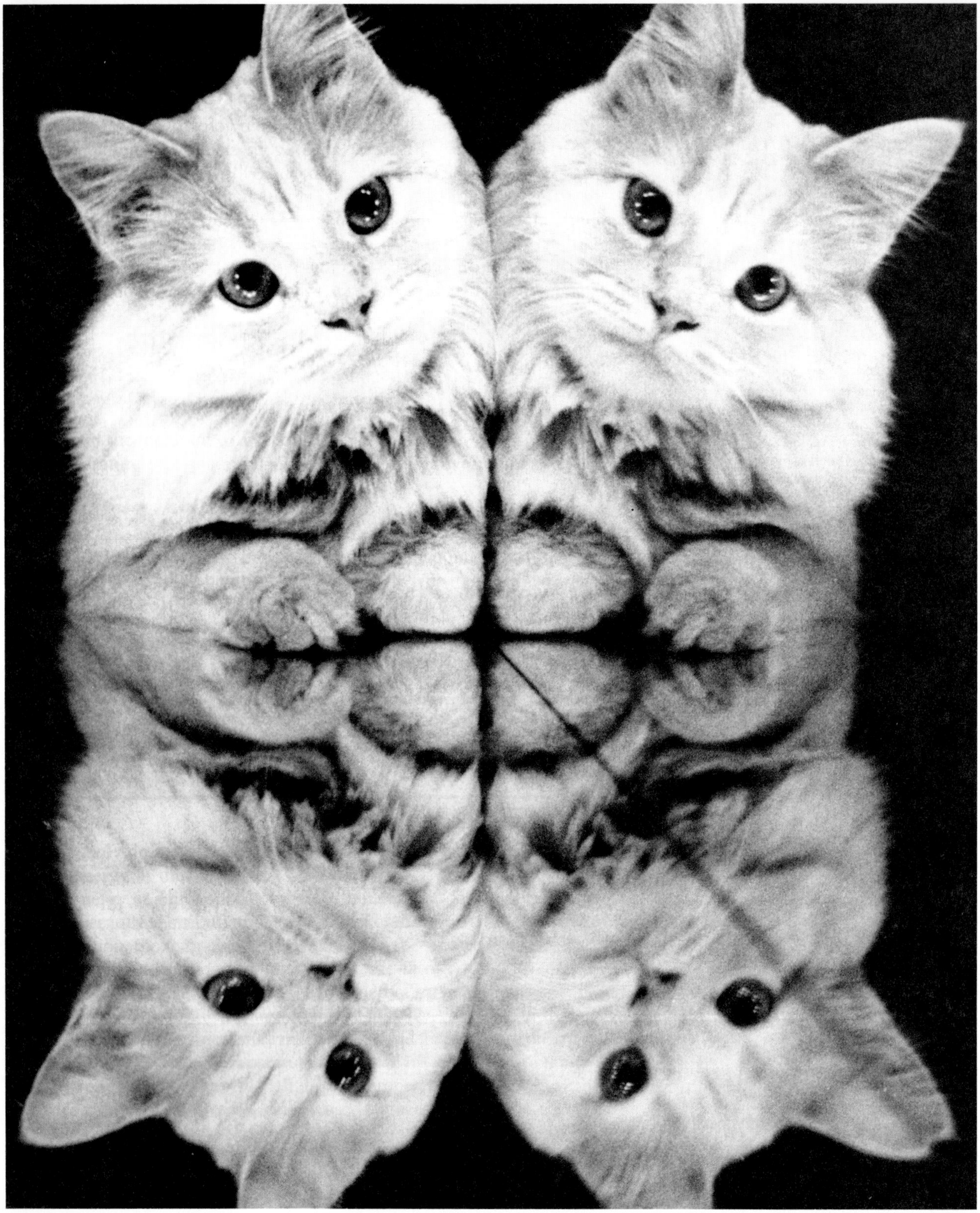

24 Light and Geometric Optics

24-1 Introduction

We use the word *optical* to describe things which are related to vision. Hence, light and its interaction with matter may be categorized as optical phenomena. You already know quite a bit about optical phenomena since you deal with many of the properites of light on a daily basis. We need light in order to see things. We know that light from a flashlight, for example, must bounce off an object and return to our eyes for us to be able to see the object. Flashlights must be aimed in a specific direction and will not illuminate the area behind them.

Most objects reflect light in all directions and may be seen from any angle when illuminated (Figure 24-1a). We also know light may reflect in a different way from shiny surfaces such as the hood of a car, a window or a mirror (Figure 24-1b). Furthermore, there are materials through which light will pass almost as though the material were not there – window glass for instance (Figure 24-1c), and materials through which light passes but do not allow you to see a clear image of what is on the other side – frosted glass, for example (Figure 24-1d).

Opaque materials pass no light at all. Light directed at an opaque object casts a sharply outlined dark shadow in the shape of the object on a surface beyond the object (Figure 24-1e). This leads us to believe that light travels in

(a)   (b)   (c)

(d)   (e)

Figure 24-1 (*a*) Diffuse reflection. Each point on the object reflects light in essentially all directions. (*b*) Specular reflection. Light falling on the surface is reflected in a specific direction depending on its angle of approach to the surface. (*c*) The cat is standing on a **transparent** surface, and the camera is positioned below the surface. All the light is transmitted through such a material. (*d*) Light passes through a **translucent** material but is randomly directed such that there is no clear image on the other side. (*e*) Light directed at the opaque cat casts a shadow on the wall beyond. (D. Riban)

straight lines and not around corners. This description of the shadow phenomenon and the inability of light to go around corners is only approximately true; but, because it was believed, the acceptance of light as having a wave nature may have been delayed for centuries.

To predict the behavior of light quantitatively requires that we obtain analytical relationships. Eyeglasses to correct defective vision, camera lenses, telescopes, and microscopes all require this analytical knowledge for their design. Along the way to obtaining this analytical knowledge, we will see that consistent optical measurements help lead us to a better understanding of the world of atoms and other areas of modern physics.

24-2 Nature of Electromagnetic Waves

Light is the visible part of the electromagnetic spectrum, the part that can be seen. In 1865, James Clerk Maxwell, a Scottish physicist, published what is now known as the electromagnetic theory, showing that the experimental results of Faraday, Gauss, and Ampere could be formulated into four mathematical equations, known forever after as Maxwell's equations. These may be manipulated simultaneously to yield a form any physicist would recognize as the equation of a wave. Maxwell calculated the speed of the waves predicted by the equations and found it to be precisely the known value for the speed of light. Experimental confirmation was needed that varying electric and magnetic fields would indeed produce such a wave, and Heinrich Hertz, in 1888, produced what we would now call short radio waves and measured many of their behaviors, confirming the electromagnetic theory.

Figure 24-2 depicts the production of an electromagnetic wave. An alternating voltage source is connected to small metal rods (an antenna) separated by a short distance but lying along the same straight line, with the whole apparatus called a transmitter. In Figure 24-2a, the switch is closed, and conventional current is directed upward in the circuit. In Figure 24-2b, the growing separation of charge gives rise to an electric field (the colored line) while at the same time the current is generating a magnetic field surrounding the antenna perpendicular to the plane of the page (the crossed circle). Only the fields to the right of the antenna are shown, but the magnetic field completely surrounds the

24-2 Nature of Electromagnetic Waves

As the switch is thrown for the circuit of Figure 24-2, imagine an isolated test charge well to the right. The imbalanced charge developing on the antenna in parts *b* and *c* should have an effect on this charge. When? At the instant an imbalanced charge appears at the antenna, the test charge does not respond to it. Instead, the electric field caused by the imbalanced charge must propagate to the right until it reaches the test charge.

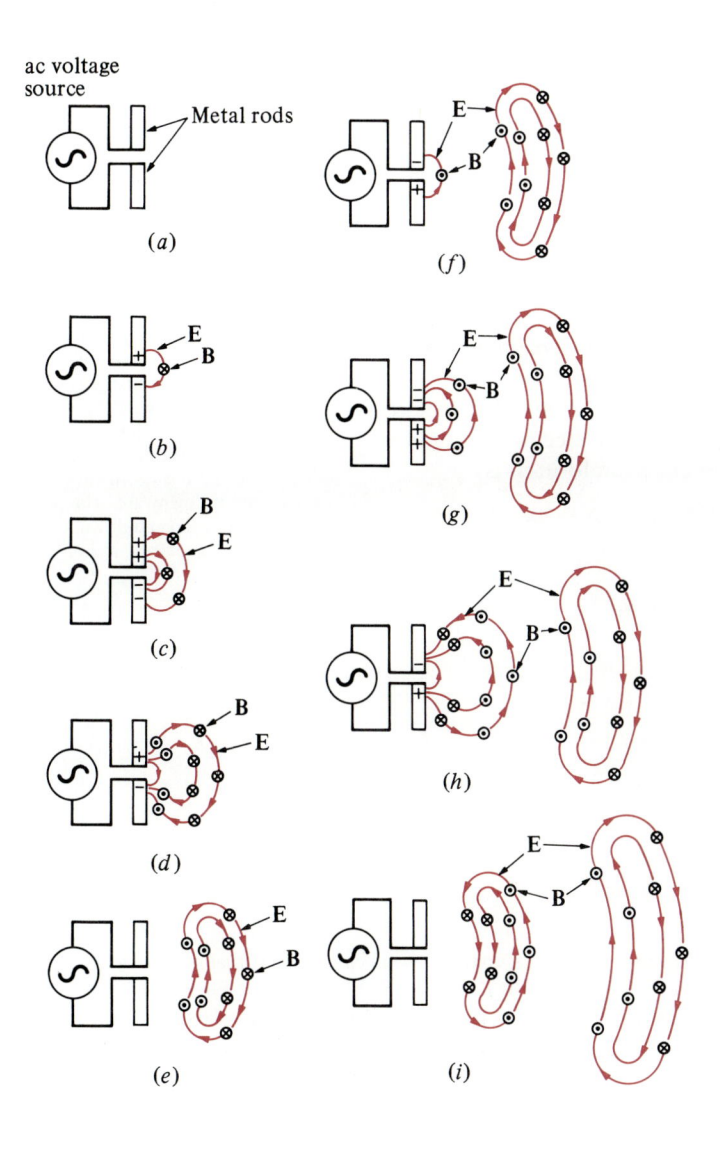

Figure 24-2 The generation of an electromagnetic wave by a simple dipole antenna.

antenna, each line closing on itself. Note that the magnetic and electric field lines are perpendicular to each other.

By Figure 24-2c, the top rod has acquired its maximum positive charge, and the current is about to reverse direction. As the charges are stopped, the current is zero, and hence no magnetic field lines are shown on the electric field line nearest the antenna. The magnetic field lines already produced have not disappeared, however, but continue to move away from the antenna at the speed of light along with the expanding electric field lines. By Figure 24-2d, current has started downward and the magnetic field lines to the right of the antenna are now out of the page. By the time of Figure 24-2e, there is no longer a separation of charge, and hence the electric field near the antenna has collapsed. The electric and magnetic fields generated during the first part of the cycle (farthest from the antenna) continue to move outward and by now have formed closed loops propagating outward. Figure 24-2f through i continues the other half of the current cycle.

As time goes on, additional loops are generated, but loops already made continue to propagate outward no matter what happens at the source. The electric and magnetic fields are changing in both space and time, are perpendicular to the direction of propagation as well as to each other, and have maximum values at the same place in space; that is, they are in phase. The changing magnetic field gives rise to the changing electric field, which gives rise to the changing magnetic field, and so on. These changing field lines constitute an **electromagnetic wave.** Suppose we pick a direction of propagation in the plane of the page and perpendicular to the antenna to plot the magnitude and directions of the fields. After the transmitter has been switched on for a time, our plot would look like Figure 24-3, where the electric field varies in the xy plane and the magnetic field varies in the xz plane. Since the variations of the fields are perpendicular to the direction of wave propagation, an electromagnetic wave is a transverse wave.

Maxwell's solution of his equations indicated that the speed of electromagnetic waves in free space should depend only upon the electric permittivity and the magnetic permeability of free space. In equation form, it is

$$c = \frac{1}{\sqrt{\epsilon_0 \mu_0}}$$

where c = speed of electromagnetic waves in free space
ϵ_0 = permittivity
μ_0 = permeability

Using the known values of these quantities, we have

$$c = \frac{1}{\sqrt{[8.85 \times 10^{-12} \text{ C}^2/(\text{N} \cdot \text{m}^2)](4\pi \times 10^{-7} \text{ T} \cdot \text{m/A})}}$$
$$= 3.00 \times 10^8 \text{ m/s}$$

The transverse nature is also made clear by picturing what an electric field is, namely, the force that acts per unit charge at a given position. Picture a positive charge to the right of the antenna. As the first part of the first loop arrives, the electric field is directed upward. The positive charge experiences an upward force. A little later, the back of the loop arrives and **E** is downward, so the force reverses and is downward. The positive charge oscillates up and down for a left-to-right-moving wave; thus, a transverse wave.

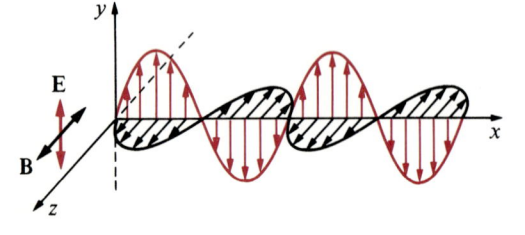

Figure 24-3 The instantaneous electric and magnetic field variations along the central axis to the right of Figure 24-2. Electric field variation is confined to a vertical plane (color), while magnetic field variation is in phase with it and in a horizontal plane (black).

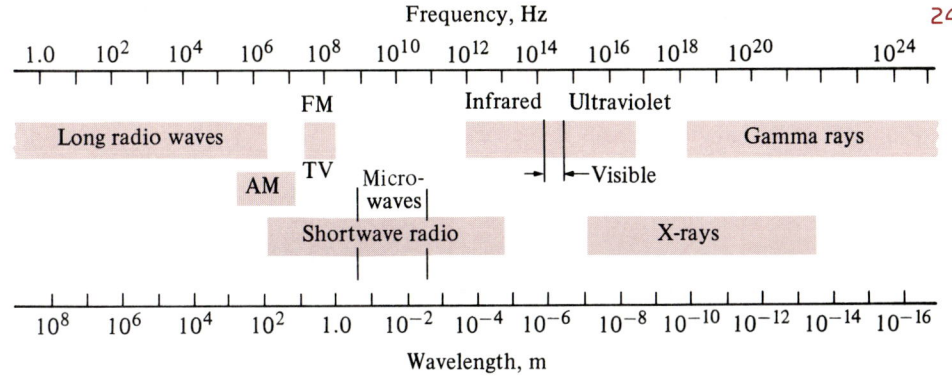

Figure 24-4 Frequencies and wavelengths of the principal divisions of the electromagnetic spectrum.

This is indeed the speed which had been measured for light traveling in free space and led Maxwell to believe that light was an electromagnetic wave.

Figure 24-4 reillustrates the spectrum of electromagnetic waves, with visible light seen to occupy the wavelength range of about 700 nm (red) down to about half this at 400 nm (violet). Within this range, the **visible spectrum,** color is determined by the wavelength of the light.

24-3 Light Wavefronts and Rays

Since light is an electromagnetic wave, we may use much of what we have already learned about wave motion to analyze its properties. You will recall from our study of waves that a wave's velocity of propagation, wavelength, and frequency are related by the equation

$$v = \lambda f$$

For free space, using c as the symbol for the speed of light, we would write

$$c = \lambda f \qquad (24\text{-}1)$$

Example 24-1 Determine the broadcast frequency for the signal for channel 2 of a television receiver if the wavelength is 5.50 m.

Solution We may apply Equation 24-1 since all electromagnetic radiation travels at the same speed, $c = 3.00 \times 10^8$ m/s.

$$c = \lambda f$$

or

$$f = \frac{c}{\lambda} = \frac{3.00 \times 10^8 \text{ m/s}}{5.50 \text{ m}} = 5.45 \times 10^7 \text{ Hz}$$

This is often stated as 54.5 megahertz (MHz).

As light (or any wave) travels through space, we find two geometric constructs very useful in our analysis. The first construct is that of a **wavefront,** a mathematical surface joining all the points of a set of waves which have equal phase. One can visualize wavefronts as the line of crests which we see moving as an unbroken water ridge toward the shore at the beach. Figure 24-5 shows outwardly moving waves caused by water drops striking the surface. We see that the wavefronts form concentric circles. We might call this disturbance **circular waves** in reference to the shape of the wavefronts. This is an example of a point source of waves from which the energy propagates along a surface in two dimensions only.

If we choose a point source of light from which the energy propagates in *three* dimensions, wavefronts form concentric spheres. These then are called **spherical waves.** Finally, if the source of waves is every point along a line (or along a plane in three dimensions), we call the resultant **plane waves** since the wavefronts form parallel planes. Drawn in two dimensions, spherical wavefronts become circles or parts of circles and plane wavefronts become straight lines, as in Figure 24-6a and b. Here, we have chosen to connect only the crests of the waves as the points of equal phase. This choice means there is exactly one wavelength's distance between adjacent fronts. This was not necessary. We could have connected the troughs of equal phase *and* the crests of equal phase, yielding a one-half wavelength's distance between adjacent fronts. It is important to note that the surface of a wavefront is perpendicular to the direction of propagation of the wave at every point. As spherical wavefronts expand very far from the source, a small section of the front looks like a plane. Thus, a point source a very large distance away is an approximate source of plane waves. For instance, light from a star or the sun may be treated as plane waves, or even the light from a very small bulb as seen by an observer at a distance of a few meters.

The second construct we find useful in our analysis of light is that of rays. A **ray** is an imaginary line drawn in the direction of the propagation of the wavefronts. As such, rays are perpendicular to fronts at every point. We often think of rays as very thin beams of light. Figure 24-7 shows the wavefronts and the rays associated with both spherical and plane waves. You can see that rays from a point source spread out, or diverge, whereas the rays associated with plane waves are parallel to each other. Sometimes we will use wavefronts to help analyze the light propagation in a real situation, and sometimes we will use rays. Occasionally it may be convenient to use both constructs.

Light rays spread out, or **diverge,** from point sources. Everything we see can be considered to be made up of a set of points forming surfaces. In general, light is diverging from all points in all directions. Our eyes intercept small cones of these diverging rays and, as we will see, form images which the brain then

Figure 24-5 Waves on a water surface caused by periodic disturbance of a single point propagate from that point in all directions. If we mark the crests of the radiating disturbances, we see a set of concentric, circular wavefronts advancing away from the point of disturbance. (R. Megna/Fundamental Photographs)

If you remember the last time you splashed in the waves at a beach, you will recognize that the motion of the wavefront was perpendicular to the line of crests. In a photograph of a large wavefront near a beach, you know intuitively which way the wavefront is moving.

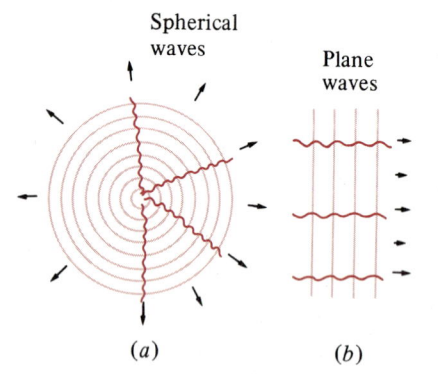

Figure 24-6 A representation of advancing wavefronts of (a) spherical and (b) plane waves. Small arrows indicate the direction of the wavefronts. Several "slices" through the wave pattern indicate the variation of wave phase with position throughout the disturbance. In actuality, the magnitude of the wave disturbance (amplitude) should decrease with increasing distance from the source in part *a*.

Spherical waves

Plane waves

(a) (b)

Figure 24-7 Instead of a wavefront, a ray construct may be used to follow the progress of waves. The ray traces the progress of a tiny piece of the wavefront as it propagates. Wavefronts (light color) and some corresponding rays (dark color) for (a) spherical and (b) plane waves.

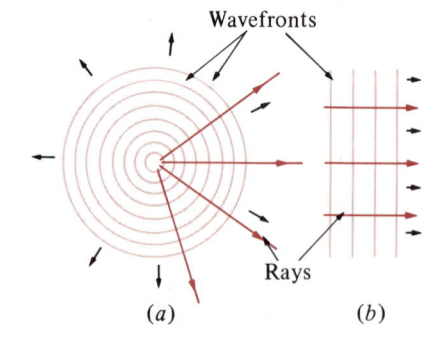

Wavefronts

Rays

(a) (b)

interprets. The *angle of divergence* is a measure of the spread of rays from a particular point as they enter the eye. Figure 24-8 shows two point objects from which light is diverging. The pupil of the eye (greatly exaggerated) intercepts a much larger angle cone of diverging rays from a point on the nearer object than from the farther object and must accommodate this varying angle to form crisp images of both objects.

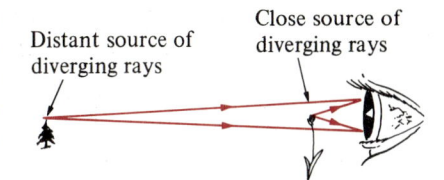

Figure 24-8 Light diverges from a source in all directions. The eye intercepts a diverging cone of light from every point on an object. The rays from the distant object are spreading apart with a much smaller angle of divergence as they reach the eye than the rays for the nearby object.

Example 24-2 Calculate the angle of divergence of the light rays which enter a 2.00-mm-diameter pupil of a human eye after they leave a point on an object which is (*a*) 20.0 cm away and (*b*) 10.0 m away.

Solution For both parts, we simply apply the definition of radian measure, assuming that the distance to the object is large enough that we may use the diameter of the pupil as the arc length (it is really the chord) of a circle centered at the point in question.

(*a*) $$\Delta\theta = \frac{\Delta s}{r} = \frac{2.00 \text{ mm}}{200 \text{ mm}} = 0.0100 \text{ rad}$$

(*b*) $$\Delta\theta = \frac{\Delta s}{r} = \frac{2.00 \text{ mm}}{1.00 \times 10^4 \text{ mm}} = 2.00 \times 10^{-4} \text{ rad}$$

Although our ability to perceive distance is primarily the result of viewing objects from two positions (one from each of the two eyes), a little of the information used by the brain is this angle of divergence.

The direction the rays are traveling when they enter the eye indicates the direction to the object. Geometric optics assumes that light travels in straight lines until it encounters a boundary between two different media. Our own optical system (eyes, optic nerve, and brain, in combination) has been trained by experience from birth to accept that the diverging light which enters the eye has traveled a straight-line path from its source. As we will see, this is not always the case.

24-4 Reflection

If you are gazing at a mirror while holding your toothbrush above your head, you might say you can see your toothbrush in the mirror. Is this really true? Not if taken literally, since the toothbrush certainly is not inside the mirror. You might really wish to say that you see the light from your toothbrush reflected from the surface of the mirror. Even then your optical system might wish to put up an argument. The light from a particular point on your toothbrush enters the eye, diverging as though the origin of the light was a point *behind* the surface of the mirror, as in Figure 24-9. In fact the rays are reflected at the surface of the mirror and only appear to originate behind it.

There is a fixed relationship between the direction of the rays before and after they strike the mirror and the orientation of the mirror. The angle an incoming, or **incident, ray** makes with an imaginary line drawn perpendicular (**normal**) to the mirror just where the ray strikes (the point of incidence) is

Figure 24-9 Light from each point on the toothbrush is striking every part of the mirror. However, only the cone of rays from a specific point striking in one small area is reflected to the eye. The reflection is such that the rate at which the rays are spreading is preserved after reflection. Extending this divergence backward, the eye "sees" each cone of rays as originating at a single point behind the mirror.

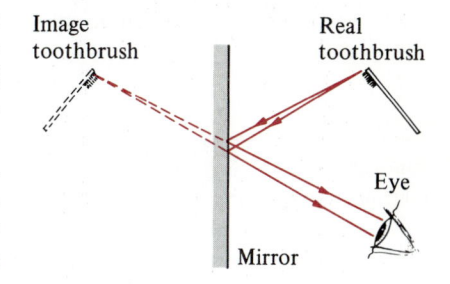

equal to the angle the ray makes on the other side of the normal after it has been reflected. This is the **law of reflection**:

> The **angle of reflection** of a light ray is equal to the **angle of incidence**.

Consider Figure 24-10, where a light ray originating at point P strikes the mirror at an angle of incidence θ_1 with the normal. After reflection, the ray makes an angle of reflection θ_1' with the same normal. The law of reflection may be stated in equation form as

$$\theta_1 = \theta_1' \qquad (24\text{-}2)$$

Note also that these angles are on opposite sides of the normal, and that the incident ray, the reflected ray, and the normal are all in the same plane. Equation 24-2 is found to be true for all reflected rays of light. Note that this means nearly every light ray we see! Most everything that we are able to see is seen by means of reflected light. The pages of this text are surely not the origin of the light with which you view them. Nor are your pencil and notebook sources of light. As you think about it, you will see that there are very few light sources. All objects reflect light to a greater or lesser extent, and most are seen by this reflected light.

At this point, you might be thinking, "Wait a minute! If all objects reflect light, why doesn't the page of this text look like a mirror?" The answer is that most surfaces are rough, having many irregularities (Figure 24-11a). A cone of rays leaving a source and being reflected from a small area on such a surface will not maintain a spreading cone shape upon reflection as it does from a mirror. Instead, the rays of the original cone will be reflected in many different directions. Rays intercepted at the eye from that part of the surface will thus have many different origins, and there is no way the eye could interpret them as coming from a single point other than a point right at the surface.

Reflection such as shown in Figure 24-11a is called diffuse reflection. You will note that each ray still obeys the basic law of reflection. However, the irregularity of the surface ensures that the normals at the points of incidence of individual rays (even those very close together before reflection) will not be parallel to each other. Thus the rays take very different directions after reflection. Figure 24-11b shows, by contrast, reflection from a mirror. Light striking the mirror as a diverging cone continues as a diverging cone after reflection, and upon interception by the eye is interpreted as originating at a point behind the mirror. This is an example of specular reflection. It occurs from surfaces that are so smooth that the irregularities which do exist are very small compared to the wavelength of light. Hence, normals to the surface at various adjacent points are parallel. Metal surfaces, glass, even polished, painted surfaces may display specular reflection. On occasion, we see both diffuse and specular reflection from the same surface, such as natural finished or stained wood which has been waxed. For most of the phenomena with which we will be concerned, we will be dealing with specular reflection only.

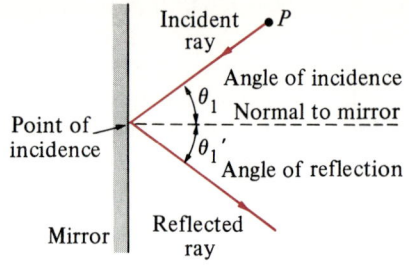

Figure 24-10 The law of reflection. The angle between the incident ray and a normal to the surface at the point of incidence equals the angle on the opposite side of the normal at which the reflected ray will emerge, or $\theta_1 = \theta_1'$.

Figure 24-11 (a) Diffuse reflection. Each ray reflects by the law of reflection, but the irregularities are such that the normals for all rays are not parallel. The rays are widely dispersed upon reflection, and only one of them moves in the direction of the eye. (b) Specular reflection. All normals to the surface of the mirror are parallel to each other, and geometric relationships between the rays are preserved upon reflection. Thus, the eye intercepts the cone of rays as though it had diverged from a point on the image bulb.

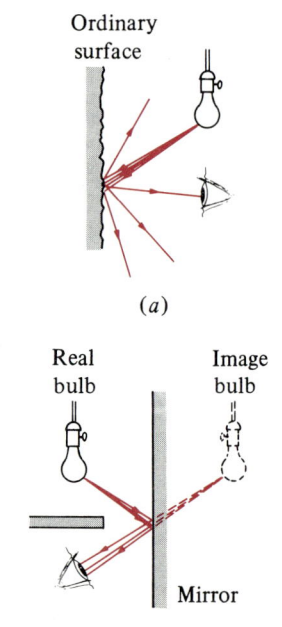

Example 24-3 Two plane mirrors are touching along one edge and oriented so that their surfaces make an angle of 50° with each other. A beam of light is directed toward one of the mirrors at an angle of incidence of 30°. What will be the angle of reflection off the second mirror?

588

Solution This problem is primarily geometrical. Thus it is a good idea to draw a diagram, as we have in Figure 24-12. Note that all the angles are labeled, and we should be able to determine their magnitudes from the law of reflection and simple geometry. We see that α is the angle of reflection at the first surface and is thus equal to the angle of incidence, or 30°. Angle β is the complement of α and is thus equal to $90° - \alpha$, or $\beta = 60°$. Because they are the interior angles of a triangle, the sum of β, γ, and 50° must equal 180°. Thus

$$\beta + \gamma + 50° = 180°$$

or

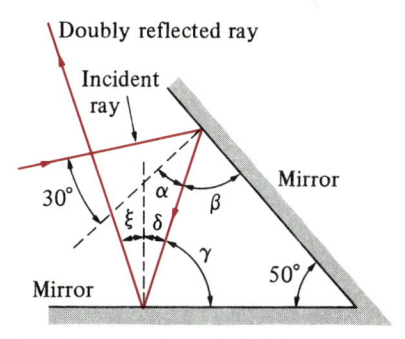

$$\gamma = 130° - \beta = 130° - 60° = 70°$$

Figure 24-12 Example 24-3.

Now δ is the complement of γ and is therefore equal to 20°. This is the angle of incidence at the second mirror. Finally, the angle of reflection at the second mirror is equal to the angle of incidence. Thus

$$\xi = 20°$$

We call what we see when we look in a mirror an **image**. We say that the eye of Figure 24-9 is seeing an image of the toothbrush, and the eye of Figure 24-11b is seeing an image of the light bulb. Plane mirrors may be used to generate simultaneously multiple images of the same object, as shown in Figure 24-13. We will deal more fully with images in Section 24-7 and Chapter 25.

24-5 Refraction

Light is not always reflected when it strikes a boundary between two media. For example, light from the sun seems to pass right through window glass. Further, you can see fish in an aquarium and swimmers under water in a clear pool. Light

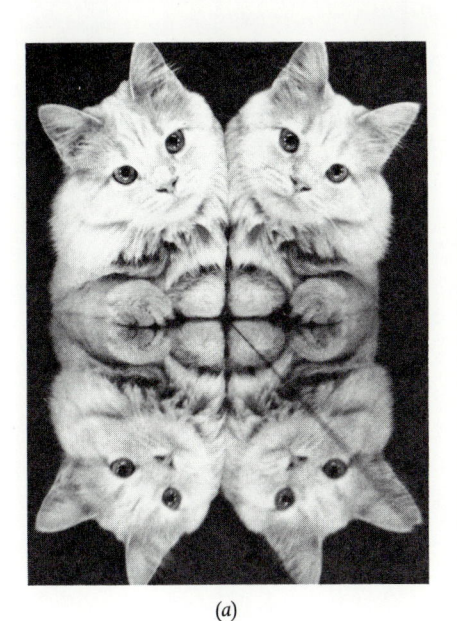

Figure 24-13 (a) Two mirrors at right angles to each other produce three images of the cat. Can you describe how the third image is produced? (b) Two mirrors may generate additional images beyond the three of part a. (c) How would you generate these eleven images (remember, the twelfth is the real cat) with two mirrors? (D. Riban)

(a) (b) (c)

does pass through materials we call transparent, but it is not necessarily unmodified. Consider Figure 24-14, which shows a beam of light directed at a thick piece of clear plastic. We can make some important observations concerning this figure. First, not all the light passes through; some is reflected at the boundary between the two media (air and plastic). Second, and more important to this section, the light transmitted through the boundary abruptly changes direction at each boundary. We call this phenomenon **refraction.** Refraction is the bending of light which takes place as it passes from one transparent medium to another. It is caused by light having different speeds in the two media. The light beam makes a smaller angle with the normal to the surface after it enters the plastic than it does while in air. We call the angle with the normal after the refraction the **angle of refraction.**

The angle of incidence, the angle of refraction, and the speeds of light in the two transparent media may be related by an equation. We find it more convenient to derive the equation by using wavefronts, as in Figure 24-15, where we show a beam of light traveling in medium 1, entering medium 2, and being refracted at the boundary. Shown are the plane wavefronts before they strike the boundary and after they leave the boundary. Shown also are three rays and the normal line, with the angle of incidence and angle of refraction labeled. Note that because the rays are perpendicular to the wavefronts and the normal line is constructed perpendicular to the surface, the angle between the wavefront and boundary surface at point P is equal to θ_1, the angle of incidence. Also, the angle between the wavefront and the boundary surface after refraction at point Q is equal to θ_2, the angle of refraction. Part of the ray on the right, the distance AQ, may be used to complete a right triangle APQ, and part of the ray on the left, the distance PB, completes the right triangle BQP. As drawn, either distance represents two wavelengths of travel for the light. Thus if the light in medium 1 travels the distance AQ in some time interval Δt, then it travels PB in the same Δt interval. Therefore, $AQ = v_1 \Delta t$ and $PB = v_2 \Delta t$. The two right triangles may be related because they share a common hypotenuse. We may write

$$\sin \theta_1 = \frac{AQ}{PQ} = \frac{v_1 \Delta t}{PQ}$$

and

$$\sin \theta_2 = \frac{PB}{PQ} = \frac{v_2 \Delta t}{PQ}$$

Solving for PQ and setting the results equal, we have

$$PQ = \frac{v_1 \Delta t}{\sin \theta_1}$$

and

$$PQ = \frac{v_2 \Delta t}{\sin \theta_2}$$

Therefore

$$\frac{v_1 \Delta t}{\sin \theta_1} = \frac{v_2 \Delta t}{\sin \theta_2}$$

Figure 24-14 A beam of light striking a thick block of plastic changes direction as it enters the plastic and reverts to its original direction upon leaving. Note that some light is reflected at both the air-to-plastic interface and the plastic-to-air interface. (D. C. Heath/Educational Development Center)

Recall that when a wave moves from one medium to another, the property which is preserved is its timing. Thus, if one point is 0.1 s behind another in the first medium, it will remain 0.1 s behind it in the second, or any successive media. Practically, for a periodic wave, this means that wavelength and velocity may change while frequency may not.

Dividing through by Δt and rearranging yields

$$\frac{\sin \theta_1}{\sin \theta_2} = \frac{v_1}{v_2} \quad (24\text{-}3)$$

Equation 24-3 is one form of what is known as **Snell's law of refraction**, named after the Dutch astronomer Willebrord Snell:

> The ratio of the sine of the angle of incidence to the sine of the angle of refraction is equal to the ratio of the speeds of light in the two media involved.

Snell's law is often used in a form other than Equation 24-3. We define a quantity associated with each transparent medium which we call the **index of refraction** of the medium. It is equal to the ratio of the speed of light in a vacuum to the speed of light in the medium. For instance, we could write the indices of refraction for the two media in Figure 24-15 as

$$n_1 = \frac{c}{v_1} \quad \text{and} \quad n_2 = \frac{c}{v_2}$$

Solving for the v's and substituting into Equation 24-3 yields

$$\frac{\sin \theta_1}{\sin \theta_2} = \frac{c/n_1}{c/n_2}$$

Finally, canceling the c's and rearranging yields

$$n_1 \sin \theta_1 = n_2 \sin \theta_2 \quad (24\text{-}4)$$

Equation 24-4 is the form of Snell's law of refraction most often used, but both forms can be important.

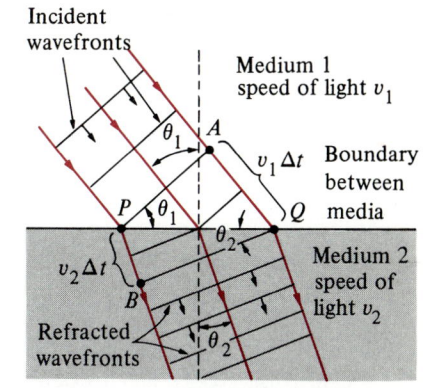

Figure 24-15 Light refracting at the boundary between two transparent media. The left edge of the incident wavefront encounters the boundary first. Since light travels more slowly in medium 2, the wavefront is slowed differentially from left to right as it enters medium 2, and its direction of propagation is pivoted.

Example 24-4 A beam of light traveling in air is incident at an angle of 41° to the normal of a surface of transparent plastic. It is refracted at the interface and makes an angle of 26.5° to the normal in the plastic. Assuming the speed of light in air is 3.00×10^8 m/s, calculate (a) the speed of light in the clear plastic and (b) the index of refraction of air and of the plastic.

Solution (a) We require only a straightforward application of Snell's law in the form of Equation 24-3.

$$\frac{\sin \theta_1}{\sin \theta_2} = \frac{v_1}{v_2}$$

or

$$v_2 = \frac{v_1 \sin \theta_2}{\sin \theta_1} = \frac{(3.00 \times 10^8 \text{ m/s})(\sin 26.5°)}{\sin 41°}$$
$$= 2.04 \times 10^8 \text{ m/s}$$

(b) We use the definition of the index of refraction, $n = c/v$. For air, we assume $c = v$; thus $n_a = 1.00$. For the plastic, we have

$$n_p = \frac{3.00 \times 10^8 \text{ m/s}}{2.04 \times 10^8 \text{ m/s}} = 1.47$$

We saw in Figure 24-14 that light is bent toward the normal when it travels from a medium of smaller index of refraction (air) to a medium of larger index of refraction (plastic). The converse should be true since Snell's law does not specify a direction of travel. When light travels from a medium of larger index of refraction to a medium of smaller index of refraction, it is refracted *away* from the normal. This, too, shows in Figure 24-14 as the beam emerges from the plastic at the second surface. Both points are made mathematically in Equation 24-4. If the index of refraction of the second medium is greater, the sine of the angle (and hence the angle) must be smaller in accordance with Snell's law. However, if the index of refraction of the second medium is smaller, Snell's law dictates that the angle of refraction must be larger than the angle of incidence.

Table 24-1 gives the index of refraction for a selected number of transparent materials. The index of refraction is sometimes called a measure of the optical density of the material. Materials with larger indices of refraction are said to be optically denser. We can see from Equation 24-4 that the larger the ratio of the index of refraction of one medium with respect to the index of refraction of another, the greater will be the bending at the interface between the two media. This is depicted in Figure 24-16. The index of refraction of the material through which the ray is passing in part *a* is much smaller than the index of refraction of the medium in part *b*.

We often see refraction effects when we look at light coming from objects after it passes through a glass of water (Figure 24-17). We also see the effects when we view objects partially submerged in water. Figure 24-18 shows that our optical system can be fooled by refraction effects just as it can be fooled by reflection. Indeed, a fish under water is not located where our optical system tells us. Figure 24-19 shows the cone of light leaving a point on the fish which will eventually enter the eye. It is bent away from the normal at the surface and appears to the eye to be diverging from a point displaced both to the right and not quite so deep as the real point. Once again we say we see an image of the fish. As in the case of the mirror, the light is not really diverging from the point our eye tells us it is.

We may solve mathematically for the apparent depth of the image fish. Suppose two observers are viewing a fish from different positions, as in Figure 24-20. The observer directly above receives a ray of light from the fish which has not been deviated from its original direction. The ray of light viewed by the other observer has been refracted at the surface, and Snell's law applies.

$$n_1 \sin \theta_1 = n_2 \sin \theta_2$$

If the angles are small enough ($<15°$), the sine may be replaced by the tangent.

TABLE 24-1 The Index of Refraction (c/v) of Selected Materials in Light of Wavelength 5.89×10^{-7} m

Material	c/v
Vacuum	1.00000
Air	1.00029
Ice	1.31
Water	1.3330
Ethyl alcohol	1.3617
Glycerine	1.4730
75% sucrose in H_2O	1.4774
Acrylic plastic	1.5100
Rock salt	1.544
Crown glass (dense)	1.5880
Flint glass (heavy)	1.7470
Calcspar:*	
Ordinary ray	1.658
Extraordinary ray	1.486
Zirconium orthosilicate	1.96
Diamond	2.4173

* The significance of two indices for some materials will be discussed in Chapter 26.

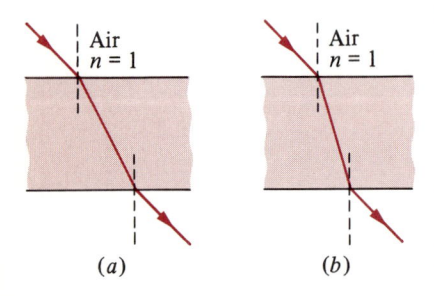

Figure 24-16 (*a*) A ray striking a transparent block is refracted on both entering and leaving the block. The refraction shows that the index of refraction of the block is greater than that of the surrounding air. (*b*) A parallel ray entering a similar block shows much greater refraction. The index of refraction of the block in this part is greater than that in part *a*.

(a)   (b)

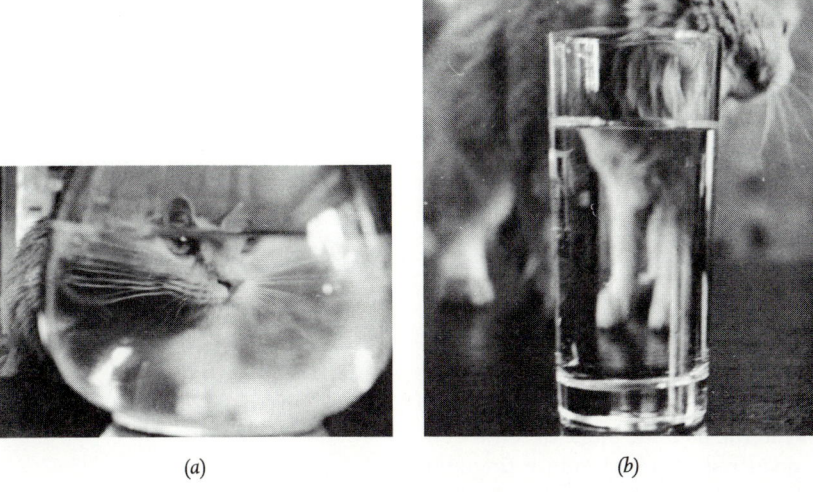

Figure 24-17 Nonplanar transparent materials distort the images of objects seen through them. (a) A transparent sphere with an index of refraction higher than that of air enlarges the image vertically and horizontally. (b) A transparent cylinder distorts the image perpendicular to its axis only. (D. Riban)

Thus
$$n_1 \tan \theta_1 = n_2 \tan \theta_2$$

But
$$\tan \theta_1 = \frac{l}{h} \quad \text{and} \quad \tan \theta_2 = \frac{l}{h'}$$

Substituting and using 1.00 for n_2 since the second medium is air and n for n_1, we have, after rearranging,

$$h' = \frac{h}{n} \qquad (24\text{-}5)$$

Equation 24-5 gives the relationship between the apparent depth (h') and the real depth (h) of the fish. Note that the apparent depth under water as viewed

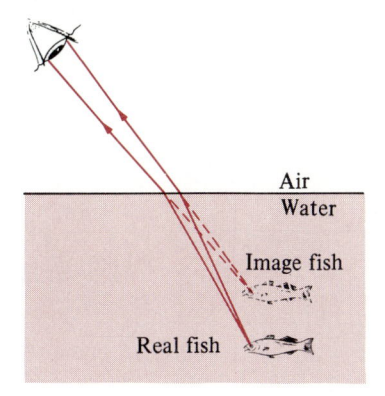

Figure 24-19 A diverging cone of rays from a point on a fish under water moves to the eye above water. The rays refract at the water-air boundary, and the eye interprets the spreading cone as coming from a point in space higher than the real source.

(a)

(b)

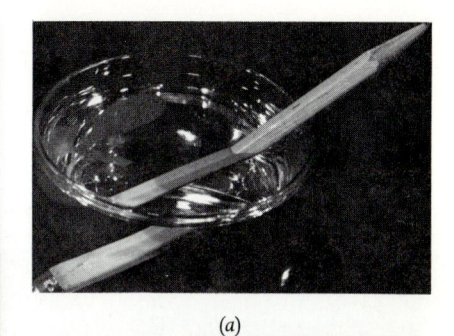

Figure 24-18 (a) The pencil appears bent. This "bending" is such that the submerged part looks closer to the surface than it actually is. (b) The coin at the left is essentially the same distance from the camera as the coin at the right, which is sitting at the bottom of a glass of water. The underwater coin looks nearly 25 percent larger in the picture, and this gives it an apparent depth much less than its real depth. Thus it appears to float in space closer to the camera. (D. Riban)

Figure 24-20 Refraction causes objects under water to appear closer to the surface than they really are.

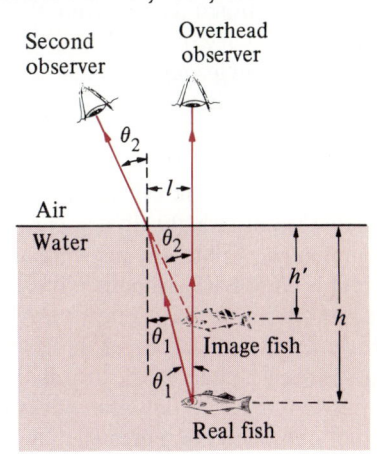

593

from air is smaller than the actual depth. This can be seen in Figure 24-18b, where the quarter which is under water and the one lying on the table are actually at the same level.

Example 24-5 The quarter in Figure 24-18b appears to be 7.00 cm below the surface of the water. How deep is the water?

Solution This only requires an application of Equation 24-5.

$$h' = \frac{h}{n}$$

or

$$h = nh' = (1.33)(7.00 \text{ cm}) = 9.31 \text{ cm}$$

When light originates under water, say at the fish, the angle of refraction is larger than the angle of incidence. This gives rise to a very interesting and useful phenomenon. Consider the rays in Figure 24-21. One can see that the rays striking the surface at larger and larger angles of incidence are refracted more and more from ray 1 to 2 to 3 until ray 4 is at an angle of refraction which barely skims the surface of the water. This is in accordance with Snell's law. At some angle of incidence, the angle of refraction will be 90°. The incident angle for which this happens is called the **critical angle.** We may write Equation 24-4 as

$$n_1 \sin \theta_c = n_2 \sin 90°$$

or

$$\sin \theta_c = \frac{n_2}{n_1} \qquad (24\text{-}6)$$

Notice that this makes physical and mathematical sense only if n_2 is less than n_1, since the sine cannot be greater than 1.

What happens if light is incident on a boundary at an angle greater than the critical angle? The answer is that the light is completely reflected at the interface, remains in the original material, and none is transmitted. This is called **total internal reflection.** Figure 24-14 shows that some light is reflected at each boundary between media. For small angles of incidence, this is only a small percentage. As the angle of incidence gets larger and larger, a bigger percentage of the incident light is reflected. Beyond the critical angle, 100 percent of the incident light is reflected.

Figure 24-22 shows what a scuba diver sees as she looks up at objects above the water. The critical angle for water is $\theta_c = \sin^{-1} \frac{1.00}{1.33} = 48.8°$. Circling about a point straight up is a cone of angular radius 48.8° which contains a complete view of the hemisphere above the water, provided the surface of the water is completely smooth. Objects from horizon to horizon are theoretically visible, although they appear more distant and rotated toward the zenith (directly overhead point). Looking at an angle greater than 48.8° away from straight up, no part of the air-filled world is in view. Because of total internal reflection, one sees only the reflection off the undersurface of the water when looking in a direction greater than the critical angle.

Figure 24-21 As the angle of incidence to the surface increases, the refraction is greater. By ray 4, the refraction is so great that the emerging ray barely skims the surface of the water. This represents the greatest angle of incidence at which light can escape the water into air, the critical angle.

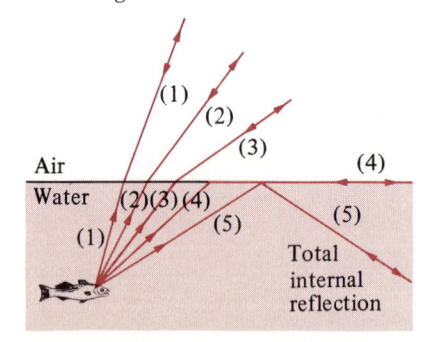

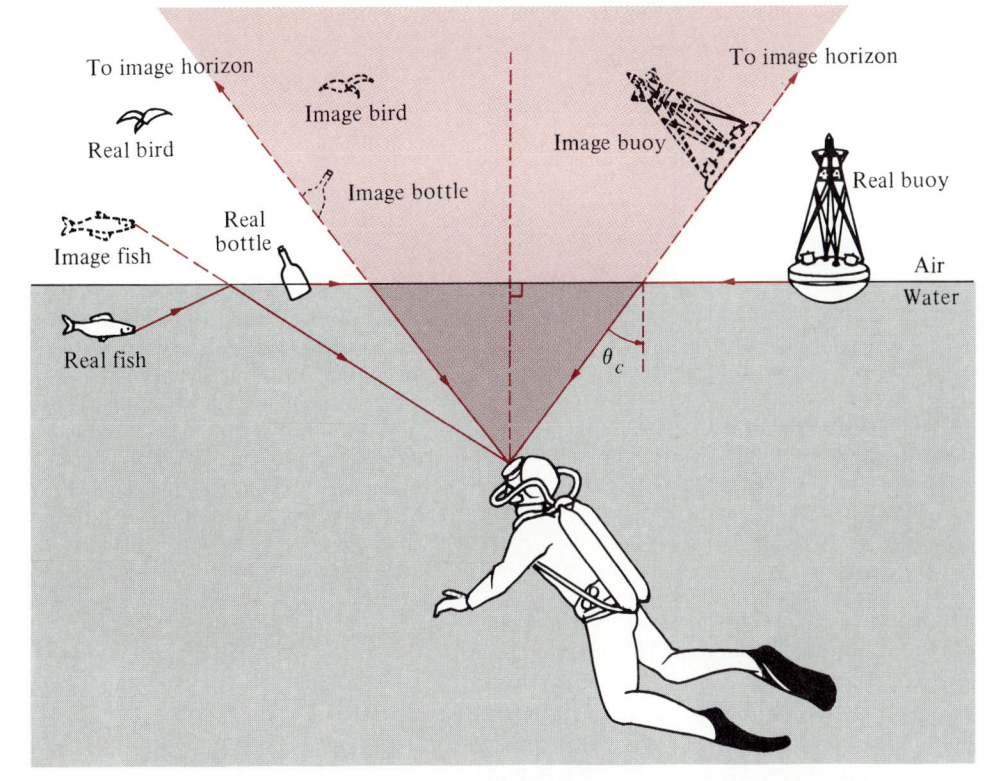

Figure 24-22 The scuba diver theoretically can see the entire scene above. Since all the light entering the water is refracted downward, the entire scene from the air is contained within a cone of angular radius θ_c (about 48.8°) from directly overhead (shaded area). The image fish is produced by total internal reflection. In practice, ripples on the water surface will distort any image that results from total internal reflection.

Example 24-6 Determine the critical angle for acrylic plastic in air. Take the index of refraction for acrylic plastic from Table 24-1 to be 1.51.

Solution This is a straightforward application of Equation 24-6, where $n_1 = 1.51$. Assume that n_2 (for air) may be taken as 1.00 and substitute into Equation 24-6.

$$\sin \theta_c = \frac{n_2}{n_1} = \frac{1.00}{1.51} \qquad \theta_c = 41.5°$$

Many fine optical instruments use the internal reflection phenomenon by employing a **prism** in place of a mirror to change the direction of light or simply to extend the optical path of rays. Prisms are used since total internal reflection really is indeed *total*, which is better than any mirrored surface, and the reflecting surfaces are not affected by tarnishing as are metallic surfaces. Figure 24-23 shows several possibilities.

595

(a)   (b)   (c)

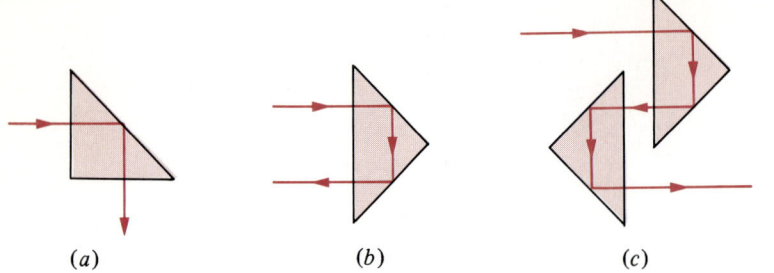

Figure 24-23 (a) A simple triangular prism may be used to divert light through a right angle. (b) The same shape prism may be used to return light back to its original direction. For example, an array of such prisms placed on the moon reflects laser beams back to earth, allowing precise determination of the moon's position. (c) Coupled prisms shorten the path of light and are used in binoculars to prevent their becoming too long and unwieldy. One of the most common applications is the pentaprism in all single-lens reflex cameras that produces an erect image for the eye.

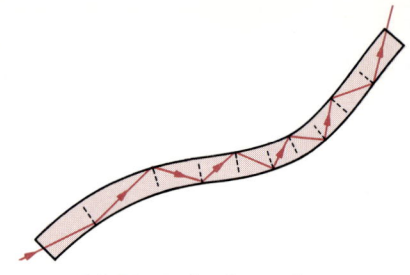

Figure 24-24 Light directed into a glass rod can be trapped. The light always strikes the walls at an angle greater than the critical angle and undergoes total internal reflection. Typically, a ray may reflect many thousands of times per meter in a fine glass fiber, but the loss of intensity is negligible if the glass is free of impurities.

Light entering the end of a transparent rod, as shown in Figure 24-24, may become trapped within the rod by total internal reflection. This trapped light will then follow the rod all the way to the other end even if the rod is curved, provided the curvature is not too great. Such a rod is often called a **light pipe** since it provides a pathway for light much as water pipes do for water. Very thin glass or plastic fibers may be used as light pipes. Many such fibers bundled together may transmit images and also have other advantages over single light pipes in certain applications. Such bundles are the basis of fiber optics in the field of applied optics (see the Special Topic at the end of the chapter).

Example 24-7 A beam of light is to be directed onto the side of a rectangular block of ice so that it will undergo total internal reflection off the interior of the bottom face. What is the largest angle of incidence for which this is possible?

Solution This is an application of Snell's law and the definition of the critical angle for total internal reflection. It is easier to solve if we draw a sketch, as in Figure 24-25. From this figure, we can see that we seek the value of θ_1 whenever θ_3 is equal to the critical angle θ_c for ice in air. We first determine θ_3 from Equation 24-6. Then, recognizing that θ_2 is the complement of θ_3, we may determine the value of θ_2. This is then used along with Snell's law to determine θ_1.

$$\sin \theta_3 = \sin \theta_c = \frac{n_{air}}{n_{ice}} = \frac{1}{1.31} \qquad \theta_3 = 49.8°$$

From $\theta_2 + \theta_3 = 90°$, we have that $\theta_2 = 40.2°$. Thus

$$n_1 \sin \theta_1 = n_2 \sin \theta_2$$

$$(1)(\sin \theta_1)_{max} = (1.31)(\sin 40.2°)$$

$$(\sin \theta_1)_{max} = 0.846$$

$$(\theta_1)_{max} = 57.8°$$

Figure 24-25 Example 24-7.

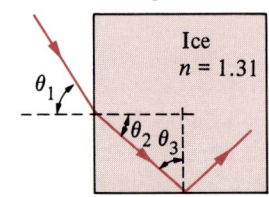

24-6 Dispersion

You may have noted that the heading to Table 24-1 indicates that the index of refraction is given for a particular wavelength of light. This could cause you to wonder whether the index of refraction depends on the wavelength. The answer is that it does! It is found experimentally that the speed of light in a medium is smaller for larger frequencies, hence, for shorter wavelengths of light. In ordinary glass, for instance, violet light is only traveling at about 99 percent the speed of red light. This means the index of refraction is larger for shorter wavelengths. Violet light has the shortest wavelength of visible light and is refracted more than red light at the same boundary between two media. A beam of light which contains a mixture of wavelengths will be separated, or dispersed, into its constituents upon refraction. The effect is called **dispersion** and is most easily demonstrated by shining a beam of white light onto a prism of glass or plastic, as depicted in Figure 24-26 and Color Plate I.

The most spectacular natural manifestation of dispersion is the **rainbow.** When conditions are proper, sunlight may be refracted, dispersed, and totally internally reflected by drops of rain. In many instances, two bows may be seen. The inner bow, called the **primary bow,** is brighter and has red on the outside and violet on the inside. The outer bow, the **secondary bow,** has the colors reversed. Figure 24-27 shows the path of light through a raindrop for both the primary bow and the secondary bow. You will note in the primary bow that the dispersed red light makes a larger angle with the incident ray of sunlight than does violet light. Actually, in the primary bow, red is observed at an angle of 42° with the incident light, while violet light is at 40°. In the secondary bow, red light occurs at an angle of 50.5°, whereas violet light is at 54°.

Figure 24-28 shows some representative raindrops. In this figure, we see that we observe the different colors of the primary bow from different drops of rain. Each drop disperses all the colors, but for the primary bow (shown), the colors are dispersed through an angle of 2°. Thus, if the eye is seeing the color red from one drop, the violet from that drop passes above it. The eye must look at successively lower drops to see orange, yellow, and so on, to violet. For the secondary bow, the colors are reversed, and the bow comes from a larger angle. The natural angles involved require that the sun not be very high in the sky. For

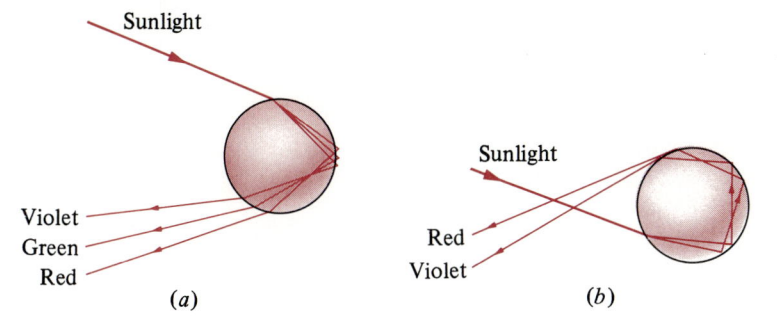

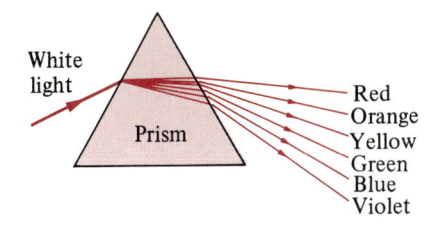

Figure 24-26 The dispersion of white light through an equilateral prism.

Figure 24-27 Formation of a rainbow. (a) Primary bow. Sunlight entering high on the drop is dispersed by refraction into the drop, reflects once at the back of the drop, and emerges with colors separated by a 2° spread centered 41° away from the path of the incident sunray. (b) A higher, less bright secondary bow originates in a ray striking farther from the central axis of the drop. These rays undergo a double internal reflection before emerging with a greater spread and at a steeper angle to the incident sunlight.

Figure 24-28 A full rainbow is seen when drops at different angles to the eye are at the right position to send a single color to the eye. The highest drop cluster here breaks the light into a spectrum, but none of these rays are seen. The medium drop cluster is just low enough that its lowest rays reach the eye, and the lowest cluster sends its violet rays to the eye. The eye thus perceives the contributions of the huge number of drops as a primary bow, with the red highest and the violet lowest.

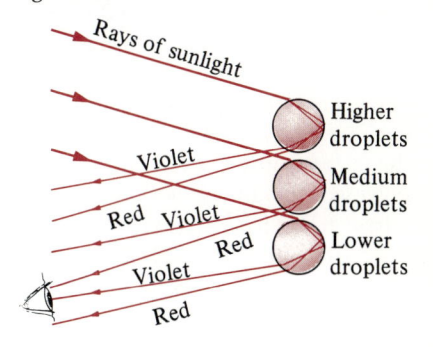

597

instance, if sunlight arrives at the earth at an angle greater than 42°, a primary rainbow cannot be observed from ground level.

24-7 Mirrors and Reflected Images

The position at which our eyes perceive the image in a mirror does not really have light diverging from it. The reflection only makes it appear that way. Such images, perceived to be at a source of divergence that the actual rays never approach, are called **virtual images.** The ordinary images we see in plane mirrors are virtual images — the light does not come from a source behind the mirror. Observation of an object and its image, as in Figure 24-9, coupled with a little geometry indicates that an object and its virtual image formed by a plane mirror have the same dimensions. Each point on the image is directly opposite the corresponding point of the object. This point-by-point transfer ensures that the image's vertical orientation is the same as the object's, and we say the image is **erect** rather than **inverted.** This same point-by-point transfer ensures that the horizontal orientation is reversed. For instance, when you raise your right hand, your image in a plane mirror appears to raise its left hand. An image for which right and left are switched is called **perverted** rather than **normal.**

Images from mirrors need not be virtual (or erect or perverted). The direction of the reflected ray depends on the orientation of the surface at the point of reflection. A continuously curved surface may reflect parallel rays to make them **converge** (come together) to a point or make them appear as though they are diverging from a point. A common and very useful curved mirrored surface is that of a sphere.

Figure 24-29 shows two cutaway views of small sections of spherical surfaces. The mirrored side in Figure 24-29a is the inside of the spherical surface. This is called a spherical **concave** mirror. Figure 24-29b shows a mirrored surface on the outside of the spherical section. This is known as a spherical **convex** mirror. Shown also for each mirror is the **center of curvature** of the spherical surface labeled C and a dashed line drawn through the center of curvature and the center of the mirror. We call this line the **principal axis,** or simply the **axis,** of the mirror.

A set of light rays parallel to the axis is shown reflected from each mirror. After reflection from the concave mirror, the rays all converge and pass through a single point on the axis from which they then diverge. The rays reflected from the convex mirror diverge, but all appear to diverge from a single point on the axis behind the mirror. These points are called the **focal points.** The distance from the focal point to the mirror surface is called the **focal length** for the mirror. It can be shown (see Problem 24-30) that the focal length for a spherical mirror is equal to one-half its radius of curvature, which is the distance from the center of curvature to the surface of the mirror. The rays from an object placed in front of either type of spherical mirror may give rise to an image. Because the convex mirror always causes rays diverging from an object to further diverge, the image from a convex mirror is always virtual. Under certain circumstances, however, the diverging rays from an object may be reconverged by a concave mirror to form a real image.

Consider Figure 24-30. Our object is an arrow perpendicular to the axis of the mirror, and we draw several selected rays from its tip to see if we can determine the position of the image of the tip. We are aware that rays leave the

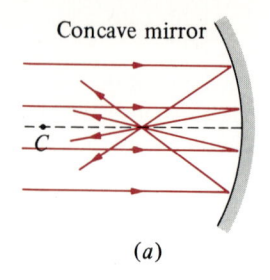

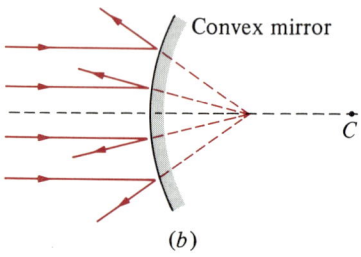

Figure 24-29 (a) A concave mirror reflects rays of light parallel to its central axis through a single point in space called its focal point. (b) A convex mirror causes light rays approaching parallel to its central axis to diverge upon reflection but in such a way that all appear to diverge from its focal point behind the convex surface.

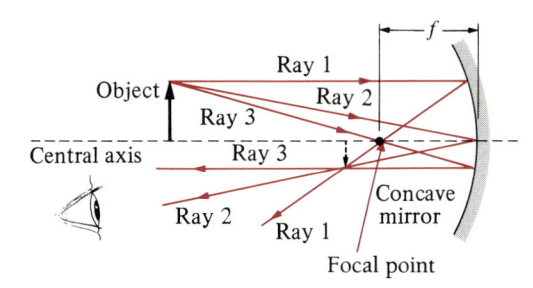

Figure 24-30 Image formation by a concave mirror. Although only three rays are drawn, all rays striking any portion of the mirror near its center would likewise reflect through the image point shown.

object in all directions, so we can select specific rays whose directions after reflection are well known. The selected rays are numbered to make each one easier to follow after reflection. Ray 1 is drawn parallel to the axis because we already know it will be reflected to pass through the focal point of the mirror. Ray 2 is drawn to strike the mirror at the point where the axis intersects the mirror. Since the axis is perpendicular to the mirror, the angle between ray 2 and the axis is the angle of incidence. By the law of reflection, ray 2 must be reflected to make the same angle with the axis. Ray 3 is drawn to pass through the focal point on its way to the mirror. By a concept called reversibility, ray 3 must be reflected on a path parallel to the axis. After all, the only way we know which ray in any of these cases is the incident ray and which the reflected ray is by the arrows we draw on them. The reflection at the surface works just the same if incident and reflected are reversed. Thus, in parallel to the axis and out through the focal point (ray 1) reverses the situation of ray 3.

We see that the three rays converge to a point and then diverge. In principle, every other ray which leaves the tip of the arrow and is reflected from the mirror will also converge to the same point before diverging. The eye shown will thus intercept a cone of rays diverging from the point where they had converged and interpret that point as the position of the tip of the arrow. The image of the arrow tip is therefore at that point from which the rays diverge. Also, a ray from the bottom of the arrow headed toward the mirror goes along the axis and strikes the mirror at a 90° angle before being reflected back along the axis. Thus we expect an image of the tail of the arrow to be on the axis and have drawn a dashed arrow to represent the total image of the arrow. We call this a **real image** because light rays are actually diverging from the position where our eye interprets the image to be (see Figure 24-31). We could place a paper or a movie screen at that position, and the image would be projected onto

An additional ray which is easy to draw is sometimes useful for spherical mirrors. Any ray passing through the center of curvature and striking the surface will do so on the normal. Recall that any line from the center of a circle (or sphere) is a radius and strikes the surface perpendicularly. Such a ray on striking the mirror must reflect right back on itself and return along its original path. By the same reasoning, a ray heading for *C* and striking the convex face will return along its original path.

Figure 24-31 In no danger of an electric shock, the finger is sharing a disconnected socket with the real image of a lit bulb. The image is formed by a spherical concave mirror placed approximately 1 m behind the socket. As long as it is viewed from near the axis of the mirror, the bulb appears real right down to the lettering and occupies the same space above the socket shared by the finger. (D. Riban)

the screen. This is sometimes mentioned as the test of whether an image is real or virtual. A real image will be displayed on a screen placed at the image position, while a virtual image will not.

We may use the rays to obtain an equation which links the distance of the objects from the mirror **(object distance)**, the image distance, and the focal length. Figure 24-32 shows an object and its real image formed by a concave mirror as determined by two of the rays. Object distance, image distance, and focal length are labeled s, s', and f, respectively. The magnitude of the heights of the object and image are labeled h and h', respectively. Several angles which the rays make with the axis are also labeled. From the definition of the tangent and a little algebra, we can determine the relationship we seek. Note from the figure that we can write both

$$\tan \theta = \frac{h}{s} \quad \text{and} \quad \tan \theta = \frac{h'}{s'}$$

Thus

$$\frac{h}{s} = \frac{h'}{s'}$$

or

$$\frac{h'}{h} = \frac{s'}{s} \tag{24-7}$$

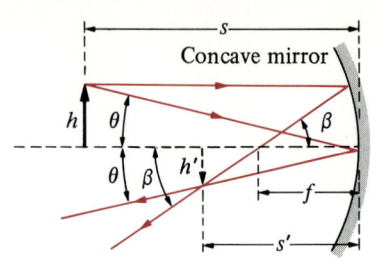

Figure 24-32 The object of height h is at an object distance s from the mirror. The dashed real image is of height h' and at an image distance s' from the mirror. The ray of light from the tip of the object reflects to cross the central axis at the focal point, which is the focal length f from the mirror. This ray makes an angle of β with the central axis. The second ray strikes the midpoint of the mirror, making an angle of θ with the central axis.

We may also write

$$\tan \beta = \frac{h}{f} \quad \text{and} \quad \tan \beta = \frac{h'}{s' - f}$$

This yields

$$\frac{h}{f} = \frac{h'}{s' - f}$$

or

$$\frac{h'}{h} = \frac{s' - f}{f} \tag{24-8}$$

Equating the right sides of Equations 24-7 and 24-8 yields

$$\frac{s'}{s} = \frac{s' - f}{f}$$

This can be written as

$$\frac{s'}{s} = \frac{s'}{f} - 1$$

Finally, dividing each term by s' and rearranging yields

$$\frac{1}{s} + \frac{1}{s'} = \frac{1}{f} \tag{24-9}$$

This is often called the **mirror equation,** but we will see that the same equation may also be applied to lenses. Although we derived the equation for a concave

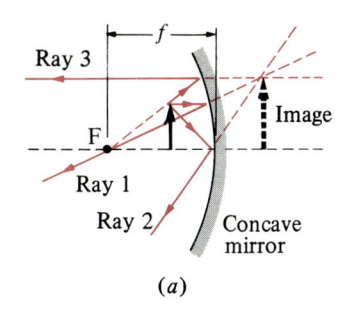

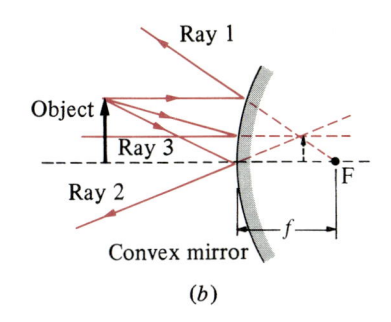

Figure 24-33 (*a*) The rays from the object reflect such that they all appear to diverge from an image point behind the mirror. The image is virtual, erect, and enlarged. (*b*) Rays to a convex mirror likewise diverge after reflection as though they all came from a single point behind the mirror. In this case the image is also virtual and erect but smaller than the object.

mirror forming a real image, it is also applicable to concave mirrors forming virtual images as well as convex mirrors (see Problem 24-32). The latter form only virtual images from real objects.

Figure 24-33 shows a virtual image formed by a concave mirror and a virtual image formed by a convex mirror. You will note that all three rays are used with each mirror. This is just to show that they all work, and, of course, the third ray always provides a check to be sure that a mistake was not made in drawing one or both of the first two. We see that to obtain the position of both these virtual images, the rays have to be extrapolated backward to the point from which they appear to diverge. This is exactly what had to be done to determine the position of the virtual images associated with a plane mirror. This is summarized in Table 24-2.

To apply Equation 24-9 to any object at any position with respect to any spherical mirror, we must agree to use some consistent convention for the signs of the focal lengths, image distance, etc. Several good possibilities have been devised, along with a few ways to describe each possibility. We will use the following convention:

1. The object distance (s) is always a positive number for a real object.
2. The image distance (s') is positive if the image is on the same side of the mirror as the object (real image) and negative if it is on the side opposite the object (virtual image).
3. The focal length (f) is positive for a converging (concave) mirror and negative for a diverging (convex) mirror.

Example 24-8   An object 3.00 cm in height is placed 20.0 cm away from a spherical mirror which has a focal length of magnitude 15.0 cm. Calculate the position of the image if the mirror is (*a*) concave and (*b*) convex.

Solution   For both parts, we simply apply the mirror equation. However, we

---

TABLE 24-2   Convenient Rays for Ray Tracing

1. Drawn from the object parallel to the axis toward the mirror. After reflection, it is directed to pass through the focal point for a concave mirror and to diverge from the focal point of a convex mirror.
2. Drawn from the object to be reflected from the mirror at the point where the axis intersects the mirror. After reflection, it makes the same angle with the axis as it did upon incidence.
3. Drawn from the object to the mirror along the line which passes through the focal point. After reflection, it is directed parallel to the axis.

(a)
$$\frac{1}{s} + \frac{1}{s'} = \frac{1}{f}$$

$$\frac{1}{s'} = \frac{1}{f} - \frac{1}{s} = \frac{1}{15.0 \text{ cm}} - \frac{1}{20.0 \text{ cm}} = \frac{1}{60.0 \text{ cm}}$$

$$s' = 60.0 \text{ cm}$$

(b)
$$\frac{1}{s} + \frac{1}{s'} = \frac{1}{f}$$

$$\frac{1}{s'} = \frac{1}{f} - \frac{1}{s} = \frac{1}{-15.0 \text{ cm}} - \frac{1}{20.0 \text{ cm}}$$

$$= \frac{4}{-60.0 \text{ cm}} - \frac{3}{60.0 \text{ cm}} = -\frac{7}{60.0 \text{ cm}}$$

$$s' = \frac{-60.0 \text{ cm}}{7} = -8.57 \text{ cm}$$

One feature we can see in Figure 24-33 is that the image size is not always the same as that of the object when a mirror is spherical. The virtual image from the concave mirror is larger than the object, while the virtual image from the convex mirror is smaller than the object. We describe this by defining **magnification** as the ratio of the image size to the object size. Using m as the symbol for magnification and h and h' as the heights of the object and image, respectively, as a measure of size, we have

$$m = \frac{h'}{h} \qquad (24\text{-}10)$$

In most instances, there is also agreement to use the sign of m to convey whether the image is erect or inverted. We agree that a distance measured upward from the axis is positive and one downward is negative. Hence, by this convention, an inverted image has a negative number for h'. To be consistent with our sign convention for image and object distance, we should rewrite Equation 24-7 as

$$\frac{h'}{h} = -\frac{s'}{s} \qquad (24\text{-}11)$$

You will recall that Equation 24-7 related only the magnitudes of those quantities. Using Equation 24-11, we may rewrite the magnification as

$$m = -\frac{s'}{s} \qquad (24\text{-}12)$$

Note that if m is positive, the image is erect; if m is negative, it is inverted. Also, if the magnitude of m is less than 1, the image is smaller than the object, while the image is larger than the object if m is greater than 1.

A concave mirror may give rise to either a virtual (Figure 24-34a) or a real

(a)   (b)   (c)

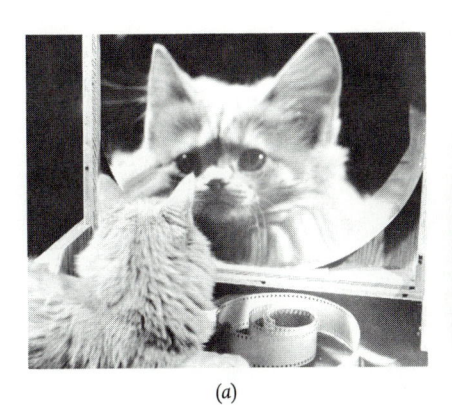

Figure 24-34 (a) The erect, enlarged image corresponds to the situation in Figure 24-33a. Such mirrors are frequently used as shaving or cosmetic mirrors. (b) The erect, diminished image corresponds to the situation of Figure 24-33b. The small image is compensated for by the wide angle of view in this typical antishoplifting mirror. (c) Harder to see than the others, this image requires careful study. It is a real, inverted image in a concave mirror as drawn in Figure 24-30. An appropriately positioned screen could record the image cat since the light from each point on the real cat, after striking the mirror, is reassembled at a single image point and then diverges to be recorded by the camera. (D. Riban)

(Figure 24-34c) image depending on the position of the object, while a convex mirror yields only a virtual image (Figure 24-34b). As we expect, this is consistent with Equation 24-9 and our sign convention. When f is positive (concave mirror), s' will be positive if s is greater than f and will be negative if s is less than f. Can you determine what happens to the image when the object is placed exactly at the focal point of a concave mirror? We see that for a negative value of f (convex mirror) and a positive value of s, the image distance s' will always be negative.

Example 24-9 Determine the magnification and the image height for both the spherical mirrors of Example 24-8.

Solution We may determine the magnification by using Equation 24-12 and then determine the image heights from the definition of magnification.

For the concave mirror,

$$m = -\frac{s'}{s} = -\frac{60.0 \text{ cm}}{15.0 \text{ cm}} = -4.00$$

and

$$m = \frac{h'}{h}$$

or

$$h' = mh = (-4.00)(3.00 \text{ cm}) = -12.0 \text{ cm}$$

Remember that the minus sign means the image is inverted.
For the convex mirror,

$$m = -\frac{s'}{s} = -\frac{-8.67 \text{ cm}}{15.0 \text{ cm}} = 0.578$$

Also

$$h' = mh = (0.578)(3.00 \text{ cm}) = 1.73$$

SPECIAL TOPIC
Fiber Optics

Total internal reflection provides a means to direct light through a complex pathway while retaining its intensity. It might be thought that such guidance of a beam could be engineered by the use of mirrors, but there is a serious problem—even the best mirrored surface absorbs some of the incident light. For example, a freshly silvered mirror surface will reflect up to 99 percent of incident light. Unfortunately, the silver tarnishes in air and rapidly drops to about 93 percent reflectivity and lower after that. Mirrors for critical uses, like astronomical telescopes, have switched to aluminum coatings with 95 percent reflectivity. The advantage is that the aluminum surface is self-protective chemically; that is, it develops a thin oxide coating which then protects the metal underneath from further contact with the air. The eye is relatively insensitive to a small loss in intensity. For example, the image in a half-silvered, or see-through, mirror may appear normal even though only half the light is reflected by the mirror. For a single reflection, a 5 percent loss may not be serious, but it limits multiple reflection use. The intensity drops to below 50 percent after only 14 reflections; that is, $(0.95)^{14} = 0.488$. After 45 reflections, the remaining intensity is less than 1 percent of the original. We can see the cumulative loss in Figure 24-35, where a laser beam is shown reflected repeatedly between two mirrors.

The method to overcome the loss of repeated reflection is to use internal reflection, which is indeed *total*. If light is introduced at one end into a solid tube of some transparent material, it can encounter the side walls at angles greater than the critical angle and be retained inside the tube. The tube may even be bent into a complex shape as long as tight turns or kinks are avoided, and the light will be conducted to the far end. In a thin optical fiber, the light may reflect a thousand times per meter; however, the only loss in this system is by absorption due to the imperfect transparency of the material itself. Such tubes could be used to deliver light to positions where it is inconvenient to mount a light bulb, as on some automobile instrument panels. Fine branching arrays of such light pipes radiating from a light bulb in a lamp base are also used for ornamentation (Figure 24-36).

A much more interesting possibility arises when the tube conducting light is a very fine fiber. These may be collected into a **fiber optical bundle** containing thousands of optical fibers, each of which can act as a separate channel for conducting light in either direction. In Figure 18-20 we saw that a television picture was composed of an array of dots of varying brightness. Viewing such an array from a distance provides a

Figure 24-35 Cumulative loss of intensity in a laser beam reflected between two mirrors. (D. Riban)

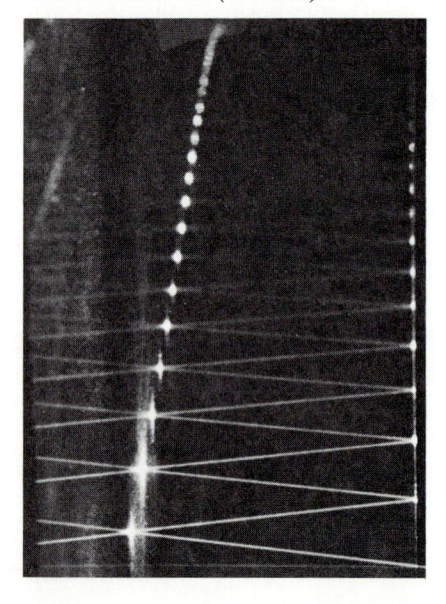

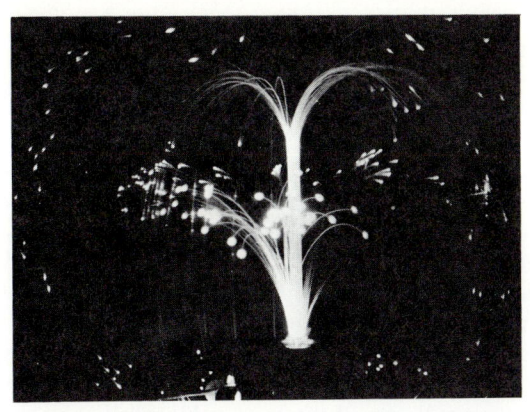

Figure 24-36 A fiber optics fountain. (D. Riban)

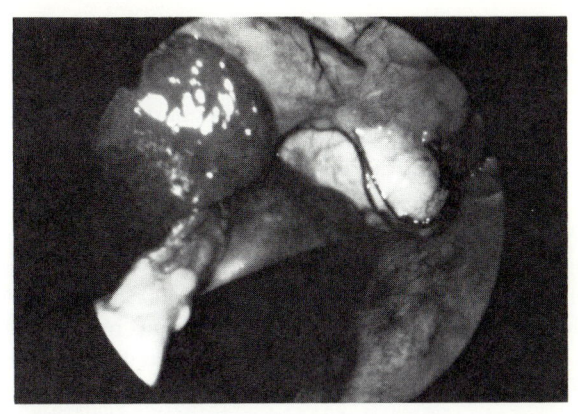

Figure 24-37 Fiber optics used in medicine: view of a gall bladder stone. (A. Tsiaras/Science Source)

seemingly continuous picture. The fiber optical bundle provides the same ability with conducted light. If a sharp image is focused on the end of such a bundle, each fiber will conduct the light falling on its end back to the opposite end, and viewing the pattern of bright and dark fiber ends will provide the complete image.

This technique is particularly useful in medicine. Many techniques exist for the indirect imaging of the body's interior, but when a visual inspection is called for, the only option used to be major surgery. The fiber optical bundle offers an alternative. Each glass fiber is extremely flexible and may be as small as 10^{-6} m, or 1 μm, in diameter. Thus, a 1-mm bundle could contain about 8×10^5 separate fibers. By contrast, a color TV screen has some 1.20×10^6 phosphor dots to make an image. Individual fibers are collected into a bundle, have their ends polished to a flat surface, and are fitted with a lens system to focus an image on the bundle end. Half the fibers are diverted to a light source to provide the illumination to view the body interior, while the other half conduct the image to a polished bundle end viewed by a microscope. While this image may be viewed directly, it is frequently displayed on a TV screen. To duplicate the sharpness of the best color TV picture, a fiber optical cable of only 3.00 mm in thickness would be required. The end of the bundle to view the body interior is usually made to fit through a standard hypodermic needle. After the needle is injected, the bundle is fed through to the site of inspection. Thus, knee injuries may be routinely viewed for physical damage, or a developing fetus may be viewed in the uterus. More complex is the observation of heart valves in action, where the bundle is inserted through the arteries of the neck and guided through to the heart. This direct optical imaging has proved very useful in medicine and has been the source of many of the body interior scenes you have see on TV or in instructional films (Figure 24-37 and Color Plate VIII).

The most widespread future use of optical fibers will be in communications. Already we are in the first phases of conversion from metal wires to optical fiber cables for information transfer. Telephone messages convert sound to a varying electric signal that is a model, or **analog,** of the variation in frequencies and intensities of the original sound. To be intelligible at the receiver, such a message must be kept free from interference, not easy since a long-distance communication requires many amplifications along its path. The alternative means of communications used in the telegraph or in flashing lights between ships is fundamentally different because messages are delivered in a three-part code (Morse code) consisting of a short burst (a dot), a long burst (a dash), or nothing. This is near to a true **digital** signal of on or off, as we use in computers. The advantage of a digital signal is that purity of signal is not as critical. As long as something can be distinguished from nothing, the information is transferred.

You may think such a system would be slow compared to a conversation. This would be true if a person were turning the light on and off. A modern microchip laser can be turned on and off in a fraction of a millionth of a second. Communicating a million on or off bursts a second would be adequate to transmit about 40 pages of a typical textbook per second, where each character was indicated by an eight-digit binary sequence of 0s and 1s. A small fraction of this capacity is needed to encode the frequencies and intensities of a spoken message for decoding at the other end to produce sound. Some encoding already occurs in telephone communications. A single microwave channel

605

can easily handle all the information for a television transmission—color picture and sound. It would be wasteful to use the capacity of such a channel for a single telephone message, and this is not done. Many conversations are encoded and transmitted simultaneously on each channel.

Light offers a tremendous advantage in digital communications. How much of a signal is "some"? Suppose that we decide that one full wave is enough for a bit of information. The human voice at 100 to 400 Hz would thus need a minimum of $\frac{1}{400}$ s to transmit a single bit of information. By the same standard, red light with a wavelength of 0.7 μm requires two millionth-billionths of a second (2×10^{-15} s) to transmit a single wave. A one-millionth of a second burst of such light would contain a half-billion waves!

This difference in the ability to compact information onto light channels as opposed to sound analogs will be most appreciated in computer communications. Today, a small computer may communicate over phone lines which convert its information into a high-speed varying sound tone. Such systems will be regarded as Stone Age electronics by the end of your life span. An optical system capable of 10^6 bits/s could transmit the entire memory of a 64K home computer from coast to coast in 0.5 s. The use of optical fibers in communications will grow until the information exchange capacity of society will exceed any reasonable guess at present.

Minimum Learning Objectives

After studying this chapter, you should be able to:
1. Define:
   - angle of incidence
   - angle of reflection
   - angle of refraction
   - center of curvature
   - concave
   - converge
   - convex
   - critical angle
   - dispersion
   - diverge
   - electromagnetic wave
   - focal length
   - focal point
   - image
   - image distance
   - incident ray
   - index of refraction
   - law of reflection
   - law of refraction
   - magnification
   - normal (line)
   - object distance
   - plane mirror
   - plane wave
   - principal axis
   - prism
   - ray
   - real image
   - reflection
   - refraction
   - Snell's law
   - total internal reflection
   - virtual image
   - wavefront

2. Understand the origin of light as electromagnetic waves.
3. Calculate frequency, wavelength, or the speed of propagation from any two of the three.
4. Interpret diagrams involving light in terms of either wavefronts or of rays.
5. Calculate the position of an image for a plane mirror from the position of the object and the orientation of the mirror.
6. Calculate the speed of light in a material from its index of refraction, or vice versa.
7. Calculate the direction of travel of a ray of light after encountering the surface of a transparent material of known index of refraction at a specified angle.
8. Calculate the critical angle for rays from an optically denser material into a less-dense material, given the indices of refraction.
9. Describe the appearance of the image in a spherical mirror of known geometry, given the size and location of the object.
10. Calculate the image position for a spherical mirror of known geometry, given the position of the object.
11. Calculate the magnification of a spherical mirror of known geometry, given the object position.

Problems

24-1 Lightning strikes a tree on a mountaintop 10.0 km away from you. Take the speed of sound to be 345 m/s. (a) How long after you see the flash will you hear the sound? (b) How long does it take for light to travel from the moon to the earth?

24-2 An optimum-length receiving antenna for electromagnetic waves should be one-half wavelength long. Calculate the best length of the antenna required for the extreme ends of the VHF television channels (channel 2 = 54.0 MHz, and channel 13 = 216 MHz).

24-3 Calculate the wavelengths of the electromagnetic waves at the extreme ends of the FM radio band, which extends from 88.0 to 108 MHz.

24-4 (a) Calculate the frequency of electromagnetic waves which have a wavelength of 1.00 μm in a vacuum. (b) What is the period of the wave motion? (c) Where does it fall in the electromagnetic spectrum?

24-5 The extreme ends of the UHF television band have wavelengths of 0.640 and 0.340 m. Calculate the range of frequencies.

24-6 Two plane mirrors intersect at an angle of 75°. A ray of light strikes the first mirror at an angle of incidence of 25°. Determine (a) the angle of reflection off the second surface and (b) the angle the reflected ray (after two reflections) makes with the incident ray.

24-7 Quality cameras require precise focusing to produce a clear picture. Suppose you were taking a picture of yourself in a plane mirror 1.00 m from you and the camera. (a) For what distance would you focus the camera lens to produce an in-focus picture of yourself? (b) Some automatic focusing cameras send out a tight beam of waves in the direction the camera is focused and time its return to get the distance to the object. Would this give you a clear picture of yourself in the mirror? (c) The most expensive autofocus cameras use different techniques. In one, the image is reflected onto a tiny bank of light-measuring photocells. When an image is out of focus, the light smears out and blends with the light from neighboring portions of the image. The autofocus computer adjusts the focus until the contrast in light readings between nearby photocells is the greatest possible, producing the sharpest image. Would the picture of your image in the mirror be in focus with such a system? (d) Suppose you had written a caption on the mirror in lipstick to show up on the photo. What would the appearance of the caption be with each of the above systems, in focus or blurred?

24-8 Two plane mirrors arranged so that the planes are at 90° to one another and intersecting along a line form what is called a corner reflector. Show that a beam of light entering the corner reflector will be reflected along the direction of incidence.

24-9 A 1.75-m-tall person with eyes 1.45 m above floor level desires to buy a plane mirror for the back of his bathroom door such that he may view himself from top of head to toe while dressing. He will view himself between a close position of 0.500 m and a greatest distance of 3.00 m from the mirror. (a) What is the minimum height of the mirror to serve this person's requirements? (b) How far above the floor must the bottom edge of this minimal full-view mirror be mounted? (c) What information in the problem as stated is irrelevant?

24-10 (a) Figure 24-13a uses two plane mirrors. Explain by the use of a careful ray diagram how the third image of the cat in the mirrors is formed. Call the vertical mirror the y plane and the horizontal mirror the x plane. The cat's nose was 4.50 cm from M_y and 10.0 cm from M_x for the photograph; thus it would have coordinates of $(-4.50$ cm, $+10.0$ cm). Give the coordinates for each image of the cat's nose. (b) Figure 24-13b is similar to Figure 24-13a, but here there are five images (plus the real cat) instead of the three of part a. Sketch the photograph and write on each image a set of letters to describe it: R or V for real or virtual; E or I for erect or inverted; N or P for normal or perverted. (c) Figure 24-13c also shows a multiple-image photograph of the cat. As in part a, only two mirrors were used. Including the real cat, take the number of images to be n. As in part b, make a sketch and identify each image as R or V, E or I, N or P. What was the angle between the two mirrors (θ) to produce this photograph? Call $360°/\theta = m$. State the relationship between m and the number of images n.

24-11 A hollow cube 30.0 cm on an edge is made of acrylic plastic 5.00 cm thick. It is filled with water, and a small bead is suspended with its position exactly in the center of the cube. Determine how far away from the side of the cube the bead appears to be when viewed along a perpendicular to the side surface of the cube.

24-12 A drain at the bottom of a 3.00-m-deep swimming pool is viewed along a direction making an angle of 60° with the normal. At what angle does the light leaving the drain strike the surface before being refracted to the intercepting eye?

24-13 Show that the index of refraction is approximately equal to the square root of the dielectric constant for other than ferromagnetic materials. Use the relationships that $\epsilon = K\epsilon_0$ and $\mu = K_m\mu_0$. Make the approximation that $K_m \approx 1$.

24-14 The bottom of a mug full of water appears to be 6.50 cm below the surface. What is the real depth of the water?

24-15 A ray of light traveling in air strikes a glass plate 3.00 cm thick at a 50° angle of incidence. The angle of refraction is 29.6°. Calculate (a) the index of refraction of the glass, (b) the speed of light in the glass, and (c) the displacement of the ray from its original path upon emerging from the plate.

24-16 A scuba diver looks up from under water and observes the sun at an angle of 25° to the vertical. (a) In what direction (i.e., at what angle to the vertical) is the sun then observed by someone on dry land? (b) At what angle to the vertical will the diver have to look as the sun sets?

24-17 A sophisticated autofocus camera described in Problem 24-7c is used to take a photograph under water. After the picture is taken, it is noted that the camera set itself for 2.25 m to take the picture. How far was the underwater object from the camera?

24-18 The source of a light determines its frequency, which will not change as the light travels through different media. However, since the speed of light depends on the media and changes on refraction, the wavelength must also change as light crosses a boundary between media. (a) Show that the wavelength of light in a medium of index of refraction n is equal to the wavelength of the same light in a vacuum divided by n. (b) Calculate the wavelengths in water of light which has wavelengths in air of 6.86×10^{-7} m (red) and 4.58×10^{-7} m (blue).

24-19 The index of refraction given for air, 1.00029, is at 1 atm of pressure. In reality, the earth's atmosphere keeps thinning out as you rise above the surface until it reaches any approximation of vacuum that is desired. For simplicity, imagine instead that the air is a uniform-density shell about the surface, 50.0 km thick with a discrete edge to the vacuum of space. (a) If you look up in the nighttime sky and see a star directly overhead, what is its real direction in space from you? (b) If you see a star at 45° from the horizon, what would be its true elevation if the earth had no atmosphere? (c) If you see the sun just setting on the horizon, you may assume its rays approached the top of the atmosphere at grazing incidence, or 90° to the normal, and were refracted downward into your eyes. By what angle is the sun really below the horizon at the instant you see it set? (d) Since the sun appears to move around us some 360° in 24.0 h, its apparent motion is 15.0°/h. How many minutes before the sun was seen to set would it have really set on an airless earth? (e) The sun's apparent size in the sky is very nearly $\frac{1}{2}$°. How many sun's diameters lower in the sky is the real sun at sunset than the image we see? (f) Ponder for a moment the problems this differential refraction of the image with position in the sky causes astronomers, who measure image position routinely to an accuracy of 0.01 second of arc. (Note: requires 6 significant figures throughout.)

24-20 An underwater swimmer stops and lies on her back on the bottom of a swimming pool so that her eyes are 3.00 m from the surface of the water. What is the diameter of the circle of light that she sees?

24-21 A prism made of crown glass of index of refraction 1.59 has angles of 45°, 90°, and 45° and is to be used to extend the optical path length by total internal reflection in a monocular. Calculate (a) the critical angle in air and (b) the critical angle if the prism is placed under water.

24-22 A clear plastic cube 8.00 cm on an edge is placed on a table. (a) What is the maximum index of refraction the plastic may have such that one can see something under the cube on the table by looking in the side? (b) How far below the top of the cube does the table appear to be if viewed from above?

24-23 When a ray of light is incident on a glass plate at an angle of 58.8°, it is found that the refracted part of the ray and the reflected part of the ray are perpendicular to each other. Calculate (a) the index of refraction of the glass and (b) the critical angle for total internal reflection for this glass.

24-24 A glass prism is determined to have a critical angle for total internal reflection of 37.6° in air. When a film of cooking oil covers the prism, measurement shows that the critical angle changes to 55°. Calculate (a) the index of refraction of the glass and (b) the index of refraction of the oil.

24-25 All types of waves reflect and refract. In particular, earthquake waves or explosion waves on the earth's surface race downward into the earth. An abrupt change of material causes both reflection and refraction. While a large disturbance may be detected completely across the earth from the site of its occurrence, it will have a closer shadow zone (a band, actually) where the wave cannot be detected. This makes great sense in terms of refraction. Suppose a wave in a lower layer, material 2, is approaching the boundary to a higher layer, material 1. If the ratio of indices of refraction for earthquake waves for the two materials, n_1/n_2, is 0.5000, what is the shallowest angle at which the wave may approach the surface and still have some of its energy cross to material 1 and reach the detectors at the surface?

24-26 A ray containing both red light of wavelength 6.86×10^{-7} m and blue light of wavelength 4.58×10^{-7} m is incident on one side of an 8.00-cm-thick glass plate at an angle of 60°. If the indices of refraction of the glass are 1.64 for the red light and 1.66 for the blue light, determine how far the two colors will be separated when they strike the other side of the plate. Use four significant figures.

24-27 Diamond has the largest known index of refraction for light. This, coupled with the dispersion effect, cause its brilliant sparkling. Diamond has an index of refraction of 2.4104 for red light and 2.4368 for blue light. Calculate (a) the critical angle for total internal reflection for each of these colors and (b) the angular separation between red and blue light which has been refracted after an angle of incidence of 60°. Use four significant figures.

24-28 Complex molecules are very specific as to which wavelengths of light they absorb and which they do not, a phenomenon used in spectrophotometry to detect complex molecules in solution. White light is broken into a spectrum, and a slit passes a narrow-wavelength band to shine through the solution and be measured by a photocell. The entire spectrum is scanned by the slit and the absorption measured for all wavelengths. In a simple spectrophotometer, an equilateral prism (each vertex angle 60°) 5.00 cm on a side is used. It is made of the same glass as in Problem 24-26. A beam of light falls on the middle of the front face such that the red ray inside the glass travels parallel to the base of the prism. (a) What is the angle of incidence of the beam of light to the glass? (b) At what angle to the base (horizontal) does the blue ray travel inside the glass? (c) At what angle to the base does the red ray travel when it reemerges into the air? (d) At what angle are red and blue rays spreading apart after reemerging into the air? (e) If the slit is located 10.0 cm from the middle of the exiting face of the prism and is 0.0100 cm wide, what is the range of wavelengths of light passed by the slit at one time?

24-29 The most common use of a spherical convex mirror is as an antishoplifting device since its wide-angle view allows clerks to see around corners and down aisles. Suppose such a mirror is cut from a sphere of radius 60.0 cm and has a diameter of 80.0 cm. If the clerk is on the central axis of the mirror and 3.00 m distant, what is the angular width of the cone of the image provided by the mirror?

24-30 Show that the focal length of a spherical mirror is approximately equal to one-half the mirror's radius of curvature R. Choose a concave mirror, and show the axis passing through the center of curvature and the focal point. Choose a ray parallel to the axis and reflected through the focal point. Show that $f = R/2$ when the angle of incidence the ray makes with the mirror is small.

24-31 An automobile headlamp consists of one or two filaments placed in front of a concave mirror and sealed in a glass enclosure. The desire is to have the high beam from the headlights produce a sharp beam of light pointing in the forward direction with very little spreading (this keeps the light intensity at considerable distance close to the intensity near the headlamp). The low beam of the headlights, in contrast, must not blind an oncoming driver. The first requirement therefore is that the reflected light be kept below the plane of the filament as it proceeds far away from the lamp. The second requirement for the low beam is that it diverge to illuminate the shoulder of the road and the opposing lane of the highway. In terms of these requirements and the geometry of the concave reflector, describe carefully the correct position in which to place (a) the high-beam filament and (b) the low-beam filament.

24-32 Use the derivation of the mirror equation given in Section 24-7 as a guide. Show that the same equation may be derived for a convex mirror provided one uses the sign convention given. That is, show that one obtains $1/s - 1/s' = -1/f$ for the equation of a convex mirror, where s' and f are the magnitudes of the distances of the image and focal point from the mirror.

24-33 An object is placed 50.0 cm in front of a concave spherical mirror, and a real image is formed on the same side of the mirror as the object at an image distance of 75.0 cm. (a) Calculate the focal length of the mirror. (b) Sketch a ray diagram showing the image, object, and focal point.

24-34 Figure 24-34c shows a cat with a spherical mirror. Take the image cat's size to be exactly one half that of the real cat. The focal length of the mirror is 40.0 cm. From the photograph, answer the following (a) Is the image real or virtual, erect or inverted, normal or perverted? (b) What is the magnification? (c) What is the distance of the cat's head from the mirror? (d) If the image is real, at what distance from the mirror would a screen have to be placed to display an in-focus image of the cat?

24-35 Use a ray diagram and the mirror equation to locate the image of an object placed (a) 30.0 cm and (b) 10.0 cm in front of a concave mirror having a focal length of 20.0 cm.

24-36 Use a ray diagram and the mirror equation to determine the distance at which an object is placed in front of a convex mirror of focal length -50.0 cm if the image appears to be on the side of the mirror opposite to that of the object and at a distance of (a) 25.0 cm and (b) 10.0 cm.

24-37 A pencil 10.0 cm long is held upright at a distance of 40.0 cm in front of a concave mirror of 50.0-cm focal length. Draw a ray diagram for this situation, and then calculate (a) the image position, (b) the magnification, and (c) the height of the image. (d) Is the image real or virtual, erect or inverted, normal or perverted?

24-38 An integrated circuit 2.00 cm high is placed 12.0 cm in front of a convex mirror of 10.0 cm focal length. (a) Draw a ray diagram to determine the position and character of the image. (b) Is the image real or virtual, erect or inverted, normal or perverted? (c) Calculate the image position. (d) What is the magnification for this situation? (e) How high is the image?

24-39 An object is placed in front of a concave mirror of focal length 30.0 cm such that the magnification is equal to -3. Draw a ray diagram, and determine whether (a) the image is real or virtual, erect or inverted, normal or perverted. (b) Calculate the object distance and the image distance.

24-40 Repeat Problem 24-39 for a magnification of $+3$.

24-41 An object 3.00 cm high is placed at a distance of 20.0 cm from a convex mirror. The magnification for this situation is determined to be 0.500. (a) Draw a ray diagram to show the situation, and determine whether the image is real or virtual, erect or inverted. (b) What is the image distance? (c) Calculate the focal length.

24-42 A spherical mirror in a fun house produces an erect image three times as large as anyone standing 2.00 m away. (a) Is the image real or virtual? (b) Is the mirror concave or convex? (c) Calculate the focal length of the mirror.

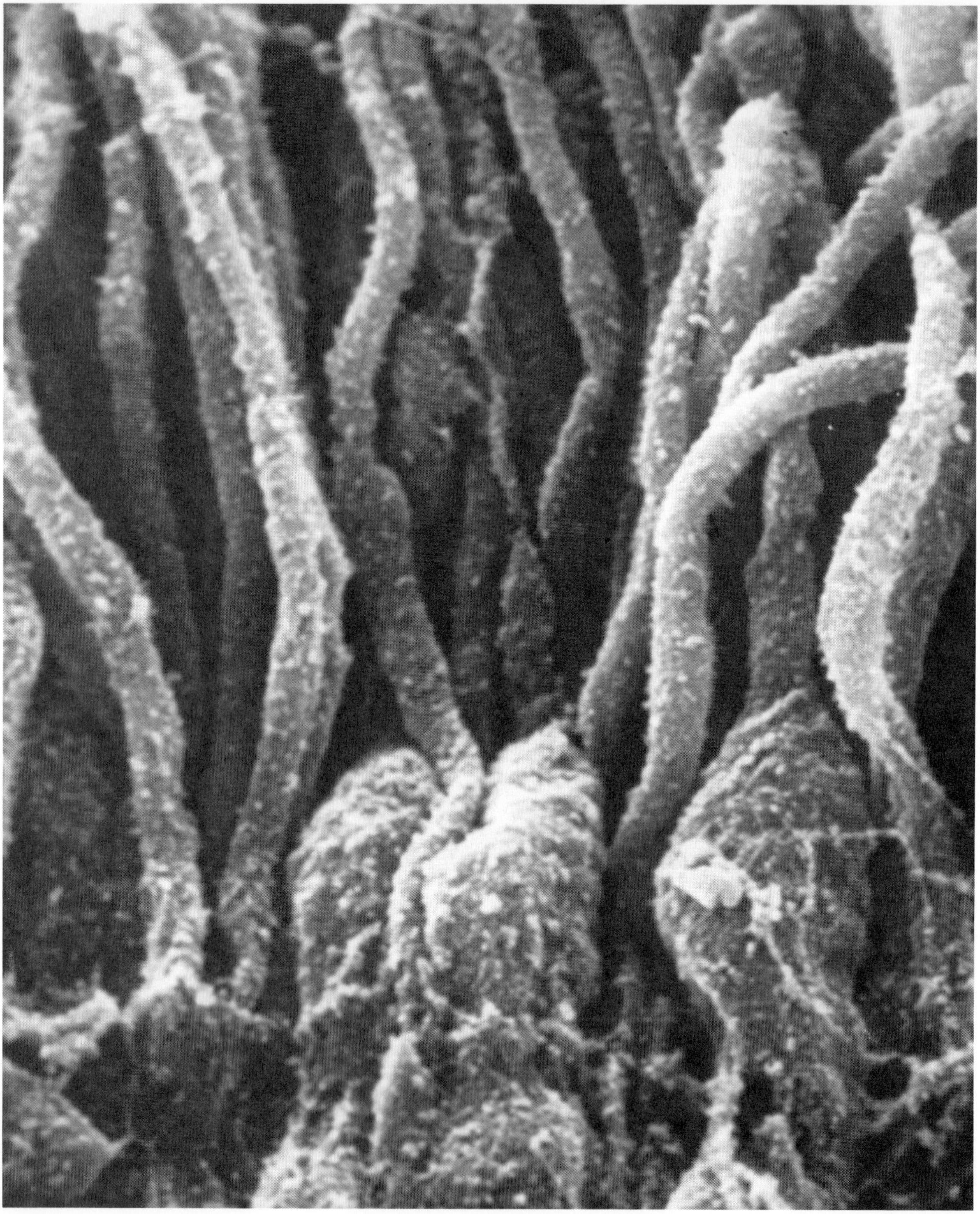

25 Lenses and Optical Instruments

25-1 Introduction

The direction of a refracted ray depends on the orientation of the surface at the point where the ray meets the second medium. A continuously curved surface may refract a set of parallel rays in such a way as to cause them to converge to a point or to make them appear to diverge from a point. Figure 25-1 shows what happens to a set of parallel rays traveling in air and incident on spherical glass surfaces. Each ray incident on either surface still obeys Snell's law. However, because the surfaces are curved, the rays, which were parallel before striking the interface, go in different directions after refraction. After being refracted at the concave surface in Figure 25-1a, the rays now diverge as if they had originated at a single point. Likewise, the parallel rays which are refracted at the convex surface in Figure 25-1b converge to a point before diverging again. Similarities with parallel rays reflected from spherical mirror surfaces are apparent.

A piece of transparent material with a curved surface to control the refraction of light is called a **lens.** Most lenses have at least one spherically shaped surface, since it is far easier to produce a spherical surface than any other, including an optically flat surface. This is because spherical surfaces automatically result when two glass blocks are slid back and forth across each

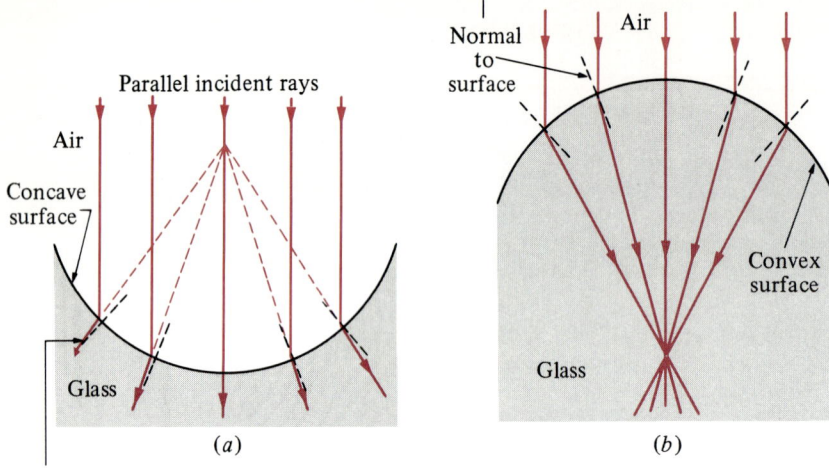

Figure 25-1 (a) Parallel incident rays falling on a concave glass surface diverge such that an eye at the bottom would interpret them as all spreading from a single point. (b) Rays falling on a convex glass surface converge to a single point. In both diagrams, the central ray strikes the surface on the normal and is undeviated.

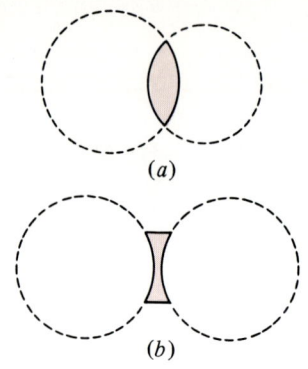

Figure 25-2 (a) The surfaces of a double-convex lens are formed from spheres of different radii. (b) The material between the two nonintersecting spherical surfaces is a double-concave lens.

other while slowly counterrotating each, with a fine abrasive between their surfaces. Nonspherical lenses are works of great skill, usually produced once and then precisely reproduced by molding in plastic if they need to be mass-produced (e.g., the Fresnel lens in the next section). In this chapter we will study image formation by lenses and then consider combined lenses and/or mirrors in common optical systems.

25-2 Types of Lenses

Figure 25-2a shows a side view of a shaded area formed by the intersection of two spheres; such an area corresponds to a double-convex lens. Figure 25-2b shows the shaded area as a space between two slightly separated spherical surfaces, corresponding to a double-concave lens.

In Figure 25-3 we see the refraction of a ray at each surface of sections of (a) a double-convex lens and (b) a double-concave lens. In each case we note that the ray is deviated from its original path by refraction at both surfaces. That the deviation is increased by refraction at the second surface is true only because these lenses are double-convex and double-concave, respectively. These are not the only shapes possible for lenses. For example, one side of each of the lenses shown could be flat. In that case the lens would be called a *planoconvex* lens or a *planoconcave* lens. Figure 25-4 shows side views of various possibilities for lens shapes. Unlike a piece of plate glass or plastic, such as we saw in Figure 24-14, each of the lenses shown here may cause a net deviation of an incident beam of light. The magnitude of the refraction at each surface depends upon the orientation of the surface with respect to the direction of the incident ray. Thus the net deviation of light will be toward the thicker part of the lens. Therefore the convex lenses in parts *a*, *b*, and *c* will cause a set of

Figure 25-3 (a) A ray striking the first surface of a double-convex lens is refracted downward relative to its original direction. At the back surface, the ray is again refracted downward. (b) The front and back surfaces of a double-concave lens likewise both deviate the ray in the same sense from its original direction. In this case, both refractions rotate the ray upward from its original path.

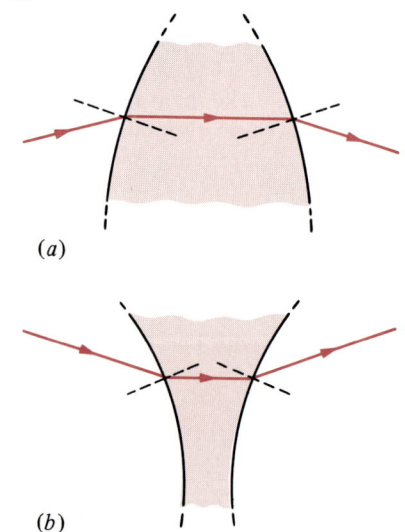

parallel rays to converge toward the center and are called **converging lenses.** The last three lens shapes of Figure 25-4 are called **diverging lenses** because they cause a set of parallel rays to diverge.

To produce a very strong convergence of light, as whenever the object or image must be close to the lens compared to its diameter, a double-convex lens must be much thicker in the center than toward its edges (Figure 25-5a). This creates problems with the bulk of glass, its mass, and also possible image distortion. One solution, developed by the French physicist Augustin Fresnel (1788–1827) for use in lighthouses, recognized that all the useful light bending occurs at the lens surfaces. He retained the surfaces but deleted much of the interior (Figure 25-5c). Figure 25-5d shows a cat viewed partially through a Fresnel lens, and Figure 25-5e shows another practical example. The cover design of this book is from a photograph similar to this one.

Most lenses are reasonably thin compared to their diameter and will be the ones considered in this chapter.

Figure 25-5 (a) A double-convex lens. (b) The interior of the lens (light shading) does not influence the light; bending occurs at the surface (dark shaded blocks). (c) A Fresnel lens. The dark blocks of part b are kept and the interior glass is discarded. Once joined, these lens segments preserve the refraction of the original thick block without its mass. (d) The transparent base of an overhead projector is a Fresnel lens. Close examination will reveal the dozens of separately curved rings. The photograph shows the ability of the lens to produce an image of an object quite close to its rear surface. (e) The most common use of the Fresnel lens is in automobile taillights. Each ring is a band of lens surface with a curvature slightly different from that of its neighbors, all producing a backward-directed parallel beam of intense light.

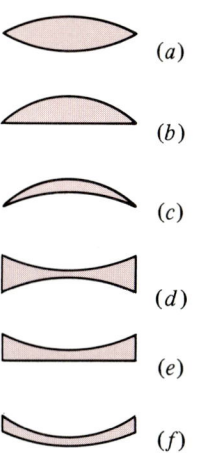

Figure 25-4 A lens thicker at its center than at its edges will converge parallel rays of light and is some form of convex lens. If thicker at the edges than at the center, the lens will diverge parallel rays and is some form of concave lens. (a) Double-convex, (b) planoconvex, (c) convex meniscus, (d) double-concave, (e) planoconcave, and (f) concave meniscus.

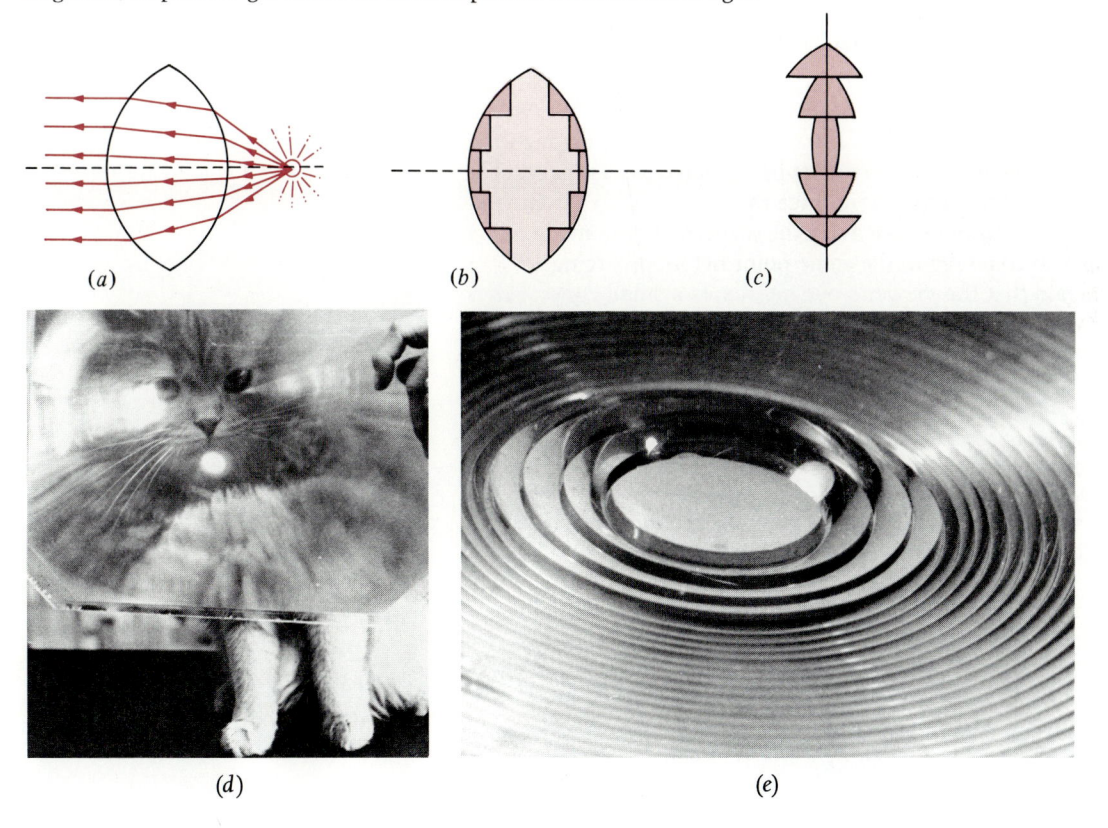

613

25-3 Thin Lenses

In many cases, the thickness of a lens is so small that we may neglect it and assume the net bending of a ray takes place at a plane through the center of the lens. Such a lens is called a **thin lens,** and it is the type on which we will concentrate. Unless otherwise stated, we will assume that the lenses are made of some transparent material with an index of refraction greater than 1 and that the lenses are immersed in air. We will use the shape of a double-convex lens to represent all converging lenses and the shape of the double-concave lens to represent all diverging lenses.

Figure 25-6 shows a set of parallel rays incident from the left on both (*a*) a converging and (*b*) a diverging lens. In each case, the dashed line perpendicular to the lens at its center represents the *principal axis* of the lens. After passing through the converging lens, the rays converge to a single point on the axis which they pass through and then diverge. The rays refracted by the diverging lens spread out but appear to diverge from a single point on the axis. In analogy with similar points for spherical mirrors, these points are called the **focal points** of the lenses, and the distance of each point from its respective lens is called the **focal length.** It does not matter whether the parallel rays are incident from the right or left of the lens. The converging lens will still cause them to converge toward a focal point, and the diverging lens will still cause the rays to diverge as if from a focal point. Thus each lens has two focal points positioned symmetrically on either side of the lens.

Just as mirrors may form images, so can lenses, as illustrated in Figure 25-7. An arrow-shaped object is placed on the axis of a lens. We select several of the rays leaving the tip of the arrow whose direction after passing through the lens we can construct from what has been said so far. The selected rays are numbered. Ray 1 is parallel to the axis, and thus it will be refracted by the lens to pass through the focal point. Ray 2 is drawn through the center of the lens. This ray should be practically undeviated by the lens since the two surfaces near the center are parallel if the lens is very thin. Ray 3 is drawn to pass through the focal point on the way to the lens. This ray must be refracted to pass parallel to the axis on the other side of the lens. These three rays converge to a point and then diverge. In principle, all other rays leaving the tip of the arrow and heading toward the lens will also converge to the same point before diverging.

We see once again that the viewer's eye intercepts a small cone of rays diverging from that point and interprets that point as the position of the tip of the arrow. Further, we know that a ray from the bottom of the arrow passes

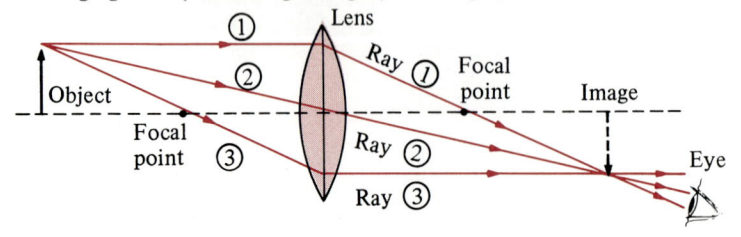

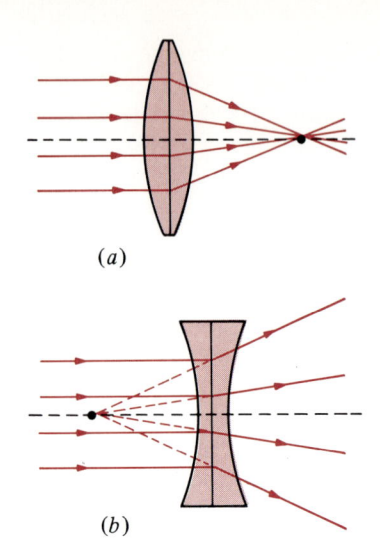

Figure 25-6 (*a*) The double-convex lens converges rays to a single point called the focal point of the lens. (*b*) The double-concave lens diverges rays such that they all appear to have diverged from a single point, on the same side of the lens as the incident rays, also called the focal point of the lens.

Figure 25-7 The behavior of certain special rays allows us to see how a lens affects light without having to calculate Snell's law at each surface for each ray. These three rays (as well as all others from the object point striking the lens) are physically reassembled in space at the image point. The eye beyond this point interprets the rays as diverging from (hence, originating at) the image point.

25-3 Thin Lenses

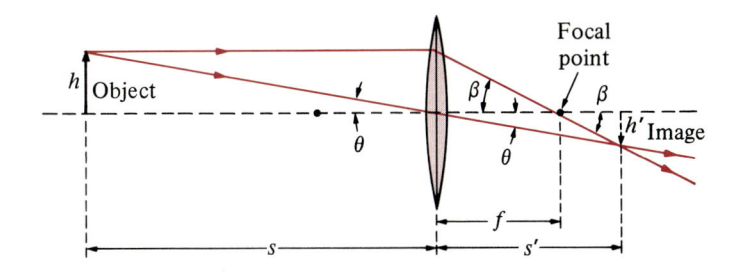

Figure 25-8 A labeled diagram for image formation by a lens: object distance s, image distance s′, focal length f, object height h, and image height h′.

along the axis and strikes the lens perpendicular to its surface. Thus it proceeds along the axis undeviated, assuring us that the eye will interpret the bottom of the arrow as being on the axis. The image of the arrow is a real image because light is actually diverging from the position where the viewer's eye interprets the arrow to be. As with all real images, this image may be projected onto a screen placed at the image position.

We can use two rays to derive an equation relating the image distance, object distance, and focal length as we did for mirrors. Figure 25-8 shows object distance, image distance, and focal length. The magnitudes of the heights of the object and image are labeled h and h', respectively, and several angles that the chosen rays make with the axis are labeled. From the right triangles in which the labeled angles appear in the figure, we may write

$$\tan \theta = \frac{h}{s} \quad \text{and} \quad \tan \theta = \frac{h'}{s'}$$

Thus

$$\frac{h}{s} = \frac{h'}{s'}$$

or

$$\frac{h'}{h} = \frac{s'}{s} \tag{25-1}$$

We have also that

$$\tan \beta = \frac{h}{f} \quad \text{and} \quad \tan \beta = \frac{h'}{s' - f}$$

which yields

$$\frac{h}{f} = \frac{h'}{s' - f}$$

or

$$\frac{h'}{h} = \frac{s' - f}{f} \tag{25-2}$$

Equating the right sides of Equations 25-1 and 25-2,

$$\frac{s'}{s} = \frac{s' - f}{f}$$

This can be written as

$$\frac{s'}{s} = \frac{s'}{f} - 1$$

Finally, dividing each term by s' and rearranging yields

$$\frac{1}{s} + \frac{1}{s'} = \frac{1}{f} \tag{25-3}$$

Used with lenses, this may be called the **lens equation,** but since it is the same as the mirror equation, it is often simply called the *law of optics.*

The lens equation is also applicable to converging lenses that form virtual images and to diverging lenses, which form *only* virtual images for real objects (see Problem 25-12). Figure 25-9 shows virtual images formed by both a converging lens and a diverging lens. In each case, we have used all three rays both to show that they trace to or from the same point and to provide a check. In both cases we see that rays from the tip of the arrow are diverging after leaving the lens. They must be extrapolated backward to the point from which they appear to diverge in order to locate the image. Thus, to see the virtual images, the viewer's eye must be on the right side of the lens. The image is viewed through the lens, and the object appears to be on the side of the lens opposite the eye. The three convenient rays used for ray tracing are summarized in Table 25-1.

As with the mirror equation, we must use a consistent sign convention when we apply Equation 25-3. The convention we will use is as follows:

1. The object distance (s) is always positive for a real object.
2. The image distance (s') is positive for a real image (side of lens opposite to that of object) and negative for a virtual image (same side of lens as object).
3. The focal length (f) is positive for converging lenses and negative for diverging lenses.

The sign convention for focal lengths has prompted people who work with lenses (optometrists, oculists, etc.) to always refer to converging lenses as **positive lenses** and to diverging lenses as **negative lenses.**

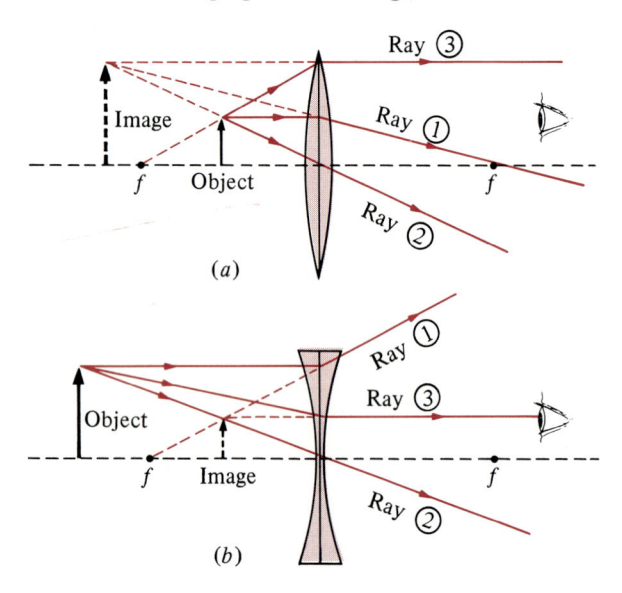

Figure 25-9 (*a*) A convex lens forms a virtual image if the object is located between the focal point and the lens. (*b*) Diverging lenses always form virtual images. Convince yourself for both drawings that if the image were replaced by a real object of the same size, its image would *not* be located at the object positions in the drawings. See if you can locate the new image positions by drawing the situation.

TABLE 25-1 Rays for Ray Tracing

1. Drawn from the object parallel to the axis of the lens. After refraction, it passes through the second focal point for a converging lens and diverges as if from the first focal point of a diverging lens.
2. Drawn from the object through the center of the lens. For thin lenses of either type, it is drawn to pass undeviated.
3. Drawn from the object along a line from the first focal point of a converging lens and toward the second focal point of a diverging lens. After refraction, it is directed parallel to the axis.

Example 25-1 What is the focal length of a converging lens which forms a real image at an image distance of 20.0 cm when the object is placed at 30.0 cm?

Solution The object distance is always a positive number for a real object, and the image distance is a positive number if the image is on the side of the lens opposite to the object (real image). Applying Equation 25-3,

$$\frac{1}{s} + \frac{1}{s'} = \frac{1}{f}$$

or

$$\frac{1}{f} = \frac{1}{30.0 \text{ cm}} + \frac{1}{20.0 \text{ cm}}$$

$$= \frac{2}{60.0 \text{ cm}} + \frac{3}{60.0 \text{ cm}} = \frac{5}{60.0 \text{ cm}}$$

$$f = 12.0 \text{ cm}$$

Example 25-2 Calculate the position of the image for an object placed at (a) 75.0 cm and (b) 25.0 cm from a converging lens of 50.0-cm focal length.

Solution Each part of this problem requires a straightforward application of Equation 25-3. We must recognize from the sign convention that the object distance s is positive in both parts and that the focal length of the lens is positive.

(a)
$$\frac{1}{s} + \frac{1}{s'} = \frac{1}{f}$$

or

$$\frac{1}{s'} = \frac{1}{f} - \frac{1}{s}$$

$$= \frac{1}{50.0 \text{ cm}} - \frac{1}{75.0 \text{ cm}} = \frac{1}{150.0 \text{ cm}}$$

$$s' = +150.0 \text{ cm}$$

This means that the image is real and on the side of the lens opposite to that of the object.

(b)
$$\frac{1}{s'} = \frac{1}{f} - \frac{1}{s}$$

$$= \frac{1}{50.0 \text{ cm}} - \frac{1}{25.0 \text{ cm}} = -\frac{1}{50.0 \text{ cm}}$$

$$s' = -50.0 \text{ cm}$$

The minus sign means that the image is virtual and on the same side of the lens as the object.

Example 25-3 (a) At what distance from a diverging lens of focal length $f = -10.0$ cm must one place an object so that its image will be at a distance of 8.00 cm? (b) Find the image distance for the same lens if the object is placed 15.0 cm from the lens.

Solution (a) Both parts are applications of Equation 25-3, but we must, as always, be careful with the signs. For this part, we must realize that the image of a diverging lens is virtual for a real object, and thus $s' = -8.00$ cm.

$$\frac{1}{s} + \frac{1}{s'} = \frac{1}{f}$$

or

$$\frac{1}{s} = \frac{1}{-10.0 \text{ cm}} - \frac{1}{-8.00 \text{ cm}}$$

$$= -\frac{1}{10.0 \text{ cm}} + \frac{1}{8.00 \text{ cm}} = \frac{1}{40.0 \text{ cm}}$$

$$s = 40.0 \text{ cm}$$

(b) Here, we must be aware that the object distance is positive, and therefore $s = +15.0$ cm.

$$\frac{1}{s'} = \frac{1}{f} - \frac{1}{s}$$

$$= \frac{1}{-10.0 \text{ cm}} - \frac{1}{15.0 \text{ cm}} = -\frac{5}{30.0 \text{ cm}}$$

$$s' = -6.00 \text{ cm}$$

Magnification for lenses is also defined in the same way as that for mirrors. That is, for lenses, we have

$$m = \frac{h'}{h} \qquad (25\text{-}4)$$

where m is the magnification and h' and h are the heights of the image and object, respectively. For the magnification of a lens, we also agree that the sign will be used to indicate the vertical orientation of the image with respect to the object. Thus, if the object is erect and the image is inverted, as in Figure 25-8, h and h' are of opposite sign. Choosing upward as the positive direction dictates that h' for that figure is a negative number. To be consistent with our sign convention for image and object distances, we should rewrite Equation 25-1 as

$$\frac{h'}{h} = -\frac{s'}{s} \qquad (25\text{-}5)$$

Note that Equation 25-1 related only the magnitudes of those quantities. Using Equation 25-5, we may rewrite the magnification as

$$m = -\frac{s'}{s} \qquad (25\text{-}6)$$

From our sign convention, we can see that the magnification is negative for a real image and positive for a virtual image.

Example 25-4 Determine the position and height of an object whose image formed by a lens is 4.00 cm high and at a distance of 25.0 cm from the lens if (*a*) the lens has a focal length of $+20.0$ cm and the image is real and (*b*) the lens has a -40.0-cm focal length and the image is virtual.

Solution Keeping our sign convention in mind, we must apply Equation 25-3 to determine the object distance for both parts *a* and *b*. Then we may apply Equation 25-6, realizing that the image distance is positive for part *a* and negative for part *b*. Finally, we solve Equation 25-4 for the height of the object for each part. The sign of *m* tells us the vertical orientation of the image with respect to that of the object.

(a)
$$\frac{1}{s} + \frac{1}{s'} = \frac{1}{f}$$

or

$$\frac{1}{s} = \frac{1}{+20.0 \text{ cm}} - \frac{1}{25.0 \text{ cm}} = \frac{1}{+100.0 \text{ cm}}$$

$$s = +100.0 \text{ cm}$$

$$m = -\frac{s'}{s} = -\frac{25.0 \text{ cm}}{100.0 \text{ cm}} = -0.250$$

$$m = \frac{h'}{h} \quad \text{or} \quad h = \frac{h'}{m} = \frac{4.00 \text{ cm}}{-0.250} = -16.0 \text{ cm}$$

The minus sign simply tells us that the vertical orientations of the image and the object are opposite each other. We already knew this since *m* is negative.

(b)
$$\frac{1}{s} = \frac{1}{-40.0 \text{ cm}} - \frac{1}{-25.0 \text{ cm}} = +\frac{3}{200 \text{ cm}}$$

$$s = 66.7 \text{ cm}$$

$$m = -\frac{s'}{s} = -\frac{-25.0 \text{ cm}}{66.7 \text{ cm}} = +0.375$$

$$h = \frac{h'}{m} = \frac{4.00 \text{ cm}}{0.375} = 10.7 \text{ cm}$$

We have assumed that the lenses are immersed in air. However, we know that the amount of bending of light which takes place at a surface between two media depends on the *ratio* of the indices of refraction of the media. Thus, remember that a glass or plastic lens will behave differently under water or if it has different media on its two surfaces.

The law of optics will still work for a lens placed under water, for example, but the lens will have a longer focal length in the water than in the air. A plastic lens with an index of refraction equal to that of water could serve as a lens in air, but it would have no effect, converging or diverging, under water.

25-4 Lens Aberrations

The derivation of the lens equation really required that we assume that all angles that focusing light rays make with the axis of the lens are small. Such rays are nearly parallel to the lens axis and are hence called **paraxial rays.** Thus the application of the lens equation is limited to paraxial rays to a certain extent. Nonparaxial rays from a point on an object will not necessarily converge to the same point (or diverge as if from the same point) after refraction by the lens. When this occurs, a point on the object does not image as a sharp point. Additionally, we recall that the index of refraction and hence the focal length of a lens vary with the wavelength of light being refracted.

Departures of an image from perfection as predicted by the lens equation are called **aberrations.** These are not the result of imperfections in the fabrication of the lens but are simply due to the nature of refraction.

Chromatic aberration is due to the variation of the index of refraction with wavelength (color). Parallel rays of white light incident on a converging lens from a distant point, for instance, will be refracted as shown in Figure 25-10a (exaggerated somewhat to clarify the effect). Violet light is refracted more strongly, and the focal point for violet light lies closer to the lens than the focal point for red light. The focal points for the other colors lie between those for violet and red. Chromatic aberrations are corrected by fabricating compound lenses of different materials, as shown in Figure 25-10b. The diverging portion of the compound lens has a bigger effect on the violet light than does the converging element. If the indices of refraction of the two parts are carefully selected and their curvatures properly designed, a compound lens may be fabricated of two single lenses to cause any two colors (wavelengths) to have the same focal length. Wavelengths other than the two chosen will not necessarily have the same focal length for the compound lens, but this type of lens may reduce all chromatic aberration to an acceptable level.

Other aberrations occur even if light of only one wavelength is used and are sometimes called **monochromatic** (single-color) **aberrations.**

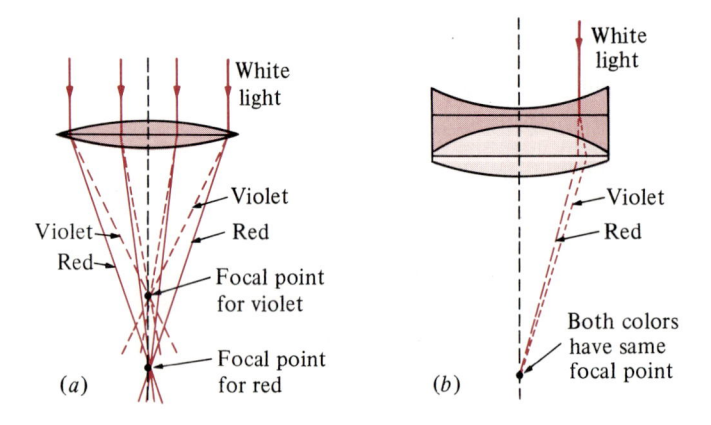

Figure 25-10 Chromatic aberration. (a) Since each wavelength is refracted through a slightly different angle, the lens will have a different focal length for each color. Images will be sharp only if a single color of light is used. (b) An anachromatic lens corrects for color separation by using both a converging and a diverging lens made of two types of glass with differing indices of refraction.

Spherical aberration is the failure of rays from a point source on the axis to converge to a point image, as in Figure 25-11. Rays which strike the lens farthest from the axis are refracted more than those which strike the lens near the axis. Thus there is no single point where the rays converge. The position along the axis where the rays form the smallest circle is taken as the best focus. The effects of spherical aberration may be minimized if only those rays near the axis are used to form the image. This can be done by placing a small opening, or **aperture**, in front of the lens which passes only rays near the center of the lens. The same defect causes point sources off the axis to image as streaked, comet-shaped figures (called **coma**) rather than as circles.

Astigmatism may be caused by a lens having different curvatures in different planes through its center. It results in the point being imaged as two perpendicular lines slightly separated along the direction of the axis. That is, the two line images of the point object are at different distances from the lens. Between the two lines is a **circle of least confusion** which gives the best image for the point object.

Curvature of field is an aberration causing points which form a plane perpendicular to the axis to be imaged as a curved surface rather than as a plane.

Distortion describes the situation in which the lens causes a straight line on the object to be imaged as a curve rather than as a line. The result is that a square grid object may be bulged out (barrel distortion) or pinched in (pincushion distortion).

It is not possible to correct a single lens for all these aberrations, but compound lenses of several elements as described for chromatic aberration and spherical aberration can greatly reduce their effects. In fine optics applications, one simply decides which aberrations will cause the most difficulty and designs to correct for those problems.

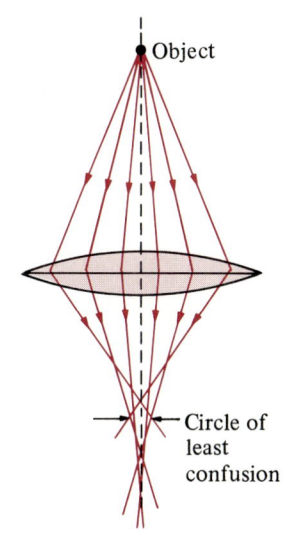

Figure 25-11 Spherical aberration is caused by rays far from the central axis being focused at a different distance than those closer to the center. It could be avoided by using only the central portion of the lens. If this is not possible, a best image-point position, called the circle of least confusion, yields the clearest image.

25-5 Single-Lens Applications

There are at least two common applications of single lenses which we have all used at one time or another: the magnifying glass and the simple camera.

A magnifying glass, sometimes called a simple microscope or a **magnifier**, forms an enlarged image so that we may see finer detail of a small object. In its simplest form, a magnifier is nothing more than a converging lens with a rather short focal length. The object to be viewed is placed closer to the magnifier than the focal length in order that a virtual image may be formed at a comfortable viewing distance for the eye (Figure 25-12).

We will obtain the greatest magnification for a single lens when the ratio of image distance to object distance is largest. From the lens equation, we know that the image distance becomes very large, approaching infinity, as the object approaches the focal point. Thus for most magnification, we wish to place the object very near the focal point, but closer to the lens than the focal length. This also places the image at a very large distance ($\sim \infty$), which is a comfortable viewing distance for the normal eye. Thus, to determine the magnification, we cannot easily compare the image size or image distance with that of the object. Instead, it is customary to compare the size of the angle subtended by the image at the eye with the angle subtended by the object when it is held at the nearest comfortable viewing distance.

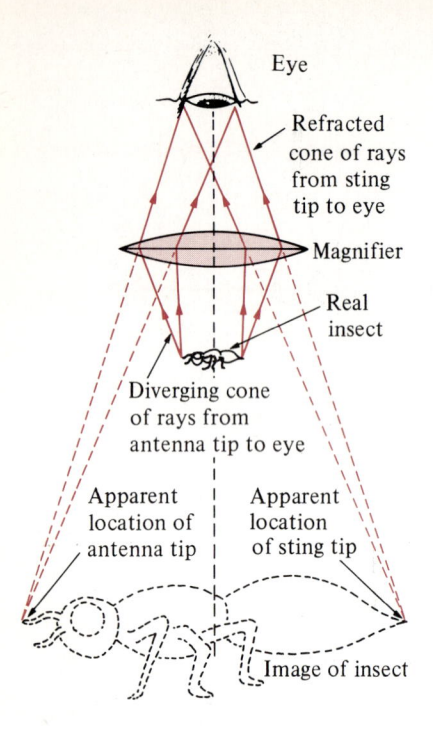

Figure 25-12 Image formation by a magnifying glass. Light from the tip of the antenna strikes all parts of the lens surface, but only the diverging cone which reaches the eye is shown. This cone is converged and bent back toward the central axis to reach the eye. The smaller angular spread of the rays gives the eye the impression that they have come from a much greater distance. The changed direction of the rays causes the eye to locate their source at a greater angle from the central axis.

The nearest comfortable viewing distance for the normal eye is about 25 cm. Figure 25-13a shows an object placed at 25 cm and the angle it subtends. In Figure 25-13b, we now see a magnifier as used with that same object and the new angle subtended by the image. Note that the object is placed very near the focal point. If we assume that the angles are small in each case, we may make the approximation that the tangent of the angle is equal to the angle measured in radians. From Figure 25-13a, we see that

$$\theta \approx \tan \theta = \frac{h}{25}$$

and from Figure 25-13b, we have

$$\theta' \approx \tan \theta' \approx \frac{h}{f}$$

Thus the angular magnification becomes

$$M = \frac{\theta'}{\theta} = \frac{h/f}{h/25}$$

or

$$M = \frac{25}{f} \quad (f \text{ in centimeters}) \quad (25\text{-}7)$$

We see that the shorter the focal length, the greater will be the angular magnification. Actually, aberrations come into play, and a single-lens magnifier gives a very poor image if f is less than a few centimeters. Thus usable M is limited to about 10 times (written 10×).

Figure 25-13 (a) An object of height h, when at a normal close viewing distance of 25 cm, subtends an angle of θ at the eye. (b) When the object is placed closer to the eye, just inside the focal length of the magnifier, an enlarged image is produced.

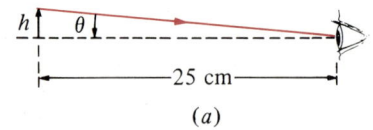

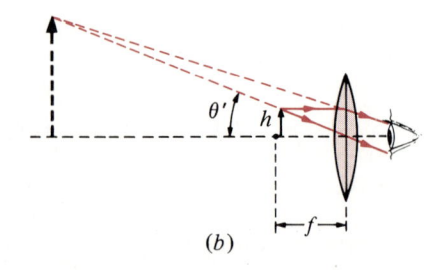

(a)

(b)

I. The Production of Color from White Light

White light is composed of all the colors of the visible spectrum mixed together. A beam of white light can be separated into its individual colors by several physical processes, the most commonly used being dispersion and diffraction. (a) Dispersion depends on the fact that the index of refraction of an object is slightly different for each wavelength of light. Here, a beam of white light is allowed to fall on one face of a triangular prism. Refraction at the first surface produces a slightly different path for each color through the glass, and refraction at the second surface enhances the separation. (*Bausch & Lomb*) (b) The finely spaced, regular grooves of a phonograph record will separate white light by diffraction. Here, light shining through the venting slits of an ordinary desk lamp falls on a record placed on the desk. Diffraction from the ridges produced by the grooves reinforces particular colors at specific angles, breaking the reflected light into a spectrum. (*D. Riban*)

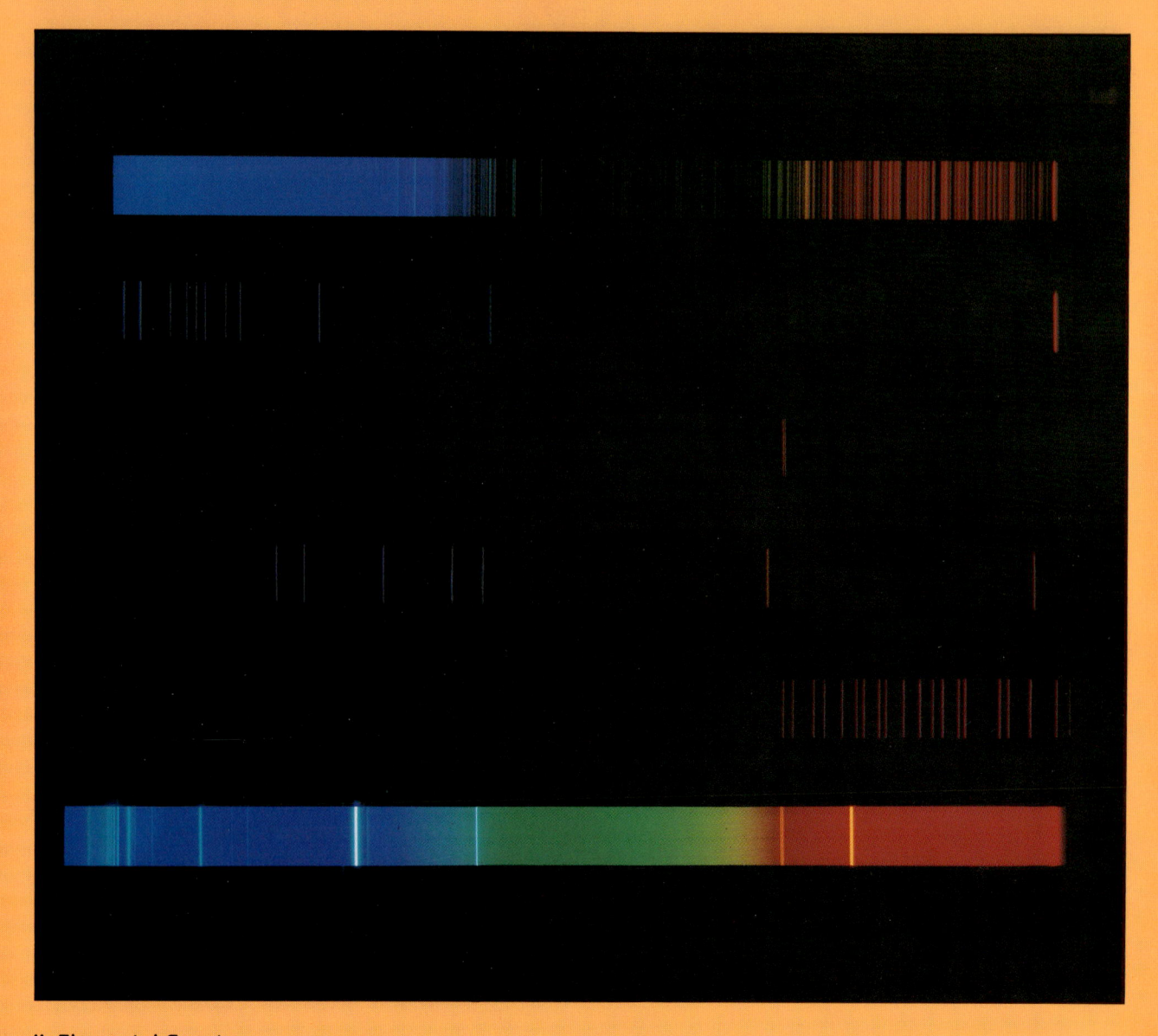

II. Elemental Spectra

The first-order spectra in the visible region for various pure substances. Such colors can be produced by either diffraction or dispersion. The substances are, from the top, molecular hydrogen, atomic hydrogen, a sodium lamp, helium, neon, and lithium. (*Bausch & Lomb*)

III. Color Mixing

Most individuals are familiar from grade-school days with how paints are mixed to produce various colors—yellow plus blue gives green, for example. This same concept holds for light, and different colors can be generated by either an additive process or a subtractive process. (*a*) The light is mixed additively where these three circles of colored light overlap. Where all three overlap, white is seen. Where two of the three overlap, other colors are seen. If the primary colors red, blue, and green are used as the source colors, the overlap areas are yellow, cyan (pale blue), and magenta, which are the three colors used in photographic films and color printing. (*Fritz Goro, Life Magazine*) (*b*) Subtractive mixing occurs when white light passes through filters that allow only certain colors to pass. For example, in this photograph, white light passing through the magenta filter and then through the yellow one has every color except red removed from it, and so red is the color we see. Where the light is passed through all three of these complementary filters (that is, filters whose colors add to produce white), all colors are absorbed and the result is black, as the center of the photograph shows. (*Fritz Goro, Life Magazine*)

(*a*)

(*b*)

(a)

(b)

(c)

(d)

(e)

IV. Light Balance and Color Film

Color photographic film must be balanced for a particular light source. For example, the film used here is balanced for use with a standard photoflood lamp emitting light as a blackbody at 3200 K. Such light is rich in red and has relatively low intensity in blue and violet. Thus, the film must have decreased sensitivity to red light and enhanced sensitivity to violet. (*a*) When the film is used with sunlight, which is much richer in shorter wavelengths than is a photoflood lamp, the image takes on a violet cast. Shadow areas of the skin are bluish, and the red flowers and lips are pale and unsaturated. The blue and violet flowers appear bright and full of detail, as do the eyes. (*b*) Fluorescent lights vary in the intensity emitted in each color. Older tubes were rich in violet and deficient in red, yellow, and orange. Modern "daylight" tubes use phosphors which simulate sunlight but are richer in longer wavelengths than earlier tubes. Note that the reds are more fully developed here than in the daylight shot, while the blue and violet flowers are still detailed. (*c*) The incandescent light used here is from two ordinary lightbulbs operating well below the 3200 K of a photoflood lamp. Hence the peak emission is shifted more to the red end of the spectrum than is intended for this film. This difference shows up mainly in the flesh tones, which are redder, or "warmer," than natural. Color values in the flowers are fully developed, with saturated reds and yellows and a natural tone in the blues and violets. (*d*) Two candles provided the illumination here. Peak emission is in the infrared, with only the long-wavelength (red and yellow) portion of the curve providing visible light. The absence of blue and violet light provides tones warmer than those from the incandescent light. White areas appear yellow due to the lack of short wavelengths in the light. Blue and violet flowers look almost black, as do the eyes. Note that the skin tone is almost the same as the light reflected from the lips. (*e*) The light source for this shot was a mercury vapor lamp. The skin and white flowers appear violet, and the lips purple. The violet flowers are quite bright, and the eyes are pale and transparent. How the red flowers look depends on whether or not they reflect the one wavelength of red light present. Note that the tulip at the center, which does not reflect this wavelength, is almost black. (*D. Riban*)

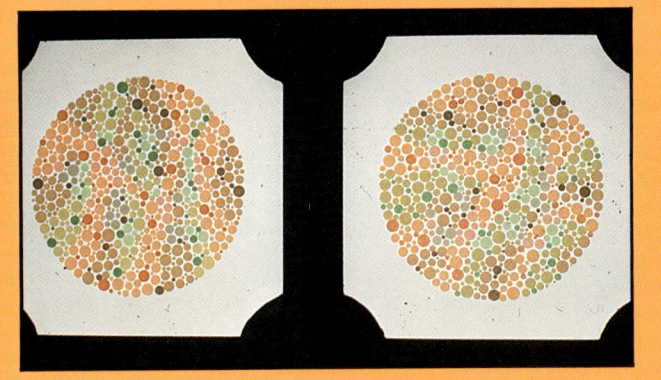

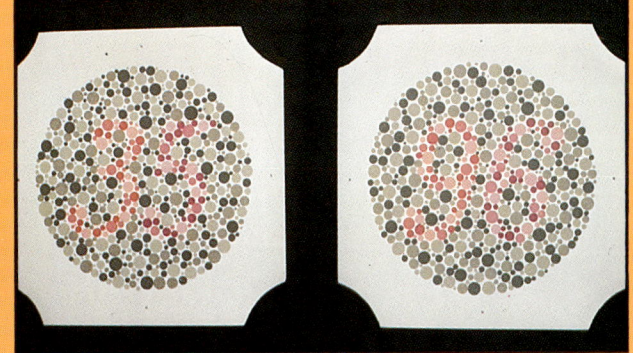

Va. Color Blindness

Color blindness affects 9 percent of males and 2 percent of females to various degrees. First described in 1774 by John Dalton, who also proposed the modern atomic theory, color blindness may take many forms. The inability to distinguish between red and green is the most common and accounts for almost 50 percent of reported cases. Persons suffering from this form of the disorder cannot see the green patterns in the photograph on the left. Persons suffering from red-brown color blindness cannot see the red numbers in the photograph on the right. (*Photo Researchers*)

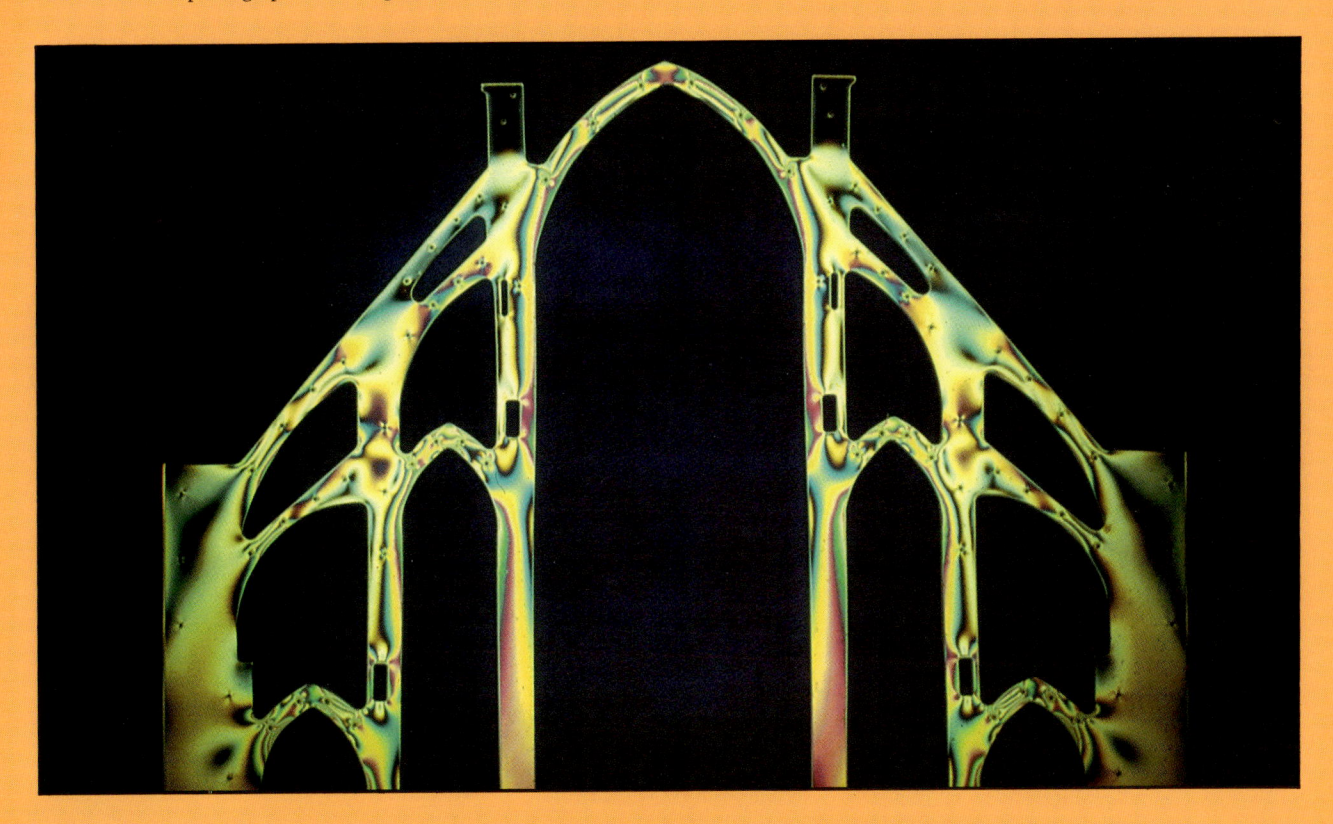

Vb. Polarization

Polarized light passes through a uniform material either with no rotation of its plane of polarization or with a uniform rotation. If the material is stressed, however, the highly stressed areas increase rotation. This phenomenon is very useful in structural analysis. Here, a plastic cutout simulates a cross section of Notre Dame Cathedral. When subjected to forces, the plastic becomes highly stressed at those places showing the greatest color variation over a small area, allowing engineers to infer positions of potential weakness and failure under stress. (*Sepp Seitz, Woodfin Camp*)

(a)

(b)

VI. Thermal Analysis by Color Imaging
An aerial photograph and a thermogram of a portion of the authors' campus reveal very different types of information. (a) This photograph shows what the eye sees. (b) This is a thermogram of the area shown in part a. Black, violet, and blue are the zones of lowest temperature, and yellow and red are the zones of highest temperature. The temperature in the white areas is off the temperature chart being used (from −8 to +2°C). Note that the grove of trees at the center is warmer on this near-freezing day than are the open areas. The diagonal white line at the upper center represents an underground steam line.

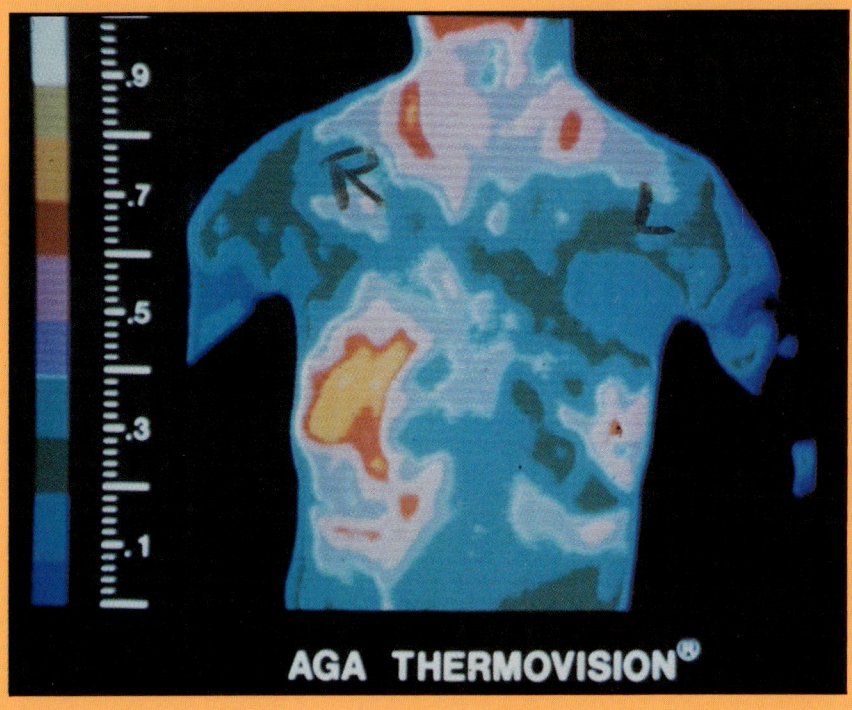

VII. Thermal Analysis by Color Imaging

(a) Infrared imaging of structures such as these houses can disclose areas of heat loss. Each color represents a temperature band. The yellow, red, and white indicate that heat loss is substantial at the windows and doors and through uninsulated walls. (*Daedalus*) (b) Infrared imaging of the human body is a simple procedure and quite useful in medical diagnosis. Here, the yellow areas on a scan of the upper torso of a woman indicate zones of elevated temperature. The prominent yellow area on the right breast could indicate the presence of an active cancer. (*Photo Researchers*)

(a) (b)

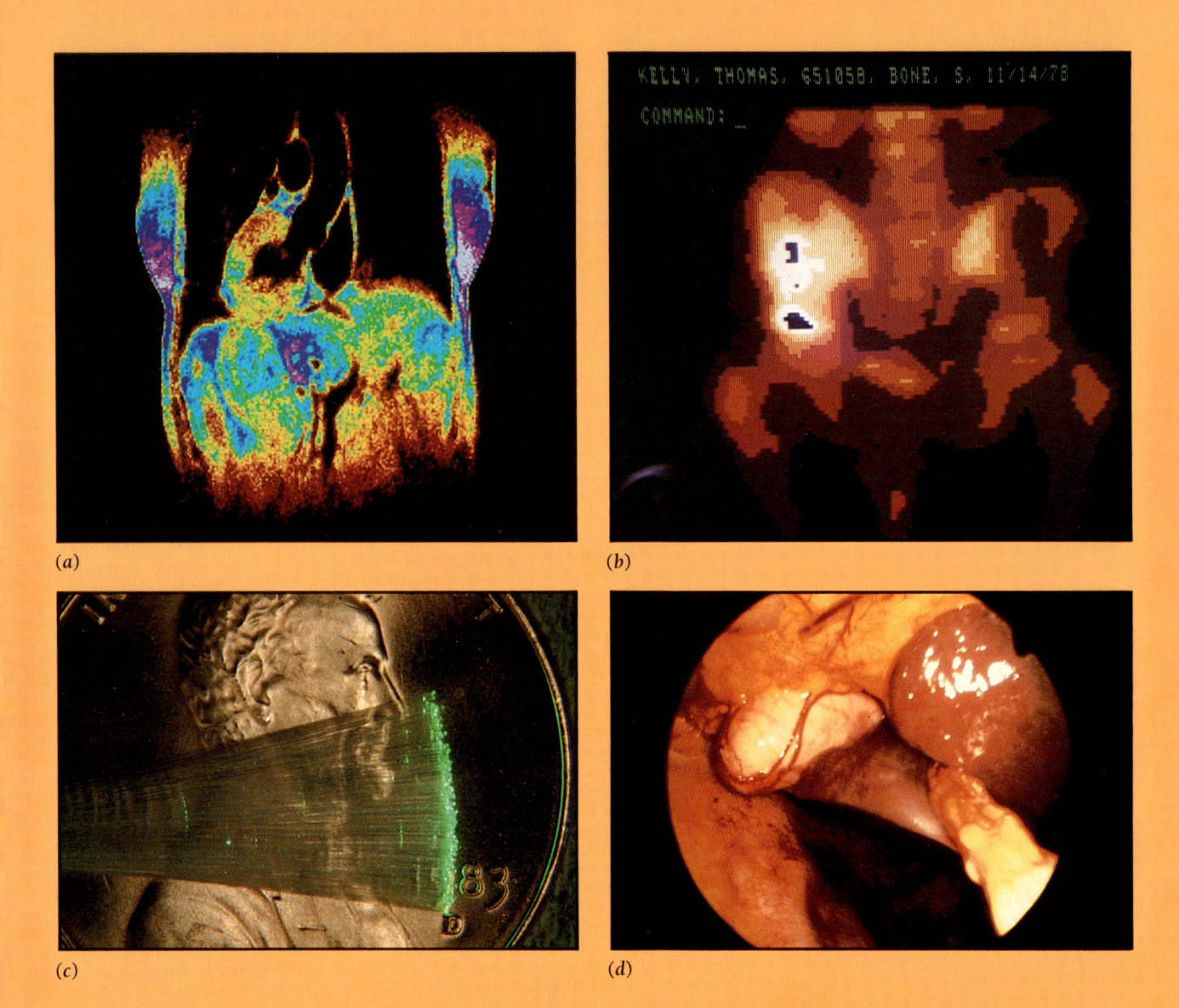

(c) (d)

VIII. Human Body Imaging

(a) An NMR (nuclear magnetic resonance) scan of the descending aorta in the human body. The term "NMR" is misleading since the patient is not exposed to any external nuclear radiation. The technology is ideal for revealing detail of soft tissues in the body and, as seen here, for examining the interior of blood vessels. Blood, being in motion, leaves the area being imaged before radiating to the detectors. This causes the image of the vessel interiors to be black and reveals constrictions and deposits on the interior walls. (*C.J. Sheldon, Phototake*)   (b) A gamma-camera scan of a hip reveals the concentration of a radioactive isotope as white "hot" spots in the bone. If the isotope has chemically concentrated in areas of bone cancer, centers of the disease may be diagnosed. (*Dan McCoy*) (c) A fiber optical bundle may be quite small and still contain a very large number of separate glass fibers to conduct light. The compact bundle shown here is small enough to be injected into the body through a hypodermic needle. Half of the fibers end in a light source that illuminates the body area being examined, and the other half conduct separate pieces of the image back to be viewed. (d) A fiber optical view of the interior of a gall bladder, showing both a gall stone and a cancerous area on the tissue.

25-5 Single-Lens Applications

Example 25-5 A single lens of focal length $f = +5.00$ cm is used as a magnifier. If a user views a textbook page on which the letters are 2.00 mm high, how high is the image of one of the letters? Assume that the image is formed at infinity.

Solution The fact that we assume the image to be at infinity means that the lens is held so the letters are at the focal point of the lens, and Equation 25-7 applies.

$$\frac{h'}{h} = M = \frac{25}{f}$$

or

$$h' = Mh = \frac{25.0 \text{ cm}}{5.00 \text{ cm}} \times 2.00 \text{ mm} = 10.0 \text{ mm} = 1.00 \text{ cm}$$

Actually, the magnification depends somewhat on the image distance, and it can be shown (see Problem 25-13) that

$$M = 1 + \frac{25}{f} \qquad (f \text{ in centimeters})$$

if the image is placed at the normal closest comfortable viewing distance of 25.0 cm.

Even a fairly simple camera presents a number of optical problems. It must have a light-tight enclosure with an area at the back to be covered with film and a lens to form a real image on the film. The lens must have a large-enough opening to allow short exposure times while still collecting enough light to expose the film. Furthermore, the lens must image a fairly wide angle, 45° or more, to simulate the scene as viewed by the eye. Thus, even the most inexpensive camera lenses are not simple lenses. They are compound lenses corrected for aberrations, at least chromatic aberration and curvature of field.

It is the job of the lens to produce a sharp image of the object on the film. The more that rays from the object diverge as they approach the camera, the farther behind the lens this image will be formed. **Focusing** the camera consists of moving the lens farther from the film for nearby objects or closer to the film for distant objects.

In addition, the camera must control the amount of light falling on the photosensitive surface of the film. To accomplish this, there are two adjustments. The *shutter* controls the duration that the film will be exposed to light, typically a fraction of a second. The **iris**, or diaphragm, controls how much of the surface of the lens will be used to collect light for the film; that is, it varies the size of the lens opening. The iris is calibrated in units called **f-stops**, corresponding to the distance from lens to film divided by the diameter of the portion of the lens used to admit light (Figure 25-14). For example, at f/2.0, it is twice as far from the lens to film as it is across the used portion of the lens. At f/4.0, the lens opening would be reduced (stopped down) to be only one-fourth as wide as the distance to the film. Thus, as f-numbers increase, the lens opening gets smaller and less light is admitted. Since the amount of light allowed into the camera varies as the area of the lens surface used rather than lens diameter, at f/4.0 the used portion of the lens would have only one-fourth the area as at f/2.0.

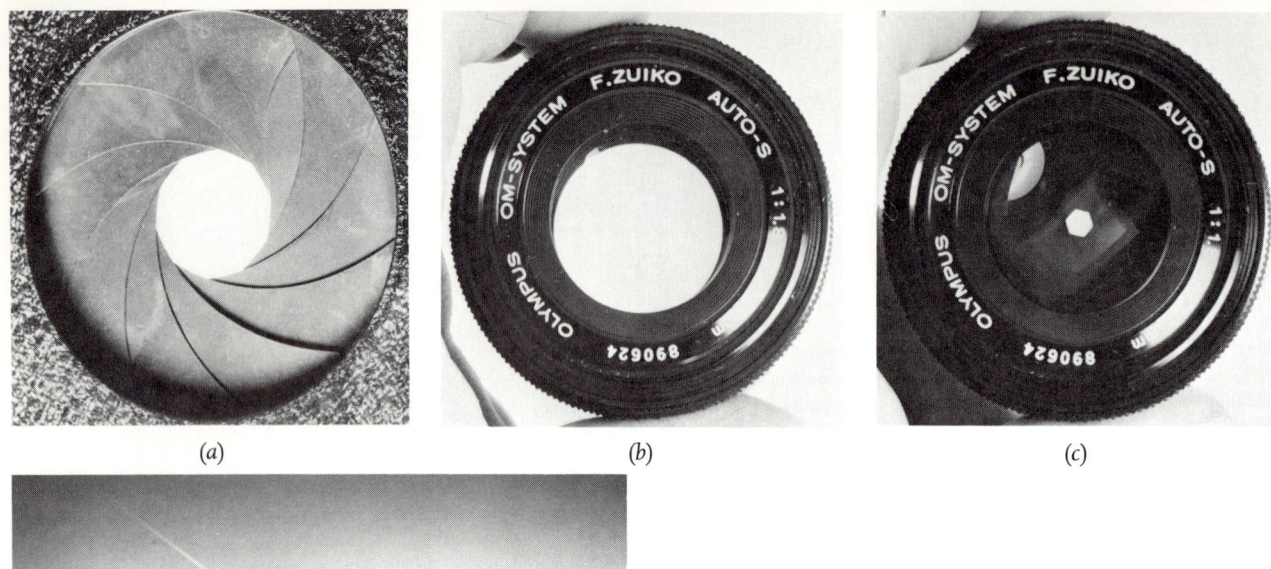

Figure 25-14 (a) The iris, or diaphragm, of a camera. (b) This lens opening is set at f/2.0. (c) Stopped down to f/11, the central opening is reduced to 1/32 of the area used at f/2.0. (d) A cutaway view of a modern 35-mm camera. The iris is between the lens elements. The 45° mirror reflects light upward into a pentaprism which provides the eye with an erect image of the scene entering the lens. At the instant a picture is taken, the mirror rotates upward out of the optical path, and a double curtain, the focal plane shutter, immediately in front of the film opens to expose the film to light. (*Olympus Corporation*)

Therefore, for equal time intervals, the camera at f/2.0 would admit four times as much light as at f/4.0.

Table 25-2 gives the customary values of full f-stops used on cameras. Moving one stop downward cuts the light reaching the film in half, while moving upward to the next stop (to a smaller f-number) doubles the light. Some cameras may include half-stops between these values. A lens with an iris operating across the entire range shown would be unusual. Typically, camera lenses can vary by about four or five full stops. A low f-stop admits more light and shortens the time required to take a picture, frequently important in dim light. However, a low f-stop also requires much more precise focusing since the light is converging to an image point through a much broader cone than for a higher f-stop. The smaller the lens opening, the more latitude the camera has in distance for correct focus, or the greater is the **depth of field** (Figure 25-15). For example, a pinhole camera at about f/128 is in perfect focus for objects at distances from zero to infinity but requires a very long time to take a picture.

Camera lenses are described in terms of their focal length and lowest

TABLE 25-2 Customary Values for Full Stops on Cameras

	f-Stop	Relative Light Intensity, f/5.6 = 1
More light	1.4	16
	2.0	8
	2.8	4
	4.0	2
	5.6	1
	8.0	$\frac{1}{2}$
	11	$\frac{1}{4}$
	16	$\frac{1}{8}$
	22	$\frac{1}{16}$
Less light	32	$\frac{1}{32}$

624

(a) (b)

Figure 25-15 Depth of field. (*a*) This scene was photographed at f/2.0 and $\frac{1}{1000}$ s. The central part of the bicycle is in precise focus, but anything much closer or farther from the camera is blurred. (*b*) The same scene photographed at f/16 and $\frac{1}{15}$ s on the same roll of film with the same lens and camera. The depth of field has been extended so that much closer objects as well as more distant ones are in sharp focus. (D. Riban)

f-stop. A typical lens for a 35-mm camera (so-named from the film size) might be a 50-mm, f/1.8 lens. The 50-mm focal-length lens is standard because its angle of view approximates the perceived angle for normal seeing. A shorter focal-length lens is a wide-angle lens (most common values, 35 and 28 mm), while a longer focal-length lens is a telephoto lens (135 mm, most common). Zoom lenses allow the continuous variation of the angle of view between two extremes by changing the spacing between groups of elements inside the lens.

A camera does not need a lens at all. A pinhole camera replaces the lens with a tiny opening. Effectively, there is no divergence of light to be compensated for since only one ray of light from a given point on the object passes through the pinhole, and this will strike the film at only one place. In practice, the pinhole admits so little light that taking a picture requires a very long time exposure.

25-6 Lens Combinations

Many applications of lenses in optical instruments require two or more lenses in combination. In most cases where two lenses are used, light from an object strikes one lens and an image is formed. The second lens is then used to magnify and position the image for further viewing. Thus, in essence, the image formed by the first lens becomes the object for the second lens which then forms the final image for the combination. Before considering several common optical instruments, we will explore the relationships between the object distance, the distance between two lenses, their focal lengths, and the ultimate image distance.

The sequence of Figure 25-16 traces the rays from an object through a two-lens system to the final image position. Part *a* shows two converging lenses separated by a distance *d*. Their focal points are shown, with focal lengths designated f_1 and f_2, respectively, while the object distance from the first lens is labeled s_1. Part *b* shows the image formed by the first lens at an image distance of s_1'. After the light from the object has converged to the position of this first image, it then diverges as it proceeds toward the second lens. The second lens thus will act on the light from the first image position as though it were a real object. Therefore, in part *c*, we start at that first image position just as if it were

625

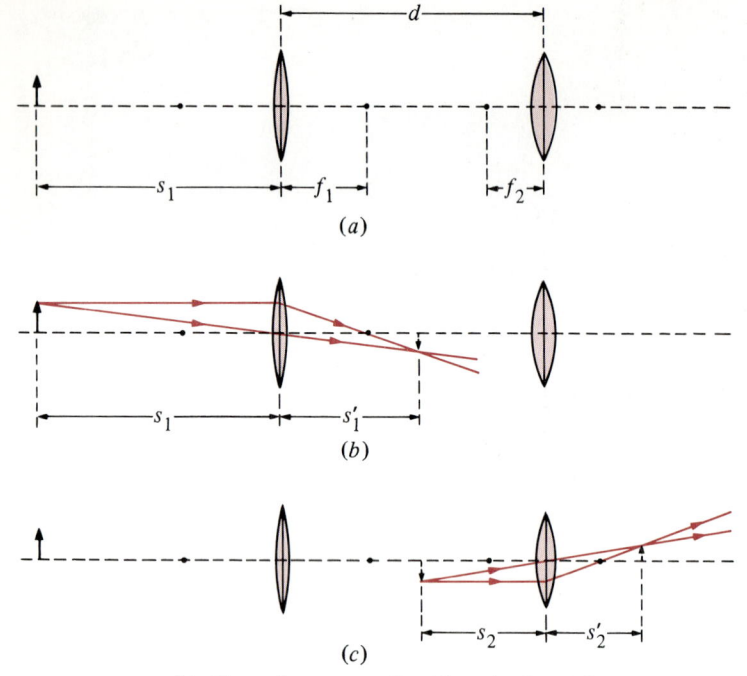

Figure 25-16 (a) Object distance s_1, focal lengths f_1 and f_2, and lens separation d are defined. (b) The location and size of the image produced by the first lens is determined by ray tracing or calculation. (c) The image produced by the first lens is now treated as the object for the second lens, and the final image is determined.

an object and trace rays through the second lens to locate the final image of the system. The illustration we have chosen is particularly simple to carry out, but the same procedure holds for any two thin lenses with any distance of separation and any object distance. For any conditions, we must keep in mind that the image position of the first lens becomes the object position for the second lens.

It is no more difficult to determine the position of the final image by applying the lens equation to each lens in turn. However, we must take particular care to follow the sign convention. Also, we need to remember that s_1', s_2, and d are related by

$$s_2 = d - s_1' \tag{25-8}$$

Example 25-6 An object 2.00 cm high is placed 15.0 cm to the left of a lens of focal length $+10.0$ cm. A second lens is placed 25.0 cm to the right of the first lens. Calculate the position of the final image of the system of two lenses if the focal length of the second lens is (a) $+20.0$ cm and (b) -10.0 cm.

Solution Both parts may be solved by simply identifying the quantities involved and applying the lens equation to each lens in succession. Note that after we find s_1' for the first lens, we utilize Equation 25-8 to obtain s_2.

(a)
$$\frac{1}{s_1} + \frac{1}{s_1'} = \frac{1}{f_1}$$

or

$$\frac{1}{s_1'} = \frac{1}{f_1} - \frac{1}{s_1} = \frac{1}{+10.0 \text{ cm}} - \frac{1}{15.0 \text{ cm}}$$

$$= \frac{3}{30.0 \text{ cm}} - \frac{2}{30.0 \text{ cm}}$$

$$s_1' = +30.0 \text{ cm}$$

$$s_2 = d - s_1' = 25.0 \text{ cm} - 30.0 \text{ cm} = -5.00 \text{ cm}$$

The minus sign on s_2 means that the image of the first lens does not act as a real object for the second lens. That is, light does not diverge from that position toward the second lens but converges toward that position when it is intercepted by the second lens. Such a situation is often described by saying the second lens has a **virtual object.** The convention is sometimes then extended to say that the sign of the object distance is negative for a virtual object. Continuing,

$$\frac{1}{s_2} + \frac{1}{s_2'} = \frac{1}{f_2}$$

or

$$\frac{1}{s_2'} = \frac{1}{f_2} - \frac{1}{s_2} = \frac{1}{+20.0 \text{ cm}} - \frac{1}{-5.00 \text{ cm}}$$

$$s_2' = 4.00 \text{ cm}$$

(b) This is the same as part *a* up to the point where we find $s_2 = -5.00$ cm. Then,

$$\frac{1}{s_2'} = \frac{1}{f_2} - \frac{1}{s_2} = \frac{1}{-10.0 \text{ cm}} - \frac{1}{-5.00 \text{ cm}}$$

$$= -\frac{1}{10.0 \text{ cm}} + \frac{2}{10.0 \text{ cm}}$$

$$s_2' = 10.0 \text{ cm}$$

Note that even the diverging lens formed a real image for a virtual object. Could you trace rays to show the image positions for these cases?

Since the object of the second lens is the image of the first, the magnification of a combination of lenses is equal to the product of the individual magnifications. This is true for any number of lenses in combination, but for two lenses we have

$$m = m_1 m_2 \qquad (25\text{-}9)$$

where m is the overall magnification and m_1 and m_2 are the magnifications of the individual lenses.

Example 25-7 Determine the overall magnification of the two systems of lenses of Example 25-6. How large is the final image in each case?

Solution For each lens, we write the equation for magnification given in Equation 25-6. Then, applying Equation 25-9, we may obtain the overall magnification.
(a) For the first system,

$$m_1 = -\frac{s_1'}{s_1} = -\frac{30.0 \text{ cm}}{15.0 \text{ cm}} = -2.00$$

$$m_2 = -\frac{s_2'}{s_2} = -\frac{4.00 \text{ cm}}{-5.00 \text{ cm}} = 0.800$$

$$m = m_1 m_2 = (-2.00)(0.800) = -1.60$$

Thus for the first system, the image is real, inverted, and enlarged relative to the object. Equation 25-4 may be used to obtain the final image height.

$$m = \frac{h'}{h}$$

yields

$$h' = mh = (-1.60)(2.00 \text{ cm}) = -3.20 \text{ cm}$$

(b) For the second system, note that the first lens is the same, and therefore

$$m_1 = -2.00$$

For the second lens,

$$m_2 = -\frac{s_2'}{s_2} = -\frac{-10.0 \text{ cm}}{-5.00 \text{ cm}} = 2.00$$

Thus

$$m = m_1 m_2 = (-2.00)(2.00) = -4.00$$

The image is again real, inverted, and enlarged.

$$h' = mh = (-4.00)(2.00 \text{ cm}) = -8.00 \text{ cm}$$

If two thin lenses are used in combination with the distance between them made very small (approximately zero), we have from Equation 25-8 that

$$s_2 \approx -s_1'$$

The thin-lens equation applied to each lens then becomes

$$\frac{1}{s_1} + \frac{1}{s_1'} = \frac{1}{f_1} \quad \text{and} \quad \frac{1}{-s_1'} + \frac{1}{s_2'} = \frac{1}{f_2}$$

When these two equations are added, we obtain

$$\frac{1}{s_1} + \frac{1}{s_2'} = \frac{1}{f_1} + \frac{1}{f_2}$$

Notice that the left side of this equation is the sum of the reciprocals of the object distance for the combination and the image distance for the combination. This equation for the combination would be just like the thin-lens equation for a single lens provided we identify the right side as the reciprocal of the combination's focal length. That is,

$$\frac{1}{f} = \frac{1}{f_1} + \frac{1}{f_2} \qquad (25\text{-}10)$$

where f is the focal length of the combination of two thin lenses in contact which have individual focal lengths of f_1 and f_2.

Equation 25-10 had led opticians to define and use a convenient unit to describe the **refractive power** of a lens. The unit is the **diopter**, and the defining equation is

$$P = \frac{1}{f \text{ in meters}} \qquad (25\text{-}11)$$

where P is the optical power of a lens expressed in diopters when f is the lens' focal length measured in meters. Thus, a converging lens of focal length $+20.0$ cm ($=0.200$ m) has a power equal to $+5.00$ diopters. Also, a diverging lens of focal length -50.0 cm ($=-0.500$ m) has a power equal to -2.00 diopters. The convenience this definition of refractive power yields is easy to see if one expresses Equation 25-10 in refractive power. It becomes

$$P = P_1 + P_2 \qquad (25\text{-}12)$$

Thus, if these two lenses were used in close combination, the power of the combination would be $+3.00$ diopters.

25-7 Optical Instruments

Several common optical instruments make use of more than one lens in combination, such as microscopes, telescopes, and binoculars. We will examine the first two systems.

As we mentioned, a magnifying glass is sometimes called a simple microscope. Thus the instrument we are about to describe, which includes more than one lens, is called a **compound microscope**. A compound microscope is used to provide higher magnifying power without the large, troublesome aberrations that a single lens would introduce. Figure 25-17 depicts a typical compound microscope, with rays traced from an object to form the image. Both lenses are converging lenses of very short focal length. The lens nearest the object is called, appropriately, the **objective lens** and has a focal length typically less than 1 cm. The lens nearest the eye is called the **eyepiece,** or sometimes the **ocular.** The eyepiece generally has a focal length of 2 to 3 cm and really acts as a magnifier. As we see in the figure, the object is placed just beyond but very near the focal point of the objective lens. The objective lens forms a real, inverted image very far from the objective but close to the eyepiece, inside the focal point of the eyepiece. This real image of the objective becomes a real object for the eyepiece, which forms a virtual image, still inverted, at a comfortable viewing distance, say 25 cm from the eye. The length L of the microscope is compressed in the figure but is usually 20 cm or more.

We may obtain the approximate overall magnification of a compound microscope rather easily. As usual with a combination of lenses, we use the fact that the overall magnification is the product of the magnifications of the individual lenses. The magnification of the object is given as usual for a lens as

$$m_o = -\frac{s'_o}{s_o}$$

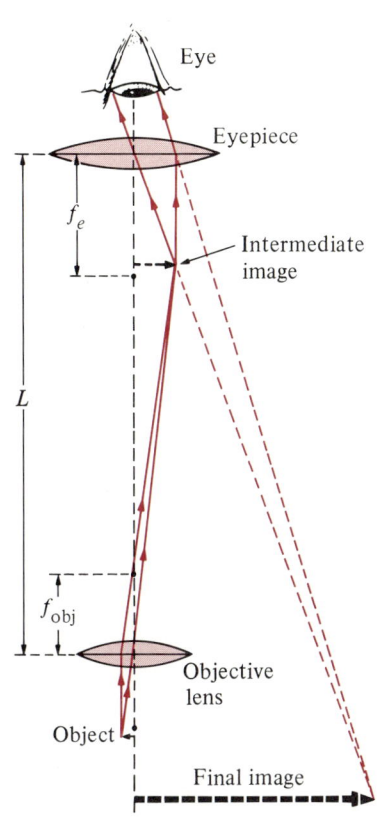

Figure 25-17 Image formation in the compound microscope. If each lens produces an image three times as large as its object, the combined magnification would be 9×.

Again, we treat both the eyepiece and the objective lenses of the microscope as single lenses optically. However, no microscope objective is a simple piece of glass. A really high-quality objective lens might have a dozen separate elements made of various types of glass, presenting two dozen precisely shaped glass surfaces (carefully spaced) which act as a unit. This complexity is necessary to correct for various aberrations. Since the combination has a fixed geometry in use, it may be treated as a single lens with a single combined focal length.

But the object distance is very near the focal length of the objective, and the image distance is almost equal to the length of the microscope. That is, $s_o \approx f_o$ and $s'_o \approx L$. Therefore, to a good approximation, we have

$$m_o \approx -\frac{L}{f_o}$$

The eyepiece is simply a magnifier, with angular magnification for normal viewing distance given approximately by Equation 25-7 as

$$M_e \approx \frac{25}{f_e}$$

The overall magnification of the compound microscope becomes

$$M = m_o M_e = -\frac{L}{f_o}\frac{25}{f_e} = -\frac{25L}{f_o f_e} \quad (25\text{-}13)$$

Example 25-8 A compound microscope has an objective with a focal length of 0.800 cm and an eyepiece of focal length 2.00 cm. The object to be viewed is placed 0.825 cm from the objective. Calculate (a) the length of the microscope and (b) its magnification for this object.

Solution (a) We must first apply the lens equation to find the image distance for the objective. This image must be formed slightly inside the focal length of the eyepiece. Remember that the eyepiece is a magnifier, and its object will be very near its focal point. The length of the compound microscope thus will be taken as the sum of the image distance of the objective and the focal length of the eyepiece.

$$\frac{1}{s} + \frac{1}{s'} = \frac{1}{f}$$

or

$$\frac{1}{s'} = \frac{1}{0.800 \text{ cm}} - \frac{1}{0.825 \text{ cm}} = \frac{0.0379}{\text{cm}}$$

$$s' = 26.4 \text{ cm}$$

$$L = s' + f_e = 26.4 \text{ cm} + 2.00 \text{ cm} = 28.4 \text{ cm}$$

(b) This is a straightforward application of Equation 25-13.

$$M = -\frac{25L}{f_o f_e} = -\frac{(25 \text{ cm})(28.4 \text{ cm})}{(0.800 \text{ cm})(2.00 \text{ cm})} = -444\times$$

A refracting **astronomical telescope** is illustrated in Figure 25-18. The lenses have the same names as those for the compound microscope because their functions are similar. An objective lens forms a real image which becomes a real object for an eyepiece. The eyepiece is a magnifier with a very short focal length which forms a greatly magnified virtual image. For objects at great distance, the image formed by the objective is practically at the focal point of the objective. In order for the eyepiece to form an enlarged virtual image, its object

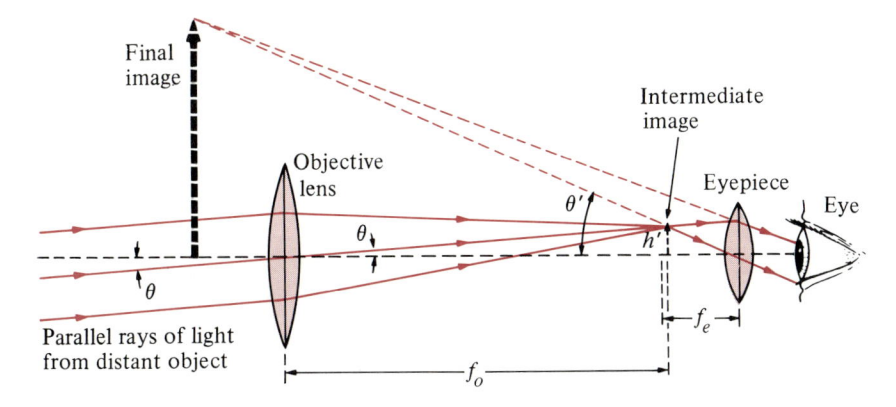

25-7 Optical Instruments

Figure 25-18 In the astronomical telescope using an objective lens, the object is very distant, essentially infinitely far. Thus, rays of light from an object point arrive at the telescope as parallel rays. The telescope objective must have a very long focal length, and the real image will be formed at or beyond this distance. As with the compound microscope, the eyepiece is simply a magnifier. Note that the incident rays are coming from below the axis but the (inverted) image is formed above the axis.

should be very near but slightly inside its focal point. Thus the focal points of the objective and the eyepiece of an astronomical telescope practically coincide, and the distance between the two lenses is approximately equal to the sum of their focal lengths.

The angular magnification of a telescope is the ratio of the angle the final image subtends at the eye to the angle subtended by the object. This would be the ratio of θ' to θ, as labeled in Figure 25-18. If we assume small angles, we may again make the approximation that the tangent of the angle is equal to the angle measured in radians. Careful examination of the figure shows we may obtain expressions for the tangent of both angles in terms of the height of the image formed by the objective and the focal lengths of the lenses. We may write

$$\theta' \approx \tan \theta' = \frac{h'}{f_e} \quad \text{and} \quad \theta \approx \tan \theta = -\frac{h'}{f_o}$$

The angular magnification becomes

$$M = \frac{h'/f_e}{-h'/f_o}$$

or

$$M = -\frac{f_o}{f_e} \quad (25\text{-}14)$$

The minus sign, as usual, indicates that the image is inverted.

The fact that the image of the telescope illustrated in Figure 25-18 is inverted limits its use primarily to astronomical observation, since for those observations the orientation of the image is unimportant. It is important, however, that a telescope used for terrestrial observations provides an erect image. This is accomplished in the opera glass or galilean telescope (Figure 25-19) by using a diverging lens as the eyepiece placed inside the focal point of the objective. The image distance of the objective for distant objects is approximately equal to the objective's focal length. Its image becomes a virtual object for the eyepiece. The eyepiece is positioned so that its object is very near but slightly outside the focal point of the lens.

The angular magnification, defined in the usual way, may be obtained rather quickly:

$$M = \frac{\theta'}{\theta}$$

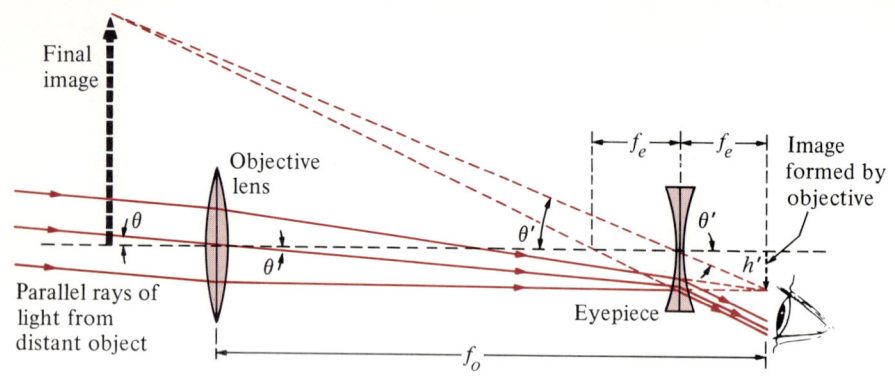

Figure 25-19 A galilean telescope. A diverging eyepiece is placed closer to the objective than with the astronomical telescope. The positioning is such that the image formed by the objective lens is just beyond the eyeward focal point of the eyepiece. The rays never converge to form a real image but are diverged to the eye to produce an erect, virtual image.

If we assume the angles are small, we see from Figure 25-19 that

$$\theta' \approx \tan \theta' = \frac{h'}{f_e}$$

where both h' and f_e are negative numbers. Further,

$$\theta \approx \tan \theta = \frac{-h'}{f_o}$$

Therefore,

$$M = \frac{h'/f_e}{-h'/f_o}$$

or

$$M = -\frac{f_o}{f_e} \tag{25-15}$$

which is the same as that for the astronomical telescope. Because the final image is erect, the galilean telescope design is used in opera glasses. It has the further advantage that it is more compact. The minimum length of the telescope is $f_o - f_e$ rather than the $f_o + f_e$ of the astronomical telescope.

An erect image is also required of binoculars and monoculars. In these instruments, the erect image is achieved by causing one more inversion. Two prisms are positioned similar to those shown in Figure 24-23c. This arrangement placed between the lenses causes one more inversion for an astronomical-type telescope so that the final image is erect. Further, it allows a greater magnification or power in a shorter distance since it extends the optical path length, allowing a longer focal-length objective lens.

Most astronomical amateur and professional telescopes are not refracting telescopes at all but are **reflectors** employing mirrors (Figure 25-20). For amateur instruments, this is because a reasonable-diameter mirror (15 to 30 cm in diameter) can be produced much more inexpensively than the same-size lens (at home, if necessary). Professional instruments require a very large collecting surface to gather enough light to produce images of very faint sources. Really

Figure 25-20 A reflecting telescope. In addition to the advantage over lenses in lower cost and possible larger size, use of mirrors eliminates chromatic aberrations in the objective. The method of viewing the image shown here uses a smaller plane mirror to reflect the image out the side and through an eyepiece.

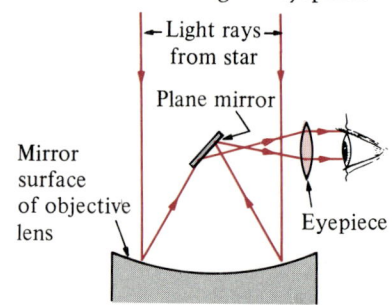

large lenses weigh a very great amount yet must be supported by their edges. After a point, the warping of the glass under its own weight as it is pointed to various positions in the sky begins to distort the image. Unlike lenses, large astronomical mirrors may be braced from underneath all across their unused surface to prevent such warping. Thus, the largest astronomical refractor, at Yerkes Observatory in Wisconsin, has a lens diameter of about 1 m, while the reflector on Mount Palomar has a 5-m diameter mirror and that on Mount Pashtukov in the Soviet Caucasus has a 6-m diameter. Astronomical mirrors must correct for edge distortions in forming an image and cannot be spherical. Starting from a spherical shape, they must be polished so that a cross section through the center of the mirror is parabolic in shape.

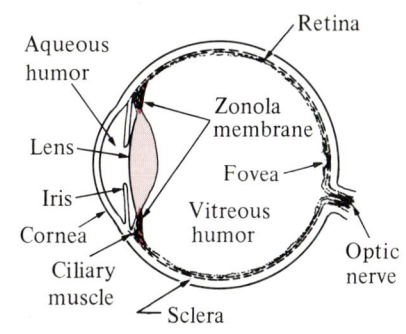

Figure 25-21 Cross section of the human eye.

25-8 The Human Eye

The human eye is very nearly spherical in shape, as illustrated by Figure 25-21. It is a jellylike mass covered by a rough fibrous membrane, the **sclera.** The front part of the eye bulges with a slightly larger curvature than the rest of the eyeball. This is covered by the **cornea,** a transparent membrane, behind which the volume is filled with the **aqueous humor,** a watery transparent liquid. The **crystalline lens,** immediately behind the aqueous humor, is a small, somewhat flexible converging lens made of many very thin layers. Behind this, the largest part of the volume of the eye is a chamber filled with a transparent, jellylike substance called the **vitreous humor.** The interior surface of this chamber is covered with the light-receptor nerve cells of the **retina.** These cells are at the ends of a fine network of nerves branching out from the optic nerve. There are two types of light receptors, **rods** and **cones,** so-named because of their shapes (Figure 25-22). Photopigments in the rods and cones convert light energy into nerve signals which are then transmitted to the brain via the optic nerve. Best studied of these pigments is the rhodopsin of the rods, a bluish pigment that takes on a purple hue when the eye is dark-adapted. For this reason it is commonly called **visual purple.** The rods are about 1000 times more sensitive to light than cones and are very important to vision in low-intensity light, providing so-called night vision. However, rods do not distinguish colors. In bright light, cones provide day, or bright-light, vision and can also distinguish colors.

Near the axis of the eye's lens on the retina there is a tiny depression about 0.250 mm in diameter called the **fovea,** containing only densely packed cones. The fovea yields our sharpest vision, while the rest of the retina provides only a somewhat hazy version of the field of view. The eye muscles involuntarily rotate the eyeball until the image on which our attention is directed falls on the fovea. Where the optic nerve enters the eye there are neither rods nor cones, and we cannot detect an image formed at this point; hence it is called the **blind spot.**

The performance of the human eye is similar to that of a camera in that light from an object is refracted by a converging type of lens system to form a real image on a light-sensitive surface. A normal adult eyeball is approximately 2.4 cm in diameter. However, the radius of curvature of the cornea is about 0.77 cm rather than 1.2 cm as one might calculate by assuming that the eye were a perfect sphere. The outer surface of the cornea is the primary refracting

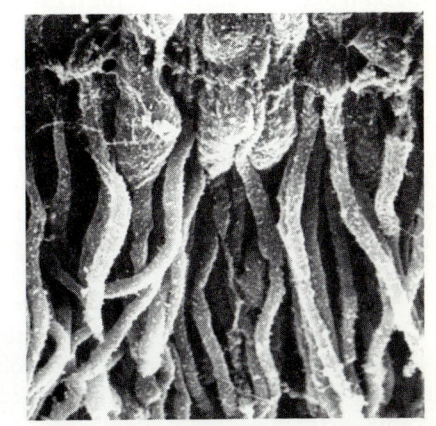

Figure 25-22 Electron micrograph image of the retina, showing rods and cones. *(R. Eagle/Photo Researchers)*

You can easily detect the blind spot of either eye. On a sheet of white paper, draw two dots, each about 1 mm in diameter and separated by a horizontal distance of 10 cm. Hold the paper 15 cm in front of your face so that the *right* dot is in front of your *left* eye. Close your *right* eye, stare at the right dot with your left eye, and slowly move the paper away from your face while continuing to stare at the right dot. At some distance, the left dot will disappear. This locates the blind spot for your left eye. Reversing the procedure locates that of your right eye.

surface of the eye. The cornea membrane has an index of refraction of 1.376, while the index of refraction of the aqueous humor is 1.336. Thus, very little bending of light takes place at the interface between the cornea and the aqueous humor.

The crystalline lens is a double-convex lens with an average index of refraction of about 1.41 (Figure 25-21). Since the refractive index of the vitreous humor is 1.336, both surfaces of the lens cause additional convergence to light rays already converged by the cornea. The additional convergence caused by the lens is not large. If we think of the eye as replaced by a single lens, this hypothetical lens would have a focal length of about 1.5 cm or a power of about 67 diopters. About 45 diopters would be due to the refraction at the air-cornea interface, while the rest would be due to the crystalline lens.

The size of the **pupil,** the lens aperture, is controlled by the **iris,** which also provides eye color. The diameter of the pupil may be varied from about 1 to 5 mm, varying the area by a factor of 25.

When light comes from objects close to the eye, the rays are not parallel and the lens changes shape to focus these rays. For the relaxed eye, focused at infinity, the lens is held under tension by the **zonula membrane,** which is attached to the ring of **ciliary muscle.** When viewing nearby objects, the ciliary muscle contracts, decreasing the tension on the zonula, which allows the lens to bulge and become more converging. This process is called **accommodation.** The crystalline lens thickens and becomes less flexible with age. For very young people, accommodation may increase the optical power of the eye by as much as 15 diopters. By age 30, the additional accommodation is reduced to about 8 to 10 diopters, while by age 50, it is further reduced to between 1 and 3 diopters for the normal eye.

In normal vision, a sharp, real image is formed on the retina. Accommodation allows the normal eye to focus on objects placed anywhere from infinity to some **near point,** the closest distance for which the eye can form a sharp image of an object. Varying from less than 10 cm for a typical teenager to about 200 cm for an average older person, the near point recedes with age. For a reference, we take the normal near point to be 25 cm. The **far point** is the farthest distance an object may be placed from the eye and still have the eye form a sharp image on the retina. For the normal eye, the far point is at infinity.

Many people's eyes fail to form focused images. There are several important types of failures, but they are all lumped together as vision defects. In **myopia,** or nearsightedness, the eyeball is too long compared to the radius of curvature of the cornea, and the image from a distant object is formed before the light rays reach the retina. Thus, the far point of an uncorrected myopic eye is not at infinity. Figure 25-23a shows a normal eye, and Figure 25-23b shows an uncorrected myopic eye. Figure 25-23c shows how a lens can correct for myopia. The myopic eye can form a sharp image if the rays from a point on an object have a large-enough divergence when they reach the cornea. To make this happen, either the object is brought closer to the eye or a diverging lens in the form of eyeglasses or contact lenses is placed between the eye and the object.

In **hypermetropia,** or farsightedness, the eyeball is too short compared to the radius of curvature of the cornea, and the light rays from a point on a close object converge toward an image point beyond the retina (Figure 25-23d). A hypermetropic eye with a corrective lens is also shown (Figure 25-23e). A hypermetropic eye can form a sharp image if the rays from a point on an object are not diverging too much when they reach the cornea. For this to happen,

Figure 25-23 (a) The normal eye. (b) The myopic (nearsighted) eye. (c) Since the problem with the myopic eye is excessive convergence of rays, the solution is to place a diverging lens in front of the eye. (d) The hypermetropic (farsighted) eye. (e) The correction for the inadequate convergence of farsightedness is to add a converging lens before the rays reach the cornea.

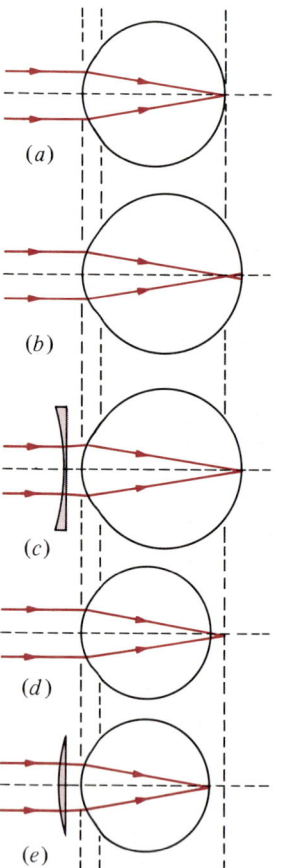

either the object is moved farther from the eye or a converging lens (eyeglass or contact) is placed between the eye and the object.

Another rather common vision defect is called **astigmatism,** which occurs if the cornea is not spherical but has different curvatures in, for example, the horizontal and vertical planes. For instance, the cornea may be shaped like a part of the surface of a football. The result is that an astigmatic eye has a different focal length in the horizontal plane than in the vertical plane, making it impossible to focus on both of two perpendicular lines at the same time (Figure 25-24). An astigmatic eye will view some set of lines as significantly fainter and less-focused than those which are perpendicular to that set. The astigmatism may be oriented along any axis and is compensated for by using a carefully oriented cylindrical lens between the eye and object.

As the lens thickens and hardens with age, the near point of the eye recedes, giving rise to a condition known as **presbyopia.** Some people do not like to consider this to be a defect, as it happens to all normal eyes and proceeds at about the same rate for everyone. It does, however, cause a decrease in vision which requires correction. Presbyopia is treated very much like farsightedness. Either the object to be viewed is moved out beyond the near point or a converging lens is used between the cornea and the object to start rays converging before they reach the cornea.

There is a simple way to think of what **corrective lenses** do for defects of vision. Essentially, the corrective lens must have a focal length such that it forms an image at a comfortable viewing distance for the uncorrected eye. Thus the corrective lens for a myopic eye must be designed so that it forms an image of very distant objects which is at or inside the eye's far point. Similarly, the lens correction for a hypermetropic eye must form an image of a nearby object which is at or beyond the eye's near point.

Figure 25-24 An astigmatic eye may see some of the bars as very dark and sharp, while those perpendicular to the sharp ones will be fainter and out of focus.

Example 25-9 A professor afflicted with hypermetropia in both eyes has near points of 90.0 cm for each. Determine the (a) focal length and (b) optical power of the corrective lenses which will allow him to read comfortably when he holds a typed page at 25.0 cm from the eye.

Solution This requires an application of the lens equation and the definition of the diopter. As usual, we must be careful with signs. Recognize that the object is the typewritten page, and therefore $s = +25.0$ cm. The image must be formed at the near point (or a greater distance), and since it is formed on the same side of the lens as the object, the image distance is negative. Thus $s' = -90.0$ cm.

(a)
$$\frac{1}{f} = \frac{1}{s} + \frac{1}{s'}$$

$$= \frac{1}{+25.0 \text{ cm}} + \frac{1}{-90.0 \text{ cm}} = \frac{1}{+34.6 \text{ cm}}$$

$$f = +34.6 \text{ cm}$$

(b)
$$P = \frac{1}{f \text{ m}} = \frac{1}{0.346 \text{ m}} = +2.89 \text{ diopters}$$

Example 25-10 A student with myopia in both eyes has a near point for each eye (uncorrected) at 10.0 cm. The student wears corrective lenses of power $P = -2.00$ diopters. Determine (a) the far point of the student's eyes and (b) the distance to the near point when the student is wearing the corrective lenses.

Solution (a) Both parts of this problem will yield to the lens equation. For part a, the corrective lenses must form an image of a very distant object at the far point. Thus we use $s = \infty$ and determine f from the power of the lenses. Then we may apply the lens equation.

$$f = \frac{1}{P} = \frac{1}{-2.00 \text{ diopters}} = -0.500 \text{ m} = -50.0 \text{ cm}$$

$$\frac{1}{s} + \frac{1}{s'} = \frac{1}{f}$$

or

$$\frac{1}{s'} = \frac{1}{f} - \frac{1}{s} = \frac{1}{-50.0 \text{ cm}} - \frac{1}{\infty}$$

$$s' = -50.0 \text{ cm}$$

and the far point is 50.0 cm away from the eye.

(b) Here we assume that an object is placed at the near point and determine its position when the image is at -10.0 cm.

$$\frac{1}{s} = \frac{1}{f} - \frac{1}{s'}$$

$$= \frac{1}{-50.0 \text{ cm}} - \frac{1}{-10.0 \text{ cm}} = \frac{4}{50.0 \text{ cm}}$$

$$s = 12.5 \text{ cm}$$

Thus the corrective lenses change the closest distance of good focus from 10.0 to 12.5 cm. This increase is not very much of a sacrifice in order to gain good vision from 50.0 cm out to infinity.

Minimum Learning Objectives

After studying this chapter, you should be able to:
1. Define:
    - aberration
    - accommodation
    - aperture
    - astigmatism
    - blind spot
    - chromatic aberration
    - circle of least confusion
    - coma
    - compound microscope
    - converging lenses
    - cornea
    - corrective lens
    - crystalline lens
    - curvature of field
    - depth of field
    - diopter [unit]
    - distortion
    - diverging lenses
    - eyepiece
    - far point
    - f-stop
    - focal point
    - hypermetropia
    - iris
    - lens
    - lens equation
    - magnification
    - magnifier
    - monochromatic
    - myopia
    - near point
    - negative lens
    - objective lens
    - positive lens
    - presbyopia
    - pupil
    - refractive power
    - retina
    - spherical aberration
    - thin lens
    - virtual object

2. Identify types of lenses.
3. Construct three rays from an object through a lens and beyond based on the definition of focal point and general behavior of light without the explicit application of Snell's law.
4. Locate the image of a given object formed by a single lens of known focal length by construction and by calculation.
5. Determine the focal length of a lens, given object and image distances from the lens.
6. Apply consistently the sign conventions for calculations involving lenses.
7. Determine the magnification for a single lens.
8. Identify common lens aberrations.
9. Determine the change in f-stop and/or shutter speed to maintain a constant exposure of the film to light in a camera.
10. Calculate the image position and size for simple combinations of lenses of known characteristics and given geometry.
11. Calculate the refractive power of combinations of known lenses.
12. Specify the appropriate lens to produce a given correction to an eye with a specified, simple visual problem.

Problems

25-1 Calculate the distance from a lens of focal length $+15.0$ cm where one should place an object such that an image is formed at (a) $+60.0$ cm, (b) $+30.0$ cm, and (c) $+20.0$ cm.

25-2 An object is placed at a distance of 100 cm from a converging lens. Determine the focal length of the lens if a real image is formed at a distance from the lens of (a) 25.0 cm, (b) 40.0 cm, and (c) 400 cm.

25-3 Determine the image distance for a lens of focal length $+30.0$ cm whenever an object is placed at a distance from the lens of (a) 120 cm, (b) 300 cm, and (c) 35.0 cm. Sketch a ray diagram for each case.

25-4 An object is placed 40.0 cm in front of a lens of focal length $+30.0$ cm. (a) Sketch a ray diagram showing the position of the image. (b) Calculate the image position by using the lens equation.

25-5 A lens of focal length -60.0 cm is used with an object which is first placed at a distance of 90.0 cm from the lens and then moved to a distance of 30.0 cm. (a) Sketch a ray diagram showing the position of the image for each object position. (b) Calculate the image positions by using the lens equation.

25-6 For a lens of focal length $+12.0$ cm, determine the object position which will yield (a) a real image at $+36.0$ cm and (b) a virtual image at -36.0 cm.

25-7 An object 1.80 cm high is placed 30.0 cm from a $+20.0$-cm focal-length lens. (a) Sketch a ray diagram showing the position and size of the image. (b) Calculate the position of the image. (c) Calculate the magnification of the lens for this situation and the height of the image.

25-8 Calculate the object position and image position for a lens of focal length $+25.0$ cm if its magnification for this situation is (a) -4.00 and (b) $+4.00$.

25-9 A single-lens slide projector is to project a real image 175 cm wide on a screen 8.00 m away from the lens for a slide 3.50 cm wide. Calculate (a) the magnification expected, (b) the position of the slide, and (c) the focal length of the lens.

25-10 A small light bulb is placed 80.0 cm from a screen and serves as an object for a $+15.0$-cm focal-length lens. (a) At what distances from the screen may the lens be placed in order that a real focused image be projected on the screen? (b) Calculate the magnification for each of the two possible cases.

25-11 How far from a lens of focal length $+12.0$ cm should an object be placed in order that the image be four times as large as the object if the image is to be (a) real and (b) virtual?

25-12 Use the derivation of the lens equation given in Section 25-3 as a guide. Show that the same equation may be derived for a diverging lens provided one uses the sign convention given. That is, show that one obtains

$$\frac{1}{s} - \frac{1}{s'} = -\frac{1}{f}$$

for the equation of a diverging lens if s' and f are the magnitudes of the distances of the image and focal point from the lens.

25-13 Show that if a magnifier is adjusted to place the image at -25.0 cm, then the angular magnification is given by $M = 1 + (25/f)$. Refer to Figure 25-13 and use the small-angle approximation that $\theta = h/25$ and $\theta' = h/s$. Then show that $M = 25/s$ and substitute for s the expression which is obtained from applying the lens equation.

25-14 Calculate the focal length of a magnifier which gives a magnification of $9\times$ if (a) the image is at infinity or (b) the image is at the normal near point of 25.0 cm (see Example 25-5).

25-15 The lens for a 35-mm camera has a focal length of 5.50 cm (usually stated as 55 mm). The lens may be moved toward or away from the film to focus images on the film for objects anywhere in the range from 1.20 m to infinity. Calculate the extremes of distance from the lens to the film.

25-16 A 35-mm camera has a lens of focal length 50 mm. What is the diameter of the lens opening at a setting of f/5.6?

637

25-17 The effective diameter of a 35-mm camera lens set at f/2.8 is 3.00 cm. Calculate (a) the focal length of the lens and (b) the exposure time at a setting of f/4 if the proper exposure time at f/2.8 is $\frac{1}{250}$ s.

25-18 A 35-mm camera is used to photograph the moon, which appears $\frac{1}{2}°$ wide in the sky. What would be the diameter of the moon's image on the film for (a) a 50-mm lens, (b) a 135-mm lens, and (c) a 400-mm lens? (d) The 35-mm camera gets its designation from its film width. Actually, sprocket holes for the winding mechanism along both edges reduce the actual picture area to 24 mm high (by 36 mm wide). What focal-length lens would be needed to produce an image of the moon on the film which just reached across the vertical picture area of the film?

25-19 A pinhole camera is made from a shoebox 32.0 cm long with a 0.400-mm hole in one end and film at the other. A camera light meter indicated that for the same film, a camera at f/8.0 would require an exposure of $\frac{1}{50}$ s. How long did it take to expose the picture with the pinhole camera?

25-20 A photographer taking a picture of a person against a background of bright snow walks up to 1.00 m from the person and takes a light-meter reading through the camera. The camera indicates that the correct exposure is f/5.6 at $\frac{1}{100}$ s. The photographer now walks to a distance of 4.00 m from where the picture will be taken. Light falls off in intensity with the inverse square of the distance. Thus, at the new distance, only one-sixteenth as much light will reach the camera from the person as at the distance of the measurement. What camera setting should the photographer use at this new distance? (Be careful: This is not a problem in computation but in perception of the process involved.)

25-21 A camera has a fixed-size shutter opening that gives it the equivalent of an f stop of f/22. When the lens-film distance is 9.00 cm, the sharpest image is obtained if the subject is placed 3.00 m from the camera. Calculate (a) the focal length of the lens, (b) the effective diameter of the lens opening, and (c) the image height on the film of a 2.00-m-tall subject standing 3.00 m in front of the lens.

25-22 A macrolens for a camera is designed for focusing on a very close object so that the image on the film will approach the size of the real object. Almost any detachable camera lens may be used to produce such close-ups, but the lens must be moved farther from the film than its mechanism normally allows. (a) A 50-mm lens may be used for objects from 0.50 m to infinity. How far is the optical center of the lens from the film in each case? (b) A penny is 19.0 mm in diameter. How far must the lens be from the film to produce the same-size image on the film? (c) If the lens is removed from the camera and extended forward, the f-stop settings on the lens are no longer valid and must be recomputed. With the lens in the position of part b, the lens is set at f/4.0. What is the real f-stop for taking the picture?

25-23 Two lenses, one of focal length +25.0 cm and the second of focal length −30.0 cm, are separated by 50.0 cm along a common axis. Calculate the position of the final image of an object placed at a distance from the converging lens of (a) 200 cm and (b) 40.0 cm. (c) What is the magnification of the system in each case?

25-24 An object is 30.0 cm from a converging lens of focal length 12.0 cm. On the other side of the lens at a distance of 45.0 cm is a second converging lens of 10.0 cm focal length. Calculate (a) the final image position and (b) the magnification of the system.

25-25 Two thin lenses in contact form a real image 60.0 cm along the axis for an object positioned 30.0 cm distant. The optical power of one of the lenses is known to be −2.00 diopters. Calculate (a) the focal length of the combination, (b) the optical power of the combination, (c) the optical power of the unknown lens, and (d) the focal length of the unknown lens.

25-26 Two lenses, each of optical power +5.00 diopters, are to share a common axis. How far apart should they be placed if an object at infinity is to form a virtual image 30.0 cm before the second lens?

25-27 A lens of optical power +1.25 diopters is 70.0 cm to the left of a second lens of optical power −10.0 diopters. An object is placed 400 cm to the left of the positive lens. Calculate (a) the final image distance of the combination and (b) the size of the object if the image is 20.0 cm tall.

25-28 A microscope has an eyepiece of magnification 12× and an objective lens of magnification 22×. The two lenses are 18.0 cm apart. Calculate (a) the overall magnification and (b) the focal length of each lens.

25-29 Many good microscopes have two or three different objective lenses which may be rotated into place over a specimen. They also may have several different eyepieces to allow a number of different possible magnifications—increasing magnification with decreasing field of view. Consider a microscope with objective lenses of focal lengths 1.80, 0.500, and 0.200 cm. It has eyepieces available with angular magnifications of 3×, 6×, and 9×. Determine the nine possible values of overall magnification for this microscope if it is 25.0 cm long.

25-30 A microscope has an objective of focal length 0.500 cm and an eyepiece of focal length 1.80 cm. The distance between the lenses is 20.0 cm. Calculate (a) the object distance and (b) the overall magnification.

25-31 An astronomical telescope has an eyepiece of focal length 8.00 cm and has a distance of 2.10 m between the objective lens and the eyepiece. Calculate (a) the focal length of the objective lens and (b) the overall magnification of the telescope.

25-32 Calculate (a) the magnification of a galilean telescope which has an objective lens of focal length 15.0 cm and an eyepiece of focal length −3.00 cm and (b) the approximate overall length of the telescope.

25-33 Because of the ease with which they are made, many amateur astronomical telescopes are constructed with a spherical concave mirror as the objective rather than a lens. Recalling that the focal length of a spherical concave mirror is one-half its radius of curvature, determine (a) the magnification of an astronomical telescope whose mirror has a radius of curvature of 5.00 m and an eyepiece of focal length 1.50 cm and (b) the diameter of the image of the moon which this objective produces if the moon subtends an angle of exactly 0.500° at the earth.

25-34 The general public is often excessively interested in the magnification of a telescope. However, a faint, fuzzy image twice the size improves nothing. The primary job of a telescope is to present a large surface area for gathering light to render the unseen visible. Astronomers rate the visible brightness of stars in magnitudes. One magnitude lower means a star appears 2.5 times as bright. Sirius, the brightest star, has a magnitude of -1.47, while the sun as seen from the earth has $m = -26.8$. Of course, the sun is much closer than Sirius. To compare real brightness, astronomers use the absolute magnitudes (M) of stars, or how bright they would appear from a distance of 10 parsecs (32.6 light-years). The sun has an M of $+5$. Since the unaided eye can see a star to magnitude $+7$, it would be visible at this distance. Light intensity falls off as the inverse square of distance; thus the sun could be about 2.5 times as far (80 light-years) and just be visible. Now, the eye has a light-gathering surface with a maximum iris diameter of 5.00 mm. Light falling on the retina has about $\frac{1}{16}$ s to accumulate before the retina is cleared for the next image. Assume that the sensitivity of film equals that of the eye. The telescope on Mount Palomar has a diameter of 5.08 m. Assume that it may collect light for 6.00 h before scattered atmospheric light fogs its photographic plate. How far away could the sun be and still be detected by the telescope on Mount Palomar?

25-35 The professor wears bifocals, the lower part of which have an optical power of $+2.50$ diopters and the upper part, a power of -0.800 diopter. The lower part is used for close work and the upper part for distance vision. Determine the professor's near point and far point.

25-36 Determine (a) the optical power of the corrective lenses necessary for a hypermetropic eye which has a near point of 150 cm and (b) the image distance for an object placed at 75.0 cm from the corrected eye.

25-37 A myopic eye has a far point of 40.0 cm and near point of 8.00 cm. Calculate (a) the power of the corrective lenses required to image a very distant point at 40.0 cm from the eye and (b) the position of the near point of the corrected eye.

25-38 At age 45 years, the professor obtains eyeglasses of optical power $+2.00$ diopters in order to read comfortably when he holds the reading material 25.0 cm away. Five years later, he finds that even while wearing the glasses he must now hold reading material 40.0 cm from his eye in order to focus sharply. What power glasses should the optometrist prescribe so that the professor can once again focus on reading material held 25 cm away?

25-39 An optometrist discovers that her patient has a near point of 90.0 cm due to lack of accommodation and a far point of 150 cm due to a naturally shortened eyeball. What should be the optical power of the two parts of the bifocals she prescribes?

25-40 A student has a far point of 40.0 cm for both eyes. (a) Calculate the power of the required corrective lens if she is to see very distant objects clearly. (b) If her power of accommodation is $+8.00$ diopters, determine her uncorrected near-point distance. (c) What is her near-point distance when wearing the lenses?

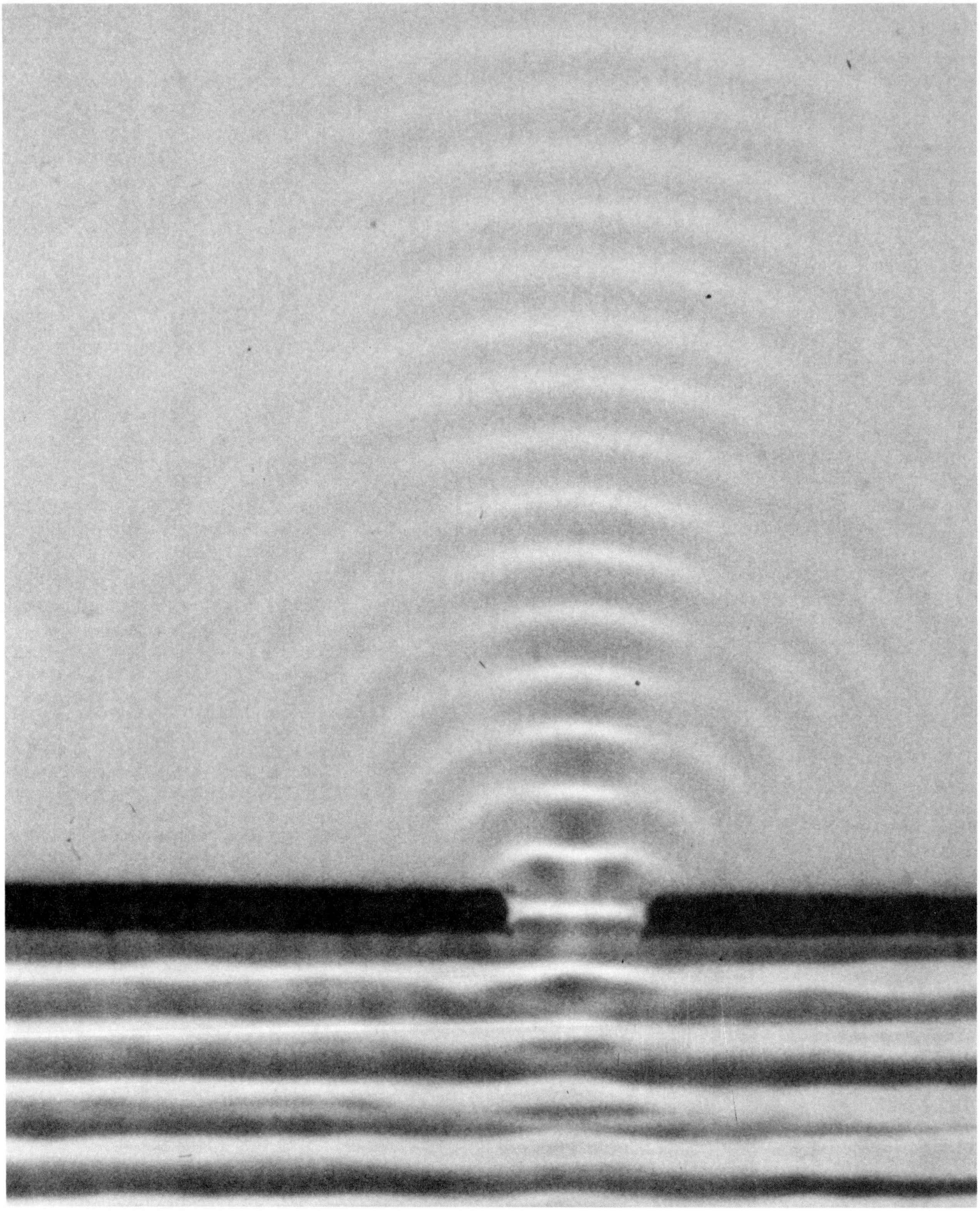

26 Physical Optics

26-1 Introduction

If we shine a light on the gap between two opaque objects (say two metal plates), the light passes through the gap and shows up as a bright band between the shadows of the plates on the wall beyond. If we push the plates closer together, the gap narrows, and, as we might expect, the bright band on the wall does, too. However, as we repeat this process, we reach a point where a further narrowing of the gap does not narrow the bright zone on the wall. In fact, further narrowing of the gap increasingly *widens* the bright zone on the wall. True, this occurs only when we reach a very narrow gap, or **slit,** but it does occur, invariably, when that fine a slit is reached. This marks the limit to treating light, as we have in the last few chapters, as inevitably traveling in straight lines in the form of fine rays. From this point on, we are obliged to consider the wave properties of light.

In geometric optics, light was assumed to travel perpendicular to wavefronts in straight lines called rays. When it encountered a boundary between two media, light was either reflected or refracted, and the rays and therefore the wavefront were redirected. Phenomena such as the behavior of light passing through a narrow slit cannot be described by geometric optics. Instead, interpretation requires that we consider fully the wave nature of light, and this treatment is called **physical optics.**

26-2 Development of Physical Optics

Newton's theory of optics provided much of the material already covered and is the base on which geometric optics is built. Newton accepted the postulate that light was made of moving small particles (called corpuscles). This model fit very well with the "known fact" that light did not bend around corners, softening shadows, as waves would be expected to do. Newton had experimented widely and made many observations which led him to forms of the laws of reflection and refraction. However, his publication *Opticks* included reports of observations for which he had no satisfactory explanation. One was his work with thin films of transparent material — soap bubbles, films of oil floating on water, etc. He observed that light reflected from these thin films exhibited seemingly random bands of color or bright and dark bands. Although these bands were not contradictory to the laws of reflection and refraction, there was nothing in Newton's optical theory to explain them.

One of Newton's contemporaries, Dutch physicist Christian Huygens (1629–1695), argued in his *Treatise on Light* that light is a form of wave motion. Huygens' wave theory did not receive wide acceptance, however, partly because most experiments reported at the time did not require the wave theory for explanation. Also, the corpuscular theory probably was most popular among scientists because it was advanced by Newton.

Others began to duplicate Newton's experiments, including those for which there was no satisfactory explanation. At the end of the eighteenth century, Thomas Young (1773–1829), an English physician, postulated that light had a wave nature and that the color fringing of sunlight observed through a pinhole, one of Newton's unexplained experiments, was an interference phenomenon. In a subsequent experiment, described in the next section, Young demonstrated the interference of light very convincingly. As other scientists reproduced Young's results, the corpuscular theory of light was gradually abandoned as the wave theory gained broader acceptance. Actually, we will see in Chapter 28 that light displays both a particle and a wave nature.

Light waves and indeed all electromagnetic waves obey the superposition principle first encountered in Chapter 11. You will recall that this means that the disturbance at any point due to two or more waves passing through that point at the same time is equal to the vector resultant of the individual disturbances. The physical effects of the superposition of two or more waves is called **interference**. Figure 26-1 is an example of interference of water waves. When the crests or the troughs of two different waves are superimposed at a point, the resultant disturbance is larger than that of either of the waves individually. This is called **constructive interference**. If, however, the crest of one wave is superimposed on the trough of a second wave, the resultant disturbance is smaller than that of the larger-amplitude wave, and we have what is known as **destructive interference**. If the two interfering waves have the same amplitude, constructive interference results in a wave of twice the amplitude of either, while destructive interference results in no disturbance at all (Figure 26-2).

Light waves from two sources incident on the same point of a wall or projection screen will interfere with each other, since the principle of superposition applies. However, for the interference to be observed, it must remain constant for a long-enough period of time, perhaps a second or more. This means that the two disturbances must maintain a constant phase relationship at

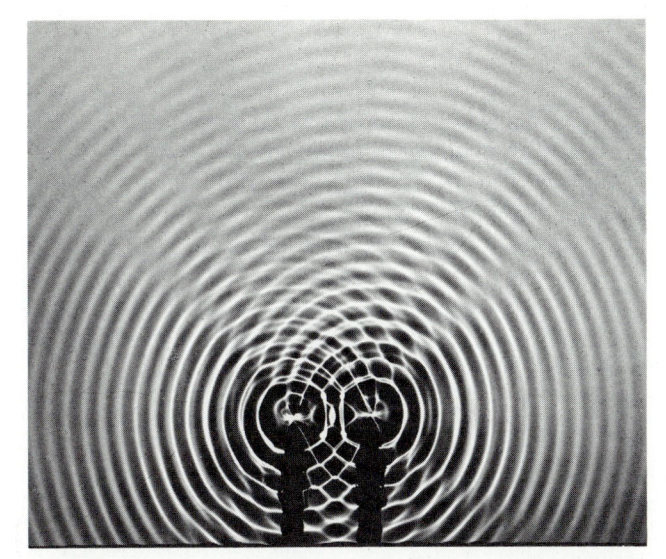

Figure 26-1 Circular waves spread from two centers of disturbance in a ripple tank. Where the patterns overlap, interference occurs. *(B. Abbott/Photo Researchers)*

the point where the interference takes place and is observed. Two waves which maintain a constant phase difference are said to be **coherent.** An essential condition for coherence is that the two waves involved have the same frequency (and wavelength). Thus, for many of the interference phenomena we describe, we will postulate light of a single frequency or color. You will recall that this is called monochromatic light.

26-3 Young's Double-Slit Experiment

The best-known demonstration of the interference of light, hence, a wave nature to light, is the double-slit experiment of Thomas Young. One of the major problems encountered in attempting to show interference of two light waves is finding or fabricating coherent sources. Young solved this problem by using only one source placed in front of an opaque material containing a single narrow slit. After passing through the very narrow slit, the light spreads out as described in Section 26-1.[1] Young then placed a second piece of opaque material with two closely spaced narrow slits between the single slit and a screen. A top view of Young's experiment is illustrated in Figure 26-3. We see that a given wavefront of light from the single slit arrives at each of the double slits at the same time. Thus the light leaving the two slits is always in phase, and the slits act as two coherent sources. Once again, the slits cause the light to spread out so that beyond the slits in the direction of the screen, light from each slit crosses over light from the other, and the principle of superposition applies.

To reach the position on the screen directly opposite the midpoint be-

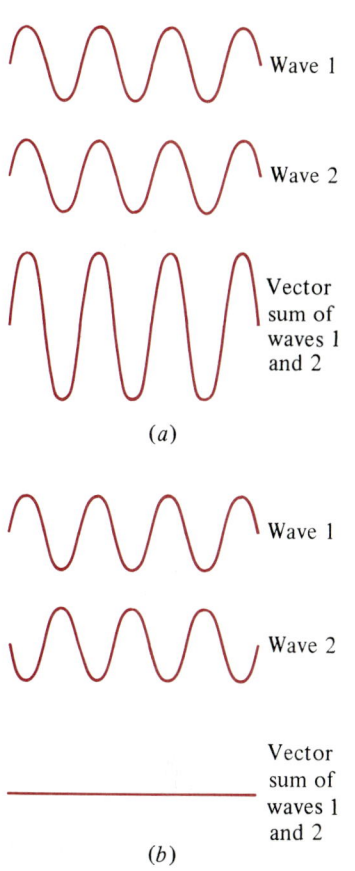

Figure 26-2 *(a)* Total constructive interference: waves of the same frequency and in phase with each other add to produce a disturbance which is the sum of the amplitudes of the waves. *(b)* Total destructive interference: waves agreeing in frequency and having the same amplitude but being 180° out of phase completely cancel each other to leave an undisturbed medium.

---

[1] This phenomenon is known as **diffraction,** or bending and spreading of a wave as it passes the edge of a barrier. We will deal more fully with diffraction in Section 26-5.

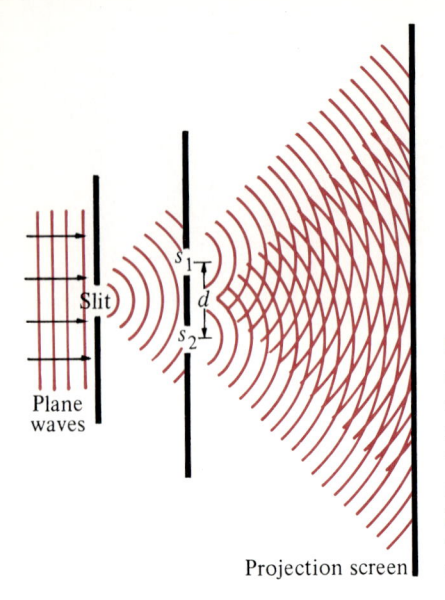

Figure 26-3 In Young's experiment, the spreading waves of light from two slits cross each other much as the waves in Figure 26-1. At places on the screen where the waves arriving from each slit agree in phase, constructive interference produces a bright spot. Where arriving waves are 180° out of phase, destructive interference occurs.

Figure 26-4 A ray diagram from Young's experiment. The difference in path length Δs defines a small right triangle near the slits. The hypotenuse is d, the separation of the slits; the smaller acute angle opposite side Δs is θ. This triangle is similar to that formed by the central axis, the screen segment over to P, and the dashed line joining P with the central point between the slits.

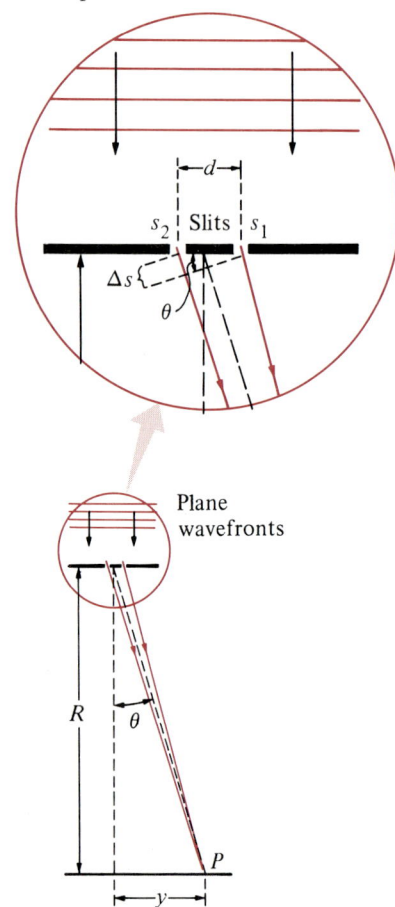

tween the two slits, the light from each slit must travel the same distance. Thus wave crests from each arrive at the central position at the same time, troughs arrive at the same time, and so on. In other words, light waves from the two slits are always in phase when they arrive at that position, and constructive interference results. At any point along the screen in either direction from the central position, light travels a different distance from one slit than from the other. This path difference may cause waves from the two slits to be in phase or out of phase when they arrive at the point in question, and constructive or destructive interference may occur.

Figure 26-4 shows rays proceeding from each of two slits of Young's experiment to a point P on a projection screen situated a distance R below the slits. The ray from slit s_2 is longer than that from slit s_1 by a distance Δs. A line drawn to point P from the midpoint between the slits makes an angle θ with the perpendicular to the screen. From the geometry, we see that the other labeled angle is also θ. Hence a little trigonometry gives us the relationship between the path difference of the rays at the angle θ and separation between slits d. We obtain

$$\Delta s = d \sin \theta$$

Let us assume that monochromatic light of the single wavelength λ (Greek lowercase lambda) is being used. (Actually, Young used sunlight, which is essentially all visible wavelengths, and so obtained multicolored fringes.) If Δs is exactly one wavelength long, then the waves from both slits arrive at point P in phase, and constructive interference results. Of course, they will be in phase if Δs is exactly two wavelengths long, or three wavelengths, or for that matter, any integral (whole) number of wavelengths m. Thus the condition for constructive interference at P is that $\Delta s = m\lambda$, or

$$m\lambda = d \sin \theta_m \qquad m = 0, 1, 2, 3, \ldots \qquad (26\text{-}1)$$

We have placed the subscript m on the angle θ as a reminder that θ will have a different value for constructive interference for each value of m. Therefore, in addition to obtaining constructive interference at the central position, we have constructive interference on either side at $\theta_1, \theta_2, \ldots, \theta_m$, where θ_m is defined by Equation 26-1. The integer m is called the **order number** of the interference. Thus first-order constructive interference occurs at $m = 1$, second order at $m = 2$, and so on. (The central constructive interference band, or **central maximum**, is called the zeroth-order interference maximum.)

Using vertical slits, we obtain vertical bright bands, images of the slits, spread out horizontally on the screen. The pattern on the screen looks somewhat like that of Figure 26-5. The physically inescapable fact is that light adds to light, at some points producing brightness and at other points producing darkness.

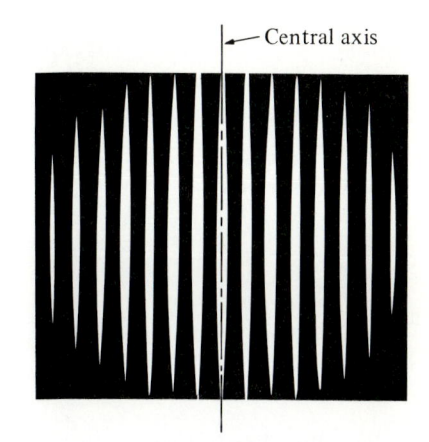

Figure 26-5 The pattern cast by a double slit in Young's experiment. The bright band on the central axis is called the central maximum ($m = 0$). If the light leaving the slits is in phase, this position will always be bright since it is equidistant from the two slits. Left and right from the central maximum are bands of constructive interference ($m = 1, 2, 3, \ldots$) separated by dark zones of destructive interference.

Example 26-1 Yellow light of wavelength 5.90×10^{-7} m is to be used in a double-slit experiment. Fourth-order constructive interference is to occur at an angle of $6.00°$. Calculate (a) the slit separation which should be used and (b) the angle at which third-order constructive interference will occur if these same slits are used with red light of wavelength 6.50×10^{-7} m.

Solution Both parts are straightforward applications of Equation 26-1.

(a) $$m\lambda = d \sin \theta_m$$

or

$$d = \frac{m\lambda}{\sin \theta_m} = \frac{4(5.90 \times 10^{-7} \text{ m})}{\sin 6.00°} = 2.26 \times 10^{-5} \text{ m}$$

(b) This requires that we use the slit separation calculated in part (a).

$$\sin \theta_m = \frac{m\lambda}{d}$$

$$\sin \theta_3 = \frac{3(6.50 \times 10^{-7} \text{ m})}{2.26 \times 10^{-5} \text{ m}} = 8.63 \times 10^{-2}$$

$$\theta_3 = 4.95°$$

Example 26-1 shows that the slit separation is quite small, less than 0.03 mm for that case. Note also the small angles involved. Neither the values of slit separation nor the angles are untypical. We may obtain a relationship which includes the distance of the mth band along the screen. Take the distance between the middle of the central maximum and the middle of the mth constructive interference band to be y_m and the distance from the slits to the screen as R. Then we may write

$$\tan \theta_m = \frac{y_m}{R}$$

Since for small angles we may make the approximation

$$\sin \theta_m \approx \tan \theta_m$$

Equation 26-1 becomes

$$m\lambda = d \frac{y_m}{R} \qquad (26\text{-}2)$$

Example 26-2 In a double-slit experiment, the distance of the slits from the projection screen is 2.00 m. Calculate the distance from the central maximum to each of the bands of interest in Example 26-1.

Solution Equation 26-2 is first solved for y_m and then applied to each of the situations described.

$$m\lambda = d\frac{y_m}{R}$$

yields

$$y_m = \frac{m\lambda R}{d}$$

For the yellow line in fourth order, we have

$$y_4 = \frac{4(5.90 \times 10^{-7}\text{ m})(2.00\text{ m})}{2.26 \times 10^{-5}\text{ m}} = 0.209\text{ m}$$

For the red line in third order,

$$y_3 = \frac{3(6.50 \times 10^{-7}\text{ m})(2.00\text{ m})}{2.26 \times 10^{-5}\text{ m}} = 0.173\text{ m}$$

A quick analysis also gives us the condition for destructive interference, which produces a dark band. Looking once again at Figure 26-4, we see that if Δs is exactly one-half wavelength, the waves from the two slits will be out of phase and will cancel each other. We also expect that the two waves will always be out of phase when they arrive at point P if the path difference Δs is exactly $1\frac{1}{2}$ wavelengths or exactly $2\frac{1}{2}$ wavelengths. In fact, destructive interference will occur whenever the path difference is any positive integer minus one-half wavelength. Thus for destructive interference in a double-slit experiment, we have

$$(m - \tfrac{1}{2})\lambda = d\sin\theta_m \qquad m = 1,2,3,\ldots \qquad (26\text{-}3)$$

We write this equation as we do so that m may designate the order number of the dark band (the destructive interference). There is no zeroth-order dark band, and m may not take on the value zero.

If we use the small-angle approximation once more, we also obtain

$$(m - \tfrac{1}{2})\lambda = d\frac{y_m}{R} \qquad (26\text{-}4)$$

where y_m is the distance from the middle of the central maximum to the middle of the mth-order dark band and R is the distance from the slits to the screen.

The most convincing feature of Young's experiment as regards the wave nature of light might be the destructive interference. If one fixes one's eyes on a dark band during a double-slit experiment while one slit is being covered, one can see the space where the dark band had been become illuminated. Uncovering the closed slit causes the dark band to return. In essence, two beams of light are added together to produce darkness. This certainly cannot be explained by the corpuscular theory of light, and thus it is not surprising that the wave theory grew in acceptance.

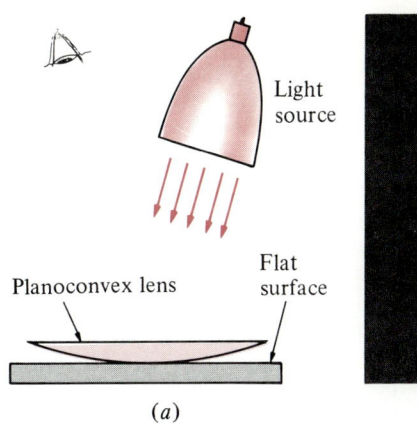

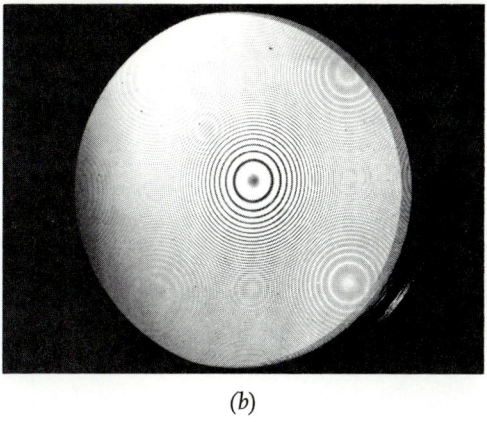

Figure 26-6 (*a*) The air wedge of varying thickness between the surfaces produces Newton's rings. (*b*) The resulting interference pattern. (*Bausch and Lomb*)

26-4 Thin-Film Interference

We have all seen the rainbowlike pattern of light reflected from a thin film of oil floating on a puddle of water. In his optics experiments, Newton recorded his observations of such colored patterns from soap bubbles, oil films, and air wedges. He refined the air wedge experiment by placing a planoconvex lens in contact with a very flat surface to obtain a distinctive pattern of rings from reflected light. These are now known as Newton's rings (Figure 26-6).

Newton's attempts to explain these thin-film phenomena were based on the behavior of particles. His suggestion that perhaps the particles of light excite vibrations in the material as they pass through, while inventive, was never really convincing. An acceptable explanation awaited the wave theory.

Figure 26-7 shows light incident on a thin film and partially reflected from each of the film's two surfaces. The light which reaches the eye is made up of two parts: light reflected from the top film surface and light which passed through the top film surface, traveled through the film, was reflected at the bottom film surface, and returned through the film to the eye. These two parts, which we consider as separate rays or waves, are superimposed at the eye and produce a **thin-film interference** phenomenon. The interference may be constructive or destructive or something in between, depending on the wavelength and the length of the path difference. The type of interference depends on the phase difference between the waves when they reach the eye. In addition to wavelength and path-length difference, there is another factor which affects the phase difference.

You may recall from Chapter 11 that when a traveling mechanical wave encounters a medium of higher density, the reflected wave changes phase by 180°, but a traveling wave reflected from a medium of lower density will not undergo a phase change. The same behavior occurs with light waves, but it is the *optical density,* or index of refraction, of the two media which determines whether or not a phase change takes place. That is, if the medium from which the light reflects has a larger index of refraction than the medium in which the light is traveling, there will be a 180° phase change. If the light is reflected from a medium which has a smaller index of refraction, there will be no phase change.

The most general case to be considered for a thin film is one for which the three media involved—the medium on one side of the film, the film itself, and

Figure 26-7 Part of the incident ray is reflected to the eye and part is refracted into the film. At the back surface of the film, part of the refracted ray is reflected back to the upper surface and refracted toward the eye parallel to the original reflection. These two reflected rays are involved in thin-film interference.

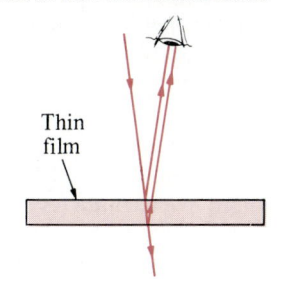

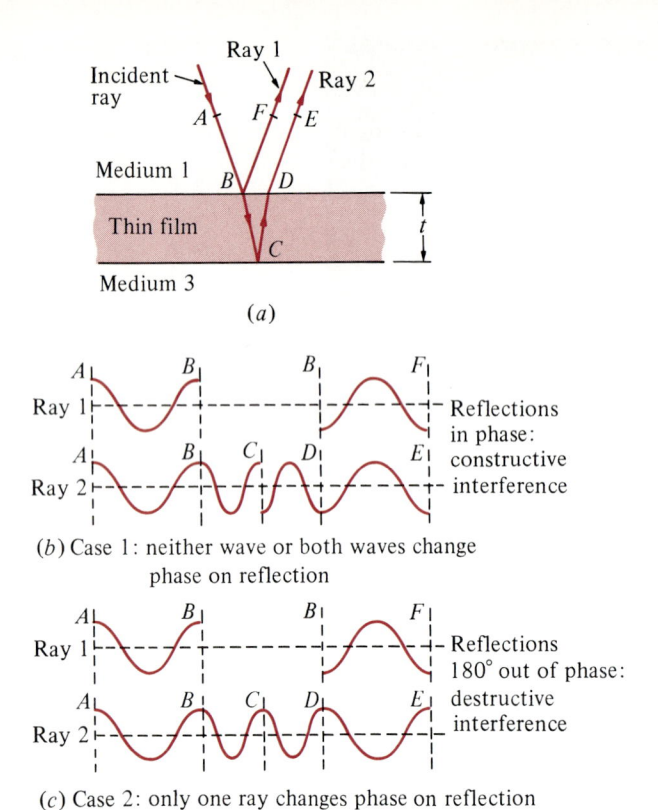

Figure 26-8 (a) Thin-film interference. From points F and E the two rays travel an identical distance to the eye, and so whatever phase relationship occurs at these points will be that observed. (b) The phases of the waves along their respective paths at the instant a crest is at A. AB is common to both rays, and at point B, ray 1 reflects and proceeds toward F. Since the film has a higher index of refraction than the air, ray 1 inverts 180° at B, with the arriving crest becoming a trough. Ray 2 enters the film, acquires a new wavelength, proceeds to C, reflects, inverts, and proceeds to D, is refracted, and moves to E. The rays at F and E are in phase, both troughs, and would be for any phase frozen at A. (c) Because medium 3 has a lower index of refraction than the film, the reflection at C does not invert in phase and so the rays at E and F are 180° out of phase.

the medium on the other side of the film—all have different indices of refraction. A film of oil floating on water and a thin layer of air trapped between glass and water are both good examples. Figure 26-8 depicts such a situation. The thickness of the film is t, and we take its index of refraction to be n. We have ignored possible multiple reflections and concern ourselves only with the interference of the first reflected rays at each side of the film. For a thin film of uniform thickness, there are only two possible interference cases to consider. The first case occurs whenever the phase difference between the two rays is entirely due to the increased path length of the ray in the film medium over the other ray. This happens if neither ray changes phase on reflection or if both change phase. The second interference case occurs whenever the phase difference is due to the path difference of the two rays plus 180° (or one-half wavelength). This happens whenever only one of the two rays changes phase on reflection.

Let us consider the first case (neither or both change phase) and refer to Figure 26-8a and b. The additional path traveled by ray 2 (B to C to D) is about equal to twice the thickness, or $\Delta s = 2t$. If that path difference is an integral multiple of wavelengths, then the two rays will be in phase as they leave the top surface, and constructive interference will result. We have a slight additional complication, however. The wavelength of light in the film is not equal to its wavelength in a vacuum (or air). However, we know that light of wavelength λ in a vacuum has a wavelength of λ/n in a medium with an index of refraction of n. (Part of Problem 24-18 required that this be proved.) We write this as

$$\lambda_n = \frac{\lambda}{n}$$

where λ_n is the wavelength of light in a medium with index of refraction equal to n and λ is its wavelength in a vacuum or air. Thus the condition of constructive interference for the first case is

$$\Delta s = m\lambda_n \qquad m = 0,1,2,3, \ldots$$

or

$$2t = m\frac{\lambda}{n} \qquad m = 0,1,2,3, \ldots \qquad (26\text{-}5)$$

Equation 26-5 is the condition for constructive interference when neither ray undergoes a phase change on reflection or both do.

The condition for destructive interference follows rather quickly. If the path difference is an integral multiple plus one-half a wavelength, the two rays will be out of phase, and destructive interference occurs. That is

$$\Delta s = (m + \tfrac{1}{2})\lambda_n \qquad m = 0,1,2, \ldots$$

Thus

$$2t = (m + \tfrac{1}{2})\frac{\lambda}{n} \qquad m = 0,1,2, \ldots \qquad (26\text{-}6)$$

is the condition for destructive thin-film interference when neither or both rays undergo phase changes on reflection. Can you justify including $m = 0$ as a possibility for both Equations 26-5 and 26-6?

The second interference case occurs when one or the other (but not both) of the rays changes phase on reflection (Figure 26-8a and c). Constructive interference requires that the path difference be one-half wavelength or three-halves wavelength or any odd integral multiple of $\lambda/2$. Thus

$$2t = (m + \tfrac{1}{2})\frac{\lambda}{n} \qquad m = 0,1,2, \ldots \qquad (26\text{-}7)$$

is the condition for constructive thin-film interference when only one of the rays changes phase on reflection. Note that this condition is identical to that for destructive interference when neither or both rays change phase as given by Equation 26-6. Similarly,

$$2t = m\frac{\lambda}{n} \qquad m = 0,1,2, \ldots \qquad (26\text{-}8)$$

is the condition for destructive thin-film interference when only one of the rays changes phase and is identical with Equation 26-5.

Example 26-3 Calculate the possible thickness of a coating of material of index of refraction 1.40 which when applied to a flint glass lens ($n = 1.65$) will cause destructive interference on reflection of yellow light ($\lambda = 5.90 \times 10^{-7}$ m).

Solution We first determine the case with which we are dealing and then apply the destructive interference condition for that case.

We note that the light is traveling in air, and part of it is reflected at the interface between air and the coating. Since the index of refraction of the coating is greater than that of air, that reflected part of the light changes phase. Some of the light which continues in the coating is reflected at the interface between the coating material and the lens. Since the flint glass lens has a larger index of refraction than the coating, this reflection causes a phase change also. Thus, both reflected rays will have changed phase, and we are dealing with the first case. We therefore apply Equation 26-6, the condition for destructive thin-film interference when both rays undergo phase changes.

$$2t = (m + \tfrac{1}{2}) \frac{\lambda}{n} \qquad m = 0, 1, 2, \ldots$$

Almost all fine lenses have many elements and are coated as in Example 26-3. This coating acts to reduce reflections from the surfaces of the individual elements, hence multiple images from repeated reflections. Equally important, it acts to preserve the light energy transmitted at each surface.

Solving for t, we obtain

$$t = (m + \tfrac{1}{2}) \frac{\lambda}{2n}$$

or

$$t = (2m + 1) \frac{\lambda}{4n} = (2m + 1) \frac{5.90 \times 10^{-7} \text{ m}}{4(1.40)}$$

$$= (2m + 1)(1.05 \times 10^{-7} \text{ m})$$

The smallest thickness of coating material is thus 1.05×10^{-7} m. Larger thicknesses which will work are 3.15×10^{-7} m, 5.25×10^{-7} m, ...

26-5 Diffraction (Single-Aperture)

The word *diffraction* as applied to wave motion means the bending of the wave as it passes an obstacle. With water waves, this bending is easy to see, since the spreading out of waves after they pass through a small opening is obvious (see Figure 26-9). Nor is it difficult to accept the diffraction of sound waves. If someone talks to you from another room, you do not have to be standing in front of the doorway to hear the sound.

The diffraction of light is not nearly so easy to detect because the bending is not very pronounced. However, under many circumstances, diffraction of light gives rise to interference effects which can be detected. The bending of light as it passes an obstacle or passes through an aperture may cause light from two different parts of the same source to give rise to a detectable interference pattern.

You may observe this type of interference effect due to diffraction rather easily. Hold your hand a few centimeters from your face, and look with one eye toward a light source through a tiny crack between your fingers. When your fingers are touching at some points along the crack, you can see the fringe effects in the spaces which are not touching. The dark lines parallel to your fingers in the slit constitute the interference effect due to the diffraction of light at the slit.

We may analyze diffraction by using Figure 26-10, which shows a top view of an opaque material in which a thin slit of width w allows light to pass. Plane waves of monochromatic light are incident on the slit. Consider each point in the aperture as a separate source of light. Thus light proceeds from each point (actually lines perpendicular to the page) of the aperture in all directions toward the screen. Although only three rays are shown to an arbitrary point P on the screen, we could draw rays from any point in the aperture. Consider two points, one very near the edge of the slit and one very near its center. The path difference of light from these two points to P is

$$\Delta s = \frac{w}{2} \sin \theta$$

If this path difference happens to be exactly equal to $\lambda/2$, light from these two points will be out of phase at point P, and destructive interference will occur. For another point near the edge of the slit but slightly closer to the center, there is another corresponding point with the same displacement from the center which will produce the same path difference of rays, hence, the same phase relationship as the first two points. In fact, for every point along one-half the aperture, there is a corresponding point along the other half which will yield the same path difference. Thus, if the original path difference happens to be $\lambda/2$, there will be a complete cancellation at P, and we observe a dark band. When this occurs, we may write

$$\frac{\lambda}{2} = \frac{w}{2} \sin \theta$$

$$\lambda = w \sin \theta \qquad (26\text{-}9)$$

is a condition for destructive interference due to the diffraction at a single slit.

The slit could be divided into fourths, or sixths, etc., and the same argument would lead to the conclusion that destructive interference occurs whenever $w \sin \theta = 2\lambda$, or 3λ, or any integral number of wavelengths. The general condition for destructive interference due to diffraction by a single slit is

$$m\lambda = w \sin \theta_m \qquad m = 1,2,3,\ldots \qquad (26\text{-}10)$$

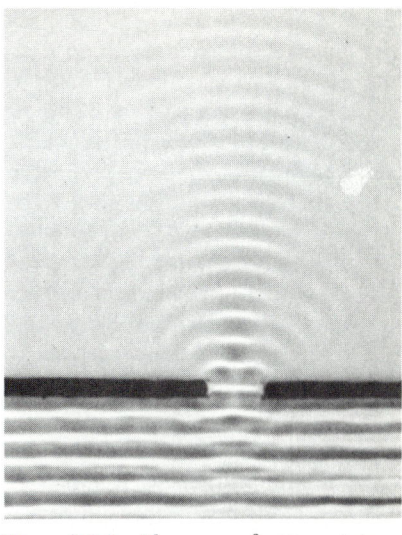

Figure 26-9 Plane wavefronts arriving at a slit are "bent" around into the shadow area by diffraction. (D. C. Heath/Education Development Center)

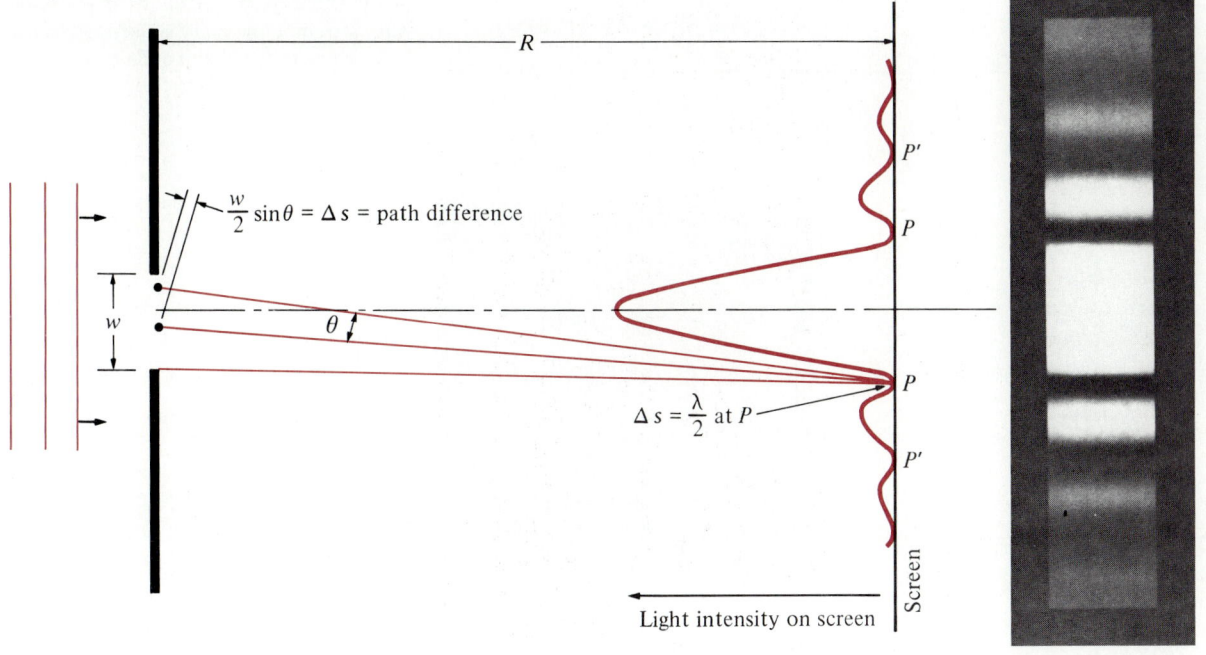

The integer m is the order number of the dark band and may take on only positive values. Note that an edge view of a projection screen is represented in

Figure 26-10 A single slit of width w may be treated as though it were a large number of separate, closely spaced slits. Such a single slit produces a pattern of alternate bright and dark areas on a screen. Although this pattern is due to interference, it is most often called a diffraction pattern. The pattern is depicted as an intensity-vs.-position curve drawn with intensity increasing toward the slit and with the horizontal position of the screen representing zero intensity. On the other side of the screen is a photograph of the light pattern formed.

651

Figure 26-10 by a vertical line a distance R away from the slit. The curved pattern touching the screen at the six points shown and drawn toward the slit represents the light intensity at a position on the screen versus the vertical coordinate of that position. To the right of the screen is a photograph of such a single-slit diffraction pattern obtained using monochromatic light. The points P and P' on the screen represent the positions of the first-order and second-order dark bands. There is a bright band at $\theta = 0$. This is not too surprising, as we might expect all the light from the slit to reach the center essentially in phase. The position of the minima, or dark bands, depends strongly on the ratio of the wavelength of light to the slit width. This is more easily seen when Equation 26-10 is solved for $\sin \theta_m$. Thus

$$\sin \theta_m = \frac{m\lambda}{w}$$

Note that there will be no dark bands when the width is less than or equal to the wavelength. Further, we see that bright bands must appear between the dark bands. Figure 26-11a through c shows the relative intensity versus angle from the central axis of the light patterns displayed on a screen by light shining through single slits of different widths.

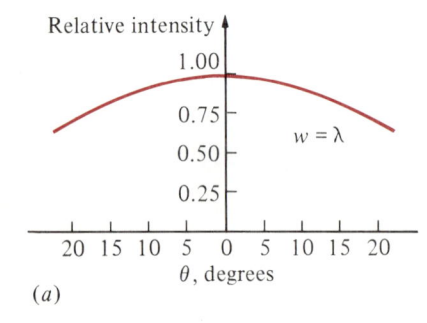

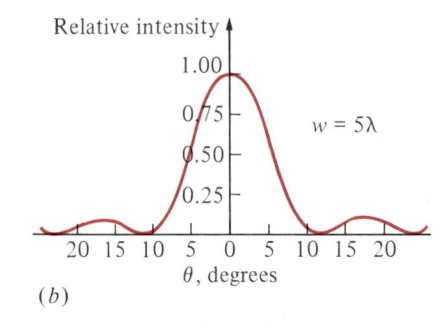

Figure 26-11 Single-slit interference patterns. Relative intensity of monochromatic light projected on a screen after passing through a single slit plotted versus angle from central axis: (a) $w = \lambda$; (b) $w = 5\lambda$; (c) $w = 10\lambda$. (d) If a pinhole rather than a linear slit is used, interference rings rather than bands result.

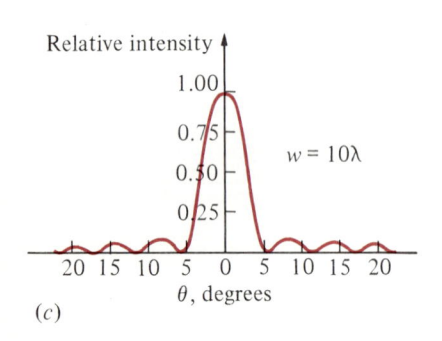

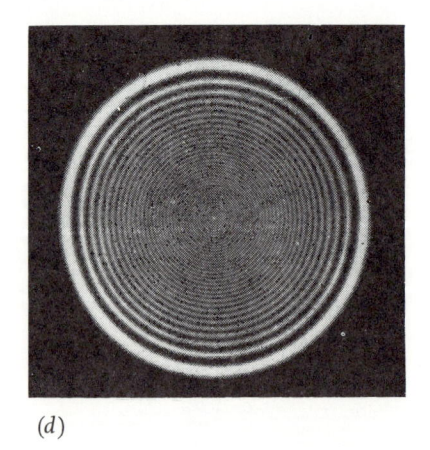

Example 26-4 Using monochromatic light of wavelength 4.50×10^{-7} m, calculate (a) the single-slit width necessary so that the first minimum will occur at an angle of $10°$ and (b) the wavelength of light which causes the first minimum to occur at $15°$ when using this same slit.

Solution Both parts may be solved by a straightforward application of Equation 26-10 with m equal to 1. Part b requires that we use the answer from part a.

(a) $$m\lambda = w \sin \theta_m$$

or

$$w = \frac{\lambda}{\sin \theta_1} = \frac{4.50 \times 10^{-7} \text{ m}}{\sin 10°} = 2.59 \times 10^{-6} \text{ m}$$

(b) $$\lambda = w \sin \theta_1 = (2.59 \times 10^{-6} \text{ m})(\sin 15°) = 6.70 \times 10^{-7} \text{ m}$$

In Figure 26-11d, we see that the light from a distant point source after diffraction at a circular opening forms a circular pattern surrounded by circular fringes rather than being imaged as a point. It can be shown that the first minimum occurs at

$$\sin \theta = 1.22 \frac{\lambda}{d} \qquad \text{(1st minimum, circular)}$$

where d is the diameter of the opening. Note the similarity to the equation for the first minimum due to the diffraction at a single slit, which is

$$\sin \theta = \frac{\lambda}{d} \qquad \text{(1st minimum, slit)}$$

If two stars, separated by a very small angle, are viewed through a telescope (which is a circular aperture), the interference patterns due to diffraction may overlap so much as to make the viewer unable to distinguish (**resolve**) two stars. Figure 26-12 shows an image for which we have no trouble making the

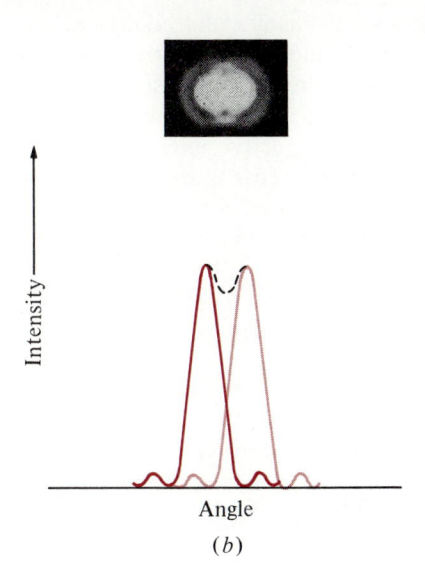

Figure 26-12 Separate circular diffraction images may be resolved if they are far enough apart. (a) If the bright central maxima of the two images are separate, the spots are easily resolved. (b) Rayleigh's criterion: the limit of resolution occurs when the central maximum of one image coincides with the first dark minimum of the second image. (c) Closer-spaced images cannot be resolved because the diffraction limit of resolution of the optical device has been exceeded.

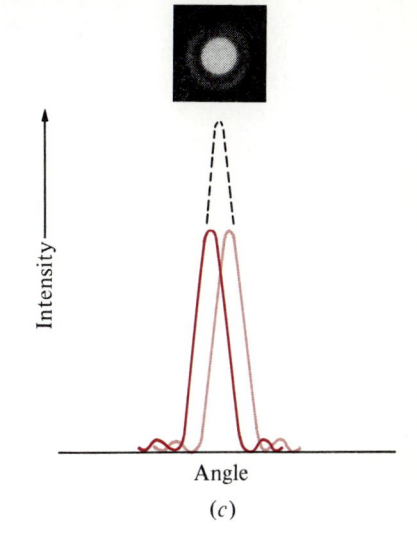

653

resolution as well as an image where resolution is impossible. In the image for which the two points are just barely resolvable, the central maximum of the pattern due to one point source falls on the first minimum of the pattern due to the other. Lord Rayleigh, an English physicist (1842–1919), chose this somewhat arbitrary criterion for deciding when two point objects are just resolved, and it is generally accepted. The angular separation is always so small that we may replace the sine of the angle with the angle itself (in radians). Thus **Rayleigh's criterion** for minimum angular separation for resolution of two point objects after circular diffraction is

$$\theta_R = 1.22 \frac{\lambda}{d}$$

Rayleigh's criterion essentially defines the limit of what is separately "seeable" with an optical instrument of "perfect" design and construction. For example, the 5-m telescope on Mount Palomar could record the light of two candles 15 cm apart and placed 10,000 km distant as coming from two independent light sources. If the candles were three times as distant, the telescope could still collect enough light to detect them, but it could not resolve the image to determine that there were two sources since the diffraction maxima would have merged into each other.

26-6 The Plane Diffraction Grating

Interference effects do not only result from the diffraction effects of one slit or from two slits in a double-slit experiment. Suppose we place many equally spaced slits side by side in a piece of opaque material and illuminate them with plane-wave monochromatic light. Such an arrangement of slits is called a **plane diffraction grating**, and its operation is illustrated in Figure 26-13. Only six slits are shown in the diagram, but an actual laboratory grating may have hundreds or thousands of slits. Initially we assume that the slits are so narrow that each pattern due to diffraction spreads out enough to interfere with patterns from all the other slits.

A converging lens is placed one focal distance from the screen near the grating to ensure that all rays which meet at a point on the screen are parallel when they leave the grating. That is, the **optical path length**[1] from the line AA' to point P is the same for all rays. Since the grating spacing, or the distance between adjacent slits, is d, the path difference of rays from adjacent slits is

$$\Delta s = d \sin \theta$$

If this path difference is an integral number of wavelengths, all rays will be in phase when they reach point P, and a bright band of constructive interference will result. If, on the other hand, this path difference is almost anything else, rays from nonadjacent slits will arrive out of phase, and we obtain a darker band of approximately destructive interference. Destructive interference of light from adjacent slits occurs whenever the path difference is an odd half-integral multiple of the wavelength used. Thus the conditions of constructive and destructive interference for a diffraction grating are

Constructive: $m\lambda = d \sin \theta_m$ $m = 0, 1, 2, \ldots$ (26-11)

Destructive: $(m - \frac{1}{2})\lambda = d \sin \theta_m$ $m = 1, 2, \ldots$ (26-12)

Figure 26-13 A diffraction grating. From the line AA', all rays have equal optical path lengths to point P and arrive at the same time. The angle θ between the grating and the line AA' is the same as the angle the undeviated ray through the optical center of the lens makes with the central axis.

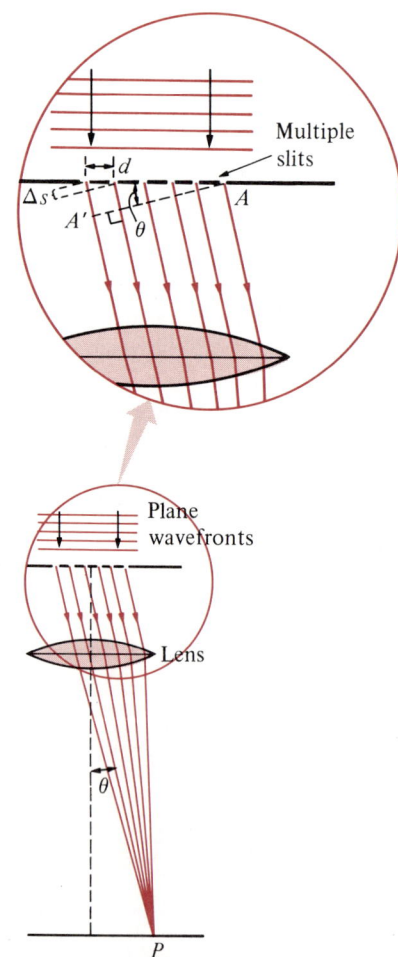

---

[1] The ray from AA' through the optical center of the lens has the shortest **geometrical path length** to point P. But, this ray spends more time in the glass of the lens than does, for example, the ray at the extreme right. Recall that in the glass, the speed of light is smaller than in air, and the wavelength is reduced as well. Thus, measured *either* by counting the number of wavelengths traveled *or* the time of flight, all rays leaving AA' at the same time along the paths shown will arrive at P together. Therefore we say they have the same optical path length.

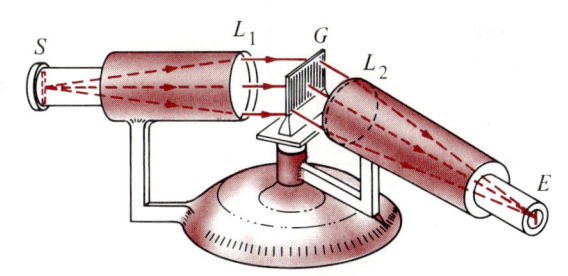

Figure 26-14 A spectroscope is used to determine the wavelengths of light. Light from the source falls on the variable-width slit S placed at the focal point of lens L_1. The lens produces plane wavefronts illuminating the diffraction grating G, with the grating lines parallel to the slit. The movable telescope rotates about the axis of the grating center and uses a lens L_2, which produces a sharply focused image of the slit in each frequency of light present. Fine cross hairs in the eyepiece E allow the telescope axis to be centered on a given line to measure the angle of deviation θ of that line, from which value the wavelength is calculated.

It is not surprising that these equations are the same as those for a double-slit experiment. A double slit may be considered as a special case of a diffraction grating with just two slits.

We see from Equation 26-11 that for a given grating spacing, the **angle of deviation** θ_m for a bright band increases with increasing wavelength. Thus if the light used with a grating has several component wavelengths, the bright bands for the different components will appear at different angles. In other words, the grating will separate multicolored light into its component colors. When a diffraction grating is used in conjunction with an instrument called a spectroscope, the angle of deviation is measured with good precision, and the wavelengths of such component colors may be determined (Figure 26-14).

When substances are heated to incandescence, they often give off light of many colors. When this light is separated by a diffraction grating, it is displayed as a set of different colored bands or lines side by side. As incandescence begins, long-wavelength red rays are the first color observed. Eventually, at high-enough temperature, all colors are present for an incandescent solid or liquid, giving a **continuous spectrum**. Incandescent, low-pressure gases are very different, giving off only certain very specific colors of light. Such a display for a given substance represents what is called an **emission spectrum** for that substance (see Color Plate II). At low density, each element has its own characteristic spectrum of emitted frequencies. Similarly, if white light, representing all colors, is passed through a low-density vapor of the same element, only those frequencies that would be emitted by the incandescent vapor are absorbed. This results in a dark-line spectrum, or **absorption spectrum**. Figure 26-15 is an absorption spectrum of the sun's light. These spectra can often be used to identify the element or elements in an unknown substance.[1]

Example 26-5 A plane diffraction grating has 5000 lines per centimeter of grating width and is used to display the emission spectrum of helium. Calculate

---

[1] The element *helium* is named after Helios, Greek god of the sun. Bright lines in the solar spectrum indicated the presence of an element which had not yet been discovered on earth. Helium was later isolated from pockets of gas trapped deep in the earth.

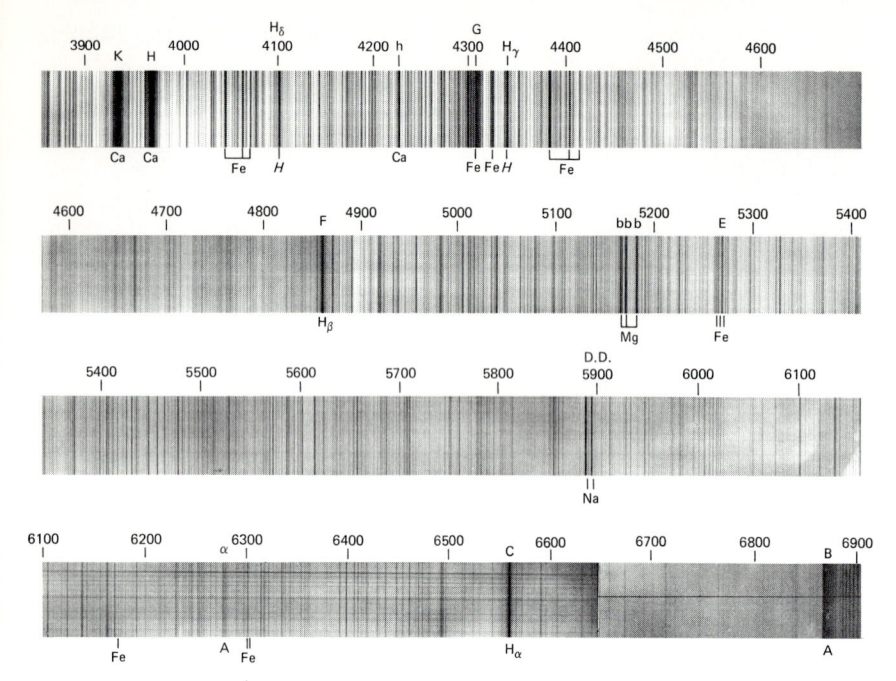

Figure 26-15 Absorption spectrum of the sun, showing dark lines representing "missing" frequencies of light (Fraunhofer lines). Light at the sun's surface presumably contains all colors; the dark lines represent frequencies absorbed by elements in the sun's atmosphere. Letters below each strip identify elements associated with some of the lines. (Hale Observatories)

the angular separation expected in second order between the red line of wavelength 6.68×10^{-7} m and the violet line of wavelength 4.03×10^{-7} m.

Solution We must first determine the grating spacing, which may be found rather easily as it is simply the reciprocal of the number of lines per unit length.

$$d = \frac{1}{5000 \text{ cm}^{-1}} = 2.00 \times 10^{-4} \text{ cm} = 2.00 \times 10^{-6} \text{ m}$$

We now apply Equation 26-11 to both wavelengths for $m = 2$.

$$m\lambda = d \sin \theta_m$$

$$2(6.68 \times 10^{-7} \text{ m}) = (2.00 \times 10^{-6} \text{ m})(\sin \theta_2)$$

$$\sin \theta_2 = 0.668 \qquad \theta_2 = 41.9°$$

$$2(4.03 \times 10^{-7} \text{ m}) = (2.00 \times 10^{-6} \text{ m})(\sin \theta'_2)$$

$$\sin \theta'_2 = 0.403 \qquad \theta'_2 = 23.8°$$

The difference of the angles of deviation is determined by subtraction.

$$\Delta \theta = \theta_2 - \theta'_2 = 18.1°$$

If the widths of the slits in a diffraction grating are not so narrow as to completely spread out the light from each slit, an interesting display occurs: diffraction and interference effects are combined. The diffraction patterns from all slits are superimposed on the screen to form a pattern varying in maximum brightness like the diffraction pattern of a single slit (see Figure 26-11). In the absence of interference effects, light intensity vs. position would vary, as in Figure 26-16a, due to diffraction alone, with minima determined by the ratio of λ to w.

26-6 The Plane Diffraction Grating

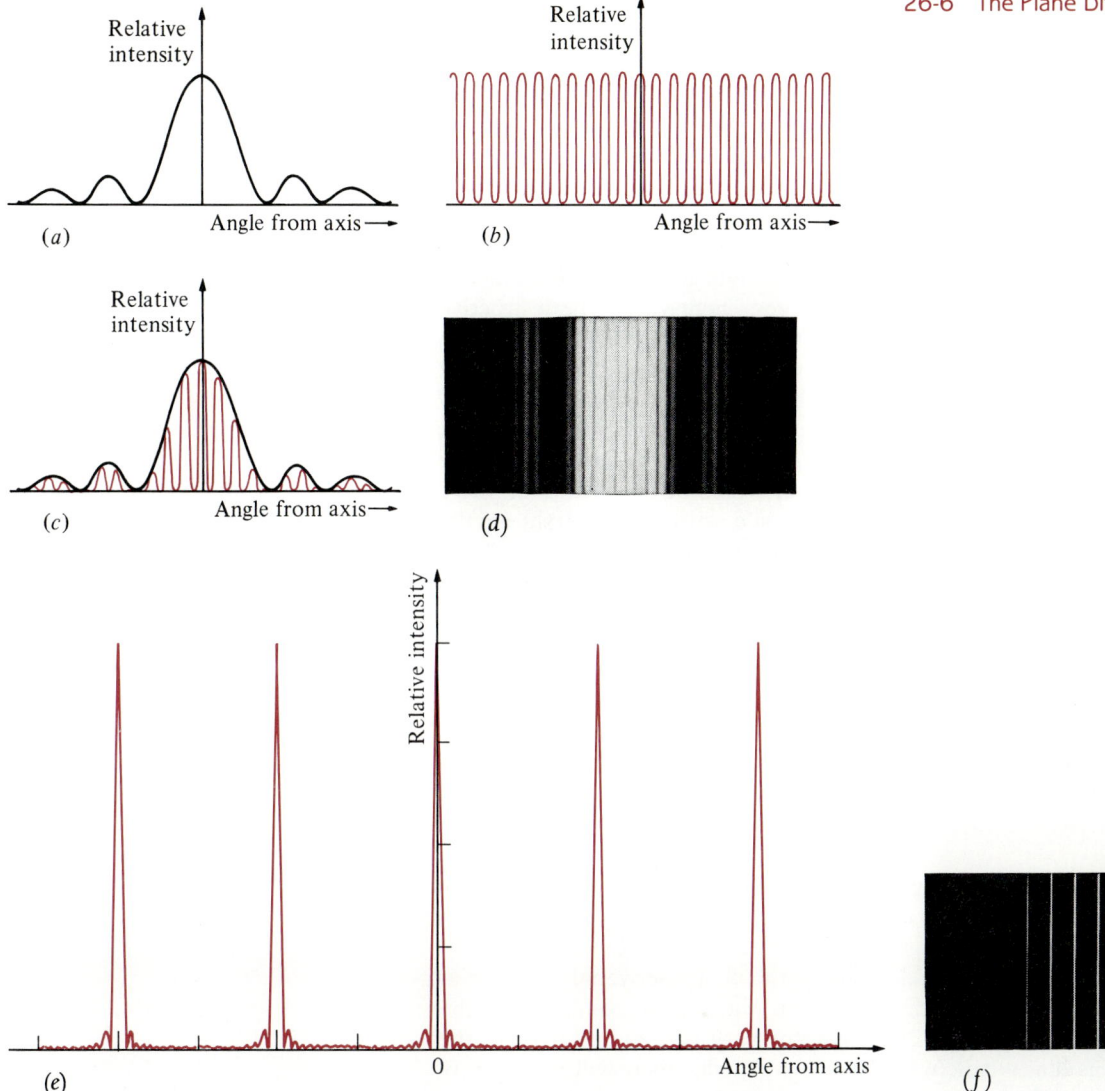

Figure 26-16 Intensity vs. position along a screen for light from (a) a single slit more than several wavelengths wide, (b) a diffraction grating with very narrow slits, and (c) a diffraction grating for which the slits are as wide as those in part a. (d) Photograph of the image produced by (c). (e) Intensity vs position for a diffraction grating of 20 slits. The very high, sharp peaks are the constructive interference maxima described by Equation 26-11. The second peaks are much smaller. (f) Diffraction pattern from a 20-slit grating. Note that the secondary maxima are not even visible.

Let us first consider a grating with only two very narrow slits. For example, take the width of the slit to be much less that the wavelength of the light to be used. The superimposed diffraction pattern for $\lambda > w$ would be so spread out that no diffraction minima would be formed and the variation in light intensity would be due to interference only (Figure 26-16b). Suppose, however, that the two slits of our double-slit grating are wider—perhaps $w = 10\lambda$. Then the

657

interference and diffraction effects similar to those of Figure 26-16a and b would be combined to form the pattern of Figure 26-16c. Figure 26-16d is a photograph of the light pattern produced on a projection screen by such a situation. Note that at each of the single-slit diffraction minima, all light is absent. The single-slit diffraction pattern acts as an envelope for the interference pattern.

As we have already indicated, a typical diffraction grating has many more than two slits, and thus the interference pattern is even further modified. For a grating with many slits, constructive interference maxima still appear at angles given by Equation 26-11 and total destructive interference occurs at angles given by Equation 26-12. Those equations are derived using the interference of light emanating from *adjacent* slits only, however. There will also be destructive interference when light from nonadjacent slits combines in pairs to cancel each other. For example, you can calculate that a grating with four slits spaced at $d = 2.00 \times 10^{-6}$ m and using monochromatic light of wavelength $\lambda = 5.58 \times 10^{-7}$ m will have first-order destructive interference at an angle of 8°. At that angle, light from, say, slit 2 would travel one-half wavelength farther than light from slit 1. They would thus cancel since they would be exactly out of phase when they reached a screen. Light from slit 3 would cancel that from slit 4 in the same manner. Consider, however, what happens at an angle of 4°. Realizing that slits 1 and 3 are separated along the grating by $2d$, we can calculate the path difference for light from these two slits at 4° to be $2(2.00 \times 10^{-6} \text{ m})(\sin 4°) = 2.79 \times 10^{-7}$ m. This is exactly 0.5λ, and therefore light from slit 3 will cancel that from slit 1. The light from slits 2 and 4 will likewise cancel each other at an angle of 4°. Therefore, for a four-slit diffraction grating, we have destructive interference occuring at an angle *smaller* than even that of the *first*-order, by Equation 26-12. The result of adding more and more slits to make a many-slit diffraction grating is to place more and more positions of destructive interference between adjacent constructive interference maxima described by Equation 26-11.

In order to illustrate this, we consider a diffraction grating with 12 slits. We can determine that destructive interference occurs for positions at which light from alternate slits OR from every third slit OR from every fourth slit OR from every sixth slit has a path difference of 0.5λ. These would all occur at angles smaller than the first-order minimum. The general result is that the maxima described by Equation 26-11 become very narrow and most of the space between maxima is dark (with the recognition that there will be some tiny secondary maxima). Figure 26-16e shows the intensity-versus-position pattern for a 20-slit diffraction grating. We see that the maxima have a very large intensity (in other words, they are brighter) as well as being very narrow (sharp). Figure 26-16f is a photograph of such a pattern made with a 20-slit grating. With the thousands of lines typical of a laboratory grating, the minima between these sharp constructive interference lines are so close together that the secondary maxima are essentially nonexistent. Gratings are thus part (with spectrometers) of an excellent tool for determining the characteristic wavelengths associated with materials in the gaseous state. Further, the more lines on the grating, the sharper the image and thus the more precision with which a wavelength may be determined.

Example 26-6 A plane diffraction grating of unknown spacing and slit width is used to examine the mercury spectrum. It is found that the red line of wavelength 6.23×10^{-7} m is deviated at an angle of 20° in second order and the green line of wavelength 5.46×10^{-7} m is missing in the third order. Calculate (a) the grating spacing, (b) the number of lines per centimeter, and (c) the minimum slit width.

Solution (a) This part may be solved by direct application of Equation 26-11, the condition for constructive interference for the grating.

$$m\lambda = d \sin \theta_m$$

or

$$d = \frac{2(6.23 \times 10^{-7} \text{ m})}{\sin 20°} = 3.64 \times 10^{-6} \text{ m} = 3.64 \times 10^{-4} \text{ cm}$$

(b) This is simply the reciprocal of the grating spacing in centimeters.

$$N = \frac{1}{d} = \frac{1}{3.64 \times 10^{-4} \text{ cm}} = 2747 \text{ lines/cm}$$

(c) For this part, we assume that the missing line is at the same angle as the diffraction minimum. We must first calculate the angular deviation of the green line in third order. That angle is then used with the single-slit destructive interference condition to obtain the slit width.

$$\sin \theta_3 = \frac{3(5.46 \times 10^{-7} \text{ m})}{3.64 \times 10^{-6} \text{ m}} = 0.45 \qquad \theta_3 = 26.7°$$

$$\lambda = w \sin \theta$$

$$w = \frac{5.46 \times 10^{-7} \text{ m}}{\sin 26.7°} = \frac{5.46 \times 10^{-7} \text{ m}}{0.45} = 1.21 \times 10^{-6} \text{ m}$$

26-7 Polarized Light

In Chapter 24 we saw that all electromagnetic radiation, including light, is made up of transverse waves. Diffraction and interference phenomena are testimony to the wave nature of light but do not distinguish the type of wave. However, **polarization** selects a particular plane of vibration for a wave perpendicular to the direction of propagation and can only be displayed by transverse waves. Originally used to study the properties of light, polarization is now primarily used in such applications as stereoscopic projection (3-D), polarizing microscopes, the dermascope, eye examinations, photography, colored advertising displays, polarizing ring gunsights, and liquid crystal displays.

You will recall from Section 24-2 that in an electromagnetic wave, both the electric field vector and the magnetic field vector are perpendicular to the direction of propagation. Further, the variation of the electric field vector of the wave is parallel to the axis of the transmitting antenna. If we look along the negative x direction of Figure 24-3, we see the electric field vector increase and decrease in magnitude in the y direction. As the wave propagates, the electric field is confined to the xy plane, and thus this wave is said to be **plane-polarized**. We might depict the electric field vector as viewed from the x axis by the diagram of Figure 26-17a. (The magnetic field vector is not shown, and we will continue this practice to avoid confusion.)

Ordinary light is not polarized. Actually, it is made up of a very large number of waves which are polarized in all planes perpendicular to the direc-

Figure 26-17 (a) A plane-polarized light wave is indicated by a dot showing the approaching wave and arrows in the plane of electric field vibration. (b) An unpolarized approaching ray is indicated by electric field vibration arrows representing all possible planes of vibration perpendicular to the line of the wave.

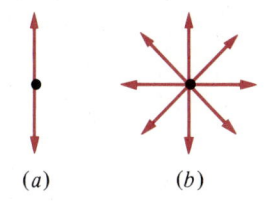

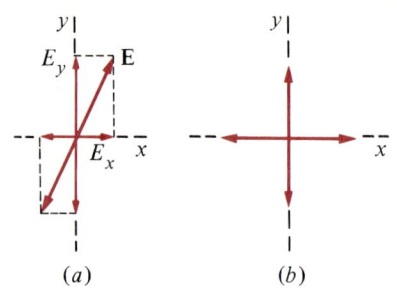

Figure 26-18 (*a*) A light ray with electric field vector vibration in some arbitrary plane may be thought to have components E_x and E_y along each of a set of perpendicular axes *x* and *y*. (*b*) An unpolarized beam may be summarized as being composed of vibration along *x* and *y* coordinate axes only.

tion of propagation. Figure 26-17*b* represents randomly polarized light, or unpolarized light, which may also be represented by two perpendicular components. Consider any one of the arbitrary electric field vectors making up unpolarized light. It may be resolved into two perpendicular components, as shown in Figure 26-18*a*. Therefore the unpolarized beam of Figure 26-17*b* may be represented by the diagram of Figure 26-18*b*.[1]

Plane-polarized light generally must be produced from a beam of unpolarized light. This is accomplished by eliminating all the components of vibration in one plane and passing the components in the plane perpendicular to this. There are several important methods by which this is done.

Selective Absorption Certain crystals (e.g., the mineral tourmaline) possess a property called **dichroism**, which means they strongly absorb components of light polarized in one plane but transmit with little loss those components that are perpendicular to this plane. In the late 1920s, Edwin H. Land, who was then a graduate student, developed an inexpensive plastic sheet which exhibits this selective-absorption property. The material, called Polaroid, was fabricated by embedding a thin layer of tiny needlelike dichroic crystals of herapathite (quinine iodosulfate) in a plastic sheet and stretching the plastic. This caused the crystals to align along the direction of stress, and the sheet became a large-area selective absorber. The sheet was protected by covering both sides with thin transparent plates. In 1938, Land's Polaroid Corporation began manufacturing the improved present-day version in which the herapathite crystals are replaced by long polymeric molecules of polyvinyl alcohol (PVA) stained with iodine. Figure 26-19 depicts how a Polaroid sheet affects a beam of unpolarized light.

Reflection When unpolarized light strikes a dielectric material such as glass or plastic, the reflected light may be partially or completely polarized. Whenever the reflected part of the beam is perpendicular to the portion which is refracted, then the reflected light is completely polarized. Figure 26-20 shows a

Figure 26-19 The **polarizer** allows only those vibration components parallel to one axis to pass. In this case, vibration parallel to the *z* axis is transmitted, while components perpendicular to this direction are absorbed. Note that the crystal axis of the polarizer defines the portion of the vibrations to be absorbed.

The electric field vector of an electromagnetic wave passing a dichroic crystal defines a variable electric force on charges. "Washed" by the wave, electrons on the crystal experience this force. If *E* varies along the crystal axis, the electrons, having some ability to rearrange in response to the force, accelerate and thus absorb energy from the wave. If *E* varies perpendicular to the crystal axis, the electrons cannot leave the crystal in response and hence do not absorb the wave.

---

[1] We remind you that the perpendicular components are each made up of a tremendous number of individual polarized waves which have arbitrary phase relations with one another. Thus these two components have an arbitrary phase difference which is changing in both time and space. Otherwise, we might think these two vectors could be added to yield a constant resultant in a single plane.

beam of unpolarized light striking a flint glass surface at the proper angle such that the reflected and refracted light will be perpendicular to each other. The incident beam is unpolarized, but its electric field vectors are shown as already resolved into components in the plane of the page and those in a plane perpendicular to the page. The small arrows shown represent **E** in the plane of the page, while the heavy dots are to represent the components perpendicular to the page. We see that the reflected beam is 100 percent plane-polarized perpendicular to the page, while the refracted beam is partially polarized in the plane of the page. Figure 26-21 shows how a polarizing sheet may be used to eliminate a specific part of reflected light responsible for glare.

The angle of incidence for which the reflected beam is completely polarized is called the polarizing angle. By using Snell's law and a little trigonometry, we may obtain an expression for this angle. To be most general, we assume light is traveling in a medium of index of refraction n_1 and encounters a medium of index of refraction n_2. Since $\theta_1 = \theta_p$, the polarizing angle, we have

$$n_1 \sin \theta_p = n_2 \sin \theta_2$$

By the law of reflection, $\theta_1' = \theta_1$, or $\theta_1' = \theta_p$. Since the condition for total polarization of the reflected beam may be written as $\theta_1' + \theta_2 = 90°$, then

$$\theta_2 = 90° - \theta_1' \quad \text{or} \quad \theta_2 = 90° - \theta_p$$

Substituting into Snell's law yields

$$n_1 \sin \theta_p = n_2 \sin (90° - \theta_p)$$

But, $\sin (90° - \theta_p) = \cos \theta_p$. Therefore,

$$n_1 \sin \theta_p = n_2 \cos \theta_p$$

Figure 26-21 (a) A waxed wooden surface with back lighting causing a glare. (b) A Polaroid sheet placed on the surface almost completely eliminates the glare, demonstrating that such reflections consist of polarized light. Orientation is important; if the Polaroid sheet is rotated 90°, it will transmit the polarized reflection.

The fact that some or all of the light in a reflection from a shiny surface is polarized explains the major use of Polaroid material in sunglasses. Ordinary sunglasses are darkened filters which absorb a fraction of any light passing through. Polaroid sunglasses are oriented so that when the user's head is held erect, the horizontal component of the light is absorbed. This reduces nonpolarized light by half but almost completely eliminates the glare off a road, a water surface, or the hood of a car.

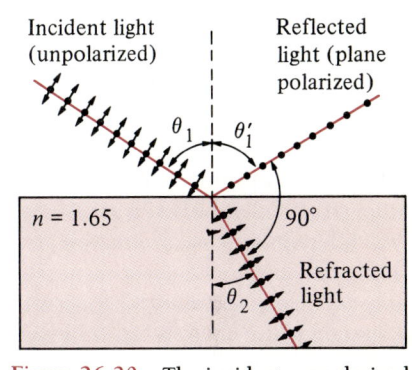

Figure 26-20 The incident, unpolarized ray has electric field vibration as shown as well as into and out of the diagram. At the polarizing (or Brewster) angle θ_1, only the one electric field component is reflected. This occurs when the reflected ray is precisely 90° from the refracted ray in the transparent substance.

661

or

$$\tan \theta_p = \frac{n_2}{n_1} \qquad (26\text{-}13)$$

Equation 26-13 is known as **Brewster's law** and is named after the Scottish physicist Sir David Brewster (1782–1868), who deduced it empirically. Actually, the partial polarization of reflected light was first noticed by the Frenchman Etienne Louis Malus (1775–1812) several years before Brewster's announcement.

Example 26-7 What is the polarizing angle for light traveling in air and incident on flint glass as shown in Figure 26-20?

Solution We simply apply Brewster's law, recognizing that $n_2 = 1.65$ and $n_1 = 1.00$.

$$\tan \theta_p = \frac{1.65}{1.00} = 1.65 \qquad \theta_p = 58.8°$$

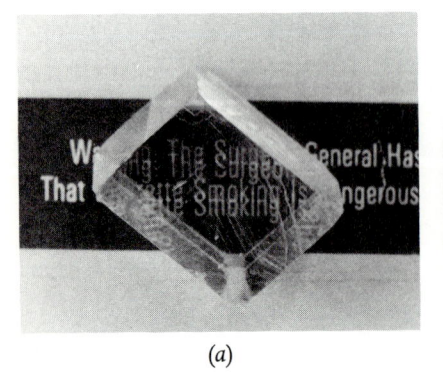

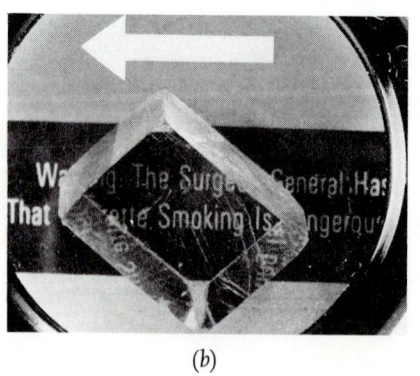

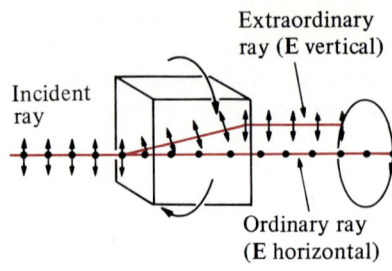

Figure 26-22 A birefringent crystal separates unpolarized light incident perpendicular to its face into an undeviated ordinary ray and a refracted extraordinary ray, each polarized in mutually perpendicular directions. If the crystal is rotated about an axis along the incident beam, the extraordinary ray revolves about the ordinary ray.

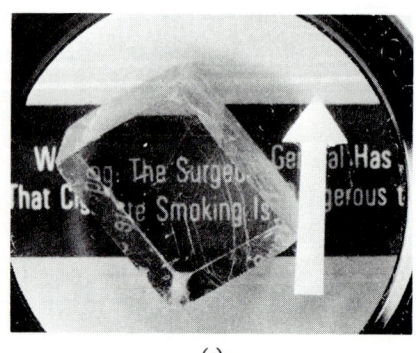

Double Refraction A number of transparent crystals (e.g., calcite) display a property called **double refraction,** or **birefringence,** meaning the material has two different indices of refraction, transmitting the components of light polarized in two perpendicular planes at different speeds. A beam of light incident on a properly oriented birefringent crystal will be split into two beams polarized in perpendicular planes and traveling in different directions in the crystal (Figure 26-22). The two beams are called the **ordinary ray** and the **extraordinary ray,** and the indices for each are so identified in Table 24-1. When a beam strikes the crystal normal to the surface, the ordinary ray proceeds undeviated, while the extraordinary ray may be refracted (Figure 26-23).

Plane-polarized light may be obtained by using a birefringent crystal and separating the two beams. One technique, invented by the Scottish physicist William Nicol (1768–1851) and vastly improved in the late 1960s, involves cementing two prisms, one of glass, the other of calcite, together. If the index of refraction of the glass is the same as that of the ordinary ray in calcite and the

Figure 26-23 (a) The double image visible through a calcite crystal can be distinguished by rotating the crystal; the ordinary image remains in place, while the extraordinary image rotates about it. (b) A polarizer, with its orientation marked, completely passes the ordinary image while the extraordinary one disappears. (c) If the polarizer is rotated 90°, the ordinary image disappears and the extraordinary one is visible.

(a)   (b)   (c)

angles of the prisms are proper, separation can occur. The extraordinary ray will be totally internally reflected at the interface, while the ordinary ray passes undeviated if the unpolarized beam is first incident on the glass part. This device is still most often called a Nicol prism and is essential for those experiments in which the beam must be 100 percent polarized.

Scattering When light passes through a transparent medium, some of the electrons in the material absorb energy from the oscillating electric field and vibrate with the same frequency as the light. The electrons then act as tiny antennas reradiating this energy as light traveling in the medium. This process is called **scattering**. A transmitting antenna does not radiate energy along the direction of oscillation of the charge. That is, if an electron is oscillating in the vertical direction, it will not radiate that energy vertically. This scattering mechanism causes skylight to be partially polarized.

Figure 26-24 shows a beam of unpolarized light passing through a gas. The energy absorbed by electrons that causes them to vibrate along the z direction may be radiated in the positive or negative x direction or in the positive or negative y direction but *not* along either z axis. Likewise, the absorbed energy that causes electrons to vibrate along the x direction may be scattered in the yz plane but not perpendicular to it. Thus, scattered light will be enhanced in light polarized in one plane and diminished in light polarized in the perpendicular plane.

Once plane-polarized light has been produced by one of the above methods, a piece of Polaroid sheet allows us to quickly determine that it is plane-polarized. If, for example, a beam of light is polarized in the vertical plane, a sheet of Polaroid oriented to absorb the vertical component will not pass any light when placed in the beam's path. We often use two sheets of Polaroid with an unpolarized beam. The first sheet becomes the polarizer, and the second is the analyzer (Figure 26-25). The lines do not really appear on the

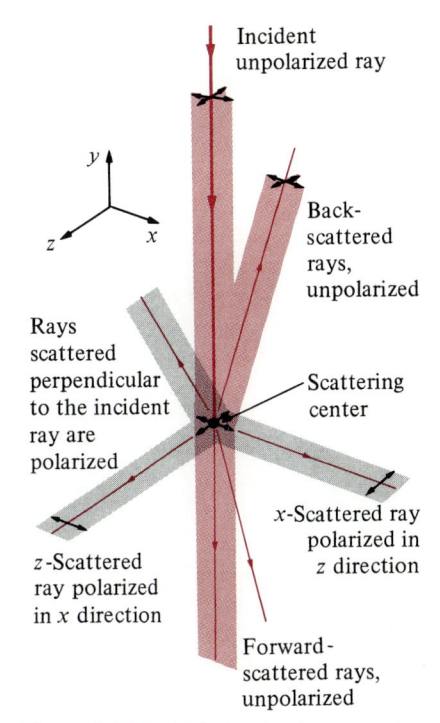

Figure 26-24 Polarization by scattering is an absorption and reemission of energy by an electron in a scattering center. Back-scattered and forward-scattered rays are unpolarized, but rays scattered perpendicular to the incident ray are polarized.

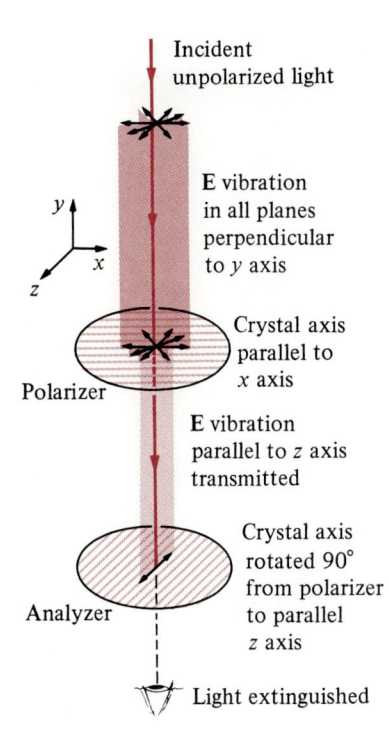

Figure 26-25 A set of polarizing sheets. The polarizer produces light polarized in a plane perpendicular to its crystal direction. The analyzer is rotated to a crystal orientation 90° from the polarizer to extinguish the light.

663

Figure 26-26 (a) A Polaroid sheet passes about 50 percent of the light incident on it. If two sheets are held crossing each other with their polarizing axes parallel, there is negligible additional loss of light due to lack of transparency. (b) If the sheets are crossed with polarizing axes perpendicular, the transmitted light is extinguished.

polarizing sheets. We have simply shown them to indicate the direction of alignment of the dichroic crystals and hence the absorbing direction. In the diagram, the absorbing directions of the sheets are perpendicular, or crossed. If the sheets had been aligned in the same direction, the analyzer would pass nearly all the light incident on it (Figure 26-26).

Whenever the direction for which an analyzer will pass polarized light makes an arbitrary angle, say θ, with the plane of polarization of incident light, only part of the light is absorbed and part is passed. In Figure 26-27, we see a head-on view of this situation. The crystals of the sheet are aligned in the horizontal direction, and thus the sheet will pass light which is polarized in the vertical plane. If the amplitude of the incident beam is E_0, then the amplitude of the transmitted beam is

$$E_y = E_0 \cos \theta$$

Since wave intensity is proportional to the square of the amplitude, the transmitted beam intensity is given by

$$I = I_0 \cos^2 \theta \qquad (26\text{-}14)$$

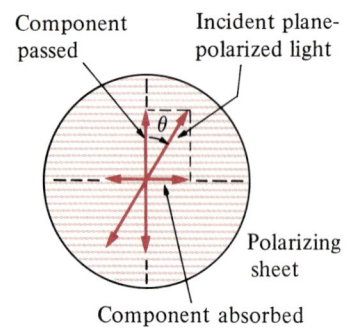

Figure 26-27 A wave plane-polarized at an arbitrary angle θ to the vertical encounters a polarizer with crystals aligned in the horizontal direction.

664

where I_0 is the intensity of the incident plane-polarized beam. Equation 26-14 was also discovered experimentally by Malus and is now known as the **law of Malus.**

The blue color we see in the sky is the result of scattered sunlight. Whenever light is scattered by molecules, or even by dust particles small compared to the wavelength of light, the scattering is proportional to the reciprocal of the fourth power of the wavelength. Thus, shorter-wavelength blue light is scattered much more effectively than longer-wavelength red. This not only gives the sky its blue color but also makes the sun redder in color when it is low in the sky or seen through a dust layer. At sunrise and sunset, the sunlight must pass through a much greater length of air to reach the eye. More of the blue is scattered out, leaving the direct light richer in red.

Because it results from scattering, the blue of the sky is partially polarized. The maximum effect is detected when the sky is viewed perpendicular to the direction of the sun's rays. Aviators determine the direction of the sun from the polarization of the blue sky, particularly near the magnetic poles where an ordinary magnetic compass is not much help.

Experimental evidence is conclusive that bees and other insects depend heavily on the polarizing angle of blue skylight (along with an internal clock) for navigation. Bees can navigate when the sun is cloud-covered provided some blue sky is visible, while placing a polarizing sheet over the hive causes bees to orient themselves by the polarization angle as seen through the polarizer. Bees' eyes are made up in part of birefringent material arranged in segments.

Many transparent materials have the property, called **optical activity,** of rotating the direction of the plane of polarization of light traversing the material. (See the Special Topic at the end of this chapter.) An optically active substance placed between crossed polarizers will rotate the plane and cause light to be transmitted where it had been extinguished. The second polarizer must then be rotated to a new position to extinguish the light passing through the optically active material. The amount of rotation depends on the concentration of optically active molecules and the path length of plane-polarized light in a material. Thus, this property can be used to test for the amount of sugar in syrups or sugar solutions. **Dextrorotary compounds** cause the plane of polarization to rotate clockwise as the beam approaches the eye. **Levulorotary compounds** rotate in the left-handed sense. The sugars dextrose (glucose) and levulose (fructose) derive their names from their optical activity and the latin words for right and left. Many kinds of glass and transparent plastics become optically active when placed under stress. Further, the greater the stress, the more concentrated the optical activity. Plastic models of objects may be placed between crossed polarizers and intentionally stressed in order to determine possible points of mechanical failure (Figure 26-28).

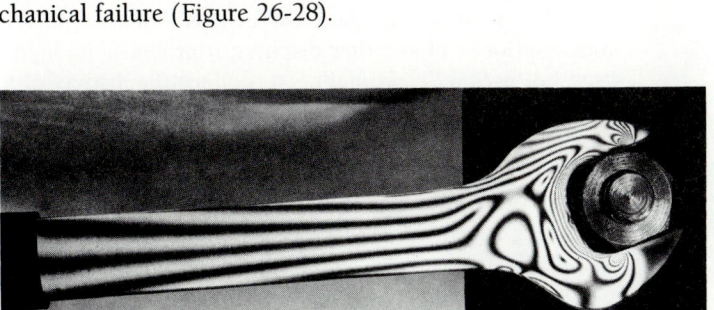

When we look at the sky, we're looking only at light that has been scattered, and this is primarily blue. When we view the sun directly, we're looking at a source of light that includes virtually all colors. As the source light passes through the atmosphere, however, much more of the blue is scattered so that when the light reaches our eyes, there is little blue left and we see the sun on the horizon as red.

Figure 26-28 The unequal forces caused by stressing a material produce the conditions necessary in a transparent material to rotate the plane of polarization. Wherever color bands are most closely packed, stresses are greatest.

SPECIAL TOPIC
Liquid Crystal Displays: A Pervasive Use of Polarization

Liquid crystal displays (LCDs) are an increasingly common part of microelectronic circuits and have come to dominate electronic watches and calculators (Figure 26-29). The very term *liquid crystal* (LC) seems to be self-contradictory. A crystal is an orderly arrangement of atoms or molecules where a particular pattern repeats again and again in space — like, for example, an ordered stack of cannonballs on a courthouse lawn. To be a liquid, by contrast, molecules must preserve the ability to rearrange in space. Toward the end of the last century, it was discovered that some organic molecules pass through an intermediate state for a certain temperature range in which they have the ability to rearrange separately; nevertheless, they maintain an overall order characteristic of the crystal. A crude analogy might be to imagine a pound of ambitious earthworms who have each swallowed a stick of chalk (presuming this would not be fatal). The worms preserve the ability to wiggle, writhe, and move somewhat as well as rotate on their long axis. However, if confined to a small space, the assemblage would inevitably work itself into a lined-up state, much as pencils in a box will if shaken.

While there are three types of LCs known, liquid crystals in use for displays are called **twisted-nematic** (Figure 26-30). For these, the earthworm model is appropriate. To produce the display, a glass plate is stroked in one direction with a fine abrasive to create channels or grooves in the glass. If the LC material is placed on the glass, its molecules will tend to line up with the grooves. A second stroked glass plate with parallel grooves is used to trap the LC layer, which is a few micrometers thick. At this point, one plate is rotated 90° while maintaining contact with the other plate through the LC material. The LC molecules in contact with the lower plate will remain parallel to its grooves, while those next to the top plate will parallel its grooves. The layers of LC material between these extremes tend to orient to parallel their upper and lower neighbors. As a result, the crystal orientation will slowly rotate through 90° from the bottom plate to the top. In this orientation, the LCs are optically active. A beam of polarized light entering the LC layers will be slowly rotated through 90° as it passes the distance between the plates. A polarizer above the top plate selects one plane of polarization to pass. Below the bottom plate, another polarizer is placed at 90° to the first. Thus, a ray of light can proceed as shown in Figure 26-30 from top to bottom, where it reflects and returns, reversing the rotation to pass the top polarizer and emerge.

Note that the light can complete the round trip only because it has the proper polarization for each of the crossed polarizers it passes. Anything that disturbs the orderly orientation of the layers of LC molecules could destroy this, resulting in the light being blocked. Since each of the LC molecules has a dipole moment, an electric field can be used to disorient them. On the grooved surfaces of the glass plates, a transparent conducting coating is placed as a fine film in patterned blocks with separate electrical connections. When these two films are connected to a battery, a strong vertical electric field is established across the LC layers. Each LC molecule experiences a force tending to align its long axis vertically. The layers along the plates maintain their orientation, but the central molecules rotate to the vertical such that their long axis is along the light path. In such a position, the LCs have no optical activity, the plane of polarization is not rotated, and the light is absorbed by the second polarizer, making the display segment opaque.

The great advantage of the LCD is in its low power consumption. Unlike other displays, it produces no light itself but uses reflected light. To produce the display of a conventional TV screen requires several milliwatts for each square centimeter of screen. The red light-emitting diodes (LEDs) of some calculators require less than a tenth of this power but still deplete small batteries with disturbing speed. The LCD requires about 1 μW/cm². If a single flashlight battery could light a 1-W bulb for 1 h,

Figure 26-29 The liquid crystal display of a calculator. An electric potential applied to any segment causes polarizers above and below the display to absorb light and turn the segment dark.

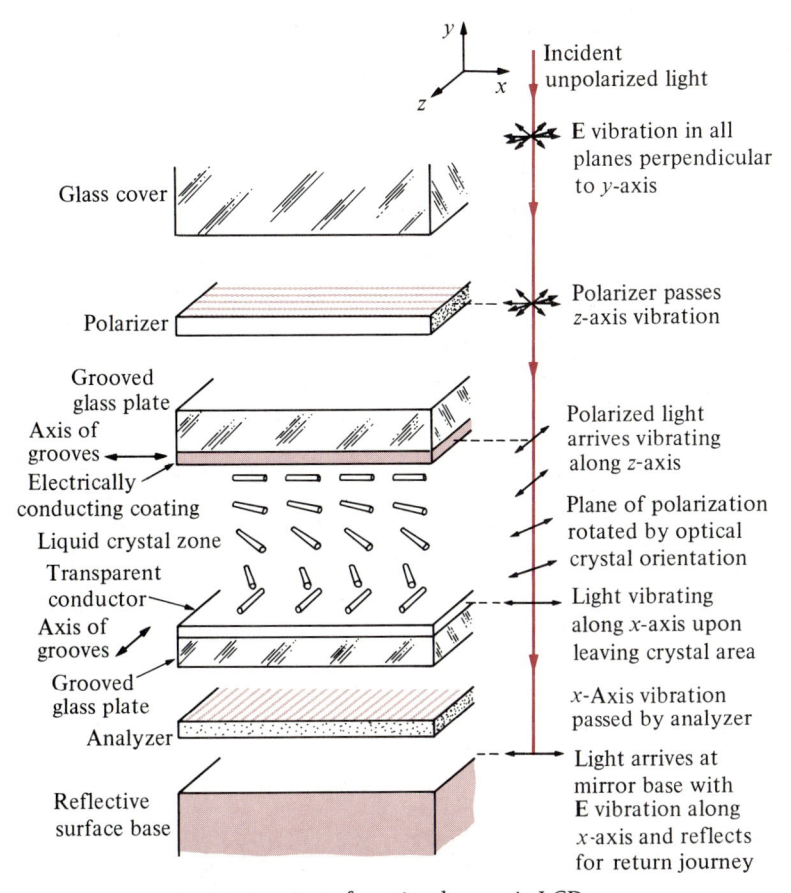

Figure 26-30 A cutaway view of a twisted-nematic LCD.

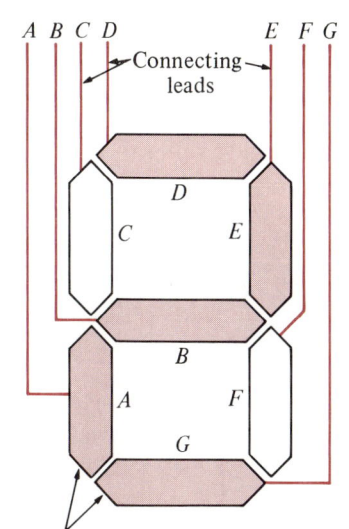

Figure 26-31 The standard seven-element LCD for numerals. The transparent, electrically conducting layer is coated on the glass in the pattern of areas shown, each with a transparent conducting lead wire to the electronics. To display the numeral 2, leads A, B, D, E, and G at the top must be connected to the potential of the battery.

it could power a 1-cm² calculator display for 114 years! The same battery could power a 4 × 4-inch LCD television display in continuous use for over a year.

While it has been almost a century since their discovery, the first commercially available LCs for experimental use appeared in the early 1970s. Since that time, progress has been rapid. Current LCDs are third-generation devices, using both microminiature circuit elements and LCD segments to produce a display. Earlier devices were limited to simple multisegment displays of numerals or letters. The limiting factor was the spacing required of transparent lead wires to energize display segments (Figure 26-31). For a complex display, like a television picture, there was not enough room between segments for all the conducting pathways. Some current devices couple *each* display segment with a fine-film electronics "package" to decode signals. In this way, many dozens of display segments may be energized by a single lead conductor. This has made LCD TV pictures possible. Devices are now available in pocket-size or even wrist-worn displays to provide a completely portable TV (Figure 26-32).

Figure 26-32 A television image produced by an LCD.

667

Minimum Learning Objectives

After studying this chapter, you should be able to:
1. Define:
   - angle of deviation
   - birefringence
   - Brewster's law
   - central maximum
   - coherent
   - dichroism
   - diffraction
   - diffraction grating
   - double refraction
   - emission spectrum
   - extraordinary ray
   - optical activity
   - order number
   - ordinary ray
   - physical optics
   - polarization
   - polarizer
   - Rayleigh's criterion
   - resolve
   - scattering
   - slit
   - thin-film interference
   - Young's experiment
2. Calculate positions of constructive or destructive interference, given the geometry of a double-slit system.
3. Assess light-wave phase change upon reflection at a boundary between media, given relative optical densities.
4. Calculate the thickness of a thin film required for constructive or destructive interference of light of a particular wavelength.
5. Calculate positions of light minima produced by a single slit of given width for a known wavelength of light.
6. Calculate the limitation on resolution of separate sources posed by diffraction.
7. Calculate the angles for maxima of light of a particular wavelength produced by a grating of known characteristics.
8. Calculate the wavelengths of light given off by a particular low-density, excited element in its spectrum from the angle at which each spectral line is formed from the central maximum of the grating.
9. Calculate the angle at which a beam reflected from a surface will be completely polarized.
10. Calculate the percentage of light polarized at a given angle passed by an analyzer at a specified angle to the plane of polarization.
11. Understand the use of optical activity with polarized light to determine concentrations of solutes or to analyze stresses in materials.

Problems

26-1 Young's double-slit experiment is duplicated by using light of wavelength 6.62×10^{-7} m. The slit separation is 3.00×10^{-5} m. Calculate (a) the angular deviation of the first-order constructive interference band and (b) the linear separation of bands on a projection screen which is 4.00 m from the slits.

26-2 The double slits in a Young's experiment are separated by 7.80×10^{-6} m and form a third-order dark band at an angle of 9°. Calculate (a) the wavelength of the light in use and (b) the linear separations of the two second-order bright bands on the screen 5.00 m from the slits.

26-3 The slits for a Young's experiment are separated by 1.20×10^{-5} m and illuminated with light of wavelength 5.25×10^{-7} m. Determine the distance away from the slits that the experimenter must place the projection screen so that the second-order dark band will be separated from the central maximum by 16.0 cm.

26-4 In a Young's experiment, the two slits are separated by 0.180 mm and placed 6.00 m from the screen. If the wavelength used is 4.50×10^{-7} m, calculate the (a) angular and (b) linear separations between the third and fourth bright fringes.

26-5 Determine the wavelength necessary for a Young's experiment such that the separation between adjacent fringes is 1.00 cm when viewed on a screen 3.50 m from the slits and the slits are separated by 2.00×10^{-4} m.

26-6 Two speakers are on a stage 1.00 m left and right from its center line. The first row of seats is 7.00 m from the line joining the speakers, and each successive row is 1.00 m farther back. The center seat of each row is on the hall's center line, with the middle of adjacent seats every 0.900 m right and left. The velocity of sound in the hall is 340 m/s. (a) An oscillator sends to an amplifier an 850-Hz signal that is amplified and passed simultaneously to both speakers. What is the closest seat to the center line in the thirtieth row where this sound intensity at the center of the seat will be zero (ignoring reflections)? (b) A person is sitting in a seat on the center line when a sound of 10,625 Hz is produced by both speakers. How many rows back from the speakers must this person be for the intensity at each ear to be zero? Use the average distance of 19.2 cm between human ears.

26-7 A harried Ms. Hepple is teaching school in a converted maximum security prison. Two tall slit windows 15 cm wide and 2.00 m apart open onto the playground and are 12.0 m from the former prison wall. To call the class in from recess, Ms. H. shrieks in C above high C (1047 Hz). All but two of the class come in at the call. The speed of

668

sound this day is 340 m/s. (a) The first missing student is found along the wall playing mumblety-peg and claims not to have heard the call. Where along the wall from the point opposite the windows would the student have to be to lend some scientific believability to his claim? (b) The second missing student went home with a cut hand. She claims to have been standing near the barred door in the wall when the shriek shattered her soft drink bottle. How far from the point on the wall opposite the windows would the barred door have to be to get the school board's lawyers worried?

26-8 Two flat glass slides 5.00 cm long are used to form an air wedge. They are touching along one edge and separated at the other edge by a 0.0425-mm-thick hair between them. When viewed by reflected light of wavelength 6.00×10^{-7} m, (a) what is the separation of adjacent fringes? (b) How many dark fringes are seen? (c) Is the line of contact a bright or dark fringe? Explain.

26-9 A film of oil (index of refraction = 1.25) in air creates constructive interference for reflected light of wavelength 5.00×10^{-7} m. Calculate (a) the thinnest the film can be and (b) the other thicknesses the film can have.

26-10 A thin piece of paper is placed between the edge of an 8.00-cm glass plate and a front surface mirror on which it is lying. This forms a wedge of air which shows 30 total dark interference fringes when viewed by reflected light of wavelength 5.50×10^{-7} m. Calculate the thickness of the paper.

26-11 Light reflected from a soap bubble shows constructive interference for light of wavelength 4.50×10^{-7} m and destructive interference for light of wavelength 6.75×10^{-7} m. Calculate the soap film thickness. Take $n = 1.33$.

26-12 What minimum thickness of coating of refractive index 1.72 is necessary on a camera lens if light of wavelength 5.75×10^{-7} m is completely eliminated from reflected light by destructive interference? ($n_{lens} = 1.59$.)

26-13 A single slit of width 3.00×10^{-6} m is illuminated with yellow light of wavelength 5.89×10^{-7} m. Calculate (a) the angular separation of the first-order dark band on either side of the central maximum and (b) the angular width of the first-order bright band.

26-14 When light of wavelength 5.50×10^{-7} m illuminates a single slit, the first dark bands on either side of the central maximum are separated by 30°. Calculate the slit width.

26-15 Assume small angles and prove that any two adjacent dark bands of a projection interference pattern due to diffraction by a single slit are separated on the screen by a distance Δy given by $\Delta y = \lambda R/w$, where λ is the wavelength of the light, R is the distance from the slit to the projection screen, and w is the slit width.

26-16 A helium-neon laser with a wavelength of 6.33×10^{-7} m illuminates a single slit 3.80×10^{-5} m wide. How wide is the central maximum bright band on a screen 2.00 m away from the slit?

26-17 Calculate (a) the angular separation of two points which are just barely resolvable at a wavelength of 5.50×10^{-7} m through the telescope on Mount Pasthukov in the Soviet Caucasus, largest in the world with an aperture of 6.00 m, and (b) the distance two points on the moon must be separated in order that they be resolvable by that telescope.

26-18 Signal gathering power is not a serious problem for radio telescopes, since even a modest one will have a larger objective area than the largest optical telescopes. Resolving power is a serious problem. Most molecules of interest radiate radio waves with lengths from a few millimeters to a few centimeters, the longest being neutral hydrogen radiation at 21.0 cm. For each of the following three telescopes, calculate (1) the minimum angular resolution when operating with waves 1.80 mm in wavelength and (2) the minimum distance between two resolvable sources of radio emission at 1.40 cm on Mars some 20.0×10^6 km away during closest approach. (a) The largest fully steerable single-dish telescope in the world at the Max Planck Institute in Bonn, Germany, with a 100-m diameter. (b) The largest fixed single-dish telescope at Arecibo, Puerto Rico, with a 305-m-diameter objective. (c) The 27 separate dishes (each 26 m in diameter) of the very large array (VLA) spread across the desert in a Y shape near Socorro, New Mexico, and electronically linked to provide the equivalent resolution of a single dish 27.0 km in diameter.

26-19 The sun is in a spiral arm of the Milky Way galaxy some 10 kpc (kiloparsecs: 1 pc = 3.26 light-years = 3.09×10^{13} km) from the center. The next spiral arm inward is about 4.00 kpc from the sun. How far apart would two gas clouds radiating neutral hydrogen waves ($\lambda =$ 21.0 cm) have to be in this inward arm to be resolved by the VLA (Problem 26-18) as two sources?

26-20 You are to design an astronomical telescope through which one can see the planet Neptune as a disk. The objective lens must be large enough to resolve by the Rayleigh criterion the edges of Neptune. Take the diameter of Neptune to be 4.54×10^7 m and its largest distance from Earth as 4.64×10^{12} m. Use 5.40×10^{-7} m as the average wavelength of light reflected from Neptune.

26-21 A ship's radar uses 3.00-cm-wavelength microwaves. If its antenna is 1.50 m in diameter, what is the closest distance to each other that two objects which are 5.00×10^3 m away from the ship may be and still be resolved by the radar?

26-22 Sunlight falls on an 0.0800-mm-diameter hole in a room-darkening window blind. It crosses the room and makes a pattern on the opposite wall some 5.50 m from the pinhole. If the sun subtends an angle of 0.53°, calculate (a) the geometrical size of the sun's image on the wall and (b) the diameter of the diffraction pattern to the first minimum. Use 5.10×10^{-7} m as the wavelength of sunlight since this is the wavelength of normal eye maximum visibility.

669

26-23 (a) What is the largest order-number image that you may view with a diffraction grating of 600 lines/cm by using green light of wavelength 5.50×10^{-7} m? (b) Answer the same question if there are 6000 lines/cm in your grating.

26-24 Light made up of two different wavelengths 4.30×10^{-7} m and 6.70×10^{-7} m is incident on a 2500 line/cm plane diffraction grating. How far apart are the lines formed by these wavelengths on a screen 3.00 m away from the grating in (a) first order and (b) second order?

26-25 What is the longest wavelength that can be observed for a transmission grating having 8000 lines/cm in (a) the second order and (b) the third order?

26-26 A plane diffraction grating using monochromatic light of wavelength 5.89×10^{-7} m produces a second-order line at $12°$. Calculate (a) the grating spacing and (b) the number of lines per centimeter.

26-27 White light containing wavelengths from 4.00×10^{-7} to 7.00×10^{-7} m is directed at a plane diffraction grating of 5000 lines/cm. (a) At what angle and (b) in what order number do the spectra overlap? (c) Would either or both answers change if the grating had a different number of lines per centimeter?

26-28 A plane transmission grating has 6000 lines/cm and produces a second-order line for certain monochromatic light at an angle of $30°$. Determine (a) the grating spacing, (b) the wavelength of the light, (c) the angle of the first-order image, and (d) the number of orders which can be seen.

26-29 At what angle of incidence will reflected light be completely polarized if it is traveling in water and reflected off (a) a water-to-air interface and (b) a water-to-flint glass interface, as at the side of an aquarium?

20-30 What is the polarizing angle for light reflected from the surface of a swimming pool?

26-31 Unpolarized light of intensity I_0 is incident on a crossed polarizer-analyzer combination, and no light is passed. (a) Through what angle must the analyzer be turned in order that light of intensity $I_0/8$ be passed? (b) With the polarizer and analyzer crossed, a third piece of polarizing sheet is placed between them and makes an angle of $45°$ with each. What intensity (as a fraction of I_0) is now passed by the combination?

26-32 A polarizer and analyzer are originally crossed, and a beam of unpolarized light is incident on the polarizer. (a) What percentage of the incident light intensity is transmitted by the polarizer? By the analyzer? (b) The analyzer is rotated through an angle of $30°$ around the beam. Now what percentage of the light incident on the polarizer is transmitted by the analyzer?

26-33 English physicist Charles Barkla (1877–1944) devised an elegant experiment to determine whether x-rays were transverse waves or not. He shot a beam of x-rays in the positive x direction to a scattering crystal. While x-rays were scattered in all directions, they were stopped by absorbers except for a beam in the positive z direction. These were allowed to travel to encounter a second scattering crystal. Barkla then measured the x-rays scattered from this second crystal in the xy plane with a detector, taking four measurements in the $+x$, $-x$, $+y$, and $-y$ directions from the crystal. What result for these four measurements demonstrates that x-rays are transverse waves? (It will help to sketch the situation.)

26-34 A beam of light reflected from a glass surface in air at an angle of $56.5°$ from the normal is completely polarized. (a) Calculate the index of refraction of the glass. (b) What is the polarizing angle for light traveling in water and incident on this glass?

26-35 A polarimeter is an instrument used to determine the concentration of certain kinds of sugar in solutions or in syrups. A cylindrical tube with transparent ends is placed between crossed polarizers. Whenever the cylinder is filled with a solution, light directed toward the cylinder is plane-polarized by one of the polarizers. As the light passes through the solution, the plane is rotated, and the beam is not extinguished at the other polarizer. This polarizer must be turned through the same angle that the plane of polarization was rotated by the sugar in order to stop all light. The angle of rotation is measured, and comparison with previously constructed tables yields the concentration. Suppose the specific rotation of a sugar solution is $50°/$dm for a concentration of 1.00 g/cm^3. What is the concentration of sugar if the net rotation of a beam is $165°$ in a polarimeter tube that is 15.0 cm (1.50 dm) long?

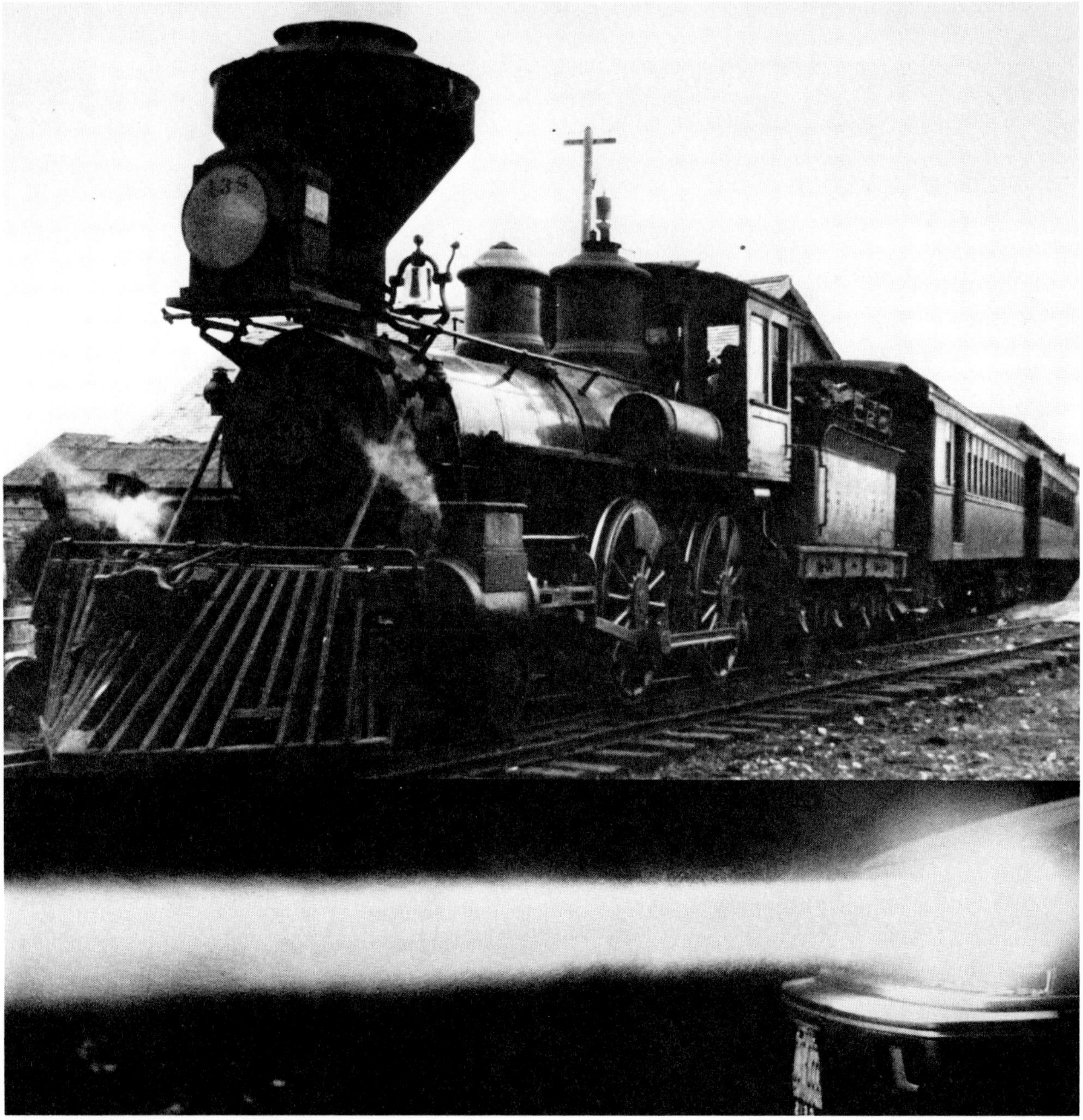

27 Theory of Relativity

27-1 Introduction

In the latter part of the nineteenth century, many scientists believed that most of physics was well known and the theories pretty much complete. In this view, physics from then on would be a mopping-up operation, tying up loose ends as it were. Little did they know that the seeds of a scientific revolution had already been sown, and the revolution was really growing in several places. To understand the origin of the first revolution, the theory of relativity, we must consider the attempts to answer a simple question: "If light is a wave, what is the medium that is wiggling to transmit it?"

27-2 The Ether: Historical Precursor to Relativity

By the 1820s, experiments by Young and others had established the wave nature of light. Every previous attempt to pose a wave theory for light raised the question "What is the elastic medium which causes light to propagate?" Before acceptance of the wave theory, this question was only one of many problems and perhaps not the most important one. Now the answer to the question had become very important. All known wave motion required a medium to provide the elastic restoring force on the wave. It was believed that this would have to be true of all waves.

The conviction that a medium was necessary was so strong that a medium was postulated to exist and called the **luminiferous ether.** It was assumed to be perfectly elastic, perfectly transparent, have vanishingly small density, and to pervade all space in the universe, even vacuums and the space between atoms. The ether concept became firmly entrenched, and it was taught to students, discussed in scientific meetings, and even had many papers and whole books written about it. However, until later in the nineteenth century, the concept remained theoretical because no one had devised an experiment to test for the ether's presence. During that period, it was generally treated as simply one of those facts of nature that "we do not fully understand."

One of the questions pondered was whether this ether in the universe is stationary and bodies which are moving simply move through it or whether the ether moves with the body it happens to pervade. Based on the hypothesis of a stationary ether, Fresnel predicted that if the speed of light was measured in a moving medium, its value would be partially altered by the motion of the medium. In other words, Fresnel stated that light would be partially "dragged" along and even derived a theoretical equation for the drag, assuming that the ether was stationary. In 1851, Armand Fizeau tested this prediction with a series of measurements of the speed of light in moving water and confirmed the equation. Data from other observations, including that from the aberration of starlight (Section 27-4), gave additional corroboration to the view that the ether was stationary.

Convinced that the earth moves through a stationary ether, A. A. Michelson (1852–1931) expected that the motion gives rise to an ether wind similar to the apparent wind you might feel while riding a bicycle through still air. If he could measure the speed of this ether wind, he could determine the absolute speed of the earth. The ether could then be used as an absolute stationary coordinate frame to which all motion could be referenced. In actuality, velocity vectors measured in a coordinate system fixed to the earth could be transformed mathematically to this **absolute system** to determine the absolute velocity of an object. In 1887, in the key experiment on the ether, Michelson and E. W. Morley used an extremely sensitive optical interferometer, invented by Michelson, to make the absolute velocity determination (Section 27-4). They expected to detect the absolute speed of the earth and hence prove the existence of a unique, stationary reference frame. Their measurement was to take place when the interferometer was rotated, and a shift of the position of interference fringes was expected. If there was a fringe shift, it would prove that the earth is moving with respect to the ether, and its speed could be calculated from the magnitude of the shift. But they detected no fringe shift at all. Again and again they conducted the experiment at different times of the day and different times of the year, always obtaining a null (zero) result.

From the Michelson-Morley experimental null result, one might conclude that the earth and ether are moving together or that the ether was dragged around by the earth. However, this had already been rejected due to the results of the Fizeau experiment repeated by Michelson himself. Other possible explanations emerged as scientists racked their brains. Perhaps the ether is not a good frame of reference from which to measure the speed of light! Perhaps light takes on the velocity of its source, as does a ball thrown from a moving automobile. This is called the emission, or ballistic, theory of light propagation, and it was also duly rejected by experiments.

> The supposed properties of the ether follow from the observed properties of light. For example, light is a transverse wave, as demonstrated by polarization, and yet only solids can transmit a transverse wave. Further, the speed of a wave is proportional to the square root of the elasticity over the density of the medium. Light is hundreds of thousands of times as fast as a wave in steel, for example. Thus the ether needed hundreds of times the springiness of steel and less than thousandths of its density to be able to spring back fast enough to transmit a light wave at this speed. We could go on. The set of required properties of ether is fascinating.

A last-gasp effort to preserve the absolute ether frame and still be consistent with the null result of Michelson and Morley was made in 1892. G. F. FitzGerald and H. A. Lorentz hypothesized that all bodies shrink, or contract, in their directions of motion with respect to the stationary ether by exactly the right amount to yield a null result. In addition to the fact that its basis was very complex, such a contraction struck scientists as a rather contrived scheme.

The various conclusions to which these different ether-measuring experiments apparently lead contradict one another. Many scientists recognized that there was something fundamentally wrong but did not know what it was. Was the concept of ether incorrect? Did our view of reference frames have a fundamental error? What is the proper reference for measuring the speed of light? With these and other questions, the scientific community seemed to be entering the twentieth century with a large number of question marks. Maybe physics was not yet complete! Maybe some of the accepted principles were not true!

Many of the questions were very hard for scientists to ask, much less answer. The most knowledgeable had lived all their lives with a set of scientific rules which had performed very well. It is not surprising that most preferred to preserve the known principles while refusing to think very hard about searching for new ones. It is as though an important part of what you know and on which you plan your future is declared (perhaps proven) to be wrong.

In 1905, Albert Einstein (1879–1955), then an obscure clerk and technical officer in the Patent Office in Berne, Switzerland, published four articles in the *Annalen der Physik,* all on different topics, which were to shake the very foundations of science. One of these articles advanced the quantum hypothesis (Chapter 28) and led to Einstein's Nobel Prize. The one most important for our present considerations provided answers to the questions about absolute reference frames by removing the apparent contradictions. The instrument of this removal required not simply the alteration of principles but a true reexamination of the fundamental quantities of measurement: mass, length, and time. We will proceed slowly through this reexamination.

27-3 Frames of Reference

The term **frame of reference,** or **reference frame,** is defined to be a vantage point or position with respect to which motion may be described or measured. This definition is really talking about a coordinate system, as we have used from the start of the book. A frame of reference is essentially the location of a coordinate system. Two bodies which are at rest with respect to each other are in the same rest frame of reference. If a third body is moving with respect to the first two, it is in a different rest reference frame. This is important, because on many occasions measurement of a physical quantity will yield different values for observers in different reference frames. For example, suppose you are riding in an automobile at a speed of 15.0 m/s when a second automobile is coming up behind you and is getting closer. We know that if an observer in a reference frame at roadside measures the speed of the second car to be 25.0 m/s, then it is catching up to you at 10.0 m/s. This is a simple example of the treatment of relative velocities first encountered in Chapter 3 (sometimes called galilean relativity). However, it is also an illustration of measurements of the same event by observers in two different reference frames.

We have no doubt that we can use the value which is measured in one

In setting up a physics problem, we define an origin, the x, y, and z axes, and decide when to start our clock. In so doing, we have defined our reference frame in space and time.

frame to calculate the value which is measured in the second frame. Such a calculation is called a **transformation.** If we know something about two different frames, we may, in principle, transform any quantity or even whole equations and laws from one frame to the other. For instance, we can make the transformation of any quantities between two coordinate systems which are moving with respect to one another, provided we know the relative velocity between them *and* the applicable transformation laws. In the example, the two frames are a coordinate system fixed to your auto and one fixed to the side of the road. The relative velocity between the two has a magnitude of 15.0 m/s. We transformed the speed of the second auto as measured at roadside to the system traveling with your auto.

It is not obvious, but we did make an assumption about the applicable transformation law. We assumed that the velocities are added algebraically (really vectorially). This seems like a rather commonsense assumption and certainly one which we have used before. The difficulty is (as we will see) that it is not applicable when the relative velocity between frames is very large.

You may remember that Maxwell's equations lead to a value of $c \approx 3.00 \times 10^8$ m/s for the speed of light in a vacuum. Using our previous assumption that velocities add vectorially, in a second frame moving at 1.00×10^8 m/s, we might expect that light could have a speed of 4.00×10^8 m/s. This would mean that observers in two different reference frames would *not* get the same result if they each measured the same electromagnetic effect. In fact, Maxwell's equations do not even retain their same form when transformed between two reference frames in the classical way.

This seems very odd! We certainly expect any law of physics to be the same for two observers if the only difference between them is that they are moving at constant velocity with respect to each other. Several scientists, including Einstein, came to the conclusion that the problem may arise from the reference frame we choose to use in measuring the speed of light. The experiments described seem to rule out the medium (ether) as a good reference frame, and others seem to rule out the source. The only remaining possibility is the observer. Perhaps the speed of light must always be measured with respect to the observer. Let us explore this possibility further by carrying out a **gedanken** (thought) **experiment,** one that you never really carry out but only think about.

For simplicity, we assume that our coordinate systems are moving with respect to one another but are not accelerating. Such a frame of reference which is either stationary or moving with a constant velocity is called an **inertial frame of reference** because Newton's law of inertia is obeyed. Will observers in different frames agree on the value of the same quantity that they both measure? For instance, consider the way in which a length measurement is made. We often compare the distance between two ends with a well-known length. As an example, we may compare the distance between the top and bottom of this page with the distance between marks on a meterstick. This works very well provided both the book and meterstick are stationary. We can read a value at one end of the book and take our time as we move our eyes to read the value at the other end.

Suppose, however, that the book is moving, as depicted in Figure 27-1. This measurement may still not be too difficult. We could simply make marks on a stationary table as the book slides by and measure the distance between the marks at our leisure. One problem we can see in the figure is that the marks must be made simultaneously. Otherwise, the mark at the bottom (P) might be made before the mark at the top (Q), and we would have a longer distance

Figure 27-1 To measure the length of a book, we read where one edge P lines up with a scale and then do the same with the other edge Q. If the book is moving with respect to the scale, we must arrange for the positions of P and Q to be read simultaneously.

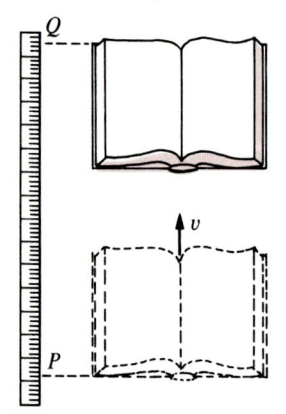

between marks than the actual book length. This is still not too hard. It could be done by two people, one assigned to each end of the book, who agree to make their respective marks simultaneously by using a common signal. Suppose, however, that the two people are to measure something much longer than a book, the length of a moving train, for example. The time required for the signal to reach one person after initiation by the other could cause a great deal of error. Sound could take many seconds to travel the length of a long train. Even light has a finite speed and can introduce an end error if the distance between the people is large enough. A final method is to have them use synchronized clocks and agree to make the marks simultaneously at a time agreed upon beforehand. There could still be a problem in synchronizing clocks.

Consider the gedanken experiment in which two people, say Sar and Sak, in the same reference frame are separated by 1.80×10^{11} m (Figure 27-2a). This distance is about six-tenths the diameter of the earth's orbit and is chosen because at the speed of light it would take the signal 10 min to go from one to the other. They synchronize their watches as follows: Sar sends a radio message telling Sak that at exactly 12:00 noon Sar will initiate an encoded radio signal toward Sak. Since they know they are separated by 1.80×10^{11} m, each knows that Sak will receive the coded signal at exactly 12:10 P.M. Sak's clock is set

Figure 27-2 (a) Sar and Sak, separated by 1.80×10^{11} m, synchronize their clocks. (b) Tar and Tak, in a reference frame moving to the left at a speed v with respect to Sar and Sak, synchronize their clocks. (c) To Tar and Tak, the Sak-Sar reference frame appears to be moving to the right at a speed v.

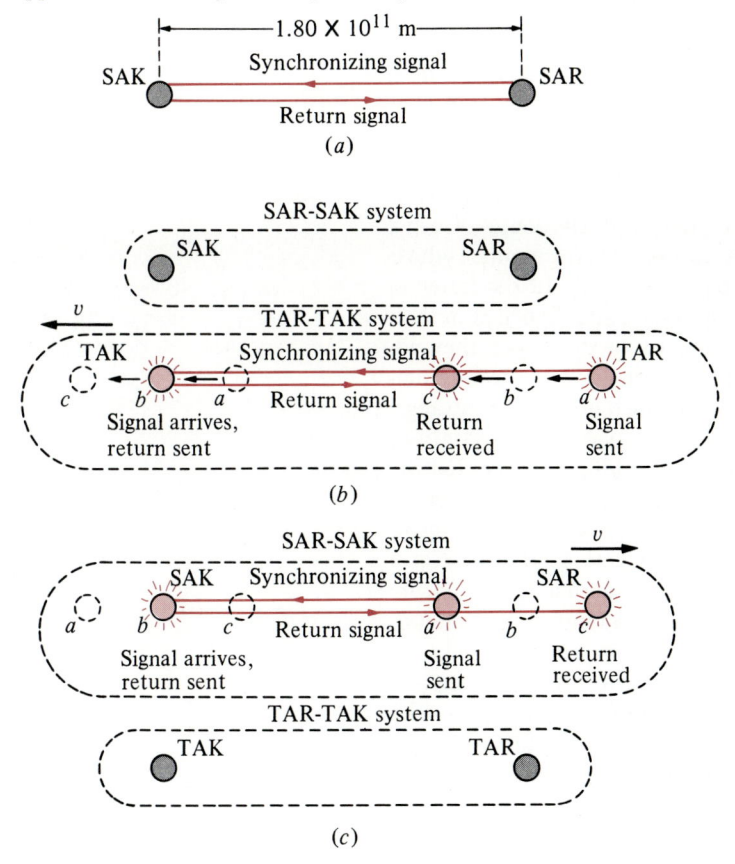

27-3 Frames of Reference

automatically to 12:10 by the signal, and a coded return signal is initiated. Upon receiving the return signal at exactly 12:20 P.M., Sar is convinced that their clocks are synchronized. They use radio signals for their synchronizing process, because traveling at the speed of light, radio signals are the fastest possible means of communication.

Now that their clocks are synchronized, they can use them to measure the distance between two other people, Tar and Tak, who are conducting their own synchronizing operation in their own reference frame. The relative velocity of the second reference frame is parallel to the line joining Sar and Sak and is nearly three-tenths the speed of light. As Tak and Tar pass, Sar points out that Tar's clock is not synchronized with Tak's clock because the Tar-Tak frame is moving at nearly three-tenths the speed of light. Sar asserts that although Tar's clock shows that exactly 20 min pass until the return signal from Tak is received by Tar, it really took 11 min to get there and only 9 min to return. This, Sar says, is because the signal must catch up with Tak, who is moving away from Tar's original position, while the return signal is intercepted by Tar 3.60×10^{10} m closer to Tak's original position (Figure 27-2b). In reply, Tar states that in reality, it is Tar and Tak who are stationary, and so it is Sar's and Sak's clocks which are not really synchronized. Tar insists that the frame of Sar and Sak is moving at nearly three-tenths the speed of light in the *other* direction (Figure 27-2c).

Now here is a problem! At first, it might seem that this disagreement can be resolved fairly easily. After all, at least one of the frames of reference (maybe both) must be moving in order that there be a relative velocity between the two. We can settle this disagreement by measuring the speed of each frame with respect to something which is absolutely stationary and determine which of the first two frames, if either, is stationary. As we think about this possible solution, we know we are in trouble. *There is no absolute frame.* The search for it is what raised all the original questions. The occupants of any third inertial frame would insist that theirs was the stationary frame and the other two frames were moving. From their own points of view, each set of observers is correct, *and* it is very important to recognize that *those* are the only available points of view. There is *no* frame in which an impartial observer can stand and announce who is moving and who is stationary. Each measurement is correct for the frame of reference in which it is made. Sar's and Sak's clocks are simultaneous in their frame. Likewise, Tar and Tak can claim simultaneity for the clocks they adjusted. No one will agree that the clocks in another reference frame are simultaneous, and this leads to the predictions of relativity which might seem strange.

Also, when Sar and Sak make the marks (or observations) necessary to measure the distance between Tar and Tak, the others will not agree that the marks were made simultaneously. Thus people outside Sar and Sak's frame of reference will insist that the length they measure is wrong. Therefore, relative motion between frames of reference precludes agreement in the setting of clocks and, as a result, prevents agreement in measurement of length. It should not surprise us that disagreement about the values of the fundamental quantities of time and distance leads also to different values of measured speeds and velocities. Conservation of linear momentum leads in turn to disagreements in inertia—mass, itself—because of motion between frames of reference. The balance of this chapter will revise carefully our understanding of these fundamental quantities.

27-4 Michelson-Morley and Other Ether Experiments

Most treatments of relativity begin with the Michelson-Morley experiment as a jumping-off point. Even though Einstein stated that he was unaware of the Michelson-Morley experiment, its null result is still required of his theory. Also, close examination of this key experiment introduces the issues in a quantitative way.

We begin with the analogy of a motorboat which can travel at a speed c in a river with a flow rate v. Consider the boat in Figure 27-3a which must travel a distance l upstream and return to the starting point. Since the boat's speed is measured with respect to the water, its speed with respect to the earth is $c - v$ as it travels upstream and $c + v$ as it travels downstream. The total time up and back is equal to the time up plus the time back, or

$$t_{ub} = t_u + t_b$$
$$= \frac{l}{c - v} + \frac{l}{c + v} = \frac{l(c + v) + l(c - v)}{c^2 - v^2}$$
$$= \frac{2lc}{c^2 - v^2}$$

We wish to compare this with the time required for the boat to go a distance across the river and return to its starting point.

In Figure 27-3b, we see the boat crossing the stream and recall from Chapter 3 that to make headway straight across, the boat must be aimed partially upstream. The speed with respect to the earth with which the boat crosses the stream is then $\sqrt{c^2 - v^2}$. On the way back, the boat must again head partially upstream and makes the same speed with respect to the earth. Thus, the time required for the boat to go across and return is equal to twice the time required to cross, and we have

$$t_{ar} = 2t_c = 2\frac{l}{\sqrt{c^2 - v^2}} = \frac{2l}{\sqrt{c^2 - v^2}}$$

The two time intervals can be compared more easily if we rearrange their forms to obtain

$$t_{ub} = \frac{2l}{c} \frac{1}{1 - v^2/c^2}$$

and

$$t_{ar} = \frac{2l}{c} \frac{1}{\sqrt{1 - v^2/c^2}}$$

To simplify these equations, we define the term γ (Greek lowercase gamma) as

$$\gamma = \frac{1}{\sqrt{1 - v^2/c^2}} \qquad (27\text{-}1)$$

We then have

$$t_{ub} = \frac{2l}{c}\gamma^2 \quad \text{and} \quad t_{ar} = \frac{2l}{c}\gamma$$

The Michelson-Morley experiment is the key experimental idea from which we may obtain the form of the equations of the theory of relativity. It is also historically a coming of age of American science of sorts. Michelson was the first American to receive a Nobel Prize in science.

Figure 27-3 (a) The boat moves with a speed c relative to the water, while the water moves at speed v relative to the shore. To an observer on shore, the boat's speed is $c - v$ moving upstream and $c + v$ moving downstream. (b) Moving across the river, part of the boat's velocity must be used to combat the current and produce a resultant directed straight across the river of magnitude $\sqrt{c^2 - v^2}$.

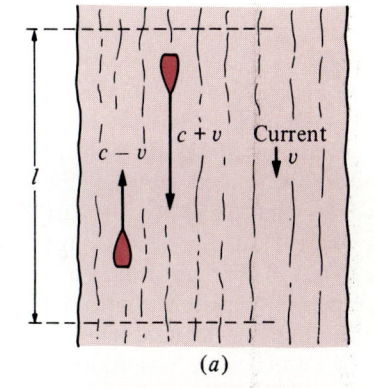

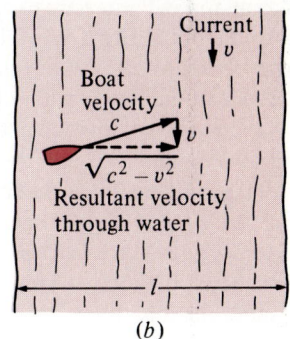

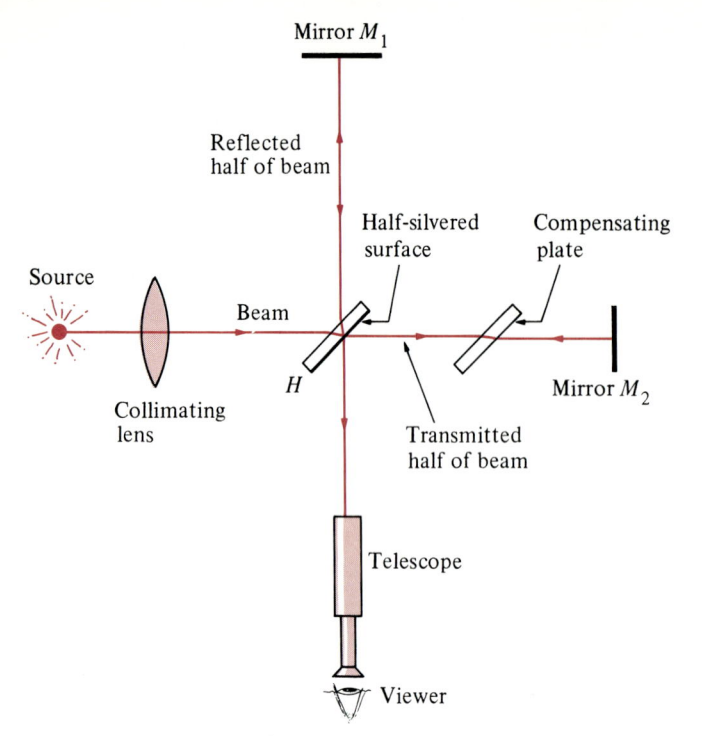

Figure 27-4 The Michelson interferometer.

From the definition, we see that γ is greater than 1, and therefore t_{ub} is greater than t_{ar}.

Figure 27-4 is a simplified representation of the Michelson interferometer. The source of light is placed at the focal point of a collimating lens which delivers a parallel beam to the half-silvered surface H. This surface divides the incident beam, reflecting half toward the mirror M_1 and allowing the other half to be transmitted to M_2. The part of the light delivered to M_1 is reflected back to H, where it is divided, half going to the source and the other half combining with that part of the light from the first path to be viewed by the telescope. The compensating plate is simply a piece of plate glass of the same type and thickness as that on which the half-silvered surface is deposited. This is to ensure that beam 1 and beam 2 have equivalent optical paths. The two combined beams are coherent and will interfere to produce a pattern like that of Figure 27-5a. If one of the mirrors is tilted very slightly, the field of view will show the vertical fringes of Figure 27-5b. Once the pattern is established, any change in the phase difference of the two beams will cause the fringes to shift one way or the other.

Figure 27-6 shows an exaggerated view of the interferometer fixed to the earth and moving through the ether. The ether wind is analogous to the flow rate of the river, whereas the speed of light with respect to the ether is analogous to the boat speed. We see that the beam to M_1 must go upstream and back, while that to M_2 goes across stream and returns. Thus, if the paths are of the same length, say l, the expressions for the times required for the beams' trips are of the same form as those for the boat moving in the river. We can write the difference in the two times as

$$\Delta t = t_{ub} - t_{ar}$$

Figure 27-5 (a) Circular interference pattern from a Michelson interferometer. (b) The edges, or **fringes**, of a pattern such as that of part *a* are the part watched as an indicator of some variation of travel time for one of the beams. For example, if one path length were changed by just one-half the wavelength of the light used, each bright fringe would move to the position of its neighboring fringe.

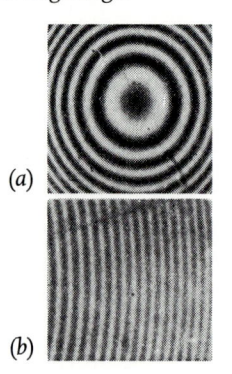

(a)

(b)

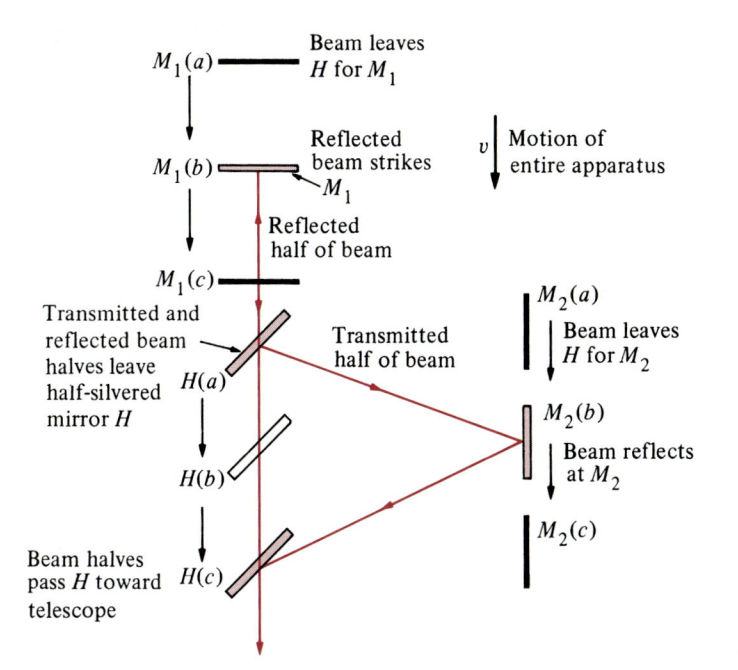

Figure 27-6 The apparatus of Figure 27-4 aligned with the axis from M_1 to the telescope pointing in the direction of the earth's motion about the sun. Thus the entire experiment is moving from top to bottom. The beam arriving at H when it is located at its position (a) is split, with half moving toward M_1 and half toward M_2. By the time the half-beams arrive at the mirrors, the mirrors have moved to the positions labeled (b). The half-beams are reflected and return to H at position (c). If we imagine the apparatus as fixed in space and the ether flowing past with velocity $-v$ (from bottom to top), the situation is the exact analog of the boats moving equivalent distances up, then downstream, vs. across and back.

or

$$\Delta t = \frac{2l}{c}\left(\frac{1}{1-v^2/c^2} - \frac{1}{\sqrt{1-v^2/c^2}}\right)$$

We expect v^2/c^2 to be much less than 1. The actual orbital speed of the earth is about 3×10^4 m/s, which makes $v/c \approx 10^{-4}$ and $v^2/c^2 \approx 10^{-8}$. Thus each of the fractions inside the parentheses may be expanded in a binomial series (Appendix A) and all terms higher than second-order be dropped. We have

$$\frac{1}{1-v^2/c^2} = \left(1-\frac{v^2}{c^2}\right)^{-1} \approx 1 + \frac{v^2}{c^2}$$

and

$$\frac{1}{\sqrt{1-v^2/c^2}} = \left(1-\frac{v^2}{c^2}\right)^{-1/2} \approx 1 + \frac{1}{2}\frac{v^2}{c^2}$$

This yields

$$\Delta t = \frac{2l}{c}\left(\frac{1}{2}\frac{v^2}{c^2}\right) = \frac{l}{c}\frac{v^2}{c^2}$$

This time difference would give rise to a phase difference of the two beams, causing the interference pattern formed when they combine to be different than it would be if the two transit times were the same. This difference in pattern is not detectable if Δt for the two beams remains constant. However, when the interferometer is turned through 90° about a vertical axis, the two beams, one with and against the ether flow and the other across it, exchange roles. There would then be a shift between the two observations of twice the time difference. The net time shift would then be

$$\Delta t_{net} = 2\,\Delta t = \frac{2l}{c}\frac{v^2}{c^2}$$

This corresponds to a path difference

$$\Delta(\text{path}) = c\,\Delta t_{net} = 2l\frac{v^2}{c^2}$$

In the first Michelson-Morley experiment, each beam was reflected back and forth many times by auxiliary mirrors to increase the effective path length. The effective path length for each beam was about 11.0 m. This gives rise to an expected path difference

$$\Delta(\text{path}) = 2(11.0\text{ m})(10^{-8}) = 2.2 \times 10^{-7}\text{ m}$$

Calculations show that this path difference would yield a shift of about four-tenths of one fringe and be easily detectable. Their final report, published in 1887 as "On the Relative Motion of the Earth and the Luminiferous Ether," showed no fringe shift of this magnitude. They concluded that it was impossible to detect the motion of the earth with respect to the ether.

But, we do have ample evidence that the earth moves through space. Figure 27-7a depicts the observation of a star in a direction perpendicular to the plane of the earth's orbit. It is assumed that no relative motion exists between the ether and the telescope. The earth could be at rest in the ether, or the ether could be dragged along with the telescope as it moves. Light entering at the center of the objective moves down the telescope axis to its eyepiece. In Figure 27-7b, we see a different situation in which the telescope is moving to the right at a speed v through the ether. The light still moves vertically downward at a speed c since the ether is its reference frame. But, as the light is moving from the objective, the telescope itself is moving to the right. We must tilt the telescope in order that the ray remains on the central axis to the eyepiece and does not appear to drift off to hit the side. We see the star in an apparent direction as determined by the orientation of the telescope. Taking the time required for the light to traverse the telescope to be t, we may determine the angle of tilt from

$$\tan\theta = \frac{vt}{ct} = \frac{v}{c}$$

Using $v = 3.00 \times 10^4$ m/s and $c = 3.00 \times 10^8$ m/s, we have

$$\tan\theta = 1.00 \times 10^{-4}$$

which yields $\theta = 5.73 \times 10^{-3}$ degrees, or $\theta = 20.6$ seconds of arc. As the earth moves in orbit, the telescope must consistently be tilted in the direction of the earth's motion (Figure 27-7c).

This is called the **aberration of starlight** (or **stellar aberration**), a well-known effect. During the course of a year, this makes the star appear to sweep out a circular path, and to follow it the telescope must sweep out a cone of angular diameter 2θ. Observations yield a cone equal to 41 seconds of arc, which is in nearly perfect agreement with the derivation. The conclusion is that the earth moves and does not drag the ether along with it.[1]

---

[1] Incredibly, the fine effect of stellar aberration was discovered by James Bradley in 1727, a half-century before the American Revolution. The angle of 20.6 seconds is the apparent width of a dime as seen across a distance of almost two football-field lengths.

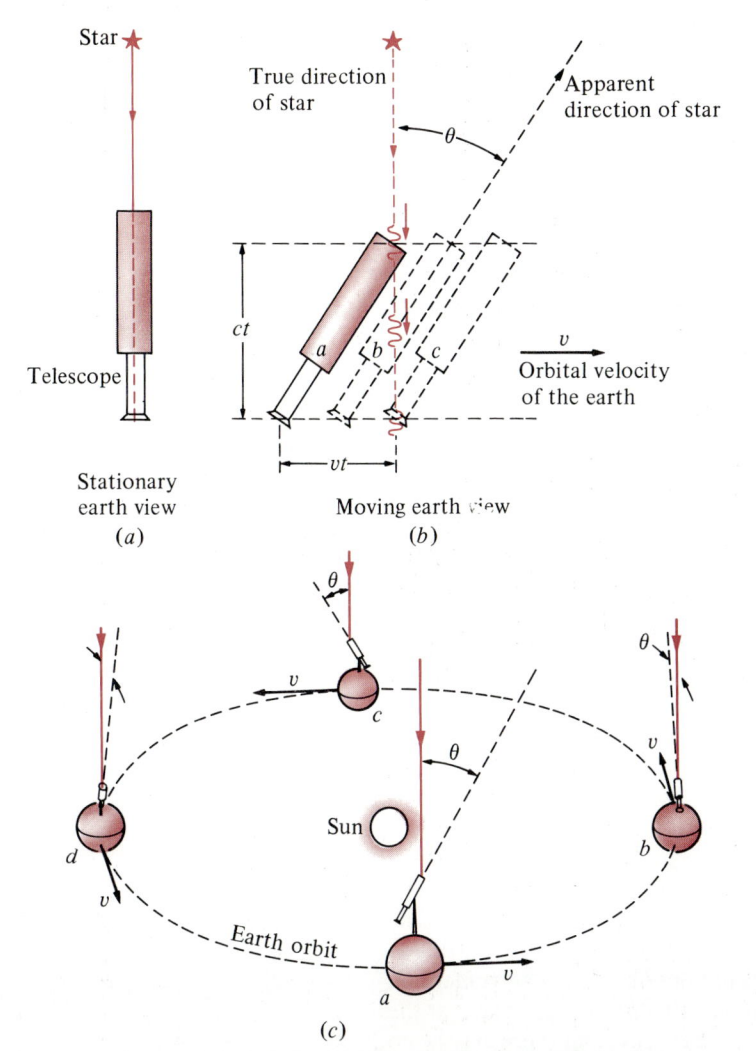

Figure 27-7 Aberration of starlight. (*a*) With a stationary earth, a telescope must be pointed at the position of the star being viewed. (*b*) With a moving earth, this may change. Viewing a star in a direction perpendicular to the earth's motion requires the telescope to be tilted in the direction of the earth's motion. (*c*) Observing the same star throughout the year requires a continuous correction of the tilt of the telescope. Watched throughout the year, the star will appear to describe a small circle in the sky of angular radius θ.

Figure 27-8 Albert Einstein (1879–1955), perhaps the most famous scientist of this century, was born in Germany and educated there through high school. After graduation from college in Switzerland, he worked at the Swiss Patent Office, where he produced some of his most famous work in his spare time, leading to the publications of 1905. By 1913, he was at the University of Berlin, where he completed the general theory of relativity. By 1933, when Hitler came to power, Einstein was in America and, being Jewish, he just stayed here, accepting a position at the Institute for Advanced Study at Princeton, where he remained for the rest of his career. (*AIP Niels Bohr Library*)

Apparently, theoretical physics was missing something. There is no ether wind (Michelson-Morley), *but* the ether does not move and the earth does (stellar aberration). Later measurements and observations demolished other hypotheses and hence set the stage for Albert Einstein (Figure 27-8).

27-5 Einstein's Postulates and the Special Theory

One of the famous Einstein articles of 1905 is titled "On the Electrodynamics of Moving Bodies." This article contains what is now known as the **special theory**

683

of relativity,[1] based on two **postulates** (assumptions) **of relativity** stated by Einstein and rephrased as follows:

1. The laws of physics are the same in all inertial systems. *No* preferred inertial system exists.
2. The speed of light in a vacuum has the same value c in all inertial systems.

Newton had stated a postulate similar to the first one of Einstein's but limited it to the laws of mechanics. Einstein extended it to include *all* the laws of physics.

Einstein's second postulate is interesting and different but not bizarre. While it is at odds with results expected by using previous methods for transforming velocities, it is consistent with Michelson-Morley, Fizeau's experiment, stellar aberration, and all the rest of the experiments so far mentioned. The test of its validity is, as usual, whether it leads to predictions which can be verified and whether there is any experimental evidence which contradicts it.

Using only the two postulates, Einstein derived a set of equations to apply when transforming variables between two reference frames in relative motion. These transformations had been first put forward by Lorentz and are called Lorentz transformations, but he had assumed them a priori in order to preserve Maxwell's equations. They give rise to some fascinating relationships of time intervals, length measurements, and mass measurements as viewed by observers in two different reference frames. Without deriving the whole set of transformation equations, we may obtain some of the more interesting relationships from gedanken experiments.

Time Dilation

In an interview in later years, Einstein stated that he had worked on the special theory for 10 years and thought his efforts fruitless "until at last it came to me that time was suspect!" We have already seen that when we are attempting to communicate between reference frames, it is not possible to get agreement on simultaneity. In general, it is only possible to say that two events occur at the same time in a single reference frame. Einstein was saying that there may be no such thing as universal time. For this reason, we use different symbols to represent the variable time in different reference frames.

Consider a gedanken experiment performed with special clocks, like the one shown in Figure 27-9 that consists of a tiny, flashing light source S in a special circuit with a light detector D which is shielded from direct light. The circuit is separated from a plane mirror M by a distance l. A flash of light by the source is reflected by the mirror back to the detector. Upon receiving the light, the detector triggers the circuit to cause the light source to flash again. The rate of flashing is then used as a measure of the passage of time. Suppose we have such a clock at rest in our reference frame and an identical clock in a different frame, say a spaceship, which is moving at a speed v with respect to our frame. The interval between flashes of our clock is equal to the time it takes for light to travel the distance l to the mirror and return to the detector. This is simply

Figure 27-9 A light clock consisting of a source S emitting light pulses, a mirror M at a distance l, and a detector D next to the source. A light pulse moves from S to M and back to D.

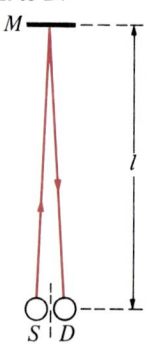

---

[1] The *special* theory applies to inertial reference frames and is really a special case of the **general theory of relativity**, developed later, which is applicable to reference frames accelerating with respect to each other.

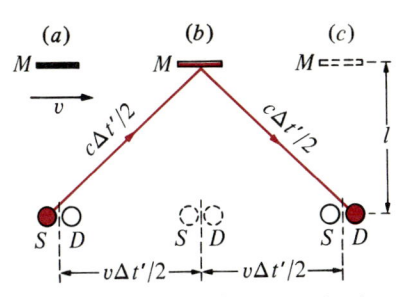

Figure 27-10 A light clock moving at speed v perpendicular to its length as seen by a stationary observer emits a light pulse at position a. Only the light emitted in a forward direction arrives at the mirror when the mirror is at position b. This light reflects and moves diagonally to the detector at position c.

$$\Delta t = \frac{2l}{c} \quad (27\text{-}2)$$

This is the proper time for our clock. **Proper time** is the time observed on a clock which is in the same reference frame as the observer.

Now we wish to determine the time interval at which the clock in the moving spaceship will flash *as we observe it*. Assume that the clock is moving perpendicular to its length, as in Figure 27-10. Using $\Delta t'$ as the symbol for time between flashes, we see that the flash takes half that time to reach the mirror and the other half to return to the detector. From the pythagorean theorem, we have

$$\left(c\frac{\Delta t'}{2}\right)^2 = l^2 + \left(v\frac{\Delta t'}{2}\right)^2$$

Solving for $\Delta t'$, we obtain

$$\Delta t' = \frac{2l}{\sqrt{c^2 - v^2}}$$

$$\Delta t' = \frac{2l/c}{\sqrt{1 - v^2/c^2}} \quad (27\text{-}3)$$

Substituting from Equation 27-2, we obtain

$$\Delta t' = \frac{\Delta t}{\sqrt{1 - v^2/c^2}} = \gamma \, \Delta t \quad (27\text{-}4)$$

The interval $\Delta t'$ is the time between flashes of the moving clock as measured in our stationary reference frame. We recognize the factor γ from Equation 27-1, and we know that γ is greater than 1. Thus $\Delta t'$ is greater than Δt, which tells us that the moving clock takes longer to tick or that the moving clock has slowed. This effect is known as **time dilation**.

We should point out that time dilation is a real effect and is not valid only for light-pulse clocks. All clocks which are moving with respect to an observer will run more slowly than if they were stationary. This includes biological clocks for growth, aging, and decay. Let us also note that the effect is reciprocal. After all, with respect to an observer in the spaceship, it is we and our clock which are moving. Observation made from the ship shows that our clock is running slow and the ship's clock shows its proper time.

27-5 Einstein's Postulates and the Special Theory

685

Example 27-1 In a spaceship moving at exactly nine-tenths the speed of light (0.9c) with respect to you, there is a pendulum clock. The occupants of the spaceship have adjusted the pendulum to have a period of exactly 2.00 s in its own frame. What would be the period of the pendulum as measured by you?

Solution We must apply Equation 27-4 for time dilation, recognizing that 2.00 s is the proper time Δt and that $v = 0.9c$.

$$\Delta t' = \frac{\Delta t}{\sqrt{1 - v^2/c^2}}$$

$$= \frac{2.00 \text{ s}}{\sqrt{1 - (0.9c)^2/c^2}} = \frac{2.00 \text{ s}}{\sqrt{1 - (0.9)^2}}$$

$$= \frac{2.00 \text{ s}}{\sqrt{1 - 0.81}} = \frac{2.00 \text{ s}}{\sqrt{0.19}} = 4.59 \text{ s}$$

Thus, you conclude from your measurement that the pendulum takes 4.59 s to complete one cycle although only 2.00 s passes in the spaceship.

Length Contraction

Suppose the light clock in the spaceship is turned through 90° so that it is aligned in the direction of motion. It will still keep the same time since the time interval between flashes will not have changed. However, what we observe from our stationary system will be much different, as we see in Figure 27-11. From our perspective, the mirror is moving away from the source and the detector is moving toward the mirror. The light flash takes a longer time t'_1 (as we measure it) to reach the mirror, because after the flash, the mirror moves a distance vt'_1 before the light catches it. Further, we observe that the light takes a shorter time t'_2 to make it back to the detector after reflection, as the detector will have moved toward the mirror by a distance vt'_2. This is just like that part of the beam in the Michelson-Morley experiment directed along the arm of the interferometer which moves parallel to the direction of motion. Note also that we have designated the length as l' rather than l because that distance is parallel to the direction of relative motion. We anticipate that we may obtain a different value for it than that measured by observers in the spaceship. The time interval we observe between ticks of the clock is equal to the sum of t'_1 and t'_2. We have for those variables that

$$t'_1 = \frac{l'}{c - v} \quad \text{and} \quad t'_2 = \frac{l'}{c + v}$$

Therefore

$$\Delta t' = \frac{l'}{c - v} + \frac{l'}{c + v}$$

which yields

$$\Delta t' = \frac{2l'/c}{1 - v^2/c^2} \qquad (27\text{-}5)$$

Figure 27-11 (a) The light clock moving in the direction of its length as seen by a stationary observer emits a light pulse when the mirror is at position 1, but the light does not arrive at the mirror until it is at position 2. (b) The return pulse is reflected from the mirror at the instant ending part a, but the apparatus advances rightward to meet it after time t'_2.

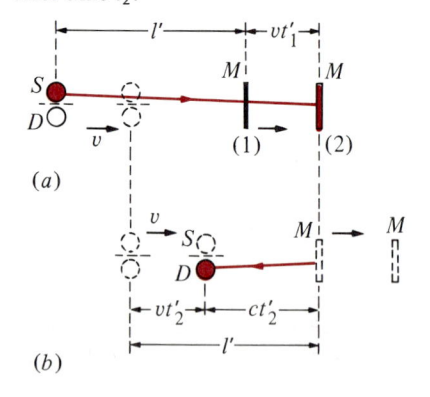

Since the expression on the right represents the time interval between flashes of the moving clock as measured in our stationary reference frame, it must be equal to that on the right side of Equation 27-3:

$$\frac{2l/c}{\sqrt{1 - v^2/c^2}} = \frac{2l'/c}{1 - v^2/c^2}$$

Solving for l' yields

$$l' = l\sqrt{1 - \frac{v^2}{c^2}} = \frac{l}{\gamma} \tag{27-6}$$

Thus, the apparent length in the direction of motion has decreased. We call this a relativistic **length contraction**.

Equation 27-6 is precisely of the form of the contraction hypothesized by FitzGerald and Lorentz to explain the null result of Michelson and Morley. It is still called the **FitzGerald contraction** or the **Lorentz contraction** or the **Lorentz-FitzGerald contraction**. Einstein derived the relationship from the simplest and most general postulates, whereas FitzGerald had assumed it. Also, Lorentz and FitzGerald had retained the concept of absolute time, whereas Einstein abandoned it. The speed of light is far removed from the realm of ordinary human experience. In Figure 27-12, we show a fanciful view in an imagined universe where the speed of light is 30 km/h.

The length of an object as measured in a frame of reference in which it is at rest is called its **proper length**. It is important when approaching a problem in special relativity to sort out who is in which reference frame, who is doing the measuring, and what is the proper frame for the proper value of the quantity.

Example 27-2 A moving meterstick is measured and determined to be exactly 0.600 m long. (a) What is the speed of relative motion between the observer's frame of reference and the frame of the meterstick? (b) Can you tell which is moving?

Solution (a) This is a straightforward application of Equation 27-6, recognizing that the proper length of the meterstick is 1.00 m.

$$l' = l\sqrt{1 - \frac{v^2}{c^2}}$$

$$\frac{l'}{l} = \sqrt{1 - \frac{v^2}{c^2}}$$

$$\left(\frac{l'}{l}\right)^2 = 1 - \frac{v^2}{c^2}$$

$$v = c\sqrt{1 - \left(\frac{l'}{l}\right)^2} = c\sqrt{1 - \left(\frac{0.600}{1.00}\right)^2}$$

$$= c\sqrt{1 - 0.36} = \sqrt{0.64}\, c = 0.8c$$

(b) A major point of the special theory is that no physical measurement can be made which will determine absolute motion. Thus, not only can *you* not tell which is moving, there is *no* way to determine which is moving.

(a)

(b)

(c)

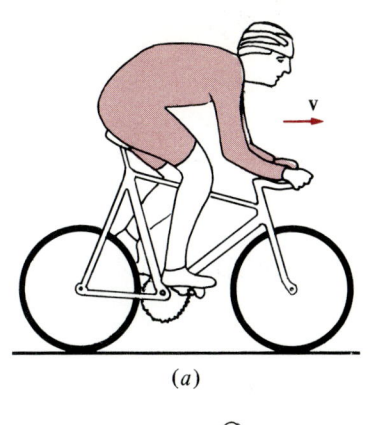

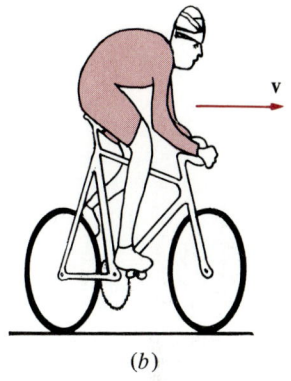

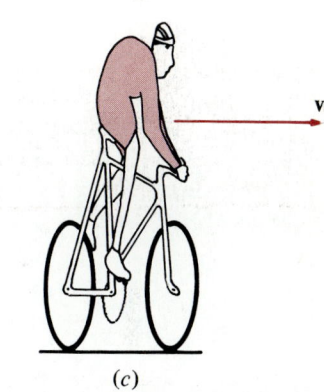

Figure 27-12 A glimpse of a universe in which the speed of light is 30.0 km/h. The racing bicycle and cyclist are being viewed by observers who are stationary with respect to the road. In these three views, the speed of the bicycle along the road is (a) 1.50 km/h, (b) 16.0 km/h, and (c) 22.0 km/h. Can you imagine what the cyclist sees when he looks at the stationary observers?

Experimental Confirmation

Cosmic radiation provides some evidence for confirmation of both time dilation and length contraction. High-speed particles from space (primary cosmic rays) interact near the top of the atmosphere, over 3.00×10^4 m from the ground, to produce muons (among other things). Muons are unstable subatomic particles which have a mass slightly more than 200 times the mass of the electron. Muons produced in the laboratory decay with a half-life of about 1.50×10^{-6} s. That is, half of all muons decay in this time, half the remainder in the next 1.50×10^{-6} s, and so on. At the speed of light (3.00×10^8 m/s), a muon would travel only 450 m in 1.50×10^{-6} s. Suppose a particular muon lasted for 20 times its half-life (a probability of less than one in a million). You can easily show that even if it is traveling at the speed of light, it will only travel 9 km and not the 30 km required to reach the surface of the earth. In fact, without involving relativity, you can calculate that the probability a muon will travel 30 km before decaying is less than 1×10^{-20}. However, if time dilates for a muon approaching the speed of light, the muon should last much longer for a stationary observer, perhaps long enough to reach the lower atmosphere.

A crucial experiment using cosmic muons was conducted in 1941 by Bruno Rossi and D. B. Hall. They measured the numbers of cosmic muons at an altitude of about 1900 m above sea level and compared them with the numbers measured at sea level. If there were no time dilation, they would expect to measure about 20 times more muons at the higher altitude. The actual result was about 1.4 times as many at the higher altitude. This is in excellent agreement with time dilation if the muons are moving at speeds greater than 99 percent that of light.

Example 27-3 Suppose a muon is traveling at a speed of $0.9999c$ with respect to the earth and decays in 1.50×10^{-6} s in its own reference frame. (a) How long will it last in the earth's frame? (b) How far will it travel during that time?

Solution (a) We recognize that the time interval is dilated and apply Equation 27-4.

$$\Delta t' = \frac{\Delta t}{\sqrt{1 - v^2/c^2}}$$

$$= \frac{1.50 \times 10^{-6} \text{ s}}{\sqrt{1 - (0.9999c/c)^2}}$$

$$= 1.06 \times 10^{-4} \text{ s}$$

(b) This solution is simply the product of the answer to part *a* times the speed of the muon (assumed to be *c* to three significant figures).

$$d = c \, \Delta t' = (3 \times 10^8 \text{ m/s})(1.06 \times 10^{-4} \text{ s})$$
$$= 3.18 \times 10^4 \text{ m}$$

From the point of view of an observer in the muon's reference frame, the muon has its proper decay time. However, that observer sees the muon at rest and the earth rushing to meet it. Thus, for that observer the distance has been shortened, and the muon can last the journey because it does not have to go nearly as far as we would measure.

Example 27-4 How far is the 3.00×10^4 m distance to the earth as measured by an observer in the muon's reference frame?

Solution We remind ourselves that 3.00×10^4 m is the proper length and apply Equation 27-6.

$$l' = l\sqrt{1 - \frac{v^2}{c^2}} = 3.00 \times 10^4 \text{ m} \sqrt{1 - \left(\frac{0.9999c}{c}\right)^2}$$

$$= 424 \text{ m}$$

27-5 Einstein's Postulates and the Special Theory

In 1971, U.S. physicists J. C. Hafele and R. E. Keating conducted an experiment with four cesium atomic clocks in an effort to confirm relativistic effects on measured time. The clocks were flown around the world in commercial airplanes, spending approximately 45 h in easterly flight and 45 h in westerly flight. Careful records were kept of speeds, altitudes, and directions of flight. The clocks were compared to reference clocks at the U.S. Naval Observatory both before and after the flights. When the flight data were analyzed, calculations were carried out to determine what changes, if any, the clocks should show to be consistent with special relativity. The predictions derived from special relativity were that during the eastward flight the clocks should have lost 40 ± 23 nanoseconds (ns) and during the westward flight they should have gained 275 ± 21 ns. Comparison with the reference clocks showed that during the eastward flight the clocks actually lost 59 ± 10 ns and during the westward flight they actually gained 273 ± 7 ns. Agreement between prediction and observation presents excellent evidence for time dilation, which is 100 percent consistent with the special theory. The magnitude of the effect for normal velocities, however, may give you some idea why it is not commonly known or used.

Example 27-5 Table 5-3 lists the average distance to the moon as 3.84×10^8 m. Suppose a lunar spaceflight travels at a rate of 3.00 km/s. Compared to a clock on earth, (a) how much time would a clock on the spaceship lose during the flight? (b) How long would it have to fly at that speed in order to lose 1.00 s?

Solution We first determine the time interval of the flight as measured on earth and then determine the time dilation. We must be careful. The proper time is that which the spaceflight clock shows in its own frame. Therefore, the time interval as measured on earth is $\Delta t'$, as in Equation 27-4.
(a) The loss in this part is just $\Delta t' - \Delta t$.

$$\Delta t' = \frac{d}{v} = \frac{3.84 \times 10^8 \text{ m}}{3.00 \times 10^3 \text{ m/s}} = 1.28 \times 10^5 \text{ s}$$

This is about $1\frac{1}{2}$ days.

$$\Delta t' = \frac{\Delta t}{\sqrt{1 - v^2/c^2}} = \frac{\Delta t}{(1 - v^2/c^2)^{\frac{1}{2}}}$$

or

$$\Delta t = \Delta t' \left(1 - \frac{v^2}{c^2}\right)^{1/2}$$

Note that v is small compared to c. The ratio of v to c is 10^{-5}, and the ratio of v^2 to c^2 is only 10^{-10}. Thus we may expand the square-root term in a binomial series and ignore all but the first two terms:

$$\Delta t = \Delta t' \left(1 - \frac{1}{2}\frac{v^2}{c^2}\right)$$

The loss is

$$\Delta t' - \Delta t = \frac{\Delta t'\, v^2}{2\, c^2} = \frac{1.28 \times 10^5 \text{ s}}{2} \times 10^{-10}$$

$$= 6.40 \times 10^{-6} \text{ s} \quad \text{or} \quad 6.40 \ \mu\text{s}$$

(b) Here we simply calculate how many times the lost interval divides into 1.00 s and multiply this result by the time required for the flight. To lose a full second, the flight must go on for N times as long as this flight, where

$$N = \frac{1.00 \text{ s}}{6.40 \times 10^{-6} \text{ s}}$$

The time the flight must go on is then

$$T = \frac{1.00 \text{ s}}{6.40 \times 10^{-6} \text{ s}} (1.28 \times 10^5 \text{ s}) = 2.00 \times 10^{10} \text{ s}$$

This is more than 630 years, by which time the spaceship would be more than 10 times farther from the sun than the planet Pluto. It is no wonder that time dilation is not obvious.

Any number of experiments—many of which are related to the decay of unstable particles moving at high speeds—have been explained by time dilation. In fact, there has never been an experimental contradiction of relativistic time dilation or length contraction.

27-6 Velocities and Masses in Relativity

At this point, we are convinced that time intervals and lengths as measured by observers in two different reference frames do not agree. Also we believe that the measured values are related by transformation equations depending on the relative velocity of the two frames and the speed of light. It is not difficult, therefore, to accept that the velocities of a body as measured by observers in two different reference frames are likewise related by a transformation equation which depends on the relative velocity and the speed of light. After all, to obtain an object's velocity, we must divide a displacement which the body undergoes by the time interval required for the displacement.

Figure 27-13 shows two different reference frames and a moving object whose velocity can be measured in either frame. We have designated the frames as the unprimed frame and the primed frame, and we have labeled all corresponding coordinates and variables with the same letter, adding a prime (') to each variable in the primed system. We take the primed system to be moving at a speed v in the positive x direction with respect to the unprimed frame. The components of velocity of the moving object may be measured and designated

Figure 27-13 Two inertial reference frames with a relative velocity **v** along the x and x' direction. The velocity **V** of the object may be measured by observers in either frame.

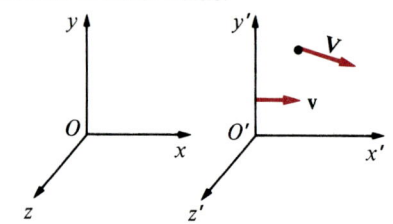

as V_x, V_y, V_z and V'_x, V'_y, V'_z in the unprimed and primed systems, respectively. At low speeds, where we may neglect relativistic effects, we would write the relationship between these components as

$$V'_x = V_x - v \qquad V'_y = V_y \qquad V'_z = V_z$$

When the special relativity conditions are applied, derivation yields the following transformation equations:

$$V'_x = \frac{V_x - v}{1 - V_x v/c^2}$$

$$V'_y = \frac{V_y \sqrt{1 - v^2/c^2}}{1 - V_x v/c^2} \qquad (27\text{-}7)$$

$$V'_z = \frac{V_z \sqrt{1 - v^2/c^2}}{1 - V_x v/c^2}$$

These transformation equations are obtained for two systems in relative motion along the x direction (and x' direction). There is no loss of generality in this because one can always choose the coordinates such that the x and x' directions are along the direction of motion. However, if coordinates had been chosen so that the relative motion was in an arbitrary direction, the transformation equations would have slightly different forms.

Example 27-6 A spaceship is passing the earth with a speed of $0.6c$ when an observer on the earth sees a missile catching up to the spaceship and measures its speed at $0.9c$. How fast is the missile traveling according to an observer on the spaceship?

Solution We designate the earth as the unprimed system and the spaceship as the primed system moving in the x direction. Then $v = 0.6c$ and $V_x = 0.9c$, and we apply the first of Equations 27-7.

$$V'_x = \frac{V_x - v}{1 - V_x v/c^2} = \frac{0.9c - 0.6c}{1 - (0.9c)(0.6c)/c^2}$$

$$= \frac{0.3c}{1 - 0.54} = 0.652c$$

Since velocities as measured in different reference frames do not follow the classical transformation, we find that we are compelled to redefine the concept of mass in order to preserve the form of the law of conservation of linear momentum. The mass transformation equation takes the form

$$m = \frac{m_0}{\sqrt{1 - v^2/c^2}} \qquad (27\text{-}8)$$

where m is the mass of an object as measured by an observer moving at a speed v with respect to the object and m_0 is the object's mass as measured in its own frame. The quantity m_0 is most often called the **rest mass**. Equation 27-8 is interpreted as saying that inertial mass (resistance to acceleration) increases as the speed of an object increases. We can see that the mass approaches infinity as the object approaches the speed of light.

The increase of mass with speed is actually the easiest of the relativity predictions to test for confirmation. While measured speeds of macroscopic objects (even spaceships) have rarely exceeded one ten-thousandth the speed of light, subatomic particles are another story. Particles from naturally occurring radioactivity often are traveling at speeds approaching that of light. Most of the measurable cosmic-ray particles have speeds very near the speed of light. In the laboratory electrons, protons, and other particles are routinely accelerated to speeds greater than 90 percent the speed of light. This gives us a way to check Equation 27-8.

In charged-particle accelerators, the number of complete circles traced out per unit time by a particle of charge q and mass m moving perpendicular to a magnetic field B is

$$f = \frac{Bq}{2\pi m}$$

This then is the frequency of the accelerating voltage required for a cyclotron. This frequency can be kept constant provided the quantities on which it depends do not change. While B and q remain constant, special relativity predicts that the mass becomes larger as the speed increases, according to Equation 27-8. If this is the case, we must begin to decrease the frequency of the accelerating voltage when the speed becomes large enough that the increase in mass is significant. The alternative is to increase the magnetic field at exactly the same rate as m is increasing in order to keep the frequency constant. The first method is characteristic of the synchrocyclotron, while the second method is used in the synchrotron. Both changes must be made exactly in accordance with Equation 27-8 or the accelerator simply will not work. This is more evidence in support of the special theory of relativity.

Example 27-7 Calculate the speed required so that an electron's mass is (a) double and (b) 10 times its rest mass.

Solution First we note that it is not necessary that we use the mass of the electron. The same solution will apply to any particle or object. We simply want to find v from Equation 27-8 whenever m is twice and 10 times m_0. Thus we solve that equation for v as a function of m/m_0.

$$m = \frac{m_0}{\sqrt{1 - v^2/c^2}}$$

$$\frac{m_0}{m} = \sqrt{1 - \frac{v^2}{c^2}}$$

$$\left(\frac{m_0}{m}\right)^2 = 1 - \frac{v^2}{c^2}$$

$$\frac{v^2}{c^2} = 1 - \left(\frac{m_0}{m}\right)^2$$

$$v = \sqrt{1 - \left(\frac{m_0}{m}\right)^2}\, c$$

(a) $$v = \sqrt{1-\left(\frac{m_0}{m}\right)^2}\,c = \sqrt{\frac{3}{4}}\,c = 0.866c$$

(b) $$v = \sqrt{1-\left(\frac{m_0}{m}\right)^2}\,c = \sqrt{1-\left(\frac{1}{10}\right)^2}\,c$$
$$= \sqrt{0.99}\,c = 0.995c$$

As you could gather from Equation 27-8 or extrapolate from the last example, a particle whose rest mass is greater than zero will never be measured as traveling at the speed of light or greater by an observer in an inertial frame. Special relativity leads to the conclusion that it is not possible to travel at the speed of light and that the speed of light constitutes an upper limit on speeds for material bodies.

27-7 Work, Energy, and Momentum in Special Relativity

Since we first introduced the concept of energy in Chapter 7, we have adhered to the work-energy principle: a net force doing work on a body or system gives rise to a change in its kinetic energy. The magnitude of the change in kinetic energy is equal to the work done. Further, we assumed that the change manifests itself by a change in the square of the body or system speed. In symbols, we write the classical relationship as

$$W_{net} = \Delta KE = \tfrac{1}{2}mv_f^2 - \tfrac{1}{2}mv_i^2$$

where v_f and v_i are the final and initial values of the body's speed. The kinetic energy is defined as $mv^2/2$. According to Equation 27-8, the mass also increases with increasing speed. Thus, at least part of the work is required to increase the mass rather than the speed. Furthermore, as the body's speed becomes very near that of light, virtually all of any additional work increases the mass. Interesting! Perhaps mass itself is a form of energy. Of course, it is!

The equation $E = mc^2$ has appeared in so many places and has become so famous that it is recognized by virtually every literate person. It first appeared in the fourth of Einstein's 1905 papers, titled "Does the Inertia of a Body Depend on Its Energy Content?" In that paper Einstein considered the radioactive decay by gamma-ray (high-energy electromagnetic radiation) emission of a particle initially at rest. With no external forces acting on the particle, both momentum and energy must be conserved during the decay. The result of Einstein's calculations is that to preserve the conservation laws, mass in the amount Δm must be lost during a gamma-ray decay. The decrease in mass is given by

$$\Delta m = \frac{\Delta E}{c^2}$$

where ΔE is the amount of energy given off in the form of radiation. This is found to be exactly correct for gamma decay. Moreover, it is even valid for decays in which a particle breaks up into smaller particles that carry away kinetic energy. The kinetic energy gained by the smaller particles is the result of a net decrease in mass. In fact, the grand conclusion to which the paper leads is that all mass is a form of energy and that a body of mass m contains an amount of energy E given by

$$E = mc^2 \qquad (27\text{-}9)$$

We carry Equation 27-9 one step further and state that whenever the body is at rest it contains an amount of energy E_0, where we have

$$E_0 = m_0 c^2 \qquad (27\text{-}10)$$

and m_0 is the body's rest mass and E_0 is called its **rest energy.**

Equations 27-9 and 27-10 lead to an expression for the kinetic energy of a body. Excluding mechanical potential energy, the mass of Equation 27-9 is greater than m_0 because of the body's speed. Then the difference between the total energy and the rest energy is equal to the kinetic energy of the body, or

$$KE = mc^2 - m_0 c^2 \qquad (27\text{-}11)$$

This is the relativistic expression for kinetic energy, where m is given by Equation 27-8.

To show Equation 27-11 consistent with our previous expression for kinetic energy, let us consider Equation 27-11 whenever v is small with respect to c. We write

$$KE = \frac{m_0 c^2}{\sqrt{1 - v^2/c^2}} - m_0 c^2$$

$$= m_0 c^2 \left[\left(1 - \frac{v^2}{c^2}\right)^{-1/2} - 1 \right]$$

We recall that for small v we may apply the binomial series expansion, keeping only the first two terms.

$$KE = m_0 c^2 \left[1 + \frac{1}{2}\left(\frac{v^2}{c^2}\right) - 1 \right]$$

Finally, we obtain

$$KE = \tfrac{1}{2} m_0 v^2$$

Example 27-8 Calculate the kinetic energy of a proton moving at $0.2c$, using (a) the nonrelativistic expression and (b) the relativistic expression.

Solution We simply apply the nonrelativistic expression and the relativistic expression in succession to solve both (a) and (b).

(a) $KE_N = \tfrac{1}{2} m_0 v^2 = \tfrac{1}{2}(1.66 \times 10^{-27} \text{ kg})[(0.2)(3.00 \times 10^8 \text{ m/s})]^2$

$\qquad = 2.99 \times 10^{-12}$ J

(b) $KE_R = m_0 c^2 \left[\left(1 - \frac{v^2}{c^2}\right)^{-1/2} - 1 \right]$

$\qquad = (1.66 \times 10^{-27} \text{ kg})(3.00 \times 10^8 \text{ m/s})^2 \left[\left(1 - \frac{0.04 c^2}{c^2}\right)^{-1/2} - 1 \right]$

$\qquad = 3.08 \times 10^{-12}$ J

The nonrelativistic expression results in an error confirmable by experimental measurement. Note, however, that even at this tremendous speed, the error is quite small.

One more relationship which is often useful includes the relationship between the magnitude of linear momentum and total energy. If we use the symbol p to represent the magnitude of the linear momentum of a particle of mass m and speed v, we have

$$p = mv \qquad (27\text{-}12)$$

This form is retained in special relativity, but m represents the relativistic mass given by Equation 27-8. Using only that equation and substituting the definitions of momentum and total energy (see Problem 27-22), we obtain

$$E^2 = p^2c^2 + m_0^2c^4 \qquad (27\text{-}13)$$

27-8 Additional Thoughts on Relativity

Einstein's postulates from which the consequences of special relativity are derived are similar to many other postulates in that they may be considered to be very well educated guesses. However, special relativity is overwhelmingly supported by experimental evidence. It is used routinely as the proper description of the physical world in many areas of physics. The lifetime of radioactive particles in flight, the energy release in all nuclear reactions, and many other real measurable events confirm the time dilation and relativistic mass equation. We have already mentioned that there is no experimental evidence which contradicts time dilation or length contraction. Indeed, there is no experimental contradiction of *any* conclusion of relativity.

Another important point is that the equations of relativity reduce to the well-known, well-established equations of newtonian mechanics whenever the speed involved is not a significant fraction of the speed of light. That is reassuring because we know newtonian mechanics performs very well in its sphere. Thus, newtonian mechanics is not incorrect, only limited, whereas relativity yields something more general.

Even given all that we have said, many people still have trouble accepting special relativity. It is certainly not due to the mathematics. The mathematics we have used in the chapter is no more complex than that used with many other concepts we have already covered. Most students find that special relativity is not so much difficult to understand as it is hard to believe. Part of the difficulty is that its conclusions and predictions are outside the realm of common, everyday experience. Nothing that we encounter moves at speeds which are significant with respect to the speed of light. If we consistently moved at speeds near that of light (either because we and our machines moved much, much faster or because the speed of light were much slower), relativity would describe commonplace mechanics, and newtonian, nonrelativistic mechanics would be considered applicable only to the growth of vines and perhaps the movement of snails and a few of the slower turtles.

And yet, many people are still made very uncomfortable by some of the implications (real or apparent) of special relativity. A partial understanding sometimes leads people to question their values, speculate that it changes the meaning of life, and lunge into areas of endeavor outside physics or even science. For example, one reaction is that "if special relativity is accurate, then everything is relative." *That is decidedly not the case.* Relativity is a theory of measurement. It concludes only that the *measured* values of time, length, and mass will be different in different reference frames or relatively moving coordinate systems. The proper or rest values of each of those quantities are absolutes. Further, the speed of light is itself absolute. In point of fact, special relativity is based on the laws of physics being absolutes. Without special relativity, the form of all those laws as well as the value of the speed of light *would* be relative, depending on the observer's state of motion.

There would be no special relativity if the speed of light were infinite or if it were possible to transmit signals to any other place at an infinite speed. Looked at in that way, we conclude that disagreements about simultaneity and all of relativity stem from this limitation of finite speed of signal transmission. Lack of an infinite transmission speed does not offend common sense. In fact, most common sense would be more offended by a theory based on being able to transmit a signal to any part of the universe in zero time.

Your reaction might be "So what? What about length contraction? Do things really shrink in the direction of motion? Is length contraction real?" In order to answer, let us consider a more familiar physical occurrence. When there is relative motion between a source of sound and an observer, the observer measures a different frequency than that of the source. Is this measured frequency of the Doppler effect real? We know that if the observer were brought to rest with respect to the source, the observer would measure the rest (proper) frequency. Whenever there is relative motion, the observer measures a frequency shift. Both effects, frequency shift and length contraction, may be viewed in the same way. They are real in the same sense that measurements are real. They are apparent in the sense that the proper quantities have not changed.

Minimum Learning Objectives

After studying this chapter, you should be able to:
1. Define:

   aberration of starlight
   absolute system
   accelerating frame of reference
   cosmic radiation
   ether drag
   fringe
   gedanken experiment
   general theory of relativity
   inertial frame of reference
   interferometer
   length contraction
   Lorentz-Fitzgerald contraction
   luminiferous ether
   mass-energy equivalence
   Michelson-Morley experiment
   postulates of relativity
   proper length
   proper time
   reference frame
   relativistic effect
   rest energy
   rest mass
   special theory of relativity
   time dilation
   transformation

2. Describe the scientific concerns and evidence leading up to the formulation of the theory of relativity.
3. Distinguish differing inertial frames of reference.
4. Discuss the significance of the Michelson-Morley experimental results, including a description of its apparatus.
5. Quantitatively describe the aberration of starlight.
6. Explain why clocks in separate inertial frames moving with high relative velocity will not agree.
7. Calculate the duration of a time interval specified for frame *A* as measured in frame *B*, knowing the relative velocity of the frames.
8. Calculate the length of an object specified for frame *A* as measured in frame *B*, knowing the relative velocity of the frames.
9. Calculate the velocity of an object whose motion is specified in frame *A* as measured in frame *B*, knowing the relative velocity of the frames.

10. Calculate the mass of an object specified in frame A as measured in frame B, knowing the relative velocity of the frames.
11. Calculate the amount of mass difference involved in an energy change, and vice versa.
12. Cite and discuss several experimental confirmations of the predictions of relativity.
13. Calculate the kinetic energy of a moving body from its measured mass and its rest mass.
14. Understand that relativity guarantees that physics is *not* relative.

Problems

27-1 A plane can fly at a speed c with respect to the air. The pilot is assigned to travel due east a distance l and return to the starting point. Give an expression (in terms of l, c, and wind speed) for the total time required for the round-trip if (a) there is no wind, (b) the wind is due east at a speed v, and (c) the wind is due south at a speed v.

27-2 The planet Mercury moves in an orbit of average radius 5.79×10^7 km with a period of 87.99 days. What would be the angular size of stellar aberration on Mercury?

27-3 A K meson (kaon) has an average lifetime in its own reference frame of 1.20×10^{-8} s. (a) How fast must it be traveling in order that the average kaon formed in the cosmic rays at an altitude of 1.00×10^4 m reach the earth? (Use six significant figures.) (b) What distance does the kaon travel as measured in its own frame?

27-4 A radio signal is beamed from earth to a colony on Mars whenever earth and Mars are separated by 8.00×10^{10} m as measured by an observer on earth. (a) How much time would the earth observer say the signal takes to reach Mars? (b) How much time would an observer traveling at $0.75c$ parallel to the beam say the signal took?

27-5 At what speed will the time interval as measured by a moving observer differ from that measured by a stationary observer by (a) 1 percent and (b) 10 percent?

27-6 Suppose the speed of light were 500 m/s and you were in an airplane traveling between two cities on opposite coasts of the United States at a speed of 325 m/s. (a) If the proper distance between the cities is 4.00×10^3 km, how long would it appear to you? (b) How much time passes on your clock during the trip? (c) How much time passes during your trip for someone on earth? (Assume the earth is an inertial frame of reference.)

27-7 A space traveler is in a ship moving at $0.99c$ for 1 year as determined on the earth. Determine how much older the traveler is.

27-8 Calculate the speed at which a meterstick must be moving parallel to its length in order that an observer at rest measures it to be 0.200 m long.

27-9 The average lifetime of pi mesons (pions) is 2.60×10^{-8} s. (a) Calculate the average lifetime as measured in the laboratory through which a beam of pions is traveling at a speed of $0.9c$. (b) Determine the distance the pions travel during this time.

27-10 Suppose the speed of light were 50.0 m/s. How much time would a runner believe it takes him to run a 1500-m race if the official timer says it takes him 250 s?

27-11 Photographs taken in a particle-detection experiment show unknown particles forming and decaying after leaving an average track length of 4.60 cm. Other considerations indicate that their speed was $0.999c$. Calculate the proper lifetime of the particles.

27-12 A missile is launched from a spaceship in a direction opposite to the motion of the spaceship. The missile's speed as measured by an observer on the spaceship is $0.8c$ but is only $0.5c$ as measured by an observer on earth. What is the speed of the spaceship with respect to the earth?

27-13 (a) Determine the mass of a proton traveling at $0.99999c$. (b) How long would a 20.0-m laboratory at rest measure to an observer in the moving proton's reference frame?

27-14 Take the escape velocity of the earth as 1.12×10^4 m/s, and determine the percentage increase in the mass of a rocket moving at that speed. (*Hint:* Use the binomial series expansion.)

27-15 Obtain an equation by using the correct transformations showing how the density of a body depends on its speed.

27-16 Two primary cosmic rays approach the earth from opposite directions at speeds of $0.75c$ each. How fast are they moving with respect to each other?

27-17 A wave that has a frequency of 500 kHz is being broadcast from a planet toward a spaceship. (a) Assuming a continuous wave, what is the spacing of wave crests as measured by an observer on the planet? (b) Consider a particular wave crest spaced exactly 1.0000 s of wave travel as seen from the earth from the instantaneous position of the spaceship. If the spaceship is approaching the earth at $0.200c$, in what elapsed time on the ship will the wave crest be encountered? (c) At what speed will an observer inside the ship measure the arriving crest? (d) How many wave crests will be counted as arriving at the ship between the initial instant in part b and the arrival of the particular crest? (e) What will be the frequency of the signal as measured on the ship?

27-18 A radioactive particle is moving at a speed of $0.5c$ as measured by a laboratory observer. It decays by emitting

697

an electron which moves at 0.9c with respect to the particle. What should be its speed as measured by the laboratory observer if the electron is emitted (a) in the direction of motion and (b) opposite to the direction of motion?

27-19 Two particles in the laboratory approach each other with velocities of 0.8c and −0.8c as measured by a laboratory observer. Calculate the velocity of one particle with respect to the other.

27-20 Use the velocity transformation to show that a light beam sent from a station on earth will be measured as moving at c by an observer in a rocket moving away from the earth at 0.6c.

27-21 A tiny spaceship which has a rest mass including its payload of 1.00×10^4 kg is moving at a speed of 0.9c. Calculate its (a) relativistic mass, (b) total energy, and (c) kinetic energy.

27-22 Use Equation 27-9 for relativistic energy and Equation 27-12 for momentum including relativistic mass to derive $E^2 = p^2c^2 + m_0^2c^4$.

27-23 An electron is accelerated through a potential difference of 2.20×10^6 V. Calculate (a) its kinetic energy in electronvolts and in joules, (b) its relativistic mass, and (c) its speed.

27-24 Calculate the rest mass energy in joules and in megaelectronvolts of (a) an electron and (b) a proton. (c) How fast is a proton moving when its kinetic energy is 150 MeV?

27-25 What is the magnitude of the magnetic field required to keep (a) an electron of kinetic energy 3.00×10^5 eV in a circle of radius 1.50 m and (b) a proton of kinetic energy 250 MeV in a circle of radius 4.00 m?

27-26 A spaceship is built to visit the stars. It has a rest mass of about 10 times the *Apollo* spacecraft we sent to the moon, or 3.00×10^4 kg. The craft is to accelerate to 0.900 times the speed of light for its journey. Energy to produce this acceleration is to be diverted from all domestic energy use in the United States. All transportation, heating, electric generation, industrial energy, etc., will be cut to half for the duration of the acceleration. Assume 100 percent efficient conversion. Daily energy use in the United States is about 2.25×10^{17} J. (a) How many days would the nation have to endure short energy use to accelerate the ship? (b) Expressed differently: if a generation is 20 years, about how many generations of Americans would have to shiver through cold, dark winters to accelerate the craft?

27-27 A 1000-MW nuclear power plant operating at full power delivers 1.00×10^9 J of electric energy and 2.00×10^9 J of waste heat every second. This energy comes from the conversion of mass. (a) Calculate the rate at which mass is converted to energy. (b) How much mass is converted, or burned, in 1 year? (Compare your answers to those of Problem 27-28.)

27-28 A 1000-MW coal-fired power plant operating at full power delivers 1.00×10^9 J of electric energy and 2.00×10^9 J of waste heat every second. This energy comes from burning coal, which provides 2.80×10^7 J for every kilogram burned. (a) Calculate the amount of coal which must be burned every second. (b) How much coal is burned in a year? (c) How much mass was converted to energy during the year's operation? (Compare your answers to those of Problem 27-27.)

27-29 The sun is considered to radiate energy as a blackbody, and from its surface temperature of 5.8×10^3 K, it is calculated to emit about 3.90×10^{26} J/s. (a) Calculate the amount of mass converted to energy in the sun every second. (b) If the sun has existed for 20 billion years (2.00×10^{10} years) and has radiated energy at the same rate all those years, how much mass has it lost? (c) What percentage of its present mass of 1.99×10^{30} kg has it lost?

27-30 Show that the speed of a relativistic particle may be written as $v = pc^2/E$.

27-31 Calculate the mass and momentum of a particle of kinetic energy 100 MeV if the particle is (a) a proton and (b) an electron.

27-32 A ship in deep space, far from any gravitational disturbance, has a gigantic scoop in front to sweep up interstellar dust and feed it to the engines. This matter is converted to energy, which is then used to accelerate the ship. (a) Ignore relativity and assume an acceleration of 9.80 m/s². How long would it take the ship to reach the speed of light starting from rest? (b) What is the ship's actual speed after the time interval calculated in part a? (Hint: The kinetic energy added to the ship during part a is $\frac{1}{2}m_0c^2$. Why?) (c) Assume energy is added at a constant rate. How long will it take the ship to reach a speed of 0.995c?

27-33 A particle of rest mass m_0 and speed 0.5c makes a perfectly inelastic collision with a stationary particle of rest mass $4m_0$ (that is, they stick together after the collision). Calculate (a) the initial momentum of the system, (b) the initial kinetic energy, (c) the final speed of the system, and (d) the final rest mass of the system.

27-34 Suppose a K meson (kaon) which has a rest-mass energy of 498 MeV decays at rest into two pi mesons (pions) of rest-mass energy 135 MeV each. Calculate (a) the kinetic energy of each pion, (b) the speed of each pion, and (c) the momentum of each pion.

27-35 A lambda particle of rest-mass energy equal to 1.12×10^3 MeV traveling at a speed of 0.5c decays into a proton (rest-mass energy = 938 MeV) and a pion (rest-mass energy = 140 MeV). Calculate the maximum speed the pion can have after the decay.

27-36 Protons moving perpendicular to a magnetic field of 0.480 T trace out a circle of radius 7.00 m in a charged-particle accelerator. Calculate the magnitude of the proton's (a) speed, (b) linear momentum, (c) kinetic energy, (d) total energy, and (e) mass.

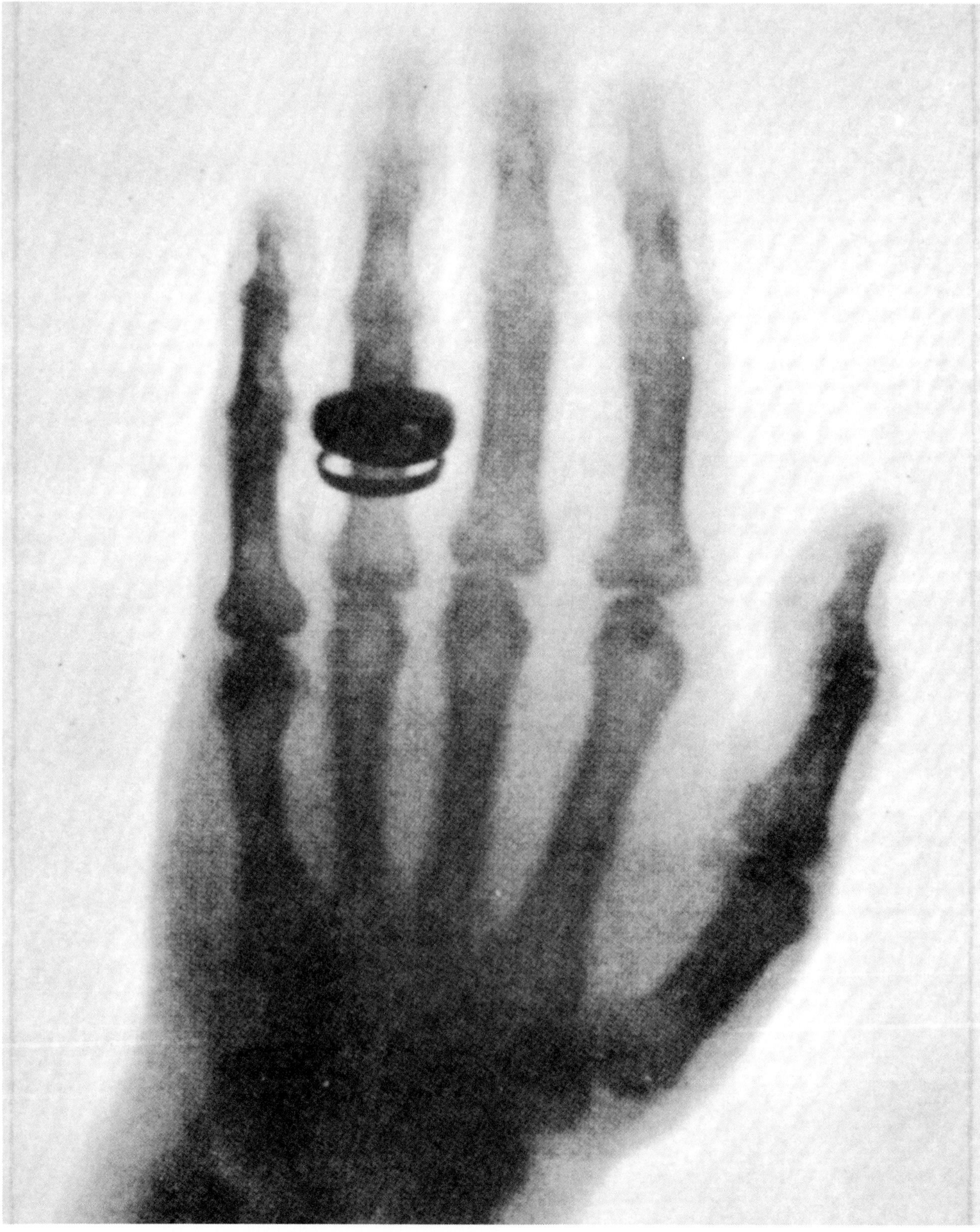

28 Birth of Quantum Physics

28-1 Introduction

Physics to date has had four great theoretical revolutions, two in the twentieth century. Difficulty with absolute reference frames was resolved in the theory of relativity (1905). In the same time period, research on another problem was evolving toward a whole new understanding of nature. The theory of how energy was radiated and absorbed was clearly incomplete. An electron in a light bulb filament, for example, was pictured as absorbing more and more energy until it was in violent oscillation. But, accelerating charge generates electromagnetic waves. Thus, the electron was pictured as radiating a continuous string of waves with the same frequency as its oscillation, losing energy in the process. This was clearly wrong, but what was a more accurate picture? The answer produced a fundamental new insight into nature in the quantum theory.

The four theoretical revolutions in physics were newtonian mechanics, Maxwell's electromagnetic theory, Einstein's relativity, and now, quantum theory. Unlike its predecessors, quantum theory is not ascribed in its final form to a single individual nor did it appear at a single moment. Still, physics, with four such reworkings of fundamental understandings, leads the sciences in such revolutions of thought.

28-2 The Nature of Radiant Energy

You will recall that any object at a temperature greater than absolute zero radiates energy. Any object which can absorb all the radiant energy falling on it at any wavelength (and can thus also radiate freely at any wavelength) is called

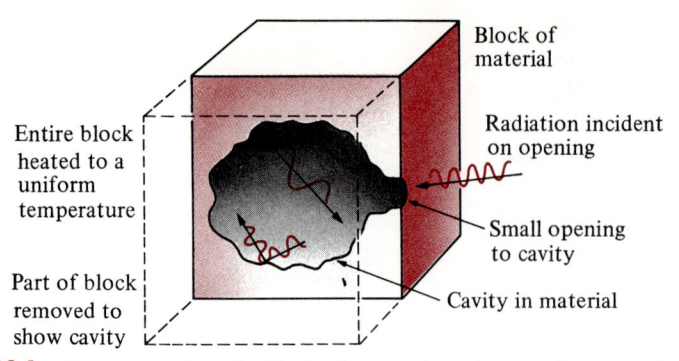

Figure 28-1 A cutaway view of a block of material used to simulate a perfect blackbody. Rays escaping through the opening sample the frequencies and intensities of the equilibrium radiations of the cavity and may be analyzed by devices like the spectroscope to yield a blackbody radiation curve for a particular temperature.

a perfect **blackbody.** For our purposes, the sun is approximately a blackbody, although in balance it is losing energy and therefore is not in thermal equilibrium. The best approximation to a perfect blackbody in a laboratory is to create a cavity in a large block of material and to treat a small opening into this cavity as the surface of a blackbody (Figure 28-1). Radiation entering the opening will reflect from the inner walls until it is absorbed, while the radiations emitted by the walls will attain an equilibrium mixture of frequencies and intensities sampled by the radiations emerging from the opening.

This emerging energy had been studied extensively before 1900 and found to produce the intensity vs. wavelength curve of Figure 28-2. As the temperature is increased, the curve increases in intensity for all wavelengths, but the peak of the curve, the most intensely radiated wavelength λ_{max}, shifts to a smaller value. The curve always retains the same shape, and, with a suitable change of scales, the curve for any one temperature can be made coincident with that for any other. The area under the curve is a measure of the total energy radiated and thus increases rapidly with increasing temperature. Stefan's measurements in 1879 showed that the energy radiated per unit time (the power) was proportional to the fourth power of the body's absolute temperature. Later, using his recently discovered statistical physics, Boltzmann derived the same expression theoretically from basic physical principles. (We encountered the Stefan-Boltzmann law, $\Delta Q/\Delta t = \sigma A T^4$, in Section 15-6; you may want to review it.)

Wilhelm Wien (1864–1928), a German physicist, discovered a simple empirical relationship, now known as the **Wien displacement law,** between the most intensely radiated wavelength λ_{max} and the absolute temperature T:

$$\lambda_{max} T = \text{constant} = 2.898 \times 10^{-3} \text{ m} \cdot \text{K} \qquad (28\text{-}1)$$

For example, at room temperature of 20°C, or 293 K, the metal heating element of an electric range will radiate most strongly at a wavelength of 9.89 μm. This is in the deep-infrared and invisible to the eye. If we switch on the element, its temperature increases. Watching the element, we notice its surface appearance changing from shades of gray to a bright red, and even orange, where it stops since the coil has reached its maximum power delivery. We are seeing a part of the short-wavelength tail on the left of the curve of Figure 28-2 as the entire curve moves to shorter wavelengths. Even at its highest temperature, λ_{max}

Figure 28-2 Intensity vs. wavelength curves for perfect blackbodies. As temperature increases, λ_{max} moves to a shorter wavelength as described by the Wien displacement law. The comparative energy radiated for similar bodies is given by the area under each curve and increases as the fourth power of absolute temperature according to the Stefan-Boltzmann law.

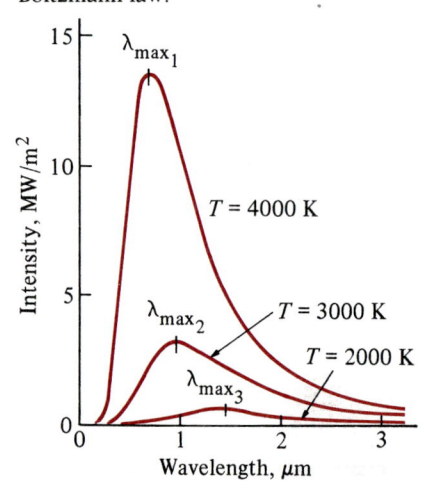

remains in the infrared portion of the spectrum for the coil. Indeed, even the filament of an incandescent light bulb never has λ_{max} reach wavelengths as short as those of visible light. To do so would require $\lambda_{max} \approx 0.70$ μm for the longest-wavelength red visible to the eye. This would correspond to a temperature of the body of about 4100 K, which exceeds the melting point of most substances.

Example 28-1 The radiation of the sun peaks at a wavelength of 503 nm. (a) What is the surface temperature of the sun? (b) Assuming it to be a perfect blackbody, what power is radiated by each square meter of the sun's surface? (c) The **solar constant** (the measured intensity of solar radiation in space above the earth's atmosphere) is 1.353 kW/m². From the known distance of the sun to the earth, what total area does this predict for the sun's surface? What radius? (d) Does this value of the sun's radius compare well enough with the accepted value for our assumption in part b to be correct?

Solution (a) We may calculate the surface temperature by the use of the Wien displacement law:

$$T = \frac{2.898 \times 10^{-3} \text{ m} \cdot \text{K}}{5.03 \times 10^{-7} \text{ m}} = 5760 \text{ K}$$

(b) Now, using this temperature with Stefan's law, we calculate

$$I = \epsilon\sigma T^4 = (1)[5.670 \times 10^{-8} \text{ W}/(\text{m}^2 \cdot \text{K}^4)](5760\text{K})^4$$
$$= 6.24 \times 10^7 \text{ W/m}^2$$

as the power radiated per square meter of surface area of the sun.

(c) The radius of the earth's orbit is 1.495×10^8 km, and a surface in space at this distance oriented perpendicular to the sun's rays receives 1.353 kW/m² of radiant power. If we enclosed the sun with a spherical shell of this radius, it would intercept the entire power output of the sun. This sphere would have an area of

$$A = 4\pi r^2 = 4\pi (1.495 \times 10^{11} \text{ m})^2 = 2.809 \times 10^{23} \text{ m}^2$$

and each of these square meters of area would receive the solar constant of power; thus

$$P = AI = (2.809 \times 10^{23} \text{ m}^2)(1.353 \times 10^3 \text{ W/m}^2) = 3.80 \times 10^{26} \text{ W}$$

which is the calculated output of the entire sun. In part b we calculated the power radiated per square meter of surface; thus

$$A_{calc} = \frac{P}{I} = \frac{3.80 \times 10^{26} \text{ W}}{6.24 \times 10^7 \text{ W/m}^2} = 6.09 \times 10^{18} \text{ m}^2$$

which is the calculated area of the sun, or, using $A = 4\pi r^2$,

$$r_{calc} = \sqrt{\frac{A}{4\pi}} = \sqrt{\frac{6.09 \times 10^{18} \text{ m}^2}{4\pi}} = 6.96 \times 10^8 \text{ m}$$

(d) The calculated radius of the sun in part c is precisely the measured value (to three significant figures), and we may conclude that it is reasonable to assume that the sun behaves like a perfect blackbody, since that assumption led to this result.

Toward the end of the nineteenth century, the Stefan-Boltzmann law was found to predict experimental results quite well. Encouraged by the success of a theoretical equation for the power radiated by a blackbody, physicists anticipated that theoretical work would yield an equation to predict the exact shape of the intensity vs. wavelength curve of Figure 28-2, starting with known physical principles. This was not to be as straightforward as was expected.

28-3 Blackbody Radiation

In 1893, Wien offered an equation for the curve of Figure 28-2 based on curve-fitting techniques as well as on some physical arguments that were not universally accepted. Still, the equation agreed fairly well with the experimental curve. His final equation was

$$I = C_1 f^3 e^{-C_2 f/T} \tag{28-2}$$

where I is the radiated energy per unit time per unit area of a blackbody at temperature T as a function of the variable frequency f of the radiation, and C_1 and C_2 are constants. The prediction of Wien's formula is plotted as a dashed line on Figure 28-3, along with the experimental curve (in black). At high frequencies, the prediction fits the curve well, but at lower frequencies, it strays from the experimental result.

Near the turn of the century, the English physicist Lord Rayleigh (1842–1919) derived an expression for the curve from basic principles. His work, modified by Sir James Jeans (1877–1946), another English physicist, became known as the **Rayleigh-Jeans law,** which is given by

$$I = C_3 f^2 T \tag{28-3}$$

where, again, I, T, and f are as defined in Equation 28-2 and C_3 is a (different) constant. This expression was derived from agreed-upon, basic relationships in physics, using no steps or assumptions with which workers in the field could disagree. Thus its impact was serious. It is clearly wrong (as shown in the solid color line of Figure 28-3). At low frequencies, the agreement with the experimental curve is perfect. However, approaching and exceeding λ_{max}, the curve does not reverse but continues to grow without bound. Since the area under the curve is the power radiated per unit area of blackbody, this predicts that any object above absolute zero radiates infinite power! To the physics community, this was worse than just impossible or wrong (which it clearly was); it was a blow to the physical theories from which the Rayleigh-Jeans law was derived. The crisis in predicting high-frequency behavior was labeled the **ultraviolet catastrophe,** and to many it demonstrated that there must be a basic flaw in physical theory as then understood. That view ultimately was proved to be correct.

Max Planck (Figure 28-4) was aware of both the Wien formula and the Rayleigh-Jeans law. He reasoned that if one equation fits exactly at high frequencies and the other fits exactly at low frequencies, then there must be a formula which incorporates both. Planck obtained an expression that was consistent with both equations. This expression was then solved for the blackbody radiation and yielded

$$I = \frac{C_1 f^3}{e^{C_2 f/T} - 1} \tag{28-4}$$

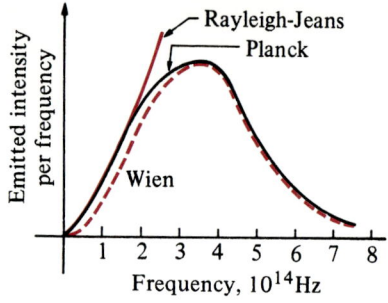

Figure 28-3 A blackbody radiation curve (black), plotting intensity vs. frequency (rather than wavelength). Wien's formula for this curve (dashed color line) matches the curve well at high frequencies but strays at lower frequencies. The Rayleigh-Jeans law (solid color) fits the curve exactly for frequencies below its peak but continues to infinity beyond the peak of the real curve.

Planck worked from the laws of thermodynamics and obtained an expression for the second time rate of change of entropy (S) with internal energy (U), or $\Delta(\Delta S/\Delta U)/\Delta U$, where ΔU is allowed to approach zero. This yielded a form for I as a function of f, as shown in Equation 28-4.

Not only does Equation 28-4 agree with the experimental curve at all frequencies, but we can also show that it reduces to Wien's formula at high frequencies and to the Rayleigh-Jeans law at low frequencies. At high frequencies, the exponential term in the denominator of Planck's equation becomes very large and the -1 is negligible. The immediate result of dropping the -1 is that we obtain Wien's formula. At very low frequencies, the exponent $C_2 f/T$ becomes very small. The exponential may then be expanded in a power series with all terms negligible except the first two. Thus

$$e^{C_2 f/T} = 1 + \frac{C_2 f}{T} + \frac{1}{2!}\left(\frac{C_2 f}{T}\right)^2 + \cdots \approx 1 + \frac{C_2 f}{T}$$

Using this in Planck's equation, we have

$$I = \frac{C_1 f^3}{(1 + C_2 f/T) - 1} = \frac{C_1}{C_2} f^2 T = C_3 f^2 T$$

which is the Rayleigh-Jeans law.

At this point, Planck knew he had the proper equation to describe blackbody radiation. However, he still had not obtained it in a derivation from fundamental principles. He had only used the mathematics associated with thermodynamics and two known but not completely accurate formulas to obtain that equation. It has been said that at this point Planck felt somewhat like the student who had looked in the back of the textbook to obtain the answer to a problem but did not know quite how to get it from fundamental principles.

In developing his derivation, Rayleigh had made the classical assumption that energy in the cavities associated with a blackbody radiator exists as standing electromagnetic waves between every two points in the cavity (Figure 28-5). Furthermore, the radiation emerging through the aperture was postulated to be a representative sample of all standing waves. However, assuming that the standing waves have nodes at each end, classical theory predicts many more short-wavelength waves than long-wavelength ones. Waves can stand between two points only when the points are separated by at least one-half a wavelength. Short wavelengths can form standing waves with nodes at each end in short distances, medium distances (more loops), and long distances (many loops). Long wavelengths have more limited possibilities. Therefore, in any cavity, theory predicts many more probabilities for short-wavelength (high-frequency) standing waves. If the radiation from the aperture is a representative sample of all standing waves in the cavity, there would be more at higher frequencies (shorter wavelength). Planck needed a model (physical

Figure 28-4 Max Planck (1858–1957), winner of the 1918 Nobel Prize in physics for his work on the quantum theory. The idea that energy could be radiated only in "packages" with an energy proportional to the frequency of the radiation was such a radical departure from classical ideas that even Planck had trouble believing it at first, but he lived to see this idea become the cornerstone of twentieth-century physics. Although anti-Nazi, he was easily the most renowned scientist remaining in Germany during World War II. Planck had been stripped of official positions and titles for trying to intercede with the Nazis for Jewish scientists prior to the war. His son was executed as part of the group which tried to assassinate Hitler. (*Culver Pictures*)

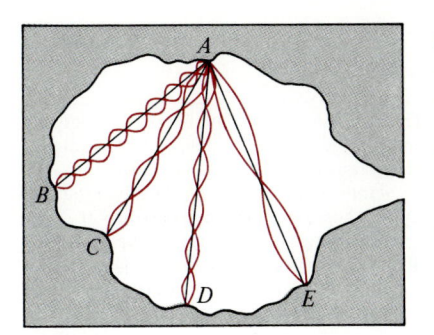

Figure 28-5 Classical theory predicts that energy resides as standing waves between any two points inside a blackbody cavity. Thus there would be standing waves between point *A* and points *B, C, D,* and *E* at (perhaps) different frequencies. Each endpoint must be a node with an integral number of half-wavelengths between it and point *A*. A moment's reflection should convince you that geometry greatly favors two points being the correct separation for a given high frequency than for a low frequency, where nodes must be farther apart. Thus, classical theory predicts the "runaway" radiation at high frequencies of the ultraviolet catastrophe.

and/or mathematical) which would limit the number of high-frequency standing waves or at least reduce their contribution to the radiation.

Planck postulated that the standing waves in a cavity were absorbed and quickly reemitted by atoms of the inside wall of the cavity. According to Maxwell's electromagnetic theory, a charged particle oscillating at a frequency f radiates electromagnetic waves of the same frequency. Also, the energy of any oscillating system depends on its frequency. Things must move faster to go back and forth at higher frequencies, and they thus have more energy. All this means is that one would expect the energy available for radiation at a given frequency to be proportional to that frequency. Even using all this, Planck chose an equal distribution of energies and derived . . . the ultraviolet catastrophe again! In an effort to obtain something different, he decided to use a technique of calculus. He assumed that the amount of energy *emitted* by a given oscillator of frequency f would be proportional to f. That is, he wrote

$$\epsilon = hf \qquad (28\text{-}5)$$

where the Greek lowercase epsilon (ϵ) is the amount of energy emitted by an oscillator of frequency f and h is the proportionality constant which makes the proportionality an equation.

Planck wrote that the energy stored in the cavity and emitted at a particular frequency would be considered to be whole-number multiples of ϵ. That is, energy could be emitted in amounts of hf, $2hf$, $3hf$, etc., but no fraction in between. Planck was essentially saying that the atom oscillators could only absorb and emit energy in discrete bundles, or jumps, and not just any amount. These discrete bundles have now been named after the Latin word **quantum**, which means "how much?" Planck expected to be able to allow h to approach zero later in the derivation to eliminate it. He discovered when he tried to do so that his equation immediately became the Rayleigh-Jeans law. However, if he allowed h to remain finite, he obtained

$$I = \frac{2hf^3}{c^2(e^{hf/kT} - 1)} \qquad (28\text{-}6)$$

which is identical to Equation 28-4 provided we identify C_1 as $2h/c^2$ and C_2 as h/k. The quantities c and k are the speed of light and the Boltzmann constant. The unknown constant h has now become known as **Planck's constant** and has the value in SI units of

$$h = 6.63 \times 10^{-34} \text{ J} \cdot \text{s}$$

Admittedly, this is a very small number, but it is still finite and has a very definite value which can be determined from fitting Planck's law, Equation 28-6, to the experimental curve.

The conclusion that Planck felt obligated to make was extremely unorthodox. In fact, it is said that when he delivered his paper on the subject, he appeared to be almost apologetic. He admitted that he was forced to conclude, in what is now known as the **quantum hypothesis**, that energy is absorbed and emitted by atomic oscillators in discrete amounts. Planck left no doubt that he still believed electromagnetic radiation existed in continuous waves. He simply stated that the proper equation required the assumption that the energy of the continuous waves could only change in discrete amounts. This was in 1900, nearly 5 years before Albert Einstein's revolutionary papers, including the one on special relativity, appeared.

The postulate that radiant energy is absorbed and emitted only in "chunks" of a magnitude proportional to frequency may not seem revolutionary at first glance. This is certainly true of those with limited experience in contemplating and dealing with the processes of emission-absorption of electromagnetic waves. Had this been your job, you would find this a new and startling view, difficult to accept.

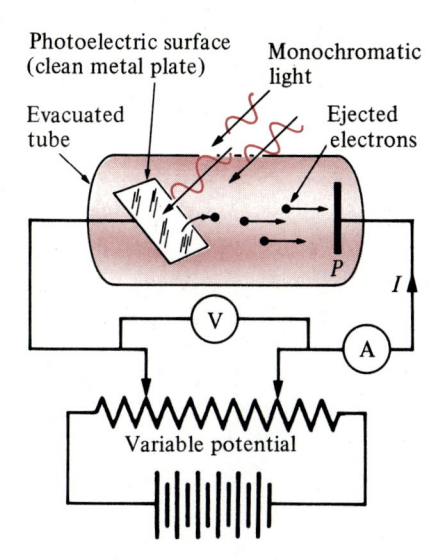

Figure 28-6 In this photoelectric apparatus, monochromatic light falls on a clean metal surface in a vacuum. Absorbed light may eject electrons from the metal, and a conventional current I is recorded by the ammeter.

28-4 The Photoelectric Effect

The quantum hypothesis was not compatible with classical physics. Nevertheless, it was readily accepted by many scientists because it was so successful in explaining and predicting experimental facts. No one, including Planck, believed that another front had opened in the ongoing scientific revolution. They felt that some explanation compatible with classical theory would emerge.

By the end of the nineteenth century, it was well known that electromagnetic waves, like light, could release electrons from a metal surface. Electrons in the metal absorb energy from the radiation and jump loose. Since light frees the electrons, this phenomenon is called the **photoelectric effect,** and Hertz is credited with its discovery.[1]

Figure 28-6 depicts an apparatus used to study the photoelectric effect. When the metal plate is illuminated by monochromatic light of constant intensity, electrons are ejected from the metal, and a current, called a photoelectric current, is detected by the ammeter. The accelerating voltage may be varied, and then the current produced vs. voltage used may be plotted to obtain curves like those shown in Figure 28-7a. The curves labeled 1, 2, and 3 are those obtained for a lower, an intermediate, and a higher incident radiation intensity, respectively. Note that each curve quickly levels off to a **saturation current** when the accelerating voltage is positive, but that they all have some finite value at zero voltage. It requires a negative voltage ($-V_0$) to stop current completely. The quantity V_0 is called **stopping potential** and is a measure of the maximum kinetic energy of the photoelectrons as they emerge from the metal. From conservation of energy, we have

$$KE_{max} = eV_0 \quad (28\text{-}7)$$

Figure 28-7 (a) Photoelectric current vs. voltage for monochromatic light of constant frequency but three different intensities. (b) The equivalent curves for the same surface exposed to light of two different frequencies of the same intensity are quite distinct.

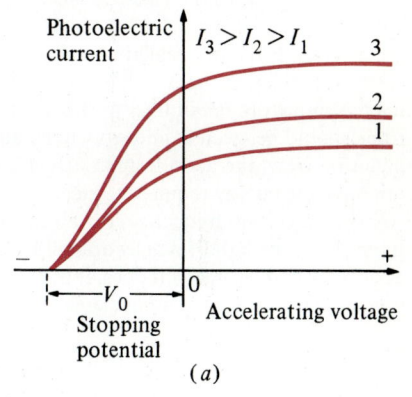

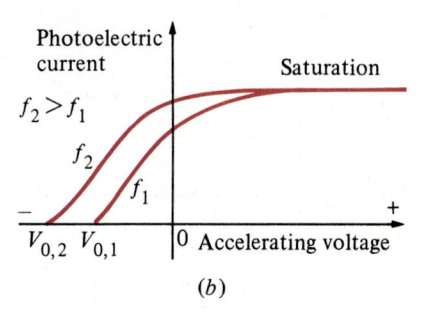

---

[1] Hertz first observed it while conducting the experiments which concluded in verifying Maxwell's theory of electromagnetism. He apparently noticed and recorded that the sparks he used to generate microwaves jumped more readily when the spark apparatus was bathed in sunlight. As other scientists became aware of this discovery, many began to investigate it. Phillip Lenard (1862–1947), a German, was the early leader, having the advantage of being a student of Hertz.

When two light sources of different frequency but the same intensity are used in succession, curves such as those of Figure 28-7b are the result. The same number of electrons is emitted, but the stopping potential is larger (more negative) for the higher frequency.

Many such photoelectric experiments yielded the following information:

1. The photocurrent is directly proportional to the intensity of the incident light.
2. The saturation current is independent of the frequency of light above a certain frequency. Below that frequency, the current gradually decreases and reaches zero at some **threshold frequency** f_0.
3. The value of f_0 depends on the type of metal but *not* on the intensity of light. A light source of frequency greater than f_0 will produce photoelectrons no matter how weak the source. When the frequency is below f_0, no photoelectrons are emitted even for the most intense incident light.
4. Above the threshold frequency, the maximum kinetic energy of the photoelectrons increases linearly with increasing frequency but is independent of intensity.
5. No matter how weak the light source, there is no noticeable time delay between the turning on of the source and the emission of the first electron.

Most of these details are inconsistent with classical electromagnetic theory. It is true that classically we would expect more electrons to be emitted for a greater incident intensity since more total energy is available. However,

Figure 28-8 (a) In the classical view of an electron being ejected from an atom by light, the plane of the light wave's electric field vector lies along a direction in which the electron is free to oscillate. As the wave with its varying **E** "washes" it, the electron accelerates back and forth, acquiring energy from the wave, eventually maybe enough to break free. In this view, the more intense the wave, the faster electrons should be ejected, and essentially all frequencies should work. (b) In the quantum view, the atom is approached by a quantum of light that either does or does not have the energy required to eject an electron. Energy depends on frequency, and so below a certain frequency, the electron cannot be ejected regardless of the intensity of the light. The electron either accepts the energy of the quantum, all of it, at once, or it does not. If it does, it instantaneously has the energy required to break free and current begins immediately.

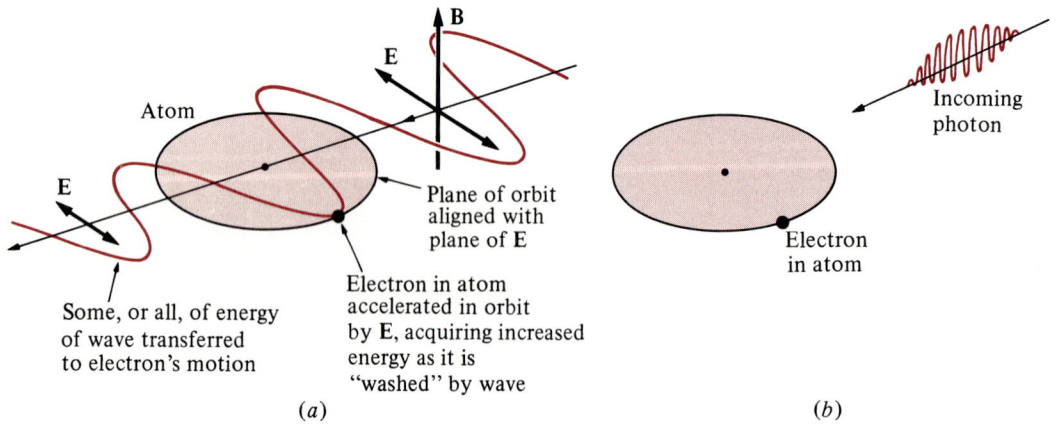

the other details on the list cannot be explained by classical theory and, in fact, directly contradict it. Classically, there should be no threshold frequency. As long as the light has enough intensity, there should always be electrons emitted. The maximum kinetic energy of the electrons should depend on the intensity rather than the frequency. Finally, we would expect a measurable time interval to pass between the instant the light is switched on and the instant when the first electron had accumulated enough energy from the arriving wave to be emitted (Figure 28-8a and b).

Example 28-2 When ultraviolet light of wavelength 1.00×10^{-7} m falls on the surface of aluminum in a photoelectric apparatus, it requires a negative voltage of 8.38 V to stop all photocurrent. That is, the stopping potential is $V_0 = 8.38$ V. Calculate the maximum kinetic energy of the photoelectrons.

Solution This is a straightforward application of Equation 28-7, which gives the value of the maximum kinetic energy in joules.

$$KE_{max} = eV_0 = (1.60 \times 10^{-19} \text{ C})(8.38 \text{ J/C}) = 1.34 \times 10^{-18} \text{ J}$$

This is also equal to 8.38 eV (electronvolts). Do you remember why?

Example 28-3 Suppose the beam from the ultraviolet light source of Example 28-2 is 4.00×10^{-3} m in diameter and delivers a power of 0.1 mW (1.00×10^{-4} J/s). Assuming classical theory is correct, calculate the minimum time necessary for an aluminum atom to absorb the energy required to eject an electron with the maximum kinetic energy. Take the diameter of the aluminum atom to be 1.10×10^{-10} m.

Solution We first calculate the energy per unit time falling on a unit area. This number multiplied by the area presented to the beam by one atom yields the energy per unit time being absorbed by the atom. The necessary time is then just the maximum kinetic energy divided by the energy per unit time.

$$\text{Cross section} = \pi r^2 = \pi(2.00 \times 10^{-3} \text{ m})^2 = 1.26 \times 10^{-5} \text{ m}^2$$

$$\text{Intensity} = \frac{\text{energy/time}}{\text{area}} = \frac{\text{power}}{\text{cross section}}$$

$$= \frac{1.00 \times 10^{-4} \text{ J/s}}{1.26 \times 10^{-5} \text{ m}^2} = 7.94 \text{ J/(s} \cdot \text{m}^2)$$

$$\text{Area of atom} = \pi r_A^2 = \pi(5.50 \times 10^{-11} \text{ m})^2 = 9.50 \times 10^{-21} \text{ m}^2$$

$$\frac{\text{Atom energy}}{\text{time}} = \left(\frac{\text{energy/time}}{\text{area}}\right)(\text{area of atom})$$

$$= [7.94 \text{ J/(s} \cdot \text{m}^2)](9.50 \times 10^{-21} \text{ m}^2)$$

$$= 7.54 \times 10^{-20} \text{ J/s}$$

$$\text{Minimum time required} = \frac{KE_{max}}{\text{energy/time}}$$

$$= \frac{1.34 \times 10^{-18} \text{ J}}{7.54 \times 10^{-20} \text{ J/s}} = 17.8 \text{ s}$$

Although it is clear that this time lag would certainly be discernible, we would actually obtain a longer time if the wave theory were a correct description of the phenomenon. We have neglected the energy necessary to break the electron free from the metal. This is not inconsiderable, as we will see.

One of the famous articles by Einstein in the 1905 issue of the *Annalen der Physik* was on the generation and transformation of light. In that article, Einstein put forth the explanation for the photoelectric effect which describes it accurately and is consistent with all known photoelectric experiments. Einstein had accepted Planck's hypothesis for blackbody radiation, but he carried it one step further. Instead of simply accepting that energy was absorbed and emitted only in discrete amounts by atomic oscillators, Einstein proposed that the quantum effect was a property of the radiation itself. That is, not only is radiation absorbed and emitted in discrete bundles, it travels in discrete bundles as well.

Consider a metallic surface irradiated by monochromatic light of frequency f. We call the difference in potential energy of an electron just outside the metal from that of an electron just inside the metal the **work function** W. Einstein reasoned that since the smallest bundle of energy has a value equal to hf, the radiation will eject electrons only if the quantum energy is greater than the work function. That is, the photoelectric effect will occur only if

$$hf > W$$

Further, he concluded that the maximum kinetic energy that an emitted electron can have is

$$KE_{max} = hf - W \qquad (28\text{-}8)$$

assuming that the energy of the quantum is completely absorbed by the electron. Equation 28-8 is known as Einsteins's **photoelectric equation**. When this is combined with Equation 28-7, we obtain

$$eV_0 = hf - W \qquad (28\text{-}9)$$

When a photoelectric experiment is conducted with a variable frequency source and the stopping energy eV_0 plotted as a function of frequency, we obtain a graph similar to Figure 28-9. Note that the slope of the line is equal to Planck's constant and that the intercept on the ordinate axis is equal to the negative of the work function of the metal. In general, the value of the work function is a different constant for each metal but is constant. Thus, if the maximum kinetic energy vs. frequency were plotted for photoelectrons of different metals, we would obtain a family of parallel straight lines. They would all have the same slope (Planck's constant) but different intercepts (their individual work functions). Einstein's equation is based on the assumption that a single quantum of electromagnetic energy, now called a **photon**, is completely absorbed in one photoelectric event. That is, the photon disappears, and the energy all goes into an increase in the potential and kinetic energy of a single electron associated with the metal.

Many photoelectrons are emitted with less than the maximum kinetic energy. The explanation is that the photoelectric effect takes place with an electron at some distance inside the metal. We think of this situation as requiring a certain amount of work to get the electron to the surface and then requiring

28 Birth of Quantum Physics

Figure 28-9 Stopping energy vs. frequency for photoelectrons. None are observed below the threshold frequency f_0, and the graph is a straight line with an x intercept f_0. The projected y intercept is the energy binding the photoelectric surface W, the work function of the metal. The slope of the line yields an experimental value of Planck's constant.

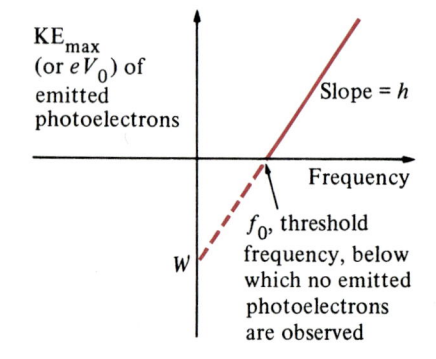

energy equal to the work function to "break it loose." In any case, the evidence is consistent with Einstein's interpretation and is so overwhelming that his explanation is accepted as fact. Several measured values of photoelectric work functions are given in Table 28-1.

TABLE 28-1 Photoelectric Work Functions for Some Selected Metals

Metal	W, eV	Metal	W, eV
Ag	4.73	K	2.24
Al	4.08	Mo	4.15
Au	4.82	Na	2.28
Cd	4.07	Ni	5.01
Co	3.90	Pb	4.14
Cs	1.98	W	4.52
Cu	4.70		

Example 28-4 Calculate (a) the energy of the photons of the ultraviolet light of Examples 28-2 and 28-3 and (b) the work function of aluminum.

Solution (a) The photon energy is simply the product of Planck's constant with the frequency:

$$\epsilon = hf = h\frac{c}{\lambda}$$

$$= (6.63 \times 10^{-34} \text{ J} \cdot \text{s}) \left(\frac{3.00 \times 10^8 \text{ m/s}}{1.00 \times 10^{-7} \text{ m}}\right)$$

$$= 1.99 \times 10^{-18} \text{ J} = \frac{1.99 \times 10^{-18} \text{ J}}{1.60 \times 10^{-19} \text{ C}} = 12.4 \text{ eV}$$

(b) The work function requires the application of Equation 28-8:

$$KE_{max} = hf - W$$

or

$$W = hf - KE_{max} = 12.4 \text{ eV} - 8.38 \text{ eV} = 4.05 \text{ eV}$$

This is in substantial agreement with the value given in Table 28-1.

The photoelectric effect shows that the quantum hypothesis of Planck and Planck's constant are not isolated curiosities associated with cavity radiation only. Indeed, Planck's constant is now accepted as one of the few universal constants of nature. It is interesting that the quanta of electromagnetic energy are thought of as little bundles of energy and remind us of the corpuscular theory of light. The quantum hypothesis indicates that light seems to act like a stream of particles. Many other phenomena which had been left unexplained surfaced in the decades after Einstein's paper. Explanations for many aspects of particle physics, atomic spectra, statistical mechanics, and even photochemistry required the quantum hypothesis.

28-5 Diffraction of X-Rays

In 1838, Michael Faraday conducted an experiment using a partially evacuated glass tube with electrodes (metallic conductors) embedded in each end. When he connected the leads from an electrostatic generator to the electrodes and applied a large voltage, he observed a purple glow around the cathode (the negative electrode). Later experimenters discovered that the voltage caused some unidentified rays to leave the cathode and travel in straight lines toward the anode (the positive electrode). Experiment showed that these rays, called **cathode rays,** could cause certain materials to glow, or fluoresce, and they could travel only a few centimeters in air. Cathode rays were subsequently shown to be electrons.

In 1895, Wilhelm Roentgen (1834–1923), a German physicist, was studying the fluorescence of certain minerals induced by cathode rays. To observe the fluorescence, Roentgen worked in a darkened room and enclosed the glowing tube in a light-tight black box. When he switched on the voltage to the tube, he was startled to see a glow from a fluorescent screen placed several meters away outside the box. Roentgen knew that cathode rays could not traverse that distance in air, and he concluded that some other unknown ray originated in the tube and penetrated the light-tight box to reach the screen. To distinguish these rays from other rays, he called them **x-rays.** Roentgen conducted additional experiments with x-rays and very shortly thereafter announced their discovery. The x-ray photograph he had made of the bones in his wife's hand caused a real sensation. Figure 28-10 shows how x-rays are presently produced.

Within a few years, it was postulated that x-rays were nothing more than electromagnetic waves of very short wavelength. Roentgen's own attempts to verify the wave nature of x-rays by showing the necessary interference and diffraction effects failed because of the very short wavelengths involved. Early experiments indicated that if x-rays are waves, the wavelength must be on the order of 1.00×10^{-10} m or less. Recall from Chapter 26 that the single-slit diffraction minimum condition is

$$m\lambda = w \sin \theta_m \qquad m = 1,2,3, \ldots \qquad (26\text{-}10)$$

Also, the condition for constructive interference by using a diffraction grating is

$$m\lambda = d \sin \theta_m \qquad m = 0,1,2, \ldots \qquad (26\text{-}11)$$

Either effect provides convincing evidence for a wave nature, and both can be observed rather readily by using visible light. However, to get reasonable values for the angle θ, the width of the slit w and the grating spacing d must not be very much larger than the wavelength λ of the radiation being observed. For example, if the grating spacing is 20 times the wavelength, the first-order constructive interference maximum occurs at an angle less than 3°. Even if $d = 10\lambda$, θ_1 is less than 6°. Thus it appears that to observe the constructive interference effect with x-rays of wavelength 1.00×10^{-10} m, a diffraction grating with spacing less than 1.00×10^{-9} m (less than 10 times the wavelength) is required. Such a grating would be impossible to construct since 1.00×10^{-9} m is on the same order as the spacing between atoms in most solids. It is not hard to imagine why Roentgen had trouble.

The problem did not go without a solution for very long, however. Max Von Laue (1879–1960), a German physicist, suggested that the atoms themselves had the proper spacing and could be used as the diffraction grating. In 1912, he produced interference patterns with x-rays passing through a sodium chloride crystal and recorded on a photographic plate. William H. Bragg (1862–1942), an English physicist, immediately recognized the significance and constructed a **crystal spectrometer** to study this phenomenon. Collaborating with him was his son Lawrence Bragg (1890–1971), who worked out a way to analyze the patterns produced.

A crystal can be considered to be a three-dimensional diffraction grating which works with reflected rather than transmitted waves. After x-rays incident on a crystal are reflected by the individual atoms, they may combine to form

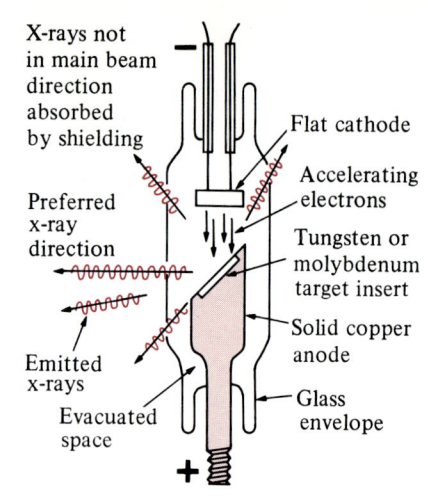

Figure 28-10 The modern Coolidge tube for producing x-rays. When a very high voltage is applied across the ends of any evacuated tube, x-rays may result at the anode if any atoms of the atomic number of aluminum or higher are present. An electron from the cathode is accelerated to high energy to strike the target, where its abrupt deceleration causes it to emit its energy as an x-ray.

28-5 Diffraction of X-Rays

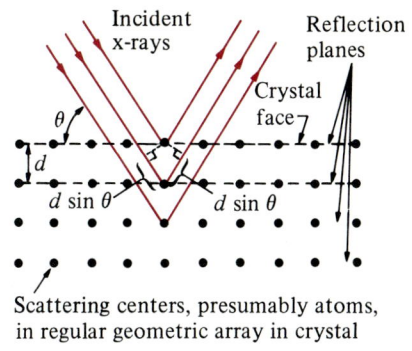

Figure 28-11 Regularly spaced planes of atoms in a crystal act for x-rays much as a diffraction grating acts for light. X-rays incident on the planes at an angle θ show positive interference at a reflection angle of θ if and only if the additional distance traveled to reach a lower plane and get back is a whole number of wavelengths.

interference patterns. Figure 28-11 represents a two-dimensional slice of a crystal with x-rays incident from above. Although the x-rays are reflected, or scattered, from the individual atoms in all directions, constructive interference will occur only for specific directions. The regularly spaced atoms form sets of planes from which we may consider the x-rays to be reflected. We see in the figure that part of the incident beam is reflected from the top plane, a portion from the second plane, and so on. Constructive interference will occur whenever the difference in path length of waves reflected at adjacent planes is a whole number of wavelengths. We show that the atoms and hence the planes are separated by a distance d. Careful examination of the figure should convince you that the path difference is equal to $2d \sin \theta$. Note that the angle of incidence θ is measured from the beam of x-rays to the crystal face rather than from the normal as we did with all our optic measurements. While it may seem confusing, we accept it since it is conventional and standard in crystallography. Therefore, the condition for constructive interference of x-rays is

$$m\lambda = 2d \sin \theta_m \qquad m = 1,2,3, \ldots \qquad (28\text{-}10)$$

Equation 28-10 is often called the **Bragg diffraction equation,** and the phenomenon which gives rise to these interference patterns is called **Bragg diffraction.**

Example 28-5 The x-rays from a cathode-ray source using molybdenum as the anode have a frequency of 4.13×10^{18} Hz. When used with a sodium chloride crystal, first-order constructive interference occurs at an angle $\theta = 7.4°$. Calculate (a) the wavelength of the x-rays and (b) the spacing between reflection planes (atoms) in the crystal.

Solution (a) We use the relationship between the speed, frequency, and wavelength of a wave.

$$c = f\lambda$$

or

$$\lambda = \frac{c}{f} = \frac{3.00 \times 10^8 \text{ m/s}}{4.13 \times 10^{18} \text{ Hz}} = 7.26 \times 10^{-11} \text{ m}$$

(b) This part requires a straightforward application of the Bragg diffraction equation.

$$m\lambda = 2d \sin \theta_m$$

or

$$d = \frac{\lambda}{2 \sin \theta_1} = \frac{7.26 \times 10^{-11} \text{ m}}{2 \sin 7.4°} = 2.82 \times 10^{-10} \text{ m}$$

The reflection planes chosen for Figure 28-11 are not the only parallel planes which can be drawn. Figure 28-12 shows several other possibilities for the same type of array. Using incident x-rays of known wavelength, crystallographers can analyze the patterns produced to determine the distance between many sets of planes (called **Bragg planes**). From these data, they can then reconstruct the arrangement and spacing of atoms in the crystal. Bragg diffraction leaves little room for doubt of the wave nature of x-rays.

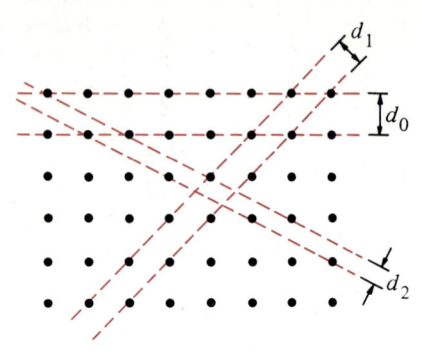

Figure 28-12 Bragg planes include many sets of parallel reflection planes, not just those parallel to the crystal faces. Here, three possible sets of parallel planes are indicated, each with a distinct value of d.

28-6 The Compton Effect

In the early 1920s, the American physicist Arthur H. Compton (1892–1962) wished to confirm the quantum, or particle, nature of electromagnetic radiation. He wished to send a beam of quanta to interact with electrons and treated the interactions as that of two kinds of particles. Although he could easily obtain a good beam of electrons (cathode rays), the number of electrons per unit volume is so tiny that the probability of a single collision between a photon and an electron is vanishingly small. Compton knew, however, that the density of electrons in matter is high enough to provide a good probability of collision of photons with electrons. Unfortunately, the electrons in matter are bound to atoms (or at least the material as a whole), and photons of ordinary light do not have enough energy to break the electrons loose and still show collision phenomena. Compton solved the problem by using x-rays rather than ordinary light. With their very high frequencies, x-rays have such large energies that the small amount required to break the electron loose from an atom can be neglected.

Compton directed a beam of x-rays from a molybdenum x-ray tube onto a very thin carbon (graphite) target and measured the resultant scattered x-rays at various angles (Figure 28-13). He plotted the relative intensity of the scat-

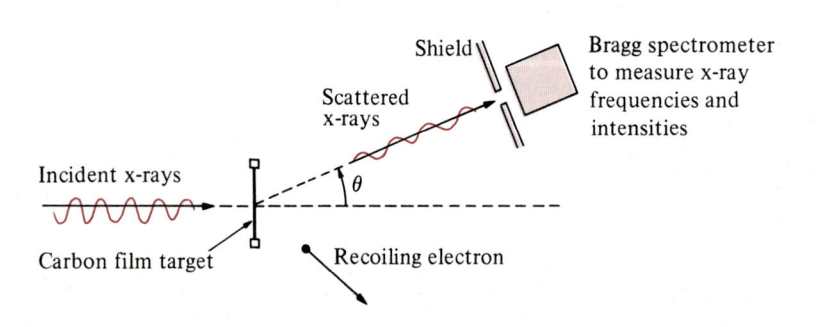

Figure 28-13 In the Compton experiment, an x-ray incident on the carbon foil may encounter an electron and be scattered at an angle of θ from the direction of incidence. The scattered x-ray has lower energy, hence, a lower frequency, than the incident x-ray. The missing energy resides with the electron, which recoils from the encounter. Energy and momentum are both conserved in the "collision" (which is better viewed as an absorption and reemission of the x-ray at reduced frequency).

tered x-rays vs. wavelength at various angles of scattering (Figure 28-14). We can see that at all angles, Compton found some of the original x-rays from molybdenum, but as the angle of observation is increased, he also saw a second set of x-rays shifted in wavelength. To explain these shifted x-rays, Compton postulated that the x-rays scatter exactly like particles. He derived an equation for this wavelength shift by using what amounts to "billiard ball physics." He treated both the x-rays and electrons as particles, using the Planck and Einstein expression for the energy of the x-ray photon. After applying the principles of conservation of energy and linear momentum, he solved the equations simultaneously (see Problem 28-27) to obtain

$$\lambda' - \lambda = \frac{h}{m_0 c}(1 - \cos\theta) \quad (28\text{-}11)$$

where λ' and λ represent the wavelengths of the shifted and original x-rays. The other quantities in the equation are Planck's constant h, the electron rest mass m_0, the speed of light c, and the angle of scattering θ.

The value of the shift in wavelengths as measured by Compton is in complete agreement with Equation 28-11. Thus his experiments confirm that x-rays may scatter from electrons as if the x-rays were particles. While there is no doubting the data and the conclusion that x-rays have particle properties, there seems to be some kind of contradiction here. Bragg diffraction confirms the wave nature of x-rays, while the Compton effect requires the particle nature for its explanation. The irony of this did not escape Compton, who had to use a crystal spectrometer to obtain the data for the curves of Figure 28-14. He is quoted as saying that "the crystal spectrometer measured a wavelike property, the wavelength, and measured it by a characteristic wavelike phenomenon, interference. But the effect of the graphite scatterer on the value of that wavelike property could be understood only in terms of a particlelike behavior." This apparent dual behavior presented another dilemma for the world of physics. It became the real difficulty of physics in the 1920s. The question, spoken and unspoken, for and by all scientists was "Is light a wave or a particle?" The twofold behavior became known as **wave-particle duality**.

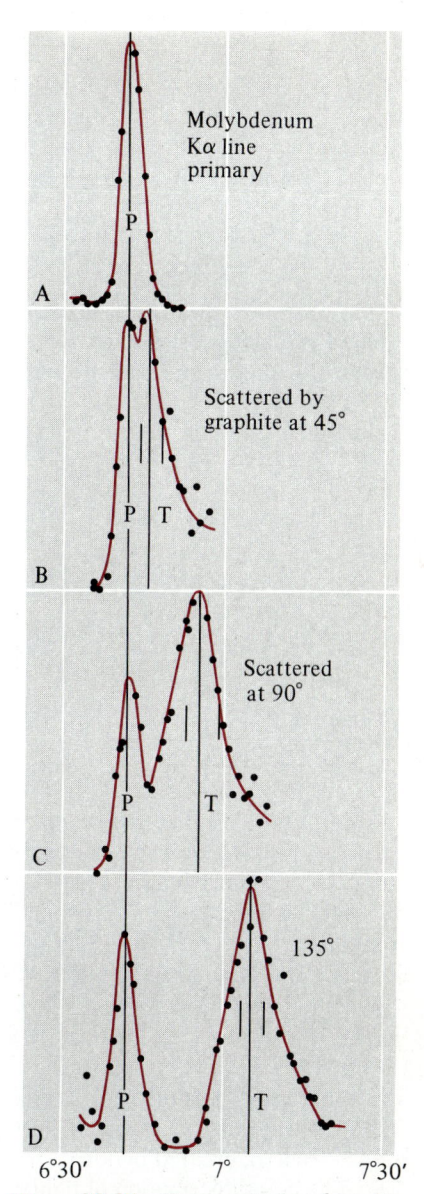

Figure 28-14 Compton's data from his 1923 publication show the original peak of x-rays from the molybdenum source at the x-ray tube. At higher angles, there is an additional peak of x-rays shifted in frequency, revealing a systematically greater energy loss for higher scattering angles. Compton's interpretation of the data earned him the Nobel Prize in physics in 1927.

Example 28-6 In Example 28-5, we were given that the frequency of the molybdenum x-rays like those used by Compton is 4.13×10^{18} Hz, and we calculated their wavelength to be 7.26×10^{-11} m. For Compton scattering, determine (a) the energy of the incident x-rays, (b) the new wavelength of a photon which is scattered at 90° to its original direction, and (c) the energy transferred to the electron.

Solution (a) The energy for this part is simply the product of Planck's constant with the frequency.

$$\epsilon = hf = (6.63 \times 10^{-34} \text{ J}\cdot\text{s})(4.13 \times 10^{18} \text{ Hz})$$
$$= 2.74 \times 10^{-15} \text{ J} \quad \text{or} \quad 1.71 \times 10^4 \text{ eV}$$

(b) Here we apply Equation 28-11 and solve for λ'.

$$\lambda' - \lambda = \frac{h}{m_0 c}(1 - \cos\theta)$$

or

$$\lambda' = \lambda + \frac{h}{m_0 c}(1 - \cos\theta)$$

$$= 7.26 \times 10^{-11} \text{ m} + \frac{6.63 \times 10^{-34} \text{ J·s}}{(9.11 \times 10^{-31} \text{ kg})(3.00 \times 10^8 \text{ m/s})}(1 - \cos 90°)$$

$$= 7.26 \times 10^{-11} \text{ m} + 2.43 \times 10^{-12} \text{ m}$$

$$= 7.50 \times 10^{-11} \text{ m}$$

(c) To solve this, we must realize that the energy transferred to the electron is equal to the energy lost by the photon. Thus we determine the energy of the new photon and subtract it from the answer to part *a*.

$$\epsilon' = hf' = \frac{hc}{\lambda'}$$

$$= \frac{(6.63 \times 10^{-34} \text{ J·s})(3.00 \times 10^8 \text{ m/s})}{7.50 \times 10^{-11} \text{ m}}$$

$$= 2.65 \times 10^{-15} \text{ J} \quad \text{or} \quad 1.66 \times 10^4 \text{ eV}$$

$$KE_{electron} = \epsilon - \epsilon' = 2.74 \times 10^{-15} \text{ J} - 2.65 \times 10^{-15} \text{ J}$$

$$= 9.00 \times 10^{-17} \text{ J}$$

This is over 500 eV. Thus the electron gains energy equivalent to being accelerated through a 500-V (or more) potential difference.

In Compton's data, we saw that, at every angle, he observed the original x-rays as well as the wavelength-shifted ones. The presence of the original x-rays is explained by recognizing that some x-rays will be reflected from electrons which are so tightly bound that they will not break loose from the atom. In this case, very little energy is transferred, and for these x-rays there is no perceptible wavelength shift.

28-7 Ordering Atomic Spectra

The quantum revolution was not related only to those events discussed so far. Scientists were also concerned with other unexplained events. One of those was the line spectra emitted by low-density incandescent materials.

The color distribution of light emitted by incandescent gases was somewhat of a puzzle right from the beginning. After Young's double-slit experiment, many investigators used slits and, later, diffraction gratings to study the light emitted by various materials. Energy was provided to the material under study by heating it or by causing an electric arc to discharge through it. The subsequent light emitted was collimated and passed through a diffraction grating. The resultant pattern was a **spectrum** for the material. In general, solid materials give rise to a continuous rainbowlike spectrum, whereas atomic gases in electric discharge tubes give off a line spectrum. Recall that the lines are simply sharp images of the slit through which the light from the source is passed. The light colored area in Figure 28-15 represents the field of view as seen in a spectrometer showing the spectrum of atomic hydrogen. (See also Color Plate II.)

In the nineteenth century, there was no theory to predict or even explain this discrete line spectrum. It was well known, however, that any given line

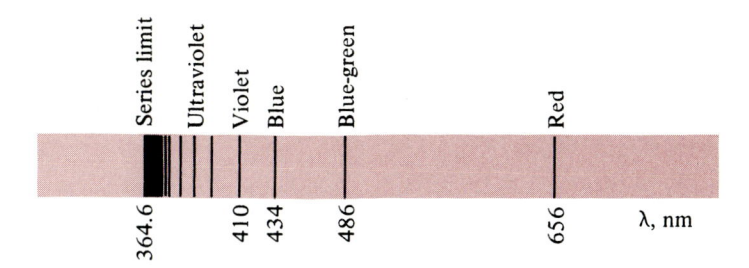

Figure 28-15 The spectrum of atomic hydrogen. The increasingly close lines beyond human vision in the ultraviolet ending at 364.6 nm are shown for completeness.

spectrum is uniquely characteristic of the element involved, and spectra can be used to identify unknowns. Precisely measured and very reproducible data in the absence of a theory often lead people to attempt to find some mathematical function which fits the data, almost as a mathematical game. This was one of the first steps taken by Wien and later by Planck with blackbody radiation.

In 1885, a Swiss schoolteacher, Johannes J. Balmer, determined that the wavelengths of the visible lines of the hydrogen spectrum fit the formula

$$\lambda = 364.56 \frac{n^2}{n^2 - 4} \qquad n = 3, 4, 5, 6 \qquad (28\text{-}12)$$

In this equation, the wavelength λ is in nanometers (nm) and n is an integer. For example, if $n = 3$, the formula yields $\lambda = 656.21$ nm. Equation 28-12 fits the visible region very well, and we can see that if n is allowed to take on larger-integer values, it predicts lines in the near-ultraviolet range. In fact, as n is allowed to get larger and larger, approaching infinity, the wavelength approaches 364.56 nm, which is observed as the series limit. This series is now known as the **Balmer series.**

The Swedish spectroscopist J. R. Rydberg rewrote the Balmer formula in the form

$$\frac{1}{\lambda} = R\left(\frac{1}{2^2} - \frac{1}{n^2}\right) \qquad n = 3, 4, 5, \ldots \qquad (28\text{-}13)$$

where R (now called the **Rydberg constant**) is 1.0974×10^7/m. Balmer went on to predict other series of spectral lines of hydrogen outside the visible range, following the formulas

$$\frac{1}{\lambda} = R\left(\frac{1}{3^2} - \frac{1}{n^2}\right) \qquad n = 4, 5, 6, \ldots \qquad (28\text{-}14)$$

and

$$\frac{1}{\lambda} = R\left(\frac{1}{4^2} - \frac{1}{n^2}\right) \qquad n = 5, 6, 7, \ldots \qquad (28\text{-}15)$$

We can see that these along with the Balmer series fit a general formula of the form

$$\frac{1}{\lambda} = R\left(\frac{1}{n_1^2} - \frac{1}{n_2^2}\right) \qquad n_1, n_2 \text{ are integers, } n_2 > n_1 \qquad (28\text{-}16)$$

A number of these **spectral series,** found later outside the visible region, were named after the scientist who first observed and reported each of them. Today, the other series are known as Lyman ($n_1 = 1$), Paschen ($n_1 = 3$), Brackett ($n_1 = 4$), and Pfund ($n_1 = 5$) series.

Example 28-7 Determine (a) the wavelengths and (b) the frequencies for $n_2 = 3$ and 4 in the Balmer series and $n_2 = 4$ in the Paschen series.

Solution (a) This requires that we apply Equation 28-16 for the Balmer series ($n_1 = 2$) twice and the Paschen series ($n_1 = 3$) once and solve for the wavelengths in each case. Designate the three wavelengths as λ_A, λ_B, and λ_C. Then

$$\frac{1}{\lambda} = R\left(\frac{1}{n_1^2} - \frac{1}{n_2^2}\right)$$

$$\frac{1}{\lambda_A} = (1.0974 \times 10^7/\text{m})(\tfrac{1}{4} - \tfrac{1}{9})$$

$$\lambda_A = 6.56 \times 10^{-7} \text{ m}$$

$$\frac{1}{\lambda_B} = (1.0974 \times 10^7/\text{m})(\tfrac{1}{4} - \tfrac{1}{16})$$

$$\lambda_B = 4.86 \times 10^{-7} \text{ m}$$

$$\frac{1}{\lambda_C} = (1.0974 \times 10^7/\text{m})(\tfrac{1}{9} - \tfrac{1}{16})$$

$$\lambda_C = 1.87 \times 10^{-6} \text{ m}$$

(b) Here we apply the relationship between the speed, wavelength, and frequency of the waves ($c = \lambda f$).

$$f_A = \frac{c}{\lambda_A} = \frac{3.00 \times 10^8 \text{ m/s}}{6.56 \times 10^{-7} \text{ m}} = 4.57 \times 10^{14} \text{ Hz}$$

$$f_B = \frac{3.00 \times 10^8}{4.86 \times 10^{-7}} = 6.17 \times 10^{14} \text{ Hz}$$

$$f_C = \frac{3.00 \times 10^8}{1.87 \times 10^{-6}} = 1.60 \times 10^{14} \text{ Hz}$$

In the early twentieth century, there was no explanation for the line spectra of atoms. By this we mean there was no model of the atom to which established physical principles could be applied to derive the discrete frequencies emitted by atoms. Bright-line spectra are produced only by low-density gases or vapors, where individual atoms are too far apart to interact rapidly enough to explain the emissions as a function of atomic collisions. Instead, spectral lines represent energy adjustments within individual, isolated atoms.

Much was known, however, and even more was speculated. Several scientists believed that line spectra and the quantum theory were somehow tied together. The reasoning was that since the excited atom gives off sharply defined electromagnetic energies, the energy differences between possible excited states (states of greater energy than the normal atom) of the atom must be sharply defined. From the principle of conservation of energy, we would expect that the energy carried away by the electromagnetic radiation would be equal to the energy difference between atomic states. What was needed was a model of the atom from which the regularities in the spectra and the arithmetic regularities determined by Balmer and others could be derived. Perhaps such a model would lead to a description that would allow us to be comfortable with the apparent wave-particle duality.

Minimum Learning Objectives

After studying this chapter, you should be able to:

1. Define:
   - Balmer series
   - blackbody
   - Bragg diffraction
   - cathode rays
   - Compton scattering
   - crystal spectrometer
   - photoelectric effect
   - photon
   - Planck's constant
   - Planck equation
   - quantum
   - quantum hypothesis
   - Rayleigh-Jeans law
   - Rydberg constant
   - spectral series
   - spectrum
   - stopping potential
   - threshold frequency
   - ultraviolet catastrophe
   - wave-particle duality
   - Wien displacement law
   - work function
   - x-rays

2. Describe the radiation emitted by a perfect blackbody in terms of the variation of intensity by frequency or wavelength and the effect of temperature variation on the curve.
3. Calculate the temperature of a blackbody from its peak radiated wavelength.
4. Calculate radiator area from the power of radiation received at a known distance from the radiator.
5. Calculate the energy of a quantum of radiated energy from its frequency, or vice versa.
6. Describe how the quantum hypothesis changes our viewpoint on radiant energy and leads toward wave-particle dualism.
7. Describe the behavior of electrons in metals exposed to light in the photoelectric effect, and tell how this behavior differed from that expected by treating light as a continuous wave.
8. Calculate the threshold frequency for a particular metal, given its work function, or vice versa.
9. Calculate the stopping potential for a photoelectric cell by using a specified metal which is exposed to light of known frequency.
10. Calculate the spacing of planes of atoms in a crystal, given the angle of constructive interference of x-rays of a known wavelength.
11. Calculate the frequency of an x-ray photon scattered from an electron at a specified angle from its direction of incidence, given its initial frequency, and determine the energy transferred to the electron.
12. Calculate several wavelengths in a spectral series for hydrogen, given the value of n_1.

Problems

28-1 The second-closest star system (5.9 light-years) to the solar system is that of Barnard's star, so dim it may only be seen with a telescope. The sun placed at the distance of Barnard's star would become the fifth brightest star in the earth's nighttime sky since it radiates 16,700 times as much power. Barnard's star radiates most strongly at a wavelength of 9.63×10^{-7} m. (a) What is the surface temperature of Barnard's star? (b) What percentage of the sun's diameter is the diameter of Barnard's star?

28-2 The star Sirius is a brilliant blue-white star which is the brightest star in the nighttime sky. Telescopic examination reveals it to be a double star, with Sirius A, as described, some 2900 times as bright as the white dwarf Sirius B of the same color (same λ_{max}). (a) Calculate the ratio of diameter of Sirius A to diameter of Sirius B. (b) If Sirius A were the size of the sun, how would Sirius B compare in size to the earth? (c) The surface temperature of Sirius A is 10,800 K; what is the most intensely radiated wavelength for this star?

28-3 Betelgeuse is the ninth brightest star in the nighttime sky on earth. Some 520 light-years distant, Betelgeuse is radiating 10,000 times as much power as the sun, with the peak of its radiations in the infrared at 905.6 nm. (a) How many times the sun's diameter is the diameter of Betelgeuse? (b) How does the radius of the star compare with the radius of the earth's orbit?

28-4 Calculate the energy of photons of wavelength (a) 4.00×10^{-7} m and (b) 7.00×10^{-7} m at the extremes of the visible spectrum.

28-5 Suppose a photon existed which had an energy of 1.00 J. What would be its (a) frequency and (b) wavelength?

28-6 The solar constant, given in Example 28-1, is 1.353 kW/m². Consider the power delivered to a 1-cm² area above the earth's atmosphere aimed perpendicular to the

719

sun's rays. (a) What is the power arriving at the 1-cm² surface? (b) Assuming that all the energy arrives at the wavelength of the sun's maximum intensity (504 nm in the yellow portion of the spectrum), how many photons arrive at the surface each second?

28-7 For an oscillating system consisting of a mass of 0.500 kg and a spring of constant 15.0 N/m vibrating at an amplitude of 5.00 cm, the frequency is 0.872 Hz and its energy is 1.88×10^{-2} J. (a) What is its quantum number n if, according to Planck's hypothesis, its energy must be equal to nhf? (b) What is the percentage change in the energy if n is increased by 1000?

28-8 Following Planck's hypothesis, we expect the energy of an oscillator to be given by nhf, where n is a positive integer, h is Planck's constant, and f is the frequency of oscillation. Suppose an oscillator system consists of a mass of 0.150 kg attached to a spring of constant $k = 25.0$ N/m. Calculate (a) the frequency of the system, (b) the energy of the system when its vibrational amplitude is 4.00 cm, and (c) the quantum number n of the system with this energy.

28-9 A candle radiates in all directions at 4.00 W, of which 2.00 percent is visible light (assume it is all yellow of wavelength 510 nm). All but the visible light is absorbed by a filter. (a) What power falls on an area of 1.00 cm² placed perpendicular to the rays from the candle at a distance of 100 m? (Recall the inverse square law for light intensity and distance.) (b) How many photons per second fall on this area?

28-10 Calculate the rate of emission of photons by a 100-W incandescent light bulb. Assume the average wavelength is equal to 5.50×10^{-7} m.

28-11 The filament of an incandescent light bulb has a surface area of 2.87×10^{-1} cm² and radiates 100 W of power. (a) What is the temperature of the filament, assuming it radiates as a blackbody? (b) What is the frequency of the most intensely radiated wavelength λ_{max}? (c) Visible light has a wavelength range of approximately 0.700 μm (red) to 0.400 μm (violet). What is the intensity of the radiated light at each of these wavelengths? (d) How does this compare with the intensity at λ_{max}?

28-12 Assuming an average wavelength of 5.50×10^{-7} m, determine the distance from a 100-W incandescent lamp at which the average number of photons striking a perpendicular surface is 1000 per square meter per second. The lamp is assumed to be emitting radiation uniformly in all directions.

28-13 Assume an average wavelength of 5.10×10^{-7} m for solar radiation and an intensity at the earth's surface of 920 W/m². Calculate the number of photons of solar light per square meter falling on the earth in 1 s.

28-14 Ultraviolet light of wavelength 1.50×10^{-7} m falls on a silver (Ag) surface. Calculate (a) the maximum kinetic energy of the photoelectrons, (b) the stopping potential, and (c) the threshold frequency for silver.

28-15 The speed of the fastest photoelectron emitted by gold (Au) in a photoelectric experiment is 1.15×10^6 m/s. Calculate (a) the stopping potential of these electrons and (b) the frequency of the light being used.

28-16 In a photoelectric experiment, it is determined that the threshold frequency is 8.00×10^{14} Hz for a particular metal. Calculate (a) the work function of the metal and (b) the frequency of the radiation required to release photoelectrons with a maximum speed of 1.00×10^6 m/s.

28-17 The stopping potential for photoelectrons emitted by potassium (K) when a particular light shines on it is 0.540 V. Determine (a) the maximum kinetic energy (in joules) of the photoelectrons, (b) the energy of the incident photons, (c) the frequency of the light, and (d) the wavelength.

28-18 For tungsten (W), calculate (a) the longest wavelength of incident radiation which will cause electron emission and (b) the stopping potential for photoelectrons emitted when tungsten is irradiated with ultraviolet light of wavelength 2.25×10^{-7} m.

28-19 What happens to the stopping potential in a photoelectric experiment for which the wavelength of the light used is decreased from 4.50×10^{-7} to 1.50×10^{-7} m?

28-20 When ultraviolet light of wavelength 2.08×10^{-7} falls on the surface of an unknown metal in a photoelectric experiment, the stopping potential is 1.20 V. Calculate the stopping potential for photoelectrons from the same surface if the wavelength of the light is reduced to 9.50×10^{-8} m.

28-21 Electrons are accelerated through a potential difference of 5.00×10^4 V. Calculate (a) the electrons' energy and (b) the maximum frequency of x-rays produced when these electrons strike a metal surface. Can you ignore the work function of the metal?

28-22 The picture tube of a 1980 model color television receiver is rated at 25 kV (2.50×10^4 V). This means that the electrons are accelerated through this much potential difference on their way to the phosphor on the inside of the tube face. Calculate the minimum wavelength of the x-rays produced when the electrons are stopped in the phosphor.

28-23 In an x-ray diffraction experiment, a first-order maximum is produced at an angle $\theta = 25°$. If the x-rays have a wavelength of 4.00×10^{-10} m, what is the spacing between reflection planes?

28-24 Two different x-ray sources are used with the same crystal in a Bragg diffraction experiment. The first x-ray has a wavelength of 5.50×10^{-10} m and yields a second-order maximum at an angle of 65°. The second x-ray gives a first-order maximum at 20°. Calculate the frequency of the second x-ray.

28-25 A crystal has a spacing between one set of reflection planes equal to 2.70×10^{-10} m. At what angles of incidence with these planes will x-rays of wavelength 9.50×10^{-11} m yield constructive interference?

28-26 An optical reflection grating (designed for visible light) may be used with x-rays of low energy if the angle of incidence is small enough. If a grating with a 2.00×10^{-6} m spacing is used with x-rays of wavelength 8.00×10^{-10} m, at what angle will one expect the first- and tenth-order maxima?

28-27 Prove that the Compton scattering equation (28-11) follows from treating the Compton effect as an elastic collision between particles. Take the initial and final energies of the photon to be hc/λ and hc/λ'. The initial and final magnitudes of linear momentum of the photon are h/λ and h/λ' (see Chapter 30). Use the relativistic expressions for the momentum and kinetic energy of the recoil electron.

28-28 X-rays of wavelength 1.50×10^{-11} m are used in a Compton effect experiment. Determine (a) the difference in wavelength of scattered photons from that of the original after they have been scattered through 45° and (b) the kinetic energy obtained by the recoil electrons.

28-29 For the Compton effect, determine the percentage of energy transferred to the recoil electrons that acquire the maximum kinetic energy when $\lambda = 1.21 \times 10^{-10}$ m.

28-30 X-ray photons of wavelength 2.50×10^{-11} m are incident on a carbon target, and compton-scattered photons are observed at 135°. Calculate (a) the wavelength of the scattered photons, (b) the magnitude of the linear momentum of the incident and scattered photons (see Problem 28-27), (c) the kinetic energy of the recoil electrons, and (d) the linear momentum (both magnitude and direction) of the recoil electrons.

28-31 Gamma rays of energy 0.511 MeV are compton-scattered. Calculate (a) the energy of the photons observed at a scattering angle of 45° and (b) the kinetic energy of the corresponding recoil electrons.

28-32 Calculate the shortest and longest wavelengths in the Paschen series of hydrogen.

28-33 Calculate the Rydberg constant by using the knowledge that the longest-wavelength line in the Balmer series of hydrogen is 6.5628×10^{-7} m.

28-34 Calculate the (a) wavelength, (b) frequency, and (c) energy of the longest-wavelength photon of the Brackett series of hydrogen.

28-35 Determine the (a) wavelength, (b) frequency, and (c) energy of the shortest-wavelength photon in the Lyman series of hydrogen.

28-36 Determine the value of n_1 for the hydrogen series which has all wavelengths greater than 4.4×10^{-6} m.

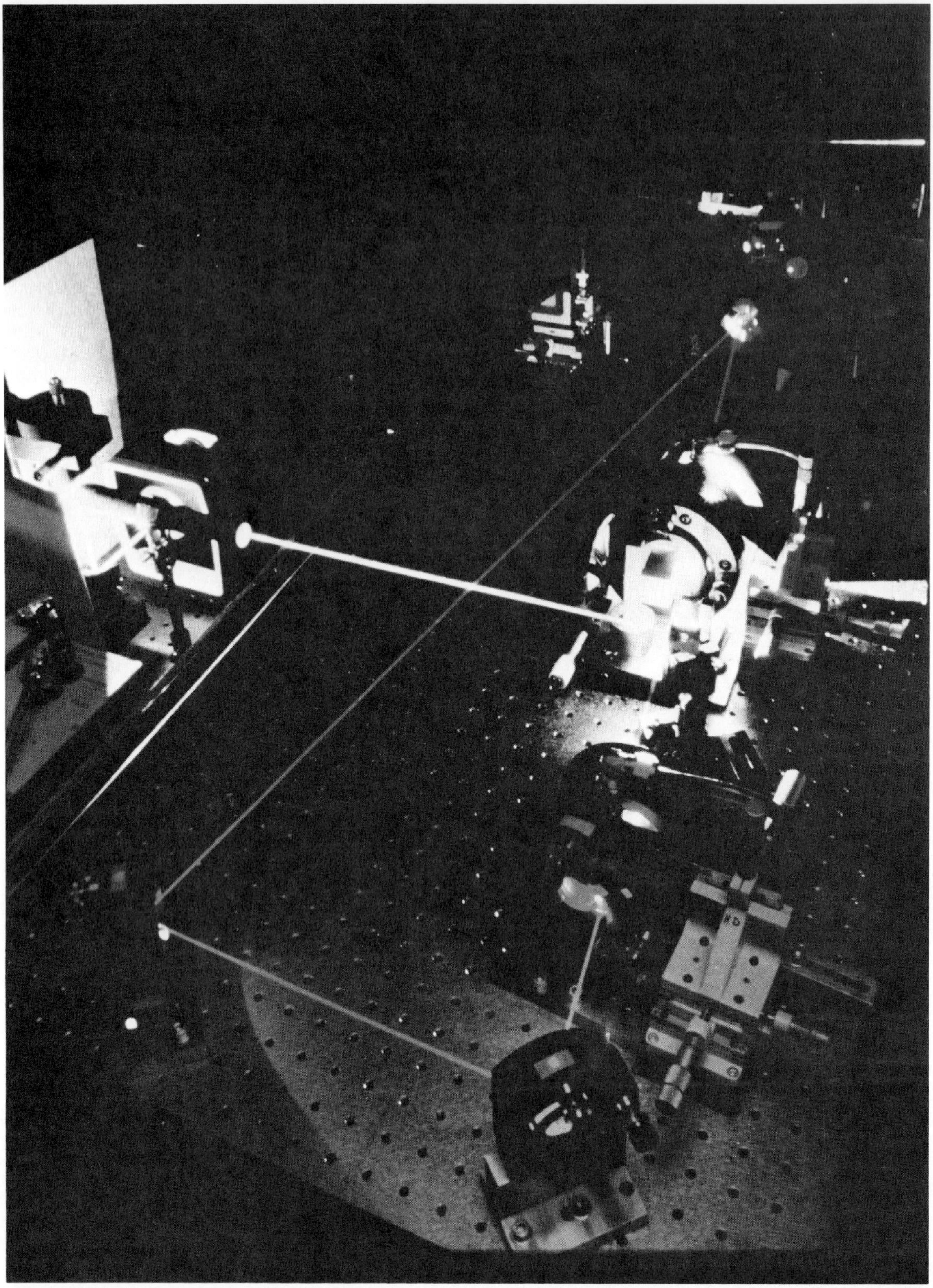

29 Atomic Physics

29-1 Introduction

The word *atom* is derived from the Greek *atomos* or the Latin *atomus,* both of which may be translated as "indivisible." Today we think of an atom as the smallest amount of an element which retains all the chemical properties of that element, but we acknowledge that the atom may be divided into smaller parts. In order that this position be accepted, some proof was required and a model was needed. We also needed a model of the atom which could be used to help explain the line spectra discussed in Section 28-7.

    Roentgen's announcement of the discovery of x-rays attracted a great deal of attention among scientists. J. J. Thomson (1856–1940), director of the Cavendish Laboratory in England, initiated a series of experiments in an attempt to identify cathode rays. By directing a beam of cathode rays through an area in which there were crossed electric and magnetic fields, Thomson and his colleagues had determined the amount of charge carried by a cathode-ray particle per unit of mass, the **charge-to-mass ratio** e/m. This ratio was much larger than the same ratio for any charged particle yet studied, implying either a large charge or an incredibly small mass, much smaller than that of the atom. Thomson concluded the latter and viewed the cathode-ray particles, later called **electrons,** as the basic carrier of electrical effects, such as currents in wires. The apparatus used was somewhat similar to the mass spectrograph we encountered in Chapter 21. Thomson's results were, within experimental error, the charge-to-mass ratio now accepted for the electron. His report is viewed (perhaps in retrospect) as both the beginning of our acceptance of the idea that atoms are *not* indivisible and of the idea that *all* matter is composed of electrically charged constituents. It was assumed (and is now known) that cathode rays are electrons which have been ripped from atoms. Further, because electrons have a negative charge, whereas atoms are electrically neutral, it means that some positive charge must be left behind when the electrons are freed.

> Cathode rays were determined to be particles with mass, transferring momentum on impact with a target. Further, whatever conducting material the cathode was made from—copper, aluminum, zinc, whatever—identical cathode rays were produced in the tube. Thus, it is a reasonable conclusion that these rays, our electrons, are an ordinary, subatomic constituent of all matter.

29-2 The Atomic Model

The facts available at the end of the nineteenth century seemed to support the idea that atoms were tiny objects constituted of positively and negatively charged parts. Thomson began to postulate a model of the atom which was consistent with the known facts. The value of such a **model** is that it suggests properties which may be experimentally tested. In his round **raisin cake model** of the atom, the fluffy cake was positively charged and distributed throughout the whole atom, and the negatively charged electrons were the raisins embedded in the cake. The electrons were thought to be held in equilibrium by electrostatic forces. Electrons were repelled by each other and attracted by and toward the center of positive charge. Thomson postulated that if this equilibrium was disturbed by a collision with another atom or particle, the electrons vibrated about their equilibrium positions. According to classical electromagnetic theory, they would thus emit radiation until they came to rest in their equilibrium positions.

While it may not sound particularly valid, this really was a pretty good model. From classical theory, Thomson "knew" that the electrons had to be stationary, because if they were moving and also confined to a limited volume, they would have to be accelerated. (Remember that even bodies moving in uniform circular motion are undergoing centripetal acceleration continuously.) In accordance with classical electromagnetic theory, accelerated charges radiate energy. Thus classical theory asserts that moving but confined electrons will radiate energy and eventually stop moving anyway. Much theoretical work was done by Thomson and his students to calculate the distribution of the electrons in the atom. From the electron distribution, they then computed vibrational frequencies to attempt to correlate theoretical values with the measured line spectra. As each attempt failed, it began to look as if the problem was with the raisin cake model.

In 1911, Ernest Rutherford (1871–1937) directed a group of researchers at the University of Manchester in England whose experiments yielded results which led to a new model of the atom. His group performed a number of experiments using **alpha (α) particles** as projectiles and gold foils as targets (Figure 29-1). These α particles were emitted by the recently discovered radioactive elements (see Chapter 31) and had previously been identified as positively charged atoms (ions) of helium by Rutherford. Rutherford recorded that most of the α particles passed through the foil without being deflected, but some

29-2 The Atomic Model

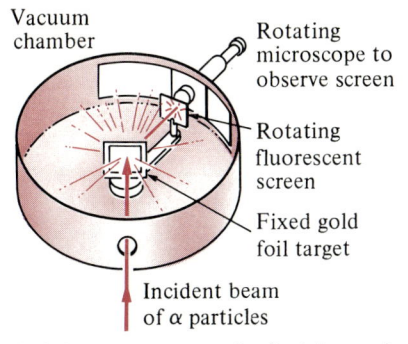

Figure 29-1 In the gold foil experiment, Rutherford directed a stream of rapid, massive alpha particles into a fine foil of gold in an evacuated chamber. The path of the particles was determined by registering their arrival at a fluorescent screen, where they produced tiny flashes of light.

of them were deflected at very large angles. In fact, some of the α particles even bounced backward in the gold foil experiment.[1]

The Thomson model would not predict large deflection angles for scattered α particles nor even allow them. The α particles were known to have nearly 10,000 times as much mass as the electron and surely would not be scattered backward even in several successive collisions with the "raisins." Consider, if you will, a fast-moving bowling ball colliding with some stationary marbles. Also, since the positively charged material of the cake had to be quite tenuous in the Thomson model, it would deflect a fast α particle about as well as a quantity of loose cotton would deflect a bowling ball.

To explain the deflection of the α particles, Rutherford postulated that all the charge of one sign (presumably the positive charge) in the atom was concentrated in a very tiny volume of the atom in which the atom's center of mass resided. This volume is called the **nucleus.** It was thus assumed that the large-value charge of the nucleus could be treated as a point charge as could the charge of the α particle. The expected angle of scattering of α particles was then calculated by assuming that the point nuclear charge was attached to a very large magnitude mass and was being approached by a fast-moving α particle. Coulomb's law and the conservation laws of momentum and energy were applied and yielded a theoretical formula which fit perfectly with the experimental data. With the new modifications, they then had what became known as the Rutherford nuclear model of the atom.

Example 29-1 The density of liquid helium at 4.20 K and 1 atm is 1.251×10^5 g/m³, and its atomic mass is 4.003 u. Avogadro's number is 6.024×10^{23} atoms/mol. (a) Calculate the volume occupied by a helium atom under these

---

[1] To record the track of the α particles, the tiny flashes they made on impact with a fluorescent screen, like that of a TV, had to be seen and counted. Rutherford's graduate students had to spend long hours each day sitting in a dark room watching a screen in a vacuum chamber through a microscope and recording flashes with a hand counter. It may be no coincidence that one of these students was Hans Geiger, who perfected the first electric particle counter which still bears his name (Chapter 31).

conditions. *(b)* Assume that the helium nucleus is a sphere of radius 2.22×10^{-15} m. Determine the fraction of the volume of the atom which is occupied by the nucleus.

Solution *(a)* We recall that the mass measured in atomic mass units (u) is equal in magnitude to the mass of 1 mol in grams. The number of atoms per mole is then found as the ratio of the density to the atomic mass times Avogadro's number. Then we can determine the volume per atom as the reciprocal of the number of atoms per cubic meter.

$$\frac{\text{Number}}{\text{Volume}} = \frac{N}{V} = \left(\frac{1.251 \times 10^5 \text{ g/m}^3}{4.003 \text{ g/mol}}\right)(6.024 \times 10^{23} \text{ atoms/mol})$$

$$= 1.883 \times 10^{28} \text{ atoms/m}^3$$

$$\text{Vol per atom} = 5.312 \times 10^{-29} \text{ m}^3$$

(b) For this part, we simply determine the volume occupied by the nucleus and divide it by the answer to part *a*.

$$\text{Vol nucleus} = \tfrac{4}{3}\pi r^3 = \tfrac{4}{3}\pi(2.22 \times 10^{-15} \text{ m})^3 = 4.583 \times 10^{-44} \text{ m}^3$$

$$f = \frac{\text{vol nucleus}}{\text{vol per atom}} = 8.63 \times 10^{-16}$$

Thus the nucleus occupies a very tiny portion of the atom's space.

29-3 The Bohr Atom

In 1912, a young physicist from Denmark, Niels Bohr (1885–1962), joined the Rutherford group. Bohr was convinced of the quantum theory of light and believed that the existence of reproducible line spectra dictated that atoms existed only in discrete energy states. He argued that the principle of conservation of energy requires that when an atom emits a photon of definite energy hf, it decreases its internal energy by that same amount. Stated another way, an atom with an initial energy E_i changes to a final (lower) energy E_f by emitting a photon of energy hf:

$$E_i - E_f = hf \tag{29-1}$$

Bohr visualized an atomic model in which all the mass (except that of the electrons) was contained in the nucleus and the electrons traveled in circular orbits about the nucleus similar to the planets orbiting about the sun. However, in circular orbits the electrons have a continuous acceleration and, according to classical electromagnetic theory, should radiate away all their energy, thus causing the atom to collapse. Bohr was apparently so convinced that his model was right that he concluded that classical theory must be incorrect on this point.

Bohr made the bold assertion that the mechanical energy of the atom is **quantized** (exists in discrete amounts only) just as electromagnetic energy is quantized. He then worked out the necessary conditions which led to a derivation of the wavelengths and frequencies of line spectra. From 1913 through 1915, Bohr presented his quantum theory of atomic structure in a series of published papers. For the moment, we confine our discussion to the hydrogen

The quantization of energy levels for an electron in an atom asserts that atoms are much more like stairways than ramps for electrons. The electron can be stable only at certain well-defined levels, just as you may rest only at certain well-defined heights on a staircase, although energy levels in atoms are not evenly spaced in energy as stairs normally are.

atom, which has only one proton in the nucleus and one external electron. Bohr postulated the following:

1. The electron revolves about the nucleus in certain circular orbits due to the attraction of the Coulomb force.
2. The only **allowed orbits** are those in which the angular momentum of the electron is an integral multiple of Planck's constant divided by 2π.
3. The atom does not radiate energy while the electron is in one of the allowed orbits.
4. The atom emits a photon of energy hf when the electron drops from an orbit of energy E_i to a lower orbit of energy E_f, where

$$hf = E_i - E_f \qquad E_i > E_f$$

Further, if the atom absorbs a photon of energy hf, the electron jumps from an orbit of energy E_i to a higher orbit of energy E_f, where

$$hf = E_f - E_i \qquad E_f > E_i$$

Using only these postulates and known physical principles, we may derive the possible energy states of the hydrogen atom and hence the energies, frequencies, and wavelengths of photons emitted by incandescent hydrogen.

Consider a hydrogen atom with the electron in uniform circular motion about the proton. The only force acting is the Coulomb force, and it is a centripetal force. Since the charge on both particles has the same magnitude e, we write for the Coulomb force

$$F = \frac{kq_1q_2}{r^2} = \frac{ke^2}{r^2}$$

Equating this to the required centripetal force yields

$$\frac{ke^2}{r^2} = \frac{mv^2}{r}$$

where m is the mass of the electron and v is its orbital speed.[1] From this we may obtain the kinetic energy as

$$KE = \tfrac{1}{2}mv^2 = \frac{ke^2}{2r} \qquad (29\text{-}2)$$

Since we know that the potential energy is

$$PE = -\frac{ke^2}{r} \qquad (29\text{-}3)$$

we have the total mechanical energy from

$$E = KE + PE = \frac{ke^2}{2r} - \frac{ke^2}{r} = -\frac{ke^2}{2r} \qquad (29\text{-}4)$$

---

[1] We make the assumption that the mass of the proton is so much larger than that of the electron that the proton does not move. Otherwise a very tiny correction (less than 0.06 percent) is required to account for the proton and electron revolving about their common center of mass.

In equation form, the second postulate quantizing the angular momentum L can be written

$$L = \frac{nh}{2\pi} \qquad n = 1, 2, \ldots \qquad (29\text{-}5)$$

For a body of mass m moving at a speed v in a circle of radius r, the angular momentum is

$$L = mvr$$

Thus

$$mvr = \frac{nh}{2\pi} \quad \text{or} \quad v = \frac{nh}{2\pi mr}$$

The kinetic energy is then

$$KE = \tfrac{1}{2}mv^2 = \tfrac{1}{2}m\left(\frac{nh}{2\pi mr}\right)^2$$

$$KE = \frac{n^2 h^2}{8\pi^2 m r^2} \qquad (29\text{-}6)$$

which, when combined with Equation 29-2, yields

$$r = r_n = \frac{n^2 h^2}{4\pi^2 m k e^2} \qquad (29\text{-}7)$$

We attach the subscript n to r to indicate that this is the equation for the radius of the nth allowed orbit (Figure 29-2). Substituting this expression for r_n in Equation 29-4 yields

$$E_n = -\frac{2\pi^2 m k^2 e^4}{h^2} \frac{1}{n^2} \qquad (29\text{-}8)$$

Equation 29-8 gives the expression for the total energy of the hydrogen atom as a function of the integer n which determines the various allowed energy states. The integer n is called a **quantum number.** The energy of the atom is negative because we chose the zero level for potential energy at infinite separation of the proton and electron.

According to the fourth postulate, if the atom decays to a lower energy state, the energy of the emitted photon is

$$hf = E_i - E_f = \frac{-2\pi^2 m k^2 e^4}{h^2}\left(\frac{1}{n_i^2} - \frac{1}{n_f^2}\right)$$

where n_i and n_f are the quantum numbers defining the initial and final energy states. This can be written as

$$hf = \frac{2\pi^2 m k^2 e^4}{h^2}\left(\frac{1}{n_f^2} - \frac{1}{n_i^2}\right) \qquad (29\text{-}9)$$

which can be solved for the reciprocal of the emitted photon's wavelength to yield

$$\frac{1}{\lambda} = \frac{2\pi^2 m k^2 e^4}{h^3 c}\left(\frac{1}{n_f^2} - \frac{1}{n_i^2}\right) \qquad (29\text{-}10)$$

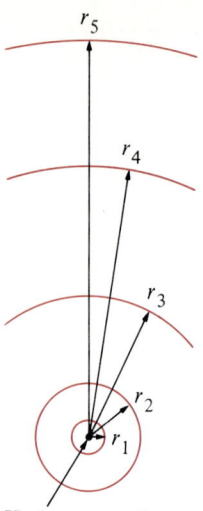

Figure 29-2 The Bohr model of the atom had the electron occupying only certain allowed orbits. The hydrogen electron normally occupies the orbit closest to the nucleus. The atom can absorb energy and the electron move to any orbit farther out, but it cannot be found between allowed orbits.

Equation 29-10 is identical in form to Equation 28-15, which is a generalization of the various formulas describing the spectral lines of hydrogen. Furthermore, when the correct numerical values of the constants are used, we find that

$$\frac{2\pi^2 mk^2 e^4}{h^3 c} = 1.0974 \times 10^7 \text{ m}^{-1}$$

This is exactly equal to the measured value of the Rydberg constant (see Problem 29-10).

This derivation of the equation for the hydrogen spectral wavelengths and especially the Rydberg constant spelled great success for the Bohr model. Because of this, support for the model was tremendous throughout the scientific community, and the concept of stationary orbits gained a high level of acceptance.

We may calculate the radii of the allowed orbits. For $n = 1$, we see from Equation 29-8 that the atom is in its lowest (most negative) energy state. This is called the normal state, or **ground state,** of the atom. We also see from Equation 29-7 that when $n = 1$, the electron is in the allowed orbit of smallest radius. Substituting the proper numerical values for the constants in Equation 29-7 gives us

$$r_1 = \frac{(1)^2 (6.63 \times 10^{-34} \text{ J}\cdot\text{s})^2}{4\pi^2 (9.11 \times 10^{-31} \text{ kg})(9 \times 10^9 \, \frac{\text{N}\cdot\text{m}^2}{\text{C}^2})(1.60 \times 10^{-19} \text{ C})^2}$$

$$= 5.30 \times 10^{-11} \text{ m}$$

Thus, the atom *should* have a particular size, determined by the values of fundamental physical constants. This is called the **Bohr radius** and is indeed the measured size of a normal hydrogen atom. This represented the first time any physical theory predicted that atoms should be of particular size.

Examining Equation 29-7 once again shows us that for any value of n, we have

$$r_n = r_1 n^2 \tag{29-11}$$

Figure 29-3 is a pictorial of the first six allowed orbits of the Bohr model and the various transitions an electron must make when photons associated with particular series are seen. For example, the first (lowest frequency, longest wave-

Figure 29-3 Electrons dropping from an outer orbit to one closer to the nucleus must lose an exact amount of energy equal to the energy difference between the two orbits.

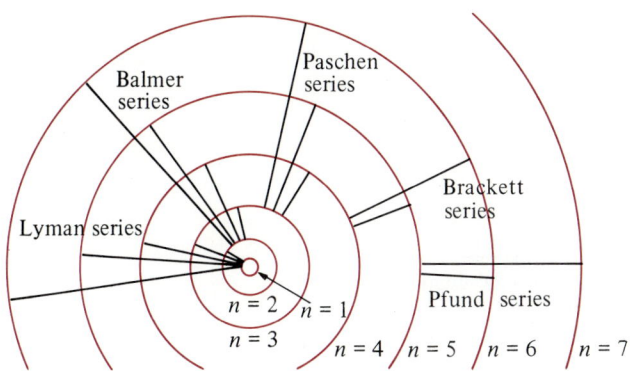

length) line in the Balmer series occurs whenever the electron drops from the $n = 3$ to the $n = 2$ orbit. Indeed, in this model, every electron which drops to the $n = 1$ orbit from any higher n orbit gives rise to a photon in the Lyman (ultraviolet) series. Those electrons which drop to the $n = 2$ orbit give rise to photons in the visible Balmer series. Of course, these electrons eventually drop to the $n = 1$ state, and a photon associated with the first line of the Lyman series is then released. The near-infrared Paschen series arises from transitions which end at the $n = 3$ state, Brackett (far-infrared) series at the $n = 4$ state, etc.

We could apply the same analysis used with the hydrogen atom to obtain the energy levels, wavelengths, and frequencies of photons emitted by any atom with a single electron. Such atoms are called **hydrogenlike atoms,** and some examples are singly ionized helium, doubly ionized lithium, triply ionized beryllium, and so on. The only difference from our original derivation is that the charge on the nucleus is $+2e$ for helium, $+3e$ for lithium, and, in general, $+Ze$ for the nucleus of an atom of atomic number Z.[1] For the general case, then, Equation 29-2 becomes

$$KE = \tfrac{1}{2}mv^2 = \frac{kZe^2}{2r}$$

and Equation 29-4 changes to

$$E = -\frac{kZe^2}{2r}$$

If we carry this through the derivation, we obtain the general equation analogous to Equation 29-9 as

$$hf = \frac{2\pi^2 m k^2 Z^2 e^4}{h^2}\left(\frac{1}{n_f^2} - \frac{1}{n_i^2}\right) \tag{29-12}$$

Example 29-2 Calculate the (a) frequency and (b) wavelength of the first line (longest wavelength) of the Brackett series for singly ionized helium.

Solution (a) We may determine the frequency by direct application of Equation 29-12. We know that for helium $Z = 2$, and we also know that the Brackett series applies to the electron transitions to the $n = 4$ state. Thus we use $n_f = 4$. We also want the longest wavelength or lowest frequency, so we must use $n_i = 5$.

$$f = \frac{2\pi^2 m k^2 Z^2 e^4}{h^3}\left(\frac{1}{n_f^2} - \frac{1}{n_i^2}\right)$$

$$= \frac{2\pi^2 (9.11 \times 10^{-31} \text{ kg})\left(9 \times 10^9 \,\frac{\text{N}\cdot\text{m}^2}{\text{C}^2}\right)^2 (2)^2 (1.60 \times 10^{-19} \text{ C})^4 \left(\frac{1}{4^2} - \frac{1}{5^2}\right)}{(6.63 \times 10^{-34} \text{ J}\cdot\text{s})^3}$$

$$= 2.95 \times 10^{14} \text{ Hz}$$

---

[1] Henry Moseley (1887–1915), another of Rutherford's students, showed that the **atomic number** Z of each atom, previously used to order the elements by increasing mass, was really the integral value of the positive charge each element possessed in units of the magnitude of the charge on the electron, $+Ze$.

(b) For the wavelength, we simply divide the speed of light by the frequency:

$$c = f\lambda$$

or

$$\lambda = \frac{c}{f} = \frac{3.00 \times 10^8 \text{ m/s}}{2.95 \times 10^{14} \text{ Hz}} = 1.02 \times 10^{-6} \text{ m}$$

29-4 Energy Levels in the Bohr Atom and Beyond

Equation 29-8 gives the value of the energy of the hydrogen atom in its nth energy state. In the ground state ($n = 1$), we calculate that

$$E_1 = -\frac{2\pi^2(9.11 \times 10^{-31} \text{ kg})\left(9 \times 10^9 \frac{\text{N} \cdot \text{m}^2}{\text{C}^2}\right)^2 (1.60 \times 10^{-19} \text{ C})^4}{(6.63 \times 10^{-34} \text{ J} \cdot \text{s})^2}$$

$$= -2.17 \times 10^{-18} \text{ J} = -13.6 \text{ eV}$$

The use of electronvolts rather than joules is both convenient and practical. While the electronvolt is not an SI unit, its use allows us to avoid carrying along large negative powers of 10. We see that when the hydrogen atom is in its ground state, it would require 13.6 eV to completely separate the electron from the proton. Thus we say that in its ground state, the electron is bound to the proton, and the **binding energy** is 13.6 eV. This is also known as the ionization energy, but we will use the term binding energy because we wish to carry it over to nuclear physics.

From Equation 29-8, we have for any value n that

$$E_n = E_1 \frac{1}{n^2} = -\frac{13.6}{n^2} \text{ eV} \quad (29\text{-}13)$$

Therefore $E_2 = -13.6/4 = -3.4$ eV, $E_3 = -13.6/9 = -1.51$ eV, and so on. We find it useful to arrange the possible energy values in increasing order and to draw lines representing these **energy levels**. The distances of separation of the levels are selected to some scale so that the separation between levels is proportional to the energy difference between them. Such a drawing is called an **energy-level diagram**, and one for the Bohr model of hydrogen is shown in Figure 29-4. The energy transitions undergone by the atoms as they emit photons which correspond to several of the series could also be shown. For example, a drop from the $n = 4$ to the $n = 2$ level is accompanied by the emission of the second Balmer line.

Such energy-level diagrams could just as easily be constructed for other hydrogenlike atoms by using the equations derived from the Bohr model to obtain numerical values. Another procedure to obtain an energy-level diagram for a certain substance is to use the experimentally determined wavelengths of the line spectra emitted by the substance. From these, all the energy differences and hence energy levels can be calculated. Energy-level diagrams are very useful to interpret and explain spectral phenomena as well as to give a better overall understanding of atomic and molecular structure.

Sometimes the characteristic spectrum of a substance is rather simple,

Figure 29-4 (a) The hydrogen spectrum. (b) The transition of an electron between two allowed orbits gives rise to one line in the spectrum shown in part a. (c) An energy-level diagram shows allowed positions, with separations proportional to energy differences. A potential of 13.6 V is required to free electrons from normal hydrogen atoms. Thus, the $n = 1$ orbit is at a potential of -13.6 eV relative to a free electron. The 486-nm line in (a) represents the electron transition from the $n = 4$ to the $n = 2$ orbit in (b), or an energy-level transition from -0.85- to -3.40-eV levels in (c).

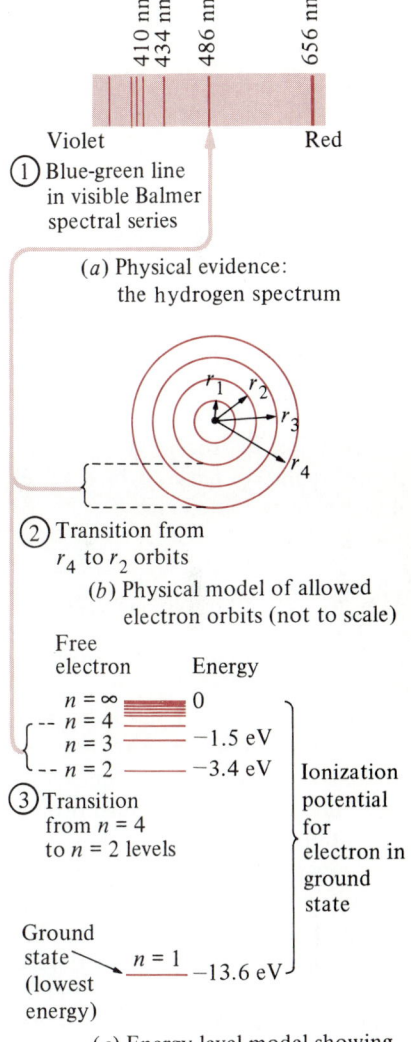

① Blue-green line in visible Balmer spectral series

(a) Physical evidence: the hydrogen spectrum

② Transition from r_4 to r_2 orbits

(b) Physical model of allowed electron orbits (not to scale)

③ Transition from $n = 4$ to $n = 2$ levels

(c) Energy level model showing relative energy at each orbit

29 Atomic Physics

(a)

(b)

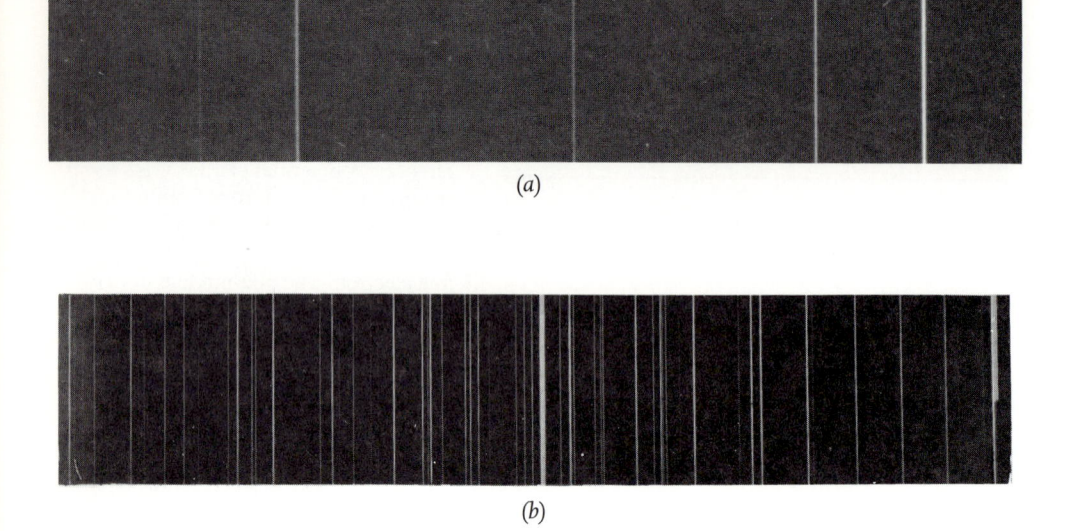

Figure 29-5 The visible spectrums of (a) helium and (b) iron.

and the energy levels may be readily determined. On the other hand, there are many molecules which have very complex spectra. Figure 29-5a shows the visible spectrum of helium, while Figure 29-5b is the iron spectrum. You can imagine that it would be much easier to obtain the energy levels from the energy differences of the helium spectrum than it would from that of iron. For the present, most energy-level diagrams must be obtained by using experimental data. To date there is *no* model (Bohr or any other) which can be used to solve analytically for all expected frequencies emitted from an atom or molecule more complex than hydrogenlike. The Bohr model provides a partial glimpse at some greater reality of nature which yet eludes us in its entirety. Yet, even greater glimpses were to come.

The key idea introduced by Bohr and forever more held to be correct is that normal atomic behavior is **quantized,** or occurs in jumps of discrete sizes only. The internal mechanics of atoms are jerky, not smooth. Electrons are either in one state or in another allowed state, but never between states somewhere.

29-5 A Modification to the Bohr Model

Although the original Bohr theory was widely supported because of its agreement with experiment, it left many questions unanswered and actually created several new ones. For instance, quantization of mechanical energy was now accepted, but there seemed to be no physical justification for it other than that it worked. There was also the question of why electrons could be accelerated (as they must be in their allowed orbits) without radiating energy. Furthermore, early attempts to apply the Bohr model to atoms with more than one electron were unsuccessful. Nonetheless, for the decade following its announcement, the Bohr model of the atom was the best available. Much theoretical work was done in an attempt to answer some of the questions, and refinements were made in the hope that further predictions could be made.

In all the work that Bohr and his colleagues undertook, he urged them to be guided by what has become known as the **correspondence principle:**

> We know in advance that any new theory in physics — whatever its character or details — must reduce to the well established classical

theory to which it corresponds when the new theory is applied to the circumstances for which the less general theory is known to hold.

This is an excellent principle which can be used to test any new more-encompassing theory. We have already seen it applied to special relativity to show that the relativistic expressions for kinetic energy and other quantities reduce to the classical expressions which apply for bodies moving at speeds which are small with respect to that of light.

The German physicist Arnold Sommerfeld (1868–1951) made the first acceptable modifications to the Bohr model by including elliptical orbits (Figure 29-6) which are expected to result from the $1/r^2$ force law. Because the position of an electron in an elliptical orbit must be specified by two variables (radial distance and angle), Sommerfeld's treatment required two related quantum conditions.[1]

Labeling the quantum numbers associated with the two variables as n_r and n_θ, Sommerfeld derived for the hydrogen atom

$$E_{n_r, n_\theta} = -\frac{2\pi^2 m k^2 e^4}{h^2} \frac{1}{(n_r + n_\theta)^2} \quad (29\text{-}14)$$

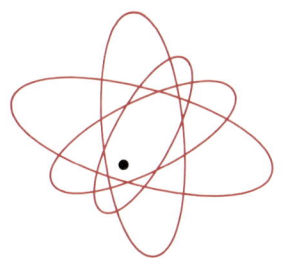

Figure 29-6 The attractive $1/r^2$ force law (Coulomb's, in this case) predicts possible elliptical orbits for the electron with the center of force at one focus. The circle is simply an ellipse with both foci at the same point.

where n_r and n_θ are both integers. Note that Equation 29-14 gives the same energy levels as those of Bohr shown in Equation 29-8. The only difference is that instead of the square of an integer n equal to 1 or more, we have the square of the sum of n_r and n_θ. This sum is also an integer equal to 1 or greater. When $n_r = 0$, we have the special case of circular orbits of Bohr's original postulates. (The correspondence principle, it seems, had already been applied.) The sum $n_r + n_\theta$ is usually denoted by n and is called the **principal quantum number.** The **orbital quantum number** Bohr associated with the angular momentum is now designated as l.

An additional extension on the Bohr model recognizes that the allowed orbits may not necessarily be in the same plane but may have many different orientations in space. This requires the defining of an additional quantum number, say m, to specify the motion of an electron in an atom. That is, at this point the three quantum numbers used to specify the electrons' motion were (1) the principal quantum number n associated with the energy of the atom, (2) the orbital quantum number l associated with the angular momentum due to orbital motion, and (3) the orientation quantum number m associated with the component of angular momentum about an arbitrarily chosen Z axis.[2]

Theoretical physicists, especially Bohr and his coworkers, struggled with other puzzling aspects of the atom. For example, the radius of the orbits of hydrogenlike atoms should be inversely proportional to Z. That is, for the hydrogenlike atom of atomic number Z, the Bohr model predicts that

$$r_n = \frac{n^2 h^2}{4\pi^2 m k Z e^2} \quad (29\text{-}15)$$

This is the same as Equation 29-7 for the Bohr radii of hydrogen except that it has a Z in the denominator. This says that for atoms of larger and larger atomic

---

[1] In a circular orbit, the radius is constant, while the angle the radius vector (the line between the electron and nucleus) makes with an arbitrary axis is variable. For an elliptical orbit, both the distance r of the electron from the nucleus and the angle θ between the radius vector and an arbitrary axis are variables.

[2] m is now called the **magnetic quantum number** because its value determines the extent to which the atom interacts with an external magnetic field along the Z axis.

number, the allowed orbits have smaller and smaller radii. In particular, the ground-state orbit becomes very small for atoms of large atomic number.

In hydrogen, after the electron moves to a larger-radius orbit, the atom gives up that excitation energy by photon emission as the electron eventually moves to the allowed orbit of smallest radius, its ground state. It seems natural to assume that if more electrons are added to a hydrogenlike atom, they too would drop to the lowest energy state corresponding to the smallest allowed radius. If this were to occur, there would be two effects, both of which could be tested. First, atoms of high atomic number should be much smaller than those of low atomic number. Second, the binding energy, or ionization energy, should be much greater for atoms of high atomic number. Neither effect is observed! Measurement indicates all atoms are within a few times of being the same size, with hydrogen being among the smallest. Furthermore, binding energy does not increase with atomic number but actually tends to decline, with only 8 of the 92 elements exceeding the 13.6 eV of hydrogen.

Perhaps the assumption that all the electrons in an atom go to the lowest allowed orbit is not true. This could lead to the apparent contradiction. However, if they do not go to the lowest level, what is it that prevents them from doing so?

29-6 The Pauli Exclusion Principle

Wolfgang Pauli (1900–1958), an Austrian theoretical physicist, worked with Bohr in the early 1920s. Pauli hypothesized that the apparent contradiction described at the end of the last section could be resolved if each possible energy state were limited in the number of electrons which could occupy it. For example, suppose only two electrons could occupy the lowest energy state. A third electron added to the atom would be excluded from that state and the lowest level available to this third electron would be the second lowest state. Pauli's hypotheses (since modified) have now taken on the respectability of a scientific principle — the **Pauli exclusion principle:**

> When attached to a single nucleus, only two electrons may occupy any quantum state and have the same three quantum numbers n, l, and m. Further, n must be a positive integer, l must be a nonnegative integer which may take on any value from zero up to $n-1$, and m may take on any integer value from $-l$ to $+l$, whatever l may be.

For example, when $n = 2$, l may be either 0 or 1. For $l = 1$, m may be $+1$, 0, or -1, and for $l = 0$, m must be 0. There are four possible quantum states for $n = 2$, as shown in Table 29-1. According to the Pauli principle, only two electrons may occupy each of these states. Thus we would expect that an atom may have at most eight electrons with principal quantum number $n = 2$.

The Pauli principle gave the prescription necessary to construct an atomic model for any element. This resolved the atomic size and binding energy problems. The models also explained many other chemical properties, including valences, the reason why certain gases are chemically "inert," and so on. In fact, the Pauli principle provided an explanation for the **periodic table** (Appendix C) designed empirically by D. I. Mendeleev 40 years earlier. All in all, the application of the Pauli principle yielded great advances for atomic physics.

TABLE 29-1
Possible Combinations of n, l, and m for $n = 2$

n	l	m
2	1	1
2	1	0
2	1	−1
2	0	0

29-6 The Pauli Exclusion Principle

In spite of the successes, problems remained. The model still did not explain all known experimental facts concerning atomic spectra. One glaring problem had to do with the apparent splitting of spectral lines. Under close examination by high-resolution spectroscopes, spectral lines were revealed to be two distinct but very closely spaced lines. This apparent split is known as **fine structure.** The existence of fine structure led to the belief that there was some additional motion, or form of energy, associated with the atom.

Arthur Compton is credited with the original suggestion that the electron may be a spinning particle. However, it was not until 1925 when Samuel Goudsmit and George Uhlenbeck, who were then graduate students at Leiden University in the Netherlands, used **electron spin** to explain fine structure. Goudsmit and Uhlenbeck postulated that the electron had rotational motion about an axis within itself and determined the conditions necessary to account for the fine structure. We might visualize this motion of the electron by analogy to the earth-sun system. The earth spins, or rotates, on its axis and simultaneously revolves in orbit about the sun. The earth thus has angular momentum due to its spinning motion as well as the angular momentum due to its orbital motion. The electron's spin should also give it a component of intrinsic angular momentum in addition to its orbital angular momentum.

An important difference between these two systems is that the electron and proton have electric charge. The spinning motion of a charged sphere gives rise to a magnetic field which makes the electron act like a tiny bar magnet. In addition, a measurement from a frame of reference attached to the spinning electron would show the proton revolving about the electron. (Just as it appears that the sun orbits the earth.) The charged proton thus causes a magnetic field at the site of the electron. From this point of view, the electron acts like a tiny magnet situated in an external magnetic field due to the proton's apparent motion.

It is postulated that the energy due to this magnetic interaction must be added to the orbital kinetic energy and the Coulomb potential energy to obtain the total energy of a given atomic state. The energy of interaction between a bar magnet and a magnetic field depends on the bar's orientation in the field. If any possible orientation is allowed, then the magnetic energy could take on any value between some maximum and some minimum. Being aware that the fine structure often shows only two closely spaced lines, Goudsmit and Uhlenbeck proposed that the electron spin could take on one of only two possible orientations. According to their model, the axis of electron spin must be perpendicular to its orbital plane, and the spin is either in the same direction or in the direction opposite to orbital motion.[1]

Goudsmit and Uhlenbeck obtained an expression for the strength of the spin-induced magnetic moment. From this expression, they concluded that the spin angular momentum had the value $\frac{1}{2}(h/2\pi)$. Thus the spin angular momentum along the Z axis is

$$S_z = \pm \frac{1}{2}\left(\frac{h}{2\pi}\right) \quad (29\text{-}16)$$

The quantity S_z is the Z component of electron-spin angular momentum, where

This discussion gives the historical reasoning leading to the concept of spin. That concept has continued to have great utility, but the mental picture leading to it has not. Picturing the electron as a charged sphere of definite size, spinning as it orbits the nucleus, does not accord with later experimentation. For example, no experiment ever has been able to define a finite physical size for an electron. As usual, we are dealing with a model, but we must remember that the reality is only what we obtain from measurement.

---

[1] These two orientations were used only to explain the facts. We will later see that additional facts require that this could be more reasonably interpreted by referring only to the time average of the spin.

the Z axis is determined rather arbitrarily. The limitation of two possible orientations for the spin leads to the choice of the **spin magnetic quantum number** as a fourth number of the group needed to define the state of an electron in an atom. The Pauli principle takes spin into account with only a slight modification:

> When attached to a single nucleus, no two electrons may occupy the same quantum state defined by the four quantum numbers n, l, m, and m_s.

The numbers n, l, and m are defined and limited as before. The number m_s may take on the value $\pm\frac{1}{2}$ only. This modification is indeed minor. We might simply have accepted the original statement but added that the two electrons allowed in a given state must have opposite spin.

The Pauli principle provided the modified Bohr model of the atom with one of its last big successes. The theorists of the early twenties were becoming increasingly aware that the Bohr model was of limited usefulness. Atomic behavior was clearly quantized, but the model did not provide the reasons for mechanical quantization. It was beginning to look like an add-on theory, where something new is appended every time trouble is encountered. A new theory was needed, one which would make the same predictions as the Bohr theory but not be bound by its limitations.

29-7 Application: The Laser

Energy-level diagrams let us understand one of the most important developments in recent technology, the laser, an acronym for light *a*mplification by *s*timulated *e*mission of *r*adiation. Introduced in 1960 amid sensational claims of all the wonderful applications which would soon be available, the laser may be one of the few such inventions to live up to its early promises. The laser produces a very intense beam of light with a very small cross section which now has a variety of uses. Different types of lasers are used to drill tiny holes through diamonds or bore larger holes through steel plate. They are being developed for use with fusion reactors which one day may produce a significant portion of our electric energy (Chapter 33). Medical science uses lasers to weld a torn retina, control bleeding in surgery, and treat skin cancer. They are used in surveying and even in determining the distance to the moon. There are laser printers and laser video disk pickups. They are used in "painting" moving light sculptures, in creating three-dimensional images, and to carry many thousands of communication signals (telephone and/or television) over fiberglass cables. In the laboratory, lasers are used to separate isotopes, study molecular structure, or provide an intense coherent beam for physical optics. Inexpensive lasers provide a good source for demonstrations of interference and diffraction effects in education.

When an atom or molecule has an energy other than that of its ground state, or normal state, it is said to be in an **excited state** E_h (where E_h means higher energy). The molecules do not radiate while they are in these states, but they usually do not remain in them very long either. Generally, an atom stays in an excited state for only about 1.00×10^{-8} s before decaying to a lower energy state E_l, emitting a photon of energy equal to the energy difference of the two states (Figure 29-7a). This is called **spontaneous emission,** because it occurs

Figure 29-7 Electron transitions. (*a*) Spontaneous emission: the electron drops from a higher to a lower orbit and the atom emits a photon equal to the energy difference between the two orbits. (*b*) Absorption: the atom absorbs a photon with the precise energy required to raise the electron to some higher orbit. (*c*) Stimulated emission: an excited atom is induced to emit a photon by a passing photon of precisely the correct energy for the electron transition to a lower state.

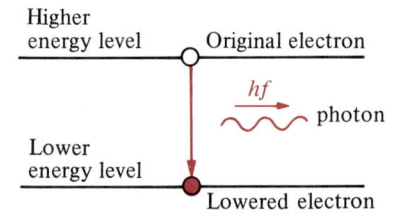

(*a*) Spontaneous emission

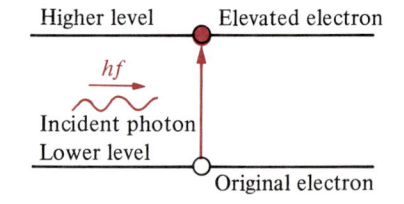

(*b*) Spontaneous absorption

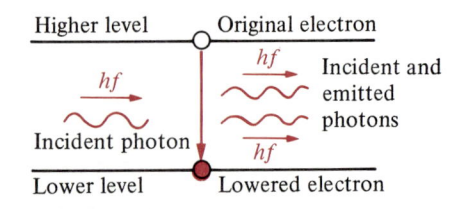

(*c*) Stimulated emission

29-7 Application: The Laser

by itself without apparent influences from outside the atom. In the laser, however, the emission that takes place is stimulated by just such an outside influence. Whenever a photon with energy equal to the energy difference of two allowed states encounters an atom in an excited state, it may cause the atom to emit another photon. The atom makes the transition to the lower state and emits a photon (1) with the same frequency, (2) with the same phase, (3) with the same polarization, and (4) going in the same direction, as the photon which stimulated it (Figure 29-7c). This process is called **stimulated emission** and was first proposed by Einstein in 1919. Alternatively, if the atom is in the lower energy state when the photon encounters it, the atom may *absorb* the photon and go to the higher state (Figure 29-7b).

Suppose we have a material whose atoms have allowed energy states E_l and E_h. Under most circumstances, if light of frequency corresponding to the energy difference is directed through the material, more photons will be absorbed than will stimulate emission. This is because, ordinarily, many more atoms will be in the E_l state and only a few in the E_h state in which they are capable of being stimulated to radiate. However, under special circumstances, it is possible to have more atoms of a material in the E_h state than in the E_l state. This condition is known as a **population inversion.** When a population inversion occurs in a material, photons of the proper frequency directed through it may stimulate emissions much more often than be absorbed. Further, the emitted photons can stimulate additional emissions, causing an avalanche. Such a system is then a source of radiation of photons with energy equal to the difference between the two states involved. A very important property of this light stems from the fact that it is the result of stimulated emission. The photons are all of the same wavelength, or *monochromatic,* and all in phase, or *coherent,* and all moving in the same direction, forming a tight beam. The system just described is a **laser,** and when the coherent beam is produced, the system is said to be **lasing.**

There are a number of ways to achieve population inversion. Most often the laser material consists of one or several types of atoms or molecules which have at least one **metastable energy state.** A metastable state is an unusual excited state of an atom or molecule in which the molecule may exist for a very long time before it decays.[1] Stimulated emission may cause an atom in a metastable state to decay, and thus a system of atoms or molecules in a metastable state may **lase.** To get the atoms of a material into an excited state, energy must be supplied or "pumped" into the system. This may be done with a flash lamp or an electric discharge in a gas discharge tube.

Figure 29-8 shows a common laser device used for physics demonstrations. This helium-neon (He-Ne) gas laser is pumped electrically and employs a mixture of helium and neon gases at very low pressure in a glass tube. When a high voltage is applied between two electrodes embedded in the tube, a stream

Figure 29-8 A helium-neon laser. The lasing material is mounted between parallel mirrors, allowing the beam to bounce back and forth along the axis of the device. Light emitted along other axes is absorbed inside the device, while axial emission is encouraged and grows.

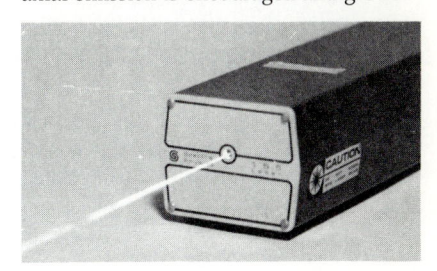

---

[1] The existence of metastable states is the basis for photography. When light strikes a silver halide molecule in the film, it provides the energy to raise an electron to a metastable state. Having already received the energy of the photon, this electron is easier to remove from the molecule than an equivalent electron never struck by light. Thus, a very weak chemical reaction (developing) can convert those molecules in a metastable state, and only those molecules, to pure metallic silver, producing a record of where light struck the film. If this were not a metastable excited state of the electron, the film would have to be developed in less than 10^{-8} s, before the electron could drop down in energy to its ground state.

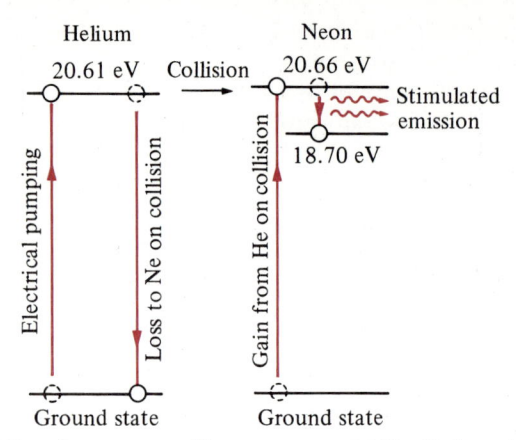

Figure 29-9 Helium has a metastable energy state 20.61 eV above its ground state, and neon has an excited state 20.66 eV above its ground state. An excited helium atom may lose its excitation energy to a neon atom in collision. The helium atom returns to its ground state, and the newly excited neon atom quickly decays to a lower state. One transition available to the excited neon atom is to a state 1.96 eV lower; since this is in the red portion of the spectrum, the light of the He-Ne laser is red.

of electrons is generated across the tube. As the electrons collide with the He and Ne atoms, they raise the atoms to excited states. One of the excited states of He, at about 20.61 eV above its ground state, is metastable. Although the He atom does not decay from that state, it may give up its excess energy in a collision with a Ne atom. Ne has an excited state at 20.66 eV above its ground state, so the excited He atom must also give up a little kinetic energy in order to raise the Ne to that state. Ne also has an excited state at 18.70 eV above its ground state to which an atom in its 20.66-eV state may decay. Thus we may have the required population inversion between the 18.70- and 20.66-eV states of Ne if collisions with He atoms can produce them in the higher state faster than they decay spontaneously to the lower state.[1] The stimulated photons are emitted with approximately 1.96 eV, corresponding to a wavelength of 6.328×10^{-7} m in the red portion of the spectrum as depicted in Figure 29-9.

Mirrors are usually placed at both ends of the laser material to reflect the light back and forth many times to stimulate as many of the excited atoms as possible. As long as there is a population inversion, more photons will be added to the beam than are absorbed from it. One of the end mirrors is only partially silvered so that part of the beam may escape and be used externally.

Example 29-3 The metastable state responsible for the lasing action of the chromium atoms in a ruby rod is about 1.79 eV above its ground state. Calculate the wavelength of the laser light of these atoms when they are stimulated to decay to the ground state.

---

[1] Actually, the higher energy state of Ne is really four closely spaced states very near to 20.66 eV, and the lower state is really ten closely spaced states at about 18.70 eV. Both atoms have many more excited states not shown in the diagram. These other states are often occupied during the process, and while photons are emitted as the excited atoms decay to their ground states, they do not contribute to the laser action.

Solution After converting the energy to joules, we may calculate the laser frequency from energy $= hf$. We then simply divide the speed of light by the frequency to obtain the wavelength.

$$\epsilon = (1.79 \text{ eV})(1.60 \times 10^{-19} \text{ J/eV}) = 2.86 \times 10^{-19} \text{ J}$$

$$f = \frac{\epsilon}{h} = \frac{2.86 \times 10^{-19} \text{ J}}{6.63 \times 10^{-34} \text{ J} \cdot \text{s}} = 4.31 \times 10^{14} \text{ Hz}$$

$$\lambda = \frac{c}{f} = \frac{3.00 \times 10^8 \text{ m/s}}{4.31 \times 10^{14} \text{ Hz}} = 6.96 \times 10^{-7} \text{ m}$$

Actually the measured wavelength of the light from a ruby-rod laser is 6.943×10^{-7} m. The difference results primarily from rounding off the numbers. This is another example where precision would require use of more than three significant figures.

In the first two decades after its invention, the laser progressed from being a solution looking for a problem to an indispensable tool of research, industry, and medicine. Lasers are constructed from many different materials. Solid ruby rods (chromium in aluminum oxide) and neodymium in glass rods are examples of solid lasers. These types are usually pumped with flash lamps and emit a pulse of laser light just after each flash of the lamp. There are many kinds of gas lasers in addition to the He-Ne just described. There are even so-called solid-state lasers in the form of tiny diodes no bigger than a pinhead. Both gas and solid-state lasers are pumped electrically and emit a continuous beam.

The output power of commercial lasers varies over a wide range. The diodes used in laser pickup from video disks have outputs near 0.001 W (1 mW), whereas a carbon dioxide gas laser used for cutting steel has a continuous 25,000-W output. Pulsed lasers can deliver up to 50 MW for intervals of 10 ns. In physical size, lasers range from the pinhead diode to the CO_2 steel cutter which takes as much space as an automobile.

The wavelengths of light one may obtain by using lasers cover the complete visible region and extend well into the ultraviolet and infrared. There are even tunable lasers used in advanced research. The wavelength of some of these lasers may be tuned continuously from 2.00×10^{-7} m to above 1.00×10^{-6} m, covering the complete visible range and then some. For many applications in science and industry, this tunability feature is crucial.

For the most part, laboratory lasers are quite safe, but the viewing of a laser beam directly may do permanent damage to the retina. Because the cornea and the crystalline lens of the eye provide additional focusing, a tight beam of parallel light entering the eye may be further focused to a point or at least a very small circle. If the power per unit area is sufficient and the time interval is long enough, a small lesion on the retina may result. The photoreceptors in that small circle are permanently destroyed if burned, and that small area becomes an additional blind spot for the eye. A 0.1-s exposure to a laser of 30 mW produces a lesion which can be seen by an ophthalmologist. Fortunately, there is a normal tiny motion of the eye including a tremor at a frequency near 50 Hz. These motions assure that a point image really sweeps out a circular area on the retina which is about 0.100 mm in diameter. Further, the blink reflex time is typically about 0.01 s. The short viewing time combines with the natural eye motion to provide some protection for the normal eye. However, there is still a certain amount of danger. There is every reason to believe that a laser with a power output from 3 to 10 mW should be considered a *possible* hazard even if not probable. In any instance, it is good practice and training to treat all lasers with respect. This even includes the common He-Ne laser used for education, which has an output of a fraction of a milliwatt.

Example 29-4 A laboratory worker views directly the beam from a He-Ne laser of 0.200 mW output power. Assume that the beam is focused to a point on the retina and that the natural eye movement causes the energy to be distributed over a circle of radius 5.00×10^{-5} m. If the worker blinks quickly so that the light is on the retina for only 0.0100 s, determine (a) the total energy deposited and (b) the energy per unit area on the circular area.

Solution (a) We may solve this part from the knowledge that the power is equal to the energy per unit time:

$$P = \frac{\epsilon}{t}$$

or

$$\epsilon = Pt = (2.00 \times 10^{-4} \text{ J/s})(0.0100 \text{ s}) = 2.00 \times 10^{-6} \text{ J}$$

(b) This requires that we calculate the area and divide it into the answer to part a.

$$A = \pi r^2 = \pi(5.00 \times 10^{-5} \text{m})^2 = 7.85 \times 10^{-9} \text{ m}^2$$

$$\frac{\epsilon}{A} = \frac{2.00 \times 10^{-6} \text{ J}}{7.85 \times 10^{-9} \text{ m}^2} = 255 \text{ J/m}^2$$

By way of comparison, this is approximately one-quarter of the solar energy which falls on a 1-m² area at the earth's surface in 1 s (although the sun, too, is an optical hazard if viewed directly). This is also less than 0.1 percent of the energy per unit area which is known to produce an observable lesion.

Minimum Learning Objectives

After studying this chapter, you should be able to:
1. Define:
   - allowed orbit
   - alpha particles
   - atomic number
   - Bohr radius
   - charge-to-mass ratio
   - correspondence principle
   - electron
   - electron spin
   - energy level
   - energy-level diagram
   - excited state
   - fine structure
   - ground state
   - hydrogenlike atom
   - lase
   - laser
   - magnetic quantum number
   - metastable energy state
   - model
   - nucleus
   - orbital quantum number
   - Pauli exclusion principle
   - periodic table
   - population inversion
   - principal quantum number
   - quantized
   - quantum number
   - raisin cake model
   - spin magnetic quantum number
   - spontaneous emission
   - stimulated emission
2. State the experimental evidence leading to a belief in subatomic structure, in particular to the existence of the electron and to the charge and mass concentration of the nucleus.
3. State the objection to orbiting electrons in atoms from physical theory.

4. State the postulates leading to the Bohr model of the atom.
5. State several quantitative successes of the Bohr atomic model.
6. Calculate the frequency and wavelength of any or all lines in the spectral series of hydrogenlike atoms, given Z and the values of fundamental constants.
7. Construct an energy-level diagram for a hydrogenlike atom, given Z and the binding (ionization) energy.
8. State the correspondence principle.
9. Define the four quantum numbers specifying the state of a particular electron in an atom.
10. State the exclusion principle, and discuss its implications in explaining the organization of the periodic table of the elements.
11. List and discuss the limitations of the Bohr model of electron structure in atoms.
12. Explain the operation of lasers.
13. Calculate the energy that will be delivered to a given area in a given time, knowing the diameter of a laser beam and its power output.

Problems

29-1 Rutherford's group often used radium as the source of alpha particles in their scattering experiments with gold foils. These particles have an energy of 4.78 MeV and a mass of 6.64×10^{-27} kg. The helium nucleus (alpha) has a charge of $+3.20 \times 10^{-19}$ C, and the gold nucleus has a charge of $+1.26 \times 10^{-17}$ C. Calculate (a) the speed of the alpha particle, (b) the distance of closest approach to the gold nucleus for a head-on collision, and (c) the energy an alpha particle would have to have to approach within a distance of 1.00×10^{-14} m to the gold nucleus.

29-2 Determine the kinetic energy of an alpha particle with a speed equal to one-twentieth the speed of light. Ignore relativity and express your answer in both joules and megaelectronvolts.

29-3 Suppose a hydrogen atom were increased in size so that the nucleus ($r = 1.40 \times 10^{-15}$ m) were as big as a bowling ball ($r = 0.109$ m). (a) How far away would the electron be in its smallest orbit? (b) How many of these sized atoms would be required to cover the state of Texas (6.92×10^5 km^2)?

29-4 The radius of a gold nucleus is about 8.15×10^{-15} m, whereas the atom has a radius of about 6.00×10^{-11} m. A beam of alpha particles is made incident on a gold foil 20 atoms thick. Assuming all gold nuclei are exposed to the beam (that is, no nuclei are hiding behind others), determine the fraction of the area of the foil which is covered by gold nuclei.

29-5 A desktop is 1.00 m wide. Embedded halfway down in its middle is a bowling ball 25.0 cm in diameter. One thousand marbles, each 1.00 cm in diameter, are rolled across the desk parallel to its length, with one started in sequence at each millimeter mark on a meterstick at the edge of the desk. (a) What number of marbles roll to the other end of the desk undeviated from their original path? What percentage? (b) In a collision with the bowling ball, the law of reflection from optics ($\theta_i = \theta_r$) applies. What number and percentage of marbles are scattered back toward the direction of their origin within (i) 10°

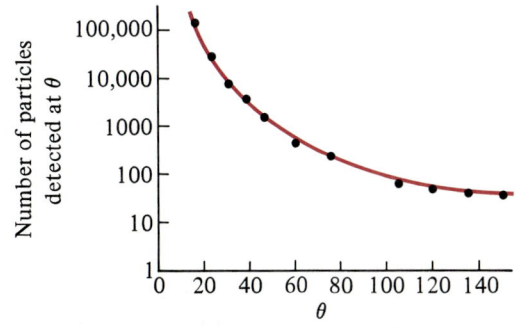

Figure 29-10 Problem 29-5.

(scattering of 170° or more), (ii) 30° (scattering of 150° or more), and (iii) 60° (scattering of 120° or more)? (c) What percentage of marbles are scattered from their original direction of motion but by less than (i) 10° and (ii) 90°? (d) The numbers calculated in the first three parts are for hard-sphere scattering. Figure 29-10 shows the actual scattering results for a gold foil experiment. Note that the scattering at 20° is about 10^4 times as frequent as the scattering at 140°. Does the result of the gold foil experiment illustrate hard-sphere scattering?

29-6 According to the Bohr theory, (a) what is the angular momentum of the hydrogen electron in the $n = 4$ state? (b) Applying conservation of angular momentum, what must be the angular momentum of the photon emitted in the $n = 4$ to $n = 2$ transition of the hydrogen atom?

29-7 Calculate the energy difference between (a) the $n = 3$ and $n = 1$ states of singly ionized helium and (b) the $n = 4$ and $n = 2$ states of doubly ionized lithium.

29-8 What is the ratio of the radius of the $n = 1$ Bohr orbit of hydrogen to the radius of the $n = 1$ Bohr orbit of triply ionized beryllium?

29-9 In the first Bohr orbit, the electron's speed is 2.19×10^6 m/s and the radius is 5.30×10^{-11} m. Calculate (a) the centripetal acceleration and (b) centripetal force.

29-10 Using the expression for the Rydberg constant from Equation 29-10 and the proper numerical values of the constants involved, calculate the value of the Rydberg constant.

29-11 Calculate the frequencies and wavelengths of the first four lines in the Balmer series of hydrogen. These correspond to transitions from the $n = 3, 4, 5$, and 6 states to the $n = 2$ state.

29-12 (a) Calculate the ratio of the speed of the electron in the $n = 1$ state of hydrogen to the speed of light. (b) Would this ratio be larger or smaller for n greater than 1? (c) Was it justified to ignore relativity in the Bohr model?

29-13 Calculate the energies of the photons of the first three Lyman lines of hydrogen. These correspond to transitions from the $n = 2, 3$, and 4 states to the $n = 1$ state.

29-14 Show that according to the Bohr atom, the frequency of orbital revolution for an electron in the nth state of hydrogen is given by $f = (4\pi^2 m k^2 e^4/h^3)(1/n^3)$.

29-15 Use the expression for the orbital revolution frequency given in Problem 29-14 and calculate (a) the orbital frequency for an electron in the $n = 1000$ state of hydrogen and (b) the frequency of the photon emitted when the electron decays from the $n = 1000$ state to the $n = 999$ state. Carry along six significant figures on your calculator. The answers to (a) and (b) should differ by less than $\frac{1}{2}$ percent from each other. This is an example of the way the correspondence principle works. Classical physics predicts that the electron should emit radiation at its orbital revolution frequency. For extremely large quantum numbers, the correspondence principle asserts that quantum theory should reduce to the classical theory.

29-16 Use the expression for the orbital revolution frequency given in Problem 29-14 and calculate (a) the orbital frequency for an electron in the $n = 1$ state of hydrogen, (b) the equivalent current to which the electron's motion corresponds, and (c) the magnetic field generated by this current loop at the site of the proton.

29-17 Calculate the binding energy (ionization energy) for singly ionized helium and doubly ionized lithium in their ground states according to the Bohr model of the atom. Express your answers in both joules and electronvolts.

29-18 An electron in a hydrogen atom decays from the $n = 6$ state to the $n = 1$ state. (a) Calculate the wavelength of the emitted photon. (b) Using the expression $p = h/\lambda$ for the linear momentum of the photon, calculate the speed of the recoil atom, assuming it was initially at rest. (c) Determine the ratio of the kinetic energy of the recoil atom to the energy of the photon.

29-19 Determine how many times an electron in the $n = 2$ state of hydrogen orbits the proton before it decays. Assume it decays to the $n = 1$ state in 1.00×10^{-8} s.

29-20 Determine the wavelength of the smallest-energy photon which can be absorbed by a hydrogen atom in its ground state.

29-21 A hydrogen atom in an excited state of binding energy 0.378 eV decays to a state of binding energy 1.511 eV. Determine (a) the energy and wavelength of the emitted photon, (b) the Bohr quantum numbers for these states, and (c) the series to which this line belongs.

29-22 Hydrogen gas in a discharge tube is excited by collisions with free electrons. The maximum energy the hydrogen atoms can absorb from the electrons is 12.75 eV. Determine which wavelengths may be emitted by the hydrogen as the atoms decay to the ground state.

29-23 A hydrogen atom, originally in its ground state, absorbs a 12.8-eV photon. (a) What is the quantum number of the atom's final state? (b) What must the energy of an absorbed photon be in order to raise the ground-state hydrogen atom to its $n = 2$ state? From the $n = 2$ to the $n = 3$ state?

29-24 Assume the Bohr model is correct. Calculate the radius of the first Bohr orbit ($n = 1$) for uranium ($Z = 92$).

29-25 A hydrogen atom is in an $n = 5$ state. List the possible combinations of values of l and m for that state.

29-26 An electron is in a state with its angular momentum quantum number $l = 3$. (a) What are the restrictions on n? (b) List the possible values of m for that state. (c) How many electrons can have that same l?

29-27 Apply the Pauli exclusion principle to construct a table of quantum numbers showing the electron configuration of (a) oxygen and (b) neon.

29-28 Uranium is the largest naturally occurring atom and has electrons filling all sites by the exclusion principle out to the seventh orbit in the Bohr model ($n = 7$). (a) What does this model predict for the radius of uranium's seventh orbit? (b) The measured radius of the uranium atom is 1.42×10^{-10} m. What is the percentage difference between this and the prediction of the Bohr model?

29-29 A pulsed ruby-rod laser emits 8.00 J during a 5.00×10^{-8} s interval. Calculate (a) the power delivered by the laser beam during that time interval and (b) the intensity of the beam (power per unit area) if the beam has a width of 0.300 mm.

29-30 A He-Ne laser is pulsed by pumping it with electric pulses rather than continuous current. Suppose each laser pulse emitted is 2.50×10^{-7} s long and there are exactly eight pulses each second, and each pulse delivers 0.300 J. Calculate (a) the average power delivered by the laser and (b) the number of photons of wavelength 6.328×10^{-7} m emitted during each pulse.

29-31 Assume a He-Ne laser has equal numbers of each type of the two atoms, both at atmospheric pressure and 273 K (the pressure is really reduced). The tube is 30.0 cm long and 1.00 cm in internal diameter. The continuous power output of red light is 0.100 mW. Neglect losses and assume that the transition producing red light has a 50 percent probability compared to other transitions. (a) How many He atoms are reduced from the metastable state each second to maintain this power output? (b) What fraction of the available He atoms is this each second? (Recall that 1 mol of any gas occupies 22.4 L at standard temperature and pressure and that 1 mol contains 6.02×10^{23} particles.) *(c)* What fraction of the output of the 50-W power source in the laser goes to pumping helium atoms to their metastable state?

29-32 An argon laser with an output wavelength of 4.88×10^{-7} m delivers 2.00 mW. The beam has a diameter of 1.50 mm at the aperture and diverges at an angle of 5.00×10^{-4} rad. Calculate *(a)* the intensity (power per unit area) of the beam at the aperture, *(b)* the intensity 5.00 m away, and *(c)* the energy deposited by the beam per unit area at a distance of 5.00 m during a blink time of 0.0100 s.

29-33 A krypton laser can be pumped by electric pulses to emit light of wavelength 6.471×10^{-8} m in 0.250-mJ pulses of duration 1.00×10^{-11} s. *(a)* What is the physical length of the pulse in a vacuum? *(b)* How many wavelengths are in the pulse (according to wave theory)? *(c)* How many photons are in the pulse?

29-34 A pulsed ruby-rod laser emits a beam 1.00 mm in diameter with a divergence of 1.80×10^{-5} rad. *(a)* Determine the diameter of the beam 200 m away from the opening. *(b)* Calculate the size of the spot which could be made on the moon 3.84×10^8 m away.

SN74LS10N
RQ8411E
PORTUGAL 8414B
ON74S00N
T74LS04BI
88415
PORTUGAL 8406A
SN74LS273N
AM95
8402
NEC
JAPAN
8407KD
D765AC
74LS08 PC
8403 SINGAPORE
PORTUGAL 8410A
ON74LS670N
74LS32 PC
8413 SINGAPORE
PORTUGAL 8410B
SN74S163N
PORTUGAL 8401C
SN74LS195AN
MALAYSIA 406C
SN74LS11N
PORTUGAL 8406A
SN74LS393N
MALAYSIA 414A
SN74LS74AN
1.Ø 5/84
L
PORTUGAL 8401E
SN74S74N
PORTUGAL 8401A
SN74S174N
T74LS00BI
88406
H

30 Quantum Mechanics

30-1 Introduction

Though the Bohr model of the atom had served very well in leading theoretical physicists to understand much of atomic behavior, it was clearly doomed. A new idea was needed which could describe a more cohesive model.

In 1925, Louis de Broglie of France provided the start of a new approach to the problem. He suggested that perhaps electrons have an associated wave nature, just as light, whose wave nature is well established, displays particle properties. De Broglie used his expression for the wavelength of moving electrons to show that the Bohr quantum condition follows. His only additional assumption was that electrons in the allowed orbits of Bohr must be represented as standing waves. What de Broglie proposed was that moving matter constitutes matter waves.

In a short time, evidence began to build for de Broglie waves, but the question we have had since Planck's derivation of the blackbody radiation equation comes back in full force. Are we dealing with *waves* or *particles*? However, the question now incorporated more components, and it somehow seemed more urgent.

About a year after de Broglie's first presentation, Erwin Schrödinger, an Austrian physicist, began publishing a series of articles on quantum mechanics. In his first paper, Schrödinger wrote a differential equation for de Broglie's matter waves. He defined a quantum wave function associated with the electron and used his equation to predict electron behavior and to solve several other important problems.

At the same time Schrödinger's papers were appearing, Werner Heisenberg, a German physicist, was publishing the results of his attack on the same problem using a completely different approach but obtaining the same results. That their results agreed was really a little startling. They used different physical assumptions and very different mathematical methods. Schrödinger later proved that the two approaches were mathematically equivalent and that either one could be shown to follow from the other. Schrödinger's approach became known as **wave mechanics,** whereas Heisenberg's approach is now called **matrix mechanics** after the kind of mathematics used.

Both wave mechanics and matrix mechanics are a little abstract. Wave mechanics requires a mathematical function descriptive of a wave, and Schrödinger showed that the matrix elements in matrix mechanics are derivable by using that wave function. However, there was not a good physical interpretation of the wave function available. It was not clear just what was doing the waving. Attempts to arrive at an interpretation led in several directions and often bore other fruit. For example, Heisenberg's attempt led him to a conclusion that it was impossible to know with infinite precision both the linear position coordinate and the linear momentum of a body at the same time. This extends to angular quantities and energy/time as well.

The new quantum mechanics went on to replace the Bohr theory of the atom. It also led to deeper scientific understanding as well as many technological advances.

30-2 The de Broglie Hypothesis and Matter Waves

The wave nature of matter as proposed by the Frenchman Louis de Broglie in 1925 was presented as part of his doctoral thesis at the University of Paris. He speculated about the symmetry of nature, suggesting that particles should have a wave nature, arguing somewhat as follows.

The photon acts somewhat like a particle, and we may obtain an expression for its linear momentum in terms of its wavelength. The same relationship should then hold between a particle's linear momentum and its wavelength.

From relativity we have that the total energy of a particle is equal to the product of its mass and the square of the speed of light. Thus for a photon of frequency f, we may write

$$E = m'c^2 = hf \tag{30-1}$$

where m' is the **effective mass of the photon**[1] and hf is the Planck expression for the photon energy. Solving for m' gives

$$m' = \frac{hf}{c^2}$$

Electromagnetic waves display the particlelike behavior of coming in "chunks"—photons—and being absorbed and radiated in packages of given energy which depends on wavelength. Why shouldn't particles display a similar versatility by having wavelike properties, such as wavelength, or even show diffraction or interference? The **de Broglie hypothesis** asserts that they do.

---

[1] The rest mass m_0 of a photon is, of course, zero. Photons travel at the speed of light, and if they had nonzero rest mass, relativity would predict an infinite mass for a photon in flight. Further, photons cannot be at rest; they move at c until they are absorbed.

and since the photon's speed is c, the linear momentum is

$$p = m'c = \frac{hf}{c}$$

But $f/c = 1/\lambda$, and we have

$$p = \frac{h}{\lambda} \tag{30-2}$$

Thus the linear momentum is inversely proportional to the wavelength. Or, we could write

$$\lambda = \frac{h}{p} \tag{30-3}$$

Applying Equation 30-3 to a particle of mass m moving with speed v, we obtain

$$\lambda = \frac{h}{mv} \tag{30-4}$$

since $p = mv$.

Equation 30-4 is the de Broglie equation and is to be used to calculate the **de Broglie wavelength** λ associated with matter waves.

Example 30-1 Calculate the de Broglie wavelength associated with the matter wave of (a) a tennis ball ($m = 5.68 \times 10^{-2}$ kg) moving at 25.0 m/s and (b) an electron ($m = 9.11 \times 10^{-31}$ kg) moving at 6.00×10^5 m/s.

Solution Both parts require straightforward applications of the de Broglie equation.

(a) $$\lambda = \frac{h}{mv} = \frac{6.63 \times 10^{-34} \text{ J} \cdot \text{s}}{(5.68 \times 10^{-2} \text{ kg})(25.0 \text{ m/s})}$$
$$= 4.67 \times 10^{-34} \text{ m}$$

It almost doesn't matter whether this is correct or not. It is not possible experimentally to measure a length so small.

(b) $$\lambda = \frac{h}{mv} = \frac{6.63 \times 10^{-34} \text{ J} \cdot \text{s}}{(9.11 \times 10^{-31} \text{ kg})(6.00 \times 10^5 \text{ m/s})}$$
$$= 1.21 \times 10^{-9} \text{ m}$$

This is about the same wavelength as a soft (very low energy) x-ray, and we will see that it may be detected and measured in the same manner that x-rays are observed.

The de Broglie hypothesis suggested a reason for the mechanical quantization of Bohr orbits in the atom. De Broglie postulated that the allowed Bohr orbits were those for which an integer number of wavelengths of matter waves would fit exactly. In other words, the allowed orbits are those in which the electron matter waves were represented by standing waves—one wave in the first orbit, two in the second, three in the third, and so on. This is represented schematically in Figure 30-1. All the patterns shown may be visualized as the envelopes of standing waves on a string which have been wrapped around in a

30-2 The de Broglie Hypothesis and Matter Waves

Figure 30-1 The de Broglie hypothesis explains fixed electron orbits as the only locations where the wavelength of the electron allows a standing wave to form.

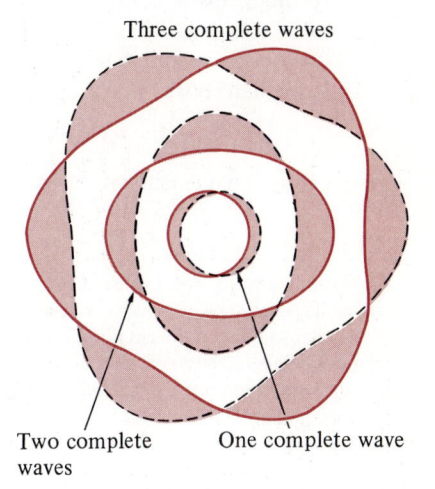

Three complete waves

Two complete waves

One complete wave

circle. Note that the smallest has one complete wave, the next smallest has two complete waves, etc. In symbols, we could write this condition as

$$n\lambda = 2\pi r_n \tag{30-5}$$

When the de Broglie wavelength is used for λ, we have

$$n\frac{h}{mv} = 2\pi r_n$$

This may be rearranged to give

$$mvr_n = \frac{nh}{2\pi} \tag{30-6}$$

Equation 30-6 is exactly the Bohr condition.

Beyond this, there was no direct addition to atomic structure from the de Broglie hypothesis. However, two important ideas had emerged. Matter waves appear to offer some justification for the mechanical quantization of atoms. Second, if these matter waves are real, we should be able to detect them experimentally.

There was no experimental evidence for the existence of matter waves when de Broglie made his hypothesis. Soon, however, the evidence was available. In 1927, the results of two experiments provided conclusive proof of the wave nature associated with moving electrons.

One experiment was conducted by two American physicists, Clinton J. Davisson and his associate Lester Germer, at Bell Telephone Laboratories. They used electrons which had been accelerated through a 54-V potential difference and treated them much like the x-rays that Bragg and Bragg had used in their crystal spectrometer experiments. Figure 30-2 depicts Davisson and Germer's apparatus, and Figure 30-3 shows their data. There is a maximum in the detector current whenever the incident and reflected beam make an angle of 50° with each other. This maximum is interpreted as constructive interference of the matter waves.

The crystal for the experiment was made of nickel which had been heated for a very long time.[1] Note that the beam of electrons is incident on the block at an angle of 90° to the crystal face. Yet the detector measured a maximum corresponding to constructive interference of the matter waves at an angle of 50°. You will recall that in a regular crystal there are many Bragg planes. Those which resulted in the constructive maximum shown must be oriented at an angle of 25° to the surface. We see this in the two-dimensional view of Figure 30-4. You may also recall that the Bragg x-ray equation uses the angle measured from the incident beam to the appropriate Bragg plane. The Bragg equation is

$$m\lambda = 2d \sin\theta \quad m = 1, 2, 3, \ldots \tag{28-10}$$

For the Bragg planes shown, x-ray spectroscopy determines that $d = 9.09 \times 10^{-11}$ m. For $m = 1$ and $\theta = 65°$, we obtain

---

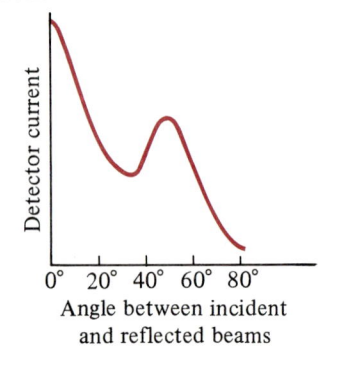

[1] This heating was initiated because of an accident with their apparatus which allowed air to get to their highly polished block of nickel, causing it to oxidize and blacken. This is a famous example of experimental serendipity. Since ordinary nickel is polycrystalline (having many small crystals with random orientation), the wave effect probably would not have been noticed except for the accident. The long, slow heating caused the small crystals to rearrange, forming one large crystal or at least a very small number of large crystals in the block. The regular cubic spacing of the atoms in the large crystal was then available to display the effect.

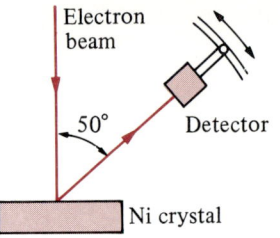

Figure 30-2 The **Davisson-Germer experiment** sends an electron beam at a nickel crystal and measures reflected electrons at various angles. If electrons were like BBs shot into an ordered stack of bowling balls, the reflected BBs would show no pronounced preferred direction of scatter. Instead, the reflected electrons display the sharp angular preference consistent with wave reflection and their de Broglie wavelengths.

Figure 30-3 The scatter measured from the Davisson-Germer nickel crystal. The pronounced peak near 50° from the direction of the incident beam of electrons follows the mathematical behavior of a constructive interference peak for waves even though made by reflected electrons.

30-2 The de Broglie Hypothesis and Matter Waves

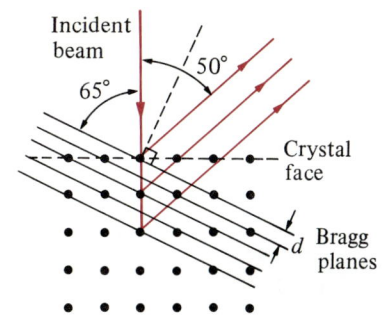

Figure 30-4 The Davisson-Germer result interpreted by the Bragg equation. The incident beam is perpendicular to the crystal face and encounters Bragg planes at an angle of 65° to the beam direction. The incident beam is thus 25° from the perpendicular to the planes and is reinforced in a direction 25° on the other side of the perpendicular.

$$\lambda = 2(9.09 \times 10^{-11} \text{ m})(\sin 65°) = 1.65 \times 10^{-10} \text{ m}$$

To confirm the wave nature of electrons, this value of the wavelength should compare favorably with the wavelength as calculated by applying the de Broglie condition. To make this calculation, first we must determine the magnitude of the linear momentum of the electrons which have been accelerated through a 54-V potential difference. Assuming the electrons start essentially from rest and applying the conservation of mechanical energy, we have

$$\Delta KE + \Delta PE = 0$$

$$\tfrac{1}{2}mv^2 - 0 - eV = 0$$

$$\tfrac{1}{2}mv^2 = eV$$

Multiplying through by twice the mass of the electron yields

$$m^2v^2 = 2meV$$

Now taking the square root, we obtain

$$mv = \sqrt{2meV} \tag{30-7}$$

Thus, application of the de Broglie condition $\lambda = h/mv$ to electrons which have been accelerated through a potential difference V reduces to

$$\lambda = \frac{h}{\sqrt{2meV}}$$

Applying this with $V = 54.0$ V,

$$\lambda = \frac{6.63 \times 10^{-34} \text{ J·s}}{\sqrt{2(9.11 \times 10^{-31} \text{ kg})(1.60 \times 10^{-19} \text{ C})(54.0 \text{ V})}}$$

and we have

$$\lambda = 1.67 \times 10^{-10} \text{ m}$$

This value is in remarkable agreement with the experimental value obtained by inserting the measured angle in the Bragg equation.

Using different voltages, different Bragg planes, and measuring at different angles, Davisson and Germer showed a large number of cases where their

(a)   (b)

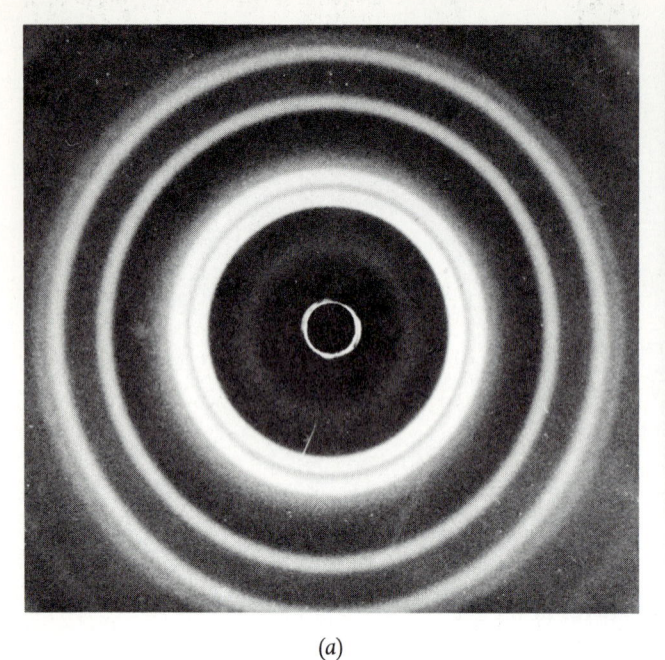

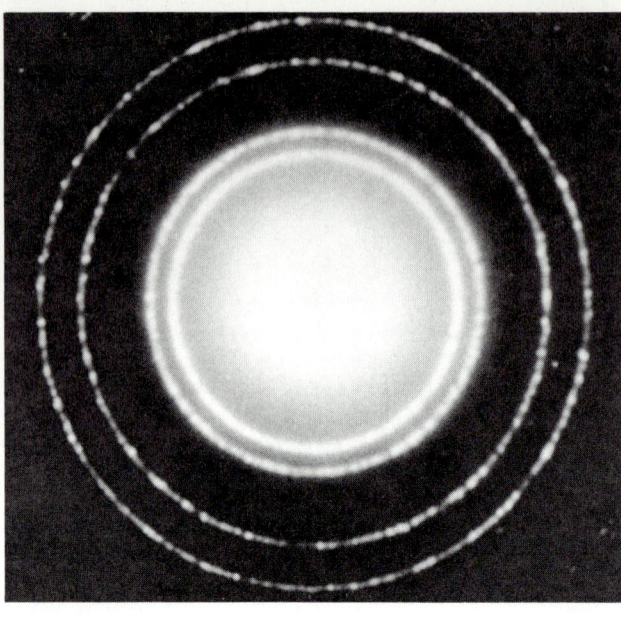

Figure 30-5 (a) X-rays shot at a foil produce this diffraction pattern on a photographic film. The symmetry is circular because microcrystals in the metal foil are found at all orientations. (b) Using the same geometry, electrons are accelerated to the same de Broglie wavelength as the x-rays and bombard the identical foil. The resulting diffraction pattern is completely explained by wave calculations for the electron. *(PSSC Physics, Heath/EDC, 1965)*

experimental data yielded Bragg wavelengths for the electron which agreed with the value calculated from the de Broglie formula. Thus the de Broglie matter waves became a part of accepted reality for physics.

The other confirming experiment was conducted by the English physicist George P. Thomson (son of J. J. Thomson) at the University of Aberdeen. Thomson accelerated electrons through a potential difference of 600 V, passed the beam through a gold foil, and obtained circular diffraction patterns (Figure 30-5) similar to those that M. von Laue obtained with x-rays. These two independent confirmations left no doubts that there is a wave nature associated with moving matter.

Example 30-2 The thin gold foil used by Thomson to obtain the circular rings was polycrystalline, with the numerous tiny crystals oriented randomly so that there were always some crystals oriented to reflect, or scatter, the waves of a given wavelength at a particular angle to the direction of the beam. These reflected waves will interfere constructively to form concentric cones about the incident beam and hence show as concentric circles on a screen. The angle of the scattered beam from the original direction is two times the Bragg angle, and the Bragg condition that $m\lambda = 2d \sin \theta$ for constructive interference applies.

For an experiment similar to Thomson's, the third ring (third order) occurs at an angle of 8.50° from the incident beam. If the foil is aluminum with a crystal lattice spacing of 4.05×10^{-10} m, calculate (a) the wavelength associated with the electron beam, (b) the momentum of the electrons, and (c) the kinetic energy of the electrons in electronvolts.

Solution (a) This is a straightforward application of the Bragg condition, although it requires some care. We must remember to use $n = 3$ and that θ is one-half the angle given, or $\theta = 4.25°$.

$$n\lambda = 2d \sin \theta$$

$$3\lambda = 2(4.05 \times 10^{-10} \text{ m})(\sin 4.25)$$

$$\lambda = 2.00 \times 10^{-11} \text{ m}$$

(b) Here we apply Equation 30-3 or the de Broglie equation.

$$\lambda = \frac{h}{mv} = \frac{h}{p}$$

or

$$p = \frac{h}{\lambda} = \frac{6.63 \times 10^{-34} \text{ J} \cdot \text{s}}{2.00 \times 10^{-11} \text{ m}}$$

$$= 3.31 \times 10^{-23} \text{ kg} \cdot \text{m/s}$$

(c) We may obtain the kinetic energy by recognizing that $KE = p^2/2m$ and then converting to electronvolts.

$$KE = \frac{p^2}{2m} = \frac{(3.31 \times 10^{-23} \text{ kg} \cdot \text{m/s})^2}{2(9.11 \times 10^{-31} \text{ kg})}$$

$$= 6.01 \times 10^{-16} \text{ J}$$

Then

$$KE = \frac{6.01 \times 10^{-16} \text{ J}}{1.60 \times 10^{-19} \text{ J/eV}} = 3.76 \times 10^3 \text{ eV}$$

Thus, this electron beam was accelerated through 3760 V.

30-3 Wave Mechanics

In 1926, Erwin Schrödinger, then a professor of physics at the University of Zurich, published the first of a series of papers on the quantum theory in which he presented a differential equation to be solved for matter waves. In deriving this equation, he had been influenced by the de Broglie hypothesis that electrons have both a particle nature and a wave nature. Schrödinger recognized that the standard wave equation could not be used to represent a particle since a pure sinusoidal wave has infinite extent. That is, the mathematical description of a single-frequency sine wave extends from minus infinity to plus infinity. On the other hand, a particle such as we have dealt with in mechanics is localized at some place. Its position is represented by a single point in space and time. The challenge, as Schrödinger saw it, was to obtain a mathematical description of a wave with finite extent, a **localized wave.** He then wished to write the differential equation for which the localized wave is a solution.

As we saw in Chapter 11, the solution which finally describes a monochromatic wave is an equation for the wave amplitude as a function of space coordinates and time. Equation 11-4 is the one-dimensional amplitude function equation we obtained.

$$y = y_{max} \sin\left[\frac{2\pi}{\lambda}(x - vt)\right] \quad (11\text{-}4)$$

where y is the amplitude of the wave at any displacement x at any time t, while v and λ are the speed of propagation and wavelength, respectively. We call y the amplitude function, or the **wave function.**

In the wave mechanics of Schrödinger, we most often use a Greek uppercase psi (Ψ) to represent the wave function and call it the **quantum wave function.** If Ψ is to represent a matter wave, it must be vibrating back and forth for the wave nature, but the vibrations must go to zero at a short distance in both directions in order that it be localized. Figure 30-6 represents such a function in one dimension. We call this a **wave packet.** This reminds one of the amplitude function which describes the superposition of two waves of slightly different frequencies yielding beats (Chapter 11). In fact, the wave packet is constructed mathematically in much the same way. However, to make the packet go to zero in both directions, a very large (actually infinite) number of slightly different frequency waves must be superimposed. This frequency spread becomes very important in the next section.

Schrödinger then solved the differential equation for the wave packet under several conditions. For a free (unbound) particle, such as one of the electrons in a beam moving at constant speed, Schrödinger's solution yielded the de Broglie condition. This lends immediate respectability to his formalism (à la the correspondence principle). Schrödinger then showed how the equation (now known as the **Schrödinger equation**) must be extended whenever the particle is bound by a variable potential energy and applied it to solve the hydrogen atom. His solution was a real triumph: the expression he obtained for the energy states was identical to that of Bohr. In addition, he showed that the angular momentum quantum number l and the magnetic quantum number m emerge quite naturally from a combination of physical and mathematical conditions which the solution must obey if it is to represent physical reality.[1] This reality requirement also provides the relationships between the quantum numbers, such as n must be greater than l, and so on.

There were some differences between Schrödinger's conclusions and the Bohr theory. For example, the Bohr theory does not predict that the atom could exist in states with zero angular momentum, whereas Schrödinger's solution allows them. Further, in the Bohr theory, the magnitude of the angular momen-

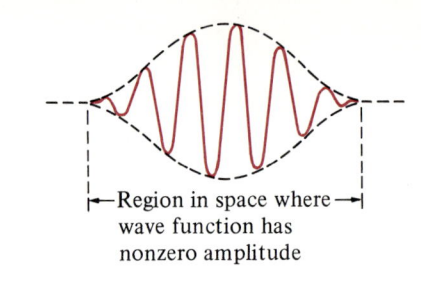

Figure 30-6 Schrödinger's wave packet. The greater the range allowed about a given frequency, the more tightly defined the spatial extent of the packet.

---

[1] For example, l is chosen originally as an arbitrary constant in the solution for the wave function. We find, however, that unless l is an integer, the wave function becomes infinite. Furthermore, if l is a negative integer, the wave function as used to represent the position of a particle goes to zero. Finally, l must be greater than or equal in magnitude to m or the solution will not exist. Since the wave function is to represent a real particle, it can neither be infinite nor zero and must exist. Thus we insist that l is a nonnegative integer which is greater than or equal to the absolute value of m. Similar restrictions must be placed on other parts of the solution to the Schrödinger equation for much the same reasons. These limit the value of n to a positive integer greater than l and also require that m be an integer.

30-3 Wave Mechanics

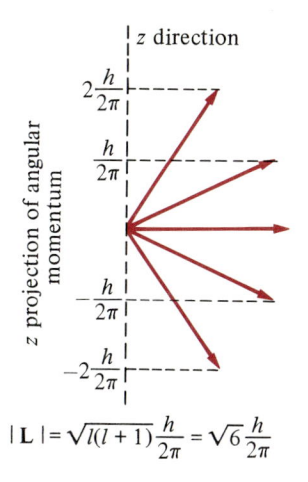

Figure 30-7 The z component of the angular momentum L_z can never be as large as **L** shown but takes on the values $mh/2\pi$ available to it. Shown is the case where $l = 2$; thus $m = 0, \pm 1, \pm 2$. The angular momentum is of constant magnitude $\sqrt{6}\, h/2\pi$, with a z component of one of the five illustrated values.

tum L due to the electron's orbital motion is equal to an integer times Planck's constant divided by 2π, while Schrödinger's solution[1] gives

$$L = \sqrt{l(l+1)}\,\frac{h}{2\pi} \qquad l = 0, 1, 2, \ldots \qquad (30\text{-}8)$$

This is interesting, because we also have that the magnitude of the z component of angular momentum must be

$$L_z = \frac{mh}{2\pi} \qquad m = 0, \pm 1, \pm 2, \ldots, \pm l \qquad (30\text{-}9)$$

Consider Equations 30-8 and 30-9 in light of the fact that both l and m are integers and remember that the angular momentum has a vector nature. We conclude that since the magnitude of L_z is an integral multiple of $h/2\pi$ and that the magnitude of L can never be a positive integral multiple of $h/2\pi$, the two can never coincide. That is, the angular momentum vector can never point along the z axis but must make an angle with that axis such that its z component is an integral multiple of $h/2\pi$. Possibilities which fulfill this requirement are shown in Figure 30-7 for $l = 2$. Actually, the angular momentum vector does not stay constant in direction, as it has a projection of $mh/2\pi$ on the z axis. It precesses about the z axis, making a constant angle such that the projection is always the constant proper value. This is shown for the $l = 2$, $m = +2$, and $m = +1$ states in Figure 30-8. In each of the two cases[2] shown, we see that while **L** is changing direction, it maintains a constant magnitude and a constant projection on the z axis.

Figure 30-8 Having chosen the z component of **L** arbitrarily and determined the possible values of L_z, we find that L_x and L_y cannot be uniquely determined. We interpret this as the **L** vector precessing continually about the defined z direction so that L_x and L_y are constantly changing.

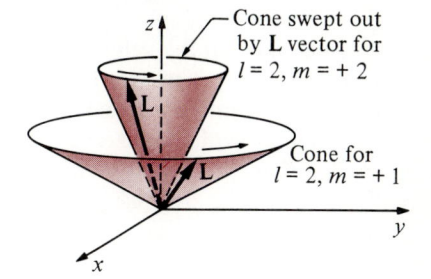

---

[1] Equation 30-8 holds as restrictions on the angular momentum of a hydrogen atom with net negative energy. If its energy is positive, any angular momentum is allowed. However, we would be then talking about a free electron and a free proton rather than a hydrogen atom.

[2] You might be wondering what is special about the z axis? The answer is "nothing!" In the next section, we will see that we can only know one component of L exactly at any given time and that if we do know one component, we can know nothing about the other two. We simply choose to know the z component. Or we might say that we choose to orient our coordinate axes so that the component of angular momentum which is known is along the z axis. The vector **L** then precesses about the z axis so that L_z remains constant but L_x and L_y are variable.

753

Example 30-3 For a hydrogen atom in an $n = 4$, $l = 3$, $m = +3$ state, calculate (a) the magnitude of the orbital angular momentum and (b) the angle the orbital angular momentum makes with the z axis.

Solution (a) This is a straightforward application of Equation 30-8.

$$L = \frac{\sqrt{l(l+1)}h}{2\pi} = \frac{\sqrt{12}(6.63 \times 10^{-34} \text{ J} \cdot \text{s})}{2\pi}$$

$$= 3.66 \times 10^{-34} \text{ J} \cdot \text{s}$$

(b) We may deduce from Figure 30-7 that the cosine of the angle between the angular momentum and the z axis is equal to the ratio of m to $\sqrt{l(l+1)}$.

$$\cos\theta = \frac{m}{\sqrt{l(l+1)}} = \frac{3}{\sqrt{12}} = 0.866$$

$$\theta = 30°$$

The Schrödinger theory had come at a propitious time. The Bohr theory was showing signs of having provided all it could to the atomic model. The Schrödinger theory was part of a twofold thrust on a badly needed new theory. It extended the atomic model far beyond that of Bohr. While we must still provide some interpretation and must still deal with spin, it became obvious that the Schrödinger theory would supplant the Bohr theory.

In 1928, the British physicist P. A. M. Dirac (1902–1984) extended Schrödinger's equation to include spin and relativity. The relativity was necessary when spin was added since the known magnetic moment of the electron would require very high angular speeds. In fact, if classical mechanics were applied, spin requires that points on the electron move at speeds greater than that of light. Dirac's solution shows that the equation describing the intrinsic, or spin, angular momentum parallels that of the orbital angular momentum. That is,

$$S = \frac{\sqrt{s(s+1)}h}{2\pi} \qquad (30\text{-}10)$$

where S is the magnitude of the spin angular momentum and $s = \frac{1}{2}$. Also, we have

$$S_z = m_s \frac{h}{2\pi} = \pm\frac{1}{2}\left(\frac{h}{2\pi}\right) \qquad (30\text{-}11)$$

for the z component of spin. Thus we see that the spin quantum number is $\frac{1}{2}$ only and m_s may take on the values $\pm\frac{1}{2}$. This is perfectly consistent with the Pauli exclusion principle. The two electrons in an atom which can have the same set of n, l, and m quantum numbers have different m_s quantum numbers ($+\frac{1}{2}$ and $-\frac{1}{2}$). Although we often differentiate the electrons with values of $m_s = +\frac{1}{2}$ and $m_s = -\frac{1}{2}$ as having **"up" spin** and **"down" spin**, respectively, their spins are really thought to be precessing about the z axis. This is depicted in Figure 30-9.

All the apparent add-on pieces of the Bohr atom emerged naturally from Dirac's analysis. With his contribution, wave mechanics was essentially complete.

Figure 30-9 Dirac's solution for the spinning electron's angular momentum. The spin angular momentum S can never fully align with the chosen z axis. Instead, S_z is either $\pm\frac{1}{2}(h/2\pi)$, or $m_s(h/2\pi)$, where $m_s = \pm\frac{1}{2}$.

(a) Up spin

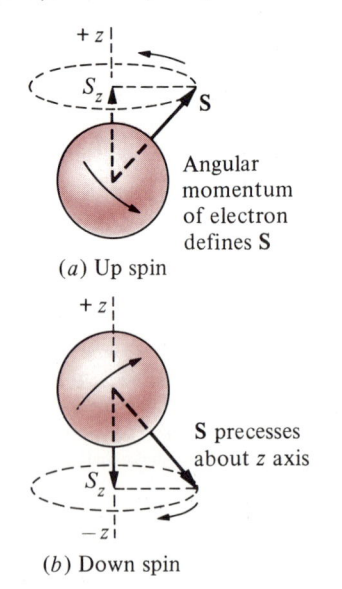

(b) Down spin

30-4 Matrix Mechanics and the Uncertainty Principle

Werner Heisenberg was at Gottingen University in 1926 when he published his first paper on quantum mechanics. As already mentioned, it appeared at the same time as Schrödinger's first paper on the subject but constituted a very different approach. Heisenberg considered the possible physical quantities which we might observe as elements of a matrix.[1] For example, the elements of a particular matrix might be what Heisenberg labeled the possible natural frequencies of vibration of an atom. The elements of another matrix might be the possible energy states of the atom, and so on. These matrices were then mathematically manipulated to yield the real possible experimentally observable values of the quantities they were to represent. Much to everyone's surprise, when the Heisenberg formulation was applied to the hydrogen atom, the results were identical with those of Schrödinger and his wave mechanics. Fortunately, in a very short time, Schrödinger showed, in one of his papers, that wave mechanics and matrix mechanics are mathematically equivalent and that either formulation can be derived from the other. Both wave mechanics and matrix mechanics are in use today. One chooses wave mechanics, matrix mechanics, or a combination of the two, whichever is most convenient to solve the problem at hand.

Both wave mechanics and matrix mechanics provide an excellent mathematical description of the hydrogen atom. Both formalisms go far beyond the Bohr theory, but there still seemed to be a weakness. Both seemed to be somewhat vague, and neither added much in the way of physical interpretation. For example, what was doing the waving in these quantum wave functions and how are these matrix elements related to the real world?

Heisenberg set out to obtain a physical interpretation of matter waves, and in the process of his investigations he concluded that the wave nature of matter imposed a fundamental limit on measurement. You will recall that to create the wave packet, Schrödinger had to superimpose an infinite number of waves of slightly different frequency. Thus, by its very nature, if a wave packet is to represent a particle, it must have a spread of frequencies or a fundamental **uncertainty** in its frequency, Δf. This uncertainty can be made to be very small by allowing the wave packet to spread out in space. In the limit of infinite spread, the packet can be a pure sine wave with zero uncertainty in its frequency ($\Delta f = 0$). But what does this mean? If we know a wave packet's frequency and hence its linear momentum exactly, we have absolutely no knowledge of its whereabouts ($\Delta x = \infty$). It goes from plus infinity to minus infinity, and its position is 100 percent uncertain. To reduce this uncertainty in position, we must localize the wave by adding different frequencies together. This then increases the uncertainty in linear momentum. Thus we may reduce the uncertainty in a wave packet's linear momentum only by increasing the uncertainty in its position, and vice versa.

---

[1] A matrix (plural, matrices) is a rectangular array of elements, usually numbers or letters which stand for numbers. For example, a square matrix may represent the elements of a determinant, or the elements of a determinant may be transformed into a matrix. In any case, matrices follow definite laws of addition and multiplication, forming a small branch of mathematics called matrix algebra.

Another way to look at this uncertainty is by a gedanken experiment. Consider matter waves incident on a single slit, as shown in Figure 30-10. We know from Chapter 26 that waves diffracted by a single slit of width w give rise to a pattern in which destructive interference occurs at an angle θ_m given by

$$m\lambda = w \sin \theta_m \qquad m = 1,2,3, \ldots \qquad (26\text{-}10)$$

Any wave (matter or otherwise) so diffracted and which strikes the screen must have passed through the slit, and thus its x coordinate is uncertain by an amount equal to the slit width w. Or, we have

$$\Delta x = w$$

Applying the de Broglie condition to Equation 26-10 for $m = 1$, we obtain

$$\sin \theta_1 = \frac{\lambda}{w} = \frac{h}{p_y \, \Delta x}$$

The matter waves which strike the screen within the central maximum may have a component of momentum in the x direction with an uncertainty as large as

$$\Delta p_x = p_y \tan \theta_1 \approx p_y \sin \theta_1$$

Thus we have

$$\Delta p_x = p_y \frac{h}{p_y \, \Delta x}$$

or

$$\Delta x \, \Delta p_x = h$$

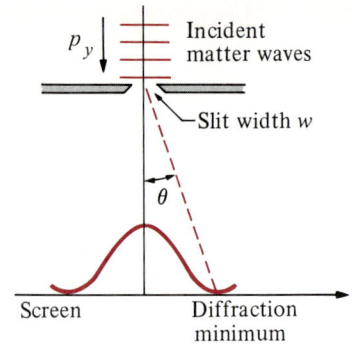

Figure 30-10 Matter waves of momentum p_y strike a slit of width w. The uncertainty in the wave's x momentum, Δp_x, is the spread of the pattern on the screen. As the slit is narrowed, Δx becomes smaller but the interference pattern spreads out. Conversely, if the slit is widened, the pattern narrows. Thus Δx and Δp_x vary inversely such that $\Delta x \, \Delta p_x = $ a constant, the Heisenberg uncertainty principle.

Any attempt to decrease the uncertainty in x, (Δx)—by decreasing the slit width, for example—will cause the diffraction pattern to spread, thereby increasing the uncertainty in p_x (Δp_x). Many gedanken experiments of this type yield similar results. There seems to be a natural conspiracy to stop us from obtaining precise measurements of both momentum and position simultaneously. It is at this point where quantum mechanics makes the irreconcilable break with classical mechanics. Classical mechanics would assert that the precision of these measurements is limited only by the instrumentation and the skill of the experimenter. Quantum mechanics is saying that the limitation is due to nature itself.

Guided by this kind of thinking, Heisenberg formulated and put forth what is now known as the **Heisenberg uncertainty principle.** It is shown to apply to all three coordinates, and, in fact, a more rigorous derivation yields a smaller lower limit than h.

$$\Delta x \, \Delta p_x \geq \frac{1}{2} \left(\frac{h}{2\pi} \right)$$

$$\Delta y \, \Delta p_y \geq \frac{1}{2} \left(\frac{h}{2\pi} \right) \qquad (30\text{-}12)$$

$$\Delta z \, \Delta p_z \geq \frac{1}{2} \left(\frac{h}{2\pi} \right)$$

The uncertainty principle also applies to the simultaneous measurement of angular coordinates and angular momentum.

$$\Delta L_i \, \Delta \theta_i \geq \frac{1}{2}\left(\frac{h}{2\pi}\right) \qquad i = x, y, z \qquad (30\text{-}13)$$

A similar relationship holds for energy and time, provided we interpret ΔE as the uncertainty in a measurement of energy which occurs during a time interval Δt.

$$\Delta E \, \Delta t \geq \frac{1}{2}\left(\frac{h}{2\pi}\right) \qquad (30\text{-}14)$$

The principle of uncertainty is now well accepted, but it was not always so. For a number of years, theoretical physicists differed about it. Even Einstein and Bohr had lively discussions concerning its validity. So far, no one has been able to invent even a gedanken experiment in which the uncertainty principle is violated. It apparently applies to any coordinate and its associated momentum.

In the last section, we stated that we could know exactly only one of the three components of angular momentum at any one time. This is in accordance with the Heisenberg uncertainty principle. If we know two of the components of the angular momentum and the magnitude of the net angular momentum, we can calculate the third component from the Pythagorean theorem. Knowing all three components means we know both the direction of the vector and the magnitude exactly, thus violating the Heisenberg uncertainty relation.

$$\Delta L \, \Delta \theta \geq \frac{1}{2}\left(\frac{h}{2\pi}\right)$$

where $\Delta \theta$ is the uncertainty in the direction of the angular momentum vector and ΔL is the uncertainty in its magnitude.

Example 30-4 At one time (before the discovery of the neutron), it was believed that the nucleus of an atom held both protons and electrons. For example, it was postulated that deuterium (hydrogen 2) had two protons and one electron in the nucleus along with one electron in orbit. Suppose that this were true and that the diameter of the deuterium nucleus is 6.00×10^{-15} m. Calculate (a) the minimum uncertainty in momentum of the trapped electron and (b) the minimum kinetic energy in electronvolts of the electron, assuming that the minimum value of momentum is equal to the uncertainty in momentum.

Solution (a) This is a straightforward application of the uncertainty principle.

$$\Delta x \, \Delta p_x \geq \frac{1}{2}\left(\frac{h}{2\pi}\right)$$

$$\Delta p_x \geq \frac{1}{2}\left[\frac{6.63 \times 10^{-34} \text{ J}\cdot\text{s}}{2\pi(6.00 \times 10^{-15} \text{ m})}\right]$$

$$\geq 8.79 \times 10^{-21} \text{ kg}\cdot\text{m/s}$$

(b) For this, we simply use our answer from the first part to obtain a value for the kinetic energy and then convert it to electronvolts.

$$KE = \frac{p^2}{2m} = \frac{\Delta p^2}{2m} = \frac{(8.79 \times 10^{-21} \text{ kg} \cdot \text{m/s})^2}{2(9.11 \times 10^{-31} \text{ kg})}$$
$$= 4.24 \times 10^{-11} \text{ J}$$
$$= 2.65 \times 10^8 \text{ eV} = 265 \text{ MeV}$$

This is a very large kinetic energy for the electron, and it seems highly unlikely that an electron is trapped in the nucleus. We should note that we are not justified in ignoring relativistic effects in the calculation of kinetic energy. You should calculate the kinetic energy by using the relativistic expression. The answer is really 16.0 MeV, and $v = 0.9995c$.

30-5 Probability, Duality, and Complementarity

While the uncertainty principle gave us some new insights into the behavior of matter waves, there was still no real interpretation of the wave function. Heisenberg was working with Max Born, another German theoretical physicist, at the University of Gottingen during the middle and late 1920s. Born researched this interpretation problem and was guided by the notion that the uncertainty principle seemed to say that electrons cannot exist in specific orbits, or at least we cannot predict exactly where they will go. In fact, the uncertainty principle seems to preclude a particle like an electron even being confined to a specific path anywhere. Born suggested that if the wave function of a particle is to be somehow related to the particle's location in space, we should be dealing with the wave's intensity. After all, the particle has a larger probability of being in those positions where the wave intensity is greater. Perhaps you will recall that the intensity of a wave is proportional to the square of the amplitude. Born knew this and proposed that for a particle with wave function Ψ, the probability of finding the particle within an element of volume ΔV should be given by[1]

$$\Psi^2 \, \Delta V \tag{30-15}$$

Born's probablistic interpretation was really saying that the uncertainty principle compels us to abandon the rigid determinism of classical mechanics. Since we are unable simultaneously to measure the position and velocity of a particle with absolute precision, we cannot predict its future history with precision. The best we can do is evaluate the probability of finding the particle in a given volume at some future time. Under these circumstances it makes no sense to talk about an electron's trajectory or orbit. Hence the Bohr orbits are interpreted as the most probable paths of the electron in a hydrogen atom.[2] Figure 30-11 shows the radial distribution of an electron in the $n = 1, l = 0$ and $n = 2, l = 0$ states of hydrogen.

Figure 30-11 The probability of finding the electron at various distances from the proton in a hydrogen atom for two quantum states ($n = 1, l = 0$ and $n = 2, l = 0$). The most probable distance of the electron in the $n = 1$, $l = 0$ state is represented by the peak of the black curve and is indeed the Bohr radius of the hydrogen atom.

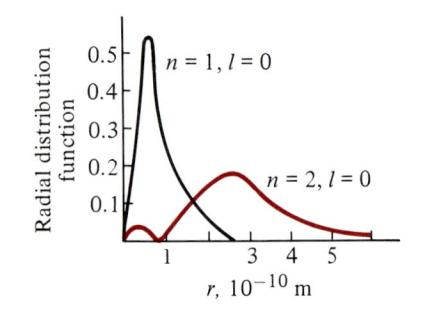

---

[1] We probably should not be surprised that the wave function must be squared if it is to be used to represent the probability of finding the particle. We could not simply use the wave function. Remember that waves take on both positive and negative values, and thus the wave function could be negative at any given position. A negative probability would not really mean anything.

[2] Even this suggests that it may be valid to picture the electron as a little, hard, planetlike sphere in some discrete but indeterminable orbit. Be aware that no physical experiment has ever revealed the electron to be of any discrete size or shape.

Born's suggestion was eventually accepted, just as the uncertainty principle was. However, the world of theoretical physics had not solved the wave-particle dilemma. The wave-particle duality had become a fact of life but was still accepted only with a certain amount of reluctance. The problem often went beyond physics and became a philosophical discussion among physicists. It eventually became evident that in asking the questions "Is light a wave or a particle?" and "Are electrons waves or particles?" we were looking for a model. Furthermore, we expected that the model would be a faithful replica of reality. Earlier, it was believed that light was neither wave nor particle but something else which combined the appropriate properties of both. This "something else" was sometimes dubbed "wavicle." This idea never really satisfied anyone, since the two natures (wave and particle) are so divergent that such a combination seemed beyond imagination.

It appears that the difficulty arises from the way we think. We have trained ourselves to want to form mental pictures of the objects we think about. Furthermore, we believe that the mental picture should be accurate in every sense of the word. For example, we expect that our model of the electron should mimic the actual electron in every respect and also be capable of a verbal description. Try as they might, no one was able to put forth such a model, and the feeling began to grow that such a model was impossible. Many physicists believed that we should simply accept it. Bohr himself advocated a point of view which he called **complementarity.** He stated that wave nature and particle nature are complementary aspects. That is, they do not merge but complement one another. Electrons, photons, and so on, exhibit wave properties or particle properties depending upon the kind of measurement we make. When subjected to diffraction experiments, photons exhibit their wave nature, but when made incident upon electrons in a Compton experiment, they show their particle nature. If an experiment is designed to measure the wave nature, that is what we see. Likewise, the particle nature shows up if that is what we are looking for. Although it is not without misgivings, it is now virtually universally accepted that nature does not always allow us to construct true models. This realization does get rid of the problem and allow physicists to get on with their work. We might close this discussion by noting that from an abstract point of view, there is really no problem. If today a theoretical physicist were asked "Is an electron a wave or a particle?", the reply might be that it is neither *and* it is both! The physicist may say that electrons obey a mathematical formulation which in one limit reduces to the mathematics we associate with the undulations on the surface of water and in another limit reduces to the mathematics we associate with the motion of tennis balls.

Because quantum mechanics is abstract and hard to form nonmathematical models for, this does not imply it is so esoteric as to be impractical. Its predictions open up vast new ranges of practical applications — transistors, lasers, the computer age, understandings of chemical bonding, and much else.

30-6 Solid-State Band Theory and Semiconductors

Because electrons carry charge in electric circuits, it seems natural to investigate their interactions in light of the quantum theory. We expect that whenever atoms are very far apart, they have negligible effect on one another. We are also aware that electrons in widely separated atoms obey the Pauli exclusion principle within their own atoms, but two electrons in two different atoms may have the same set of quantum numbers. When two identical atoms are brought closer together, the associated electron wave functions begin to overlap, distorting one another and changing the energy levels of each. It is found that the common levels divide, or split, and that the amount of splitting is dependent on the distance of separation of the atoms. Figure 30-12a illustrates the energy level of the outer electron of two identical multielectron atoms. Note that in addition to the splitting, the magnitude of the energy levels increases as the atoms get closer together. Can you suggest a reason why this is so?

Whenever a larger number of atoms are brought together to form a solid, the same type of splitting occurs. When the atoms are far apart, the electrons which are in different atoms but have the same quantum numbers have the same energy. As the atoms are moved together, the energy levels split. Figure 30-12b could represent this situation for the single outer electrons of six atoms being brought together. We see that as the atoms are moved closer together, the common energy level divides into six closely spaced levels due to the overlapping of the electron wave functions.

As we increase to very large numbers of atoms (there are more than 10^{20} atoms in even a pinhead-sized piece of metal), the common level splits into a number so large with the difference between the splits so small that we cannot distinguish them as separate. We can think of this as forming an almost continuous band of energy levels. Figure 30-12c depicts the energy band for the outer electrons of some solid.

Although we show only the energy levels of the outer electrons in Figure 30-12, every energy level associated with the atoms of a solid will, in principle, undergo a similar splitting into bands. Each band has a total number of individual levels equal to the number of electrons which *could* occupy the band according to the Pauli exclusion principle. For a single atom, a level with associated angular momentum quantum number l could hold $2(2l + 1)$ electrons corresponding to 2 orientations of the spin angular momentum and $2l + 1$ orientations of the orbital angular momentum. Therefore the capacity of the band into which that level splits is equal to $N(2)(2l + 1)$ electrons, where N is the number of atoms in the solid.

For many atoms in the ground state, the inner levels are completely full, and thus the corresponding inner bands of a solid of these atoms are filled to capacity. It happens that many kinds of atoms have an outer ground-state energy level which is *not* full. In this case, the corresponding band is not full, either. A schematic of the bands of sodium is shown in Figure 30-13. The shaded portions show the extent to which the band is filled in the ground state. We see that the three lower levels are filled to capacity, whereas the $n = 3, l = 0$ state is only half-full and the higher states are empty. Note also that for sodium there is a gap between the bands. These **energy gaps** include energy states in which the electron may not exist. This band structure gives a clue as to why sodium is a good conductor of electricity. The highest ground-state energy band which contains electrons has many unoccupied states. This means that if even a tiny amount of energy is added to the electrons in this band (by, for example, applying a fraction of a volt potential difference), they may accelerate with no apparent impediment. The electrons are essentially free to move since there are numerous unoccupied states available and within easily accessible energy range. This is the mark of a good conductor of electricity.

Not all solids are good conductors, and the band structure of such poor conductors is not like that of sodium. In general, the highest energy band of a material which in the ground state contains some electrons is called the **valence band**. Also, the lowest energy band which is not filled to capacity is called the

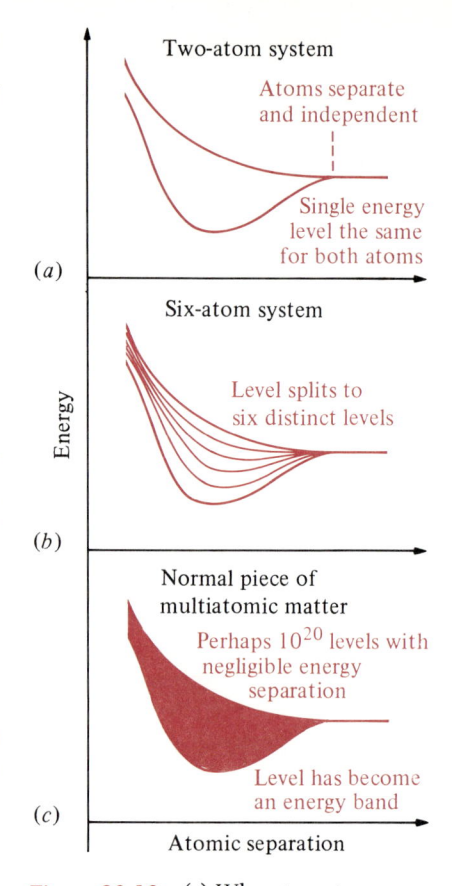

Figure 30-12 (a) When two atoms are far apart, they are independent of each other and the energy level is the same for both. As the atoms approach and interact, these levels split. (b) In a six-atom system there is a sixfold separation of levels between the limits established for the two-atom system. (c) In any real solid, billions of billions of atoms are present, splitting the allowed energy range into that many sublevels separated by negligible energy differences. We say that a particular energy level has become an **energy band**.

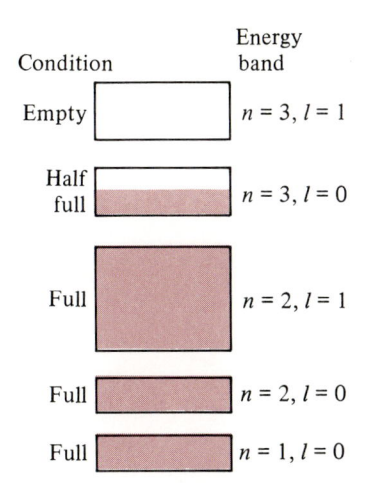

30-6 Solid-State Band Theory and Semiconductors

Figure 30-13 The energy-band structure of sodium metal. For a 1-g piece, the lowest energy band has about 10^{22} levels of negligible separation, and other bands at least as many. The outermost band ($n = 3, l = 0$) is only half-filled with electrons. Thus, at least $\frac{1}{2} \times 10^{22}$ levels are available for any electron to move into with a negligible energy addition.

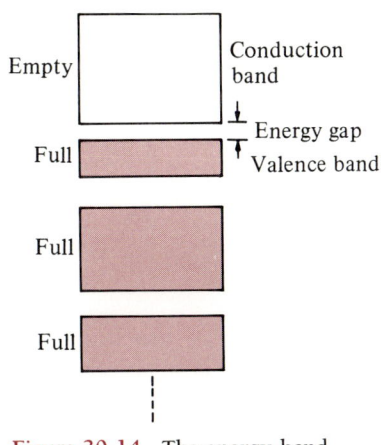

Figure 30-14 The energy-band structure for a poor conductor.

conduction band. We saw that for sodium (and for other good conductors), the valence and conduction bands are the same because that band has some electrons (valence) but is not filled (conduction). Consider, however, the band structure of Figure 30-14. We see that the valence band of this material is filled to capacity, whereas the conduction band is unoccupied. Furthermore, there is a distinct gap between these two bands. Although there are many electrons in the valence band which are available for conduction, there are no unoccupied states through which they can move. Hence, these electrons will not contribute to the conductivity (reciprocal of resistivity) of this material. Furthermore, although there are many unoccupied states in the conduction band, there are normally no electrons to conduct. Therefore, unless electrons can gain enough energy to jump the gap, the material of Figure 30-14 will be a poor conductor.

If the gap between the valence band and the conduction band is fairly large (more than several electronvolts), the material is an insulator. However, if the gap is small (1 eV or less), the material is called a **semiconductor.**

Because of the small gap between the valence band and the conduction band in semiconductors, thermal energy often is enough to raise a fair number of electrons from the valence band to the conduction band. This gives a semiconductor at room temperature a much higher conductivity than insulators. The small gap also makes the conductivity of semiconductors much more temperature-dependent than that of conductors or insulators.

Semiconductors, such as silicon and germanium, are the basis of much of today's technology in the electronics industry. The usefulness of semiconductors in electronic devices was enhanced with the discovery that their properties

761

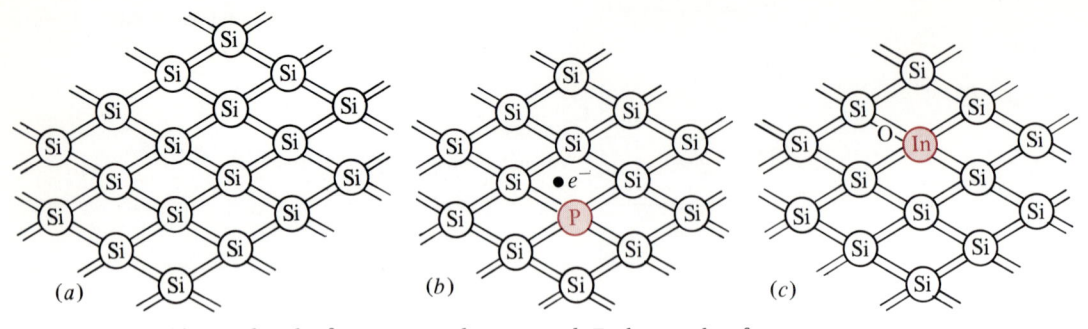

Figure 30-15 (a) A gridwork of atoms in a silicon crystal. Each atom has four valence electrons and is thus capable of four bonds. (b) An atom of phosphorus in a silicon lattice structure has five valence electrons. After bonding, the "extra" electron can easily be promoted to the conduction band of the silicon atoms. (c) An atom of indium in the silicon structure has only three valence electrons and can bond to three neighboring silicon atoms. A nearby electron may move into the hole left at the fourth bond site but in so doing will leave another bonding pattern incomplete. This hole may wander through the structure, much like a moving positive structure.

can be altered by adding certain impurities in very low concentrations. This process is known as **doping,** and the useful impurities are called **dopants.** The process enhances the conductivity of the semiconductor because of the chemistry of the materials involved.

Silicon and germanium atoms both have four outer electrons. In the states of lowest energy, two of those outer electrons have $l = 0$ and two have $l = 1$. When the atoms of silicon or germanium are brought together to form a solid, those two groups first begin to form overlapping bands which separate as the atoms move closer. A valence band is formed which is filled and separated in energy by a small gap from the conduction band (empty). In this configuration, the pure silicon or pure germanium is a very poor conductor. Figure 30-15a represents a two-dimensional view of a pure silicon crystal. Note that every silicon atom is bonded to four others. Each bond is covalent, in which the two atoms of the bond share two electrons. Figure 30-15b shows the same structure in which one of the silicon atoms has been replaced by an atom of phosphorus having five outer electrons. All the covalent bonds can still be formed, but now there is one electron left over. This "extra" electron easily moves to the conduction band, and thus one "free" electron for conduction results every time a silicon atom is replaced by a phosphorus atom. In Figure 30-15c, one of the silicon atoms has been replaced by an atom of indium which has only three outer electrons. In the covalent bonding process, we see that one of the bonds will have only one electron and an empty space rather than two electrons. The bond can be completed by moving a nearby electron to that site which effectively moves the "empty" space to a new location. This process can repeat, and the empty space, called a **hole,** moves through the crystal. In other words, the capturing of an electron to complete the bond creates a hole in the valence band. This hole, wandering about, acts like a positive particle and contributes to a greater value of conductivity.

Impurities which have five outer electrons (e.g., phosphorus, arsenic, and antimony) are called **donors** because they donate an electron to the conduction band. The impurities which have three outer electrons (e.g., boron, aluminum, gallium, and indium) are called **acceptors** because they form a state which accepts electrons from the valence band.

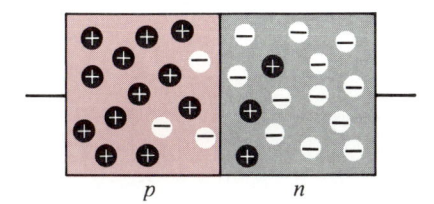

Figure 30-16 A junction diode. The deficiency of electrons on the *p* side makes electron flow from *p* to *n* very difficult even if an applied voltage encouraged such a migration. The opposite field, encouraging flow from *n* to *p*, deposits excess electrons in the holes of the *p* side, allowing new electrons to enter the *n* side from the right and holes to deliver their electrons to the wire at the left.

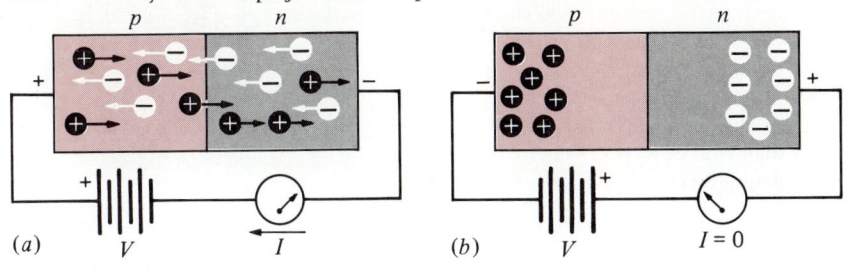

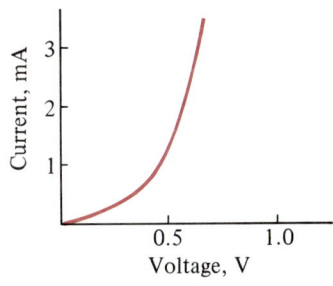

Figure 30-17 Flow of charge vs. applied voltage across a junction diode. Reverse (negative) voltages produce negligible current until very high breakdown voltages are reached. Voltages in the direction of natural flow produce first a trickle and then substantial currents across the diode as voltage is increased.

A semiconductor doped with a donor is called an **n-type semiconductor**, because the conductivity is due to negatively charged electrons. Those semiconductors which are doped with an acceptor are called **p-type semiconductors**, because the conductivity is due to holes, which contribute to the conductivity like positively charged particles.

Solid-state electronic devices such as diodes, transistors, and integrated circuits may be fabricated by joining several different types of semiconductors in particular ways. For example, Figure 30-16 depicts a *p*-type semiconductor joined with an *n*-type semiconductor to form a diode. The boundary along which the two semiconductor materials are joined is called a **pn junction**, and this device is called a **junction diode**. The most useful property of a diode is that it will easily pass current in one direction but presents a very high resistance to current in the other direction. Figure 30-17 shows the current through a silicon diode vs. the voltage across the diode. We may deduce why the diode conducts in one way and not the other. When the *p*-type material is connected to high positive voltage and the *n*-type connected to negative voltage, we are attempting to drive conventional current through the junction from *p*-type to *n*-type material (Figure 30-18*a*). Note that the electrons from the *n*-type and the holes from the *p*-type material are driven toward the junction, where they may combine immediately or drift into the other type of material before they do. When the applied voltage is driving the conventional current in this direction (called the forward direction), very large currents may result.

When the voltage is reversed, as in Figure 30-18*b*, both electrons and holes are driven away from the junction. The junction now has few charge

Figure 30-18 (*a*) Excess electrons from the wire move into the *n* side at the right, where they migrate under the influence of the field to the center boundary. Here, an electron jumps across into a hole, jumping farther leftward every time a hole approaches. (*b*) The reversed polarity of the battery tends to pull excess electrons from the *n* side and supply electrons to the *p* side. This pulls both mobile electrons and holes away from the *pn* junction and produces no flow across the material.

carriers, and the region in the vicinity of the junction has a very small value of conductivity. The junction thus acts as an insulator, allowing only a tiny current to flow.

The **transistor** is a very important development of **solid-state physics** which employs semiconductors joined together. Figure 30-19 illustrates the most common kinds of transistor, an *npn* and a *pnp*, both named after the type of material used to make the junctions. The central region is usually called the **base**, and the two outside regions are called the **emitter** and the **collector**. The base is normally very narrow, so that the probability is high that charge carriers from the other two regions may drift across it. By varying the voltage on the base, one can control the flow of charge carriers between the emitter and the collector. Figure 30-20 shows a typical circuit in which a transistor may be used. In this configuration the transistor acts as a voltage amplifier. We see that the applied voltage is in the forward direction across the emitter-base junction. Thus large numbers of holes are incident on the junction from the emitter. Most of these holes will pass right through the thin-base *n*-type material to the collector. The number of holes being produced in the emitter per unit time depends on the magnitude of the input current. Thus the output-circuit current changes with the input-circuit current. Because the base-collector junction has a reverse voltage, there may be a much higher voltage in the output circuit than in the input circuit. Finally, since power is equal to the product of current and voltage, the configuration shown can represent either a voltage amplifier or a power amplifier.

Advances in technology have come so rapidly in the years since the invention of the transistor that even the most optimistic of scientists and engineers find it difficult to fathom. New designs for creating transistors as well as the development of techniques for incorporating large numbers of transistors in very small spaces have contributed a great deal to the technology explosion. Development of what is called the **integrated circuit** has led the way in this explosion. Conducting and semiconducting material as well as insulating material may be deposited successively in layers to create many interconnected diodes, transistors, and passive circuit elements (resistors, capacitors). Carefully designed and controlled etching then defines the paths for current, making a complete and perhaps very complex circuit (see Figure 30-21). The wonder of the integrated circuit is not so much its capability but its tiny size. The principles of quantum mechanics which were developed to explain the submicroscopic part of the universe generated each step toward miniaturized technology.

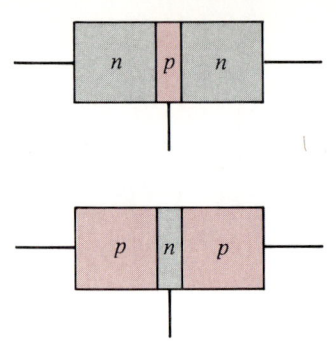

Figure 30-19 The transistor uses a thin center layer between blocks of opposite material, either *npn* or *pnp*. In either case, control of the voltage applied to the central section controls the flow of charge between the ends of the device.

The first commercial computer appeared in the 1950s (the Univac I), filled a room, and needed another small room for its air conditioner. It employed vacuum tubes, each a bit like the electron gun assembly and grid described in Chapter 18. Each vacuum tube may be replaced by a transistor, thousands of which can be produced on a microchip of silicon. As a result, you can hold more calculating power in your hand calculator today than the Univac I at a hundred-thousandth the cost. All this is the result of a quantum mechanical view of matter.

Figure 30-21 An integrated circuit positioned atop the end of a paper clip to indicate scale. (*Bell Laboratories*)

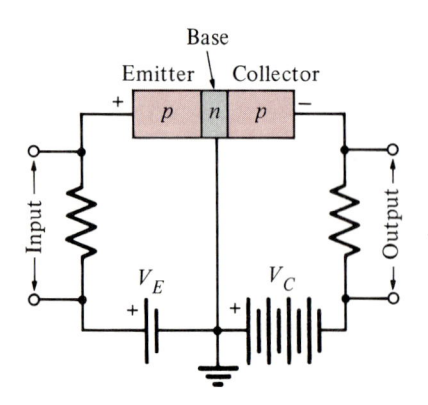

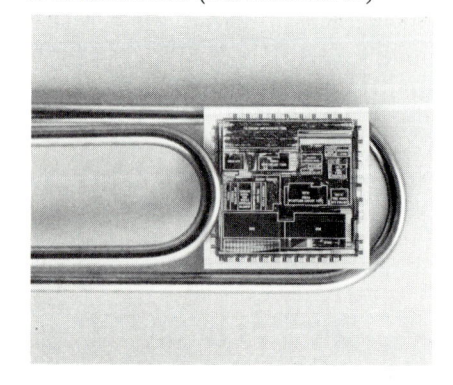

Figure 30-20 A circuit containing a transistor. By varying the voltage of the base relative to the emitter and collector, charge migration may be encouraged or discouraged. In the configuration shown, current out would be proportional to current in, but migrating charges would receive an energy boost crossing the junctions. Thus, voltage out would be greater than voltage in, and the circuit is a voltage amplifier.

SPECIAL TOPIC

"Seeing" with Electrons

A wave cannot reflect from something much smaller than its own wavelength. Dust grains in air, for example, will not reflect normal sound waves, nor will a flying insect. The insect will, however, reflect the much shorter ultrasound waves generated by a bat to detect the insect. A rule of thumb is that a wave detector must be at least $\frac{1}{4}\lambda$ in length for good reception.[1] One implication of this rule is that, in the light microscope, there is a seeing limit imposed by the light itself. Visible light ranges from 700 nm (red) to about 400 nm (violet) in wavelength. An atom is on the order of 1 nm in diameter; thus only large aggregates of atoms can reflect light and be detected visually. Figure 30-22 shows the sizes of microscopic subjects along with the wavelength range of visible light. This visible-light range is approximately coincident with the size range of most bacteria, which accounts for their appearance in the microscope as hazy shapes revealing little detail. Viruses are completely submicroscopic, with no hope of being seen individually with visible light.

One solution is to use shorter-wavelength ultraviolet light to "see" finer detail with detectors other than the eye. In biology, this extends the ability to resolve fine structures within the cell by a small amount, but not nearly enough to do more than infer that still-finer structures exist. X-rays have still shorter wavelengths but cannot be refracted and reflected to control their paths by blocks of matter (lenses) as light is.

Following the confirmation of the de Broglie hypothesis that matter in motion has wave properties, a new possibility became apparent. An **electron microscope** might be constructed, using electrons much as a normal microscope uses light. Beams of electrons may be produced with little more difficulty than producing light. Furthermore, the charge on the electron allows its path to be controlled by a shaped magnetic (or electric) field. Above an accelerating potential of about 50 kV, the effects of relativity on the electron begins to present serious problems. Using this as an upper potential limit, we can calculate the shortest de Broglie wavelength we may use to examine fine structures. Equating the final kinetic energy acquired by the electron with the maximum energy obtained in accelerating across this potential difference, we obtain

$$\tfrac{1}{2}mv^2 = eV$$

Figure 30-22 Sizes of some very small but common objects along with the range of wavelengths of visible light.

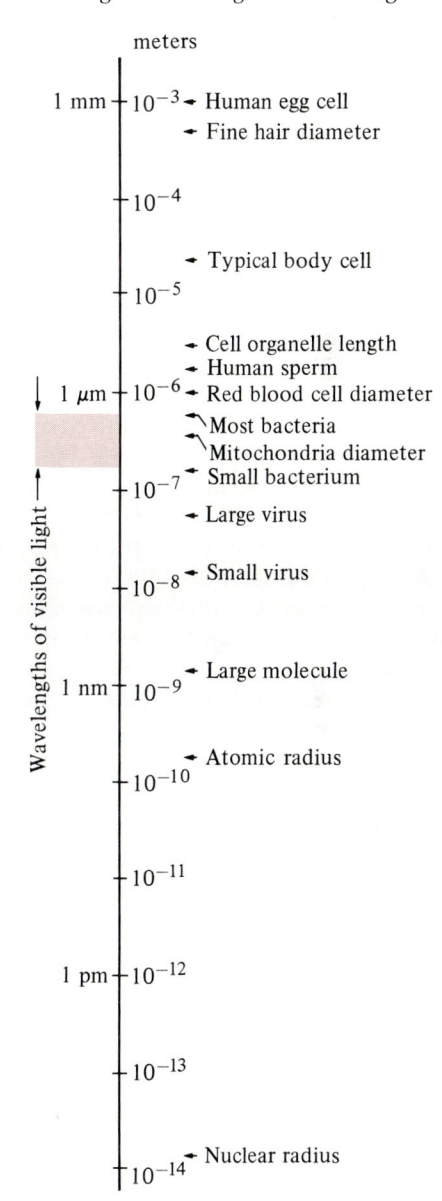

meters

1 mm	10^{-3}	→ Human egg cell
		→ Fine hair diameter
	10^{-4}	
		→ Typical body cell
	10^{-5}	
		→ Cell organelle length
		→ Human sperm
1 μm	10^{-6}	→ Red blood cell diameter
		Most bacteria
		Mitochondria diameter
	10^{-7}	→ Small bacterium
		→ Large virus
	10^{-8}	→ Small virus
		→ Large molecule
1 nm	10^{-9}	
		→ Atomic radius
	10^{-10}	
	10^{-11}	
1 pm	10^{-12}	
	10^{-13}	
	10^{-14}	→ Nuclear radius

Wavelengths of visible light

---

[1] TV waves for channel 2 are 6.0 m long, and this rule would imply that an antenna 1.5 m across is needed for good reception. Some TVs do not come with an antenna 1.5 m long (most do). If this is the case with your set and you have poor reception of channel 2, simply extend the antenna and try for a better picture. A flattened wire coat hanger, bent at the hook to extend from the antenna end, works well.

or, solving for v, we have

$$v = \sqrt{\frac{2eV}{m}}$$

Substituting this into the de Broglie equation (30-4),

$$\lambda = \frac{h}{mv} = \frac{h}{m\sqrt{\frac{2eV}{m}}} = \frac{h}{\sqrt{2meV}}$$

$$= \frac{6.63 \times 10^{-34} \text{ J} \cdot \text{s}}{\sqrt{2(9.11 \times 10^{-31} \text{ kg})(1.60 \times 10^{-19} \text{ C})(50.0 \times 10^3 \text{ J/C})}}$$

$$= 5.49 \times 10^{-12} \text{ m}$$

Comparing this wavelength with Figure 30-22 would seem to imply that the wave properties of the moving electron could be used to explore even the structure of the atom in some detail. Many problems intervene to prevent this; still, the best electron microscopes have attained resolutions down to nearly atomic diameters under the best of operating conditions (and operating beyond 50 kV). Expressed differently, while the light microscope can provide an 800× magnification, or almost 1000× with ultraviolet light, the electron microscope can attain magnifications greater than 100,000×.

Magnetic fields cannot be shaped to have the subtleties of a finely curved piece of glass; thus, spherical aberration is a serious problem for all electron microscopes, requiring long, thin beam paths to keep electrons close to the central axis. The typical **transmission electron microscope** (Figure 30-23) is a long evacuated column containing a specimen holder between magnetic lenses. Other lenses shape the electron path moving through the specimen to produce an image on a fluorescent screen at the bottom. Typically, the specimen must be very thin to allow substantial electron penetration and prevent energy losses (hence wavelength changes) which produce the equivalent of chromatic aberration. A specimen sliced to less than one one-hundredth the thickness of the typical cell is placed on a microfine carbon film deposited over a supporting screen of wires and freeze-dried to remove moisture. Really delicate specimens require even more preparation since they are destroyed in the electron bombardment before a stable image is produced. Such specimens may be mounted in a vacuum chamber and a heavy metal evaporated onto them from the side. It is the metal ghost of the structure which is then seen, improved in contrast by the heavy, slanted shadows resulting from the evaporation technique.

The **scanning electron microscope,** developed in the 1960s, produces an almost three-dimensional image

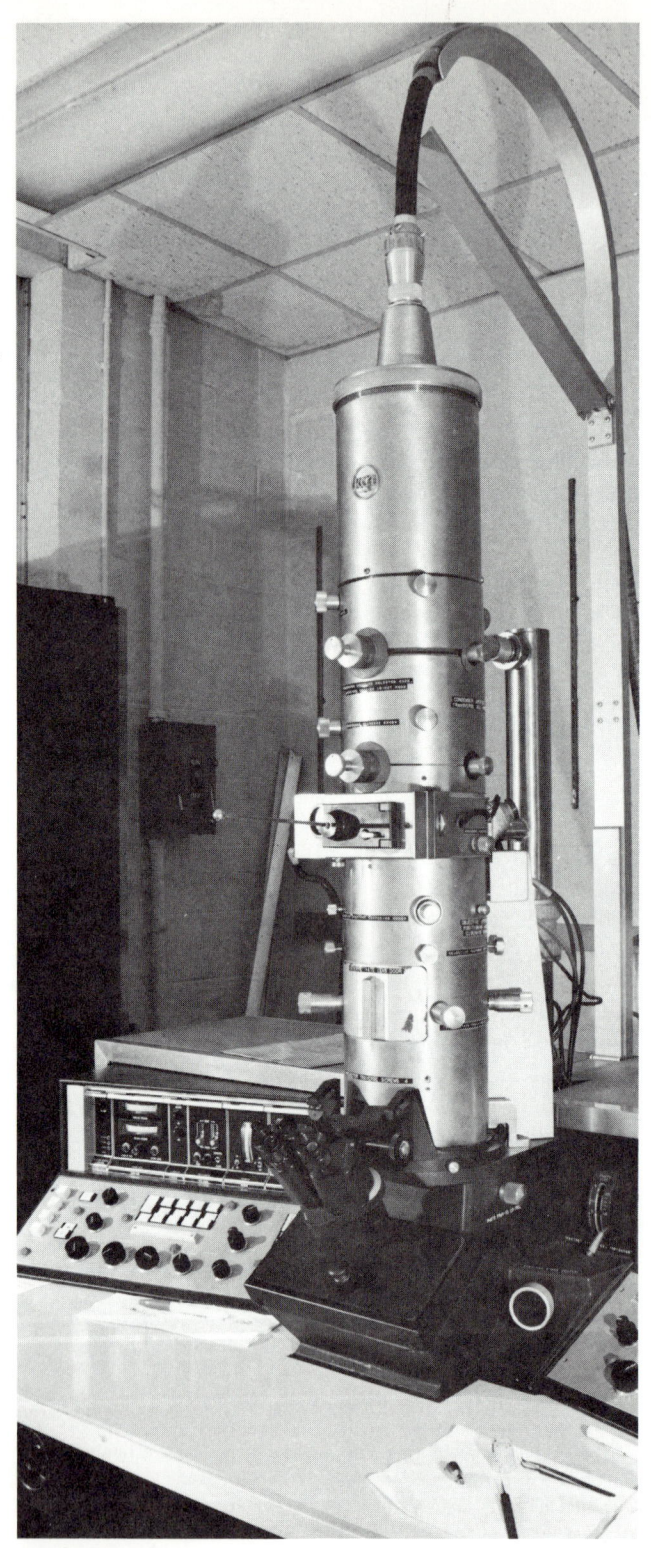

Figure 30-23 A transmission electron microscope. (G. Heilman/Runk-Schoenberger)

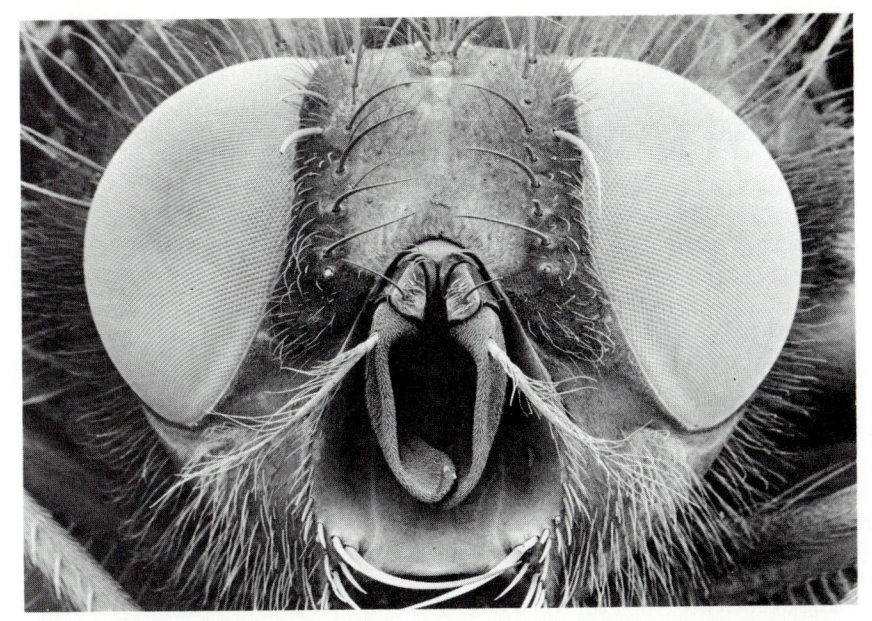

Figure 30-24 Scanning electron micrograph of the head of a bluebottle fly, a familiar house pest. *(Photo Researchers)*

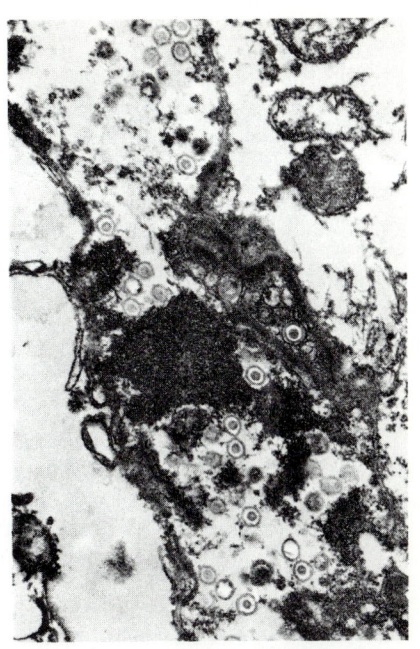

Figure 30-25 Scanning electron micrograph of the herpes simplex I virus in brain tissue. *(M. Rotker/Taurus)*

with incredible depth of field but loses some resolution. Since its image is produced by reflected electrons, entire thick objects (for example, an insect) may be examined. An extremely fine ($\approx$ 10 nm) beam of electrons is scanned across the area being examined, much as an electron beam scans the face of a TV picture tube.

Electrons reflected to the side toward a detector are measured instant by instant, and their intensity is fed to control the point-by-point brightness of the image on a TV screen scanning in synchronization with the beam in the microscope. Figures 30-24 and 30-25 are photomicrographs from a scanning electron microscope.

Minimum Learning Objectives

After studying this chapter, you should be able to:

1. Define:
   - complementarity
   - conduction band
   - Davisson-Germer experiment
   - de Broglie hypothesis
   - de Broglie wavelength
   - doping
   - effective mass of a photon
   - energy band
   - energy gap
   - Heisenberg uncertainty principle
   - hole
   - integrated circuit
   - junction diode
   - localized wave
   - n-type semiconductor
   - p-type semiconductor
   - pn junction
   - quantum mechanics
   - quantum wave function
   - Schrödinger equation
   - semiconductor
   - transistor
   - "up" or "down" spin
   - valence band
   - wave function
   - wave mechanics
   - wave packet

2. Use the de Broglie hypothesis to calculate the wavelength of a particle with a specified motion.
3. Determine the wavelength of electrons accelerated through a given voltage.
4. Describe allowed orbits in terms of standing waves.
5. Discuss the experimental evidence supporting the wave nature of electrons.
6. For electrons accelerated across a known voltage, calculate the angle of a diffraction maximum from planes of atomic centers of known separation in a crystal.

7. Discuss how a quantum wave function must differ from the classical description of a wave.
8. Calculate the orbital angular momentum of an electron in a given energy state in an atom.
9. Calculate the amount of uncertainty in the momentum of a particle from a knowledge of how well its position is known, or vice versa.
10. Calculate the amount of uncertainty in the measured energy of a particle from a knowledge of how well its time of measurement is known, or vice versa.
11. Interpret atomic orbitals in terms of the probability of localizing an electron in space in time.
12. Discuss the formation of energy bands from electron energy levels in multiatom structures.
13. Interpret differences in the electric conductivity of various substances in terms of energy-band structure.
14. Explain the advantage gained by doping a semiconductor.
15. Explain the operation of a junction diode of *npn* or *pnp* transistors.

Problems

30-1 Calculate the de Broglie wavelength of (a) a 5000-kg truck moving at 6.00 m/s, (b) a 2.86×10^{-5} kg ball bearing moving at 0.100 m/s, and (c) a 1.00×10^{-12} kg amoeba moving at 1.15×10^{-5} m/s.

30-2 An electron is accelerated through a potential difference of 900 V. Calculate its (a) kinetic energy in joules, (b) linear momentum, and (c) de Broglie wavelength.

30-3 Calculate the de Broglie wavelength of 10.0-keV (a) x-rays, (b) electrons, and (c) neutrons.

30-4 (a) Calculate the voltage through which a proton would have to be accelerated to have the same de Broglie wavelength as an 800-eV electron. (b) Repeat for an alpha particle.

30-5 What must be the velocity and energy of (a) a proton of de Broglie wavelength 1.00×10^{-12} m and (b) an electron of de Broglie wavelength 2.50×10^{-9} m?

30-6 Calculate the ratio of the de Broglie wavelengths for a proton and an electron of the same (nonrelativistic) kinetic energy.

30-7 Calculate the energy of a proton with a de Broglie wavelength of (a) 1.00×10^{-10} m and (b) 1.00×10^{-13} m.

30-8 Neutrons are the neutral constituents of the atomic nucleus which we will encounter in the next chapter. Because they are uncharged, neutrons do not respond to electric or uniform magnetic fields, and thus another way must be employed to obtain a beam of neutrons with a single velocity. One method which works for very low energies is to make use of their de Broglie wavelengths. Consider a beam of low-energy neutrons incident normally on a crystal of rock salt which has a grating spacing of 2.814×10^{-10} m. (a) Calculate the Bragg angle at which first-order constructive interference would occur for 0.0250-eV neutrons. (b) What angle is this with the crystal face? (Take the mass of the neutron as 1.67×10^{-27} kg.)

30-9 The electron microscope was developed to attain much higher resolution than is possible with visible light (Special Topic, this chapter) because of the shorter wavelengths attainable. Calculate the wavelength of electrons after they have been accelerated through 50.0 kV in an electron microscope, (a) ignoring special relativity and (b) using the special relativity expression for linear momentum.

30-10 A neutral phosphorus atom has 15 electrons. Construct a table showing the ground-state quantum numbers of phosphorus.

30-11 Construct a table showing the possible ground-state quantum numbers for argon ($Z = 18$).

30-12 Show in tabular form the electron configuration of a neutral copper atom in the ground state. Copper has 29 electrons.

30-13 Wave mechanics predicts that an electron confined to a one-dimensional box of length l may have only those wavelengths associated with standing waves. That is, the only wavelengths possible are given by $\lambda_n = 2l/n$. Obtain the expression for the (a) linear momentum and (b) kinetic energy, assuming the potential energy is zero.

30-14 Show that the total number N of electrons which can occupy a state of principal quantum number n is given by $N = 2n^2$.

30-15 Calculate the minimum possible error in measuring the speed of an electron in the first Bohr radius; that is, $d = 1.06 \times 10^{-10}$ m.

30-16 Determine the linear width of the central maximum due to a beam of electrons of speed 1.00×10^7 m/s diffracted by a single slit 2.00×10^{-6} m wide. The screen is 2.50 m away from the slit.

30-17 Calculate (a) the uncertainty in linear momentum and (b) the percentage uncertainty in linear momentum of an electron moving at a speed of 1.00×10^6 m/s if its position is known to within 1.00×10^{-9} m.

30-18 To conduct a precision experiment in the laboratory, you must make a measurement of the position of a diffraction grating to an accuracy of 1.00×10^{-3} mm. (a) Calculate the minimum uncertainty in the momentum of the grating. (b) Assume that the grating has a mass of 0.0250 kg, and determine the minimum uncertainty in the grating's speed. (c) Assume that the grating is moving at a speed equal to the uncertainty. How long would it take the grating to move 1.00×10^{-3} mm? For comparison, the greatest estimate of the age of the universe is approximately 7.00×10^{17} s.

30-19 Given the possible energy that must be associated with an electron, which we calculated in Example 30-4, we concluded that it seemed unlikely that an electron was trapped in the nucleus. For comparison, (a) calculate the minimum uncertainty in linear momentum of a proton ($m = 1.67 \times 10^{-27}$ kg) trapped in a deuterium nucleus of diameter 6.00×10^{-15} m, and (b) calculate the minimum kinetic energy, assuming the minimum value of momentum is equal to the uncertainty in momentum. (c) Are you justified in ignoring relativistic effects?

30-20 An electron microscope has a resolution of $\Delta y = 4.00 \times 10^{-10}$ m and an angular aperture of $\alpha = 0.150$ rad. The electrons have an energy of 50.0 keV and are scattered off the target atoms into an angle of ± 0.150 rad. The uncertainty in their momentum is given by $\Delta p_y = p_y \sin \alpha$. Calculate (a) the momentum of the electrons, (b) the uncertainty in momentum of the electrons, and (c) the product of uncertainty in the y components of position and momentum.

30-21 Suppose a radioactive nucleus (Chapter 31) remains in the state for a time interval on the average of 1.00×10^{-13} s. (a) What is the minimum uncertainty of the energy of the system (in electronvolts)? (b) What is the minimum uncertainty of the frequency of the photon emitted? Take the emitted energy to be 500 eV.

30-22 You are to measure the position of an electron without causing a change in its kinetic energy by more than 10.0 eV. Determine the maximum accuracy with which the position can be determined if its speed is 2.30×10^7 m/s.

30-23 In Chapter 29, we saw that an atom typically exists in an excited state for a time interval of 1.00×10^{-8} s or less. With what minimum uncertainty can one expect to measure (a) the energy of such an excited state and (b) the frequency of the photon emitted when the atom decays to its ground state?

30-24 (a) How much time is needed to measure the kinetic energy of an electron moving at 9.38×10^4 m/s to an accuracy of 0.100 percent? (b) Calculate the distance traveled by the electron during this time interval.

30-25 Suppose there exists another universe in which Planck's constant is 1.00×10^4 J·s but other fundamental constants are the same as ours. For that universe, calculate (a) the wavelength of a 1000-kg automobile moving at 20.0 m/s, (b) the uncertainty in position of the auto if you can measure the speed to within 0.1 m/s, and (c) the angle of first-order diffraction if the auto passes through a garage door of width 3.00 m.

30-26 In a mythical universe where Planck's constant is 80.0 J·s, a tennis player serves a ball toward an opponent at 40.0 m/s. If the tennis ball has a mass of 0.0568 kg, and its position must be measured to within 0.0600 m, calculate (a) the uncertainty in the speed of the ball and (b) the uncertainty in the time it will pass the opponent, who is 30.0 m away.

30-27 Determine the number of free electrons in a piece of copper wire 1.00 m long and 2.00×10^{-3} m in diameter. Take the atomic mass of copper to be 63.5 u and its density to be 8.29×10^3 kg/m³. Assume one free electron per atom.

30-28 Figure 30-20 shows a *pnp* transistor connected so that the base lead is the common wire between a loop containing the input and V_E and another loop containing the output and V_C. In this common-base wiring scheme, collector current and emitter current are approximately equal, and voltage is amplified. However, the transistor is inherently a current-amplifying device. Used in the common-emitter connection shown in Figure 30-26a, it generates the transistor characteristic curve shown in Figure 30-26b. (a) From the curve, what is the amplification of the circuit operating at 4.5 V for a base current I_B of 40 and 20 μA? (b) All methods of connecting the transistor into the circuit of Figure 30-26a would provide a power gain (more power out than in). If the connections were switched so that the lowest wire were connected to the base and not to the emitter as shown, we have the circuit of Figure 30-20 with a common base, no current gain, but power and voltage gains. If the lowest wire were connected to the collector, the common-collector circuit would amplify power and current but not voltage. In each case, where does this power gain come from without violating the law of conservation of energy?

30-29 A transistor is a 0.200-cm cube, essentially half collector, half emitter with a vapor-film base of negligible thickness separating them. (a) If the density of silicon is 2.33×10^3 kg/m³ and it has approximately 28.1 g/mol, how many silicon atoms are present in the cube? (b) If the *npn* transistor is doped with phosphorus atoms with five valence electrons at a rate of one part per million, how many unbonded electrons are available in the cube? (c) If each phosphorus atom contributed its unbonded electron to the electron flow across the transistor in 1.00 s, what would be the current?

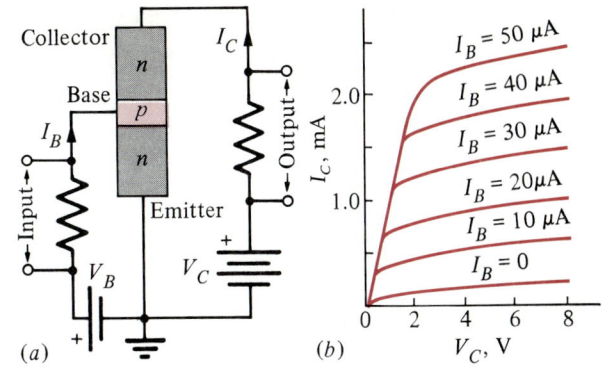

Figure 30-26 Problem 30-28.

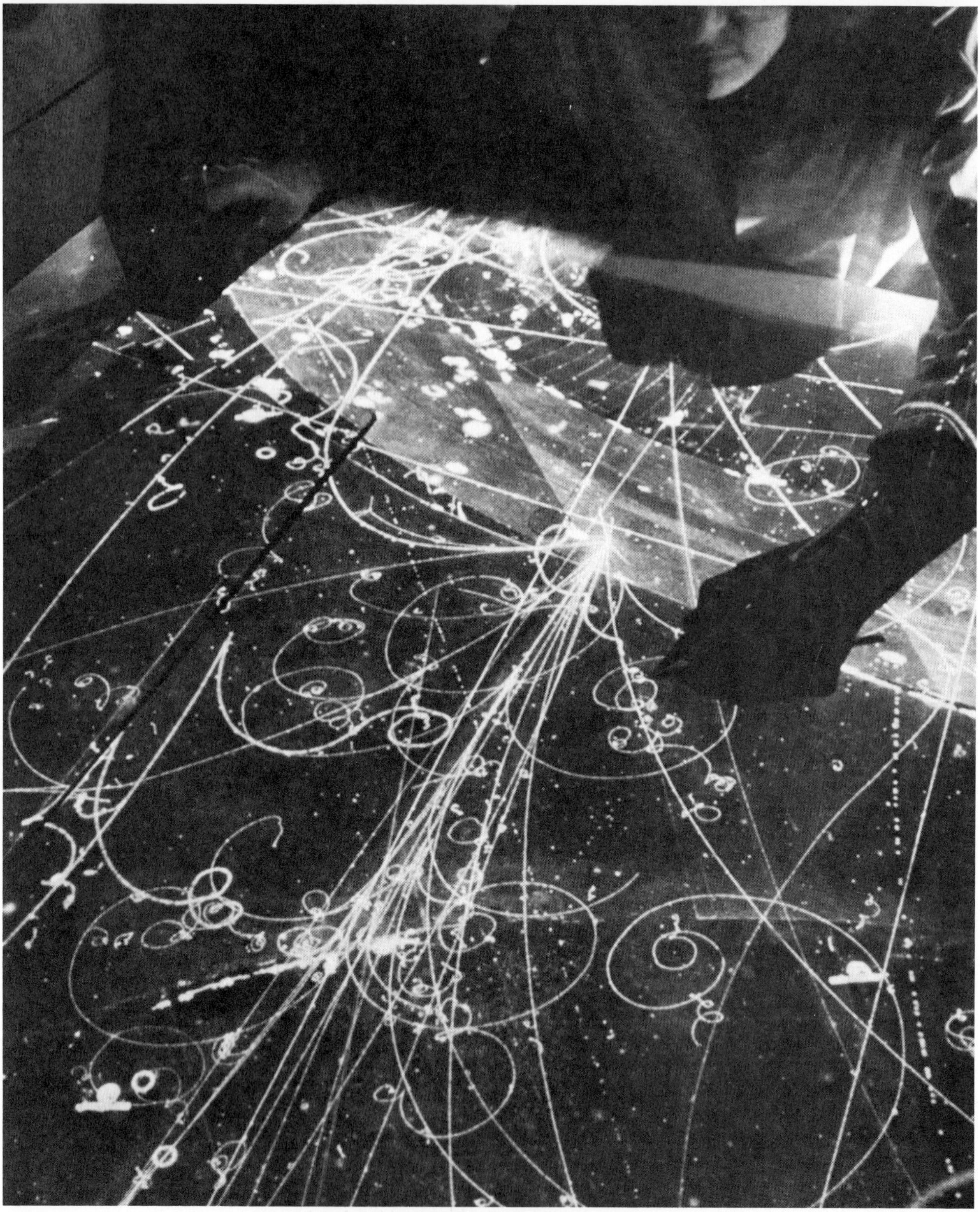

31 The Nucleus

31-1 Introduction

The discovery of the nucleus from Rutherford's alpha particle scattering experiments (Chapter 29) was announced in 1911, but nuclear physics really began earlier with the discovery of radioactivity. Within a short time after Roentgen's announcement of his discovery of x-rays, many scientists began investigating their properties. Henri Becquerel (1852–1908), a professor of chemistry at the University of Paris, in what many consider to be a series of lucky accidents which went far beyond serendipity, discovered what we call radioactivity.

When Becquerel learned of the green glow observed by Roentgen in the x-ray tube glass (Section 28-5), he wondered whether glowing phosphorescent substances might emit x-rays (or other penetrating rays). To test this, he wrapped an unexposed photographic plate in black paper, placed a phosphorescent material on top of the wrapped plate, and put them both in the sunlight. After a few hours in the sun, Becquerel developed the plate to determine if any rays from the material had penetrated the paper. Just as he had hoped, there sometimes appeared on the developed plate a silhouette of the phosphorescent material. During a period of sunless days in late February 1896, he put his wrapped photographic plate and phosphorescent sample away in a dark drawer. When the sun did not come out again for a few days, he developed the plate, expecting to see only a faint image, if any. Much to his surprise, the silhouettes appeared with even greater intensity than they had for those materials which were exposed to sunlight. Becquerel reports that he immediately concluded "that the action might be able to go on in the dark." He continued his experiments with the same substance (which happened to be a uranium compound) as well as with some other phosphorescent substances. He found that the original substance continued to emit rays and expose photographic plates kept light-tight even after being in the dark for 2 months. Furthermore, Becquerel determined that it is the presence of uranium rather than the phosphorescent property which is important to exposing the plates (Figure 31-1). After many experiments, using different concentrations of uranium both in solid form and in solution, he concluded that the uranium was emitting some type of rays and that the intensity depended on the amount of uranium. He also determined that when uranium was brought near a charged electroscope, the leaves immediately collapsed. Evidently, the rays from uranium made the air conductive. These rays were originally called Becquerel rays. Later, however, one of Becquerel's doctoral students invented the term **radioactivity** when she saw radium glowing in the dark. Apparently, she combined the words *radiance* and *activity*. Becquerel had discovered natural radioactivity.[1]

The student credited with the term radioactivity, Marie Curie (1867–1934), and her husband, Pierre (1850–1906), began a systematic study of both the chemical compounds of uranium and all the other known chemical elements for radioactivity. They found that the element thorium was also radioactive. Following this, they expected to find radioactivity only in those samples which contained uranium or thorium. Further, they believed that the level of activity would depend on how much of those elements were present. When they tested pitchblende, an ore which contains uranium, they found the activity to be much greater than could be attributed to its uranium concentration. They concluded it must contain an additional, undiscovered radioactive element. Thus, they began the long, arduous task of chemically separating tons of pitchblende in order to isolate the source of the high radioactivity. Along the way, they discovered not one, but two new elements: polonium, named after Marie's native Poland, and radium. The Curies and Becquerel shared the 1903 Nobel Prize in physics for the discovery of these two elements.

It was Rutherford along with a coworker, English chemist Frederick Soddy (1877–1956), who first concluded that radioactivity is the result of atoms breaking into smaller pieces and flying apart. Rutherford is also credited with recognizing that there are several different kinds of rays coming from uranium. He showed this by stacking very thin sheets of aluminum foil between the uranium and the detector. A few sheets caused a large drop in the amount of radioactivity detected, but some rays made it through the foil. After about five sheets, further layers caused little reduction in the rays reaching the detector.

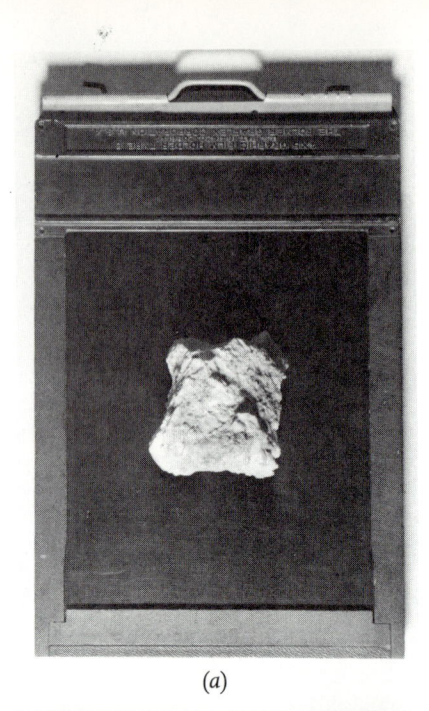

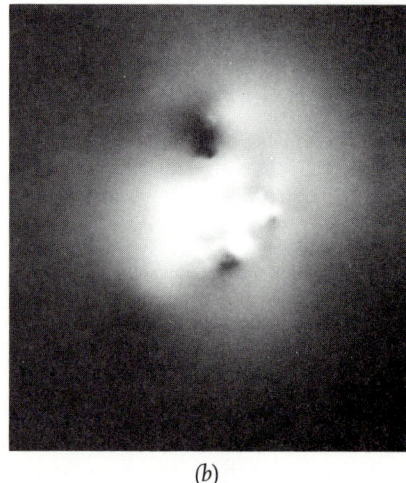

Figure 31-1 An autoradiogram is a picture taken in the absence of light by radioactive emission. (*a*) A sample of uraninite, the principal ore of uranium, rests on a light-tight film holder and exposes the film inside. (*b*) A positive print of the film after 4 days. (*Fundamental Photographs*)

---

[1] That good fortune smiled on Henri Becquerel is evident. First, he was looking for something else, which does make it serendipitous. In addition, he happened to choose a salt of uranium as the phosphorescent substance. (Of the numerous substances available which phosphoresce, very few are naturally radioactive.) Also, it happened to be cloudy when he prepared an experiment and decided to leave the materials together for awhile. Finally, he developed the plate anyway in spite of his expectation that he would "find the images very feeble." Although it was indeed a series of lucky happenstances, it should be noted that Becquerel was intelligent enough, educated enough, and dedicated enough to carry out the experiments and correctly to interpret the results.

Apparently there were at least two types of radiation: an easily absorbed kind which he called **alpha (α) rays** and a very penetrating kind which he called **beta (β) rays**. Rutherford subsequently showed that the alpha rays were really positively charged helium atoms (Figure 31-2) and that the beta rays were negative. Becquerel used a mass spectrometer type of technique to determine their charge-to-mass ratio and identified beta rays as electrons. Not long afterward, a third type of radiation was detected. This new kind is uncharged and hence very penetrating. It was (and is) called **gamma (γ) radiation** and was eventually shown to be very high energy electromagnetic radiation.

All three rays are emitted by radium and may be distinguished quite nicely. Figure 31-3 depicts a lead block with a narrow hole drilled part way through it, forming a well. A small sample of radium is placed in the bottom of the well, and the block is then placed in an evacuated chamber. A magnetic field whose direction is into the plane of the page is imposed in the chamber. The three types of rays leave the block in the same direction along the well. A photographic plate, placed perpendicular to the original beam, would record that alphas were deflected in the direction expected for positively charged particles, betas were deflected strongly in the opposite direction, and gammas were undeflected.

After Rutherford's discovery of the nucleus and its acceptance, it was immediately realized that radioactivity originated in the nucleus and is hence a nuclear property.

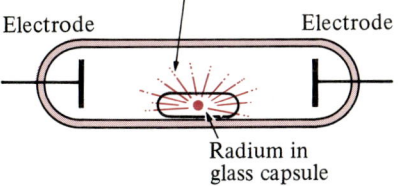

Figure 31-2 Rutherford's experiment that established the nature of alpha particles. A capsule containing radium was sealed inside a glass tube. High voltage across the ends of the tube caused the low-pressure gas inside to glow with its spectrum, which was recorded immediately. As the tube sat, the alpha particles given off by the radium penetrated the thin glass walls of the capsule but not the outer walls of the tube and accumulated in the tube. After a time, a second spectrum of the gas in the tube revealed that helium was now present, establishing alpha particles as helium nuclei, $^4_2$He.

31-2 The Composition and Properties of the Nucleus

According to the Rutherford-Soddy interpretation of radioactive disintegration, we would expect alpha-particle decay to result in a new element. That is, we expect that a nucleus of some element (a **parent nucleus**) which disintegrates by alpha decay is really splitting into a helium nucleus and a nucleus of some different element (now called the **daughter nucleus**). The new element created belongs somewhere in the periodic table, but Rutherford and Soddy found far too many radioactive substances to have all their daughters fit into new spaces in the periodic table. In addition, Soddy observed that two substances with very different radioactivities had identical chemical properties and suggested that perhaps many of the daughters fit into spaces that are already occupied. Since the periodic table orders elements according to their chemical properties, he was suggesting that there are atoms of different mass which have the same chemical properties. He was quite correct. Atoms with different masses but the same chemical properties are called **isotopes,** the existence of which was quickly confirmed by J. J. Thomson in 1913 and several others shortly thereafter. Scientists have identified more than one isotope for almost all the elements. For example, hydrogen has three known isotopes, with masses of 1.007825, 2.014102, and 3.016049 u (often called normal hydrogen, **deuterium,** and **tritium,** respectively). Some elements, like tin, have more than 15 known isotopes.

The chemical properties of a neutral atom are determined by the number (and configuration) of electrons outside the nucleus. Because the atom is electrically neutral, the nucleus must possess a net positive charge equal to the combined negative charge of all the atom's electrons. Thus, the nucleus of each

Figure 31-3 The nature of radium radiations was demonstrated by a piece of radium contained in a well in a lead block.

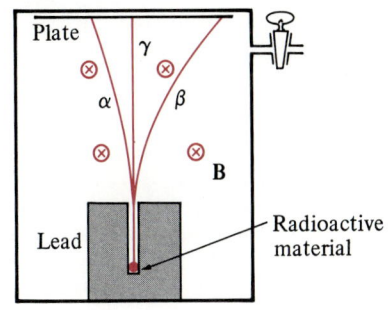

773

element has a positive charge equal to an integral multiple of $+e$. We call this integer the **atomic number** of the element (designated Z). All the isotopes of all the elements have masses very near to integral numbers of atomic mass units (u). Rounded to this nearest integer, this is the **atomic mass number** (designated A). Therefore, to represent a particular isotope with a chemical symbol, say X, we write

$$^A_Z X$$

For instance, we would designate the most common isotope of hydrogen as ^1_1H, because both its atomic number and atomic mass number are 1. The other two isotopes of hydrogen are designated ^2_1H and ^3_1H, respectively. Incidentally, the most abundant isotope of carbon is used to determine the value of the **atomic mass unit**. The mass of $^{12}_6C$ is defined to be exactly 12.000000 u.[1]

Rutherford and Soddy eventually formulated what became known as the **displacement laws** of radioactivity decay.

1. When an element decays by alpha-particle emission, the daughter nucleus has an atomic number which is 2 less than the parent nucleus and an atomic mass number which is 4 less than the parent.

   Suppose, for example, that a uranium 238 nucleus decays by alpha emission. Since the atomic number of uranium is 92, the daughter nucleus must have an atomic number of 90 (making it thorium) and an atomic mass number of 234. We write this as a decay reaction:

   $$^{238}_{92}U \rightarrow {}^{234}_{90}Th + {}^4_2He$$

Note that the sum of the atomic numbers on the right equals that on the left, as does the atomic mass numbers.

2. When the element decays by beta-particle emission, the daughter nucleus has an atomic number which is 1 *more* than the parent nucleus and an atomic mass number which is the same as the parent.

   In beta decay, an electron leaves the nucleus. The charge of the electron is -1, and its mass is only $\frac{1}{1836}$ u, or effectively zero when compared with a unit of atomic mass. Thus we may indicate the beta particle as $_{-1}^0e$, as in the reaction

   $$^{14}_6C \rightarrow {}^{14}_7N + {}^{0}_{-1}e$$

We see that the negative charge leaving the nucleus leaves it more positive and Z increases by 1, while the loss of negligible mass leaves the atomic mass number unchanged.

Both decay examples are in accordance with the displacement laws, which are really statements of conservation of nuclear charge and conservation of mass number. For completeness, we note that when an element decays by gamma-ray emission, the energy of the gamma ray represents an internal energy adjustment and leaves the atomic number and atomic mass number unchanged.

$$^{60}_{27}Co \rightarrow {}^{60}_{27}Co + \gamma$$

---

[1] Some people may think that this representation is overkill. After all, the chemical symbol uniquely determines the atomic number; all carbon atoms must have $Z = 6$ to be carbon. Thus you will often see the symbol as ^{12}C or written out as carbon 12. We will see, however, that the designation showing both A and Z ($^{12}_6C$) comes in handy in nuclear reactions when we must keep track of total charge and total mass.

Example 31-1 A nucleus of bismuth 210 decays first by alpha-particle emission and then by beta-particle emission. Use the displacement laws and the table of isotopes in Appendix B to determine the successive daughter nuclei. Show the decay reaction in each case.

Solution Bismuth (Bi) has an atomic number of 83. Thus for the alpha decay, the daughter nucleus must have an atomic number of 81 and an atomic mass number of 206. This makes the first daughter thallium (Tl) 206, and the decay reaction is

$$^{210}_{83}\text{Bi} \rightarrow {}^{206}_{81}\text{Tl} + {}^{4}_{2}\text{He}$$

Now when the thallium decays by beta emission, its atomic number must increase by 1, making the second daughter (granddaughter of bismuth) equal to lead 206. The decay reaction is

$$^{206}_{81}\text{Tl} \rightarrow {}^{206}_{82}\text{Pb} + {}^{0}_{-1}e$$

31-2 The Composition and Properties of the Nucleus

Nearly a decade after the discovery of the nucleus, it was still not known what was in the nucleus—what gave it its mass and charge. Many suspected that hydrogen nuclei (**protons**) provided the mass as well as combined with the proper number of electrons to provide the correct net positive charge. A neutral atom of atomic mass number A and atomic number Z would have to have A protons in its nucleus along with $A - Z$ electrons. The presence of electrons in the nucleus would partially explain beta decay.

In 1919, Rutherford announced the results of another history-making experiment. By bombarding nitrogen gas with alpha particles, he had induced the first artificial *transmutation,* converting one element into another, and as a by-product provided convincing evidence that the proton was a constituent of the nucleus since one was emitted as a product. The reaction equation for Rutherford's experiment is

$$^{4}_{2}\text{He} + {}^{14}_{7}\text{N} \rightarrow {}^{17}_{8}\text{O} + {}^{1}_{1}\text{H}$$

Note again that the sum of the atomic numbers on the left equals the sum of atomic numbers on the right and likewise for atomic mass numbers. Of course, energy is required (supplied by the kinetic energy of the alpha particle) for the alpha particle to approach the nitrogen nucleus closely because both are positively charged. Also, the final products of the reaction have kinetic energy.

Example 31-2 The alpha particles used by Rutherford in the first transmutation experiment had kinetic energies of 7.70 MeV. Assuming that both the alpha particles and the nitrogen nucleus are point masses, determine how close the alpha particles could approach the nucleus before being stopped due to Coulomb repulsion.

Solution This is a straightforward application of the conservation of energy, if we recall that the alpha particle approaches the nitrogen from a very large distance and comes to a stop at the point of closest approach.

$$\Delta KE + \Delta PE = 0$$

$$0 - KE_i + \frac{kq_1q_2}{r_f} - 0 = 0$$

775

An alpha particle is a helium nucleus, or $_2^4$He. Recognizing that $Z_\alpha = 2$ and $Z_N = 7$, we have

$$r_f = \frac{k(2e)(7e)}{KE_i} = \frac{14ke^2}{KE_i}$$

$$= \frac{14(9.00 \times 10^9 \text{ N} \cdot \text{m}^2/\text{C}^2)(1.60 \times 10^{-19} \text{ C})^2}{(7.70 \text{ MeV})(1.60 \times 10^{-13} \text{ J/MeV})}$$

$$= 2.62 \times 10^{-15} \text{ m}$$

Actually, the approximate radius of the alpha particle is 1.90×10^{-15} m and that of the nitrogen nucleus is 2.90×10^{-15} m.

In his paper, Rutherford suggested that the nucleus might also contain a neutral particle. Although, as we mentioned, beta decay could be explained in part by the presence of electrons in the nucleus, there seemed to be several reasons to suspect that this was not possible. The value of intrinsic angular momentum of the nucleus (nuclear spin) was much too small to include electrons, as was the nuclear magnetic moment. Further, we saw in Example 30-4 that the Heisenberg uncertainty principle makes it highly unlikely that the electron would be confined to so small a volume.

By 1930, it became well known that whenever beryllium and boron were bombarded by alpha particles, radiation with very great penetrating power resulted. While this was originally assumed to be gamma radiation, it was shown to be capable of ejecting protons from materials containing hydrogen. It was also known that this radiation could transfer a great deal of kinetic energy to gaseous nitrogen nuclei. The velocities of the ejected protons and the nitrogen nuclei were then measured. James Chadwick (1891–1974), a coworker of Rutherford, expected to be able, using a Compton-scattering type of calculation, to determine the energy and thus the wavelength of the "gamma rays." When he discovered that the two sets of data yielded very inconsistent results, he slowly became convinced that the alpha-on-beryllium source was not producing gamma rays. Chadwick proposed that the radiation consisted "of neutral particles of mass very nearly equal to that of the proton." If this were true, the energy inconsistency would go away. Chadwick calculated a value for the mass of these particles which is within a few percent of present-day accepted value. The existence of the new particle was accepted immediately, and it was called a **neutron**. Chadwick, who received the 1935 Nobel Prize in physics for his work, even provided the reaction equation he used for the production of these neutrons:

$$_2^4\text{He} + {}_4^9\text{Be} \rightarrow {}_6^{12}\text{C} + {}_0^1\text{n}$$

where we use the symbol $_0^1$n for the neutron to represent the fact that it has no charge and a mass of about 1 u. The present accepted mass of the neutron is

$$m_n = 1.008665 \text{ u}$$

The announcement of the neutron made theorists happy. It removed the necessity of having electrons in the nucleus with all its attendant inconsistencies. From Chadwick's announcement to the present time, we believe the nucleus to consist of Z protons and $A - Z$ neutrons. Neutrons and protons in the nucleus are given the combined name **nucleons**.

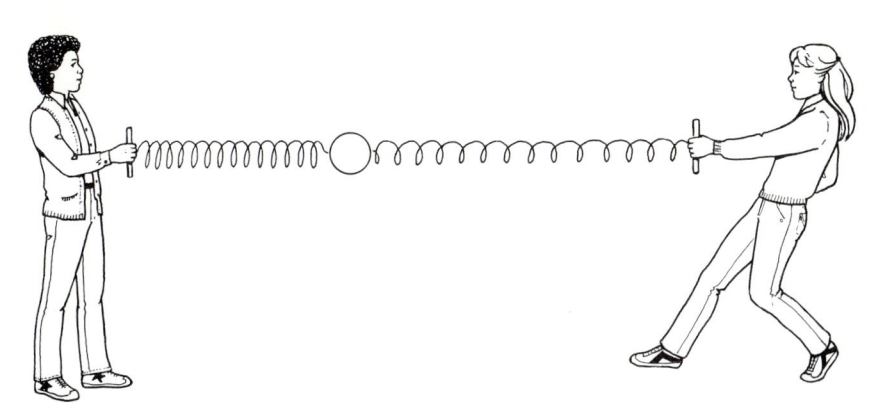

Figure 31-4 An analogy to the effects of an **exchange particle** has two individuals playing catch with a ball to which they are both attached by springs. The rapid back-and-forth motion of the ball causes a net attractive force between the players.

31-3 Nuclear Forces and Binding Energy

Even though the proton-neutron model of the nucleus cleared up some problems, there were still a number of unanswered questions. Since protons are positively charged and repel each other, why doesn't the nucleus fly apart? There must be some other attractive force acting between nucleons which is much greater in magnitude than the Coulomb repulsion. This force is now known as the **nuclear strong force.** In addition to having a huge magnitude, the nuclear strong force must also have a very short range and drop to zero as the distance between nucleons becomes greater than the diameter of one nucleon (~ 2 to 3×10^{-15} m). If this force were not very short range, the alpha particles in the Rutherford scattering experiment, for example, would be absorbed by the gold nuclei in spite of the Coulomb repulsion. Backscattering would never be possible and would not have been observed. Furthermore, there does seem to be a Coulomb repulsive effect which shows up in larger nuclei having many protons in close proximity.

Our knowledge of the nuclear strong force is primarily empirical. There are models, but none has been completely successful in explaining the properties of nucleon interaction. One model that has had some success in explaining the origin and short range uses what is called the **exchange force.** In this model, the attraction between two nucleons is caused by the two exchanging (passing back and forth) a smaller particle (called an **exchange particle**) to which each is strongly attracted. This might be compared loosely to two people playing catch with a ball to which each is attached by a spring (Figure 31-4).[1] If a nucleus were struck hard enough in a nuclear reaction, such an exchange particle may be temporarily separated as a real, free particle. The exchange particles, called **pi mesons,** or **pions,** were first detected in cosmic rays. Now, however, they are routinely produced in the laboratory. Mesons and other exotic particles will be considered in Section 31-7.

When we compare the masses of the initial and final products of a nuclear reaction and the energy liberated or absorbed, we see that mass-energy is conserved by Einstein's equivalence. Thus, from the masses involved, we may predict whether the reaction will consume or release energy and how much. You will recall that when an electron and proton are brought together, there is an energy release (the hydrogen spectrum) and that net amount of energy must be resupplied in order to separate the two. This is true as well for nucleons in a

---

[1] Remember, this is only a model and is not entirely correct.

nucleus. For example, when a proton and a neutron are brought together to form a deuterium nucleus, $^2_1$H, there is a net release of slightly more than 2.20 MeV of energy. The proton and neutron are then bound together, and it requires this same amount of energy to break them apart. We call this the **binding energy** of the nucleus.

We may calculate this binding energy from the masses of the particles involved: the proton, the neutron, and the $^2_1$H nucleus.

$$m_p + m_n - m\,^2_1H = \Delta m$$

$$1.007825 \text{ u} + 1.008665 \text{ u} - 2.014102 \text{ u} = 0.002388 \text{ u}$$

From $E = mc^2$, we obtain

$$E = (0.002388 \text{ u})(1.66 \times 10^{-27} \text{ kg/u})(3.00 \times 10^8 \text{ m/s})^2$$
$$= 3.57 \times 10^{-13} \text{ J}$$

or

$$E = (3.57 \times 10^{-13} \text{ J})\left(\frac{1 \text{ eV}}{1.60 \times 10^{-19} \text{ J}}\right) = 2.22 \times 10^6 \text{ eV} = 2.22 \text{ MeV}$$

The assembled hydrogen 2 nucleus has less mass than the sum of the masses of its constituent particles. This lost mass represents the energy binding the particles together by Einstein's mass-energy equivalence formula.

Every nucleus has its own binding energy, and this increases with increasing atomic mass number. The total binding energy of any nucleus is equal to the mass loss which would occur if its constituents were assembled from separate particles times the speed of light squared. We may generalize this relationship. Consider a nucleus of atomic number Z and atomic mass number A having a mass of M. The nucleus contains Z protons and $A - Z$ neutrons. Thus, we might expect its mass to be

$$Zm_p + (A - Z)m_n$$

where m_p and m_n are the mass of a proton and neutron, respectively. In reality, this will exceed the measured mass of the nucleus, M. The difference in mass ΔM times the speed of light squared is the binding energy BE of the nucleus.

$$\text{BE} = (\Delta M)c^2 = [Zm_p + (A - Z)m_n - M]c^2 \tag{31-1}$$

Figure 31-5 should help make this clear.

In practice, we work with atoms, not nuclei. If we replace m_p with the mass of the hydrogen atom, $^1_1$H, this adds the mass of Z electrons to the first term on the right. This is subtracted out immediately if we take M to be the mass of the atom rather than its nucleus.[1] Use of the atomic mass M and the mass of the hydrogen atom rather than the proton is standard practice which we will follow.

Binding energy per nucleon in a nucleus is more informative than net binding energy. This is just the net binding energy BE divided by the atomic mass number A. Figure 31-6 shows binding energy per nucleon vs. the atomic mass number for naturally occurring nuclei. Initially, for small nuclei, as nucleons are added, the binding energy per nucleon increases. Above atomic mass number 30 or so, it remains essentially constant for a range as additional

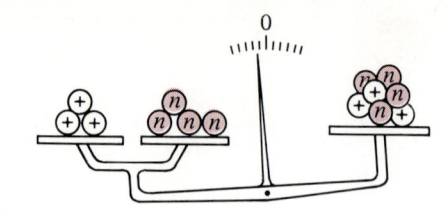

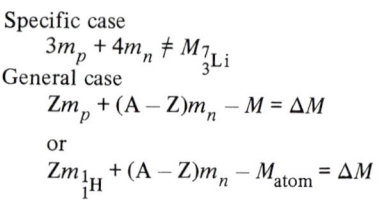

Specific case
$$3m_p + 4m_n \neq M\,^7_3\text{Li}$$
General case
$$Zm_p + (A - Z)m_n - M = \Delta M$$
or
$$Zm_{^1_1\text{H}} + (A - Z)m_n - M_{\text{atom}} = \Delta M$$

Figure 31-5 When nucleons are bound together, the resulting nucleus has less mass than the component nucleons. For example, three protons and four neutrons bind to form the nucleus of $^7_3$Li, resulting in a lost mass of ΔM, the amount of mass converted to the nuclear binding energy.

Figure 31-6 Binding energy per nucleon vs. atomic mass number. Nuclei in the plateau region (e.g., iron) have the greatest binding energy per nucleon and hence are the most stable. A nuclear reaction to break such nuclei in smaller pieces or a reaction to assemble them into larger ones would require additional energy. Dots representing specific individual isotopes display a small variation about the smooth curve. Most notable of these is the dot for $A = 4$, the helium nucleus, clearly an unusually stable, well bound nucleus for its region of the chart.

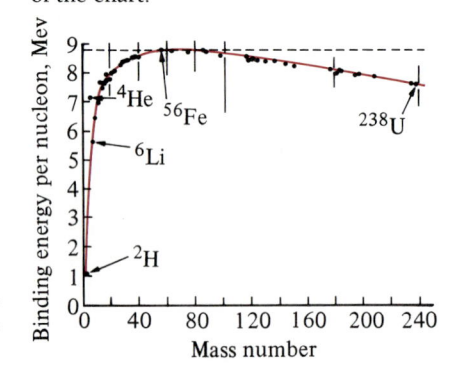

---

[1] When using atomic masses, we ignore the tiny differences between the ionization energies of the electrons in the Z hydrogen atoms and that of the Z electrons in the single atom of mass M.

nucleons are acquired. Finally, at about $A = 100$, the curve shows a consistent downward slope with increasing mass number.

The near-constant value of BE/A for most of the range can be explained if the binding force on each nucleon is exerted only by its nearest-neighbor nucleons. Remember, the nuclear strong force is a very short range force. For small nuclei, each new nucleon represents additional binding for every nucleon present. At some level, this binding becomes saturated, with each nucleon already binding to as many nucleons as possible. Above this level, the binding energy per nucleon due to the nuclear strong force remains essentially constant. However, as we get to larger nuclei, additional protons with their positive charges are present. The increasing Coulomb repulsion is *not* a short-range force experienced by neighboring nucleons only. As a result, for large atoms, each additional proton represents a fixed amount of increased nucleon binding due to the strong force but an ever-increasing amount of Coulomb repulsion. This could account for the downward trend of the curve at high atomic mass number.

31-3 Nuclear Forces and Binding Energy

Example 31-3 Determine the (*a*) total binding energy and (*b*) the binding energy per nucleon of $^{11}_{5}$B in joules and in megaelectronvolts. Use the *atomic* masses given in Appendix B, and recall that we should use the mass of $^{1}_{1}$H instead of the mass of the proton.

Solution (*a*) This is a straightforward application of Equation 31-1, substituting the mass of hydrogen 1 for the mass of the proton.

$$\Delta M = 5(1.007825 \text{ u}) + 6(1.008665 \text{ u}) - 11.009305 \text{ u}$$
$$= 0.08181 \text{ u} = 1.36 \times 10^{-28} \text{ kg}$$

$$\text{BE} = (\Delta M)c^2 = (1.36 \times 10^{-28} \text{ kg})(3.00 \times 10^8 \text{ m/s})^2$$
$$= 1.22 \times 10^{-11} \text{ J} = 76.2 \text{ MeV}$$

(*b*) This is obtained by dividing the answer to part *a* by the atomic mass number (11) of the nucleus in question.

$$\frac{\text{BE}}{A} = \frac{76.2 \text{ MeV}}{11 \text{ nucleons}} = 6.93 \text{ MeV/nucleon}$$

Since most of the mass differences with which we deal in nuclear physics are in atomic mass units (u) and as a matter of convenience we use megaelectronvolts for energies, it is worthwhile to obtain a relationship between them. That is, we wish to determine the energy equivalence in megaelectronvolts of a mass of 1 u.

$$mc^2 = (1 \text{ u})(c^2) = (1.660566 \times 10^{-27} \text{ kg})(299{,}792{,}458 \text{ m/s})^2$$
$$= 1.4924 \times 10^{-10} \text{ J} = 931.5 \text{ MeV}$$

Thus

$$1 \text{ u} = 931.5 \text{ MeV}/c^2$$

where we have carried along all the significant figures available in the known constants to avoid round-off errors.

To see how convenient it is to have this conversion to use, consider the previous example. As soon as we obtained $\Delta M = 0.08181$ u, we could have converted to energy directly, as

$$BE = (\Delta M)c^2 = (0.08181 \text{ u})\left(\frac{931.5 \text{ MeV}/c^2}{1 \text{ u}}\right)(c^2) = 76.2 \text{ MeV}$$

We will find this conversion quite useful in this and the next chapters.

31-4 Beta Decay and Neutrinos

One problem not cleared up by the discovery of the neutron had to do with beta decay. Whenever a nucleus decays by beta emission, we detect only the beta particle emitted, and we have by the second displacement law that the daughter nucleus has an atomic number which is 1 greater than that of the parent. The masses of both daughter and parent can often be measured with great precision, and their difference is a *single* numerical value. Thus, by conservation of energy, we would expect the kinetic energy of the ejected beta particle to have a definite, single value. This does not happen.

Consider, for example, the beta decay of strontium 90 to yttrium.

$$^{90}_{38}\text{Sr} \rightarrow ^{90}_{39}\text{Y} + ^{\ 0}_{-1}e$$

We may determine the energy available for the beta particle simply by using the difference of the parent and daughter atomic masses.

$$\text{Mass } ^{90}\text{Sr} = 89.907746 \text{ u}$$
$$\text{Mass } ^{90}\text{Y} = 89.907160 \text{ u}$$
$$\Delta M = 0.000586 \text{ u}$$

and

$$E = (\Delta M)c^2$$
$$= (0.000586 \text{ u})\left(\frac{931.5 \text{ MeV}/c^2}{\text{u}}\right)(c^2)$$
$$= 0.546 \text{ MeV}$$

If the beta particle receives virtually all the energy, we would expect each one emitted from strontium 90 to have a kinetic energy of 0.546 MeV, but, in fact, a distribution of energies is obtained (Figure 31-7). The curve intersects the kinetic energy axis at just about the point we expected for the total kinetic

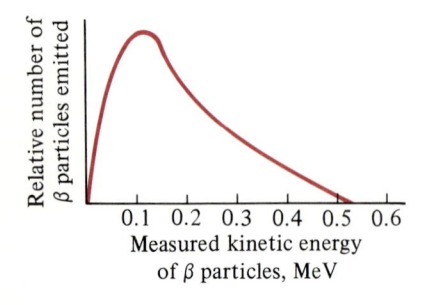

Figure 31-7 Beta decay of strontium 90, showing the wide range of energies. Since the energy converted in this decay is 0.546 MeV, we might expect all the particles to be ejected with this kinetic energy. Instead, 0.546 MeV appears to represent the upper limit of beta-particle energy.

energy of the beta particle. Thus, no beta particles exceed the calculated kinetic energy, but most do not have this much. Is the principle of conservation of energy violated?

After Chadwick first observed this energy problem in 1914, other anomalies arose. With detectors in which the direction of the electron and that of the recoil nucleus could both be observed, it was found that the two particles often did not go in opposite directions even when the parent nucleus could be considered at rest. This would seem to violate the principle of conservation of linear momentum. Also, it is known that the intrinsic angular momentum (spin) of nucleons (protons and neutrons) is $\pm \frac{1}{2}(h/2\pi)$. Thus, since the parent and daughter nuclei have the same number of nucleons, they must have the same spin (or possibly change by an integral number of $h/2\pi$ if one or more nucleons flip over during the beta decay). Yet the beta particle is an electron with half-integer spin, and we might expect it would take $\pm \frac{1}{2}(h/2\pi)$ spin away and change the nuclear spin by $\frac{1}{2}$ rather than an integer times $h/2\pi$. Therefore, beta decay seemed to be inconsistent with the law of conservation of angular momentum.

Thus, beta decay appeared to violate three conservation laws: energy, linear momentum, and angular momentum. In 1930, Pauli proposed that these laws would not be violated if there was another particle emitted during beta decay with exactly the right properties. The particle has to have no charge, vanishingly small mass (or none), spin of $\frac{1}{2}(h/2\pi)$, and be extremely penetrating, reacting practically not at all as it passes through matter. If it did interact with high probability, it would surely have already been detected by direct methods, and it had not. In spite of this, the particle's existence was fairly well accepted.[1] In 1934, Enrico Fermi (1901–1954), an Italian, developed a complete theory of beta decay and called Pauli's particle the **neutrino,** which is Italian for "little neutral one." According to Fermi's theory, which is still taken to be correct, there is another very short range interaction between nucleons, electrons, and neutrinos (now known as the **weak interaction**) which transforms a neutron into a proton in the nucleus and simultaneously causes the emission of an electron and a neutrino. Direct evidence for the existence of the neutrino was obtained and reported by Clyde Cowan (1919–1974) and Frederick Reines (b. 1918) in a series of experiments beginning in 1952 and concluding in 1960.

Actually, there is another kind of beta particle. Dirac's theory of quantum mechanics had predicted the existence of a particle with the same mass and spin as the electron but with a positive charge. It was discovered by Carl Anderson (b. 1905), an American, in cosmic rays but is often found in nuclear decay. This particle is called the **positron** and is the electron's **antiparticle.** That is, when an electron and a positron get together, the masses annihilate with the release of electromagnetic energy.[2] Fermi's theory states that during positive beta decay, a proton in the nucleus is transformed into a neutron with the simultaneous emission of a positron and a neutrino. The Pauli particle emitted with negative

---

[1] It is interesting to note that physicists believe so strongly in the conservation laws that they will accept the existence of something they cannot detect in order to preserve them.

[2] We call the ordinary kinds of things of which we are made **matter,** and the other kind **antimatter.** We will see in Section 31-7 that antiparticles exist for all particles, although in specific cases a particle may be identical to its antiparticle. In fact, we may speculate that there are antiatoms and even whole galaxies of antimatter. There is no way to detect this by looking at the light from a distant source, since the photon (light) is one of the special cases for which particle and antiparticle are the same.

beta decay is actually an antineutrino. We indicate the negative and positive beta-decay reactions in the nucleus as

$$^1_0n \rightarrow {}^1_1p + {}^{\ 0}_{-1}e + \bar{v}_e$$

and

$$^1_1p \rightarrow {}^1_0n + {}^0_1e + v_e$$

where v_e represents an electron neutrino and $\bar{v}_e$ is an antineutrino.

31-5 Radioactive Decay

The decay of any radioactive nucleus is spontaneous and random but occurs with a constant probability. That is, the probability of a particular nucleus decaying in the next second is independent of the age of the nucleus or of any environmental condition, such as temperature or pressure. Suppose, for example, we have 1000 radioactive nuclei of a particular type which have a probability of decay of 0.100 per second. At the end of 1 s, we would expect to have had 100 decays and 900 of the original nuclei remaining. At the end of the next second, we would expect to have had an additional 90 decays and 810 originals remaining. At the end of one more second, we would expect to have had 81 more decays and still have 729 originals. Because the decay probability is constant, the number of decays per second depends directly on the number of radioactive nuclei available.* The time rate of change of radioactive nuclei is proportional to the number of nuclei present:

$$\frac{\Delta N}{\Delta t} \propto N$$

where N is the number of radioactive nuclei existing at time t. As usual, the insertion of the proper constant will make this proportionality an equation. We do this as

$$\frac{\Delta N}{\Delta t} = -\lambda N \tag{31-2}$$

Since N is decreasing with time, $\Delta N/\Delta t$ must be a negative quantity. The proportionality constant is negative, with the minus sign shown explicitly. Note that λ is positive. The quantity λ is known as the **decay constant** and physically is the probability per unit time for any particular nucleus to decay, as in the 0.100 per second of the example.

We have seen equations like Equation 31-2 before (for example, the charge remaining on a discharging capacitor vs. time, Equation 20-24), where the time rate of change of some quantity is equal to a negative constant times the quantity itself. The solution of such an equation is an exponential decay law, which in our present case is

$$N = N_0 e^{-\lambda t} \tag{31-3}$$

---

* Actually, because the decay is random, we may see 105 decays in the first second and 87 in the next second for such a small sample. In practice, we most often deal with 10^{20} or more radioactive nuclei at once, and the statistical treatment is then sound.

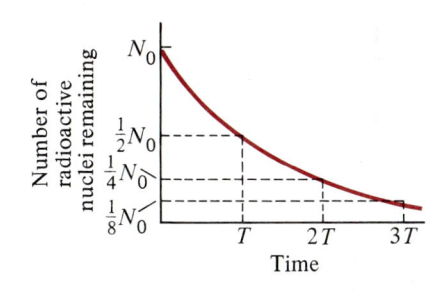

Figure 31-8 Rate of radioactive decay depends upon the number of radioactive atoms present at any time. If we start with N_0 atoms of a particular radioactive isotope, the number declines in time exponentially. Such a curve is characterized by a half-life T, the constant time period for half the atoms of that isotope to decay.

where N_0 is the number of radioactive nuclei present at time $t = 0$. Equation 31-3 could also be written as

$$N = N_0 e^{-t/\tau} \qquad (31\text{-}4)$$

to put it in the same form as the solution to Equation 20-24. By inspection, we see that the time constant τ is related to the decay constant λ as

$$\lambda = \frac{1}{\tau} \qquad (31\text{-}5)$$

Figure 31-8 shows the number of radioactive nuclei vs. time, a plot of either Equation 31-3 or 31-4.

From the equations or the graph, we see that, in principle, a radioactive substance is never all gone. In effect then, it is not possible to talk about the lifetime of a radioactive substance. In spite of this, we must differentiate among radioactive substances in a concise and easily understood way. Thus, we use the concept of **half-life.** The half-life of a radioactive substance is the time interval during which the number of nuclei has been reduced by one-half. If we designate the half-life as T, we can write

$$t = T \qquad \text{when } N = \frac{N_0}{2}$$

Substituting this condition into Equation 31-3, we have

$$\frac{N_0}{2} = N_0 e^{-\lambda T}$$

or

$$\tfrac{1}{2} = e^{-\lambda T}$$

which, if we take the reciprocal of both sides, becomes

$$2 = e^{\lambda T}$$

Taking the natural logarithm of both sides, we obtain

$$\ln 2 = \lambda T$$

Finally, we have

$$T = \frac{\ln 2}{\lambda} = \tau \ln 2 \qquad (31\text{-}6)$$

The concept of half-life means that if we start with N_0 nuclei, at the end of T we will have $N_0/2$ remaining. After another half-life, we have $N_0/4$ (one-half of one-half) remaining. In general, after n half-lives, we have $N_0/2^n$ remaining. After seven half-lives, a radioactive substance will have decayed to $\frac{1}{128}$, or 0.78 percent, of its original number of nuclei. Also, after a time interval equal to five time constants has passed, we have

$$N = N_0 e^{-t/\tau} = N_0 e^{-5\tau/\tau} = N_0 e^{-5} = 0.0067 N_0$$

In both cases, less than 1 percent of the original number of nuclei remains. Thus, it is common engineering practice to say that an exponential decay process is essentially complete after seven half-lives or after five time constants.

Example 31-4 Iodine 131 has a half-life of 8.05 days. (a) Calculate the decay constant in units of s^{-1}. (b) If the director of the nuclear physics laboratory obtained a sample of $^{131}_{53}$I containing 1.00×10^{15} atoms near the beginning of the academic year, how many of the radioactive nuclei would be remaining at the end of the term (90 days later)? (c) How many decays per second would there be at the end of the term?

Solution We must convert the half-life to seconds for parts a and c.
(a) We apply Equation 31-6, with the half-life in seconds:

$$T = (8.05 \text{ days})\left(\frac{8.64 \times 10^4 \text{ s}}{\text{day}}\right) = 6.96 \times 10^5 \text{ s}$$

Then,

$$\lambda = \frac{\ln 2}{T} = \frac{\ln 2}{6.96 \times 10^5 \text{ s}} = 9.96 \times 10^{-7} \text{ s}^{-1}$$

(b) This is an application of Equation 31-3. However, we must either obtain λ in units of $days^{-1}$ or convert 90 days to seconds. Choosing the latter, we have

$$t = (90 \text{ days})\left(\frac{8.64 \times 10^4 \text{ s}}{\text{day}}\right) = 7.78 \times 10^6 \text{ s}$$

$$N = N_0 e^{-\lambda t}$$
$$= (1.00 \times 10^{15} \text{ atoms}) e^{-(9.96 \times 10^{-7} \text{ s}^{-1})(7.78 \times 10^6 \text{ s})}$$
$$= 4.33 \times 10^{11} \text{ atoms left}$$

(c) From the right side of Equation 31-2, this is (in magnitude) simply the product of λ and N, provided that $N \approx$ constant for that second.

$$\frac{\Delta N}{\Delta t} = \lambda N = (9.96 \times 10^{-7} \text{ s}^{-1})(4.33 \times 10^{11} \text{ atoms})$$
$$= 4.31 \times 10^5 \text{ atoms/s decaying}$$

The answer to part c of Example 31-4, the number of decays per unit time, is called the **activity** of a radioactive substance.[1] Since the activity is nothing more than a constant times the number of nuclei remaining (see Equation 31-2),

---

[1] The activity may be calculated in this way only if Δt is small with respect to the half-life. One could not, for example, find the decay rate per week in this way if the half-life were 8.00 days, because N would change significantly during Δt.

it follows the same exponential equation as the decaying nuclei. Thus in magnitude we have

$$A = \frac{\Delta N}{\Delta t} = \lambda N = \lambda N_0 e^{-\lambda t}$$
$$A = A_0 e^{-\lambda t} \tag{31-7}$$

where we have used A to represent the activity and A_0 for the activity at $t = 0$.

Note that we could also write the activity as

$$A = \frac{\ln 2}{T} N$$

since $\lambda = (\ln 2)/T$ from Equation 31-6. This is an important consideration. It says that for the same amount of radioactive material, the longer the half-life, the smaller is the activity. Further, if a substance is very active, it must necessarily be short-lived. In other words, either a radioactive substance is very active *or* it is very long lived. It cannot be both.

31-6 Detection of Ionizing Radiation

Ionizing radiation includes all the natural radiation (α, β, and γ) we have discussed plus any moving particle or photon that can create ions when it passes through and interacts with matter. Ultraviolet light, x-rays, cosmic rays (Chapter 32), and even neutrons are ionizing radiation.

A number of methods and devices are used to detect ionizing radiation. Each type has its advantages and disadvantages. Thus, it is important to choose the right detector for the application at hand.

The first detectors of ionizing radiation were *photographic plates,* which have evolved a great deal since Becquerel. The trail of ions left in a thick emulsion may be seen as a track in the developed film, indicating the direction in which the particle moved through the emulsion. Also, the number of ionizations per unit path length on the film indicates the amount of energy deposited by the ray. Comparing this with other tracks may even identify the type and speed of an unknown particle.

The *leaf electroscope* was a second kind of detector, where half the ions in the trail caused by the radiation were attracted to the charged leaves which then collapsed. This quickly gave rise to early **ionization chambers.** These were usually two metal plates separated by several centimeters of air with 100 V or more potential difference between them connected in a circuit with a meter to measure the current. Detectors similar to this were used by the Curies, Rutherford's group, and other pioneers in detection of radiation.

This eventually evolved to become an instrument that detects individual ionizing particles and is called a **Geiger-Mueller counter** or a **G-M counter** or sometimes just a **Geiger counter** (Figure 31-9). Instead of using air at atmospheric pressure between the electrodes, the G-M tube uses an inert gas at reduced pressure. One electrode is a conducting cylinder, while the other electrode is a stiff wire situated along the cylinder's axis. A potential difference on the order of 1000 V is placed between the electrodes. When a ray or particle enters, it may leave an ion (or more likely, a trail of ions) in the low-pressure gas. These ions are charged particles in a very strong electric field, and the electric force accelerates them toward the electrodes. As the positive ions are acceler-

Figure 31-9 The Geiger-Mueller counter consists of a conducting cylinder containing an inert gas with a conducting wire down its center.

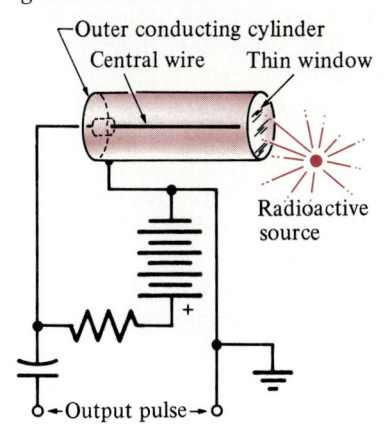

ated toward the negative electrode and the negative ions accelerated toward the positive electrode, they gain energy and collide with gas molecules, creating secondary ions. These secondary ions will also be accelerated and collide with gas molecules, causing more ionizations. In this way, an **avalanche** occurs, and a significant pulse of current may pass through the resistor, creating a significant voltage pulse which may be amplified and then used to advance a counter, ring a bell, flash a light, or whatever we wish. Each trail of ions created by a particle or ray gives rise to a pulse which we then call a count. The time for one avalanche to clear and leave the tube ready for another count is called the **resolving time** of the counter.

The end of the cylinder is covered by a very thin material called the **window,** which keeps the gas confined but allows radioactivity to pass through. It passes beta particles and gamma rays fairly well, but most alpha particles are stopped before they reach the gas. Beta particles create ions all along their path by collisions until their kinetic energy is expended. Gamma rays, by contrast, may interact only once, and a Geiger counter will usually detect the secondary electrons due either to the photoelectric effect or to the Compton effect. Thus, beta particles provide a much more substantial pulse and are counted more easily. Geiger counters are probably the most common radiation detectors in use. They are sturdy and may be made to be portable. Hence, they are very good as survey meters. In many models, a speaker or an earphone is activated by each count, allowing the user to hear the count rate as a series of clicks.

Many substances will emit a **scintillation** (small burst of light) following the deposition of energy by radiation or a fast-moving atomic particle. This fact led to the **scintillation detector.** Rutherford's group embedded tiny natural scintillating crystals of zinc sulfide in small screens, placed them in an evacuated chamber with the source to be counted, and viewed them under a microscope.

Scintillation detectors, too, have evolved. Scientists have learned to "grow" their own scintillation crystals in much larger sizes than those used by Rutherford and to make liquid scintillators to be used with transparent containers of any size and shape they wish. Coupled with a photomultiplier tube, these scintillators are part of present-day scintillation detectors. The photomultiplier tube takes a flash of light, converts it to an electric current pulse, and amplifies it. Its output pulse is then used very much like that from a Geiger tube. However, scintillation detectors with photomultiplier tubes are 10,000 or more times faster than a Geiger counter. They also have the advantage that the output pulse is proportional to the amount of energy deposited.

The **solid-state detector,** depicted in Figure 31-10, is basically a *pn* junction diode with reverse voltage which has been altered somewhat to increase its sensitivity to ionizing-particle detection. Typically, the incident particles enter the diode on the *n*-type side and stop in the region in the vicinity of the junction. This is known as the **depletion region,** or **depletion layer,** because with the reverse voltage, it is somewhat depleted of charge carriers. The incident particles produce electrons and holes which then move toward the *n*-type layer and *p*-type layer, respectively. The arrival of these charge carriers at the two layers causes a voltage drop across the junction and hence a voltage pulse in the output circuit. The pulse, which is proportional to the energy deposited, may then be amplified to be used in much the same way as the voltage pulses from other detectors. Solid-state detectors are normally used

---

The voltage pulse produced by a radiation detector often requires that more be done to it than simply making it bigger. The light (or bell, counter, speaker) might require that the pulse be of shorter (or longer) duration, reach its maximum sooner (or later), and/or decay less quickly (or more quickly). In these cases, a shaping circuit must be included, either separately or as part of the amplifier. The pulse is then shaped (in addition to being amplified) to accommodate the next step in its use.

Figure 31-10 A solid-state detector of radiation uses a *pn* junction diode with the voltage applied in reverse to its normal conducting direction.

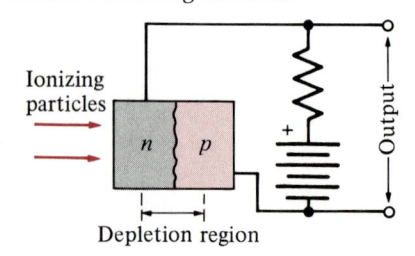

with radiation which is not very penetrating and for which energy information is desired.

Another group of detectors provides us with the visible track of the particle. The **cloud chamber** was developed by the English physicist C. T. R. Wilson (1869–1959) as the first radiation detection device which showed the actual path of the detected particle. Wilson developed a method in which he quickly released the pressure in a glass chamber containing air saturated with alcohol vapor. Because of its speed, the expansion is essentially adiabatic. Thus the system's temperature drops and the air becomes supersaturated with alcohol vapor. Electrically charged ions become centers of condensation on which small droplets of alcohol form. If the ions are part of a trail produced by the recent passage of a charged particle, the droplets form in a track marking this passage. Because it takes ions from 10 to more than 20 s to recombine in the expansion cloud chamber, tracks may show for particles which passed through anytime in the 20 s previous to the expansion. Continuously sensitive cloud chambers have also been designed and are often used for demonstrations. In these chambers, there is a very large temperature gradient built in so that there will be a portion of a closed volume of an air-alcohol vapor mixture which is always supersaturated.

Another detector which shows tracks but absorbs energy more rapidly and stops particles in a shorter distance is the **bubble chamber.** In 1952, Donald Glaser (b. 1926), an American, recognized the need for seeing detail of the whole tracks of very high energy particles. This is not possible in the cloud chamber, as very high energy particles lose very little energy per unit path length in gas and usually pass all the way through the chamber long before they stop. The bubble chamber uses a volume of superheated liquid created by a slight adiabatic expansion of a liquid which is originally just below the boiling point. The ions in the liquid then become centers of evaporation (or boiling). A track of bubbles is thus formed along a track of ions made by a recently passing charged particle. To date, no one has created a continuously sensitive bubble chamber.

An even more recent development than the bubble chamber is the **spark chamber.** This is a set of parallel conducting plates alternately connected to a source of high dc voltage and ground. High electric fields placed across the gaps by the sudden application of high voltage cause a breakdown anywhere a trail of ions exists between plates. The path of a particle which recently passed through the system shows as a series of sparks. The gas in the system and the value of the electric field are important considerations. An inert gas is usually chosen and may be saturated with an alcohol vapor. The electric field is approximately 1×10^6 V/m but must be adjusted for most efficient operation.

To obtain good directional information, all three detectors—the cloud chamber, the bubble chamber, and the spark chamber—are photographed stereoscopically and reprojected for measurement.

31-7 Elementary Particles

The nuclear reactions which we have seen so far might be called low-energy reactions, since the particles involved have energies of perhaps a few to as many as 10 MeV. The products of these reactions are alpha, beta, and gamma rays as well as neutrons, protons, and neutrinos. At low energies, those are the only particles generally produced. Until the advent of charged-particle accelerators, experimenters were limited to the low energies of nuclear-decay particles. However, with accelerators, very high energy particles became available to bombard the nucleus, and many new particles were discovered in high-energy reactions. We mentioned the exchange force particle, the pion, in Section 31-3, but many more particles with masses anywhere from about 200 times the electron mass to well beyond the mass of the proton are also seen.

After a few particles were discovered, we began to call them **elementary particles,** and a new branch of science, elementary particle physics, began to grow. As more particles were discovered, it became obvious that a method of classification was necessary. The particles are all distinguishable by unique combinations of rest mass, charge and spin, and/or some other properties. Particles were originally grouped according to mass. Photons are by themselves, **leptons** are the smaller-mass particles (including electrons), **mesons** are the intermediate-mass particles (including pions), and **baryons** are the larger-mass particles (including the proton and neutron). The names *lepton, meson,* and *baryon* are derived from Greek words meaning "lightweight," "middleweight," and "heavyweight," respectively. It was subsequently shown that other properties are more important than mass, and several recently discovered mesons are more massive than many of the baryons.[1]

We have seen throughout the text that scientists rely heavily on conservation principles. Linear momentum, angular momentum, mass-energy, and charge are examples of quantities conserved under a wide variety of conditions. In each case of a conservation law, some type of change to a system (under conditions for which we might expect the change) is not observed to occur and seems to be forbidden. For example, provided a photon has enough energy, we would expect that sometimes it would simply disappear and that an electron and a positron appear in its place. In fact, this does occur, and we call it **pair production,** or **pair creation.** However, it never occurs when the photon is isolated but only when the photon is in the presence of a massive particle like a nucleus. Calculations show that this is necessary for conservation of linear momentum. It is true that some of our conservation laws are derived beginning with Newton's laws, but others, like conservation of charge, are determined by induction. In every interaction we observe, the net charge before is the same as the net charge after. It seems that interactions in which charge is created are forbidden, and thus we conclude that charge is always conserved.

Certain interactions, especially expected decay schemes, simply do not take place for a number of elementary particles. Scientists conclude that those interactions must violate some conservation principle even if they seem to obey all those which are known. They assert that there must be some other property whose conservation is necessary and which would be violated if the expected interaction took place. For example, when the expected decay schemes of K mesons (kaons) took longer than expected by a factor of 10^{14} or so, it seemed to be very strange. From this was defined the quantity **strangeness,** an additional quantum number associated with elementary particles. All the leptons and pions have a strangeness equal to zero, but kaons have strangeness other than zero. Positive and neutral kaons are assigned a strangeness of $+1$, and the negative kaon has a strangeness of -1. It seems that strangeness is conserved in

---

[1] Sometimes it is convenient to group baryons and mesons together under the name **hadrons** as including all particles (and only those particles) which interact via the nuclear strong force.

TABLE 31-1   Elementary Particles

Family	Particle	Symbol	Rest Energy, MeV	Charge Units of e	Spin Quantum No.	Strangeness	Antiparticle	Half-Life, s
Photon	Photon	γ	0	0	1	0	Self	Stable
Leptons	Electron	e^-	0.511	-1	$\tfrac{1}{2}$	0	e^+	Stable
	Muon	μ^-	105.7	-1	$\tfrac{1}{2}$	0	μ^+	1.5×10^{-6}
	Electron neutrino	ν_e	0	0	$\tfrac{1}{2}$	0	$\bar{\nu}_e$	Stable
	Muon neutrino	ν_μ	0	0	$\tfrac{1}{2}$	0	$\bar{\nu}_\mu$	Stable
Mesons	Pion	$\pi^\pm$	139.6	± 1	0	0	$\pi^\mp$	1.8×10^{-8}
		π^0	135.0	0	0	0	Self	5.8×10^{-17}
	Kaon	$K^\pm$	493.7	± 1	0	± 1	$K^\mp$	8.3×10^{-9}
		K^0	497.7	0	0	$+1$	$\bar{K}_0$	6.2×10^{-11}
	Eta	η	548.8	0	0	0	Self	5.3×10^{-19}
	Eta'	η'	957.6	0	0	0	Self	1.7×10^{-21}
Baryons	Proton	p	938.3	$+1$	$\tfrac{1}{2}$	0	$\bar{p}$	Stable
	Neutron	n	939.6	0	$\tfrac{1}{2}$	0	$\bar{n}$	920
	Lambda	Λ^0	1116	0	$\tfrac{1}{2}$	-1	$\bar{\Lambda}^0$	1.8×10^{-10}
	Sigma plus	Σ^+	1189	$+1$	$\tfrac{1}{2}$	-1	$\bar{\Sigma}^+$	5.5×10^{-11}
	Sigma zero	Σ^0	1193	0	$\tfrac{1}{2}$	-1	$\bar{\Sigma}^0$	4.0×10^{-20}
	Sigma minus	Σ^-	1197	-1	$\tfrac{1}{2}$	-1	$\bar{\Sigma}^-$	1.0×10^{-10}
	Delta star	Δ^*	1232	$+2, +1, 0, -1$	$\tfrac{3}{2}$	0	$\bar{\Delta}^*$	4×10^{-24}
	Xi zero	Ξ^0	1315	0	$\tfrac{1}{2}$	-2	$\bar{\Xi}^0$	2.0×10^{-10}
	Xi minus	Ξ^-	1321	-1	$\tfrac{1}{2}$	-2	$\bar{\Xi}^-$	1.1×10^{-10}
	Sigma star	Σ^*	1385	$+1, 0, -1$	$\tfrac{3}{2}$	-1	$\bar{\Sigma}^*$	1.4×10^{-23}
	Xi star	Ξ^*	1530	$-1, 0$	$\tfrac{3}{2}$	-2	$\bar{\Xi}^*$	4×10^{-23}
	Omega minus	Ω^-	1672	-1	$\tfrac{3}{2}$	-3	$\bar{\Omega}^-$	5.7×10^{-11}

the nuclear strong interaction and the electromagnetic interaction but is not conserved in the weak interaction.

There are several other quantities which are apparently conserved in nuclear interactions: The **lepton number,** which is assigned as $+1$ to electrons, muons, and their neutrinos and as -1 to their antiparticles, is conserved. All particles which are not leptons have a zero lepton number. The baryon number, which is assigned as $+1$ for protons, neutrons, and the other "ordinary" baryons, as -1 for their antiparticles, and as zero for nonbaryons, is also conserved. There is no meson number conservation law. Table 31-1 shows some of the members of the families of elementary particles, with enough of their properties to determine whether a particular interaction is allowed by the conservation laws. Note that this is not a complete list of elementary particles. Other leptons, mesons, and baryons have been detected and have had important impacts on theory.

You can see just with the particles shown in Table 31-1 that the total number seems to be getting out of hand. When the many other particles which have been discovered are added (two more leptons and over sixty more mesons and baryons), it becomes very difficult to continue to think of them all as truly elementary. This is especially so when one becomes aware of electron-scattering experiments which indicate that mesons and baryons have internal structure and all except the proton decay very quickly into something more fundamental.[1]

To make some sense out of what has been called the "nuclear zoo of particles" and also account for the evidence of structure, theorists have pro-

---

[1] It has even been proposed that the proton is unstable with a very long half-life (10^{31} years), and large experiments have been under way since 1983 to attempt to detect its proposed decay.

TABLE 31-2 Family of Quarks

Quark Name	Symbol	Charge	Baryon Number	Strangeness	Charm	Antiparticle
Up	u	$+\frac{2}{3}e$	$\frac{1}{3}$	0	0	$\bar{u}$
Down	d	$-\frac{1}{3}e$	$\frac{1}{3}$	0	0	$\bar{d}$
Strange	s	$-\frac{1}{3}e$	$\frac{1}{3}$	-1	0	$\bar{s}$
Charm	c	$+\frac{2}{3}e$	$\frac{1}{3}$	0	1	$\bar{c}$
Top (truth)	t	$+\frac{2}{3}e$	$\frac{1}{3}$	0	0	$\bar{t}$
Bottom (beauty)	b	$-\frac{1}{3}e$	$\frac{1}{3}$	0	0	$\bar{b}$

posed a different classification. In 1963, Murray Gell-Mann (b. 1929) and George Zweig (b. 1937) published independent papers suggesting that mesons and baryons are made up of more fundamental particles. These fundamental particles have become known as **quarks,** as proposed by Gell-Mann, who took a word from a line in James Joyce's *Finnegan's Wake*. These quarks are essentially points, carry fractional charge, and have other properties which allow them in combinations of twos and threes to make up the known mesons and baryons.

The quark model is not universally accepted, but it does represent a consensus of opinion based on theoretical developments and new particle discoveries. The original proposal called for only three quarks, but the fourth, fifth, and now sixth were added to maintain consistency with theory while explaining properties of newly discovered particles. Table 31-2 shows the presently accepted family of quarks. Note that each has an antiparticle with opposite charge, baryon number, strangeness, and charm.

Table 31-3 shows several of the known mesons and baryons and their quark constituents. For example, the proton is seen as composed of two up quarks and one down quark.

The quark called charm became necessary to explain the particle we presently call J/Ψ, which was discovered in simultaneous experiments in 1974 and was given different names by the two groups. Its very large mass and relatively long life made its being a combination of any two of the three then-accepted quarks an impossibility. When the upsilon particle was discovered in 1977, it was interpreted as being composed of a new quark and its antiparticle in close combination. This is now known as the bottom-antibottom pair. Theory dictates that there is yet another quark, a partner to bottom, which has been dubbed top.[1]

Quarks have not yet been detected in direct experiment. Although many experiments have been designed to detect fractional charge and hence give some direct evidence for quarks, most have reported negative results. One notable experiment which began in 1977 and continued through the 1980s has reported the detection of charges of $+\frac{1}{3}e$ and $-\frac{1}{3}e$. However, the data were so sparse that even the experimenters themselves reported their results as being inconclusive as late as 1984. Many theoretical physicists believe that quarks must be so tightly bound that they can never exist in a free state. A very few even doubt their existence at all, claiming that the quark model is simply a good "bookkeeping" device. As we acquire the ability to impart larger and larger energies in accelerators, we may remove or confirm some of the doubts. In one way, elementary particle physics is a very new field, with scientists groping to find their way. The future in this area is bound to be exciting.

TABLE 31-3 Quark Composition of Particles

Particle	Quark Combination
π^+	$u\bar{d}$
π^-	$\bar{u}d$
K^+	$u\bar{s}$
K^-	$\bar{u}s$
J/Ψ	$c\bar{c}$
Y	$b\bar{b}$
p	uud
n	udd
Λ^0	uds
Σ^+	uus
Σ^0	uds
Σ^-	dds
Ξ^0	uss
Ξ^-	dss

---

[1] Some people prefer to call these last two quarks **truth** and **beauty** rather than top and bottom.

Minimum Learning Objectives

After studying this chapter, you should be able to:
1. Define:

    activity
    alpha particle
    antimatter
    antiparticle
    atomic mass number
    atomic mass unit (u)
    atomic number
    beta particle
    binding energy
    bubble chamber
    cloud chamber
    daughter nucleus
    decay constant
    deuterium
    elementary particle
    exchange force
    exchange particle
    gamma radiation
    Geiger-Mueller (G-M) counter
    half-life
    ionization chamber
    ionizing radiation
    isotopes
    neutrino
    neutron
    nuclear strong force
    nucleon
    pair production
    parent nucleus
    pion
    positron
    proton
    quark
    radioactivity
    scintillation
    scintillation detector
    solid-state detector
    spark chamber
    tritium
    weak interaction

2. List techniques and devices for the detection of radioactivity, summarizing their operating principles.
3. List the types of radiations emitted by a radioactive substance, and state the properties of each.
4. Describe a given nucleus in terms of atomic number and atomic mass number.
5. Determine the daughter nucleus produced in the specified decay of a given parent nucleus.
6. Calculate the binding energy of a specified nucleus.
7. Calculate the half-life of a radioactive material, given its decay constant.
8. Calculate the amount of a radioactive sample remaining after any given time interval, knowing either its decay constant or its half-life and its original amount.
9. Calculate the activity of a sample, given its mass and either its half-life or its decay constant.
10. Calculate the future activity of a sample at any specified time, given its current activity and its decay constant.
11. Describe the subdivisions used to categorize elementary particles.
12. Describe the quark model for elementary particles.

Problems

(The masses needed for these problems are listed in Appendix B.)

31-1 Actinium 227 decays by either alpha or beta emission. (a) Determine the daughter nucleus in each case. (b) Write the decay reaction which produces each of the daughter nuclei in part a. (c) Write the decay reaction which each daughter must undergo if both produce radium 223 as granddaughters of actinium.

31-2 There is no way, at least at present, to make a direct measurement of the size of the nucleus. However, many different indirect methods combine experimental evidence with nuclear theory to give a reasonable number for the nuclear radius. The value of the radius obtained depends somewhat on the type of experiment and the assumptions made. Present analysis of any of a number of methods indicates that we may express the nuclear radius as $r_A = R_0 A^{1/3}$, where R_0 may be as small as 1.00×10^{-15} m or as large as 1.50×10^{-15} m, depending on the method of measurement. Use $R_0 = 1.20 \times 10^{-15}$ m, and calculate the radius of the (a) $^4_2$He nucleus and (b) $^{238}_{92}$U nucleus.

31-3 Calculate the distance of closest approach to an iron nucleus ($Z = 26$) of a proton of energy 2.00 MeV.

31-4 Complete the following nuclear reactions:

$^4_2$He + $^9_4$Be → ? + $^1_0$n

$^1_0$n + $^{238}_{92}$U → $^{239}_{93}$Np + ?

$^{15}_7$N + $^1_1$H → ? + $^{12}_6$C

$^4_2$He + $^{14}_7$N → $^{17}_8$O + ?

$^4_2$He + $^{60}_{28}$Ni → ? + $^1_0$n

$^1_0$n + $^{14}_7$N → ? + $^1_1$H

31-5 Complete the following nuclear decay reactions:

$^{235}_{92}$U → ? + $^4_2$He

$^{239}_{93}$Np → $^{239}_{94}$Pu + ?

$^{131}_{53}$I → ? + $^{131}_{54}$Xe

? → $^{60}_{28}$Ni + $^0_{-1}$e

? → $^{206}_{82}$Pb + $^4_2$He

$^{14}_6$C → $^{14}_7$N + ?

31-6 Calculate the kinetic energy (in megaelectronvolts) that an alpha particle must have initially if it is to approach a target nucleus to within 2.50×10^{-14} m if the target nucleus is (a) lead ($Z = 82$), and (b) beryllium ($Z = 4$).

791

31-7 Determine how many (a) alpha and (b) negative beta decays are required in order for a ^{238}U nucleus to decay to a ^{206}Pb nucleus. (c) Are these numbers unique?

31-8 Every naturally occurring radioactive isotope will eventually decay, frequently to another radioactive isotope. As decays continue, some stable isotope will be reached. The sequence of steps by which a particular radioactive isotope decays to eventually form a particular stable isotope is called a **decay chain**. For example, ^{238}U nuclei α decay in the first of 14 steps which eventually lead to an atom of ^{206}Pb. (Here, we designate α decay by α, β decay by β, and either α then β or β then α by δ.) Write the reactions in the decay chain of (a) ^{238}U to produce ^{206}Pb (in order, these reactions are $\alpha, \beta, \beta, \alpha, \alpha, \alpha, \alpha, \delta, \delta, \beta, \delta$) and (b) ^{235}U to produce ^{207}Pb (in order, these reactions are $\alpha, \beta, \alpha, \delta, \alpha, \alpha, \delta, \delta$).

31-9 Promethium 145 decays by alpha emission to praseodymium 141. (a) Calculate the energy release in the decay. (b) Explain why promethium 145 does not decay to promethium 144 (143.91270 u), emitting a neutron.

31-10 Tritium ($^3_1$H) decays to the light isotope of helium ($^3_2$He) by emitting a negative beta particle. (a) Write the decay reaction equation. (b) Calculate the maximum kinetic energy of the beta.

31-11 Determine the binding energy per nucleon of (a) carbon 14 and (b) nitrogen 14.

31-12 Calculate the binding energy per nucleon for sodium (Na) isotopes (a) ^{22}Na, (b) ^{23}Na, and (c) ^{24}Na. (Masses: 21.994437 u; 22.989771 u, 23.990964 u.)

31-13 Calculate the binding energy of (a) the earth-moon system and (b) the sun-earth system.

31-14 How much energy is released when ^{14}C decays to ^{14}N with the emission of a negative beta particle?

31-15 Determine the net energy release in the reaction $^1_1$H + $^2_1$H $\rightarrow$ $^3_2$He.

31-16 Potassium 40 may undergo either positive beta decay, becoming argon 40, or negative beta decay, becoming calcium 40. (a) Calculate the total binding energy and binding energy per nucleon of all three isotopes. (b) On the basis of only those numbers, which decay would you expect to be more probable?

31-17 (a) Calculate the change in binding energy of a bowling ball–earth system if the bowling ball is rocketed to the moon. (b) Calculate the change in mass of the system. Take the bowling ball's original mass to be 7.00 kg.

31-18 How much time is required for a sample of polonium 210 (T = 138 days) to be reduced by (a) 90 percent and (b) 99 percent?

31-19 A sample of radioactive material shows a measured activity of 5.40×10^3 disintegrations per minute when first measured. If measurement 1 h later shows 3.80×10^3 disintegrations per minute, calculate the sample's (a) decay constant, (b) time constant, and (c) half-life.

31-20 Phosphorus 32 has a decay constant of 8.12×10^{-7} s^{-1}. Determine its (a) half-life in days, (b) time constant in days, and (c) the activity of a sample 125 days after its activity was measured to be 350 disintegrations per second.

31-21 Cesium 137, with a half-life of about 30.2 years, is the longest-lived of the radioactive waste products of nuclear power. (a) How many years must pass before the activity of cesium 137 in a sample of waste is reduced to 1 percent of its original value? (b) To what percentage of the original activity will the cesium 137 have decayed after 600 years has passed?

31-22 Sodium 24 decays by negative beta emission with a half-life of 15.0 h. When measured in a laboratory experiment on a Thursday afternoon, a sample of sodium 24 has an activity of 360 disintegrations per second. (a) What activity will the samples show at the same time of day on the following Monday? (b) About when will the sample have an activity of 1.00 disintegration per second?

31-23 A certain radioactive state of technicium 99 is used in nuclear medicine and has a half-life of 6.02 h. Determine (a) the activity (in disintegrations per second) of a sample which is produced at 4:00 A.M. if it is to be used at noon and must have 8.50×10^6 disintegrations per second when injected and (b) the number of atoms of technicium still to be left at noon.

31-24 A sensitive Geiger counter which detects only 40 percent of the particles emanating from a radioactive source registers 180 counts per second from a tritium source. The half-life of tritium is 12.3 years. Calculate (a) the activity of the source, (b) the decay constant of tritium, (c) the number of tritium nuclei in the measured source, and (d) the mass of tritium present.

31-25 Silicon 31 has a half-life of 2.62 h and is to be used in a radiation exercise in a physics laboratory course. The most active sample which can be prepared by the professor has 4.65×10^4 disintegrations per minute. What is the earliest that the professor may prepare the sample to be used at 10:30 A.M. on a Friday morning if it must have an activity of at least 120 disintegrations per minute at the beginning of the exercise?

31-26 The activity of a radioactive sample is directly proportional to the number of radioactive nuclei present, which, in turn, determines the mass of the sample. How much time must pass before 1.00 g of radium ($T = 1.60 \times 10^3$ years) is reduced to (a) 0.800 g and (b) 0.100 g?

31-27 The half-life of cobalt 60 is 5.27 years. Calculate (a) the decay constant, (b) the number of cobalt 60 atoms present in a radioactive sample with an activity of 5.00×10^3 decays per second, and (c) the mass of the sample.

31-28 The nucleus $^{20}_{10}$Ne is formed in an excited (unstable) state but is essentially at rest when it immediately decays to an alpha particle and an oxygen nucleus. Assume that the alpha and oxygen nuclei share 5.30 MeV. (a) How is it divided between them? (b) What is the speed of the alpha particle?

31-29 Naturally occurring uranium is 99.27 percent ^{238}U and 0.72 percent ^{235}U (with a trace of ^{234}U). Assume that the ratio of isotopes is 99.3 percent ^{238}U to 0.7 percent ^{235}U. The half-life of ^{235}U is 7.10×10^8 years, thus shorter than that of ^{238}U, which is 4.51×10^9 years. This would imply that the percentage of naturally occurring uranium composed of ^{235}U was higher in the past than at present. Many lines of evidence indicate that the earth is approximately 1 half-life of ^{238}U in age. What would the percentage of ^{235}U in the uranium present at the earth's formation have been?

31-30 Carbon 14 is a radioactive isotope which decays by beta emission with a half-life of 5.73×10^3 years. Because it is produced by secondary cosmic rays in our atmosphere ($_0^1 n + {}_7^{14}N \rightarrow {}_6^{14}C + {}_1^1H$) at a fairly constant rate, it constitutes a minute percentage of the carbon in our environment all the time. Plants and animals metabolize carbon, and since, chemically, there is no distinction, carbon 14 makes up a small part of all these living beings. However, when a plant or animal dies, the carbon 14 continues to decay and is not replaced. An average sample of 1.00 g of natural carbon contains enough carbon 14 to give it an activity of 15.1 disintegrations per minute. How old is a tree found under a glacier if an average sample of 1.00 g of carbon from the tree has an activity of 6.40 disintegrations per minute?

31-31 The half-life of ^{235}U is 7.10×10^8 years, and the half-life of ^{238}U is 4.51×10^9 years. Natural uranium now contains about 99.3 percent ^{238}U and 0.7 percent ^{235}U. Suppose natural uranium contained 50 percent of each of these isotopes when the earth was "born." What would be the age of the earth?

31-32 When molten rock cools, lead ions become incorporated in several common minerals which solidify early (at a high temperature), but uranium ions do not. Thus, slow cooling separates lead from uranium. The uranium is frequently frozen into other mineral structures, such as zircon, which may thus be substantially lead-free. Most (assume all) of the uranium is ^{238}U, which chain-decays to the stable isotope ^{206}Pb (^{238}U $\rightarrow$ ^{206}Pb + 8 ^{4}He + 6β). While this decay requires many steps, the relatively long half-life for ^{238}U (4.51×10^9 years) ensures that at any given moment, a negligible fraction of atoms are in other than the beginning or ending state. (a) For an isolated, lead-free crystal containing uranium, calculate and plot the ^{206}Pb/^{238}U ratio from time zero, using intervals of 10^9 years up to age 6×10^9 years. (b) The Canadian Shield formation is a large area of exposed granites many times the size of (and surrounding) Hudson Bay. Isolated crystals from a sample of these rocks yield a ^{206}Pb/^{238}U ratio of 0.536. Meanwhile, similar samples from Greenland yield a ratio of 0.634. Which sample is older? (c) Using your graph, what are the approximate ages of the samples in (b)?

31-33 Determine which of the following reactions violates one or more of the conservation laws (charge, baryon number, lepton number, strangeness), and name the law or laws violated.
(a) $\nu_e + n \rightarrow p + e^-$
(b) $\pi^- + p \rightarrow \Xi^0 + K^0 + K^0$
(c) $\pi^- + n \rightarrow K^- + \Lambda^0$
(d) $\nu_e + p \rightarrow n + e^+$
(e) $p + p \rightarrow p + p + \Lambda^0 + K^0$

31-34 Determine which of the following decay reactions of elementary particles are (a) allowed or (b) forbidden due to a violation of one of the laws of conservation (mass-energy, charge, baryon number, lepton number, strangeness). For those which are forbidden, indicate which conservation law is violated.
(a) $\Sigma^0 \rightarrow \Lambda^0 + \gamma$
(b) $\Lambda^0 \rightarrow p + K^-$
(c) $\Xi^0 \rightarrow \Sigma^0 + K^0$
(d) $K^0 \rightarrow \pi^+ + \pi^-$
(e) $\nu_e \rightarrow \gamma + \gamma$

31-35 Confirm that the charge, baryon number, and strangeness are proper according to the quark makeup of (a) K^+, (b) Λ^0, (c) Ξ^0.

31-36 Confirm that the charge, baryon number, and strangeness are proper according to the quark makeup of (a) π^-, (b) n, (c) Σ^0, and (d) Ξ^-.

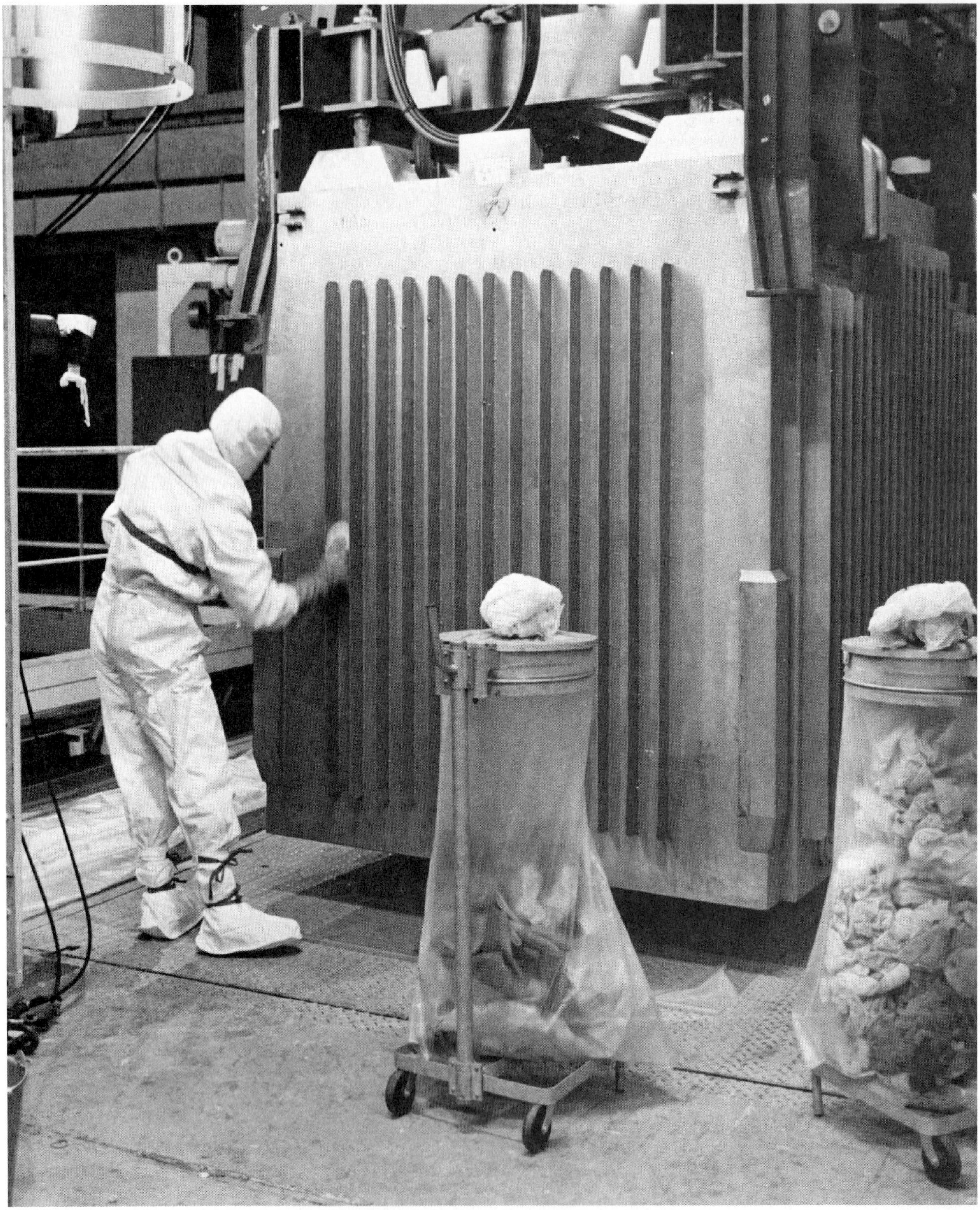

32 Ionizing Radiation, Safety, and Nuclear Medicine

32-1 Introduction

Movies frequently show the use of a Geiger counter to detect a radiation source. The meter has a speaker clicking to indicate the arrival of a particle. Far from the source, it clicks only once in a while, somewhat randomly spaced in time but usually several seconds apart. As the meter gets closer to the source, the clicks get closer together, and when the meter is supposedly at the source, the clicks become a continuous buzz.

Such a movie scene is fairly authentic. While we easily understand the increase in frequency of detection as the meter approaches the source, you might wonder what triggers the randomly spaced clicks when the meter is very far from the source. The answer is that there is ionizing radiation all over the place, coming from all directions and present all the time. We call this, collectively, the **background radiation.** Much of it is cosmic radiation, which we consider in the next section, but the rest comes from the atmosphere, the earth, and objects in and on the earth. Naturally occurring radioactive isotopes of uranium, thorium, radon, potassium, and so on, are the sources of the non-cosmic-ray natural background. Other radiation to which we are exposed is the result of human activity, and we will label it **artificial background radiation.** It includes medical and dental x-rays, x-radiation emitted by television receivers, radiation from isotopes used in nuclear medicine, global fallout from nuclear weapons testing, and discharges from nuclear power plants. In this chapter, we will study the units of activity and absorbed radiation, the relative quantities we receive, and their possible biological effects, as well as safety standards and protection methods.

32-2 Cosmic Rays

The discovery that radioactivity caused charged electroscopes to discharge was thought to provide the solution to a then-existing puzzle. In the nineteenth century, it was well known that a charged electroscope would always lose its charge spontaneously if left alone. Even a well designed electroscope for which charge leakage is reduced to practically zero will not retain its charge indefinitely. Turn-of-the-century scientists believed that the radioactivity discovered by Becquerel was ionizing the air in the vicinity of the electroscope's leaves and thus causing the discharge. It was postulated that small traces of radioactive substances were present in the material of which the electroscopes were made. This proved to be correct in part but could not account for all the discharge. Several experimenters showed that the rate of discharge could be decreased if the electroscope were surrounded with lead or water. It was therefore assumed that a good part of the radiation discharging electroscopes must come from radioactive isotopes in the earth's crust. This opinion prevailed for more than 10 years. This hypothesis could easily be tested by observing the rate of discharge of electroscopes at different altitudes. If the earth is the source of the radiation, the electroscope should discharge fastest at the earth's surface and progressively slower with increasing altitude.

In 1912, Victor F. Hess (1883–1964), an Austrian physicist, published measurements made with electroscopes in a balloon at altitudes up to 16,000 ft. Contrary to the expectation that electroscopes would discharge faster at lower altitudes, Hess actually observed that they discharged about four times faster at 16,000 ft than they had at sea level. He summarized his paper with "The results of my observations are best explained by the assumption that a radiation of very great penetrating power enters our atmosphere from above."

Many groups began to investigate this radiation from above. There was much speculation concerning its origin, and experiments were designed to test theories. Finally, in 1926, Robert Millikan (1868–1953), an American, announced the results which convinced the scientific community that this radiation did come from beyond the earth's atmosphere. In the process, Millikan gave this radiation the name **cosmic rays** because he believed they came from the cosmos (outside the earth's atmosphere). Millikan was initially very skeptical that radiation came from beyond the earth's atmosphere. His experiments were conducted at two lakes in southern California, Muir and Arrowhead. Muir Lake is 11,800 ft above sea level, and Arrowhead is 5100 ft (Figure 32-1). The 6700-ft difference of atmosphere above Arrowhead is equivalent to about 6 ft of water in its ability to absorb radiation. Millikan reasoned that if cosmic rays were created in the air layer between the two lakes, the number of rays reaching an electroscope in Arrowhead would be larger. Millikan measured the rate of discharge of his electroscope sunk to various depths in the two lakes. He

32 Ionizing Radiation, Safety, and Nuclear Medicine

32-2 Cosmic Rays

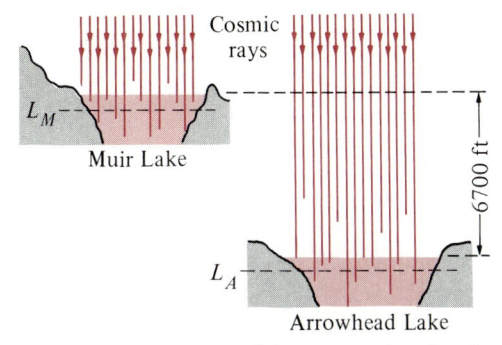

Figure 32-1 Millikan used two mountain lakes to test the idea that cosmic rays are made in the atmosphere. Note that while we have drawn the cosmic-ray tracks as vertical, they really approach the earth's surface from all directions from outside the atmosphere.

reported that "within limits of observational error, every reading in Arrowhead corresponded to a reading six feet farther under water in Muir Lake, thus showing that the rays definitely do come in from above and that their origin is entirely outside the layer of atmosphere between the levels of the two lakes."

No serious attempt was made to identify the type of ray in the cosmic rays until 16 years after Hess' discovery, partly because of the relatively crude and cumbersome detectors and techniques available in 1912 and partly because most scientists believed they already knew what the cosmic rays were. Recognize that the rays had to be extremely penetrating to get through the whole atmosphere (the equivalent of a 10-m depth of water) to the earth. The only really penetrating rays which were then known were gamma rays. Thus, most scientists believed that cosmic rays were simply photons of higher energy than any previously encountered. On the basis of absorption curves, Millikan hypothesized that the cosmic rays consisted of three distinct groups of photons of energies about 26, 110, and 220 MeV.

In 1929, Hans Geiger and his student Wilhelm Mueller made modifications to Geiger's point counter and created the G-M counter (Section 31-6). At the same time, electronics technology was beginning to come into its own, and it became possible, using vacuum tubes, to construct a **coincidence circuit.** In its simplest form, a coincidence circuit might well be considered a black box with two or more separate connections which will accept input pulses and one connection for an output pulse. The circuit is such that a pulse appears at the output connector *only* if pulses appear simultaneously (within the resolving time) at the input connectors. The conclusion is that the simultaneous pulses represent the same ray. When G-M counters are used with coincidence circuits, they may show penetrating power, as in Figure 32-2a, or even the actual path, as in Figure 32-2b. Over the next 10 years, various experiments which applied these developments were carried out. Scientists slowly became convinced that the local component of cosmic rays was not necessarily the same as the cosmic rays which struck the upper atmosphere from the cosmos. The question was split into two: What is the composition of the local radiation (**secondaries**) in the atmosphere? What is the nature of the **primary cosmic radiation** which falls on the atmosphere from outer space?

Evidence began to accumulate that the local radiation included electrons and photons of energies up to 1 or 2 MeV. The consensus was that these were secondaries created during some process or interaction which included the very

Figure 32-2 (*a*) Three G-M tubes with space for variable-thickness absorbers between them. Such an arrangement can measure the penetrating power of radiation while neglecting random background counts. (*b*) Two or more detectors can be used to determine the radiation path. The coincidence circuit ensures that only those particles which move through both detectors within a very short time period are counted. Such schemes allow us to reject irrelevant data when we wish to examine a specific case.

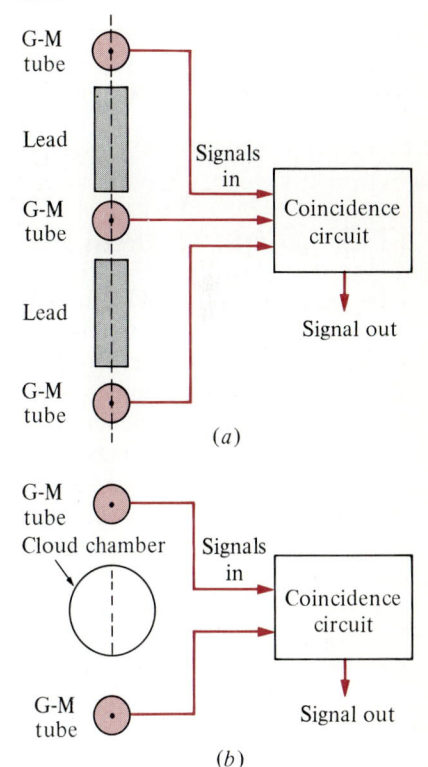

penetrating component of local radiation. In 1932, Carl D. Anderson (b. 1905), an American physicist, obtained the first photograph of a cloud chamber track which could be identified as a positron. Five years later, Anderson and a colleague, Seth Neddermeyer, published the results of new experiments from which they concluded that the very penetrating local radiation had a charge of magnitude equal to the electron and a mass between that of the electron and proton. It was subsequently shown to have a mass about 207 times that of the electron and was eventually called the **muon.**

The primary cosmic radiation was tentatively identified during balloon experiments in which the balloons ascended to altitudes of up to 70,000 ft. These experiments indicated that the rays were "probably protons," and this eventually was proved to be correct, although we now know that there are also alpha particles (~6 percent) and other bare nuclei (slightly less than 1 percent).

Pure research was interrupted by World War II but resumed immediately afterward with many new discoveries in the cosmic rays. In 1947, the pions were discovered in the cosmic rays, and it was determined that they were the parents of the muons by the reactions

$$\pi^- \to \mu^- + \bar{v}_\mu$$

and

$$\pi^+ \to \mu^+ + v_\mu$$

Since pions have a half-life of 1.80×10^{-8} s, it is rare that they travel more than a few tens of meters before they decay. Further, since they are created in the nuclear strong interaction of the primary protons with nuclei, they are usually "born" and subsequently decay near the top of the atmosphere.

In the next few years, the kaons, the sigma particles, and the xi particles were all discovered in the cosmic rays. These were exciting times for cosmic-ray physicists. Elementary particle physics was born, first as a subfield of cosmic rays and then as a separate area.

32-3 Other Natural Background Radiation

Uranium, a constituent of the rock pitchblende and of certain salts, is one of those ubiquitous elements — it is virtually everywhere on earth. Tiny amounts are present in every bit of soil, rock, and so on. In some areas its concentration is unusually high, and it is mined in those places. It is radioactive and provides some of the background radiation from the ground, building materials, and so on. It is even in coal in concentrations averaging about 1 part per million (1 ppm). In addition, uranium 238 undergoes about 14 generations of decay in series until it reaches the stable isotope lead 206. Along the way, one of the generations is radium 226, which decays to radon 222, an inert gas, which in turn decays to polonium. All these generations provide part of the natural background radiation, but as we will see, radon is one of the most hazardous to human health.

Thorium 232, another heavy element which begins a radioactive series, is actually more abundant in the earth's crust than uranium. However, it is less active than uranium 238 by a factor of 3, and the radon isotope in its decay series has a very short half-life, thus decaying too fast to concentrate as a gas.

Potassium 40 is a naturally occurring isotope which makes up about 0.0118 percent of the available potassium, though its chemistry is the same as any other isotope of potassium. Thus potassium 40 is present in foods, table salt substitutes, animals, the human body (especially the blood), and so on. Some background radiation, therefore, is due to potassium in our own bodies.

Carbon 14 and hydrogen 3 are radioactive isotopes of those elements which are products of secondary cosmic-ray nuclear reactions. They are both a part of the life cycle and are, therefore, found in every living thing as well as anything else which contains carbon and hydrogen. All water, food, paper, wood, and so on, is radioactive.

A number of other radioactive nuclei either occur naturally or are created in nuclear reactions initiated by cosmic rays. Altogether, however, these others contribute very little additional radiation to the background.

Approximately one-third of what we call background radiation is artificial, as we mentioned in Section 32-1. We should emphasize that exposure to artificial radiation may vary much more than exposure to natural radiation (although that, too, is quite variable). Some people may never receive diagnostic x-rays or nuclear medicine procedures, hence no radiation from these sources, while others might undergo extensive x-ray examination and radiation therapy. In most cases, we will discuss average exposures. To understand these, we must first define units for exposure.

All living systems contain radioactive ^{14}C, ^{3}H, etc., and always have from the earliest life. The amount is not large, of course, but it is easily measurable. This may present problems for people who have not studied science. For example, one of the U.S. state legislatures almost passed a law totally banning the disposal of any radioactive waste in the state. Taken literally, all human wastes would have to be collected and shipped across state lines, and it would have become illegal to exhale.

32-4 Units of Ionizing Radiation

Many different types of radiation units are of practical interest. Each click of a counter records a single radiation, representing the decay of a single atom of a radioactive substance. If we are interested in the rate at which atoms are decaying in a sample, we need a measure of **activity.** However, two different radiations may each produce one click in a counter but differ widely in their ability to produce the ions which give ionizing radiation its name. Ion-producing ability gives a measure of **exposure.** Of great interest is the amount of radiation energy released in the absorbing tissue, or the **absorbed dose.** Finally, the biological effects of a given absorbed dose vary with the type of ray or particle being absorbed, requiring a unit of **effective dose** or **dose equivalent** to adjust for these differences. We will discuss each type of unit in this order.

Activity of Source

When we detect radiation from a radioactive sample with a G-M counter, we are simply determining the number of ionizing particles. If we divide this number by the time interval during which we employ the detector, we obtain a measure of the activity. To compare the strength of different samples without the use of a lot of powers of 10, a standard unit of radioactivity is defined: one **curie** (Ci) is defined as exactly 3.70×10^{10} disintegrations per second (s^{-1}). The curie was named in honor of Marie and Pierre Curie and was intended to be equal to the activity of a freshly prepared gram of radium 226.

Example 32-1 Tritium ($^{3}_{1}$H) decays by beta emission with a half-life of 12.33 years. What mass of pure tritium will have an activity of 1.00 Ci?

Solution We use the given activity (1.00 Ci) in disintegrations per second to obtain the number of radioactive nuclei present. We then use Avogadro's number to obtain the number of moles of tritium. Finally, we calculate the number of grams from the atomic mass of tritium.

$$A = \lambda N = 1.00 \text{ Ci} = 3.70 \times 10^{10} \text{ s}^{-1}$$

but

$$\lambda = \frac{\ln 2}{T}$$

Thus

$$N = \frac{A}{\lambda} = \frac{3.70 \times 10^{10} \text{ s}^{-1}}{(\ln 2)/T}$$

Converting T to seconds, we have

$$T = (12.33 \text{ yr})(365 \text{ days/yr})(86{,}400 \text{ s/day})$$
$$= 3.89 \times 10^8 \text{ s}$$

Thus

$$N = \frac{(3.70 \times 10^{10} \text{ s}^{-1})(3.89 \times 10^8 \text{ s})}{\ln 2}$$
$$= 2.08 \times 10^{19} \text{ nuclei}$$

Then

$$\text{No. mol} = \frac{2.08 \times 10^{19} \text{ nuclei}}{6.023 \times 10^{23} \text{ nuclei/mol}} = 3.45 \times 10^{-5} \text{ mol}$$

and

$$m = (\text{no. mol})(\text{atomic mass})$$
$$= (3.45 \times 10^{-5} \text{ mol})(3.016049 \text{ g/mol})$$
$$= 1.04 \times 10^{-4} \text{ g}$$

This is a very small mass for such a large activity.

The curie is not an SI unit of radioactivity. The 1975 General International Conference on Weights and Measures formally named the **becquerel** (Bq), which is equal to one disintegration per second, or 1 s^{-1}, as the SI unit of activity. But they also approved the interim use of the curie, recognizing that it may be many years before the becquerel replaces it in the literature.

Exposure: Ion Pairs Produced

The **roentgen** (R) is a unit of exposure and is defined as the amount of x-radiation or gamma radiation which will produce 2.082×10^9 ion pairs in one cubic centimeter of dry air. This is equivalent to producing 1.61×10^{12} ion pairs per gram of air or an ion charge per unit mass of 2.58×10^{-4} C/kg. Radiologic physicists are gradually attempting to change from roentgens to the SI unit, coulombs per kilogram of dry air, as a unit of exposure.

Exposure is the easiest to measure of the properties associated with radiation. It may be measured by ion chambers similar to those used by the Curies and Rutherford. Another device using the ion-chamber method is a pocket **dosimeter**, like that shown in Figure 32-3. The tiny capacitor-like ion chamber has a flexible fiber attached to a metal frame which is in electrical contact with one of the capacitor plates. The chamber is originally charged and the optics adjusted so that the position of the fiber reads zero. As radiation causes ions to form in the chamber and discharge the capacitor, the fiber moves back toward the frame, indicating the accumulated dose. Dosimeters are also made to monitor efficiently neutron radiation. This is done by placing in the ion chamber some material with which neutrons react with good probability to produce high-energy charged particles. For this purpose, neutron dosimeters often have their ion chambers filled with boron trifluoride (BF_3), a gas. The neutrons then react with boron 10 to produce an alpha particle in the reaction

$$ {}^1_0n + {}^{10}_5B \rightarrow {}^7_3Li + {}^4_2He $$

Absorbed Energy

Another important unit is the **rad,** which stands for *r*adiation *a*bsorbed *d*ose. The rad is defined as the amount of radiation necessary to deposit 100 ergs of energy (1.00×10^{-5} J) in one gram (1.00×10^{-3} kg) of material. An exposure of 1 R absorbed in tissue gives a dose of between 0.83 and 0.93 rad, depending on the energy of the photons (x-ray or gamma).

The SI unit of absorbed radiation dose is the **gray** (Gy), the amount of radiation which will deposit one joule of energy in a kilogram of material. From these definitions, we have that

$$ 1 \text{ rad} = 100 \text{ ergs/g} = 1.00 \times 10^{-2} \text{ J/kg} = 1.00 \times 10^{-2} \text{ Gy} $$

Biological Effect

Most of the biological effects of radiation depend on the rate at which energy is deposited in living tissue. Since different types of radiation lose energy to ionizations at different rates along their paths, we define the term **quality factor** (QF) to relate the relative biological effectiveness of various types of radiation.[1] For some kinds of radiation, quality factor changes as the energy of the ionizing radiation changes, because a faster-moving charged particle ionizes less effectively than those which are slower moving, and high-energy gammas interact differently than those of lower energy.

The unit of effective dose, or dose equivalent of radiation exposure, is the **rem** (*r*oentgen *e*quivalent *m*an). Effective dose in rems is equal to the product of the absorbed dose in rads and the quality factor of the radiation involved in the exposure. The terms *rem* and *millirem* may be familiar to you from news reports of radiation absorbed. However, the rem will eventually be displaced by the approved SI unit, the **sievert** (Sv). The effective dose in sieverts is equal to the absorbed dose in grays times the quality factor. Thus

$$ 1 \text{ rem} = 1 \text{ rad} \times \text{QF} = 10^{-2} \text{ Gy} \times \text{QF} = 10^{-2} \text{ Sv} $$

Figure 32-3 A dosimeter (left) and a film badge. One end of the dosimeter houses a lens and the other a window, allowing viewing along the axis to read the position of a movable fiber indicating exposure. The instrument is essentially a well insulated capacitor charged to high voltage. Ion pairs generated in the dosimeter by penetrating radiation progressively discharge the capacitor and thus alter the position of the fiber. The typical film badge is a small piece of radiation-sensitive film wrapped in light-tight paper and used to monitor accumulated dose. Film badges are further described in Section 32-7.

---

[1] Quality factor seems to be the designation used by radiologic scientists and health physicists, but others often call this quantity the **relative biological effectiveness** (RBE).

TABLE 32-1 Radiation Units

Quantity	Common Unit	SI Unit	Value
Activity	Ci	Bq	1 Ci = 3.70 × 10^{10} Bq = 3.70 × 10^{10} s^{-1}
Exposure	R	C/kg	1 R = 2.58 × 10^{-4} C/kg
Absorbed dose	rad	Gy	1 rad = 10^{-2} Gy = 10^{-2} J/kg
Dose equivalent	rem	Sv	1 rem = 1 rad × QF = 10^{-2} Sv

TABLE 32-2 Quality Factors of Selected Types of Radiation

Radiation	Quality Factor*
X-rays, gammas < 4 MeV	1
Gammas > 4 MeV	0.7
Beta particles < 30 keV	1.7
Beta particles > 30 keV	1
Slow and thermal neutrons	4 or 5
Fast neutrons	10
Protons	10
Alpha particles	10
Heavy ions	20

* Note that these values are only approximate. For the most part, they are used for determining occupational exposures. For nonoccupational exposures, we usually deal in millirems (mrem).

Table 32-1 gives the four units associated with radiation measurement and their SI counterparts.

The quality factors for several different kinds of radiation are given in Table 32-2. Typically, for most medical diagnostic exposures, one makes the assumption that 1 R = 1 rad = 1 rem, a pretty good assumption. Most medical exposures are due to x-rays or gamma rays of energy less than 4 MeV, and we see in Table 32-2 that the quality factor for those types of radiation is 1.

32-5 Levels of Human Exposure

What levels of radiation are of human concern? We know that human beings who are exposed to a whole-body dose of 50 rem (50,000 mrem) or less display *no* clinically observable effects. Certain kinds of radiation, x-rays for example, have even higher limits. We also know that exposures of 250 rem on a large number of people lead to few or no immediate deaths (within a month or two). However, most people so exposed show acute radiation sickness from which they recover in a few weeks to a month or two. About 50 percent of the people who receive a whole-body dose of 600 rems all at once die within a few days to a few weeks after exposure (without treatment).[1] The rest recover from the radiation sickness with significant probability of a shorter life span. The details on effects of radiation are discussed in Section 32-6.

Table 32-3 shows the effective radiation dose one might expect to receive from the background (including artificial radiation). You might note that the dose rates given are in millirems per year, *not* in rems per year. Further, we emphasize that these numbers are best-estimate averages for people living in the United States, and numbers may vary widely from person to person.

Some of the individual variations in effective radiation dose which we receive are obvious. The only source which is nearly constant is that which we receive from radioactive nuclei inside our bodies. About half of that is from potassium 40, and most of the rest is from uranium or its decay daughters. The dose due to global fallout from nuclear weapons testing is primarily from the isotopes strontium 90 and cesium 137, which both have half-lives near 30 years. Barring the resumption of atmospheric testing, this source should eventually become negligible. At present, however, it is virtually the same value for everyone.

TABLE 32-3 Average Effective Ionizing Radiation Doses Received by the U.S. Population

Source	Average Effective Dose, mrem/yr
Cosmic rays	45
Internal	35
Building materials	40
Ground	11
Air	5
Medical	70
Global fallout	4
Color TV	1
Nuclear power	0.003
Total	211

---

[1] If a dentist x-rays a single tooth, the absorbed dose is a certain number of joules per kilogram to a portion of the head. This is clearly less of an insult to the body than if every kilogram of body mass had been given the same dose, a whole-body dose. Likewise, 1 rem/day for nearly 2 years would be a 600-rem cumulative dose, but less of a trauma than the same 600 rem delivered in a single, short period of time.

32-5 Levels of Human Exposure

As we are now aware, the shielding of the atmosphere decreases the dose due to cosmic rays at lower altitudes. At sea level, the cosmic-ray dose is about 40 mrem/yr and increases by approximately 1 mrem/yr for every 100 ft of altitude.

Although there is not a great deal of reliable data on radiation dose levels inside U.S. buildings, we know that all building materials come from the earth's crust and thus contain radioactive nuclei. Careful readings taken in Swedish buildings showed dose rates ranging from a low of 48 mrem/yr in one wooden building to as high as 202 mrem/yr in a building made of concrete composed of aluminum shale. A radiation survey conducted by the Environmental Protection Agency showed dose rates of 525 mrem/yr in New York's Grand Central Station, due to the radioactive isotopes present in the granite of which the station is built.

Because uranium is everywhere, its decay chain, including radon, is everywhere also. Radon 222 is the isotope in the decay chain of uranium 238 and makes up virtually 100 percent of the radon present in our surroundings. It decays with a half-life of 3.8 days and emits both alpha particles and gamma rays. Since it is an inert gas, it diffuses through materials without being trapped in a chemical reaction. Thus, once it is formed as a decay product of uranium in the soil, rocks, bricks, and concrete, radon diffuses upward and escapes into the atmosphere. Being a gas, radon has easy access to the human body by inhalation and may cause damage to lung tissue, resulting in lung cancer. This can be a hazard wherever its concentrations are high. In particular, radon diffusing from walls and foundations may be trapped in unventilated homes and office buildings. In 1977, the United Nations Scientific Committee on Effects of Atomic Radiation released a report showing that radon is the most serious radioactive health hazard in the natural environment, causing 10,000 cancer deaths per year in the United States. The thrust to conserve money spent for heating and cooling homes in the last several decades has resulted in many homes which are so air-tight that radon concentrations greatly exceed the maximum permissible levels in uranium mines. Although the issue is not clear-cut, there appears to be a consensus developing in the building industry that infiltration rates of one-half an air change per hour will keep pollutants, including radon, at acceptable levels in residences and other buildings.

The average effective dose received from medical sources is practically all due to diagnostic x-rays, with 1 or 2 percent due to radiopharmaceuticals. This average has decreased in recent years because of great improvements in x-ray equipment and techniques. Changes in attitude of radiologic scientists should further reduce this average over the next several decades. For instance, it is now believed that periodic chest x-rays among the general population are not sufficiently productive as a screening procedure for the detection of tuberculosis, pulmonary disease, or heart disease and should not be done. Even routine hospital admission chest x-ray exams have been questioned as a procedure. The attitude seems to be that x-rays should be taken only if ordered by a physician trained to make the risk vs. benefit decision.

The x-rays which result from the stopping of very high energy electrons in color television receivers can be avoided only by not watching, or if there is a breakthrough in the production of color receivers using liquid crystals. This dose is much smaller for an average year of watching than the additional cosmic-ray dose received during one coast-to-coast airplane flight.

The effective average dose due to nuclear power is insignificant compared to other sources and is included only so that we may keep the numbers in their proper perspective.

Example 32-2 Calculate (a) the total energy deposited in the body of an 80.0-kg person who absorbs 100 rad of ionizing radiation and (b) the total effective dose the person received if the radiation was from beta particles of energy 25 keV.

Solution (a) This part is simply an application of the definition of the rad converted to joules per kilogram.

$$\text{Absorbed energy} = \text{absorbed dose} \times \text{mass}$$
$$= (100 \text{ rad})[1.00 \times 10^{-2} \text{ J/(kg} \cdot \text{rad)}](80.0 \text{ kg})$$
$$= 80.0 \text{ J}$$

(b) The effective dose in rems equals the absorbed dose in rads times the quality factor. For low-energy betas, Table 32-2 gives a quality factor of 1.7. Thus

$$\text{Dose equivalent} = (100 \text{ rad})(1.7 \text{ rem/rad}) = 170 \text{ rem}$$

Note that only a small amount of energy was absorbed—about the same amount the person would receive by spending a minute or so in the sun in a bathing suit. Energy is not the measure of what does the damage.

32-6 Biological Effects of Ionizing Radiation

When ionizing radiation passes through living tissue, the important effect is the damage which may be done to cells due to the ionization of one or more molecules within the cell. It is possible that unusual ions or free radicals may be produced, initiating chemical reactions which interfere with the cell's normal function. If this happens, the cell may (1) repair itself and go back to its normal function, (2) be unable to reproduce and die, or (3) live damaged and reproduce more copies of damaged or defective cells.

If the number of dead cells caused by a particular irradiation is not very large, it may be that no problem is created for the organism. The dead cells are simply replaced by new ones, and the organism goes on as if nothing happened. On the other hand, a large number of cell deaths may affect the whole organism, causing, for example, radiation sickness in human beings. A severe case may bring death to the whole organism.

The defective cells which reproduce copies of themselves may be the start of a cancerous growth. This is normally the biggest concern for radiation, the fact that it may cause cancer. The risk for both cancer and radiation-induced cell death is higher in cells that are dividing rapidly, such as those in bone marrow, the origin of red blood cells. Apparently, rapid growth means less time between divisions, and it is therefore less probable that the time interval necessary for cell repair is available before the cell must divide. Thus cell death or the reproduction of a defective cell is more likely.

The other possibility of a defective cell is a genetic effect. Radiation damage may affect the genes, resulting in a mutation. If the mutation happens to occur in egg or sperm cells, it is possible that it will be passed along to offspring.

It should be noted that more is known about the radiation hazard than we know about almost any other hazard, possible exceptions being specific chemical poisons. In fact, we have more information on the effect of radiation on human beings than we have on the effect of most of today's pollutants and pesticides in our environment. In addition to the vast amount of detailed research carried out irradiating plants and animals and the numerous experiments confirming and extending general laws, a great deal of data is available on the effects of radiation on human beings. As examples: (1) The 24,000 Japanese survivors of the nuclear bombings of Hiroshima and Nagasaki were followed very closely for evidence of any statistically significant effects. Many of these survivors received more than 200 rem, and the average effective dose was about 130 rem. (2) Nearly 15,000 victims of the disease ankylosing spondylitis, an arthriticlike disease of the spine, were given massive doses of x-rays in England as therapy for their condition. The spinal doses they received ranged from 375 to 2750 rad. This group was the subject of many follow-up studies. (3) Thousands of miners (primarily uranium miners) inhaled radon, some receiving doses to the lung of nearly 5000 rem. (4) For a 20-year period, beginning in 1915, there were 775 American women employed to paint radium numerals on watch dials. These workers would lick the ends of their brushes to point them and thus ingest quantities of radium.

Genetic defects have been shown to be passed along from irradiated parents to their offspring in experiments with animals, particularly fruitflys, so we know they exist. However, inherited genetic defects due to ionizing radiation have never been observed in human beings, not even in the offspring of the survivors of Hiroshima and Nagasaki. While it is believed that these effects do exist, it is probable that they are too small to be observed above the "noise." Many genetic (congenital) effects due to spontaneous mutations in the sex cells occur even without artificial radiation. Fully 10 percent of all live births show such congenital deformities, which include extra fingers and toes, double elbows, cleft palates, etc., as well as other defects which do not show up until later in life. With so many "natural" genetic defects, it would require a very large artificial-radiation-induced effect to be observable. Nonetheless, an enormous number of research efforts have been made and are still being made, all without success to date.

Even artificial-radiation-induced cancers are difficult to detect with confidence. There is already a very high risk of contracting cancer. In the United States, the present chance of dying of cancer is 16.8 percent. That is, approximately one person out of every six dies of cancer. Thus, identifying radiation-induced cancer requires very good statistics among a large group of exposed people. Even then, all we can conclude is that the group contracted a certain number of *excess* cancers over what we would expect them to get "naturally." There seems to be no method by which the cause of a particular cancer can be determined. However, painstaking research of the records and follow-up studies of the exposed groups listed above have been made. The 24,000 survivors of the Hiroshima and Nagasaki bombings have had about 100 excess cancer deaths. Of the 15,000 ankylosing spondylitis patients who received large x-ray doses to the spine, 60 developed leukemia over a 25-year period. This is 11 times the incidence of leukemia that is expected in a similar but unradiated population. We could go on, but it is not necessary. The statistics are solid enough so that it is clear that there is a relationship between large radiation doses and excess cancer incidence.

It is not at all clear what the effects of low doses of radiation are. However, the assumption in the United States is to be most conservative and use the

32-6 Biological Effects of Ionizing Radiation

Radioactive substances are incredibly easy to trace and follow in extremely small amounts. Each click of a counter records the decay of a *single* atom. Measurement of the energy of the absorbed radiation often reveals what isotope of what element that atom was.

so-called **linear hypothesis.** By this we mean that if studies of radiation victims show that an acute 100-rem dose increases the probability of contracting cancer by 1.8 percent (i.e., from 16.8 to 18.6 percent), then a 1-rem exposure increases the cancer-contracting probability by one-hundredth of this, or 0.018 percent. By extension, an additional 1-mrem exposure is linearly hypothesized to increase the probability of contracting cancer by 0.000018 percent. There are many scientists who insist that the linear hypothesis is a gross overestimate. They point out that there must be a "threshold" value of effective dose below which there will be no long-term effect whatsoever. To reinforce this suggestion, they note that there are densely populated areas of the world where the natural background radiation averages more than 20 times greater than the United States average. In the Kerala region of India and in a portion of Brazil along the Atlantic coast, there are many towns and villages built on monazite sand which is rich in thorium. Surveys showed a mean dose rate of 2164 mrem/yr in one town in India and dose rates up to 2 mrem/h (17.5 rem/yr) in a town in Brazil. Studies of these populations have revealed no unusual effects.

The possibility of a threshold seems to have some merit. After all, tissue which has been slightly damaged will most often heal itself if it has enough time. If the damage is due to low doses of ionizing radiation, the time will be available. In any event, it is irrelevant in the United States whether or not a threshold exists. Although it is not accepted by international standard-setting institutions, the linear hypothesis for low radiation levels is used by the U.S. National Council on Radiation Protection, which sets the standards for maximum permissible exposure. The United Nations Scientific Committee on the Effects of Atomic Radiation and the International Atomic Energy Agency both accept thresholds.

Finally, it must be noted that there are some beneficial effects of ionizing radiation beyond standard diagnostic x-rays and radiation therapy. Three years after the discovery of x-rays, a radiation biology experiment with algae found that algae which had been x-irradiated grew faster than unirradiated controls. Stimulated growth in trees was noted in 1908, then increased life span in invertebrates (1918) and insects (1919). Increased life span was shown to be the result of irradiation of houseflys, rats, dogs, cats, and even human beings. Hiroshima and Nagasaki survivors who received doses between 10 and 120 rem not only have a longer life span than those who received greater doses but also than those who received lower doses as well as those who received no bomb radiation at all.[1]

32 Ionizing Radiation, Safety, and Nuclear Medicine

By the linear hypothesis, the nuclear release at the Three Mile Island power plant would generate less than one excess cancer case among the exposed population over their life spans. Advocates of extreme positions argue the number from zero to 10. What is clear is that we can never know the actual number due to that accident, since the exposed population statistically will develop over 320,000 other cases of cancer over their life spans.

32-7 Radiation Safety and Standards for Protection

When x-rays and radioactivity were discovered, it was not known that they could produce harmful biological effects. Within weeks of Roentgen's announcement, a manufacturer of glass tubes in Chicago, Illinois, retooled to make x-ray tubes. In the process of testing, he overexposed the fingers of his left hand, suffering an observable burn. Sometime later, he developed dermatitis, and the fingers of his left hand and eventually the whole hand had to be

---

[1] From T. D. Luckey, *Hormesis with Ionizing Radiation*, CRC Publishing Company, Boca Raton, Fla., 1980.

amputated. Even so, x-ray machines could be built very inexpensively, and many were sold and, subsequently, used by physicians indiscriminately. Unfortunately, these machines had essentially no shielding; x-rays were given off in all directions, and no one had any idea of limiting exposure levels and times. As a result, many doctors (and their patients as well) were overexposed to x-radiation, many developing leukemia and other cancers which led to early deaths.

The overuse of x-rays finally prompted the formation in 1921 of professional groups to establish recommended limits of radiation exposure. Original limits were outrageously high by today's standards, but they were set to be a small fraction of the dose which would give a known biological response — reddening of the skin. As more and more was learned, the limits have been reduced and controls made more strict. At present in the United States, limits are enforced by the U.S. Nuclear Regulatory Commission and are promulgated in the *Code of Federal Regulations* dealing with energy. Table 32-4 gives the permissible effective doses from those regulations. You might note that these limits are the **maximum permissible effective doses** from artificial radiation and are a tiny fraction of the *smallest* dose *known* to cause health problems. Virtually no one receives the maximum dose.

Many people work in occupations which require that they be routinely exposed to ionizing radiation, including x-ray technicians, radiologists, nuclear medicine employees, and nuclear reactor personnel. To ensure that radiation standards are maintained, the dose these people receive is monitored, usually by a **film badge** (Figure 32-3), a small piece of radiation-sensitive film wrapped in a light-tight cover. The badge is periodically developed in order to determine the accumulated exposure for that period from the darkness of the developed film. Badges are supplied and developed for a fee by commercial film badge services which duly report the exposures. These records are then kept by Radiation Safety Officers at the occupation site, and total accumulated doses may be obtained by qualified authorities at any time. Pocket dosimeters are also available and used routinely in these occupations. Even visitors to the sites are sometimes required to wear film badges and/or clip-on dosimeters. Normally, these sites also have a number of survey meters and radiation monitors in operation to detect the various types of radiation, including neutrons. Studies

32-7 Radiation Safety and Standards for Protection

TABLE 32-4 U.S. Government Maximum Above-Background-Radiation Dose Limits

Occupational	
Whole body, head and trunk, active blood-forming organs, lens of the eye, or gonads	5 rem/yr or 3 rem/calendar quarter
Retrospective limit	10–15 rem/yr
Hands and forearms, feet and ankles	$18\frac{3}{4}$ rem/calendar quarter
Skin of body only	$7\frac{1}{2}$ rem/calendar quarter
Long-term accumulation to age N years	$(N - 18) \times 5$ rem
Nonoccupationally Exposed (General Population)	
Population average	0.17 rem/yr
Individual	0.5 rem/yr
Students	0.1 rem/yr
Family of radioactive therapy patients:	
Under age 45	0.5 rem/yr
Over age 45	5 rem/yr

are conducted very often with these workers to determine if there are any statistically significant health effects resulting from their exposures.

Three important parameters may be varied to *minimize* the amount of radiation absorbed: time of exposure, distance from the source, and shielding. These three are the basis of any radiation safety effort. Since the total radiation dose received is directly proportional to the time of exposure, keeping that time as short as possible is very important in minimizing the effective dose.

A physically small source of unshielded radiation generally radiates equally in all directions. Thus, in a vacuum, at a distance r from the source, the radiation is spread over a spherical area equal to $4\pi r^2$. You can see that if the distance is doubled to $2r$, the same radiation is spread over an area four times as large $[4\pi(2r)^2 = 16\pi r^2]$. Therefore, a small area on a sphere of radius $2r$ is exposed to only one-fourth the radiation to which it would be exposed on a sphere of radius r. Thus exposure decreases as the reciprocal of the square of the distance from a point source of radiation. This is known as the "one over r-squared" law and is applicable to all point sources from which the radiation spreads out in all directions without being absorbed. From it we conclude that we can minimize exposures by keeping as much distance as possible between ourselves and radioactive sources.[1]

If one cannot decrease the time of exposure or increase the distance from the source, the absorbed radiation may be reduced by providing **shielding,** either around the source or around the exposed people. For alpha particles from a source of radioactive isotopes, the shielding may be a sheet of paper or several centimeters of air, since alpha particles are not very penetrating. Naturally occurring beta radiation is also not very penetrating (though more so than alpha) and may be stopped by a 1-mm thickness of aluminum. Gamma radiation is the most penetrating of that coming from radioactive isotopes and may require thicknesses of several inches of lead or several feet of water to absorb it and act as a shield. Where appropriate, sheets of lead or lead bricks are often used in this capacity.

In addition to time and distance, the dose from an unshielded radiation source depends on the activity of the source as well as the type of radiation and its energy. Since alpha and beta particles are stopped in the skin, they are hazardous only if their sources are ingested or inhaled. Gamma sources are often used in radiation therapy, because their greater penetrating power makes them able to penetrate deeply into the body where a tumor might be growing. For use in determining dose rates from exposures to different isotopes, a **specific equivalent dose rate,** labeled the Rhm, has been defined and measured for a number of commonly used isotopes. The Rhm for a particular isotope is the absorbed dose rate in rads per hour at a distance of one meter from a one-curie source. The SI unit of the Rhm is $(\text{rad/h})(\text{m}^2/\text{Ci})$. Thus, the dose rate in rads per hour received by a person at a distance r from an isotope of activity A is

$$\text{Dose rate} = \text{Rhm}\left(\frac{A}{r^2}\right)$$

Table 32-5 gives Rhm values for several commonly used isotopes.

---

[1] There is one caution! This does not apply to particle beams or scattered radiation from beams or x-ray tubes, since in those cases the radiation does not spread out equally in all directions.

TABLE 32-5 Specific Equivalent Dose Rate (Rhm) for Selected Isotopes

Isotope	Rhm Value rad·m²/(h·Ci)
Cobalt 60	1.3
Technetium 99m	0.059
Iodine 131	0.21
Cesium 137	0.32
Radium 226	0.83

Example 32-3 (a) What is the rate of radiation absorbed by a cancer patient placed for radiation therapy at 0.800 m from a 600-Ci cesium 137 source? (b) How long must the patient remain in position to receive a total dose of 75 rad?

Solution (a) This requires applying the definition of the Rhm value, using the value for cesium 137 given in Table 32-5.

$$\text{Dose rate} = \text{Rhm}\left(\frac{A}{r^2}\right)$$

$$= \left(0.32 \, \frac{\text{rad} \cdot \text{m}^2}{\text{h} \cdot \text{Ci}}\right)\left[\frac{600 \text{ Ci}}{(0.800)^2 \text{ m}^2}\right]$$

$$= 300 \text{ rad/h} = 5.00 \text{ rad/min}$$

(b) The time required is simply the total dose desired divided by the dose rate.

$$\text{Time} = \frac{75 \text{ rad}}{5 \text{ rad/min}} = 15 \text{ min}$$

32-8 Radiation in Medicine

In addition to the x-ray films (sometimes called radiographs) used in medical and dental diagnosis, there are many other medical applications of ionizing radiation for both diagnosis and therapy. These applications have become so commonplace that most hospitals in the United States have a separate department of nuclear medicine.

Most radiodiagnostic procedures employ a chemical compound which, when ingested or given intravenously, concentrates in the area or organ of interest. The compound is prepared such that one of its ordinary atoms is replaced by an appropriate radioactive isotope, and the compound is then said to be radioactively labeled. After the compound has been administered, the concentration in the organ of interest is determined by a technique called **imaging**. Either a radiation-scanning scintillation detector or a gamma camera is used. The scanning device, shown schematically in Figure 32-4, is a typical scintillation detector employing a scintillation crystal optically coupled to a photomultiplier tube. The lead collimator has long tapered holes which allow only those gammas emitted from a small area to enter the crystal. The detector is then moved uniformly in a series of passes (scanned) over the area of interest, and the intensity of the radiation from each area is recorded on film or video tape.

Figure 32-5 depicts a **gamma camera** which has a large time advantage over a scanning device and is in very widespread use. The gamma camera has a very large area scintillating crystal with a flat lead collimator which has as many as 10,000 holes. Only the gamma rays emitted perpendicular to the collimator and headed toward the holes will enter the crystal to be detected. The crystal is optically coupled to a selected array of photomultiplier tubes which detect the light generated in the crystal and convert it to an electric pulse. For a particular gamma ray, the amount of light entering each photomultiplier tube, and hence the size of the pulse that tube generates, depend on the tube's distance from the

Figure 32-4 A scanning scintillation detector has a lead collimator with numerous tubes to allow gamma rays coming from a particular zone within the patient to penetrate to a scintillation crystal. The apparatus scans across the patient to sample the radiation from a series of strips which are merged to form a picture of radiation intensity vs. position.

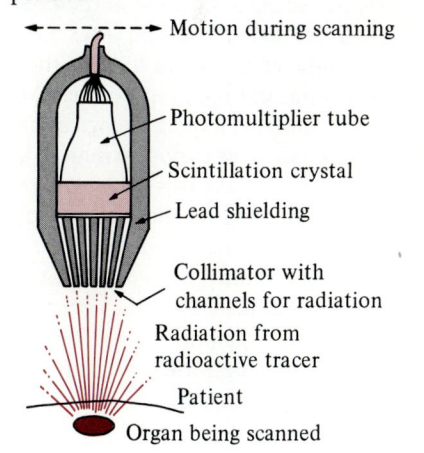

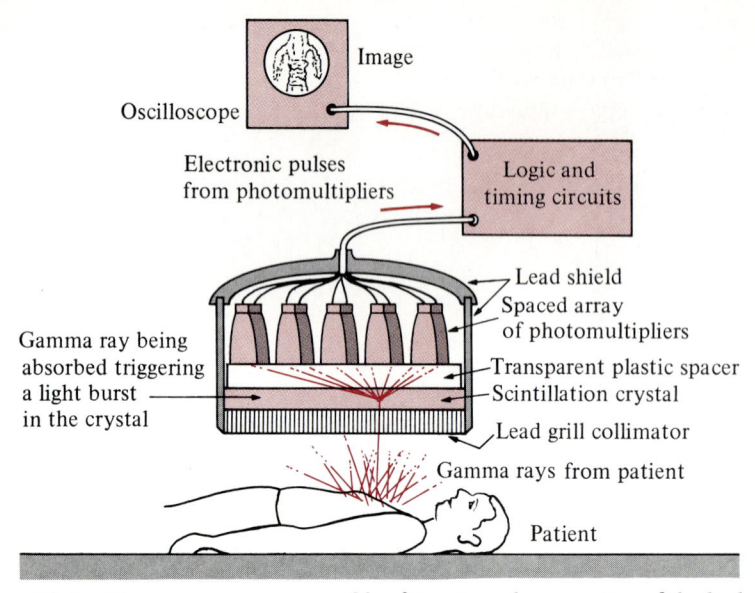

32 Ionizing Radiation, Safety, and Nuclear Medicine

Figure 32-5 The gamma camera, capable of imaging a large portion of the body at once, uses a lead collimator plate with thousands of vertical holes to pass gamma rays.

point where the gamma ray strikes the crystal. Using the relative pulse sizes from all the photomultiplier tubes in the array, electronic circuitry (actually a microcomputer) identifies the position of the point and positions and triggers the oscilloscope beam. As a result, the oscilloscope screen displays a two-dimensional array of dots whose intensity is proportional to the level of radioactivity at the corresponding point in the organ. A photograph of the screen can then be analyzed at any time. Because a gamma camera detects and records all the gammas from a large area at the same time, examinations may be completed in a minute or so. Some scanning devices may take as much as an hour. Figure 32-6 shows the image result of using a gamma camera for medical diagnosis or training.

Table 32-6 gives several isotopes which are in routine use for diagnostics. They are selected because of their gamma energies as well as their physical or **biological half-lives.** The latter is the time required for half the radioactivity of an administered isotope to clear the system either by decay or by elimination through natural bodily processes. Many possible radioactive isotopes are not used for imaging if their biological half-life is so long that their use can contribute more radiation exposure than is deemed desirable for the diagnosis. Further, except for special cases, isotopes which emit beta particles as well as gamma rays are not used, as the beta particles add to the radiation absorbed without contributing to the diagnosis.

Technetium 99m and indium 113m are simply excited states of their daughter products. Thus they decay to a lower energy state of the same isotope by emitting a gamma ray. (The *m* identifies a so-called metastable excited nuclear state.) Because most of these isotopes have such short half-lives, they must be produced and delivered to the hospital very shortly before they are to be used. Hospitals in many areas receive their daily shipment of radioisotopes

Figure 32-6 Images resulting from a gamma camera. *(Photo Researchers)*

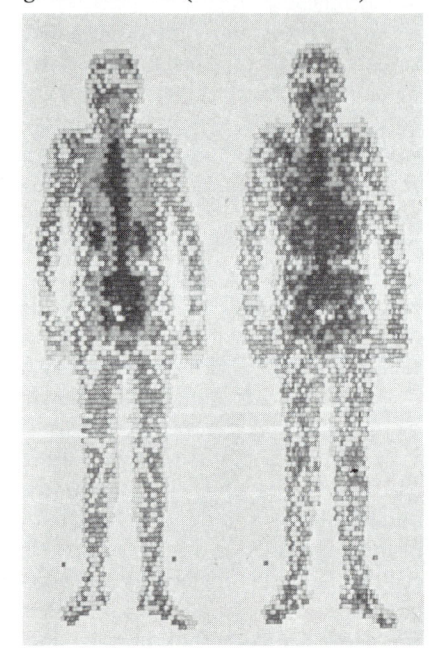

32-8 Radiation in Medicine

TABLE 32-6 Some Radioisotopes Used in Diagnostic Medicine

Isotope	Half-Life	Biological Half-Life	Imaging Use
^{18}F (fluorine 18)	110 min	110 min	Bone
^{67}Ga (gallium 67)	78 h	46.5 h	Lymphomas and malignant tumors
^{99m}Tc (technetium 99m)	6 h	5.92 h	Brain, thyroid, kidney, liver, spleen, heart, blood flow
^{113m}In (indium 113m)	1.7 h	1.45 h	Liver, lung, brain blood pool
^{125}I (iodine 125)	1.8 h	1.8 h	Thyroid
^{131}I (iodine 131)	8.1 days	7.65 days	Thyroid imaging and treatment, blood flow, kidney, liver
^{198}Au (gold 198)	2.7 days	2.67 days	Liver structure

in the very early hours of the morning. Many of these isotopes are the product nucleus of some nuclear reaction. For instance, gold 198 is produced when ordinary gold 197 absorbs a neutron. This is done on a large scale in the neutron flux of a commercial production nuclear reactor.

Radiation in therapy is used primarily to kill cancerous tissue. As an example, iodine 131 is used for thyroid imaging and treatment. Each iodine 131 atom emits both a beta and a gamma radiation, and when taken up by the thyroid, it concentrates in any cancerous tissue there. The betas, therefore, selectively kill the cancerous cells. Because they are growing and dividing rapidly as compared to most normal cells, cancer cells are much more vulnerable to being killed by radiation. One way which has been used to destroy tumors is to implant some radioactive isotope directly in them. A favorite used to be radium 226 encapsulated in gold needles. Various types of "seeds," wires, or ribbons containing such isotopes as cobalt 60 and iridium 192 have also been used.

The therapy radiation may also be administered externally by employing the gamma rays from very intense, very well shielded sources of cobalt 60 or cesium 137 (Figure 32-7). With recent advances in technology, x-ray machines are now available which produce beams with energies up to 4 MeV. These are higher energy than even the gammas from cobalt 60 (1.17 and 1.33 MeV) and can thus penetrate more deeply. In recent years, beams of high-energy electrons generated in small accelerators have also been used for external radiation therapy. These are very efficient if they penetrate deeply enough, since they can be focused to a very tight beam to exactly the small area desired with minimal damage to surrounding tissue. In use, such devices are rotated so that the beam passes through the tumor at all times, but surrounding healthy tissues receive a much smaller, partial dose. At present, medical specialists are experimenting with high-energy beams of alpha particles and other heavy ions. These particles deposit most of their energies in a small distance near the ends of their ranges. Hence they should be the most efficient of all in destroying malignancies with minimal damage to healthy tissue. They may never come into common usage, however, because of the very high cost of heavy-ion accelerators. It is ironic that cancer which may be caused by ionizing radiation is also treated by ionizing radiation.

Figure 32-7 A cobalt 60 gamma ray teletherapy unit being prepared for use. (*M. Rotker/Taurus Photos*)

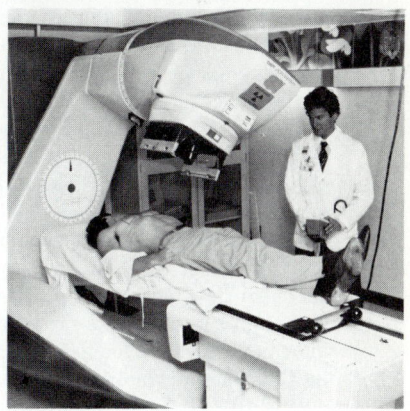

Minimum Learning Objectives

After studying this chapter, you should be able to:
1. Define:
   - absorbed dose
   - activity
   - artificial background radiation
   - background radiation
   - becquerel (Bq) [unit]
   - biological half-life
   - coincidence circuit
   - cosmic ray
   - curie (Ci) [non-SI unit]
   - dosimeter
   - effective dose (dose equivalent)
   - exposure
   - film badge
   - gamma camera
   - gray (Gy) [unit]
   - imaging
   - linear hypothesis
   - maximum permissible effective dose
   - muon
   - occupational dose limit
   - primary cosmic radiation
   - quality factor (QF)
   - rad [non-SI unit]
   - rem [non-SI unit]
   - roentgen (R) [non-SI unit]
   - secondaries
   - shielding
   - sievert (Sv) [unit]
   - specific equivalent dose rate

2. Give the approximate effective dose of background radiation exposure and roughly apportion it to its proper sources.
3. Calculate the mass of a sample from its activity and its known half-life or decay constant.
4. Distinguish the concepts of activity, exposure, absorbed dose, and dose equivalent, giving approximate units in which each is measured.
5. Calculate the absorbed dose and the dose equivalent from the energy deposited by ionizing radiation in a given mass of tissue and a knowledge of the type of radiation.
6. Summarize the evidence for a genetic effect and a cancer-inducing effect of ionizing radiation on human beings.
7. Describe strategies to minimize exposure from a radioactive source.
8. Calculate the dose rate for a given isotope of known Rhm and activity at a given distance from the isotope.
9. Describe the use of ionizing radiation in medical diagnosis and therapy.

Problems

32-1 If a radiation detector which counts 50 percent of the radiation from a certain sample reads 782 counts in 10 s, determine the activity of the sample in (a) becquerels, (b) curies, (c) microcuries, and (d) picocuries.

32-2 A G-M detector which is 12 percent efficient in counting radiation from a ^{99m}Tc sample registers 1.42×10^4 counts/min. Determine the activity of the sample in (a) becquerels, (b) curies, and (c) microcuries.

32-3 A sample of radium 226 is counted by a 40 percent efficient detector and gives rise to 4.18×10^4 counts/min. Determine (a) the activity of the source in microcuries and (b) the number of micrograms of radium present.

32-4 Determine the activity in becquerels and curies of a 1.00-g sample of ^{210}Po, which has a half-life of 138 days.

32-5 A 2.00-μCi cobalt 60 source is to be detected by a scintillation counter of 30 percent efficiency. Calculate (a) the activity of the source in becquerels and (b) the number of counts per minute which should be expected to register on the detector.

32-6 A diagnostic x-ray exposes a patient's broken leg to 4.50 rad. (a) What is the effective dose to the exposed tissue? (b) How much energy is absorbed per kilogram of tissue? (c) How many x-ray photons are absorbed per kilogram if each photon is 70.0 keV?

32-7 In a medical diagnostic procedure, a patient absorbs 6.80×10^{-3} Gy in one forearm and hand. A mass of 2.50 kg received radiation. Calculate (a) the total amount of energy absorbed, (b) the absorbed dose in rads, and (c) the effective dose in rems and mrems if the radiation was due to low-energy beta particles.

32-8 Calculate the absorbed dose of radiation necessary to raise the temperature of water by 0.500 K.

32-9 How large a radiation dose must be deposited in lead to increase the temperature by 1°C? [Specific heat of lead = 128 J/(kg·K).]

32-10 A cancer patient receives an exposure of 50.0 rem to the head during treatment of a tumor. The head has a mass of 3.80 kg. Calculate (a) the total energy absorbed by the head and (b) the increase in the temperature of the head due to the exposure. [Assume that the specific heat of the head is about the same as that of water, 4.18×10^3 J/(kg·K).]

32-11 Calculate the energy required to produce an ion pair in dry air. Use the definition of roentgen and rad.

32-12 There are approximately 5.00×10^3 kg of uranium and 8.00×10^3 kg of thorium in the first meter of the earth's crust for every square kilometer on the earth's surface. Neglect the contribution due to ^{235}U, and calculate the activity in microcuries of a 1-km^2 area due to just the two isotopes ^{238}U and ^{232}Th.

32-13 Calculate the exposure and the exposure rate for an x-ray beam which produces 2.80×10^{10} ion pairs in 5.00 s in 5.16×10^{-4} kg of dry air. ($\rho_{air} = 1.29$ kg/m^3.)

32-14 A nuclear physics professor's film badge indicates an exposure of 42.0 mR over a 1-month period. What is the equivalent dose in mrems if the radiation was (a) low-energy (<30 keV) beta particles, (b) slow neutrons, and (c) alpha particles?

32-15 Determine the absorbed dose rate for a layer of air 1.40 m thick from a low-energy x-ray beam. The intensity of the x-rays is 8.40×10^{-3} W/m². ($\rho_{air} = 1.29$ kg/m³.)

32-16 In a radiology laboratory, an ionization chamber containing 8.25×10^{-5} kg of dry air is used as a monitor. If it shows a current of 1.20×10^{-10} A when the x-ray machine is on, and an exposure requires 4.00 s, determine the absorbed dose by someone in the room.

32-17 Airline passengers receive an annual excess dose of approximately 330 million rem due to high-altitude flying. Using the linear hypothesis, how many excess cancers may we expect this to induce among the flying population?

32-18 Airlines used to have a rule requiring stewardesses to be unmarried (thus presumably not pregnant), and they lost their in-flight jobs when wed. Women's groups successfully argued that this was an example of sexual discrimination. Actually, there was a scientific rationale behind the rule. Occupational radiation standards limit exposure for a developing fetus to 0.500 rem for the entire 9 months of gestation. The background radiation at 30,000 ft in an airplane averages 4000 mrem/yr, but at 60,000 ft this rises to 13 rem/yr. (a) How many hours would have to be spent flying at 60,000 ft to absorb a 0.500-rem dose? (b) If the stewardess works 6 h/day at this altitude, how many flying days would be required to absorb the maximum permissible dose (MPD)? (c) Assuming she becomes aware of her pregnancy 2 months after conception, is it likely for the fetus to have absorbed the MPD in this time? (d) During periods after a solar flare, the background radiation at high altitudes can rise to over 1000 rem/yr and remain there for several hours. How many hours of flying under such condition will exceed the MPD for the fetus? (e) Under conditions with frequent solar flares, is it probable that a stewardess could exceed the MPD before she knew she was pregnant?

32-19 A 200-Ci cobalt 60 source is to provide a dose rate of 5.00 Gy/h to a cancer patient. Calculate (a) the time necessary for an 80.0-rad exposure and (b) the distance the source should be from the patient.

32-20 Calculate the radiation exposure rate at a distance of (a) 0.500 m, (b) 1.00 m, and (c) 100 m from a 2.00-Ci source of radium 226.

32-21 Absorption of gamma rays passing through matter is a statistical process, much like radioactive decay itself and mathematically similar in its treatment. If an absorber of a given thickness reduces gamma intensity by one-half, another second layer of the same thickness would reduce it to one-quarter its original intensity. Analogous to the half-life of decay, the thickness $X_{1/2}$ would be called a half-value layer of absorber. The relationship between incident radiation intensity on an absorber I_0 and the intensity emerging unabsorbed I is

$$I = I_0 e^{-\mu x}$$

where X is the thickness of the absorber and μ is called the attenuation coefficient of the material. (a) Gamma intensity is measured at 18,000 s⁻¹ without an absorber. When a 2.00-mm-thick lead absorber is placed between counter and source, the count drops to 7000 s⁻¹. What is the attenuation coefficient for lead for gamma rays of this energy? (b) What is the half-value-layer thickness for the lead? (c) What would the intensity become with a 1.00-cm lead absorber in place?

32-22 How long would it take to get a 50.0-mrad absorbed dose of radiation at 2.00 m away from a 5.00-Ci source of iodine 131?

32-23 It is desired to deliver a dose of 75.0 rad to a radiation therapy patient in a 10.0-min time interval. If a cesium 137 source is to be used and placed 1.40 m from the patient, what is the (a) required activity of the source and (b) the mass of the source?

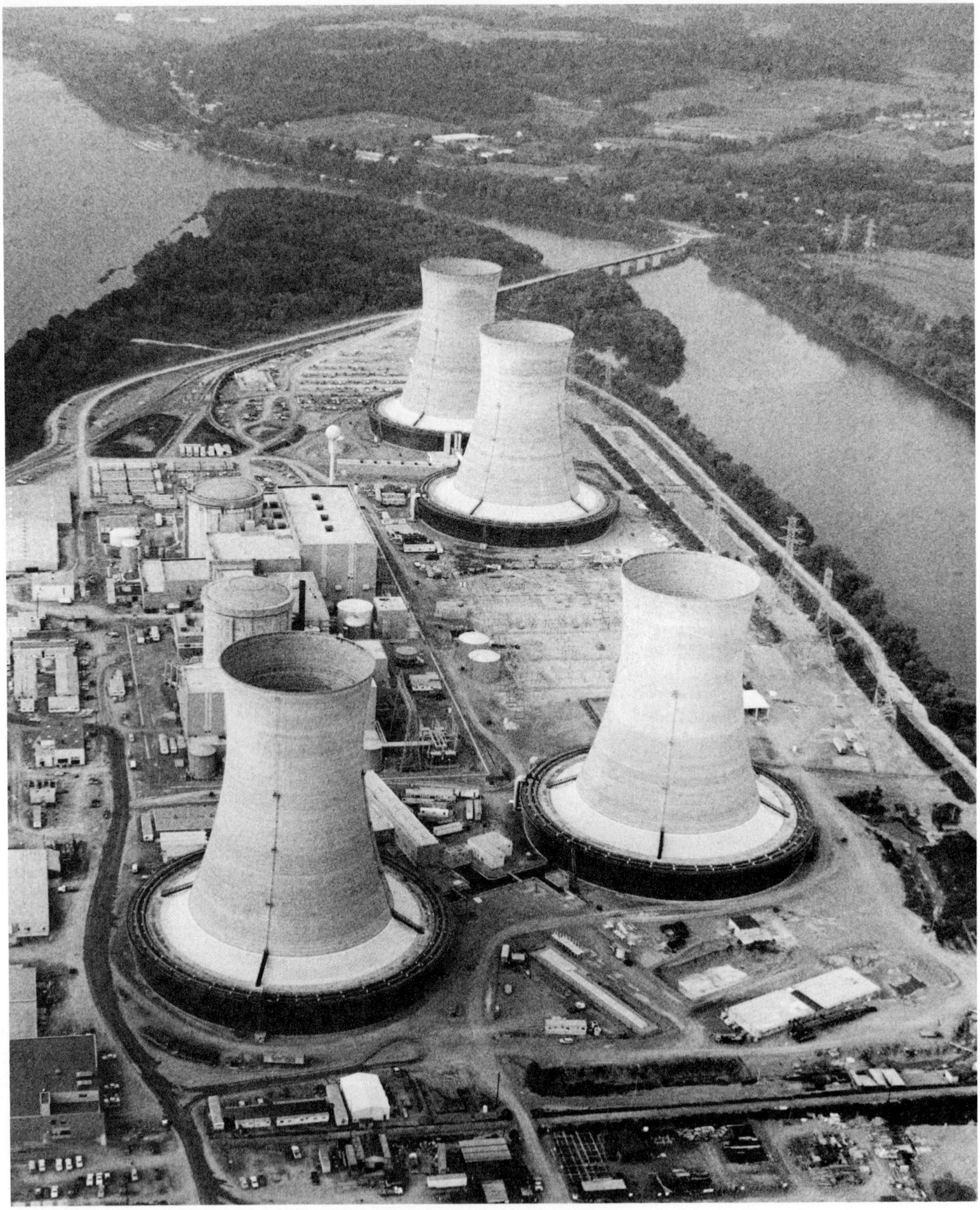

33 Nuclear Fission and Fusion

33-1 Introduction

Many of the ideas of modern physics considered in the last six chapters have already been adapted for applications by present-day technology. A prime example is the use of semiconductors to fabricate diodes, transistors, and whole integrated circuits. Their expanded application in electronic equipment has had a very important effect on our daily lives. In some ways, the discovery of nuclear fission and fusion and their subsequent development have had an even more profound impact. The possession of nuclear weapons heightens tension among nations and even threatens the future of life as we know it. The use of nuclear power to generate electricity, in nuclear reactors, has both proponents and opponents. Scientific data are not always easily interpreted; hence, factual and speculative assertions are often confused. Under these circumstances, reason often gives way to emotion. To avoid this trap, it requires knowledgeable individuals to make informed decisions. Whether you approve or disapprove of any technological development, your position should be based firmly on fact. In this chapter, we consider nuclear energy and its applications.

33-2 The Discovery of Nuclear Fission

After artificial radioactivity was discovered by bombarding target nuclei with alpha particles, many groups of scientists began to conduct similar experiments. They would bombard a target with alpha particles and then test the target for radioactivity after the alpha source was removed. Fermi and his students at the University of Rome tried the same kind of experiment with neutrons as the projectiles rather than alpha particles. Fermi believed it should be advantageous to use neutrons in spite of the fact that much larger sources of alpha particles were available. He reasoned, quite correctly, that the absence of charge on the neutron would allow it to penetrate to the nucleus without being repelled by the Coulomb force as were most alpha particles. Fermi's group used neutrons produced by alpha particles striking beryllium in the reaction

$$^{9}_{4}\text{Be} + ^{4}_{2}\text{He} \rightarrow ^{12}_{6}\text{C} + ^{1}_{0}\text{n}$$

They planned to bombard many of the elements in order, beginning with hydrogen. Their first observable success in inducing radioactivity occurred when fluorine was the target. To prove it, they had to develop a new technique to determine what element was responsible for the radioactivity. Ordinary chemical separation was unsuccessful because not enough radioactive atoms were produced in an activated sample. However, they suspected that some element close in the periodic table to their target nuclei was responsible for the radioactivity. Thus they would dissolve the activated sample in acid, add small amounts of neighboring elements, and selectively precipitate the elements including the target out of solution. They then checked the separated precipitates with a detector such as a G-M counter to determine which element the radioactivity followed. For instance, when iron ($Z = 26$) was the target and became radioactive, they dissolved the activated iron in nitric acid and added small amounts of chromium ($Z = 24$), manganese ($Z = 25$), and cobalt ($Z = 27$). Upon precipitation, they found that the radioactivity followed the manganese. They eventually concluded that they had caused the reaction

$$^{56}_{26}\text{Fe} + ^{1}_{0}\text{n} \rightarrow ^{56}_{25}\text{Mn} + ^{1}_{1}\text{H}$$

The resulting isotope of manganese then decays by beta emission as

$$^{56}_{25}\text{Mn} \rightarrow ^{56}_{26}\text{Fe} + ^{0}_{-1}e + \overline{\nu}_e$$

This identification technique was probably an important part of why Fermi was awarded the Nobel Prize in physics in 1938.

As their experiments continued up the periodic table, Fermi's group eventually reached the element with the largest known atomic number, uranium ($Z = 92$), which also showed increased radioactivity upon neutron bombardment. They were unable to identify the element responsible by their chemical separation method, and Fermi suggested that perhaps it was a "new" element of $Z = 93$. This could result if uranium 238 absorbed a neutron, becoming uranium 239, and then emitted a beta particle. Fermi's group also discovered that the induced activity for many target elements was greater if the neutrons were slowed before bombarding the target. Fermi concluded that a slow-moving neutron spends more time in the immediate vicinity of a target nucleus and thus has a greater probability of being captured.

Lise Meitner (1878–1968), an Austrian physicist, and Otto Hahn (1879–1968), a German chemist, duplicated the neutron activation experiments of Fermi and eliminated all the elements near uranium as responsible for the large

---

When exposed to neutrons, many types of nuclei may be activated, that is, absorb the neutron to become some new, radioactive isotope. This new isotope will decay in a known way with a given energy. Thus, **neutron activation analysis** is a powerful tool. There is no chemical method to detect even a million atoms of one type in a sample. However, the precise energy of a radioactive emission can record the disintegration of a single atom of a particular type, making very fine traces of an element detectable.

increase in radiation.[1] They tentatively reached the same conclusion as Fermi that the element $Z = 93$ was produced. After Meitner fled Germany for Stockholm to avoid Hitler and the Nazis, Hahn wrote her a letter stating that he and Fritz Strassmann, another German scientist, had tried lighter and lighter elements in their chemical separation with the uranium-activated target. They discovered that at least part of the increased radioactivity was due to an isotope of barium. The conclusion that barium ($Z = 56$) was produced by bombarding uranium ($Z = 92$) with neutrons had been so startling to Hahn and Strassmann that they performed the experiment and chemical separations again and again to rule out any errors. After considering all possibilities, Meitner and her nephew, Otto Frisch (b. 1904), also an Austrian physicist, finally concluded that the process involved was not a small chipping of the uranium nucleus but a division into two separate, fairly large nuclei. It could best be explained by using Bohr's idea that the nucleus was somewhat like a liquid drop which elongated and divided in pieces after absorbing a neutron (Figure 33-1). On the basis of the calculated Coulomb repulsion of the residual fragments, Meitner and Frisch even calculated that approximately 200 MeV of energy should be released in the process.

Hahn and Strassmann published their results soon after the letter to Meitner. In their publication, they wrote of the "bursting" of uranium. This dividing, or splitting, process is now called nuclear **fission**. They also speculated on what the other product of this splitting of uranium might be. By application of conservation of charge, we conclude that it must be krypton ($Z = 36$), as shown in Figure 33-1.

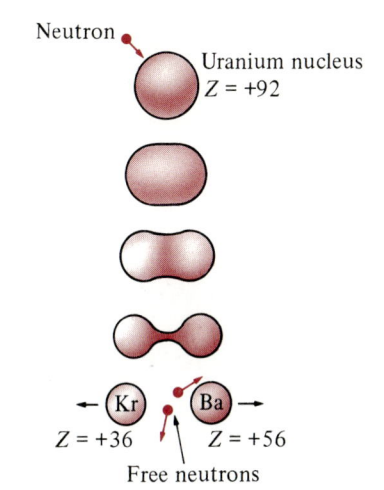

Figure 33-1 The **liquid-drop model** of the nucleus treats the nucleus as analogous to a deformable drop of a liquid. Neutron absorption by a U-235 nucleus (but not a U-238 nucleus) causes the nucleus to elongate, then separate into two more or less equal-sized pieces (**nuclear fission**). The two new nuclei together must have the same total number of protons as the original nucleus, but other than that, the new nuclei are not uniquely specified in the model.

Example 33-1 Assume that in a fission of uranium the isotopes of krypton ($Z = 36$) and barium ($Z = 56$) are fully formed at a separation of 1.45×10^{-14} m between their centers. Calculate their mutual Coulomb potential energy (in megaelectronvolts) with respect to infinite separation and hence the kinetic energy they will expend in stopping.

Solution We must apply the relationship to obtain the Coulomb potential energy, recognizing that the charges are $q_1 = +36(1.60 \times 10^{-19}$ C) and $q_2 = +56(1.60 \times 10^{-19}$ C). This answer in joules is then converted to electronvolts by dividing it by 1.60×10^{-19} J/eV.

$$PE = \frac{kq_1q_2}{r}$$

$$= \frac{\left(9.00 \times 10^9 \frac{\text{N} \cdot \text{m}^2}{\text{C}^2}\right)(+36)(+56)(1.60 \times 10^{-19} \text{ C})^2}{1.45 \times 10^{-14} \text{ m}}$$

$$= 3.20 \times 10^{-11} \text{ J}$$

Thus

$$PE = \frac{3.20 \times 10^{-11} \text{ J}}{1.60 \times 10^{-19} \text{ J/eV}} = 2.00 \times 10^8 \text{ eV} = 200 \text{ MeV}$$

Apparently, Meitner and Frisch made a similar calculation.

---

[1] This was something Fermi was unable to do as he had no sample of protactinium ($Z = 91$). Meitner and Hahn had discovered protactinium and, of course, had samples available.

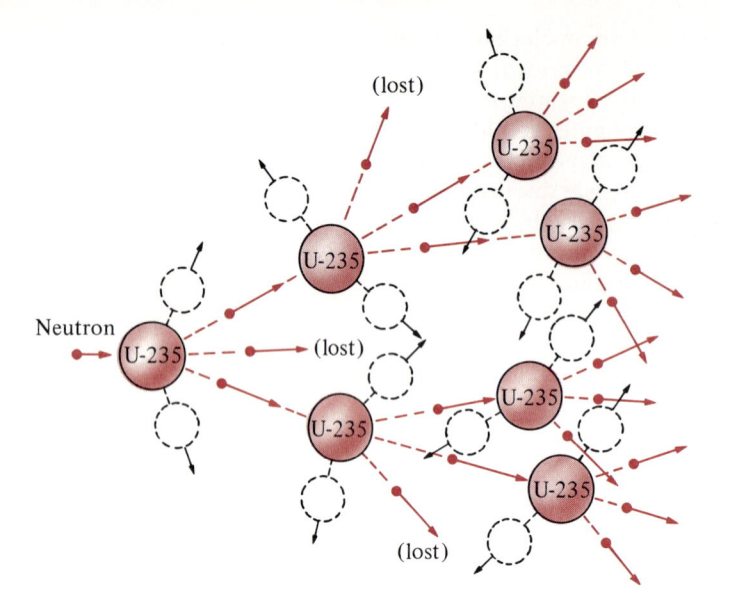

Figure 33-2 A chain reaction is possible whenever one product of a reaction is the same as the particle which began the reaction. Here, a neutron encounters a U-235 nucleus, is absorbed, and causes fission. The several free neutrons liberated can encounter other U-235 nuclei, causing further fissions and releasing still more neutrons, and the process repeats. If more than one neutron per fission triggers another fission, reaction rate increases — dramatically so in the case of a nuclear bomb.

33-3 Nuclear Fission and Its Applications

Frisch passed this idea of nuclear fission along to Niels Bohr, who was just leaving Copenhagen for Princeton. Bohr shared the news with Fermi on arriving in the United States. In 1938, Fermi had moved permanently to America with his family after being allowed to leave Mussolini's Italy to accept the Nobel Prize in physics. At Columbia University, Fermi and a graduate student, H. L. Anderson (b. 1914), began experiments to determine the energy release in the fission of uranium. At a conference on theoretical physics in Washington, D.C., Fermi mentioned the possibility that neutrons might be emitted during the splitting of uranium since the fission product isotopes should have a large excess of neutrons relative to their stable counterparts. This raised the possibility of a **chain reaction,** in which one or more neutrons from a fission might cause other fissions to occur, which, in turn, may release neutrons, causing yet more fissions, and so on (Figure 33-2).

Example 33-2 How many **excess neutrons** are in the fission products when a uranium 235 nucleus absorbs a neutron and fissions to cesium and rubidium? The stable isotopes of these elements are cesium 133 and rubidium 85.

Solution Since the protons have already been taken into account, we need only subtract the sum of the mass numbers of the stable isotopes (133 + 85) from the total number of nucleons of the reactants (1 + 235). Thus,

$$\text{Neutron excess} = 236 - (133 + 85) = 18 \text{ neutrons}$$

Most of these are transformed to protons during beta decays, but a few are released as neutrons.

33-3 Nuclear Fission and Its Applications

The chain reaction possibility created a great deal of excitement at the conference when the amount of energy available was considered. If the figures were correct, 1 metric ton (1000 kg) of uranium could possibly supply more energy than would be used by a city of a million people in a whole year! Fermi immediately returned to Columbia to see if he could determine whether or not neutrons are ejected during a fission. He designed an experiment which was successful in showing that they were. Fermi and his group concluded that approximately two neutrons are produced per fission.[1]

The next step was to determine whether a chain reaction is feasible and, if so, to design and construct a demonstration. However, there were still a number of questions to be answered and problems to be solved before considering such a demonstration. The release of neutrons during the fission process was obviously necessary. More to the point, however, is that for a sustained chain reaction, conditions must be such that an average of one neutron from each fission is absorbed and causes an additional fission. The question is "What are the conditions?" Fermi already knew from his earlier experiments that many more fissions resulted whenever the neutrons were slowed before encountering uranium. Also, to this point, natural uranium had been used in all the experimental work on fission. This naturally occurring uranium is composed of approximately 99.3 percent of the ^{238}U isotope and 0.7 percent of the ^{235}U isotope (with a trace of ^{234}U). Bohr pointed out that based on the liquid-drop model of the nucleus, slow neutrons should cause fissions in ^{235}U but *not* in ^{238}U. In short order, two experiments were conducted which confirmed Bohr's hypothesis. Using a mass spectrometer, a fraction of a microgram of naturally occurring uranium was separated into very nearly pure samples of ^{235}U and ^{238}U. Each sample was then irradiated with slow neutrons. Many fissions were seen with the ^{235}U sample as a target and practically none with ^{238}U.

The problems of sustaining a chain reaction were: (1) An average of at least one neutron per fission must cause another fission. (2) Neutrons which leave the uranium without being absorbed or neutrons which are absorbed without causing a fission to occur are lost to the chain. (3) The fission neutrons must first be slowed to increase the probability of a fission capture. (4) The probability of fission with ^{235}U is high for slow neutrons, while the probability with ^{238}U is quite low. Also, both isotopes of uranium may capture a neutron without fissioning. Many physics problems had to be solved to produce a sustained chain reaction. However, it was believed that technological problems might be more difficult, perhaps making it impossible.

As the design for the first nuclear reactor developed, it was planned to embed lumps of uranium in a **moderator,** the material used to slow the neutrons. After a fission has occurred, a neutron leaves the uranium and gives up most of its energy to the moderator nuclei. In its subsequent random motion, the slowed neutron enters a lump of uranium with a good probability of being captured by a ^{235}U nucleus. The idea was to construct a pile of uranium and graphite and make it large enough to sustain a reaction. The larger the pile, the better would be the probability that a neutron would eventually encounter a ^{235}U nucleus and cause a fission before it escaped at the pile's boundary. Because of its geometry, a reactor was at first called an **atomic pile** and often still is.

---

[1] Experimental evidence is now available which shows that the *average* number of neutrons produced with each fission of uranium is more nearly 2.5.

The difficulty and expense involved in separating the isotopes led Fermi to choose natural uranium as the fuel. The moderator had to be a light element in order to cause the neutrons to slow. You will recall that conservation of energy and momentum dictates that a small mass will not lose very much energy during an elastic collision with a large mass. In this regard, the best possible moderator would be hydrogen nuclei (protons) which have nearly the same mass as the neutron. Unfortunately, protons often capture neutrons, creating $^{2}_{1}H$ (deuterium), and those neutrons are lost to the reaction. Other light elements were considered, but those below carbon all have some disadvantages which make them unsuitable as moderators. At that time the reasonable choice was graphite (carbon), because its natural isotopes have a very small probability for neutron capture, and also it is solid, machinable, and plentiful.[1]

Fermi had agreed to head the project to produce a controlled chain reaction, and it was begun at Columbia University but later moved to the University of Chicago. The final design included many detectors and other instruments. Open slots were built into the pile, and cadmium (a good neutron absorber) bars were inserted as **control rods.** (Every nuclear reactor has provision for inserting rods of some material which captures neutrons with a very high probability. After enough fuel and moderator have been assembled to sustain a chain reaction, the reaction can be started by removing the rods partway and is further controlled by proper positioning of the rods.)

On Dec. 2, 1942, the first artificial nuclear reactor, housed in a squash court under the stands of Stagg Field at the University of Chicago, sustained the first chain reaction. This ushered in what has been called the atomic age but might be more accurately named the nuclear age.

A tremendous amount of experimental research on the topic of nuclear fission has been conducted since it was discovered in 1938. Early on, it was discovered that neutrons incident on heavy nuclei did indeed sometimes result in the production of isotopes with atomic numbers greater than that of uranium. These are the so-called **transuranic elements,** usually produced whenever the heavy element absorbs a neutron and then undergoes one or more beta decays, which increase the atomic number of the residual nucleus. An important transuranic element, **plutonium** ($Z = 94$), is the result of uranium 238 absorbing a neutron and decaying by beta emission to neptunium 239 ($Z = 93$), which, in turn, decays by beta emission to become plutonium 239. This and another isotope of plutonium, ^{241}Pu, will undergo fission similar to that of uranium 235. In fact, after a uranium reactor has been in operation for some time, a significant percentage of the fissions is due to plutonium. A fourth isotope which will fission with slow neutrons is uranium 233. It may be produced with neutrons bombarding thorium 232, which then undergoes two beta decays.[2]

---

[1] In the fall of 1939, Leo Szilard (1898–1964), a Hungarian physicist, wrote a letter to President Roosevelt and with Edward Teller (b. 1908) persuaded Albert Einstein to sign it. The letter advised the President of the chain reaction and the possibility of very powerful new bombs. It also notes that Germany had taken control of the Czechoslovakian mines and stopped the sale of uranium. As a result, a meeting of several scientists was called at the Bureau of Standards by its director, and a plea was made for U.S. government involvement. Eventually funds were provided to purchase the pure graphite needed for the moderator for the first nuclear reactor.

[2] Actually, a significant number of other nuclei will fission when they absorb fast neutrons, but their probability for fission is so small that they are unable to sustain a chain reaction. The isotopes ^{233}U, ^{235}U, ^{239}Pu, and ^{241}Pu will fission with either fast or slow neutrons, and these are the nuclei we call **fissionable.**

Although the first identified fission product, or fragment, was barium, it was quickly discovered that fission may occur in many different ways. When Fermi's group first bombarded uranium, they found increased radioactivity with what appeared to be a number of different half-lives. Later it was determined that in addition to producing isotopes of barium and krypton, the uranium nuclei may fission to produce isotopes of lanthanum and bromine or of antimony and niobium. In fact, well over 100 different elements, or fission fragments, have been identified. The reactions of the three examples mentioned are

$$^{235}_{92}\text{U} + ^{1}_{0}\text{n} \rightarrow ^{236}_{92}\text{U} \rightarrow ^{139}_{56}\text{Ba} + ^{95}_{36}\text{Kr} + 2\,^{1}_{0}\text{n}$$

$$^{235}_{92}\text{U} + ^{1}_{0}\text{n} \rightarrow ^{236}_{92}\text{U} \rightarrow ^{149}_{57}\text{La} + ^{85}_{35}\text{Br} + 2\,^{1}_{0}\text{n}$$

$$^{235}_{92}\text{U} + ^{1}_{0}\text{n} \rightarrow ^{236}_{92}\text{U} \rightarrow ^{132}_{51}\text{Sb} + ^{100}_{41}\text{Nb} + 4\,^{1}_{0}\text{n}$$

The way a particular fission occurs is no doubt related to the conservation laws and depends in large part on the energy levels of the temporary ^{236}U nucleus. The mass distribution of fission fragments from a large number of fissions is shown in Figure 33-3. You will note that very few nuclei split into equal parts. The most probable values of A for the fragments are 141 and 95. Virtually all fission fragments are isotopes with many more neutrons than their naturally occurring counterparts and are hence unstable to beta decay. Occasionally after the first beta decay, the daughter emits a neutron. These are called **delayed neutrons** to distinguish them from those released during the fission, which are called **prompt neutrons.** Delayed neutrons comprise about 1 percent of the neutrons in the fission process and make control of the chain reaction much easier.

Reactors release their energies over a relatively long period of time, and it was quickly deduced that constructing a bomb which is to deliver a tremendous amount of energy in a very short time interval is a much different problem. For a bomb, succeeding generations of fissions must occur much more quickly than is required just to sustain a chain reaction. Thus, a bomb requires that the proportion of fissionable material be considerably greater than that in natural uranium. Instead of 0.7 percent ^{235}U and 99.3 percent ^{238}U, a reasonable uranium bomb must be **enriched** so that the proportion of uranium isotopes is over 90 percent ^{235}U and less than 10 percent ^{238}U. Although many techniques to accomplish this enrichment are now available or under development, the earliest successful large-scale method involved a gaseous diffusion cascade. Natural uranium, as the gas uranium hexaflouride, UF$_6$, is used, and molecules containing ^{235}U, having smaller mass, move faster at a given temperature than those which contain ^{238}U. This allows diffusion to separate the isotopes. Uranium for most American reactors, including those in electric power plants, is also enriched but only to about 3.5 percent ^{235}U, which is inadequate to sustain the rapid chain reaction necessary for a bomb.

Nuclear bombs can also be made with plutonium. In fact, the first nuclear weapon exploded in a test was plutonium-based. The bomb at Hiroshima used uranium, while that at Nagasaki used plutonium. Plutonium may be obtained by chemical separation after a quantity of ^{238}U has been irradiated with neutrons in a **production reactor** (a reactor designed to produce transmutation products). The plutonium produced in a **power reactor** (a reactor designed to produce useful power) would not work very well since its isotopic content includes much ^{240}Pu, which is nonfissionable but absorbs neutrons. Thus, a bomb made with plutonium of power-reactor grade would be very inefficient if it exploded at all.

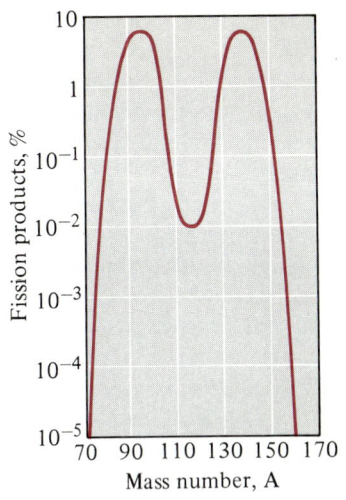

Figure 33-3 The product nuclei from U-235 fission include all nuclei with mass numbers from about 70 to 160. Fragments of equal mass are not favored in the fission but do occur. The logarithmic scale of percentage emphasizes the symmetry of the curve but disguises the fact that the bulk of the product nuclei are found in the twin peaks.

Until recently, the need to produce material enriched in ^{235}U to make a uranium bomb prevented the spread of such weapons. Gaseous diffusion enrichment plants are large, costly, difficult to operate, and more difficult to conceal. The U.S. plant built at Oak Ridge, Tennessee, for this purpose, was the largest industrial installation of any kind to the 1940s. Recent innovations have produced cheaper, smaller isotope-separation techniques. A result is that smaller nations may become capable of producing weapons-grade uranium.

For a nuclear bomb to explode, two pieces of fissionable material, each more than half a **critical mass** (the amount necessary to sustain a chain reaction), must be brought together very rapidly and held for a time interval equal to many generations of fissions. The technological and engineering construction problems for a bomb are enormous, requiring very many skilled and knowledgeable workers. The concern that some mad scientist will build such a bomb in a basement or garage is unrealistic.

Many nuclear reactors are now used for experimental research or isotope production. Both uses take advantage of the large number of neutrons available in a reactor, and many are designed with access to an open space near the center of the fuel or reactor core. In research reactors, the neutrons are used as projectiles to help determine reaction probabilities, yields, and so on. Their primary purpose is to investigate fundamental laws of nature. In production reactors, the neutrons are most often used to produce nonnaturally occurring isotopes of various elements for medicine, research, and education. There are hundreds of research and production reactors in the United States.

Perhaps the most well known applications of nuclear reactors are those used to power oceangoing ships and to generate commercial electric power. These reactors are designed to take advantage of the energy emitted in fission. In the most common types of electric power reactors, this energy is used to produce steam, which then drives a turbine generator. The reactor simply acts as a large source of heat much like the burning of fossil fuels in a coal, oil, or natural gas power plant. After the steam is produced, all these plants generate the same type of electricity in exactly the same way. By 1985, there were over 90 licensed nuclear power electric generating plants operating in the United States and about 200 in the world.

The largest number of power reactors are used in naval vessels: aircraft carriers, cruisers, but mainly submarines. The steam generated is used for propulsion as well as for generating electricity. By 1985, the U.S. Navy had accumulated over 3000 reactor-years of experience with these power sources.

33-4 Electricity Generated from Nuclear Power

Most of the kinetic energy which results from a fission goes to the neutrons and in a reactor is eventually absorbed by the moderator. This additional kinetic energy of the material is really thermal energy and is transferred as heat. In a power reactor, the intent is to use the heat to produce the steam required to drive the machinery involved. Thus the heat must be taken from the core by a **coolant** and delivered to water somewhere. Careful design is obviously necessary, and a judicious choice of materials, especially for the moderator, may offer physical or economic advantages.

In the United States, most electric power reactors use ordinary water (so-called light water) as both a moderator and a medium to deliver heat from the core to wherever it is to be utilized. Since the ^{1}H in the water absorbs some neutrons, natural uranium cannot be used because too many neutrons would be lost. Thus, the uranium in power reactors is usually enriched to about 3.5 percent ^{235}U. We repeat that *at this percentage enrichment, the uranium cannot be made into an explosive device no matter how much of it one has*. Therefore, *a commercial nuclear power reactor cannot become a nuclear bomb under any circumstances*.[1] In Canada, power reactors use heavy water (deuterium oxide)

---

[1] Polls continue to show that despite the nuclear power debate of recent years, the majority of people do not know this fact; hence, simple education demands some emphasis.

as both moderator and primary coolant. **Heavy water** is water in which ^1_1H has been replaced by ^2_1H (deuterium). Because deuterium has a much smaller probability of absorbing neutrons than ordinary hydrogen, fewer neutrons are lost in the moderating process. Thus these reactors, called **CanDU** (acronym for *Canadian-deuterium-uranium*) **reactors,** may use natural uranium as a fuel and do not require expensive enrichment facilities.

Figure 33-4 shows several types of power reactors in use in the United States. Shown also is a fossil-fuel-fired plant for comparison.

Nuclear power generation of electricity is seen by many people as both a supplement and an alternative to fossil-fuel power plants. Enormous amounts of electricity are required to run any industrialized society, and projections for the foreseeable future show that the need for electricity in the United States and the rest of the world will continue to grow. Nuclear power has emerged as a factor in the production of electric power for a number of reasons. Coal, oil, and natural gas are in limited supply, and curbing our present use should make them last for a longer time. Also, fossil fuels have other uses which may be more valuable and for which they will be unavailable if they are burned to produce electricity. Presently, fossil fuels are used for home heating, for vehicular fuel, and as a feedstock for the chemicals necessary for drugs, fertilizers, plastics, lubricants, and so on. Unless sources of abundant, economical substitutes for these fossils are found, it seems wise to limit their use in those areas where substitutes are available. Economics, initially, played a positive role in the growth of nuclear power in the United States and still does elsewhere. The lower cost of fuel for nuclear power plants kept the average cost of nuclear-generated electric energy less than that of oil or gas and comparable to that of coal. Electricity from existing nuclear plants is still less expensive than that from coal-fired plants. However, dramatic increases in construction costs and historically unprecedented interest rates have seriously reduced new nuclear plant construction in the United States (but not elsewhere in the world). Finally, there are some adverse side effects to the environment in the obtaining of and the burning of fossil fuels which we might wish to minimize by employing other alternatives.

Nuclear power has been a topic of some concern during the last few decades throughout the world. One difficulty is that the estimated risks associated with the generation of nuclear power are publicized but rarely compared with the risks of the possible alternatives. A number of books that do make these comparisons are available. To read and truly understand these books, however, requires more time and background information than the average person has. It is impossible to discuss here all the issues that have been raised concerning nuclear power, but the following are chosen from the most common. Reference numbers refer to sources listed at the end of the section.

33-4 Electricity Generated from Nuclear Power

Nuclear power plants require longer to build than a coal-fired plant, for example. Furthermore, the United States takes about twice as long to produce an operating nuclear plant as most other nations. This time is a serious problem. Need for power must be forecast as much as 20 years ahead of time. Also, no installation is as sensitive to interest rates as a nuclear plant. With construction requiring well more than a decade before any income can be generated, interest on the money for construction can more than triple the cost of the plant itself.

Safety

Because a nuclear explosion from a reactor is impossible, the largest safety concern is a release to the environment of radioactive material with all its attendant dangers. Since the Atoms for Peace program was initiated just after World War II, the charge of safety engineers has been to postulate the worst possible accident they can imagine which could result in human exposure to

33 Nuclear Fission and Fusion

(a) Fossil fuel combustion

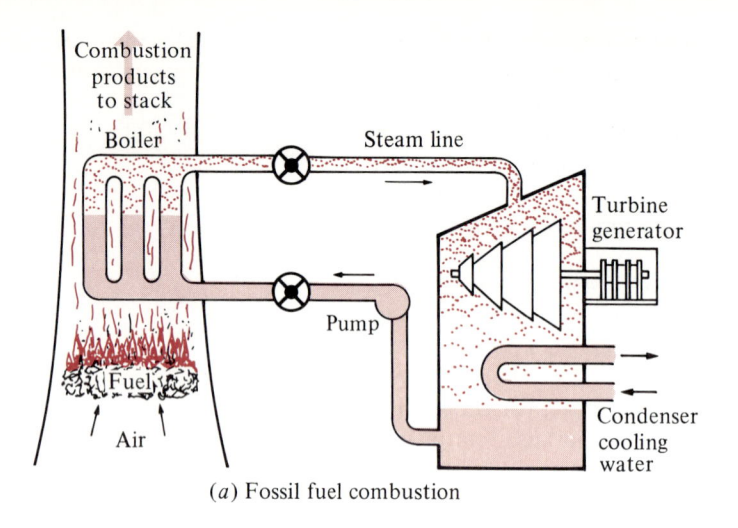

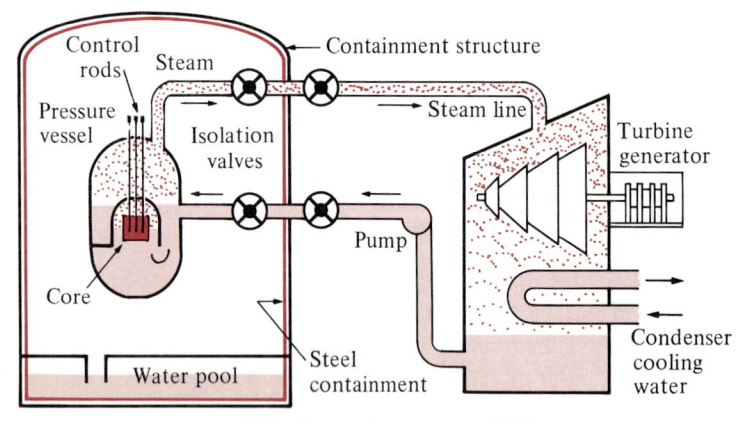

(b) Boiling water reactor (BWR)

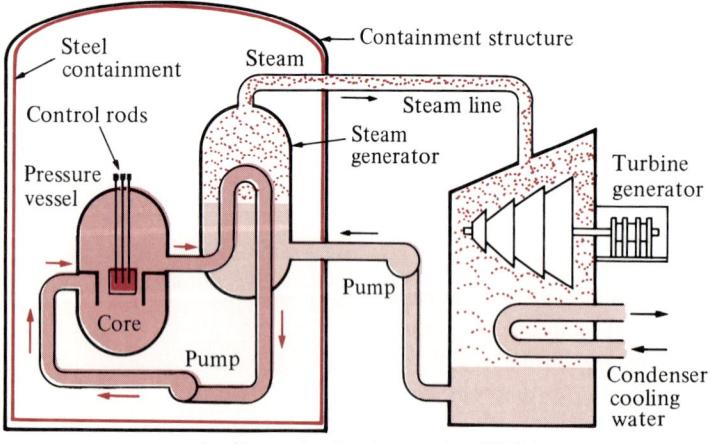

(c) Pressurized water reactor (PWR)

Figure 33-4 (a) In a **fossil-fuel plant**, the burning fuel heats water in the boilers. The steam produced turns a steam turbogenerator. At the condenser, the steam is changed to liquid water, maintaining the pressure difference across the turbine. The condensed water is returned to the boiler for reuse. (b) In a **boiling-water reactor**, the core separates the pressure vessel into upper and lower parts. Water passing the core flashes into steam, which then occupies the upper section, flows to the turbine, and is condensed and pumped back to the pressure vessel. (c) The **pressurized-water reactor** keeps the water as a superheated, high-pressure liquid and circulates it to **heat exchangers** inside the steam generator. Here, external water is boiled to provide the steam for the turbines. One advantage of this design is that none of the water circulating around the nuclear fuel leaves the containment.

33-4 Electricity Generated from Nuclear Power

radioactivity. They are then to design in such a way as to avoid that exposure occurring. From this charge, there evolved a concept called *defense in depth*. Built into the design of a nuclear reactor is a series of physical barriers that inhibit or prevent the release of fission products from the reactor's core. Furthermore, there are many safety systems which are designed to keep the plant operating normally or to prevent it from operating abnormally. The reactor is shut down and will not operate unless all systems and subsystems indicate normal operation. This has been compared to the way in which the safety system of an elevator operates. Considering the fact that millions of people ride in passenger elevators every day, we might find it remarkably good fortune that none ever plunge to their death. The reason is that every elevator is equipped with powerful jaws which may grip the guiderails of the elevator cabin and stop it from dropping. The safety feature is that the jaws are not activated if something goes wrong but are kept inoperative only if everything is going right. Similarly, nuclear reactors are not designed to activate safety measures when something goes wrong but to stay inoperative unless everything works normally. For example, control rods are brought into the core from above, withdrawn vertically upward to increase power, and suspended by electromagnets. If power to the electromagnets is lost for any reason, the control rods drop back to their natural position, immediately shutting off the reactor.

Some of the specific safety devices and barriers in the design include the following:

1. The fuel is formed into very small cylindrical pellets of hard, dense ceramic with a very high melting point ($>5200°F$). The uranium and solid radioactive waste material from fission fragments are thus bound together and should remain in place.
2. The pellets are inserted into long zirconium tubes to make a fuel rod. This tube provides an additional barrier to radioactive release.
3. Bundles of the fuel rods are placed vertically in the moderator-coolant (water, a good shield) with control rods interspersed, inside a pressure vessel or reactor vessel.
4. The reactor vessel is made of 9-in-thick steel fed by steel pipes with walls which are 3 to 4 in thick.
5. The reactor vessel is encircled by a concrete wall 7 to 10 ft thick to provide additional shielding.
6. The reactor and all reactor system components are enclosed in a large leak-tight steel shell to contain any radioactive materials in the event that some escape the core.
7. The containment building of which the steel containment forms the inner surface is concrete and steel, $3\frac{1}{2}$ to $4\frac{1}{2}$ ft thick, providing an additional barrier to possible radiation leaks. The containment building is required to be able to withstand the direct impact of the largest passenger jetliner, fully fueled, without rupture.
8. A designated area surrounding the plant is of restricted access and separates the plant from the public.
9. An emergency core cooling system is always on standby to provide a backup supply of coolant if the primary water supply is reduced.

In addition, backups and redundancies are built into the system. Air and water to be released to the outside environment are passed through various filters and ion-exchange devices to remove possible radioactive isotopes. Releases which include radioactive isotopes of the inert gases xenon and krypton are made in a controlled manner at amounts far lower than the strict standards set by federal authorities. Parallel electric power sources are available which are not dependent on the power delivered by the reactor. Radiation detectors are placed all over the plant and in the surrounding countryside. This monitoring is reviewed on a continuing basis by federal authorities, who also make regular unannounced safety inspections.

The reasoning for the emergency core cooling system is to prevent a core melt, or **meltdown.** After a power reactor has been in operation for some time, there will have been a buildup of fission fragments embedded in the fuel pellets. These fragments are generally radioactive, and this radiation incident on the atoms in the fuel and rods ionizes and increases the kinetic energy (the temperature) of the core. The core is then thermally hot due to the radioactivity even in the absence of nuclear fissions. Thus, when a reactor is shut down, water must continue to circulate or the temperature may increase, causing core damage. In the worst case anyone can imagine, if the cooling water is removed, the rods may melt, forming a molten mass which continues to melt through the pressure vessel. The most pessimistic calculations have the molten core continuing through the containment building floor some 15 ft before it stops.[1] Of course, the removal of the water would cause the reactor to shut down if it were not already off, because the water is also the moderator for neutrons. In any case, both knowledgeable opponents and proponents of nuclear power agree that if a meltdown should occur, the most probable number of deaths resulting is zero (References 2 and 3).

Many studies have been done comparing the safety of alternative power sources. In 1977, the results of what is probably the most nearly objective of those studies was released in the form of a 400-page book called *Nuclear Power Issues and Choices* (Reference 4). The study was sponsored by the Ford Foundation, which also sponsors one of the most fervently antinuclear organizations in the United States. To obtain the objective report desired, the following ground rules were established before the study group was selected:

1. The participants in the study must be recognized as highly qualified in their own fields of investigation and analysis.
2. As a group, they must be — and must be recognized as being — essentially open-minded on the general debate concerning nuclear power.
3. The participants themselves, as a group, must have sole and complete responsibility for their findings.

A group of 21 highly regarded individuals was assembled, and was subsequently referred to as the Nuclear Energy Policy Study Group. They included economists, political scientists, and attorneys as well as physicists, chemists,

The accident at the Three Mile Island plant was certainly the worst incident to date at a U.S. nuclear plant. You are urged to read the summary report of the Kemeny commission charged with investigating this accident (Reference 1). It is clear that there was sufficient blame for everyone involved; but, oddly enough, both sides of the argument draw some support from the accident. Antinuclear groups conclude, "See, with all your reassurances and safety systems, major accidents can still happen." Pronuclear groups conclude, "See, with every conceivable wrong action being taken and many safety systems being prevented from acting, the multiple lines of defense still held and prevented a disaster." Read about it and see what you conclude.

A major nuclear accident occurred at Chernobyl, U.S.S.R., in April 1986. Although available information is limited, enough is known to underscore the fact that the release of a large amount of radioactive material is much more dangerous than a meltdown. The Chernobyl reactor used graphite as a moderator and was designed both to produce electrical power and to breed weapons-grade plutonium. It had no containment building, and the power density at which it was run was high enough to ignite the graphite if the core was not cooled continuously. There are no comparable dual-purpose reactors outside the Soviet Union that operate in this way. Also, physical laws do not allow for this type of accident in pressurized or boiling water reactors.

---

[1] In fact, a number of experts believe a meltdown is impossible under any circumstances. The geometry inside a pressure vessel is very complex, and the assumptions which must be made are, for safety's sake, necessarily the most pessimistic. At the Three Mile Island nuclear plant accident in 1979, a reactor core was partially uncovered for several hours, and evidence indicates that the temperature remained several thousand degrees below the melting point of the fuel, although it did cause damage to the zirconium cladding. However, until it is proven otherwise, it is assumed that a meltdown is possible, and we act accordingly.

biologists, geologists, and engineers. The presidents of the California Institute of Technology and New York University were among the group. The group members took more than a year to complete their study. Their conclusions compared the risks associated with the use of nuclear power to generate electricity with those associated with the use of coal. This particular comparison was made because it is widely accepted that these are the only viable alternatives for large-scale electric energy production in the next several decades. In their concluding paragraph in the chapter on "Health Effects," they state that "the general conclusion is that on the average, *new* coal-fueled power plants meeting *new* source standards will probably exact a considerably higher cost in life and health than new nuclear plants" (Reference 4, page 196). Every study recognizes that there are risks associated with every method of electrical production and that they should be minimized. However, the risks of any given course of action must be compared with the risks of the possible alternatives.

33-4 Electricity Generated from Nuclear Power

Environmental Impact

The environmental effects of the whole of the nuclear fuel cycle have been compared with those of the fossil fuels. While all methods use and disrupt land areas, at present much more land is disturbed by extraction of fossil fuels than by uranium mining *for the same amount of power generated*.

Thermal discharges to the environment, sometimes called thermal pollution, are common to all heat engines, as you know. The amount of heat discharged for a given amount of electric energy output depends on the efficiency of the heat engine. The average efficiency of existing fossil-fueled plants and nuclear power plants is about 33 percent, although modern fossil plants may have efficiencies up to 40 percent. The amount of heat per unit of electric energy output is thus comparable for all these plants. Therefore, if thermal pollution is a problem, it is shared by all alternatives.

In addition, there are environmental effects which are unique to each cycle — acid runoff from coal mining, possible oil spills, perhaps acid rain and CO_2 buildup from burning of all fossil fuels. In balance, nuclear power seems to create fewer problems in this regard than possible alternatives. In fact, the conclusion of the "Environmental Effects" chapter in *Nuclear Power Issues and Choices* is that "on balance, the local environmental consequences of the nuclear power cycle in normal operation are not as serious as those from fossil-fuel power generation" (Reference 4, page 211).

The health effects of alternatives to nuclear power are rarely compared to those of reactors. The most thorough evaluation of the associated risks of coal-burning electric generating plants was in a study by the government conducted at the request of Congress (Reference 5). It estimated that the United States during 1985 had, due to coal-burning, some 49,500 excess deaths due to lung disease. This is slightly more than the mortality induced by automobile accidents.

Waste Disposal

A large number of possible disposal methods of **high-level radioactive waste** have been considered and found to be feasible. Several methods have been endorsed by the National Academy of Sciences and the American Physical Society. The conclusion of the Nuclear Energy Policy Study Group at the end of the chapter on "Radioactive Waste" is that "disposal of waste in stable geological formations which are isolated from appreciable contact with groundwater appears to provide adequate assurance against the escape of consequential amounts of radioactivity even over long periods of time" (Reference 4, page 266).

After about 3 years of service in a power reactor, fuel rods are replaced. They still contain nearly 2 percent ^{235}U and about 95 percent ^{238}U as well as some fissionable plutonium, but the inventory of fission fragments has increased to a point where they are absorbing a significant number of neutrons. By law, the federal government took over the responsibility for nuclear waste disposal in the late 1940s. The plan devised to manage the spent fuel rods is as follows: After removal from the reactor, the rods are stored for some time in cooling ponds at the plant site. This allows their temperature to decrease and allows the short-lived components of fission fragments to die away. The rods are then taken to a reprocessing plant, where they are cut into small pieces and dissolved in a very strong acid. From this solution, virtually all the uranium and plutonium is chemically separated for recycle as fresh fuel. The remaining solution is then vitrified into a glassy or ceramic compound, molded into solid cylinders, and encased in stainless steel. Finally, these cylinders are to be lowered through drill holes into salt domes or unfractured granite 1800 ft or more below the surface of the earth. This method, published by the American Physical Society in 1978, is already in use in France and Sweden. An important consideration is that if it is handled in this way, the high-level waste from a 1000-MW nuclear power plant is very small in volume, amounting to about 2 m^3/year.[1] Waste disposal (site selection, demonstration, etc.) is progressing very slowly in the United States. Some people say this deliberate approach has the advantage of being able to profit from others' research, and besides, reprocessing and early disposal are not critical since spent fuel can be stored in cooling ponds for decades. Others insist that the delay is politically motivated. It is understandable that many people are not enthusiastic about having nuclear waste depositories nearby regardless of how potentially benign they are said to be. However, federal laws have now mandated dates by which some of the hard decisions must be made.

High-level radioactive waste must be distinguished from low-level waste. Low-level wastes are defined as having less than 10 mCi/kg and comprise worker's gloves, paper, plastic, glass, etc., that *may* have been contaminated. Nobel laureate Dr. Rosalyn Yalow has noted that if the radioactivity currently in your body were injected into a laboratory rat, its body would become a low-level radioactive waste. By law, it could not be burned or buried but would have to be sealed in multiple plastic layers, then enclosed in steel for deposit in a low-level radioactive waste landfill.

Numbered References

1. *Report of the President's Commission on the Accident at Three Mile Island,* Oct. 30, 1979. John G. Kemeny, Chairman. Library of Congress Catalog Card #79-25694, ISBN 0-935758-00-3.

2. Bernard L. Cohen, "Impacts of the Nuclear Energy Industry on Human Health and Safety," *American Scientist,* vol. 64, September–October 1976.

3. H. Kendall (spokesman for the Union of Concerned Scientists) in testimony before the U.S. Congressional Committee on Energy and the Environment, *Congressional Record,* April 28–May 2, 1975.

4. *Nuclear Power Issues and Choices—Report of the Nuclear Energy Policy Study Group,* Ballinger Publishing Company, Cambridge, Mass., 1977. (Copyright by the Ford Foundation.)

5. *The Direct Use of Coal,* report by the Congressional Office of Technological Assessment, 1979, U.S. Government Printing Office, Stock No. 052-003-00664-2.

---

[1] For comparison, a 1000-MW coal-fired plant produces nearly a quarter of a million tons of ash each year, plus a similar amount of sludge if the plant is equipped with scrubbers. This amounts to about 80,000 truckloads of waste each year.

33-5 Breeder Reactors

The idea of using a nuclear power reactor to "breed" additional fuel to be used for its own cycle or that of other power reactors developed in the 1940s. As we have seen, it became known that other fissionable nuclei could be obtained by bombarding certain nonfissionable nuclei with neutrons and that the average number of neutrons per fission was more than needed to sustain a chain reaction. The breeding idea followed naturally with the expectation that the excess neutrons in a power reactor could be used to convert **fertile nuclei** like ^{238}U to fissionable ^{239}Pu. A big advantage of using this type of reactor is that virtually all the uranium can be used to supply energy rather than just the ^{235}U isotope. In principle, we could increase our nuclear fuel supply a hundredfold.

The first demonstration breeder reactor which successfully produced both electricity and additional fuel operated in the United States in 1951. It was fueled with enriched uranium and cooled by liquid sodium and potassium. Breeders are now in operation in France, Great Britain, and the Soviet Union and have been employed successfully since the mid-1970s. Many professional groups believe that breeders are needed as soon as possible "to conserve our nuclear fuel supply, to reduce the requirements for enriching facilities, and to reduce (eventually) the cost of power." Other groups disagree and recommend a more relaxed timetable, stating that "introduction of the breeder may be deferred for ten, twenty or more years without seriously affecting the economic health or energy security of the United States." All expert study groups, however, appear to conclude that breeder reactors will become an important part of our energy mix in the future.

The most well known breeder is the **liquid-metal fast breeder reactor** (LMFBR) used with the ^{238}U to ^{239}Pu cycle. There is no moderator, as the core is cooled with liquid sodium or potassium and the fissions are induced by fast neutrons. Whenever ^{239}Pu absorbs a slow neutron, the probability is only about 70 percent that a fission will occur. In 30 percent of the absorptions, the nucleus is converted to ^{240}Pu, which also absorbs slow neutrons without fission. Thus both isotopes act as **neutron poisons** for the chain reaction, depleting slow neutrons from the fuel. For this reason the design was made to utilize fast neutrons. Fast breeder reactors do provide high fuel economy but have the disadvantage of requiring more sophisticated control techniques. Thus exceptional precautions, including a very skilled technical staff, are required.

Another breeder possibility, which many experts believe to be more desirable, is one which utilizes a **thorium-uranium cycle.** In this cycle, thorium 232 absorbs slow neutrons, converting to thorium 233. This decays by beta emission to protactinium 233, which, in turn, decays to uranium 233. The uranium 233 is fissionable with the absorption of slow neutrons in the same way as uranium 235. Several reactor designs can operate with this cycle, including the CanDU reactor and the high-temperature gas-cooled reactor like that which is in commercial operation in Platteville, Colorado.

One difference between the two cycles is that uranium 233 emits fewer neutrons on the average than does plutonium. Every 10 plutonium fissions in a fast breeder reactor can produce 10 new fissions plus 13 or 14 additional plutonium nuclei. However, every 10 uranium fissions in a slow breeder can produce 10 new fissions plus only 10 additional uranium 233 nuclei. But, once it is started (using uranium 235 since uranium 233 is not found in nature), a thorium slow breeder need never be fueled with mined uranium. It will breed the uranium it needs indefinitely from thorium. An advantage of the thorium cycle is that there is more thorium in the earth's crust than there is uranium. Further, it occurs naturally only in the isotope thorium 232 and need not be enriched, only refined.

33-6 Controlled Fusion

Nuclear energy can be obtained from fusion as well as from fission. In nuclear **fusion,** several light nuclei merge, or fuse together, to form one larger nucleus, with the release of energy. Perhaps the energy release can be most easily understood by reference to the binding-energy curve of Figure 33-5. Note (as we did in Section 31-3) that those naturally occurring nuclei with large nucleon numbers and small nucleon numbers have less binding energy per nucleon than those with nucleon numbers intermediate between the two extremes. You will recall that the binding energy per nucleon for a particular nucleus is equal to the energy per nucleon which would be released if that particular nucleus were assembled from its constituent protons and neutrons. Thus, if a larger nucleus were disassembled and reassembled into several smaller intermediate nuclei, we would expect a net release of energy. This is essentially what takes place during the fission reaction. A large (^{235}U, for example) nucleus absorbs a neutron and may fission into two intermediate nuclei, with the release of energy. In fusion, something like the reverse of fission takes place. Two smaller nuclei may fuse together, releasing the net difference in binding energy in the process. Consider, for example, a possible reaction in which two deuterium ($^{2}_{1}$H) nuclei fuse to form a helium 4 nucleus ($^{4}_{2}$He) plus energy. The reaction equation is

$$^{2}_{1}\text{H} + ^{2}_{1}\text{H} \rightarrow ^{4}_{2}\text{He} + \gamma \text{ (or several } \gamma\text{'s)}$$

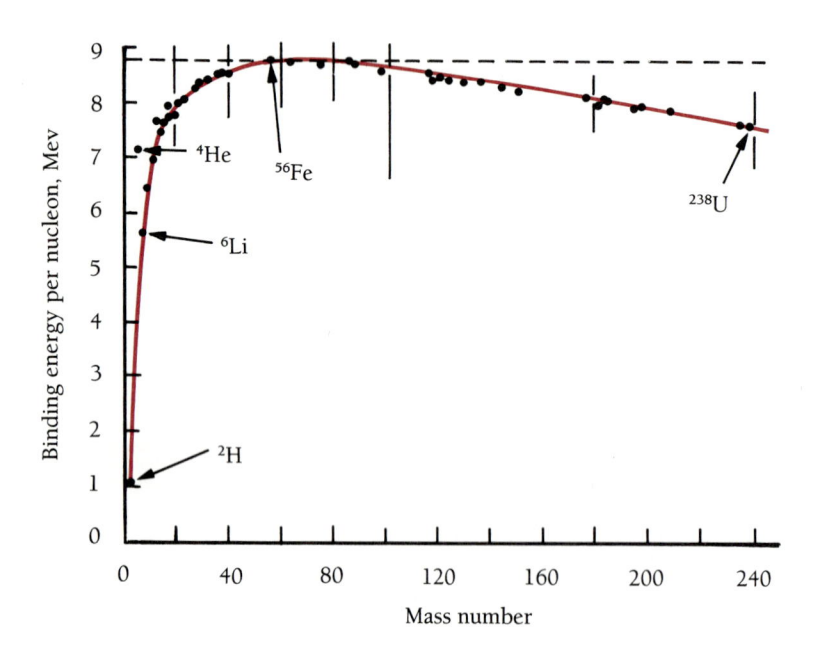

Figure 33-5 Binding energy per nucleon vs. mass number for the naturally occurring isotopes. A net release of energy occurs when nuclei in the midrange are produced, either by the breaking apart of larger nuclei (fission) or by the bringing together of smaller nuclei (fusion).

33-6 Controlled Fusion

Example 33-3 Given that the binding energy of deuterium is 2.22 MeV and the binding energy of helium 4 is 28.3 MeV, determine how much energy is released in the fusion of two deuterium nuclei to form helium.

Solution We simply determine the difference between the binding energy per nucleon before and after the fusion. This number is then multiplied by the number of nucleons involved (four in this case).

$$\frac{BE}{A} \text{ before} = \frac{2.22 \text{ MeV}}{2 \text{ nucleons}} = 1.11 \text{ MeV/nucleon}$$

$$\frac{BE}{A} \text{ after} = \frac{28.3 \text{ MeV}}{4 \text{ nucleons}} = 7.08 \text{ MeV/nucleon}$$

$$\Delta \frac{BE}{A} = 7.08 \text{ MeV/nucleon} - 1.11 \text{ MeV/nucleon} = 5.97 \text{ MeV/nucleon}$$

$$\text{Energy release} = 4 \Delta \left(\frac{BE}{A}\right) = 4 \text{ nucleons } (5.97 \text{ MeV/nucleon})$$
$$= 23.9 \text{ MeV}$$

In fact, the reaction of Example 33-3, if it occurs at all, is very rare. More likely, if two deuterium nuclei fuse, the reactions are

$$^2_1H + ^2_1H \rightarrow ^3_1H + ^1_1H$$

and

$$^2_1H + ^2_1H \rightarrow ^3_2He + ^1_0n$$

The net energy release in these reactions is 4.0 and 3.3 MeV, respectively (see Problem 33-17).

Many other fusion reactions and combinations of reactions which have a net energy release are possible. Several are included in the end-of-chapter problems.

Two well-known examples incorporate fusion reactions in their energy releases: one is the sun (and all other stars), and the other is thermonuclear weapons — hydrogen bombs. However, neither of these is an example of controlled fusion, which is the topic of this section.

We have been emphasizing that in a fusion reaction there may be a *net* energy release, because the reactants must have a great deal of energy originally in order to get close enough to fuse in the first place. Remember, nuclei are positively charged and will repel each other. Initial energy is needed to overcome this Coulomb repulsion. Once the reactants fuse, there is a net energy release if there is more energy from the fusion than had to be supplied initially to the reactants. We may calculate the approximate value of the energy which must be supplied. Consider two identical single-charged nuclei at some large distance of separation and having equal speeds v while aiming for a head-on collision. We assume their separation is such that their initial potential energy is zero and that they come to rest just as they fuse at a separation r_f. Applying the conservation of mechanical energy, we have

$$\Delta KE + \Delta PE = 0$$

or

$$0 - \text{KE}_i + \left(\frac{ke^2}{r_f} - 0\right) = 0$$

Thus the sum of their initial kinetic energies is

$$\text{KE}_i = \frac{ke^2}{r_f}$$

Making the assumption that r_f must be on the order of a typical nuclear radius, or about 1.00×10^{-14} m, we have

$$\text{KE}_i = \frac{\left(9.00 \times 10^9 \, \frac{\text{N} \cdot \text{m}^2}{\text{C}^2}\right)(1.60 \times 10^{-19} \, \text{C})^2}{1.00 \times 10^{-14} \, \text{m}}$$

$$= 2.30 \times 10^{-14} \, \text{J}$$

Thus, each of the initial nuclei would have to have about 1.15×10^{-14} J of kinetic energy (approximately 70 keV) in order that they approach each other to within a distance of 1.00×10^{-14} m.

Suppose we wish to confine a quantity of these nuclei in a form similar to a gas and heat them to a high-enough temperature so that they would have an average kinetic energy of 1.15×10^{-14} J. Normal collisions would then cause many, many fusions to occur. Let us calculate the temperature to which we would have to heat this nuclear "gas."[1]

In Chapter 14, we learned that temperature is defined in terms of the average random kinetic energy of the particles in a system as

$$\text{KE}_{av} = \tfrac{3}{2}kT$$

where k is the Boltzmann constant and T is the absolute temperature in kelvins. Using this, we can calculate the absolute temperature of a system of single-charged nuclei such that their average kinetic energy would be enough so that they could approach within 1.00×10^{-14} m in head-on collisions.

$$T = \frac{2}{3} \times \frac{\text{KE}}{k} = \frac{2}{3} \times \frac{1.15 \times 10^{-14} \, \text{J}}{1.38 \times 10^{-23} \, \text{J/K}} = 5.56 \times 10^8 \, \text{K}$$

This is 556 million kelvins! Even if we recognize that the *average* kinetic energy need not be so large that a significant number of nuclei in a system have enough energy to fuse, we must realize we are dealing with energies equivalent to temperatures in the vicinity of millions of degrees. (It hardly matters whether we are talking of kelvins or degrees Celsius at this point.) The hydrogen in the sun is confined by the gravitational field. With its enormous amount of matter, the core temperature is raised to values near 5 million kelvins. This is enough to allow at least a small percentage of most of the proposed fusion reactions to proceed. These reactions produce yet more energy, some of which provides an outward pressure necessary to prevent the gravitational collapse of the sun and some of which eventually radiates into space. In a hydrogen bomb, the equivalent temperatures required for fusion are generated by using a fission explosion as an energy source.

---

[1] This system would not really be a gas. At the energies necessary, all matter breaks down into separate nuclei and electrons. This highly ionized state is known as a **plasma** and is the topic of a very important area of physics.

33-6 Controlled Fusion

Immediately after the first successful test of the hydrogen bomb, the scientists who worked on the project were prodded to solve the problem of controlled fusion. The peaceful use of this enormous energy source seemed to be uppermost in everyone's mind. The key word is *controlled,* since fusion is certain at sufficient kinetic energies.

There were (and are) a very large number of very difficult problems in science to be overcome before physicists could be convinced that practical application of controlled fusion is even feasible. Many of these problems have been solved for two possible methods, but there are still enormous engineering difficulties to overcome.

The two possibilities for accomplishing controlled fusion which are presently being pursued are:

1. Slow the rate of the reaction by decreasing the density of the reactants.
2. Work with tiny amounts of a very dense plasma, creating a succession of microminiature hydrogen bombs.

Either of the two methods could in principle use hydrogen, deuterium, or tritium (or some combination of the three) as fuel. Extensive research showed that the reaction fueled by deuterium and tritium releases more energy and thus can achieve the same power level at much lower equivalent temperatures. Therefore, in the interest of obtaining a self-heating, net power-producing fusion reactor in the shortest time period, controlled-fusion research groups have been concentrating on the deuterium-tritium reaction. This reaction is

$$ {}^2_1H + {}^3_1H \rightarrow {}^4_2He + {}^1_0n $$

About 17.6 MeV is released in this reaction, but slightly more than 14 MeV is carried by the neutron (see Problem 33-21). A number of scientists and engineers express doubts about using this reaction, stating that the high-energy neutrons will create substantial engineering difficulties. In the 1980s the magnitude of these difficulties began to be obvious, but work continues with this reaction because many are confident that the engineering problems of neutron damage and induced radioactivity can be solved. The proponents of this reaction point out that any of the possible reactions will create many free neutrons (although not so energetic), and thus the induced radioactivity and neutron damage problems must be solved in any event.

The first method for controlling fusion is being pursued at several places in the United States, Great Britain, West Germany, France, Japan, and the Soviet Union. In this method, the density of the fuel is reduced to about one hundred-millionth of normal liquid density, and the energy of the particles is increased.[1] The first difficulty encountered was to design a container for the rarefied, high-energy plasma. In such a tenuous system, the particles may travel relatively long distances in straight lines between collisions. (They are said to have a very long mean free path.) Under these circumstances, most particles will very quickly strike the walls of any material container and give up some or all of their energies to the wall. They will then be unavailable to collide with each other, and the desired reaction will not proceed. One container that can work, some people call a **magnetic bottle.** Under certain circumstances, a

---

[1] Dr. Edward Teller calls this system a "high-pressure vacuum." He states that it is a vacuum because any given volume contains very few particles but that it has high pressure because the few particles which are there have tremendous energies.

properly shaped magnetic field can provide the forces to confine moving charged particles to a limited volume. In this way, the magnetic field might act like a bottle which contains gas. Several designs are possible and are being pursued, but they all rely on the fact that a magnetic field exerts forces on fast-moving charged particles. You will recall from Chapter 21 that the magnitude of the force **F** exerted by a magnetic field on a charge q moving with a velocity **v** in the field is given by

$$F = qvB \sin \theta \qquad (21\text{-}1)$$

where θ is the angle between **v** and **B**. Further, the direction of the force is perpendicular to the plane formed by **v** and **B**. If the charged particle's initial velocity is perpendicular to **B**, the force will be centripetal in nature, causing the particle to move in a circle. However, if the particle has an initial velocity with components which are both perpendicular and parallel to **B**, only the perpendicular component will be affected. The particle's parallel component of velocity will be unaffected by the magnetic field, and the perpendicular component will continuously change direction. The particle thus traces a helical path along the magnetic field lines. Figure 33-6 depicts the motion of a charged particle in a uniform magnetic field. The particle is shown progressing in a direction opposite to that of the magnetic field, but it may trace its path in either direction depending on the direction of its velocity when it enters the field. In either case, the helix is wrapped around the field lines, and if the lines are curved, the helix is also.

One type of magnetic bottle depends on the phenomenon called a **magnetic mirror.** This reflection effect can take place in a magnetic field where the lines are caused to converge by an increase in magnetic induction. The cause of this reflection is shown in Figure 33-7. As the particle approaches the stronger region of the field, the radius of the curvature gets smaller as the force increases in magnitude. Also, the circular paths get closer together since the force (which must be perpendicular to both **v** and **B**) now has a backward component. If this backward component of force becomes large enough, the particle's motion in the direction of the field is slowed to a stop and the particle acquires a component of velocity in the direction opposite to that of the field. A magnetic bottle can be formed by "pinching" a formerly uniform field at two points, forming two such magnetic mirrors, as shown in Figure 33-8. This type of magnetic bottle is somewhat unstable and also leaks particles out the ends if they happen to be moving parallel to the field near the mirrors.[1] An advanced version of this is known as the tandem mirror experiment (TMX) and is being used at the Lawrence Livermore Laboratory.

The greatest amount of time and money for controlled fusion development in the United States and elsewhere have been given to another form of magnetic containment called the **tokamak,** a fusion reactor of toroidal shape (Figure 33-9). Much of the early development of this system to its present form is credited to Russian experiments conducted in the late 1960s. The tokamak

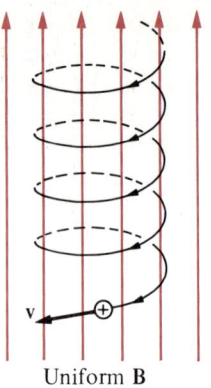

Figure 33-6 The force on a proton in a uniform magnetic field is always perpendicular to its velocity and to **B**. In this uniform field shown, this force deflects the proton to its right as it advances. Note that since **F** is at every instant perpendicular to **v**, the magnitude of **v** cannot change. The force can neither speed up nor slow down the particle, only deflect its path.

Figure 33-7 As the proton approaches the compressed region of the magnetic field, the force on it is increasing and acquiring a larger and larger downward component, which reduces the proton's upward motion loop by loop. The proton will eventually stop entirely and be accelerated downward until the uniform-field region at the bottom is reached and **F** loses any vertical component.

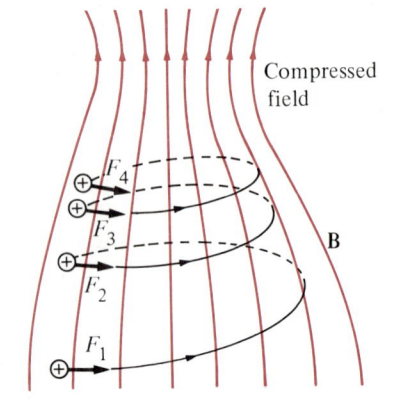

---

[1] The earth's magnetic field provides a natural set of magnetic mirrors for electrons and protons injected by the sun. These particles are trapped in what are known as the **Van Allen belts** and reflect near the poles where the magnetic field is stronger. When the direction or energy of the particles is such that they mirror too low in the atmosphere, they excite oxygen and nitrogen. These then emit their characteristic colored lines as they return to their ground states. This is seen as the northern lights (aurora borealis) in the northern latitudes and as the southern lights (aurora australis) in the southern latitudes.

has a doughnut-shaped magnetic field similar to any toroid but suitably modified to avoid certain instabilities. The magnetic bottle is composed of the toroidal-type field lines which close on themselves while the plasma particles trace bent helical paths around these lines. The plasma acts as a single-loop conductor and receives energy, acting as a single-turn secondary of an iron-core transformer. A current in the primary windings produces a very large current in the secondary (the plasma). This current, in addition to providing the energy to the fuel, also has its own associated magnetic field which helps to compress the plasma. The tokamak fusion test reactor (TFTR) in Princeton, New Jersey, began operating in 1983 and was projected to reach the **breakeven stage** (where as much energy is produced by fusion as is required to run the reactor) by 1986 or 1987.

The second possibility under active investigation in an attempt to produce controlled fusion is called **inertial confinement fusion.** In this approach, a tiny glass sphere containing a mixture of deuterium and tritium at pressures of 10 to 100 atm is bombarded by a burst of high-intensity laser light or particle beams from many sides simultaneously. The pellet's surface is heated and explodes outward, while the reaction force establishes a shock front (a pressure wave) inward in the fuel, confining it by its own inertia. Once the conditions for fusion are established on the surface of the fuel, a thermonuclear burn front propagates inward, causing the fuel's density and rate of fusion reaction to increase. The fusion explosion that results is to take place in a reactor chamber, where the heat generated is absorbed by some liquid and can be used to produce electricity just as with the heat from the tokamak.

For both magnetic and inertial confinement fusion, there must be an assured supply of fuel. Deuterium is readily available, as it constitutes about 1 part in 6700 naturally occurring hydrogen atoms. The oceans would contain enough deuterium to last for many millions of years without seriously depleting the supply. Further, there are relatively inexpensive processes for extracting it. Tritium has a different story. It is radioactive with a half-life of about 12.3 years. Thus, virtually none is found in nature, and it must be obtained from a nuclear

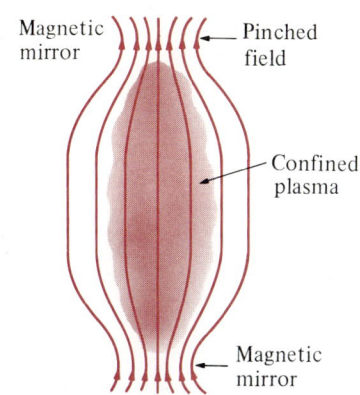

Figure 33-8 A **magnetic bottle** to confine a plasma consists of a uniform magnetic field constricted at each end. These pinched zones act as magnetic mirrors to reflect charged particles back into the uniform-field section. Magnetic confinement is necessary for a plasma because its effective temperature would vaporize any physical walls it contacted.

Figure 33-9 A tokamak fusion reactor confines the plasma in a toroidal magnetic "doughnut" rather than pinching a linear field to produce a magnetic bottle. (*Princeton Plasma Physics Laboratory*)

Figure 33-10 Target chamber of the Nova laser fusion system. (*Lawrence Livermore National Laboratory*)

reaction. It is intended that the neutrons released in the fission be used for this purpose by assuring that they react with lithium inside the fusion reactor. The reactions would be

$$^1_0n + ^7_3Li \rightarrow ^3_1H + ^4_2He + ^1_0n$$

and

$$^1_0n + ^6_3Li \rightarrow ^3_1H + ^4_2He$$

The reaction with lithium 7 requires an energetic neutron (2.47 MeV) and yields a slow neutron. The slow neutron then reacts with lithium 6.

The design and construction of these inertial confinement fusion reactors require great attention to detail, and they become very complex instruments. The first large laser fusion device at Lawrence Livermore Laboratory was named after Shiva, the Hindu goddess of love and destruction. This unit has now been superseded by the Nova laser fusion system (Figure 33-10), which will yield 20 to 30 times as much energy.

Figure 33-11 is a conceptual drawing of a nuclear fusion electric power plant. Such an installation is many decades in the future. It is even unlikely that a demonstration plant will be in operation very early in the next century. The neutron damage and activation problem along with the much higher heat loads means there will be stricter requirements for structural integrity than for an equivalent fission reactor. Also, beyond the energy breakeven point, there must be a great deal of additional energy generated before the fusion reactor plant becomes economically practical. We are optimistic, however, that all these problems will eventually be solved, and that in future centuries a significant fraction of our electric energy will be generated by nuclear fusion power plants.

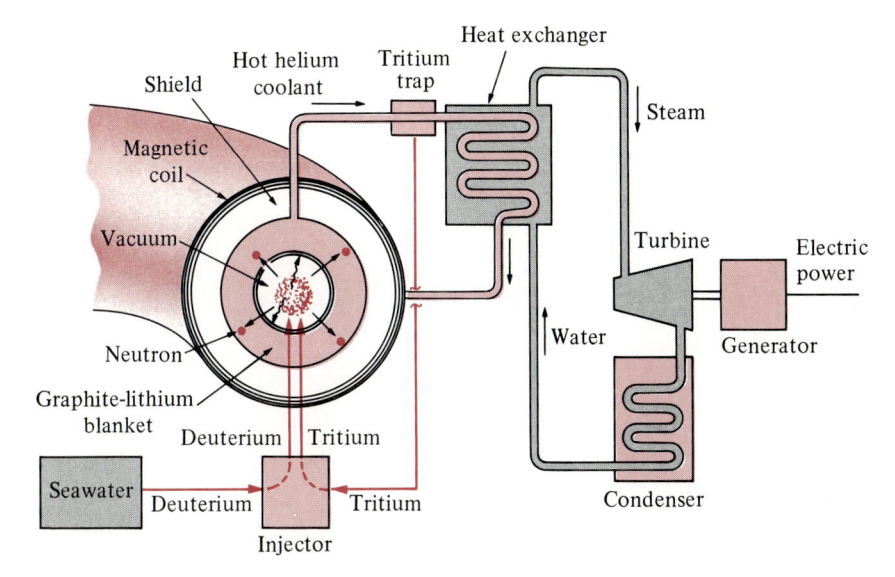

Figure 33-11 A continuous fusion plant, showing a magnetically confined plasma fusing to release energy and neutrons. The energy is absorbed and carried off by a helium coolant circulated through the graphite-lithium blanket. Lithium atoms absorbing a neutron produce tritium nuclei, and this gas is flushed from the blanket by the helium coolant and later condensed out by cooling. The tritium is then used as part of the plasma. Heat from the process runs through a heat exchanger, boiling external water to power a conventional steam-generating station.

Minimum Learning Objectives

After studying this chapter, you should be able to:

1. Define:

   atomic pile
   breakeven stage
   breeder reactor
   CanDU reactor
   chain reaction
   control rod
   coolant
   critical mass
   delayed neutrons
   enriched
   excess neutrons
   fertile nucleus
   fission
   fusion
   heavy water
   high-level radioactive waste
   inertial confinement fusion
   liquid-metal fast breeder reactor (LMFBR)
   magnetic bottle
   magnetic mirror
   meltdown
   moderator
   neutron activation analysis
   neutron poison
   plasma
   plutonium
   power reactor
   production reactor
   prompt neutrons
   reactor
   thorium-uranium cycle
   tokamak
   transuranic element
   Van Allen belt

2. Calculate several nuclei likely to result from neutron activation of a given nucleus.

3. Calculate the Coulomb potential energy for a positively charged particle just expelled from a given nucleus.
4. Describe the conditions necessary for a sustained nuclear chain reaction.
5. Calculate the excess neutrons for given fission fragments from a known nucleus.
6. Explain the need for a moderator in a nuclear reactor, and identify appropriate substances to serve as a moderator.
7. Describe the nuclear details of controlled fission in a nuclear reactor.
8. Distinguish several types of fission reactors, explaining their differences.
9. Summarize areas of contention in the debate on the widespread use of nuclear reactors.
10. Define breeder reactor and the nuclear requirements for producing one.
11. Calculate the energy release in specified nuclear fission or fusion reactions.
12. Describe the strategies for producing controlled nuclear fusion and the problems each must overcome.

Problems

33-1 Calculate the net energy released in the fission reaction

$$^{1}_{0}n + ^{235}_{92}U \rightarrow ^{139}_{56}Ba + ^{95}_{36}Kr + 2\,^{1}_{0}n$$

by determining the amount of mass converted to energy. Take the mass of ^{139}Ba as 138.908830 u and the mass of ^{95}Kr as 94.897331 u. Compare your answer with that of Example 33-1.

33-2 Calculate the net energy released in the fission reaction

$$^{1}_{0}n + ^{235}_{92}U \rightarrow ^{140}_{55}Cs + ^{92}_{37}Rb + 4\,^{1}_{0}n$$

837

Take the mass of ^{140}Cs as 139.917140 u and the mass of ^{92}Rb as 91.919088 u.

33-3 Calculate the net energy release in the fission reaction

$$^1_0n + ^{235}_{92}U \rightarrow ^{134}_{51}Sb + ^{95}_{39}Yt + ^4_2He + 3\,^1_0n$$

Take the mass of ^{134}Sb as 133.896857 u and the mass of ^{95}Yt as 94.912540 u.

33-4 If the molecular mass of an average gasoline molecule is 104 u and gasoline releases 4.40×10^7 J/kg when burned, calculate (a) the energy release per molecule in megaelectronvolts and (b) the ratio of the average energy released per fission (200 MeV) to that released per combustion of one gasoline molecule.

33-5 If the average energy release per fission is 200 MeV = 3.20×10^{-11} J, how much (a) energy is released if 1.00 kg of ^{235}U fissions and (b) mass is converted to energy when 1.00 kg of ^{235}U fissions?

33-6 The nuclear power plant for a submarine must deliver 6.50×10^7 W when the submarine is traveling at its fastest possible speed. Assuming 200 MeV per fission, calculate (a) the number of fissions required when the submarine travels at top speed for $2\frac{1}{2}$ days and (b) the amount of mass converted to energy during those $2\frac{1}{2}$ days.

33-7 Determine (a) the random kinetic energy and (b) the root mean square speed of an average neutron in a group whose equivalent temperature is 20°C. (Recall from the kinetic theory of gases that $KE_{av} \approx \frac{3}{2}kT$, where k is the Boltzmann constant, 1.38×10^{-23} J/K.) Neutrons of this average energy are called thermal neutrons.

33-8 Assuming that a neutron will on the average lose half its kinetic energy in each collision with a proton, how many collisions must a 14.0-MeV neutron make in order that its energy be reduced to 3.80×10^{-2} eV?

33-9 It is estimated that the annual energy consumption on earth is 4.00×10^{20} J. Calculate the mass of uranium which must fission to provide this energy.

33-10 (a) Write the reaction equation and (b) calculate the fission energy release when a ^{235}U nucleus fission produces two equal-mass fission fragments and six neutrons. Take the mass of $^{115}_{46}$Pd as 114.91269 u.

33-11 (a) Determine the average output power of a nuclear reactor if it uses 12.0 kg of ^{235}U each week. (b) At 33 percent efficiency, how much thermal energy must be dissipated per second?

33-12 When coal is burned, it releases about 2.80×10^7 J/kg. Assuming 33 percent efficiency, how much coal is burned per week to generate 550 MW in an electric power plant? Compare your answer to the data given in Problem 33-11.

33-13 (a) Calculate the mass of ^{235}U which fissions each year in a 1000-MW electric power plant operating at full power 70 percent of the time. Assume 33 percent efficiency. (b) What volume of ^{235}U is this if the density of uranium is 1.89×10^4 kg/m^3?

33-14 Calculate the binding energy of the last neutron added to (a) ^{236}U and (b) ^{239}U. (Hint: These isotopes are formed by adding a neutron to ^{235}U and ^{238}U, respectively.)

33-15 Calculate the binding energy of the last neutron added to (a) ^{234}U and (b) ^{240}Pu. (Hint: These isotopes are formed by adding a neutron to ^{233}U and ^{239}Pu, respectively.)

33-16 Calculate the energy necessary to separate a neutron from $^4_2$He.

33-17 Write the reaction equations and calculate the amount of energy released when two deuterium ($^2_1$H) nuclei fuse to form (a) a tritium ($^3_1$H) nucleus and a proton and (b) helium 3 and a neutron.

33-18 Two deuterium nuclei can fuse if they have energies of 10 keV or more. To what temperature is this energy equivalent?

33-19 One of the series of reactions believed to produce fusion energy in the stars is called the carbon-nitrogen cycle, which proceeds in order as

$$^1_1H + ^{12}_6C \rightarrow ^{13}_7N$$

$$^{13}_7N \rightarrow ^{13}_6C + ^0_1e$$

$$^1_1H + ^{13}_6C \rightarrow ^{14}_7N$$

$$^1_1H + ^{14}_7N \rightarrow ^{15}_8O$$

$$^{15}_8O \rightarrow ^{15}_7N + ^0_1e$$

$$^1_1H + ^{15}_7N \rightarrow ^{12}_6C + ^4_2He$$

Calculate the energy released in each step of the cycle. The mass of $^{15}_8$O is 15.003070 u.

33-20 Show that the net energy release when deuterium ($^2_1$H) fuses with tritium ($^3_1$H) to form helium 4 and a neutron is 17.6 MeV.

33-21 After the fusion of a deuterium ($^2_1$H) nucleus and a tritium nucleus ($^3_1$H), determine (a) the ratio of the kinetic energies delivered to the product nuclei (^{4}He and $^1_0$n) and (b) the kinetic energy of the neutron if the net energy of the reaction is 17.6 MeV. (Hint: Assume that the center of mass of the products is stationary. Apply the conservation of momentum and the definition of kinetic energy.)

33-22 A series of reactions called the proton-proton cycle may be responsible for much of the fusion energy of stars:

$$^1_1H + ^1_1H \rightarrow ^2_1H + ^0_1e$$

$$^2_1H + ^1_1H \rightarrow ^3_2He$$

$$^3_2He + ^3_2He \rightarrow ^4_2He + ^1_1H + ^1_1H$$

Calculate the net energy release during each step of the cycle.

33-23 The energy release per deuterium nucleus in deuterium fusion is about 7.00 MeV. (a) Calculate the number of deuterium nuclei in 1.00 m^3 of water. (Recall that one of every 6700 hydrogen atoms is deuterium.) (b) Calculate the energy available for release by fusing the deuterium in 1.00 m^3 of water. (c) When burned, a barrel of oil releases about 5.90×10^9 J, and a ton of coal when burned releases 2.76×10^{10} J. Determine the energy content equivalent for both oil and coal from the fusion of the deuterium in 1.00 m^3 of water.

33-24 If the annual energy consumption on earth of about 4.00×10^{20} J is to be provided by the fusion of deuterium, (a) how many deuterium nuclei (at 7.00 MeV per fusion) must be fused? (b) What volume of water contains this number of deuterium nuclei?

33-25 How much energy would be produced if three ^{4}He nuclei fused to form a ^{12}C nucleus? (This is the reaction believed to provide some of the energy of older stars.)

33-26 Calculate the net energy released by the nuclear reaction

$$^1_0 n + {}^6_3 Li \rightarrow {}^4_2 He + {}^3_1 H$$

Appendix A
Mathematical Review

If you are a typical student in a college physics course, you have already been exposed to most of the material in this appendix. Your exposure to some of the topics may have been brief, however, and it may have been a number of years since you applied very much of this material. Thus it would be wise to read the whole appendix very carefully and work through the examples. The division by topics is intended as an aid if you need to refer to a specific point in the future.

A-1 Algebra

Algebra is defined as a generalization of arithmetic in which letters (or other symbols representing numbers) are combined according to the rules of arithmetic. Thus algebra plays a central role in physics in that we may use it to state general relationships among quantities without stating (or even knowing) the numerical value of those quantities. We may have relationships among numbers and symbols in which the symbols represent unknowns. We most often state these algebraic relationships in equation form, in which the most important symbol is the equal sign:

$$x + 4 = 9 \quad \text{and} \quad x + y = 3$$

In these and other true equations which we will encounter, the equal sign (=) means that the expression on its right has the same mathematical and physical value as the expression on its left. When we wish to manipulate an equation (perhaps to solve for an unknown), we must remember that the equality will be retained if any operation we carry out is conducted on *both* sides of the equation. That is, any equation remains an equation whenever any quantity (letters and/or numbers) is added to or subtracted from *both* sides or is used to multiply or to divide *both* sides. The only restrictions in manipulative strategies are that you may *not* multiply by infinity or divide by zero.

Solving Equations

One Unknown: Linear In a linear equation, the variable appears only to the first power. It is solved simply by manipulating the equation to isolate the unknown on one side of the equal sign and by collecting all the numbers and other letters on the other side.

Example A-1 Solve for x in the equation $6x - 9 = 2x + 3$.

Solution First, subtract $2x$ from both sides of the equation, then add 9 to both sides. Finally, divide both sides by the coefficient of x (which should be 4).

$$(6x - 9) - 2x = (2x + 3) - 2x$$

Thus

$$4x - 9 = 3$$

$$(4x - 9) + 9 = 3 + 9$$

or

$$4x = 12$$

Finally,

$$\frac{4x}{4} = \frac{12}{4}$$

$$x = 3$$

Example A-2 Solve for Δs in terms of the known quantities v, v_0, and a in the equation $v^2 = 2a\,\Delta s + v_0^2$.

Solution Isolate Δs without writing all the intermediate steps. Subtracting v_0^2 from both sides yields

$$2a\,\Delta s = v^2 - v_0^2$$

Now dividing both sides by $2a$ gives

$$\Delta s = \frac{v^2 - v_0^2}{2a}$$

It is very often useful to solve equations in terms of symbols or letters before inserting numbers.

One Unknown: Quadratic A general quadratic equation in one unknown contains both first and second powers of the unknown quantity as well as a known term. In standard form, the quadratic equation is written

$$ax^2 + bx + c = 0$$

where x is the unknown and a, b, and c are known constants. It can be shown that there are two possible values of x for which this equation is solved. Dividing by a yields

$$x^2 + \frac{b}{a}x + \frac{c}{a} = 0$$

> You are encouraged to solve equations for unknowns in symbols before inserting numbers. It will save you calculating time and also give you practice with the techniques used in derivations. Furthermore, an equation with symbols is more generally applicable than one with numbers. We rely on symbols when we attempt to describe general laws.

Adding and subtracting $b^2/4a^2$ (which is the square of one-half the coefficient of the linear term) to complete a square gives

$$x^2 + \frac{b}{a}x + \frac{b^2}{4a^2} + \frac{c}{a} - \frac{b^2}{4a^2} = 0$$

$$x^2 + \frac{b}{a}x + \frac{b^2}{4a^2} = \frac{b^2 - 4ac}{4a^2}$$

$$\left(x + \frac{b}{2a}\right)^2 = \frac{b^2 - 4ac}{4a^2}$$

Extracting the square root of both sides gives

$$x + \frac{b}{2a} = \pm\frac{\sqrt{b^2 - 4ac}}{2a}$$

Finally,

$$x = \frac{-b \pm \sqrt{b^2 - 4ac}}{2a}$$

where the plus sign is used for one of the solutions and the minus sign is used for the other. The quadratic equation often arises in physics problems. It is usually first encountered in kinematics.

Example A-3 A tennis ball is batted straight upward at an initial speed of 30.0 m/s and obeys the equation

$$\Delta s = v_0 t + \tfrac{1}{2}at^2$$

where Δs = distance of ball above its initial position at time t
v_0 = its initial upward speed
a = its upward acceleration

Noting that the acceleration is really downward at the rate of 9.80 m/s², we have $a = -9.80$ m/s², and our equation (dropping the units) becomes

$$\Delta s = 30t - 4.9t^2$$

Determine the value of the time when Δs is 20.0 m above the initial position.

Solution Substituting and rearranging yields

$$4.9t^2 - 30t + 20 = 0$$

Applying the quadratic formula, we have

$$t = \frac{-(-30) \pm \sqrt{(-30)^2 - 4(4.9)(20)}}{2(4.9)}$$

$$= 0.761 \text{ s} \quad \text{and} \quad 5.36 \text{ s}$$

The ball is at that distance once on the way up and again on the way down.

Two Unknowns: Linear A solution again requires isolating the unknowns, but for two unknowns, this must be done in separate equations. Thus to solve for two unknowns, you must have two independent equations which are satisfied simultaneously. These are appropriately called **simultaneous equations,** and there are several good techniques which may be used to isolate the unknowns.

One technique is to solve each equation for one of the unknowns in terms of the other and then set them equal to each other. This yields an equation in one unknown which may be manipulated in a way similar to that in Example A-1. Its solution may then be substituted into either of the original equations to solve for the other unknown. Finally, the solution can be checked by substituting the values of both unknowns into the other original equation to see if it reduces to an identity.

Example A-4   Solve the following simultaneous equations for x and y:

$$6y + 4x = 3y + 4$$

$$5y - 2x = 3y + 4x + 20$$

Solution   We solve each equation for y and set them equal. The first equation becomes $3y = 4 - 4x$, or

$$y = \frac{4 - 4x}{3}$$

The second equation is $2y = 6x + 20$, or

$$y = 3x + 10$$

Thus

$$\frac{4 - 4x}{3} = 3x + 10$$

$$4 - 4x = 9x + 30$$

yielding

$$13x = -26$$

$$x = -2$$

Substituting this value into the first of the original equations gives

$$6y + 4(-2) = 3y + 4$$

which yields

$$3y = 8 + 4 = 12$$

$$y = 4$$

If we substitute these values into the second of the original equations, we obtain

$$5(4) - 2(-2) = 3(4) + 4(-2) + 20$$

or

$$20 + 4 = 12 - 8 + 20$$

which reduces to an identity. Therefore the solution to the simultaneous equations is $x = -2$ and $y = +4$.

A variation of the technique employed in the previous example is also in common usage. Here, one of the original equations is solved for one of the unknowns in terms of the other unknown. This is then substituted into the second equation, yielding a single equation in one unknown, and its solution proceeds along the lines previously discussed.

Example A-5 Show that the solution to the simultaneous equations of Example A-4 may be obtained by using the second technique.

Solution The first equation, $6y + 4x = 3y + 4$, yields

$$y = \frac{4 - 4x}{3}$$

Substituting this into the second equation, $5y - 2x = 3y + 4x + 20$, gives us

$$5\left(\frac{4 - 4x}{3}\right) - 2x = 3\left(\frac{4 - 4x}{3}\right) + 4x + 20$$

Expanding and clearing fractions yields

$$20 - 20x - 6x = 12 - 12x + 12x + 60$$

giving $x = -2$ as before. Using this with either of the original equations obviously gives $y = +4$.

A third commonly used technique requires that we gather the terms in each of the unknowns. Then we multiply each of the equations by an appropriate constant so that coefficients of one of the unknowns have the same magnitude in each equation. Finally, we add or subtract one equation to or from the other to eliminate that unknown, yielding an equation in one unknown.

Example A-6 Apply the third technique to the simultaneous equations of Example A-4.

Solution First, we must rewrite the equations by gathering the unknowns. This yields

$$3y + 4x = 4$$
$$2y - 6x = 20$$

If we multiply each term in the first equation by 2 and each term in the second by 3, the equalities will be retained and the coefficients of y will be the same in each equation.

$$6y + 8x = 8$$
$$6y - 18x = 60$$

Now if we subtract the left side of the second equation from the left side of the first and the right side of the second equation from the right side of the first, we have

$$26x = -52$$

which will give us $x = -2$ and (when combined with either of the original equations) $y = +4$.

More than Two Unknowns: Linear Any of the techniques employed to solve two simultaneous equations for two unknowns may be extended to solve simultaneous linear equations in more than two unknowns. We must remember, however, that in every case a set of simultaneous equations cannot be solved completely unless the set includes as many *independent* equations as there are unknowns. When this condition is met, one can manipulate the set of equations by eliminating one of the unknowns while creating a new set of simultaneous equations (one fewer in number). This is done in succession until there is only one equation in one unknown. This is then solved and its solution used with the intermediate and finally the original equations to obtain a complete solution.

Example A-7 Solve the following equations for x, y, and z.

$$x + y + z = 3 \quad \text{(A)}$$
$$3x - 3y - 2z = 1 \quad \text{(B)}$$
$$4x + 5y + z = -2 \quad \text{(C)}$$

Solution We will eliminate the unknown z between Equations A and B and then between B and C. This will yield two equations in two unknowns, which are then solved by using a technique of the previous examples. Solving both Equations A and B for z and setting them equal yields

$$3 - x - y = \tfrac{3}{2}x - \tfrac{3}{2}y - \tfrac{1}{2}$$

or

$$6 - 2x - 2y = 3x - 3y - 1$$
$$5x - y = 7 \quad \text{(D)}$$

Solving Equations B and C for z and setting them equal yields

$$\tfrac{3}{2}x - \tfrac{3}{2}y - \tfrac{1}{2} = -2 - 4x - 5y$$

or

$$3x - 3y - 1 = -4 - 8x - 10y$$
$$11x + 7y = -3 \quad \text{(E)}$$

It is left as an exercise for you to solve Equations D and E simultaneously and with their solution determine the value of z from any of the original equations. You should obtain $x = 1$, $y = -2$, and $z = 4$.

A-2 Exponents and Logarithms

Exponents and logarithms comprise shorthand methods for writing certain mathematical operations often used in physics. You will recall that the rules of exponents (or powers) are:

1. A quantity's positive exponent shows the number of times the quantity must be taken as a multiplicative factor. For instance, $4^3 = 4 \times 4 \times 4$. Thus it follows that:

(a) In multiplying, we add the exponents of the common factor to find the exponent of the product:

$$4^3 \times 4^2 = 4^5$$

(b) In dividing, we subtract the exponent of the divisor from that of the dividend to find the exponent of the quotient.

$$4^5 \div 4^2 = 4^3$$

Consistency requires that any number raised to the zero power equals 1. Thus

$$\frac{4^5}{4^5} = 4^0 = 1$$

2. A quantity's negative exponent shows the reciprocal of that number with a positive exponent.

$$4^{-3} = \frac{1}{4^3} = \frac{1}{64}$$

You may also recall that a logarithm of a number is associated with a base number and that the logarithm of a number to a base is the power to which the base must be raised to equal that number. For example, the logarithm to the base 10 of 1000 is 3 since 10 to the third power is 1000:

$$\log_{10} 1000 = 3 \quad \text{since } 10^3 = 1000$$

The number 10 is the base of what are called the common logarithms. We often drop the 10 from the notation, and the base 10 is assumed or understood in the expression log x (logarithm to the base 10 of the number x). Another base which is even more popular is the base of what are called the natural logarithms. This base is often designated by a lowercase e and has the value 2.71828.... We sometimes write $\log_e x$, but more often we will denote this as ln x, which means natural logarithm of x (or logarithm to the base e of x). Several rules of operating with logarithms are useful and apply no matter what the base:

1. The logarithm of the product of two numbers is equal to the sum of the logarithms of the two numbers.

$$\log xy = \log x + \log y$$

2. The logarithm of the ratio of two numbers is equal to the difference of the logarithms of the two numbers.

$$\log \frac{x}{y} = \log x - \log y$$

3. The logarithm of a number to a power is equal to the power times the logarithm of the number.

$$\log x^n = n \log x$$

Each of these rules follows from the definitions of logarithms and the rules of exponents. It would be a good exercise for you to prove them all.

A-3 Series Expansions

In many physical problems, we must evaluate (or at least employ) a binomial, say $a + b$, to some power. It can be shown that the general expression is given by

$$(a+b)^n = a^n + \frac{n}{1} a^{n-1}b + \frac{n(n-1)}{2!} a^{n-2}b^2 + \frac{n(n-1)(n-2)}{3!} a^{n-3}b^3 + \cdots$$

where the exclamation point (!) means **factorial,** and we have, for example, that $3! = (3)(2)(1) = 6$ and $5! = (5)(4)(3)(2)(1) = 120$. The expression above is known as the binomial series and is valid for negative and fractional values of n as well as for positive integers. It is very valuable, and we can use it to make a good approximation whenever b is very small with respect to a. This is most easily seen if we write

$$(a+b)^n = a^n \left(1 + \frac{b}{a}\right)^n$$

From the binomial series, we have

$$\left(1 + \frac{b}{a}\right)^n = 1 + \frac{n}{1}\left(\frac{b}{a}\right) + \frac{n(n-1)}{2!}\left(\frac{b}{a}\right)^2 + \cdots$$

If $b \ll a$, then $b/a \ll 1$ and $(b/a)^2 \ll b/a$. Therefore, to a good approximation, we may write

$$\left(1 + \frac{b}{a}\right)^n \approx 1 + n\frac{b}{a}$$

where we have dropped the $(b/a)^2$ and higher-order terms as being negligible with respect to b/a. Note also that either b or n or both can be negative numbers and the approximation will still be valid.

Example A-8 Show that $(1-x)^{-1/2}$ can be approximated to four significant figures by

$$(1-x)^{-1/2} \approx 1 - (-\tfrac{1}{2})(x) = 1 + \frac{x}{2} \qquad \text{for } x = 0.02$$

Solution We carry out the calculation on the left to five significant figures (using an electronic calculator). Then we compare the value of the quantity as calculated by the approximation to five significant figures.

$$(1-x)^{-1/2} = (0.98)^{-1/2} = 1.0102$$

$$1 + \frac{x}{2} = 1 + \frac{0.02}{2} = 1.0100$$

Thus the two values are the same to four significant figures. Hence this is a good approximation.

Several other series expansions are important:

$$e^x = 1 + \frac{x}{1!} + \frac{x^2}{2!} + \frac{x^3}{3!} + \cdots$$

$$\sin\theta = \theta - \frac{\theta^3}{3!} + \frac{\theta^5}{5!} - \cdots$$

$$\cos\theta = 1 - \frac{\theta^2}{2!} + \frac{\theta^4}{4!} - \cdots$$

For these last two series, it is required that θ be in radians. One can see immediately that for small values of θ, $\sin\theta \approx \theta$ and $\cos\theta \approx 1$.

A-4 Geometry and Some Common Shapes

We must often use some of the results of plane geometry as follows:

1. A triangle is a closed figure with three straight sides and for any triangle the sum of the internal angles is 180°
2. Two angles may be equal if:
   (a) Their sides are parallel (Figure A-1)
   (b) Their sides are mutually perpendicular (Figure A-2)
   (c) They are vertical angles (Figure A-3).

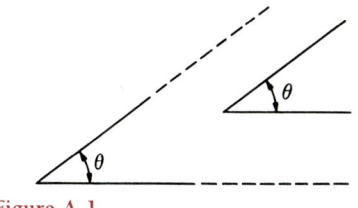

Figure A-1

Areas and Volumes of Some Common Figures

	Area	Volume
Square of side a	a^2	
Triangle of base b, height h	$\frac{1}{2}bh$	
Circle:		
Of radius r	πr^2	
Of diameter $d = 2r$	$\pi d^2/4$	
Cube of side a	$6a^2$	a^3
Sphere of radius r	$4\pi r^2$	$\frac{4}{3}\pi r^3$
Cylinder of radius r, height h	$2\pi rh$ (area of curved surface)	$\pi r^2 h$

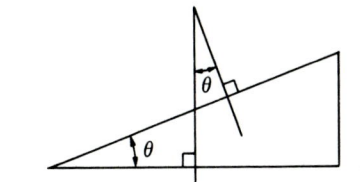

Figure A-2

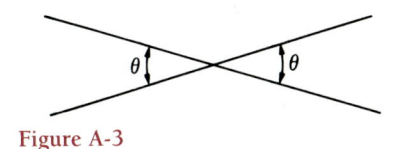

Figure A-3

A-5 Trigonometry and Trigonometric Functions

Trigonometry is the study of the properties of triangles. For most purposes, however, we will confine our applications of trigonometry to a specific triangle, the right triangle. Such a triangle has two of its sides perpendicular to each other, forming a 90° angle, or right angle, between them (Figure A-4). Note that the right angle is between the two sides of lengths a and b.

To define the trigonometric functions, we designate the lengths of the sides as a, b, and c and the spread between the lines b and c as the angle θ. We see that a is the side opposite to the angle θ and b is the side adjacent to θ. The side c, opposite the right angle, is called the hypotenuse. We use the ratios of the lengths of the sides to define the three basic functions: the sine (sin), cosine (cos), and tangent (tan). In terms of θ, these are

$$\sin\theta = \frac{\text{side opposite}}{\text{hypotenuse}} = \frac{a}{c}$$

$$\cos\theta = \frac{\text{side adjacent}}{\text{hypotenuse}} = \frac{b}{c}$$

$$\tan\theta = \frac{\text{side opposite}}{\text{side adjacent}} = \frac{a}{b}$$

Figure A-4

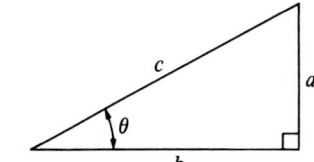

If the triangle is larger or smaller than that of Figure A-4 but the angles are the same, then these ratios remain the same. That is, Figure A-5 yields

$$\frac{a''}{c''} = \frac{a}{c} = \frac{a'}{c'}$$

The lengths of the sides of any right triangle are related by the **pythagorean theorem:**

$$c^2 = a^2 + b^2$$

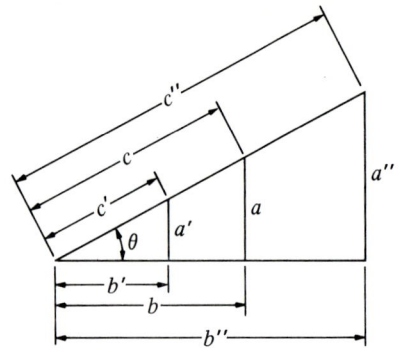

We often wish to indicate an angle, say θ, as the angle whose sine is (for example) a/c. This is written either as

$$\theta = \arcsin \frac{a}{c} \quad \text{or} \quad \theta = \sin^{-1} \frac{a}{c}$$

Figure A-5

This is sometimes known as the inverse-sine function, and the notation "inv" is used on some calculators. That is, pressing inv and then sin after entering a number will cause the calculator to present the angle whose sine is the number you entered. Similar designations and statements are true for cosine and tangent, yielding arccos ($\cos^{-1}$) and arctan ($\tan^{-1}$).

There are many useful relationships known as **trigonometric identities.** The following are most common:

For any angle θ, we have

$$\sin^2 \theta + \cos^2 \theta = 1$$

For any two angles θ and ϕ, we have

$$\sin(\theta \pm \phi) = \sin \theta \cos \phi \pm \cos \theta \sin \phi$$
$$\cos(\theta \pm \phi) = \cos \theta \cos \phi \mp \sin \theta \sin \phi$$

Most other identities may be derived rather easily by redesignating angles in the above and using the definitions given previously. For example, we can see that if we redesignate ϕ as θ in $\sin(\theta + \phi)$, we obtain $\sin(\theta + \theta) = \sin \theta \cos \theta + \cos \theta \sin \theta$ or $\sin 2\theta = 2 \sin \theta \cos \theta$.

Finally, for *any* triangle, not just right triangles, with sides of lengths a, b, and c opposite angles α, β, and γ, respectively, we have two laws:

Law of sines:
$$\frac{a}{\sin \alpha} = \frac{b}{\sin \beta} = \frac{c}{\sin \gamma}$$

Law of cosines:
$$c^2 = a^2 + b^2 - 2ab \cos \gamma$$

Appendix B
Table of Selected Isotopes

Atomic Number Z	Symbol	Name	Mass Number A	Atomic Mass, u*	Percent Natural Abundance and/or Half-Life	Decay Mode (if Radioactive)
0	n	Neutron	1	1.0086652	11 min	β^-
1	H	Hydrogen	1	1.0078252	99.985%	
		Deuterium	2	2.0141022	0.015%	
		Tritium	3	3.0160497	12.3 y	β^-
2	He	Helium	3	3.0160297	0.00013%	
			4	4.0026030	100%	
3	Li	Lithium	6	6.015123	7.42%	
			7	7.016004	92.58%	
4	Be	Beryllium	7	7.016929	53.37 d	EC, γ
			9	9.012183	100%	
5	B	Boron	10	10.0129385	19.78%	
			11	11.0093050	80.22%	
6	C	Carbon	11	11.011432	20.4 min	β^+, EC
			12	12.0000000	98.89%	
			13	13.003354	1.11%	
			14	14.0032420	5730 y	β^-
7	N	Nitrogen	13	13.005738	10 min	β^+
			14	14.0030744	99.63%	
			15	15.000108	0.37%	

* Includes electrons in the neutral atom.

Source: American Institute of Physics Handbook. 3d ed. McGraw-Hill, New York, 1972.

Atomic Number Z	Symbol	Name	Mass Number A	Atomic Mass, u*	Percent Natural Abundance and/or Half-Life	Decay Mode (if Radioactive)
8	O	Oxygen	16	15.9949150	99.759%	
			17	16.9991330	0.037%	
			18	17.9991600	0.204%	
9	F	Fluorine	19	18.998405	100%	
10	Ne	Neon	20	19.992440	90.92%	
			21	20.993847	0.257%	
			22	21.991385	8.82%	
11	Na	Sodium	23	22.989771	100%	
12	Mg	Magnesium	24	23.985044	78.70%	
13	Al	Aluminum	27	26.981541	100%	
14	Si	Silicon	28	27.976929	92.21%	
15	P	Phosphorus	31	30.973765	100%	
			32	31.973909	14.29 d	β^-
16	S	Sulfur	32	31.972074	95.0%	
17	Cl	Chlorine	35	34.968854	75.53%	
			37	36.965903	24.47%	
18	Ar	Argon	40	39.962383	99.60%	
19	K	Potassium	39	38.963710	93.10%	
			40	39.964000	0.0118%, 1.28×10^9 y	β^-
			41	40.961827	6.88%	
20	Ca	Calcium	40	39.962592	96.97%	
21	Sc	Scandium	45	44.955917	100%	
22	Ti	Titanium	48	47.947949	73.94%	
23	V	Vanadium	51	50.943964	99.76%	
24	Cr	Chromium	52	51.940510	83.76%	
25	Mn	Manganese	55	54.938046	100%	
26	Fe	Iron	56	55.934934	91.66%	
27	Co	Cobalt	59	58.933189	100%	
			60m	59.933816	10.47 min	γ, (IT)
			60	59.933811	5.26 y	β^-
28	Ni	Nickel	58	57.935336	67.88%	
			60	59.930780	26.23%	
29	Cu	Copper	63	62.929590	60.09%	
			65	64.927790	30.91%	
30	Zn	Zinc	64	63.929140	48.89%	
			66	65.926040	27.18%	
31	Ga	Gallium	69	68.925580	60.4%	
			71	70.924706	39.6%	
32	Ge	Germanium	72	71.922082	27.43%	
			74	73.921179	36.54%	
33	As	Arsenic	75	74.921600	100%	
34	Se	Selenium	78	77.917309	23.52%	
			80	79.916525	49.82%	
35	Br	Bromine	79	78.918332	50.54%	
			81	80.916292	49.46%	
36	Kr	Krypton	84	83.911506	56.90%	
			85	84.912537	10.76 y	β^-
37	Rb	Rubidium	85	84.911800	72.15%	
38	Sr	Strontium	88	87.905625	82.56%	
			90	89.907746	28.1 y	β^-
39	Y	Yttrium	89	88.905856	100%	
40	Zr	Zirconium	90	89.904708	51.46%	
41	Nb	Niobium	93	92.906378	100%	

* Includes electrons in the neutral atom.

Source: *American Institute of Physics Handbook*. 3d ed. McGraw-Hill, New York, 1972.

Atomic Number Z	Symbol	Name	Mass Number A	Atomic Mass, u*	Percent Natural Abundance and/or Half-Life	Decay Mode (if Radioactive)
42	Mo	Molybdenum	98	97.905405	23.78%	
43	Tc	Technetium	97	96.906362	2.6×10^6 y	EC
			99m	98.906403	6.0 h	γ (IT)
44	Ru	Ruthenium	102	101.904348	31.61%	
45	Rh	Rhodium	103	102.905503	100%	
46	Pd	Palladium	106	105.903475	27.33%	
47	Ag	Silver	107	106.905095	51.82%	
			109	108.904754	48.18%	
48	Cd	Cadmium	114	113.903361	28.86%	
49	In	Indium	115	114.903875	95.76%	
50	Sn	Tin	120	119.902199	32.85%	
51	Sb	Antimony	121	120.903824	57.25%	
52	Te	Tellurium	130	129.906229	34.48%	
53	I	Iodine	127	126.904477	100%	
			131	130.905076	8.07 d	β^-, γ
54	Xe	Xenon	131m	130.905252	11.8 d	γ (IT)
			132	131.904148	26.89%	
55	Cs	Cesium	133	132.905433	100%	
			137	136.907070	30.0 y	β^-, γ
56	Ba	Barium	137m	136.906526	2.558 min	γ (IT)
			138	137.905236	71.66%	
57	La	Lanthanum	139	138.906355	99.91%	
58	Ce	Cerium	140	139.905442	88.48%	
59	Pr	Praseodymium	141	140.907657	100%	
60	Nd	Neodymium	142	141.907731	27.11%	
61	Pm	Promethium	145	144.912754	17.7 y	EC
62	Sm	Samarium	152	151.919741	26.72%	
63	Eu	Europium	153	152.921243	52.18%	
64	Gd	Gadolinium	158	157.924111	24.87%	
65	Tb	Terbium	159	158.925350	100%	
66	Dy	Dysprosium	164	163.929183	28.15%	
67	Ho	Holmium	165	164.930332	100%	
68	Er	Erbium	166	165.930305	33.41%	
69	Tm	Thulium	169	168.934225	100%	
70	Yb	Ytterbium	174	173.938873	31.84%	
71	Lu	Lutetium	175	174.940785	97.41%	
72	Hf	Hafnium	180	179.946561	35.24%	
73	Ta	Tantalum	181	180.948014	99.988%	
74	W	Tungsten	184	183.950953	30.64%	
75	Re	Rhenium	185	184.952977	37.07%	
			187	186.955765	62.93%, 5×10^{10} y	β^-
76	Os	Osmium	192	191.961487	41.0%	
77	Ir	Iridium	191	190.960603	37.30%	
			193	192.962942	62.70%	
78	Pt	Platinum	195	194.964785	33.8%	
79	Au	Gold	197	196.966560	100%	
80	Hg	Mercury	202	201.970632	29.80%	
81	Tl	Thalium	203	202.972336	29.50%	
			205	204.974410	70.50%	
82	Pb	Lead	204	203.973037	1.48%	
			206	205.974455	23.6%	
			207	206.975885	22.6%	
			208	207.976641	52.3%	

* Includes electrons in the neutral atom.

Source: American Institute of Physics Handbook. 3d ed. McGraw-Hill, New York, 1972.

Atomic Number Z	Symbol	Name	Mass Number A	Atomic Mass, u*	Percent Natural Abundance and/or Half-Life	Decay Mode (if Radioactive)
83	Bi	Bismuth	209	208.930388	100%	
84	Po	Polonium	210	209.982864	138.4 d	α
85	At	Astatine	211	210.987490	7.21 h	EC
86	Rn	Radon	222	222.017574	3.824 d	α
87	Fr	Francium	223	223.019734	22 min	β^-
88	Ra	Radium	226	226.025406	1600 y	α, γ
			228	228.031100	5.75 y	β^-
89	Ac	Actinium	227	227.027751	21.8 y	α, β^-, γ
90	Th	Thorium	230	230.033131	8.0×10^4 y	α
			232	232.038054	100%, 1.41×10^{10} y	α, γ
91	Pa	Protactinium	231	231.035881	3.24×10^4 y	α, γ
92	U	Uranium	233	233.039629	1.62×10^5 y	α, γ
			234	234.040947	0.0057%, 2.5×10^5 y	α, γ
			235	235.043925	0.72%, 7.1×10^8 y	α, γ
			236	236.045563	2.39×10^7 y	α, γ
			238	238.050786	99.27%, 4.5×10^9 y	α, γ
			239	239.054291	23.5 min	β^-
93	Np	Neptunium	239	239.052950	2.35 d	β^-
94	Pu	Plutonium	238	238.049555	86 y	α, γ
			239	239.052158	2.44×10^4 y	α, γ
			240	240.053809	6580 y	α, γ
			241	241.056847	14 y	α, β^-, γ
			243	243.062030	4.98 h	β^-
95	Am	Americium	243	243.061374	7.37×10^3 y	α, γ
96	Cm	Curium	245	245.065510	9.3×10^3 y	α, γ
97	Bk	Berkelium	247	247.070300	1.4×10^3 y	α, γ
98	Cf	Californium	249	249.079581	360 y	α, γ
99	Es	Einsteinium	254	254.08282	276 d	α, β^-, γ
100	Fm	Fermium	253	253.095103	3 d	EC, α, γ
101	Md	Mendelevium	258	258.09857	54 d	α, γ
102	No	Nobelium	255	255.100941	180 s	EC, α, γ
103	Lr	Lawrencium	257	257.10536	35 s	α
104	Rf	Rutherfordium	261	261.10869	1.1 min	α
105	Ha	Hahnium	262	262.11384	0.7 min	α
106			263	263.1184	0.9 s	α
107			261	261.	1–2 ms	α
108						
109						
110						

* Includes electrons in the neutral atom.

Source: American Institute of Physics Handbook. 3d ed. McGraw-Hill, New York, 1972.

Appendix C
Periodic Table of the Elements

The values listed are based on $^{12}_{6}C = 12$ u exactly. For radioactive elements, the approximate atomic weight of the most stable isotope is given in brackets.

Period	I_A	II_A	III_B	IV_B	V_B	VI_B	VII_B	VIII			I_B	II_B	III_A	IV_A	V_A	VI_A	VII_A	0
1	1 H 1.00797																	2 He 4.003
2	3 Li 6.939	4 Be 9.012											5 B 10.81	6 C 12.011	7 N 14.007	8 O 15.9994	9 F 19.00	10 Ne 20.183
3	11 Na 22.990	12 Mg 24.31											13 Al 26.98	14 Si 28.09	15 P 30.974	16 S 32.064	17 Cl 35.453	18 Ar 39.948
4	19 K 39.102	20 Ca 40.08	21 Sc 44.96	22 Ti 47.90	23 V 50.94	24 Cr 52.00	25 Mn 54.94	26 Fe 55.85	27 Co 58.93	28 Ni 58.71	29 Cu 63.54	30 Zn 65.37	31 Ga 69.72	32 Ge 72.59	33 As 74.92	34 Se 78.96	35 Br 79.909	36 Kr 83.80
5	37 Rb 85.47	38 Sr 87.62	39 Y 88.905	40 Zr 91.22	41 Nb 92.91	42 Mo 95.94	43 Tc [99]	44 Ru 101.1	45 Rh 102.905	46 Pd 106.4	47 Ag 107.870	48 Cd 112.40	49 In 114.82	50 Sn 118.69	51 Sb 121.75	52 Te 127.60	53 I 126.90	54 Xe 131.30
6	55 Cs 132.905	56 Ba 137.34	†	72 Hf 178.49	73 Ta 180.95	74 W 183.85	75 Re 186.2	76 Os 190.2	77 Ir 192.2	78 Pt 195.09	79 Au 196.97	80 Hg 200.59	81 Tl 204.37	82 Pb 207.19	83 Bi 208.98	84 Po [210]	85 At [210]	86 Rn [222]
7	87 Fr [223]	88 Ra [226]	‡															

† Lanthanide series

57 La 138.91	58 Ce 140.12	59 Pr 140.91	60 Nd 144.24	61 Pm [147]	62 Sm 150.35	63 Eu 152.0	64 Gd 157.25	65 Tb 158.92	66 Dy 162.50	67 Ho 164.93	68 Er 167.26	69 Tm 168.93	70 Yb 173.04	71 Lu 174.97

‡ Actinide series

89 Ac [227]	90 Th 232.04	91 Pa [231]	92 U 238.03	93 Np [237]	94 Pu [242]	95 Am [243]	96 Cm [247]	97 Bk [247]	98 Cf [251]	99 Es [254]	100 Fm [253]	101 Md [256]	102 No [254]	103 Lw [257]

Appendix D
Answers to Odd-Numbered Problems

Chapter 1

1-1 (a) m² (b) m³/s (c) kg/m³
1-3 (a) m/s² (b) kg · m/s² (c) kg/m·s²
   (d) kg·m²/s²
1-5 (a) 1115 ft/s (b) 339.8 m/s (c) 1223 km/h
1-7 (a) 27 ft³ (b) 35.3 ft³ (c) 0.0164 liters
1-9 (a) 86,400 s (b) 604,800 s (c) 31,622,400 s
1-11 (a) 171.5 mi (b) No; by 5 p.m. you will have traveled only (3.33 h)(45.0 mi/h) = 150 mi.
1-13 (a) 170.2 cm (b) 1.702 m
1-15 1.93×10^4 kg/m³
1-17 (a) 63.0 yd² (b) 169 m³ (c) Each dimension has four significant figures. Therefore the answer should be reported to at most *four* figures.
1-19 (a) 102.1 cm (b) 644 cm²
1-21 (a) Five (b) Five (c) Three (d) Three
1-23 (a) 7116 in² (b) 347.62 in
1-25 (a) 5.216×10^3 m (b) 4.17×10^{-3} kg
(c) 1.121×10^1 s (d) 2.13×10^{-6} s
(e) 2.99792458×10^8 m/s
1-27 1.3×10^{-10} m³/s

Chapter 2

2-1 4.69 m/s
2-3 4.62×10^{-3} m/s
2-5 5.00×10^5 m/s
2-7 (a) 1.47×10^3 km/h (b) (i) The rotation of the earth is to the east and gives the rocket added eastward motion. (ii) Over the ocean for safety (not true in California).
2-9 (a) 0.200 h (b) Car 1 27.0 km, car 2 24.0 km
2-11 (a) 1.74 m/s (b) 5.28 km
2-13 (a) 25.3 s (b) 2.66×10^3 m
2-15 (a) -0.300 m/s² (b) 20.0 s (c) 9.00 m/s
2-17 0.322 cm/day²
2-19 (a) 238 m/s (b) 2.61×10^5 m

A16

2-21 (a) -15.0 m/s² (b) 4.00 s
2-23 (a) 5.60×10^3 km (b) 2800 km (c) Answer (a) is more likely.
2-25 3.96×10^3 people/h
2-27 (a) Yes, you will have 3.06 m to spare. (b) No, collision will take place 30.0 m before you can stop.
2-29 (a) 22.1 m/s (b) 2.26 s
2-31 (a) 100 m (b) 184 m (c) 6.12 s
2-33 6.26 m/s
2-35 (a) — (b) $v_0 = 23.3$ cm/s; $a \approx 114$ cm/s². We can obtain the acceleration due to gravity but only after we study combined rotational and translational motion in Chapter 9. These data yield 922 cm/s² = 9.22 m/s², about a 6% error.
2-37 (a) 2.30 s (b) 7.94 m (c) Stone 22.6 m/s, elevator 6.90 m/s

Chapter 3

3-1 6.43 blocks, 51.2° W of S
3-3 Approximately 18.2 m/s due east, 10.5 m/s due north
3-5 4.58 km at 40.9° N of W
3-7 $\mathbf{r} = 5.44$ m at $\theta = 41°$, $\mathbf{R} = 8.69$ m at 66.2°
3-9 6.16×10^{-6} N at 52° with respect to either O–H bond
3-11 (a) 12.6 m (b) 5.03 m/s
3-13 11.4 N at 13.3° below positive x axis
3-15 131 m
3-17 105 m
3-19 (a) 75.0 m (b) 14.7 m/s (c) 30.5°
3-21 (a) 0.589 s (b) 1.19 m/s
3-23 (a) 21.3 m/s (b) 5.78 m (c) 2.17 s
3-25 0.373 s
3-27 (a) 2.82 s (b) 79.5 m
3-29 19.6 m
3-31 (a) 32.9 m/s (b) 33.4 m/s
3-33 833 km/h at 5.16° N of W
3-35 (a) 3.00 km/h (b) 36.9° upstream with respect to a line directed across stream (c) 22.5 min
3-37 13.6 m/s

Chapter 4

4-1 1.68×10^3 N
4-3 2.50×10^3 N
4-5 0.420 N
4-7 479 N against motion of ball
4-9 340 N
4-11 (a) 1.60 m/s (b) 20.0 kg/s
4-13 -3.80 m/s² (downward)
4-15 (a) 60.1 N (b) 0.700 cm
4-17 (a) 3.92×10^4 N (b) 0.0891 (c) 7.84×10^3 N
4-19 (a) 1.07×10^4 N (b) 1.50 s (c) 7.13×10^7 N/m², not large enough to shatter bone
4-21 (a) 163 N (b) 74.0 N
4-23 (a) 200 m/s² (b) 1.00×10^3 N (c) 12.8 cm²
4-25 (a) 0.563 kg (b) 2.11 N
4-27 Wt in N = wt in lb $\left(\dfrac{9.80 \text{ m/s}^2}{2.20 \text{ lb/kg}} \right)$

Mass in kg = $\dfrac{\text{wt in lb}}{2.20 \text{ lb/kg}}$

Wt on moon = (mass in kg)(1.62 m/s²)
4-29 (a) 1.14×10^4 N (b) 3.04×10^3 N (c) 2.54 m/s² (d) 123 m
4-31 14.3°
4-33 Tension in cord, weight of water in bucket, weight of bucket, pull by boy, . . .
4-35 Normal force, weight force
4-37 (a) 88.0 N directed opposite the motion (b) 0.408
4-39 40.8 m
4-41 0.588
4-43 (a) 52.0 N (b) 48.0 N directed up the plane
4-45 $a = 1.31$ m/s², $T_A = 24.8$ N, $T_B = 102$ N

Chapter 5

5-1 (a) 22.0 rad (b) 7.70 m
5-3 23.6 min
5-5 77.8 rad/s
5-7 (a) I, VIII (b) III (c) IV, VII (d) II, IX (e) V (f) VI
5-9 (a) 8.17 rad/s (b) 0.409 rad/s²
5-11 (a) -31.5 rad/s² (b) 62.5 rev
5-13 (a) -0.166 rad/s² (b) 4.83 rad/s
5-15 (a) 3.00 rad/s² (b) 15.0 s (c) 338 rad
5-17 (a) 1.29 m/s² (b) 162 m/s (c) 126 s
5-19 (a) 0.165 rad/s (b) 0.578 m/s²
5-21 (a) 0.571 rad/s (b) 2.57 m/s (c) 1.47 m/s² (d) 44.1 N
5-23 (a) 2.00×10^7 rad/s (b) 8.00×10^{14} m/s² (c) 7.29×10^{-16} N
5-25 2.86 rad/s
5-27 211 rad/s
5-29 (a) 669 N (b) 899 N
5-31 1.22×10^{-9} N
5-33 3.46×10^8 m
5-35 (a) 122 kg (b) 532 N (no) (c) 6.46×10^3 m/s (d) 6.76×10^{-4} rad/s
5-37 (a) 28.8 m/s (forward) (b) 1.50 m/s (backward) (c) No
5-39 (a) 1.26×10^{22} N (b) 4.78×10^4 m/s (c) 7.62×10^6 s

5-41 (a) 4.16×10^{41} kg (b) 2.09×10^{11}
5-43 (a) 4.98×10^3 m/s aphelion, 3.85×10^4 m/s perihelion (b) 5.44×10^9 km

Chapter 6

6-1 112 N at 26.6°
6-3 (a) 3.82 N, 3.82 N (b) 3.82 N, 0.989 N, 3.69 N
6-5 2.87×10^3 N
6-7 (a) 750 N (b) 0.300
6-9 0.286 m²
6-11 —
6-13 28.0 N·m about end nearer to the force, 92.4 N·m about farther end
6-15 5.60 N
6-17 37.5 N
6-19 435 N
6-21 (a) 295 N·m clockwise (b) 0.492 m
6-23 1.20 m from the 14.0-N weight
6-25 1.04×10^3 N, 781 N
6-27 1.85 m from the left end
6-29 630 N
6-31 Top: 296 N up, 295 N away from door; bottom: 296 N up, 295 N toward door. Must make some assumption about the vertical forces—for example, that the two vertical forces are equal to each other.
6-33 0.848 m
6-35 (a) 0.656 m to the right of left end (b) Yes, the center of mass and center of gravity are at the same point in a constant gravitational field.
6-37 7.84×10^4 N (nearly 9 tons)
6-39 (a) 624 N (b) 15.6 N·m (c) 312 N, 195 N, 173 N

Chapter 7

7-1 2.40×10^5 J
7-3 (a) 600 J (b) -600 J (c) -600 J
7-5 (a) 426 N (b) 114 N (c) 34.1 N (d) 148 N (e) 2220 J (f) -512 J (g) 0
7-7 (a) 17.7 N (b) -4.43 J (c) -23.4 J (d) 27.8 J
7-9 4.48×10^4 J
7-11 833 W, 1.12 hp
7-13 1.49 m/s
7-15 (a) 9.48×10^7 J (b) 191 s
7-17 53.0 N
7-19 (a) 852 J (b) -852 J
7-21 (a) 492 J (b) 6.32 m/s
7-23 3.77×10^3 m
7-25 273 W
7-27 (a) 5.63×10^3 J (b) 626 W
7-29 (a) Yes (b) 14.0°
7-31 (a) 1.70×10^{18} J (b) 2.88×10^8 barrels (c) 4.05×10^4 bombs (d) 20.2 bombs (e) 5.39×10^6 W, 7.23×10^3 hp (f) 8.40×10^{20} W, 1.13×10^{18} hp
7-33 (a) -11.7 J, -33.3 J, 21.6 J (b) 0.237 m
7-35 (a) 57.2 m/s (b) 232 J (c) 774 N
7-37 (a) 3.35×10^4 J (b) 42.0 m/s
7-39 3.69 m

Chapter 8

8-1 (a) 133 N (b) 26.6 N·s
8-3 132 kg·m/s due west
8-5 (a) -1.80×10^4 kg·m/s (b) 3.60×10^3 N
8-7 (a) 24.0 N·s in direction of motion (b) 0.0300 kg
8-9 (a) 1.83×10^4 kg·m/s at 20.3° W of S (b) 2.29×10^3 N at 20.3° W of S
8-11 (a) 27.0 N·s (b) 145 m/s in direction opposite initial motion
8-13 (a) $A - 19.5$ kg·m/s, $B - 10.0$ kg·m/s (b) A 163 N, B 83.3 N
8-15 3.60×10^4 kg
8-17 Be 1.91×10^6 m/s, D 4.91×10^6 m/s
8-19 Cue ball 5.18 m/s, ten ball 1.34 m/s
8-21 (a) 22.0 m/s (b) -5.25×10^4 J
8-23 (a) 34.6° (b) 9.12 kg·m/s away from wall
8-25 (a) 604 m/s (b) -1.41×10^4 J
8-27 1200-kg automobile 15.5 m/s, 1000-kg automobile 11.5 m/s
8-29 (a) 7.03 m/s at 47.4° (b) -3.75×10^3 J
8-31 2.40 m/s
8-33 9.60 m/s²
8-35 0.639 N
8-37 (a) 5.00 m/s (b) 11.0 m/s
8-39 22.2 N
8-41 90.0 N

Chapter 9

9-1 21.0 rad/s²
9-3 0.435 kg·m²
9-5 1.08 kg·m²
9-7 165 N·m
9-9 (a) -85.5 N·m (b) 24.0 rad
9-11 (a) 216 kg·m² (b) 43.2 N·m (c) 123 rad
9-13 0.918 s
9-15 —
9-17 (a) 0.520 N·m (b) 65.3 W
9-19 (a) 84.1 rad/s (b) 412 J
9-21 $T_1 = 3.84$ N, $T_2 = 5.93$ N, 1.25×10^{-2} kg·m²
9-23 (a) 5.03 N·m (b) 15.8 J
9-25 (a) 0.656 J (b) 2.30 J (c) 615 J (d) 0.656 W, 2.30 W, 615 W

9-27 (a) Orange 1.67×10^{-4} kg·m², can 2.08×10^{-4} kg·m² (b) Orange 2.96 m/s, can 2.85 m/s (c) Orange 2.03 s, can 2.10 s

9-29 (a) 5.50 rad/s (b) 145 J (c) 145 J (d) 290 J

9-31 (a) 54.2 rad/s, 5.42 m/s (b) 62.6 rad/s, 6.26 m/s (c) Trans KE: cyl > hoop, rot KE: hoop > cyl, total KE: the same for both

9-33 2.78×10^4 kg·m²/s

9-35 (a) 0.938 kg·m² (b) 4.69 kg·m²/s (c) 11.7 J (d) 7.46 N

9-37 2.49 rad/s

9-39 95.5%

9-41 (a) 15.7 kg·m²/s (b) 31.4 J (c) 7.73 rad/s (d) 60.7 J

9-43 —

9-45 (a) 25.0 kg·m²/s (b) 22.1 N·m (c) 0.882 rad/s

Chapter 10

10-1 $3.82 \cos[(2\pi/3)t + (\pi/6)]$

10-3 $0.120 \cos[6\pi t + (\pi/2)]$

10-5 Approximately 4.47×10^5 km to the right

10-7 (a) 55.6 N/m (b) 10.5/s (c) 0.504 kg

10-9 (a) 1.67×10^{-2} Hz (b) 4.32×10^4 s, 2.31×10^{-5} Hz

10-11 536 N

10-13 (a) 62.8 m/s (b) 3.95×10^{14} m/s²

10-15 0.444

10-17 (a) 18.0 rad/s (b) 0.349 s (c) 0.167 m

10-19 87.7 J

10-21 (a) KE 75.0%, PE 25.0% (b) $\pm\sqrt{2}/2 A$

10-23 (a) 4.49 s (b) 0.223 Hz (c) 20.0 m

10-25 (a) 2.24 s (b) 3.07 s

10-27 9.7464 m/s²

10-29 (a) $ml^2/3$ (b) $2\pi\sqrt{2l/3g}$ (c) 0.280 m

10-31 (a) $2mr^2$ (b) $2\pi\sqrt{2r/g}$

10-33 0.990 s

10-35 $k_{\text{eff}} = \dfrac{k_1 k_2}{k_1 + k_2}, f = \dfrac{1}{2\pi}\sqrt{\dfrac{k_1 k_2}{(k_1 + k_2)m}}$

10-37 $k_{\text{eff}} = k_1 + k_2 + \dfrac{k_3 k_4}{k_3 + k_4},$

$f = \dfrac{1}{2\pi}\sqrt{\dfrac{(k_1 + k_2)(k_3 + k_4) + k_3 k_4}{(k_3 + k_4)m}}$

10-39 (a) 3.48 s (b) 25.0 pushes

Chapter 11

11-1 3.41 m, 2.78 m

11-3 8.96×10^{-3} m

11-5 (a) 357 Hz (b) 300 Hz

11-7 —

11-9 0.750 m

11-11 1:2

11-13 (a) 3.52×10^3 m/s (b) 2.15×10^3 m/s

11-15 (a) $(n - \frac{1}{8})$ cm; $n = 0, 1, 2, \ldots$ (b) $[(2n + \frac{1}{2}) - \frac{1}{8}]$ cm; $n = 0, 1, 2, \ldots$

11-17 (a) Yes (b) No (c) 3.33×10^{-7} m

11-19 (a) Plot of $y = 0.400 \cos[10\pi t + (\pi/3)]\sin[2\pi x + (\pi/3)]$ at times indicated (b) 1.00 cm

11-21 (a) 4.00 m (b) 113 Hz

11-23 (a) 1.98×10^3 N (b) 0.0421 m

11-25 246 Hz

11-27 (a) 0.165 m (b) 0.018 m, 0.041 m, 0.055 m, 0.0672 m, 0.0825 m

11-29 (a) 18.9 Hz (b) 37.8 Hz

11-31 Source of vibration is difference in frequency of engines (a resonance phenomenon). It probably can be eliminated by making certain the two engines have exactly the same frequency.

11-33 3.16×10^{-2} W/m², 0.316 W/m²

11-35 (a) 513 Hz (b) 3850 rpm (c) 39.2 m/s

11-37 15.4 m/s due west

Chapter 12

12-1 (a) 3.60×10^{-6} m² (b) 5.56×10^5 N/m²

12-3 1.43×10^{-2} m

12-5 (a) 3.40 m/s (b) 0.995 N (c) 0.0729 m

12-7 (a) 2.00×10^6 N/m² (b) 3.00×10^{-5} (c) 6.67×10^{10} N/m²

12-9 7.16×10^{-2} m

12-11 8.66×10^9 N/m²

12-13 (a) 1.42×10^5 N (b) 946 people (473 on each side) (c) No (d) 43.8 m (e) 3.11×10^6 J (f) 3.26 sticks (g) 5.68×10^3 m/s² (h) 56.8 m/s (i) 588 ft

12-15 (a) 46.9 N/m² (b) 0.384 (c) 122 N/m²

12-17 (a) 1.09×10^{-7} m (b) -8.60×10^{-5} m

12-19 (a) 36.3 m², 20.6 m³ (b) 20.6 m³ (c) 4.72×10^{-4} m³

12-21 (a) 2.23×10^3 N (b) 31.8 m² (c) 2.23×10^7 N (d) 2.48 (e) 0.0248

Chapter 13

13-1 2.69 kg

13-3 1.65×10^{-3} m³

13-5 1.55×10^6 N

13-7 (a) 1.80 m (b) 1.13 m

13-9 1.60×10^4 N

13-11 (a) 8.01×10^3 m (b) 10.3 m

13-13 17.1 N

13-15 (a) 5.21×10^3 kg/m³ (b) 7.43 N

13-17 (a) 4.50 m/s (b) 3.42×10^5 N/m²

A19

13-19 (a) 2.04 m/s, 22.6 m/s (b) 2.53×10^5 N/m²
13-21 0.179 m
13-23 (a) $R = 2\sqrt{h(H-h)}$ (b) 6.93 m, 8.00 m, 6.93 m
13-25 6.22 m
13-27 (a) — (b) 3.12×10^4 N/m², 228 m/s
13-29 (a) 6.66×10^4 N/m² (b) 2.03 m/s (c) 2.30×10^{-2} m³/s
13-31 98.2 m/s
13-33 4.54×10^4 N/m²
13-35 (a) Clockwise (b) Right to left (c) Open-faced
13-37 (a) $v_1 A_1 = v_2 A_2$; thus if $v_2 > v_1$, then $A_2 < A_1$. (b) 9.26×10^{-3} m
13-39 (a) 0.564 m/s (b) 2.61×10^{-2} m/s (c) 44.8 m/s (d) 1.32×10^{-4} m/s
13-41 60.0 m/s
13-43 0.0174 m/s
13-45 0.406 m

Chapter 14

14-1 (a) $-297.33°F$ (b) $787.24°F$
14-3 $136.4°F = 58.0°C = 331$ K, $-126.9°F = -88.3°C = 185$ K
14-5 (a) $20 + 14.1 t_A$ (b) $14.1 t_A - 253$ (c) $(t_C + 253)/14.1$ (d) $25.0°A$
14-7 15.11 cm
14-9 6.30×10^{-2} m
14-11 1.80 cm³
14-13 2.5396 cm
14-15 30.0078 cm
14-17 —
14-19 18.2
14-21 53.0 cm
14-23 5.39×10^{-2} m
14-25 8.29 N·m/mol·K
14-27 (a) 19.6 mol (b) 0.486 m³
14-29 1.09
14-31 2.47×10^4 molecules
14-33 (a) 232°C (b) 281°C
14-35 (a) 10.0 atm (b) 37.8 atm (c) 47.6 atm (d) 8.34×10^4 N (e) Answers (c) and (d) would change because the number of moles would increase by a factor smaller than 34/27.
14-37 (a) 478 m/s (b) 1350 m/s (c) 218 m/s
14-39 (a) 6.36×10^3 K for nitrogen, 7.27×10^3 K for oxygen (b) 1.41×10^5 K for nitrogen, 1.61×10^5 K for oxygen

Chapter 15

15-1 5.23×10^5 cal
15-3 (a) 78.4 J (b) 4.69×10^{-2} K
15-5 (a) 1.02 C° (b) 0.162%
15-7 (a) 4.18×10^6 J (b) 1.85 kg
15-9 1.41×10^2 m/s
15-11 (a) 1.05×10^5 J (b) 3.35×10^5 J (c) 4.18×10^5 J (d) 2.26×10^6 J (e) 2.01×10^5 J, Total $= 3.32 \times 10^6$ J
15-13 1.44×10^3 kg, 0.576 m³
15-15 10.5 kg
15-17 (a) 95.8°C (b) 8.65×10^{-3} kg (c) 74.3°C
15-19 (a) 9.93°C (b) 3.24
15-21 (a) 0.333 kg (b) 3.63 kg
15-23 (a) 5.12×10^8 J (b) 6.47×10^9 J (c) $239.41 (d) $1890 (e) $1122 (f) $4200
15-25 1.43×10^7 J
15-27 (a) 2.95×10^9 J (b) $13.50
15-29 (a) 1.35×10^9 J (b) 6.75×10^8 J (c) 1.81×10^7 J
15-31 (a) 0.584 W/m·°C (b) 27.8 W/m²·°C at 1.00 m/s, 62.5 W/m²·°C at 10 m/s
15-33 (a) 9.93×10^{-8} W/m²·°C (b) 0.162 J
15-35 (a) 174 W (b) 103 W (c) 71.6 W (d) 6.19×10^6 J
15-37 (a) 6.42×10^7 W/m² (b) 3.91×10^{26} W (c) 1.38×10^3 W/m²
15-39 (a) 2.64×10^{-5} kg/s (b) 0.474 kg
15-41 1.66×10^4 g

Chapter 16

16-1 (a) 2.00×10^4 J (b) 1.00×10^4 J (c) 3.00×10^4 J
16-3 (a) 2.27×10^3 J (b) 1.03×10^4 J
16-5 (a) 0, 6.39×10^3 J (b) 1.12×10^3 K, 1.03×10^3 K
16-7 (a) 1.57×10^7 J (b) 1.60×10^3 m (1.00 mile)
16-9 (a) 9.95% (b) 1.09×10^3 kg/m³ (c) 65.0% (d) 8.04×10^5 N/m²
16-11 7.00×10^4 J
16-13 (a) 9.52 atm $= 9.64 \times 10^5$ N/m² (b) 244 K, 464 K (c) -9.15×10^3 J (d) 9.15×10^3 J
16-15 (a) 2.98×10^{-4} K/m (b) About 2.20×10^3 K
16-17 1.42×10^2 m, not reasonable.
16-19 (a) 0.9886 atm (b) 1.008 m³ (c) 292.0 K (1°C drop) (d) 17.12 g, 16.22 g (e) Energy is tranferred from the water to the air as heat of vaporization; 2.034×10^3 J; lower (f) As moist air rises, it cools. If saturated, it must release moisture as water and transfer the heat of vaporization (or condensation) to the air. When the dry air drops, it is warmed at the dry lapse rate. The heat released stays with it, and the result is air at a higher temperature than the original.

16-21 (a) 2.00×10^4 J (b) $0.400 = 40.0\%$
16-23 (a) 1.00×10^5 J (b) 7.00×10^4 J
16-25 (a) 6.00×10^4 J (b) 1.50
16-27 2.58 to 1.49
16-29 435 J/s
16-31 (a) 0.602 (b) 32
16-33 1.62×10^5 J
16-35 (a) 0.446 (b) 1.24
16-37 $8.58 \times 10^{-2} = 8.58\%$
16-39 (a) 1.68×10^7 J (b) 11.9 (c) 1.82×10^7 J
  (d) 1.40×10^6 J
16-41 (a) 0.625 (b) 300 K
16-43 (a) 1.87×10^4 J (b) 3.70×10^3 J
16-45 (a) -9.56 J/K (b) $+10.2$ J/K (c) $+0.640$ J/K

Chapter 17

17-1 118 N
17-3 1.80×10^{-8} N
17-5 1.81×10^{13} C
17-7 (a) Electric repulsion (b) 1.71×10^{18} N
  (c) 1.71×10^{18} m/s² (d) 1.75×10^{-10} s
  (e) Electrostatic forces are much stronger than gravitational forces. (f) The force required and hence the work (or energy) to collect them in one central location and hold them there are impractical to achieve.
17-9 0.427 m
17-11 1.82×10^{-3} N, $-34.9°$
17-13 1.02×10^7 N/C
17-15 (a) 7.03×10^{15} m/s² (b) 3.83×10^{12} m/s²
  (c) 2.00 m/s²
17-17 $qEx^2/2mv_0^2$
17-19 4.00×10^{-8} C
17-21 106 m
17-23 -3.36×10^{-3} C
17-25 (a) 1.44×10^3 N/C (b) 1.38×10^3 N/C
17-27 —
17-29 (a) 2.68×10^{-7} m (b) 4.29×10^{-25} C·m
17-31 2.88×10^{-26} N·m

Chapter 18

18-1 (a) 2.88×10^5 J (b) 9.80 m
18-3 (a) 694 V (b) -1.39×10^{-3} J (c) $+1.39 \times 10^{-3}$ J
18-5 4.00×10^5 J
18-7 (a) To a distance 3.00 m from the second charge. (b) 3.60×10^3 V
18-9 -2.78×10^{25} J
18-11 (a) 217 V (b) 3.47×10^{-17} J (c) 2.60×10^{-17} J
18-13 (a) 0.300 m (b) 1.50 m
18-15 (a) 1.64×10^{-16} J (b) 6.84×10^{-17} J
  (c) -225 V
18-17 9.17×10^5 V
18-19 (a) 3.02×10^{-19} J (b) 1.63×10^{-18} J
18-21 5.81 C/s
18-23 1.12 s
18-25 (a) -400 V (b) 5.00×10^3 V/m directed from P to Q
18-27 6.93×10^3 V
18-29 (a) 22.6 m² (b) 2.50×10^{-7} F
18-31 (a) 1.07×10^{-5} C (b) 1.67×10^{-3} C
  (c) 1.05×10^{-3} C
18-33 (a) 0.984 V (b) 3.94 V (c) 6.56×10^3 V/m, 6.56×10^3 V/m
18-35 (a) 44.7 V (b) 2.24×10^{-3} C
18-37 (a) 7.07×10^3 V (b) 0.113 C
18-39 (a) 9.00×10^{-4} C (b) 2.03×10^{-2} J
  (c) 9.00×10^{-4} C (d) 0.122 J (e) Work was done to remove the dielectric, which goes to strengthen the electric field.
18-41 (a) 7.20×10^{-8} F (b) 9.00×10^{-5} J
  (c) 3.60×10^{-7} F (d) 10.0 V (e) 1.80×10^{-5} J
18-43 6.67×10^6 V/m

Chapter 19

19-1 2.40×10^5 A
19-3 (a) 2.90×10^{-6} m/s (b) 1.45×10^5 s = 40.2 h
19-5 (a) 0.222 h = 13.3 min (b) 2.70×10^4 C
19-7 (a) 6.00×10^{-3} s (b) 3.00×10^{-2} A
19-9 (a) 0.139 A (b) 5.85×10^{-5} m/s (c) 0.105 m
19-11 (a) 0.150 V (b) 3.00 V
19-13 (a) 1.15 m (b) 20.8 A
19-15 4.48 h
19-17 (a) 0.692 Ω (b) 17.3 V
19-19 (a) 22.2 V (b) 17.0 V
19-21 (a) 4.05 Ω (b) 162 V (c) 211 V
19-23 (a) 6.07×10^{-5} Ω (b) 0.0535 Ω (c) 6.88×10^7 Ω (d) 1.91×10^{21} Ω
19-25 (a) 7.59×10^{-2} Ω (b) 0.307 Ω (c) 2.58 mm
19-27 (a) 80.0 Ω (b) 761°C (c) 180 W (d) 1.08×10^4 J
19-29 (a) 144 Ω (b) 8.07×10^{-7} Ω·m (c) 4.48×10^{-10} m² (d) 10.0 Ω
19-31 (a) 7.23×10^{-5} m/s (b) 2.77×10^4 s
  (c) 150 Ω (d) 96.0 W
19-33 (a) 7.20 V (b) 0.200 Ω (c) 173 W (d) 115 W
19-35 180 J
19-37 (a) 12.0 Ω (b) 10.0 A (c) 1.20×10^3 J/s
  (d) 0.574 C°/s (e) 60.0 V/m
19-39 (a) $I = \mathscr{E}/(R+r)$ (b) $P = \mathscr{E}^2R/(R+r)^2$
  (c) $P_{1.10} = 24.9$ W, $P_{1.00} = 25.0$ W, $P_{0.900} = 24.9$ W

A21

Chapter 20

20-1 (a) 0.0400 A (b) 0.200 A (c) 0.480 W, 2.40 W (d) 0.160 W, 4.00 W
20-3 (a) 80.0 Ω (b) 5.00 W (c) 5.00 V, 0.167 A
20-5 (a) 0.500 A (b) 240 Ω (c) 16.0 Ω (d) 8.00 V
20-7 (a) 576 Ω (b) 0.208 A (c) 8.31 W (d) — (e) 288 Ω (f) 0.417 A (g) 0.208 A (h) 50.0 W (i) 66.7%
20-9 R_1: 0.0525 A, 0.630 V, 0.0331 W
R_2: 0.0525 A, 0.630 V, 0.0331 W
R_3: 0.0350 A, 0.420 V, 0.0147 W
R_4: 0.140 A, 1.68 V, 0.235 W
R_5: 0.0175 A, 0.210 V, 0.00368 W
R_6: 0.0175 A, 0.210 V, 0.00368 W
R_7: 0.193 A, 2.32 V, 0.448 W
20-11 (a) 5.71 h (b) 1.82 h
20-13 (a) 60.0 Ω (b) 120 Ω (c) 420 Ω
20-15 —
20-17 (a) $\dfrac{\mathscr{E}_2(R_1 + R_3) - \mathscr{E}_1 R_3}{R_1 R_2 + R_1 R_3 + R_2 R_3}$ (b) —
20-19 0.927 A, 2.13 A, 2.67 V
20-21 $\mathscr{E} \dfrac{R_1(R_3 + R_4) - R_3(R_1 + R_2)}{(R_1 + R_2)(R_3 R_4) + R_1 R_2(R_3 + R_4) + R_m(R_1 + R_2)(R_3 + R_4)}$
20-23 (a) 30.0 A, 10.0 A, 20.0 A (b) 6.00 A, 8.57 A, 10.0 A
20-25 (a) 1.25×10^5 Ω (b) 4.50×10^{-4} Ω
20-27 (a) 3.33×10^{-2} A (b) 2.96×10^3 Ω
20-29 (a) 90.0 μC, 54.0 μC, 18.0 μC; 9.00 V (b) 11.7 μC; 1.17 V, 1.95 V, 5.85 V
20-31 (a) 6.00 μF (b) 192 μC (c) 96.0 μC (d) 7.68×10^{-4} J
20-33 (a) 22.3 μF in series with the battery or 8.57 μF in series with the 30.0-μF and 60.0-μF capacitors. (b) 4.80×10^{-4} C, 5.17×10^{-3} J (240 μC, 3.36×10^{-3} J)
20-35 (a) 3.50×10^{-3} C, 7.00×10^{-3} C (b) 350 V (c) 0.613 J, 1.23 J
20-37 (a) 16.0 s (b) 73.7 s
29-39 2.67×10^{-7} F
20-41 (a) Joule (b) 133 C (c) 2.66×10^5 A (d) 0.668 s (e) 46.8 F (f) 0.0143 Ω (g) 0.250 V/cell (h) 0.0356 Ω (i) 1.67×10^4 J (j) $51.84/kWh

Chapter 21

21-1 5.30×10^{-5} T upward
21-3 (a) 1.96×10^7 m/s (b) 0.818 T
21-5 (a) 8.00×10^5 m/s (b) 1.00×10^6 m/s
21-7 (a) 4.61×10^{-17} N (b) West (c) 5.06×10^{13} m/s²
21-9 (a) 0.130 m/s (b) 6.12×10^{-4} m³
21-11 (a) 1.47×10^{-26} kg (b) 1.73×10^{-26} kg (c) 6.00×10^5 m/s
21-13 (a) 1.05×10^{-3} A (b) 12.4 T
21-15 (a) 3.20×10^{-17} N toward wire (b) 3.20×10^{-17} N away from wire (c) Opposite the direction of the current as the proton approaches the wire and in the direction of the current as proton departs.
21-17 0.524 T
21-19 (a) 3.00×10^{-3} T (b) This field is about 50 times as strong as the earth's. (c) (i) Points north, (ii) Abruptly swings to point west, (iii) Stays pointing west
21-21 0.928 A
21-23 (a) 1.20×10^{-3} N/m (b) 447 A (c) Repulsive
21-25 (a) 4.00×10^{-3} N/m (b) Repulsive (c) 3.50×10^{-3} N/m (d) 0.571 m (e) Western
21-27 1.71 N repulsive
21-29 (a) 0.202 T (b) 5.71 m/s
21-31 (a) 2.51 m/s (b) 3.77×10^{-3} V
21-33 83.3 rad/s
21-35 (a) $NBlw\omega \sin \omega t$ (b) 8.15×10^{-2} V
21-37 —
21-39 (a) 1.84×10^{10} A (b) Counterclockwise as viewed from the north pole

Chapter 22

22-1 (a) 0.200 H (b) 2.50×10^4 A/s
22-3 (a) 24.0 H (b) 4.20×10^4 V
22-5 (a) 2.79×10^{-3} H (b) 4.18×10^{-4} V (c) 0.314 J (d) 624 J/m³
22-7 Plot using $V = IR + L(\Delta I/\Delta t)$ and taking data from Figure 22-19. For example, in the interval 0 to 2.00 s, we calculate $\Delta I/\Delta t = 4.00$ A/s.
22-9 (a) 75.0 A/s (b) 3.77 A (c) 201 V
22-11 (a) 1.39 A (b) 1.01 A/s (c) 1.12 W (d) 0.966 W (e) 2.09 W
22-13 (a) 8.00 A (b) 640 J (c) 12.0 s
22-15 (a) 1.80 A (b) 4.50 A/s (c) 0.400 s (d) 0.400 s
22-17 (a) 91.9 Hz (b) 8.00×10^{-5} F
22-19 (a) 2.16×10^{-5} N·m (b) 46.0°
22-21 (a) 48.0° (b) 1.88×10^{-3} A
22-23 (a) 30.0 V (b) 32.4 V (c) 162 A
22-25 (a) 45.0 V (b) 0.250 Ω (c) 7.36×10^{-2} T (d) 60.0 rad/s
22-27 (a) 200 turns (b) 0.159 N·m
22-29 (a) 188 turns (b) 60.0 Hz
22-31 21.1 A

22-33 (a) 18.0 turns (b) 0.333 A, 5.00 A
22-35 (a) 1000 turns (b) 16.0 A (c) 2.27×10^{-2} m (d) Small primary current yields small joule heating.
22-37 (a) 0.2 (or 1 to 5) (b) 48.0 V (c) 7.20×10^3 W

Chapter 23

23-1 (a) 150 V (b) 5.00 Hz (c) 0.200 s (d) 75.0 V
23-3 (a) $V = 40 \sin(50\pi t - \pi/6)$ (b) $I = 2.5 \sin(50\pi t - 5\pi/6)$; need sign of $\Delta V/\Delta t$ or $\Delta I/\Delta t$ at $t = 0$.
23-5 (a) 4.42×10^{-4} F (b) 165 Hz
23-7 (a) 4.71 A (b) 212 Hz
23-9 (a) 34.0 Ω (b) 119 V (c) 121 V (d) 7.67×10^{-5} F
23-11 (a) 11.4 V (b) 978 Hz (c) $-49.3°$
23-13 (a) 24.9 Hz (b) 50.1 V
23-15 —
23-17 (a) 6.36 H (b) 2.30×10^3 Ω
23-19 (a) 400 Ω (b) 1.08 H (c) 45.6°
23-21 (a) 2.00 A (b) 45.0 Ω, 25.0 Ω (c) 134 Hz
23-23 (a) 265 Ω, 100 Ω (b) 244 Ω (c) 1.64 A (d) $-42.5°$
23-25 (a) 98.0 Ω (b) 35.3° (c) 1.88×10^{-6} F
23-27 (a) 1.57×10^5 Ω, 7.96×10^4 Ω (b) 1.26×10^5 Ω (c) 1.59×10^{-3} A (d) 37.7° (e) 123 V
23-29 (a) 0.870 (b) 29.5° (c) 9.19 Ω (d) 5.86×10^{-4} F (e) 1.32
23-31 —
23-33 (a) 2.39×10^{-2} H (b) 0.707 (c) Capacitor (d) 2.94×10^{-4} F (e) It would act as an open (broken) circuit.
23-35 (a) 219 Ω (b) 6.86×10^{-2} A (c) 15.0 V (d) 0.0311 A (e) 11.7 V
23-37 (a) 0.17 s, 5.9 Hz (b) 4.6 V (c) 0.067 s, 15 Hz, 0.65 V for both
23-39 (a) 3.90×10^5 Ω (b) 12.0 mA, will notice but cannot let go (c) (i) 0.0800 mA, no sensation (ii) 80.0 mA, labored breathing, cannot let go (borders on electrocution)

Chapter 24

24-1 (a) 29.0 s (b) 1.28 s
24-3 3.41 m, 2.78 m
24-5 4.69×10^8 to 8.82×10^8 Hz
24-7 (a) 2.00 m (b) No (c) Yes (d) Out of focus for (a), in focus for (b), out of focus for (c)
24-9 (a) 0.875 m (b) 0.725 m (c) Distance to mirror
24-11 10.8 cm
24-13 —
24-15 (a) 1.55 (b) 1.94×10^8 m/s (c) 1.20 cm
24-17 3.00 m
24-19 (a) Directly overhead (b) 44.9834° (c) 1.38° (d) 5.52 min (e) 2.76 diameters
24-21 (a) 39.0° (b) 56.8°
24-23 (a) 1.65 (b) 37.3°
24-25 30.0°
24-27 (a) 24.52°, 24.24° (b) 0.24°
24-29 182°
24-31 (a) Approximately at focal point (b) Slightly above and behind focal point
24-33 (a) $+30.0$ cm (b) —
24-35 (a) $+60.0$ cm (b) -20.0 cm
24-37 (a) -200 cm (b) $+5.00$ (c) $+50.0$ cm (d) Virtual, erect, perverted
24-39 (a) Real, inverted, normal (b) $+40.0$ cm, $+120$ cm
24-41 (a) Virtual, erect (b) -10.0 cm (c) -20.0 cm

Chapter 25

25-1 (a) 20.0 cm (b) 30.0 cm (c) 60.0 cm
25-3 (a) 40.0 cm (b) 33.3 cm (c) 210 cm
25-5 (a) — (b) -36.0 cm, -20.0 cm
25-7 (a) — (b) 60.0 cm (c) $-2.00, -3.60$ cm
25-9 (a) -50.0 (b) 0.160 m from the lens (c) 0.157 m
25-11 (a) 15.0 cm (b) 9.00 cm
25-13 —
25-15 55.0 to 57.6 mm
25-17 (a) 8.40 cm (b) 1/125 s
25-19 200 s
25-21 (a) $+8.74$ cm (b) 0.409 cm (c) 6.00 cm
25-23 (a) -12.5 cm (before second lens) (b) $+37.5$ cm (after second lens) (c) $-0.0833, -3.75$
25-25 (a) 20.0 cm (b) 5.00 diopters (c) 7.00 diopters (d) 14.3 cm
25-27 (a) -15.0 cm (before second lens) (b) 160 cm
25-29 41.7, 83.3, 125, 150, 300, 450, 375, 750, 1125
25-31 (a) 202 cm (b) -25.3
25-33 (a) -167 (b) 2.18 cm
25-35 66.7 cm, 125 cm
25-37 (a) -2.50 diopters (b) 10.0 cm
25-39 -0.667 diopters, $+2.89$ diopters

Chapter 26

26-1 (a) 1.26° (b) 8.83 cm
26-3 2.44 m
26-5 5.71×10^{-7} m
26-7 (a) 0.975 m, 2.93 m, . . . (b) 1.95 m, 3.90 m, . . .
26-9 (a) 1.00×10^{-7} m (b) 3.00×10^{-7} m, 5.00×10^{-7} m, . . .
26-11 2.54×10^{-7} m

26-13 (a) 22.6° (b) 11.8°
26-15 —
26-17 (a) 1.12×10^{-7} rad (b) 43.0 m
26-19 1.17×10^{12} km
26-21 122 m
26-23 (a) 30 (b) 3
26-25 (a) 6.25×10^{-7} m (b) 4.17×10^{-7} m
26-27 (a) 36.9° (b) Order 2 for 6.00×10^{-7} m, order 3 for 4.00×10^{-7} m (c) Angle would change but order number would not.
26-29 (a) 36.9° (b) 52.7°
26-31 (a) 30.0° from perpendicular (b) 1/8
26-33 X-rays observed in both positive and negative x directions, none in positive and negative y directions.
26-35 2.20 g/cm^3

Chapter 27

27-1 (a) $2l/c$ (b) $\dfrac{2l/c}{1-(v^2/c^2)}$ (c) $\dfrac{2l/c}{\sqrt{1-(v^2/c^2)}}$
27-3 (a) $0.999999c$ (b) 3.60 m
27-5 (a) 4.23×10^7 m/s ($0.141c$) (b) 1.25×10^8 m/s ($0.417c$)
27-7 0.141 years
27-9 (a) 5.96×10^{-8} s (b) 16.1 m
27-11 6.86×10^{-12} s
27-13 (a) 3.73×10^{-25} kg (b) 8.94×10^{-2} m
27-15 $\rho = \dfrac{\rho_0}{1-(v^2/c^2)}$
27-17 (a) 600 m (b) 0.816 s (c) 3.00×10^8 m/s (d) 5.00×10^5 crests (e) 613 kHz
27-19 $0.976c$
27-21 (a) 2.29×10^4 kg (b) 2.06×10^{21} J (c) 1.16×10^{21} J
27-23 (a) 3.52×10^{-13} J (b) 4.82×10^{-30} kg (c) 2.95×10^8 m/s
27-25 (a) 1.40×10^{-3} T (b) 0.607 T
27-27 (a) 3.33×10^{-8} kg/s (b) 1.05 kg
27-29 (a) 4.33×10^9 kg (b) 2.73×10^{27} kg (c) 0.137%
27-31 (a) 1.85×10^{-27} kg, 2.37×10^{-19} kg·m/s (b) 1.79×10^{-28} kg, 5.36×10^{-20} kg·m/s
27-33 (a) $0.577 m_0 c$ (b) $0.155 m_0 c^2$ (c) $0.115c$ (d) $5.12 m_0$
27-35 $0.919c = 2.76 \times 10^8$ m/s

Chapter 28

28-1 (a) 3.01×10^3 K (b) 2.83%
28-3 (a) $d_B = 324 d_S$ (b) $r_B = 1.51 r_{EO}$
28-5 (a) 1.51×10^{33} Hz (b) 1.99×10^{-25} m
28-7 (a) 3.25×10^{31} (b) 3.08×10^{-27}%
28-9 (a) 6.37×10^{-11} W (b) 1.63×10^8 photons/s
28-11 (a) 2.80×10^3 K (b) 2.90×10^{14} Hz (c) 7.40×10^{-10} W/m^2 (red), 1.60×10^{-11} W/m^2 (violet) (d) $0.296 I_{max}$ (red), $(6.40 \times 10^{-3}) I_{max}$ (violet)
28-13 2.36×10^{21} photons/m^2·s
28-15 (a) 3.76 V (b) 2.07×10^{15} Hz
28-17 (a) 8.64×10^{-20} J (b) 4.45×10^{-19} J (c) 6.71×10^{14} Hz (d) 4.47×10^{-7} m
28-19 The magnitude increases by 5.53 V.
28-21 (a) 8.00×10^{-15} J (b) 1.21×10^{19} Hz; yes
28-23 4.73×10^{-10} m
28-25 10.1°, 20.6°, 31.9°, 44.7°, 61.6°
28-27 —
28-29 3.93%
28-31 (a) 0.395 MeV (b) 0.116 MeV
28-33 1.0971×10^7/m
28-35 (a) 9.11×10^{-8} m (b) 3.29×10^{15} Hz (c) 2.18×10^{-18} J

Chapter 29

29-1 (a) 1.52×10^7 m/s (b) 4.74×10^{-14} m (c) 3.63×10^{-12} J
29-3 (a) 4.13×10^3 m (b) 1.29×10^4 atoms
29-5 (a) 740, 74.0%
 (b) (i) 23, 2.30%
  (ii) 67, 6.70%
  (iii) 131, 13.1%
 (c) (i) 0.200%
  (ii) 7.80%
 (d) No
29-7 (a) 7.72×10^{-18} J (b) 3.66×10^{-18} J
29-9 (a) 9.05×10^{22} m/s^2 (b) 8.24×10^{-8} N
29-11 4.572×10^{14} Hz, 6.173×10^{14} Hz, 6.914×10^{14} Hz, 7.316×10^{14} Hz; 6.561×10^{-7} m, 4.860×10^{-7} m, 4.339×10^{-7} m, 4.101×10^{-7} m
29-13 1.637×10^{-18} J, 1.940×10^{-18} J, 2.046×10^{-18} J
29-15 (a) 6.55092×10^6 Hz (b) 6.56076×10^6 Hz
29-17 8.69×10^{-18} J, 54.3 eV; 1.95×10^{-17} J, 122 eV
29-19 8.19×10^6 times
29-21 (a) 1.13 eV = 1.81×10^{-19} J, 1.10×10^{-6} m (b) 6 and 3 (c) Paschen series
29-23 (a) 4 (b) 1.63×10^{-18} J, 3.02×10^{-19} J
29-25

l	m
0	0
1	$0, \pm 1$
2	$0, \pm 1, \pm 2$
3	$0, \pm 1, \pm 2, \pm 3$
4	$0, \pm 1, \pm 2, \pm 3, \pm 4$

29-27

n	l	m	m_s
1	0	0	$\pm\tfrac{1}{2}$
2	1	1	$\pm\tfrac{1}{2}$
2	1	0	$\pm\tfrac{1}{2}$
2	1	-1	$\pm\tfrac{1}{2}$
2	0	0	$\pm\tfrac{1}{2}$

Only four of the six possibilities are filled in the $l=1$ state for oxygen. Same table for neon, but all possibilities are filled.

29-29 (a) 1.60×10^8 W (b) 2.26×10^{15} W/m²
29-31 (a) 6.36×10^{14} atoms/s (b) 2.01×10^{-6}/s
 (c) 4.19×10^{-5}
29-33 (a) 3.00×10^{-3} m (b) 4.64×10^4 waves/pulse
 (c) 8.13×10^{14} photons/pulse

Chapter 30

30-1 (a) 2.21×10^{-38} m (b) 2.32×10^{-28} m
 (c) 5.77×10^{-17} m
30-3 (a) 1.24×10^{-10} m (b) 1.23×10^{-11} m
 (c) 2.87×10^{-13} m
30-5 (a) 3.97×10^5 m/s, 1.32×10^{-16} J (b) 2.91×10^5 m/s, 3.86×10^{-20} J
30-7 (a) 1.32×10^{-20} J (b) 1.32×10^{-14} J
30-9 (a) 5.49×10^{-12} m (b) 5.36×10^{-12} m

30-11

n	l	m	m_s
1	0	0	$\pm\tfrac{1}{2}$
2	0	0	$\pm\tfrac{1}{2}$
2	1	-1	$\pm\tfrac{1}{2}$
2	1	0	$\pm\tfrac{1}{2}$
2	1	$+1$	$\pm\tfrac{1}{2}$
3	0	0	$\pm\tfrac{1}{2}$
3	1	-1	$\pm\tfrac{1}{2}$
3	1	0	$\pm\tfrac{1}{2}$
3	1	$+1$	$\pm\tfrac{1}{2}$

30-13 (a) $nh/2l$ (b) $n^2h^2/8ml^2$
30-15 5.46×10^5 m/s
30-17 (a) 5.28×10^{-26} kg·m/s (b) 5.80%
30-19 (a) 8.79×10^{-21} kg·m/s (b) 2.31×10^{-14} J
 (c) Yes
30-21 (a) 3.30×10^{-3} eV (b) 7.96×10^{11} Hz
30-23 (a) 5.28×10^{-27} J (b) 7.96×10^6 Hz
30-25 (a) 0.500 m (b) 7.96 m (c) 9.59°
30-27 2.47×10^{23} free electrons
30-29 (a) 3.99×10^{20} atoms (b) 3.99×10^{14} electrons
 (c) 6.38×10^{-5} A

Chapter 31

31-1 (a) $^{223}_{87}$Fr, $^{227}_{90}$Th
 (b) $^{227}_{89}$Ac $\rightarrow$ $^{223}_{87}$Fr + $^4_2$He
  $^{227}_{89}$Ac $\rightarrow$ $^{227}_{90}$Th + $^{\ 0}_{-1}$e
 (c) $^{223}_{87}$Fr $\rightarrow$ $^{223}_{88}$Ra + $^{\ 0}_{-1}$e
  $^{227}_{90}$Th $\rightarrow$ $^{223}_{88}$Ra + $^4_2$He
31-3 1.87×10^{-14} m
31-5 $^{231}_{90}$Th, $^{\ 0}_{-1}$e, $^{\ 0}_{-1}$e, $^{60}_{27}$Co, $^{210}_{84}$Po, $^{\ 0}_{-1}$e
31-7 (a) 8 (b) 6 (c) Yes
31-9 (a) 2.30 MeV (b) The mass difference is too small.
31-11 (a) 7.52 MeV/nucleon (b) 7.48 MeV/nucleon
31-13 (a) 3.82×10^{28} J (b) 2.66×10^{33} J
31-15 5.49 MeV
31-17 (a) -4.31×10^8 J (b) 4.79×10^{-9} kg
31-19 (a) 0.351/h (b) 2.85 h (c) 1.97 h
31-21 (a) 201 years (b) 1.05×10^{-4}%
31-23 (a) 2.13×10^7/s (b) 2.66×10^{11} atoms
31-25 Thursday at noon
31-27 (a) 0.132/year, 4.17×10^{-9}/s (b) 1.20×10^{12} atoms (c) 1.20×10^{-10} g
31-29 22.4%
31-31 6.02×10^9 years
31-33 (a) No violation (b) No violation
 (c) Strangeness (d) Lepton number
 (e) Baryon number
31-35 —

Chapter 32

32-1 (a) 156 Bq (b) 4.22×10^{-9} Ci (c) 4.22×10^{-3} µCi (d) 4.22×10^3 pCi
32-3 (a) 4.71×10^{-2} µCi (b) 4.71×10^{-2} µg
32-5 (a) 7.40×10^4 Bq (b) 1.33×10^6 counts/min
32-7 (a) 1.70×10^{-2} J (b) 0.680 rad (c) 1.16 rem, 1.16×10^3 mrem
32-9 1.28×10^4 rad
32-11 5.15×10^{-18} J
32-13 3.37×10^{-2} R, 6.74×10^{-3} R/s
32-15 0.465 rad/s
32-17 5.94×10^4 cancers
32-19 (a) 9.60 min (b) 0.721 m
32-21 (a) 0.472/mm (b) 1.47 mm (c) 160/s
32-23 (a) 2.76×10^3 Ci (b) 31.8 g

Chapter 33

33-1 213 MeV, 6.3% difference
33-3 200 MeV
33-5 (a) 8.20×10^{13} J (b) 9.11×10^{-4} kg
33-7 (a) 6.07×10^{-21} J (b) 2.69×10^3 m/s

33-9 4.88×10^6 kg
33-11 (a) 1.63×10^9 W (b) 1.09×10^9 J/s
33-13 (a) 816 kg (b) 4.32×10^{-2} m³ (a cube about 35 cm on a side)
33-15 (a) 6.84 MeV (b) 6.53 MeV
33-17 (a) $^2_1\text{H} + ^2_1\text{H} \rightarrow ^3_1\text{H} + ^1_1\text{H}$, 4.03 MeV (b) $^2_1\text{H} + ^2_1\text{H} \rightarrow ^3_2\text{He} + ^1_0\text{n}$, 3.27 MeV
33-19 1.94 MeV, 1.20 MeV, 7.55 MeV, 7.29 MeV, 1.74 MeV, 4.96 MeV
33-21 (a) 0.252 (b) 14.1 MeV
33-23 (a) 9.98×10^{24} deuterium nuclei/m³ (b) 6.99×10^{25} MeV $= 1.12 \times 10^{13}$ J (c) 1900 barrels of oil, 405 tons of coal
33-25 7.27 MeV

Index

Aberration:
  of lenses, 620, 621
  of starlight, 674, 682–683
Absolute frame of reference, 678
Absolute magnitude of star, (prob.) 639
Absolute potential, 428–430
Absolute pressure, 290–291
Absolute system of measure, 674
Absolute temperature, 317
Absolute zero, 317, 325
Absorbed dose of radiation, units, 799
Absorption of radiant energy, 355
Absorption curves and cosmic rays, 797
Absorption spectrum, 655
AC (alternating current), 551–579
  and capacitor, 554–556
  current vs. time, 551
  distinguished from dc, 471–472, 551–552
  and inductor, 558–559
  power delivered, 567–570
  production of, 551
  RCL circuit, 562–567
  and resistance, 553–554
  RL circuit, 560–562
  simple circuits, 553–562
  standard frequency requirements, 540, 541
AC circuits, 551–579
AC motors, 540–541
Accelerating voltage, 707
Acceleration:
  centripetal, 90–92, 724
  defined, 21
  from graph of v vs. t, 22
  gravitational, 24–27
  multidimensional, 44–45
  negative, 23
  radial, 91–92
  of rocket, 172
  rotational, 88–90
  in SHM, 210, 214
  tangential, 90
  uniform, 21–24

Acceleration due to gravity (g):
  defined, 24–27
  and height, 96–97
  from law of gravitation, 96–97
  measurement of, 24–25
  on moon, 63
  on other planets, (table) 96
  and weight, 62–63
Accelerator:
  particle, 434, 692, 788, 811
  in radiation therapy, 811
  Van de Graaff, (illus.) 434
Acceptor of electrons, 762
Action potential of heart, 573
Action-reaction pair, 64
Activity (radioactive):
  and dose, 808–809
  and longevity, 785
  of sample, 784–785
  units of, 799, 800
Adiabat, (illus.) 381
Adiabatic enclosure, cloud chamber as, 787
Adiabatic expansion in bubble chamber, 787
Adiabatic lapse rate:
  of air, (prob.) 398
  in earth, (prob.) 397
Adiabatic process, 380–382
Airplane, 302
Airplane flight and radiation, (probs.) 813
Air resistance (curve), 107
Air track, (illus.) 160
Alcohol in cloud chamber, 787
Allowed orbits of electron, 727–731
Alpha bombardment:
  to produce neutrons, 776
  and transmutation, 775–776
Alpha particle, 724–726
  emission, 774
  and G-M counter, 786
  identified, 773
  scattering, 771
Alternating current (*see* ac)

Alternator, 539*n*.
Altitude and radiation, 796–797
Amber, 403
Ammeter, 484–485
  circuit symbol, 484
  internal circuit, (illus.) 485
  resistance of, 484–485
  use of, 484–485
Ampere, Andre, 453, 509–510, 582
Ampere:
  interpretation of magnetism, 509–510
  operational, defined, 508–509
  unit (A), 453
Amplification, 446
Amplifier, 786
Amplitude, 214
Analog:
  vs. digital meters, (illus.) 483–484
  vs. digital signals, 605
Analyzer, 663
Anderson, Carl, 798
Anderson, H. L., 818
Anemometer, (prob.) 547
Angle:
  critical, 594
  of deviation, 655
  of divergence, 587
  natural measure of, 80
  of reflection, 588
  of refraction, 590
  of shear, 275
  subtended by image, 621–622
Angular acceleration, 88–90
Angular dynamics, 179–198
Angular impulse, 193
Angular magnification, 631
Angular measure, 80–81
Angular momentum:
  of atom, 752
  conservation of, 194–195
  defined, 193
  direction of, in 3-dimensions, 196–198
  and helicopters, 195
  quantized, 727, 733

I-1

Angular momentum (*Cont.*):
   in sports, 194–195
   and stability of motorcycle, 196–197
   z component, 753
Angular momentum quantum number, 733
Angular velocity, 81–88
   and arc length, 85
   average, 89
   defined, 83
   and linear distance, 85
   and linear velocity, 85
   units, 83
Annalen der Physik, 675, 710
Annihilation of matter, 781
Anode, 711
Antenna, 583–584
Antimatter, 781
Antineutrino in beta decay, 782
Antinode of wave, 247, (illus.) 248
Antiparticle, 781
Aperture, 621
Apparent depth in refraction, 592–593
Aqueous humor, 633, 634
Archimedes, 294
Archimedes' principle, 294–296
Armature, 537
Artificial background radiation, 795–796, 802–804
   variability, 799
Astigmatism, 621
   test chart (illus.), 635
Atmosphere, shielding cosmic rays, 803
Atmospheric pressure, 290
   measurement, value of, 293
Atom:
   hydrogenlike, 730
   simple model of, 406
   size of, 734
      from theory, 729
   stationary electron model, 724
   structure, 723–730
Atomic mass number, 774–776
   and binding energy, 778–779
Atomic mass unit, 774
   to MeV, 779–780
Atomic model:
   Bohr, 726–733
   need for, 718
Atomic number, 730, 774–776
   and atomic properties, 734
Atomic oscillator, energy emission, 710
Atomic physics, 723–743
Atomic pile, 819
Atomic spectra, 716–718
Atomic theory of matter, 327
Atomizer, 301–302
Attraction:
   electric, 404–405
   of magnets, 498
   of uncharged object, 418–419
Aurora, 834

Automobile and air resistance, 107–108
Avalanche of ions in G-M tube, 786
Average speed and velocity, 16–18
Average velocity, 16–18, 42
   of gas molecule, 329
Avogadro hypotheses, (prob.) 337
Avogadro's number, 323
Axis of rotation:
   choice arbitrary for static equilibrium, 113–114
   fulcrum, 113
   and moment of inertia, 182–186
   and torques, 109

B (*see* Magnetic field)
Back emf, 538
Back torque, 539
Background radiation, 795–796, 798–799, 802
   high-dose areas, 806
Bacteria, 765
Ball, height of bounce, 169
Ballistic pendulum, 166–168
Ballistic theory of light propagation, 674
Balloon experiments with cosmic rays, 796, 798
Balmer, Johannes, 717
Balmer series, 717–718, 729–731
Band theory, solid-state, 759–764
Banked roadways, 94–95
Bar (unit), 293
Barium as uranium product, 817
Barkla, Charles, (prob.) 670
Barnard's star, (prob.) 719
Barometer, 293
Barrel distortion, 621
Baryon, (table) 788–789
Basal metabolic rate, 362
Base of transistor, 764
Batteries:
   of different emf, 473
   in series and parallel, 472–473
Battery, 457
   charging of, 460
   construction from cells, 472
   current and plate area, 472
   internal resistance, 459–460
   power dissipated in, 464
   and work, 427
Beam, distortion of, 278–280
Beam size, scanning electron microscope, 767
Beats, 255–256
Becquerel, Henri, 771–772
Becquerel (Bq) (unit), 800
Bees, vision of, 665
Bending strength, 279
Bernoulli, Daniel, 298
Bernoulli's equation, 297–299
Bernoulli's principle, 299–302

Beta decay, 780–782
   theory of, 781–782
Beta emission and conservation of energy, 780–781
Beta particle:
   emission, 774
   and G-M counter, 786
   identified, 773
Betelgeuse, (prob.) 719
Billiard balls, collisions of, 165–166
Binary code in computer, 605
Binding energy, 731, 734
   nuclear, 777–779
   for nucleon, 830
   and stability, 778–779
Binoculars, 596
Binomial expansion, 694
Biological clocks, 685
Biological effects:
   of currents, 574–575
   of ionizing radiation, 804–806
   of types of radiation, 801–802
Biological half-life, 810–811
Biological signals, electrical, 572–573
Biological specimens in electron microscope, 766–767
Birefringence, (illus.) 662
Bit of digital data, (prob.) 521
Bitter cold, defined, (prob.) 368
Blackbody, 355, 702
Blackbody radiation curve, 702, 704–706
Blind spot of eye, 633
Blink reflex of eye, 739–740
Block and tackle, 129–130
Blood flow in fingers, 331–332
Blood pressure, measurement and values, 308
Body temperature, 317, 331–332
   variation of, 331
Bohr, Neils, 726–733
   and complementarity, 759
   and fissionability, 819
   and liquid-drop model of nucleus, 817
Bohr condition, 748
Bohr model:
   limitations, 732
   modifications, 732–734
   and quantum mechanics, 745
Bohr orbits, 758
Bohr postulates, 727, 733
Bohr radius, 729, 733
Boiling point, 345
   of water, 317, 326, 334
Boiling water reactor (BWR), 824
Boltzmann, Ludwig, 702
Boltzmann constant, 330, 706
Bombardment of nucleus, 788
Bombs, nuclear, 821
Bones, strength of, 279–281
Boron trifluoride, 801
Bottom-antibottom pair, 790
Boundaries and waves, 244–245

Boyle, Robert, 320
Boyle's law, 320
Brackett series, 717, 729–730
Bradley, James, 682
Bragg, Lawrence, 712
Bragg, William H., 712
Bragg diffraction, 713
Bragg diffraction equation, 713, 748–750
Bragg planes, 714, 748–749
Bragg spectrometer, 712, 714
Brake system, (illus.) 309
Breakdown strength of dielectric, (table) 440
Breakeven in fusion, 835
Breaking point, 271
Breeder reactors, 829–830
Brewster, Sir David, 662
Brewster angle, 661–662
Bridge, resonance of, 225
Bright-line spectra, 718
British engineering system of units, 5
British thermal unit (Btu) (non-SI unit), 342
Brittle materials, 270–271
Brushes of motor, 537
Bubble chamber, 787
Buckling of structure, 279
Building materials, 278–280
Buildings, radiation dose in, 803
Bulk modulus, (table) 277
 and speed of wave, 242–243
Buoyant force in fluid, 294
Byte of digital data, (prob.) 521

Cadmium as control rod, 820
Calcite, birefringence in, (illus.) 662
Caloric, 341
Calorie (cal) (non-SI unit), 341
Calorie (Cal) (non-SI unit), 342
Calorimeter, 347
Calorimetry, 346–348
Camera, 621, 623–625
 pinhole, 625
Camera lenses, description of, 624–625
Cancer:
 current death rate, 805
 and radiation, 804
 radiation therapy, 811
Capacitance:
 defined, 435
 equivalent, 486–488
 from geometry, 436
Capacitive reactance, 562–565
Capacitor, 435–441
 and ac, 554–556
  vs. with dc, 552
 ac charge-discharge cycle, 555
 charge remaining on, 490
 charging, 438
  and discharging, 488–491
 common sizes, 436–437

Capacitor (*Cont.*):
 construction, (illus.) 436–437
 current during charging, 489–490
 in dc circuits, 486–491
 and dielectrics, 439–441
 electric field inside, (illus.) 435
 in electronic flash, 488–489
 energy per unit volume, 439
 energy stored, 437–439
 and inductor in oscillation, 532–534
 instantaneous charge, 489
 in pacemaker, 491
 in parallel, 487
 parallel-plate, 435–436
 potential difference of, 435
 in series, 487–488
 time constant in circuit, 489–491
 variable, (illus.) 437
Carbon-14, 799
Carbon-14 dating, (prob.) 793
Carnot, Sadi, 386
Carnot cycle, 386–388
 efficiency of, 387
 as most efficient, 386
 P vs. V diagram of, 386
CAT scan, 225
Cathode, 711
Cathode rays, 711–714, 723–724
Cavendish, Lord Henry, 97–98
Cavendish Laboratory, 723
Cell, electrochemical, 457–458
Cell damage from radiation, 804
Cell growth and scaling, 282
Cell sizes, 765
Celsius temperature scale, (illus.) 317
Center of curvature of mirror, 598
Center of gravity, 114–119
 used, 81–82
Center of mass, 118
 of body parts, (illus.) 119
Center of moments, 116
Central maximum, 645
Centripetal acceleration, 90–92
 of electron, 724
Centripetal force, 92–94
 on charged particle in magnetic field, 500–501
Chadwick, James, 776, 781
Chain reaction, 818
Charge, 403–412
 balanced, 406
 conservation of, 474, 788
 discovery, 403–404
 drift velocity, 453–454
 on electron, 409
 forces between, 405
 fractional, 790
 magnetic force on, 498–503
 motion of, 452–454
 potential energy of, 424–428
 production of, 404–406

Charge (*Cont.*):
 on surface of conductor, 433
 types, 405
Charge density, 432*n*.
Charge-to-mass ratio, 723–724
Charged object, atomic interpretation, 406
Charged sphere:
 electric field of, (illus.) 432
 potential difference due to, (illus.) 432
Charging, 404–406
 atomic interpretation, 405–406
 of battery, 457, 460
 of capacitor, 488–491
 by conduction, 407
 by induction, 407
 by rubbing insulators, 406
Charles, Jacques, 321
Charles' law, 321
Charm, quark, 790
Chart recorder, 573
Chop of water, 245
Chromatic aberration, 620
 in electron microscope, 766
Circle of least confusion, 621
Circuit:
 closed, 458
 coincidence, 797
 integrated, 764
 LCR series, 532–534
 open, 458
 RC (*see RC* circuit)
 RCL (*see RCL* circuit)
 redrawing schematic, 477
 schematic of, 458
 simple, (illus.) 458
 time constant of, 489–491
Circuit analysis:
 complex dc circuits, 477–483
 simple dc, strategy, 476
Circuit breaker, 575
Circuit elements (symbols), 458
 ohmic vs. nonlinear, 456
 variable (symbol), 458
Circuit phasor, 553
Circuit symbol, rectifier, 570
Circular motion as SHM, (illus.) 208, 210–216
Circular waves, 586
Classical mechanics, 756
Clausius, Rudolf, 388–392
Clocks:
 biological, 685
 confirmation of relativity, 689
 light-pulse, 684–687
Closed circuit, 458
Closed end of medium, 243
Clothes, insulating value of, 363–364
Cloud chamber, 787
Coal burning, health effects, 827
Cobalt 60, 811
Coefficient of friction, (table) 69

I-3

Coefficient of performance (COP):
  Carnot refrigerator, 388
  ideal refrigerator, 395
  of refrigerator, 384
Coefficient of restitution, (table) 168–169
Coefficient of thermal expansion, (tables) 319–320
  units of, 320
Coefficient of viscosity, (table) 303
Coherent, 643
Coherent beam, 737
Coincidence circuit, 797
Collector of transistor, 764
Collimator, 809–810
Collision, 155–174
  of billiard balls, 165–166
  elastic, 161–166, 327–328
  of electron and x-ray, 714–716
  of equal masses, 165–166
  inelastic, 166–171
  and kinetic energy, 161–162, 166, 168
  one-dimensional, 161–164
  partially elastic, 168–169
  perfectly inelastic, 168
  reproducibility of, 159
  two-dimensional, 164–171
Color TV picture, 446
Coma, 621
Combinations of lenses, 625–628
Combustion of gasoline, (prob.) 338
Commutator, 537
  split-ring, 539
Compensating plate, 680
Complementarity, 759
Compound lens, 623–624
Compound microscope, 629–630
Compressibility, (table) 277
Compression, 269, 271
  of sound wave, 239
  table of Young's modulus for, 272
Compression phase in refrigerator, 392–394
Compression ratio, 386
Compression stroke of engine, (illus.) 385
Compton, Arthur, 714–716
Compton effect, 714–716
  and G-M counter, 786
Compton scattering, 776
Computer, data transmission, 605–606
Concave mirror, 598–603
Condensation in cloud chamber, 787
Conduction:
  charging by, 407
  of heat, 348–352
Conduction band, 760–761
Conductivity and band structure, 760–761
Conductor:
  electric, 405
  in electric field, 414
  of heat, 349
  resistivity of, (table) 461

Conductor (*Cont.*):
  work to move charge in, 433
Cones of eye, 633
Confirmation of relativity, 688–690
Congenital effects, 805
Conservation:
  of charge, 474
  of energy in atom, 726–727
  of linear momentum, 160
    in explosions, 171
    and pair production, 788
    and relativity, 678
    vector nature of, 160, 161
  of mass-energy, 777–778
  of mechanical energy, 143–147
  and pair production, 788
  of spin, 781
Conservation laws and beta decay, 780–781
Conservative force, 142
  electrostatic force as, 424
Constant-volume gas thermometer, (illus.) 324
Constructive interference, 245, 642
Containment of reactor, 824–826
Continuous spectra, 655
Control rods, 820
Convection, 348
  forced, 352–353
  natural, 353
Convection coefficient, 353–354
  at various speeds, 363
Conventional current, 453
Convergence, 598
  in lens, 612–613
Converging lens, 613
Convex mirror, 598, 601–603
Coolidge tube for x-rays, 712
Cooling of canteen, 357
Cooling meat in refrigerator, calculation of, 395
Cooling phase of refrigerator, 394
Coordinate system, choice of, 675–678
Core of reactor, 824–826
Cornea, 633, 634
Corpuscular theory of light, 642, 646
Corrective lenses for eye, 635–636
Correspondence principle, 732–733
Cosine curve, 206–207, 210
Cosmic radiation, 688, 795–798, 802–803
Cosmic rays:
  dose and altitude, 803
  source, 796–798
Cost:
  of freezing meat, 396
  of operating refrigerator, 396
Coulomb, Charles Augustin de, 408
Coulomb (C) (unit), 409
Coulomb force, 424, 425
  in nucleus, 777, 779
Coulomb's law, 407–412
  and particle scattering, 725

Count Rumford (Benjamin Thompson), 341
Cowan, Clyde, 781
Crest of wave, 237
Critical angle, 594
Critical isotherm for real gas, 333
Critical mass, 822
Critical point, 333
Critical temperature of superconductor, 462
CRT (oscilloscope), 571–572
  for biological signals, 572–573
Crystal, 666, 712–713
  in scintillation detector, 786
Crystal spectrometer, 712
Crystalline lens of eye, 633, 634
Crystallography, 713
Curie, Marie, 772
Curie, Pierre, 772
Curie (Ci) (non-SI unit), 799–800
Curie point for magnet, 518
Current, 451–454
  effective ac, 568
  vs. frequency, 566
  magnetic field around, 504–505
  maximum ac, 562
  measurement of, 484–485
  photoelectric, 707
  or sum of drift of charges, 454
Current amplitude, 553
Current direction:
  conventional, 453
  in Kirchhoff's rules, 479
Current loop in magnetic field, torque on, 534–536
Curvature of field, 621–623
Curve ball, 302
Cycle, Otto, 384–386
Cycle:
  of motion, 205
  thermodynamic, 382–388
Cyclotron, 692
Cylinder, hollow vs. solid, strength of, 279–280
Cylindrical lens, (illus.) 593

Dalton, John, 327
Damped, driven harmonic oscillator, ac analog, 566
Damped harmonic oscillator, 223
  circuit, 532–534
Dark-line spectrum, 655
d'Arsonval, Arsene, 535
d'Arsonval galvanometer, 535–536
Dating, radioactive, (probs.) 793
Daughter nucleus, 773
Davisson, Clinton J., 748
Davisson-Germer experiment, 748–750
DC (direct current) vs. ac, 471–472, 551–552
DC circuits, 471–495

DC Circuits (*Cont.*):
  examples, (illus.) 471–472
  sources, 471
DC motors, 536–538
de Broglie, Louis, 745–747
de Broglie condition, 749, 756
  hypothesis, 746–751
  wavelength, 747
    and electron microscope, 766
Decay:
  of elementary particle, 788
  radioactive, laws of, 773–774, 782–785
Decay constant, 782
Decibel (dB) (unit), 257
Declination, magnetic, 516–517
Defense-in-depth concept, 825
Deformation of objects, 267–282
Delayed neutrons, 821
Delta (Δ) as prefix, 16
Density, 288–289
  and floating, 295
  linear mass, 241–242
  of planets, (table) 96
  and speed of wave, 241–243
  table of, 288
  units of, 288
  volume mass, 242–243
  of water vapor in saturated air, (table) 359
Depletion layer, 786
Depletion region, 786
Depth of field, photographic, 623–625
Derived quantity, 2
Derived units, 5
Description of motion, 13–14
Destructive interference, 245, 642
  conditions, 646
Detectors:
  for ionizing radiation, 785–787
  relative speed, 786
Deuterium, 773
  availability, 835
Deuterium-tritium fusion, 833
Deviation, angle of, 655
Dextrorotary compounds, 665
Diamagnetism, 510
Diaphragm, camera, 623
Diastolic pressure of blood, 308
Dichroism, 660
Dielectric, 439
  breakdown strength, (table) 440
  permittivity of, 441
  and voltage of capacitor, 440–441
Dielectric coefficient, (table) 439–440
Dielectric constant (*see* Dielectric coefficient)
Dielectric strength, (table) 440
Diffraction, 643, 650–654
  circular aperature, 653
  of electrons, 750–751
  and interference, 656–659
  of x-rays, 711–714

Diffraction grating, 654–659
  for x-rays, 712–713
Diffraction pattern, 645, 651–653, 656
Diffuse reflection, 581–582, 588
Digital vs. analog signals, 605
Digital communications, 605–606
Dilation of time, 684–686
Dimensions, 3–4
Diode, junction, 763
Diopter (unit), 629
Dip, magnetic, 516–517
Dipole:
  electric, 416–419
  heart as, 573
Dipole moment, 417
  induced, 417–419
  permanent, 417
  of water, 417, 419
Dirac, P. A. M., 754
Dirac and positron, 781
Direct current (*see* DC)
Direction:
  of magnetic field about wire, 504–505
  of magnetic force on current, 508
Discharge of capacitor, 490–491
Disorder and entropy, 319
Dispersion, 597–598
Displacement:
  defined, 15
  directional property, 15
  of fluid, 294–296
Displacement laws of radioactive decay, 774
Display segment in LCD, 666–667
Disposal of nuclear wastes, 827–828
Distance:
  vs. displacement, 14–15
  and dose of radiation, 808
  and force, 130
Distortion:
  of beam, 278–280
  of lens, 621
Divergence:
  angle of, 587
  in lens, 612–613
  of waves, 586
Diverging lens, 613
Domains, magnetic, 511
Donor of electrons, 762
Dopant, 762
Doping of semiconductor, 762
Doppler, Christian, 259
Doppler effect, 259–260
  sign convention, 259
  stars and galaxies, 260
Dose of radiation:
  average, 802
  U.S. standards, 807
Dose equivalent of radiation, units, 799
Dosimeter, (illus.) 801
Double-concave lens, 614
Double-convex lens, 614

Double image, (illus.) 662
Double refraction, (illus.) 662–663
Double-slit interference, 643–646
Drag coefficient, 307
Drag force, 307
Drift velocity of charges, 453–454
Driven harmonic oscillator, 224
Duality, wave-particle, 758–759
Ductile materials, 270–271
Dynamics:
  angular, 179–198
  defined, 14, 56
  equilibrium, 105, 107

Earth:
  absolute motion of, 682
  absolute speed, 674
  circumnavigating and time dilation, 689
  magnetic field, of, 516–518
  motion through space, 682–683
  size and orbit, 96
Ebonite, 403
Edison, Thomas, 541
Effective current, ac, 568
Effective dose of radiation, units, 799
Effective frontal area, 107
Effective mass of photon, 746
Effective spring constant, 223
Effective voltage, ac, 568–569
Efficiency:
  of Carnot cycle, 387
  of heat engine, 383
  of Otto cycle, (prob.) 399
Einstein, Albert, 675, 679, (illus.) 683, 684, 693, 695, 706, 711, 777, 820
Einstein's photoelectric equation, 710
  postulates of relativity, 683–684, 695
Elastic collisions, 161–166
  of gas molecules, 327–328
Elastic limit, 270
Elastic modulus, 271
Elasticity and collisions, 169
Electric charge, 403–412
  (*See also* Charge)
Electric current (*see* Current)
Electric dipole, 416–419
Electric field, 412–415
  and cathode rays, 723
  inside conductor, 414–415
  and electromagnetic waves, 583–584
  induced, 418
  magnetic, 498–510
  and polarization, 659
  and potential difference, 431–435
  shown by lines of force, 415–416
  in spark chamber, 787
  at surface of conductor, 415
  in vicinity of point charge, 413–414
  and voltage, 432

I-5

Electric field (Cont.):
  in wire, 451
  work done in, 423–425
Electric field intensity, 413
Electric force (see Electrostatic force)
Electric oscillations in LCR circuit, 532–534
Electric permittivity, 436
Electric potential (see Potential electric)
Electric shock, 574–575
Electric signals, transmission, 605–606
Electrical codes for wiring, 575
Electrical meters (see Meters, electrical)
Electrical potential energy, 424–428
Electrical resistance, 455–456
Electrical safety, 573–576
Electrical wiring, safety, 575–576
Electricity, 404
  ac vs. dc use, 541
  commercial, 544–545
  measurement of, 483–486
Electrocardiogram (EKG), 572–573
Electrochemical cell, 457–458
Electrocution, accidental, rate, 575
Electrode, 711
  biological, 572
  of G-M counter, 785
Electroencephalogram (EEG), 572
Electrolyte, 457
Electromagnet, 511
Electromagnetic spectrum, (illus.) 355, 585
Electromagnetic theory and orbiting electrons, 724, 726
Electromagnetic waves, 260, 355, 582–585
  experimental production, 583
  speed of, 584–585
  transverse nature, 584
Electromotive force (see emf)
Electron, 406
  allowed orbits, 727–731
  charge of, 409
  constructive interference, 748–750
  deflection in TV set, 441–443
  diffraction, 750–751
  discovery of, 723–724
  drift velocity in wire, 453–454
  elliptical orbits, 733
  free, 406, 762
  incomplete picture of, 758n.
  in metals, 406
  no measured size, 735
  orbital, 726–733
  in nucleus, unlikely, 776
  spinning, magnetic field of, 735
  tendency to lose on rubbing, 406
  thermal ejection from filament, 441
  wave nature, 747–751
  and x-rays, 714–716
Electron gun, (illus.) 441–442
Electron jump, 727

Electron microscope, 765–767
Electron spin, 735–736
  and magnetism, 510
Electron transitions, 729
Electronic flash, capacitor in, 488–489
Electronics, 797
Electronics industry, 761
Electronvolt (eV) (non-SI unit), 430
Electroscope, 406–407
  and cosmic rays, 796
  as detector, 785
  as source of current, 451–452
Electrostatic energy, 423–441
Electrostatic force, 403–421
  in atom, 724
  as conservative force, 424
Elementary particles, (table) 787–790
  in cosmic rays, 798
Elements, identified by spectra, 655
Elevators and safety, 825
Elliptical electron orbits, 733
Elongation under stress, 268–272
Emergency core cooling system, 825
emf (electromotive force), 456–459
  back, 538
  equivalent, 473
  of generator, 538–540
  motional, (illus.) 512–514
Emission of radiant energy, 355
Emission spectrum, 655, 718
Emission theory of light propagation, 674
Emissivity, 356
  of human body, 356
Emitter of transistor, 764
Empirical relationship, 72
End behavior of waves, 243–245
Energy, 128–148
  of capacitor, 437–439
  of configuration, 139
  conservation of mechanical, 143–147
  in electric circuit, 463–465
  forms of, 147–148
  in fusion, 830–832
  gravitational potential, 139–141
  kinetic (see Kinetic energy)
  mass as, 693–694
  measurement of, 143–147
  mechanical, 143–147, 216–219, 726–728
  of motion, 139
  nuclear binding, 777–779
  of pendulum, 143–144
  of pole vault, 145–147
  of position, 139
  and position in SHM, 218
  potential (see Potential energy)
  radiant, 355, 701–704
  of radiation, detection, 786–787
  of ray, 785
  in RCL circuit, 566
  and relativity, 693–694
  release of, (table) 148

Energy (Cont.):
  rest, 694
  simplest definition, 139
  stored by inductor, 529
  thermal (see Thermal energy)
  translational and rotational, 190–192
  units of, (table) 131
  of wave, 233, 235
Energy band from energy levels, 760–761
Energy content of fuels, (table) 148
Energy conversion of waterfall, 330
Energy density of solenoid, 529
Energy efficiency ratio (EER) of air conditioners, (prob.) 399
Energy gap, 760
Energy-level diagram, 731
Energy levels:
  helium, 738
  hydrogen, 731–732
  of multiatom systems, 760–761
  neon, 738
Energy release in fission, 817
Energy states:
  of atom, 727–731
  discrete, 726
Engine, four-stroke cycle, (illus.) 384–385
Enriched uranium, 821
Enrichment for reactor fuel, 821, 822
Entropy:
  defined, 389
  and disorder, 319
  and lost opportunity, 390
  Planck's use, 704
  statistical nature, 390
  units, 389
Entropy change in refrigerator, 395
Envelope of standing waves, 247–248
Environmental impact of nuclear power, 827
Environmental Protection Agency, 803
Equation of line, use of, in kinematics, 17
Equations summarizing two-dimensional motion, 45
Equilibrium, 105–121
  and center of gravity, 115
  first condition of, 105–106
  and human body, 118–121
  second condition of, 113–121
  static vs. dynamic, 105
  thermal, 315–316
Equipotentials, 433
Equivalent capacitance, 486–488
Equivalent emf, 473
Equivalent resistance, 473–477
Erect image, 598
Escape velocity of the earth, 173
  and gravitational pull, (prob.) 338
Ether, luminiferous, 673–675, 682–683
Ether drag, 674
Ether experiments, 679–683
Evaporation, 357–362, 365

I-6

Evaporation (*Cont.*):
   body heat loss through, 346, 365
Excess cancers, 805
Excess deaths, 827
Excess neutrons, 818
Exchange coupling, 511
Exchange force, 777
Exchange particle, 777
Excited state, 736, 810
Exhaust stroke of engine, (illus.) 385
Expanding universe, 260
Expansion:
   coefficient of thermal, (table) 319
   thermal, 319–320
Expansion phase in refrigerator, 394
Experiment, role of, 2, 56
Explosions and momentum, 171
Exponential decay law, 782
Exposure to radiation, units, 799–801
Extraordinary ray, 662–665
Extrapolation:
   to absolute zero, 325
   work function, 710
Eye, 633–636
   defects, 634–636
   and divergence, 587
   high-frequency tremor, 739–740
   and laser damage, 739–740
Eyeball, size of, 633–634
Eyepiece, 629–631

f-stops, (table) 623–625
Fahrenheit, Gabriel, 316–317
Fahrenheit temperature scale, (illus.) 317
Falling objects, 24–27
Fallout, nuclear, 796, 802
Far point of eye, 634
Farad (F) (unit), 435
Faraday, Michael, 415, 433, (illus.) 514–515, 582, 711
Faraday's ice pail experiment, (illus.) 433–434
Faraday's law, 514–515, 537
Farsightedness, 634–635
Fatigue of structure, 279
Feeling, threshold of, for sound, 257
Fermi, Enrico, 781
   and first reactor, 818–820
   and neutron activation, 816
Ferromagnet, 511–512
Ferromagnetism, 511–512
Fertile nuclei, 829
Fiber optical bundle, 596, 604
Fiber optics, 604–606
Fibrillation of heart, 575
Field:
   electric, 412–415
   induced, 418, 515
   magnetic (*see* Magnetic field)
   shown by lines of force, 415–416
Filament, 441

Film, camera, 623
Film badge, (illus.) 801, 807
Fires, electrical, 575
First condition of equilibrium, 105–106
First law of thermodynamics, 373–374
   as conservation of energy, 373
   sign convention, 373
Fission, nuclear, 816–822
Fission products, 821
Fissionable isotopes, 820
FitzGerald, G. F., 675, 687
FitzGerald contraction, 684
Fixed end of medium, 243
Fizeau, Armand, 674
Floating and density, 295
Floating object, 295–296
Fluid flow, 296–308
   of blood, 308
   continuity, 296–297
   with friction, 302–305
   frictionless, 297–302
   laminar, 306–307
   measurement, 300–301
   from tank, 299–300
   in tubes and pipes, 302–305
   turbulent, 306–307
Fluids, defined, 287
Fluorescence, 711–712
Fluorescent screen, 725
Flux:
   changing, 512, 525–527
   magnetic, 512, 514–516
Flux linkage, 525
Focal length:
   of lens, 614
   of mirror, 598–603
Focal point:
   of lens, 614
   of mirror, 598–603
Focusing, camera, 623
Forces:
   average, in impact, 274
   buoyant, 294
   centrifugal, 93
   centripetal, 92–94, 500–501
   between charges, 407–412
   conservative vs. nonconservatives, 142
   exchange, 777
   exerted by a confined gas, 327
   and impulse, 156–158
   intuitive definition, 55
   with laminar flow through fluid, 305–306
   line of action of, 109–111
   lines of, 415–416, 498
   and machines, 128
   magnetic: on current, 507–509
      between current-carrying wires, 508–509
      on moving charge, 498–503
   between magnets, 498
   moment arm of, 109–110

Forces (*Cont.*):
   on moving object in air, 307
   of muscles, 119
   normal, defined, 69
   nuclear, 777–782
   nuclear weak interaction, 781
   parallel component and work, 132–133
   perpendicular component of, 110–111
   qualitative definition, 58
   quantitative definitions, 60
   restoring, 209
   and stretched spring, 133
   thrust of rocket, 171–173
   with turbulent flow through fluid, 307
   unbalanced, 57
   vector nature of, 57–58
Forced-air heating, (illus.) 353
Fossil fuels, 823
Four-stroke cycle engine, (illus.) 384–385
Fovea of eye, 633
Fracture of bones, 273–274
Frame of reference, 675–678
Franklin, Benjamin, 405
Free electron, 406, 762
Free end, 243
Free fall, 99
Free radicals, 804
Freezing point, 317, 334
Frequency:
   of beats, 256
   of closed-end pipe, 253
   of cyclotron, 692
   of doppler shift, 259–260
   and hearing, (illus.) 256–257
   of *LCR* circuit, 534
   of open-end pipe, 254
   of pendulum, 220
   and period, 84
   and pitch, 254
   of precession, 198
   of rotation, 83
   in SHM, 216
   threshold, 708–709
   unit of, 216
   and wavelength, 238
   of waves on strings, 250–251
Frequency dependance of ac circuit, 562–565
Frequency response of hearing, (illus.) 257
Fresnel, Augustin, 613, 674
Fresnel lens, (illus.) 613
Friction, 68–72
   coefficient of, 69–72
   and contact area, 68
   defined, 68
   empirical nature of, 72
   factors involved in, 68–69
   and fluid flow, 302–307
   kinetic, 68–70
   as nonconservative force, 142

I-7

Friction (Cont.):
  and normal force, 69
  rolling, 72
  and shortest distance to stop, 70–71
  static, 67–69
  and thermal energy, 148
  types of, 68
  and work in rolling, 191–192
  work done by, 132
Fringes, 680
Fuels, (table) 148
Fulcrum, 113
Full-scale reading, 485
Fundamental mode of vibration, 248, 250
Fundamental quantity, 3
Fundamental standard, 3
Fuses, 575
Fusion, 830–837
  confinement, 832
  controlled, 833–837
  current research, 833
  temperature of, 832
Fusion reactors, 833–837

g (see Acceleration due to gravity)
Galilean relativity, 675
Galilean telescope, 631
Galileo Galilei, (illus.) 56
  and falling objects, 24
  and nature of force, 55
  and projectile, 47
  and range of cannon, 13–14
Galvanometer, d'Arsonval, 535–536
Gamma camera, 809–811
Gamma decay, 693
Gamma emission, 774
Gamma radiation, discovery of, 773
Gamma ray, 585, 693
  absorption, (prob.) 813
  in G-M counter, 786
Gas laws, 320–331
Gaseous diffusion, 821
Gases, 288
  work done by, 374
Gauge pressure, 292
Gauges, 292–293
Gauss, 582
Gedanken experiment, 676, 684, 756
Geiger, Hans, 725, 797
Geiger counter, 785–786
Geiger-Mueller (G-M) counter, 785–786, 797
Gell-Mann, Murray, 790
General theory of relativity, 684n.
Generator, 538–540
  back torque, 539
  commercial, 540
  emf produced, 539
  vs. motor, 540
  Van de Graaff, (illus.) 434
Genetic effect of radiation, 804–805

Geomagnetic reversal, 518
Geometric constructs for waves, 585–587
Geometric optics, 581–639
Geometrical path length, 654
Germanium as semiconductor, 761–762
Germer, Lester, 748
Gilbert, Sir William, 516
Glaser, Donald, 787
Glasses for eye, 635–636
Goddard, Robert H., (illus.) 173
Gold foil experiment, 724–725
G-M (Geiger-Mueller) counter, 785–786, 797
Goudsmit, Samuel, 735
Graphite as moderator, 819–820
Grating, diffraction, 654–659, 712–713
Gravitation:
  law of, 95–99
    and g, 96–97
    and motions of moon, 98–99
    original statement, 95
    and satellites, 98–99
    summary, 96
  work done by, 140
Gravitational constant, 97–98
Gravitational force, 62
Gravitational potential energy, 139–141
Gravity, acceleration due to (see Acceleration due to gravity)
Gray (Gy) (unit), 801
Grid, 445
Ground, electrical, 451
Ground potential, 428
Ground state of atom, 729
Gyroscope:
  and precession, 198
  use in rocket, 173

Hafele, J. C., 689
Hahn, Otto, 816–817
Half-life, 783
  biological, 810–811
Half-value layer, (prob.) 813
Hall, D. B., 688
Harmonic frequencies in sound, (illus.) 249, 251–254
Harmonic function, 206
Harmonic oscillation of LCR circuit, 532–534
Hearing:
  range of, (illus.) 257
  threshold of, 257
Heartbeat, human, 573
Heat, 315–316
  mechanical equivalent of, 342
Heat of fusion, 344
Heat of vaporization, 345
  of water, 346
Heat absorption phase, 395
Heat capacity, molar, 374–375
Heat engines, 382–388

Heat engines (Cont.):
  defined, 371
  efficiency of, 383
  strategy of, 374
  work done in cycle, 383
Heat loss, detection, 356–357
Heat regulation of body, 331–332
Heat reservoir, 378, 387
Heat transfer, 348–365
Heating curve for water, (illus.) 345
Heating systems, 352–354
Heavy water, 823
Henry, Joseph, 527
Henry (H) (unit), 527
Heisenberg, Werner, 746, 755
Helicopter, 195
Helium, 655
Helium ions (alpha particles), 724–726, 773
Helium-neon laser, 737–740
Hertz, H., 707
Hertz (unit), 216
Hess, Victor, 796
High-level radioactive waste, 827–828
Hiroshima survivors, 805–806
Hitler, Adolph, 705
Hole in crystal, 762
Hollow vs. solid cylinder, 279
Home computer, 606
Home heating systems, 352–354
Home insulation, 351
Hooke, Robert, 271
Hooke's law, 271
  applied to springs, 278
Horizontal motion, 43
Horsepower, 136
Hot-air furnace, (illus.) 353
Hot-water heating system, (illus.) 354
Human auditory response, (illus.) 257
Human body:
  area of, 363
  dimensions, 119
  electrical resistance, 574
  heating and cooling, 362–365
  mass of parts, (illus.) 119
  metabolic rates, 362–363
  perspiration rate, 365
  pivot points, (illus.) 119
  response to cold, 331–332
  temperature of, 331–332
Human exposure to radiation:
  biological effect, 804–806
  levels, 802–804
  medical, 809–811
  safety standards, 806–809
Human eye, 633–636
Humidifying air, 359–360
Humidity, 358–362
  indoors, 359–360
Hurricanes, energy of, (prob.) 366
Huygens, Christian, 642
Hydraulic press, 291

Hydrodynamics, 296–308
(See also Fluid flow)
Hydrogen:
  isotopes, 773
  as moderator, 820
  spectral lines, 729–732, 777
Hydrogen atom, 726–730
Hydrogen bomb, 831, 833
Hydrogenlike atom, 730
Hydrostatic paradox, 291
Hydrostatics, 288
Hypermetropia, 634–635
Hypertension, 308
Hysteresis, 512

Ice caps, volume of, (prob.) 366
Ideal gas, 323–324
  P vs. T curves, 332
  vs. real gas, 323–324
Ideal gas law, 323–324
  from Newton's laws, 327–329
Ignition sequence of space shuttle, 173
Image, 589
  angle subtended by, 621–622
  in different medium, 592–595
  erect, 598
  inverted, 598
  in mirrors, 598–604
  multiple, (illus.) 589
  perverted, 598
  real, (illus.) 599
  in spherical mirrors, (illus.) 603
  virtual, 598
Image distance, 600
Image formation:
  in lenses, 614–620
  of magnifier, (illus.) 622
Image horizon, underwater, 596
Imaging in nuclear medicine, 809–811
Impact load of structures, 279–280
Impedance, 564–565
Impedance matching and transformer, 545
Impending motion, 69
Impulse, 156–159
  angular, 193
  as area under curve, 157–158
  of gas molecule, 327–328
  to generate wave motion, 241–242
  and rocket, 171–172
  units of, 157
  vector nature, 157
Impulse-momentum theorem, 156
  and collisions, 159–160
  and wave motion, 241–242
Incandescence, 655
Incidence, angle of, 588
Incident ray, 587
Incident wave, 246
Index of refraction, (table) 591–596
  and interference, 647–648

Index of refraction, (table) (Cont.):
  and parts of eye, 633–634
Induced current, 513–514
Induced emf, 512–516
  between coils, 515
  direction of, 514
  motional, 512–513
Induced field, 418, 515
Inductance, 525–548
  mutual, 525–527
  self, 525, 527–530
  stray, 530
Induction, charging by, 407
Induction motor, 541
Inductive reactance, 562–565
Inductor, 528
  and ac, 558–559
  and capacitor, 532–534
  charge-discharge cycle, 559
  current vs. time in, 531–534
  in electric circuits, 530–534
  energy stored by, 528–529
  and shorted circuit, 531
Inelastic collisions, 166–171
Inertia, 57
  law of, 57
  moment of (see Moment of inertia)
Inertial confinement fusion, 833, 835–836
Inertial frame of reference, 676
Inertial mass, 691
Inertial system, 684
Infiltration rate for homes, 803
Infrared in medicine, 356
Infrared waves, 585
Ingestion of radioisotopes, 805, 809–811
Initial phase, 556–557
Initial phase angle, 552
Instantaneous speed and velocity, 18–21
Insulators:
  electric, 405
  resistivity, (table) 461
  thermal, 349, (table) 352
Intake stroke, (illus.) 385
Integrated circuit, 764
Intensity:
  loss in reflection, (illus.) 604
  passing polarizer, 664–665
  of radiant energy, 355, 706
  of sound, 256–258
  vs. wavelength, (illus.) 356
Interactions, 155–174
  of elementary particles, 788
Interface, 244
Interference:
  of electrons, 748–750
  of light, 642–659
    and diffraction, 656–659
    double-slit, 643–646
    single-slit, 650–654
    thin-film, 647–650
  of waves, 245

Interference (Cont.):
  x-ray, 713
Interferometer, 680–682
Internal energy, 373–374
Internal resistance:
  of battery, 459–460
  power dissipated by, 464
Inverted image, 598
Ion pairs produced, 800
Ionization chamber, 785
Ionization energy, 734
Ionization potential, 731
Ionizing radiation, 785–787
  and humans, 795–813
  units of, 799–802
Ions:
  mass determination, 503
  trail of, in G-M counter, 785
Iris, camera, 623–625
Isobar, (illus.) 376–377
Isobaric process, 376–378
Isolating bodies (analytical technique), 66
Isomet, (illus.) 376–377
Isotherm, 379, (illus.) 381
Isothermal process, 378–380
Isotopes, 773–774
  fissionable, 820
  in medicine, 809–811
  representation of, 774
  separation of, 819
Isovolumetric process, 376

Joule, James Prescott, 131, (illus.) 342
Joule (unit), 131
Joule heating, 463
  in ac circuits, 568
  in power transmission, 542
  and safety, 573
Joule's law, 463
Junction diode, 763

Keating, R. E., 689
Kelvin, Lord, 388, 390–391
Kelvin temperature scale, 317, (illus.) 323–325
Kelvins (absolute temperature unit), 317
Kilogram (unit), 4
Kinematics, 14
  angular, (table) 89
  rotational, 79–95
Kinetic energy, 138–139, 141–148
  and absolute temperature, 329
  in beta emission, 780–781
  of charges and potential difference, 429–430
  of fluid, 298
  as limit of relativistic energy, 694
  lost to friction, 315
  of molecule, 329
  of pendulum, (illus.) 143–144

I-9

Kinetic energy (*Cont.*):
  and relativity, 693–694
  rotational, 187–193
  in SHM, 217–219
Kinetic friction, 68–70
Kinetic theory:
  of gases, 326–331
  of matter, 327
Kirchhoff, Gustav, 478
Kirchhoff's rules, 477–483
  independence of equations, 480
  for inductors, 530–531
  loop rule, 478
  point rule, 478
  prescription for application, 480–481
  in *RC* circuit, 556
    ac, 561
  sign conventions, 479

Lags, 552
Lake Superior, (prob.) 367
Laminar flow, 306–307
Land, Edwin, 660
Lase, 737
Laser, 736–740
  as acronym, 736
  in communications, 605–606
  and eye damage, 739–740
  in fusion, 835–836
  medicine and (*see* Medicine and Laser)
  tunable, 739
  types of, 739
  uses of, 736
Latent heats, 344–348
Law of inertia, 57
Law of Malus, 664–665
Law of optics, 600, 616
Laws, nature of, 72
LCDs (*see* Liquid-crystal displays)
LCR series circuit, 532–534
Leads, 552
LEDs (*see* Light-emitting diodes)
Lenard, Phillip, 707
Length:
  as fundamental quantity, 3
  relativistic, 686–687
  standard of, 4
  vs. weight in animals, 282
Length contraction, 686–687
Lens, 611–639
  aberrations, 620–621
  applications, 632–636
  camera, 623–625
  combinations, 625–628
  compound, 623–624
  corrective, for eye, 635–636
  cylindrical, (illus.) 593
  divergence and convergence, 612
  of eye, 633, 634
  Fresnel, 613
  image formation, 614–620

Lens (*Cont.*):
  manufacture, 612
  meniscus, 613
  nonspherical, 612
  positive and negative, 616
  spherical, (illus.) 593, 611–612
  thin, 614–620
  types of, 612–613
  underwater, 620
Lens equation, 616
Lenz, Heinrich, 514
Lenz's law, 514–515, 527, 537
Lepton, (table) 788–789
Lepton number, 789
Levers, 119
Levulorotary compounds, 665
Life span of nuclear bomb survivors, 806
Lifetime of muon, 688–689
Light, 581–671
  as electromagnetic wave, 584–585
  medium of propagation, 673–674
  polarization of, 659–667
  speed of, 584–585, 676, 684, 693
  travel in straight lines, 587
  wave properties, 641–667
Light-emitting diodes (LEDs), 666
Light intensity and f-stop, (table) 624
Light pipe, 596
Lighthouses, 613
Limitation of measurement, 756
Limiting process (mathematical procedure), 19–21
Limiting velocity, 693
Linear hypothesis for radiation effect, 805–806
Linear mass density, 241–242
Linear momentum (*see* Momentum, linear)
Linear motion, 13–30
Lines of force, 415–416, 498
Liquid crystal (LC), 666
Liquid-crystal displays (LCDs), 666–667
Liquid-drop model of nucleus, 817, 819
Liquids, 288
Load:
  electrical, 472
  on structures, 279
Localized wave, 751
Lodestone, 497
Longevity of radioactive sample, 785
Longitudinal waves, 235
  depicted as transverse wave, (illus.) 239
Loop rule, Kirchhoff's, 478
Lorentz, H. A., 675, 684, 687
Lorentz contraction, 687
Lorentz transformations, 684
Low-level radioactive wastes, 828
Lyman series, 717, 729–730

Machine, 128

Magnet:
  effect of breaking, 510
  electromagnet, 511
  in generator, 539
  loss of magnetism, 512
  natural, 497–498
  permanent, 497–498
Magnetic bottle, 833–835
Magnetic confinement fusion, 833–835
Magnetic declination, 516–517
Magnetic deflection in TV set, 441
Magnetic dip, 516–517
Magnetic domains, 511–512
Magnetic energy of inductor, 529
Magnetic field:
  axial, of loop, (prob.) 522
  and cathode rays, 723
  cause, 504
  around current-carrying wire, 504–505
  of current loop, 505
  direction of, 498–500
  of the earth, 516–518
    changes in, 516–518
    measurement of, (prob.) 548
    origin, 518
  and electromagnetic waves, 583–584
  in iron, 511–512
  "pinched," 834–835
  and polarization, 659
  representation on page, 501
  of solenoid, 506
  in space around earth, (illus.) 517
  of spinning electron, 735
  of toroid, 506
Magnetic field intensity, 499n.
Magnetic field lines, 498
Magnetic flux, 512, 514–516
Magnetic flux density, 499
Magnetic force:
  on current, 507–509
  between current-carrying wires, 508–509
  on moving charge, 498–503
Magnetic induction, 499
  of the earth, 516, 518
Magnetic lines of force, 497–498
Magnetic mirror, 834
Magnetic moment of electron spin, 735
Magnetic permeability, 505
Magnetic poles, 498
Magnetic properties of materials, 509–512
Magnetic quantum number, 733
Magnetic torque on current loop, 534–536
Magnetism, 497–522
  discovery, 497
  of materials, types of, 510–512
  permanent, interpretation, 509–512
  residual, 512
  and rocks, 518
  and temperature, 518

Magnetopause, 517
Magnification, 602
   angular, 631
   of combined lenses, 627–628
   for lenses, 618–620
   of magnifier, 621–623
   of microscope, 630
   of telescope, 631
Magnifier, 621–623
   image formation, (illus.) 622
Magnitudes:
   of stars, (prob.) 639
   of vector, 35
Malus, Etienne Louis, 662
Malus, law of, 664–665
Manometer, 292–293
Mass:
   in collisions, 163
   distribution of, and rotation, 179–187
   as energy, 693–694
   as fundamental quantity, 3
   in gravitation, 95–98
   of human body parts, (illus.) 119
   of ions, determined, 503
   as measure of inertia, 57
   of planets, (table) 96
   relativistic, 691–692
   rest, 691
   of rocket as payload, 172–173
   of space shuttle, 173
   on spring, motion of, 209–210
   unit of, 4
   vs. weight, 62–63
Mass change in accelerator, 692
Mass-energy conservation, 777–778
Mass number, atomic, 774–776
   and binding energy, 778–779
Mass ratio, electron and alpha particle, 725
Mass spectrograph, 502–503
Mass spectrometer, 819
Materials, uranium concentration in, 798
Mathematical techniques and applications:
   area under curve, 379
      impulse, 157–159
      work, 132–133
   binomial expansion, relativity, 694
   calculus alluded to, 96, 172, 183, 379, 432
   common logarithm, sound intensity level, 257–258
   confirmation of physical idea, 218
   equation of parabola, projectile, 47
   examination of special solutions (billiard balls), 163, 166
   exponential equation: decay of inductor field, 531
      radioactive decay, 782–784
      time constant, 489–491
   extrapolation, work function, 710
   limiting process, instantaneous velocity, 20–21

Mathematic techniques and applications (*Cont.*):
   logarithmic equation, 379
   quadratic equation, interpretation of both roots, 26–27
   matrix described, 755
   natural logarithm as general solution, 172
   parabolic velocity profile, fluid flow, 304
   plausibility calculation, moment of inertia of hoop, 185
   simultaneous equations:
      collisions, 162–163
      2-dimensional collision, 164–168
      equilibrium, 120–121
      Kirchhoff's rules, 481–483
      projectile, 46–47
   slope of curve at a point, 20–21
   slope-intercept form, 710
   small-angle approximation: centripetal acceleration, 91
      pendulum, 219
      physical pendulum, 221
   summation notation, 58
   summation process, moment of inertia, 183
   trigonometric identities: double-angle formula, 47
      $\sin^2\theta + \cos^2\theta = 1$, 120–121, 215, 217
      sum and difference of angles applied to standing waves, 246
   trigonometry, law of cosines, 165
   verification of form of equation, 217
Matrix mechanics, 746, 755
Matter, wave nature of, 745–769
Matter waves:
   evidence, 748–750
   physical interpretation, 755
   wavelength, 747
Maximum permissible effective doses, 807
Maxwell, James Clerk, 582–585
Maxwell's electromagnetic theory, 706, 707
Maxwell's equations, 582, 676, 684
Mean free path in fusion, 833
Measure, standard of, 2
Measurement, 2–3
   of electricity, 483–486
   and reference frame, 676–678
   relativity as a system of, 696
Mechanical energy, 143–147
   of atom, quantized, 726–728
   in SHM, 216–219
Mechanical equivalent of heat, 342
Mechanics, wave and matrix, equivalent, 755
Media in interference, 647–648
Medical and dental x-rays, 796, 802–803, 805–807
Medical radiation dose, 803
Medicine and laser, 736, 739–740

Medicine and laser (*Cont.*):
   imaging with fiber optics, 605
   nuclear, 809–811
   use of infrared, 356–357
Meitner, Lise, 816–817
Meltdown, 826
Melting point of ice, 344
Mendeleev, D. I., 734
Meniscus lens, 613
Mercurial barometer, 293
Meson, (table) 788–789
   pi, 777
Metabolic rate for various activities, 362–363
Metabolism, 362–363
Metastable energy state, 737, 738
Meter (unit), defined, 4
Meter movement, 485
   construction, (illus.) 535–536
Meters, electrical, 483–486
   ac, 570–571
   analog vs. digital, (illus.) 483–484
   internal circuit, (illus.) 485–486
   nuclear survey, 786
   reading scales, 485
Metric system of measure, 4
MeV to u conversion, 779–780
Michelson, Albert A., 674, 679–683
Michelson interferometer, 680–682
Michelson-Morley experiment, 679–683, 686
Microscope, 629–630
   electron, 765–767
   limits, 765
Microscopic objects, size, (illus.) 765
Microstates and entropy, 390–391
Microwave transmission, 605–606
Microwaves, 585
Milky Way galaxy, (prob.) 669
Millikan, Robert, 796
Miners and radiation dose, 805
Mirrors, 587–589
   concave, 598–603
   convex, 598–603
   equation, 600
   and images, 598–604
   in laser, 738
   parabolic, 632–633
   plane, 582, 587–589
   vs. prisms, 595–596
   spherical, 598–604
Model:
   atomic, Thomson, 724–725
   for mechanical waves, 234–235
Moderator, 819
Modulus:
   bulk, 277–278
   elastic, 271
   shear, 275
   Young's, 271
Molar heat capacity, 374–375
   at constant pressure (C_P), 375

I-11

Molar heat capacity (Cont.):
   at constant volume (C_V), 375
   difference derived, 377
   ratio of (γ), (table) 381
   table, 377
   units, 375
Molar volume of a gas, 323
Mole, 323
Molecular mass, 323
Molecules and entropy, 390–391
Molybdenum, 714
Moment arm of a force, 109–111
Moments of force, 118
Moments of inertia, 180–187
   and axis of rotation, 182–186
   defined, 181
   experimental determination, 183
   general form, 183
   of irregular body, 182–183
   parallel axis theorem, 186
   of point mass, 180–181
   of regular objects, (table) 184, (illus.) 185
   rotation not through center of mass, 186
   as rotational inertia, 181n.
   table of, 184–185
   and torque, 183
   units of, 181
Moments of mass, 118
Momentum:
   angular (see Angular momentum)
   linear, 156–174
      conservation of (see Conservation, of linear momentum)
      and rocket, 171–172
      units of, 157
      vector nature, 157, 165
   of photon, 746–747
   and position of wave packet, 755
   in relativity, 695
Monochromatic aberrations, 620–621, 623
Monochromatic radiation, 643, 710
Month, calculation of, 98
Moon:
   data on, (table) 96
   motion of, 98
   period of, 98
   rocket to, (illus.) 171
Morley, E. W., 674, 679–683
Morse code, 605
Moseley, Henry, 730
Motion:
   changes in and force, 58
   equations of, (table) 22
   first law of, 56–58
      and natural state of motion, 56
      original statement of, 56
      summary of points, 58
   of fluids, 296–308
   impending, 69

Motion (Cont.):
   of mass on spring, (illus.) 206–207, 209
   as natural state, 58
   Newton's defined, 59
   of pendulum in swing, (illus.) 207
   rotational and translational, 80
   second law of, 59–62
      most common statement, 60
      original statement, 59
   simple harmonic (see Simple harmonic motion)
   third law of, 63–65
      forces in matched sets, 64
      forces act on different bodies, 64
      original statement, 64
   through viscous fluid, 305–307
   (See also specific type, for example: Simple harmonic motion)
Motional emf, (illus.) 512–514
Motors, 536–538
   vs. generators, 540
   torque produced, 537
Mt. Palomar telescope, (prob.) 639, 654
Mt. Pashtukov telescope, 633
Moving sources and waves, 259–260
Mueller, Wilhelm, 797
Multidimensional motion, 35–51
Multimeter:
   full-scale reading, 485
   ranges of, 486
Multiple images, (illus.) 589
Multiple reflection, intensity in, (illus.) 604
Muon, 688–689
   discovery, 798
Muscles, 119
Musical instruments, sound spectrum, (illus.) 250, 254
Mutation:
   and radiation, 804–805
   spontaneous rate, 805
Myopia, 634

n-type semiconductor, 763
Natural convection, 353
Natural uranium in reactor, 823
Near point of eye, 634
Nearsightedness, 634
Neddermeyer, Seth, 798
Negative charge, 405
Negative lens, 616
Neon, properties of, (table) 393
Net energy in fusion, 831–832
Neutrino, 781–782
   evidence for, 781
Neutron, 406
   availability in reactor, 822
   decay of, 781–782
   delayed, 821
   discovery, 776
   excess, 818

Neutron (Cont.):
   production of, 816
   prompt, 821
   slow, 819
Neutron activation analysis, 816
Neutron bombardment, 816
Neutron capture, probability, 820
Neutron poisons, 829
Newton, Sir Isaac, 57
   and gravitation, 95–96, 99
   and optics, 642, 647
   role in prediction of motion, 14
Newton (unit), 60
   equivalence in pounds, 62
Newton's laws of motion (see Motion, laws of)
Newton's rings, 647
Nicol, William, 662
Nicol prism, 662–663
NMR scan (see Nuclear magnetic resonance)
Nobel prize, 657, 679, 715, 772, 816
Node of waves, 246, (illus.) 248
   position on string, 248–249
Nonconservative forces, 142
Nonfissionable plutonium, 821
Nonlinear circuit elements, 456
Normal, 587
Normal force, 69
Normal image, 598
North-seeking pole, 498
Notation:
   change in, 16
   force on body due to other body, 65
   proportional to, 59
   rate of change, 16
   relative velocities, 49
   sum of, 58
   vectors, 36
Nova laser fusion system, 836
Nuclear binding energy, 777–779
Nuclear bombardment and momentum, 163
Nuclear equations, symbolism, 774
Nuclear fallout, 796, 802
Nuclear fission, 816–822
Nuclear forces, 777–782
Nuclear fusion, 831–837
Nuclear magnetic resonance (NMR), (illus.) 225–226
Nuclear medicine, 809–811
Nuclear power:
   economics, 823
   environmental impact, 827
   knowledge of public, 822–823
   plants, radiation, 796, 802–804
   radiation dose, 803–804
   safety, 823–827
   waste disposal, 827–828
Nuclear reactors, 815–839
   isotope production, 811
   (See also Reactors)

I-12

Nuclear spin, 776
　and magnetism, 510
Nuclear stability and binding energy, 778–779
Nuclear strong force, 777–780
Nuclear weapons, 815
Nuclear "zoo" of particles, 789–790
Nucleon, 776
　finding energy per, 778–779
Nucleus (atomic), 406, 771–793
　composition and properties, 773–776
　discovery, 724–725
　liquid-drop model of, 817, 819
Null result, 674, 679, 682
Nutcracker as lever, 127

Object, 599
　distance, 600
Objective lens, 629–631
Observation, role of, in science, 2
Observer frame of reference, 677
Oceans, volume of, (prob.) 366
Ocular, 629
Ohm, Georg Simon, 455
Ohm (Ω) (unit), 455
Ohmic circuit elements, 456
Ohmmeter, 486
　accuracy of, 486
　internal circuit, 486
Ohm's law, 455–456
　ac analog, 565
　limitation in analysis, 477
One-dimensional motion, 13–30
Opaque, 581–582
Open circuit, 458
Open end, 243
Open-tube manometer, 292–293
Opera glass, 631
Operational definition, 508–509
Optical activity, 665
　density, 592, 647
　instruments, 611–639
　path length, 654
Optics:
　geometrical, 581–639
　physical, 641–670
Orbit:
　being in, 99
　electron, elliptical, 733
　radius of, for planets, (table) 96
Orbital quantum number, 733
Order number, 645
　single-slit, 651
Ordinary ray, 662–663
Organ pipes, 251–255
Organ transplants and temperature, 331
Orientation quantum number, 733
Origin of reference frame, 675
Oscillation, 205
　of LCR circuit, 532–534

Oscillator:
　atomic, 710
　energy emitted, 706
Oscillatory motion, 205
Oscilloscope, 571–573, 810
Oscilloscope traces, 572
Otto, Nicholas, 384
Otto cycle:
　P vs. V diagram, 385
　strokes of, (illus.) 385
　theoretical efficiency, (prob.) 399
Out-of-phase voltage, 552
Outlets, electrical, 575
Overhead projector lens, 613
Overlap of wave functions, 759–760
Overtones, (illus.) 249, 250–254

p-type semiconductor, 763
Pacemaker, capacitor in, 491
Pain, threshold of, 257
Pair creation, 788
Parabolic reflector, 633
Parabolic velocity profile of fluid in tube, 304
Parallel:
　batteries in, 472–473
　capacitors in, 487
Parallel-axis theorem, 186
Parallel circuit, simple, 473
Parallel-plate capacitor, 435–436
Paramagnetism, 511
Paraxial rays, 620
Parent nucleus, 773
Partial pressures:
　Dalton's law of, (prob.) 369
　of gas, 359
Particle physics, state of, 790
Particles:
　elementary, (table) 787–790
　　in cosmic rays, 798
　　"nuclear zoo" of, 789–790
　　(See also specific particle)
　of gas in given volume, 323
Pascal, Blaise, 275, 291
Pascal (Pa) (unit), 275
Pascal's principle, 291–292
Paschen series, 717–718, 729–730
Passing truck, 301
Path of particle, 787
Path difference, 651
Pattern:
　double-slit diffraction, 645
　single-slit diffraction, 651
Pauli, Wolfgang, 734
　exclusion principle, 734–736
　　for multiatomic systems, 759–760
　and neutrino, 781
Payload of rocket, 172–173
Pendulum:
　ballistic, 166–168
　energy in swing, (illus.) 143–144

Pendulum (Cont.):
　motion as SHM, 219–220
　period of, 220
　physical, 221–223
　simple, 219–220
Penetrating power of radiation, 772, 776, 808
Pentaprism, 596, 624
Perfectly inelastic collisions, 168
Period, 205
　and frequency, 84
　of pendulum, 220
　of physical pendulum, 222
　of rotation, 84
　in SHM, 215–216
　sidereal vs. synodic, 98
Periodic motion, 205
Periodic table, 734
Periodic waves, 235
Permeability:
　magnetic, 505
　and speed of light, 584
Permittivity:
　of dielectric, 441
　electric, of free space, 436
　and speed of light, 584
Perspiration, 357, 365
Perverted image, 598
Pfund series, 717, 729
Phase:
　and ac circuit, 552
　initial, 214
　and interference, 644, 647–650
　inversion of, in reflection, 243–245
　of matter, 325
Phase angle, 214
Phase-angle difference, 552–553
Phase change:
　latent heat of, 344–348
　on reflection, 647–650
Phase difference, 552–553
　of capacitor, 555
　of inductor, 558–559
　and power delivered, 567–570
Phasor, 553
Phasor diagrams, 552–553
　ac capacitor, 556
　inductor, 559
　instantaneous values, 553
　RC circuit, 557
　RCL circuit, 564–565
　RL circuit, 560
Phosphor, 441, 446, 571
Phosphor dots, number, 605
Phosphorescence, 771
Photoelectric apparatus, 707
　current, 707
　effect, 707–711
　equation, 710
Photoelectrons, 710
Photoemission:
　classical view, 708–709

I-13

Photoemission (*Cont.*):
  quantum view, 708
Photograph, x-ray, 700, 712
Photographic data:
  ball on ramp, 33
  bouncing ball, 169
  bowling ball drop, 24
  pendulum swing, 143
  pole vault and energy, 146
  rate of rotation and turntable, 25
  rotation of gymnast, 82–84
  tennis ball rising and falling, 19
  vertical and horizontal ball drop, 43
Photographic plate:
  as detector, 785
  and discovery of radioactivity, 771–772
  trail of ions on, 785
Photography:
  camera, 623–625
  and metastable states, 737
  stereoscopic, 787
Photomultiplier, 809–810
  tube, 786
Photon, 710, 726–727
  emission, 736–737
  linear momentum, 746–747
  as particle, 746–747
Physical barriers in reactors, 825
Physical laws, nature of, 72
Physical optics, 641–670
Physical pendulum, 221–223
Physical quantity, 2
Physics, 2, 14
Physiological effects of currents, (table) 575
Pi meson, 777
Pincushion distortion, 621
Pinhole camera, 625
Pion, 777
  in cosmic rays, 798
Pipes or tubes and waves, 251–255
Pitch (musical) and frequency, 254
Pitchblende, 798
Pitot tube, (illus.) 309
Pivot, 113
  location on human body, (illus.) 119
Planck, Max, 704–707, (illus.) 705, 710, 711
Planck's constant, 706
Planck's expression, 704–706
Plane diffraction grating, 654–659
  polarized, 659
  wavefronts, 644
  waves, 586
Planetary retention of atmosphere, (prob.) 338
Planoconcave lens, 612–613
Planoconvex lens, 612–613
Plasma, 832–835
  confinement, 833–835
Plastic flow, 271
Plutonium, 820

pn junction, 763
Point masses, 95–96
  and rotations, 180–181
Point rule, Kirchhoff, 478
Poiseuille, Jean, 303
Poiseuille (Pl) (unit), 303
Poiseuille's law, 302–305
Polarimeter, (prob.) 670
Polarization, 659–667
  and LCD's, 666–667
  and stress analysis, 665
Polarizer, 663
Polaroid, 660–665
  sunglasses, 661
Pole vault, energy of, 145–147
Poles, magnetic, 498
Polonium, 772
Polyvinyl alcohol (PVA), 660
Population inversion, 737
Positive charge, 405
Positive lens, 616
Positron, 781, 798
Positron-electron pair, 788
Postulates:
  Bohr, 727
  de Broglie, 747
Potassium 40, 799
Potential, electric:
  absolute, 428–430
  defined, 426
  ground, 428
  and KE of charges, 429–430
  of many charges, 429–430
  zero, 428
Potential difference, 431–435
  between ends of wires, 451
  across heart, 431
  terminal, 459–460
  (*see also* Voltage)
Potential drop, 456
Potential energy, 139–148
  electrical, 424–428
  independent of path, 140
  of pendulum, (illus.) 143–144
  in SHM, 217–219
  of spring, 140
Potential gradient, 431
Potential limit, electron microscope, 765
Pound (non-SI unit), 62
Power, 136–137
  in ac circuits, 567–570
  average ac, 568
  of dc motor, 538
  in electric circuit, 463–465
  nuclear, 815–839
  refractive, 629
  of eye, 634
  and torque, 137
  and velocity, 136–137
Power consumption, 666–667
Power factor, 568–570
Power plant wastes, volume, 828

Power stroke of engine, (illus.) 385
Power transfer, maximum, 545
Powers of ten, prefixes, (table) 8–9
Precession, 197–198
  frequency of, 198
Prefix, metric, (table) 9
Presbyopia, 635
Pressure:
  absolute, 290–291
  atmospheric, 290
    measurement of, 293
  of blood, 308
  defined, 275
  at depth in liquid, 289–290
  drop in atomizer, 301–302
  of gas and volume, 320–321
  gauge, 292
  kinetic model for gases, 327
  lack of direction, 290
  due to liquid, 290
  non-SI units, 293
  transmission in fluid, 291–292
  variation in sound waves, 239–240
  and velocity of fluid, 299–302
  vessel, 825
Pressurized water reactor (PWR), 824
Primary bow of rainbow, 597
Primary coil, 541
Primary cosmic radiation, 797
Primary standard of measure, 3
Principal axis of mirror, 598
Principal quantum number, 733
Principia, 56
Principle of superposition, 244–245
Prisms, 595–596
  vs. mirrors for reflection, 595–596
  Nicol, 662–663
  and spectrum, 597
Probability:
  of localized electron, 758
  of nuclear decay, 782
Problems:
  coding by difficulty, 10
  solving techniques, 29–30
Product nuclei from fission, 821
Projectile, 45
  mass of, and collisions, 163
  maximum height, 47
  optimum angle of projection, 47–48
  parabolic path, 47
  range of, 46–48
  symmetry of trajectory, 47
  trajectory of, 45, 46
Projectile motion, 45–48
Prompt neutrons, 821
Propagation:
  of light, theories, 674
  of waves, 233–235, 241–245
Proper time, 685
Proportion (mathematical), 59
Proportional limit, 270
Proportionality constant, 59

I-14

Proton, 406, 775
  from neutron, 781–782
  stability of, 789n.
Pulley, 128–130
Pulse (wave), 243
Pumping of laser, 737
Pupil of eye, 634, 635
PVT surface for water, 335
Pyrometer, optical, 316, 326

Quality factor, 801, (table) 802
Quanta, 711
Quantization, de Broglie interpretation, 747–748
Quantize, 726, 732
Quantum, 706
Quantum hypothesis, 706, 711
Quantum mechanics, 745–769
  overview, 745–746
  practical nature, 759
Quantum number, 728
  magnetic, 733
  orbital, 733
  orientation, 733
  principal, 733
  spin magnetic, 736
  in wave mechanics, 752
Quantum physics, 701–721
Quantum theory, 701, 726
Quantum wave function, 752
Quark model, 790
Quarks, (table) 790

R-value, 351
  table of, 352
Rad (non-SI unit), 801
Radian, 80
  degree conversion, (table) 81
  measure, 80–81
Radiant energy, 355, 706
  nature, 701–704
Radiation:
  average dose, 802
  background, 795–796, 798–799, 802, 806
  beneficial effects, 806
  biological effectiveness, 801–802
  blackbody, 702–706
  classical theory, 705
  cosmic, 688, 795–798, 802–803
  detection of, 785–787
  effect of: on cells, 805
    large doses, 805
    low doses, 805–806
  of energy, early theories, 701
  from electric stove, 702
  factors to minimize dose, 808
  from fusion reactor, 826
  ionizing, 785–787, 795–813
  in medicine, 809–811

Radiation (Cont.):
  occupational dose, 807
  quality factor, 801
  record keeping requirements, 807
  relative ease of detection, 805
  release from nuclear plant, 826
  thermal, 348, 355–357
    and human body, 374
  units of measure, (table) 802
Radiation dose:
  with altitude, 803
  in buildings, 803
  ion pairs produced, 800–801
  U.S. standards, 807
Radiation level and altitude, 796–797
Radiation safety, 806–809
Radiation sickness, 804
Radiation therapy, 811
Radio telescope, (prob.) 669
Radio waves, 585
Radioactive dating, (prob.) 797
Radioactive decay, 773–774, 782–785
Radioactivity, 771–776, 782–785
  in building materials, 803
  discovery, 771–772
  in environment, 798–799
  in living systems, 799
  natural, 795
  types, 772–773
Radioisotopes:
  diagnostic, (table) 811
  ingestion of, 805, 809–811
Radiopharmaceuticals, 803, 809–811
Radium, 772, 773
  ingestion, 805
Radon, 798
  in buildings, 803
Radius:
  of curvature of mirror, 598
  of electron orbit, 728–730
  of planetary orbit, (table) 96
  of planets, (table) 96
Rainbow, 597–598
Raindrops and rainbow, 597
Raisin cake model of atom, 724
Ramp as machine, 129–130
Randomized motion and temperature, 330
Range of cannon experiments, 16th Century, 13
Rarefaction of sound wave, 239
Rate, defined, 16
Ray tracing:
  curved mirrors, 599–601
  hints for lenses, (table) 617
  hints for mirrors, (table) 601
  for lenses, 614–617
Rayleigh, Lord, 654, 704–705
Rayleigh criterion, 653–654
Rayleigh-Jeans law, 704–705
Rays, 586
  extraordinary, 662–665

Rays (Cont.):
  incident, 587
  paraxial, 620
  vs. wavefronts, 585–587
RC circuit:
  ac, 556–557
  dc, 488–491
RCL circuit, 562–567
  analog to damped, driven oscillator, 566
  energy in, 566
Reactance, 562–565
Reactor:
  breeder, 829–830
  CanDU, 823
  first, 819–820
  fusion, 833–837
  high-temperature gas-cooled, 829
  liquid-metal fast breeder, 829
  number operating, 823
  power, 821
  production, 821
  research, 822
  in ships, 822
  thorium-uranium cycle breeder, 829
  safety, 823–827
Real gas:
  vs. ideal gas, 323–324
  P vs. T curves, 333–334
  states of, 332–335
Real image, (illus.) 599
Rectifier, 570
Red shift of galaxies, 260
Reference, frame of, 675–678
Reference circle for SHM, 210–216
Reference points for temperature, (table) 326
References on nuclear power, 828
Reflected ray, 588
Reflection:
  angle of, 588
  diffuse, 581–582, 588
  intensity loss, (illus.) 604
  law of, 588
  phase change on, 647–650
  polarization by, 660–662
  specular, 581–582, 588–589
  total internal, 594–596
  of waves, 243–245
Reflectors, 632–633
Refraction, 589–596
  index of (see Index of refraction)
  of lenses, 611–612
  Snell's law of, 590–591
  wave interpretation, (illus.) 591
Refractive power, 629
  of eye, 634
Refrigerants, 392
Refrigerator, 384
  calculation of cycle, 392–396
  Carnot, 388
  operating cycle, 392

I-15

Refrigerator (*Cont.*):
   real vs. ideal, 392
Reines, Frederick, 781
Relative, everything is not, 696
Relative biological effectiveness, 801
Relative humidity, 358–362
Relative velocities, 49–51
   in relativity, 677–678
Relativistic velocities, 690–691
Relativity, 673–698
   acceptance, 695–696
   and common experience, 695–696
   experimental contraction, 688–690
   general theory of, 684n.
   history of, 673–675
   and length, 686–687
   and mass, 691–693
   momentum in, 695
   special theory of, 683–695
   as a system of measurement, 696
   and time, 684–686
   velocities in, 690–691
Releases from nuclear plants, 826
Rem (non-SI unit), 801–802
Reprocessing nuclear fuel, 828
Repulsion:
   electric, 404–405
   of magnets, 498
Resistance:
   electrical, 455–456
   equivalent, 473–477
   of human body, 574
   internal, 459–460, 464
   measurement of, 486
   stray, 530
Resistivity, 460–463
   table of, 461
Resistor, (illus.) 456
   and ac, 553–554
   in parallel, 474–475
   in series, 474
   series and parallel combined, 475–477
Resolution, 653–654
   of electron microscope, 766–767
   limits of, 765
Resolving time of G-M counter, 786
Resonance, 225–226
   in ac circuit, 566–567
   and structures, 225
Resonant frequency, 225, 556–567
Rest energy, 694
Rest mass, 691
   of photon, 746
Restoring force, 209
Resultant, 36
Retarding force in fluid, 305–307
Retina, (illus.) 633
Reversible thermodynamic processes, 375–382
Revolutions in physics, 701
Reynolds, Osborne, 307
Reynolds number, 307

Rhm, 808
Right-hand rule:
   and three-dimensional rotations, 196–198
   for torques, 112
Right-hand screw rule for magnetism, 499–500
Risk vs. benefit decision, 803
RL circuit (ac), 560–561
rms (root mean square), 568
   current, 568
   values, 568–569
   velocity, 330
   voltage, 568
   for capacitor and inductor, 569
Rocket, 171–174
   final velocity, 172–173
   stability of, 173
   thrust factors, 172
Rods of eye, 633
Roentgen, Wilhelm, 712, 723, 771
Roentgen (R) (non-SI unit), 800
Rolling object, 189–192
   alternate ways to treat, 189–190
   frictional work in, 191–192
   need for friction, 191–192
   vs. slipping, 189
Room temperature, 317
Root mean square (*see* rms)
Rossi, Bruno, 688
Rotation, 79
   axis of (*see* Axis of rotation)
   with changing geometry, 192–193
   with changing mass, 192–193
   distinguished from vector, 86–88
   not commutative, 86–88
   of point mass, 180–181
   sense of, 82
   and sign, 111–112
   sign convention for, 82
Rotational dynamics, 179–198
Rotational inertia (*see* Moment of inertia)
Rotational kinematics, 79–95
   defined, 79
Rotational kinetic energy, 187–193
Rotational motion, changes in, causes of, 79, 179–180
Rotational work, 135, 187–189, 193
   sign of, 188
Rounding numbers, rules, 7–8
Rutherford, Ernest:
   and discovery of nucleus, 724–726, 771
   and nature of alpha particle, 773
   suggestion of neutron, 776
   and transmutation, 775
   and types of radiation, 772
Rutherford nuclear model of the atom, 724–725
Rutherford-Soddy interpretation of radioactivity, 772–774
Rydberg, J. R., 717

Rydberg constant, 717
   from theory, 729

Safety:
   electrical, 573–576
   with laser, 739–740
   nuclear vs. non-nuclear, 826–827
   nuclear power, 823–827
   radiation, 806–809
Safety factor, (table) 279
Safety ground in electrical outlets, 575
Sailboat, 302
Satellites, 98–99
Saturated air, 358–359
   water vapor density, (table) 359
Saturation current, 707
Saturation curve for real gas, 333
Sawtooth wave, 552
   of CRT, 571
Scalar, defined, 35
Scaling factor, 281–282
   laws, 280–282
Scanning electron microscope, 766–767
Scanning rate of TV, 444
Scattering:
   of alpha particles, 724–725
   and color of sky, 665
   polarization by, 663
   of x-rays, 714–716
Schematic, 477
Schrodinger, Erwin, 746, 751–754
Schrodinger equation, 752
Science, 2, 815
Scientific notation, 8
Scintillation, 786
Scintillation detector, 786
   scanning, 809–810
Sclera, 633
Second of time (unit), 4
Second condition of equilibrium, 113–121
Second law of thermodynamics, 388–392
   in terms of entropy, 390
Secondaries (cosmic rays), 797
Secondary bow of rainbow, 597
Secondary coil, 541
Secondary standard of measure, 3
Selective absorption, polarization by, 660
Semiconductor, 759, 761–764
   conductivities, (table) 461–462
   doped, 762
   p-type and *n*-type, 763
   temperature dependence, 761
Sensations of electric shock, 574–575
Sense of rotation, 82, 111–112
Serendipity, 748n., 771, 772
Series:
   batteries in, 472
   capacitors in, 487–488
Series circuit, 473
Series-wound motor, 540–541

I-16

Service box, home ac, 575
Shadow, 582
Shaped field, 765
Shear, angle of, 275
Shear forces, 274
Shear modulus, 275, (table) 277
Shear stress, 269
Shearing, 274–275
Shielding and radiation, 808
SHM (see Simple harmonic motion)
Shock, electric, 574–575
Shock wave in water, 287
Shorted circuit, 531
SI (Système International d'Unites), 3
Sievert (Sv) (unit), 801–802
Sign convention:
   images in mirrors, 601
   Kirchhoff's rules, 479
Signal, TV, 445
Signal purity in communication, 605–606
Significant figures, 7–8
SIL (see Sound intensity level)
Silicon as semiconductor, 761–762
Simple harmonic motion (SHM), 205–227
   acceleration in, 210, 214
   adding energy to, 224–225
   amplitude, 214
   and angular velocity, 212
   damped, 223
   defining condition, 210
   described, 209
   displacement equation, 212
   driven, 224
   energy in, 216–219
     and position, 218
   frequency of, 214
   initial phase, 214
   KE vs. PE, time variation, 217
   of mass on spring, 209–210
   period of, 215–216
   phase, 214
   production of, 206–208
   reference circle, 210–216
   and resonance, 224–225
   velocity equation, 214
   x, v, and A related, (illus.) 214
Simple pendulum, 219–220
Simultaneous pulses, 797
Sine curve, 206–207, 210
Sine wave in ac, 552
Single-lens reflex camera, operation, (illus.) 624
Single-slit diffraction, 650–654
Sinusoidal waveform, 236
Siphon, 311
Sirius, (prob.) 639, 719
Skin, dry vs. wet, resistance, 574
Sky:
   color of, 665
   polarization, 665

Slit for diffraction, 641
Slow neutrons, 819
Small-angle approximation:
   justification, for physical pendulum, 221
   for pendulum, 219
Snell, Willebord, 591
Snell's law, 590–591, 611
   and polarization, 661–662
Soddy, Frederick, 772
Sodium, energy bands of, 760–761
Solar constant, (prob.) 703, 719
Solenoid, 505–506
   energy density of, 529
   energy stored by, 529
Solid-state detector, 786
Solid-state physics, 759–764
Solids:
   defined, 287
   energy band structure in, 760–761
Sommerfeld, Arnold, 733
Sound:
   beats, 255
   as longitudinal wave, 238–240
Sound intensity level, (SIL), 256–258
Sound waves, 238–240
South-seeking pole, 498
Space shuttle, 173–174
Spark chamber, 787
Special relativity and correspondence principle, 733
Special theory of relativity, 683–695
   (See also Relativity)
Special topics:
   Fiber Optics, 604–606
   How Hot Are You?, 331–332
   Keeping Your Cool, 363–365
   Liquid Crystal Displays, 666–667
   Magnetic Field of the Earth, 516–518
   Measuring Blood Pressure, 308
   Problem Solving in Physics, 27–30
   The Rocket, 171–174
   "Seeing" with Electrons, 765–767
   States of a Real Gas, 332–335
   Strength and Scaling Laws, 280–282
Specific equivalent dose rate (Rhm), 808
Specific heat:
   table of, 343
   units of, 343
Specific heat capacity, 343–344
Spectra, 655, 723, 726
   atomic, 716, 718
   fine structure of, 735
   types, 655
   (See also specific spectrum)
Spectral series, 717–718
Spectrophotometry, (prob.) 608
Spectroscope, 655
Spectrum hydrogen, 717
   solar, (illus.) 656
   visible, 585
Specular reflection, 581–582, 588–589

Speed:
   instantaneous, 18–21
   of light, 584–585, 676, 684
     as limit, 693
   and slope of s vs. t curve, 17
   vs. velocity, 16
   of waves, 241–243
     in fluid, 242
     and frequency, 238
     on string, 242
   (See also Velocity)
Spherical aberration, 621
   in electron microscope, 766
Spherical lens, (illus.) 593, 611–612
Spherical mirrors, 598–604
Spherical surface of lenses, 611–612
Spherical waves, 586
Sphygmomanometer, 308
Spin:
   conservation of, 781
   electron, 735–736
   and relativity, 754
   "up" and "down," 754
Spin angular momentum, 754
   of electron, 735
Spin magnetic quantum number, 736
Spontaneous absorption, 736–737
Spontaneous emission, 736–737
Spring:
   bent pole as, 146
   force vs. stretch, 133
   and Hooke's law, 278
   potential energy of, 140
   and work, 133–134
Spring constant, 134
   of bent pole, 146
   effective, 223
Square wave, 552
   on CRT, 572
Stability, nuclear, 778–779
Standing waves, 245–255
   in blackbody cavity, 705
   electrons as, 745–748
   envelope, 247–248
   equations for, 246
   longitudinal, 251–255
   time-dependent portion, 247–248
State of a gas, 373
Static electricity, 403
Static equilibrium, 105
Static friction, 67–69
Stationary orbits, 729
Steam, heat of condensation, 346
Stefan, Josef, 702
Stefan-Boltzmann constant, 355
Stefan-Boltzmann law, 355, 702–704
Stefan's law, 356
Stellar aberration, 674, 682–683
Step-down transformer, 542
Step-up transformer, 542
Stimulated emission, 736–737
Stokes, Sir George, 305

I-17

Stokes' law, 305–306
Stopping potential, 707
Stops on camera, (table) 623–625
STP (standard temperature and pressure), 323
Strain:
  defined, 269
  hardening, 271
Strangeness, 789–790
Strassmann, Fritz, 817
Stray inductance and resistance, 530
Streamline flow, 306–307
Streamlining, 307
Strength:
  of hollow vs. solid cylinder, 279
  of materials, 278–280
  and scaling, 280–282
  of steel, 279
Stress, 268
Stress vs. strain curves, 270
Stress analysis with polarizer, 665
Strings, vibration of, 247–251
Strokes of engine, (illus.) 385
Strong force, nuclear, 777–780
Sublimation, 334
Sugars, optical activity, 665
Summation notation, 58
Sun:
  energy delivered at distance of the earth, (prob.) 369
  fusion in, 831, 832
  heat absorbed from, 364
  radiation from, 703
  spectrum of, (illus.) 656
  surface temperature, (prob.) 369
Sunlight:
  and discovery of radioactivity, 771–772
  and rainbow, 597–598
  in Young's experiment, 644
Superconductors, 462–463
Superposition and wave packet, 752
Superposition principle, 244–245
  for light, 642
Surface charge density, 432n.
Survey meter, nuclear, 786
Sustained chain reaction, 819–820
Sweep:
  of beam on CRT, 571
  of TV picture, 444
Symbolism in nuclear equations, 774
Synchronization of clocks, 677–678
Synchronous motor, 541
System, thermodynamic, 372
Systolic pressure of blood, 308
Szilard, Leo, 820

Tables:
  Table 1-1 Powers of 10 used in SI, 9
  Table 2-4 Equations Describing Motion at Constant Acceleration, 22
  Table 4-1 Coefficients of Friction, 69
  Table 5-2 Translational and Rotational Equations of Motion, 89
  Table 5-3 Data for Bodies in the Solar System, 96
  Table 7-2 Energy Release by Reaction of Various Substances, 148
  Table 8-1 Coefficient of Restitution of Various Balls, 169
  Table 9-1 Moments of Inertia, 184
  Table 9-2 Translational and Rotational Analogs, 193
  Table 12-1 Young's Modulus and Ultimate Strengths, 272
  Table 12-2 Shear Modulus, Bulk Modulus, Compressibility, and Ultimate Shear Strength, 277
  Table 12-3 Safety Factors in Building Design, 279
  Table 13-1 Densities, 288
  Table 13-2 Coefficient of Viscosity, 303
  Table 14-1 Linear Coefficient of Thermal Expansion, 319
  Table 14-2 Volume Coefficient of Thermal Expansion, 319
  Table 14-3 Temperature Reference Points, 326
  Table 15-1 Specific Heat, 343
  Table 15-2 Thermal Conductivity, 350
  Table 15-3 R-Values, 352
  Table 15-4 Water Vapor Density and Partial Pressure by Temperature, 359
  Table 16-1 Molar Heat Capacities, 377
  Table 16-2 Ratio of Molar Heat Capacities, 381
  Table 16-3 Summary of Thermodynamic Processes, 382
  Table 16-4 Properties of Neon, 393
  Table 18-1 Dielectric Constant, 440
  Table 19-1 Circuit Element Symbols, 458
  Table 19-2 Resistivity, 461
  Table 19-3 Thermal Coefficient of Resistivity, 462
  Table 23-1 Physiological Effects of Currents, 575
  Table 24-1 Index of Refraction, 592
  Table 24-2 Convenient Rays for Ray Tracing, 601
  Table 25-1 Rays for Ray Tracing (lenses), 617
  Table 25-2 Customary Full Stops, 624
  Table 28-1 Work Function for Metals, 711
  Table 31-1 Elementary Particles, 789
  Table 31-2 Family of Quarks, 790
  Table 31-3 Quark Composition of Particles, 790
  Table 32-1 Radiation Units, 802
  Table 32-2 Quality Factor of Radiations, 802
  Table 32-3 Average Radiation Dose, 802
  Table 32-4 Radiation Dose, U.S. Standards, 807
  Table 32-5 Specific Equivalent Dose Rates, 808
Tacoma Narrows Bridge, 225
Tandem mirror experiment, 834
Tangential acceleration, 90
Tartaglia, 13–14, 47
Telephone messages, 605–606
Telephoto lens, 625
Telescope, 630–632
  largest, 633
  reflecting, 632–633
Television picture tube, operation, 441–446
Teller, Edward, 820, 833
Temperature:
  absolute, and radiation, 355
  body, 317, 331–332
  of fusion, 832
  of gas and volume, 321
  and internal energy, 376
  kinetic definition, 330
  measurement of, 316–318
  as random motion, 330
  and thrust of rocket, 172
Temperature conversions, 317–318
Temperature scales, 316–319
Tension, 65–68, 269–271
  and speed of wave, 242
  table of Young's modulus for, 272
Terminal of battery, 457
Terminal potential difference, 459–460
Terminal speed, 306
Terminal voltage, 459–460
Tesla, Nikola, 499
Tesla (T) (unit), 499
Test charge, 412
Thales of Miletus, 403
Theory, modification of, 732–733
Thermal coefficient of resistivity, (table) 462
Thermal conductivity, 349–352
  combined materials, 351–352
  table of, 350
Thermal energy, 315
  and friction, 148
  to mechanical energy, 371–400
  net flow in calorimeter, 347
Thermal equilibrium, 315–316
Thermal expansion, 319–320
Thermal gradient in cloud chamber, 787
Thermal pollution, 388
Thermistor, 316, 326
Thermocouple, 316, 326
Thermodynamic processes, 374–382
  extended example, 392–396
  listed, 376
  summary table for, 382
Thermodynamics, 371–400

Thermodynamics (*Cont.*):
   first law of, 373–374
   second law of, 388–392
   third law of, 391
   zeroth law of, 316
Thermograms, 356
Thermometer:
   calibration, 317, 325–326
   constant-volume gas, (illus.) 324
   construction of, 316
   gas, 316
   liquid-in-glass, 316, 326
   and "lost" KE, 315
   operating principles, 316
   ranges of, (illus.) 325–326
Thermonuclear weapons, 831
Thin-film interference, 647–650
Thin lenses, 614–620
Third law of thermodynamics, 391
35-mm camera, 624–625
Thomson, George, 750
Thomson, J. J., 723–724
   and confirmation of isotopes, 773
Thomson model of atom, 724–725
Thorium, abundance, 798
Thorium-uranium cycle, 829
Thought (gedanken) experiment, 677, 684, 756
Three Mile Island, 806, 826
Threshold(s):
   of effects of radiation, 806
   of pain and feeling, 257
Threshold frequency, 708–710
Thrust:
   of rockets, 171–173
   of Saturn (moon) rocket, (illus.) 170
   of space shuttle, 173
Thyroid and radiation, 811
Time:
   of exposure to radiation, 808
   as fundamental quantity, 3
   proper, 685
   relativistic, 682–684
   unit of, 4
   and work (power), 136
Time constant, LR circuit, 531
Time dilation, 684–686
Time rate of change:
   defined, 16
   of radioactive nuclei, 782
Tokamak, 834–835
Tokamak fusion test reactor, 835
Toroid:
   in fusion, 834–835
   magnetic field, 506
Torque, 108–121
   back, of generators, 539
   on current loop in field, 534–536
   direction of, 111–112
      in three dimensions, 196–197
   magnetic, 535
   and moment of inertia, 183

Torque (*Cont.*):
   and moment arm, 109
   of motor, 537
   net, 188
   perpendicular to angular momentum, 196–198
   and physical pendulum, 221
   positive and negative, 112
   and power, 136
   and rotational acceleration, 179
   and rotational work, 135
   unbalanced, 179
   uniquely defined direction, 112
   units of, 110
Torr (non-SI unit), 293
Torsional stress, 269
Torsional waves, 235
Total internal reflection, 594–596
Total mechanical energy of atom, 727
Tourmaline, 660
Track of particle, 787
Trail of ions on photographic plate, 785
Transformation:
   mathematical, 676
   velocity, 691
Transformer, 541–545
   commercial, 544–545
Transistor, 764
   current amplifier circuit, (prob.) 769
   voltage amplifier circuit, 764
Transistor characteristic curve, (prob.) 769
Transition temperature of superconductor, 462
Translational and rotational equations summarized, (table) 89, 193
Translational motion, 79
   separated from rotational motion, 81–82
Translucent, 582
Transmission electron microscope, (illus.) 766
Transmutation, 775–776
Transparent, 582
Transuranic elements, 820
Transverse waves, 235
Traveling wave, (illus.) 235–240
Triangle wave, 552
   on CRT, 572
Triboelectric series, 406
Triple points of water, 325–326, (table) 326, 334
Tritium, 773, 799
   availability, 835
Trough of wave, 237
Truth and beauty, quarks, 790
Tubes of pipes and waves, 251–255
Turbulence, 306–307
   and Reynolds number, 307
Turbulent flow, 306–307
TV, color:
   LCD screen, 666–667

TV, color (*Cont.*):
   lines per scan, 444
   picture production, 441–446
   power consumption, 666–667
   and radiation, 803
   reception and wavelength, 765*n*.
Twisted-nematic LCD, 666–667

U-233 in reactors, 829
Uhlenbeck, George, 735
Ultimate shear strength, (table) 277
Ultimate strength, 271, (table) 272
Ultraviolet catastrophe, 704, 706
Ultraviolet microscope, 766
Ultraviolet waves, 585
Uncertainty:
   of electron in nucleus, 757–758
   energy and time, 757
   momentum and position, 755–756
   of wave packet, 755
Uncertainty principle, 755–758
Underwater image, 595
Uniform circular motion, 91
   acceleration in, 91–92
   force to produce, 91–92
   as SHM, 210–216
Unit conversion, 5–7
Units:
   atomic mass, to MeV, 779–780
   for ionizing radiation, 799–802
   use in checking problems, 5
Universal gas constant, 323
   derived from molar heat capacities, 377
Universal gravitational constant, 95–98
   measurement of, 97–98
Unpolarized light, 659–660
Unventilated homes, radon hazard, 803
Uranium:
   dating, (prob.) 793
   hexafluoride, 821
   natural isotopes, 819
   neutron bombardment of, 816–817
   occurrence, 798

Valence, 734
Valence band, 760–761
Van Allen belts, 834
Van de Graaff, Robert, 434
Van de Graaff accelerator, (illus.) 434
Vapor vs. gas, 333
Vapor density, 358–359
Varying load on structures, 279–280
Vector:
   in angular dynamics, 196–198
   components of, 38–39
   defined, 35
   direction from components, 40
   magnitude of, 35
      from components, 40
   need for, 35–36

Vector (*Cont.*):
   rectangular components, 38–40
   representation of, 36
   resolution of, 38–42
   resultant, 36
   subtraction, 38
Vector addition:
   graphical method, 36–38
   order of addition, 37
Vector equation, agreement of sides in direction, 59
Vector resultant, ac voltage as, 561
Velocity:
   addition of, 675–678
   average, 16–18, 42, 329
   from curve of s vs. t, 19–20
   defined, 16
   drift, of charges, 453–454
   of ejected fuel of rocket, 172
   of escape from earth, 173
   independence of horizontal and vertical, 43–44
   instantaneous, 18–21, 43
   and pressure of fluid, 299–302
   relative, 49–51, 677–678
   in relativity, 690–691
   of rocket, (illus.) 172
   root-mean-square (rms), 330
   in SHM, 214
   in two dimensions, 42–44
   (*See also* Speed)
Velocity gradient of fluid, 303
Velocity selector, 502
Venturi meter, 300–301
Vibrating string, 247–251
Vibration, 205
Vibratory motion, 205
Virtual image, 598
Virtual object, 627
Viruses, 765, 767
Viscosity, 302–307
   table, 303
   units of, 303
Viscous fluid, 305–307
Visible spectrum, 585, 597
Vision defects, 634–636
Visual purple, 633
Vitreous humor, 633, 634
Volt (V) (unit), 426
   intuitive definition, 426–427
Volta, Alessandro, 426
Voltage:
   constancy of, 427
   to houses, 524
   maximum ac, 562
   measurement of, 483–485
   as potential difference, 431
   rate of change of, 432
   sources, 456–459
   vs. time on CRT, 571
   and work, 427
Voltage amplifier circuit, 764

Voltage amplitude, 552
Voltage drop, 456
Voltmeter, 483–485
   circuit symbol, 484
   connection to circuit, 483–484
   internal circuit, (illus.) 485
   internal resistance, 484–485
Volume:
   of fluid, and pressure, 276–277
   of gas and pressure, 320–321
   of gas and temperature, 321
   and phase change, 326–327
Volume expansion, thermal, 319
Von Laue, Max, 712, 750

Water, heating curve, (illus.) 345
Water pump, hand, (illus.) 310
Watt, James, 136
Watt (unit), 136
Wave function, 752
Wave mechanics, 746, 751–754
   probability interpretation, 748–759
Wave motion, model for, 234–235
Wave nature of matter, 745–769
Wave-particle duality, 715, 759
Wave pocket, 752
   uncertainty, 755
Wave train, 236
Wavefront, 585
   and diffraction, 651
   plane, 644
   vs. ray, 585–587
Wavelength, (illus.) 237
   of emitted photon, 728
   of emitted radiation, (illus.) 356
   and frequency of waves, 238
   and gratings, 658–659
   of He-Ne laser, 738
   maximum, 702
   and resolution, 765
   of waves on strings, 249
   of x-rays, 712
Wavelength distribution of radiation, 702–706
Wavelength shift, 715
Waves:
   beats, 255
   boundary behavior, 243–245
   in closed pipes, 251–253
   defined, 233
   electromagnetic (*see* Electromagnetic waves)
   and energy, 233, 235
   equation for, 238
   frequency and period, 237
   interference, 245
   localized, 751
   longitudinal, 235, (illus.) 239
   mechanical, 233–260
   and moving source, 259–260
   in open pipes, 253–255

Waves (*Cont.*):
   and particle motion, 235
   parts of, 237
   periodic, 235
   phase of, 238
   phase inversion, 243–245
   propagation, 233–235, 241–245
   reflections, 243–245
   speed of propagation, 241–245
   speed and wavelength, 238
   standing, 245–255, 705, 745–748
   strategy to describe, 236
   superposition, 244–245
   torsional, 235
   transverse, 235
   traveling, 235–240
   types of, 235
   on water surface, (illus.) 233–234
   wavelength of, 237
Weak interaction, 781
Weight:
   defined, 62
   and immersion in fluid, 294–296
   vs. length of animals, 282
   vs. mass, 62–63
   and scaling, 280–282
   units of: in British engineering system, 62
   in SI, 62
Whole-body dose, 802
Wide-angle lens, 625
Wien, Wilhelm, 702–705
Wien displacement law, 702–703
Wien formula, 704–705
Wilson, C. T. R., 787
Windchill index, (illus.) 363–364
Window of G-M tube, 786
Work, 130–148
   as area under curve, 133
   to change speed, 138
   and changing force, 132–133
   of charge in field, 423–425
   defined, 130–131
   done by charges, 426–427
   done by gas, 374, 379
   done by a system, 373
   frictional, 132
   moving charge along conductor, 433
   on moving fluid, 297–299
   in refrigerator, 394
   and relativity, 693
   rotational, 135, 187–189, 193
   and rotations, 134–135
   as scalar, 131
   and springs, 133–134
   units of, (table) 131
Work-energy principle and fluids, 297–299
Work-energy theorem, 138
   modified for PE, 141
   rotational, 187
Work function, (table) 710–711

X-ray, 585, 711–715, 771
  beam energy, 811
  diagnostic, dose, 803
  and electrons, 714–716
  medical and dental, 796, 802–803, 805–807
  overexposure, 806–807
  wave nature, 712
X-ray scans, 225

X-ray scattering, 714–716
X-ray tube, 712

Yalow, Roslyn, 828
Yerkes Observatory, 633
Yield point, 271
Young, Thomas, 642–646, 673, 716
Young's experiment, 643–646

Young's modulus, 271–272
  table of, 272

Zeppelin, (prob.) 337
Zeroth law of thermodynamics, 316
Zinc sulfide screen, 786
Zweig, George, 790

Chapter Opening Photographs

1. Adler Planetarium; 2. Photo Researchers; 3. H. Schafer; 4. Rapho/Photo; 5. NASA; 6. J. Cooke; 7. D. Riban; 8. NASA; 9. Photo Researchers; 10. Rapho/Photo; 11. G. Heilman; 12. Photo Researchers; 13. Photo Researchers; 14. Photo Researchers; 15. Photo Researchers; 16. P. Arnold; 17. Photo Researchers; 18. T. Pix; 19. Chattanooga Area Regional Transportation Authority; 20. R. Ellis; 21. Fundamental Photographs; 22. D. Riban; 23. D. Riban; 24. D. Riban; 25. R. Eagle/Photo Researchers; 26. D. Riban; 27. S. Tenney/Bettmann/B. Rogers; 28. Burndy Library; 29. D. Quat/Phototake; 30. B. Rogers; 31. P. Arnold; 32. Rapho/Photo; 33. Photo Researchers